McGraw-Hill

Dictionary of
Geology and
Mineralogy

Second
Edition

D0011914

McGraw-Hill

New York   Chicago   San Francisco   Lisbon   London   Madrid
Mexico City   Milan   New Delhi   San Juan   Seoul   Singapore
Sydney   Toronto

This book is printed on recycled, acid-free paper containing a minimum of 50% recycled, de-inked fiber.

This book was set in Helvetica Bold and Novarese Book by the Clarinda Company, Clarinda, Iowa. It was printed and bound by RR Donnelley, The Lakeside Press.

McGraw-Hill books are available at special quantity discounts to use as premiums and sales promotions, or for use in corporate training programs. For more information, please write to the Director of Special Sales, McGraw-Hill, Professional Publishing, Two Penn Plaza, New York, NY 10121-2298. Or contact your local bookstore.

Library of Congress Cataloging-in-Publication Data

McGraw-Hill dictionary of geology and mineralogy — 2nd. ed.
    p.   cm.
   "All text in this dictionary was published previously in the McGraw-Hill dictionary of scientific and technical terms, sixth edition,
— T.p. verso.
   ISBN 0-07-141044-9 (alk. paper)
    1. Geology—Dictionaries. 2. Mineralogy—Dictionaries. I. Title: Dictionary of geology and mineralogy. II. McGraw-Hill dictionary of scientific and technical terms. 6th ed.

QE5.M3654   2003
550'.3—dc21   2002033173

Contents

Preface

The *McGraw-Hill Dictionary of Geology and Mineralogy* provides a compendium of more than 9000 terms that are central to a broad range of geological sciences and related fields. The coverage in this Second Edition is focused on the areas of geochemistry, geology, geophysics, mineralogy, paleobotany, paleontology, and petrology, with new terms added and others revised as necessary.

Geology deals with the solid earth and the processes that formed and modified it as it evolved. Related disciplines include the study of the physics of the earth (geophysics); earth chemistry, composition, and chemical changes (geochemistry); the composition, properties, and structure of minerals (mineralogy); the description, classification, origin, and evolution of rocks (petrology); and the study of ancient life (paleontology).

All of the definitions are drawn from the *McGraw-Hill Dictionary of Scientific and Technical Terms*, Sixth Edition (2003). Each definition is classified according to the field with which it is primarily associated; if it is used in more than one area; it is identified by the general label [GEOLOGY]. The pronunciation of each term is provided along with synonyms, acronyms, and abbreviations where appropriate. A guide to the use of the Dictionary appears on pages vii-viii, explaining the alphabetical organization of terms, the format of the book, cross referencing, and how synonyms, variant spellings, abbreviations, mineral formulas, and similar information are handled. The Pronunciation Key is provided on page x. The Appendix provides conversion tables for commonly used scientific units as well as revised geologic time scale, periodic table, historical information, and useful listings of geological and mineralogical data.

It is the editors' hope that the Second Edition of the *McGraw-Hill Dictionary of Geology and Mineralogy* will serve the needs of scientists, engineers, students, teachers, librarians, and writers for high-quality information, and that it will contribute to scientific literacy and communication.

Mark D. Licker
Publisher

Staff

How to Use the Dictionary

ALPHABETIZATION. The terms in the McGraw-Hill *Dictionary of Geology and Mineralogy*, Second Edition, are alphabetized on a letter-by-letter basis; word spacing, hyphen, comma, solidus, and apostrophe in a term are ignored in the sequencing. For example, an ordering of terms would be:

abnormal fold	**acre-yield**
a-b plane	**Agassiz orogeny**
ACF diagram	**Age of Fishes**

FORMAT. The basic format for a defining entry provides the term in boldface, the field is small capitals, and the single definition in lightface:

> **term** [FIELD]   Definition.

A field may be followed by multiple definitions, each introduced by a boldface number:

> **term** [FIELD]   **1.** Definition. **2.** Definition. **3.** Definition.

A simple cross-reference entry appears as:

> **term**   See another term.

A cross reference may also appear in combination with definitions:

> **term** [FIELD]   **1.** Definition.   **2.** See another term.

CROSS REFERENCING. A cross-reference entry directs the user to the defining entry. For example, the user looking up "abyssal" finds:

> **abyssal**   See plutonic.

The user then turns to the "P" terms for the definition. Cross references are also made from variant spellings, acronyms, abbreviations, and symbols.

> **aenigmatite**   See enigmatite.
> **aggradation**   See accretion.
> **barkhan**   See barchan.

ALSO KNOWN AS . . . , etc. A definition may conclude with a mention of a synonym of the term, a variant spelling, an abbreviation for the term, or other such information, introduced by "Also known as . . . ," "Also spelled . . . ," "Abbreviated . . . ," "Symbolized . . . ," "Derived from . . . ." When a term has more than one definition, the positioning of any of these phrases conveys the extent of applicability. For example:

term [FIELD] **1.** Definition. Also known as synonym. **2.** Definition. Symbolized T.

In the above arrangement, "Also known as . . ." applies only to the first definition; "Symbolized . . ." applies only to the second definition.

term [FIELD] Also known as synonym. **1.** Definition. **2.** Definition.

In the above arrangement, "Also known as . . ." applies to both definitions.

MINERAL FORMULAS. Mineral definitions may include a formula indicating the composition.

Fields and Their Scope

[GEOCHEM] **geochemistry**—The field that encompasses the investigation of the chemical composition of the earth, other planets, and the solar system and universe as a whole, as well as the chemical processes that occur within them.

[GEOL] **geology**—The study or science of earth, its history, and its life as recorded in the rocks; includes the study of the geologic features of an area, such as the geometry of rock formations, weathering and erosion, and sedimentation.

[GEOPHYS] **geophysics**—The branch of geology in which the principles and practices of physics are used to study the earth and its environment, that is, earth, air, and (by extension) space.

[MINERAL] **mineralogy**—The study of naturally occurring inorganic substances, called minerals, whether of terrestrial or extraterrestrial origin.

[PALEOBOT] **paleobotany**—The study of fossil plants and vegetation of the geologic past.

[PALEON] **paleontology**—The study of life in the geologic past as recorded by fossil remains.

[PETR] **petrology**—The branch of geology dealing with the origin, occurrence, structure, and history of rocks, especially igneous and metamorphic rocks.

Pronunciation Key

Vowels

a   as in bat, that
ā   as in bait, crate
ä   as in bother, father
e   as in bet, net
ē   as in beet, treat
i   as in bit, skit
ī   as in bite, light
ō   as in boat, note
ȯ   as in bought, taut
u̇   as in book, pull
ü   as in boot, pool
ə   as in but, sofa
au̇   as in crowd, power
ȯi   as in boil, spoil
yə   as in formula, spectacular
yü   as in fuel, mule

Semivowels/Semiconsonants

w   as in wind, twin
y   as in yet, onion

Stress (Accent)

ˈ   precedes syllable with primary stress

ˌ   precedes syllable with secondary stress

ǀ   precedes syllable with variable or indeterminate primary/secondary stress

Consonants

b   as in bib, dribble
ch   as in charge, stretch
d   as in dog, bad
f   as in fix, safe
g   as in good, signal
h   as in hand, behind
j   as in joint, digit
k   as in cast, brick
k̲   as in Bach (used rarely)
l   as in loud, bell
m   as in mild, summer
n   as in new, dent
n̲   indicates nasalization of preceding vowel
ŋ   as in ring, single
p   as in pier, slip
r   as in red, scar
s   as in sign, post
sh   as in sugar, shoe
t   as in timid, cat
th   as in thin, breath
t̲h̲   as in then, breathe
v   as in veil, weave
z   as in zoo, cruise
zh   as in beige, treasure

Syllabication

·   Indicates syllable boundary when following syllable is unstressed

A

aa channel [GEOL] A narrow, sinuous channel in which a lava river moves down and away from a central vent to feed an aa lava flow. { 'ä'ä 'chan·əl }

aa lava *See* block lava. { 'ä'ä 'lä·və }

Aalenian [GEOL] Lowermost Middle or uppermost Lower Jurassic geologic time. { ȯ'lēn·ē‚ən }

a axis [GEOL] The direction of movement or transport in a tectonite. { 'ā 'ak‚sis }

abandoned channel *See* oxbow. { ə'ban·dənd 'chan·əl }

ABC system [GEOPHYS] A procedure in seismic surveying to determine the effect of irregular weathering thickness. { 'ā'bē'sē 'sis·təm }

ablation [GEOL] The wearing away of rocks, as by erosion or weathering. { ə'blā·shən }

ablation moraine [GEOL] **1.** A layer of rock particles overlying ice in the ablation of a glacier. **2.** Drift deposited from a superglacial position through the melting of underlying stagnant ice. { ə'blā·shən mə'rān }

abnormal anticlinorium [GEOL] An anticlinorium with axial planes of subsidiary folds diverging upward. { ab'nȯr·məl ‚an·tə·kli'nȯ·rē·əm }

abnormal fold [GEOL] An anticlinorium in which there is an upward convergence of the axial surfaces of the subsidiary folds. { ab'nȯr·məl 'fōld }

abnormal magnetic variation [GEOPHYS] The anomalous value in magnetic compass readings made in some local areas containing unknown sources that deflect the compass needle from the magnetic meridian. { ab'nȯr·məl mag'ned·ik ve·rē'ā·shən }

abnormal synclinorium [GEOL] A synclinorium with axial planes of subsidiary folds converging downward. { ab'nȯr·məl ‚sin·kli'nȯ·rē·əm }

a-b plane [GEOL] The surface along which differential movement takes place. { ā‚bē ‚plän }

abrade [GEOL] To wear away by abrasion or friction. { ə'brād }

abrasion [GEOL] Wearing away of sedimentary rock chiefly by currents of water laden with sand and other rock debris and by glaciers. { ə'brā·zhən }

abrasion platform [GEOL] An uplifted marine peneplain or plain, according to the smoothness of the surface produced by wave erosion, which is of large area. { ə'brā·zhən 'plat·fȯrm }

abrasive [GEOL] A small, hard, sharp-cornered rock fragment, used by natural agents in abrading rock material or land surfaces. Also known as abrasive ground. { ə'brās·əv }

absarokite [PETR] An alkalic basalt of about equal portions of olivine, augite, labradorite, and sanidine with accessory biotite, apatite, and opaque oxides; leucite is occasionally present in small amounts. { ab'sä·rə·kīt }

absolute age [GEOL] The geologic age of a fossil, or a geologic event or structure expressed in units of time, usually years. Also known as actual age. { 'ab·sə‚lüt 'āj }

absolute geopotential topography *See* geopotential topography. { 'ab·sə‚lüt jē·ō·pə'ten·shəl tə'päg·rə·fē }

absolute time [GEOL] Geologic time measured in years, as determined by radioactive decay of elements. { 'ab·sə‚lüt 'tīm }

Abukuma-type facies [PETR] A type of dynathermal regional metamorphism characterized by low pressure. { ab·ə'kü·mə ‚tīp 'fā·shēz }

abundance

abundance [GEOCHEM] The relative amount of a given element among other elements. { ə'bən·dəns }

abyssal *See* plutonic. { ə'bis·əl }

abyssal cave *See* submarine fan. { ə'bis·əl 'kāv }

abyssal fan *See* submarine fan. { ə'bis·əl 'fan }

abyssal floor [GEOL] The ocean floor, or bottom of the abyssal zone. { ə'bis·əl 'flȯr }

abyssal gap [GEOL] A gap in a sill, ridge, or rise that lies between two abyssal plains. { ə'bis·əl 'gap }

abyssal hill [GEOL] A hill 2000 to 3000 feet (600 to 900 meters) high and a few miles wide within the deep ocean. { ə'bis·əl 'hil }

abyssal injection [GEOL] The process of driving magmas, originating at considerable depths, up through deep-seated contraction fissures in the earth's crust. { ə'bis·əl in'jek·shən }

abyssal plain [GEOL] A flat, almost level area occupying the deepest parts of many of the ocean basins. { ə'bis·əl 'plān }

abyssal rock [GEOL] Plutonic, or deep-seated, igneous rocks. { ə'bis·əl 'räk }

abyssal theory [GEOL] A theory of the origin of ores involving the separation of ore silicates from the liquid stage during the cooling of the earth. { ə'bis·əl 'thē·ə·rē }

abyssolith [GEOL] A molten mass of eruptive material passing up without a break from the zone of permanently molten rock within the earth. { ə'bis·ō,lith }

Acadian orogeny [GEOL] The period of formation accompanied by igneous intrusion that took place during the Middle and Late Devonian in the Appalachian Mountains. { ə'kād·ē·ən ȯr'äj·ə·nē }

acanthite [MINERAL] Ag₂S A blackish to lead-gray silver sulfide mineral, crystallizing in the orthorhombic system. { ə'kan·thīt }

Acanthodes [PALEON] A genus of Carboniferous and Lower Permian eellike acanthodian fishes of the family Acanthodidae. { ə,kan'thō·dēz }

Acanthodidae [PALEON] A family of extinct acanthodian fishes in the order Acanthodiformes. { ə,kan'thō·də,dē }

Acanthodiformes [PALEON] An order of extinct fishes in the class Acanthodii having scales of acellular bone and dentine, one dorsal fin, and no teeth. { ə,kan·thō·də'fȯr,mēz }

Acanthodii [PALEON] A class of extinct fusiform fishes, the first jaw-bearing vertebrates in the fossil record. { ə,kan'thō·dē,ī }

acanthopore [PALEON] A tubular spine in some fossil bryozoans. { ə'kan·thə,pȯr }

acaustobiolith [PETR] A noncombustible organic rock, or one formed by organic accumulation of minerals. { ¦ā,kȯs·tə'bī·ə·lith }

acaustophytolith [PETR] An acaustobiolith resulting from plant activity, such as a pelagic ooze that contains diatoms. { ¦ā,kȯs·tə'fīd·ə,lith }

accelerated erosion [GEOL] Soil erosion that occurs more rapidly than soil horizons can form from the parent regolith. { ak'sel·ər,ā·dəd i'rō·zhən }

acceptable risk [GEOPHYS] In seismology, that level of earthquake effects which is judged to be of sufficiently low social and economic consequence, and which is useful for determining design requirements in structures or for taking certain actions. { ak¦sep·tə·bəl 'risk }

accessory ejecta [GEOL] Pyroclastic material formed from solidified volcanic rocks that are from the same volcano as the ejecta. { ak'ses·ə·rē i'jek·tə }

accessory element *See* trace element. { ak'ses·ə·rē 'el·ə·mənt }

accessory mineral [MINERAL] A minor mineral in an igneous rock that does not affect its general character. { ak'ses·ə·rē ,min·rəl }

accidental ejecta [GEOL] Pyroclastic rock formed from preexisting nonvolcanic rocks or from volcanic rocks unrelated to the erupting volcano. { ¦ak·sə¦den·təl i'jek·tə }

accidental inclusion *See* xenolith. { ¦ak·sə¦den·təl in'klü·zhən }

accident block [GEOL] A solid chip of rock broken off from the subvolcanic basement and ejected from a volcano. { 'ak·sə,dent ,bläk }

acclivity [GEOL] A slope that is ascending from a reference point. { ə'kliv·əd·ē }

accordant |GEOL| Pertaining to topographic features that have nearly the same elevation. { ə'körd·ənt }

accordant fold |GEOL| One of several folds that are similarly oriented. { ə'körd·ənt ,föld }

accordant summit level |GEOL| A hypothetical horizontal plane that can be drawn over a broad region connecting mountain summits of similar elevation. { ə'körd·ənt 'səm·ət ,lev·əl }

accretion |GEOL| **1.** Gradual buildup of land on a shore due to wave action, tides, currents, airborne material, or alluvial deposits. **2.** The process whereby stones or other inorganic masses add to their bulk by adding particles to their surfaces. Also known as aggradation. **3.** *See* accretion tectonics. { ə'krē·shən }

accretionary lapilli *See* mud ball. { ə'krē·shən,er·ē lə'pi·lē }

accretionary lava ball |GEOL| A rounded ball of lava that occurs on the surface of an aa lava flow. { ə'krē·shən,er·ē 'lä·və ,böl }

accretionary limestone |PETR| A type of limestone formed by the slow accumulation of organic remains. { ə'krē·shən,er·ē 'līm·stōn }

accretionary ridge |GEOL| A beach ridge located inland from the modern beach, indicating that the coast has been built seaward. { ə'krē·shən,er·ē ,rij }

accretion tectonics |GEOL| The bringing together, or suturing, of terranes; regarded by many geologists as an important mechanism of continental growth. Also known as accretion. { ə'krē·shən tek'tän·iks }

accretion topography |GEOL| Topographic features built by accumulation of sediment. { ə'krē·shən tä'päg·rə·fē }

accretion vein |GEOL| A type of vein formed by the repeated filling of channels followed by their opening because of the development of fractures in the zone undergoing mineralization. { ə'krē·shən ,vān }

accretion zone |GEOL| Any beach area undergoing accretion. { ə'krē·shən ,zōn }

accumulation zone |GEOL| The area where the bulk of the snow contributing to an avalanche was originally deposited. { ə·kyü·myə'lā·shən ,zōn }

ACF diagram |PETR| A triangular diagram showing the chemical character of a metamorphic rock; the three components plotted are A $=Al_2O_3 + Fe_2O_3 - (Na_2O + K_2O)$, $C= CaO$, $F = FeO + MgO + MnO$. { ,ā,sē'ef 'dī·ə,gram }

a-c girdle |GEOL| A girdle of points in a petrofabric diagram that have a tread parallel with the plane of the a and c fabric axes. { 'a'sē 'gərd·əl }

Achaenodontidae |PALEON| A family of Eocene dichobunoids, piglike mammals belonging to the suborder Palaeodonta. { ə,kēn·ə'dän·tə·dē }

achondrite |GEOL| A stony meteorite that contains no chondrules. { ¦ā'kän,drīt }

achroite |MINERAL| A colorless variety of tourmalines found in Malagasy. { 'ak·rō,īt }

acid clay |GEOL| A type of clay that gives off hydrogen ions when it dissolves in water. { 'as·əd 'klā }

acidic lava |GEOL| Extruded felsic igneous magma which is rich in silica (SiO_2 content exceeds 65). { ə'sid·ik 'lä·və }

acidic rock |PETR| Igneous rock containing more than 66% SiO_2, making it silicic. { ə'sid·ik 'räk }

acidity coefficient |GEOCHEM| The ratio of the oxygen content of the bases in a rock to the oxygen content in the silica. Also known as oxygen ratio. { ə'sid·ə·tē ,kō·ə'fish·ənt }

acid soil |GEOL| A soil with pH less than 7; results from presence of exchangeable hydrogen and aluminum ions. { 'as·əd 'söil }

acid spar |MINERAL| A grade of fluorspar containing over 98% CaF_2 and no more than 1% SiO_2; produced by flotation; used for the production of hydrofluoric acid. { 'as·əd ,spär }

aclinal |GEOL| Without dip; horizontal. { ¦ā'klīn·əl }

aclinic |GEOPHYS| Referring to a situation where a freely suspended magnetic needle remains in a horizontal position. { a'klin·ik }

aclinic line *See* magnetic equator. { a'klin·ik 'līn }

3

acme

acme [PALEON] The time of largest abundance or variety of a fossil taxon; the taxon may be either general or local. { 'ak·mē }

acmite [MINERAL] $NaFeSi_2O_6$ A brown or green silicate mineral of the pyroxene group, often in long, pointed prismatic crystals; hardness is 6–6.5 on Mohs scale, and specific gravity is 3.50–3.55; found in igneous and metamorphic rocks. { 'ak,mīt }

acre-yield [GEOL] The average amount of oil, gas, or water taken from one acre of a reservoir. { 'ā·kər ¦yēld }

acritarch [PALEON] A unicellular microfossil of unknown or uncertain biological origin that occurs abundantly in strata from the Precambrian and Paleozoic. { 'ak·rə,tark }

acrobatholithic [GEOL] A stage in batholithic erosion where summits of cupolas and stocks are exposed without any exposure of the surface separating the barren interior of the batholith from the mineralized upper part. { ,ak·rə¦bath·ə¦lith·ik }

acromorph [GEOL] A salt dome. { 'ak·rō,mȯrf }

Acrosaleniidae [PALEON] A family of Jurassic and Cretaceous echinoderms in the order Salenoida. { ¦ak·rō,sal·ə'nī·ə·dē }

Acrotretacea [PALEON] A family of Cambrian and Ordovician inarticulate brachiopods of the suborder Acrotretidina. { ,ak·rō·tre'tās·ē·ə }

acrozone See range zone. { 'ak·rō,zōn }

actinolite [MINERAL] $Ca_2(Mg,Fe)_5Si_8O_{22}(OH)_2$ A green, monoclinic rock-forming amphibole; a variety of asbestos occurring in needlelike crystals and in fibrous or columnar forms; specific gravity 3–3.2. { ,ak'tin·ə,līt }

Actinostromariidae [PALEON] A sphaeractinoid family of extinct marine hydrozoans. { ,ak·tə·nō,strō·mə'rī·ə,dē }

active layer [GEOL] That part of the soil which is within the suprapermafrost layer and which usually freezes in winter and thaws in summer. Also known as frost zone. { 'ak·tiv 'lā·ər }

active margin [GEOL] A continental margin that is characterized by earthquakes, volcanic activity, and orogeny resulting from movement of tectonic plates. { 'ak·təv 'mär·jən }

active permafrost [GEOL] Permanently frozen ground (permafrost) which, after thawing by artificial or unusual natural means, reverts to permafrost under normal climatic conditions. { 'ak·tiv 'pər·mə,frȯst }

active volcano [GEOL] A volcano capable of venting lava, pyroclastic material, or gases. { 'ak·tiv ,väl'kā·nō }

activity ratio [GEOL] The ratio of plasticity index to percentage of clay-sized minerals in sediment. { ,ak'tiv·əd·ē ,rā·shō }

actual age See absolute age. { 'ak·chə·wəl āj }

actualism See uniformitarianism. { 'ak·chü·ə,liz·əm }

actual relative movement See slip. { 'ak·chə·wəl 'rel·ə·tiv 'müv·mənt }

acute angle block [GEOL] A fault block in which the strike of strata on the down-dip side meets a diagonal fault at an acute angle. { ə'kyüt ¦aŋ·gəl 'bläk }

acute bisectrix [MINERAL] A bisecting line of the acute angle of the optic axes of biaxial minerals. { ə'kyüt ,bī'sek·triks }

adakites [GEOL] Rocks formed from lavas that melted from subducting slabs associated with other volcanic arcs or arc/continent collision zones; they were first described from Adak Island in the Aleutians. { 'a·də,kīts }

adamantine spar [MINERAL] A silky brown variety of corundum. { ,ad·ə'man,tēn 'spär }

adamellite See quartz monzonite. { ə'dam·ə,līt }

adamite [MINERAL] $Zn_2(AsO_4)(OH)$ A colorless, white, or yellow mineral consisting of basic zinc arsenate, crystallizing in the orthorhombic system; hardness is 3.5 on Mohs scale, and specific gravity is 4.34–4.35. { 'ad·ə,mīt }

adamsite [MINERAL] Greenish-black mica. { 'a·dəm,zīt }

adcumulus [PETR] Pertaining to the growth of a cumulus crystal so as to exclude the growth of other phases; results in a monomineralic rock. { ad'kyü·myə·ləs }

adelite [MINERAL] $CaMg(AsO_4)(OH,F)$ A colorless to gray, bluish-gray, yellowish-gray, yellow, or light green orthorhombic mineral consisting of a basic arsenate of calcium and magnesium; usually occurs in massive form.   { 'ad·əl,īt }

ader wax *See* ozocerite.   { 'ad·ər ‚waks }

adiagnostic   [PETR]   Pertaining to a rock texture in which identification of individual components is not possible macroscopically or microscopically; applied especially to igneous rock.   { ¦ā,dī·əg'näs·tik }

adinole   [GEOL]   An argillaceous sediment that has undergone albitization at the margin of a basic intrusion.   { 'ad·ən,ōl }

adipocerite   *See* hatchettite.   { ‚ad·ə'päs·ə,rīt }

adipocire   *See* hatchettite.   { ‚ad·ə'pä,sir }

admixture   [GEOL]   One of the lesser or subordinate grades of sediment.   { ¦ad¦miks·chər }

adobe   [GEOL]   Heavy-textured clay soil found in the southwestern United States and in Mexico.   { ə'dō·bē }

adobe flats   [GEOL]   Broad flats that are floored with sandy clay and have been formed from sheet floods.   { ə'dō·bē 'flats }

adolescence   [GEOL]   Stage in the cycle of erosion following youth and preceding maturity.   { ‚ad·əl'es·əns }

adolescent coast   [GEOL]   A type of shoreline characterized by low but nearly continuous sea cliffs.   { ‚ad·əl'es·ənt ‚kōst }

adularia   [MINERAL]   A weakly triclinic form of the mineral orthoclase occurring in transparent, colorless to milky-white pseudo-orthorhombic crystals.   { ‚aj·ə'la·rē·ə }

adularization   [GEOL]   Replacement by or introduction of the mineral adularia.   { ə,jül·ə·rə'zā·shən }

advance   [GEOL]   **1.** A continuing movement of a shoreline toward the sea.   **2.** A net movement over a specified period of time of a shoreline toward the sea.   { əd'vans }

adventive cone   [GEOL]   A volcanic cone that is on the flank of and subsidiary to a larger volcano.   Also known as lateral cone; parasitic cone.   { ad'ven·tiv 'kōn }

adventive crater   [GEOL]   A crater opened on the flank of a large volcanic cone.   { ad'ven·tiv 'krāt·ər }

Aechminidae   [PALEON]   A family of extinct ostracodes in the order Paleocopa in which the hollow central spine is larger than the valve.   { ēk'min·ə,dē }

Aeduellidae   [PALEON]   A family of Lower Permian palaeoniscoid fishes in the order Palaeonisciformes.   { ‚ē·dü'el·ə,dī }

aegirine   [MINERAL] $NaFe(SiO_3)_2$ A brown or green clinopyroxene occurring in alkali-rich igneous rocks.   Also known as aegirite.   { 'ā·gə,rēn }

aegirite   *See* aegirine.   { 'ā·gə,rīt }

Aegyptopithecus   [PALEON]   A primitive primate that is thought to represent the common ancestor of both the human and ape families.   { ə,jip·tō'pith·e,kəs }

aenigmatite   *See* enigmatite.   { ə'nig·mə,tīt }

Aepyornis   [PALEON]   A genus of extinct ratite birds representing the family Aepyornithidae.   { ‚ē·pē'órn·əs }

Aepyornithidae   [PALEON]   The single family of the extinct avian order Aepyornithiformes.   { ‚ē·pē,ór'nith·ə,dē }

Aepyornithiformes   [PALEON]   The elephant birds, an extinct order of ratite birds in the superorder Neognathae.   { ‚ē·pē,ór,nith·ə'fór,mēz }

aerogeology   [GEOL]   The geologic study of earth features by means of aerial observations and aerial photography.   { ‚e·rō·jē'äl·ə·jē }

aerohydrous mineral   [MINERAL]   A mineral containing water in small cavities.   { ¦e·rō¦hī·drəs 'min·rəl }

aerolite   *See* stony meteorite.   { 'e·rō,līt }

aeromagnetic surveying   [GEOPHYS]   The mapping of the magnetic field of the earth through the use of electronic magnetometers suspended from aircraft.   { ‚e·rō·mag'ned·ik sər'vā·iŋ }

aeropalynology   [PALEOBOT]   A branch of palynology that focuses on the study of pollen grains and spores that are dispersed into the atmosphere.   { ‚er·ō,pal·ə'näl·ə·jē }

aerosiderite   [GEOL]   A meteorite composed principally of iron.   { ‚e·rō'sīd·ə,rīt }

5

affine deformation

affine deformation [GEOL] A type of deformation in which very thin layers slip against each other so that each moves equally with respect to its neighbors; generally does not result in folding. { ə'fīn ,dē-fòr'mä·shən }

affine strain [GEOPHYS] A strain in the earth that does not differ from place to place. { ə'fīn 'strān }

African superplume [GEOPHYS] A large, discrete, slowly rising plume of heated material in the earth's mantle, beneath southern Africa, believed by some to contribute to the movement of tectonic plates. { ¦af·ri·kən 'sü·pər,plüm }

aftershock [GEOPHYS] A small earthquake following a larger earthquake and originating at or near the larger earthquake's epicenter. { 'af·tər,shäk }

Aftonian interglacial [GEOL] Post-Nebraska interglacial geologic time. { ,af'ton·ē·ən ,in·tər'glā·shəl }

afwillite [MINERAL] $Ca_3Si_2O_4(OH)_6$ A colorless mineral consisting of a hydrous calcium silicate and occurring in monoclinic crystals; specific gravity is 2.6. { 'af·wə,līt }

agalite [MINERAL] A mineral with the same composition as talc but with a less soapy feel; used as a filler in writing paper. { 'a·gə,līt }

agalmatolite [GEOL] A soft, waxy, gray, green, yellow, or brown mineral or stone, such as pinite and steatite; used by the Chinese for carving images. Also known as figure stone; lardite; pagodite. { ,a·gəl'mad·əl,īt }

agaric mineral See rock milk. { ə'gar·ik 'min·rəl }

Agassiz orogeny [GEOL] A phase of diastrophism confined to North America Cordillera occurring at the boundary between the Middle and Late Jurassic. { 'ag·ə·sē ò'räj·ə·nē }

Agassiz Valleys [GEOL] Undersea valleys in the Gulf of Mexico between Cuba and Key West. { 'ag·ə·sē 'val·ēz }

agate [MINERAL] SiO_2 A fine-grained, fibrous variety of chalcedony with color banding or irregular clouding. { 'ag·ət }

agate jasper [MINERAL] An impure variety of quartz consisting of jasper and agate. Also known as jaspagate. { 'ag·ət 'jas·pər }

agatized wood See silicified wood. { 'ag·ə·tīzd 'wúd }

age [GEOL] **1.** Any one of the named epochs in the history of the earth marked by specific phases of physical conditions or organic evolution, such as the Age of Mammals. **2.** One of the smaller subdivisions of the epoch as geologic time, corresponding to the stage or the formation, such as the Lockport Age in the Niagara Epoch. { āj }

aged [GEOL] Of a ground configuration, having been reduced to base level. { 'ā·jəd }

age determination [GEOL] Identification of the geologic age of a biological or geological specimen by using the methods of dendrochronology or radiometric dating. { 'āj di,tər·mə'nā·shən }

aged shore [GEOL] A shore long established at a constant level and adjusted to the waves and currents of the sea. { 'ā·jəd 'shòr }

Age of Fishes [GEOL] An informal designation of the Silurian and Devonian periods of geologic time. { 'āj əv 'fish·əz }

Age of Mammals [GEOL] An informal designation of the Cenozoic era of geologic time. { 'āj əv 'mam·əlz }

Age of Man [GEOL] An informal designation of the Quaternary period of geologic time. { 'āj əv 'man }

age ratio [GEOL] The ratio of the amount of daughter to parent isotope in a mineral being dated radiometrically. { 'āj ,rā·shō }

agglomerate [GEOL] A pyroclastic rock composed of angular rock fragments in a matrix of volcanic ash; typically occurs in volcanic vents. { ə'gläm·ə·rət }

agglutinate cone See spatter cone. { ə'glüt·ən,āt ,kōn }

aggradation See accretion. { ,ag·rə'dā·shən }

aggradation recrystallization [GEOL] Recrystallization resulting in the enlargement of crystals. { ,ag·rə'dā·shən rē,kris·tə·lə'zā·shən }

aggraded valley floor [GEOL] The surface of a flat deposit of alluvium which is thicker than the stream channel's depth and is formed where a stream has aggraded its valley. { ə'grād·əd 'val·ē 'flȯr }

aggraded valley plain See alluvial plain. { ə'grād·əd 'val·ē 'plān }

aggregate |GEOL| A collection of soil grains or particles gathered into a mass. { 'ag·rə·gət }

aggregate structure |GEOL| A mass composed of separate small crystals, scales, and grains that, under a microscope, extinguish at different intervals during the rotation of the stage. { 'ag·rə·gət 'strək·chər }

aggressive magma |GEOL| A magma that forces itself into place. { ə'gres·iv 'mag·mə }

Aglaspida |PALEON| An order of Cambrian and Ordovician merostome arthropods in the subclass Xiphosurida characterized by a phosphatic exoskeleton and vaguely trilobed body form. { ə'glas·pə·də }

agmatite |PETR| 1. A migmatite that contains xenoliths. 2. Fragmental plutonic rock with granitic cement. { 'ag·mə,tīt }

agonic line |GEOPHYS| The imaginary line through all points on the earth's surface at which the magnetic declination is zero; that is, the locus of all points at which magnetic north and true north coincide. { ā'gän·ik līn }

agravic |GEOPHYS| Of or pertaining to a condition of no gravitation. { ,ā'grav·ik }

agpaite |PETR| A group of igneous rocks containing feldspathoids; includes naujaite, lujavrite, and kakortokite. { 'ag·pə,īt }

agricere |GEOL| A waxy or resinous organic coating on soil particles. { 'ag·rə,sir }

agricolite See eulytite. { ə'grik·ə,līt }

agricultural geology |GEOL| A branch of geology that deals with the nature and distribution of soils, the occurrence of mineral fertilizers, and the behavior of underground water. { ¦ag·rə¦kəl·chə·rəl jē'äl·ə·jē }

Agriochoeridae |PALEON| A family of extinct tylopod ruminants in the superfamily Merycoidodontoidea. { ,ag·rē·ō'kir·ə,dē }

aguilarite |MINERAL| Ag$_4$SeS An iron-black mineral associated with argentite and silver in Mexico. { äg·ə'lä,rīt }

ahlfeldite |MINERAL| (Ni,Co)SeO$_3$·2H$_2$O A triclinic mineral identified as green to yellow crystals with a reddish-brown coating, consisting of a hydrous selenite of nickel. { äl'fel,dīt }

aiguille |GEOL| The needle-top of the summit of certain glaciated mountains, such as near Mont Blanc. { ,ā'gwēl }

aikinite |MINERAL| PbCuBiS$_3$ A mineral crystallizing in the orthorhombic system and occurring massive and in gray needle-shaped crystals; hardness is 2 on Mohs scale, and specific gravity is 7.07. Also known as needle ore. { 'ā·kə,nīt }

ailsyte |PETR| An alkalic microgranite containing a considerable amount of riebeckite. Also known as paisanite. { 'āl,sīt }

air current |GEOPHYS| See air-earth conduction current. { 'er ,kər·ənt }

air gap See wind gap. { 'er ,gap }

air heave |GEOL| Deformation of plastic sediments on a tidal flat as a result of the growth of air pockets in them; the growth occurs by accretion of smaller air bubbles oozing through the sediment. { 'er ,hēv }

air sac See vesicle. { 'er ,sak }

air shooting |GEOPHYS| In seismic prospecting, the technique of applying a seismic pulse to the earth by detonating a charge or charges in the air. { 'er ,shüd·iŋ }

air volcano |GEOL| An eruptive opening in the earth from which large volumes of gas emanate, in addition to mud and stones; a variety of mud volcano. { ¦er ,väl¦kā·nō }

Airy isostasy |GEOPHYS| A theory of hydrostatic equilibrium of the earth's surface which contends that mountains are floating on a fluid lava of higher density, and that higher mountains have a greater mass and deeper roots. { ¦er·ē i'säs·tə·sē }

Aistopoda |PALEON| An order of Upper Carboniferous amphibians in the subclass Lepospondyli characterized by reduced or absent limbs and an elongate, snakelike body. { ,ā·ə'stäp·ə·də }

akaganeite [MINERAL] β-FeO(OH) A mineral found in meteorites and considered to be formed in flight or by alteration. { ˌa·kə'gan·ēˌīt }

akenobeite [PETR] A form of aplite composed of orthoclase and oligoclase with quartz in the interstices. { ˌa·kə'nōb·ēˌit }

akerite [PETR] A rock composed of quartz syenite containing soda microcline, oligoclase, and augite. { 'ō·kəˌrīt }

akermanite [MINERAL] $Ca_2MgSi_2O_7$ Anhydrous calcium-magnesium silicate found in igneous rocks; a melilite. { 'ō·kər·məˌnīt }

AKF diagram [PETR] A triangular diagram showing the chemical character of a metamorphic rock in which the three components plotted are A = Al_2O_3 + Fe_2O_3 + (CaO + Na_2O), K = K_2O, and F = FeO + MgO + MnO. { ¦ā¦kā¦ef 'dī·əˌgram }

akrochordite [MINERAL] $Mn_4Mg(AsO_4)_2(OH)_4·4H_2O$ Mineral consisting of a hydrous basic manganese magnesium arsenate and occurring in reddish-brown rounded aggregates; hardness is 3 on Mohs scale, and specific gravity is 3.2. { ˌak·rō'kórˌdit }

aktological [GEOL] Nearshore shallow-water areas, conditions, sediments, or life. { ˌak·tə'läj·ə·kəl }

alabandite [MINERAL] MnS A complex sulfide mineral that is a component of meteorites and usually occurs in iron-black massive or granular form. Also known as manganblende. { ˌal·ə'banˌdīt }

alabaster [MINERAL] **1.** $CaSO_4·2H_2O$ A fine-grained, colorless gypsum. **2.** See onyx marble. { 'al·əˌbas·tər }

alamosite [MINERAL] $PbSiO_3$ A white or colorless monoclinic mineral consisting of lead silicate and occurring in radiating fibers; hardness is 4.5 on Mohs scale, and specific gravity is 6.5. { ˌal·ə'mōˌsīt }

alaskaite [MINERAL] A light lead-gray sulfide mineral consisting of a mixture of lead, silver, copper, and bismuth. { ə'las·kəˌīt }

alaskite [PETR] A granitic rock composed mainly of quartz and alkali feldspar, with few dark mineral components. { ə'lasˌkīt }

albafite [MINERAL] Greenish to brownish bitumen which becomes white when exposed to air; contains up to 15% oxygen; fusible; insoluble in organic solvents; varies from soft to hard, porous to compact; atomic ratio H/C 1.75–2.25. { 'al·bəˌfīt }

albanite [PETR] A melanocratic leucitite found near Rome, Italy. { 'al·bəˌnīt }

albertite [MINERAL] Jet-black, brittle natural hydrocarbon with conchoidal fracture, hardness of 1–2, and specific gravity of approximately 1.1. Also known as asphaltite coal. { 'al·bərˌtīt }

Albertosaurus [PALEON] A carnivorous therapod dinosaur, 30 feet (9 meters) long, from the Late Cretaceous Period that had long muscular hindlimbs, comparatively weak forelimbs (with two-fingered hands), and powerful jaws lined with sharp teeth; related to Tyrannosaurus. { alˌber·də'sór·əs }

Albian [GEOL] Uppermost Lower Cretaceous geologic time. { 'al·bē·ən }

albic horizon [GEOL] A soil horizon from which clay and free iron oxides have been removed or in which the iron oxides have been segregated. { 'al·bik hə'rīz·ən }

Albionian [GEOL] Lower Silurian geologic time. { ˌal·bē'ōn·ē·ən }

albite [MINERAL] $NaAlSi_3O_8$ A colorless or milky-white variety of plagioclase of the feldspar group found in granite and various igneous and metamorphic rocks. Also known as sodaclase; sodium feldspar; white feldspar; white schorl. { 'alˌbīt }

albite-epidote-amphibolite facies [PETR] Rocks of metamorphic type formed under intermediate temperature and pressure conditions by regional metamorphism or in the outer contact metamorphic zone. { 'alˌbīt 'ep·əˌdōt ˌam'fib·əˌlīt 'fāˌshēz }

albitite [PETR] A porphyritic dike rock that is coarse-grained and composed almost wholly of albite; common accessory minerals are muscovite, garnet, apatite, quartz, and opaque oxides. { 'al·bəˌtīt }

albitization [PETR] The formation of albite in a rock as a secondary mineral. { ˌal·bəd·ə'zā·shən }

albitophyre [PETR] A porphyritic rock that contains albite phenocrysts in a groundmass composed mostly of albite. { al'bid·əˌfīr }

alkali feldspar

Alboll [GEOL] A suborder of the soil order Mollisol with distinct horizons, wet for some part of the year; occurs mostly on upland flats and in shallow depressions. { 'al,bȯl }

alboranite [PETR] Olivine-free hypersthene basalt. { ‚al·bə'ra,nīt }

alcove [GEOL] A large niche formed by a stream in a face of horizontal strata. { 'al,kōv }

alcove lands [GEOL] Terrain where the mud rocks or sandy clays and shales that compose the hills (badlands) are interstratified by occasional harder beds; the slopes are terraced. { 'al,kōv ,lanz }

alee basin [GEOL] A basin formed in the deep sea by turbidity currents aggrading courses where the currents were deflected around a submarine ridge. { ə'lē ,bās·ən }

aleishtite [GEOL] A bluish or greenish mixture of dickite and other clay minerals. { ə'lē·ish,tīt }

Alexandrian [GEOL] Lower Silurian geologic time. { ‚al·ig'zan·dre·ən }

alexandrite [MINERAL] A gem variety of chrysoberyl; emerald green in natural light but red in transmitted or artificial light. { ‚al·ig'zan,drīt }

Alfisol [GEOL] An order of soils with gray to brown surface horizons, a medium-to-high base supply, and horizons of clay accumulation. { 'al·fə,sōl }

algal [GEOL] Formed from or by algae. { 'al·gəl }

algal biscuit [GEOL] A disk-shaped or spherical mass, up to 20 centimeters in diameter, made up of carbonate that is probably the result of precipitation by algae. { ¦al·gəl ¦bis·kət }

algal coal [GEOL] Coal formed mainly from algal remains. { 'al·gəl ,kōl }

algal limestone [PETR] A type of limestone either formed from the remains of calcium-secreting algae or formed when algae bind together the fragments of other lime-secreting organisms. { 'al·gəl 'līm,stōn }

algal pit [GEOL] An ablation depression that is small and contains algae. { 'al·gəl ,pit }

algal reef [GEOL] An organic reef which has been formed largely of algal remains and in which algae are or were the main lime-secreting organisms. { 'al·gəl ,rēf }

algal ridge [GEOL] Elevated margin of a windward coral reef built by actively growing calcareous algae. { 'al·gəl ,rij }

algal rim [GEOL] Low rim built by actively growing calcareous algae on the lagoonal side of a leeward reef or on the windward side of a patch reef in a lagoon. { 'al·gəl ,rim }

algal structure [GEOL] A deposit, most frequently calcareous, with banding, irregular concentric structures, crusts, and pseudo-pisolites or pseudo-concretionary forms resulting from organic, colonial secretion and precipitation. { ¦al·gəl ¦strək·chər }

Algerian onyx See onyx marble. { al'jer·ē·ən 'än·iks }

alginite See algite. { 'al·jə,nīt }

algite [PETR] The petrological unit that constitutes algal material present in considerable amounts in algal or boghead coal. Also known as alginite. { 'al,jīt }

algodonite [MINERAL] Cu_6As A steel gray to silver white mineral consisting of copper arsenide and occurring as minute hexagonal crystals or in massive and granular form. { al'gäd·ə,nīt }

Algoman orogeny [GEOL] Orogenic episode affecting Archean rocks of Canada about 2.4 billion years ago. Also known as Kenoran orogeny. { al'gōm·ən ȯ'räj·ə·nē }

Algonkian See Proterozoic. { al'gäŋ·kē·ən }

alkali See alkalic. { 'al·kə,lī }

alkalic Also known as alkali. [PETR] **1.** Of igneous rock, containing more than average alkali (K_2O and Na_2O) for that clan in which they are found. **2.** Of igneous rock, having feldspathoids or other minerals, such as acmite, so that the molecular ratio of alkali to silica is greater than 1:6. **3.** Of igneous rock, having a low alkali-lime index (51 or less). { ‚al'kal·ik }

alkali-calcic series [PETR] The series of igneous rocks with weight percentage of silica in the range 51–55, and weight percentages of CaO and $K_2O + Na_2O$ equal. { ¦al·kə,lī ¦kal,sik ,sir·ēz }

alkali emission [GEOPHYS] Light emission from free lithium, potassium, and especially sodium in the upper atmosphere. { 'al·kə,lī i'mish·ən }

alkali feldspar [MINERAL] A feldspar composed of potassium feldspar and sodium

9

alkali flat

feldspar, such as orthoclase, microcline, albite, and anorthoclase; all are considered alkali-rich. { 'al·kə,lī 'feld,spar }

alkali flat [GEOL] A level lakelike plain formed by the evaporation of water in a depression and deposition of its fine sediment and dissolved minerals. { 'al·kə,lī ,flat }

alkali-lime index [PETR] The percentage by weight of silica in a sequence of igneous rocks on a variation diagram where the weight percentages of CaO and of K_2O and Na_2O are equal. { 'al·kə,lī 'līm ,in·deks }

alkaline soil [GEOL] Soil containing soluble salts of magnesium, sodium, or the like, and having a pH value between 7.3 and 8.5. { 'al·kə,līn 'sȯil }

alkali soil [GEOL] A soil, with salts injurious to plant life, having a pH value of 8.5 or higher. { 'al·kə,lī ,sȯil }

alkenones [GEOL] Long-chain (37–39 carbon atoms) di-, tri-, and tetraunsaturated methyl and ethyl ketones produced by certain phytoplankton (coccolithophorids), which biosynthetically control the degree of unsaturation (number of carbon-carbon double bonds) in response to the water temperature; the survival of this temperature signal in marine sediment sequences provides a temporal record of sea surface temperatures that reflect past climates. { 'al·kə,nōnz }

allactite [MINERAL] $Mn_7(AsO_4)_2(OH)_8$ A brownish-red mineral consisting of a basic manganese arsenate. { ə'lak,tīt }

allalinite [PETR] An altered gabbro with original texture and euhedral pseudomorphs. { ə'lal·ə,nīt }

allanite [MINERAL] $(Ca,Ce,La,Y)_2(Al,Fe)_3Si_3O_{12}(OH)$ Monoclinic mineral distinguished from all other members of the epidote group of silicates by a relatively high content of rare earths. Also known as bucklandite; cerine; orthite; treanorite. { 'al·ə,nīt }

allcharite [MINERAL] A lead gray mineral, supposed to be a lead arsenic sulfide and known only crystallographically as orthorhombic crystals. { 'ȯl·kə,rīt }

alleghanyite [MINERAL] $Mn_5(SiO_4)_2(OH)_2$ A pink mineral consisting of basic manganese silicate. { ¦al·ə¦gā·nē,īt }

Alleghenian [GEOL] Lower Middle Pennsylvanian geologic time. { ¦al·ə¦gān·ē·ən }

Alleghenian orogeny [GEOL] Pennsylvanian and Early Permian orogenic episode which deformed the rocks of the Appalachian Valley and the Ridge and Plateau provinces. { ¦al·ə¦gān·ē·ən ȯ'räj·ə·nē }

allemontite [MINERAL] AsSb Rhombohedric, gray or reddish, native antimony aresenide occurring in reniform masses. Also known as arsenical antimony. { ,al·ə'män,tīt }

Allende meteorite [GEOL] A meteorite that fell in Mexico in 1969 and contains inclusions that have been radiometrically dated at 4.56×10^9 years, the oldest found so far, presumably indicating the time of formation of the first solid bodies in the solar system. { ai¦yen·de 'mēd·ē·ə,rīt }

allevardite See rectorite. { ,al·ə'vär,dīt }

allivalite [PETR] A form of gabbro composed of anorthite and olivine; accessories are augite, apatite, and opaque iron oxides. { 'al·ə·və,līt }

allochem [GEOL] Sediment formed by chemical or biochemical precipitation within a depositional basin; includes intraclasts, oolites, fossils, and pellets. { 'a·lō,kem }

allochemical metamorphism [PETR] Metamorphism accompanied by addition or removal of material so that the bulk chemical composition of the rock is changed. { ,a·lō'kem·ə·kəl ,med·ə'mȯr,fiz·əm }

allochetite [PETR] A porphyritic igneous rock composed of phenocrysts of labradorite, orthoclase, titanaugite, nepheline, magnetite, and apatite in a groundmass of augite, biotite, magnetite, hornblende, nepheline, and orthoclase. { ,a·lə'ked,īt }

allochthon [GEOL] A rock that was transported a great distance from its original deposition by some tectonic process, generally related to overthrusting, recumbent folding, or gravity sliding. { ə'läk·thən }

allochthonous [PETR] Of rocks whose primary constituents have not been formed in situ. { ə'läk·thə·nəs }

allochthonous coal [GEOL] A type of coal arising from accumulations of plant debris moved from their place of growth and deposited elsewhere. { ə'läk·thə·nəs ,kōl }

10

allogene [GEOL] A mineral or rock that has been moved to the site of deposition. Also known as allothigene; allothogene { 'a·lə,jēn }

allogenic See allothogenic. { |a·lə|jen·ik }

allomorphism See paramorphism. { ,a·lə'mȯr,fiz·əm }

allomorphite [MINERAL] A mineral consisting of barite that is pseudomorphous after anhydrite. { ,a·lə'mȯr,fīt }

allophane [GEOL] $Al_2O_3 \cdot SiO_2 \cdot nH_2O$ A clay mineral composed of hydrated aluminosilicate gel of variable composition; P_2O_5 may be present in appreciable quantity. { 'a·lə,fān }

Allosaurus [PALEON] A carnivorous theropod dinosaur, 40 feet (12 meters) long, and weighing 1.5 tons, from the Late Jurassic Period that had muscular hindlimbs, small forelimbs (with three-fingered hands), and sharp teeth; similar to but smaller than Tyrannosaurus. { ,al·ə'sȯr·əs }

Allotheria [PALEON] A subclass of Mammalia that appeared in the Upper Jurassic and became extinct in the Cenozoic. { ,a·lō'thir·ē·ə }

allothigene See allogene. { ə'läth·ə,jēn }

allothimorph [GEOL] A metamorphic rock constituent which retains its original crystal outlines in the new rock. { ə'läth·ə,mȯrf }

allothogene See allogene. { ə'läth·ə,jēn }

allothogenic [GEOL] Formed from preexisting rocks which have been transported from another location. Also known as allogenic. { ə|läth·ə|jen·ik }

allotrioblast See xenoblast. { ,a·lə'trē·ə,blast }

allotriomorphic [MINERAL] Of minerals in igneous rock not bounded by their own crystal faces but having their outlines impressed on them by the adjacent minerals. Also known as anhedral; xenomorphic. { ə|lä·trē·ə|mȯr·fik }

alluvial [GEOL] **1.** Of a placer, or its associated valuable mineral, formed by the action of running water. **2.** Pertaining to or consisting of alluvium, or deposited by running water. { ə'lüv·ē·əl }

alluvial cone [GEOL] An alluvial fan with steep slopes formed of loose material washed down the slopes of mountains by ephemeral streams and deposited as a conical mass of low slope at the mouth of a gorge. Also known as cone delta; cone of dejection; cone of detritus; debris cone; dry delta; hemicone; wash. { ə'lüv·ē·əl 'kōn }

alluvial dam [GEOL] A sedimentary deposit which is built by an overloaded stream and dams its channel; especially characteristic of distributaries on alluvial fans. { ə'lüv·ē·əl 'dam }

alluvial deposit See alluvium. { ə'lüv·ē·əl di'päz·ət }

alluvial fan [GEOL] A fan-shaped deposit formed by a stream either where it issues from a narrow moutain valley onto a plain or broad valley, or where a tributary stream joins a main stream. { ə'lüv·ē·əl 'fan }

alluvial flat [GEOL] A small alluvial plain having a slope of about 5 to 20 feet per mile (1.5 to 6 meters per 1600 meters) and built of fine sandy clay or adobe deposited during flood. { ə'lüv·ē·əl 'flat }

alluvial ore deposit [GEOL] A deposit in which the valuable mineral particles have been transported and left by a stream. { ə'lüv·ē·əl |ȯr di|päz·ət }

alluvial plain [GEOL] A plain formed from the deposition of alluvium usually adjacent to a river that periodically overflows. Also known as aggraded valley plain; river plain; wash plain; waste plain. { ə'lüv·ē·əl 'plān }

alluvial slope [GEOL] A surface of alluvium which slopes down from mountainsides and merges with the plain or broad valley floor. { ə'lüv·ē·əl 'slōp }

alluvial soil [GEOL] A soil deposit developed on floodplain and delta deposits. { ə'lüv·ē·əl 'sȯil }

alluvial terrace [GEOL] A terraced embankment of loose material adjacent to the sides of a river valley. Also known as built terrace; drift terrace; fill terrace; stream-built terrace; wave-built platform; wave-built terrace. { ə'lüv·ē·əl 'ter·əs }

alluvial valley [GEOL] A valley filled with a stream deposit. { ə'lüv·ē·əl 'val·ē }

alluviation [GEOL] The deposition of sediment by a river. { ə,lüv·ē'ā·shən }

alluvion See alluvium. { ə'lüv·ē·ən }

alluvium [GEOL] The detrital materials that are eroded, transported, and deposited by streams; an important constituent of shelf deposits. Also known as alluvial deposit; alluvion. { ə'lüv·ē·əm }

almandine [MINERAL] $Fe_3Al_2(SiO_4)_3$ A variety of garnet, deep red to brownish red, found in igneous and metamorphic rocks in many parts of world; used as a gemstone and an abrasive. Also known as almandite. { 'al·mən,dēn }

almandite See almandine. { 'al·mən,dīt }

almeriite See natroalunite. { ,al·mə'rē,īt }

alnoite [PETR] A variety of biotite lamprophyres characterized by lepidomelane phenocrysts; it is feldspar-free but contains melitite, perovskite, olivine, and carbonate in the matrix. { 'al·nə,wit }

aloisite [MINERAL] A brown to violet mineral consisting of a hydrous subsilicate of calcium, iron, magnesium, and sodium, and occurring in amorphous masses. { ,a·lə'wis·ē,īt }

Alpides [GEOL] Great east-west structural belt including the Alps of Europe and the Himalayas and related mountains of Asia; mostly folded in Tertiary times. { 'al·pə,dēz }

alpine [GEOL] Similar to or characteristic of a lofty mountain or mountain system. { 'al,pīn }

Alpine orogeny [GEOL] Jurassic through Tertiary orogeny which affected the Alpides. { 'al,pīn ó'räj·ə·nē }

alpine-type facies [PETR] High-pressure, low-temperature (150–400°C) dynamothermal metamorphism characterized by the presence of the pumpellyite and glaucophane schist facies. { ¦al,pīn¦tīp 'fā,shez }

alpinotype tectonics [GEOL] Tectonics of the alpine-type geosynclinal mountain belts characterized by deep-seated plastic folding, plutonism, and lateral thrusting. { al'pē·nō,tīp ,tek'tän·iks }

alsbachite [PETR] A plutonic rock of sodic plagioclase, quartz, and subordinate orthoclase and accessory garnet, biotite, and muscovite; a variety of porphyritic granodiorite. { 'ólz·bä,kīt }

alstonite See bromlite. { 'ólz·tə,nīt }

Altaid orogeny [GEOL] Mountain building in Central Europe and Asia that occurred from the late Carboniferous to the Permian. { ¦al,tād ó'räj·ə·nē }

altaite [MINERAL] PbTe A tin-white lead-tellurium mineral occurring as isometric crystals with tin ores in central Asia. { al'tā,īt }

alteration [PETR] A change in a rock's mineral composition. { ,ól·tə'rā·shən }

altiplanation [GEOL] A phase of solifluction that may be seen as terracelike forms, flattened summits, and passes that are mainly accumulations of loose rock. { ,al·tə·plā'nā·shən }

altiplanation surface [GEOL] A flat area fronted by scarps a few to hundreds of feet in height; the area ranges from several square rods to hundreds of acres. Also known as altiplanation terrace. { ,al·tə·plā'nā·shən ,sər·fəs }

altiplanation terrace See altiplanation surface. { ,al·tə·plā'nā·shən ,ter·əs }

altithermal [GEOPHYS] Period of high temperature, particularly the postglacial thermal optimum. { ¦al·tə¦thər·məl }

Altithermal [GEOL] A dry postglacial interval centered about 5500 years ago during which temperatures were warmer than at present. Also known as Hypsithermal. { ¦al·tə¦thər·məl }

altithermal soil [GEOL] Soil recording a period of rising or high temperature. { ¦al·tə¦thər·məl 'sóil }

alum [MINERAL] $KAl(SO_4)_2 \cdot 12H_2O$ A colorless, white, astringent-tasting evaporite mineral. { 'al·əm }

alum coal [GEOL] Argillaceous brown coal rich in pyrite in which alum is formed on weathering. { 'al·əm ,kōl }

aluminite [MINERAL] $Al_2(SO_4)(OH)_4 \cdot 7H_2O$ Native monoclinic hydrous aluminum sulfate; used in tanning, papermaking, and water purification. Also known as websterite. { ə'lüm·ə,nīt }

12

aluminum ore [GEOL] A natural material from which aluminum may be economically extracted. { ə'lüm·ə·nəm 'ȯr }

alumite See alunite. { 'al·ə,mīt }

alum rock See alunite. { 'al·əm ,räk }

alum schist See alum shale. { 'al·əm ,shist }

alum shale [PETR] A shale containing pyrite that is decomposed by weathering to form sulfuric acid, which acts on potash and alumina constituents to form alum. Also known as alum schist; alum slate. { 'al·əm ,shāl }

alum slate See alum shale. { 'al·əm ,slāt }

alumstone See alunite. { 'al·əm,stōn }

alunite [MINERAL] KAl₃(SO₄)₂(OH)₆ A mineral composed of a basic potassium aluminum sulfate; it occurs as a hydrothermal-alteration product in feldspathic igneous rocks and is used in the manufacture of alum. Also known as alumite; alum rock; alumstone. { 'al·yə,nīt }

alunitization [GEOL] Introduction of or replacement by alunite. { ,al·yə·nə·tə'zā·shən }

alunogen [MINERAL] Al₂(SO₄)₃·18H₂O A white mineral occurring as a fibrous incrustation of hydrated aluminum sulfate by volcanic action or decomposition of pyrite. Also known as feather alum; hair salt. { ə'lün·ə·jən }

alurgite [MINERAL] A purple manganiferous variety of muscovite mica. { ə'lür,jīt }

alyphite [GEOL] Bitumen that yields a high percentage of open-chain aliphatic hydrocarbons upon distillation. { 'al·ə,fīt }

amalgam [MINERAL] A silver mercury alloy occurring in nature. { ə'mal·gəm }

amarantite [MINERAL] Fe(SO₄)(OH)·3H₂O An amaranth red to brownish- or orange-red triclinic mineral consisting of a hydrated basic sulfate of ferric iron. { ,a·mə'ran,tīt }

amarillite [MINERAL] NaFe(SO₄)₂·6H₂O A pale greenish-yellow mineral consisting of a hydrous sodium ferric sulfate. { ,a·mə'ri,līt }

amazonite [MINERAL] An apple-green, bright-green, or blue-green variety of microcline found in the United States and the former Soviet Union; sometimes used as a gemstone. Also known as amazon stone. { ¦a·mə¦zō,nīt }

amazon stone See amazonite. { 'a·mə·zän ,stōn }

ambatoarinite [MINERAL] A mineral consisting of a carbonate of cerium metals and strontium. { ,am·bə,tō'ä·rə,nīt }

amber [MINERAL] A transparent yellow, orange, or reddish-brown fossil resin derived from a coniferous tree; used for ornamental purposes; it is amorphous, has a specific gravity of 1.05–1.10, and a hardness of 2–2.5 on Mohs scale. { 'am·bər }

amberoid [MINERAL] A gem-quality mineral composed of small fragments of amber that have been reunited by heat or pressure. { 'am·bə,rȯid }

ambient stress field [GEOPHYS] The distribution and numerical value of the stresses present in a rock environment prior to its disturbance by man. Also known as in-place stress field; primary stress field; residual stress field. { 'am·bē·ənt 'stres ,fēld }

amblygonite [MINERAL] (Li,Na)AlPO₄(F,OH) A mineral occurring in white or greenish cleavable masses and found in the United States and Europe; important ore of lithium. { am'bli·gə,nīt }

ambonite [PETR] Any of a group of hornblende-biotite andesites and dacites containing cordierite. { 'am·bə,nīt }

ambrite [MINERAL] A yellow-gray, semitransparent fossil resin resembling amber; found in large masses in New Zealand coal fields and regarded as a semiprecious stone. { 'am,brīt }

ambrosine [MINERAL] A yellowish to clove-brown variety of amber rich in succinic acid; occurs as rounded masses in phosphate beds near Charleston, South Carolina. { 'am·brə,zēn }

Amebelodontinae [PALEON] A subfamily of extinct elephantoid proboscideans in the family Gomphotheriidae. { ,a·mə,bel·ə'dän·tə,nē }

amemolite [GEOL] A stalactite with one or more changes in its axis of growth. { ə'mem·ə,līt }

American jade See californite. { ə'mer·ə·kən 'jād }

amesite [MINERAL] $(Mg,Fe)_4Al_4Si_2O_{10}(OH)_8$ An apple-green phyllosilicate mineral occurring in foliated hexagonal plates. { 'am,zīt }

amethyst [MINERAL] The transparent purple to violet variety of the mineral quartz; used as a jeweler's stone. { 'am·ə,thist }

amherstite [PETR] A syenodiorite containing andesine and antiperthite. { 'a·mər,stīt }

amianthus [MINERAL] A fine, silky variety of asbestos, such as chrysotile. { ,a·mē'an·thəs }

amino acid dating [GEOCHEM] Relative or absolute age determination of materials by measuring the degree of racemization of certain amino acids, which generally increases with geologic age. { ə,mē·nō ¦as·əd ¦dā·diŋ }

Ammanian [GEOL] Middle Upper Cretaceous geologic time. { ,ä'man·ē·ən }

ammonioborite [MINERAL] $(NH_4)_2B_{10}O_{16}\cdot5H_2O$ A white mineral consisting of a hydrous ammonium borite and occurring as aggregates of minute plates. { ə,mōn·ē·ō'bȯr,īt }

ammoniojarosite [MINERAL] $(NH_4)Fe_3(SO_4)_2(OH)_6$ Pale-yellow mineral consisting of basic ferric ammonium sulfate. { ə,mōn·ē·ō·jə'rō,sīt }

ammonite [PALEON] A fossil shell of the cephalopod order Ammonoidea. { 'a·mə,nīt }

ammonoid [PALEON] A cephalopod of the order Ammonoidea. { 'a·mə,nȯid }

Ammonoidea [PALEON] An order of extinct cephalopod mollusks in the subclass Tetrabranchia; important as index fossils. { ,a·mə'nȯid·ē·ə }

amoeboid fold [GEOL] A fold or structure, such as an anticline, having no prevailing trend or definite shape. { ə'mē,bȯid 'fōld }

amorphous mineral [MINERAL] A mineral without definite crystalline structure. { ə'mȯr·fəs 'min·rəl }

amorphous peat [GEOL] Peat composed of fine grains of organic matter; it is plastic like wet, heavy soil, with all original plant structures destroyed by decomposition of cellulosic matter. { ə'mȯr·fəs 'pēt }

amosite [MINERAL] A monoclinic amphibole form of asbestos having long fibers and a high iron content; used in insulation. { 'am·ə,zīt }

ampangabeite See samarskite. { ,äm,päŋ'gä·bē,īt }

ampelite [PETR] A graphite schist containing silica, alumina, and sulfur; used as a refractory. { 'am·pə,līt }

amphibole [MINERAL] Any of a group of rock-forming, ferromagnesian silicate minerals commonly found in igneous and metamorphic rocks; includes hornblende, anthophyllite, tremolite, and actinolite (asbestos minerals). { 'am·fə,bōl }

amphibolite [PETR] A crystalloblastic metamorphic rock composed mainly of amphibole and plagioclase; quartz may be present in small quantities. { am'fib·ə,līt }

amphibolite facies [PETR] Rocks produced by medium- to high-grade regional metamorphism. { am'fib·ə,līt 'fā,shēz }

amphibolization [PETR] Formation of amphibole in a rock as a secondary mineral. { am,fib·ə·lə'zā·shən }

Amphichelydia [PALEON] A suborder of Triassic to Eocene anapsid reptiles in the order Chelonia; these turtles did not have a retractable neck. { ,am·fə·kə'lid·ē·ə }

Amphicyonidae [PALEON] A family of extinct giant predatory carnivores placed in the infraorder Miacoidea by some authorities. { ¦am·fə·sī¦än·ə,dē }

amphigene See leucite. { 'am·fə,jēn }

Amphilestidae [PALEON] A family of Jurassic triconodont mammals whose subclass is uncertain. { ,am·fə'les·tə,dē }

Amphimerycidae [PALEON] A family of late Eocene to early Oligocene tylopod ruminants in the superfamily Amphimerycoidea. { ,am·fə·mə'ris·ə,dē }

Amphimerycoidea [PALEON] A superfamily of extinct ruminant artiodactyls in the infraorder Tylopoda. { ,am·fə,mir·ə'kȯid·ē·ə }

amphimorphic [GEOL] A rock or mineral formed by two geologic processes. { ,am·fə'mȯr·fik }

amphisapropel [GEOL] Cellulosic ooze containing coarse plant debris. { ,am¦fīz·ə¦prō,pel }

Amphissitidae [PALEON] A family of extinct ostracods in the suborder Beyrichicopina. { ¦am·fə¦sid·ə,dē }

14

Amphitheriidae [PALEON] A family of Jurassic therian mammals in the infraclass Pantotheria. { ‚am·tə·thə'rī·ə‚dē }

amphoterite [GEOL] A stony meteorite containing bronzite and olivine with some oligoclase and nickel-rich iron. { am'fäd·ə‚rīt }

amygdaloid [GEOL] Lava rock containing amygdules. Also known as amygdaloidal lava. { ə'mig·də‚lȯid }

amygdaloidal lava *See* amygdaloid. { ə'mig·də‚lȯid·əl'läv·ə }

amygdule [GEOL] **1.** A mineral filling formed in vesicles (cavities) of lava flows; it may be chalcedony, opal, calcite, chlorite, or prehnite. **2.** An agate pebble. { ə'mig‚dyül }

Amynodontidae [PALEON] A family of extinct hippopotamuslike perissodactyl mammals in the superfamily Rhinoceratoidea. { ‚a·mə·nə'dän·tə‚dē }

anabohitsite [PETR] A variety of olivine-pyroxenite containing hornblende and hypersthene and a high proportion (about 30%) of magnetite and ilmenite. { ‚an·ə·bō'hit‚sīt }

anaclinal [GEOL] Having a downward inclination opposite to that of a stratum. { ¦an·ə¦klīn·əl }

anaerobic sediment [GEOL] A highly organic sediment formed in the absence or near absence of oxygen in water that is rich in hydrogen sulfide. { ¦an·ə¦rōb·ik 'sed·ə·mənt }

analbite [MINERAL] A triclinic albite which is not stable and becomes monoclinic at about 700°C. { ə'nal‚bīt }

analcime [MINERAL] NaAlSi$_2$O$_6$·H$_2$O A white or slightly colored isometric zeolite found in diabase and in alkali-rich basalts. Also known as analcite. { ə'nal‚sēm }

analcimite [PETR] An extrusive or hypabyssal rock that consists primarily of pyroxene and analcime. { ə'nal·sə‚mīt }

analcimization [GEOL] The replacement in igneous rock of feldspars or feldspathoids by analcime. { ə¦nal·sə·mə¦zā·shən }

analcite *See* analcime. { ə'nal‚sīt }

analytical geomorphology *See* dynamic geomorphology. { ‚an·əl'id·ə·kəl ‚jē·ō‚mȯr'fäl·ə·jē }

anamigmatism [GEOL] A process of high-temperature, high-pressure remelting of sediment to yield magma. { ‚an·ə'mig·mə‚tiz·əm }

anamorphic zone [GEOL] The zone of rock flow, as indicated by reactions that may involve decarbonation, dehydration, and deoxidation; silicates are built up, and the formation of denser minerals and of compact crystalline structure takes place. { ¦an·ə¦mȯr·fik 'zōn }

anamorphism [GEOL] A kind of metamorphism at considerable depth in the earth's crust and under great pressure, resulting in the formation of complex minerals from simple ones. { ‚an·ə'mȯr·fiz·əm }

Anancinae [PALEON] A subfamily of extinct proboscidean placental mammals in the family Gomphotheriidae. { ə'nan·sə‚nē }

anapaite [MINERAL] Ca$_2$Fe(PO$_4$)$_2$·4H$_2$O A pale-green or greenish-white triclinic mineral consisting of a ferrous iron hydrous phosphate and occurring in crystals and massive forms; hardness is 3–4 on Mohs scale, and specific gravity is 3.81. { ə'nap·ə‚īt }

anapeirean *See* Pacific suite. { ‚an·ə'pir·ē·ən }

Anaplotheriidae [PALEON] A family of extinct tylopod ruminants in the superfamily Anaplotherioidea. { ‚an·ə‚pläth·ə'rī·ə‚dē }

Anaplotherioidea [PALEON] A superfamily of extinct ruminant artiodactyls in the infraorder Tylopoda. { ‚an·ə‚pläth·ə‚rē'ȯid·ē·ə }

Anasca [PALEON] A suborder of extinct bryozoans in the order Cheilostomata. { ə'nas·kə }

anaseism [GEOPHYS] Movement of the earth in a direction away from the focus of an earthquake. { ¦an·ə¦sīz·əm }

Anaspida [PALEON] An order of extinct fresh- or brackish-water vertebrates in the class Agnatha. { ə'nas·pə·də }

anatase [MINERAL] The brown, dark-blue, or black tetragonal crystalline form of titanium dioxide, TiO_2; used to make a white pigment. Also known as octahedrite. { 'an·ə,tās }

anatexis [GEOL] A high-temperature process of metamorphosis by which plutonic rock in the lowest levels of the crust is melted and regenerated as a magma. { ,an·ə'tek·səs }

anathermal [GEOL] A period of time between the age of other strata or units of reference in which the temperature is increasing. { ,an·ə'thər·məl }

anauxite [MINERAL] $Al_2(SiO_7)(OH)_4$ A clay mineral that is a mixture of kaolinite and quartz. Also known as ionite. { ə'nȯk,sīt }

anchieutectic [GEOL] A type of magma which is incapable of undergoing further notable main-stage differentiation because its mineral composition is practically in eutectic proportions. { ¦aŋ·kē·yü¦tek·tik }

anchimonomineralic [PETR] Of rock composed mostly of one kind of mineral. { ¦aŋ·kē,män·ō,min·ə¦ral·ik }

anchored dune [GEOL] A sand dune stabilized by growth of vegetation. { 'aŋ·kərd 'dün }

anchorite [PETR] A variety of diorite having nodules of mafic minerals and veins of felsic minerals. { 'aŋ·kə,rīt }

anchor stone [GEOL] A rock or pebble that has marine plants attached to it. { 'aŋ·kər ,stōn }

ancylite [MINERAL] $SrCe(CO_3)_2(OH)·H_2O$ A mineral consisting of hydrous basic carbonate of cerium and strontium. { 'an·sə,līt }

ancylopoda [PALEON] A suborder of extinct herbivorous mammals in the order Perissodactyla. { ,an·sə'lä·pə·də }

andalusite [MINERAL] Al_2SiO_5 A brown, yellow, green, red, or gray neosilicate mineral crystallizing in the orthorhombic system, usually found in metamorphic rocks. { ¦an·də'lü,sīt }

Andean-type continental margin [GEOL] A continental margin, as along the Pacific coast of South America, where oceanic lithosphere descends beneath an adjacent continent producing andesitic continental margin volcanism. { 'an·dē·ən ,tīp ,känt·ən'ent·əl 'mär·jən }

Andept [GEOL] A suborder of the soil order Inceptisol, formed chiefly in volcanic ash or in regoliths with high components of ash. { ¦an¦dept }

andersonite [MINERAL] $Na_2Ca(UO_2)(CO_3)_3·6H_2O$ Bright yellow-green secondary mineral consisting of a hydrous sodium calcium uranium carbonate. { 'an·dər·sən,īt }

andesine [MINERAL] A plagioclase feldspar with a composition ranging from $Ab_{70}An_{30}$ to $Ab_{50}An_{50}$, where $Ab = NaAlSi_3O_8$ and $An = CaAl_2Si_2O_8$; it is a primary constituent of intermediate igneous rocks, such as andesites. { 'an·də,zēn }

andesite [PETR] Very finely crystalline extrusive rock of volcanic origin composed largely of plagioclase feldspar (oligoclase or andesine) with smaller amounts of dark-colored mineral (hornblende, biotite, or pyroxene), the extrusive equivalent of diorite. { 'an·də,zīt }

andesite line [GEOL] The postulated geographic and petrographic boundary between the andesite-dacite-rhyolite rock association of the margin of the Pacific Ocean and the olivine-basalt-trachyte rock association of the Pacific Ocean basin. { 'an·də,zīt ,līn }

andesitic glass [GEOL] A natural glass that is chemically equivalent to andesite. { 'an·də,zīt·ik ,glas }

andorite [MINERAL] $AgPbSb_3S_6$ A dark-gray or black orthorhombic mineral. Also known as sundtite. { 'an·də,rīt }

andradite [MINERAL] The calcium-iron end member of the garnet group. { an'drä,dīt }

andrewsite [MINERAL] $(Cu,Fe^{2+})Fe_3^{3+}(PO_4)_3(OH)_2$ A bluish-green mineral consisting of a basic phosphate of iron and copper. { 'an·drü,zīt }

andrite [GEOL] A meteorite composed principally of augite with some olivine and troilite. { 'an,drīt }

anemoclast [GEOL] A clastic rock that was fragmented and rounded by wind. { ¦a·nə·mō¦klast }

anemoclastic [GEOL] Referring to rock that was broken by wind erosion and rounded by wind action. { ¦a·nə·mō¦klas·tik }

angaralite [MINERAL] $Mg_2(Al,Fe)_{10}Si_6O_{29}$ A mineral of the chlorite group, occurring in thin black plates. { an'gar·ə,līt }

Angara Shield [GEOL] A shield area of crystalline rock in Siberia. { ¸äŋ·gə'rä ¸shēld }

angle of dip See dip. { 'aŋ·gəl əv 'dip }

angle of shear [GEOL] The angle between the planes of maximum shear which is bisected by the axis of greatest compression. { 'aŋ·gəl əv 'shēr }

anglesite [MINERAL] $PbSO_4$ A mineral occurring in white or gray, tabular or prismatic orthorhombic crystals or compact masses. Also known as lead spar; lead vitriol. { 'aŋ·glə,sīt }

Angoumian [GEOL] Upper middle Upper Cretaceous (Upper Turonian) geologic time. { ¸än'güm·ē·ən }

angrite [GEOL] An achondrite stony meteorite composed principally of augite with a little olivine and troilite. { 'aŋ,grīt }

anguclast [GEOL] An angular phenoclast. { 'aŋ·gyü,klast }

angular unconformity [GEOL] An unconformity in which the older strata dip at a different angle (usually steeper) than the younger strata. { 'aŋ·gyə·lər ¸ən·kən'fȯrm·əd·ē }

anhedral See allotriomorphic. { an'hēd·rəl }

anhedron [PETR] Rock that has the organized internal structure of a crystal without the external geometric form of a crystal. { an'hēd·rən }

anhydrite [MINERAL] $CaSO_4$ A mineral that represents gypsum without its water of crystallization, occurring commonly in white and grayish granular to compact masses; the hardness is 3–3.5 on Mohs scale, and specific gravity is 2.90–2.99. Also known as cube spar. { an'hī,drīt }

anhydrite evaporite [PETR] $CuSO_4$ A sedimentary rock composed chiefly of copper sulfate in compact granular form deposited by evaporation of water; resembles marble and differs from gypsum in lack of water of hydration and hardness. { an'hī,drīt i'vap·ə,rīt }

anhydrock [PETR] A sedimentary rock chiefly made of anhydrite. { an'hi,dräk }

Animikean [GEOL] The middle subdivision of Proterozoic geologic time. Also known as Penokean; Upper Huronian. { ə¦nim·ə¦kē·ən }

animikite [GEOL] An ore of silver, composed of a mixture of sulfides, arsenides, and antimonides, and containing nickel and lead; occurs in white or gray granular masses. { ə'nim·ə,kīt }

Anisian [GEOL] Lower Middle Triassic geologic time. { ə'nis·ē·ən }

anisodesmic [MINERAL] Pertaining to crystals or compounds in which the ionic bonds are unequal in strength. { ¸a,nis·ə'dez·mik }

ankaramite [PETR] A mafic olivine basalt primarily composed of pyroxene with smaller amounts of olivine and plagioclase and accessory biotite, apatite, and opaque oxides. { 'aŋ·kə'rä,mīt }

ankaratrite See olivine nephelinite. { ¸aŋ·kə'rä,trīt }

ankerite [MINERAL] $Ca(Fe,Mg,Mn)(CO_3)_2$ A white, red, or gray iron-rich carbonate mineral associated with iron ores and found in thin veins in coal seams; specific gravity is 2.95–3.1. Also known as cleat spar. { 'aŋ·kə,rīt }

Ankylosauria [PALEON] A suborder of Cretaceous dinosaurs in the reptilian order Ornithischia characterized by short legs and flattened, heavily armored bodies. { ¦aŋ·kə·lə'sȯr·ē·ə }

annabergite [MINERAL] $(Ni,Co)_3(AsO_4)_2·8H_2O$ A monoclinic mineral usually found as apple-green incrustations as an alteration product of nickel arsenides; it is isomorphous with erythrite. Also known as nickel bloom; nickel ocher. { 'a·nə,bər,gīt }

annual layer [GEOL] **1.** A sedimentary layer deposited, or presumed to have been deposited, during the course of a year; for example, a glacial varve. **2.** A dark layer in a stratified salt deposit containing disseminated anhydrite. { 'an·yə·wəl 'lā·ər }

annual magnetic change See magnetic annual change. { 'an·yə·wəl ¸mag'ned·ik 'chānj }

17

annual magnetic variation *See* magnetic annual variation. { 'an·yə·wəl ˌmag'ned·ik ver·ē'ā·shən }

annual variation [GEOPHYS] A component in the change with time in the earth's magnetic field at a specified location that has a period of 1 year. { 'an·yə·wəl ver·ē'ā·shən }

anomalous magma [GEOL] Magma formed or obviously changed by assimilation. { ə'näm·ə·ləs 'mag·mə }

anomaly [GEOL] A local deviation from the general geological properties of a region. { ə'näm·ə·lē }

anomite [MINERAL] A variety of biotite different only in optical orientation. { 'an·ə‚mīt }

Anomphalacea [PALEON] A superfamily of extinct gastropod mollusks in the order Aspidobranchia. { ə‚näm·fə' lāsh·ə }

anorogenic [GEOL] Of a feature, forming during tectonic quiescence between orogenic periods, that is, lacking in tectonic disturbance. { ¦a‚nó·rō¦jen·ik }

anorogenic time [GEOL] Geologic time when no significant deformation of the crust occurred. { ¦a‚nó·rō¦jen·ik 'tīm }

anorthite [MINERAL] The white, grayish, or reddish calcium-rich end member of the plagioclase feldspar series; composition ranges from $Ab_{10}An_{90}$ to Ab_0An_{100}, where $Ab = NaAlSi_3O_8$ and $An = CaAl_2Si_2O_8$. Also known as calciclase; calcium feldspar. { ə'nór‚thīt }

anorthite-basalt [PETR] A rock composed of a basic variety of basalt with anorthite instead of labradorite. { ə'nór‚thīt bə'sólt }

anorthoclase [MINERAL] A triclinic alkali feldspar having a chemical composition ranging from $Or_{40}Ab_{60}$ to $Or_{10}Ab_{90}$ to about 20 mole % An, where $Or = KAlSi_3O_8$, $Ab = NaAlSi_3O_8$, and $An = CaAl_2Si_2O_8$. Also known as anorthose; soda microcline. { ə'nór·thə‚klās }

anorthose *See* anorthoclase. { ə'nór‚thōs }

anorthosite [PETR] A visibly crystalline plutonic rock composed almost entirely of plagioclase feldspar (andesine to anorthite) with minor amounts of pyroxene and olivine. { ə'nór·thə‚sīt }

anorthositization [GEOL] A process of anorthosite formation by replacement or metasomatism. { ə¦nór·thə‚sid·ə'zā·shən }

antecedent platform [GEOL] A submarine platform 165 feet (50 meters) or more below sea level from which barrier reefs and atolls are postulated to grow toward the water's surface. { ‚ant·ə'sēd·ənt 'plat‚fórm }

antecedent valley [GEOL] A stream valley that existed before uplift, faulting, or folding occurred and which has maintained itself during and after these events. { ‚ant·ə'sēd·ənt 'val·ē }

antediluvial [GEOL] Formerly referred to time or deposits antedating Noah's flood. { ¦an·tē·də¦lüv·ē·əl }

antetheca [PALEON] The last or exposed septum at any stage of fusulinid growth. { ¦an·tē¦thek·ə }

Anthocyathea [PALEON] A class of extinct marine organisms in the phylum Archaeocyatha characterized by skeletal tissue in the central cavity. { ‚an·thə‚sī'ā·thē·ə }

anthodite [GEOL] Gypsum or aragonite growing in clumps of long needle- or hairlike crystals on the roof or wall of a cave. { 'an·thə‚dīt }

anthoinite [MINERAL] $Al_2W_2O_9·3H_2O$ A white mineral consisting of a hydrous basic aluminum tungstate. { ‚an'thói‚nīt }

anthophyllite [MINERAL] A clove-brown orthorhombic mineral of the amphibole group, a variety of asbestos occurring as lamellae, radiations, fibers, or massive in metamorphic rocks. Also known as bidalotite. { ‚an·thō'fi‚līt }

anthracite [MINERAL] A high-grade metamorphic coal having a semimetallic luster, high content of fixed carbon, and high density, and burning with a short blue flame and little smoke or odor. Also known as hard coal; Kilkenny coal; stone coal. { 'an·thrə‚sīt }

anthracitization |GEOCHEM| The natural process by which bituminous coal is transformed into anthracite coal. { ,an·thrə,sīd·ə'zā·chən }

Anthracosauria |PALEON| An order of Carboniferous and Permian labyrinthodont amphibians that includes the ancestors of living reptiles. { ,an·thrə·kə'sȯr,ē·ə }

Anthracotheriidae |PALEON| A family of middle Eocene and early Pleistocene artiodactyl mammals in the superfamily Anthracotherioidea. { ,an·thrə·kə·thə'rī·ə,dē }

Anthracotherioidea |PALEON| A superfamily of extinct artiodactyl mammals in the suborder Paleodonta. { 'an·thrə·kə·thə,rī'ȯid·ē·ə }

anthracoxene |GEOL| A brownish resin that occurs in brown coal; in ether it dissolves into an insoluble portion, anthrocoxenite, and a soluble portion, schlanite. { ,an·thrə'käk,sēn }

anthraxolite |GEOL| Anthracite-like asphaltic material occurring in veins in Precambrian slate of Sudbury District, Ontario. { an'thrak·sə,līt }

anthraxylon |GEOL| The vitreous-appearing components of coal that are derived from the woody tissues of plants. { an'thrak·sə,län }

Antiarchi |PALEON| A division of highly specialized placoderms restricted to freshwater sediments of the Middle and Upper Devonian. { ,an·tē'är,kī }

anticenter |GEOL| The point on the surface of the earth that is diametrically opposite the epicenter of an earthquake. Also known as antiepicenter. { ¦an·tē'sent·ər }

anticlinal |GEOL| Folded as in an anticline. { ¦an·tē¦klīn·əl }

anticlinal axis |GEOL| The median line of a folded structure from which the strata dip on either side. { ¦an·tē¦klīn·əl 'ak·səs }

anticlinal bend |GEOL| An upwardly convex flexure of rock strata in which one limb dips gently toward the apex of the strata and the other dips steeply away from it. { ¦an·tē¦klīn·əl 'bend }

anticlinal mountain |GEOL| Ridges formed by a convex flexure of the strata. { ¦an·tē¦klīn·əl 'maun·tən }

anticlinal theory |GEOL| A theory relating trapped underground oil accumulation to anticlinal structures. { ¦an·tē¦klīn·əl 'thē·ə·rē }

anticlinal trap |GEOL| A formation in the top of an anticline in which petroleum has accumulated. { ¦ant·i¦klīn·əl 'trap }

anticlinal valley |GEOL| A valley that follows an anticlinal axis. { ¦an·tē¦klīn·əl 'val·ē }

anticline |GEOL| A fold in which layered strata are inclined down and away from the axes. { 'an·ti,klīn }

anticlinorium |GEOL| A series of anticlines and synclines that form a general arch or anticline. { ,an·ti,klī'nȯr·ē·əm }

antidune |GEOL| A temporary form of ripple on a stream bed analogous to a sand dune but migrating upcurrent. { 'an·tē,dün }

antiepicenter See anticenter. { ,an·tē'ep·i,sent·ər }

antiform |GEOL| An anticline-like structure whose stratigraphic sequence is not known. { 'an·tē,fȯrm }

antigorite |MINERAL| $Mg_3Si_2O_5(OH)_4$ Brownish-green variety of the mineral serpentine. Also known as baltimorite; picrolite. { an'tig·ə,rīt }

antimonite |MINERAL| Sb_2S_3 A lead-gray antimony sulfide mineral, the primary source of antimony; sometimes contains gold or silver; has a brilliant metallic luster, and occurs as prismatic orthorhombic crystals in massive forms. Also known as antimony glance; gray antimony; stibium; stibnite. { 'an·tə·mə,nīt }

antimony |MINERAL| A very brittle, tin-white, hexagonal mineral, the native form of the element. { 'an·tə,mō·nē }

antimony blende See kermesite. { 'an·tə,mō·nē 'blend }

antimony glance See antimonite. { 'an·tə,mō·nē 'glans }

antiperthite |GEOL| Natural intergrowth of feldspars formed by separation of sodium feldspar (albite) and potassium feldspar (orthoclase) during slow cooling of molten mixtures; the potassium-rich phase is evolved in a plagioclase host, exactly the inverse of perthite. { ,an·ti'pər,thīt }

antistress mineral |MINERAL| Minerals such as leucite, nepheline, alkalic feldspar, andalusite, and cordierite which cannot form or are unstable in an environment of high shearing stress, and hence are not found in highly deformed rocks. { ¦an·tē¦stres ˌmin·ə·rəl }

antlerite [MINERAL] $Cu_3SO_4(OH)_4$ Emerald- to blackish-green mineral occurring in aggregates of needlelike crystals; an ore of copper. Also known as vernadskite. { 'ant·ləˌrīt }

Antler orogeny [GEOL] Late Devonian and Early Mississippian orogeny in Nevada, resulting in the structural emplacement of eugeosynclinal rocks over microgeosynclinal rocks. { 'ant·lər ȯ'räj·ə·nē }

Ao horizon [GEOL] That portion of the A horizon of a soil profile which is composed of pure humus. { ¦ā¦ō hə'rīz·ən }

Aoo horizon [GEOL] Uppermost portion of the A horizon of a soil profile which consists of undecomposed vegetable litter. { ¦ā¦ō¦ō hə'rīz·ən }

Apatemyidae [PALEON] A family of extinct rodentlike insectivorous mammals belonging to the Proteutheria. { əˌpad·ə'mī·əˌdē }

apachite [PETR] A phonolite consisting of enigmatite and hornblende in about the same quantity as the pyroxene, but of a later crystallization phase. { ə'paˌchīt }

Apathornithidae [PALEON] A family of Cretaceous birds, with two species, belonging to the order Ichthyornithiformes. { ˌa·pəˌthȯr'nith·əˌdē }

apatite [MINERAL] A group of phosphate minerals that includes 10 mineral species and has the general formula $X_5(YO_4)_3Z$, where X is usually Ca^{2+} or Pb^{3+}, Y is P^{5+} or As^{5+}, and Z is F^-, Cl^-, or OH^-. { 'ap·əˌtīt }

Apatosaurus [PALEON] A herbivorous sauropod dinosaur, approximately 70 feet (21 meters) long and weighing 30 tons, from the Jurassic Period that had much longer hindlimbs than forelimbs. Also known as Brontosaurus. { əˌpad·ə'sȯr·əs }

apex [GEOL] The part of a mineral vein nearest the surface of the earth. { 'āˌpeks }

aphaniphyric [PETR] Denoting a texture of porphyritic rocks with microaphanitic groundmasses. Also known as felsophyric. { ˌaf·ə·nə'fīr·ik }

aphanite [PETR] **1.** A general term applied to dense, homogeneous rocks whose constituents are too small to be distinguished by the unaided eye. **2.** A rock having aphanitic texture. { 'af·əˌnīt }

aphanitic [PETR] Referring to the texture of an igneous rock in which the crystalline components are not distinguishable by the unaided eye. { ˌaf·ə'nid·ik }

Aphrosalpingoidea [PALEON] A group of middle Paleozoic invertebrates classified with the calcareous sponges. { ¦af·rōˌsalˌpiŋ'gȯid·ē·ə }

aphrosiderite See ripidolite. { ˌaf·rō'sid·əˌrīt }

aphthitalite [MINERAL] $(K,Na)_3Na(SO_4)_2$ A white mineral crystallizing in the rhombohedral system and occurring massively or in crystals. { ˌaf'thid·əlˌīt }

aphyric [PETR] Of the texture of fine-grained igneous rocks, showing two generations of the same mineral but without phenocrysts. { ā'fir·ik }

apjohnite [MINERAL] $MnAl_2(SO_4)_4 \cdot 22H_2O$ A white, rose-green, or yellow mineral containing water and occurring in crusts, fibrous masses, or efflorescences. { 'apˌjäˌnīt }

aplite [PETR] Fine-grained granitic dike rock made up of light-colored mineral constituents, mostly quartz and feldspar; used to manufacture glass and enamel. { 'aˌplīt }

apophyllite [MINERAL] A hydrous calcium potassium silicate containing fluorine and occurring as a secondary mineral with zeolites with geodes and other igneous rocks; the composition is variable but approximates $KFCa_4(Si_2O_5)_4 \cdot 8H_2O$. Also known as fish-eye stone. { ə'päf·əˌlīt }

Appalachia [GEOL] Proposed borderland along the southeastern side of North America, seaward of the Appalachian geosyncline in Paleozoic time. { ¦ap·ə¦lā·chə }

Appalachian orogeny [GEOL] An obsolete term referring to Late Paleozoic diastrophism beginning perhaps in the Late Devonian and continuing until the end of the Permian; now replaced by Alleghenian orogeny. { ¦ap·ə¦lā·chən ȯ'räj·ə·nē }

apparent cohesion [GEOL] In soil mechanics, the resistance of particles to being pulled apart due to the surface tension of the moisture film surrounding each particle. Also known as film cohesion. { ə'pa·rənt ˌkō'hē·zhən }

apparent dip [GEOL] Dip of a rock layer as it is exposed in any section not at a right angle to the strike. { ə'pa·rənt 'dip }

20

apparent movement of faults [GEOL] The apparent motion observed to have occurred in any chance section across a fault. { ə'pa·rənt ¦müv·mənt əv ¦fólts }

apparent plunge [GEOL] Inclination of a normal projection of lineation in the plane of a vertical cross section. { ə'pa·rənt 'plənj }

apparent precession *See* apparent wander. { ə'pa·rənt pri'sesh·ən }

apparent vertical [GEOPHYS] The direction of the resultant of gravitational and all other accelerations. Also known as dynamic vertical. { ə'pa·rənt 'verd·ə·kəl }

apparent wander [GEOPHYS] Apparent change in the direction of the axis of rotation of a spinning body, such as a gyroscope, due to rotation of the earth. Also known as apparent precession; wander. { ə'pa·rənt 'wän·dər }

appinite [PETR] Hornblende-rich plutonic rock with high feldspar content. { 'ap·ə,nīt }

apple coal [GEOL] Easily mined soft coal that breaks into small pieces the size of apples. { 'ap·əl ,kōl }

apposition beach [GEOL] One of a series of parallel beaches formed on the seaward side of an older beach. { ,ap·ə'zish·ən ,bēch }

apposition fabric [PETR] A primary orientation of the elements of a sedimentary rock that is developed or formed at time of deposition of the material; fabrics of most sedimentary rocks belong to this type. Also known as primary fabric. { ,ap·ə'zish·ən ,fab·rik }

apron *See* outwash plain. { 'ā·prən }

Aptian [GEOL] Lower Cretaceous geologic time, between Barremian and Albian. Also known as Vectian. { 'ap·tē·ən }

aquagene tuff *See* hyaloclastite. { 'ak·wə,jēn 'təf }

aqualf [GEOL] A suborder of the soil order Alfisol, seasonally wet and marked by gray or mottled colors; occurs in depressions or on wide flats in local landscapes. { 'ak·wəlf }

aquamarine [MINERAL] A pale-blue or greenish-blue transparent gem variety of the mineral beryl. { ,ak·wə·mə'rēn }

Aquent [GEOL] A suborder of the soil order Entisol, bluish gray or greenish gray in color; under water until very recent times; located at the margins of oceans, lakes, or seas. { 'ā·kwənt }

aqueous lava [GEOL] Mud lava produced by the mixing of volcanic ash with condensing volcanic vapor or other water. { 'āk·wē·əs 'läv·ə }

aqueous rock [PETR] A sedimentary rock deposited by or in water. Also known as hydrogenic rock. { 'āk·wē·əs 'räk }

Aquept [GEOL] A suborder of the soil order Inceptisol, wet or drained, which lacks silicate clay accumulation in the soil profiles; surface horizon varies in thickness. { 'ak·wəpt }

aquiclude [GEOL] A porous formation that absorbs water slowly but will not transmit it fast enough to furnish an appreciable supply for a well or spring. { 'ak·wə,klüd }

aquifer [GEOL] A permeable body of rock capable of yielding quantities of groundwater to wells and springs. { 'ak·wə·fər }

aquifuge [GEOL] An impermeable body of rock which contains no interconnected openings or interstices and therefore neither absorbs nor transmits water. { 'ak·wə,fyüj }

Aquitanian [GEOL] Lower lower Miocene or uppermost Oligocene geologic time. { ,ak·wə'tān·ē·ən }

aquitard [GEOL] A bed of low permeability adjacent to an aquifer; may serve as a storage unit for groundwater, although it does not yield water readily. { 'ak·wə,tärd }

Aquod [GEOL] A suborder of the soil order Spodosol, with a black or dark brown horizon just below the surface horizon; seasonally wet, it occupies depressions or wide flats from which water cannot escape easily. { 'ak·wəd }

Aquoll [GEOL] A suborder of the soil order Mollisol, with thick surface horizons; formed under wet conditions, it may be under water at times, but is seasonally rather than continually wet. { 'ak·wól }

Aquox [GEOL] A suborder of the soil order Oxisol, seasonally wet, found chiefly in shallow depressions; deeper soil profiles are predominantly gray, sometimes mottled, and contain nodules or sheets of iron and aluminum oxides. { 'ak·wəks }

Aquult [GEOL] A suborder of the soil order Ultisol; seasonally wet, it is saturated with water a significant part of the year unless drained; surface horizon of the soil profile is dark and varies in thickness, grading to gray in the deeper portions; it occurs in depressions or on wide upland flats from which water drains very slowly. { 'ak·wəlt }

Araeoscelidia [PALEON] A provisional order of extinct reptiles in the subclass Euryapsida. { ə¦rē·ə·sə'lid·ē·ə }

aragonite [MINERAL] $CaCO_3$ A white, yellowish, or gray orthorhombic mineral species of calcium carbonate but with a crystal structure different from those of vaterite and calcite, the other two polymorphs of the same composition. Also known as Aragon spar. { ə'räg·ə,nīt }

Aragon spar *See* aragonite. { 'ar·ə,gän ,spär }

aramayoite [MINERAL] $Ag(Sb,Bi)S_2$ An iron-black mineral consisting of silver antimony bismuth sulfide. { ,ar·ə'mī·ə,wīt }

arapahite [PETR] A dark-colored, porous, fine-grained basic basalt consisting of magnetite, bytownite, and augite. { ə'rap·ə,hīt }

Arbuckle orogeny [GEOL] Mid-Pennsylvanian episode of diastrophism in the Wichita and Arbuckle Mountains of Oklahoma. { 'är·bək·əl ȯ'räj·ə·nē }

arc [GEOL] A geologic or topographic feature that is repeated along a curved line on the surface of the earth. { ärk }

arcanite [MINERAL] K_2SO_4 A colorless, vitreous orthorhombic sulfate mineral. Also known as glaserite. { 'är·kə,nīt }

Archaeoceti [PALEON] The zeuglodonts, a suborder of aquatic Eocene mammals in the order Cetacea; the oldest known cetaceans. { ,ärk·ē·ə'sē,tī }

Archaeocidaridae [PALEON] A family of Carboniferous echinoderms in the order Cidaroida characterized by a flexible test and more than two columns of interambulacral plates. { ,ärk·ē·ə,sə'dar·ə,dē }

Archaeocopida [PALEON] An order of Cambrian crustaceans in the subclass Ostracoda characterized by only slight calcification of the carapace. { ,ärk·ē·ə'käp·ə·də }

Archaeopteridales [PALEOBOT] An order of Upper Devonian sporebearing plants in the class Polypodiopsida characterized by woody trunks and simple leaves. { ,ärk·ē,äp·tə'rīd·ə·lēz }

Archaeopteris [PALEOBOT] A genus of fossil plants in the order Archaeopteridales; used sometimes as an index fossil of the Upper Devonian. { ,ärk·ē'äp·tə·rəs }

Archaeopterygiformes [PALEON] The single order of the extinct avian subclass Archaeornithes. { ,ärk·ē,äp·tə,rij·ə'fȯr,mēz }

Archaeopteryx [PALEON] The earliest known bird; a genus of fossil birds in the order Archaeopterygiformes characterized by flight feathers like those of modern birds. { ,ärk·ē'äp·tə·riks }

Archaeornithes [PALEON] A subclass of Upper Jurassic birds comprising the oldest fossil birds. { ,ärk·ē'ȯr·nə,thēz }

Archanthropinae [PALEON] A subfamily of the Hominidae, set up by F. Weidenreich, which is no longer used. { ,ärk·ən'thräp·ə,nē }

Archean [GEOL] A term, meaning ancient, which has been applied to the oldest rocks of the Precambrian; as more physical measurements of geologic time are made, the usage is changing; the term Early Precambrian is preferred. { är'kē·ən }

archeomagnetic dating [GEOPHYS] An absolute dating method based on the earth's shifting magnetic poles. When clays and other rock and soil materials are fired to approximately 1300°F (700°C) and allowed to cool in the earth's magnetic field, they retain a weak magnetism which is aligned with the position of the poles at the time of firing. This allows for dating, for example, of when a fire pit was used, based on the reconstruction of pole position for earlier times. { ¦är·kē·ō,mag¦ned·ik 'dā·diŋ }

Archeozoic [GEOL] **1.** The era during which, or during the latter part of which, the oldest system of rocks was made. **2.** The last of three subdivisions of Archean time, when the lowest forms of life probably existed; as more physical measurements of geologic time are made, the usage is changing; it is now considered part of the Early Precambrian. { ¦är·kē·ə¦zō·ik }

Archeria [PALEON] Genus of amphibians, order Embolomeri, in early Permian in Texas; fish eaters. { ,är'kir·ē·ə }

arching [GEOL] The folding of schists, gneisses, or sediments into anticlines. { 'ärch·iŋ }

archipelagic apron [GEOL] A fan-shaped slope around an oceanic island differing from deep-sea fans in having little, if any, sediment cover. { ¦är·kə·pə¦laj·ik 'ā·prən }

architectonic [GEOL] Of forces that determine structure. { ¦är·kə,tek¦tän·ik }

Arctic suite [PETR] A group of basic igneous rocks intermediate in composition between Atlantic and Pacific suites. { 'ärd·ik 'swēt }

Arctocyonidae [PALEON] A family of extinct carnivore-like mammals in the order Condylarthra. { 'ärk·tō,sī'än·ə,dē }

Arctolepiformes [PALEON] A group of the extinct joint-necked fishes belonging to the Arthrodira. { ,ärk·tō,lep·ə'fór,mēz }

arcuate delta [GEOL] A bowed or curved delta with the convex margin facing the body of water. Also known as fan-shaped delta. { 'ärk·yə·wət 'del·tə }

arcuation [GEOL] Production of an arc, as in rock flowage where movement proceeded in a fanlike manner. { ,ärk·yə'wā·shən }

Arcyzonidae [PALEON] A family of Devonian paleocopan ostracods in the superfamily Kirkbyacea characterized by valves with a large central pit. { ¦är,sī'zän·ə,dē }

ardealite [MINERAL] $Ca_2(HPO_4)(SO_4)\cdot4H_2O$ A white or light-yellow mineral consisting of a hydrous acid calcium phosphate-sulfate. { ,är·dē'ä,līt }

Ardennian orogeny [GEOL] A short-lived orogeny during the Ludlovian stage of the Silurian period of geologic time. { är'den·ē·ən ó'räj·ə·nē }

ardennite [MINERAL] $Mn_5Al_5(VO_4)(SiO_4)_5(OH)_2\cdot2H_2O$ A yellow to yellowish-brown mineral consisting of a hydrous silicate vanadate and arsenate of manganese and aluminum. { är'den,īt }

arduinite See mordenite. { är'dwin,īt }

areal eruption [GEOL] Volcanic eruption resulting from collapse of the roof of a batholith; the volcanic rocks grade into parent plutonic rocks. { 'er·e·əl i'rəp·shən }

areal geology [GEOL] Distribution and form of rocks or geologic units of any relatively large area of the earth's surface. { 'er·e·əl jē'äl·ə·jē }

arenaceous [GEOL] Of sediment or sedimentary rocks that have been derived from sand or that contain sand. Also known as arenarious; psammitic; sabulous. { ¦a·rə¦nāsh·əs }

arenarious See arenaceous. { ¦a·rə¦ner·ē·əs }

arendalite [MINERAL] A dark-green variety of epidote found in Arendal, Norway. { ə'rend·əl,īt }

arenicolite [GEOL] A hole, groove, or other mark in a sedimentary rock, generally sandstone, interpreted as a burrow made by an arenicolous marine worm or a trail of a mollusk or crustacean. { ,a·rə'nik·ə,līt }

Arenigian [GEOL] A European stage including Lower Ordovician geologic time (above Tremadocian, below Llanvirnian). Also known as Skiddavian. { ,a·rə'nij·ē·ən }

arenite [PETR] Consolidated sand-texture sedimentary rock of any composition. Also known as arenyte; psammite. { 'a·rə,nīt }

Arent [GEOL] A suborder of the soil order Entisol, consisting of soils formerly of other classifications that have been severely disturbed, completely disrupting the sequence of horizons. { 'a·rənt }

arenyte See arenite. { 'a·rə,nīt }

arête [GEOL] Narrow, jagged ridge produced by the merging of glacial cirques. Also known as arris; crib; serrate ridge. { a'rāt }

arfvedsonite [MINERAL] A black monoclinic amphibole, containing sodium and silicon trioxide with occluded water and some calcium. Also known as soda hornblende. { 'är·vəd·sə,nīt }

argentite [MINERAL] Ag_2S A lustrous, lead-gray ore of silver; it is a monoclinic mineral and is dimorphous with acanthite. Also known as argyrite; silver glance; vitreous silver. { 'är·jən,tīt }

argentojarosite |MINERAL| AgFe$_3$(SO$_4$)$_2$(OH)$_6$ A yellow or brownish mineral consisting of basic silver ferric sulfate. { är,jen·tō'jär·ə,sīt }

Argid |GEOL| A suborder of the soil order Aridisol, well drained, having a characteristically brown or red color and a silicate accumulation below the surface horizon; occupies older land surfaces in deserts. { 'är·jəd }

argillaceous |GEOL| Of rocks or sediments made of or largely composed of clay-size particles or clay minerals. { ,är·jə'lā·shəs }

argillation |GEOL| Development of clay minerals by weathering of aluminum silicates. { ,är·jə'lā·shən }

argillic alteration |GEOL| A rock alteration in which certain minerals are converted to minerals of the clay group. { är'jil·ik ,ȯl·tə'rā·shən }

argilliferous |GEOL| Abounding in or producing clay. { ,är·jə¦lif·ə·rəs }

argillite |PETR| A compact rock formed from siltstone, shale, or claystone but intermediate in degree of induration and structure between them and slate; argillite is more indurated than mudstone but lacks the fissility of shale. { 'är·jə,līt }

Argovian |GEOL| Upper Jurassic (lower Lusitanian), a substage of geologic time in Great Britain. { är'gōv·ē·ən }

argyrite See argentite. { är'jir,īt }

argyrodite |MINERAL| Ag$_8$GeS$_6$ A steel-gray mineral, one of two germanium minerals and a source for germanium; crystallizes in the isometric system and is isomorphous with canfieldite. { är'jir·ə,dīt }

arid erosion |GEOL| Erosion or wearing away of rock that occurs in arid regions, due largely to the wind. { 'ar·əd i'rō·zhən }

Aridisol |GEOL| A soil order characterized by pedogenic horizons; low in organic matter and nitrogen and high in calcium, magnesium, and more soluble elements; usually dry. { a'rid·ə,sȯl }

ariegite |PETR| A group of pyroxenites composed principally of clinopyroxene, orthopyroxene, and spinel. { ,ar·ē'ā,zhīt }

Arikareean |GEOL| Lower Miocene geologic time. { ə,rik·ə'rē·ən }

Arizona ruby |MINERAL| A ruby-red pyrope garnet of igneous origin found in the southwestern United States. { ¦ar·ə¦zōn·ə 'rü·bē }

arizonite |MINERAL| Fe$_2$Ti$_3$O$_9$ A steel-gray mineral containing iron and titanium and found in irregular masses in pegmatite. |PETR| A dike rock composed of mostly quartz, some orthoclase, and accessory mica and apatite. { ,ar·ə'zō,nīt }

Arkansas stone |PETR| A variety of novaculite quarried in Arkansas. { 'är·kən,sȯ ,stōn }

arkite |PETR| A feldspathoid-rich rock consisting largely of pseudoleucite and nepheline, subordinate melanite and pyroxene, and accessory orthoclase, apatite, and sphene. { 'är,kīt }

arkose |PETR| A sedimentary rock composed of sand-size fragments that contain a high proportion of feldspar in addition to quartz and other detrital minerals. { 'är,kōs }

arkose quartzite See arkosite. { 'är,kōs 'kwȯrt,sīt }

arkosic |PETR| Having wholly or partly the character of arkose. { är'kōs·ik }

arkosic bentonite |PETR| Bentonite derived from volcanic ash which contains 25–75% sandy impurities and whose detrital crystalline grains remain essentially unaltered. Also known as sandy bentonite. { är'kōs·ik 'ben·tə,nīt }

arkosic limestone |PETR| An impure clastic limestone composed of a relatively high proportion of grains or crystals of feldspar. { är'kōs·ik 'līm,stōn }

arkosic sandstone |PETR| A sandstone in which much feldspar is present, ranging from unassorted products of granular disintegration of granite to partly sorted riverlaid or even marine deposits. { är'kōs·ik 'san,stōn }

arkosic wacke See feldspathic graywacke. { är'kōs·ik¦wak·ə }

arkosite |PETR| A quartzite with a high proportion of feldspar. Also known as arkose quartzite. { är'kō,sīt }

arksutite See chiolite. { ärk'sü,tīt }

arm |GEOL| A ridge or a spur that extends from a mountain. { ärm }

armangite [MINERAL] $Mn_3(AsO_3)_2$ A black mineral crystallizing in the rhombohedral system and consisting of manganese arsenite. { är'man,gīt }

armenite [MINERAL] $BaCa_2Al_6Si_8O_{28} \cdot 2H_2O$ Mineral composed of a hydrous calcium barium aluminosilicate. { är'mē,nīt }

armored mud ball [GEOL] A large (0.4–20 inches or 1–50 centimeters in diameter) subspherical mass of silt or clay coated with coarse sand and fine gravel. Also known as pudding ball. { 'är·mərd 'məd ,bȯl }

Armorican orogeny [GEOL] Little-used term, now replaced by Hercynian or Variscan orogeny. { är'mȯr·ə·kən ȯ'räj·ə·nē }

arnimite [MINERAL] $Cu_5(SO_4)_2(OH)_6 \cdot 3H_2O$ Mineral consisting of a hydrous copper sulfate. { 'ärn·ə,mīt }

arquerite [MINERAL] A mineral consisting of a soft, malleable, silver-rich variety of amalgam, containing about 87% silver and 13% mercury. { är'kē,rīt }

arrested decay [GEOL] A stage in coal formation where biochemical action ceases. { ə'res·təd di'kā }

arrhenite [MINERAL] A variety of fergusonite. { ə'rā,nīt }

arris See arête. { 'ar·əs }

arrival time [GEOPHYS] In seismological measurements, the time at which a given wave phase is detected by a seismic recorder. { ə'rī·vəl ,tīm }

arrojadite [MINERAL] $Na_2(Fe,Mn)_5(PO_4)_4$ Dark-green mineral crystallizing in the monoclinic system, being isostructural with dickinsonite and occurring in masses. { ,ar·ə'jä,dīt }

arroyo [GEOL] Small, deep gully produced by flash flooding in arid and semiarid regions of the southwestern United State. { ə'rȯi·ō }

arsenic [MINERAL] A brittle, steel-gray hexagonal mineral, the native form of the element. { 'ärs·ən·ik }

arsenical antimony See allemontite. { ar'sen·ə·kəl 'ant·ə,mō·nē }

arsenical nickel See niccolite. { ar'sen·ə·kəl 'nik·əl }

arsenic bloom See arsenolite. { 'ärs·ən·ik ,blüm }

arseniopleite [MINERAL] A reddish-brown mineral consisting of a basic arsenate of manganese, calcium, iron, lead, and magnesium and occurring in cleavable masses. { är'sēn·ē·ō'plē,īt }

arseniosiderite [MINERAL] $Ca_3Fe_4(AsO_4)_4(OH)_4 \cdot 4H_2O$ A yellowish-brown mineral consisting of a basic iron calcium arsenate and occurring as concretions. { är'sēn·ē·ō'sid·ə,rīt }

arsenobismite [MINERAL] $Bi_2(AsO_4)(OH)_3$ A yellowish-green mineral consisting of a basic bismuth arsenate and occurring in aggregates. { ,ärs·ən·ō'biz,mīt }

arsenoclasite [MINERAL] $Mn_5(AsO_4)_2(OH)_4$ A red mineral consisting of a basic manganese arsenate. Also spelled arsenoklasite. { ,ärs·ən·ō'klā,sīt }

arsenoklasite See arsenoclasite. { ,ärs·ən·ō'klä,sīt }

arsenolamprite [MINERAL] FeAsS A lead gray mineral consisting of nearly pure arsenic; occurs in masses with a fibrous foliated structure. { ,ärs·ən·ō'lam,prīt }

arsenolite [MINERAL] As_2O_3 A mineral crystallizing in the isometric system and usually occurring as a white bloom or crust. Also known as arsenic bloom. { är'sen·əl,īt }

arsenopyrite [MINERAL] FeAsS A white to steel-gray mineral crystallizing in the monoclinic system with pseudo-orthorhombic symmetry because of twinning; occurs in crystalline rock and is the principal ore of arsenic. Also known as mispickel. { ,ärs·ən·ō'pī,rīt }

arsoite [PETR] An olivine-bearing diopside trachyte. { 'är·sō,īt }

arterite [PETR] **1.** A migmatite produced as a result of regional contact metamorphism during which residual magmas were injected into the host rock. **2.** Gneisses characterized by veins formed from the solution given off by deep-seated intrusions of molten granite. **3.** A veined gneiss in which the vein material was injected from a magma. { är'tir,īt }

arteritic migmatite [GEOL] Injection gneiss supposedly produced by introduction of pegmatite, granite, or aplite into schist parallel to the foliation. { ¦ard·ə¦rid·ik 'mig·mə,tīt }

Arthrodira [PALEON] The joint-necked fishes, an Upper Silurian and Devonian order of the Placodermi. { ,är·thrō'dī·rə }

articulite *See* itacolumite. { är'tik·yə,līt }

artinite [MINERAL] $Mg_2CO_3(OH)_2 \cdot 3H_2O$ A snow-white mineral crystallizing in the orthorhombic system and occurring in crystals or fibrous aggregates. { är'tē,nīt }

Artinskian [GEOL] A European stage of geologic time including Lower Permian (above Sakmarian, below Kungurian). { är'tin·skē·ən }

arzrunite [MINERAL] A bluish-green mineral consisting of a basic copper sulfate with copper chloride and lead, and occurring as incrustations. { ärz'rü,nīt }

asar *See* esker. { 'a·sər }

asbestos [MINERAL] A general name for the useful, fibrous varieties of a number of rock-forming silicate minerals that are heat-resistant and chemically inert; two varieties exist: amphibole asbestos, the best grade of which approaches the composition $Ca_2Mg_5(OH)_2Si_8O_{22}$ (tremolite), and serpentine asbestos, usually chrysotile, Mg_3Si_2-$(OH)_4O_5$. { as'bes·təs }

asbolane *See* asbolite. { 'az·bə,lān }

asbolite [MINERAL] A black, earthy mineral aggregate containing hydrated oxides of manganese and cobalt. Also known as asbolane; black cobalt; earthy cobalt. { 'az·bə,līt }

aschistic [GEOL] Pertaining to rocks of minor igneous intrusions that have not been differentiated into light and dark portions but that have essentially the same composition as the larger intrusions with which they are associated. { ā'skis·tik }

aseismic [GEOPHYS] Not subject to the occurrence or destructive effects of earthquakes. { ā'sīz·mik }

ash [GEOL] Volcanic dust and particles less than 4 millimeters in diameter. { ash }

Ashby [GEOL] A North American stage of Middle Ordovician geologic time, forming the upper subdivision of Chazyan, and lying above Marmor and below Porterfield. { 'ash·bē }

ash cone [GEOL] A volcanic cone built primarily of unconsolidated ash and generally shaped somewhat like a saucer, with a rim in the form of a wide circle and a broad central depression often nearly at the same elevation as the surrounding country. { 'ash ,kōn }

ash fall [GEOL] **1.** A fall of airborne volcanic ash from an eruption cloud; characteristic of Vulcanian eruptions. Also known as ash shower. **2.** Volcanic ash resulting from an ash fall and lying on the ground surface. { 'ash ,fól }

ash field [GEOL] A thick, extensive deposit of volcanic ash. Also known as ash plain. { 'ash ,fēld }

ash flow [GEOL] **1.** An avalanche of volcanic ash, generally a highly heated mixture of volcanic gases and ash, traveling down the flanks of a volcano or along the surface of the ground. Also known as glowing avalanche; incandescent tuff flow. **2.** A deposit of volcanic ash and other debris resulting from such a flow and lying on the surface of the ground. { 'ash ,flō }

ash-flow tuff *See* ignimbrite. { 'ash,flō ,təf }

ash fusibility [GEOL] The gradual softening and melting of coal ash that takes place with increase in temperature as a result of the melting of the constituents and chemical reactions. { 'ash ,fyüz·ə'bil·əd·ē }

Ashgillian [GEOL] A European stage of geologic time in the Upper Orodovician (above Upper Caradocian, below Llandoverian of Silurian). { ash'gil·yən }

ash plain *See* ash field. { 'ash ,plān }

ash rock [GEOL] The material of arenaceous texture produced by volcanic explosions. { 'ash ,räk }

ash shower *See* ash fall. { 'ash ,shaú·ər }

ashstone [PETR] A rock composed of fine volcanic ash; particles are less than 0.06 millimeter in diameter. { 'ash,stōn }

ashtonite *See* mordenite. { 'ash·tə,nīt }

ash viscosity [GEOL] The ratio of shearing stress to velocity gradient of molten ash;

indicates the suitability of a coal ash for use in a slag-tap-type boiler furnace. { 'ash vɪs'kas·əd·ē }

ashy grit [GEOL] **1.** Pyroclastic material of sand and smaller size. **2.** Mixture of ordinary sand and volcanic ash. { 'ash·ē 'grit }

asiderite See stony meteorite. { ə'sīd·ə,rīt }

Aso lava [GEOL] A type of indurated pyroclastic deposit produced during the explosive eruptions that formed the Aso Caldera of Kyushu, Japan. { 'äs·ō 'läv·ə }

asparagolite See asparagus stone. { ,as·pə'rag·ə,līt }

asparagus stone [MINERAL] A yellow-green variety of apatite occurring in crystals. Also known as asparagolite. { ə'spar·ə·gəs ,stōn }

aspect [GEOL] **1.** The general appearance of a specific geologic entity or fossil assemblage as considered more or less apart from relations in time and space. **2.** The direction toward which a valley side or slope faces with respect to the compass or rays of the sun. { 'a,spekt }

aspect angle [GEOL] The angle between the aspect of a slope and the geographic south (Northern Hemisphere) or the geographic north (Southern Hemisphere). { 'a,spekt ,aŋ·gəl }

asperity [GEOL] A type of surface roughness appearing along the interface of two faults. { a'sper·ə·dē }

asphaltic sand [GEOL] Deposits of sand grains cemented together with soft, natural asphalt. { a'sfólt·ik 'sand }

asphaltite [GEOL] Any of the dark-colored, solid, naturally occurring bitumens that are insoluble in water, but more or less completely soluble in carbon disulfide, benzol, and so on, with melting points between 250 and 600°F (121 and 316°C); examples are gilsonite and grahamite. { a'sfól,tīt }

asphaltite coal See albertite. { a'sfól,tīt ,kōl }

asphalt rock [GEOL] Natural asphalt-containing sandstone or dolomite. Also known as asphalt stone; bituminous rock; rock asphalt. { 'a,sfólt 'räk }

asphalt stone See asphalt rock. { 'a,sfólt 'stōn }

Aspidorhynchidae [PALEON] The single family of the Aspidorhynchiformes, an extinct order of holostean fishes. { ¦as·pə,dō'riŋ·kə,dē }

Aspidorhynchiformes [PALEON] A small, extinct order of specialized holostean fishes. { ¦as·pə,dō,riŋk·ə'fór,mēz }

Aspinothoracida [PALEON] The equivalent name for Brachythoraci. { a,spīn·ō·thə'ras·əd·ə }

aspite [GEOL] A cratered volcano with the base wide in relation to the height; for example, Mauna Loa. { 'as,pīt }

assemblage [GEOL] **1.** A group of fossils that, appearing together, characterize a particular stratum. **2.** A group of minerals that compose a rock. [PALEON] A group of fossils occurring together at one stratigraphic level. { ə'sem·blij }

assemblage zone [PALEON] A biostratigraphic unit defined and identified by a group of associated fossils rather than by a single index fossil. { ə'sem·blij ,zōn }

assimilation [GEOL] Incorporation of solid or fluid material that was originally in the rock wall into a magma. { ə,sim·ə'lā·shən }

assyntite [PETR] A plutonic rock consisting largely of orthoclase and pyroxene, lesser amounts of sodalite and nepheline, and accessory biotite, sphene, apatite, and opaque oxides. { ə'sin,tīt }

Astartian See Sequanian. { ə'stär·shən }

asthenolith [GEOL] A body of magma locally melted at any time within any solid portion of the earth. { as'then·ə,lith }

asthenosphere [GEOL] That portion of the upper mantle beneath the rigid lithosphere which is plastic enough for rock flowage to occur; extends from a depth of 30–60 miles (50–100 kilometers) to about 240 miles (400 kilometers) and is seismically equivalent to the low velocity zone. { as'then·ə,sfir }

Astian [GEOL] A European stage of geologic time: upper Pliocene, above Plaisancian, below the Pleistocene stage known as Villafranchian, Calabrian, or Günz. { 'as·tē·ən }

27

astrakanite *See* bloedite. { 'as·trə·kə,nīt }

Astrapotheria |PALEON| A relatively small order of large, extinct South American mammals in the infraclass Eutheria. { ,as·trə·pə'thir·ē·ə }

Astrapotheroidea |PALEON| A suborder of extinct mammals in the order Astrapotheria, ranging from early Eocene to late Miocene. { ,as·trə·pə·thə'rȯid·ē·ə }

astrobleme |GEOL| A circular-shaped depression on the earth's surface produced by the impact of a cosmic body. { 'as·trō,blēm }

astrochanite *See* bloedite. { ə'sträk·ə,nīt }

astrophyllite |MINERAL| $(K,Na)_3(Fe,Mn)_7Ti_2Si_8O_{24}(O,OH)_7$ A mineral composed of a basic silicate of potassium or sodium, iron or manganese, and titanium. { ,as·trə'fī,līt }

Asturian orogeny |GEOL| Mid-Upper Carboniferous diastrophism. { ə'stur·ē·ən ȯ'räj·ə·nē }

asymmetrical bedding |GEOL| An order in which lithologic types or facies follow one another in a circuitous arrangement so that, for example, the sequence of types 1-2-3-1-2-3-1-2-3 indicates asymmetry (while the sequence 1-2-3-2-1-2-3-2-1 indicates symmetrical bedding). { ¦ā·sə¦me·tri·kəl 'bed·iŋ }

asymmetrical fold |GEOL| A fold in which one limb dips more steeply than the other. { ¦ā·sə¦me·tri·kəl 'fōld }

asymmetrical laccolith |GEOL| A laccolith in which the beds dip at conspicuously different angles in different sectors. { ¦ā·sə¦me·tri·kəl 'lak·ə,lith }

asymmetrical ripple mark |GEOL| The normal form of ripple mark, with short downstream slopes and comparatively long, gentle upstream slopes. { ¦ā·sə¦me·tri·kəl 'rip·əl ,märk }

asymmetrical vein |GEOL| A crustified vein of geologic material with unlike layers on each side. { ¦ā·sə¦me·tri·kəl 'vān }

atacamite |MINERAL| $Cu_2Cl(OH)_3$ Native, green hydrouscopper oxychloride crystallizing in the orthorhombic system. { ,ad·ə'kam,īt }

ataxic |GEOL| Pertaining to unstratified ore deposits. { ə'tak·sik }

ataxite |GEOL| An iron meteorite that lacks the structure of either hexahedrite or octahedrite and contains more than 10% nickel. |PETR| A taxitic rock whose components are arranged in a breccialike manner, that is, there is no specific arrangement. { ə'tak,sīt }

atectonic |GEOL| Of an event that occurs when orogeny is not taking place. { ¦ā·tek'tän·ik }

atectonic pluton |GEOL| A pluton that is emplaced when orogeny is not occurring. { ¦ā·tek'tän·ik 'plü,tän }

atelestite |MINERAL| $Bi_8(AsO_4)_3O_5(OH)_5$ A yellow mineral consisting of basic bismuth arsenate and occurring in minute crystals; specific gravity is 6.82. { ,ad·əl'e,stīt }

athrogenic |PETR| Of or pertaining to pyroclastics. { ¦ath·rə¦jen·ik }

Athyrididina |PALEON| A suborder of fossil articulate brachiopods in the order Spiriferida characterized by laterally or, more rarely, ventrally directed spires. { ,ath·ə·rə'də'dī·nə }

Atlantic series |PETR| A great group of igneous rocks, based on tectonic setting, found in nonorogenic areas, often associated with block sinking and great crustal instability, and erupted along faults and fissures or through explosion vents. Also known as Atlantic suite. { ət'lan·tik 'sir·ēz }

Atlantic suite *See* Atlantic series. { ət'lan·tik 'swēt }

Atlantic-type continental margin |GEOL| A continental margin typified by that of the Atlantic which is aseismic because oceanic and continental lithospheres are coupled. { ət'lan·tik ,tīp ,känt·ən'ent·əl 'mär·jən }

atlantite |PETR| An olivine-bearing nepheline tephrite. { ət'lan,tīt }

atmoclast |GEOL| A fragment of rock broken off in place by atmospheric weathering. { 'at·mə,klast }

atmoclastic |PETR| Of a clastic rock, composed of atmoclasts that have been recemented without rearrangement. { ¦at·mə¦klas·tik }

atmogenic |GEOL| Of rocks, minerals, and other deposits derived directly from the atmosphere by condensation, wind action, or deposition from volcanic vapors; for example, snow. { ¦at·mə¦jeír·ik }

atmolith |GEOL| A rock precipitated from the atmosphere, that is, an atmogenic rock. { 'at·mə,lith }

Atokan |GEOL| A North American provincial series in lower Middle Pennsylvanian geologic time, above Morrowan, below Desmoinesian. { ə'tō·kən }

atoll texture |GEOL| The surrounding of a ring of one mineral with another mineral, or minerals, within and without the ring. Also known as core texture. { 'a,tȯl ,teks·chər }

atopite |MINERAL| A yellow or brown variety of romeite that contains fluorine. { 'ad·ə,pīt }

Atrypidina |PALEON| A suborder of fossil articulate brachiopods in the order Spiriferida. { a·trī'pid·ə·nə }

attached dune |GEOL| A dune that has formed around a rock or other geological feature in the path of windblown sand. { ə'tacht 'dün }

attapulgite |MINERAL| $(Mg,Al)_2Si_4O_{10}(OH)·4H_2O$ A clay mineral with a needlelike shape from Georgia and Florida; active ingredient in most fuller's earth, and used as a suspending agent, as an oil well drilling fluid, and as a thickener in latex paint. { ,ad·ə'pəl,jīt }

Atterberg scale |GEOL| A geometric and decimal grade scale for classification of particles in sediments based on the unit value of 2 millimeters and involving a fixed ratio of 10 for each successive grade; subdivisions are geometric means of the limits of each grade. { 'at·ər,bərg ,skāl }

Attican orogeny |GEOL| Late Miocene diastrophism. { 'ad·ə·kən ȯ'räj·ə·nē }

attitude |GEOL| The position of a structural surface feature in relation to the horizontal. { 'ad·ə,tüd }

attrital coal |GEOL| A bright coal composed of anthraxylon and of attritus in which the translucent cell-wall degradation matter or translucent humic matter predominates, with the ratio of anthraxylon to attritus being less than 1:3. { ə'trīd·əl 'kōl }

attrition |GEOL| The act of wearing and smoothing of rock surfaces by the flow of water charged with sand and gravel, by the passage of sand drifts, or by the movement of glaciers. { ə'trish·ən }

attritus |GEOL| **1.** Visible-to-ultramicroscopic particles of vegetable matter produced by microscopic and other organisms in vegetable deposits, particularly in swamps and bogs. **2.** The dull gray to nearly black, frequently striped portion of material that makes up the bulk of some coals and alternate bands of bright anthraxylon in well-banded coals. { ə'trīd·əs }

aubrite |GEOL| An enstatite achondrite (meteorite) consisting almost wholly of crystalline-granular enstatite (and clinoenstatite) poor in lime and practically free from ferrous oxide, with accessory oligoclase. Also known as bustite. { 'ō,brīt }

auganite |PETR| An olivine-free basalt (calcic plagioclase and augite are the essential mineral components) or an augite-bearing andesite. { 'ȯg·ə,nīt }

augelite |MINERAL| Natural, basic aluminum phosphate. { 'ȯj·ə,līt }

augen |PETR| Large, lenticular eye-shaped mineral grain or mineral aggregate visible in some metamorphic rocks. { 'ȯg·ən or 'aů·gən }

augen kohle See eye coal. { 'aů·gən ,kōl·ə }

augen schist |PETR| A mylonitic rock characterized by the presence of recrystallization. { 'aů·gən ,shist }

augen structure |PETR| A structure found in some gneisses and granites in which certain of the constituents are squeezed into elliptic or lens-shaped forms and, especially if surrounded by parallel flakes of mica, resemble eyes. { 'aů·gən ,strək·chər }

augite |MINERAL| $(Ca,Mg,Fe)(Mg,Fe,Al)(Al,Si)_2O_6$ A general name for the monoclinic pyroxenes; occurs as dark green to black, short, stubby, prismatic crystals, often of octagonal outline. { 'ȯ,jīt }

augitite |PETR| A volcanic rock consisting of abundant phenocrysts of augite in a glassy groundmass containing microlites of nepheline and plagioclase, with accessory biotite, apatite, and opaque oxides. { 'ȯ·jə,tīr }

augitophyre [PETR] A porphyritic rock in which the phenocrysts are augite and the groundmass is potash feldspar. { ȯ'jid·ə,fī·ər }

aulacogen [GEOL] A major fault-bounded trough considered to be one part of a three-rayed fault system on the domes above mantle hot spots; the other two rays open as proto-ocean basins. { ,au̇'läk·ə·jən }

Aulolepidae [PALEON] A family of marine fossil teleostean fishes in the order Ctenothrissiformes. { ,ȯl·ə'lep·ə,dē }

Auloporidae [PALEON] A family of Paleozoic corals in the order Tabulata. { ,ȯl·ə'pȯr·ə,dē }

aureole [GEOL] A ring-shaped contact zone surrounding an igneous intrusion. Also known as contact aureole; contact zone; exomorphic zone; metamorphic aureole; metamorphic zone; thermal aureole. { 'ȯr·ē,ōl }

aurichalcite [MINERAL] $(Zn,Cu)_5(CO_3)_2(OH)_6$ Pale-green or pale-blue mineral consisting of a basic copper zinc carbonate and occurring in crystalline incrustations. Also known as brass ore. { ,ȯr·ə'kal,sīt }

auriferous [GEOL] Of a substance, especially a mineral deposit, bearing gold. { ȯ'rif·ə·rəs }

aurora [GEOPHYS] The most intense of the several lights emitted by the earth's upper atmosphere, seen most often along the outer realms of the Arctic and Antarctic, where it is called the aurora borealis and aurora australis, respectively; excited by charged particles from space. { ə'rȯr·ə }

aurosmiridium [MINERAL] A brittle, silver-white, isometric mineral consisting of a solid solution of gold and osmium in iridium. { ¦ȯr·ō·smə'rid·ē·əm }

austinite [MINERAL] $CaZnAsO_4(OH)$ A colorless or yellowish mineral crystallizing in the orthorhombic system; consists of a basic calcium zinc arsenate; hardness is 4.5 on Mohs scale, and specific gravity is 4.13. { 'ȯs·tə,nīt }

austral axis pole [GEOPHYS] The southern intersection of the geomagnetic axis with the earth's surface. { 'ȯs·trəl ¦ak·səs ,pōl }

australite [GEOL] A tektite found in southern Australia, occurring as glass balls and spheroidal dumbbell forms of green and black, similar to obsidian and probably of cosmic origin. { 'ȯs·trə,līt }

Australopithecinae [PALEON] The near-men, a subfamily of the family Hominidae composed of the single genus *Australopithecus*. { ȯ,strā·lō,pith·ə'sī·nē }

Australopithecus [PALEON] A genus of near-men in the subfamily Australopithecinae representing a side branch of human evolution. { ȯ,strā·lō'pith·ə·kəs }

Austrian orogeny [GEOL] A short-lived orogeny during the end of the Early Cretaceous. { 'ȯs·trē·ən ȯ'räj·ə·nē }

autallotriomorphic [PETR] Pertaining to an aplitic texture in which all mineral constituents crystallized simultaneously, preventing the development of euhedral crystals. { ¦au̇d·ə¦lä·trē·ə¦mȯr·fik }

authigene [MINERAL] A mineral which has not been transported but has been formed in place. Also known as authigenic mineral. { 'ȯ·thə,jēn }

authigenic [GEOL] Of constituents that came into existence with or after the formation of the rock of which they constitute a part; for example, the primary and secondary minerals of igneous rocks. { ¦ȯ·thə¦jen·ik }

authigenic mineral *See* authigene. { ¦ȯ·thə¦jen·ik 'min·rəl }

authigenic sediment [GEOL] Sediment occurring in the place where it was originally formed. { ¦ȯ·thə¦jen·ik 'sed·ə·mənt }

autobrecciation [GEOL] The process whereby portions of the first consolidated crust of a lava flow are incorporated into the still-fluid portion. { ¦ȯd·ō,brech·ē'ā·shən }

autochthon [GEOL] A succession of rock beds that have been moved comparatively little from their original site of formation, although they may be folded and faulted extensively. [PALEON] A fossil occurring where the organism once lived. { ȯ'täk·thən }

autochthonous [GEOL] Having been formed or occurring in the place where found. { ȯ'täk·thə·nas }

autochthonous coal [GEOL] Coal believed to have originated from accumulations of plant debris at the place where the plants grew. Also known as indigenous coal. { ȯ'täk·thə·nas 'kōl }

autochthonous sediment [GEOL] A residual soil deposit formed in place through decomposition. { ȯ'täk·thə·nas 'sed·ə·mənt }

autoclastic [GEOL] Of rock, fragmented in place by folding due to orogenic forces when the rock is not so heavily loaded as to render it plastic. { ¦ȯd·ō¦klas·tik }

autoclastic schist [GEOL] Schist formed in place from massive rocks by crushing and squeezing. { ¦ȯd·ō¦klas·tik 'shist }

autogenetic topography [GEOL] Conformation of land due to the physical action of rain and streams. { ¦ȯd·ō·jə¦ned·ik tə'päg·rə·fē }

autogeosyncline [GEOL] A parageosyncline that subsides as an elliptical basin or trough nearly without associated highlands. Also known as intracratonic basin. { ¦ȯd·ō¦jē·ō'sin,klīn }

autoinjection See autointrusion. { ¦ȯd·ō,in'jek·shən }

autointrusion [GEOL] A process wherein the residual liquid of a differentiating magma is drawn into rifts formed in the crystal mesh at a late stage by deformation of unspecified origin. Also known as autoinjection. { ¦ȯd·ō·in'trü·zhən }

autolith [PETR] 1. A fragment of igneous rock enclosed in another igneous rock of later consolidation, each being regarded as a derivative from a common parent magma. 2. A round, oval, or elongated accumulation of iron-magnesium minerals of uncertain origin in granitoid rock. { 'ȯd·ō,lith }

autolysis [GEOCHEM] Return of a substance to solution, as of phosphate removed from seawater by plankton and returned when these organisms die and decay. { ȯ'täl· ə·səs }

autometamorphism [PETR] Metamorphism of an igneous rock by the action of its own volatile fluids. Also known as autometasomatism. { ¦ȯd·ō,med·ə'mȯr,fiz·əm }

autometasomatism See autometamorphism. { ¦ȯd·ō,med·ə'sō·mə,tiz·əm }

automorphic [PETR] Of minerals in igneous rock bounded by their own crystal faces. Also known as euhedral; idiomorphic. { ¦ȯd·ō¦mȯr·fik }

automorphosis [PETR] Metamorphosis of solidified igneous rock by solutions from its heated interior. { ,ȯd·ə'mȯr·fə·səs }

autophytograph [GEOL] An imprint on a rock surface made by chemical activity of a plant or plant part. { ,ȯd·ə'fīd·ə,graf }

autopneumatolysis [GEOL] The occurrence of metamorphic changes at the pneumatolytic stage of a cooling magma when temperatures are approximately 400–600°C. { ¦ȯd·ō,nü·mə'täl·ə·səs }

Autunian [GEOL] A European stage of Lower Permian geologic time, above the Stephanian of the Carboniferous and below the Saxonian. { ,ō'tün·ē·ən }

autunite [MINERAL] $Ca(UO_2)_2(PO_4)_2 \cdot 10H_2O$ A common fluorescent mineral that occurs as yellow tetragonal plates in uranium deposits; minor ore of uranium. { ō'tə,nīt }

Auversian See Ledian. { ,ō'vərzh·ən }

auxiliary fault [GEOL] A branch fault; a minor fault ending against a major one. { ȯg'zil·yə·rē 'fȯlt }

auxiliary mineral [MINERAL] A light-colored, relatively rare or unimportant mineral in an igneous rock; examples are apatite, muscovite, corundrum, fluorite, and topaz. { ȯg'zil·yə·rē 'min·rəl }

auxiliary plane [GEOL] A plane at right angles to the net slip on a fault plane as determined from analysis of seismic data for an earthquake. { ȯg'zil·yə·rē 'plān }

available relief [GEOL] The vertical distance after uplift between the altitude of the original surface and the level at which grade is first attained. { ə'vāl·ə·bəl ri'lēf }

aven See pothole. { 'av·ən }

aventurine [MINERAL] 1. A glass or mineral containing sparkling gold-colored particles, usually copper or chromic oxide. 2. A shiny red or green translucent quartz having

31

small, but microscopically visible, exsolved hematite or included mica particles.
{ ə'vench·ə,rēn }

average igneous rock [PETR] A hypothetical rock whose composition is thought to be similar to the average chemical composition of the outermost 10-mile (16-kilometer) shell of the earth. { 'av·rij 'ig·nē·əs 'räk }

aviolite [PETR] A mica-cordierite-hornfels. { ā'vī·ə,līt }

avogadrite [MINERAL] (K,Cs)BF₄ An orthorhombic fluoborate mineral occurring in small crystals on Vesuvian lava. { ,a·və'gäd,rīt }

Avonian See Dinantian. { ə'vōn·ē·ən }

awaruite [MINERAL] Native nickel-iron alloy containing 57.7% nickel. { ,a·wä'rü,īt }

axial compression [GEOL] A compression applied parallel with the cylinder axis in experimental work involving rock cylinders. { 'ak·sē·əl kəm'presh·ən }

axial culmination [GEOL] Distortion of the fold axis upward in a form similar to an anticline. { 'ak·sē·əl ,kəl·mə'nā·shən }

axial dipole field [GEOPHYS] A postulated magnetic field for the earth, consisting of a dipolar field centered at the earth's center, with its axis coincident with the earth's rotational axis. { 'ak·sē·əl 'di,pōl ,fēld }

axial plane [GEOL] A plane that intersects the crest or trough in such a manner that the limbs or sides of the fold are more or less symmetrically arranged with reference to it. Also known as axial surface. { 'ak·sē·əl 'plān }

axial-plane cleavage [GEOL] Rock cleavage essentially parallel to the axial plane of a fold. { 'ak·sē·əl ¦plān ,klē·vij }

axial-plane foliation [GEOL] Foliation developed in rocks parallel to the axial plane of a fold and perpendicular to the chief deformational pressure. { 'ak·sē·əl ¦plān ,fō·lē'ā·shən }

axial-plane schistosity [GEOL] Schistosity developed parallel to the axial planes of folds. { 'ak·sē·əl ¦plān ,shis'täs·əd·ē }

axial-plane separation [GEOL] The distance between axial planes of adjacent anticline and syncline. { 'ak·sē·əl ¦plān sep·ə'rā·shən }

axial surface See axial plane. { 'ak·sē·əl 'sər·fəs }

axial trace [GEOL] The intersection of the axial plane of a fold with the surface of the earth or any other specified surface; sometimes such a line is loosely and incorrectly called the axis. { 'ak·sē·əl 'trās }

axial trough [GEOL] Distortion of a fold axis downward into a form similar to a syncline. { 'ak·sē·əl 'tróf }

axinite [MINERAL] H₂(Ca,Fe,Mn)₄(BO)Al₂(SiO₄)₅ Brown, blue, green, gray, or purplish gem mineral that commonly forms glassy triclinic crystals. Also known as glass schorl. { 'ak·sə,nīt }

axinitization [GEOL] The replacement of rocks by axinite, as in the border zones of some granites. { ak,zin·ə·tə'zā·shən }

axiolite [MINERAL] A variety of elongated spherulite in which there is an aggregation of minute acicular crystals arranged at right angles to a central axis. { 'ak·sē·ə,līt }

axis [GEOL] **1.** A line where a folded bed has maximum curvature. **2.** The central portion of a mountain chain. { 'ak·səs }

Azoic [GEOL] That portion of the earlier Precambrian time in which there is no trace of life. { ā'zō·ik }

azonal soil [GEOL] Any group of soils without well-developed profile characteristics, owing to their youth, conditions of parent material, or relief that prevents development of normal soil-profile characteristics. Also known as immature soil. { 'ā,zōn·əl 'sóil }

azulite [MINERAL] A translucent pale-blue variety of smithsonite found in large masses in Arizona and Greece. { 'azh·ə,līt }

azurite [MINERAL] Cu₃(CO₃)₂(OH)₂ A blue monoclinic mineral consisting of a basic carbonate of copper; an ore of copper. Also known as blue copper ore; blue malachite; chessylite. { 'azh·ə,rīt }

azurmalachite [MINERAL] A mixture of azurite and malachite, usually occurring massive with concentric banding; used as an ornamental stone. { ¦a·zhər'mal·ə,kīt }

B

back-arc basin [GEOL] The region (small ocean basin) between an island arc and the continental mainland formed during oceanic plate subduction, containing sediment eroded from both. { 'bak,ärk ,bās·ən }

back beach *See* backshore. { 'bak ,bēch }

backbone [GEOL] **1.** A ridge forming the principal axis of a mountain. **2.** The principal mountain ridge, range, or system of a region. { 'bak,bōn }

backdeep [GEOL] An epieugeosynclinal basin; a nonvolcanic postorogenic geosynclinal basin whose sediments are derived from an uplifted eugeosyncline. { 'bak,dēp }

backfolding [GEOL] Process in mountain forming in which the folds are overturned toward the interior of an orogenic belt. Also known as backward folding. { 'bak ,fōld·iŋ }

backlands [GEOL] A section of a river floodplain lying behind a natural levee. { 'bak,lanz }

backlimb [GEOL] Of the two limbs of an asymmetrical anticline, the one that is more gently dipping. { 'bak,lim }

back-set bed [GEOL] Cross bedding that dips in a direction against the flow of a depositing current. { 'bak ,set ,bed }

backshore [GEOL] The upper shore zone that is beyond the advance of the usual waves and tides. Also known as back beach; backshore beach. { 'bak,shȯr }

backshore beach *See* backshore. { 'bak,shȯr ,bēch }

backshore terrace *See* berm. { 'bak,shȯr 'ter·əs }

back slope *See* dip slope. { 'bak ,slōp }

backswamp [GEOL] Swampy depressed area of a floodplain between the natural levees and the edge of the floodplain. { 'bak,swamp }

backthrusting [GEOL] The thrusting in the direction of the interior of an orogenic belt, opposite the general structural trend. { 'bak,thrəst·iŋ }

backward folding *See* backfolding. { 'bak·wərd ¦fōld·iŋ }

backwash mark [GEOL] A crisscross ridge pattern in beach sand, caused by backwash. { 'bak,wäsh ,märk }

backwash ripple mark [GEOL] Ripple marks that are broad and flat and parallel to the shoreline, with narrow, shallow troughs and crests about 30 centimeters apart; formed by backwash above the maximum wave retreat level. { 'bak,wäsh 'rip·əl ,märk }

baculite [GEOL] A crystallite that looks like a dark rod. { 'bak·yə,līt }

baddeleyite [MINERAL] ZrO_2 A colorless, yellow, brown, or black monoclinic zirconium oxide mineral found in Brazil and Ceylon; used as heat- and corrosion-resistant linings for furnaces and muffles. { 'bad·əl·ē,īt }

bahada *See* bajada. { bə'häd·ə }

bahamite [PETR] A consolidated limestone formed of sediment similar to a type currently found accumulating in the Bahamas. { bə'ham,īt }

bahiaite [PETR] Holocrystalline igneous rock formed mainly of hypersthene with subordinate hornblende and sometimes minor amounts of other minerals. { bə'hī·yə,īt }

baikerite [MINERAL] A waxlike mineral from the vicinity of Lake Baikal, Siberia; apparently about 60% ozocerite with other tarry, waxy, and resinous hydrocarbons. { 'bī·kə,rīt }

bajada [GEOL] An alluvial plain formed as a result of lateral growth of adjacent alluvial fans until they finally coalesce to form a continuous inclined deposit along a mountain front. Also spelled bahada. { bə'häd·ə }

bajada breccia [PETR] An imperfectly stratified accumulation of coarse, angular rock fragments mixed with mud that formed in arid climates and results from a mudflow containing considerable water. { bə'häd·ə 'brech·ə }

Bajocian [GEOL] A European stage: the middle Middle or lower Middle Jurassic geologic time; above Toarcian, below Bathonian. { bə'jō·shən }

bakerite [MINERAL] $8CaO·5B_2O_3·6SiO_2·6H_2O$ White mineral, occurring in fine-grained, nodular masses, resembling marble and unglazed porcelain, and consisting of hydrous calcium borosilicate. { 'bāk·ə,rīt }

balanced rock See perched block. { 'bal·ənst ,räk }

baldheaded anticline [GEOL] An upfold with a crest that has been deeply eroded before later deposition. { 'bȯld,hed·əd 'an·ti,klīn }

ball [GEOL] **1.** A low sand ridge, underwater by high tide, which extends generally parallel with the shoreline; usually separated by an intervening trough from the beach. **2.** A spheroidal mass of sedimentary material. **3.** Common name for a nodule, especially of ironstone. { bȯl }

ball-and-socket joint [GEOL] See cup-and-ball joint. { ¦bȯl ən 'säk·ət ,jȯint }

ballas [MINERAL] A spherical aggregate of small diamond crystals; used in diamond drill bits and other diamond tools. { 'bal·əs }

ball coal [GEOL] A variety of coal occurring in spheroidal masses. { 'bȯl ,kōl }

ballstone [GEOL] **1.** Large mass or concretion of fine, unstratified limestone resulting from growth of coral colonies. **2.** A nodule of rock, especially ironstone, in a stratified unit. { 'bȯl,stōn }

balm [GEOL] A concave cliff or precipice that forms a shelter. { bäm }

banakite [PETR] An alkalic basalt made up of plagioclase, sanidine, and biotite, with small quantities of analcime, augite, and olivine; quartz or leucite may be present. { 'ban·ə,kīt }

band [GEOL] A thin layer or stratum of rock that is noticeable because its color is different from the colors of adjacent layers. { band }

bandaite [PETR] A dacite type of extrusive rock composed of hypersthene and labradorite. { 'ban·də,īt }

banded [PETR] Pertaining to the appearance of rocks that have thin and nearly parallel bands of different textures, colors, and minerals. { 'ban·dəd }

banded coal [GEOL] A variety of bituminous and subbituminous coal made up of a sequence of thin lenses of highly lustrous coalified wood or bark interspersed with layers of more or less striated bright or dull coal. { 'ban·dəd 'kōl }

banded differentiate [PETR] A type of igneous rock made up of bands of different composition, frequently alternating between two varieties as in a layered intrusion. { 'ban·dəd ,dif·ə'ren·chē,āt }

banded iron formation [GEOL] A sedimentary mineral deposit consisting of alternate silica-rich (chert or quartz) and iron-rich layers formed 2.5–3.5 billion years ago; the major source of iron ore. { ¦band·əd 'ī·ərn fȯr,mā·shən }

banded ore [GEOL] Ore made up of layered bands composed either of the same minerals that differ from band to band in color or textures or proportion, or of different minerals. { 'ban·dəd 'ȯr }

banded peat [GEOL] Peat formed of alternate layers of vegetable debris. { 'ban·dəd 'pēt }

banded structure [PETR] An outcrop feature in igneous and metamorphic rocks due to alternation of layers, stripes, flat lenses, or streaks that obviously differ in mineral composition or texture. { 'ban·dəd 'strək·chər }

banded vein [GEOL] A vein composed of layers of different minerals that lie parallel to the walls. Also known as ribbon vein. { 'ban·dəd 'vān }

banding [PETR] **1.** The series of layers occurring in a banded structure. **2.** In sedimentary rocks, the thin bedding of alternate layers of different materials. { 'band·iŋ }

bandylite [MINERAL] $CuB_2O_4 \cdot CuCl_2 \cdot 4H_2O$ A tetragonal mineral that is deep blue with greenish lights and consists of a hydrated copper borate-chloride. { 'ban·də,līt }
bank [GEOL] **1.** The edge of a waterway. **2.** The rising ground bordering a body of water. **3.** A steep slope or face, generally consisting of unconsolidated material. { baŋk }
bank deposit [GEOL] Mounds, ridges, and terraces of sediment rising above and about the surrounding sea bottom. { 'haŋk di'päz·ət }
banket [GEOL] A conglomerate containing valuable metal to be exploited. { baŋ'ket }
bank-inset reef [GEOL] A coral reef situated on island or continental shelves well inside the outer edges. { 'baŋk 'in,set ,rēf }
bank reef [GEOL] A reef which rises at a distance back from the outer margin of rimless shoals. { 'baŋk ,rēf }
bank-run gravel [GEOL] A natural deposit comprising gravel or sand. { 'baŋk ,rən 'grav·əl }
bank sand [GEOL] Deposits occurring in banks or pits and containing a low percentage of clay; used in core making. { 'baŋk ,sand }
bar [GEOL] **1.** Any of the various submerged or partially submerged ridges, banks, or mounds of sand, gravel, or other unconsolidated sediment built up by waves or currents within stream channels, at estuary mouths, and along coasts. **2.** Any band of hard rock, for example, a vein or dike, that extends across a lode. { bär }
baraboo [GEOL] A monadnock buried by a series of strata and then reexposed by the partial erosion of these younger strata. { 'bär·ə,bü }
bararite [MINERAL] $(NH_4)_2SiF_6$ A white, hexagonal mineral consisting of ammonium silicon fluoride; occurs in tabular, arborescent, and mammillary forms. { bə'rä,rīt }
Barbados earth [GEOL] A deposit of fossil radiolarians. { bar'bā·dəs ,ərth }
bar beach [GEOL] A straight beach of offshore bars that are separated by shallow bodies of water from the mainland. { 'bär ,bēch }
barbertonite [MINERAL] $Mg_6Cr_2(OH)_{16}CO_3 \cdot 4H_2O$ A lilac to rose pink, hexagonal mineral consisting of a hydrated carbonate-hydroxide of magnesium and chromium; occurs in massive form or in masses of fibers or plates. { 'bär·bər·tə,nīt }
barbierite [MINERAL] $NaAlSi_3lO_8$ A hypothetical soda feldspar thought to be isomorphous with orthoclase. { bar'bi,rīt }
barchan [GEOL] A crescent-shaped dune or drift of windblown sand or snow, the arms of which point downwind; formed by winds of almost constant direction and of moderate speeds. Also known as barchane; barkhan; crescentic dune. { bär'kän }
barchane See barchan. { bär'kän }
bar finger sand [GEOL] An elongated lenticular sand body that lies beneath a distributory in a birdfoot delta. { 'bär ,fiŋ·gər ,sand }
baring See overburden. { 'ba·riŋ }
barite [MINERAL] $BaSO_4$ A white, yellow, or colorless orthorhombic mineral occurring in tabular crystals, granules, or compact masses; specific gravity is 4.5; used in paints and drilling muds and as a source of barium chemicals; the principal ore of barium. Also known as baryte; barytine; cawk; heavy spar. { 'ba,rīt }
barite dollar [MINERAL] Barite in the form of rounded disk-shaped masses; formed in a sandstone or sandy shale. { 'ba,rīt ,däl·ər }
barkevikite [MINERAL] A brown or black member of the amphibole mineral group; looks like basaltic hornblende but differs from it in its iron concentration. { 'bär·kə,vi,kīt }
barkhan See barchan. { bär'kän }
bar plain [GEOL] A plain formed by a stream without a low-water channel or an alluvial cover. { 'bär ,plān }
barranca [GEOL] A hole or deep break made by heavy rain; a ravine. { bə'raŋ·kə }
barred basin See restricted basin. { 'bärd ,bās·ən }
barred beach sequence [GEOL] A sequence comprising longshore bars, barrier beaches, and lagoons that develop when, under low-energy conditions, waves cross a broad continental shelf before impinging on a shoreline where sand-sized sediments are abundant. { 'bärd 'bēch 'sē·kwəns }
Barremian [GEOL] Lower Cretaceous geologic age, between Hauterivian and Aptian. { bə'räm·ē·ən }

35

barrier bar |GEOL| Ridges whose crests are parallel to the shore and which are usually made up of water-worn gravel put down by currents in shallow water at some distance from the shore. { 'bar·ē·ər ‚bär }

barrier basin |GEOL| A basin formed by natural damming, for example, by landslides or moraines. { 'bar·ē·ər ‚bās·ən }

barrier beach |GEOL| A single, long, narrow ridge of sand which rises slightly above the level of high tide and lies parallel to the shore, from which it is separated by a lagoon. Also known as offshore beach. { 'bar·ē·ər ‚bēch }

barrier chain |GEOL| A series of barrier spits, barrier islands, and barrier beaches extending along a coastline. { 'bar·ē·ər ‚chān }

barrier flat |GEOL| An area which is relatively flat and frequently occupied by pools of water that separate the seaward edge of the barrier from a lagoon on the landward side. { 'bar·ē·ər ‚flat }

barrier island |GEOL| An elongate accumulation of sediment formed in the shallow coastal zone and separated from the mainland by some combination of coastal bays and their associated marshes and tidal flats; barrier islands are typically several times longer than their width and are interrupted by tidal inlets. { 'bar·ē·ər ‚ī·lənd }

barrier reef |GEOL| A coral reef that runs parallel to the coast of an island or continent, from which it is separated by a lagoon. { 'bar·ē·ər ‚rēf }

barrier spit |GEOL| A barrier of sand joined at one of its ends to the mainland. { 'bar·ē·ər ‚spit }

Barrovian metamorphism |GEOL| A regional metamorphism that can be zoned into facies that are metamorphic. { bə'rōv·ē·ən ‚med·ə'mór‚fiz·əm }

Barstovian |GEOL| Upper Miocene geologic time. { ‚bär'stōv·ē·ən }

bar theory |GEOL| A theory that accounts for thick deposits of salt, gypsum, and other evaporites in terms of increased salinity of a solution in a lagoon caused by evaporation. { bär 'thē·ə·rē }

Bartonian |GEOL| A European stage: Eocene geologic time above Auversian, below Ludian. Also known as Marinesian. { bär'tōn·ē·ən }

Barychilinidae |PALEON| A family of Paleozoic crustaceans in the suborder Platycopa. { ‚bar·ə·kə'lin·ə‚dē }

Barylambdidae |PALEON| A family of late Paleocene and early Eocene aquatic mammals in the order Pantodonta. { ‚bar·ə'lam·də‚dē }

barysphere See centrosphere. { 'bar·ə‚sfir }

baryta feldspar See hyalophane. { bə'rīd·ə 'fel‚spär }

baryte See barite. { 'ba‚rīt }

Barytheriidae |PALEON| A family of extinct proboscidean mammals in the suborder Barytherioidea. { ‚bar·ə·thə'rī·ə‚dē }

Barytherioidea |PALEON| A suborder of extinct mammals of the order Proboscidea, in some systems of classification. { ‚bar·ə‚thir·ē'óid·ē·ə }

barytine See barite. { 'bar·ə‚tēn }

barytocalcite |MINERAL| $CaBa(CO_3)_2$ A colorless to white, grayish, greenish, or yellowish monoclinic mineral consisting of calcium and barium carbonate. { bə‚rīd·ə'kal‚sīt }

basal arkose |PETR| Partially reworked feldspathic residuum in the lower section of a sandstone that overlies granitic rock. { 'bā·səl 'är‚kōs }

basal complex See basement. { 'bā·səl 'käm‚pleks }

basal conglomerate |GEOL| A coarse gravelly sandstone or conglomerate forming the lowest member of a series of related strata which lie unconformably on older rocks; records the encroachment of the seabeach on dry land. { 'bā·səl kən'gläm·ə·rət }

basalt |PETR| An aphanitic crystalline rock of volcanic origin, composed largely of plagioclase feldspar (labradorite or bytownite) and dark minerals such as pyroxene and olivine; the extrusive equivalent of gabbro. { bə'sólt }

basalt glass See tachylite. { bə'sólt ‚glas }

basaltic dome See shield volcano. { bə'sól·tik 'dōm }

basaltic hornblende |PETR| A black or brown variety of hornblende rich in ferric iron

36

and occurring in basalts and other iron-rich basic igneous rocks.   Also known as basaltine; lamprobolite, oxyhornblende.   { bə'sȯl·tik 'hȯrn,blend }

basaltic lava   [PETR] A volcanic fluid rock of basaltic composition.   { bə'sȯl·tik 'lav·ə }

basaltic magma   [GEOL] Mobile rock material of basaltic composition.   { bə'sȯl·tik 'mag·mə }

basaltic rock   [PETR] Igneous rock that is fine-grained and contains basalt, diabase, and dolerite; if andesite is included the rock is dark in color.   { bə'sȯl·tik 'räk }

basaltic shell   [GEOL] The lower crystal layer of basalt underlying the oceans and beneath the sialic layer of continents.   { bə'sȯl·tik 'shel }

basaltiform   [GEOL] Similar to basalt in form.   { bə'sȯl·tə,fȯrm }

basaltine See basaltic hornblende.   { bə'sȯl,tēn }

basalt obsidian See tachylite.   { bə'sȯlt əb'sid·ē·ən }

basaluminite   [MINERAL] $Al_4(SO_4)(OH)_{10}\cdot5H_2O$ A white mineral consisting of hydrated basic aluminum sulfate; occurs in compact masses.   { ¦bäs·ə'lüm·ə,nīt }

basanite   [PETR] A basaltic extrusive rock closely allied to chert, jasper, or flint.   Also known as Lydian stone; lydite.   { 'bas·ə,nīt }

basculating fault See wrench fault.   { 'ba·skyə,lād·iŋ 'fȯlt }

base exchange   [GEOCHEM] Replacement of certain ions by others in clay.   { 'bās iks'chānj }

base level   [GEOL] That critical plane of erosion and deposition represented by river level on continents and by wave or current base in the sea.   { 'bās ,lev·əl }

base-leveled plain   [GEOL] Any land surface changed almost to a plain by subaerial erosion.   Also known as peneplain.   { 'bās ,lev·əld 'plān }

base-leveling epoch See gradation period.   { 'bās ,lev·əl·iŋ 'ep·ək }

basement   [GEOL] **1.** A complex, usually of igneous and metamorphic rocks, that is overlain unconformably by sedimentary strata.   Also known as basement rock.   **2.** A crustal layer beneath a sedimentary one and above the Mohorovičić discontinuity.   **3.** The ancient continental igneous rock base that lies beneath Precambrian rocks.   Also known as basal complex; basement complex.   { 'bās·mənt }

basement complex See basement.   { 'bās·mənt 'käm,pleks }

basement rock See basement.   { 'bās·mənt ,räk }

basic   [PETR] Of igneous rocks, having low silica content (generally less than 54%) and usually being rich in iron, magnesium, or calcium.   { 'bā·sik }

basic front   [GEOL] An advancing zone of granitization enriched in calcium, magnesium, and iron.   { 'bā·sik ¦frənt }

basic hornfels   [PETR] A type of hornfels derived from a basic igneous rock.   { 'bā·sik 'hȯrn,felz }

basic rock   [PETR] An igneous rock with a relatively low silica content, and rich in iron, magnesium, or calcium.   { 'bā·sik 'räk }

basic schist   [PETR] A schistose rock that forms from the metamorphism of a basic igneous rock.   { 'bā·sik 'shist }

basification   [GEOL] Development of a more basic rock, usually with more hornblende, biotite, and oligoclase, by contamination of a granitic magma in the assimilation of country rock.   { ,bäs·ə·fə'kā·shən }

basimesostasis   [GEOL] A process of the partial or entire enclosure of plagioclase crystals in a diabase by augite.   { ¦bā·zē,mez·ə'stā·səs }

basin   [GEOL] **1.** A low-lying area, wholly or largely surrounded by higher land, that varies from a small, nearly enclosed valley to an extensive, mountain-rimmed depression.   **2.** An entire area drained by a given stream and its tributaries.   **3.** An area in which the rock strata are inclined downward from all sides toward the center.   **4.** An area in which sediments accumulate.   { 'bās·ən }

basin-and-range structure   [GEOL] Regional structure dominated by fault-block mountains separated by basins filled with sediment.   { 'bās·ən ən 'ranj ,strək·chər }

basin fold   [GEOL] Synclinal and anticlinal folds in structural basins.   { 'bās·ən ,fōld }

basining   [GEOL] A settlement of earth in the form of basins due to the solution and transportation of underground deposits of salt and gypsum.   { 'bās·ən·iŋ }

basin length [GEOL] Length in a straight line from the mouth of a stream to the farthest point on the drainage divide of its basin. { 'bās·ən ‚leŋkth }

basin order [GEOL] A classification of basins according to stream drainage; for example, a first-order basin contains all of the drainage area of a first-order stream. { 'bās·ən ‚órd·ər }

basin peat *See* local peat. { 'bās·ən ‚pēt }

basin range [GEOL] A mountain range characteristic of the Great Basin in the western United States and formed by a faulted and tilted block of strata. { 'bās·ən ‚rānj }

basin valley [GEOL] The filled-in depression of large intermountain areas; an example is Salt Lake Valley in Utah. { 'bās·ən ‚val·ē }

bassanite [MINERAL] A white mineral consisting of hydrated calcium sulfate; a pseudomorph of gypsum. { bə'sä‚nīt }

basset [GEOL] The outcropping edge of a layer of rock exposed to the surface. { 'bas·ət }

bassetite [MINERAL] A transparent, yellow, monoclinic mineral presumably consisting of a hydrated uranium phosphate containing divalent iron; occurs in groups of thin tablets. { 'bas·əd‚īt }

bastion [GEOL] A prominent aggregation of bedrock extending from the mouth of a hanging glacial trough and reaching well into the main glacial valley. { 'bas·chən }

bastite [MINERAL] A hydrated magnesium silicate, a variety of serpentine occurring from the alteration of orthorhombic pyroxenes such as enstatite. { 'ba‚stīt }

bastnaesite [MINERAL] (Ce,La)CO_3(F,OH) A greasy yellow to reddish-brown fluorocarbonate rare-earth metal mineral; source of rare earths, for example, cerium and lanthanum. { 'bast·nə‚sīt }

batholite [GEOL] An older massive protrusion of magma that solidifies as coarse crystalline rock in the deep horizons of the earth's crust. { 'bath·ə‚līt }

batholith [GEOL] A body of igneous rock, 40 square miles (100 square kilometers) or more in area, emplaced at great or intermediate depth in the earth's crust. { 'bath·ə‚lith }

Bathonian [GEOL] A European stage of geologic time: Middle Jurassic, below Callovian, above Bajocian. Also known as Bathian. { bə'thōn·ē·ən }

Bathornithidae [PALEON] A family of Oligocene birds in the order Gruiformes. { ‚ba·thȯr'nith·ə‚dē }

bathvillite [MINERAL] An oxygenated hydrocarbon mineral, found in Tortane Hill, Scotland, that is amorphous, fawn-brown, opaque, and quite friable. { 'bath·və‚līt }

bathymetric biofacies [GEOL] The lateral distribution and character of underwater sedimentary strata. { ‚bath·ə'me·trik ¦bī·ō'fā·shēz }

battery reefs *See* Kimberley reefs. { 'bad·ə·rē ‚rēfs }

batture [GEOL] An elevation of the bed of a river under the surface of the water; sometimes used to signify the same elevation when it has risen above the surface. { ba'túr }

baumhauerite [MINERAL] $Pb_4As_6S_{13}$ A lead to steel gray, monoclinic mineral consisting of lead arsenic sulfide. { baú'maú·ə‚rīt }

bauxite [PETR] A whitish, grayish, brown, yellow, or reddish-brown rock composed of hydrous aluminum oxides and aluminum hydroxides and containing impurities such as free silica, silt, iron hydroxides, and clay minerals; the principal commercial source of aluminum. { 'bȯk‚sīt }

bauxitization [GEOL] Bauxite development from either primary aluminum silicates or secondary clay minerals. { ¦bȯk·sə·də'zā·shən }

bavenite *See* duplexite. { bə've‚nīt }

b axis [PETR] A direction in the plane of movement that is at a right angle to the tectonic transport direction. { 'bē ‚ak·səs }

bay [GEOPHYS] A simple transient magnetic disturbance, usually an hour in duration, whose appearance on a magnetic record has the shape of a V or a bay of the sea. { bā }

bay bar *See* baymouth bar. { 'bā ‚bär }

bay barrier [GEOL] A narrow shoal or small point of land projecting from the shore across the mouth of a bay and severing the bay's connection with the main body of water. { 'bā ,bar·ē·ɔr }

bay delta [GEOL] A usually triangular alluvial deposit formed at the point where the mouth of a stream enters the head of a drowned valley. { 'bā ,del·tə }

bay head [GEOL] A swampy region at the head of a bay. { ¦bā ¦hed }

bay-head bar [GEOL] A bar formed a short distance from the shore at the head of a bay. { ¦bā ¦hed ,bär }

bay-head beach [GEOL] A beach formed around a bay head by storm waves; layers of sediment cover the bay floor and bare rock benches front the headland cliffs. { ¦bā ¦hed ,bēch }

bay-head delta [GEOL] A delta at the head of an estuary or a bay into which a river discharges because of the margin of the land's late partial submergence. { ¦bā ¦hed ,del·tə }

bayldonite [MINERAL] $Cu_3(AsO_4)_2(OH)_2$ An apple green to yellowish-green monoclinic mineral consisting of a basic arsenate of copper and lead; occurs in minute mammillary concretions, in massive form, and as crusts. { 'bāl·də,nīt }

bayleyite [MINERAL] $Mg_2(UO_2)(CO_3)_3\cdot18H_2O$ A sulfur yellow monoclinic mineral consisting of a hydrated carbonate of magnesium and uranium; occurs as minute, short-prismatic crystals. { 'bā·lē,īt }

baymouth bar [GEOL] A bar extending entirely or partially across the mouth of a bay. Also known as bay bar. { 'bā,maúth ,bär }

bayside beach [GEOL] A beach formed at the side of a bay by materials eroded from nearby headlands and deposited by longshore currents. { 'bā,sīd ,bēch }

bazzite [MINERAL] $Sc_2Be_3Si_6O_{18}$ An azure-blue mineral that crystallizes in the hexagonal system; the rare scandium analog of beryl. { 'ba,zīt }

b-c fracture [GEOL] A tension fracture parallel with the fabric plane and normal to the *a* axis. { ¦bē¦sē 'frak·chɔr }

b-c plane [GEOL] A plane that is perpendicular to the plane of movement and parallel to the *b* direction in that plane. { ¦bē¦sē ,plān }

beach [GEOL] The zone of unconsolidated material that extends landward from the low-water line to where there is marked change in material or physiographic form or to the line of permanent vegetation. { bēch }

beach cusp *See* cusp. { bēch ,kɔsp }

beach cycle [GEOL] Periodic retreat and outbuilding of beaches resulting from waves and tides. { bēch ,sī·kɔl }

beach drift [GEOL] The material transported by drifting of beach. { bēch ,drift }

beach face *See* foreshore. { bēch ,fās }

beach gravel [GEOL] Gravels in which most of the particles cluster about one size. { bēch ,grav·ɔl }

beach plain [GEOL] Embankments of wave-deposited material added to a prograding shoreline. { 'bēch ,plān }

beach platform *See* wave-cut bench. { 'bēch ,plat,fórm }

beach profile [GEOL] Intersection of a beach's ground surface with a vertical plane perpendicular to the shoreline. { 'bēch ,prō,fīl }

beach ridge [GEOL] A continuous mound of beach material behind the beach that was heaped up by waves or other action. { 'bēch ,rij }

beachrock [PETR] A friable to well-cemented rock made of calcareous skeletal debris that is cemented together by calcium carbonate. { 'bēch,räk }

beach scarp [GEOL] A nearly vertical slope along the beach caused by wave erosion. { 'bēch ,skärp }

bean ore [GEOL] A lenticular, pisolitic aggregate of limonite. { 'bēn ,òr }

beaverite [MINERAL] $Pb(Cu,Fe,Al)_3(SO_4)_2(OH)_6$ A canary yellow, hexagonal mineral consisting of a basic sulfate of lead, copper, iron, and aluminum. { 'bē·vɔ,rīt }

Becke test [MINERAL] A microscope test in which indices of refraction are compared for minerals; the Becke line appears to move toward the material of higher refractivity as the tube of the microscope is raised. { 'bek·ɔ ,test }

beckerite [MINERAL] A brown variety of the fossil resin retinite with a very high oxygen content. { 'bek·ə,rīt }

becquerelite [MINERAL] $CaU_6O_{19}\cdot11H_2O$ An orthorhombic mineral consisting of a hydrated oxide of uranium; occurs in tabular, elongated, striated, and massive form. { be'kre,līt }

bed [GEOL] **1.** The smallest division of a stratified rock series, marked by a well-defined divisional plane from its neighbors above and below. **2.** An ore deposit, parallel to the stratification, constituting a regular member of the series of formations; not an intrusion. { bed }

bedded [GEOL] Pertaining to rocks exhibiting depositional layering or bedding formed from consolidated sediments. { 'bed·əd }

bedded chert [PETR] Chert of brittle, close-jointed, rhythmically layered character found over large areas in thick deposits, the usually even-bedded layers separated by partings of dark siliceous shale or by siderite layers. { 'bed·əd ,chərt }

bedded vein [GEOL] A lode occupying the position of a bed that is parallel with the enclosing rock stratification. { 'bed·əd ,vān }

bedding [GEOL] Condition where planes divide sedimentary rocks of the same or different lithology. { 'bed·iŋ }

bedding cleavage [GEOL] Cleavage parallel to the rock bedding. { 'bed·iŋ ,klēv·ij }

bedding fault [GEOL] A fault whose fault surface is parallel to the bedding plane of the constituent rocks. Also known as bedding-plane fault. { 'bed·iŋ ,fólt }

bedding fissility [GEOL] Primary foliation parallel to the bedding of sedimentary rocks. { 'bed·iŋ fi'sil·əd·ē }

bedding joint [GEOL] A joint parallel to the rock bedding. { 'bed·iŋ ,jóint }

bedding plane [GEOL] Any of the division planes which separate the individual strata or beds in sedimentary or stratified rock. { 'bed·iŋ ,plān }

bedding-plane fault See bedding fault. { 'bed·iŋ ,plān ,fólt }

bedding-plane slip See flexural slip. { 'bed·iŋ ,plān ,slip }

bedding schistosity [GEOL] Schistosity that is parallel to the rock bedding. { 'bed·iŋ ,shis'täs·əd·ē }

bedding thrust [GEOL] A thrust fault parallel to bedding. { 'bed·iŋ ,thrəst }

bedding void [GEOL] A void formed between successive batches of lava that are discharged in a single short activity of a volcano, as well as between flows made a long time apart. { 'bed·iŋ ,vóid }

Bedford limestone See spergenite. { 'bed·fərd 'līm,stōn }

bediasite [GEOL] A black to brown tektite found in Texas. { bē'dī·ə,zīt }

bed load [GEOL] Particles of sand, gravel, or soil carried by the natural flow of a stream on or immediately above its bed. Also known as bottom load. { 'bed ,lōd }

Bedoulian [GEOL] Lower Cretaceous (lower Aptian) geologic time in Switzerland. { bə'dül·ē·ən }

bedrock [GEOL] General term applied to the solid rock underlying soil or any other unconsolidated surficial cover. { 'bed,räk }

beegerite [MINERAL] $Pb_6Bi_2S_9$ A light to dark gray mineral consisting of lead bismuth sulfide; usually occurs in granular to dense massive form. { 'be·gə,rīt }

beekite [MINERAL] **1.** A concretionary form of calcite or silica that occurs in small rings on the surface of a fossil shell which has weathered out of its matrix. **2.** White, opaque accretions of silica found on silicified fossils or along joint surfaces as a replacement of organic matter. { 'bē,kīt }

beerbachite [PETR] A hornfels with large poikiloblastic crystals of olivine. { 'bir·bə,kīt }

beetle stone See septarium. { 'bēd·əl ,stōn }

beidellite [MINERAL] A clay mineral of the montmorillonite group in which Si^{4+} has been replaced by Al^{3+} and in which there is virtual absence of Mg or Fe replacing Al. { bī'de,līt }

Belemnoidea [PALEON] An order of extinct dibranchiate mollusks in the class Cephalopoda. { ,bə·ləm'nóid·ē·ə }

Belinuracea [PALEON] An extinct group of horseshoe crabs; arthropods belonging to the Limulida. { ,bel·ə·nú'rās·ē·ə }

belite *See* larnite. { 'bē,līt }

Bellerophontacea [PALEON] A superfamily of extinct gastropod mollusks in the order Aspidobranchia. { bə,ler·ə,fän'tās·ē·ə }

bellingerite [MINERAL] 3Cu(IO₃)₂·2H₂O A light green triclinic mineral consisting of hydrated copper iodate. { bə'liŋ·ə,rīt }

bell-metal ore *See* stannite. { 'bel ,med·əl ,ȯr }

belonite [GEOL] A rod- or club-shaped microscopic embryonic crystal in a glassy rock. { 'bel·ə,nīt }

belted plain [GEOL] A plain whose surface has been slowly worn down and sculptured into bands or belts of different levels. { 'bel·təd ¦plān }

belteroporic [GEOL] Of crystals in rocks whose growth was determined by the direction of easiest growth. { ,bel¦ter·ə'pȯr·ik }

belt of cementation *See* zone of cementation. { ¦belt əv ,si·men'tā·shən }

belt of soil moisture *See* belt of soil water. { ¦belt əv 'sȯil ,mȯis·chər }

belt of soil water [GEOL] The upper subdivision of the zone of aeration limited above by the land surface and below by the intermediate belt; this zone contains plant roots and water available for plant growth. Also known as belt of soil moisture; discrete film zone; soil-water belt; soil-water zone; zone of soil water. { ¦belt əv 'sȯil ,wȯd·ər }

bench [GEOL] A terrace of level earth or rock that is raised and narrow and that breaks the continuity of a declivity. { bench }

bench gravel [GEOL] Gravel beds found on the sides of valleys above the present stream bottoms, representing parts of the bed of the stream when it was at a higher level. { 'bench ,grav·əl }

bench lava [GEOL] Semiconsolidated, crusted basaltic lava forming raised platforms and crags about the edges of lava lakes. Also known as bench magma. { 'bench ,la·və }

bench magma *See* bench lava. { 'bench ,mag·mə }

bench placer [GEOL] A placer in ancient stream deposits from 50 to 300 feet (15 to 90 meters) above present streams. { 'bench ,plās·ər }

bend [GEOL] **1.** A curve or turn occurring in a stream course, bed, or channel which has not yet become a meander. **2.** The land area partly encircled by a bend or meander. { bend }

Benioff zone [GEOPHYS] A zone of earthquake hypocenters distributed on well-defined planes that dips from a shallow depth into the earth's mantle to depths as great as 420 miles (700 kilometers). Also known as Benioff-Wadati zone; Wadati-Benioff zone. { 'ben·ē·ȯf ,zōn }

Benioff-Wadati zone *See* Benioff zone. { ¦ben·ē,ȯf wə'dä·tē ,zōn }

benitoite [MINERAL] BaTi(SiO₃)₃ A blue to violet barium-titanium silicate mineral; at one time it was cut and sold as sapphire. { bə'nēd·ə,wīt }

benjaminite [MINERAL] Pb₂(Cu,Ag)₂Bi₄S₉ A gray mineral occurring in granular massive form. { 'ben·jə·mə,nīt }

Bennettitales [PALEOBOT] An equivalent name for the Cycadeoidales. { bə,ned·ə'tā,lēz }

Bennettitatae [PALEOBOT] A class of fossil gymnosperms in the order Cycadeoidales. { be,ned·ə'tā,dē }

bentonite [GEOL] A clay formed from volcanic ash decomposition and largely composed of montmorillonite and beidellite. Also known as taylorite. { 'bent·ən,īt }

beraunite [MINERAL] Fe²⁺Fe³⁺(PO₄)₃(OH)₅·3H₂O A reddish-brown to blood red, monoclinic mineral consisting of hydrated basic phosphate of ferric and ferrous iron. { bə'raú,nīt }

beresorite *See* phoenicochroite. { bə'res·ə,rīt }

berg crystal *See* rock crystal. { 'bərg ,kris·təl }

bergmehl *See* rock milk. { 'berk,mel }

berg till *See* floe till. { 'bərg ,til }

41

berkeyite See lazulite. { 'bərk·ē‚īt }

berlinite [MINERAL] $Al(PO_4)$ A colorless to gray or pale rose, hexagonal mineral consisting of aluminum orthophosphate; occurs in massive form. { 'bər·lə‚nīt }

berm [GEOL] **1.** A narrow terrace which originates from the interruption of an erosion cycle with rejuvenation of a stream in the mature stage of its development and renewed dissection. **2.** A horizontal portion of a beach or backshore formed by deposit of material as a result of wave action. Also known as backshore terrace; coastal berm. { bərm }

bermanite [MINERAL] $Mn^{2+}Mn_2^{3+}(PO_4)_2(OH)_2·4H_2O$ A reddish-brown, orthorhombic mineral consisting of a hydrated basic phosphate of manganese; occurs in crystal aggregates and as lamellar masses. { 'bər·mə‚nīt }

berm crest [GEOL] The seaward limit and usually the highest spot on a coastal berm. Also known as berm edge. { 'bərm ‚krest }

berm edge See berm crest. { 'bərm ‚ej }

bernalite [MINERAL] $Fe(OH)_3$ An iron hydroxide, yellow-green or dark green in color. { 'bərn·ə‚līt }

Berriasian [GEOL] Part of or the underlying stage of the Valanginian at the base of the Cretaceous. { ‚ber·ē'ā·zhən }

berthierite [MINERAL] $FeSb_2S_4$ A dark steel gray, orthorhombic mineral consisting of iron antimony sulfide. { 'bər·thē·ə‚rīt }

berthonite See bournonite. { 'bər·thə‚nīt }

bertrandite [MINERAL] $Be_4Si_2O_7(OH)_2$ A colorless or pale-yellow mineral consisting of a beryllium silicate occurring in prismatic crystals; hardness is 6–7 on Mohs scale, and specific gravity is 2.59–2.60. { 'bər·trən‚dīt }

beryllonite [MINERAL] $NaBe(PO_4)$ A colorless or yellow mineral occurring in short, prismatic or tabular, monoclinic crystals with two good pinacoidal cleavages at right angles; hardness is 5.5–6 on Mohs scale, and specific gravity is 2.85. { bə'ril·ə‚nīt }

berzelianite [MINERAL] Cu_2Se A silver-white mineral composed of copper selenide and found in igneous rock; specific gravity is 4.03. { ‚bər'zēl·yə‚nīt }

beta chalcocite See chalcocite. { 'bād·ə 'chal·kə‚sīt }

betafite See ellsworthite. { 'bed·ə‚fīt }

beta plane [GEOPHYS] The model, introduced by C.G. Rossby, of the spherical earth as a plane whose rate of rotation (corresponding to the Coriolis parameter) varies linearly with the north-south direction. { 'bād·ə ‚plān }

betrunked river [GEOL] A river that is shorn of its lower course as a result of submergence of the land margin by the sea. { bē'trəŋkt 'riv·ər }

betwixt mountains See median mass. { bə'twikst ‚maúnt·ənz }

beudantite [MINERAL] $PbFe_3(AsO_4)(SO_4)(OH)_6$ A black, dark green, or brown, hexagonal mineral consisting of a basic sulfate-arsenate of lead and ferric iron; occurs as rhombohedral crystals. { 'byüd·ən‚īt }

beveling [GEOL] Planing by erosion of the outcropping edges of strata. { 'bev·ə·liŋ }

beyerite [MINERAL] $(Ca,Pb)Bi_2(CO_3)_2O_2$ A bright yellow to lemon yellow, tetragonal mineral consisting of bismuth and calcium carbonate; occurs as thin plates and compact earthy masses. { 'bī·ə‚rīt }

Beyrichacea [PALEON] A superfamily of extinct ostracodes in the suborder Beyrichicopina. { ‚bī·rə'kās·ē·ə }

Beyrichicopina [PALEON] A suborder of extinct ostracodes in the order Paleocopa. { ‚bī·rə·kə‚kō'pī·nə }

Beyrichiidae [PALEON] A family of extinct ostracodes in the superfamily Beyrichacea. { ‚bī·rə'kī·ə‚dē }

B girdle [PETR] A circular pattern in petrofabric diagrams that indicates a B axis. { 'bē ‚gərd·əl }

B horizon [GEOL] The zone of accumulation in soil below the A horizon (zone of leaching). Also known as illuvial horizon; subsoil; zone of accumulation; zone of illuviation. { 'bē hə'rīz·ən }

bianchite [MINERAL] $(Fe,Zn)SO_4·6H_2O$ A white, monoclinic mineral consisting of iron and zinc sulfate hexahydrate; occurs in crusts of indistinct crystals. { bē'aŋ‚kīt }

bidalotite See anthophyllite. { bə'däl·ə,tīt }

bieberite |MINERAL| $CoSO_4·7I_2O$ A rose red or flesh red, monoclinic mineral consisting of cobalt sulfate heptahydrate; occurs as crusts and stalactites. { 'bē·bə,rīt }

bight |GEOL| **1.** A long, gradual bend or recess in the coastline which forms a large, open receding bay. **2.** A bend in a river or mountain range. { bīt }

bigwoodite |PETR| A medium-grained plutonic rock consisting of microcline, microcline-microperthite, sodic plagioclase, and hornblende, aegirine-augite, or biotite. { big'wù,dīt }

bilinite |MINERAL| $Fe^{2+}Fe^{3+}(SO_4)_4·22H_2O$ A white to yellowish mineral consisting of a hydrated sulfate of divalent and trivalent iron; occurs in radial-fibrous aggregates. { 'bil·ə,nīt }

Billingsellacea |PALEON| A group of extinct articulate brachiopods in the order Orthida. { ,bil·iŋ·sə'lās·ē·ə }

bimaceral |GEOL| A coal microlithotype that consists of a mixture of two macerals. { bī'mas·ə·rəl }

binary granite |PETR| **1.** A granite made up of quartz and feldspar. **2.** A granite containing muscovite mica and biotite. { 'bīn·ə·rē 'gran·ət }

bindheimite |MINERAL| $Pb_2Sb_2O_6(O,OH)$ A hydrous lead antimonate mineral produced from natural oxidation of jamesonite; found in Nevada. { 'bint,hī,mīt }

binding coal See caking coal. { 'bīn·diŋ ,kōl }

bing ore |GEOL| The purest lead ore, with the largest crystals of galena. { 'biŋ ,òr }

biochemical deposit |GEOL| A precipitated deposit formed directly or indirectly from vital activities of organisms, such as bacterial iron ore and limestone. { ¦bī·ō'kem·ə·kəl di'päz·ət }

biochemical rock |PETR| A type of sedimentary rock primarily comprising deposits resulting directly or indirectly from processes and activities of living organisms. { ¦bī·ō'kem·i·kəl 'räk }

biochron |PALEON| A fossil of relatively short range of time. { 'bī·ō,krän }

biochronology |GEOL| The relative age dating of rock units based on their fossil content. { ,bī·ō·krə'näl·ə·jē }

bioclastic rock |PETR| Rock formed from material broken or arranged by animals, humans, or sometimes plants; a rock composed of broken calcareous remains of organisms. { ¦bī·ō¦klas·tik 'räk }

biofacies |GEOL| **1.** A rock unit differing in biologic aspect from laterally equivalent biotic groups. **2.** Lateral variation in the biologic aspect of a stratigraphic unit. { 'bī·ō,fā·shēz }

biogenic chert |PETR| Chert derived from the tests of pelagic silica-secreting organisms, particularly diatoms and radiolarians. { ¦bī·ō¦jen·ik 'chərt }

biogenic mineral |MINERAL| A mineral in sediments or sedimentary rock which represents the hard parts of dead organisms. { ¦bī·ō¦jen·ik 'min·rəl }

biogenic reef |GEOL| A mass consisting of the hard parts of organisms, or of a biogenically constructed frame enclosing detrital particles, in a body of water; most biogenic reefs are made of corals or associated organisms. { ¦bī·ō¦jen·ik 'rēf }

biogenic sediment |GEOL| A deposit resulting from the physiological activities of organisms. { ¦bī·ō¦jen·ik 'sed·ə·mənt }

biogeochemical cycle |GEOCHEM| The chemical interactions that exist between the atmosphere, hydrosphere, lithosphere, and biosphere. { ,bī·ō,jē·ō'kem·ə·kəl 'sīkəl }

biogeochemical prospecting |GEOCHEM| A prospecting technique for subsurface ore deposits based on interpretation of the growth of certain plants which reflect subsoil concentrations of some elements. { ,bī·ō,jē·ō'kem·ə·kəl 'präs,pek·tiŋ }

biogeochemistry |GEOCHEM| A branch of geochemistry that is concerned with biologic materials and their relation to earth chemicals in an area. { ,bī·ō,jē·ō'kem·ə·strē }

bioherm |GEOL| A circumscribed mass of rock exclusively or mainly constructed by marine sedimentary organisms such as corals, algae, and stromatoporoids. Also known as organic mound. { 'bī·ō,hərm }

biohermal limestone |PETR| Reefs or reeflike mounds of carbonate that accumulated

43

much in the same fashion as modern reefs and atolls of the Pacific Ocean. { ¦bī·ō¦hər·məl 'līm,stōn }

biohermite [PETR] Limestone formed of debris from a bioherm. { ¦bī·ō'hər,mīt }

biolite [GEOL] A concretion formed of concentric layers through the action of living organisms. [PETR] See biolith. { 'bī·ō,līt }

biolith [PETR] A rock formed from or by organic material. Also known as biolite. { 'bī·ō,lith }

biolithite [PETR] An inclusive category for all organic limestone. { ,bī·ō'lith,īt }

biologic weathering See organic weathering. { ¦bī·ə¦läj·ik 'weth·ə·riŋ }

biomarkers [GEOL] Complex organic compounds found in oil, bitumen, rocks, and sediments that are linked with and distinctive of a particular source (such as algae, bacteria, or vascular plants); they are useful dating indicators in stratigraphy and molecular paleontology. Also known as chemical fossils; molecular fossils. { 'bī·ō,mär·kərz }

biomicrite [PETR] A limestone resembling biosparite except that the microcrystalline calcite matrix exceeds calcite cement. { ,bī·ə'mī,krīt }

biomicrosparite [PETR] **1.** Biomicrite in which the micrite groundmass has recrystallized to microspar. **2.** Microsparite containing fossil fragments or fossils. { ,bī·ō,mī·krō'spär,īt }

biomicrudite [PETR] Biomicrite with fossil fragments or fossils greater than 1 millimeter in diameter. { ,bī·ō'mī·krə,dīt }

biopelite See black shale. { bī'äp·ə,līt }

biopelmicrite [PETR] A limestone similar to biopelsparite but with a microcrystalline matrix that exceeds calcite cement. { bī·ə'pel·mə,krīt }

biopelsparite [PETR] A limestone similar to biosparite but with the ratio of fossils and fossil fragments to pellets between 3:1 and 1:3. { bī·ə'pel·spə,rīt }

biopyribole [MINERAL] **1.** A collective term for the rock-forming minerals pyroxene, amphibole, and mica. **2.** A chemically diverse but structurally related group of minerals that constitute substantial fractions of both the earth's crust and upper mantle; they exhibit single-chain, double-chain, triple-chain, and sheet silicate structures. { ,bī·ō'pir·ə,bōl }

biosparite [PETR] A limestone made up of less than 25% oolites and less than 25% intraclasts, with the ratio by volume of fossils and fragments to pellets being more than 3:1 and the calcite cement content being greater than the microcrystalline calcite content. { bī'äs·pə,rīt }

biostratigraphic unit [GEOL] A stratum or body of strata that is defined and identified by one or more distinctive fossil species or genera without regard to lithologic or other physical features or relations. { ¦bī·ō,strad·ə'graf·ik 'yü·nət }

biostratigraphy [PALEON] A part of paleontology concerned with the study of the conditions and deposition order of sedimentary rocks. { ¦bī·ō·strə'tig·rə·fē }

biostromal limestone [GEOL] Biogenic carbonate accumulations that are laterally uniform in thickness, in contrast to the moundlike nature of bioherms. { ¦bī·ə¦strō·məl 'līm,stōn }

biostrome [GEOL] A bedded structure or layer (bioclastic stratum) composed of calcite and dolomitized calcarenitic fossil fragments distributed over the sea bottom as fine lentils, independent of or in association with bioherms or other areas of organic growth. { 'bī·ə,strōm }

biotite [MINERAL] A black, brown, or dark green, abundant and widely distributed species of rock-forming mineral in the mica group; its chemical composition is variable: $K_2[Fe(II),Mg]_{6-4}[Fe(III),Al,Ti]_{0-2}(Si_{6-5},Al_{2,3})O_{20-22}(OH,F)_{4-2}$. Also known as black mica; iron mica; magnesia mica; magnesium-iron mica. { 'bī·ə,tīt }

biotite schist [PETR] A schist composed of biotite. { 'bī·ə,tīt 'shist }

bioturbation [GEOL] The disruption of marine sedimentary structures by the activities of benthic organisms. { ¦bī·ō·tər'bā·shən }

biozone [PALEON] The range of a single taxonomic entity in geologic time as reflected by its occurrence in fossiliferous rocks. { 'bī·ō,zōn }

bipedal dinosaur [PALEON] A dinosaur having two long, stout hindlimbs for walking and two relatively short forelimbs. { bī'ped·əl 'dīn·ə,sȯr }
bird-hipped dinosaur [PALEON] Any member of the order Ornithischia, distinguished by the birdlike arrangement of their hipbones. { 'bərd ,hipt 'dīn·ə,sȯr }
bird's-foot delta [GEOL] A delta with long, projecting distributary channels that branch outward like the toes or claws of a bird. { 'bərdz ,fút 'del·tə }
birnessite [MINERAL] A manganese oxide mineral often found as a primary constituent of manganese nodules or crusts. { bər'nes·īt }
bischofite [MINERAL] $MgCl_2·6H_2O$ A colorless to white, monoclinic mineral consisting of magnesium chloride hexahydrate. { 'bish·ə,fīt }
bisilicate [MINERAL] See metasilicate. { ,bī'sil·ə·kət }
bismite [MINERAL] Bi_2O_3 A monoclinic mineral composed of bismuth trioxide; native bismuth ore, occurring as a yellow earth. Also known as bismuth ocher. { 'biz,mīt }
bismuth [MINERAL] The brittle, rhombohedral mineral form of the native element bismuth. { 'biz·məth }
bismuth blende See eulytite. { 'biz·məth ¦blend }
bismuth glance See bismuthinite. { 'biz·məth ¦glans }
bismuthinite [MINERAL] Bi_2S_3 A mineral consisting of bismuth trisulfide, which has an orthorhombic structure and is usually found in fibrous or leafy masses that are lead gray with a yellowish tarnish and a metallic luster. Also known as bismuth glance. { 'biz·məth·ə,nīt }
bismuth ocher See bismite. { 'biz·məth 'ō·kər }
bismuth spar See bismutite. { 'biz·məth 'spär }
bismutite [MINERAL] $(BiO)_2CO_3$ A dull-white, yellowish, or gray, earthy, amorphous mineral consisting of basic bismuth carbonate. Also known as bismuth spar. { 'biz·məd,īt }
bismutotantalite [MINERAL] $Bi(Ta,Nb)O_4$ A pitch black, orthorhombic mineral consisting of an oxide of bismuth and tantalum and occurring in crystals. { ,biz·məd·ə'tan·tə,līt }
bitumenite See torbanite. { bī'tü·mə,nīt }
bituminization See coalification. { bī,tü·mə·nə'zā·shən }
bituminous [MINERAL] Of a mineral, having the odor of bitumen. { bī'tü·mə·nəs }
bituminous coal [GEOL] A dark brown to black coal that is high in carbonaceous matter and has 15–50% volatile matter. Also known as soft coal. { bī'tü·mə·nəs 'kōl }
bituminous lignite [GEOL] A brittle, lustrous bituminous coal. Also known as pitch coal. { bī'tü·mə·nəs 'lig,nīt }
bituminous rock See asphalt rock. { bī'tü·mə·nəs 'räk }
bituminous sand [GEOL] Sand containing bituminous-like material, such as the tar sands at Athabasca, Canada, from which oil is extracted commercially. { bī'tü·mə·nəs 'sand }
bituminous sandstone [PETR] A sandstone containing bituminous matter. { bī'tü·mə·nəs 'sand,stōn }
bituminous shale [PETR] A shale containing bituminous material. { bī'tü·mə·nəs 'shāl }
bituminous wood [GEOL] A variety of brown coal having the fibrous structure of wood. Also known as board coal; wood coal; woody lignite; xyloid coal; xyloid lignite. { bī'tü·mə·nəs 'wúd }
bixbyite [MINERAL] $(Mn,Fe)_2O_3$ A manganese-iron oxide mineral; black cubic crystals found in cavities in rhyolite. Also known as partridgeite; sitaparite. { 'biks·bē,īt }
black alkali [GEOL] A deposit of sodium carbonate that has formed on or near the surface in arid to semiarid areas. { ¦blak 'al·kə,lī }
black amber See jet coal. { ¦blak 'am·bər }
blackband [GEOL] An earthy carbonate of iron that is present with coal beds. { 'blak,band }
black coal See natural coke. { ¦blak 'kōl }
black cobalt See asbolite. { ¦blak 'kō·bȯlt }
black cotton soil See regur. { ¦blak ¦kat·ən 'sȯil }

black diamond

black diamond See carbonado. { ¦blak 'dī·mənd }

black durain [GEOL] A durain that has high hydrogen content and volatile matter, many microspores, and some vitrain fragments. { ¦blak 'dü,rān }

black granite See diorite. { ¦blak 'gran·ət }

black lead See graphite. { ¦blak 'led }

black lignite [GEOL] A lignite with a fixed carbon content of 35–60% and a total carbon content of 73.6–76.2% that contains between 6300 and 8300 Btu per pound; higher in rank than brown lignite. Also known as lignite A. { 'blak 'lig,nīt }

black mica See biotite. { ¦blak 'mī·kə }

black mud [GEOL] A mud formed where there is poor circulation or weak tides, such as in lagoons, sounds, or bays; the color is due to iron sulfides and organic matter. { ¦blak 'məd }

black ocher See wad. { ¦blak 'ō·kər }

black opal [MINERAL] A variety of gem-quality opal displaying internal reflections against a dark background. { ¦blak 'ō·pəl }

black sand [GEOL] Heavy, dark, sandlike minerals found on beaches and in stream beds; usually magnetite and ilmenite and sometimes gold, platinum, and monazite are present. { ¦blak 'sand }

black shale [PETR] Very thinly bedded shale rich in sulfides such as pyrite and organic material deposited under barred basin conditions so that there was an anaerobic accumulation. Also known as biopelite. { ¦blak 'shāl }

black silver See stephanite. { ¦blak 'sil·vər }

black tellurium See nagyagite. { 'blak ta'lür·ē·əm }

bladder See vesicle. { 'blad·ər }

Blaine formation [GEOL] A Permian red bed formation containing red shale and gypsum beds of marine origin in Oklahoma, Texas, and Kansas. { 'blān fȯr'mā·shən }

blairmorite [PETR] A porphyritic extrusive rock consisting mainly of analcite phenocrysts in a groundmass of sanidine, analcite, and alkalic pyroxene, with accessory sphene, melanite, and nepheline. { 'bler·mə,rīt }

blakeite [MINERAL] A deep reddish-brown to deep brown mineral consisting of anhydrous ferric tellurite; occurs in massive form, as microcrystalline crusts. { 'blā,kīt }

Blancan [GEOL] Upper Pliocene or lowermost Pleistocene geologic time. { 'blän·kən }

blanket deposit [GEOL] A flat deposit of ore; its length and width are relatively great compared with its thickness. { 'blaŋ·kət di'päz·ət }

blanket sand [GEOL] A relatively thin body of sand or sandstone covering a large area. Also known as sheet sand. { 'blaŋ·kət ,sand }

blastic deformation [GEOL] Rock deformation involving recrystallization in which space lattices are destroyed or replaced. { 'blas·tik ,dē,fȯr'mā·shən }

blasting [GEOL] Abrasion caused by movement of fine particles against a stationary fragment. { 'blas·tiŋ }

blasto- [PETR] A prefix indicating the presence in a rock of residual structures somewhat modified by metamorphism. { 'blas·tō }

blastogranitic rock [PETR] A metamorphic granitic rock which still has parts of the original granitic texture. { ¦blas·tō·grə'nid·ik 'räk }

Blastoidea [PALEON] A class of extinct pelmatozoan echinoderms in the subphylum Crinozoa. { bla'stȯid·ē·ə }

blastomylonite [PETR] Rock which has recrystallized after granulation. { ,blas·tə'mī·lə,nīt }

blastopelitic [PETR] Descriptive of the structure of metamorphosed argillaceous rocks. { ¦blas·tō·pə'lid·ik }

blastophitic [PETR] A metamorphosed rock which once contained lath-shaped crystals partly or wholly enclosed in augite and in which part of the original texture remains. { ¦blas·tō'fid·ik }

blastoporphyritic [PETR] Applied to the textures of metamorphic rocks that are derived from porphyritic rocks; the porphyritic character still remains as a relict feature. { ¦blas·tō¦pȯr·fə¦rid·ik }

blastopsammite [GEOL] A relict fragment of sandstone that is contained in a metamorphosed conglomerate. { bla'stäp·sə‚mīt }

blastopsephitic [GEOL] Descriptive of the structure of metamorphosed conglomerate or breccia. { bla¦stäp·sə¦fid·ik }

bleach spot [GEOL] A green or yellow area in red rocks formed by reduction of ferric oxide around an organic particle. Also known as deoxidation sphere. { 'blēch ‚spät }

bleb [PETR] A small, usually spherical inclusion in a rock mass. { bleb }

blende See sphalerite. { blend }

blended unconformity [GEOL] An unconformity that is not sharp because the original erosion surface was covered by a thick residual soil that graded downward into the underlying rock. { ¦blen·dəd ‚ən·kən'fȯr·məd·ē }

blind [GEOL] Referring to a mineral deposit with no surface outcrop. { blīnd }

blind coal See natural coke. { ¦blīnd ¦kōl }

blind valley [GEOL] A valley that has been made by a spring from an underground channel which emerged to form a surface stream, and that is enclosed at the head of the stream by steep walls. { 'blīnd 'val·ē }

blister [GEOL] A domelike protuberance caused by the buckling of the cooling crust of a molten lava before the flowing mass has stopped. { 'blis·tər }

blister hypothesis [GEOL] A theory of the formation of compressional mountains by a process in which radiogenic heat expands and melts a portion of the earth's crust and subcrust, causing a domed regional uplift (blister) on a foundation of molten material that has no permanent strength. { 'blis·tər hī'päth·ə·səs }

block clay See mélange. { 'bläk ‚klā }

block faulting [GEOL] A type of faulting in which fault blocks are displaced at different orientations and elevations. { 'bläk ‚fȯl·tiŋ }

block glide [GEOL] A translational landslide in which the slide mass moves outward and downward as an intact unit. { 'bläk ‚glīd }

block lava [GEOL] Lava flows which occur as a tumultuous assemblage of angular blocks. Also known as aa lava. { 'bläk ‚läv·ə }

block mountain [GEOL] A mountain formed by the combined processes of uplifting, faulting, and tilting. Also known as fault-block mountain. { 'bläk ‚maún·tən }

blödite See bloedite. { 'blō‚dīt }

bloedite [MINERAL] $MgSO_4 \cdot Na_2SO_4 \cdot 4H_2O$ A white or colorless monoclinic mineral consisting of magnesium sodium sulfate. Also spelled blödite. Also known as astrakanite; astrochanite. { 'blō‚dīt }

blomstrandine See priorite. { ‚blȯm'stran‚dēn }

bloodstone [MINERAL] **1.** A form of deep green chalcedony flecked with red jasper. Also known as heliotrope; oriental jasper. **2.** See hematite. { 'bləd ‚stōn }

bloom See blossom; efflorescence. { blüm }

blossom [GEOL] The oxidized or decomposed outcrop of a vein or coal bed. Also known as bloom. { 'bläs·əm }

blowhole [GEOL] A longitudinal tunnel opening in a sea cliff, on the upland side away from shore; columns of sea spray are thrown up through the opening, usually during storms. { 'blō‚hōl }

blowing cave [GEOL] A cave with an alternating air movement. Also known as breathing cave. { ¦blō·iŋ ¦kāv }

blowout [GEOL] Any of the various trough-, saucer-, or cuplike hollows formed by wind erosion on a dune or other sand deposit. { 'blō‚aút }

blowout dune See parabolic dune. { 'blō‚aút ‚dün }

blue asbestos See crocidolite. { ¦blü as'bes·təs }

blue band [GEOL] **1.** A layer of bubble-free, dense ice found in a glacier. **2.** A bluish clay found as a thin, persistent bed near the base of No. 6 coal everywhere in the Illinois-Indiana coal basin. { ¦blü 'band }

blue copper ore See azurite. { ¦blü ¦käp·ər 'ȯr }

blue ground [GEOL] **1.** The decomposed peridotite or kimberlite that carries the diamonds in the South African mines. **2.** Strata of the coal measures, consisting principally of beds of hard clay or shale. { ¦blü ¸graünd }

blue iron earth *See* vivianite. { ¦blü ¦ī·ərn 'ərth }

blue lead *See* galena. { ¦blü 'led }

blue magnetism [GEOPHYS] The magnetism displayed by the south-seeking end of a freely suspended magnet; this is the magnetism of the earth's north magnetic pole. { ¦blü 'mag·nə¸tiz·əm }

blue malachite *See* azurite. { ¦blü 'mal·ə¸kīt }

blue metal [GEOL] The common fine-grained blue-gray mudstone which is part of many of the coal beds of England. { 'blü ¸med·əl }

blue mud [GEOL] A combination of terrigenous and deep-sea sediments having a bluish gray color due to the presence of organic matter and finely divided iron sulfides. { 'blü ¸məd }

blue ocher *See* vivianite. { ¦blü 'ō·kər }

blueschist facies [PETR] High-pressure, low-temperature metamorphism associated with subduction zones which produces a broad mineral association including glaucophane, actinolite, jadeite, aegirine, lawsonite, and pumpellyite. { 'blü¸shist 'fā¸shēz }

blue spar *See* lazulite. { 'blü ¸spär }

bluestone [MINERAL] *See* chalcanthite. [PETR] **1.** A sandstone that is highly argillaceous and of even texture and bedding. **2.** The commercial name for a feldspathic sandstone that is dark bluish gray; it is easily split into thin slabs and used as flagstone. { 'blü¸stōn }

blue vitriol [MINERAL] *See* chalcanthite. { 'blü 'vit·rē¸ól }

board coal *See* bituminous wood. { 'bórd ¸kōl }

boart *See* bort. { bórt }

Bobasatranidae [PALEON] A family of extinct palaeonisciform fishes in the suborder Platysomoidei. { bə¸bas·ə'tran·ə¸dē }

bobierrite [MINERAL] $Mg_3(PO_4)_2 \cdot 8H_2O$ A transparent, colorless or white, monoclinic mineral consisting of octahydrated magnesium phosphate. { 'bō·bē·ə¸rīt }

bodenite [MINERAL] A metallic, steel-gray mineral consisting of cobalt, nickel, iron, arsenic, and bismuth; occurs in granular to fibrous masses. { 'bōd·ən¸īt }

bodily tide *See* earth tide. { ¦bäd·əl·ē 'tīd }

body [GEOL] An ore body, or pocket of mineral deposit. { 'bäd·ē }

body wave [GEOPHYS] A seismic wave that travels within the earth, as distinguished from one that travels along the surface. { 'bäd·ē ¸wāv }

boehmite [MINERAL] AlO(OH) Gray, brown, or red orthorhombic mineral that is a major constituent of some bauxites. { 'bā¸mīt }

boehm lamellae [GEOL] Lines or bands with dusty inclusions that are subparallel to the basal plane of quartz. { ¦bāmlə'mel·ē }

bogen structure [GEOL] The structure of vitric tuffs composed largely of shards of glass. { 'bō·gən ¸strək·chər }

boghead cannel shale [GEOL] A coaly shale that contains much waxy or fatty algae. { 'bäg¸hed ¦kan·əl ¸shäl }

boghead coal [GEOL] Bituminous or subbituminous coal containing a large proportion of algal remains and volatile matter; similar to cannel coal in appearance and combustion. { 'bäg¸hed ¸kōl }

bog iron ore [MINERAL] A soft, spongy, porous deposit of impure hydrous iron oxides formed in bogs, marshes, swamps, peat mosses, and shallow lakes by precipitation from iron-bearing waters and by the oxidation action of algae, iron bacteria, or the atmosphere. Also known as lake ore; limnite; marsh ore; meadow ore; morass ore; swamp ore. { 'bäg ¦ī·ərn ¸ór }

bog manganese *See* wad. { 'bäg 'maŋ·gə¸nēs }

bog-mine ore *See* bog ore. { 'bäg ¸mīn ¸ór }

bog ore [MINERAL] A poorly stratified accumulation of earthy metallic mineral substances, consisting mainly of oxides, that are formed in bogs, marshes, swamps, and other low-lying moist places. Also known as bog-mine ore. { 'bäg ¸ór }

Bohemian ruby *See* rose quartz. { bō'hem·ē·ən ¦rü·bē }

Bohemian topaz *See* citrine. { bō'hem·ē ən ¦tō,paz }

boiler plate [GEOL] A fairly smooth surface on a cliff, consisting of flush or overlapping slabs of rock, having little or no foothold. { 'bȯil·ər ,plāt }

bojite [PETR] **1.** A gabbro with primary hornblende substituting for augite. **2.** Hornblende diorite. { 'bō,jīt }

bole [GEOL] Any of various red, yellow, or brown earthy clays consisting chiefly of hydrous aluminum silicates. Also known as bolus; terra miraculosa. { bōl }

boleite [MINERAL] A deep Prussian blue, tetragonal mineral consisting of a hydroxide-chloride of lead, copper, and silver. { bō'lā,īt }

bolson [GEOL] In the southwestern United States, a basin or valley having no outlet. { bōl,sän }

boltwoodite [MINERAL] $K_2(UO_2)_2(SiO_3)_2(OH)_2 \cdot 5H_2O$ Yellow mineral consisting of hydrous potassium uranyl silicate. { 'bōlt·wə,dīt }

bolus *See* bole. { 'bō·ləs }

bolus alba *See* kaolin. { ¦bō·ləs 'äl·bə }

bomb [GEOL] Any large (greater than 64 millimeters) pyroclast ejected while viscous. { bäm }

bombiccite *See* hartite. { bäm'bē,chīt }

bomb sag [GEOL] Depressed and deranged laminae mainly found in beds of fine-grained ash or tuff around an included volcanic bomb or block which fell on and became buried in the deposit. { 'bäm ,sag }

bone bed [GEOL] Several thin strata or layers with many fragments of fossil bones, scales, teeth, and also organic remains. { 'bōn ,bed }

bone chert [PETR] A weathered residual chert that appears chalky and porous with a white color but may be stained red or other colors. { 'bōn ,chərt }

bone coal [GEOL] Argillaceous coal or carbonaceous shale that is found in coal seams. { 'bōn ,kōl }

boninite [PETR] An andesitic rock that contains much glass and abundant phenocrysts of bronzite and less of olivine and augite. { 'bän·ə,nīt }

Bononian [GEOL] Upper Jurassic (lower Portlandian) geologic time. { bə'nōn·ē·ən }

book *See* mica book. { bùk }

book structure [GEOL] A rock structure of numerous parallel sheets of slate alternating with quartz. { 'bùk ,strək·chər }

boothite [MINERAL] $CuSO_4 \cdot 7H_2O$ A blue, monoclinic mineral consisting of copper sulfate heptahydrate; usually occurs in massive or fibrous form. { 'bü,thīt }

boracite [MINERAL] $Mg_3B_7O_{13}Cl$ A white, yellow, green, or blue orthorhombic borate mineral occurring in crystals which appear isometric in external form; it is strongly pyroelectric, has a hardness of 7 on Mohs scale, and a specific gravity of 2.9. { 'bȯr·ə,sīt }

Boralf [GEOL] A suborder of the soil order Alfisol, dull brown or yellowish brown in color; occurs in cool or cold regions, chiefly at high latitudes or high altitudes. { 'bȯr,alf }

borate mineral [MINERAL] Any of the large and complex group of naturally occurring crystalline solids in which boron occurs in chemical combination with oxygen. { 'bȯ,rāt 'min·rəl }

borax [MINERAL] $Na_2B_4O_7 \cdot 10H_2O$ A white, yellow, blue, green, or gray borate mineral that is an ore of boron and occurs as an efflorescence or in monoclinic crystals; when pure it is used as a cleaning agent, antiseptic, and flux. Also known as diborate; pyroborate; sodium (1:2) borate; sodium tetraborate; tincal. { 'bȯ,raks }

border facies [GEOL] The outer portion of an igneous intrusion which differs in composition and texture from the main body. { 'bȯrd·ər ,fā·shēz }

borderland [GEOL] One of the crystalline, continental landmasses postulated to have existed on the exterior (oceanward) side of geosynclines. { 'bȯrd·ər,land }

borderland slope [GEOL] A declivity which indicates the inner margin of the borderland of a continent. { 'bȯrd·ər,land 'slōp }

borickite [MINERAL] $CaFe_5(PO_4)_2(OH)_{11} \cdot 3H_2O$ Reddish-brown, isotropic mineral consisting of a hydrated basic phosphate of calcium and iron; occurs in compact reniform masses. { 'bȯr·ə‚kīt }

bornhardt [GEOL] A large dome-shaped granite-gneiss outcrop having the characteristics of an inselberg. { 'bȯrn‚härt }

bornite [MINERAL] Cu_5FeS_4 A primary mineral in many copper ore deposits; specific gravity 5.07; the metallic and brassy color of a fresh surface rapidly tarnishes upon exposure to air to an iridescent purple. { 'bȯr‚nīt }

boroarsenate [MINERAL] One of a group of borate minerals containing arsenic; cahnite is an example. { ‚bȯr·ō'ar·sə‚nāt }

borolanite [PETR] A hypabyssal rock that is essentially orthoclase and melanite with subordinate nepheline, biotite, and pyroxene. { bə'räl·ə‚nīt }

Boroll [GEOL] A suborder of the soil order Mollisol, characterized by a mean annual soil temperature of less than 8°C and by never being dry for 60 consecutive days during the 90-day period following the summer solstice. { 'bȯ‚rȯl }

boronatrocalcite See ulexite. { ‚bȯr·ō‚na·trō'kal‚sīt }

borosilicate [MINERAL] A salt of boric and silicic acids which occurs in the natural minerals tourmaline, datolite, and dumortierite. { ‚bȯr·ō'sil·i·kət }

bort [MINERAL] Imperfectly crystallized diamond material unsuitable for gems because of its shape, size, or color and because of flaws or inclusions; used for abrasive and cutting purposes. Also spelled boart. { bȯrt }

Boryhaenid [PALEON] A carnivorous marsupial from the Miocene Epoch that resembled the wolf. { ‚bȯr·ē'han·əd }

boss [GEOL] A large, irregular mass of crystalline igneous rock that formed some distance below the surface but is now exposed by denudation. { bȯs }

bostonite [PETR] A rock with coarse trachytic texture formed almost wholly of albite and microcline and with accessory pyroxene. { 'bȯs·tə‚nīt }

botallackite [MINERAL] $Cu_2(OH)_3Cl \cdot 3H_2O$ A pale bluish-green to green, orthorhombic mineral consisting of a basic copper chloride; occurs as crusts of crystals. { bə'tal·ə‚kīt }

Bothriocidaroida [PALEON] An order of extinct echinoderms in the subclass Perischoechinoidea in which the ambulacra consist of two columns of plates, the interambulacra of one column, and the madreporite is placed radially. { ‚bä·thrē·ō‚sik·ə'rȯid‚ē·ə }

botryogen [MINERAL] $MgFe(SO_4)_2(OH) \cdot 7H_2O$ Orange-red, monoclinic mineral consisting of a hydrated basic sulfate of magnesium and trivalent iron. { 'bä·trē·ə‚jen }

botryoid [GEOL] **1.** A mineral formation shaped like a bunch of grapes. **2.** Specifically, such a formation of calcium carbonate occurring in a cave. Also known as clusterite. { 'bä·trē‚ȯid }

bottom [GEOL] **1.** The bed of a body of running or still water. **2.** See root. { 'bäd·əm }

bottomland [GEOL] A lowland formed by alluvial deposit about a lake basin or a stream. { 'bäd·əm‚land }

bottom load See bed load. { 'bäd·əm ‚lōd }

bottom moraine See ground moraine. { 'bäd·əm mə'rān }

bottomset beds [GEOL] Horizontal or gently inclined layers of finer material carried out and deposited on the bottom of a lake or sea in front of a delta. { 'bäd·əm‚set ‚bedz }

bottom terrace [GEOL] A landform deposited by streams with moderate or small bottom loads of coarse sand and gravel, and characterized by a broad, sloping surface in the direction of flow and a steep escarpment facing downstream. { 'bäd·əm ‚ter·əs }

boudin [GEOL] One of a series of sausage-shaped segments found in a boudinage. { bü'dan }

boudinage [GEOL] A structure in which beds set in a softer matrix are divided by cross fractures into segments resembling pillows. { ‚büd·ən‚äzh }

Bouguer correction See Bouguer reduction. { bü'ger kə'rek·shən }

Bouguer gravity anomaly [GEOPHYS] A value that corrects the observed gravity for latitude and elevation variations, as in the free-air gravity anomaly, plus the mass of material above some datum (usually sea level) within the earth and topography. { bü'ger 'grav·əd·ē ə'näm·ə·lē }

Bouguer reduction [GEOL] A correction made in gravity work to take account of the station's altitude and the rock between the station and sea level. Also known as Bouguer correction. { bü'ger ri'dək·shən }

boulangerite [MINERAL] $Pb_5Sb_4S_{11}$ A bluish-lead-gray, monoclinic mineral consisting of lead antimony sulfide. { bü'lan·jə,rīt }

boulder [GEOL] A worn rock with a diameter exceeding 256 millimeters. Also spelled bowlder. { 'bōl·dər }

boulder barricade [GEOL] An accumulation of large boulders that is visible along a coast between low and half tide. { 'bōl·dər ,bar·ə,kād }

boulder belt [GEOL] A long, narrow accumulation of boulders elongately transverse to the direction of glacier movement. { 'bōl·dər ,belt }

boulder clay See till. { 'bōl·dər ,klā }

boulder pavement [GEOL] A surface of till with boulders; the till has been abraded to flatness by glacier movement. { 'bōl·dər ,pāv·mənt }

boulder train [GEOL] Glacial boulders derived from one locality and arranged in a right-angled line or lines leading off in the direction in which the drift agency operated. { 'bōl·dər ,trān }

bounce cast [GEOL] A short ridge underneath a stratum fading out gradually in both directions. { 'baůns ,kast }

boundary [GEOL] A line between areas occupied by rocks or formations of different type and age. { 'baůn·drē }

boundary wave [GEOPHYS] A seismic wave that propagates along a free surface or an interface between defined layers. { 'baůn·drē ,wāv }

bournonite [MINERAL] $PbCuSbS_3$ Steel-gray to black orthorhombic crystals; mined as an ore of copper, lead, and antimony. Also known as berthonite; cogwheel ore. { 'bůr·nə,nīt }

boussingaultite [MINERAL] $(NH_4)_2Mg(SO_4)_2 \cdot 6H_2O$ A colorless to yellowish-pink, monoclinic mineral consisting of a hydrated sulfate of ammonium and magnesium; usually occurs in massive form, as crusts or stalactites. { ,büs·ən'gȯl,tīt }

Bowen reaction series [MINERAL] A series of minerals wherein any early-formed phase will react with the melt later in the differentiation to yield a new mineral further in the series. { 'bō·ən rē'ak·shən ,sir·ēz }

Bowie formula [GEOPHYS] A correction used for calculation of the local gravity anomaly on earth. { 'bō·ē ,fȯrm·yə·lə }

bowlder See boulder. { 'bōl·dər }

bowlingite See saponite. { 'bō·liŋ,gīt }

box fold [GEOL] A fold in which the broad, flat top of an anticline or the broad, flat bottom of a syncline is bordered by steeply dipping limbs. { 'bäks ,fōld }

Box Hole [GEOL] A meteorite crater in central Australia, 575 feet (175 meters) in diameter. { 'bäks ,hōl }

boxwork [GEOL] Limonite and other minerals which formed at one time as blades or plates along cleavage or fracture planes, after which the intervening material dissolved, leaving the intersecting blades or plates as a network. { 'bäks,wərk }

Brachiosaurus [PALEON] A herbivorous sauropod dinosaur, 90 feet (27 meters) long and weighing 85–110 tons, from the Late Jurassic that had a very long neck. { ,brā·kē·ə'sȯr·əs }

brachypinacoid [GEOL] A pinacoid parallel to the vertical and the shorter lateral axis. { ,brak·i'pin·ə,kȯid }

brachysyncline [GEOL] A broad, short syncline. { ,brak·i'sin,klīn }

Brachythoraci [PALEON] An order of the joint-neckfishes, now extinct. { ,brak·i'thȯr·ə·sī }

brackebuschite [MINERAL] $Pb_4MnFe(VO_4)_4 \cdot 2H_2O$ Dark brown to black, monoclinic mineral consisting of a hydrated vanadate of lead, manganese, and iron. { 'bra·kə,bů,shīt }

51

Bradfordian

Bradfordian |GEOL| Uppermost Devonian geologic time. { ˌbrad'fȯrd·ē·ən }

bradleyite |MINERAL| $Na_3Mg(PO_4)(CO_3)$ A light gray mineral consisting of a phosphate-carbonate of sodium and magnesium; occurs as fine-grained masses. { 'brad·lē͵īt }

Bradyodonti |PALEON| An order of Paleozoic cartilaginous fishes (Chondrichthyes), presumably derived from primitive sharks. { ˌbrä·dē·ō'dän͵tī }

braggite |MINERAL| PtS A steel-gray platinum sulfide mineral with tetragonal crystals. { 'bra͵gīt }

brammalite |MINERAL| A mica-type clay mineral that is different from illite because it has soda instead of potash; it is the sodium analog of illite. Also known as sodium illite. { 'bram·ə͵līt }

branchite See hartite. { 'bran͵chīt }

brandtite |MINERAL| $Ca_2Mn(AsO_4)_2·2H_2O$ A colorless to white, monoclinic mineral consisting of a hydrated arsenate of calcium and manganese. { 'brant͵īt }

brannerite |MINERAL| A complex, black, opaque titanite of uranium and other elements in which the weight of uranium exceeds the weight of titanium; monoclinic and possibly $(U,Ca,Fe,Y,Th)_3Ti_5O_6$· { 'bran·ə͵rīt }

brass |GEOL| A British term for sulfides of iron (pyrites) in coal. Also known as brasses. { bras }

brasses See brass. { 'bras·əz }

brass ore See aurichalcite. { 'bras͵ȯr }

braunite |MINERAL| $3Mn_2O_3·MnSiO_3$ Brittle mineral that forms tetragonal crystals; commonly found as steel-gray or brown-black masses in the United States, Europe, and South America; it is an ore of manganese. { 'braủ͵nīt }

bravoite |MINERAL| $(Ni,Fe)S_2$ A yellow sulfide ore of nickel containing iron. { 'brä͵vō͵īt }

brazilianite |MINERAL| $NaAl_3(PO_4)_2(OH)_4$ A chartreuse yellow to pale yellow, monoclinic mineral consisting of a basic phosphate of sodium and aluminum. { brə'zil·yə͵nīt }

breached anticline |GEOL| An anticline that has been more deeply eroded in the center. Also known as scalped anticline. { ˌbrēcht 'an·ti͵klīn }

breached cone |GEOL| A cinder cone in which lava has broken through the sides and broken material has been carried away. { 'brēcht ͵kōn }

breadcrust |GEOL| A surficial structure resembling a crust of bread, as the concretions formed by evaporation of salt water. { 'bred͵krəst }

breadcrust bomb |GEOL| A volcanic bomb with a cracked exterior. { 'bred͵krəst ͵bäm }

break |GEOL| See knickpoint. { brāk }

breaker terrace |GEOL| A type of shore found in lakes in glacial drift; the terrace is formed from stones deposited by waves. { 'brā·kər ͵ter·əs }

break thrust |GEOL| A thrust fault cutting across one limb of a fold. { 'brāk ͵thrəst }

breathing cave See blowing cave. { 'brēth·iŋ ͵kāv }

breccia |PETR| A rock made up of very angular coarse fragments; may be sedimentary or may be formed by grinding or crushing along faults. { 'brech·ə }

breccia dike |GEOL| A dike formed of breccia injected into the country rock. { 'brech·ə ͵dīk }

breccia marble |PETR| Any marble containing angular fragments. { 'brech·ə ͵mär·bəl }

breccia pipe See pipe. { 'brech·ə ͵pīp }

breithauptite |MINERAL| NiSb A light copper red mineral consisting of nickel antimonide; commonly occurs in association with silver minerals. { 'brīt͵haủp͵tīt }

Bretonian orogeny |GEOL| Post-Devonian diastrophism that is found in Nova Scotia. { bre'tōn·ē·ən ȯ'räj·ə·nē }

Bretonian strata |GEOL| Upper Cambrian strata in Cape Breton, Nova Scotia. { bre'tōn·ē·ən 'strad·ə }

breunnerite |MINERAL| $(Mg,Fe,Mn)CO_3$ A carbonate mineral consisting of an isomorphous system of the metallic components. { 'brȯin·ə͵rīt }

brewsterite |MINERAL| $Sr(Al_2Si_6O_{18})·5H_2O$ A member of the zeolite family of minerals; crystallizes in the monoclinic system and usually contains some calcium. { 'brü·stə͵rīt }

bright-banded coal See bright coal. { 'brīt ‚ban·dəd ¦kōl }
bright coal [GEOL] A jet-black, pitchlike type of banded coal that is more compact than dull coal and breaks with a shell-shaped fracture; microscopic examination shows a consistency of more than 5% anthraxyllon and less than 20% opaque matter. Also known as bright-banded coal; brights. { 'brīt ‚kōl }
brights See bright coal. { brīts }
brimstone [MINERAL] A common or commercial name for native sulfur. { 'brim‚stōn }
britholite [MINERAL] (Na,Ce,Ca)$_5$(OH)|(P,Si)O$_4$|$_3$ A rare-earth phosphate found in carbonatites in Kola Peninsula, former Soviet Union. { 'brith·ə‚līt }
brittle mica [MINERAL] Hydrous sodium, calcium, magnesium, and aluminum silicates; a group of more or less related minerals that resemble true micas but cleave to brittle flakes and contain calcium as the essential constituent. { ¦brid·əl 'mī·kə }
brittle silver ore See stephanite. { ¦brid·əl 'sil·vər ‚òr }
brochanite See brochantite. { brō'shän‚īt }
brochanthite See brochantite. { brō'shän‚thīt }
brochantite [MINERAL] Cu$_4$(SO$_4$)(OH)$_6$ A monoclinic copper mineral, emerald to dark green, commonly found with copper sulfide deposits; a minor copper ore. Also known as brochanite; brochanthite; warringtonite. { brō'shän‚tīt }
bromellite [MINERAL] BeO A white hexagonal mineral consisting of beryllium oxide; it is harder than zincite. { brō'me‚līt }
bromlite [MINERAL] BaCa(CO$_3$)$_2$ An orthorhombic mineral composed of a carbonate of barium and calcium. Also known as alstonite. { 'brōm‚līt }
bromyrite [MINERAL] AgBr A secondary ore of silver that occurs in the oxidized zone of silver deposits; exists in crusts and coatings resembling a wax. { 'brō·mə‚rīt }
brontides [GEOPHYS] Low, rumbling, thunderlike sounds of short duration, most frequently heard in active seismic regions and believed to be of seismic origin. { 'brän‚tīdz }
Brontosaurus See Apatosaurus. { ‚brän·tə'sòr·əs }
Brontotheriidae [PALEON] The single family of the extinct mammalian superfamily Brontotherioidea. { ¦brän·tō·thə'rī·ə‚dē }
Brontotherioidea [PALEON] The titanotheres, a superfamily of large, extinct perissodactyl mammals in the suborder Hippomorpha. { ¦brän·tō‚the·rē'òid·ē·ə }
bronze mica See phlogopite. { ¦bränz 'mī·kə }
bronzite [MINERAL] (Mg,Fe)(SiO$_3$) An orthopyroxene mineral that forms metallic green orthorhombic crystals; a form of the enstatite-hypersthene series. { 'brän‚zīt }
bronzitfels See bronzitite. { 'brän·zət‚felz }
bronzitite [PETR] A pyroxenite that is composed almost entirely of bronzite. Also known as bronzitfels. { 'brän·zə‚tīt }
brookite [MINERAL] TiO$_2$ A brown, reddish, or black orthorhombic mineral; it is trimorphous with rutile and anatase, has hardness of 5.5–6 on Mohs scale, and a specific gravity of 3.87–4.08. Also known as pyromelane. { 'brú‚kīt }
brown clay See red clay. { ¦braún ¦klā }
brown clay ironstone [GEOL] Limonite in the form of concrete masses, often in concretionary nodules. { 'braún ‚klā 'ī·ərn‚stōn }
brown coal See lignite. { ¦braún ¦kōl }
brown hematite See limonite. { ¦braún 'hem·ə‚tīt }
brown iron ore See limonite. { ¦braún 'ī·ərn ‚òr }
brown lignite [GEOL] A type of lignite with a fixed carbon content ranging from 30 to 55% and total carbon from 65 to 73.6; contains 6300 Btu per pound (14.65 megajoules per kilogram). Also known as lignite B. { ¦braún 'lig‚nīt }
brown mica See phlogopite. { ¦braún 'mī·kə }
brown soil [GEOL] Any of a zonal group of soils, with a brown surface horizon which grades into a lighter-colored soil and then into a layer of carbonate accumulation. { ¦braún ¦sòil }
brown spar [GEOL] Any light-colored crystalline carbonate that contains iron, such as ankerite or dolomite, and is therefore brown. { 'braún ‚spär }
brucite [MINERAL] Mg(OH)$_2$ A hexagonal mineral; native magnesium hydroxide that

appears gray and occurs in serpentines and impure limestones; hardness is 2.5 on Mohs scale, and specific gravity is 2.38–2.40. { 'brü,sīt }

brownstone [PETR] Ferruginous sandstone with its grains coated with iron oxide. { 'braún,stōn }

brugnatellite [MINERAL] $Mg_6Fe(OH)_{13}CO_3 \cdot 4H_2O$ A flesh pink to yellowish- or brownish-white, hexagonal mineral consisting of a hydrated carbonate-hydroxide of magnesium and ferric iron; occurs in massive form. { ,brü·nyə'te,līt }

brushite [MINERAL] $CaHPO_4 \cdot 2H_2O$ A nearly colorless mineral that is a constituent of rock phosphates that crystallizes in slender or massive crystals. { 'brə,shīt }

Bruxellian [GEOL] Lower middle Eocene geologic time. { brü'sel·yən }

B tectonite [PETR] Tectonite with a fabric dominated by linear elements indicating an axial direction rather than a slip surface. { ¦bē 'tek·tə,nīt }

bubble pulse [GEOPHYS] An extraneous effect during a seismic survey caused by a bubble formed by a seismic charge, explosion, or spark fired in a body of water. { 'bəb·əl ,pəls }

bubble train [GEOL] A string or strings of vesicles in lava, indicating the path of rising gas escaping a flow of lava. { 'bəb·əl ,trān }

bubble wall fragment [GEOL] A glassy volcanic shard revealing part of a vesicle surface which may be curved or flat. { 'bəb·əl ,wól ,frag·mənt }

bucaramangite [MINERAL] A pale yellow variety of retinite that looks like amber but is insoluble in alcohol. { ,byü·kə·rə'maŋ,gīt }

buchite [PETR] A partially vitrified inclusion of sandstone in basalt. { 'bü,kīt }

buchonite [PETR] An extrusive rock formed of labradorite, titanaugite, and titaniferous hornblende, with nepheline and sodic sanidine and accessory biotite, apatite, and opaque oxides. { 'bü·kə,nīt }

bucklandite See allanite. { 'bək·lən,dīt }

buckle fold [GEOL] A double flexure of rock beds formed by compression acting in the plane of the folded beds. { 'bək·əl ,fōld }

buckwheat coal [GEOL] An anthracite coal that passes through 9/16-inch (14-millimeter) holes and over 5/16-inch (8-millimeter) holes in a screen. { 'bək,wēt ,kōl }

buetschliite [MINERAL] $K_6Ca_2(CO_3)_5 \cdot 6H_2O$ A mineral that is probably hexagonal and consists of a hydrated carbonate of potassium and calcium. { 'büch·lē,īt }

bughole See vug. { 'bəg,hōl }

buhrstone [PETR] A silicified fossiliferous limestone with abundant cavities previously occupied by fossil shells. Also known as millstone. { 'bər,stōn }

built terrace See alluvial terrace. { ¦bilt ¦ter·əs }

bunsenite [MINERAL] NiO A pistachio-green mineral consisting of nickel monoxide and occurring as octahedral crystals. { 'bən·sə,nīt }

Bunter [GEOL] Lower Triassic geologic time. Also known as Buntsandstein. { 'bún·tər }

Buntsandstein See Bunter. { 'bùnt·sən,shtīn }

burden [GEOL] All types of rock or earthy materials overlying bedrock. { 'bərd·ən }

Burdigalian [GEOL] Upper lower Miocene geologic time. { ,bərd·i'gāl·yən }

Burgess Shale [GEOL] A fossil deposit in the Canadian Rockies, British Columbia, consisting of a diverse fauna that accumulated in a clay and silt sequence during the Cambrian. { ¦bər·jəs 'shāl }

burial metamorphism [GEOL] A kind of regional metamorphism which affects sediments and interbedded volcanic rocks in a geosyncline without the factors of orogenesis or magmatic intrusions. { 'ber·ē·əl med·ə'mòr,fiz·əm }

buried hill [GEOL] A hill of resistant older rock over which later sediments are deposited. { 'ber·ēd 'hil }

buried placer [GEOL] Old deposit of a placer which has been buried beneath lava flows or other strata. { 'ber·ēd 'plās·ər }

buried river [GEOL] A river bed which has become buried beneath streams of alluvial drifts or basalt. { 'ber·ēd 'riv·ər }

buried soil See paleosol. { 'ber·ēd 'sóil }

burkeite [MINERAL] $Na_6(CO_3)(SO_4)_2$ A white to pale buff or gray mineral consisting of a carbonate-sulfate of sodium. { 'bɔr,kīt }

Bushveld Complex [GEOL] In South Africa, an enormous layered intrusion, containing over half the world's platinum, chromium, vanadium, and refractory minerals. { ¦bush,veld 'käm,pleks }

bustite See aubrite. { 'bəs,tīt }

butlerite [MINERAL] $Fe(SO_4)(OH)·2H_2O$ A deep orange, monoclinic mineral consisting of a hydrated basic ferric sulfate. Also known as parabutlerite. { 'bət·lə,rīt }

butter rock See halotrichite. { 'bəd·ər ,räk }

buttgenbachite [MINERAL] $Cu_{19}(NO_3)_2Cl_4(OH)_{32}·3H_2O$ An azure blue, hexagonal mineral consisting of a hydrated basic chloride-sulfate-nitrate of copper. { 'bət·gən,ba,kīt }

buttress [PALEON] A ridge on the inner surface of a pelecypod valve which acts as a support for part of the hinge. { 'bə·trəs }

buttress sands [GEOL] Sandstone bodies deposited above an unconformity; the upper portion rests upon the surface of the unconformity. { 'bə·trəs ,sanz }

byerite [GEOL] Bituminous coal that does not crack in fire and melts and enlarges upon heating. { 'bī·ə,rīt }

byon [GEOL] Gem-bearing gravel, particularly that with brownish-yellow clay in which corundum, rubies, sapphires, and so forth occur. { 'bī,än }

bysmalith [GEOL] A body of igneous rock that is more or less vertical and cylindrical; it crosscuts adjacent sediments. { 'biz·mə,lith }

bytownite [MINERAL] A plagioclase feldspar with a composition ranging from $Ab_{30}An_{70}$ to $Ab_{10}An_{90}$, where Ab = $NaAlSi_3O_8$ and An = $CaAl_2Si_2O_8$; occurs in basic and ultrabasic igneous rock. { 'bī·tau̇,nīt }

C

cacoxenite [MINERAL] $Fe_4(PO_4)_3(OH)_3 \cdot 12H_2O$ Yellow or brownish mineral consisting of a hydrous basic iron phosphate occurring in radiated tufts. { kə'käk·sə,nīt }

cadmium blende See greenockite. { 'kad·mē·əm ,blend }

cadmium ocher See greenockite. { 'kad·mē·əm 'ō·kər }

cadwaladerite [MINERAL] $Al(OH)_2Cl \cdot 4H_2O$ A mineral consisting of a hydrous basic aluminum chloride. { kad'wäl·ə·də,rīt }

Caenolestidae [PALEON] A family of extinct insectivorous mammals in the order Marsupialia. { ,sē·nə'les·tə,de }

cahnite [MINERAL] $Ca_2B(OH)_4(AsO_4)$ A tetragonal borate mineral occurring in white, sphenoidal crystals. { 'kä,nīt }

Cainotheriidae [PALEON] The single family of the extinct artiodactyl superfamily Cainotherioidea. { ,kān·ə·thə'rī·ə,dē }

Cainotherioidea [PALEON] A superfamily of extinct, rabbit-sized tylopod ruminants in the mammalian order Artiodactyla. { ,kān·ə·ther·ē'òid·ē·ə }

Cainozoic See Cenozoic. { ,kān·ə'zō·ik }

cairngorm See smoky quartz. { 'kern,górm }

caking coal [GEOL] A type of coal which agglomerates and softens upon heating; after volatile material has been expelled at high temperature, a hard, gray cellular mass of coke remains. Also known as binding coal. { 'kāk·iŋ ,kōl }

Calabrian [GEOL] Lower Pleistocene geologic time. { kə'läb·rē·ən }

calaite See turquoise. { kə'lā,īt }

calamine See hemimorphite; smithsonite. { 'kal·ə,mīn }

Calamitales [PALEOBOT] An extinct group of reedlike plants of the subphylum Sphenopsida characterized by horizontal rhizomes and tall, upright, grooved, articulated stems. { kə,lam·ə'tā·lēz }

calaverite [MINERAL] $AuTe_2$ A yellowish or tin-white, monoclinic mineral commonly containing gold telluride and minor amounts of silver. { kə'lav·ə,rīt }

calc-alkalic series [PETR] Series of igneous rocks in which the weight percentage of silica is 55–61. { ¦kalk ¦al'kal·ik ,sir·ēz }

calcarenite [PETR] A type of limestone or dolomite composed of coral or shell sand or of sand formed by erosion of older limestones, with particle size ranging from 1/16 to 2 millimeters. { kal·kə'rē,nīt }

calcareous crust See caliche. { kal'ker·ē·əs 'krəst }

calcareous duricrust See caliche. { kal'ker·ē·əs 'dúr·i,krəst }

calcareous ooze [GEOL] A fine-grained pelagic sediment containing undissolved sand- or silt-sized calcareous skeletal remains of small marine organisms mixed with amorphous clay-sized material. { kal'ker·ē·əs 'üz }

calcareous schist [PETR] A coarse-grained metamorphic rock derived from impure calcareous sediment. { kal'ker·ē·əs 'shist }

calcareous sinter See tufa. { kal'ker·ē·əs 'sin·tər }

calcareous soil [GEOL] A soil containing accumulations of calcium and magnesium carbonate. { kal'ker·ē·əs 'sòil }

calcareous tufa See tufa. { kal'ker·ē·əs 'tü·fə }

calciclastic [PETR] Pertaining to calcium carbonate-containing rock eroded from a

calcification

preexisting source, transported some distance, and then redeposited; for example, calciclastic limestone. { ‚kal·sə'klas·tik }

calcification [GEOCHEM] Any process of soil formation in which the soil colloids are saturated to a high degree with exchangeable calcium, thus rendering them relatively immobile and nearly neutral in reaction. { ‚kal·sə·fə'kā·shən }

calcilutite [PETR] **1.** A dolomite or limestone formed of calcareous rock flour that is typically nonsiliceous. **2.** A rock of calcium carbonate formed of grains or crystals with average diameter less than 1/16 millimeter. { ‚kal·sə'lü‚tīt }

calciocarnotite See tyuyamunite. { ‚kal·sē·ō'kär·nə‚tīt }

calcioferrite [MINERAL] $Ca_2Fe_2(PO_4)OH \cdot 7H_2O$ A yellow or green mineral consisting of a hydrous basic calcium iron phosphate and occurring in nodular masses. { ‚kal·sē·ō'fe‚rīt }

calciovolborthite [MINERAL] $CaCu(VO_4)(OH)$ Green, yellow, or gray mineral consisting of a basic vanadate of calcium and copper. Also known as tangeite. { ‚kal·sē·ō'vòl‚bór‚thīt }

calcirudite [PETR] Dolomite or limestone formed of worn or broken pieces of coral or shells or of limestone fragments coarser than sand; the interstices are filled with sand, calcite, or mud, the whole bound together with a calcareous cement. { kal'sir·ə‚dīt }

calcite [MINERAL] $CaCO_3$ One of the commonest minerals, the principal constituent of limestone; hexagonal-rhombohedral crystal structure, dimorphous with aragonite. Also known as calcspar. { 'kal‚sīt }

calcite compensation depth [GEOL] The depth in the ocean (about 5000 meters) below which solution of calcium carbonate occurs at a faster rate than its deposition. Abbreviated CCD. { 'kal‚sīt käm·pən'sā·shən ‚depth }

calcite dolomite [PETR] A carbonate rock with a composition of 10–50% calcite and 90–50% dolomite. { 'kal‚sīt 'dol·ə‚mīt }

calclacite [MINERAL] $CaCl_2Ca(C_2H_3O_2)\cdot 10H_2O$ A white mineral consisting of a hydrated chloride-acetate of calcium; occurs as hairlike efflorescences. { 'kal·klə‚sīt }

Calclamnidae [PALEON] A family of Paleozoic echinoderms of the order Dendrochirotida. { kal'klam·nə‚dē }

calclithite [PETR] Limestone with 50% or more fragments of older limestone that was redeposited after being eroded from the land. { 'kal·klə‚thīt }

calcrete [GEOL] A conglomerate of surficial gravel and sand cemented by calcium carbonate. { 'kal‚krēt }

calc-silicate [GEOL] Referring to a metamorphic rock consisting mainly of calcite and calcium-bearing silicates. { 'kalk 'sil·ə·kət }

calc-silicate hornfels [PETR] A metamorphic rock with a fine grain of calcium silicate minerals. { ‚kalk 'sil·ə‚kät 'hórn‚felz }

calc-silicate marble [PETR] Marble having conspicuous calcium silicate or magnesium silicate minerals. { ‚kalk 'sil·ə‚kät 'mär·bəl }

calcspar See calcite. { 'kalk‚spär }

calcsparite See sparry calcite. { kalk'spä‚rīt }

caldera [GEOL] A large collapse depression at a volcano summit that is typically circular to slightly elongate in shape, with dimensions many times greater than any included vent. It ranges from a few miles to 37 miles (60 kilometers) in diameter. It may resemble a volcanic crater in form, but differs in that it is a collapse rather than a constructional feature. { kal'der·ə }

Caledonian orogeny [GEOL] Deformation of the crust of the earth by a series of diastrophic movements beginning perhaps in Early Ordovician and continuing through Silurian, extending from Great Britain through Scandinavia. { ‚kal·ə‚dōn·ē·ən ò'räj·ə·nē }

Caledonides [GEOL] A mountain system formed in Late Silurian to Early Devonian time in Scotland, Ireland, and Scandinavia. { ‚kal·ə'dä‚nīdz }

caledonite [MINERAL] $Cu_2Pb_5(SO_4)_3CO_3(OH)_6$ A mineral occurring as green, orthorhombic crystals composed of basic copper lead sulfate; found in copper-lead deposits. { ‚kal·ə'dä‚nīt }

caliche [GEOL] **1.** Conglomerate of gravel, rock, soil, or alluvium cemented with sodium

58

salts in Chilean and Peruvian nitrate deposits; contains sodium nitrate, potassium nitrate, sodium iodate, sodium chloride, sodium sulfate, and sodium borate   **2.** A thin layer of clayey soil capping auriferous veins (Peruvian usage).   **3.** Whitish clay in the selvage of veins (Chilean usage).   **4.** A recently discovered mineral vein.   **5.** A secondary accumulation of opaque, reddish brown to buff or white calcareous material occurring in layers on or near the surface of stony soils in arid and semiarid regions of the southwestern United States; called hardpan, calcareous duricrust, and kanker in different geographic regions.   Also known as calcareous crust; croute calcaire; nari; sabach; tepetate.   { kə'lē·chē }

californite   [MINERAL] $Ca_{10}Al_4(Mg,Fe)_2Si_9O_{34}(OH,F)_4$ A variety of vesuvianite resembling jade; it is dark-, yellowish-, olive-, or grass-green and occurs in translucent to opaque compact or massive form.   Also known as American jade.   { ,kal·ə'fór,nīt }

callenia See stromatolite.   { kə'lēn·yə }

Callovian   [GEOL]   A stage in uppermost Middle or lowermost Upper Jurassic which marks a return to clayey sedimentation.   { kə'lōv·ē·ən }

calomel   [MINERAL] Hg_2Cl_2 A colorless, white, grayish, yellowish, or brown secondary, sectile, tetragonal mineral; used as a cathartic, insecticide, and fungicide.   Also known as calomelene; calomelite; horn quicksilver; mercurial horn ore.   { 'kal·ə·məl }

calomelene See calomel.   { kə'läm·ə,lēn }

calomelite See calomel.   { ,kal·ə'me,līt }

calving   [GEOL]   The breaking off of a mass of ice from its parent glacier, iceberg, or ice shelf.   Also known as ice calving.   { 'kav·iŋ }

Camarasaurus   [PALEON]   A herbivorous sauropod dinosaur, 60-feet (18 meters) long and weighing 20 tons, from the Late Jurassic Period that had a very long neck and tail.   { ,ka·mə·rə'sór·əs }

camber   [GEOL]   **1.** A terminal, convex shoulder of the continental shelf.   **2.** A structural feature that is caused by plastic clay beneath a bed flowing toward a valley so that the bed sags downward and seems to be draped over the sides of the valley.   { 'kam·bər }

Cambrian   [GEOL]   The lowest geologic system that contains abundant fossils of animals, and the first (earliest) geologic period of the Paleozoic era from 570 to 500 million years ago.   { 'kam·brē·ən }

Camerata   [PALEON]   A subclass of extinct stalked echinoderms of the class Crinoidea.   { ,kam·ə'räd·ə }

Campanian   [GEOL]   European stage of Upper Cretaceous.   { kam'pan·ē·ən }

camptonite   [PETR]   A lamprophyre containing pyroxene, sodic hornblende, and olivine as dark constituents and labradorite as the light constituent; sodic orthoclase may be present.   { 'kam·tə,nīt }

Canadian Shield See Laurentian Shield.   { kə'nād·ē·ən 'shēld }

Canastotan   [GEOL]   Lower Upper Silurian geologic time.   { kə'nas·tə·tən }

cancrinite   [MINERAL] $Na_3CaAl_3Si_3O_{12}CO_3(OH)_2$ A feldspathoid tectosilicate occurring in hexagonal crystals in nepheline syenites, usually in compact or disseminated masses.   { 'kaŋ·krə,nīt }

candite See ceylonite.   { 'kan,dīt }

canfieldite   [MINERAL] Ag_8SnS_6 A black mineral of the argyrodite series consisting of silver thiostannate, with a specific gravity of 6.28; found in Germany and Bolivia.   { 'kan,fēl,dīt }

cannel coal   [GEOL]   A fine-textured, highly volatile bituminous coal distinguished by a greasy luster and blocky, conchoidal fracture; burns with a steady luminous flame.   Also known as cannelite.   { 'kan·əl ,kōl }

cannelite See cannel coal.   { 'kan·əl,īt }

canneloid   [GEOL]   **1.** Coal that resembles cannel coal.   **2.** Coal intermediate between bituminous and cannel.   **3.** Durain laminae in banded coal.   **4.** Cannel coal of anthracite or semianthracite rank.   { 'kan·əl,óid }

cannel shale   [GEOL]   A black shale formed by the accumulation of an aquatic ooze rich in bituminous organic matter in association with inorganic materials such as silt and clay.   { 'kan·əl ,shāl }

canyon bench

canyon bench |GEOL| A steplike level of hard strata in the walls of deep valleys in regions of horizontal strata. { 'kan·yən ‚bench }

canyon fill |GEOL| Loose, unconsolidated material which fills a canyon to a depth of 50 feet (15 meters) or more during periods between great floods. { 'kan·yən ‚fil }

capacity of the wind |GEOL| The total weight of airborne particles (soil and rock) of given size, shape, and specific gravity, which can be carried in 1 cubic mile (4.17 cubic kilometers) of wind blowing at a given speed. { kə'pas·əd·ē əv thə ¦wind }

capillary |GEOL| A fissure or a crack in a formation which provides a route for flow of water or hydrocarbons. { 'kap·ə‚ler·ē }

capillary ejecta See Pele's hair. { 'kap·ə‚ler·ē i'jek·tə }

capillary pyrites See millerite. { 'kap·ə‚ler·ē 'pī‚rīts }

cappelenite |MINERAL| (Ba,Ca,Na)(Y,La)$_6$B$_6$Si$_{13}$(O,OH)$_{27}$ A greenish-brown hexagonal mineral consisting of a rare yttrium-barium borosilicate occurring in crystals. { 'kap·lə‚nīt }

capping |GEOL| **1.** Consolidated barren rock overlying a mineral or ore deposit. **2.** See gossan. { 'kap·iŋ }

cap rock |GEOL| **1.** An overlying, generally impervious layer or stratum of rock that overlies an oil- or gas-bearing rock. **2.** Barren vein matter, or a pinch in a vein, supposed to overlie ore. **3.** A hard layer of rock, usually sandstone, a short distance above a coal seam. **4.** An impervious body of anhydrite and gypsum in a salt dome. { 'kap ‚räk }

Captorhinomorpha |PALEON| An extinct subclass of primitive lizardlike reptiles in the order Cotylosauria. { ‚kap·tə¦rī·nə¦mȯr·fə }

capture |GEOCHEM| In a crystal structure, the substitution of a trace element for a lower-valence common element. { 'kap·chər }

caracolite |MINERAL| A rare, colorless mineral occurring as crystalline incrustations, and consisting of a sulfate and chloride of sodium and lead. { ‚kar·ə'kō‚līt }

Caradocian |GEOL| Lower Upper Ordovician geologic time. { kar·ə'dō·shən }

carapace |GEOL| The upper normal limb of a fold having an almost horizontal axial plane. { 'kar·ə‚pās }

carbohumin See ulmin. { ‚kär·bō'hyü·mən }

carbonaceous chondrite |GEOL| A chondritic meteorite that contains a relatively large amount of carbon and has a resulting dark color. Also known as carbonaceous meteorite. { ‚kär·bə'nā·shəs 'kän‚drīt }

carbonaceous meteorite See carbonaceous chondrite. { kär·bə'nā·shəs 'mēd·ē·ə‚rīt }

carbonaceous rock |PETR| Rock with carbonaceous material included. { kär·bə'nā·shəs 'räk }

carbonaceous sandstone |PETR| Sandstone rich in carbon. { kär·bə'nā·shəs 'san‚stōn }

carbonaceous shale |GEOL| Shale rich in carbon. { kär·bə'nā·shəs 'shāl }

carbonado |MINERAL| A dark-colored, fine-grained diamond aggregate; valuable for toughness and absence of cleavage planes. Also known as black diamond; carbon diamond. { kär·bə'nā·dō }

carbonate cycle |GEOCHEM| The biogeochemical carbonate pathways, involving the conversion of carbonate to CO_2 and HCO_3, the solution and deposition of carbonate, and the metabolism and regeneration of it in biological systems. { 'kär·bə·nət ‚sī·kəl }

carbonate mineral |MINERAL| A mineral containing considerable amounts of carbonates. { 'kär·bə·nət 'min·rəl }

carbonate reservoir |GEOL| An underground oil or gas trap formed in reefs, clastic limestones, chemical limestones, or dolomite. { 'kär·bə·nət 'rez·əv‚wär }

carbonate rock |PETR| A rock composed principally of carbonates, especially if at least 50% by weight. { 'kär·bə·nət 'räk }

carbonation |GEOCHEM| A process of chemical weathering whereby minerals that contain soda, lime, potash, or basic oxides are changed to carbonates by the carbonic acid in air or water. { ‚kär·bə'nā·shən }

---

carbonatite [PETR] **1.** Intrusive carbonate rock associated with alkaline igneous intrusive activity. **2.** A sedimentary rock that is composed of at least 80% calcium or magnesium. { kär'bän·ə,tīt }

carbon cycle [GEOCHEM] The cycle of carbon in the biosphere, in which plants convert carbon dioxide to organic compounds that are consumed by plants and animals, and the carbon is returned to the biosphere in the form of inorganic compounds by processes of respiration and decay. { 'kär·bən ,sī·kəl }

carbon diamond See carbonado. { ¦kär·bən 'dī·mənd }

Carboniferous [GEOL] A division of late Paleozoic rocks and geologic time including the Mississippian and Pennsylvanian periods. { ,kär·bə'nif·ə·rəs }

carbonification See coalification. { kär,bän·ə·fə'kā·shən }

carbon isotope ratio [GEOL] Ratio of carbon-12 to either of the less common isotopes, carbon-13 or carbon-14, or the reciprocal of one of these ratios; if not specified, the ratio refers to carbon-12/carbon-13. Also known as carbon ratio. { ¦kär·bən 'is·ə,tōp ,rā·shō }

carbonite See natural coke. { 'kär·bə,nīt }

carbonization [GEOCHEM] **1.** In the coalification process, the accumulation of residual carbon by changes in organic material and their decomposition products. **2.** Deposition of a thin film of carbon by slow decay of organic matter underwater. **3.** A process of converting a carbonaceous material to carbon by removal of other components. { ,kär·bə·nə'zā·shən }

carbon pool [GEOCHEM] A reservoir with the capacity to store and release carbon, such as soil, terrestrial vegetation, the ocean, and the atmosphere. { 'kär·bən ,pül }

carbon ratio [GEOL] **1.** The ratio of fixed carbon to fixed carbon plus volatile hydrocarbons in a coal. **2.** See carbon isotope ratio. { 'kär·bən ,rā·shō }

carbon-ratio theory [GEOL] The theory that the gravity of oil in any area is inversely proportional to the carbon ratio of the coal. { 'kär·bən ,rā·shō ,thē·ə·rē }

carbon sequestration [GEOCHEM] The uptake and storage of atmospheric carbon in, for example, soil and vegetation. { ,kär·bən ,sē·kwes'trā·shən }

carbon sink [GEOCHEM] A reservoir that absorbs or takes up atmospheric carbon; for example, a forest or an ocean. { 'kär·bən ,siŋk }

carminite [MINERAL] $PbFe_2(AsO_4)_2(OH)_2$ A carmine to tile-red mineral consisting of a basic arsenate of lead and iron. { 'kär·mə,nīt }

carnallite [MINERAL] $KMgCl_3 \cdot 6H_2O$ A milky-white or reddish mineral that crystallizes in the orthorhombic system and occurs in deliquescent masses; it is valuable as an ore of potassium. { 'kärn·əl,īt }

carnegieite [MINERAL] $NaAlSiO_4$ An artificial mineral similar to feldspar; it is triclinic at low temperatures, isometric at elevated temperatures. { 'kär·nə·gē,īt }

Carnian [GEOL] Lower Upper Triassic geologic time. Also spelled Karnian. { 'kärn·ē·ən }

Carnosauria [PALEON] A group of large, predacious saurischian dinosaurs in the suborder Theropoda having short necks and large heads. { ,kär·nə'sór·ē·ə }

carnotite [MINERAL] $K(UO_2)_2(VO_4)_2 \cdot nH_2O$ A canary-yellow, fine-grained hydrous vanadate of potassium and uranium having monoclinic microcrystals; an ore of radium and uranium. { 'kär·nə,tīt }

carpholite [MINERAL] $MnAl_2Si_2O_6(OH)_4$ A straw-yellow fibrous mineral consisting of a hydrous aluminum manganese silicate occurring in tufts; specific gravity is 2.93. { 'kär·fə,līt }

carphosiderite [MINERAL] A yellow mineral consisting of a basic hydrous iron sulfate occurring in masses and crusts. { ,kär·fō'sīd·ə,rīt }

Carpoidea [PALEON] Former designation for a class of extinct homalozoan echinoderms. { kär'póid·ē·ə }

carpoids [PALEON] An assemblage of three classes of enigmatic, rare Paleozoic echinoderms formerly grouped together as the class Carpoidea. { 'kär,póidz }

Carrara marble [PETR] All marble quarried near Carrara, Italy, having a prevailing white to bluish color, or white with blue veins. { kə'rä·rə 'mär·bəl }

61

caryinite |MINERAL| $(Ca,Pb,Na)_5(Mn,Mg)_4(AsO_4)_5$ A mineral consisting chiefly of a calcium manganese arsenate. { 'kar·ē·ə,nīt }

cascade |GEOL| A landform structure formed by gravity collapse, consisting of a bed that buckles into a series of folds as it slides down the flanks of an anticline. { ka'skād }

Cascadian orogeny |GEOL| Post-Tertiary deformation of the crust of the earth in western North America. { ka'skād·ē·ən ȯ'räj·ə·nē }

case hardening |GEOL| Formation of a mineral coating on the surface of porous rock by evaporation of a mineral-bearing solution. { kās ,härd·ən·iŋ }

Cassadagan |GEOL| Middle Upper Devonian geologic time, above Chemungian. { kə'sad·ə·gən }

Casselian See Chattian. { ka'sel·yən }

Cassiar orogeny |GEOL| Orogenic episode in the Canadian Cordillera during late Paleozoic time. { 'kas·ē·ər ȯ'räj·ə·nē }

cassidyite |MINERAL| $Ca_2(Ni,Mg)(PO_4)_2·2H_2O$ A mineral found in meteorites. { kə'sid·ē,īt }

cassiterite |MINERAL| SnO_2 A yellow, black, or brown mineral that crystallizes in the tetragonal system in prisms terminated by dipyramids; the most important ore of tin. Also known as tin stone. { kə'sid·ə,rīt }

cast |PALEON| A fossil reproduction of a natural object formed by infiltration of a mold of the object by waterborne minerals. { kast }

castings See fecal pellets. { 'kast·iŋz }

castorite |MINERAL| A transparent variety of petalite occurring in crystals. { 'kas·tə,rīt }

catachosis |GEOL| Fracturing or crushing of rock during metamorphism. { ,kad·ə'kō·səs }

cataclasis |GEOL| Deformation of rock by fracture and rotation of aggregates or mineral grains. { ,kad·ə'klā·səs }

cataclasite See cataclastic rock. { ,kad·ə'klā,sīt }

cataclastic metamorphism |PETR| Local metamorphism restricted to a region of faults and overthrusts involving purely mechanical forces resulting in cataclasis. { ¦kad·ə¦klas·tik ,med·ə'mȯr,fiz·əm }

cataclastic rock |PETR| Rock containing angular fragments formed by cataclasis. Also known as cataclasite. { ¦kad·ə¦klas·tik 'räk }

cataclastic structure See mortar structure. { ¦kad·ə¦klas·tik 'strək·chər }

catapleiite |MINERAL| $(Na_2,Ca)ZrSi_3O_9·2H_2O$ A yellow or yellowish-brown mineral crystallizing in the hexagonal system, consisting of a hydrous silicate of sodium, calcium, and zirconium, and occurring in thin tabular crystals; hardness is 6 on Mohs scale, and specific gravity is 2.8. { ,kad·ə'plī,īt }

catastrophism |GEOL| The theory that most features in the earth were produced by the occurrence of sudden, short-lived, worldwide events. |PALEON| The theory that the differences between fossils in successive stratigraphic horizons resulted from a general catastrophe followed by creation of the different organisms found in the next-younger beds. { kə'tas·trə,fiz·əm }

catazone |GEOL| The deepest zone of rock metamorphism where high temperatures and pressures prevail. { 'kad·ə,zōn }

catena |GEOL| A group of soils derived from uniform or similar parent material which nonetheless show variations in type because of differences in topography or drainage. { kə'tē·nə }

catoptrite |MINERAL| An iron black to jet black, monoclinic mineral consisting of a silicoantimonate of aluminum and divalent manganese. Also spelled katoptrite. { kə'täp,trīt }

cauldron subsidence |GEOL| **1.** A structure formed by the lowering along a steep ring fracture of a more or less cylindrical block, usually 1 to 10 miles (1.6 to 16 kilometers) in diameter, into a magma chamber. **2.** The process of forming such a structure. { 'kȯl·drən səb'sī·dəns }

62

caustobiolith [GEOL] Combustible organic rock formed by direct accumulation of plant materials; includes coal peat. { ¦kȯ,stō'bī·ə,lith }

cave [GEOL] A natural, hollow chamber or series of chambers and galleries beneath the earth's surface, or in the side of a mountain or hill, with an opening to the surface. { kāv }

cave breccia [GEOL] Sharp fragments of limestone debris deposited on the floor of a cave. { ¦kāv 'brech·ə }

cave formation See speleothem. { 'kāv fȯr'mā·shən }

Cavellinidae [PALEON] A family of Paleozoic ostracodes in the suborder Platycopa. { ,kav·ə'lin·ə,dē }

cave pearl [GEOL] A small, smooth, rounded concretion of calcite or aragonite, formed by concentric precipitation about a nucleus and usually found in limestone caves. { 'kāv ,pərl }

cavern [GEOL] An underground chamber or series of chambers of indefinite extent carved out by rock springs in limestone. { 'kav·ərn }

cavernous [GEOL] **1.** Having many caverns or cavities. **2.** Producing caverns. **3.** Of or pertaining to a cavern, that is, suggesting vastness. { 'kav·ər·nəs }

c axis [GEOL] The reference axis perpendicular to the plane of movement of rock or mineral strata. { 'sē ,ak·səs }

cay [GEOL] **1.** A flat coral island. **2.** A flat mound of sand built up on a reef slightly above high tide. **3.** A small, low coastal islet or emergent reef composed largely of sand or coral. { kā }

cay sandstone [GEOL] Firmly cemented or friable coral sand formed near the base of coral reef cays. { ¦kā 'san,stōn }

Caytoniales [PALEOBOT] An order of Mesozoic plants. { ,kā·tän·ē'ā,lēz }

Cayugan [GEOL] Upper Silurian geologic time. { kī'yü·gən }

Cazenovian [GEOL] Lower Middle Devonian geologic time. { kaz·ə'nōv·ē·ən }

CCD See calcite compensation depth.

Cebochoeridae [PALEON] A family of extinct palaeodont artiodactyls in the superfamily Enteledontoidae. { ¦seb·ə,kō'er·ə,dē }

cebollite [MINERAL] $H_2Ca_4Al_2Si_3O_{16}$ A greenish to white mineral consisting of hydrous calcium aluminum silicate occurring in fibrous aggregates; hardness is 5 on Mohs scale, and specific gravity is 3. { 'seb·ə,līt }

cecilite [PETR] A basaltic rock having few phenocrysts and consisting of at least 50% leucite with augite, melilite, nepheline, olivine, anorthite, magnetite, and apatite. { 'ses·əl,īt }

cedricite [MINERAL] A variety of lamproite composed principally of diopside, leucite, and phlogopite and usually containing crystals of serpentine. { 'sed·rə,sīt }

celadonite [MINERAL] A soft, green variety of mica having high iron content and containing silicates of magnesium and potassium. { 'sel·ə·də,nīt }

celestine See celestite. { 'sel·ə,stēn }

celestite [MINERAL] $SrSO_4$ A colorless or sky-blue mineral occurring in orthorhombic, tabular crystals and in compact forms; fracture is uneven and luster is vitreous; principal ore of strontium. Also known as celestine. { 'sel·ə,stīt }

cellular [PETR] Pertaining to igneous rock having a porous texture, usually with the cavities larger than pore size and smaller than caverns. { 'sel·yə·lər }

cellular soil See polygonal ground. { 'sel·yə·lər 'sȯil }

celsian [MINERAL] $BaAl_2Si_2O_8$ Colorless, monoclinic mineral consisting of barium feldspar. { 'sel·sē,an }

cement [GEOL] Any chemically precipitated material, such as carbonates, gypsum, and barite, occurring in the interstices of clastic rocks. { si'ment }

cementation [GEOL] The precipitation of a binding material around minerals or grains in rocks. { ,sē,men'tā·shən }

cement gravel [GEOL] Gravel consolidated by clay, silica, calcite, or other binding material. { si'ment ,grav·əl }

cement rock [PETR] An argillaceous limestone containing lime, silica, and alumina in

Cenomanian

variable proportions and usually some magnesia; used in the manufacture of natural hydraulic cement. { si'ment ,räk }

Cenomanian [GEOL] Lower Upper Cretaceous geologic time. { ¦sen·ə¦mān·ē·ən }

cenote See pothole. { sə'nōd·ē }

Cenozoic [GEOL] The youngest of the eras, or major subdivisions of geologic time, extending from the end of the Mesozoic Era to the present, or Recent. Also spelled as Cainozoic. { ¦sen·ə¦zō·ik }

central valley See rift valley. { 'sen·trəl 'val·ē }

centrifugal drainage pattern See radial drainage pattern. { ,sen'trif·i·gəl 'drān·ij ,pad·ərn }

centroclinal [GEOL] Referring to geologic strata dipping toward a common center, as in a structural basin. { ¦sen·trō¦klīn·əl }

Centronellidina [PALEON] A suborder of extinct articulate brachiopods in the order Terebratulida. { ¦sen·trō·nə'lid·ən·ə }

centrosphere [GEOL] The central core of the earth. Also known as the barysphere. { 'sen·trə,sfir }

Cephalaspida [PALEON] An equivalent name for the Osteostraci. { ,sef·ə'las·pə·də }

ceramicite [PETR] A porcelained pyrometamorphic rock composed of basic plagioclase and cordierite with a small amount of hypersthene and a groundmass of glass. { sə'ram·ə,sīt }

Ceramoporidae [PALEON] A family of extinct, marine bryozoans in the order Cystoporata. { sə,ram·ə'pór·ə,dē }

cerargyrite [MINERAL] AgCl A colorless to pearl-gray mineral; crystallizes in the isometric system, but crystals, usually cubic, are rare; a secondary mineral that is an ore of silver. Also known as chlorargyrite; horn silver. { sa'rär·jə,rīt }

ceratite [PALEON] A fossil ammonoid of the genus *Ceratites* distinguished by a type of suture in which the lobes are further divided into subordinate crenulations while the saddles are not divided and are smoothly rounded. { 'ser·ə,tīt }

ceratitic [PALEON] Pertaining to a ceratite. { ,ser·ə'tid·ik }

Ceratodontidae [PALEON] A family of Mesozoic lungfishes in the order Dipteriformes. { ,ser·ə·tō'dän·tə,dē }

Ceratopsia [PALEON] The horned dinosaurs, a suborder of Upper Cretaceous reptiles in the order Ornithischia. { ,ser·ə'täp·sē·ə }

Ceratosaurus [PALEON] A carnivorous therapod dinosaur, 20 feet (6 meters) long, from the Late Jurassic Period that had strong hindlimbs, short and weak forelimbs (with four-fingered hands), and massive jaws lined with enormous teeth. { sə,rad·ə'sór·əs }

cerine See allanite. { 'sir,ēn }

cerite [MINERAL] (Ca,Fe)Ce$_3$Si$_3$O$_{12}$·H$_2$O A brown rare-earth hydrous silicate of cerium and other metals found in gneiss; hardness is 5.5 on Mohs scale, and specific gravity is 4.86. { 'sir,īt }

cerolite [MINERAL] A mixture of serpentine and stevensite occurring in yellow or greenish waxlike masses. { 'sir·ə,līt }

cerussite [MINERAL] PbCO$_3$ A yellow or white member of the aragonite group occurring in orthorhombic crystals; produced by the action of carbon dioxide on lead ore. { sə'rəs,īt }

cervantite [MINERAL] Sb$_2$O$_4$ A white or yellow secondary mineral crystallizing in the orthorhombic system and formed by oxidation of antimony sulfide. { sər'van,tīt }

cesarolite [MINERAL] H$_2$PbMn$_3$O$_8$ A steel-gray mineral consisting of a hydrous lead manganate occurring in spongy masses. { ,chāz·ə'rō,līt }

ceylonite [MINERAL] A dark-green, brown, or black iron-bearing variety of spinel. Also known as candite; pleonaste; zeylanite. { sə'lä,nīt }

C figure See C index. { 'sē ,fig·yər }

chabazite [MINERAL] CaAl$_2$Si$_4$O$_{12}$·6H$_2$O A white to yellow or red member of the zeolite group occurring in glassy rhombohedral crystals; hardness is 4–5 on Mohs scale, and specific gravity is 2.08–2.16. { 'kab·ə,zīt }

64

Chaetetidae [PALEON] A family of Paleozoic corals of the order Tabulata. { ,kē'tē·də,dē }

chain [GEOL] A series of interconnected or related natural features, such as lakes, islands, or seamounts, arranged in a longitudinal sequence. { chān }

chalazoidite See mud ball. { 'kal·ə,zȯi,dīt }

chalcanthite [MINERAL] $CuSO_4 \cdot 5H_2O$ A blue to bluish-green mineral which occurs in triclinic crystals or in massive fibrous veins or stalactites. Also known as bluestone; blue vitriol. { kal'kan,thīt }

chalcedony [MINERAL] A cryptocrystalline variety of quartz; occurs as crusts with a rounded, mammillary, or botryoidal surface and as a major constituent of nodular and bedded cherts; varieties include carnelian and bloodstone. { kal'sed·ən·ē }

chalcedonyx [MINERAL] A mineral consisting of onyx with alternating gray and white bands; valued as a semiprecious stone. { ,kal·sə'dän·iks }

chalcoalumite [MINERAL] $CuAl_4(SO_4)(OH)_{12} \cdot 3H_2O$ A turquoise-green to pale-blue mineral consisting of a hydrous basic sulfate of copper and aluminum. { ¦kal·kō'al·ə,mīt }

chalcocite [MINERAL] Cu_2S A fine-grained, massive mineral with a metallic luster which tarnishes to dull black on exposure; crystallizes in the orthorhombic system, the crystals being rare and small usually with hexagonal outline as a result of twinning; hardness is 2.5–3 on Mohs scale, and specific gravity is 5.5–5.8. Also known as beta chalcocite; chalcosine; copper glance; redruthite; vitreous copper. { 'kal·kə,sīt }

chalcocyanite [MINERAL] $CuSO_4$ A white mineral consisting of copper sulfate. Also known as hydrocyanite. { ,kal·kə'sī·ə,nīt }

chalcolite See torbernite. { 'kal·kə,līt }

chalcomenite [MINERAL] $CuSeO_3 \cdot 2H_2O$ A blue mineral consisting of copper selenite occurring in crystals. { ,kal·kə'mē,nīt }

chalcophanite [MINERAL] $(Zn,Mn,Fe)Mn_2O_5 \cdot nH_2O$ A black mineral with metallic luster consisting of hydrous manganese and zinc oxide. { kal'käf·ə,nīt }

chalcophile [GEOL] Having an affinity for sulfur and therefore massing in greatest concentration in the sulfide phase of a molten mass. { 'kal·kə,fīl }

chalcophyllite [MINERAL] $Cu_{18}Al_2(AsO_4)_3(OH)_{27} \cdot 33H_2O$ A green mineral consisting of basic arsenate and sulfate of copper and aluminum occurring in tabular crystals or foliated masses. Also known as copper mica. { ,kal·kō'fi,līt }

chalcopyrite [MINERAL] $CuFeS_2$ A major ore mineral of copper; crystallizes in the tetragonal crystal system, but crystals are generally small with diphenoidal faces resembling the tetrahedron; usually massive with a metallic luster and brass-yellow color; hardness is 3.5–4 on Mohs scale, and specific gravity is 4.1–4.3. Also known as copper pyrite; yellow pyrite. { ,kal·kō'pī,rīt }

chalcopyrrohite [MINERAL] $CuFe_4S_5$ A sulfide mineral occurring in meteorites. { ,kal·kō'pī·rə,nīt }

chalcosiderite [MINERAL] $Cu(Fe,Al)_6(PO_4)_4(OH)_8 \cdot 4H_2O$ A green mineral, isomorphous with turquoise, consisting of a hydrous basic phosphate of copper, iron, and aluminum. { ¦kal·kō'sīd·ə,rīt }

chalcosine See chalcocite. { 'kal·kə,sēn }

chalcostibite [MINERAL] $CuSbS_2$ A lead-gray mineral consisting of antimony copper sulfide. { ,kal·kō'sti,bīt }

chalcotrichite [MINERAL] A capillary variety of cuprite occurring in long needlelike crystals. Also known as hair copper; plush copper ore. { ,kal·kō'tri,kīt }

Chalicotheriidae [PALEON] A family of extinct perissodactyl mammals in the superfamily Chalicotherioidea. { ,kal·ə,kō·thə'rī·ə,dē }

Chalicotherioidea [PALEON] A superfamily of extinct, specialized perissodactyls having claws rather than hooves. { ¦kal·ə,kō,thi·rē'ȯid·ē·ə }

chalk [PETR] A variety of limestone formed from pelagic organisms; it is very fine-grained, porous, and friable; white or very light-colored, it consists almost entirely of calcite. { chȯk }

chalmersite See cubanite. { 'chä·mər,zīt }

chalybite See siderite. { 'kal·ə,bīt }

chamosite [MINERAL] A greenish-gray or black mineral consisting of silicate belonging to the chlorite group and having monoclinic crystals; found in many oolitic iron ores. { 'sham·ə,zīt }

Champlainian [GEOL] Middle Ordovician geologic time. { ,sham'plān·ē·ən }

champsosaur [PALEON] A large crocodile-like reptile that lived in freshwater ponds and swamps 55–65 million years ago. { 'champ·sə,sòr }

Chandler motion See polar wandering. { 'chand·lər ,mō·shən }

channel fill [GEOL] Accumulations of sand and detritus in a stream channel where the transporting capacity of the water is insufficient to remove the material as rapidly as it is delivered. { 'chan·əl ,fil }

channel frequency See stream frequency. { 'chan·əl ,frē·kwən·sē }

channel gradient ratio See stream gradient ratio. { 'chan·əl 'grād·ē·ənt ,rā·shō }

channel-lag deposit [GEOL] Coarse residual material left as accumulations in the channel in the normal processes of the stream. { 'chan·əl ,lag di,päz·ət }

channel morphology See river morphology. { 'chan·əl ,mór'fäl·ə·jē }

channel-mouth bar [GEOL] A bar formed where moving water enters a body of still water, due to decreased velocity. { 'chan·əl ,maùth ,bär }

channel roughness [GEOL] A measure of the resistivity offered by the material constituting stream channel margins to the flow of water. { 'chan·əl ,rəf·nəs }

channel sand [GEOL] A sandstone or sand deposited in a stream bed or other channel eroded into the underlying bed. { 'chan·əl ,sand }

channel splay See floodplain splay. { 'chan·əl ,splā }

channel width [GEOL] The distance across a stream or channel as measured from bank to bank near bankful stage. { 'chan·əl ,width }

chapmanite [MINERAL] $Fe_2Sb(SiO_4)_2(OH)$ A mineral consisting of a silicate of iron and antimony. { 'chap·mə,nīt }

Charmouthian [GEOL] Middle Lower Jurassic geologic time. { chär'maùth·ē·ən }

charnockite [PETR] Any of various faintly foliated, nearly massive varieties of quartzo-feldspathic rocks containing hypersthene. { 'chär·nə,kīt }

charnockite series [GEOL] A series of plutonic rocks compositionally similar to the granitic rock series but characterized by the presence of orthopyroxene. { 'chär·nə,kīt ,sir·ēz }

chassignite [GEOL] An achondritic stony meteorite composed chiefly of olivine (95); resembles dunite. { 'shas·ən,yīt }

chatoyant [MINERAL] Of a mineral or gemstone, having a changeable luster or color marked by a band of light, resembling the eye of a cat in this respect. { shə'tói·ənt }

chatter mark [GEOL] A scar on the surface of bedrock made by the abrasive action of drift carried at the base of a glacier. { 'chad·ər ,märk }

Chattian [GEOL] Upper Oligocene geologic time. Also known as Casselian. { 'chad·ē·ən }

Chautauquan [GEOL] Upper Devonian geologic time, below Bradfordian. { shə'täk·wən }

Chazyan [GEOL] Middle Ordovician geologic time. { 'chaz·ē·ən }

Cheiracanthidae [PALEON] A family of extinct acanthodian fishes in the order Acanthodiformes. { ,kī·rə'kan·thə,dē }

chemical denudation [GEOL] Wasting of the land surface by water transport of soluble materials into the sea. { 'kem·i·kəl ,dē·nü'dā·shən }

chemical fossils See biomarkers. { ¦kem·i·kəl 'fäs·əlz }

chemical precipitates [GEOL] A sediment formed from precipitated materials as distinguished from detrital particles that have been transported and deposited. { 'kem·i·kəl pri'sip·ə,tāts }

chemical remanent magnetization [GEOPHYS] Permanent magnetization of rocks acquired when a magnetic material, such as hematite, is grown at low temperature through the oxidation of some other iron mineral, such as magnetite or goethite; the growing mineral becomes magnetized in the direction of any field which is present. Abbreviated CRM. { 'kem·i·kəl 'rem·ə·nənt ,mag·nət·ə'zā·shən }

chemical reservoir [GEOL] An underground oil or gas trap formed in limestones or dolomites deposited in quiescent geologic environments. { 'kem·i·kəl 'rez·əv,wär }

chemical rock [PETR] A type of sedimentary rock comprising material deposited directly by precipitation from solution or colloidal suspension and frequently possessing a crystalline texture. { 'kem·i·kəl 'räk }

chemical weathering [GEOCHEM] A weathering process whereby rocks and minerals are transformed into new, fairly stable chemical combinations by such chemical reactions as hydrolysis, oxidation, ion exchange, and solution. Also known as decay; decomposition. { 'kem·i·kəl 'weth·ə·riŋ }

chemostratigraphy [GEOCHEM] The correlation and dating of marine sediments and sedimentary rocks through the use of trace-element concentrations, molecular fossils, and certain isotopic ratios that can be measured on components of the rocks. { ,kē·mō·strə'tig·rə·fē }

Chemungian [GEOL] Middle Upper Devonian geologic time, below Cassodagan. { ke'mən·jē·ən }

chenevixite [MINERAL] $Cu_2Fe_2(AsO_4)_2(OH)_4 \cdot H_2O$ A dark-green to greenish-yellow mineral consisting of a hydrous copper iron arsenate occurring in masses. { ¦shen·ə¦vik,sīt }

chenier [GEOL] A continuous ridge of beach material built upon swampy deposits; often supports trees, such as pines or evergreen oaks. { 'shen·yā }

Chernozem [GEOL] One of the major groups of zonal soils, developed typically in temperate to cool, subhumid climate; the Chernozem soils in modern classification include Borolls, Ustolls, Udolls, and Xerolls. Also spelled Tchernozem. { ¦chər·nəz¦yóm }

chert [PETR] A hard, dense, sedimentary rock composed of fine-grained silica, characterized by a semivitreous to dull luster and a splintery to conchoidal fracture; commonly gray, black, reddish brown, or green. Also known as hornstone; phthanite. { chərt }

chertification [GEOL] A process of replacement by silica in limestone in the form of fine-grained quartz or chalcedony. { ,chərd·ə·fə'ka·shən }

chessylite See azurite. { 'shes·ə,līt }

Chesteran [GEOL] Upper Mississippian geologic time. { che'stir·ē·ən }

chestnut coal [GEOL] Anthracite coal small enough to pass through a round mesh of $1 \frac{5}{8}$ inches (3.1 centimeters) but too large to pass through a round mesh of $1 \frac{13}{16}$ inches (1.7 centimeters). { 'ches,nət ,kōl }

Chestnut soil [GEOL] One of the major groups of zonal soils, developed typically in temperate to cool, subhumid to semiarid climate; the Chestnut soils in modern classification include Ustolls, Borolls, and Xerolls. { 'ches,nət ¦sóil }

chevkinite [MINERAL] $(Fe,Ca)(Ce,La)_2(Si,Ti)_2O_8$ A mineral consisting of silicotitanate of iron, calcium, and rare-earth elements. { 'chef·kə,nīt }

chevron fold [GEOL] An accordionlike fold with limbs of equal length. { 'shev·rən ,fōld }

chiastolite [MINERAL] A variety of andalusite whose crystals have a cross-shaped appearance in cross section due to the arrangement of carbonaceous impurities. Also known as macle. { kī'as·tə,līt }

Chideruan [GEOL] Uppermost Permian geologic time. { chi'der·ə·wən }

childrenite [MINERAL] $(Fe,Mn)AlPO_4(OH)_2 \cdot H_2O$ A pale-yellowish to dark-brown orthorhombic mineral consisting of a hydrous basic iron aluminum phosphate occurring as translucent crystals; it is isomorphous with eosphorite; hardness is 4.5–5 on Mohs scale, and specific gravity is 3.18–3.24. { 'chil·drə,nīt }

Chile niter See Chile saltpeter. { ¦chil·ē 'nīd·ər }

Chile saltpeter [MINERAL] Also known as Chile niter. **1.** Soda niter found in large quantities in caliche in arid regions of northern Chile. **2.** Deposits of sodium nitrate. { ¦chil·ē ,sólt'pēd·ər }

chilled contact [PETR] The finer-grained portion of an igneous rock found near its contact with older rock. { 'child 'kän,takt }

Chilobolbinidae

Chilobolbinidae [PALEON] A family of extinct ostracods in the superfamily Hollinacea showing dimorphism of the velar structure. { ˌkī·lə‚bäl'bīn·ə‚dē }

chimney [GEOL] *See* pipe; spouting horn. { 'chim‚nē }

chimney rock [GEOL] **1.** A chimney-shaped remnant of a rock cliff whose sides have been cut into and carried away by waves and the gravel beach. **2.** A rock column rising above its surroundings. { 'chim‚nē ‚räk }

chiolite [MINERAL] Na₅Al₃F₁₄ A snow white mineral resembling cryolite. Also known as arksutite. { 'kī·ə‚līt }

Chirodidae [PALEON] A family of extinct chondrostean fishes in the suborder Platysomoidei. { ˌkī'räd·ə‚dē }

Chirognathidae [PALEON] A family of conodonts in the suborder Neurodontiformes. { ˌkī·rəg'näth·ə‚dē }

Chitinozoa [PALEON] An extinct group of unicellular microfossils of the kingdom Protista. { ¦kīt·ən·ə¦zō·ə }

chiviatite [MINERAL] Pb₂Bi₆S₁₁ A lead-gray mineral consisting of a lead bismuth sulfide occurring in foliated masses. { ˌchiv·ē'ä‚tīt }

chloanthite [MINERAL] NiAs₂₋₃ A white or gray mineral with metallic luster forming crystals in the isometric system; it is isomorphous with nickel-skutterudite. { klō'an‚thīt }

chloraluminite [MINERAL] AlCl₃·6H₂O A mineral consisting of hydrous aluminum chloride. { ¦klȯr·ə¦lüm·ə‚nīt }

chlorapatite [MINERAL] Ca₅(PO₄)₃Cl An apatite mineral containing chlorine. { klȯr'ap·ə‚tīt }

chlorargyrite *See* cerargyrite. { klȯr'ar·jə‚rīt }

chlorastrolite [MINERAL] A mottled, green variety of pumpellyite occurring as grains or small nodules of a stellate structure in basic igneous rock in the Lake Superior region; used as a semiprecious stone. { klȯr'as·trə‚līt }

chlorite [MINERAL] Any of a group of greenish, platyhydrous monoclinic silicates of aluminum, ferrous iron, and magnesium which are closely associated with and resemble the micas. { 'klȯr‚īt }

chlorite schist [PETR] A metamorphic rock whose composition is dominated by members of the chlorite group. { 'klȯr‚īt ‚shist }

chlorite-sericite schist [PETR] A low-grade, fine-grained variety of mica schist without biotite. { 'klȯr‚īt 'ser·ə‚sīt ‚shist }

chloritoid [MINERAL] FeAl₄Si₂O₁₀(OH)₄ A micaceous mineral related to the brittle mica group; has both monoclinic and triclinic modifications, a gray to green color, and weakly pleochroic crystals. { 'klȯr·ə‚toid }

chloritoid schist [PETR] A variety of mica schist whose composition is dominated by chloritoid. { 'klȯr·ə‚toid ‚shist }

chlormanganokalite [MINERAL] K₄MnCl₆ A wine yellow to lemon or canary yellow, hexagonal mineral consisting of potassium and manganese chloride; occurs as rhombohedrons. { 'klȯr¦maŋ·gə‚nō'kä‚līt }

chlorocalcite [MINERAL] KCaCl₃ A white mineral consisting of a chloride of potassium and calcium. Also known as hydrophilite. { ¦klȯr·ō'kal‚sīt }

chloromagnesite [MINERAL] MgCl₂ A mineral consisting of anhydrous magnesium chloride, found on the volcano Vesuvius. { ¦klȯr·ō'mag·nə‚sīt }

chloropal *See* nontronite. { 'klȯr·ə‚pal }

chlorophoenicite [MINERAL] (Mn,An)₅(AsO₄)(OH)₇ Gray-green monoclinic mineral consisting of a basic arsenate of manganese and zinc occurring in crystals. { ˌklȯr·ō'fēn·ə‚sīt }

chlorothionite [MINERAL] K₂Cu(SO₄)Cl₂ Bright-blue secondary mineral consisting of potassium copper sulfate chloride, found on the volcano Vesuvius. { ˌklȯr·ə'thī·ə‚nīt }

chloroxiphite [MINERAL] Pb₃CuCl₂(OH)₂O₂ A dull-olive or pistachio-green mineral consisting of a basic chloride of lead and copper, found in the Mendip Hills of England. { klə'räk·sə‚fīt }

68

choanate fish [PALEON] Any of the lobefins composing the subclass Crossopterygii.
{ 'kō·ə,nāt ,fish }

Choeropotamidae [PALEON] A family of extinct palaeodont artiodactyls in the super-family Entelodontoidae. { ,kir·ə·pə'täm·ə,dē }

chondrite [GEOL] A stony meteorite containing chondrules. { 'kän,drīt }

chondrodite [MINERAL] $Mg_5(SiO_4)_2(F_7OH)_2$ A monoclinic mineral of the humite group; has a resinous luster, is yellow-red in color, and occurs in contact-metamorphosed dolomites. { 'kän·drō,dīt }

Chondrostei [PALEON] The most archaic infraclass of the subclass Actinopterygii, or rayfin fishes. { kän'dräs·tē,ī }

Chondrosteidae [PALEON] A family of extinct actinopterygian fishes in the order Acipenseriformes. { ,kän·drə'stē·ə,dē }

chondrule [GEOL] A spherically shaped body consisting chiefly of pyroxene or olivine minerals embedded in the matrix of certain stony meteorites. { 'kän,drül }

Chonetidina [PALEON] A suborder of extinct articulate brachiopods in the order Strophomenida. { ,kän·ə·tə·dī·nə }

chorismite [PETR] A mixed rock whose fabric is macropolyschematic and which consists of petrologically dissimilar materials of varied origins. { kə'riz,mīt }

Choristodera [PALEON] A suborder of extinct reptiles of the order Eosuchia composed of a single genus, *Champsosaurus*. { ,kȯr·ə'städ·ə·rə }

C horizon [GEOL] The portion of the parent material in soils which has been penetrated with roots. { 'sē hə'rīz·ən }

christophite *See* marmatite. { 'kris·tə,fīt }

chromate [MINERAL] A mineral characterized by the cation $CrO_4{}^{2-}$. { 'krō,māt }

chromatic mineral [MINERAL] A mineral with color. { krō'mad·ik ,min·rəl }

chrome diopside [MINERAL] A bright green variety of diopside containing a small amount of Cr_2O_3. { 'krōm dī'äp,sīd }

chrome iron ore *See* chromite. { ¦krōm 'ī·ərn ,ȯr }

chrome spinel *See* picotite. { ¦krōm spə'nel }

chromite [MINERAL] $FeCr_2O_4$ A mineral of the spinel group; crystals and pure form are rare, and it usually is massive; the only important ore mineral of chromium. Also known as chrome iron ore. { 'krō,mīt }

chromocratic *See* melanocratic. { ,krō·mə'krad·ik }

chron [GEOL] **1.** The time unit equivalent to the stratigraphic unit, subseries, and geologic name of a division of geologic time. **2.** The geochronological equivalent of chronozone. { krän }

chronocline [PALEON] A cline shown by successive morphological changes in the members of a related group, such as a species, in successive fossiliferous strata. { 'krän·ō,klīn }

chronolith *See* time-stratigraphic unit. { 'krän·ə,lith }

chronolithologic unit *See* time-stratigraphic unit. { ¦krän·ə¦lith·ə'läj·ik 'yü·nət }

chronostratic unit *See* time-stratigraphic unit. { ¦krän·ə¦strad·ik 'yü·nət }

chronostratigraphic unit *See* time-stratigraphic unit. { ¦krän·ə¦strad·ə'graf·ik 'yü·nət }

chronostratigraphic zone *See* chronozone. { ,krän·ə,strad·ə'graf·ik 'zōn }

chronostratigraphy [GEOL] A division of stratigraphy that uses age determination and time sequence of rock strata to develop an interpretation of the earth's geologic history. { ,krän·ə·strə'tig·rə·fē }

chronozone [GEOL] **1.** A formal time-stratigraphic unit used to specify strata equivalent in time span to a zone in another type of classification, for example, a biostratigraphic zone. Also known as chronostratigraphic zone. **2.** The smallest subdivision of chronostratigraphic units, below stage, composed of rocks formed during a chron of geologic time. { krän·ə,zōn }

chrysoberyl [MINERAL] $BeAl_2O_4$ A pale green, yellow, or brown mineral that crystallizes in the orthorhombic system and is found most commonly in pegmatite dikes; used as a gem. Also known as chrysopal; gold beryl. { 'kris·ə,ber·əl }

Chrysochloridae [PALEON] The golden moles, a family of extinct lipotyphlan mammals in the order Insectivora. { ¦kris·ə'klȯr·ə,dē }

69

chrysocolla [MINERAL] $CuSiO_3 \cdot 2H_2O$ A silicate mineral ordinarily occurring in impure cryptocrystalline crusts and masses with conchoidal fracture; a minor ore of copper; luster is vitreous, and color is normally emerald green to greenish-blue. { ,kris·ə'käl·ə }

chrysolite [MINERAL] **1.** A gem characterized by light-yellowish-green hues, especially the gem varieties of olivine, but also including beryl, topaz, and spinel. **2.** A variety of olivine having a magnesium to magnesium-iron ratio of 0.90–0.70. { 'kris·ə,līt }

chrysopal See chrysoberyl. { kri'sō·pəl }

chrysoprase [MINERAL] An apple-green variety of chalcedony that contains nickel; used as a gem. Also known as green chalcedony. { 'kris·ə,prāz }

chrysotile [MINERAL] $Mg_3Si_2O_5(OH)_4$ A fibrous form of serpentine that constitutes one type of asbestos. { 'kris·ō,tīl }

Chubb [GEOL] A meteorite crater in Ungava, Quebec, Canada. { chəb }

churchite See weinschenkite. { 'chər,chīt }

churnhole See pothole. { 'chərn ,hōl }

ciminite [PETR] An extrusive rock consisting essentially of olivine with sanidine and pyroxene and basic plagioclase. { 'chīm·ə,nīt }

Cimmeria [PALEON] In the Jurassic, a narrow continent that extended east-west at the southern margin of Eurasia. The name comes from the Crimean peninsula of Russia, where there is well-displayed evidence of an intra-Jurassic orogenic disturbance, indicative of continental collision. { sə'mer·ē·ə }

cimolite [MINERAL] $2Al_2O_3 \cdot 9SiO_3 \cdot 6H_2O$ A white, grayish, or reddish mineral consisting of hydrous aluminum silicate occurring in soft, claylike masses. { 'sim·ə,līt }

Cincinnatian [GEOL] Upper Ordovician geologic time. { sin·sə'nad·ē·ən }

cinder [GEOL] Fine-grained pyroclastic material ranging in diameter from 0.16 to 1.28 inch (4 to 32 millimeters). { 'sin·dər }

cinder coal See natural coke. { 'sin·dər ,kōl }

cinder cone [GEOL] A conical elevation formed by the accumulation of volcanic debris around a vent. { 'sin·dər ,kōn }

C index [GEOPHYS] A subjectively obtained daily index of geomagnetic activity, in which each day's record is evaluated on the basis of 0 for quiet, 1 for moderately disturbed, and 2 for very disturbed. Also known as C figure; magnetic character figure. { 'sē ,in,deks }

cinnabar [MINERAL] HgS A vermilion-red mineral that crystallizes in the hexagonal system, although crystals are rare, and commonly occurs in fine, granular, massive form; the only important ore of mercury. Also known as cinnabarite; vermilion. { 'sin·ə,bär }

cinnabarite See cinnabar. { ,sin·ə'bä,rīt }

CIPW classification [PETR] A designation for the Norm system of classifying igneous rocks; from the initial letters of the names of those who devised it: Cross, Iddings, Pirsson, and Washington. { 'sē'ī'pē'dəb·əl·yü ,klas·ə·fə'kā·shən }

circle of illumination [GEOL] The edge of the sunlit hemisphere, which forms a circular boundary separating the earth into a light half and a dark half. { 'sər·kəl əv ə,lü·mə'nā·shən }

circular coal See eye coal. { 'sər·kyə·lər ,kōl }

circum-Pacific province See Pacific suite. { ,sər·kəm·pə'sif·ik 'prä·vəns }

cirque [GEOL] A steep elliptic to elongated enclave high on mountains in calcareous districts, usually forming the blunt end of a valley. Also known as corrie; cwm. { sərk }

cistern [GEOL] A hollow that holds water. { 'sis·tərn }

citrine [MINERAL] An important variety of crystalline quartz, yellow to brown in color and transparent. Also known as Bohemian topaz; false topaz; quartz topaz; topaz quartz; yellow quartz. { 'si,trēn }

cladodont [PALEON] Pertaining to sharks of the most primitive evolutionary level. { 'klad·ə,dänt }

Cladoselachii [PALEON] An order of extinct elasmobranch fishes including the oldest and most primitive of sharks. { ,klad·ō·sə'läk·ē,ī }

70

Claibornian |GEOL| Middle Eocene geologic time. { ˌklerˈbȯrn·ē·ən }

clairite See enargite. { 'kle,rīt }

clan |PETR| A category of igneous rocks defined in terms of similarities in mineralogical or chemical composition. { klan }

clarain |GEOL| A coal lithotype appearing as stratifications parallel to the bedding plane and usually having a silky luster and scattered or diffuse reflection. Also known as clarite. { 'kla,rān }

Clarendonian |GEOL| Lower Pliocene or upper Miocene geologic time. { ˌkla·rənˈdōn·ē·ən }

clarinite |MINERAL| A heterogeneous, generally translucent material making up the major micropetrological ingredient of clarain. { 'klar·ə,nīt }

clarite See clarain. { 'kla,rīt }

clarke |GEOCHEM| A unit of the average abundance of an element in the earth's crust, expressed as a percentage. Also known as crustal abundance. { klärk }

Clarkecarididae |PALEON| A family of extinct crustaceans in the order Anaspidacea. { ˌklär·kəˈrid·ə,dē }

clarkeite |MINERAL| (Na,Ca,Pb)$_2$U$_2$(O,OH)$_7$ A dark reddish-brown or dark brown mineral consisting of a hydrous or hydrated uranium oxide. { 'klär,kīt }

clarodurain |GEOL| A transitional lithotype of coal composed of vitrinite and other macerals, principally micrinite and exinite. { ¦kla·rōˈdu̇,rān }

clarofusain |GEOL| A transitional lithotype of coal composed of fusinite and vitrinite and other macerals. { ¦kla·rōˈfyü,zān }

clarovitrain |GEOL| A transitional lithotype of coal rock composed primarily of the maceral vitrinite, with lesser amounts of other macerals. { ¦kla·rōˈvi,trān }

clast |GEOL| An individual grain, fragment, or constituent of detrital sediment or sedimentary rock produced by physical breakdown of a larger mass. { klast }

clastation See weathering. { kla'stā·shən }

clastic |GEOL| Rock or sediment composed of clasts which have been transported from their place of origin, as sandstone and shale. { 'klas·tik }

clastic dike |GEOL| A tabular-shaped sedimentary dike composed of clastic material and transecting the bedding of a sedimentary formation; represents invasion by extraneous material along a crack of the containing formation. { 'klas·tik 'dīk }

clastic pipe |GEOL| A cylindrical body of clastic material having an irregular columnar or pillarlike shape, standing approximately vertically through enclosing formations (usually limestone), and measuring a few centimeters to 50 meters (165 feet) in diameter and 1 to 60 meters (3 to 200 feet) in height. { 'klas·tik 'pīp }

clastic ratio |GEOL| The ratio of the percentage of clastic rocks to that of nonclastic rocks in a geologic section. Also known as detrital ratio. { 'klas·tik 'rā·shō }

clastic reservoir |GEOL| An underground oil or gas trap formed in clastic limestone. { 'klas·tik 'rez·əv,wär }

clastic sediment |GEOL| Deposits of clastic materials transported by mechanical agents. Also known as mechanical sediment. { 'klas·tik 'sed·ə·mənt }

clastic wedge |GEOL| The sediments of the exogeosyncline, derived from the tectonic landmasses of the adjoining orthogeosyncline. { 'klas·tik 'wej }

clathrate |GEOCHEM| See gas hydrate. |PETR| Pertaining to a condition, chiefly in leucite rock, in which clear leucite crystals are surrounded by tangential leucite crystals to give the rock an appearance of a net or a section of sponge. Also known as enclosure compound. { 'klath,rāt }

clathrate hydrate See gas hydrate. { ¦klath,rāt 'hī,drāt }

claudetite |MINERAL| As$_2$O$_3$ A mineral containing arsenic that is dimorphous with arsenolite; crystallizes in the monoclinic system. { 'klȯd·ə,tīt }

clausthalite |MINERAL| PbSe A mineral consisting of lead selenide and resembling galena; specific gravity is 7.6–8.8. { 'klau̇s·tə,līt }

Clavatoraceae |PALEOBOT| A group of middle Mesozoic algae belonging to the Charophyta. { ˌklav·əd·əˈrās·ē,ē }

clay |GEOL| **1.** A natural, earthy, fine-grained material which develops plasticity when mixed with a limited amount of water; composed primarily of silica, alumina, and

71

water, often with iron, alkalies, and alkaline earths. **2.** The fraction of an earthy material containing the smallest particles, that is, finer than 3 micrometers. { klā }

Clay Belt [GEOL] A lowland area bordering on the western and southern portions of Hudson and James bays in Canada, composed of clays and silts recently deposited in large glacial lakes during the withdrawal of the continental glaciers. { 'klā ,belt }

clay gall [GEOL] A dry, curled clay shaving derived from dried, cracked mud and embedded and flattened in a sand stratum. { 'klā ,gȯl }

clay ironstone [PETR] **1.** A clayey rock containing large quantities of iron oxide, usually limonite. **2.** A clayey-looking stone occurring among carboniferous and other rocks; contains 20–30% iron. { 'klā 'ī·ərn,stōn }

clay loam [GEOL] Soil containing 27–40% clay, 20–45% sand, and the remaining portion silt. { ¦klā 'lōm }

clay marl [GEOL] A chalky clay, whitish with a smooth texture. { ¦klā 'märl }

clay mineral [MINERAL] One of a group of finely crystalline, hydrous silicates with a two-or three-layer crystal structure; the major components of clay materials; the most common minerals belong to the kaolinite, montmorillonite, attapulgite, and illite groups. { ¦klā ¦min·rəl }

claypan [GEOL] A stratum of compact, stiff, relatively impervious noncemented clay; can be worked into a soft, plastic mass if immersed in water. { 'klā,pan }

clay plug [GEOL] Sediment, with a great deal of organic muck, deposited in a cutoff river meander. { ¦klā ¦pləg }

clay shale [GEOL] **1.** Shale composed wholly or chiefly of clayey material which becomes clay again on weathering. **2.** Consolidated sediment composed of up to 10% sand and having a silt to clay ratio of less than 1:2. { ¦klā ¦shāl }

clay soil [GEOL] A fine-grained inorganic soil which forms hard lumps when dry and becomes sticky when wet. { ¦klā ¦sȯil }

claystone [GEOL] Indurated clay, consisting predominantly of fine material of which a major proportion is clay mineral. { 'klā,stōn }

clay vein [GEOL] A body of clay which is similar to an ore vein in form and fills a crevice in a coal seam. Also known as dirt slip. { 'klā ,vān }

cleat [GEOL] Vertical breakage planes found in coal. Also spelled cleet. { klēt }

cleat spar See ankerite. { 'klēt ,spär }

cleavage [GEOL] Splitting, or the tendency to split, along parallel, closely positioned planes in rock. { 'klēv·ij }

cleavage banding [GEOL] A compositional banding, usually formed from incompetent material such as argillaceous rocks, that is parallel to the cleavage rather than the bedding. { 'klēv·ij ,band·iŋ }

cleavelandite [MINERAL] A white, lamellar variety of albite that is almost pure NaAl-Si$_3$O$_8$ and has a tabular habit, with individuals often showing mosaic developments and tending to occur in fan-shaped aggregates. { 'klēv·lən,dīt }

cleet See cleat. { klēt }

cliachite [MINERAL] A group of brownish, colloidal aluminum hydroxides that constitutes most bauxite. { 'klī·ə,kīt }

cliff of displacement See fault scarp. { 'klif əv dis'plā·smənt }

Cliftonian [GEOL] Middle Middle Silurian geologic time. { klif'tän·ē·ən }

Climatiidae [PALEON] A family of archaic tooth-bearing fishes in the suborder Climatioidei. { ,klī·mə'tī·ə,dē }

Climatiiformes [PALEON] An order of extinct fishes in the class Acanthodii having two dorsal fins and large plates on the head and ventral shoulder. { ,klī·mə,tī·ə'fȯr,mēz }

Climatioidei [PALEON] A suborder of extinct fishes in the order Climatiiformes. { ,klī·mə,tī'ȯid·ē,ī }

climatochronology [GEOL] The absolute age dating of recent geologic events by using the oxygen isotope ratios in ice, shells, and so on. { klī¦mad·ō·krə'näl·ə·jē }

climbing dune [GEOL] A dune that develops on the windward side of mountains or hills. { 'klīm·iŋ 'dün }

clinker [GEOL] Burnt or vitrified stony material, as ejected by a volcano or formed in a furnace. { 'kliŋ·kər }

clinoamphibole [MINERAL] A group of amphiboles which crystallize in the monoclinic system. { ¦klī·nō'am·fə,bōl }

clinochlore [MINERAL] $(Mg,Fe,Al)_3(Si,Al)_2O_5(OH)_4$ Green mineral of the chlorite group, occurring in monoclinic crystals, in folia or scales, or massive. { 'klī·nə,klȯr }

clinoclase [MINERAL] $Cu_3(AsO_4)(OH)_3$ A dark-green mineral consisting of basic copper arsenate occurring in translucent prismatic crystals or massive. Also known as clinoclasite. { 'klī·nə,klās }

clinoclasite See clinoclase. { ¦klī·nə¦klā,sīt }

clinoenstatite [MINERAL] $Mg_2(Si_2O_6)$ A monoclinic pyroxene consisting principally of magnesium silicate; occurs frequently in stony meteorites, but is rare in terrestrial environments. { ¦klī·nō'enz·tə,tīt }

clinoferrosilite [MINERAL] $Fe_2(Si_2O_6)$ A monoclinic pyroxene consisting of iron silicate. { ¦klī·nō,fe·rō'sī,līt }

clinoform [GEOL] A subaqueous landform, such as the continental slope of the ocean or the foreset bed of a delta. { 'klī·nə,fȯrm }

clinohedrite [MINERAL] $CaZnSiO_3(OH)_2$ A colorless, white, or purplish monoclinic mineral consisting of a calcium zinc silicate occurring in crystals; hardness is 5.5 on Mohs scale, and specific gravity is 3.33. { ¦klī·nō¦hē,drīt }

clinohumite [MINERAL] $Mg_9(SiO_4)_4(F,OH)_2$ A monoclinic mineral of the humite group. { ¦klī·nō'hyü,mīt }

clinoptilolite [MINERAL] $(Na,K,Ca)_{2-3}Al_3(Al,Si)_2Si_{13}O_{36}\cdot 12H_2O$ A zeolite mineral that is considered to be a potassium-rich variety of heulandite. { 'klin·əp'til·ə,līt }

clinopyroxene [MINERAL] The general term for any of those pyroxenes that crystallize in the monoclinic system; on occasion, these pyroxenes have large amounts of calcium with or without aluminum and the alkalies. Also known as monopyroxene clinoaugite. { ¦klī·nə·pə'räk,sēn }

clinozoisite [MINERAL] $Ca_2Al_3(SiO_4)_3(OH)$ A grayish-white, pink, or green monoclinic mineral of the epidote group. { ¦klī·nə'zō·i,sīt }

clint [GEOL] A hard or flinty rock, such as a projecting rock or ledge. { klint }

Clintonian [GEOL] Lower Middle Silurian geologic time. { klin'tōn·ē·ən }

clintonite [MINERAL] $Ca(Mg,Al)_3(Al,Si)O_{10}(OH)_2$ A reddish-brown, copper-red, or yellowish monoclinic mineral of the brittle mica group occurring in crystals or foliated masses. Also known as seybertite; xanthophyllite. { 'klint·ən,īt }

closed fold [GEOL] A fold whose limbs have been compressed until they are parallel, and whose structure contour lines form a closed loop. Also known as tight fold. { ¦klōzd ¦fōld }

close-joints cleavage See slip cleavage. { ¦klōs ¦jȯins 'klē·vij }

close sand See tight sand. { ¦klōs ¦sand }

closure [GEOL] The vertical distance between the highest and lowest point on an anticline which is enclosed by contour lines. { 'klō·zhər }

cluse [GEOL] A narrow gorge, trench, or water gap with steep sides that cuts transversely through an otherwise continuous ridge. { klüz }

clusterite See botryoid. { 'klə·stə,rīt }

Coahuilan [GEOL] A North American provincial series in Lower Cretaceous geologic time, above the Upper Jurassic and below the Comanchean. { kō·ə'wēl·ən }

coal [GEOL] The natural, rocklike, brown to black derivative of forest-type plant material, usually accumulated in peat beds and progressively compressed and indurated until it is finally altered into graphite or graphite-like material. { kōl }

coal ball [GEOL] A subspherical mass containing mineral matter embedded with plant material, found in coal seams and overlying beds of the late Paleozoic. { kōl ,bȯl }

coal bed [GEOL] A seam or stratum of coal parallel to the rock stratification. Also known as coal rake; coal seam. { kōl ,bed }

coal breccia [GEOL] Angular fragments of coal within a coal bed. { 'kōl ,brech·ə }

coal clay See underclay. { kōl ,klā }

coalification [GEOL] Formation of coal from plant material by the processes of diagenesis and metamorphism. Also known as bituminization; carbonification; incarbonization; incoalation. { ,kōl·ə·fə'kā·shən }

Coal Measures

Coal Measures [GEOL] The sequence of rocks typically containing coal of the Upper Carboniferous. { 'kōl ,mezh·ərz }

coal paleobotany [PALEOBOT] A branch of the paleobotanical sciences concerned with the origin, composition, mode of occurrence, and significance of fossil plant materials that occur in or are associated with coal seams. { 'kōl ,pā·lē·ō'bät·ən·ē }

coal pebbles [GEOL] Rounded masses of coal occurring in sedimentary rock. { 'kōl ,peb·əlz }

coal petrology [GEOL] The science that deals with the origin, history, occurrence, structure, chemical composition, and classification of coal. { 'kōl pə'träl·ə·jē }

coal rake *See* coal bed. { 'kōl ,rāk }

coal seam *See* coal bed. { 'kōl ,sēm }

coal split *See* split. { 'kōl ,split }

coarse fragment [GEOL] A rock or mineral fragment in the soil with an equivalent diameter greater than 0.08 inch (2 millimeters). { 'kȯrs 'frag·mənt }

coarse-grained [PETR] *See* phaneritic. { 'kȯrs ¦grānd }

coastal berm *See* berm. { 'kōs·təl 'bərm }

coastal dune [GEOL] A mobile mound of windblown material found along many sea and lake shores. { 'kōs·təl 'dün }

coastal plain [GEOL] An extensive, low-relief area that is bounded by the sea on one side and by a high-relief province on the landward side. Its geologic province actually extends beyond the shoreline across the continental shelf; it is linked to the stable part of a continent on the trailing edge of a plate. Typically, it has strata that dip gently and uniformly toward the sea. { 'kōs·təl 'plān }

coastal sediment [GEOL] The mineral and organic deposits of deltas, lagoons, and bays, barrier islands and beaches, and the surf zone. { 'kōs·təl 'sed·ə·mənt }

coast shelf *See* submerged coastal plain. { 'kōst ,shelf }

cobalt bloom *See* erythrite. { 'kō,bȯlt ,blüm }

cobalt glance *See* cobaltite. { 'kō,bȯlt 'glans }

cobaltite [MINERAL] CoAsS A silver-white mineral with a metallic luster that crystallizes in the isometric system, resembling crystals of pyrite; it is one of the chief ores of cobalt. Also known as cobalt glance; gray cobalt; white cobalt. { kə'bȯl,tīt }

cobaltocalcite [MINERAL] A red, cobalt-bearing variety of calcite. { kə¦bȯl·tō'kal,sīt }

cobalt ocher *See* asbolite; erythrite. { 'kō,bȯlt 'ō·kər }

cobaltomenite [MINERAL] CoSeO$_3$·2H$_2$O A mineral consisting of a hydrous cobalt selenium oxide. { ,kō,bȯl'tä·mə,nīt }

cobalt pyrites *See* linnaeite. { 'kō,bȯlt 'pī,rīts }

cobble [GEOL] A rock fragment larger than a pebble and smaller than a boulder, having a diameter in the range of 64–256 millimeters (2.5–10.1 inches), somewhat rounded or otherwise modified by abrasion in the course of transport. { 'käb·əl }

cobble beach *See* shingle beach. { 'käb·əl ,bēch }

Coblentzian [GEOL] Upper Lower Devonian geologic time. { kō'blens·ē·ən }

coccolith ooze [GEOL] A fine-grained pelagic sediment containing undissolved sand- or silt-sized particles of coccoliths mixed with amorphous clay-sized material. { 'käk·ə,lith ,üz }

coccosphere [PALEOBOT] The fossilized remains of a member of Coccolithophorida. { 'käk·ə,sfir }

Coccosteomorphi [PALEON] An aberrant lineage of the joint-necked fishes. { kä¦kä·stē·ə¦mȯr·fē }

Cochliodontidae [PALEON] A family of extinct chondrichthian fishes in the order Bradyodonti. { ,kōk·lē·ō'dän·tə,dē }

cocinerite [MINERAL] Cu$_4$AgS A silver gray mineral consisting of copper and silver sulfide; occurs in massive form. { kō·sə'ne,rīt }

cockpit karst *See* cone karst. { 'käk,pit 'karst }

Coelacanthidae [PALEON] A family of extinct lobefin fishes in the order Coelacanthiformes. { ,sē·lə'kan·thə,dē }

Coelolepida [PALEON] An order of extinct jawless vertebrates (Agnatha) distinguished

74

by skin set with minute, close-fitting scales of dentine, similar to placoid scales of sharks. { ˌsē·lō'lep·ə·da }

Coelurosauria [PALEON] A group of small, lightly built saurischian dinosaurs in the suborder Theropoda having long necks and narrow, pointed skulls. { sə,lür·ə'sȯr·ē·ə }

Coenopteridales [PALEOBOT] A heterogeneous group of fernlike fossil plants belonging to the Polypodiophyta. { ˌsē·näp,ter·ə'dā·lēz }

coeruleolactite [MINERAL] $(Ca,Cu)Al_6(PO_4)_4(OH)_8 \cdot 4-5H_2O$ A milky-white to sky-blue mineral consisting of an aluminum phosphate. { sə,rül·ē·ō'lak,tīt }

coesite [MINERAL] A high-pressure polymorph of SiO_2 formed in nature only under unique physical conditions, requiring pressures of more than 20 kilobars (2 gigapascals); usually found in meteor impact craters. { 'sē,zīt }

coffinite [MINERAL] $USiO_4$ A black silicate important as a uranium ore; found in sandstone deposits and hydrothermal veins in New Mexico, Utah, and Wyoming. { 'kȯf·ə,nīt }

cognate [GEOL] Pertaining to contemporaneous fractures in a system with regard to time of origin and deformational type. { ˌkäg,nāt }

cognate ejecta [GEOL] Essential or accessory pyroclasts derived from the magmatic materials of a current volcanic eruption. { ˌkäg,nāt ē'jek·tə }

cognate inclusion See autolith. { ˌkäg,nāt in'klü·zhən }

cogwheel ore See bournonite. { 'käg,wēl ˌȯr }

cohenite [MINERAL] $(Fe,Ni,Co)_3C$ A tin-white, isometric mineral found in meteorites. { 'kō·ə,nīt }

coherent deposit [GEOL] A consolidated sedimentary deposit that is not easily shattered. { kō¦hir·ənt di'päz·ət }

cohesionless [GEOL] Referring to a soil having low shear strength when dry, and low cohesion when wet. Also known as frictional; noncohesive. { kō'hē·zhən·ləs }

cohesiveness [GEOL] Property of unconsolidated fine-grained sediments by which the particles stick together by surface forces. { kō'hē·siv·nəs }

cohesive soil [GEOL] A sticky soil, such as clay or silt; its shear strength equals about half its unconfined compressive strength. { kō'hē·siv 'sȯil }

coke coal See natural coke. { 'kōk ˌkōl }

cokeite [GEOL] Naturally occurring coke formed by the action of magma on coal or by natural combustion of coal. { 'kō,kīt }

coking coal [GEOL] A very soft bituminous coal suitable for coking. { 'kok·iŋ ˌkōl }

col [GEOL] A high, sharp-edged pass occurring in a mountain ridge, usually produced by the headward erosion of opposing cirques. { käl }

cold glacier [GEOL] A glacier whose base is at a temperature much below 32°F (0°C) and frozen to the bedrock, resulting in insignificant movement and almost no erosion. { ¦kōld 'glā·shər }

colemanite [MINERAL] $Ca_2B_6O_{11} \cdot 5H_2O$ A colorless or white hydrated borate mineral that crystallizes in the monoclinic system and occurs in massive crystals or as nodules in clay. { 'kōl·mə,nīt }

Coleodontidae [PALEON] A family of conodonts in the suborder Neurodontiformes. { ˌkō·lē·ō'dän·tə,dē }

colk See pothole. { kōk }

collapse breccia [GEOL] Angular rock fragments derived from the collapse of rock overlying a hollow space. { kə'laps ˌbrech·ə }

collapse caldera [GEOL] A caldera formed primarily as a result of collapse due to withdrawal of magmatic support. { kə'laps kal'dir·ə }

collapse sink [GEOL] A sinkhole resulting from local collapse of a cavern that has been enlarged by solution and erosion. { kə'laps ˌsiŋk }

collapse structure [GEOL] A structure resulting from rock slides under the influence of gravity. Also known as gravity-collapse structure. { kə'laps ˌstrək·chər }

collenia [PALEOBOT] A convex, slightly arched, or turbinate stromatolite produced by late Precambrian blue-green algae of the genus *Collenia*. { kə'len·ē·ə }

collinite [GEOL] The maceral, of collain consistency, of jellified plant material precipitated from solution and hardened; a variety of euvitrinite. { 'käl·ə,nīt }

collinsite [MINERAL] $Ca_2(Mg,Fe)(PO_4)_2$ A phosphate mineral occurring in concentric layers in phosphoric nodules; found in meteorites. { 'käl·ən,zīt }

colloform [GEOL] Pertaining to the rounded, globular texture of mineral formed by colloidal precipitation. { 'käl·ə,fórm }

collophane [MINERAL] A massive, cryptocrystalline, carbonate-containing variety of apatite and a principal source of phosphates for fertilizers. Also known as collophanite. { 'käl·ə,fän }

collophanite See collophane. { kə'läf·ə,nīt }

colluvium [GEOL] Loose, incoherent deposits at the foot of a slope or cliff, brought there principally by gravity. { kə'lü·vē·əm }

Collyritidae [PALEON] A family of extinct, small, ovoid, exocyclic Euechinoidea with fascioles or a plastron. { ,käl·ə'rid·ə,dē }

Coloradoan [GEOL] Middle Upper Cretaceous geologic time. { ,käl·ə'rad·ə·wən }

coloradoite [MINERAL] HgTe A grayish-black, isometric telluride mineral with a metallic luster; specific gravity is 8.6. { ,käl·ə'rad·ə,wīt }

columbite [MINERAL] $(Fe,Mn)(Cb,Ta)_2O_6$ An iron-black mineral with a submetallic luster that crystallizes in the orthorhombic system; the chief ore mineral of niobium (columbium); hardness is 6 on Mohs scale, and specific gravity is 5.4–6.5. Also known as dianite; greenlandite; niobite. { kə'ləm,bīt }

column [GEOL] See geologic column; stalacto-stalagmite. { 'käl·əm }

columnar jointing [GEOL] Parallel, prismatic columns that are formed as a result of contraction during cooling in basaltic flow and other extrusive and intrusive rocks. Also known as columnar structure; prismatic jointing; prismatic structure. { kə'ləm· nər 'jóint·iŋ }

columnar section [GEOL] A vertical strip or scale drawing of the strip taken from a given area or locality showing the sequence of the rock units and their stratigraphic relationship, and indicating the thickness, lithology, age, classification, and fossil content of the rock units. Also known as section. { kə'ləm·nər 'sek·shən }

columnar structure [GEOL] See columnar jointing. [MINERAL] Mineral structure consisting of parallel columns of slender prismatic crystals. [PETR] A primary sedimentary structure consisting of columns arranged perpendicular to the bedding. { kə'ləm·nər ,strək·chər }

colusite [MINERAL] $Cu_3(As,Sn,V,Fe,Te)S_4$ A bronze-colored mineral consisting of a sulfide of copper and arsenic with vanadium, iron, and tellurium substituting for arsenic; usually occurs in massive form. { kə'lü,sīt }

comagmatic province See petrographic province. { ¦kō·mag'mad·ik 'prä·vəns }

Comanchean [GEOL] A North American provincial series in Lower and Upper Cretaceous geologic time, above Coahuilan and below Gulfian. { kə'man·chē·ən }

combination trap [GEOL] Underground reservoir structure closure, deformation, or fault where reservoir rock covers only part of the structure. { ,käm·bə'nā·shən ¦trap }

combined water [GEOCHEM] Water attached to soil minerals by means of chemical bonds. { kəm'bīnd 'wód·ər }

combustible shale See tasmanite. { kəm'bəs·tə·bəl ,shäl }

comendite [GEOL] A white, sodic rhyolite containing alkalic amphibole or pyroxene. { kə'men,dīt }

Comleyan [GEOL] Lower Cambrian geologic time. { 'käm·lā·ən }

common feldspar See orthoclase. { ¦käm·ən 'feld,spär }

common mica See muscovite. { ¦käm·ən 'mī·kə }

common pyrite See pyrite. { ¦käm·ən 'pī,rīt }

common salt See halite. { ¦käm·ən 'sólt }

compaction [GEOL] Process by which soil and sediment mass loses pore space in response to the increasing weight of overlying material. { kəm'pak·shən }

competence [GEOL] The ability of the wind to transport solid particles either by rolling, suspension, or saltation (intermittent rolling and suspension); usually expressed in terms of the weight of a single particle. { 'käm·pəd·əns }

competent beds [GEOL] Beds or strata capable of withstanding the pressures of folding without flowing or changing in original thickness. { ¦käm·pəd·ənt ¦bedz }

complementary rocks [GEOL] Rocks which are differentiated from the same magma, and whose average composition is the same as the parent magma. { ˌkäm·plə'men· trē 'räks }

complex [GEOL] An assemblage of rocks that has been folded together, intricately mixed, involved, or otherwise complicated. [MINERAL] Composed of many ingredients. { 'käm,pleks }

complex dune [GEOL] A dune of varying forms, often very large, and produced by variable, shifting winds and the merging of various dune types. { 'käm,pleks 'dün }

complex fold [GEOL] A fold whose axial line is also folded. { 'käm,pleks 'fōld }

complex tombolo [GEOL] A system resulting when several islands and the mainland are interconnected by a complex series of tombolos. Also known as tombolo cluster; tombolo series. { 'käm,pleks 'täm·bə,lō }

composite cone [GEOL] A large volcanic cone constructed of lava and pyroclastic material in alternating layers. { kəm'päz·ət 'kōn }

composite dike [GEOL] A dike consisting of several intrusions differing in chemical and mineralogical composition. { kəm'päz·ət 'dīk }

composite fold [GEOL] A fold having smaller folds on its limbs. { kəm'päz·ət 'fōld }

composite gneiss [PETR] A banded rock formed by intimate penetration of magma into country rocks. { kəm'päz·ət 'nīs }

composite grain [GEOL] A sedimentary clast formed of two or more original particles. { kəm'päz·ət 'grān }

composite sequence [GEOL] An ideal sequence of cyclic sediments containing all the lithological types in their proper order. { kəm'päz·ət 'sē·kwəns }

composite sill [GEOL] A sill consisting of several intrusions differing in chemical and mineralogical compositions. { kəm'päz·ət 'sil }

composite topography [GEOL] A topography whose features have developed in two or more erosion cycles. { kəm'päz·ət tə'päg·rə·fē }

composite unconformity [GEOL] An unconformity that has resulted from more than one episode of nondeposition and possible erosion. { kəm'päz·ət ˌən·kən'fòr· məd·ē }

composite vein [GEOL] A large fracture zone composed of parallel ore-filled fissures and converging diagonals, whose walls and intervening country rock have been replaced to a certain degree. { kəm'päz·ət 'vān }

composite volcano See stratovolcano. { kəm'päz·ət väl'kā·nō }

compositional maturity [GEOL] Concept of a type of maturity in sedimentary rocks in which a sediment approaches the compositional end product to which formative processes drive it. { ˌkäm·pə'zish·ən·əl mə'chùr·əd·ē }

compound alluvial fan [GEOL] Structure formed by the lateral growth and merger of fans made by neighboring streams. { 'käm,paúnd ə¦lü·vē·əl ¦fan }

compound fault [GEOL] A zone or series of essentially parallel faults, closely spaced. { 'käm,paúnd 'fòlt }

compound ripple marks [GEOL] Complex ripple marks of great diversity which originate by simultaneous interference of wave oscillation with current action. { 'käm ,paúnd 'rip·əl ,märks }

compound volcano [GEOL] **1.** A volcano consisting of a complex of two or more cones. **2.** A volcano with an associated volcanic dome. { 'käm,paúnd väl'kā·nō }

compression [GEOL] A system of forces which tend to decrease the volume or shorten rocks. { kəm'presh·ən }

concentric faults [GEOL] Faults that are arranged concentrically. { kən'sen·trik 'fòlts }

concentric fold [GEOL] A fold in which the original thickness of the strata is unchanged during deformation. Also known as parallel fold. { kən'sen·trik 'fōld }

concentric fractures [GEOL] A system of fractures concentrically arranged about a center. { kən'sen·trik 'frak·chərz }

concentric weathering See spheroidal weathering. { kən'sen·trik 'weth·ə·riŋ }

conchoidal

conchoidal [GEOL] Having a smoothly curved surface; used especially to describe the fracture surface of a mineral or rock. { käŋ'kȯid·əl }

concordant body [GEOL] An intrusive igneous body whose contacts are parallel to the bedding of the country rock. Also known as concordant injection; concordant pluton. { kən 'kȯrd·ənt 'bäd·ē }

concordant coastline [GEOL] A coastline parallel to the land structures which form the margin of an ocean basin. { kən'kȯrd·ənt 'kōst,līn }

concordant injection See concordant body. { kən'kȯrd·ənt in'jek·shən }

concordant pluton See concordant body. { kən'kȯrd·ənt 'plü,tän }

concretion [GEOL] A hard, compact mass of mineral matter in the pores of sedimentary or fragmental volcanic rock; represents a concentration of a minor constituent of the enclosing rock or of cementing material. { kän'krē·shən }

concretionary [GEOL] Tending to grow together, forming concretions. { kən'krē·shə,ner·ē }

concretioning [GEOL] The process of forming concretions. { kən'krē·shən·iŋ }

concussion fracture [GEOL] Radiating system of fractures in a shock-metamorphosed rock. { kən'kəsh·ən ,frak·chər }

condensate field [GEOL] A petroleum field developed in predominantly gas-bearing reservoir rocks, but within which condensation of gas to oil commonly occurs with decreases in field pressure. { 'kän·dən,sāt ,fēld }

conductivity See permeability. { ,kän,dək'tiv·əd·ē }

conduit [GEOL] A water-filled underground passage that is always under hydrostatic pressure. { 'kän·də·wət }

Condylarthra [PALEON] A mammalian order of extinct, primitive, hoofed herbivores with five-toed plantigrade to semidigitigrade feet. { ,kän·də'lär·thrə }

cone [GEOL] A mountain, hill, or other landform having relatively steep slopes and a pointed top. { kōn }

cone delta See alluvial cone. { 'kōn ,del·tə }

cone dike See cone sheet. { 'kōn ,dīk }

cone-in-cone structure [GEOL] The structure of a concretion characterized by the development of a succession of cones one within another. { 'kōn in 'kōn 'strək·chər }

cone karst [GEOL] A type of karst, typical of tropical regions, characterized by a pattern of steep, convex sides and slightly concave floors. Also known as cockpit karst; Kegel karst. { 'kōn ,kärst }

Conemaughian [GEOL] Upper Middle Pennsylvanian geologic time. { ,kän·ə'mȯg·ē·ən }

cone of dejection See alluvial cone. { 'kōn əv di'jek·shən }

cone of detritus See alluvial cone. { 'kōn əv di'trīd·əs }

cone sheet [GEOL] An accurate dike forming part of a concentric set that dips inward toward the center of the arc. Also known as cone dike. { 'kōn ,shēt }

Conewangoan [GEOL] Upper Upper Devonian geologic time. { ,kän·ə'waŋ·gə·wən }

confining bed [GEOL] An impermeable bed adjacent to an aquifer. { kən'fīn·iŋ ,bed }

confining pressure [GEOL] An equal, all-sided pressure, such as lithostatic pressure produced by overlying rocks in the crust of the earth. { kən'fīn·iŋ ,presh·ər }

conformable [GEOL] **1.** Pertaining to the contact of an intrusive body when it is aligned with the internal structures of the intrusion. **2.** Referring to strata in which layers are formed above one another in an unbroken, parallel order. { kən'fȯr·mə·bəl }

conformity [GEOL] The shared and undisturbed correspondence between adjacent sedimentary strata that have been deposited in orderly sequence with little or no indication of time lapses. { kən'fȯr·məd·ē }

congelifluction See gelifluction. { kən,jel·ə'flək·shən }

congelifraction [GEOL] The splitting or disintegration of rocks as the result of the freezing of the water contained. Also known as frost bursting; frost riving; frost shattering; frost splitting; frost weathering; gelifraction; gelivation. { kən,jel·ə,frak·shən }

congeliturbate [GEOL] Soil or unconsolidated earth which has been moved or disturbed by frost action. { kən,jel·ə'tər·bət }

congeliturbation [GEOL] The churning and stirring of soil as a result of repeated cycles of freezing and thawing; includes frost heaving and surface subsidence during thaws. Also known as cryoturbation; frost churning; frost stirring; geliturbation. { kən‚jel· ə·tər'bā·shən }

conglomerate [GEOL] Cemented, rounded fragments of water-worn rock or pebbles, bound by a siliceous or argillaceous substance. { kən'gläm·ə·rət }

conglomeratic mudstone See paraglomerate. { kən‚gläm·ə‚rad·ik 'məd‚stōn }

congruent melting [GEOL] Melting of a solid substance to a liquid identical in composition. { kən'grü·ənt 'melt·iŋ }

Coniacian [GEOL] Lower Senonian geologic time. { ‚kän·ē'ā·shən }

conichalcite [MINERAL] CaCu(AsO₄)(OH) A grass green to yellowish-green or emerald green, orthorhombic mineral consisting of a basic arsenate of calcium and copper. { ‚kän·ə'kal‚sīt }

Coniconchia [PALEON] A class name proposed for certain extinct organisms thought to have been mollusks; distinguished by a calcareous univalve shell that is open at one end and by lack of a siphon. { ‚kän·ə'käŋ·kē·ə }

conjugate [GEOL] **1.** Pertaining to fractures in which both sets of veins or joints show the same strike but opposite dip. **2.** Pertaining to any two sets of veins or joints lying perpendicular. { 'kän·jə·gət }

conjugate joint system [GEOL] Two joint sets with a symmetrical pattern arranged about another structural feature or an inferred stress axis. { 'kän·jə·gət ‚jȯint 'sis·təm }

connarite [MINERAL] A green mineral consisting of hydrous nickel silicate occurring as small crystals or grains. { 'kän·ə‚rīt }

connate [GEOL] Referring to materials involved in sedimentary processes that are contemporaneous with surrounding materials. { kə'nāt }

connecting bar See tombolo. { kə'nekt·iŋ ‚bär }

connellite [MINERAL] Cu₁₉(SO₄)Cl₄(OH)₃₂·3H₂O A deep-blue striated copper mineral; crystals are in the hexagonal system. Also known as footeite. { 'kän·əl‚īt }

Conoclypidae [PALEON] A family of Cretaceous and Eocene exocyclic Euechinoidea in the order Holectypoida having developed aboral petals, internal partitions, and a high test. { ‚kän·ō·klə'pid·ē‚ē }

Conocyeminae [PALEON] A subfamily of Mesozoan parasites in the family Dicyemidae. { ‚kän·ə‚sī'em·ə‚nē }

conodont [PALEON] A minute, toothlike microfossil, composed of translucent amber-brown, fibrous or lamellar calcium phosphate; taxonomic identity is controversial. { 'kän·ə‚dänt }

Conodontiformes [PALEON] A suborder of conodonts from the Ordovician to the Triassic having a lamellar internal structure. { ‚kän·ə‚dän·tə'fȯr‚mēz }

Conodontophoridia [PALEON] The ordinal name for the conodonts. { ‚kän·ə‚dän·tə· fə'rid·ē·ə }

conoplain See pediment. { 'kän·ə‚plān }

Conrad discontinuity [GEOPHYS] A relatively abrupt discontinuity in the velocity of elastic waves in the earth, increasing from 6.1 to 6.4–6.7 kilometers per second; occurs at various depths and marks contact of granitic and basaltic layers. { 'kän‚rad dis‚känt·ən'ü·əd·ē }

consanguineous [GEOL] Of a natural group of sediments or sedimentary rocks, having common or related origin. { ‚kän·saŋ‚gwin·ē·əs }

consanguinity [PETR] The genetic relationship between igneous rocks in a single petrographic province which are presumably derived from a common parent magma. { ‚kän·saŋ‚gwin·əd·ē }

consequent [GEOL] Of, pertaining to, or characterizing movements of the earth resulting from the external transfer of material in the process of gradation. { 'kän· sə·kwənt }

consequent stream [GEOL] A stream whose course is determined by the slope of the land. Also known as superposed stream. { 'kän·sə·kwənt ‚strēm }

consequent valley [GEOL] **1.** A valley whose direction depends on corrugation. **2.** A

consolidation

valley formed by the widening of a trench cut by a consequent stream. { 'kän·sə·kwənt ,val·ē }

consolidation |GEOL| **1.** Processes by which loose, soft, or liquid earth become coherent and firm. **2.** Adjustment of a saturated soil in response to increased load; involves squeezing of water from the pores and a decrease in void ratio. { kən,säl·ə'dā·shən }

Constellariidae |PALEON| A family of extinct, marine bryozoans in the order Cystoporata. { ,kän·stə·lə'rī·ə,dē }

contact |GEOL| The surface between two different kinds of rocks. { 'kän,takt }

contact aureole See aureole. { 'kän,takt 'ȯr·ē,ōl }

contact breccia |PETR| Angular rock fragments resulting from shattering of wall rocks around laccolithic and other igneous masses. { 'kän,takt 'brech·ə }

contact metamorphic rock |PETR| A rock formed by the processes of contact metamorphism. { 'kän,takt ,med·ə'mȯr·fik 'räk }

contact metamorphism |PETR| Metamorphism that is genetically related to the intrusion or extrusion of magmas and takes place in rocks at or near their contact. { 'kän,takt ,med·ə'mȯr·fiz·əm }

contact metasomatism |GEOL| One of the main local processes of thermal metamorphism that is related to intrusion of magmas; takes place in rocks or near their contact with a body of igneous rock. { 'kän,takt ,med·ə'sō·mə,tiz·əm }

contact mineral |MINERAL| A mineral formed by the processes of contact metamorphism. { 'kän,takt ,min·rəl }

contact vein |GEOL| **1.** A variety of fissure vein formed by deposition of minerals in a fault fissure at a rock contact. **2.** A replacement vein formed by mineralized solutions percolating along the more permeable surface areas of the contact. { 'kän,takt ,vān }

contact zone See aureole. { 'kän,takt ,zōn }

contamination |GEOL| A process in which the chemical composition of a magma changes due to the assimilation of country rocks. { kən,tam·ə'nā·shən }

contemporaneous |GEOL| **1.** Formed, existing, or originating at the same time. **2.** Of a rock, developing during formation of the enclosing rock. { kən,tem·pə'rā·nē·əs }

continental accretion |GEOL| The theory that continents have grown by the addition of new continental material around an original nucleus, mainly through the processes of geosynclinal sedimentation and orogeny. { ¦känt·ən¦ent·əl ə'krē·shən }

continental borderland |GEOL| The area of the continental margin between the shoreline and the continental slope. { ¦känt·ən¦ent·əl 'bȯr·dər,land }

continental crust |GEOL| The basement complex of rock, that is, metamorphosed sedimentary and volcanic rock with associated igneous rocks mainly granitic, that underlies the continents and the continental shelves. { ¦känt·ən¦ent·əl 'krəst }

continental deposits |GEOL| Sedimentary deposits laid down within a general land area. { ¦känt·ən¦ent·əl di'päz·əts }

continental displacement See continental drift. { ¦känt·ən¦ent·əl di'splās·mənt }

continental divide |GEOL| A drainage divide of a continent, separating streams that flow in opposite directions; for example, the divide in North America that separates watersheds of the Pacific Ocean from those of the Atlantic Ocean. { ¦känt·ən¦ent·əl di'vīd }

continental drift |GEOL| The concept of continent formation by the fragmentation and movement of land masses on the surface of the earth. Also known as continental displacement. { ¦känt·ən¦ent·əl 'drift }

continental geosyncline |GEOL| A geosyncline filled with nonmarine sediments. { ¦känt·ən¦ent·əl ¦jē·ō'sin,klīn }

continental growth |GEOL| The processes contributing to growth of continents at the expense of ocean basins. { ¦känt·ən¦ent·əl 'grōth }

continental heat flow |GEOPHYS| The amount of thermal energy escaping from the earth through the continental crust per unit area and unit time. { ¦känt·ən¦ent·əl 'hēt ,flō }

continental margin |GEOL| Those provinces between the shoreline and the deep-sea

bottom; generally consists of the continental borderland, shelf, slope, and rise.
{ ¦känt·ən¦ent·əl 'mär·jən }

continental nucleus |GEOL| A large area of basement rock consisting of basaltic and more mafic oceanic crust and periodotitic mantle from which it is postulated that continents have grown. Also known as continental shield; cratogene; shield.
{ ¦känt·ən¦cnt·əl 'nü·klē·əs }

continental plate |GEOL| Thick continental crust. { ¦känt·ən¦ent·əl 'plāt }

continental platform See continental shelf. { ¦känt·ən¦ent·əl 'plat,fórm }

continental rise |GEOL| A transitional part of the continental margin; a gentle slope with a generally smooth surface, built up by the shedding of sediments from the continental block, and located between the continental slope and the abyssal plain.
{ ¦känt·ən¦ent·əl 'rīz }

continental shelf |GEOL| The zone around a continent, that part of the continental margin extending from the shoreline and the continental slope; composes with the continental slope the continental terrace. Also known as continental platform; shelf. { ¦känt·ən¦ent·əl 'shelf }

continental shield See shield. { ¦känt·ən¦ent·əl 'shēld }

continental slope |GEOL| The part of the continental margin consisting of the declivity from the edge of the continental shelf extending down to the continental rise.
{ ¦känt·ən¦ent·əl 'slōp }

continental shield See shield. { ¦känt·ən¦ent·əl 'shēld }

continental terrace |GEOL| The continental shelf and slope together. { ¦känt·ən¦ent·əl 'ter·əs }

continent formation |GEOL| A series of six or seven major episodes, resulting from the buildup of radioactive heat and then the melting or partial melting of the earth's interior; the molten rock melt rises to the surface, differentiating into less primitive lavas; the continent then nucleates, differentiates, and grows from oceanic crust and mantle. { ¦känt·ən¦ent·əl fər'mā·shən }

continuous permafrost zone |GEOL| Regional zone predominantly underlain by permanently frozen subsoil that is not interrupted by pockets of unfrozen ground.
{ kən¦tin·yə·wəs 'pər·mə,fróst ,zōn }

continuous profiling |GEOL| A method of shooting in seismic exploration in which uniformly placed seismometer stations along a line are shot from holes spaced along the same line so that each hole records seismic ray paths geometrically identical with those from adjacent holes. { kən¦tin·yə·wəs 'prō,fīl·iŋ }

continuous reaction series |MINERAL| A branch of Bowen's reaction series comprising the plagioclase mineral group in which reaction of early-formed crystals with water takes place continuously, without abrupt changes in crystal structure. { kən¦tin·yə·wəs rē'ak·shən ,sir,ēz }

contraction hypothesis |GEOL| Theory that shrinking of the earth is the cause of compression folding and thrusting. { kən'trak·shən hī'päth·ə·səs }

Conularida |PALEON| A small group of extinct invertebrates showing a narrow, four-sided, pyramidal-shaped test. { ,kän·əl'ar·ə·də }

Conulidae |PALEON| A family of Cretaceous exocyclic Euechinoidea characterized by a flattened oral surface. { kə'nü·lə,dē }

convection current |GEOPHYS| Mass movement of subcrustal or mantle material as the result of temperature variations. { kən'vek·shən ,kər·ənt }

convergence |GEOL| Diminution of the interval between geologic horizons. { kən'vər·jəns }

convolute bedding |GEOL| The extremely contorted laminae usually confined to a single layer of sediment, resulting from subaqueous slumping. { 'kän·və,lüt ,bed·iŋ }

convolution |GEOL| **1.** The process of developing convolute bedding. **2.** A structure resulting from a convolution process, such as a small-scale but intricate fold. { ,kän·və'lü·shən }

cooperite |MINERAL| (Pt,Pd)S A steel-gray tetragonal mineral of metallic luster consisting of a sulfide of platinum, occurring in irregular grains in igneous rock.
{ 'kü·pə,rīt }

81

coorongite

coorongite [GEOL] A boghead coal in the peat stage. { kō'ä·rən‚jīt }

copiapite [MINERAL] **1.** $Fe_5(SO_4)_6(OH)_2 \cdot 20H_2O$ A yellow mineral occurring in granular or scalar aggregates. Also known as ihleite; knoxvillite; yellow copperas. **2.** A group of minerals containing hydrous iron sulfates. { 'kō·pē·ə‚pīt }

Copodontidae [PALEON] An obscure family of Paleozoic fishes in the order Bradyodonti. { ‚kō·pə'dän·tə‚dē }

copper glance See chalcocite. { 'käp·ər 'glans }

copperite [MINERAL] An important platinum mineral, composed of platinum sulfide. { 'käp·ə‚rīt }

copper mica See chalcophyllite. { 'käp·ər 'mī·kə }

copper nickel See niccolite. { 'käp·ər 'nik·əl }

copper ore [GEOL] Rock containing copper minerals. { 'käp·ər ‚ȯr }

copper pyrite See chalcopyrite. { 'käp·ər 'pī‚rīt }

copper uranite See torbernite. { 'käp·ər 'yur·ə‚nīt }

coprolite [GEOL] Petrified excrement. { 'käp·rə‚līt }

coquimbite [MINERAL] $Fe_2(SO_4)_3 \cdot 9H_2O$ A white mineral that crystallizes in the hexagonal system; it is dimorphous with paracoquimbite. { kō'kim‚bīt }

coquina [PETR] A coarse-grained, porous, easily crumbled variety of limestone composed principally of mollusk shell and coral fragments cemented together as rock. { kō'kē·nə }

coquinoid [PETR] **1.** Of or pertaining to coquina. **2.** Lithified coquina. **3.** An autochthonous deposit of limestone made up of more or less whole mollusk shells. { 'kō·kə‚nȯid }

coracite See uraninite. { 'kȯr·ə‚sīt }

coral head [GEOL] A small reef patch of coralline material. Also known as coral knoll. { 'kä·rəl ‚hed }

coral knoll See coral head. { 'kä·rəl ‚nōl }

coral mud [GEOL] Fine-grade deposits of coral fragments formed around coral islands and coasts bordered by coral reefs. { 'kär·əl ‚məd }

coral pinnacle [GEOL] A sharply upward-projecting growth of coral rising from the floor of an atoll lagoon. { 'kär·əl 'pin·ə·kəl }

coral reef [GEOL] A ridge or mass of limestone built up of detrital material deposited around a framework of skeletal remains of mollusks, colonial coral, and massive calcareous algae. { 'kär·əl ‚rēf }

coral-reef shoreline [GEOL] A shoreline formed by reefs composed of coral polyps. Also known as coral shoreline. { 'kär·əl ‚rēf 'shȯr‚līn }

coral rock See reef limestone. { 'kär·əl ‚räk }

coral sand [GEOL] Coarse-grade deposits of coral fragments formed around coral islands and coasts bordered by coral reefs. { 'kär·əl ‚sand }

coral shoreline See coral-reef shoreline. { 'kär·əl 'shȯr‚līn }

Cordaitaceae [PALEOBOT] A family of fossil plants belonging to the Cordaitales. { ‚kȯr·dā‚ī'tās·ē‚ē }

Cordaitales [PALEOBOT] An extensive natural grouping of forest trees of the late Paleozoic. { ‚kȯr·dā‚ī'tā·lēz }

cordierite [MINERAL] $Mg_2(Al_4Si_5O_{18})$ A blue, orthorhombic magnesium aluminosilicate mineral frequently occurring associated with thermally metamorphosed rocks derived from argillaceous sediments. { 'kȯrd·ē·ə‚rīt }

cordilleran geosyncline [GEOL] The Devonian geosynclinal region of western North America. { ‚kȯrd·əl'er·ən ‚jē·ō'sin‚klīn }

cordylite [MINERAL] $(Ce,La)_2Ba(CO_3)_3F_2$ A colorless to wax-yellow mineral consisting of a carbonate and fluoride of cerium, lanthanum, and barium. { 'kȯrd·əl‚īt }

core [GEOL] **1.** Center of the earth, beginning at a depth of 2900 kilometers. Also known as earth core. **2.** A vertical, cylindrical boring of the earth from which composition and stratification may be determined; in oil or gas well exploration the presence of hydrocarbons or water are items of interest. { kȯr }

core analysis [GEOL] The use of core samples taken from the borehole during drilling

82

to give information on strata age, composition, and porosity, and the presence of hydrocarbons or water along the length of the borehole. { 'kȯr ə'nal·ə·səs }

core intersection [GEOL] **1.** The point in a borehole where an ore vein or body is encountered as shown by the core. **2.** The width or thickness of the ore body, as shown by the core. Also known as core interval. { 'kȯr ˌin·tər,sek·shən }

core interval *See* core intersection. { 'kȯr ˌin·tər·vəl }

core logging [GEOL] The analysis of the strata through which a borehole passes by the taking of core samples at predetermined depth intervals as the well is drilled. { 'kȯr ˌläg·iŋ }

core sample [GEOL] A sample of rock, soil, snow, or ice obtained by driving a hollow tube into the undisturbed medium and withdrawing it with its contained sample or core. { 'kȯr ˌsam·pəl }

corestone [GEOL] A rounded or broadly rectangular joint block of granite formed as a result of subsurface weathering in a manner similar to a tor but entirely separated from the bedrock. { 'kȯr,stōn }

corneite [GEOL] A biotite-hornfels formed during deformation of shale by folding. { 'kȯr·nē,īt }

cornetite [MINERAL] $Cu_3(PO_4)(OH)_3$ A peacock-blue mineral consisting of basic copper phosphate. { 'kȯr·nə,tīt }

cornwallite [MINERAL] $Cu_5(AsO_4)_2(OH)_4 \cdot H_2O$ A verdigris green to blackish-green mineral consisting of a hydrated basic arsenate of copper; occurs as small botryoidal crusts. { 'kȯrn,wȯ,līt }

corona [GEOL] A mineral zone that is usually radial about another mineral or at the area between two minerals. Also known as kelyphite. [MINERAL] An annular zone of minerals that is disposed either around another mineral or at the contact between two minerals. { kə'rō·nə }

coronadite [MINERAL] $Pb(Mn^{2+},Mn^{4+})_8O_{16}$ A black mineral consisting of a lead and manganese oxide, occurring in massive form with fibrous structure; an important constituent of manganese ore. { ˌkȯr·ə'nä,dīt }

corrasion [GEOL] Mechanical wearing away of rock and soil by the action of solid materials moved along by wind, waves, running water, glaciers, or gravity. Also known as mechanical erosion. { kə'rā·zhən }

correlation [GEOL] **1.** The determination of the equivalence or contemporaneity of geologic events in separated areas. **2.** As a step in seismic study, the selecting of corresponding phases, taken from two or more separated seismometer spreads, of seismic events seemingly developing at the same geologic formation boundary. { ˌkär·ə'lā·shən }

corrie *See* cirque. { kȯr·ē }

corrosion [GEOCHEM] Chemical erosion by motionless or moving agents. { kə'rō·zhən }

corrosion border *See* corrosion rim. { kə'rō·zhən ˌbȯrd·ər }

corrosion rim [MINERAL] A modification of the outlines of a porphyritic crystal due to the corrosive action of a magma on previously stable minerals. Also known as corrosion border. { kə'rō·zhən ˌrim }

corsite [PETR] A spheroidal variety of gabbro. Also known as miagite; napoleonite. { 'kȯr,sīt }

cortlandite [PETR] A peridotite consisting of large crystals of hornblende with poikilitically included crystals of olivine. Also known as hudsonite. { 'kȯrt·lən,dīt }

corundum [MINERAL] Al_2O_3 A hard mineral occurring in various colors and crystallizing in the hexagonal system; crystals are usually prismatic or in rounded barrel shapes; gem varieties are ruby and sapphire. { kə'rən·dəm }

corvusite [MINERAL] $V_2O_4 \cdot 6V_2O_5 \cdot nH_2O$ A blue-black to brown mineral consisting of a hydrous oxide of vanadium; occurs in massive form. { 'kȯr·və,sīt }

Coryphodontidae [PALEON] The single family of the Coryphodontoidea, an extinct superfamily of mammals. { ˌkȯr·ə·fə'dän·tə,dē }

Coryphodontoidea [PALEON] A superfamily of extinct mammals in the order Pantodonta. { ˌkȯr·ə·fə,dän'tȯid·ē·ə }

cosalite |MINERAL| $Pb_2Bi_2S_5$ A lead-gray or steel-gray mineral consisting of lead, bismuth, and sulfur; specific gravity is 6.39–6.75. { 'kō·zə,līt }

cosmic sediment |GEOL| Particles of extraterrestrial origin which are observed as black magnetic spherules in deep-sea sediments. { 'käz·mik 'sed·ə·mənt }

cosmic spherules |GEOCHEM| Solidified, millimeter-sized to microscopic, rounded particles of extraterrestrial materials that melted either during high-velocity entry into the atmosphere or during hypervelocity impact of large meteoroids onto the earth's surface. { ¦käz·mik 'sfe·rülz }

cosmochlore See ureyite. { 'käz·mə,klȯr }

cotton ball See ulexite. { 'kät·ən ,bȯl }

cotunnite |MINERAL| $PbCl_2$ An alteration product of galena; a soft, white to yellowish mineral that crystallizes in the orthorhombic crystal system. { kə'tə,nīt }

Cotylosauria |PALEON| An order of primitive reptiles in the subclass Anapsida, including the stem reptiles, ancestors of all of the more advanced Reptilia. { ¦käd·əl·ə¦sȯr·ē·ə }

coulee |GEOL| **1.** A thick, solidified sheet or stream of lava. **2.** A steep-sided valley or ravine, sometimes with a stream at the bottom. { kü'lā }

country rock |GEOL| **1.** Rock that surrounds and is penetrated by mineral veins. **2.** Rock that surrounds and is invaded by an igneous intrusion. { ¦kən·trē 'räk }

Couvinian |GEOL| Lower Middle Devonian geologic time. { kü'vin·ē·ən }

covellite |MINERAL| CuS An indigo-blue mineral of metallic luster that crystallizes in the hexagonal system; it is usually massive or occurs in disseminations through other copper minerals and represents an ore of copper. Also known as indigo copper. { kō've,līt }

covite |PETR| A rock of igneous origin composed of sodic orthoclase, hornblende, sodic pyroxene, nepheline, and accessory sphene, apatite, and opaque oxides. { 'kō,vīt }

crag |GEOL| A steep, rugged point or eminence of rock, as one projecting from the side of a mountain. { krag }

crandallite |MINERAL| $CaAl_3(PO_4)_2(OH)_5 \cdot H_2O$ A white to light-grayish mineral consisting of a hydrous phosphate of calcium and aluminum occurring in fine, fibrous masses. { 'krand·əl,īt }

crater |GEOL| **1.** A large, bowl-shaped topographic depression with steep sides. **2.** A rimmed structure at the summit of a volcanic cone; the floor is equal to the vent diameter. { 'krād·ər }

crater cone |GEOL| A cone built around a volcanic vent by lava extruded from the vent. { 'krād·ər ,kōn }

craton |GEOL| A large, stable portion of the continental crust. Cratons are the broad heartlands of continents with subdued topography, encompassing the largest areas of most continents. { 'krā,tän }

crednerite |MINERAL| $CuMn_2O_4$ A steel-gray to iron-black foliated mineral consisting of copper, manganese, and oxygen. { 'kred·nə,rīt }

creedite |MINERAL| $Ca_3Al_2(SO_4)(F,OH)_{10} \cdot 2H_2O$ A white or colorless monoclinic mineral consisting of hydrous calcium aluminum fluoride with calcium sulfate, occurring in grains and radiating crystalline masses; hardness is 2 on Mohs scale, and specific gravity is 2.7. { 'krē,dīt }

creep |GEOL| A slow, imperceptible downward movement of slope-forming rock or soil under sheer stress. { krēp }

crenitic |GEOL| Relating to or resulting from the raising of subterranean minerals by the action of spring water. { krə'nid·ik }

crenulation cleavage See slip cleavage. { ,kren·yə'lā·shən ,klēv·ij }

Creodonta |PALEON| A group formerly recognized as a suborder of the order Carnivora. { ,krē·ə'dän·tə }

crescent beach |GEOL| A crescent-shaped beach at the head of a bay or the mouth of a stream entering the bay, with the concave side facing the sea. { 'kres·ənt ,bēch }

crescentic dune See barchan. { krə'sen·tik 'dün }

crestal plane |GEOL| The plane formed by joining the crests of all beds of an anticline. { 'krest·əl ,plān }

crest line |GEOL| The line connecting the highest points on the same bed of an anticline in an infinite number of cross sections.   { 'krest ‚līn }

Cretaceous |GEOL| In geological time, the last period of the Mesozoic Era, preceded by the Jurassic Period and followed by the Tertiary Period; it extended from 144 million years to 65 million years before present.   { kri'tā·shəs }

crevasse |GEOL| An open, nearly vertical fissure in a glacier or other mass of land ice or the earth, especially after earthquakes.   { krə'vas }

crevasse deposit |GEOL| Kame deposited in a crevasse.   { krə'vas di'päz·ət }

crib See arête.   { krib }

crinoidal limestone |PETR| A rock composed predominantly of crystalline joints of crinoids, with foraminiferans, corals, and mollusks.   { krī'nȯid·əl 'līm‚stōn }

cristobalite |MINERAL| SiO_2 A silicate mineral that is a high-temperature form of quartz; stable above 1470°C; crystallizes in the tetragonal system at low temperatures and the isometric system at high temperatures.   { kri'stō·bə‚līt }

critical bottom slope |GEOL| The depth distribution in which depth d of an ocean increases with latitude ϕ according to an equation of the form $d = d_0 \sin \phi +$ constant.   { 'krid·ə·kəl 'bäd·əm ‚slōp }

critical density |GEOL| That degree of density of a saturated, granular material below which, as it is rapidly deformed, it will decrease in strength and above which it will increase in strength.   { 'krid·ə·kəl 'den·səd·ē }

crocidolite |MINERAL| A lavender-blue, indigo-blue, or leek-green asbestiform variety of riebeckite; occurs in fibrous, massive, and earthy forms.   Also known as blue asbestos; krokidolite.   { krō'sīd·əl‚īt }

crocoisite See crocoite.   { 'kräk·wə‚zīt }

crocoite |MINERAL| $PbCrO_4$ A yellow to orange or hyacinth-red secondary mineral occurring as monoclinic, prismatic crystals; it is also massive granular.   Also known as crocoisite; red lead ore.   { 'kräk·ə‚wīt }

Croixian |GEOL| Upper Cambrian geologic time.   { 'krȯi·ən }

Cro-Magnon man |PALEON| **1.** A race of tall, erect Caucasoid men having large skulls; identified from skeletons found in southern France.   **2.** A general term to describe all fossils resembling this race that belong to the upper Paleolithic (35,000–8000 B.C.) in Europe.   { krō'mag·nən 'man }

cromfordite See phosgenite.   { 'kräm·fər‚dīt }

cronstedtite |MINERAL| $Fe_4^{2+}Fe_2^{3+}(Fe_2^{3+}Si_2)O_{10}(OH)_8$ A black to brownish-black mineral consisting of a hydrous iron silicate crystallizing in hexagonal prisms; specific gravity is 3.34–3.35.   { 'krän‚sted‚īt }

crookesite |MINERAL| $(Cu,Tl,Ag)_2Se$ An important selenium mineral occurring in lead-gray masses and having a metallic appearance.   { 'krúk‚sīt }

crop out See outcrop.   { 'kräp ‚aút }

cross-bedding |GEOL| The condition of having laminae lying transverse to the main stratification planes of the strata; occurs only in granular sediments.   Also known as cross-lamination; cross-stratification.   { ¦krȯs 'bed·iŋ }

crosscutting relationships |GEOL| Relationships which may occur between two adjacent rock bodies, where the relative age may be determined by observing which rock "cuts" the other, for example, a granitic dike cutting across a sedimentary unit.   { 'krȯs‚kəd·iŋ ri'lā·shən‚ships }

cross fault |GEOL| **1.** A fault whose strike is perpendicular to the general trend of the regional structure.   **2.** A minor fault that intersects a major fault.   { 'krȯs ‚fȯlt }

cross fold |GEOL| A secondary fold whose axis is perpendicular or oblique to the axis of another fold.   Also known as subsequent fold; superimposed fold; transverse fold.   { 'krȯs ‚fōld }

cross joint |GEOL| A fracture in igneous rock perpendicular to the lineation caused by flow magma.   Also known as transverse joint.   { 'krȯs ‚jȯint }

cross-lamination See cross-bedding.   { ¦krȯs lam·ə'nā·shən }

Crossopterygii |PALEON| A subclass of the class Osteichthyes comprising the extinct lobefins or choanate fishes and represented by one extant species; distinguished by two separate dorsal fins.   { krä‚säp·tə'rij·ē‚ī }

cross section [GEOL] **1.** A diagram or drawing that shows the downward projection of surficial geology along a vertical plane, for example, a portion of a stream bed drawn at right angles to the mean direction of the flow of the stream.   **2.** An actual exposure or cut which reveals geological features.   { 'krós ,sek·shən }

cross-stone See harmotome; staurolite.   { 'krós ,stōn }

cross-stratification See cross-bedding.   { 'krós ,strad·ə·fə'kā·shən }

cross valley See transverse valley.   { 'krós ,val·ē }

croute calcaire See caliche.   { ,krüt kal'ker }

crude oil [GEOL] A comparatively volatile liquid bitumen composed principally of hydrocarbon, with traces of sulfur, nitrogen, or oxygen compounds; can be removed from the earth in a liquid state.   { ¦krüd 'óil }

crumb structure [GEOL] A soil condition in which the particles are crumblike aggregates; suitable for agriculture.   { 'krəm ,strək·chər }

crush breccia [GEOL] A breccia formed in place by mechanical fragmentation of rock during movements of the earth's crust.   { 'krəsh ,brech·ə }

crush conglomerate [GEOL] Beds similar to a fault breccia, except that the fragments are rounded by attrition.   Also known as tectonic conglomerate.   { 'krəsh kən'gläm·ə·rət }

crush fold [GEOL] A fold of large dimensions that may involve considerable minor folding and faulting such as would produce a mountain chain or an oceanic deep.   { 'krəsh ,fōld }

crush zone [GEOL] A zone of fault breccia on fault gouge.   { 'krəsh ,zōn }

crust [GEOL] The outermost solid layer of the earth, mostly consisting of crystalline rock and extending no more than a few miles from the surface to the Mohorovičić discontinuity.   Also known as earth crust.   { krəst }

crustal motion [GEOL] Movement of the earth's crust.   { ¦krəst·əl 'mō·shən }

crustal plate See tectonic plate.   { 'krəst·əl ,plāt }

cryoconite [GEOL] A dark, powdery dust transported by wind and deposited on the surface of snow or ice; found, however, mainly in cryoconite holes.   [MINERAL] A mixture of garnet, sillimanite, zircon, pyroxene, quartz, and various other minerals.   { krī'äk·ə,nīt }

cryoconite hole [GEOL] A cylindrical dust well filled with cryoconite; absorbs solar radiation, causing melting of glacier ice around and below it.   { krī'äk·ə,nīt ,hōl }

cryogenic period [GEOL] A time period in geologic history during which large bodies of ice appeared at or near the poles and climate favored the formation of continental glaciers.   { ,krī·ə'jen·ik ¦pir·ē·əd }

cryolaccolith See hydrolaccolith.   { ¦krī·ō'lak·ə,lith }

cryolite [MINERAL] Na_3AlF_6 A white or colorless mineral that crystallizes in the monoclinic system but has a pseudocubic aspect; found in masses of waxy luster; hardness is 2.5 on Mohs scale, and specific gravity is 2.95–3.0; used chiefly as a flux in producing aluminum from bauxite and for making salts of sodium and aluminum and porcelaneous glass.   Also known as Greenland spar; ice stone.   { 'krī·ə,līt }

cryolithionite [MINERAL] $Na_3Li_3Al_2F_{12}$ A colorless mineral that crystallizes in the isometric system; found in the Ural Mountains.   { ,krī·ō'lith·ē·ə,nīt }

cryomorphology [GEOL] The branch of geomorphology that treats the processes and topographic features of regions where the ground is permanently frozen.   { ¦krī·ō·mór'fäl·ə·jē }

cryopedology [GEOL] A branch of geology that deals with the study of intensive frost action and permanently frozen ground.   { ¦krī·ō·pə'däl·ə·jē }

cryoplanation [GEOL] Land erosion at high latitudes or elevations due to processes of intensive frost action.   { ¦krī·ō·plə'nā·shən }

cryosphere [GEOL] That region of the earth in which the surface is perennially frozen.   { 'krī·ə,sfir }

cryostatic pressure [GEOL] Hydrostatic pressure exerted on soil and rocks when soil water freezes.   { 'krī·ə,stad·ik 'presh·ər }

cryoturbation See congeliturbation.   { ¦krī·ō·tər'bā·shən }

cryptoclastic [GEOL] Composed of extremely fine, almost submicroscopic, broken or fragmental particles. { ¦krip·tə¦klas·tik }

cryptocrystalline [GEOL] Having a crystalline structure but of such a fine grain that individual components are not visible with a magnifying lens. { ¦krip·tō'krist·əl·ən }

cryptohalite [MINERAL] (NH₄)₂SiF₆ A colorless to white or gray, isometric mineral consisting of ammonium silicon fluoride; occurs in massive and arborescent forms. { ¦krip·tō'ha,līt }

cryptolite See monazite. { 'krip·tə,līt }

cryptomelane [MINERAL] KMn₈O₁₆·H₂O A usually massive mineral, common in manganese ores; contains an oxide of manganese and potassium and crystallizes in the monoclinic system. { ¦krip·tō·mə'lān }

cryptoperthite [MINERAL] A fine-grained, submicroscopic variety of perthite consisting of an intergrowth of potassic and sodic feldspar, detectable only by means of x-rays or with the aid of an electron microscope. { ¦krip·tō'pər,thīt }

Cryptostomata [PALEON] An order of extinct bryozoans in the class Gymnolaemata. { ¦krip·tə'stō·məd·ə }

cryptovolcanic [GEOL] A small, nearly circular area of highly disturbed strata in which there is no evidence of volcanic materials to confirm the origin as being volcanic. { ¦krip·tō·väl'kan·ik }

cryptozoon [PALEOBOT] A hemispherical or cabbagelike reef-forming fossil algae, probably from the Cambrian and Ordovician. { ¦krip·tō'zō·ən }

crystal See rock crystal. { 'krist·əl }

crystalline-granular texture [PETR] A primary texture of an igneous rock due to crystallization from a fluid medium. { ¦kris·tə·lən ¦gran·yə·lər 'teks·chər }

crystalline porosity [GEOL] Porosity in crystalline limestone and dolomite, making possible underground oil reservoirs. { 'kris·tə·lən pə'räs·əd·ē }

crystalline rock [PETR] **1.** Rock made up of minerals in a clearly crystalline state. **2.** Igneous and metamorphic rock, as opposed to sedimentary rock. { 'kris·tə·lən 'räk }

crystallinity [PETR] Degree of crystallization exhibited by igneous rock. { ¦kris·tə'lin·əd·ē }

crystallite [GEOL] A small, rudimentary form of crystal which is of unknown mineralogic composition and which does not polarize light. { 'kris·tə,līt }

crystallization differentiation See fractional crystallization. { ¦kris·tə·lə'zā·shən ¦dif·ə,ren·chē'ā·shən }

crystalloblast [MINERAL] A mineral crystal produced by metamorphic processes. { 'kris·tə·lō,blast }

crystalloblastic series [GEOL] A series of metamorphic minerals ordered according to decreasing formation energy, so crystals of a listed mineral have a tendency to form idioblastic outlines at surfaces of contact with simultaneously developed crystals of all minerals in lower positions. { 'kris·tə·lə'blas·tik 'sir,ēz }

crystalloblastic texture [GEOL] A crystalline texture resulting from metamorphic recrystallization under conditions of high viscosity and directed pressure. { 'kris·tə·lə'blas·tik 'teks·chər }

crystallographic texture [MINERAL] A texture of replacement or exsolution mineral deposits, with the distribution and form of the inclusions controlled by the host-mineral crystallography. { ¦kris·tə·lō¦graf·ik 'teks·chər }

crystal sandstone [GEOL] Siliceous sandstone in which deposited silica is precipitated upon the quartz grains in crystalline position. { ¦krist·əl 'sand,stōn }

crystal settling [GEOL] Sinking of crystals in magma from the liquid in which they formed, by the action of gravity. { ¦krist·əl ¦set·liŋ }

crystal tuff [GEOL] Consolidated volcanic ash in which crystals and crystal fragments predominate. { ¦krist·əl 'təf }

crystal-vitric tuff [GEOL] Consolidated volcanic ash composed of 50–75% crystal fragments and 25–50% glass fragments. { ¦krist·əl ¦vi·trik 'təf }

Ctenothrissidae [PALEON] A family of extinct teleostean fishes in the order Ctenothrissiformes. { ten·ə'thris·ə,dē }

Ctenothrissiformes [PALEON] A small order of extinct teleostean fishes; important as a group on the evolutionary line leading from the soft-rayed to the spiny-rayed fishes. { ,ten·ə,thris·ə'fȯr,mēz }

cubanite [MINERAL] $CuFe_2S_3$ Bronze-yellow mineral that crystallizes in the orthorhombic system. Also known as chalmersite. { 'kyü·bə,nīt }

cube ore *See* pharmacosiderite. { 'kyüb ,ȯr }

cube spar *See* anhydrite. { 'kyüb ,spär }

culmination [GEOL] A high point on the axis of a fold. { kəl·mə'nā·shən }

cumberlandite [PETR] A coarse-grained, ultramafic, ultrabasic rock composed principally of olivine crystals in a ground mass of magnetite and ilmenite with minor plagioclase. { 'kəm·bər·lən,dīt }

cumbraite [PETR] A variety of dacite or rhyodacite containing very calcic plagioclase and pyroxene in a glassy groundmass. { kyüm'brā,īt }

cumengite [MINERAL] $Pb_4Cu_4Cl_8(OH)_8 \cdot H_2O$ A deep-blue or light-indigo-blue tetragonal mineral consisting of a basic lead-copper chloride occurring in crystals. { kyü'men,jīt }

cummingtonite [MINERAL] $(Fe,Mg)_7Si_8O_{22}(OH)_2$ A brownish mineral that crystallizes in the monoclinic system; usually occurs as lamellae or fibers in metamorphic rocks. { 'kəm·iŋ·tə,nīt }

cumulate [PETR] Any igneous rock formed by the accumulation of crystals settling out of a magma. { 'kyü·myə,lāt }

cumulus [GEOCHEM] The accumulation of minerals which have precipitated from a liquid without having been modified by later crystallization. { 'kyü·myə·ləs }

cup-and-ball joint [GEOL] A dish-shaped transverse fracture which divides a basalt column into segments. Also known as ball-and-socket joint. { ,kəp ən 'bȯl ,jȯint }

cupola [GEOL] An isolated, upward-projecting body of plutonic rock that lies near a larger body; both bodies are presumed to unite at depth. { 'kyü·pə·lə }

cupped pebble [GEOL] A pebble fragment that has become hollow after being subjected to solution. { 'kəpt ¦peb·əl }

cuprite [MINERAL] Cu_2O A red mineral that crystallizes in the isometric system and is found in crystals and fine-grained aggregates or is massive; a widespread supergene copper ore. Also known as octahedral copper ore; red copper ore; ruby copper ore. { 'kyü,prīt }

cuprocopiapite [MINERAL] $CuFe_4(SO_4)_6(OH)_2 \cdot 20H_2O$ A sulfur yellow to orange-yellow, triclinic mineral consisting of a hydrated basic sulfate of copper and iron. { ¦kyü·prō'kō·pē·ə,pīt }

cuprodescloizite *See* mottramite. { ¦kyü·prō·des'klō·ə,zīt }

cuprotungstite [MINERAL] $Cu_2(WO_4)(OH)_2$ A green mineral that forms compact masses; soluble in acids; the crystal system is not known. { ¦kyü·prō'təŋ,stīt }

cuprouranite *See* torbernite. { ¦kyü·prō'yùr·ə,nīt }

curite [MINERAL] $Pb_2U_5O_{17} \cdot 4H_2O$ An orange-red radioactive mineral, occurring in acicular crystals, an alteration product of uraninite. { 'kyù,rīt }

current-bedding [GEOL] Cross-bedding resulting from water or air currents. { 'kər·ənt ,bed·iŋ }

current lineation *See* parting lineation. { 'kər·ənt lin·ē'ā·shən }

current mark [GEOL] Any structure formed by direct or indirect action of a water current on a sedimentary surface. { 'kər·ənt ,märk }

current ripple [GEOL] A type of ripple mark having a long, gentle slope toward the direction from which the current flows, and a shorter, steeper slope on the lee side. { 'kər·ənt ,rip·əl }

curtain [GEOL] **1.** A thin sheet of dripstone that hangs or projects from a cave wall. **2.** A rock formation connecting two adjacent bastions. { 'kərt·ən }

cusp [GEOL] One of a series of low, crescent-shaped mounds of beach material separated by smoothly curved, shallow troughs spaced at more or less regular intervals along and generally perpendicular to the beach face. Also known as beach cusp. [GEOPHYS] Any of the funnel-shaped regions in the magnetosphere extending from the front magnetopause to the polar ionosphere, and filled with solar wind plasma. { kəsp }

cuspate bar [GEOL] A crescentic bar joining with the shore at each end. { 'kə,spāt ,bär }

cuspate ripple mark *See* linguoid ripple mark. { 'kə,spāt 'rip·əl ,märk }

cut and fill [GEOL] **1.** Lateral corrosion of one side of a meander accompanied by deposition on the other. **2.** A sedimentary structure consisting of a small filled-in channel. { ¦kət ən 'fil }

cutbank [GEOL] The concave bank of a winding stream that is maintained as a steep or even overhanging cliff by the action of water at its base. { 'kət,baŋk }

cutinite [GEOL] A variety of exinite consisting of plant cuticles. { 'kyüt·ən,īt }

cutoff [GEOL] A new, relatively short channel formed when a stream cuts through the neck of an oxbow or horseshoe bend. { 'kət,öf }

cutout [GEOL] *See* horseback. { 'kət,aút }

cut platform *See* wave-cut platform. { 'kət ,plat,förm }

Cuvieroninae [PALEON] A subfamily of extinct proboscidean mammals in the family Gomphotheriidae. { küv·yə'rän·ə,nē }

cwm *See* cirque. { küm }

cyanite *See* kyanite. { 'sī·ə,nīt }

cyanochroite [MINERAL] $K_2Cu(SO_4)_2·6H_2O$ A blue mineral consisting of a hydrous sulfate of potassium and copper. { ,sī·ə'nä·krə,wīt }

cyanotrichite [MINERAL] $Cu_4Al_2(SO_4)(OH)_{12}·2H_2O$ A bright-blue or sky-blue mineral consisting of a hydrous basic copper aluminum sulfate. { ,sī·ə'nä·trə,kīt }

Cycadeoidaceae [PALEOBOT] A family of extinct plants in the order Cycadeoidales characterized by sparsely branched trunks and a terminal crown of leaves. { sī,kad·ē·öid'ās·ē,ē }

Cycadeoidales [PALEOBOT] An order of extinct plants that were abundant during the Triassic, Jurassic, and Cretaceous periods. { sī,kad·ē·öid'ā·lēz }

Cycadofilicales [PALEOBOT] The equivalent name for the extinct Pteridospermae. { ¦sī·kə·dō,fil·ə'kā·lēz }

cycle of erosion *See* geomorphic cycle. { 'sī·kəl əv i'rō·zhən }

cycle of sedimentation [GEOL] Also known as sedimentary cycle. **1.** A series of related processes and conditions appearing repeatedly in the same sequence in a sedimentary deposit. **2.** The sediments deposited from the beginning of one cycle to the beginning of a second cycle of the spread of the sea over a land area, consisting of the original land sediments, followed by those deposited by shallow water, then deep water, and then the reverse process of the receding water. { 'sī·kəl əv ,sed·ə·mən'tā·shən }

cyclic sedimentation [GEOL] Deposition of various kinds of sediment in a repeated regular sequence. { 'sīk·lik ,sed·ə·mən'tā·shən }

Cyclocystoidea [PALEON] A class of small, disk-shaped, extinct echinozoans in which the lower surface of the body probably consisted of a suction cup. { ¦sī·klō·si'stöid·ē·ə }

cyclopean *See* mosaic. { ¦sī·klə¦pē·ən }

cyclopean stairs [GEOL] The landscape that results in a glacial trough after the ice has melted away, and that consists of an irregular series of rock steps, with steep cliffs on the down-valley side and small lakes in the shallow excavated depressions of the rock steps. { ,sī·klə'pē·ən 'sterz }

cyclosilicate [MINERAL] A silicate having the SiO_4 tetrahedra linked to form rings, with a silicon-oxygen ratio of 1:3, such as $Si_3O_9^{6-}$ or $Si_6O_{18}^{12-}$. Also known as ring silicate. { ¦sī·klō'sil·ə,kāt }

Cyclosteroidea [PALEON] A class of Middle Ordovician to Middle Devonian echinoderms in the subphylum Echinozoa. { ¦sī·klō·stə'röid·ē·ə }

cyclothem [GEOL] A rock stratigraphic unit associated with unstable shelf of interior basin conditions, in which the sea has repeatedly covered the land. { 'si·klə,them }

89

cylindrite [MINERAL] $Pb_3Sn_4Sb_2S_{14}$ A blackish-gray mineral consisting of sulfur, lead, antimony, and tin, occurring in cylindrical forms that separate under pressure into distinct sheets or folia. { sə'lin₁drīt }

cymrite [MINERAL] $Ba_2Al_5Si_5O_{19}(OH) \cdot 3H_2O$ Zeolite mineral consisting of a basic aluminosilicate of barium. { 'kəm₁rīt }

Cystoidea [PALEON] A class of extinct crinozoans characterized by an ovoid body that was either sessile or attached by a short aboral stem. { si'stȯid·ē·ə }

Cystoporata [PALEON] An order of extinct, marine bryozoans characterized by cystopores and minutopores. { ₁sis·tə'pȯr·əd·ə }

D

dachiardite [MINERAL] $(Na_2Ca)_2(Al_4Si_{20}O_{48})\cdot12H_2O$ A white to colorless mineral in the mordenite group of the zeolite family that crystallizes in the monoclinic system. { ,däk·ē'är,dīt }

Dacian [GEOL] Lower upper Pliocene geologic time. { 'dā·shən }

dacite [GEOL] Very fine crystalline or glassy rock of volcanic origin, composed chiefly of sodic plagioclase and free silica with subordinate dark-colored minerals. { 'dā,sīt }

dacite glass [GEOL] A natural glass formed by rapid cooling of dacite lava. { 'dā,sīt ,glas }

dactylitic [GEOL] Of a rock texture, characterized by fingerlike projections of a mineral that penetrate another mineral. { dak·tə'lid·ik }

daily variation [GEOPHYS] Oscillation occurring in the earth's magnetic field in a 1-day period. { ¦dā·lē ,ver·ē'ā·shən }

Dakotan [GEOL] Lower Upper Cretaceous geologic time. { də'kot·ən }

damkjernite [PETR] A melanocratic dike rock composed of biotite and pyroxene phenocrysts in a groundmass of pyroxene, biotite, and magnetite. { 'dam·kyər,nīt }

danalite [MINERAL] $(Fe,Mn,Zn)_4Be_3(SiO_4)_3S$ A mineral consisting of a silicate and sulfide of iron and beryllium; it is isomorphous with helvite and genthelvite. { 'dā·nə,līt }

danburite [MINERAL] $CaB_2(SiO_4)_2$ An orange-yellow, yellowish-brown, grayish, or colorless transparent to translucent borosilicate mineral with a feldspar structure crystallizing in the orthorhombic system; it resembles topaz and is used as an ornamental stone. { 'dan·bə,rīt }

Danian [GEOL] Lowermost Paleocene or uppermost Cretaceous geologic time. { 'dān·ē·ən }

dannemorite [MINERAL] $(Fe,Mn,Mg)_7Si_8O_{22}(OH)_2$ A yellowish-brown to greenish-gray monoclinic mineral consisting of a columnar or fibrous amphibole. { ,dan·ə'mȯr,īt }

daphnite [MINERAL] $(MgFe)_3(Fe,Al)_3(Si,Al)_4O_{10}(OH)_8$ A mineral of the chlorite group consisting of a basic aluminosilicate of magnesium, iron, and aluminum. { 'daf,nīt }

Daphoenidae [PALEON] A family of extinct carnivoran mammals in the superfamily Miacoidea. { də'fēn·ə,dē }

darapskite [MINERAL] $Na_3(NO_3)(SO_4)\cdot H_2O$ A naturally occurring hydrate mineral consisting of a hydrous nitrate and sulfate of sodium. { də'rap,skīt }

dark-red silver ore See pyrargyrite. { ¦därk 'red 'sil·vər ,ȯr }

dark-ruby silver See pyrargyrite. { ¦därk ¦rü·bē ,sil·vər }

Darwin glass [GEOL] A highly siliceous, vesicular glass shaped in smooth blobs or twisted shreds, found in the Mount Darwin range in western Tasmania. Also known as queenstownite. { 'där·wən ,glas }

dashkesanite [MINERAL] $(Na,K)Ca_2(Fe,Mg)_5(Si,Al)_8O_{22}Cl_2$ A monoclinic mineral of the amphibole group consisting of a chloroaluminosilicate of sodium, potassium, iron, and magnesium. { ,dash·kə'sa,nīt }

datolite [MINERAL] $CaBSiO_4(OH)$ A mineral nesosilicate crystallizing in the monoclinic system; luster is vitreous, and crystals are colorless or white with a greenish tinge. { 'dad·əl,īt }

datum [GEOL] The top or bottom of a bed of rock on which structure contours are drawn. { 'dad·əm, 'dād·əm, or 'däd·əm }

daubreeite [MINERAL] $FeCr_2S_4$ A mineral composed of a black chromium iron sulfide; occurs in some meteors. { 'dȯ·brē,īt }

Davian [GEOL] A subdivision of the Upper Cretaceous in Europe; a limestone formation with abundant hydrocorals, bryozoans, and mollusks in Denmark; marine limestone and nonmarine rocks in southeastern France; and continental formations in the Davian of Spain and Portugal. { 'dä·vē·ən }

davidite [MINERAL] A black primary pegmatite uranium mineral of the general formula $A_6B_{15}(O,OH)_{36}$, where $A = Fe^{2+}$, rare earths, uranium, calcium, zirconium, and thorium, and $B = $ titanium, Fe^{3+}, vanadium, and chromium. { 'dā·və,dīt }

daviesite [MINERAL] An orthorhombic mineral consisting of a lead oxychloride, occurring in minute crystals. { 'dā·vē,zīt }

davisonite [MINERAL] $Ca_3Al(PO_4)_2(OH)_3 \cdot H_2O$ A white mineral consisting of a hydrous basic phosphate of calcium and aluminum. { 'dā·və·sə,nīt }

dawsonite [MINERAL] $NaAl(OH)_2CO_3$ A white, bladed mineral found in certain oil shales that contains large quantities of alumina; specific gravity is 2.40. { 'dȯs·ən,īt }

dead [GEOL] In economic geology, designating a region with no economic value. { ded }

dead cave [GEOL] A cave where there is no moisture or no growth of mineral deposits associated with moisture. { ¦ded 'kāv }

death assemblage See thanatocoenosis. { 'deth ə,sem·blij }

debris [GEOL] Large fragments arising from disintegration of rocks and strata. { də'brē }

debris avalanche [GEOL] The sudden and rapid downward movement of incoherent mixtures of rock and soil on deep slopes. { də'brē 'av·ə,lanch }

debris cone [GEOL] **1.** A mound of fine-grained debris piled atop certain boulders moved by a landslide. **2.** A mound of ice or snow on a glacier covered with a thin layer of debris. { də'brē ,kōn }

debris fall [GEOL] A relatively free downward or forward falling of unconsolidated or poorly consolidated earth or rocky debris from a cliff, cave, or arch. { də'brē ,fȯl }

debris flow [GEOL] A variety of rapid mass movement involving the downslope movement of high-density coarse clast-bearing mudflows, usually on alluvial fans. { də'brē ,flō }

debris line See swash mark. { də'brē ,līn }

debris slide [GEOL] A type of landslide involving a rapid downward sliding and forward rolling of comparatively dry, unconsolidated earth and rocky debris. { də'brē ,slīd }

debris slope See talus slope. { də'brē ,slōp }

decay See chemical weathering. { di'kā }

Deccan basalt [GEOL] Fine-grained, nonporphyritic, tholeiitic basaltic lava consisting essentially of labradorite, clinopyroxene, and iron ore; found in the Deccan region of southeastern India. Also known as Deccan trap. { 'dek·ən bə'sȯlt }

Deccan trap See Deccan basalt. { 'dek·ən 'trap }

declination [GEOPHYS] The angle between the magnetic and geographical meridians, expressed in degrees and minutes east or west to indicate the direction of magnetic north from true north. Also known as magnetic declination; variation. { ,dek·lə'nā·shən }

declivity [GEOL] **1.** A slope descending downward from a point of reference. **2.** A downward deviation from the horizontal. { də'kliv·əd·ē }

décollement [GEOL] Folding or faulting of sedimentary beds by sliding over the underlying rock. { dā'käl·mənt }

decomposition See chemical weathering. { dē,käm·pə'zish·ən }

decrepitation [GEOPHYS] Breaking up of mineral substances when exposed to heat; usually accompanied by a crackling noise. { di,krep·ə'tā·shən }

decussate structure [GEOL] A crisscross microstructure of certain minerals; most noticeable in rocks composed predominantly of minerals with a columnar habit. { 'dek·ə,sāt ,strək·chər }

dedolomitization [GEOL] Destruction of dolomite to form calcite and periclase, usually by contact metamorphism at low pressures. { dē,dō·lə,mīd·ə'zā·shən }

deep-marine sediments [GEOL] Sedimentary environments occurring in water deeper than 200 meters (660 feet), seaward of the continental shelf break, on the continental slope and the basin. { ‚dēp·mə¦rēn 'sed·ə·mins }

deep-sea basin [GEOL] A depression of the sea floor more or less equidimensional in form and of variable extent. { ¦dēp ¦sē 'bās·ən }

deep-sea channel [GEOL] A trough-shaped valley of low relief beyond the continental rise on the deep-sea floor. Also known as mid-ocean canyon. { ¦dēp ¦sē 'chan·əl }

deep-sea plain [GEOL] A broad, almost level area forming the predominant portion of the ocean floor. { ¦dēp ¦sē 'plān }

deep-seated See plutonic. { ¦dēp 'sēd·əd }

deep-sea trench [GEOL] A long, narrow depression of the deep-sea floor having steep sides and containing the greatest ocean depths; formed by depression, to several kilometers' depth, of the high-velocity crustal layer and the mantle. { ¦dēp ¦sē 'trench }

Deerparkian [GEOL] A North American stage of geologic time in the Lower Devonian, above Helderbergian and below Onesquethawan. { dir'pärk·ē·ən }

deflation [GEOL] The sweeping erosive action of the wind over the ground. { di'flā·shən }

deflation basin [GEOL] A topographic depression formed by deflation. { di'flā·shən ‚bās·ən }

deformation fabric [GEOL] The space orientation of rock elements produced by external stress on the rock. { ‚def·ər‚mā·shən ‚fab·rik }

deformation lamella [GEOL] A type of slipband in the crystalline grains of a material (particularly quartz) produced by intracrystalline slip during tectonic deformation. { ‚def·ər‚mā·shən lə‚mel·ə }

degenerative recrystallization See degradation recrystallization. { di'jen·ə·rəd·iv rē‚krist·əl·ə'zā·shən }

degradation [GEOL] The wearing down of the land surface by processes of erosion and weathering. { ‚deg·rə'dā·shən }

degradation recrystallization [GEOL] Recrystallization resulting in a decrease in the size of crystals. Also known as degenerative recrystallization; grain diminution. { ‚deg·rə'dā·shən rē‚krist·əl·ə'zā·shən }

degraded illite [MINERAL] Illite with a depleted potassium content because of prolonged leaching. Also known as stripped illite. { dē'grād·əd i'līt }

dehrnite [MINERAL] $(Ca,Na,K)_5(PO_4)_3(OH)$ A colorless to pale green, greenish-white, or gray, hexagonal mineral consisting of a basic phosphate of calcium, sodium, and potassium; occurs as botryoidal crusts and minute hexagonal prisms. { 'der‚nīt }

Deinotheriidae [PALEON] A family of extinct proboscidean mammals in the suborder Deinotherioidea; known only by the genus Deinotherium. { ‚dī·nō·thə'rī·ə‚dē }

Deinotherioidea [PALEON] A monofamilial suborder of extinct mammals in the order Proboscidea. { ‚dī·nō‚ther·ē'oid·ē·ə }

Deister phase [GEOL] A subdivision of the late Ammerian phase of the Jurassic period between the Kimmeridgian and lower Portlandian. { 'dī·stər ‚fāz }

delafossite [MINERAL] $CuFeO_2$ A mineral consisting of an oxide of copper and iron. { ‚de·lə'fȯ‚sīt }

dellenite See rhyodacite. { 'del·ə‚nīt }

Delmontian [GEOL] Upper Miocene or lower Pliocene geologic time. { del'män·chən }

delorenzite See tanteuxenite. { dē·lə'ren‚zīt }

delta [GEOL] An alluvial deposit, usually triangular in shape, at the mouth of a river, stream, or tidal inlet. { 'del·tə }

delta geosyncline See exogeosyncline. { 'del·tə ‚jē·ō'sin‚klīn }

deltaic deposits [GEOL] Sedimentary deposits in a delta. { del'tā·ik di'päz·əts }

deltaite [MINERAL] A mixture of crandallite and hydroxylapatite. { 'del·tə‚īt }

delta moraine See ice-contact delta. { 'del·tə mə'rān }

delta plain [GEOL] A plain formed by deposition of silt at the mouth of a stream or by overflow along the lower stream courses. { 'del·tə ‚plān }

Deltatheridia

Deltatheridia [PALEON] An order of mammals that includes the dominant carnivores of the early Cenozoic. { ,del·tə·thə'rid·ē·ə }

delvauxite [MINERAL] A mineral, with the approximate formula $Fe_4(PO_4)_2(OH)_6 \cdot nH_2O$, consisting of a hydrous phosphate of iron. { del'vók,sīt }

demantoid [MINERAL] A lustrous, green variety of andradite; used as a gem. { də'man,toid }

demorphism See weathering. { dē'mȯr·fiz· əm }

dendritic valleys [GEOL] Treelike extensions of the valleys in a region lying upon horizontally bedded rock. { den'drid·ik 'val·ēz }

dendrochronology [GEOL] The science of measuring time intervals and dating events and environmental changes by reading and dating growth layers of trees as demarcated by the annual rings. { ¦den·drō·krə'näl·ə·jē }

Dendroidea [PALEON] An order of extinct sessile, branched colonial animals in the class Graptolithina occurring among typical benthonic fauna. { den'drȯid·ē·ə }

densofacies See metamorphic facies. { ,den·sō'fā·shēz }

denudation [GEOL] General wearing away of the land; laying bare of subjacent lands. { ,dē·nü'dā·shən }

deoxidation sphere See bleach spot. { dē,äk·sə'dā·shən ,sfir }

Depertellidae [PALEON] A family of extinct perissodactyl mammals in the superfamily Tapiroidea. { de·pər'tel·ə,dē }

depocenter [GEOL] A site of maximum deposition. { 'dep·ə,sen·tər }

deposit [GEOL] Consolidated or unconsolidated material that has accumulated by a natural process or agent. { də'päz·ət }

deposition [GEOL] The laying, placing, or throwing down of any material; specifically, the constructive process of accumulation into beds, veins, or irregular masses of any kind of loose, solid rock material by any kind of natural agent. { ,dep·ə'zish·ən }

depositional dip See primary dip. { ,dep·ə'zish·ən·əl 'dip }

depositional fabric [PETR] Arrangement of detrital particles settled from suspension or of crystals from a differentiating magma determined by the plane of the surface on which they come to rest. { ,dep·ə'zish·ən·əl 'fab·rik }

depositional sequence [GEOL] A major but informal assemblage of formations or groups and supergroups, bounded by regionally extensive unconformities at both their base and top and extending over broad areas of continental cratons. { ,dep·ə'zish·ən·əl 'sē·kwəns }

depositional strike [GEOL] Sedimentary deposits that are continuous laterally on a gently sloping surface. { ,dep·ə'zish·ən·əl 'strīk }

depression [GEOL] **1.** A hollow of any size on a plain surface having no natural outlet for surface drainage. **2.** A structurally low area in the crust of the earth. { di'presh·ən }

depth of compensation [GEOPHYS] That depth at which density differences occurring in the earth's crust are compensated isostatically; calculated to be between 62 and 70–73 miles (100 and 113–117 kilometers). { 'depth əv ,käm·pən'sā·shən }

depth zone [GEOL] A zone within the earth giving rise to different metamorphic assemblages. { 'depth ,zōn }

derbylite [MINERAL] $Fe_6Ti_6Sb_2O_{23}$ A black or brown orthorhombic mineral occurring in cinnabar-bearing gravels. { 'dər·bē,līt }

Derbyshire spar See fluorite. { 'där·bə,shir ,spär }

derivative rock See sedimentary rock. { də'riv·əd·iv 'räk }

descendant [GEOL] A topographic feature that is formed from the mass beneath an older topographic form, now removed. { di'sen·dənt }

desert crust See desert pavement. { ¦dez·ərt ¦krəst }

desert pavement [GEOL] A mosaic of pebbles and large stones which accumulate as the finer dust and sand particles are blown away by the wind. Also known as desert crust. { ¦dez·ərt 'pāv·mənt }

desert peneplain See pediplain. { ¦dez·ərt 'pen·ə,plān }

desert plain See pediplain. { ¦dez·ərt 'plān }

desert polish [GEOL] A smooth, shining surface imparted to rocks and other hard substances by the action of windblown sand and dust of desert regions. { ¦dez·ərt 'päl·ish }

desert soil [GEOL] In early United States classification systems, a group of zonal soils that have a light-colored surface soil underlain by calcareous material and a hardpan. { ¦dez·ərt 'soil }

desert varnish See rock varnish. { ¦dez·ərt 'vär·nish }

desiccation breccia [GEOL] Fragments of a mud-cracked layer of sediment deposited with other sediments. { ¸des·ə'kā·shən ¸brech·ə }

desiccation crack See mud crack. { ¸des·ə'kā·shən ¸krak }

desilication [GEOCHEM] Removal of silica, as from rock or a magma. { dē¸sil·ə'kā·shən }

desmine See stilbite. { 'dez¸mēn }

Desmodonta [PALEON] An order of extinct bivalve, burrowing mollusks. { ¸dez·mə'dän·tə }

Des Moinesian [GEOL] Lower Middle Pennsylvanian geologic time. { də'móin·ē·ən }

Desmostylia [PALEON] An extinct order of large hippopotamuslike, amphibious, gravigrade, shellfish-eating mammals. { ¸dez·mə'stīl·ē·ə }

Desmostylidae [PALEON] A family of extinct mammals in the order Desmostylia. { ¸dez·mə'stīl·ə¸dē }

detached core [GEOL] The inner bed or beds of a fold that may become separated or pinched off from the main body of the strata due to extreme folding and compression. { di'tacht 'kòr }

detrital fan See alluvial fan. { də'trīd·əl 'fan }

detrital minerals [MINERAL] Grains of heavy minerals found in sediment, resulting from mechanical disintegration of the parent rock. { də'trīd·əl 'min·rəlz }

detrital ratio See clastic ratio. { də'trīd·əl 'rā·shō }

detrital remanent magnetization [GEOPHYS] Magnetization acquired by magnetic grains during formation of a sedimentary rock.   Abbreviated DRM. { də'trīd·əl 'rem·ə·nənt 'mag·nəd·ə'zā·shən }

detrital reservoir [GEOL] A clastic or detrital-granular reservoir, classified by rock type and other factors such as sediments (quartzose-type, graywacke, or arkose sediments). { də'trīd·əl 'rez·əv¸wär }

detrital sediment [GEOL] Accumulations of the organic and inorganic fragmental products of the weathering and erosion of land transported to the place of deposition. { də'trīd·əl 'sed·ə·mənt }

detritus [GEOL] Any loose material removed directly from rocks and minerals by mechanical means, such as disintegration or abrasion. { də'trīd·əs }

deuteric [GEOL] Of or pertaining to alterations in igneous rock during the later stages and as a direct result of consolidation of magma or lava.   Also known as epimagmatic; paulopost. { dü'tir·ik }

development [GEOL] The progression of changes in fossil groups which have succeeded one another during deposition of the strata of the earth. { də'vel·əp·mənt }

deviatoric stress [GEOL] A condition in which the stress components operating at a point in a body are not the same in every direction.   Also known as differential stress. { ¦dēv·ē·ə¦tòr·ik 'stres }

devillite [MINERAL] $Cu_4Ca(SO_4)_2(OH)_6 \cdot 3H_2O$ A dark-green mineral consisting of a hydrous basic sulfate of copper and calcium, occurring in six-sided platy crystals. { də'vē¸līt }

Devonian [GEOL] The fourth period of the Paleozoic Era, covering the geological time span between about 412 and 354×10^6 years before present. { di'vō·nē·ən }

De Vries effect [GEOCHEM] A relatively short-term oscillation, on the order of 100 years, in the radiocarbon content of the atmosphere, and the resulting variation in the apparent radiocarbon age of samples. { də'vrēz i'fekt }

deweylite [MINERAL] A mixture of clinochrysolite and stevensite.   Also known as gymnite. { 'dü·ē¸līt }

dewindtite [MINERAL] $Pb(UO_2)_2(PO_4)_2 \cdot 3H_2O$ A canary-yellow secondary mineral consisting of a hydrous phosphate of lead and uranium. { də'win¸tīt }

95

de Witte relation [GEOPHYS] Graphical plot of the relation between electrical conductivity and distance over which the conductivity is measured through reservoir rock with clay minerals, (the effect is similar to two parallel electrical circuits), the current passing through the conducting clay minerals and the water-filled pores. { də'wit rē'lā·shən }

dextral drag fold [GEOL] A drag fold in which the trace of a given surface bed is displaced to the right. { 'dek·strəl 'drag ,fōld }

dextral fault [GEOL] A strike-slip fault in which an observer approaching the fault sees the opposite block as having moved to the right. Also known as right-lateral fault; right-lateral slip fault; right-slip fault. { 'dek·strəl 'fȯlt }

dextral fold [GEOL] An asymmetric fold in which the long limb appears to be offset to the right to an observer looking along the long limb. { 'dek·strəl 'fōld }

D horizon [GEOL] A soil horizon sometimes occurring below a B or C horizon, consisting of unweathered rock. { 'dē hə'rīz·ən }

diabantite [MINERAL] $(Mg,Fe^{2+},Al)_6(Si,Al)_4O_{10}(OH)_8$ Mineral of the chlorite group consisting of a basic silicate of magnesium, iron, and aluminum, occurring in cavities in basic igneous rock. { ,dī·ə'ban,tīt }

diabase [PETR] An intrusive rock consisting principally of labradorite and pyroxene. { 'dī·ə,bās }

diabase amphibolite [PETR] Amphibolite formed by dynamic metamorphism of diabase. { 'dī·ə,bās am'fib·ə,līt }

diabasic [PETR] Denoting igneous rock in which the inter-stices between the feldspar crystals are filled with discrete crystals or grains of pyroxene. { ¦dī·ə¦bās·ik }

diablastic [PETR] Pertaining to a texture in metamorphic rock that consists of intergrown and interpenetrating rod-shaped components. { ,dī·ə'blas·tik }

diaboleite [MINERAL] $Pb_2CuCl_2(OH)_4$ A sky-blue mineral consisting of a basic chloride of lead and copper. { ¦dī·ə·bō'lā,īt }

diachronous [GEOL] Of a rock unit, varying in age in different areas or cutting across time planes or biostratigraphic zones. Also known as time-transgressive. { dī'ak·rə·nəs }

diaclinal [GEOL] Pertaining to a stream crossing a fold, perpendicular to the strike of the underlying strata it traverses. { ¦dī·ə¦klīn·əl }

Diacodectidae [PALEON] A family of extinct artiodactyl mammals in the suborder Palaeodonta. { ,dī·ə·kə'dek·tə,dē }

diadochite [MINERAL] $Fe_2(PO_4)(SO_4)(OH) \cdot 5H_2O$ A brown or yellowish mineral consisting of a basic hydrous ferric phosphate and sulfate. { dī'ad·ə,kīt }

diagenesis [GEOL] Chemical and physical changes occurring in sediments during and after their deposition but before consolidation. { ,dī·ə'jen·ə·səs }

diagonal fault [GEOL] A fault whose strike is diagonal or oblique to the strike of the adjacent strata. Also known as oblique fault. { dī'ag·ən·əl 'fȯlt }

diagonal joint [GEOL] A joint having its strike oblique to the strike of the strata of the sedimentary rock, or to the cleavage plane of the metamorphic rock in which it occurs. Also known as oblique joint. { dī'ag·ən·əl 'jȯint }

diallage [MINERAL] A green, brown, gray, or bronze-colored clinopyroxene characterized by prominent parting parallel to the front pinacoid *a* (100). { 'dī·ə·lij }

diamantine [MINERAL] Consisting of or resembling diamond. { ¦dī·ə¦man,tēn }

diamictite [PETR] A calcareous, terrigenous sedimentary rock that is not sorted or poorly sorted and contains particles of many sizes. Also known as mixtite. { dī·ə'mik,tīt }

diamicton [PETR] A nonlithified diamictite. Also known as symmicton. { dī·ə'mik,tän }

diamond [MINERAL] A colorless mineral composed entirely of carbon crystallized in the isometric system as octahedrons, dodecahedrons, and cubes; the hardest substance known; used as a gem and in cutting tools. { 'dī,mənd }

diamond matrix [GEOL] The rock material in which diamonds are formed. { 'dī·mənd 'mā·triks }

dianite *See* columbite. { 'dī·ə,nīt }

Dianulitidae [PALEON] A family of extinct, marine bryozoans in the order Cystoporata. { dī,an·yə'lid·ə,dē }

diaphorite [MINERAL] $PB_2Ag_3Sb_3S_8$ A gray-black orthorhombic mineral consisting of sulfide of lead, silver, and antimony, occurring in crystals. Also known as ultrabasite. { dī'af·ə,rīt }

diaphthoresis *See* retrograde metamorphism. { dī¦af·thə'rē·səs }

diaphthorite [PETR] Schistose rocks in which minerals have formed by retrograde metamorphism. { dī'af·thə,rīt }

diapir [GEOL] A dome or anticlinal fold in which a mobile plastic core has ruptured the more brittle overlying rock. Also known as diapiric fold; piercement dome; piercing fold. { 'dī·ə,pir }

diapiric fold *See* diapir. { ¦dī·ə¦pir·ik 'fōld }

diaspore [MINERAL] AlO(OH) A mineral composed of some bauxites occurring in white, lamellar masses; crystallizes in the orthorhombic system. { 'dī·ə,spȯr }

diastem [GEOL] A temporal break between adjacent geologic strata that represents nondeposition or local erosion but not a change in the general regimen of deposition. { 'dī·ə,stem }

diastrophism [GEOL] **1.** The general process or combination of processes by which the earth's crust is deformed. **2.** The results of this deforming action. { dī'as·trə,fiz·əm }

diatomaceous earth [GEOL] A yellow, white, or light-gray, siliceous, porous deposit made of the opaline shells of diatoms; used as a filter aid, paint filler, adsorbent, abrasive, and thermal insulator. Also known as kieselguhr; tripolite. { ¦dī·ə·tə¦mā·shəs 'ərth }

diatomaceous ooze [GEOL] A pelagic, siliceous sediment composed of more than 30% diatom tests, up to 40% calcium carbonate, and up to 25% mineral grains. { ¦dī·ə·tə¦mās·shəs 'üz }

diatomite [GEOL] Dense, chert-like, consolidated diatomaceous earth. { dī'ad·ə,mīt }

diatreme [GEOL] A circular volcanic vent produced by the explosive energy of gas-charged magmas. { 'dī·ə,trēm }

Diatrymiformes [PALEON] An order of extinct large, flightless birds having massive legs, tiny wings, and large heads and beaks. { dī,a·trə·mə'fōr,mēz }

diborate *See* borax. { dī'bȯr,āt }

Dichobunidae [PALEON] A family of extinct artiodactyl mammals in the superfamily Dichobunoidea. { ,dī·kə'byün·ə,dē }

Dichobunoidea [PALEON] A superfamily of extinct artiodactyl mammals in the suborder Paleodonta composed of small to medium-size forms with tri- to quadritubercular bunodont upper teeth. { ,dī·kə·byə'nȯid·ē·ə }

Dickinsoniidae [PALEON] A family that comprises extinct flat-bodied, multisegmented coelomates; identified as ediacaran fauna. { ,dik·ən·sə'nī·ə,dē }

dickinsonite [MINERAL] $H_2Na_6(Mn,Fe,Ca,Mg)_{14}(PO_4)_{12}{\cdot}H_2O$ A green mineral consisting of foliated hydrous acid phosphate, chiefly of manganese, iron, and sodium, and is isostructural with arrojadite; specific gravity is 3.34. { 'dik·ən·sə,nīt }

dickite [MINERAL] $Al_2Si_2O_5(OH)_4$ A mineral of the kaolin group found crystallized in clay in hydrothermal veins; it is polymorphous with kaolinite and nacrite. { 'di,kīt }

Dictyonellidina [PALEON] A suborder of extinct articulate brachiopods. { ¦dik·tē·ō·ne'lid·ən·ə }

dictyonema bed [GEOL] A thin shale bed rich in remains of graptolites of the genus *Dictyonema*. { ,dik·tē·ə'nē·mə ,bed }

Dictyospongiidae [PALEON] A family of extinct sponges in the subclass Amphidiscophora having spicules resembling a one-ended amphidisc (paraclavule). { ¦dik·tē·ō,spən'jī·ə,dē }

Didolodontidae [PALEON] A family consisting of extinct medium-sized herbivores in the order Condylarthra. { dīd·əl·ō'dänt·ə,dē }

didymolite [MINERAL] $Ca_2Al_6Si_9O_{29}$ A dark-gray monoclinic mineral consisting of a calcium aluminum silicate, occurring in twinned crystals. { dī'dim·ə,līt }

dietrichite

dietrichite [MINERAL] $(Zn,Fe,Mn)Al_2(SO_4)_4 \cdot 22H_2O$ Mineral consisting of a hydrous sulfate of aluminum and one or more of the metals zinc, iron, and manganese. { 'dē·tri,kīt }

dietzeite [MINERAL] $Ca_2(IO_3)_2(CrO_4)$ A dark-golden-yellow iodate mineral commonly in fibrous or columnar form as a component of caliche. { 'dēt·sə,īt }

differential compaction [GEOL] Compression in sediments, such as sand or limestone, as the weight of overburden causes reduction in pore space and forcing out of water. { ,dif·ə'ren·chəl kəm'pak·shən }

differential erosion [GEOL] Rapid erosion of one area of the earth's surface relative to another. { ,dif·ə'ren·chəl i'rō·zhən }

differential fault See scissors fault. { ,dif·ə'ren·chəl 'fólt }

differential stress See deviatoric stress. { ,dif·ə'ren·chəl 'stres }

digenite [MINERAL] Cu_9S_5 A blue to black mineral consisting of an isometric copper sulfide having a variable deficiency in copper. Also known as alpha chalcocite; blue chalcocite. { 'dī·jə,nīt }

digitation [GEOL] A secondary recumbent anticline emanating from a larger recumbent anticline. { ,dij·ə'tā·shən }

dike [GEOL] A tabular body of igneous rock that cuts across adjacent rocks or cuts massive rocks. { dīk }

dike ridge [GEOL] Any small wall-like ridge created by differential erosion. { 'dīk ,rij }

dike set [GEOL] A small group of dikes arranged linearly or parallel to each other. { 'dīk ,set }

dike swarm [GEOL] A large group of parallel, linear, or radially oriented dikes. { 'dīk ,swórm }

dilatancy [GEOL] Expansion of deformed masses of granular material, such as sand, due to rearrangement of the component grains. { dī'lāt·ən·sē }

dimorphite [MINERAL] As_4S_3 An orange-yellow mineral consisting of arsenic sulfide. { dī'mór,fīt }

Dimylidae [PALEON] A family of extinct lipotyphlan mammals in the order Insectivora; a side branch in the ancestry of the hedgehogs. { dī'mil·ə,dē }

Dinantian [GEOL] Lower Carboniferous geologic time. Also known as Avonian. { di'nan·chən }

Dinocerata [PALEON] An extinct order of large, herbivorous mammals having semigraviportal limbs and hoofed, five-toed feet; often called uintatheres. { ,dī·nō'ser·ə·də }

Dinornithiformes [PALEON] The moas, an order of extinct birds of New Zealand; all had strong legs with four-toed feet. { ,dīn·ór,nith·ə'fór,mēz }

dinosaur [PALEON] The name, meaning terrible lizard, applied to the fossil bones of certain large, ancient bipedal and quadripedal reptiles placed in the orders Saurischia and Ornithischia. { 'dī·nə,sór }

diogenite [MINERAL] An achondritic stony meteorite composed essentially of iron-rich pyroxene minerals. Also known as rodite. { dī'ä·jə,nīt }

diopside [MINERAL] $CaMg(SiO_3)_2$ A white to green monoclinic pyroxene mineral which forms gray to white, short, stubby, prismatic, often equidimensional crystals. Also known as malacolite. { dī'äp,sīd }

dioptase [MINERAL] $CuSiO_2(OH)_2$ A rare emerald-green mineral that forms hexagonal, hydrous crystals. { dī'äp,tās }

diorite [PETR] A phaneritic plutonic rock with granular texture composed largely of plagioclase feldspar with smaller amounts of dark-colored minerals; used occasionally as ornamental and building stone. Also known as black granite. { 'dī·ə,rīt }

dip [GEOL] **1.** The angle that a stratum or fault plane makes with the horizontal. Also known as angle of dip; formation dip; true dip. **2.** A pronounced depression in the land surface. { dip }

dip fault [GEOL] A type of fault that strikes parallel with the dip of the strata involved. { 'dip ,fólt }

dip joint [GEOL] A joint that strikes approximately at right angles to the cleavage or bedding of the constituent rock. { 'dip ,jóint }

Diplacanthidae [PALEON] A family of extinct acanthodian fishes in the suborder Diplacanthoidei. { ‚dip·lə'kan·thə‚dē }

Diplacanthoidei [PALEON] A suborder of extinct acanthodian fishes in the order Climatiiformes. { ‚dip·lə‚kan'thȯid·ē‚ī }

Diplobathrida [PALEON] An order of extinct, camerate crinoids having two circles of plates beneath the radials. { ‚dip·lō'bath·rə·də }

Diplodocus [PALEON] Herbivorous sauropod dinosaur, approximately 100 feet (30 meters) long and weighing 12 tons, from the Late Jurassic Period that had a very long neck and tail and a very small body. { di'pläd·ə·kəs }

dip log [GEOL] A log of the dips of formations traversed by boreholes. { 'dip ‚läg }

Diploporita [PALEON] An extinct order of echinoderms in the class Cystoidea in which the thecal canals were associated in pairs. { ‚dip·lə'pȯr·əd·ə }

dipmeter log [GEOL] A dip log produced by reading of the direction and angle of formation dip as analyzed from impulses from a dipmeter consisting of three electrodes 120° apart in a plane perpendicular to the borehole. { 'dip‚mēd·ər ‚läg }

dip pole See magnetic pole. { 'dip ‚pōl }

dip reversal See reversal of dip. { 'dip ri'vər·səl }

Diprotodontidae [PALEON] A family of extinct marsupial mammals. { dī‚prōd·ə'dän·tə‚dē }

dip slip [GEOL] The component of a fault parallel to the dip of the fault. Also known as normal displacement. { 'dip ‚slip }

dip slope [GEOL] A slope of the surface of the land determined by and conforming approximately to the dip of the underlying rocks. Also known as back slope; outface. { 'dip ‚slōp }

dip-strike symbol [GEOL] A geologic symbol used on maps to show the strike and dip of a planar feature. { 'dip ‚strīk ‚sim·bəl }

dipyre See mizzonite. { 'dī‚pīr }

direction See trend. { də'rek·shən }

directional structure [GEOL] Any sedimentary structure having directional significance; examples are cross-bedding and ripple marks. Also known as vectorial structure. { də'rek·shən·əl 'strək·chər }

direct stratification See primary stratification. { də'rekt ‚strad·ə·fə'kā·shən }

dirt band [GEOL] A dark layer in a glacier representing a former surface, usually a summer surface, where silt and debris accumulated. { 'dərt ‚band }

dirt bed [GEOL] A buried soil containing partially decayed organic material; sometimes occurs in glacial drift. { 'dərt ‚bed }

dirt slip See clay vein. { 'dərt ‚slip }

Disasteridae [PALEON] A family of extinct burrowing, exocyclic Euechinoidea in the order Holasteroida comprising mainly small, ovoid forms without fascioles or a plastron. { ‚dis·ə'ster·ə‚dē }

Discoidiidae [PALEON] A family of extinct conical or globular, exocyclic Euechinoidea in the order Holectypoida distinguished by the rudiments of internal skeletal partitions. { dis‚kȯi'dī·ə‚dē }

disconformity [GEOL] Unconformity between parallel beds or strata. { ‚dis·kən'fȯr·məd·ē }

discontinuity [GEOL] **1.** An interruption in sedimentation. **2.** A surface that separates unrelated groups of rocks. [GEOPHYS] A boundary at which the velocity of seismic waves changes abruptly. { dis‚känt·ən'ü·əd·ē }

discontinuous reaction series [GEOL] The branch of Bowen's reaction series that include olivine, pyroxene, amphibole, and biotite; each change in the series represents an abrupt change in phase. { ‚dis·kən'tin·yə·wəs rē'ak·shən ‚sir·ēz }

discordance [GEOL] An unconformity characterized by lack of parallelism between strata which touch without fusion. { di'skȯrd·əns }

discordant pluton [GEOL] An intrusive igneous body that cuts across the bedding or foliation of the intruded formations. { di'skȯrd·ənt 'plü‚tän }

discrete-film zone See belt of soil water. { di'skrēt ‚film ‚zōn }

disharmonic fold |GEOL| A fold in which changes in form or magnitude occur with depth. { ,dis·här'män·ik 'fōld }

disjunct endemism |PALEON| A type of regionally restricted distribution of a fossil taxon in which two or more component parts are separated by a major physical barrier and hence not readily explicable in terms of present-day geography. { 'dis,jəŋkt 'en·də,miz·əm }

dislocation |GEOL| Relative movement of rock on opposite sides of a fault. Also known as displacement. { ,dis·lō'kā·shən }

dislocation breccia See fault breccia. { ,dis·lō'kā·shən 'brech·ə }

dismicrite |GEOL| Fine-grained limestone of obscure origin, resembling micrite but containing sparry calcite bodies. { diz'mī,krīt }

dispersal pattern |GEOCHEM| Distribution pattern of metals in soil, rock, water, or vegetation. { də'spər·səl ,pad·ərn }

dispersed elements |GEOCHEM| Elements which form few or no independent minerals but are present as minor ingredients in minerals of abundant elements. { də'spərst 'el·ə·mənts }

dispersion |MINERAL| In optical mineralogy, the constant optical values at different positions on the spectrum. { də'spər·zhən }

displaced ore body |GEOL| An ore body which has been subjected to displacement or disruption after its initial deposition. { dis'plāst 'òr ,bäd·ē }

displacement |GEOL| See dislocation. { dis'plās·mənt }

dissection |GEOL| Destruction of the continuity of the land surface by erosive cutting of valleys or ravines into a relatively even surface. { də'sek·shən }

dissepiment |PALEON| One of the vertically positioned thin plates situated between the septa in extinct corals of the order Rugosa. { də'sep·ə·mənt }

dissipation constant |GEOPHYS| In atmospheric electricity, a measure of the rate at which a given electrically charged object loses its charge to the surrounding air. { ,dis·ə'pā·shən ,kän·stənt }

Distacodidae |PALEON| A family of conodonts in the suborder Conodontiformes characterized as simple curved cones with deeply excavated attachment scars. { dis·tə'käd·ə,dē }

disthene See kyanite. { 'dis,thēn }

distortional wave See S wave. { di'stòr·shən·əl 'wāv }

distributed fault See fault zone. { di'strib·yəd·əd ,fòlt }

distributive fault See step fault. { di'strib·yəd·iv 'fòlt }

disturbance |GEOL| Folding or faulting of rock or a stratum from its original position. { də'stər·bəns }

diurnal age See age of diurnal inequality. { dī'ərn·əl 'āj }

diurnal variation |GEOPHYS| Daily variations of the earth's magnetic field at a given point on the surface, with both solar and lunar periods having their source in the horizontal movements of air in the ionosphere. { dī'ərn·əl ,ver·ē'ā·shən }

divergence loss |GEOPHYS| During geophysical prospecting, the portion of the power lost in transmitting signals that is caused by the spreading of seismic or sound rays by the geometry of the geologic features. { də'vər·jəns ,lòs }

Divesian See Oxfordian. { də'vēzh·ən }

dixenite |MINERAL| $Mn_5(SiO_3)(AsO_3)(OH)_2$ A black hexagonal mineral consisting of a manganese arsenite and silicate, occurring in scales. { 'dik·sə,nīt }

djalmaite See microlite. { 'jal·mə,īt }

djerfisherite |MINERAL| $K_3CuFe_{12}S_{14}$ A sulfide mineral found only in meteorites. { jər'fish·ə,rīt }

Djulfian |GEOL| Upper upper Permian geologic time. { 'jùl·fē·ən }

D layer |GEOL| The lower mantle of the earth, between a depth of 600 and 1800 miles (1000 and 2900 kilometers). { 'dē ,lā·ər }

dneprovskite See wood tin. { ne'prōv,skīt }

Docodonta |PALEON| A primitive order of Jurassic mammals of North America and England. { ,däk·ə'dän·tə }

dogger [GEOL] Concretionary masses of calcareous sandstone or ironstone. { 'dȯg·ər }

dolerophanite [MINERAL] $Cu_2(SO_4)O$ A brown, monoclinic mineral consisting of a basic copper sulfate, occurring in crystals. { ˌdäl·ə'räf·ə,nīt }

Dolichothoraci [PALEON] A group of joint-necked fishes assigned to the Arctolepiformes in which the pectoral appendages are represented solely by large fixed spines. { ˌdäl·ə·kō'thȯr·ə,sī }

doline [GEOL] A general term for a closed depression in an area of karst topography that is formed either by solution of the surficial limestone or by collapse of underlying caves. { də'lēn }

dolocast [GEOL] The cast or impression of a dolomite crystal. { 'dō·lə,kast }

dolomite [MINERAL] $CaMg(CO_3)_2$ The carbonate mineral; white or colorless with hexagonal symmetry and a structure similar to that of calcite, but with alternate layers of calcium ions being completely replaced by magnesium. { 'dō·lə,mīt }

dolomite rock See dolomitic limestone. { 'dō·lə,mīt 'räk }

dolomitic limestone [PETR] A limestone whose carbonate fraction contains more than 50% dolomite. Also known as dolomite rock; dolostone. { ˌdō·lə¦mid·ik 'līm,stōn }

dolomitization [GEOL] Conversion of limestone to dolomite rock by replacing a portion of the calcium carbonate with magnesium carbonate. { ˌdō·lə·məd·ə'zā·shən }

dolostone See dolomitic limestone. { 'dō·lə,stōn }

dome [GEOL] **1.** A circular or elliptical, almost symmetrical upfold or anticlinal type of structural deformation. **2.** A large igneous intrusion whose surface is convex upward. { dōm }

Domerian [GEOL] Upper Charmouthian geologic time. { dō'mer·ə·ən }

domeykite [MINERAL] Cu_3As A tin-white or steel-gray mineral consisting of copper arsenide; specific gravity is 7.2–7.75. { dō'mā,kīt }

Donau glaciation [GEOL] A Pleistocene glacial time unit in the Alps region in Europe. { 'dō,naȯ glä·sē'ä·shən }

doodlebug [GEOL] Also known as douser. **1.** Any unscientific device or apparatus, such as a divining rod, used to locate subsurface water, minerals, gas, or oil. **2.** A scientific instrument used for locating minerals. { 'düd·əl,bəg }

dopplerite [GEOL] A naturally occurring gel of humic acids found in peat bags or where an aqueous extract from a low-rank coal can collect. { 'däp·lə,rīt }

Dorypteridae [PALEON] A family of Permian palaeonisciform fishes sometimes included in the suborder Platysomoidei. { dȯ,rip'ter·ə,dē }

doubly plunging fold [GEOL] A fold that plunges in opposite directions, either away from or toward a central point. { ¦dəb·lē ,plənj·iŋ 'fōld }

douglasite [MINERAL] $K_2FeCl_4·2H_2O$ Ore from Stassfurt, Germany; a member of the erythrosiderite group; orthorhombic, in the isomorphous series. { 'dəg·lə,sīt }

douser See doodlebug. { 'daȯs·ər }

down [GEOL] **1.** Hillock of sand thrown up along the coast by the sea or the wind. **2.** A flat eminence on the top of a hill or mountain. { daȯn }

downcutting [GEOL] Stream erosion in which the cutting is directed in a downward direction. { 'daȯn,kəd·iŋ }

downdip [GEOL] Pertaining to a position parallel to or in the direction of the dip of a stratum or bed. { 'daȯn,dip }

downthrow [GEOL] The side of a fault whose relative movement appears to have been downward. { 'daȯn,thrō }

downwarp [GEOL] A segment of the earth's crust that is broadly bent downward. { 'daȯn,wȯrp }

Dowtonian [GEOL] Uppermost Silurian or lowermost Devonian geologic time. { daȯ'tōn·ē·ən }

drag fold [GEOL] A minor fold formed in an incompetent bed by movement of a competent bed so as to subject it to couple; the axis is at right angles to the direction in which the beds slip. { 'drag ,fōld }

drag mark [GEOL] Long, even mark usually having longitudinal striations produced by current drag of an object across a sedimentary surface. { 'drag ,märk }

drainage divide [GEOL] **1.** The border of a drainage basin. **2.** The boundary separating adjacent drainage basins. { 'drān·ij də,vīd }

draping [GEOL] Structural concordance of the strata overlying a limestone reef or other hard core to the surface of the reef or core. { 'drāp·iŋ }

dreikanter [GEOL] A pebble with three facets shaped by sandblasting. { 'drī,kän·tər }

Drepanellacea [PALEON] A monomorphic superfamily of extinct paleocopan ostracods in the suborder Beyrichicopina having a subquadrate carapace, many with a marginal rim. { drə,pan·əl'ās·ē·ə }

Drepanellidae [PALEON] A monomorphic family of extinct ostracodes in the superfamily Drepanellacea. { ,dre·pə'nel·ə,dē }

Dresbachian [GEOL] Lower Croixan geologic time. { drez'bäk·ē·ən }

drewite [GEOL] Calcareous ooze composed of impalpable calcareous material. { 'drü,īt }

drift [GEOL] **1.** Rock material picked up and transported by a glacier and deposited elsewhere. **2.** Detrital material moved and deposited on a beach by waves and currents. { drift }

drift dam [GEOL] A dam formed by glacial drift in a stream valley. { 'drift ,dam }

drift terrace See alluvial terrace. { 'drift ,ter·əs }

dripstone [GEOL] A cave feature, such as a stalagmite, which is formed by precipitation of calcium carbonate or another mineral from dripping water. { 'drip,stōn }

DRM See detrital remanent magnetization.

drop [MINERAL] A funnel-shaped downward intrusion of sedimentary rock into the roof of a coal seam. { dräp }

dropstone [GEOL] A rock that was carried by a glacier or iceberg, and deposited as the ice melted. { 'dräp,stōn }

drowned atoll [GEOL] An atoll which has not reached the water surface. { ¦draund 'a,tol }

drowned coast [GEOL] A shoreline transformed from a hilly land surface to an archipelago of small islands by inundation by the sea. { ¦draund 'kōst }

drowned valley [GEOL] A valley whose lower part has been inundated by the sea due to submergence of the land margin. { ¦draund 'val·ē }

drumlin [GEOL] A hill of glacial drift or bedrock having a half-ellipsoidal streamline form like the inverted bowl of a spoon, with its long axis paralleling the direction of movement of the glacier that fashioned it. { 'drəm·lən }

drumlinoid See rock drumlin. { 'drəm·lə,noid }

druse [GEOL] A small cavity in a rock or vein encrusted with aggregates of crystals of the same minerals which commonly constitute the enclosing rock. { drüz }

drusy [GEOL] Of or pertaining to rocks containing numerous druses. { 'drüz·ē }

dry-bone ore See smithsonite. { 'drī ,bōn ,or }

dry delta See alluvial fan. { ¦drī 'del·tə }

dry-hot-rock geothermal system [GEOL] A water-deficient hydrothermal reservoir dominated by the presence of rocks at depths in which large quantities of heat are stored. { ¦drī 'hät ,räk jē·ō¦thər·məl 'sis·təm }

dry permafrost [GEOL] A loose and crumbly permafrost which contains little or no ice. { ¦drī 'pər·mə,frost }

dry quicksand [GEOL] An accumulation of alternate layers of firmly compacted sand and loose sand that cannot support heavy loads. { ¦drī 'kwik,sand }

dry sand [GEOL] **1.** A formation, underlying the production sand, into which oil has leaked due to careless drilling practices. **2.** A nonproductive oil sand. { ¦drī ¦sand }

drystone [GEOL] A stalagmite or stalactite formed by dropping water. { 'drī,stōn }

dry valley [GEOL] A valley, usually in a chalk or karst type of topography, that has no permanent water course along the valley floor. { ¦drī 'val·ē }

dry wash [GEOL] A wash, arroyo, or coulee whose bed lacks water. { 'drī ,wäsh }

Dst [GEOPHYS] The "storm-time" component of variation of the terrestrial magnetic field, that is, the component which correlates with the interval of time since the onset of a magnetic storm; used as an index of intensity of the ring current.

duck-billed dinosaur [PALEON] Any of several herbivorous, bipedal ornithopods having the front of the mouth widened to form a ducklike beak. { ¦dək ¦bild 'dīn·ə‚sȯr }

dufrenite [MINERAL] A blackish-green, fibrous ferric phosphate mineral; commonly massive or in nodules. { dü'frā‚nīt }

dufrenoysite [MINERAL] $Pb_2As_2S_5$ A lead gray to steel gray, monoclinic mineral consisting of lead arsenic sulfide. { ‚dü·frə'nȯi‚zīt }

duftite [MINERAL] $PbCu(AsO_4)(OH)$ Orthorhombic mineral that is composed of a basic arsenate of lead and copper. { 'dəf‚tīt }

dull coal [GEOL] A component of banded coal with a grayish color and dull appearance, consisting of small anthraxylon constituents in addition to cuticles and barklike constituents embedded in the attritus. { ¦dəl ¦kōl }

dumontite [MINERAL] $Pb_2(UO_2)_3(PO_4)_2(OH)_4·3H_2O$ Yellow orthorhombic mineral consisting of a hydrated phosphate of uranium and lead, occurring in crystals. { dü'män‚tīt }

dumortierite [MINERAL] $Al_8BSi_3O_{19}(OH)$ A pink, green, blue, or violet mineral that crystallizes in the orthorhombic system but commonly occurs in parallel or radiating fibrous aggregates; mined for the manufacture of high-grade porcelain. { dü·mȯr'tir‚īt }

dundasite [MINERAL] $PbAl_2(CO_3)_2(OH)_4·2H_2O$ A white mineral consisting of a basic lead aluminum carbonate, occurring in spherical aggregates. { 'dən·də‚sīt }

dune [GEOL] A mound or ridge of unconsolidated granular material, usually of sand size and of durable composition (such as quartz), capable of movement by transfer of individual grains entrained by a moving fluid. { dün }

dunite [PETR] An ultrabasic rock consisting almost solely of a magnesium-rich olivine with some chromite and picotite; an important source of chromium. { 'dü‚nīt }

duplexite [MINERAL] $Ca_4BeAl_2Si_9O_{24}(OH)_2$ A white fibrous mineral consisting of hydrous beryllium calcium aluminosilicate. Also known as bavenite. { 'dü‚plek‚sīt }

durain [GEOL] A hard, granular ingredient of banded coal which occurs in lenticels and shows a close, firm texture. Also known as durite. { 'dü‚rān }

durangite [MINERAL] $NaAlF(AsO_4)$ An orange-red, monoclinic mineral consisting of a fluoarsenate of sodium and aluminum; occurs in crystals. { də'ran‚jīt }

Durargid [GEOL] A great soil group constituting a subdivision of the Argids, indicating those soils with a hardpan cemented by silica and called a duripan. { dür'är·jəd }

duricrust [GEOL] The case-hardened soil crust formed in semiarid climates by precipitation of salts; contains aluminous, ferruginous, siliceous, and calcareous material. { 'dür·ə‚krəst }

durinite [GEOL] The principal maceral of durain; a heterogeneous material, semiopaque in section (including all parts of plants); micrinite, exinite, cutinite, resinite, collinite, xylinite, suberinite, and fusinite may be present. { 'dür·ə‚nīt }

duripan [GEOL] A horizon in mineral soil characterized by cementation by silica. { 'dür·ə‚pan }

durite See durain. { 'dü·rīt }

dussertite [MINERAL] $BaFe_3(AsO_4)_2(OH)_5$ A mineral consisting of a hydrous basic arsenate of barium and iron. { 'dəs·ər‚tīt }

dust [GEOL] Dry solid matter of silt and clay size (less than 1/16 millimeter). { dəst }

dust avalanche [GEOL] An avalanche of dry, loose snow. { 'dəst ‚av·ə‚lanch }

dust-devil effect [GEOPHYS] In atmospheric electricity, rather sudden and short-lived change (positive or negative) of the vertical component of the atmospheric electric field that accompanies passage of a dust devil near an instrument sensitive to the vertical gradient. { 'dəst ‚dev·əl i‚fekt }

Dwyka tillite [GEOL] A glacial Permian deposit that is widespread in South Africa. { də¦vīk·ə 'ti‚līt }

dynamic breccia See tectonic breccia. { dī¦nam·ik 'brech·ə }

dynamic geomorphology [GEOL] The quantitative analysis of steady-state, self-regulatory geomorphic processes. Also known as analytical geomorphology. { dī¦nam·ik ‚jē·ō·mȯr'fäl·ə·jē }

dynamic height [GEOPHYS] As measured from sea level, the distance above the geoid of points on the same equipotential surface, in terms of linear units measured along a plumb line at a given latitude, generally 45°. { dī¦nam·ik 'hīt }

dynamic metamorphism [GEOL] Metamorphism resulting exclusively or largely from rock deformation, principally faulting and folding. Also known as dynamometamorphism. { dī¦nam·ik ˌmed·ə'mȯr,fiz·əm }

dynamic vertical *See* apparent vertical. { dī¦nam·ik 'vərd·ə·kəl }

dynamo effect [GEOPHYS] A process in the ionosphere in which winds and the resultant movement of ionization in the geomagnetic field give rise to induced current. { 'dī·nə,mō i,fekt }

dynamometamorphism *See* dynamic metamorphism. { ¦dī·nə,mō,med·ə'mȯr,fiz·əm }

dynamo theory [GEOPHYS] The hypothesis which explains the regular daily variations in the earth's magnetic field in terms of electrical currents in the lower ionosphere, generated by tidal motions of the ionized air across the earth's magnetic field. { 'dī·nə,mō ,thē·ə·rē }

dysanalyte [MINERAL] A variety of the mineral perovskite in which Nb^{5+} substitutes for Ti^{5+}, and Na^+ for Ca^{2+} in the formula $Ca[TiO_3]$. { də'san·əl,īt }

dyscrasite [MINERAL] Ag_2Sb A gray mineral that forms rhombic crystals. { 'dis·krə,sīt }

Dysodonta [PALEON] An order of extinct bivalve mollusks with a nearly toothless hinge and a ligament in grooves or pits. { ˌdis·ə'dän·tə }

E

earlandite [MINERAL] $Ca_3(C_6H_5O_7)_2 \cdot 4H_2O$ A mineral consisting of a hydrous citrate of calcium; found in sediments in the Weddell Sea. { 'ir·lən,dīt }

earth [GEOL] **1.** Solid component of the globe, distinct from air and water. **2.** Soil; loose material composed of disintegrated solid matter. { ərth }

earth core See core. { 'ərth ,kòr }

earth crust See crust. { 'ərth ,krəst }

earth current [GEOPHYS] A current flowing through the ground and due to natural causes, such as the earth's magnetic field or auroral activity. Also known as telluric current. { 'ərth ,kə·rənt }

earth-current storm [GEOPHYS] Irregular fluctuations in an earth current in the earth's crust, often associated with electric field strengths as large as several volts per kilometer, and superimposed on the normal diurnal variation of the earth currents. { 'ərth ,kə·rənt ,stòrm }

earthflow [GEOL] A variety of mass movement involving the downslope slippage of soil and weathered rock in a series of subparallel sheets. { 'ərth,flō }

earth hummock [GEOL] A small, dome-shaped uplift of soil caused by the pressure of groundwater. Also known as earth mound. { 'ərth ,həm·ək }

earth interior [GEOL] The portion of the earth beneath the crust. { ¦ərth in¦tir·ē·ər }

earth-layer propagation [GEOPHYS] **1.** Propagation of electromagnetic waves through layers of the earth's atmosphere. **2.** Electromagnetic wave propagation through layers below the earth's surface. { 'ərth ,lā·ər ,präp·ə,gā·shən }

earth mound See earth hummock. { 'ərth ,maùnd }

earth movements [GEOPHYS] Movements of the earth, comprising revolution about the sun, rotation on the axis, precession of equinoxes, and motion of the surface of the earth relative to the core and mantle. { 'ərth ¦müv·məns }

earth oscillations [GEOPHYS] Any rhythmic deformations of the earth as an elastic body; for example, the gravitational attraction of the moon and sun excite the oscillations known as earth tides. { ¦ərth ,äs·ə'lā·shənz }

earth pillar [GEOL] A tall, conical column of earth materials, such as clay or landslide debris, that has been sheltered from erosion by a cap of hard rock. { 'ərth ¦pil·ər }

earthquake [GEOPHYS] A sudden movement of the earth caused by the abrupt release of accumulated strain along a fault in the interior. The released energy passes through the earth as seismic waves (low-frequency sound waves), which cause the shaking. { 'ərth,kwāk }

earthquake tremor See tremor. { 'ərth,kwāk ,trem·ər }

earthquake zone [GEOL] An area of the earth's crust in which movements, sometimes with associated volcanism, occur. Also known as seismic area. { 'ərth,kwāk ,zōn }

earth system [GEOPHYS] The atmosphere, oceans, biosphere, cryosphere, and geosphere, together. { 'ərth , sis·təm }

earth tide [GEOPHYS] The periodic movement of the earth's crust caused by forces of the moon and sun. Also known as bodily tide. { 'ərth ,tīd }

earth tremor See tremor. { 'ərth ,trem·ər }

earth wax See ozocerite. { 'ərth,waks }

earthy cobalt See asbolite. { ¦ərth·ē 'kō,bòlt }

earthy manganese See wad. { ¦ərth·ē 'maŋ·gə,nēs }

eastonite

eastonite [MINERAL] $K_2Mg_5AlSi_5Al_3O_{20}(OH_4)$ A mineral consisting of basic silicate of potassium, magnesium, and aluminum; it is an end member of the biotite system. { 'ē·stə,nīt }

ebb-and-flow structure [GEOL] Rock strata with alternating horizontal and cross-bedded layers, believed to have been produced by ebb and flow of tides. { ¦eb ən 'flō ¦strək·chər }

ecdemite [MINERAL] $Pb_6As_2O_7Cl_4$ A greenish-yellow to yellow, tetragonal mineral consisting of an oxychloride of lead and arsenic; occurs as coatings of small tabular crystals and as coarsely foliated masses. { 'ek·də,mīt }

echelon faults [GEOL] Separate, parallel faults having steplike trends. { 'esh·ə,län ,fȯls }

Echinocystitoida [PALEON] An order of extinct echinoderms in the subclass Perischoechinoidea. { ,ek·ə·nō,sis·tə'toid·ə }

eckermannite [MINERAL] $Na_3(Mg,Li)_4(Al,Fe)Si_8O_{22}(OH,F)_2$ Mineral of the amphibole group containing magnesium, lithium, iron, and fluorine. { 'ek·ər·mə,nīt }

eclogite [PETR] A class of metamorphic rocks distinguished by their composition, consisting essentially of omphacite and pyrope with small amounts of diopside, enstatite, olivine, kyanite, rutile, and rarely, diamond. { 'ek·lə,jīt }

eclogite facies [PETR] A type of facies composed of eclogite and formed by regional metamorphism at extremely high temperature and pressure. { 'ek·lə,jīt ,fā·shēz }

economic geology [GEOL] **1.** Application of geologic knowledge to materials usage and principles of engineering. **2.** The study of metallic ore deposits. { ,ek·ə'näm·ik jē'äl·ə·jē }

economic mineral [MINERAL] Mineral of commercial value. { ,ek·ə'näm·ik 'min·rəl }

ectinites [PETR] One of two major groups of metamorphic rocks comprising those formed with no accession or introduction of feldspathic material. { 'ek·tə,nīts }

ectohumus [GEOL] An accumulation of organic matter on the soil surface with little or no mixing with mineral material. Also known as mor; raw humus. { ¦ek·tō'hyü·məs }

Edaphosuria [PALEON] A suborder of extinct, lowland, terrestrial, herbivorous reptiles in the order Pelycosauria. { ,ed·ə·fō'sȯr·ē·ə }

eddy mill See pothole. { 'ed·ē ,mil }

Edenian [GEOL] Lower Cincinnatian geologic stage in North America, above the Mohawkian and below Maysvillian. { ,ē'dēn·ē·ən }

edge water [GEOL] In reservoir structures, the subsurface water that surrounds the gas or oil. { 'ej ,wȯd·ər }

Ediacaran fauna [PALEON] The oldest known assemblage of fossil remains of soft-bodied marine animals; first discovered in the Ediacara Hills, Australia. { ,ēd·ē·ə'kar·ən 'fȯn·ə }

edingtonite [MINERAL] $BaAl_2Si_3O_{10}·4H_2O$ Gray zeolite mineral that forms rhombic crystals; sometimes contains large amounts of calcium. { 'ed·iŋ·tə,nīt }

Edrioasteroidea [PALEON] A class of extinct Echinozoa having ambulacral radial areas bordered by tube feet, and the mouth and anus located on the upper side of the theca. { ,ed·rē·ō,as·tə'rȯid·ē·ə }

effective porosity [GEOL] A property of earth containing interconnecting interstices, expressed as a percent of bulk volume occupied by the interstices. { ə¦fek·tiv pə'räs·əd·ē }

effective pressure See effective stress. { ə¦fek·tiv 'presh·ər }

effective stress [GEOL] The average normal force per unit area transmitted directly from particle to particle of a rock or soil mass. Also known as effective pressure; intergranular pressure. { ə¦fek·tiv 'stres }

efflorescence [MINERAL] A whitish powder, consisting of one or several minerals produced as an encrustation on the surface of a rock in an arid region. Also known as bloom. { ,ef·lə'res·əns }

effusive stage [GEOL] The second cooling stage for volcanic rocks. { e'fyü·siv ,stāj }

eggstone See oolite. { 'eg,stōn }

eglestonite |MINERAL| Hg_4Cl_2O Rare mercuric oxide mineral; forms yellow-brown isometric crystals upon exposure to air. { 'eg·əl·stə,nīt }

eguëite |MINERAL| $CaFe_{14}(PO_4)_{10}(OH)_{14}·21H_2O$ A brownish-yellow mineral consisting of a hydrated basic phosphate of calcium and iron; occurs as small nodules. { e'gwā,īt }

Egyptian asphalt |GEOL| A glance pitch (bituminous mixture similar to asphalt) found in the Arabian Desert. { i'jip·shən 'as,fòlt }

einkanter |GEOL| A stone shaped by windblown sand only upon one facet. { 'īn,kän·tər }

ejecta |GEOL| Material which is discharged by a volcano. { ē'jek·tə }

elastic bitumen *See* elaterite. { i'las·tik bī'tü·mən }

elastic rebound theory |GEOL| A theory which attributes faulting to stresses (in the form of potential energy) which are being built up in the earth and which, at discrete intervals, are suddenly released as elastic energy; at the time of rupture the rocks on either side of the fault spring back to a position of little or no strain. { i'las·tik 'rē,baúnd ,thē·ə·rē }

elaterite |GEOL| A light-brown to black asphaltic pyrobitumen that is moderately soft and elastic. Also known as elastic bitumen; mineral caoutchouc. { i'lad·ə,rīt }

electric calamine *See* hemimorphite. { i¦lek·trik 'kal·ə,mīn }

electrofiltration |GEOL| Counterprocess during electrical logging of well boreholes, in which mud filtrate forced through the mud cake produces an emf in the mud cake opposite a permeable bed, positive in the direction of filtrate flow. { i,lek·trō·fil'trā·shən }

eleolite *See* nepheline. { ə'lē·ə,līt }

elephant-hide pahoehoe |GEOL| A type of pahoehoe on whose surface are innumerable tummuli, broad swells, and pressure ridges which impart the appearance of elephant hide. { 'el·ə·fənt ,hīd pa'hō·ē,hō·ē }

ellestadite |MINERAL| A pale rose, hexagonal mineral consisting of an apatite-like calcium sulfate-silicate; occurs in granular massive form. { 'el ə,stə,dīt }

ellipsoidal lava *See* pillow lava. { ə,lip'sòid·əl 'läv·ə }

ellsworthite |MINERAL| $(Ca,Na,U)_2(Nb,Ta)_2O_6(O,OH)$ A yellow, brown, greenish or black mineral of the pyrochlore group occurring in isometric crystals and consisting of an oxide of niobium, titanium, and uranium. Also known as betafite; hatchettolite. { 'elz·wər,thīt }

elpasolite |MINERAL| K_2NaAlF_6 Mineral composed of sodium potassium aluminum fluoride. { el'pas·ō,līt }

elpidite |MINERAL| $Na_2ZrSi_6O_{15}·3H_2O$ A white to brick-red mineral composed of hydrated sodium zirconium silicate. { 'el·pə,dīt }

elutriation |GEOL| The washing away of the lighter or finer particles in a soil, especially by the action of raindrops. { ē,lü·trē'ā·shən }

eluvial |GEOL| Of, composed of, or relating to eluvium. { ē'lüv·ē·əl }

eluvial placer |GEOL| A placer deposit that is concentrated near the decomposed outcrop of the source. { ē'lüv·ē·əl 'plā·sər }

eluvium |GEOL| Disintegrated rock material formed and accumulated in situ or moved by the wind alone. { ē'lü·vē·əm }

embatholithic |GEOL| Pertaining to ore deposits associated with a batholith where exposure of the batholith and country rock is about equal. { em,bath·ə'lith·ik }

embayed coastal plain |GEOL| A coastal plain that has been partly sunk beneath the sea, thereby forming a bay. { em'bād ¦kòst·əl 'plān }

embayed mountain |GEOL| A mountain that has been depressed enough for sea water to enter the bordering valleys. { em'bād 'maún,tən }

embayment |GEOL| 1. Act or process of forming a bay. 2. A reentrant of sedimentary rock into a crystalline massif. { em'bā·mənt }

embolite |MINERAL| $Ag(Cl,Br)$ A yellow-green mineral resembling cerargyrite; composed of native silver chloride and silver bromide. { 'em·bə,līt }

Embolomeri |PALEON| An extinct side branch of slender-bodied, fish-eating aquatic anthracosaurs in which intercentra as well as centra form complete rings. { ,em·bə'lä·mə,rī }

107

embouchure [GEOL] **1.** The mouth of a river. **2.** A river valley widened into a plain. { ¦äm·bə¦shùr }

embrechites [PETR] A type of migmatite in which structural features of crystalline shifts are preserved but often partially obliterated by metablastesis. { 'em·brə‚kīts }

Embrithopoda [PALEON] An order established for the unique Oligocene mammal *Arsinoitherium*, a herbivorous animal that resembled the modern rhinoceros. { ‚em·brə'thä·pə·də }

emerald [MINERAL] $Al_2(Be_3Si_6O_{18})$ A brilliant-green to grass-green gem variety of beryl that crystallizes in the hexagonal system; green color is caused by varying amounts of chromium. Also known as smaragd. { 'em·rəld }

emerged shoreline See shoreline of emergence. { ə¦mərjd 'shòr‚līn }

emergence [GEOL] **1.** Dry land which was part of the ocean floor. **2.** The act or process of becoming an emergent land mass. { ə'mər·jəns }

emery [MINERAL] A fine, granular, gray-black, impure variety of corundum containing iron oxides, either hematite or magnetite; occurs as masses in limestone and as segregations in igneous rock. { 'em·ə·rē }

emery rock [PETR] A rock that contains corundum and iron ores. { 'em·ə·rē ‚räk }

emmonsite [MINERAL] $Fe_2Te_3O_9·2H_2O$ A yellow-green mineral composed of a hydrous oxide of iron and tellurium. { 'em·ən‚zīt }

emplacement [GEOL] Intrusion of igneous rock or development of an ore body in older rocks. { em'plās·mənt }

emplectite [MINERAL] $CuBiS_3$ A grayish or white mineral that crystallizes in the orthorhombic system; occurs in masses. { em'plek‚tīt }

empressite [MINERAL] AgTe An opaque, pale-bronze mineral whose crystal system is unknown. { 'em·prə‚sīt }

Enaliornithidae [PALEON] A family of extinct birds assigned to the order Hesperornithiformes, having well-developed teeth found in grooves in the dentary and maxillary bones of the jaws. { e¦nal·ē·òr¦nith·ə‚dē }

enargite [MINERAL] A lustrous, grayish-black mineral which is found in orthorhombic crystals but is more commonly columnar, bladed, or massive; hardness is 3 on Mohs scale, specific gravity is 4.44; in some places enargite is a valuable copper ore. Also known as clairite; luzonite. { e'när‚jīt }

enclosure compound See clathrate. { in'klō·zhər ‚käm‚paùnd }

encrinal limestone [GEOL] A limestone consisting of more than 10% but less than 50% of fossil crinoidal fragments. { en'krīn·əl 'līm‚stōn }

encrinite [PALEON] One of certain fossil crinoids, especially of the genus *Encrinus*. { 'eŋ·krə‚nīt }

endellite [MINERAL] $Al_2Si_2O_5(OH)_4·4H_2O$ Term used in the United States for a clay mineral, the more hydrous form of halloysite. Also known as hydrated halloysite; hydrohalloysite; hydrokaolin. { 'en·də‚līt }

endlichite [MINERAL] A mineral similar to vanadinite, but with the vanadium replaced by arsenic. { 'end·li‚kīt }

end member [MINERAL] One of the two or more pure chemical compounds that enters into solid solution with other pure chemical compounds to make up a series of minerals of similar crystal structure (that is, an isomorphous, solid-solution series). { 'end ‚mem·bər }

end moraine [GEOL] An accumulation of drift in the form of a ridge along the border of a valley glacier or ice sheet. { 'end mə‚rān }

endobatholithic [GEOL] Pertaining to ore deposits along projecting portions of a batholith. { ¦en·dō·bath·ə'lid·ik }

endocast See steinkern. { 'en·dō‚kast }

endogenetic See endogenic. { ¦en·dō·jə'ned·ik }

endogenic [GEOL] Of or pertaining to a geologic process, or its resulting feature such as a rock, that originated within the earth. Also known as endogenetic; endogenous. { ¦en·dō¦jen·ik }

endogenous [GEOL] See endogenic. { en'däj·ə·nəs }

endometamorphism [GEOL] A phase of contact metamorphism involving changes in

108

an igneous rock due to assimilation of portions of the rocks invaded by its magma. { ¦en·dō,med·ə'mȯr,tiz·əm }

Endotheriidae [PALEON] A family of Cretaceous insectivores from China belonging to the Proteutheria. { ,en·dō·thə'rī·ə,dē }

Endothyracea [PALEON] A superfamily of extinct benthic marine foraminiferans in the suborder Fusulinina, having a granular or fibrous wall. { ,en·dō·thə'rās·ē·ə }

en echelon [GEOL] Referring to an overlapped or staggered arrangement of geologic features. { 'en ,esh·ə,län }

en echelon fault blocks [GEOL] A belt in which the individual fault blocks trend approximately 45° to the trend of the entire fault belt. { 'en ,esh·ə,län 'fȯlt ,bläks }

energy level [GEOL] The kinetic energy supplied by waves or current action in an aqueous sedimentary environment either at the interface of deposition or several meters above. { 'en·ər·jē ,lev·əl }

englishite [MINERAL] $K_2Ca_4Al_8(PO_4)_8(OH)_{10} \cdot 9H_2O$ A white mineral composed of hydrous basic phosphate of potassium, calcium, and aluminum. { 'iŋ·gli,shīt }

engysseismology [GEOPHYS] Seismology dealing with earthquake records made close to the disturbance. { ¦en·jə·sīz'mäl·ə·jē }

enigmatite [MINERAL] $Na_2Fe_5TiSi_6O_{20}$ A black amphibole mineral occurring in triclinic crystals; specific gravity is 3.14–3.80. Also spelled aenigmatite. { ə'nig·mə,tīt }

ensialic geosyncline [GEOL] A geosyncline whose geosynclinal prism accumulates on a sialic crust and contains clastics. { en·sē'al·ik ,jē·ō'sin,klīn }

ensimatic geosyncline [GEOL] A geosyncline whose geosynclinal prism accumulates on a simatic crust and is composed largely of volcanic rock or sediments of volcanic debris. { en·sə'mad·ik ,jē·ō'sin,klīn }

enstatite [MINERAL] $MgOSiO_2$ A member of the pyroxene mineral group that crystallizes in the orthorhombic system; usually yellowish gray but becomes green when a little iron is present. { 'en·stə,tīt }

enstatite chondrite [GEOL] A type of chondritic meteorite consisting almost entirely of enstatite, with metal inclusions that may be abundant and are usually low in nickel. { 'en·stə,tīt 'kän,drīt }

Enteletacea [PALEON] A group of extinct articulate brachiopods in the order Orthida. { ,en·tə·lə'tās·ē·ə }

Entelodontidae [PALEON] A family of extinct palaeodont artiodactyls in the superfamily Entelodontoidea. { ,en·tə·lə'dän·tə,dē }

Entelodontoidea [PALEON] A superfamily of extinct piglike mammals in the suborder Palaeodonta having huge skulls and enlarged incisors. { ,en·tə·lə,dän'tȯid·ē·ə }

enterolithic [GEOL] Of or pertaining to structures, such as small folds, formed in evaporites due to flowage or hydration. { ,ent·ə·rə'lith·ik }

Entisol [GEOL] An order of soil having few or faint horizons. { 'ent·ə,sȯl }

Entomoconchacea [PALEON] A superfamily of extinct marine ostracods in the suborder Myodocopa that are without a rostrum above the permanent aperture. { ,ent·ə·mō,käŋ'kās·ē·ə }

entrail pahoehoe [GEOL] A type of pahoehoe having a surface that resembles an intertwined mass of entrails. { 'en,trāl pə'hō·ē,hō·ē }

entrapment [GEOL] The underground trapping of oil or gas reserves by folds, faults, domes, asphaltic seals, unconformities, and such. { en'trap·mənt }

environment of sedimentation [GEOL] A more or less destructive geomorphologic setting in which sediments are deposited as beach environment. { in¦vī·ərn¦mənt əv ,sed·ə·men'tā·shən }

Eocambrian [GEOL] Pertaining to the thick sequences of strata conformably underlying Lower Cambrian fossils. Also known as Infracambrian. { ,ē·ō'kam·brē·ən }

Eocene [GEOL] The next to the oldest of the five major epochs of the Tertiary period (in the Cenozoic era). { 'ē·ə,sēn }

Eocrinoidea [PALEON] A class of extinct echinoderms in the subphylum Crinozoa that had biserial brachioles like those of cystoids combined with a theca like that of crinoids. { ,ē·ō·krə'nȯid·ē·ə }

Eogene See Paleogene. { 'ē·ə,jēn }

Eohippus [PALEON] The earliest, primitive horse, included in the genus *Hyracotherium*; described as a small, four-toed species. { ¸ē·ō'hip·əs }

eolation [GEOL] Any action of wind on the land. { ¸ē·ə'lā·shən }

eolian dune [GEOL] A dune resulting from entrainment of grains by the flow of moving air. { ē'ōl·yən 'dün }

eolian erosion [GEOL] Erosion due to the action of wind. { ē'ōl·yən ə'rō·zhən }

eolianite [GEOL] A sedimentary rock consisting of clastic material which has been deposited by wind. { ē'ōl·yə¸nīt }

eolian ripple mark [GEOL] A mark made in sand by the wind. { ē'ōl·yən 'rip·əl ¸märk }

eolian sand [GEOL] Deposits of sand arranged by the wind. { ē'ōl·yən 'sand }

eolian soil [GEOL] A type of soil ranging from sand dunes to loess deposits whose particles are predominantly of silt size. { ē'ōl·yən 'soil }

Eomoropidae [PALEON] A family of extinct perissodactyl mammals in the superfamily Chalicotherioidea. { ¸ē·ō·mə'räp·ə¸dē }

eonothem [GEOL] A chronostratigraphic unit, above erathem, composed of rocks formed during an eon of geologic time. { 'ēn·ə¸them }

eosphorite [MINERAL] $(Mn,Fe)Al(PO_4)(OH)_2 \cdot H_2O$ A usually rose-pink mineral composed of hydrous aluminum manganese phosphate, found massive or in prismatic crystals. { ē'äs·ə·fə¸rīt }

Eosuchia [PALEON] The oldest, most primitive, and only extinct order of lepidosaurian reptiles. { ¸ē·ō'sü·kē·ə }

eötvös [GEOPHYS] A unit of horizontal gradient of gravitational acceleration, equal to a change in gravitational acceleration of 10^{-9} galileo over a horizontal distance of 1 centimeter. { 'ət·vəsh }

epeirogeny [GEOL] Movements which affect large tracts of the earth's crust. { ¸e¸pī'räj·ə·nē }

ephemeral gully [GEOL] A channel that forms in a cultivated field when precipitation exceeds the rate of soil infiltration. { ə¦fem·ə·rəl ¦gəl·ē }

epicenter [GEOL] A point on the surface of the earth which is directly above the seismic focus of an earthquake and where the earthquake vibrations reach first. { 'ep·ə¸sen·tər }

epiclastic [GEOL] Pertaining to the texture of mechanically deposited sediments consisting of detrital material from preexistent rocks. { ¦ep·ə'klas·tik }

epicontinental [GEOL] Located upon a continental plateau or platform. { ¦ep·ə¸kant·ən'ent·əl }

epidiorite [PETR] A dioritic rock formed by alteration of pyroxenic igneous rocks. { ¦ep·ə'dī·ə¸rīt }

epidosite [PETR] A rare metamorphic rock composed of epidote and quartz. { ¸ep·ə'dō¸sīt }

epidote [MINERAL] A pistachio-green to blackish-green calcium aluminum sorosilicate mineral that crystallizes in the monoclinic system; the luster is vitreous, hardness is $6^1/_2$ on Mohs scale, and specific gravity is 3.35–3.45. { 'ep·ə¸dōt }

epidote-amphibolite facies [PETR] Metamorphic rocks formed under pressures of 3000–7000 bars and temperatures of 250–450°C with conditions intermediate between those that formed greenschist and amphibolite, or with characteristics intermediate. { ¦ep·ə¸dōt am'fib·ə¸līt ¸fā·shēz }

epidotization [GEOL] The introduction of epidote into, or the formation of epidote from, rocks. { ¸ep·ə¸dōd·ə'zā·shən }

epieugeosyncline [GEOL] Deep troughs formed by subsidence which have limited volcanic power and overlie a eugeosyncline. { ¦ep·ē¸yü¸jē·ō'sin¸klīn }

epigene [GEOL] **1.** A geologic process originating at or near the earth's surface. **2.** A structure formed at or near the earth's surface. { 'ep·ə¸jēn }

epigenesis [GEOL] Alteration of the mineral content of rock due to outside influences. { ¸ep·ə'jen·ə·səs }

epigenetic [GEOL] Produced or formed at or near the surface of the earth. { ¦ep·ə·jə¦ned·ik }

epigenite [MINERAL] $(Cu,Fe)_5AsS_6$ A steel gray, orthorhombic mineral consisting of copper and iron arsenic sulfide. { ə'pij·ə,nīt }

epimagma [GEOL] A gas-free, vesicular to semisolid magmatic residue of pasty consistency formed by cooling and loss of gas from liquid lava in a lava lake. { ,ep·ə'mag·ma }

epimagmatic *See* deuteric. { ,ep·ə·mag'mad·ik }

episode [GEOL] A distinctive event or series of events in the geologic history of a region or feature. { 'ep·ə,sōd }

epistilbite [MINERAL] $CaAl_2Si_6O_{16}·5H_2O$ A mineral of the zeolite family that contains calcium and aluminosilicate and crystallizes in the monoclinic system; occurs in white prismatic crystals or granular forms. { |ep·ə'stil,bīt }

epithermal [GEOL] Pertaining to mineral veins and ore deposits formed from warm waters at shallow depth, at temperatures ranging from 50–200°C, and generally at some distance from the magmatic source. { |ep·ə'thər·məl }

epithermal deposit [GEOL] Ore deposit formed in and along openings in rocks by deposition at shallow depths from ascending hot solutions. { |ep·ə'thər·məldə'päz·ət }

epizone [GEOL] **1.** The zone of metamorphism characterized by moderate temperature, low hydrostatic pressure, and powerful stress. **2.** The outer depth zone of metamorphic rocks. { 'ep·ə,zōn }

epoch [GEOL] A major subdivision of a period of geologic time. { 'ep·ək }

epsomite [MINERAL] $MgSO_4·7H_2O$ A mineral that occurs in clear, needlelike, orthorhombic crystals; commonly, it is massive or fibrous; luster varies from vitreous to milky, hardness is 2–2.5 on Mohs scale, and specific gravity is 1.68; it has a salty bitter taste and is soluble in water. Also known as epsom salt. { 'ep·sə,mīt }

epsom salt *See* epsomite. { 'ep·səm ,sȯlt }

equatorial electrojet [GEOPHYS] A concentration of electric current in the atmosphere found in the magnetic equator. { ,e·kwə'tȯr·ē·əl ə'lek·trə,jet }

equigranular [PETR] Pertaining to the texture of rocks whose essential minerals are all of the same order of size. { |ē·kwə'gran·yə·lər }

equilibrium profile *See* profile of equilibrium. { ,ē·kwə'lib·rē·əm 'prō,fīl }

equiphase zone [GEOPHYS] That region in space where the difference in phase of two radio signals is indistinguishable. { 'e·kwə,fāz ,zōn }

equipotential surface [GEOPHYS] A surface characterized by the potential being constant everywhere on it for the attractive forces concerned. { |e·kwə·pə'ten·chəl 'sər·fəs }

equivalent diameter *See* nominal diameter. { i'kwiv·ə·lənt dī'am·əd·ər }

equivoluminal wave *See* S wave. { |e·kwə·və|lüm·ə·nəl 'wāv }

era [GEOL] A unit of geologic time constituting a subdivision of an eon and comprising one or more periods. { 'ir·ə }

erathem [GEOL] A chronostratigraphic unit, below eonothem and above system, composed of rocks formed during an era of geologic time. { 'er·ə,them }

Erian [GEOL] Middle Devonian geologic time; a North American provincial series. { 'i·rē·ən }

Erian orogeny [GEOL] One of the orogenies during Phanerozoic geologic time, at the end of the Silurian; the last part of the Caledonian orogenic era. Also known as Hibernian orogeny. { 'i·rē·ən ȯ'räj·ə·nē }

erikite [MINERAL] A brown mineral consisting of a silicate and phosphate of cerium metals; occurs in orthorhombic crystals. { 'er·ə,kīt }

erinite [MINERAL] $Cu_5(OH)_4(AsO_4)_2$ Emerald-green mineral composed of basic copper arsenate. { 'er·ə,nīt }

erionite [MINERAL] A chabazite mineral of the zeolite family that contains calcium ions and crystallizes in the hexagonal system. { 'er·ē·ə,nīt }

eroding velocity [GEOL] The minimum average velocity required for eroding homogeneous material of a given particle size. { ə'rōd·iŋ və'läs·əd·ē }

erosion [GEOL] **1.** The loosening and transportation of rock debris at the earth's surface. **2.** The wearing away of the land, chiefly by rain and running water. { ə'rō·zhən }

111

erosional unconformity |GEOL| The surface that separates older, eroded rocks from younger, overlying sediments. { ə'rō·zhən·əl ,ən·kən'for·məd·ē }

erosion cycle |GEOL| A postulated sequence of conditions through which a new landmass proceeds as it wears down, classically the concept of youth, maturity, and old age, as stated by W.M. Davis; an original landmass is uplifted above base level, cut by canyons, gradually converted into steep hills and wide valleys, and is finally reduced to a flat lowland at or near base level. { ə'rō·zhən ,sī·kəl }

erosion pavement |GEOL| A layer of pebbles and small rocks that prevents the soil underneath from eroding. { ə'rō·zhən ,pāv·mənt }

erosion platform See wave-cut platform. { ə'rō·zhən ,plat,form }

erosion surface |GEOL| A land surface shaped by agents of erosion. { ə'rō·zhən ,sər·fəs }

erratic |GEOL| A rock fragment that has been transported a great distance, generally by glacier ice or floating ice, and differs from the bedrock on which it rests. { ə'rad·ik }

eruption |GEOL| The ejection of solid, liquid, or gaseous material from a volcano. { i'rəp·shən }

eruptive rock |PETR| 1. Rock formed from a volcanic eruption. 2. Igneous rock that reaches the earth's surface in a molten condition. { ə'rəp·tiv 'räk }

erythrine See erythrite. { 'er·ə,thrēn }

erythrite |MINERAL| $Co_3(AsO_4)_2 \cdot 8H_2O$ A crimson, peach, or pink-red secondary oxidized cobalt mineral that occurs in monoclinic crystals, in globular and reniform masses, or in earthy forms. Also known as cobalt bloom; cobalt ocher; erythrine; peachblossom ore; red cobalt. { 'er·ə,thrīt }

erythrosiderite |MINERAL| $K_2FeCl_5 \cdot H_2O$ Mineral composed of hydrous potassium iron chloride; occurs in lavas. { ə|rith·rə'sid·ə,rīt }

Erzgebirgian orogeny |GEOL| Diastrophism of the early Late Carboniferous. { 'erts·gə,bər·jən ò'räj·ə·nē }

escar See esker. { 'es·kər }

escarpment |GEOL| A cliff or steep slope of some extent, generally separating two level or gently sloping areas, and produced by erosion or by faulting. Also known as scarp. { ə'skärp·mənt }

eschar See esker. { 'es·kər }

eschwegeite See tanteuxenite. { ,esh'vä·gē,īt }

eschynite |MINERAL| $(Ce,Ca,Fe,Th)(Ti,Cb)_2O_6$ A black mineral, occurring in prismatic crystals; a rare oxide of cesium, titanium, and other metals, which is isomorphous with priorite. { 'es·kə,nīt }

eskar See esker. { 'es·kər }

eskebornite |MINERAL| $CuFeSe_2$ The selenium analog of the mineral pyrrhotite ($Fe_{1-x}S$). { ,es·kə'bȯr,nīt }

esker |GEOL| A sinuous ridge of constructional form, consisting of stratified accumulations, glacial sand, and gravel. Also known as asar; eschar; eskar; osar; serpent kame. { 'es·kər }

essexite |PETR| A rock of igneous origin composed principally of plagioclase hornblende, biotite, and titanaugite. { 'e·sik,sīt }

estuarine deposit |GEOL| A sediment deposited at the heads and floors of estuaries. { 'es·chə·wə,rēn də'päz·ət }

etch figures |MINERAL| A minute pit produced by a solvent on the crystal face of a mineral which reveals its molecular structure. { 'ech ,fig·yərz }

ethmolith |GEOL| A downward tapering, funnel-shaped, discordant intrusion of igneous rocks. { 'eth·mə,lith }

ettringite |MINERAL| $Ca_6Al_2(SO_4)_3(OH)_{12} \cdot 26H_2O$ A mineral composed of hydrous basic calcium and aluminum sulfate. { 'e·triŋ,īt }

eucairite |MINERAL| CuAgSe A white, native selenide that crystallizes in the isometric crystal system. { yü'kī,rīt }

euchlorin |MINERAL| $(K,Na)_8Cu_9(SO_4)_{10}(OH)_6$ An emerald-green mineral consisting of a basic sulfate of potassium, sodium, and copper; found in lava at Vesuvius. { yü'klȯr·ən }

euchroite [MINERAL] $Cu_2(AsO_4)(OH)\cdot 3H_2O$ An emerald green or leek green, orthorhombic mineral consisting of a hydrated basic copper arsenate. { 'yü·krō,īt }

euclase [MINERAL] $BeAlSiO_4(OH)$ A brittle, pale green, blue, yellow, or violet monoclinic mineral, occurring as prismatic crystals. { 'yü,klās }

eucrite [MINERAL] An olivine-bearing gabbro containing unusually calcic plagiocase; a meteorite component { 'yü,krīt }

eucryptite [MINERAL] $LiAlSiO_4$ A colorless or white lithium aluminum silicate mineral, crystallizing in the hexagonal system; specific gravity is 2.67. { yü'krip,tīt }

eudialite [MINERAL] $(Na,Ca,Fe)_6ZrSi_6O_{18}(OH,Cl)$ Hexagonal-crystalline silicate chloride mineral; color is red to brown. { yü'dī·ə,līt }

eudidymite [MINERAL] $NaBeSi_3O_7(OH)$ A glassy white mineral composed of sodium beryllium silicate. { yü'did·ə,mīt }

eugeosyncline [GEOL] The internal volcanic belt of an orthogeosyncline. { yü,jē·ō'sin,klīn }

euhedral See automorphic. { yü'hē·drəl }

eulytine See eulytite. { 'yü·lə,tēn }

eulytite [MINERAL] $Bi_4Si_3O_{12}$ A bismuth silicate mineral usually found as minute dark-brown or gray tetrahedral crystals; specific gravity is 6.11. Also known as agricolite; bismuth blende; eulytine. { 'yü·lə,tīt }

Euomphalacea [PALEON] A superfamily of extinct gastropod mollusks in the order Aspidobranchia characterized by shells with low spires, some approaching bivalve symmetry. { yü,äm·fə'lās·ē·ə }

eupelagic See pelagic. { yü·pə'laj·ik }

Euproopacea [PALEON] A group of Paleozoic horseshoe crabs belonging to the Limulida. { yü,prō·ə'pās·ē·ə }

Euramerica [GEOL] The continent that was composed of Europe and North America during most of the Mesozoic Era. { ,yür·ə'mer·ə·kə }

Euryapsida [PALEON] A subclass of fossil reptiles distinguished by an upper temporal opening on each side of the skull. { yür·ē'ap·sə·də }

Eurychilinidae [PALEON] A family of extinct dimorphic ostracodes in the superfamily Hollinacea. { ,yür·ə·kə'lin·ə,dē }

Eurymylidae [PALEON] A family of extinct mammals presumed to be the ancestral stock of the order Lagomorpha. { ,yür·ə'mil·ə,dē }

Eurypterida [PALEON] A group of extinct aquatic arthropods in the subphylum Chelicerata having elongate-lanceolate bodies encased in a chitinous exoskeleton. { ,yür·əp'ter·ə·də }

eutaxite [PETR] A rock exhibiting eutaxitic structure. { yü'tak,sīt }

eutaxitic [PETR] Referring to the banded appearance in certain extrusive rocks, resulting from the layering of different textures, materials, or colors. { 'yü·tak'sid·ik }

eutectofelsite See eutectophyre. { yü'tek·tō'fel,sīt }

eutectophyre [PETR] A light-colored tufflike igneous rock exhibiting a network of interlocking quartz and orthoclase crystals. Also known as eutectofelsite. { yü'tek·tə,fīr }

Euthacanthidae [PALEON] A family of extinct acanthodian fishes in the order Climatiiformes. { ,yü·thə'kan·thə,dē }

euxenite [MINERAL] A brownish-black rare-earth mineral that crystallizes in the orthorhombic system, contains oxide of calcium, cerium, columbium, tantalum, titanium, and uranium, and has a metallic luster; hardness is 6.5 on Mohs scale, and specific gravity is 4.7–5.0. { 'yük·sə,nīt }

evansite [MINERAL] $Al_3(PO_4)(OH)_6\cdot 6H_2O$ A colorless to milky white mineral consisting of a hydrated basic aluminum phosphate; occurs in massive form and as stalactites. { 'ev·ən,zīt }

evaporite [GEOL] Deposits of mineral salts from sea water or salt lakes due to evaporation of the water. { i'vap·ə,rīt }

event [GEOL] An incident of probable tectonic significance, but whose full implications are unknown. { i'vent }

113

evjite |PETR| A gabbro of hornblende in which the only light-colored mineral is labradorite or bytownite; hornblende must be primary, not uralitic. { 'ev,yīt }

evorsion |GEOL| The process of pothole formation in riverbeds; plays an important role in denudation. { ē'vȯr·shən }

evorsion hollow See pothole. { ē'vȯr·shən ,häl·ō }

exchange capacity |GEOL| The ability of a soil material to participate in ion exchange as measured by the quantity of exchangeable ions in a given unit of the material. { iks'chānj kə,pas·əd·ē }

exfoliation |GEOL| See sheeting. |PETR| The breaking off of thin concentric shells, sheets, scales, plates, and so on, from a rock mass; measuring less than a centimeter to several meters in thickness, the loosened rock is spalled, peeled, or stripped. { eks,fō·lē'ā·shən }

exfoliation dome |GEOL| A large rounded dome-shaped structure produced in massive homogeneous coarse-grained rocks (usually igneous) by exfoliation. { eks,fō·lē'ā·shən ,dōm }

exfoliation joint See sheeting structure. { eks,fō·lē'ā·shən ,jȯint }

exhalation |GEOPHYS| The process by which radioactive gases escape from the surface layers of soil or loose rock, where they are formed by decay of radioactive salts. { ,eks·ə'lā·shən }

exhumation |GEOL| The uncovering or exposure through erosion of a former surface, landscape, or feature that had been buried by subsequent deposition. { ,eks·yü'mā·shən }

exhumed See resurrected. { ig'zyümd }

exinite |GEOL| A hydrogen-rich maceral group consisting of spore exines, cuticular matter, resins, and waxes; includes sporinite, cutinite, alginite, and resinite. Also known as liptinite. { 'ek·sə,nīt }

exocline |GEOL| An inverted anticline or syncline. { 'ek·sə,klīn }

exogenous inclusion See xenolith. { ,ek'säj·ə·nəs in'klü·zhən }

exogeosyncline |GEOL| A parageosyncline that lies along the cratonal border and obtains its clastic sediments from erosion of the adjacent orthogeosynclinal belt outside the craton. Also known as delta geosyncline; foredeep; transverse basin. { ¦ek·sō,jē·ō'sin,klīn }

exomorphic zone See aureole. { ¦ek·sə¦mȯr·fik ,zōn }

exomorphism |PETR| A change in a rock mass caused by intrusion of external igneous material; in the usual sense, contact metamorphism. { ,ek·sə'mȯr,fiz·əm }

exorheic |GEOL| Referring to a basin or region characterized by external drainage. { ek·sə'rē·ik }

expansion fissures |GEOL| A system of fissures which radiate randomly and pass through feldspars and other minerals adjacent to olivine crystals that have been replaced by serpentine. { ik'span·shən ,fish·ərz }

expansion joint See sheeting structure. { ik'span·shən ,jȯint }

experimental petrology |PETR| A branch of petrology in which phenomena that occur during petrological processes are reproduced and studied in the laboratory. { ik,sper·ə'ment·əl pə'träl·ə·jē }

explosion breccia |PETR| Breccia resulting from volcanic eruption or a phreatic explosion. { ik'splō·zhən,brech·ə }

explosion crater |GEOL| A volcanic crater formed by explosion and commonly developed along rift zones on the flanks of large volcanoes. { ik'splō·zhən ,krād·ər }

explosion tuff |GEOL| A tuff whose constituent ash particles are in the place they fell after being ejected from a volcanic vent. { ik'splō·zhən ,təf }

explosive index |GEOL| The percentage of pyroclastics in the material from a volcanic eruption. { ik'splō·siv 'in,deks }

exsolution |GEOL| A phenomenon during which molten rock solutions separate when cooled. { ¦ek·sə'lü·shən }

exsolution lamellae |GEOL| Layers of sedimentary rock that solidify from solution by either precipitation or secretion. { ¦ek·sə'lü·shən lə'mel·ē }

extended valley [GEOL] **1.** A valley that is lengthened downstream either by a regression of the sea or by uplift of the coastal region. **2.** A valley formed by or containing an extended stream. { ik¦stend·əd 'val·ē }

extensional fault See tension fault. { ik'sten·chən·əl 'fȯlt }

extension fracture [GEOL] A fracture that develops perpendicular to the direction of greatest stress and parallel to the direction of compression. { ik'sten·chən ‚frak·chər }

extension joints [GEOL] Fractures that form parallel to a compressive force. { ik'sten·chən ‚jȯins }

extravasation [GEOL] The eruption of lava from a vent in the earth. { ik‚strav·ə'sā·shən }

extrusion [GEOL] Emission of magma or magmatic materials at the surface of the earth. { ek'strü·zhən }

extrusive rock See volcanic rock. { ik'strü·siv 'räk }

exudation vein See segregated vein. { ‚ek·syə'dā·shən‚vān }

eye coal [GEOL] Coal characterized by small, circular or elliptic structural disks that reflect light and are arranged in parallel planes either in or normal to the bedding. Also known as augen kohle; circular coal. { 'ī ‚kōl }

F

fabric |GEOL| The spatial orientation of the elements of a sedimentary rock. |PETR| The sum of all the structural and textural features of a rock. Also known as petrofabric; rock fabric; structural fabric. { 'fab·rik }

fabric analysis See structural petrology. { 'fab·rik ə,nal·əs· əs }

fabric diagram |PETR| In structural petrology, a graphic representation of the data of fabric elements. Also known as petrofabric diagram. { 'fab·rik 'dī·ə,gram }

fabric domain |PETR| A three-dimensional area or volume of uniform rock fabric delineated by boundaries such as structural or compositional discontinuities. { 'fab· rik də'mān }

fabric element |PETR| A surface or line of structural discontinuity in a rock fabric. { 'fab·rik 'el·ə·mənt }

face |GEOL| **1.** The main surface of a landform. **2.** The original surface of a layer of rock. { fās }

facellite See kaliophilite. { fə'se,līt }

faceted pebble |GEOL| A pebble with three or more faces naturally worn flat and meeting at sharp angles. { 'fas·əd·əd 'peb·əl }

faceted spur |GEOL| A spur or ridge with an inverted-V face resulting from faulting or from the trimming, beveling, or truncating motion of streams, waves, or glaciers. { 'fas·əd·əd 'spər }

facies |GEOL| Any observable attribute or attributes of a rock or stratigraphic unit, such as overall appearance or composition, of one part of the rock or unit as contrasted with other parts of the same rock or unit. { 'fā·shēz }

facies map |GEOL| A stratigraphic map indicating distribution of sedimentary facies within a specific geologic unit. { 'fā· shēz ,map }

fahlband |GEOL| A stratum containing metal sulfides; occurs in crystalline rock. { 'fäl,bänt }

fahlore See tetrahedrite. { 'tä,lor }

fairchildite |MINERAL| $K_2Ca(CO_3)_2$ A mineral composed of potassium calcium carbonate; occurs in partly burned trees. { 'fer,chīl,dīt }

fairfieldite |MINERAL| $Ca_2Mn(PO_4)_2·2H_2O$ A white or pale-yellow mineral composed of hydrous calcium manganese phosphate and occurring in foliated or fibrous form. { 'fer,fēl,dīt }

fairy stone See staurolite. { 'fer·ē stōn }

fallback |GEOL| Fragmented ejecta from an impact or explosion crater during formation which partly refills the true crater almost immediately. { 'fól,bak }

fall line |GEOL| **1.** The zone or boundary between resistant rocks of older land and weaker strata of plains. **2.** The line indicated by the edge over which a waterway suddenly descends, as in waterfalls. { 'fól ,līn }

false bedding |GEOL| An inclined bedding produced by currents. { ¦fóls 'bed·iŋ }

false cleavage |GEOL| **1.** A weak cleavage at an angle to the slaty cleavage. **2.** Spaced surfaces about a millimeter apart along which a rock splits. { ¦fóls 'klēv·ij }

false drumlin See rock drumlin. { ¦fóls 'drəm·lən }

false form See pseudomorph. { ¦fóls 'fórm }

false galena See sphalerite. { ¦fóls gə'lē·nə }

false lapis See lazulite. { ¦fóls 'lap·əs }

false oolith *See* pseudo-oolith. { ¦fóls 'ō,ō,līth }

false topaz *See* citrine. { ¦fóls 'tō,paz }

famatinite [MINERAL] Cu_3SbS_4 A reddish-gray mineral composed of copper antimony sulfide. { ,fam·ə'tē,nīt }

fan [GEOL] A gently sloping, fan-shaped feature usually found near the lower termination of a canyon. { fan }

fan fold [GEOL] A fold of strata in which both limbs are overturned, forming a syncline or anticline. { 'fan ,fōld }

fanglomerate [GEOL] Coarse material in an alluvial fan, with the rock fragments being only slightly worn. { fan'gläm·ə·rət }

fan-shaped delta *See* arcuate delta. { 'fan ,shapt 'del·tə }

farinaceous [GEOL] Of a rock or sediment, having a texture that is mealy, soft, and friable, for example, a limestone or a pelagic ooze. { ¦far·ə¦nā·shəs }

farringtonite [MINERAL] $Mg_3(PO_4)_2$ A colorless, wax-white, or yellow phosphate mineral known only in meteorites. { 'far·iŋ·tə,nīt }

fassaite [GEOCHEM] $Ca(Mg,Ti,Al)(Al,Si)_2O_6$ A mineral found in the millimeter-sized rocklets or refractory inclusions of carbonaceous chondrite meteorites. { 'fas·ə,yīt }

faujasite [MINERAL] $(Na_2,Ca)Al_2Si_4O_{12}·6H_2$ O Zeolite mineral of the sodalite group, crystallizing in the cubic system. { 'fō·zhə,sīt }

fault [GEOL] A fracture in rock along which the adjacent rock surfaces are differentially displaced. { fólt }

fault basin [GEOL] A region depressed in relation to surrounding regions and separated from them by faults. { 'fólt ,bās·ən }

fault block [GEOL] A rock mass that is bounded by faults; the faults may be elevated or depressed and not necessarily the same on all sides. { 'fólt ,bläk }

fault-block mountain *See* block mountain. { 'fólt ,bläk ,maúnt·ən }

fault breccia [GEOL] The assembly of angular fragments found frequently along faults. Also known as dislocation breccia. { 'fólt ,brech·ə }

fault cliff *See* fault scarp. { 'fólt ,klif }

fault escarpment *See* fault scarp. { 'fólt e,skärp·mənt }

faulting [GEOL] The fracturing and displacement processes which produce a fault. { 'fól·tiŋ }

fault ledge *See* fault scarp. { 'fólt ,lej }

fault line [GEOL] Intersection of the fault surface with the surface of the earth or any other horizontal surface of reference. Also known as fault trace. { 'fólt,līn }

fault-line scarp [GEOL] A cliff produced when a soft rock erodes against hard rock at a fault. { 'fólt,līn ,skärp }

fault plane [GEOL] A planar fault surface. { 'fólt ,plān }

fault rock [GEOL] A rock often found along a fault plane and made up of fragments formed by the crushing and grinding which accompany a dislocation. { 'fólt ,räk }

fault scarp [GEOL] A steep cliff formed by movement along one side of a fault. Also known as cliff of displacement; fault cliff; fault escarpment; fault ledge. { 'fólt ,skärp }

fault separation [GEOL] Apparent displacement of a fault measured on the basis of disrupted linear features. { 'fólt ,sep·ə,rā·shən }

fault strike [GEOL] The angular direction, with respect to north, of the intersection of the fault surface with a horizontal plane. { 'fólt ,strīk }

fault system [GEOL] Two or more fault sets which interconnect. { 'fólt ,sis·təm }

fault terrace [GEOL] A step on a slope, produced by displacement of two parallel faults. { 'fólt ,ter·əs }

fault throw [GEOL] The amount of vertical displacement of rocks due to faulting. { 'fólt ,thrō }

fault trace *See* fault line. { 'fólt ,trās }

fault trap [GEOL] Oil or gas reservoir formed by a structural trap limited in one or more directions by subterranean geological faulting. { 'fólt ,trap }

fault-trough lake *See* sag pond. { 'fólt ,tróf ,lāk }

fault vein [GEOL] A mineral vein deposited in a fault fissure. { 'fólt ,vān }

fault wall [GEOL] The mass of rock on a particular side of a fault. { 'fólt ,wól }

fault zone |GEOL| A fault expressed as an area of numerous small fractures. Also known as distributed fault. { 'fȯlt ˌzōn }

faunizone |GEOL| A bed characterized by fossils of a particular assemblage of fauna. { 'fȯn·ə,zōn }

faunule |PALEON| The localized stratigraphic and geographic distribution of a particular taxon. { 'fȯˌnyül }

Favositidae |PALEON| A family of extinct Paleozoic corals in the order Tabulata. { ˌfav ə'sid·ə,dē }

fayalite |MINERAL| Fe$_2$SiO$_4$ A brown to black mineral of the olivine group, consisting of iron silicate and found either massive or in crystals; specific gravity is 4.1. { fə'yä,līt }

feather alum See alunogen; halotrichite. { 'feth·ər ,al·əm }

feather joint |GEOL| One of a series of joints in a fault zone formed by shear and tension. Also known as pinnate joint. { 'feth·ər ,jȯint }

feather ore See jamesonite. { 'feth·ər ,ȯr }

fecal pellets |GEOL| Mainly the excreta of invertebrates occurring in marine deposits and as fossils in sedimentary rocks. Also known as castings. { 'fē·kəl 'pel·əts }

feeder |GEOL| A small ore-bearing vein which merges with a larger one. { 'fēd·ər }

feeder beach |GEOL| A beach that is artificially widened and nourishes downdrift beaches by natural littoral currents or forces. { 'fēd·ər ,bēch }

feldspar |MINERAL| A group of silicate minerals that make up about 60% of the outer 9 miles (15 kilometers) of the earth's crust; they are silicates of aluminum with the metals potassium, sodium, and calcium, and rarely, barium. { 'fel,spär }

feldspathic graywacke |PETR| Sandstone containing less than 75% quartz and chert and 15–75% detrital clay matrix, and having feldspar grains in greater abundance than rock fragments. Also known as arkosic wacke; high-rank graywacke. { fel'spath·ik 'grā,wak·ə }

feldspathic sandstone |PETR| Sandstone rich in feldspar; intermediate in composition between arkosic sandstone and quartz sandstone, made up of 10–25% feldspar and less than 20% matrix material. { fel'spath·ik 'san,stōn }

feldspathic shale |PETR| A well-laminated shale with more than 10% feldspar in the silt size and with a finer matrix of kaolinitic clay minerals. { fel'spath·ik 'shāl }

feldspathization |GEOL| Formation of feldspar in a rock usually as a result of metamorphism leading toward granitization. { ˌfel,spa·thə'zā·shən }

feldspathoid |GEOL| Aluminosilicates of sodium, potassium, or calcium that are similar in composition to feldspars but contain less silica than the corresponding feldspar. { 'fel,spa,thȯid }

felsenmeer |GEOL| A flat or gently sloping veneer of angular rock fragments occurring on moderate mountain slopes above the timber line. { 'felz·ən,mer }

felsic |MINERAL| A light-colored mineral. |PETR| Of an igneous rock, having a mode containing light-colored minerals. { 'fel·sik }

felsite |PETR| 1. A light-colored, fine-grained igneous rock composed chiefly of quartz or feldspar. 2. A rock characterized by felsitic texture. { 'fel,sīt }

felsöbányaite |MINERAL| Al$_4$(SO$_4$)(OH)$_{10}$·5H$_2$O A yellow to white, probably orthorhombic mineral consisting of a hydrated basic sulfate of aluminum; occurs as aggregates of lamellar crystals. { ˌfel·sō'ban·yə,īt }

felsophyric See aphaniphyric. { ˌfel·sə,fir·ik }

felty |GEOL| Referring to a pilotaxitic texture in which the microlites are randomly oriented. { 'fel·tē }

Fenestellidae |PALEON| A family of extinct fenestrated, cryptostomatous bryozoans which abounded during the Silurian. { ˌfen·ə'stel·ə,dē }

fen peat See low-moor peat. { 'fen ,pēt }

fenster See window. { 'fen·stər }

ferberite |MINERAL| FeNO$_4$ A black mineral of the wolframite solid-solution series occurring as monoclinic, prismatic crystals and having a submetallic luster; hardness is 4.5 on Mohs scale, and specific gravity is 7.5. { 'fər·bə,rīt }

ferghanite |MINERAL| U$_3$(VO$_4$)$_2$·6H$_2$O Sulfur-yellow mineral composed of hydrated uranium vanadate, occurring in scales. { fər'gä,nīt }

119

fergusonite |MINERAL| $Y_2O_3 \cdot (Nb,Ta)_2O_5$ Brownish-black rare-earth mineral with a tetragonal crystal form; it is isomorphous with formanite. { 'fər·gə·sə,nīt }

fermorite |MINERAL| $(Ca,Sr)_5|(As,P)O_4]_3$ A white mineral composed of arsenate, phosphate, and fluoride of calcium and strontium, occurring in crystalline masses. { 'fər·mə,rīt }

fernandinite |MINERAL| A dull green mineral composed of hydrous calcium vanadyl vanadate. { ,fər·nən'dē,nīt }

ferriamphibole |MINERAL| The ferric ion equivalent of the amphibole group of minerals. { ,fer·ē'am·fə,bōl }

ferricrete |GEOL| A conglomerate of surficial sand and gravel held together by iron oxide resulting from percolating solutions of iron salts. { 'fer·ə,krēt }

ferrierite |MINERAL| $(Na,K)_2MgAl_3Si_{15}O_{36}(OH) \cdot 9H_2O$ A zeolite mineral crystallizing in the orthorhombic system. { fə'rē·ə,rīt }

ferriferous |GEOL| Of a sedimentary rock, iron-rich. |MINERAL| Of a mineral, iron-bearing. { fə'rif·ə·rəs }

ferrimolybdite |MINERAL| $Fe_2(MoO_4)_3 \cdot 8H_2O$ A colorless to canary yellow, probably orthorhombic mineral consisting of hydrated ferric molybdate; occurs in massive form, as crusts or aggregates. { ¦fe·ri·mə'lib,dīt }

ferrinatrite |MINERAL| $Na_3Fe(SO_4)_3 \cdot 3H_2O$ A greenish or white mineral composed of sodium ferric iron double sulfate; usually occurs in spherical forms. { ,fe·ri'nā,trīt }

ferrisicklerite |MINERAL| $(Li,Fe,Mn)(PO_4)$ Mineral composed of phosphate of lithium, ferric iron, and manganese, more iron being present than manganese; it is isomorphous with sicklerite. { ¦fe·ri'sik·lə,rīt }

ferrite |PETR| Grains or scales of unidentifiable, generally transparent amorphous iron oxide in the matrix of a porphyritic rock. { 'fe,rīt }

ferritremolite |MINERAL| The ferric ion equivalent of the monoclinic amphibole, tremolite. { ¦fe·ri'trem·ə,līt }

ferritungstite |MINERAL| $Fe_2(WO_4)(OH)_4 \cdot 4H_2O$ A yellow ocher mineral composed of hydrous ferric tungstate, occurring as a powder. { ¦fer·ri'təŋ·,stīt }

ferroamphibole |MINERAL| The ferrous iron equivalent of the amphibole group of minerals. { ¦fe·rō'am·fə,bōl }

ferroan dolomite |MINERAL| A species of ankerite having less than 20% of the manganese positions occupied by iron. { 'fer·ə·wən 'dōl,mīt }

ferroaugite |MINERAL| A form of monoclinic pyroxene. { ¦fe·rō'ȯ,gīt }

Ferrod |GEOL| A suborder of the soil order Spodosol that is well drained and contains an iron accumulation with little organic matter. { 'fe,räd }

ferrodolomite |MINERAL| $CaFe(CO_3)_2$ A mineral composed of calcium iron carbonate, isomorphous with dolomite, and occurring in ankerite. { ¦fe·rō'dō·lə,mīt }

ferrogabbro |PETR| A gabbro rock in which the pyroxene and olivine constituents have an unusually high iron content. { ¦fe·rō'ga·brō }

ferrosilite |MINERAL| A mineral in the orthopyroxene group; the iron analog of enstatite; occurs in hypersthene, but is not found separately in nature. { 'fe·rō'si,līt }

ferrospinel See hercynite. { ¦fe·rō·spə'nel }

ferrotremolite |MINERAL| The ferrous iron equivalent of the monoclinic amphibole, tremolite. { ¦fe·rō'tre·mə,līt }

ferruccite |MINERAL| $NaBF_4$ An orthorhombic boron mineral consisting of sodium fluoborate. { fə'rü,chīt }

fersmanite |MINERAL| $(Na,Ca)_2(Ti,Cb)Si(O,F)_6$ A brown mineral composed of a silicate fluoride of sodium, calcium, titanium, and columbium. { 'fərz·mə,nīt }

fersmite |MINERAL| $(Ca,Ce)(Cb,Ti)_2(O,F)_6$ A black mineral composed of an oxide and fluoride of calcium and columbium with cerium and titanium. { 'fərz,mīt }

fervanite |MINERAL| $Fe_4V_4O_{16} \cdot 5H_2O$ Golden-brown mineral composed of a hydrated iron vanadate; although itself not radioactive, it occurs with radioactive minerals. { 'fər·və,nīt }

Fibrist |GEOL| A suborder of the soil order Histosol, consisting mainly of recognizable plant residues or sphagnum moss and saturated with water most of the year. { 'fī·brəst }

fibroblastic |PETR| Of a metamorphic rock, having a texture that is homeoblastic as a result of the development of minerals with a fibrous habit during recrystallization. { ¦fī·brə¦blas·tik }

fibroferrite |MINERAL| Fe(SO₄)(OH)·5H₂O A yellowish mineral composed of a hydrous basic ferric sulfate, occurring in fibrous form. { ¦fī·brō'fe,rīt }

fibrolite See sillimanite. { 'fī·brə,līt }

fiedlerite |MINERAL| Pb₃(OH)₂Cl₄ A colorless mineral composed of a hydroxychloride of lead, occurring as monoclinic crystals. { 'fēd·lə,rīt }

field |GEOL| A region or area with a particular mineral resource, for example, a gold field. |GEOPHYS| That area or space in which a particular geophysical effect, such as gravity or magnetism, occurs and can be measured. { fēld }

field focus |GEOPHYS| The total area or volume occupied by an earthquake source. { 'fēld ,fō·kəs }

field geology |GEOL| The study of rocks and rock materials in their environment and in their natural relations to one another. { 'fēld jē,äl·ə·jē }

field pressure |GEOL| The pressure of natural gas in the underground formations from which it is produced. { 'fēld ,presh·ər }

figure stone See agalmatolite. { 'fig·yər ,stōn }

filiform lapilli See Pele's hair. { 'fil·ə,fórm lə'pil·ē }

fillowite |MINERAL| H₂Na₆(Mn,Fe,Ca)₁₄(PO₄)₁₂·H₂O A brown, yellow, or colorless mineral composed of a hydrous phosphate of manganese, iron, sodium, and other metals. { 'fil·ə,wīt }

fill terrace See alluvial terrace. { 'fil ,ter·əs }

fine admixture |GEOL| The smaller size grades of a sediment of mixed size grades. { ¦fīn 'ad,miks·chər }

fine earth |GEOL| A soil which can be passed through a 2-millimeter sieve without grinding its primary particles. { ¦fīn 'ərth }

fine gravel |GEOL| Gravel consisting of particles with a diameter range of 1 to 2 millimeters. { ¦fīn 'grav·əl }

fine sand |GEOL| Sand grains between 0.25 and 0.125 millimeter in diameter. { ¦fīn 'sand }

finger |GEOL| The tendency for gas which is displacing liquid hydrocarbons in a heterogeneous reservoir rock system to move forward irregularly (in fingers), rather than on a uniform front. { 'fiŋ·gər }

finger coal See natural coke. { 'fiŋ·gər ,kōl }

finnemanite |MINERAL| Pb₅Cl(AsO₃)₃ A gray, olive-green, or black hexagonal mineral composed of arsenite and chloride of lead. { 'fin·ə·mə,nīt }

fiorite See siliceous sinter. { fē'ór,īt }

fireclay |GEOL| **1.** A clay that can resist high temperatures without becoming glassy. **2.** Soft, embedded, white or gray clay rich in hydrated aluminum silicates or silica and deficient in alkalies and iron. { 'fīr ,klā }

fire fountain See lava fountain. { 'fīr ,faúnt·ən }

fire opal |MINERAL| A translucent or transparent, orangy-yellow, brownish-orange, or red variety of opal that gives out fiery reflections in bright light and that may have a play of colors. Also known as pyrophane; sun opal. { 'fīr ,ō·pəl }

firestone See flint. { 'fīr,stōn }

firn limit See firn line. { 'fərn ,lim·ət }

firn line |GEOL| **1.** The regional snow line on a glacier. **2.** The line that divides the ablation area of a glacier from the accumulation area. Also known as firn limit. { 'fərn ,līn }

first bottom |GEOL| The floodplain of a river, below the first terrace. { ¦fərst 'bäd·əm }

fischerite |MINERAL| A green mineral composed of a basic aluminum phosphate; may be identical to wavellite. { 'fish·ə,rīt }

fish-eye stone See apophyllite. { 'fish ,ī ,stōn }

fissile |GEOL| Capable of being split along the line of the grain or cleavage plane. { 'fis·əl }

fission-track dating |GEOL| A method of dating geological specimens by counting the radiation-damage tracks produced by spontaneous fission of uranium impurities in minerals and glasses. { 'fish·ən ,trak ,dād·iŋ }

fissure |GEOL| **1.** A high, narrow cave passageway. **2.** An extensive crack in a rock. { 'fish·ər }

fissure system |GEOL| A group of fissures having the same age and generally parallel strike and dip. { 'fish·ər ,sis·təm }

fissure vein |GEOL| A mineral deposit in a cleft or crack in the rock material of the earth's crust. { 'fish·ər ,vān }

Fistuliporidae |PALEON| A diverse family of extinct marine bryozoans in the order Cystoporata. { ,fis·chə·lə'pór·ə,dē }

fizelyite |MINERAL| A metallic, lead-gray mineral composed of a lead silver antimony sulfide, occurring as prisms. { fə'zā·lē,īt }

flaggy |GEOL| **1.** Of bedding, consisting of strata 4–40 inches (10–100 centimeters) in thickness. **2.** Of rock, tending to split into layers of suitable thickness (0.4–2 inches or 1–5 centimeters) for use as flagstones. { 'flag·ē }

flagstone |GEOL| **1.** A hard, thin-bedded sandstone, firm shale, or other rock that splits easily along bedding planes or joints into flat slabs. **2.** A piece of flagstone used for making pavement or covering the side of a house. { 'flag,stōn }

flajolotite |MINERAL| $4FeSbO_4 \cdot 3H_2O$ A claylike, lemon-yellow mineral composed of a hydrous iron antimonate, occurring in nodular masses. { 'flaj·ə'lō,tīt }

flamboyant structure |GEOL| The optical continuity of crystals or grains as disturbed by a structure that is divergent. { flam'bói·ənt 'strək·chər }

flank See limb. { flaŋk }

flaser |GEOL| Streaky layer of parallel, scaly aggregates that surrounds the lenticular bodies of granular material in flaser structure; caused by pressure and shearing during metamorphism. { 'flā·zər }

flaser gabbro |GEOL| A cataclastic gabbro that contains augen of feldspar or quartz surrounded by flakes of mica or chlorite. { 'flā·zər 'ga,brō }

flaser structure |GEOL| **1.** A metamorphic structure in which small lenses and layers of granular material are surrounded by a matrix of sheared, crushed material, resembling a crude flow structure. Also known as pachoidal structure. **2.** A primary sedimentary structure consisting of fine-sand or silt lenticles that are aligned and cross-bedded. { 'flā·zər ,strək·chər }

flat |GEOL| See mud flat. |MINERAL| An inferior grade of rough diamonds. { flat }

flat-lying |GEOL| Of mineral deposits and coal seams, having a relatively flat dip, up to 5°. { 'flat ,lī·iŋ }

flaw |MINERAL| A faulty part of a gemstone, such as a crack, visible imperfect crystallization, or internal twinning or cleavage. { fló }

flaxseed ore |GEOL| Iron ore composed of disk-shaped oauolites that have been partially flattened parallel to the bedding plane. { 'flak,sēd ,ór }

Flexibilia |PALEON| A subclass of extinct stalked or creeping Crinoidea; characteristics include a flexible tegmen with open ambulacral grooves, uniserial arms, a cylindrical stem, and five conspicuous basals and radials. { ,flek·sə'bil·ē·ə }

flexible sandstone |GEOL| A variety of itacolumite that consists of fine grains and occurs in thin layers. { ,flek·sə·bəl 'san,stōn }

flexural slip |GEOL| The slipping of sedimentary strata along bedding planes during folding, producing disharmonic folding and, when extreme, découllement. Also known as bedding-plane slip. { 'flek·shə·rəl 'slip }

flexure |GEOL| **1.** A broad, domed structure. **2.** A fold. { 'flek·shər }

flinkite |MINERAL| $Mn_3(AsO_4)(OH)_4$ Greenish-brown mineral composed of basic manganese arsenate, occurring in feathery forms. { 'fliŋ,kīt }

flint |MINERAL| A black or gray, massive, hard, somewhat impure variety of chalcedony, breaking with a conchoidal fracture. Also known as firestone. { flint }

flint clay |GEOL| A hard, smooth, flintlike fireclay; when it is ground, it develops no plasticity, and it breaks with conchoidal fracture. { 'flint ,klā }

float |GEOL| An isolated, displaced rock or ore fragment. { flōt }

float coal [GEOL] Small, irregularly shaped, isolated deposits of coal embedded in sandstone or in siltstone. Also known as raft. { 'flōt ,kōl }

floating sand [PETR] A single grain of quartz sand that does not appear to touch surrounding sand grains scattered throughout the finer-grained matrix of a sedimentary rock. { ¦flōd·iŋ 'sand }

float mineral [GEOL] Small ore fragments carried from the ore bed by the action of water or by gravity; a float mineral often leads to discovery of mines. { 'flōt ,min·rəl }

floe till [GEOL] **1.** A glacial till resulting from the intact deposition of a grounded iceberg in a lake bordering an ice sheet. **2.** A lacustrine clay with boulders, stones, and other glacial matter dropped into it by melting icebergs. Also known as berg till. { 'flō ,til }

flokite See mordenite. { 'flō,kīt }

flood basalt See plateau basalt. { 'fləd bə,sȯlt }

flood basin [GEOL] **1.** The tract of land actually submerged during the highest known flood in a specific region. **2.** The flat, wide area lying between a low, sloping plain and the natural levee of a river. { 'fləd ,bās·ən }

flood fringe See pondage land. { 'fləd ,frinj }

floodplain [GEOL] The relatively smooth valley floors adjacent to and formed by alluviating rivers which are subject to overflow. { 'fləd,plān }

floodplain splay [GEOL] A small alluvial fan or other outspread deposit formed where an overloaded stream breaks through a levee (artificial or natural) and deposits its material (often coarse-grained) on the floodplain. Also known as channel splay. { 'fləd,plān ,splā }

flood tuff See ignimbrite. { 'fləd ,təf }

floor [GEOL] **1.** The rock underlying a stratified or nearly horizontal deposit, corresponding to the footwall of more steeply dipping deposits. **2.** A horizontal, flat ore body. { flȯr }

florencite [MINERAL] $CeAl_3(PO_4)_2(OH)_6$ Pale-yellow mineral composed of basic phosphate of cerium and aluminum. { 'flär·ən,sīt }

flow [GEOL] Any rock deformation that is not instantly recoverable without permanent loss of cohesion. Also known as flowage; rock flowage. { flō }

flowage See flow. { 'flō·ij }

flowage line [GEOL] A contour line at the edge of a body of water, such as a reservoir, representing a given water level. { 'flō·ij ,līn }

flow banding [GEOL] An igneous rock structure resulting from flowing of magmas or lavas and characterized by alternation of mineralogically unlike layers. { 'flō ,band·iŋ }

flow breccia [GEOL] A breccia formed with the movement of lava flow while the flow is still in motion. { 'flō ,brech·ə }

flow cast [PETR] One of a group of bedding plane structures formed in graywacke. { 'flō ,kast }

flow cleavage [GEOL] Rock cleavage in which solid flow of rock accompanies recrystallization. Also known as slaty cleavage. { 'flō ,klē·vij }

flow earth See solifluction mantle. { 'flō ,ərth }

flow fold [GEOL] Folding in beds, composed of relatively plastic rock, that assume any shape impressed upon them by the more rigid surrounding rocks or by the general stress pattern of the deformed zone; there are no apparent surfaces of slip. { 'flō ,fōld }

flow layer [PETR] In an igneous rock, a layer which is different in composition or texture from adjacent layers. { 'flō ,lā·ər }

flow line [PETR] In an igneous rock, any internal structure produced by parallel orientation of crystals, mineral streaks, or inclusions. { 'flō ,līn }

flow rock [PETR] An igneous rock that had been liquid. { 'flō ,räk }

flow slide [GEOL] A slide of waterlogged material in which the slip surface is not well defined. { 'flō ,slīd }

flowstone [GEOL] Deposits of calcium carbonate that accumulated against the walls of a cave where water flowed on the rock. { 'flō,stōn }

flow structure [GEOL] A primary sedimentary structure due to underwater slump or flow. { 'flō ‚strək·chər }

flow texture [PETR] A pattern of an igneous rock that is formed when the stream or flow lines of a once-molten material have a subparallel arrangement of prismatic or tabular cyrstals or microlites. Also known as fluidal texture. { 'flō ‚teks·chər }

flow velocity [GEOL] In soil, a vector point function used to indicate rate and direction of movement of water through soil per unit of time, perpendicular to the direction of flow. { 'flō və'läs·əd·ē }

fluellite [MINERAL] $AlF_3 \cdot H_2O$ A colorless or white mineral composed of aluminum fluoride, occurring in crystals. { 'flü·ə‚līt }

fluidal texture See flow texture. { ¦flü·əd·əl 'teks·chər }

fluid geometry [GEOL] Fluid distribution in reservoir strata controlled by rock effective pore-size distribution, rock wettability characteristics in relation to the fluids present, method of producing saturation, and rock heterogeneity. { ¦flü·əd jē'äm·ə·trē }

fluid inclusion [PETR] A tiny fluid-filled cavity in an igneous rock that forms by the entrapment of the liquid from which the rock crystallized. { ¦flü·əd in'klü·zhən }

fluid saturation [GEOL] Measure of the gross void space in a reservoir rock that is occupied by a fluid. { ¦flü·əd ‚sach·ə'rā·shən }

flume [GEOL] A ravine with a stream flowing through it. { flüm }

fluoborite [MINERAL] $Mg_3(BO_3)(F,OH)_3$ A colorless mineral composed of magnesium fluoborate; occurs in hexagonal prisms. Also known as nocerite. { ‚flü·ə'bȯr‚īt }

fluocerite [MINERAL] $(Ce,La,Nd)F_3$ A reddish-yellow mineral composed of fluoride of cerium and related elements. { ¦flü·ə'se‚rīt }

fluolite See pitchstone. { 'flü·ə‚līt }

fluor See fluorite; luminophor. { 'flü‚ȯr }

fluorapatite [MINERAL] **1.** $Ca_5(PO_4)_3F$ A mineral of the solid-solution series of the apatite group; common accessory mineral in igneous rocks. **2.** An apatite mineral in which the fluoride member dominates. { flù·'rap·ə‚tīt }

fluoridation [GEOCHEM] Formation in rocks of fluorine-containing minerals such as fluorite or topaz. { flùr·ə'dā·shən }

fluorite [MINERAL] CaF_2 A transparent to translucent, often blue or purple mineral, commonly found in crystalline cubes in veins and associated with lead, tin, and zinc ores; hardness is 4 on Mohs scale; the principal ore of fluorine. Also known as Derbyshire spar; fluor; fluorspar. { 'flùr‚īt }

fluorocummingtonite [MINERAL] Cummingtonite with a high content of fluorine. { ¦flùr·ō'kəm·iŋ·tə‚nīt }

fluorspar See fluorite. { 'flùr‚spär }

flute [GEOL] **1.** A natural groove running vertically down the face of a rock. **2.** A groove in a sedimentary structure formed by the scouring action of a turbulent, sediment-laden water current, and having a steep upcurrent end. { flüt }

flute cast [GEOL] A raised, oblong, or subconical welt on the bottom surface of a siltstone or sandstone bed formed by the filling of a flute. { flüt ‚kast }

Fluvent [GEOL] A suborder of the soil order Entisol that is well-drained with visible marks of sedimentation and no identifiable horizons; occurs in recently deposited alluvium along streams or in fans. { 'flü·vənt }

fluvial cycle of erosion See normal cycle. { 'flü·vē·əl 'sī·kəl əv ə'rō·zhən }

fluvial deposit [GEOL] A sedimentary deposit of material transported by or suspended in a river. { ¦flü·vē·əl di'päz·ət }

fluvial sand [GEOL] Sand laid down by a river or stream. { ¦flü·vē·əl 'sand }

fluvial soil [GEOL] Soil laid down by a river or stream. { ¦flü·vē·əl 'sȯil }

fluviatile [GEOL] Resulting from river action. { 'flü·vē·ə‚tīl }

fluviomorphology See river morphology. { ¦flü·vē·ō·mȯr'fäl·ə·jē }

flying veins [GEOL] A series of mineral-deposit veins which overlap or intersect in a branchlike pattern. { ¦flī·iŋ 'vānz }

flysch [GEOL] Deposits of dark, fine-grained, thinly bedded sandstone shales and of clay, thought to be deposited by turbidity currents and originally defined as rock formations on the northern and southern borders of the Alps. { flīsh }

foam See pumice. { fōm }

foam mark |GEOL| A surface sedimentary structure comprising a pattern of barely visible ridges and hollows formed where wind-driven sea foam passes over a surface of wet sand. { 'fōm ,märk }

focus |GEOPHYS| The center of an earthquake and the origin of its elastic waves within the earth. { 'fō·kəs }

fold |GEOL| A bend in rock strata or other planar structure, usually produced by deformation; folds are recognized where layered rocks have been distorted into wavelike form. { fōld }

fold belt See orogenic belt. { 'fōld ,belt }

folding |GEOL| Compression of planar structure in the formation of fold structures. { 'fōld·iŋ }

fold system |GEOL| A group of folds with common trends and characteristics. { 'fōld ,sis·təm }

folia |PETR| Thin, leaflike layers that occur in gneissic or schistose rocks. { 'fō·lē·ə }

foliaceous |GEOL| Having a leaflike or platelike structure composed of thin layers of minerals. { ,fō·lē'ā·shəs }

foliation |GEOL| A laminated structure formed by segregation of different minerals into layers that are parallel to the schistosity. { ,fō·lē'ā·shən }

Folist |GEOL| A suborder of the soil order Histosol, consisting of wet forest litter resting on rock or rubble. { 'fäl·əst }

Fontéchevade man |PALEON| A fossil man representing the third interglacial Homo sapiens and having browridges and a cranial vault similar to those of modern Homo sapiens. { fōn·te·che'väd ,man }

fool's gold See pyrite. { ¦fülz ¦gōld }

footeite See connellite. { 'fút,īt }

footwall |GEOL| The mass of rock that lies beneath a fault, an ore body, or a mine working. Also known as heading side; heading wall; lower plate. { 'fút,wòl }

forbesite |MINERAL| $H(Ni,Co)AsO_4 \cdot 3^1/_2H_2O$ A grayish-white mineral composed of hydrous nickel cobalt arsenate; occurs in fibrocrystalline form. { 'fórb,zīt }

forearc |GEOL| The area between the trench and the volcanic arc of a subduction zone. { 'fór,ärk }

forebulge |GEOL| An uplift at the edge of a glacier caused by tilting of the lithosphere. { 'fòr,bəlj }

foredeep |GEOL| 1. A long, narrow depression that borders an orogenic belt, such as an island arc, on the convex side. 2. See exogeosyncline. { 'fór,dēp }

foredune |GEOL| A coastal dune or ridge that is parallel to the shoreline of a large lake or ocean and is stabilized by vegetation. { 'fór,dün }

foreign inclusion |PETR| A fragmentary piece of country rock which is enclosed in an igneous intrusion. { ¦fär·ən in'klü·zhən }

foreland |GEOL| 1. A lowland area onto which piedmont glaciers have moved from adjacent mountains. 2. A stable part of a continent bordering an orogenic or mobile belt. { 'fór·lənd }

foreland facies See shelf facies. { 'fór·lənd ,fā·shēz }

forellenstein See troctolite. { fə'rel·ən,stīn }

foreset bed |GEOL| One of a series of inclined symmetrically arranged layers of a cross-bedding unit formed by deposition of sediments that rolled down a steep frontal slope of a delta or dune. { 'fór,set ,bed }

foreshock |GEOPHYS| A tremor which precedes a larger earthquake or main shock. { 'fór,shäk }

foreshore |GEOL| The zone that lies between the ordinary high- and low-watermarks and is daily traversed by the rise and fall of the tide. Also known as beach face. { 'fór,shòr }

formanite |MINERAL| A mineral composed of an oxide of uranium, zirconium, thorium, calcium, tantalum, and niobium with some rare-earth metals. { 'fór·mə,nīt }

formation |GEOL| Any assemblage of rocks which have some common character and are mappable as a unit. { fòr'mā·shən }

formation factor [GEOCHEM] The ratio between the conductivity of an electrolyte and that of a rock saturated with the same electrolyte. Also known as resistivity factor. [GEOL] A function of the porosity and internal geometry of a reservoir rock system, expressed as $F = \phi^{-m}$, where ϕ is the fractional porosity of the rock, and m is the cementation factor (pore-opening reduction). { fȯr'mā·shən ‚fak·tər }

formation pressure See reservoir pressure. { fȯr'mā·shən ‚presh·ər }

formation resistivity [GEOPHYS] Electrical resistivity of reservoir formations measured by electrical log sondes; used for clues to formation lithography and fluid content. { fȯr'mā·shən ri‚zis'tiv·əd·ē }

forril farina See rock milk. { 'fär·əl fə‚rēn·ē }

forsterite [MINERAL] Mg_2SiO_4 A whitish or yellowish, magnesium-rich variety of olivine. Also known as white olivine. { 'fȯr·stə‚rīt }

fortification agate See landscape agate. { ‚fȯrd·ə·fə'kā·shən 'ag·ət }

foshagite [MINERAL] $Ca_5Si_3O_{10}(OH)_2 \cdot 2H_2O$ A white mineral composed of a basic hydrous calcium silicate. { 'fō·shə‚gīt }

fossil [PALEON] The organic remains, traces, or imprint of an organism preserved in the earth's crust since some time in the geologic past. { 'fäs·əl }

fossil dune [GEOL] An ancient desert dune. { ¦fäs·əl¦'dün }

fossil fuel [GEOL] Any hydrocarbon deposit that may be used for fuel; examples are petroleum, coal, and natural gas. { ¦fäs·əl 'fyül }

fossil man [PALEON] Ancient human identified from prehistoric skeletal remains which are archeologically earlier than the Neolithic. { ¦fäs·əl 'man }

fossil permafrost See passive permafrost. { ¦fäs·əl 'pər·mə‚frȯst }

fossil reef [GEOL] An ancient reef. { ¦fäs·əl 'rēf }

fossil resin [GEOL] A natural resin in geologic deposits which is an exudate of long-buried plant life; for example, amber, retinite, and copal. { ¦fäs·əl 'rez·ən }

fossil soil See paleosol. { ¦fäs·əl 'sȯil }

fossil wax See ozocerite. { ¦fäs·əl 'waks }

foundation coefficient [GEOPHYS] A coefficient which expresses how much stronger the effect of an earthquake is on a given rock than it would be on an undisturbed crystalline rock under the same conditions. { faùn'dā·shən ‚kō·i‚fish·ənt }

founder [GEOL] To sink under water either by depression of the land or by rise of sea level, especially in reference to large crustal masses, islands, or significant portions of continents. { 'faùn·dər }

fourchite [PETR] A monchiquite that lacks feldspar and olivine. { 'fùr‚shīt }

fourmarierite [MINERAL] An orange-red to brown mineral composed of a hydrous oxide of lead and uranium. { fùr'mar·ē·ə‚rīt }

four-way dip [GEOPHYS] In seismic prospecting, dip determined by an array of geophones which are set up at points in four directions from a shot point; three of the locations are essential and the fourth serves as a control point. { 'fȯr ‚wā 'dip }

fowlerite [MINERAL] A zinc-bearing variety of rhodonite. { 'faù·lə‚rīt }

foyaite [PETR] A nepheline syenite composed chiefly of potassium feldspar. { 'fȯi·yə‚īt }

fractional crystallization [PETR] Separation of a cooling magma into multiple minerals as the different minerals cool and congeal at progressively lower temperatures. Also known as crystallization differentiation; fractionation. { ¦frak·shən·əl ‚krist·əl·ə'zā·shən }

fractionation See fractional crystallization. { ‚frak·shə'nā·shən }

fractoconformity [GEOL] The relation between conformable strata, where faulting of the older beds occurs at the same time as deposition of the newer beds. { ¦frak·tō·kən'fȯr·məd·ē }

fracture [GEOL] A crack, joint, or fault in a rock due to mechanical failure by stress. Also known as rupture. [MINERAL] A break in a mineral other than along a cleavage plane. { 'frak·shər }

fracture cleavage [GEOL] Cleavage that occurs in deformed but only slightly metamorphosed rocks along closely spaced, parallel joints and fractures. { 'frak·shər ‚klēv·ij }

fracture-plane inclination [GEOL] Gradient or inclination of the plane of fracture formed in a reservoir formation. { 'frak·shər ‚plan ‚in·klə'nā·shən }

fracture system [GEOL] A stress-related group of contemporaneous fractures. { 'frak· shər ‚sis·təm }

fracture zone [GEOL] An elongate zone on the deep-sea floor that is of irregular topography and often separates regions of different depths; frequently crosses and displaces the mid-oceanic ridge by faulting. { 'frak·shər ‚zōn }

fragipan [GEOL] A dense, natural subsurface layer of hard soil with relatively slow permeability to water, mostly because of its extreme density or compactness rather than its high clay content or cementation. { 'fraj·ə‚pan }

framboid [GEOL] A microscopic aggregate of pyrite grains, often occurring in spheroidal clusters. { 'fram‚bȯid }

framework [GEOL] **1.** In a sediment or sedimentary rock, the rigid arrangement created by particles that support one another at contact points. **2.** A fixed calcareous structure impervious to waves, built by sedentary organisms (for example, sponges, corals, and bryozoans) in a high-energy environment. { 'frām‚wərk }

framework silicate See tectosilicate. { 'frām‚wərk 'sil·ə·kət }

franckeite [MINERAL] A dark-gray or black massive mineral composed of lead antimony tin sulfide. { 'fräŋ·kə‚īt }

francolite [MINERAL] $Ca_5(PO_4,CO_3)_3(F,OH)$ Colorless fluoride-bearing carbonate-apatite. { 'fraŋ·kə‚līt }

Franconian [GEOL] A North American stage of geologic time; the middle Upper Cambrian. { fraŋ'kō·nē·ən }

franklinite [MINERAL] $ZnFe_2O_4$ Black, slightly magnetic mineral member of the spinel group; usually possesses extensive substitution of divalent manganese and iron for the divalent zinc, and limited trivalent manganese for the trivalent iron. { 'fraŋ· klə‚nīt }

free air See free atmosphere. { ¦frē 'er }

free-air anomaly See free-air gravity anomaly. { 'frē ‚er ə'näm·ə·lē }

free-air gravity anomaly [GEOPHYS] A measure of the mass excesses and deficiencies within the earth; calculated as the difference between the measured gravity and the theoretical gravity at sea level and a free-air coefficient determined by the elevation of the measuring station. Also known as free-air anomaly. { 'frē ‚er 'grav·əd·ē ə‚näm·ə·lē }

free atmosphere [GEOPHYS] That portion of the earth's atmosphere, above the planetary boundary layer, in which the effect of the earth's surface friction on the air motion is negligible and in which the air is usually treated (dynamically) as an ideal fluid. Also known as free air. { ¦frē 'at·mə‚sfir }

free-burning coal See noncaking coal. { ¦frē ‚bȯrn·iŋ 'kōl }

free face [GEOL] A vertical or steeply inclined layer of rock from which weathered material falls to form talus at its base. { ¦frē 'fās }

freestone [GEOL] Stone, particularly a thick-bedded, even-textured, fine-grained sandstone, that breaks freely and is able to be cut and dressed with equal facility in any direction without tending to split. { 'frē‚stōn }

F region [GEOPHYS] The general region of the ionosphere in which the F_1 and F_2 layers tend to form. { 'ef ‚rē·jən }

freibergite [MINERAL] A steel-gray, silver-bearing variety of tetrahedrite. { 'frī‚bər‚gīt }

freieslebenite [MINERAL] $Pb_3Ag_5Sb_5S_{12}$ A steel-gray to dark mineral composed of a sulfide of antimony, lead, and silver. { ¦frī·əs¦lā·bə‚nīt }

freirinite [MINERAL] $Na_3Cu_3(AsO_4)_2(OH)_3·H_2O$ A lavender to turquoise-blue mineral composed of a basic hydrous arsenate of sodium and copper. { frā'rē‚nīt }

fremontite See natromontebrasite. { 'frē·mən‚tīt }

fresh [GEOL] Unweathered in reference to a rock or rock surface. { fresh }

Fresnian [GEOL] A North American stage of upper Eocene geologic time, above Narizian and below Refugian. { 'frez·nē·ən }

frictional See cohesionless. { 'frik·shən·əl }

friction crack [GEOL] A short, crescent-shaped crack in glaciated rock produced by a

127

localized increase in friction between rock and ice, oriented transverse to the direction of ice flow.   { 'frik·shən ,krak }

friedelite   [MINERAL] $Mn_8Si_6O_{18}(OH,Cl)_4 \cdot 3H_2O$ A rose-red mineral composed of manganese silicate with chlorine.   { frē·de,līt }

fringe joint   [GEOL]   A small-scale joint peripheral to, and usually at a 5–25° angle from the face of, the master joint.   { 'frinj ,jóint }

fringe ore   [GEOL]   Ore located on the outer boundary of a mineralization pattern or halo.   Also known as halo ore.   { 'frinj ,ór }

fringing reef   [GEOL]   A coral reef attached directly to or bordering the shore of an island or continental landmass.   { ¦frin·jiŋ 'rēf }

frohbergite   [MINERAL] $FeTe_2$ A mineral composed of iron telluride; it is isomorphous with marcasite.   { 'frō,bər,gīt }

frondelite   [MINERAL] $MnFe_4(PO_4)_5(OH)_5$ A mineral composed of basic phosphate of manganese and iron; it is isomorphous with rockbridgeite.   { frän'de,līt }

front abutment pressure   [GEOPHYS]   The release of energy in the superincumbent strata above the seam induced by the extraction of the seam.   { ¦frənt ə'bət·mənt ,presh·ər }

frontal apron *See* outwash plain.   { ¦frənt·əl 'ā·prən }

frontal plain *See* outwash plain.   { ¦frənt·əl ¦plān }

front slope *See* scarp slope.   { 'frənt ,slōp }

frost action   [GEOL]   **1.** The weathering process caused by cycles of freezing and thawing of water in surface pores, cracks, and other openings.   **2.** Alternate or repeated cycles of freezing and thawing of water contained in materials; the term is especially applied to disruptive effects of this action.   { 'fróst ,ak·shən }

frost boil   [GEOL]   **1.** An accumulation of water and mud released from ground ice by accelerated spring thawing.   **2.** A low mound formed by local differential frost heaving at a location most favorable for the formation of segregated ice and accompanied by the absence of an insulating cover of vegetation.   { 'fróst ,bóil }

frost bursting *See* congelifraction.   { 'fróst ,bərst·iŋ }

frost churning *See* congeliturbation.   { 'fróst ,chərn·iŋ }

frost heaving   [GEOL]   The lifting and distortion of a surface due to internal action of frost resulting from subsurface ice formation; affects soil, rock, pavement, and other structures.   { 'fróst ,hēv·iŋ }

frost line   [GEOL]   **1.** The maximum depth of frozen ground during the winter.   **2.** The lower limit of the permafrost.   { 'fróst ,līn }

frost mound   [GEOL]   A hill and knoll associated with frozen ground in a permafrost region, containing a core of ice.   Also known as soffosian knob; soil blister.   { 'fróst ,maúnd }

frost riving *See* congelifraction.   { 'fróst ,rīv·iŋ }

frost shattering *See* congelifraction.   { 'fróst ,shad·ə·riŋ }

frost splitting *See* congelifraction.   { 'fróst ,splid·iŋ }

frost stirring *See* congelifraction.   { 'fróst ,stər·iŋ }

frost table   [GEOL]   An irregular surface in the ground which, at any given time, represents the penetration of thawing into seasonally frozen ground.   { 'fróst ,tā·bəl }

frost thrusting   [GEOL]   Lateral dislocation of soil and rock materials by the action of freezing and resulting expansion of soil water.   { 'fróst ,thrəst·iŋ }

frost weathering *See* congelifraction.   { 'fróst ,weth·ə·riŋ }

frost wedging *See* congelifraction.   { 'fróst ,wej·iŋ }

frost zone *See* seasonally frozen ground.   { 'fróst ,zōn }

frozen ground   [GEOL]   Soil having a temperature below freezing, generally containing water in the form of ice.   Also known as gelisol; merzlota; taele; tjaele.   { ¦frōz·ən 'graúnd }

fuchsite   [MINERAL]   A bright-green variety of muscovite rich in chromium.   { 'fyük,sīt }

fucoid   [GEOL]   A tunnellike marking on a sedimentary structure identified as a trace fossil but not referred to a described genus.   { 'fyü,kóid }

fulgurite   [GEOL]   A glassy, rootlike tube formed when a lightning stroke terminates in dry sandy soil; the intense heating of the current passing down into the soil along an irregular path fuses the sand.   { 'fúl·gə,rīt }

fuller's earth |GEOL| A natural, fine-grained earthy material, such as a clay, with high adsorptive power; consists principally of hydrated aluminum silicates; used as an adsorbent in refining and decolorizing oils, as a catalyst, and as a bleaching agent. { ˈfu̇l·ərz ˌərth }

fuloppite |MINERAL| $Pb_3Sb_8S_{15}$ A lead gray, monoclinic mineral consisting of lead antimony sulfide. { 'fu̇l·ə,pīt }

fumarole |GEOL| A hole, usually found in volcanic areas, from which vapors or gases escape. { 'fyü·mə,rōl }

fundamental complex |GEOL| An agglomeration of metamorphic rocks underlying sedimentary or unmetamorphosed rocks; specifically, an agglomeration of Archean rocks supporting a geological column. { ˌfən·dəˌment·əl 'käm,pleks }

fundamental jelly See ulmin. { ˌfən·dəˌment·əl 'jel·ē }

fundamental strength |GEOPHYS| The maximum stress that a geological structure can withstand without creep under certain conditions but without reference to time. { ˌfən·dəˌment·əl 'streŋkth }

fundamental substance See ulmin. { ˌfən·dəˌment·əl 'səb·stəns }

fusain |GEOL| The local lithotype strands or patches, characterized by silky luster, fibrous structure, friability, and black color.   Also known as mineral charcoal; mother-of-coal. { 'fyü,zān }

fusinite |GEOL| The micropetrological constituent of fusain which consists of carbonized woody tissue. { 'fyüz·ən,īt }

fusinization |GEOL| The process of formation of fusain in coal. { ˌfyüz·ən·ə'zā·shən }

fusion crust |GEOL| A thin, glassy coating, usually black and rerely more than 1 millimeter thick, which is formed by ablation on the surface of a meteorite. { 'fyü·zhən ,krəst }

Fusulinacea |PALEON| A superfamily of large, marine extinct protozoans in the order Foraminiferida characterized by a chambered calcareous shell. { ˌfyü·zə·lə'nās·ē·ə }

Fusulinidae |PALEON| A family of extinct protozoans in the superfamily Fusulinacea. { ˌfyü·zə'lin·ə,dē }

Fusulinina |PALEON| A suborder of extinct rhizopod protozoans in the order Foraminiferida having a monolamellar, microgranular calcite wall. { ˌfyü·zə·lə'nī·nə }

G

gabbro [PETR] A group of dark-colored, intrusive igneous rocks with granular texture, composed largely of basic plagioclase and clinopyroxene. { 'gab·rō }

gadolinite [MINERAL] $Be_2FeY_2Si_2O_{10}$ A black, greenish-black, or brown rare-earth mineral; hardness is 6.5–7 on Mohs scale, and specific gravity is 4–4.5. { 'gad·əl·ə,nīt }

gageite [MINERAL] $(Mn,Mg,Zn)_8Si_3O_{14}\cdot2H_2O$ (or $3H_2O$) A mineral composed of a hydrous silicate of manganese, magnesium, and zinc. { 'gā,īt }

gahnite [MINERAL] $ZnAl_2O_4$ A usually dark-green, but sometimes yellow, gray, or black spinel mineral consisting of an oxide of zinc and aluminum. Also known as zinc spinel. { 'gä,nīt }

galaxite [MINERAL] $MnAl_2O_4$ A black mineral of the spinel series composed of an oxide of manganese and aluminum. { 'gā·lak,sīt }

galena [MINERAL] PbS A bluish-gray to lead-gray mineral with brilliant metallic luster, specific gravity 7.5, and hardness 2.5 on Mohs scale; occurs in cubic or octahedral crystals, in masses, or in grains. Also known as blue lead; lead glance. { gə'lē·nə }

galenic [MINERAL] Containing galena. Also known as galenical. { gə'len·ik }

galenical See galenic. { gə'len·i·kəl }

galenobismutite [MINERAL] $PbBi_2S_4$ A lead-gray or tin-white mineral consisting of bismuth sulfide; specific gravity is 6.9. { gə'lē·nō'biz·mə,tīt }

Galeritidae [PALEON] A family of extinct exocyclic Euechinoidea in the order Holectypoida, characterized by large ambulacral plates with small, widely separated pore pairs. { ,ga·lə'rid·ə,dē }

gallery [GEOL] **1.** A horizontal, or nearly horizontal, underground passage. **2.** A subsidiary passage in a cave at a higher level than the main passage. { 'gal·rē }

galmei See hemimorphite. { gäl'mī }

Gampsonychidae [PALEON] A family of extinct crustaceans in the order Palaeocaridacea. { ,gam·sə'nī·kə,dē }

gangue [GEOL] The valueless rock or aggregates of minerals in an ore. { gaŋ }

ganister [PETR] A fine, hard quartzose sandstone; used to make refractory silica brick to line furnace reactors. { 'gan·ə·stər }

ganomalite [MINERAL] $(Ca_2)Pb_3Si_3O_{11}$ A colorless to gray silicate of lead with calcium crystallizing in the tetragonal system. { gə'näm·ə,līt }

ganophyllite [MINERAL] $(Na,K)(Mn,Fe,Al)_5(Si,Al)_6O_{15}(OH)_5\cdot2H_2O$ A brown, prismatic crystalline or foliated mineral composed of a hydrous silicate of manganese and aluminum. { ,gan·ə'fi,līt }

garnet [MINERAL] A generic name for a group of mineral silicates that are isometric in crystallization and have the general chemical formula $A_3B_2(SiO_4)_3$, where A is Fe^{2+}, Mn^{2+}, Mg, or Ca, and B is Al, Fe^{3+}, Cr^{3+}, or Ti^{3+}; used as a gemstone and as an abrasive. { 'gär·nət }

garnierite [MINERAL] $(Ni,Mg)_3Si_2O_5(OH)_4$ An apple-green or pale-green, monoclinic serpentine; a gemstone and an ore of nickel. Also known as nepuite; noumeite. { 'gär·nē·ə,rīt }

garronite [MINERAL] $Na_2Ca_5Al_{12}Si_{20}O_{64}\cdot27H_2O$ A zeolite mineral belonging to the phillipsite group; crystallizes in the tetragonal system. { 'ga·rə,nīt }

gas clathrate See gas hydrate. { ¦gas 'klath,rāt }

gas column [GEOL] The difference in elevation between the highest and lowest parts of the various producing zones of a gas-producing formation. { 'gas ,ka·ləm }

gas-condensate reservoir [GEOL] Hydrocarbon reservoir in which conditions of temperature and pressure have resulted in the condensation of the heavier hydrocarbon constituents from the reservoir gas. { ¦gas 'känd·ən,sāt ,rez·əv,wär }

gas-filled porosity [GEOL] A reservoir formation in which the pore space is filled by gas instead of liquid hydrocarbons. { 'gas ,fild pə'räs·əd·ē }

gas floor [GEOL] In a sedimentary basin, the depth below which there is no economic accumulation of gaseous hydrocarbons. { 'gas ,flȯr }

gash fracture [GEOL] Open gashes that are formed diagonally to a fault or fault zone. { 'gash ,frak·chər }

gas hydrate [GEOCHEM] A naturally occurring solid composed of crystallized water (ice) molecules, forming a rigid lattice of cages (a clathrate) with most of the cages containing a molecule of natural gas, mainly methane. Also known as clathrate hydrate, gas clathrate. { ¦gas 'hī,drāt }

gash vein [GEOL] A mineralized fissure that extends a short distance vertically. { 'gash ,vān }

gaspeite [MINERAL] $NaCO_3$ An anhydrous normal carbonate mineral with calcite structure. { ga'spē,īt }

gas pocket [GEOL] A gas-filled cavity in rocks, especially above an oil pocket. { 'gas ,päk·ət }

gas reservoir [GEOL] An accumulation of natural gas found with or near accumulations of crude oil in the earth's crust. { ¦gas ¦rez·əv,wär }

gas sand [GEOL] A stratum of sand or porous sandstone from which natural gas may be extracted. { 'gas ,sand }

gas spurt [GEOL] An accumulation of organic matter on certain strata caused by escaping gas. { 'gas ,spərt }

gas zone [GEOL] A rock formation containing gas under a pressure large enough to force the gas out if tapped from the surface. { 'gas ,zōn }

gaufrage See plaiting. { gō'fräzh }

gaylussite [MINERAL] $Na_2Ca(CO_3)_2 \cdot 5H_2O$ A translucent, yellowish-white hydrous carbonate mineral, with a vitreous luster, crystallizing in the monoclinic system; found in dry lakes. { 'gā·lə,sīt }

geanticline [GEOL] A broad land uplift; refers to the land mass from which sediments in a geosyncline are derived. { ,jē'ant·i,klīn }

gearksutite [MINERAL] $CaAl(OH)F_4 \cdot H_2O$ A clayey mineral composed of hydrous calcium aluminum fluoride, occurring with cryolite. { jē'ärk·sə,tīt }

gedanite [MINERAL] A brittle, wine-yellow variety of amber containing little succinic acid; found on the shore of the Baltic Sea. { 'ged·ən,īt }

gedrite [MINERAL] An aluminous variety of the mineral anthophyllite. { 'je,drīt }

gehlenite [MINERAL] $Ca_2Al_2SiO_7$ A mineral of the melilite group that crystallizes in the tetragonal crystal system and is isomorphous with akermanite; a green, resinous material found with spinel. { 'gā·lə,nīt }

geikielite [MINERAL] $MgTiO_3$ A bluish-black or brownish-black mineral that crystallizes in the rhombohedral system and occurs in the form of rolled pebbles; it is isomorphous with ilmenite. { 'gē·kē,līt }

gelifluction [GEOL] The slow, continuous downslope movement of rock debris and water-saturated soil that occurs above frozen ground, as in most polar regions and in many high mountain ranges. Also known as congelifluction; gelisolifluction. { ¦jel·ə¦flək·shən }

gelifraction See congeliturbation. { ¦jel·ə¦frak·shən }

gelisol See frozen ground. { 'jel·ə,sȯl }

gelisolifluction See gelifluction. { jə,las·ə'fiək·shən }

geliturbation See congeliturbation. { ,jel·ə,ter'bāsh·ən }

gelivation See congelifraction. { ¦jel·ə¦vā·shən }

gel mineral See mineraloid. { 'jel ,min·rəl }

132

geocosmogony

Gelocidae [PALEON] A family of extinct pecoran ruminants in the superfamily Tragu-loidea. { jə'läs·ə,dē }

gelose *See* ulmin. { 'je,lōs }

gem [MINERAL] A natural or artificially produced mineral or other material that has sufficient beauty and durability for use as a personal adornment. { jem }

gemology [MINERAL] The science concerned with the identification, grading, evaluation, fashioning, and other aspects of gemstones. { je'mäl·ə·jē }

gemstone [GEOL] A mineral or petrified organic matter suitable for use in jewelry. { 'jem,stōn }

Gemuendinoidei [PALEON] A suborder of extinct raylike placoderm fishes in the order Rhenanida. { je¦myü·ən·də¦nȯid·ē,ī }

generalized hydrostatic equation [GEOPHYS] The vertical component of the vector equation of motion in natural coordinates when the acceleration of gravity is replaced by the virtual gravity; for most purposes it is identical to the hydrostatic equation. { 'jen·rə,līzd ,hī·drə¦stad·ik i'kwā·zhən }

genesis rocks [GEOL] Rocks that have retained their character from nearly 4.6 × 10⁹ years ago, when planets were still occulting out of the cloud of dust and gas referred to as the solar nebula; examples are meteorites and asteroids. { 'jen·ə·səs ,räks }

genetic facies [GEOL] An ancient deposit of rocks which have been formed by similar sedimentary processes. { jə¦ned·ik 'fā·shēz }

Geniohyidae [PALEON] A family of extinct ungulate mammals in the order Hyracoidea; all members were medium to large-sized animals with long snouts. { ¦jē·nē·ō'hī·ə,dē }

gentnerite [MINERAL] Cu₈Fe₃Cr₁₁S₁₈ A sulfide mineral known only in meteorites. { 'jent·nə,rīt }

geobotanical prospecting [GEOL] The use of the distribution, appearance, and growth anomalies of plants in locating ore deposits. { ¦jē·ō·bə¦tan·ə·kəl 'präs·pek·tiŋ }

geocerite [MINERAL] A white, waxy mineral composed of carbon, oxygen, and hydrogen, occurring in brown coal. { ,jē·ō'si,rīt }

geochemical anomaly [GEOCHEM] Above-average concentration of a chemical element in a sample of rock, soil, vegetation, stream, or sediment; indicative of nearby mineral deposit. { ¦jē·ō¦kem·ə·kəl ə'näm·ə·lē }

geochemical balance [GEOCHEM] The proportional distribution, and the migration rate, in the global fractionation of elements, minerals, or compounds; for example, the distribution of quartz in igneous rocks, its liberation by weathering, and its redistribution into sediments and, in solution, into lakes, rivers, and oceans. { ¦jē·ō¦kem·ə·kəl 'bal·əns }

geochemical cycle [GEOCHEM] During geologic changes, the sequence of stages in the migration of elements between the lithosphere, hydrosphere, and atmosphere. { ¦jē·ō¦kem·ə·kəl 'sī·kəl }

geochemical evolution [GEOCHEM] **1.** A change in any constituent of a rock beyond that amount present in the parent rock. **2.** A change in chemical composition of a major segment of the earth during geologic time, as the oceans. { ¦jē·ō¦kem·ə·kəl ,ev·ə'lü·shən }

geochemistry [GEOL] The study of the chemical composition of the various phases of the earth and the physical and chemical processes which have produced the observed distribution of the elements and nuclides in these phases. { ¦jē·ō¦kem·ə·strē }

geochron *See* isochron. { 'jē·ə,krän }

geochronology [GEOL] **1.** The dating of the events in the earth's history. **2.** A system of dating developed for the purposes of study of the earth's history. { ¦jē·ō·krə'näl·ə·jē }

geochronometry [GEOL] The study of the absolute age of the rocks of the earth based on the radioactive decay of isotopes, such as ²³⁸U, ²³⁵U, ²³²Th, ⁸⁷Rb, ⁴⁰K, and ¹⁴C, present in minerals and rocks. { ¦jē·ō·krə'näm·ə·trē }

geocosmogony [GEOL] The study of the origin of the earth. { ¦jē·ō,käz'mäj·ə·nē }

133

geocronite

geocronite [MINERAL] $Pb_5(Sb,As)_2S_3$ A mineral composed of lead-gray lead antimony arsenic sulfide. { jē'äk·rə,nīt }

geode [GEOL] A roughly spheroidal, hollow body lined inside with inward-projecting, small crystals; found frequently in limestone beds but may occur in shale. { 'jē,ōd }

geodesy [GEOPHYS] A subdivision of geophysics which includes determination of the size and shape of the earth, the earth's gravitational field, and the location of points fixed to the earth's crust in an earth-referred coordinate system. { jē'äd·ə·sē }

geodynamics [GEOPHYS] The branch of geophysics concerned with measuring, modeling, and interpreting the configuration and motion of the crust, mantle, and core of the earth. { ¦jē·ō·dī¦nam·iks }

geodynamo [GEOPHYS] The self-sustaining process responsible for maintaining the earth's magnetic field in which the kinetic energy of convective motion of the earth's liquid core is converted into magnetic energy. { ,jē·ō'dī·nə·mō }

geoelectricity See terrestrial electricity. { ¦jē·ō·i,lek'tris·əd·ē }

geoflex See orocline. { 'jē·ə,fleks }

geognosy [GEOL] The science dealing with the solid body of the earth as a whole, occurrences of minerals and rocks, and the origin of these and their relations. { jē'äg·nə·sē }

geographical cycle See geomorphic cycle. { ¦jē·ə¦graf·ə·kəl 'sī·kəl }

geoisotherm [GEOPHYS] The locus of points of equal temperature in the interior of the earth; a line in two dimensions or a surface in three dimensions. Also known as geotherm; isogeotherm. { ¦jē·ō'ī·sə,thərm }

geolith See rock-stratigraphic unit. { 'jē·ə,lith }

geologic age [GEOL] **1.** Any great time period in the earth's history marked by special phases of physical conditions or organic development. **2.** A formal geologic unit of time that corresponds to a stage. **3.** An informal geologic time unit that corresponds to any stratigraphic unit. { ¦jē·ə¦läj·ik 'āj }

geological oceanography [GEOL] The study of the floors and margins of the oceans, including descriptions of topography, composition of bottom materials, interaction of sediments and rocks with air and sea water, the effects of movements in the mantle on the sea floor, and action of wave energy in the submarine crust of the earth. Also known as marine geology; submarine geology. { ¦jē·ə¦läj·ə·kəl ,ō·shə'näg·rə·fē }

geological survey [GEOL] **1.** An organization making geological surveys and studies. **2.** A systematic geologic mapping of a terrain. { ¦jē·ə¦läj·ə·kəl 'sər,vā }

geological transportation [GEOL] Shifting of material by the action of moving water, ice, or air. { ¦jē·ə¦läj·ə·kəl ,tranz·pər'tā·shən }

geologic climate See paleoclimate. { ¦jē·ə¦läj·ik 'klī·mət }

geologic column [GEOL] **1.** The vertical sequence of strata of various ages found in an area or region. Also known as column. **2.** The geologic time scale as represented by rocks. { ¦jē·ə¦läj·ik 'käl·əm }

geologic erosion See normal erosion. { ¦jē·ə¦läj·ik ə'rō·zhən }

geologic log [GEOL] A graphic presentation of the lithologic or stratigraphic units or both traversed by a borehole; used in petroleum and mining engineering as well as geological surveys. { ¦jē·ə¦läj·ik 'läg }

geologic map [GEOL] A representation of the geologic surface or subsurface features by means of signs and symbols and with an indicated means of orientation; includes nature and distribution of rock units, and the occurrence of structural features, mineral deposits, and fossil localities. { ¦jē·ə¦läj·ik 'map }

geologic noise [GEOPHYS] Disturbances in observed data caused by random inhomogeneities in surface and near-surface material. { ¦jē·ə¦läj·ik'nóiz }

geologic province [GEOL] An area in which geologic history has been the same. { ¦jē·ə¦läj·ik 'präv·əns }

geologic section [GEOL] Any succession of rock units found at the surface or below ground in an area. Also known as section. { ¦jē·ə¦läj·ik 'sek·shən }

geologic structure [GEOL] The total structural features in an area. { ¦jē·ə¦läj·ik ,strək·chər }

geologic thermometer See geothermometer. { ¦jē·ə¦läj·ik thər'mäm·əd·ər }
geologic thermometry See geothermometry. { ¦jē·ə¦läj·ik thər'mäm·ə·trē }
geologic time [GEOL] The period of time covered by historical geology, from the end of the formation of the earth as a separate planet to the beginning of written history. { ¦jē·ə¦läj·ik 'tīm }
geologic time scale [GEOL] The relative age of various geologic periods and the absolute time intervals. { ¦jē·ə¦läj·ik 'tīm ‚skāl }
geologist [GEOL] An individual who specializes in the geological sciences. { jē'äl·ə·jəst }
geomagnetic coordinates [GEOPHYS] A system of spherical coordinates based on the best fit of a centered dipole to the actual magnetic field of the earth. { ¦jē·ō·mag¦ned·ik kō'ȯrd·ən·əts }
geomagnetic cutoff [GEOPHYS] The minimum energy of a cosmic-ray particle able to reach the top of the atmosphere at a particular geomagnetic latitude. { ¦jē·ō·mag¦ned·ik 'kə‚dȯf }
geomagnetic dipole [GEOPHYS] The magnetic dipole caused by the earth's magnetic field. { ¦jē·ō·mag¦ned·ik 'dī‚pōl }
geomagnetic equator [GEOPHYS] That terrestrial great circle which is 90° from the geomagnetic poles. { ¦jē·ō·mag¦ned·ik i'kwād·ər }
geomagnetic field [GEOPHYS] The earth's magnetic field. { ¦jē·ō·mag¦ned·ik 'fēld }
geomagnetic field reversal [GEOPHYS] Reversed magnetization in sedimentary and igneous rock, that is, polarized opposite to the mean geomagnetic field. { ¦jē·ō·mag¦ned·ik 'fēld ‚ri‚vər·səl }
geomagnetic latitude [GEOPHYS] The magnetic latitude that a location would have if the field of the earth were to be replaced by a dipole field closely approximating it. { ¦jē·ō·mag¦ned·ik 'lad·ə‚tüd }
geomagnetic longitude [GEOPHYS] Longitude that is determined around the geomagnetic axis instead of around the rotation axis of the earth. { ¦jē·ō·mag¦ned·ik 'län·jə‚tüd }
geomagnetic meridian [GEOPHYS] A semicircle connecting the geomagnetic poles. { ¦jē·ō·mag¦ned·ik mə'rid·ē·ən }
geomagnetic noise [GEOPHYS] Unwanted frequencies caused by fluctuations in the geomagnetic field of the earth. { ¦jē·ō·mag¦ned·ik 'nȯiz }
geomagnetic pole [GEOPHYS] Either of two antipodal points marking the intersection of the earth's surface with the extended axis of a powerful bar magnet assumed to be located at the center of the earth and having a field approximating the actual magnetic field of the earth. { ¦jē·ō·mag¦ned·ik 'pōl }
geomagnetic reversal [GEOPHYS] Reversed magnetization of the earth's magnetic dipole. { ¦jē·ō·mag¦ned·ik ri'vər·səl }
geomagnetic secular variation See secular variation. { ¦jē·ō·mag¦ned·ik ¦sek·yə·lər ver·ē'ā·shən }
geomagnetic storm See magnetic storm. { ¦jē·ō·mag¦ned·ik 'stȯrm }
geomagnetic variation [GEOPHYS] Temporal changes in the geomagnetic field, both long-term (secular) and short-term (transient). { ¦jē·ō·mag¦ned·ik ver·ē'ā·shən }
geomagnetism [GEOPHYS] **1.** The magnetism of the earth. Also known as terrestrial magnetism. **2.** The branch of science that deals with the earth's magnetism. { ¦jē·ō'mag·nə‚tiz·əm }
geomorphic cycle [GEOL] The cycle of change in the surface configuration of the earth. Also known as cycle of erosion; geographical cycle. { ¦jē·ō¦mȯr·fik 'sī·kəl }
geomorphology [GEOL] The study of the origin of secondary topographic features which are carved by erosion in the primary elements and built up of the erosional debris. { ¦jē·ō·mȯr'fäl·ə·jē }
geopetal [PETR] Pertaining to the top-to-bottom relations in rocks at the time of formation. { ¦jē·ə¦ped·əl }
geopetal fabric [PETR] The internal structure of a rock indicating the original orientation of the top-to-bottom strata. { ¦jē·ə¦ped·əl 'fab·rik }

geophysical fluid dynamics [GEOPHYS] The study of the naturally occurring, large-scale flows in the atmosphere and oceans, such as in weather patterns, atmospheric fronts, ocean currents, coastal upwelling, and the El Niño phenomenon. { ¦jē·ō¦fiz·ə·kəl ¦flü·əd dī'nam·iks }

geophysicist [GEOPHYS] An individual who specializes in geophysics. { ¦jē·ə'fiz·ə·sist }

geophysics [GEOL] The physics of the earth and its environment, that is, earth, air, and (by extension) space. { ¦jē·ə'fiz·iks }

geopotential topography [GEOPHYS] The topography of any surface as represented by lines of equal geopotential; these lines are the contours of intersection between the actual surface and the level surfaces (which everywhere are normal to the direction of the force of gravity), and are spaced at equal intervals of dynamic height. Also known as absolute geopotential topography. { ¦jē·ō·pə'ten·chəl tə'päg·rə·fē }

geopotential unit [GEOPHYS] A unit of gravitational potential used in describing the earth's gravitational field; it is equal to the difference in gravitational potential of two points separated by a distance of 1 meter when the gravitational field has a strength of 10 meters per second squared and is directed along the line joining the points. Abbreviated gpu. { ¦jē·ō·pə'ten·chəl 'yü·nət }

geopressure [GEOPHYS] An unusually high pressure exerted by a subsurface formation. { 'jē·ō,presh·ər }

geopressurized geothermal system [GEOL] A geothermal system dominated by the presence of hot fluids under high pressure (brine plus methane) and having higher-than-normal temperatures because of their low thermal conductivity, the presence of interbedded shale layers, or the existence of local, exothermic chemical reactions. { ¦jē·ō'presh·ə,rīzd ¦jē·ō¦thər·məl 'sis·təm }

Georges Banks [GEOL] An elevation beneath the sea east of Cape Cod, Massachusetts. { ¦jór·jəz 'baŋks }

georgiadesite [MINERAL] Pb₃(AsO₄)Cl₃ A white or brownish-yellow mineral composed of lead chloroarsenate, occurring in orthorhombic crystals. { jór'jäd·ə,sīt }

georgiaite [GEOL] Any of a group of North American tektites, 134 million years of age, found in Georgia. { 'jór·jə,īt }

geosere [GEOL] A series of ecological climax communities following each other in geologic time and changing in response to changing climate and physical conditions. { 'jē·ō,sir }

geosol [GEOL] A body of sediment or rock composed of one or more soil horizons. { 'jē·ə,sōl }

geosphere [GEOL] **1.** The solid mass of earth, as distinct from the atmosphere and hydrosphere. **2.** The lithosphere, hydrosphere, and atmosphere combined. { 'jē·ō,sfir }

geostatic pressure See ground pressure. { ¦jē·ō¦stad·ik 'presh·ər }

geostatistics [GEOL] A branch of applied statistics that focuses on mathematical description and analysis of geological observations. { ,jē·ō·stə'tis·tiks }

geostrophic [GEOPHYS] Pertaining to deflecting force resulting from the earth's rotation. { ¦jē·ō¦sträf·ik }

geostrophic approximation [GEOPHYS] The assumption that the geostrophic current can represent the actual horizontal current. Also known as geostrophic assumption. { ¦jē·ō¦sträf·ik ə¦präk·sə'mā·shən }

geostrophic assumption See geostrophic approximation. { ¦jē·ō¦sträf·ik ə'səm·shən }

geostrophic current [GEOPHYS] A current defined by assuming the existence of an exact balance between the horizontal pressure gradient force and the Coriolis force. { ¦jē·ō¦sträf·ik 'kə·rənt }

geostrophic equation [GEOPHYS] An equation, used to compute geostrophic current speed, which represents a balance between the horizontal pressure gradient force and the Coriolis force. { ¦jē·ō¦sträf·ik i'kwā·shən }

geostrophic equilibrium [GEOPHYS] A state of motion of a nonviscous fluid in which the horizontal Coriolis force exactly balances the horizontal pressure force at all points of the field so described. { ¦jē·ō¦sträf·ik ,ē·kwə'lib·rē·əm }

geostrophic flow [GEOPHYS] A form of gradient flow where the Coriolis force exactly balances the horizontal pressure force. { ¦jē·ō¦sträf·ik 'flō }

geosynclinal couple See orthogeosyncline. { ¦jē·ō,sin'klīn·əl 'kəp·əl }

geosynclinal cycle See tectonic cycle. { ¦jē·ō,sin'klīn·əl 'sī·kəl }

geosynclinal facies [GEOL] A sedimentary facies marked by great thickness, a generally argillaceous character, and few carbonate rocks. { ¦jē·ō,sin'klīn·əl 'fā·shēz }

geosyncline [GEOL] A linear part of the earth's crust, hundreds of kilometers long and tens of kilometers wide, that subsided during millions of years as it received thousands of meters of sedimentary and volcanic accumulations. { ¦jē·ō'sin,klīn }

geotectogene See tectogene. { ¦jē·ō¦tek·tə,jēn }

geotectonic cycle See orogenic cycle. { ¦jē·ō·tek'tän·ik 'sī·kəl }

geotectonics See tectonics. { ¦jē·ō·tek'tän·iks }

geotherm See geoisotherm. { 'jē·ō,thərm }

geothermal [GEOPHYS] Pertaining to heat within the earth. { ¦jē·ō¦thər·məl }

geothermal energy [GEOPHYS] Thermal energy contained in the earth; can be used directly to supply heat or can be converted to mechanical or electrical energy. { ¦jē·ō,thərm·əl 'en·ər·jē }

geothermal gradient [GEOPHYS] The change in temperature with depth of the earth. { ¦jē·ō¦thər·məl 'grād·ē·ənt }

geothermal system [GEOL] Any regionally localized geological setting where naturally occurring portions of the earth's internal heat flow are transported close enough to the earth's surface by circulating steam or hot water to be readily harnessed for use; examples are the Geysers Region of northern California and the hot brine fields in the Imperial Valley of southern California. { ¦jē·ō¦thər·məl 'sis·təm }

geothermometer [GEOL] A mineral that yields information about the temperature range within which it was formed. Also known as geologic thermometer. { ¦jē·ō·thər'mäm·əd·ər }

geothermometry [GEOL] Measurement of the temperatures at which geologic processes occur or occurred. Also known as geologic thermometry. { ¦jē·ō·thər'mäm·ə·trē }

gerhardtite [MINERAL] $Cu_2(NO_3)(OH)_3$ An emerald-green mineral composed of basic copper nitrate. { 'ger,härd,īt }

germanite [MINERAL] $Cu_3(Ge,Ga,Fe)(S,As)_4$ Reddish-gray mineral occurring in massive form; an important ore of germanium. { 'jər·mə,nīt }

germination See grain growth. { ,jer·mə'nā·shən }

gersdorffite [MINERAL] NiAsS A silver-white to steel-gray mineral, crystallizing in the isometric system; resembles cobaltite and may contain some iron and cobalt. Also known as nickel glance. { 'gerz,dór,fīt }

geyserite See siliceous sinter. { 'gī·zə,rīt }

ghost [PETR] The discernible outline of the shape of a former crystal or of another rock structure that has been partly obliterated and has as its boundaries inclusions, bubbles, or other foreign matter. Also known as phantom. { gōst }

giant granite See pegmatite. { ¦jī·ənt 'gran·ət }

giant's cauldron See giant's kettle. { ¦jī·əns 'kól·drən }

giant's kettle [GEOL] A cylindrical hole bored in bedrock beneath a glacier by water falling through a deep moulin or by boulders rotating in the bed of a meltwater stream. Also known as giant's cauldron; moulin pothole; potash kettle. { ¦jī·əns 'ked·əl }

gibbsite [MINERAL] $Al(OH)_3$ A white or tinted mineral, crystallizing in the monoclinic system; a principal constituent of bauxite. Also known as hydrargillite. { 'gib,zit }

Gibraltar stone See onyx marble. { jə'bról·dər ,stōn }

gillespite [MINERAL] $BaFeSi_4O_{10}$ A micalike mineral composed of barium and iron silicate. { gə'le,spīt }

gilsonite [MINERAL] A variety of asphalt; it has black color, brilliant luster, brown streaks, and conchoidal fracture. { 'gil·sə,nīt }

ginorite [MINERAL] $Ca_2B_{14}O_{23}\cdot 8H_2O$ A white monoclinic mineral composed of hydrous borate of calcium. { 'jin·ə,rīt }

giobertite See magnesite. { 'jō·bər,tīt }

girdle [PETR] With reference to a fabric diagram or equal-area projection net, a belt showing concentration of points which is approximately coincident with a great circle of the net and which represents orientation of the fabric elements. { 'gərd·əl }

gismondite [MINERAL] $CaAl_2Si_2O_8 \cdot 4H_2O$ A light-colored mineral composed of hydrous calcium aluminum silicate, occurring in pyramidal crystals. { jiz'män,dīt }

glacial [GEOL] Pertaining to an interval of geologic time which was marked by an equatorward advance of ice during an ice age; the opposite of interglacial; these intervals are variously called glacial periods, glacial epochs, glacial stages, and so on. { 'glā·shəl }

glacial abrasion [GEOL] Alteration of portions of the earth's surface as a result of glacial flow. { ¦glā·shəl ə'brā·zhən }

glacial accretion [GEOL] Deposition of material as a result of glacial flow. { ¦glā·shəl ə'krē·shən }

glacial advance [GEOL] **1.** Increase in the thickness and area of a glacier. **2.** A time period equal to that increase. { ¦glā·shəl əd'vans }

glacial boulder [GEOL] A boulder moved to a point distant from its original site by a glacier. { ¦glā·shəl 'bōl·dər }

glacial deposit [GEOL] Material carried to a point beyond its original location by a glacier. { ¦glā·shəl di'päz·ət }

glacial drift [GEOL] All rock material in transport by glacial ice, and all deposits predominantly of glacial origin made in the sea or in bodies of glacial meltwater, including rocks rafted by icebergs. { ¦glā·shəl 'drift }

glacial epoch [GEOL] **1.** Any of the geologic epochs characterized by an ice age; thus, the Pleistocene epoch may be termed a glacial epoch. **2.** Generally, an interval of geologic time which was marked by a major equatorward advance of ice; the term has been applied to an entire ice age or (rarely) to the individual glacial stages which make up an ice age. { ¦glā·shəl 'ep·ək }

glacial erosion [GEOL] Movement of soil or rock from one point to another by the action of the moving ice of a glacier. Also known as ice erosion. { ¦glā·shəl ə'rō·zhən }

glacial flour See rock flour. { ¦glā·shəl 'flaů·ər }

glacial geology [GEOL] The study of land features resulting from glaciation. { ¦glā·shəl jē'äl·ə·jē }

glacial lake [GEOL] A lake that exists because of the effects of the glacial period. { ¦glā·shəl 'lāk }

glacial maximum [GEOL] The time or position of the greatest extent of any glaciation; most frequently applied to the greatest equatorward advance of Pleistocene glaciation. { ¦glā·shəl 'mak·sə·məm }

glacial outwash See outwash. { ¦glā·shəl 'aůt,wäsh }

glacial period [GEOL] **1.** Any of the geologic periods which embraced an ice age; for example, the Quaternary period may be called a glacial period. **2.** Generally, an interval of geologic time which was marked by a major equatorward advance of ice. { ¦glā·shəl 'pir·ē·əd }

glacial plucking See plucking. { ¦glā·shəl 'plək·iŋ }

glacial retreat [GEOL] A condition occurring when backward melting at the front of a glacier takes place at a rate exceeding forward motion. { ¦glā·shəl ri'trēt }

glacial scour [GEOL] Erosion resulting from glacial action, whereby the surface material is removed and the rock fragments carried by the glacier abrade, scratch, and polish the bedrock. Also known as scouring. { ¦glā·shəl 'skaůr }

glacial striae [GEOL] Scratches, commonly parallel, on smooth rock surfaces due to glacial abrasion. { ¦glā·shəl 'strī,ī }

glacial till See till. { ¦glā·shəl 'til }

glacial trough [GEOL] A deep U-shaped valley with steep sides that leads down from a cirque and was excavated by a glacier. { ¦glā·shəl 'trof }

glacial varve See varve. { ¦glā·shəl 'värv }

glaciated terrain [GEOL] A region that once bore great masses of glacial ice; a distinguishing feature is marks of glaciation. { 'glā·shē,ād·əd tə'rān }

glaciation [GEOL] Alteration of any part of the earth's surface by passage of a glacier, chiefly by glacial erosion or deposition. { ˌglā·chē'ā·shən }

glaciation limit [GEOPHYS] For a given locality, the lowest altitude at which glaciers can develop. { ˌglā·shē'ā·shən ˌlim·ət }

glacier table [GEOL] A stone block supported by an ice pedestal above the surface of a glacier. { 'glā·shər ˌtā·bəl }

glacioeustasy [GEOL] Changes in sea level due to storage or release of water from glacier ice. { ¦glās·ē·ō'yü·stə·sē }

glaciofluvial [GEOL] Pertaining to streams fed by melting glaciers, or to the deposits and landforms produced by such streams. { ¦glā·shē·ō¦flü·vē·əl }

glacioisostasy [GEOL] Lithospheric depression or rebound due to the weight or melting of glacier ice. { ˌglā·sē·ō·ī'sās·tə·sē }

glaciolacustrine [GEOL] Pertaining to lakes fed by melting glaciers, or to the deposits forming therein. { ¦glā·shē·ō·lə'kəs·trən }

glaciology [GEOL] A broad field encompassing all aspects of the study of ice: glaciers, the largest ice masses on earth; ice that forms on rivers, lakes, and the sea; ice in the ground, including both permafrost and seasonal ice such as that which disrupts roads; ice that crystallizes directly from the air on structures such as airplanes and antennas, and all forms of snow research, including hydrological and avalanche forecasting. { ˌglā·shē'äl·ə·jē }

gladite [MINERAL] PbCuBi₅S₉ A lead gray mineral consisting of lead and copper bismuth sulfide; occurs as prismatic crystals. { 'gla,dīt }

glance pitch [GEOL] A variety of asphaltite having brilliant conchoidal fracture, and resembling gilsonite but having higher specific gravity and percentage of fixed carbon. { 'glans ˌpich }

glaserite See arcanite. { 'gla·zə,rīt }

glass porphyry See vitrophyre. { ¦glas 'pȯr·fə·rē }

glass schorl See axinite. { 'glas ˌshȯrl }

glassy feldspar See sanidine. { ¦glas·ē 'fel,spär }

glauberite [MINERAL] Na₂Ca(SO₄)₂ A brittle, gray-yellow monoclinic mineral having vitreous luster and saline taste. { 'glau̇·bə,rīt }

glaucocerinite [MINERAL] A mineral composed of a hydrous basic sulfate of copper, zinc, and aluminum. { ˌglȯ·kō'se·rə,nīt }

glaucochroite [MINERAL] CaMnSiO₄ A bluish-green mineral that is related to monticellite, is composed of calcium manganese silicate, and occurs in prismatic crystals. { ˌglō·kə'krō,īt }

glaucodot [MINERAL] (Co,Fe)AsS A grayish-white, metallic-looking mineral composed of cobalt iron sulfarsenide, occurring in orthorhombic crystals. { 'glȯ·kə,dät }

glauconite [MINERAL] K₁₅(Fe,Mg,Al)₄₋₆(Si,Al)₈O₂₀(OH)₄ A type of clay mineral; it is dioctohedral and occurs in flakes and as pigmentary material. { 'glȯ·kə,nīt }

glauconitic sandstone [PETR] A quartz sandstone or an arkosic sandstone that has many glauconite grains. { ¦glȯ·kə¦nid·ik 'san,stōn }

glaucophane [MINERAL] Na₂Mg₃Al₂Si₈ A blue to black monoclinic sodium amphibole; blue to black coloration with marked pleochroism. { 'glȯ·kə,fān }

glaucophane schist [PETR] Metamorphic schist that contains glaucophane. { 'glȯ·kə,fān ˌshist }

glessite [GEOL] Fossil resin similar to amber. { 'gle,sīt }

gley [GEOL] A sticky subsurface layer of clay in some waterlogged soils. { glā }

glide fold See shear fold. { 'glīd ˌfōld }

globigerina ooze [GEOL] A pelagic sediment consisting of than 30% calcium carbonate in the form of foraminiferal tests of which *Globigerina* is the dominant genus. { glō,bij·ə'rī·nə ˌüz }

globular See spherulitic. { 'gläb·yə·lər }

globulite [GEOL] A small, isotropic, globular of spherulelike crystallite; usually dark in color and found in glassy extrusive rocks. { 'gläb·yə,līt }

glockerite [MINERAL] A brown, ocher yellow, black, or dull green mineral consisting of a hydrated basic sulfate of ferric iron; occurs in stalactitic, encrusting, or earthy forms. { 'glä·kə,rīt }

glossopterid flora [PALEOBOT] Permian and Triassic fossil ferns of the genus *Glossopteris*. { glä'säp·tə·rəd 'flör·ə }

gloup [GEOL] An opening in the roof of a sea cave. { glüp }

glowing avalanche *See* ash flow. { ¦glō·iŋ 'av·ə,lanch }

glowing cloud *See* nuée ardente. { ¦glō·iŋ 'klaůd }

Glyphocyphidae [PALEON] A family of extinct echinoderms in the order Temnopleuroida comprising small forms with a sculptured test, perforate crenulate tubercles, and diademoid ambulacral plates. { ,glif·ō'sīf·ə,dē }

Glyptocrinina [PALEON] A suborder of extinct crinoids in the order Monobathrida. { ¦glip·tō·krə'nī·nə }

glyptolith *See* ventifact. { 'glip·tə,lith }

gmelinite [MINERAL] $(Na_2Ca)Al_2Si_4O_{12} \cdot 6H_2O$ Zeolite mineral that is colorless or lightly colored and crystallizes in the hexagonal system. { gə'mel·ə,nīt }

Gnathobelodontinae [PALEON] A subfamily of extinct elephantoid proboscideans containing the shovel-jawed forms of the family Gomphotheriidae. { nā¦thäb·ə·lō'dän·tə,nē }

Gnathodontidae [PALEON] A family of extinct conodonts having platforms with large, cup-shaped attachment scars. { ,nā·thō'dän·tə,dē }

gneiss [PETR] A variety of rocks with a banded or coarsely foliated structure formed by regional metamorphism. { nīs }

gneissic granodiorites [PETR] Granodiorite rocks with gneissic characteristics. { 'nīs·ik ¦gra·nō'dī·ə,rīts }

gobi [GEOL] Sedimentary deposits in a synclinal basin. { 'gō·bē }

Gobiatheriinae [PALEON] A subfamily of extinct herbivorous mammals in the family Uintatheriidae known from one late Eocene genus; characterized by extreme reduction of anterior dentition and by lack of horns. { gō¦bī·ə·thə'rī·ə,nē }

goethite [MINERAL] FeO(OH) A yellow, red, or dark-brown mineral crystallizing in the orthorhombic system, although it is usually found in radiating fibrous aggregates; a common constituent of natural rust or limonite. Also known as xanthosiderite. { 'gə,tīt }

gold beryl *See* chrysoberyl. { 'gōld ,ber·əl }

goldschmidtine *See* stephanite. { 'gōl,shmid,ēn }

goldschmidtite *See* sylvanite. { 'gōl,shmid,īt }

Goldschmidt's mineralogical phase rule [GEOL] The rule that the probability of finding a system with degrees of freedom less than two is small under natural rock-forming conditions. { 'gōl,shmits ,min·ə·rə'läj·ə·kəl ¦fāz ,rül }

Gomphotheriidae [PALEON] A family of extinct proboscidean mammals in the suborder Elephantoidea consisting of species with shoveling or digging specializations of the lower tusks. { ,gäm·fō·thə'rī·ə,dē }

Gomphotheriinae [PALEON] A subfamily of extinct elephantoid proboscideans in the family Gomphotheriidae containing species with long jaws and bunomastodont teeth. { ,gäm·fō·thə'rī·ə,nē }

Gondwana [GEOL] The ancient continent that is supposed to have fragmented and drifted apart during the Triassic to form eventually the present continents. Also known as Gondwanaland. { gänd'wä·nə }

Gondwanaland *See* Gondwana. { gän'dwän·ə,land }

gonnardite [MINERAL] $Na_2CaAl_4Si_6O_{20} \cdot 7H_2O$ Zeolite mineral occurring in fibrous, radiating spherules; specific gravity is 2.3. { 'gän·ər,dīt }

goongarrite [MINERAL] $Pb_4Bi_2S_7$ A mineral composed of a sulfide of lead and bismuth. { gün'ga,rīt }

gooseberry stone *See* grossularite. { 'güs,ber·ē ,stōn }

gorceixite [MINERAL] $BaAl_3(PO_4)_2(OH)_5 \cdot H_2O$ A brown mineral composed of a hydrous basic phosphate of barium and aluminum. { 'gȯr·sək,sīt }

gordonite [MINERAL] $MgAl_2(PO_4)_2(OH)_2 \cdot 8H_2O$ A colorless mineral composed of a hydrous basic phosphate of magnesium and aluminum. { 'gȯrd·ən,īt }

goslarite [MINERAL] $ZnSO_4 \cdot 7H_2O$ A white mineral composed of hydrous zinc sulfate. { 'gäs·lə‚rīt }

gossan [GEOL] A rusty, ferruginous deposit filling the upper regions of mineral veins and overlying a sulfide deposit; formed by oxidation of pyrites. Also known as capping; gozzan; iron hat. { 'gas·ən }

Gotlandian [GEOL] A geologic time period recognized in Europe to include the Ordovician; it appears before the Devonian. { gät'lan·dē·ən }

gouge [GEOL] Soft, pulverized mixture of rock and mineral material found along shear (fault) zones and produced by the differential movement across the plane of slippage. { gaúj }

goyazite [MINERAL] $SrAl_3(PO_4)_2(OH)_5 \cdot H_2O$ A granular, yellowish-white mineral composed of a hydrous strontium aluminum phosphate. { 'gói·ə‚zīt }

gozzan See gossan. { 'gäz·ən }

gpu See geopotential unit.

graben [GEOL] A block of the earth's crust, generally with a length much greater than its width, that has dropped relative to the blocks on either side. { 'grä·bən }

gradation [GEOL] **1.** The leveling of the land, or the bringing of a land surface or area to a uniform or nearly uniform grade or slope through erosion, transportation, and deposition. **2.** Specifically, the bringing of a stream bed to a slope at which the water is just able to transport the material delivered to it. { grā'dā·shən }

gradation period [GEOL] The time during which the base level of the sea remains in one position. Also known as base-leveling epoch. { grā'dā·shən ‚pir·ē·əd }

grade [GEOL] The slope of the bed of a stream, or of a surface over which water flows, upon which the current can just transport its load without either eroding or depositing. { grād }

grade correction See slope correction. { 'grād kə‚rek·shən }

graded [GEOL] Brought to or established at grade. { 'grād·əd }

graded bedding [GEOL] A stratification in which each stratum displays a gradation in the size of grains from coarse below to fine above. { ¦grād·əd 'bed·iŋ }

graded profile See profile of equilibrium. { ¦grād·əd 'prō‚fīl }

grade scale [GEOL] A continuous scale of particle sizes divided into a series of size classes. { 'grād ‚skāl }

gradient [GEOL] The rate of descent or ascent (steepness of slope) of any topographic feature, such as streams or hillsides. { 'grād·ē·ənt }

grading [GEOL] The gradual reduction of the land to a level surface; for example, erosion of land to base level by streams. { 'grād·iŋ }

graftonite [MINERAL] $(Fe,Mn,Ca)_3(PO_4)_2$ A salmon-pink mineral, crystallizing in the monoclinic system, and found as laminated intergrowths of triphylite; hardness is 5 on Mohs scale, and specific gravity is 3.7. { 'graf·tə‚nīt }

grahamite [GEOL] See mesosiderite. [MINERAL] A solid, jet-black hydrocarbon that occurs in veinlike masses; soluble in carbon disulfide and chloroform. { 'grā·ə‚mīt }

grain [GEOL] The particles or discrete crystals that make up a sediment or rock. { grān }

grain diminution See degradation recrystallization. { 'grān dim·yə'nish·ən }

grain growth [PETR] Enlargement of some individual crystals in a monomineralic rock, producing a coarser texture. Also known as germination. { 'grān ‚grōth }

grain size [GEOL] Average size of mineral particles composing a rock or sediment. { 'grān ‚sīz }

grainstone [PETR] A mud-free (micrite-free) limestone. { 'grān‚stōn }

gramenite See nontronite. { 'gra·mə‚nīt }

grandite [MINERAL] A garnet that is intermediate in chemical composition between grossular and androdite. { 'gran‚dīt }

granite [PETR] A visibly crystalline plutonic rock with granular texture; composed of quartz and alkali feldspar with subordinate plagioclase and biotite and hornblende. { 'gran·ət }

granite-gneiss [PETR] A banded metamorphic rock derived from igneous or sedimentary rocks mineralogically equivalent to granite. { 'gran·ət ¦nīs }

granite pegmatite *See* pegmatite. { 'gran·ət 'peg·mə,tīt }

granite porphyry *See* quartz porphyry. { 'gran·ət 'pȯr·fə·rē }

granite series [GEOL] A sequence of products that evolve continuously during crustal fusion; earlier products tend to be deep-seated, syntectonic, and granodioritic, and later products tend to be shallower, late syntectonic, or postsyntectonic, and more potassic. { 'gran·ət ,sir·ēz }

granite wash [GEOL] Material eroded from granites and redeposited, forming a rock with the same major mineral constituents as the original rock. { 'gran·ət ,wäsh }

granitic batholith [GEOL] A granitic shield mass intruded as the fusion of older formations. { grə'nid·ik 'bath·ə,lith }

granitic layer *See* sial. { grə'nid·ik 'lā·ər }

granitic magma [PETR] A coarse-grained igneous rock. { grə'nid·ik 'mag·mə }

granitization [PETR] A process whereby various types of rock may be converted to granite or closely related material. { ,gran·əd·ə'zā·shən }

granoblastic fabric [PETR] The texture of metamorphic rocks composed of equidimensional elements formed during recrystallization. { ¦gra·nō¦blas·tik 'fab·rik }

granodiorite [PETR] A visibly crystalline plutonic rock composed chiefly of sodic plagioclase, alkali feldspar, quartz, and subordinate dark-colored minerals. { ¦gra·nō'dī·ə,rīt }

granofels [PETR] A medium-to coarse-grained metamorphic rock possessing a granoblastic fabric and either lacking foliation or lineation entirely or exhibiting such characteristics only indistinctly. { 'gran·ə,felz }

granogabbro [PETR] Plutonic rock composed of quartz, basic plagioclase, potash-feldspar, and at least one ferromagnesian mineral; intermediate between a granite and a gabbro, and in a strict sense, a granodiorite with more than 50% boric plagioclase. { ¦gra·nō'ga·brō }

granophyre [PETR] A quartz porphyry or fine-grained porphyritic granite. { 'gran·ə,fīr }

granularity [PETR] The feature of rock texture relating to the size of the constituent grains or crystals. { ,gran·yə'lar·əd·ē }

granule [GEOL] A somewhat rounded rock fragment ranging in diameter from 2 to 4 millimeters; larger than a coarse sand grain and smaller than a pebble. { 'gran·yül }

granulite [PETR] **1.** Granite that contains muscovite. **2.** A relatively coarse, granuloblastic rock formed at the high temperatures and pressures of the granulite facies. { 'gran·yə,līt }

granulite facies [PETR] A group of gneissic rocks characterized by a granoblastic fabric and formed by regional dynamothermal metamorphism at temperatures above 650°C and pressures of 3000–12,000 bars. { 'gran·yə,līt 'fā·shēz }

granulometry [PETR] Measurement of grain sizes of sedimentary rock. { ,gran·yə'läm·ə·trē }

grapestone [GEOL] A cluster of sand-size grains, such as calcareous pellets, held together by incipient cementation shortly after deposition; the outer surface is lumpy, resembling a bunch of grapes. { 'grāp,stōn }

graphic granite [PETR] A distinct type of pegmatite in which quartz and orthoclase crystals grew together along a parallel axis. Also known as Hebraic granite; runite. { ¦graf·ik'gran·ət }

graphic intergrowth [PETR] An intergrowth of crystals, commonly feldspar and quartz, that produces a type of poikilitic texture in which the larger crystals have a fairly regular geometric outline and orientation, and resemble cuneiform writing. { ¦graf·ik 'in·tər,grōth }

graphic tellurium *See* sylvanite. { ¦graf·ik te'lür·ē·əm }

graphic texture [GEOL] A pattern of rocks that is similar to cuneiform characters. { 'graf·ik ,teks·chər }

graphite [MINERAL] A mineral consisting of a low-pressure allotropic form of carbon; it is soft, black, and lustrous and has a greasy feeling; it occurs naturally in hexagonal crystals or massive or can be synthesized from petroleum coke; hardness is 1–2 on Mohs scale, and specific gravity is 2.09–2.23; used in pencils, crucibles, lubricants, paints, and polishes. Also known as black lead; plumbago. { 'gra,fīt }

graptolites See graptolithina. { 'grap·tə,līts }

graptolite shale [GEOL] Shale containing an abundance of extinct colonial marine organisms known as graptolites. { 'grap·tə,līt 'shāl }

Graptolithina [PALEON] A class of extinct colonial animals believed to be related to the class Pterobranchia of the Hemichordata. Also known as graptolites. { ,grap·tə·lə'thīn·ə }

Graptoloidea [PALEON] An order of extinct animals in the class Graptolithina including branched, planktonic forms described from black shales. { ,grap·tə'lòid·ē·ə }

Graptozoa [PALEON] The equivalent name for Graptolithina. { ,grap·tə'zō·ə }

gratonite [MINERAL] $Pb_9As_4S_{15}$ A mineral composed of lead arsenic sulfide, occurring in rhombohedral crystals. { 'grat·ən,īt }

gravel [GEOL] A loose or unconsolidated deposit of rounded pebbles, cobbles, or boulders. { 'grav·əl }

gravel bank [GEOL] A natural mound or exposed face of gravel, particularly such a place from which gravel is dug. { 'grav·əl ,baŋk }

gravel desert See reg. { 'grav·əl ¦dez·ərt }

gravitational settling [GEOL] A movement of sediment resulting from gravitational forces. { ,grav·ə'tā·shən·əl 'set·liŋ }

gravitational sliding [GEOL] Extensive sliding of strata down a slope of an uplifted area. Also known as sliding. { ,grav·ə'tā·shən·əl 'slīd·iŋ }

gravity anomaly [GEOPHYS] The difference between the observed gravity and the theoretical or predicted gravity. { 'grav·əd·ē ə,näm·ə·lē }

gravity-collapse structure See collapse structure. { 'grav·əd·ē kə¦laps ,strək·chər }

gravity drainage reservoir [GEOL] A reservoir in which production is significantly affected by gas, oil, and water separating under the influence of gravity while production takes place. { 'grav·əd·ē 'drān·ij ,rez·əv,wär }

gravity erosion See mass erosion. { 'grav·əd·ē i,rō·zhən }

gravity fault See normal fault. { 'grav·əd·ē ,fólt }

gravity map [GEOPHYS] A map of gravitational variations in an area displaying gravitational highs and lows. { 'grav·əd·ē ,map }

gravity slope [GEOL] The relatively steep slope on a hillside above the wash slope; usually situated at the angle of repose of the material eroded from it. { 'grav·əd·ē ,slōp }

gravity tide [GEOPHYS] Cyclic motion of the earth's surface caused by interaction of gravitational forces of the moon, sun, and earth. { 'grav·əd·ē ,tīd }

gray antimony See antimonite; jamesonite. { ¦grā 'ant·ə,mō·nē }

gray cobalt See cobaltite. { ¦grā 'kō,bólt }

gray copper ore See tetrahedrite. { 'gra 'käp·ər ,ór }

gray hematite See specularite. { ¦grā 'hē·mə,tīt }

gray manganese ore See manganite. { ¦grā 'maŋ·gə,nēs ,ór }

graywacke [PETR] An argillaceous sandstone characterized by an abundance of unstable mineral and rock fragments and a fine-grained clay matrix binding the larger, sand-size detrital fragments. { 'grā,wak·ə }

greasy quartz See milky quartz. { 'grē·sē 'kwórts }

Great Ice Age [GEOL] The Pleistocene epoch. { ¦grāt 'īs ,āj }

great soil group [GEOL] A group of soils having common internal soil characteristics; a subdivision of a soil order. { ¦grāt 'sóil ,grüp }

green chalcedony See chrysoprase. { ¦grēn kal'sed·ən·ē }

greenlandite See columbite. { 'grēn·lən,dīt }

Greenland spar See cryolite. { 'grēn·lənd 'spär }

green lead ore See pyromorphite. { 'grēn 'led ,ór }

green mud [GEOL] **1.** A fine-grained, greenish terrigenous mud or oceanic ooze found near the edge of a continental shelf at depths of 300–7500 feet (90–2300 meters). **2.** A deep-sea terrigenous deposit characterized by the presence of a considerable proportion of glauconite and calcium carbonate. { ¦grēn 'məd }

greenockite [MINERAL] CdS A green or orange mineral that crystallizes in the hexagonal

greensand

system; occurs as an earthy encrustation and is dimorphous with hawleyite. Also known as cadmium blende; cadmium ocher; xanthochroite. { 'grē·nə,kīt }

greensand [GEOL] A greenish sand consisting principally of grains of glauconite and found between the low-water mark and the inner mud line. [PETR] Sandstone composed of greensand with little or no cement. { 'grēn,sand }

greenschist [PETR] A schistose metamorphic rock with abundant chlorite, epidote, or actinolite present, giving it a green color. { 'grēn,shist }

greenschist facies [PETR] Any schistose rock containing an abundance of green minerals and produced under conditions of low to intermediate temperatures (300–500°C) and low to moderate hydrostatic pressures (3000–8000 bars). { 'grēn,shist 'fā·shēz }

greenstone [MINERAL] See nephrite. [PETR] Any altered basic igneous rock which is green due to the presence of chlorite, hornblende, or epidote. { 'grēn,stōn }

greenstone belts [GEOL] Oceanic and island arclike sequences that are similar to, and run to the south and north of, the Swaziland System. { 'grēn,stōn ,belts }

greenstone schist [PETR] Greenstone with a foliated structure. { 'grēn,stōn ,shist }

greisen [PETR] A pneumatolytically altered granite consisting of mainly quartz and a light-green mica. { 'grīz·ən }

grenatite See leucite; staurolite. { 'gren·ə,tīt }

Grenville orogeny [GEOL] A Precambrian mountain-forming epoch. { 'gren·vəl ȯ'räj·ə·nē }

griffithite [MINERAL] A micalike mineral containing magnesium, iron, calcium, and aluminosilicate. { 'grif·ə,thīt }

grike [GEOL] A vertical fissure developed along a joint in limestone by dissolution of some of the rock. Also spelled gryke. { grīk }

griphite [MINERAL] $(Na,Al,Ca,Fe)_6Mn_4(PO_4)_5(OH)_4$ Mineral composed of a basic phosphate of sodium, calcium, iron, aluminum, and manganese. { 'gri,fīt }

griquaite [PETR] A hypabyssal rock that contains garnet and diopside, and sometimes olivine or phlogopite, and is found in kimberlite pipes and dikes. { 'grē·kwə,īt }

grit [GEOL] **1.** A hard, sharp granule, as of sand. **2.** A coarse sand. [PETR] A sandstone composed of angular grains of different sizes. { grit }

Groeberiidae [PALEON] A family of extinct rodentlike marsupials. { ,grə·bə'rī·ə,dē }

groove [GEOL] Glaciated marks of large size on rock. { grüv }

groove casts [GEOL] Rounded or sharp, crested, rectilinear ridges that are a few millimeters high and a few centimeters long; found on the undersurfaces of sandstone layers lying on mudstone. { 'grüv ,kasts }

grossular See grossularite. { 'gräs·yə·lər }

grossularite [MINERAL] $Ca_3Al_2(SiO_4)_3$ The colorless or green, yellow, brown, or red end member of the garnet group, often occurring in contact-metamorphosis impure limestones. Also known as gooseberry stone; grossular. { 'gräs·yə·lə,rīt }

grothite See sphene. { 'grō,thīt }

ground [GEOL] **1.** Any rock or rock material. **2.** A mineralized deposit. **3.** Rock in which a mineral deposit occurs. { graünd }

ground ice mound [GEOL] A frost mound containing bodies of ice. Also known as ice mound. { 'graünd ,īs ,maünd }

groundmass See matrix. { 'graünd,mas }

ground moraine [GEOL] Rock material carried and deposited in the base of a glacier. Also known as bottom moraine; subglacial moraine. { 'graünd mə,rān }

ground noise [GEOPHYS] In seismic exploration, disturbance of the ground due to some cause other than the shot. { 'graünd ,nȯiz }

ground pressure [GEOPHYS] The pressure to which a rock formation is subjected by the weight of the superimposed rock and rock material or by diastrophic forces created by movements in the rocks forming the earth's crust. Also known as geostatic pressure; lithostatic pressure; rock pressure. { 'graünd ,presh·ər }

ground truth measurements [GEOPHYS] Measurements of various properties, such as temperature and land utilization, which are conducted on the ground to calibrate observations made from satellites or aircraft. { 'graünd ,trüth ,mezh·ər·məns }

group |GEOL| A lithostratigraphic material unit comprising several formations. { grüp }

groutite |MINERAL| HMnO$_2$ A mineral of the diaspore group, composed of manganese, hydrogen, and oxygen; it is polymorphous with manganite. { 'graú,tīt }

growl |GEOPHYS| Noise heard when strata are subjected to great pressure. { graúl }

growth fabric |PETR| Orientation of fabric elements independent of the influences of stress and resultant movement. { 'grōth ,fab·rik }

growth lattice |GEOL| The rigid, reef-building, inplace framework of an organic reef, consisting of skeletons of sessile organisms and excluding reef-flank and other associated fragmental deposits. Also known as organic lattice. { 'grōth ,lad·əs }

gruenlingite |MINERAL| Bi$_4$TeS$_3$ A mineral composed of sulfide and telluride of bismuth. { 'grün·liŋ,īt }

grunerite |MINERAL| (Mg,Fe)$_7$Si$_8$O$_{22}$(OH)$_2$ Variety of amphibole; forms monoclinic crystals. { 'grün·ə,rīt }

grus See gruss. { grüs }

gruss |GEOL| A loose accumulation of fragmental products formed from the weathering of granite. Also spelled grus. { grüs }

gryke See grike. { grīk }

Guadalupian |GEOL| A North American provincial series in the Lower and Upper Permian, above the Leonardian and below the Ochoan. { ˌgwäd·əl'ü·pē·ən }

guanajuatite |MINERAL| Bi$_2$Se$_3$ Bluish-gray mineral composed of bismuth selenide, occurring in crystals or masses. { ˌgwän·ə'hwä,tīt }

gudmundite |MINERAL| FeSbS A silver-white to steel-gray orthorhombic mineral composed of a sulfide and antimonide of iron. { 'gúd·mən,dīt }

guest element See trace element. { 'gest ˌel·ə·mənt }

guide fossil |PETR| A fossil used for rock correlation and age determination. { 'gīd ,fäs·əl }

guildite |MINERAL| (Cu,Fe)$_3$(Fe,Al)$_4$(SO$_4$)$_7$(OH)$_4$·15H$_2$O A dark-brown mineral composed of a basic hydrated sulfate of copper, iron, and aluminum. { 'gil,dīt }

guitermanite |MINERAL| Pb$_{10}$Ar$_6$S$_{19}$ A bluish-gray mineral composed of lead, arsenic, and sulfur, occurring in compact masses. { 'gid·ər·mə,nīt }

Gulfian |GEOL| A North American provincial series in Upper Cretaceous geologic time, above the Comanchean and below the Paleocene of the Tertiary. { 'gəlf·ē·ən }

gully erosion |GEOL| Erosion of soil by running water. { 'gəl·ē iˌrō·zhən }

gumbo |GEOL| A soil that forms a sticky mud when wet. { 'gəm·bō }

gumbotil |GEOL| Deoxidized, leached clay that contains siliceous stones. { 'gəm·bō,til }

gummite |MINERAL| Any of various yellow, orange, red, or brown secondary minerals containing hydrous oxides of uranium, thorium, and lead. Also known as uranium ocher. { 'gə,mīt }

Günz |GEOL| A European stage of geologic time, in the Pleistocene (above Astian of Pliocene, below Mindel); it is the first stage of glaciation of the Pleistocene in the Alps. { gints }

Günz-Mindel |GEOL| The first interglacial stage of the Pleistocene in the Alps, between Günz and Mindel glacial stages. { 'gints 'mind·əl }

gut |GEOL| **1.** A narrow water passage such as a strait. **2.** A channel deeper than the surrounding water; generally formed by water in motion. { gət }

guyot |GEOL| A seamount, usually deeper than 100 fathoms (180 meters), having a smooth platform top. Also known as tablemount. { gē'ō }

Gymnarthridae |PALEON| A family of extinct lepospondylous amphibians that have a skull with only a single bone representing the tabular and temporal elements of the primitive skull roof. { ˌjim'närth·rə,dē }

gymnite See deweylite. { 'jim,nīt }

Gymnocodiaceae |PALEOBOT| A family of fossil red algae. { ˌjim·nō,kō·dē'as·ē,ē }

gypcrete |GEOL| A type of duricrust composed of hydrous calcium sulfate. { 'jip,krēt }

gypsite |GEOL| A variety of gypsum consisting of dirt and sand; found as an efflorescent deposit in arid regions, overlying gypsum. Also known as gypsum earth. { 'jip,sīt }

gypsum [MINERAL] $CaSO_4 \cdot 2H_2O$ A mineral, the commonest sulfate mineral; crystals are monoclinic, clear, white to gray, yellowish, or brownish in color, with well-developed cleavages; luster is subvitreous to pearly, hardness is 2 on Mohs scale, and specific gravity is 2.3; it is calcined at 190–200°C to produce plaster of paris. { 'jip·səm }

gypsum earth See gypsite. { 'jip·səm ¦ərth }

Gyracanthididae [PALEON] A family of extinct acanthodian fishes in the suborder Diplacanthoidei. { ¸jī·rə¸kan'thid·ə¸dē }

gyrogonite [PALEOBOT] A minute, ovoid body that is the residue of the calcareous encrustation about the female sex organs of a fossil stonewort. { jī'räg·ə¸nīt }

gyttja [GEOL] A fresh-water anaerobic mud containing an abundance of organic matter; capable of supporting aerobic life. { 'yi¸chä }

H

Haanel depth rule [GEOPHYS] A rule for estimating the depth of a magnetic body, provided the body may be considered magnetically equivalent to a single pole; the depth of the pole is then equal to the horizontal distance from the point of maximum vertical magnetic intensity to the points where the intensity is one-third of the maximum value. { 'hän·əl 'depth ˌrül }

hackly fracture [MINERAL] A break in a mineral characterized by jagged irregular surfaces with sharp edges. { ˈhak·lē ˈfrak·chər }

hackmanite [MINERAL] A mineral of the sodalite family containing a small amount of sulfur; fluoresces orange or red in ultraviolet light. { 'hak·məˌnīt }

hade [GEOL] **1.** The angle of inclination of a fault as measured from the vertical. **2.** The inclination angle of a vein or lode. { hād }

Hadean [GEOL] The period (more than 3800 million years ago) extending for several hundred millions of years from the end of the accretion of the earth to the formation of the oldest recognized rocks. { 'hā·dē·Pen }

hadrosaur [PALEON] A duck-billed dinosaur. { 'had·rəˌsȯr }

haidingerite [MINERAL] HCaAsO₄·H₂O A white mineral composed of hydrous calcium arsenate. { 'hī·diŋ·əˌrīt }

hair copper See chalcotrichite. { 'her ˌkäp·ər }

hair pyrites See millerite. { 'her ˈpīˌrīts }

hair salt See alunogen. { 'her ˌsȯlt }

hairstone [GEOL] Quartz embedded with hairlike crystals of rutile, actinolite, or other mineral. { 'herˌstōn }

haldenhang See wash slope. { 'hal·dənˌhaŋ }

halite [MINERAL] NaCl Native salt; an evaporite mineral occurring as isometric crystals or in massive, granular, or compact form. Also known as common salt; rock salt. { 'haˌlīt }

Halitheriinae [PALEON] A subfamily of extinct sirenian mammals in the family Dugongidae. { həˌlith·ə'rī·əˌnē }

Hallian [GEOL] A North American stage of Pleistocene geologic time, above the Wheelerian and below the Recent. { 'hȯl·ē·ən }

halloysite [MINERAL] Al₂Si₂O₅(OH)₄·2H₂O Porcelainlike clay mineral whose composition is like that of kaolinite but contains more water and is structurally distinct; varieties are known as metahalloysites. { hə'lȯiˌsīt }

halmeic [GEOL] Referring to minerals or sediments derived directly from sea water. Also known as halmyrogenic; halogenic. { hal'mē·ik }

halmyrogenic See halmeic. { halˌmī·rə'jen·ik }

halmyrolysis [GEOCHEM] Postdepositional chemical changes that occur while sediment is on the sea floor. { ˌhal·mə'räl·ə·səs }

halo [GEOL] A ring or crescent surrounding an area of opposite sign; it is a diffusion of a high concentration of the sought mineral into surrounding ground or rock; it is encountered in mineral prospecting and in magnetic and geochemical surveys. { 'hā·lō }

halogenic See halmeic. { ˈhal·əˌjen·ik }

halogen mineral [MINERAL] Any of the naturally occurring compounds containing a halogen as the sole or principal anionic constituent. { 'hal·ə·jən ˌmin·rəl }

halokinesis

halokinesis *See* salt tectonics. { ¦hal·ə·kə'nē·səs }

halomorphic [GEOCHEM] Referring to an intrazonal soil whose features have been strongly affected by either neutral or alkali salts, or both. { ¦hal·ə¦mȯr·fik }

halo ore *See* fringe ore. { 'hā·lō ˌȯr }

halotrichite [MINERAL] **1.** FeAl$_2$(SO$_4$)$_4$·22H$_2$O A mineral composed of hydrous sulfate of iron and aluminum. Also known as butter rock; feather alum; iron alum; mountain butter. **2.** Any sulfate mineral resembling halotrichite in structure and habit. { ha'lä·trə,kīt }

Halysitidae [PALEON] A family of extinct Paleozoic corals of the order Tabulata. { ˌhal·ə'sid·ə,dē }

hamada [GEOL] A barren desert surface composed of consolidated material usually consisting of exposed bedrock, but sometimes of consolidated sedimentary material. { hə'mä·də }

hambergite [MINERAL] Be$_2$BO$_3$OH A grayish-white or colorless mineral composed of beryllium borate and occurring as prismatic crystals; hardness is 7.5 on Mohs scale, and specific gravity is 2.35. { 'ham·bər,gīt }

hammarite [MINERAL] Pb$_2$Cu$_2$Bi$_4$S$_9$ A monoclinic mineral whose color is a steel gray with red tone; consists of lead and copper bismuth sulfide. { 'ham·ə,rīt }

hancockite [MINERAL] A complex silicate mineral containing lead, calcium, strontium, and other minerals; it is isomorphous with epidote. { 'han,kä,kīt }

hanger *See* hanging wall. { 'haŋ·ər }

hanging *See* hanging wall. { 'haŋ·iŋ }

hanging side *See* hanging wall. { 'haŋ·iŋ ¦sīd }

hanging valley [GEOL] A valley whose floor is higher than the level of the shore or other valley to which it leads. { 'haŋ·iŋ ¦val·ē }

hanging wall [GEOL] The rock mass above a fault plane, vein, lode, ore body, or other structure. Also known as hanger; hanging; hanging side. { 'haŋ·iŋ ¦wȯl }

hanksite [MINERAL] Na$_{22}$K(SO$_4$)$_9$(CO$_3$)$_2$Cl A white or yellow mineral crystallizing in the hexagonal system; found in California. { 'haŋk,sīt }

hannayite [MINERAL] Mg$_3$(NH$_4$)$_2$H$_2$(PO$_4$)$_4$·8H$_2$O Mineral composed of hydrous acid ammonium magnesium phosphate; occurs as yellow crystals in guano. { 'ha·nē,īt }

Haplolepidae [PALEON] A family of Carboniferous chondrostean fishes in the suborder Palaeoniscoidei having a reduced number of fin rays and a vertical jaw suspension. { ¦ha·plō¦lep·ə,dē }

haplopore [PALEON] Any randomly distributed pore on the surface of fossil cystoid echinoderms. { 'ha·plō,pȯr }

hard coal *See* anthracite. { 'härd ¦kōl }

hardpan *See* caliche. { 'härd,pan }

hard rock [GEOL] Rock which needs drilling and blasting for removal. { 'härd ¦räk }

hardystonite [MINERAL] Ca$_2$ZnSi$_2$O$_7$ A white mineral composed of zinc calcium silicate. { 'här·dē·stə,nīt }

Harker diagram *See* variation diagram. { 'härk·ər ˌdī·ə,gram }

Harlechian [GEOL] A European stage of geologic time: Lower Cambrian. { här'lek·ē·ən }

harmonic folding [GEOL] Folding in the earth's surface, with no sharp changes with depth in the form of the folds. { här'män·ik 'fōld·iŋ }

harmotome [MINERAL] (K,Ba)(Al,Si)$_2$(Si$_6$O$_{16}$)·6H$_2$O A zeolite mineral with ion-exchange properties that forms cruciform twin crystals. Also known as cross-stone. { 'här·mə,tōm }

harstigite [MINERAL] Be$_2$Ca$_3$Si$_3$O$_{11}$ A mineral composed of silicate of beryllium and calcium. { 'härs·tə,gīt }

hartite [GEOL] A white, crystalline, fossil resin that is found in lignites. Also known as bombiccite; branchite; hofmannite; josen. { 'här,tīt }

harzburgite [PETR] A peridotite consisting principally of olivine and orthopyroxene. { 'härts,bər,gīt }

hastingsite [MINERAL] NaCa$_2$(Fe,Mg)$_5$Al$_2$Si$_6$O$_{22}$(OH)$_2$ A mineral of the amphibole group crystallizing in the monoclinic system and composed chiefly of sodium, calcium, and iron, but usually with some potassium and magnesium.   { 'hās·tiŋ,zīt }

hatchettine See hatchettite.   { 'ha·chəd,ēn }

hatchettite |MINERAL| $C_{38}H_{78}$ A yellow-white mineral paraffin wax, melting at 55–65°C in the natural state and 79°C in the pure state; occurs in masses in ironstone nodules or in cavities in limestone.   Also known as adipocerite; adipocire; hatchettine; mineral tallow; mountain tallow; naphthine.   { 'ha·chəd,ıt }

hatchettolite See ellsworthite.   { 'ha·ched·ō,līt }

hatchite |MINERAL| A lead-gray mineral composed of sulfide of lead and arsenic; occurs in triclinic crystals.   { 'ha,chīt }

hauerite |MINERAL| MnS_2 A reddish-brown or brownish-black mineral composed of native manganese sulfide; occurs massive or in octahedral or pyritohedral crystals.   { 'haù·ə,rīt }

haughtonite |PETR| A black variety of biotite that is rich in iron.   { 'hòt·ən,īt }

hausmannite |MINERAL| Mn_3O_4 Brownish-black, opaque mineral composed of manganese tetroxide.   { 'haùs·mä,nīt }

Hauterivian |GEOL| A European stage of geologic time, in the Lower Cretaceous, above Valanginian and below Barremian.   { ō·trə've̅·ən }

haüyne |MINERAL| $(Na,Ca)_{4-8}(Al_6Si_6O_{24})(SO_4,S)_{1-2}$ An isometric silicate mineral of the sodalite group occurring as grains embedded in various igneous rocks; hardness is 5.5–6 on Mohs scale, and specific gravity is 2.4–2.5.   Also known as haüynite.   { ä'wēn }

haüynite See haüyne.   { ä'we̅,nīt }

hazel sandstone |GEOL| An arkosic, iron-bearing redbed sandstone from the Precambrian found in western Texas.   { 'ha·zəl 'san,stōn }

head erosion See headward erosion.   { 'hed i,rō·zhən }

heading side See footwall.   { 'hed·iŋ ,sīd }

heading wall See footwall.   { 'hed·iŋ ,wòl }

headwall |GEOL| The steep cliff at the back of a cirque.   { 'hed,wòl }

headward erosion |GEOL| Erosion caused by water flowing at the head of a valley. Also known as head erosion; headwater erosion.   { 'hed·wərd i'rō·zhən }

headwater erosion See headward erosion.   { 'hed,wòd·ər i'rō·zhən }

heat budget |GEOPHYS| Amount of heat needed to raise a lake's water from the winter temperature to the maximum summer temperature.   { 'hēt ,bəj·ət }

heat flow province |GEOPHYS| A geographic area in which the heat flow and heat production are linearly related.   { 'hēt ,flō ,präv·əns }

heave |GEOL| The horizontal component of the slip, measured at right angles to the strike of the fault.   { hēv }

heavy mineral |MINERAL| A mineral with a density above 2.9, which is the density of bromoform, the liquid used to separate the heavy from the light minerals.   { 'hev·ē 'min·rəl }

heazelwoodite |MINERAL| Ni_3S_2 A meteorite mineral consisting of a sulfide of nickel.   { 'hē·zəl,wù,dīt }

Hebraic granite See graphic granite.   { hē'brā·ik ,gran·ət }

hecatolite See moonstone.   { hə'kat·əl,īt }

hectorite |MINERAL| $(Mg,Li)_3Si_4O_{10}(OH)_2$ A trioctohedral clay mineral of the montmorillonite group composed of a hydrous silicate of magnesium and lithium.   { 'hek·tə,rīt }

hedenbergite |MINERAL| $CaFeSi_2O_6$ A black mineral consisting of calcium-iron pyroxene and occurring at the contacts of limestone with granitic masses.   { 'hed·ən,bər,gīt }

hedleyite |MINERAL| A mineral composed of an alloy of bismuth and tellurium.   { 'hed·lē,īt }

hedreocraton |GEOL| A craton that influenced later continental development.   { ,hed·rē·ō'krā,tän }

hedyphane |MINERAL| $(Ca,Pb)_5Cl(AsO_4)_3$ Yellowish-white mineral composed of lead and calcium arsenate and chloride; occurs in monoclinic crystals.   { 'hed·ə,fān }

149

Heidelberg man [PALEON] An early type of European fossil man known from an isolated lower jaw; considered a variant of *Homo erectus* or an early stock of Neanderthal man. { 'hīd·əl·bərg ,man }

Helaletidae [PALEON] A family of extinct perissodactyl mammals in the superfamily Tapiroidea. { ,hel·ə'led·ə,dē }

Helcionellacea [PALEON] A superfamily of extinct gastropod mollusks in the order Aspidobranchia. { ¦hel·sē·ō·nə'las·ē·ə }

Helderbergian [GEOL] A North American stage of geologic time, in the lower Lower Devonian. { ,hel·dər'bərg·ē·ən }

Helicoplacoidea [PALEON] A class of free-living, spindle- or pear-shaped, plated echinozoans known only from the Lower Cambrian of California. { ,hel·ə·kō·plə'kȯid·ē·ə }

helictite [GEOL] A speleothem whose origin is similar to that of a stalactite or stalagmite but that angles or twists in an irregular fashion. { 'hē·lik,tīt }

heliolite *See* sunstone. { 'hē·lē·ə,līt }

Heliolitidae [PALEON] A family of extinct corals in the order Tabulata. { ,hē·lē·ō'lid·ə,dē }

heliophyllite [MINERAL] $Pb_6As_2O_7Cl_4$ A yellow to greenish-yellow, orthorhombic mineral consisting of an oxychloride of lead and arsenic; occurs in massive and tabular form and as crystals. { ,hē·lē·ō'fi,līt }

heliotrope *See* bloodstone. { 'hē·lē·ə,trōp }

hellandite [MINERAL] Mineral composed of silicate of metals in the cerium group with aluminum, iron, manganese, and calcium. { 'hel·ən,dīt }

Helmert's formula [GEOPHYS] A formula for the acceleration due to gravity in terms of the latitude and the altitude above sea level. { 'hel·mərts ,fȯr·myə·lə }

Helodontidae [PALEON] A family of extinct ratfishes conditionally placed in the order Bradyodonti. { ,he·lō'dänt·ə,dē }

helvine *See* helvite. { 'hel,vēn }

helvite [MINERAL] $(Mn,Fe,Zn)_4Be_3(SiO_4)_3S$ A silicate mineral isomorphous with danalite and genthelvite. Also known as helvine. { 'hel,vīt }

hemafibrite [MINERAL] $Mn_3(AsO_4)(OH)_3·H_2O$ A brownish to garnet-red mineral composed of basic manganese arsenate. { ¦hē·mə¦fī,brīt }

hematite [MINERAL] Fe_2O_3 An iron mineral crystallizing in the rhombohedral system; the most important ore of iron, it is dimorphous with maghemite, occurs in black metallic-looking crystals, in reniform masses or fibrous aggregates, or in reddish earthy forms. Also known as bloodstone; red hematite; red iron ore; red ocher; rhombohedral iron ore. { 'hē·mə,tīt }

hematolite [MINERAL] $(Mn,Mg)_4Al(AsO_4)(OH)_8$ A brownish-red mineral composed of aluminum manganese arsenate; occurs in rhombohedral crystals. { 'he·məd·ō,līt }

hematophanite [MINERAL] $Pb_5Fe_4O_{10}(Cl,OH)_2$ A mineral composed of oxychloride lead and iron. { ,hē·mə'täf·ə,nīt }

Hemicidaridae [PALEON] A family of extinct Echinacea in the order Hemicidaroida distinguished by a stirodont lantern, and ambulacra abruptly widened at the ambitus. { ¦he·mē·si'där·ə,dē }

Hemicidaroida [PALEON] An order of extinct echinoderms in the superorder Echinacea characterized by one very large tubercle on each interambulacral plate. { ¦he·mē,sid·ə'rȯid·ə }

hemicone *See* alluvial cone. { 'he·mē,kōn }

hemicrystalline *See* hypocrystalline. { ¦he·mē'krist·əl·ən }

hemimorphite [MINERAL] $Zn_4Si_2O_7(OH)_2·H_2O$ A white, colorless, pale-green, blue, or yellow mineral having an orthorhombic crystal structure; an ore of zinc. Also known as calamine; electric calamine; galmei. { ,he·mē'mȯr,fīt }

hemipelagic sediment [GEOL] Deposits containing terrestrial material and the remains of pelagic organisms, found in the ocean depths. { ¦he·mē·pə'laj·ik 'sed·ə·mənt }

Hemist [GEOL] A suborder of the soil order Histosol, consisting of partially decayed plant residues and saturated with water most of the time. { 'he·mist }

Hemizonida [PALEON] A Paleozoic order of echinoderms of the subclass Asteroidea having an ambulacral groove that is well defined by adambulacral ossicles, but with restricted or undeveloped marginal plates. { ˈheˈmēˈzänˈəˈdə }

Hercules stone *See* lodestone. { ˈhərˈkyəˌlēz ˌstōn }

Hercynian geosyncline [GEOL] A principal area of geosynclinal sediment accumulation in Devonian time; found in south-central and southern Europe and northern Africa. { hərˈsinˈēˈən ˌjēˈōˈsinˌklīn }

Hercynian orogeny *See* Variscan orogeny. { hərˈsinˈēˈən ȯˈräjˈəˈnē }

hercynite [MINERAL] $(Fe,Mg)Al_2O_4$ A black mineral of the spinel group; crystallizes in the isometric system. Also known as ferrospinel; iron spinel. { ˈhərsˈənˌīt }

herderite [MINERAL] $CaBe(PO_4)(F,OH)$ A colorless to pale-yellow or greenish-white mineral consisting of phosphate and fluoride of calcium and beryllium; hardness is 7.5–8 on Mohs scale, and specific gravity is 3.92. { ˈhərˈdəˌrīt }

hervidero *See* mud volcano. { ˈhərˈvaˌderˈō }

Hesperornithidae [PALEON] A family of extinct North American birds in the order Hesperornithiformes. { ˌhesˈpərˌȯrˈnithˈəˌdē }

Hesperornithiformes [PALEON] An order of ancient extinct birds; individuals were large, flightless, aquatic diving birds with the shoulder girdle and wings much reduced and the legs specialized for strong swimming. { hesˈpəˌrȯrˌnithˈəˈfȯrˌmēz }

hessite [MINERAL] Ag_2Te A lead-gray sectile mineral crystallizing in the isometric system; usually massive and often auriferous. { ˈheˌsīt }

hetaerolite [MINERAL] $ZnMn_2O_4$ A black mineral consisting of zinc-manganese oxide found with chalcophanite. { həˈtirˈəˌlīt }

Heteractinida [PALEON] A group of Paleozoic sponges with calcareous spicules; probably related to the Calcarea. { ˌhedˈəˈrakˈtinˈədˈə }

heteroblastic [PETR] Pertaining to rocks in which the essential constituents are of two distinct orders of magnitude of size. { ˌhedˈəˈrōˌblasˈtik }

heterochronism [GEOL] A phenomenon in which two similar geologic deposits may not be of the same age even though they underwent like processes of formation. { ˌhedˈəˈräkˈrəˌnizˈəm }

Heterocorallia [PALEON] An extinct small, monofamilial order of fossil corals with elongate skeletons; found in calcareous shales and in limestones. { ˌhedˈəˈrōˈkəˈralˈēˈə }

heterogeneous reservoir [GEOL] Formation with two or more noncommunicating sand members, each possibly with different specific- and relative-permeability characteristics. { ˌhedˈəˈrəˈjēˈnēˈəs ˈrezˈəvˌwär }

heterogenite [MINERAL] $CoO(OH)$ A black cobalt mineral, sometimes with some copper and iron, found in mammillary masses. Also known as stainierite. { ˌhedˈəˈräjˈəˌnīt }

heteromorphite [MINERAL] $Pb_7Sb_8S_{19}$ An iron black, monoclinic mineral consisting of lead antimony sulfide. { ˌhedˈəˈrōˈmȯrˌfīt }

Heterophyllidae [PALEON] The single family of the extinct cnidarian order Heterocorallia. { ˌhedˈəˈrōˈfilˈəˌdē }

heterosite [MINERAL] A mineral composed of phosphate of iron and manganese; it is isomorphous with purpurite. { ˈhedˈəˈrəˌsīt }

Heterosoricinae [PALEON] A subfamily of extinct insectivores in the family Soricidae distinguished by a short jaw and hedgehoglike teeth. { ˌhedˈəˈrōˈsəˈrisˈəˌnē }

Heterostraci [PALEON] An extinct group of ostracoderms, or armored, jawless vertebrates; armor consisted of bone lacking cavities for bone cells. { ˌhedˈəˈräsˈtrəˌsī }

Hettangian [GEOL] A stage of Lower Jurassic geologic time. { heˈtanˈjēˈən }

heulandite [MINERAL] $CaAl_2Si_6O_{16}\cdot5H_2O$ A zeolite mineral that crystallizes in the monoclinic system; often occurs as foliated masses or in crystal form in cavities of decomposed basic igneous rocks. { ˈhyüˈlənˌdīt }

hewettite [MINERAL] $CaV_6O_{16}\cdot9H_2O$ A deep-red mineral composed of hydrated calcium vanadate; found in silky orthorhombic crystal aggregates in Colorado, Utah, and Peru. { ˈhyüˈəˌtīt }

hexahedrite [GEOL] An iron meteorite composed of single crystals or aggregates of kamacite, usually containing 4–6% nickel in the metal phase. { ˌhekˈsəˈheˌdrīt }

hexahydrite

hexahydrite [MINERAL] $MgSO_4 \cdot 6H_2O$ A white or greenish-white monoclinic mineral composed of hydrous magnesium sulfate. { ,hek·sə'hī,drīt }

hiatus [GEOL] A gap in a rock sequence due to a lack of deposition of a bed or to erosion of beds. { hī'ād·əs }

Hibernian orogeny See Erian orogeny. { hī'bər·nē·ən ȯ'räj·ə·nē }

hibonite [MINERAL] $CaAl_{12}O_{19}$ Common mineral found in carbonaceous chondrite meteorites; occurs only rarely on earth. { 'hib·ə,nīt }

hiddenite [MINERAL] A transparent green or yellowish-green spodumene mineral containing chromium and valued as a gem. { 'hid·ən,īt }

hieratite [MINERAL] K_2SiF_6 A grayish mineral composed of potassium fluosilicate; occurs as deposits in volcanic holes. { 'hī·ər·ə,tīt }

hieroglyph [GEOL] Any sort of sedimentary mark or structure occurring on a bedding plane. { 'hī·rə,glif }

high-angle fault [GEOL] A fault with a dip greater than 45°. { ¦hī ,aŋ·gəl 'fȯlt }

high-energy environment [GEOL] An aqueous sedimentary environment which features a high energy level and turbulent motion, created by waves, currents, or surf, which prevents the settling and piling up of fine-grained sediment. { 'hī ,en·ər·jē in'vī·ərn·mənt }

highland [GEOL] **1.** A lofty headland, cliff, or other high platform. **2.** A dissected mountain region composed of old folded rocks. { 'hī·lənd }

high quartz [MINERAL] Quartz that was formed at high temperatures. { 'hī ¦kwȯrts }

high-rank coal [GEOL] Coal consisting of less than 4% moisture when air-dried, or more than 84% carbon. { 'hī ,raŋk 'kōl }

high-rank graywacke See feldspathic graywacke. { 'hī ,raŋk 'grā,wak·ə }

high-volatile bituminous coal [GEOL] A bituminous coal composed of more than 31% volatile matter. { 'hī ¦väl·əd·əl bə¦tü·mə·nəs 'kōl }

high-water platform See wave-cut bench. { 'hī ¦wȯd·ər 'plat,fȯrm }

hilgardite [MINERAL] $Ca_8(B_6O_{11})_3Cl_4 \cdot H_2O$ Colorless mineral composed of hydrous borate and chloride of calcium; occurs as monoclinic domatic crystals. { 'hil,gär,dīt }

hill creep [GEOL] Slow gravity movement of rock and soil waste down a steep hillside. Also known as hillside creep. { 'hil ,krēp }

hillebrandite [MINERAL] $Ca_2SiO_3(OH)_2$ A white mineral composed of hydrous calcium silicate; occurs in masses. { 'hil·ə,bran,dīt }

hillock [GEOL] A small, low hill. { 'hil·ək }

hillside creep See hill creep. { 'hil,sīd ,krēp }

Hilt's law [GEOL] The law that in a small area the deeper coals are of higher rank than those above them. { 'hilts ,lȯ }

hinge fault [GEOL] A fault whose movement is an angular or rotational one on a side of an axis that is normal to the fault plane. { 'hinj ,fȯlt }

hinge line [GEOL] **1.** The line separating the region in which a beach has been thrust upward from that in which it is horizontal. **2.** A line in the plane of a hinge fault separating the part of a fault along which thrust or reverse movement occurred from that having normal movement. { 'hinj ,līn }

hinsdalite [MINERAL] $(Pb,Sr)Al_3(PO_4)(SO_4)(OH)_6$ Dark-gray or greenish rhombohedral mineral composed of basic lead and strontium aluminum sulfate and phosphate; occurs in coarse crystals and masses. { 'hinz,dā,līt }

hinterland [GEOL] **1.** The region behind the coastal district. **2.** The terrain on the back of a folded mountain chain. **3.** The moving block which forces geosynclinal sediments toward the foreland. { 'hin·tər,land }

hisingerite [MINERAL] $Fe_2^{3-}Si_2O_5(OH)_4 \cdot 2H_2O$ A black, amorphous mineral composed of hydrous ferric silicate; an iron ore. { 'hī·siŋ·ə,rīt }

historical geology [GEOL] A branch of geology concerned with the systematic study of bedded rocks and their relations in time and the study of fossils and their locations in a sequence of bedded rocks. { hi'stär·ə·kəl jē'äl·ə·jē }

Histosol [GEOL] An order of wet soils consisting mostly of organic matter, popularly called peats and mucks. { 'his·tə,sȯl }

hitch [GEOL] **1.** A fault of strata common in coal measures, accompanied by displacement. **2.** A minor dislocation of a vein or stratum not exceeding in extent the thickness of the vein or stratum. { hich }

hjelmite [MINERAL] A black mineral containing yttrium, iron, manganese, uranium, calcium, columbium, tantalum, tin, and tungsten oxide; often occurs with crystal structure disrupted by radiation. { 'yel,mīt }

hodgkinsonite [MINERAL] $MnZnSiO_5 \cdot H_2O$ A pink to reddish-brown mineral composed of hydrous zinc manganese silicate; occurs as crystals. { 'häj·kən·sə,nīt }

hoegbomite [MINERAL] $Mg(Al,Fe,Ti)_4O_7$ A black mineral composed of an oxide of magnesium, aluminum, iron, and titanium. Also spelled högbomite. { 'hāg·bə,mīt }

hoernesite [MINERAL] $Mg_3As_2O_8 \cdot H_2O$ A white, monoclinic mineral composed of hydrous magnesium arsenate; occurs as gypsumlike crystals. { 'hər·nə,sīt }

hofmannite See hartite. { 'häf·mə,nīt }

hogback [GEOL] Alternate ridges and ravines in certain areas of mountains, caused by erosive action of mountain torrents. { 'häg,bak }

högbomite See hoegbomite. { 'hāg·bə,mīt }

hohmannite [MINERAL] $Fe_2(SO_4)_2(OH)_2 \cdot 7H_2O$ A chestnut brown to burnt orange and amaranth red, triclinic mineral consisting of a hydrated basic sulfate of iron. { 'hō·mə,nīt }

holdenite [MINERAL] A red, orthorhombic mineral composed of basic manganese zinc arsenate with a small amount of calcium, magnesium, and iron. { 'hōl·də,nīt }

Holectypidae [PALEON] A family of extinct exocyclic Euechinoidea in the order Holectypoida; individuals are hemispherical. { hō,lek'tip·ə,dē }

hollandite [MINERAL] $Ba(Mn^{2+},Mn^{4+})_8O_{16}$ A silvery-gray to black mineral composed of manganate of barium and manganese; occurs as crystals. { 'hä·lən,dīt }

Hollinacea [PALEON] A dimorphic superfamily of extinct ostracods in the suborder Beyrichicopina including forms with sulci, lobation, and some form of velar structure. { ,häl·ə'nās·ē·ə }

Hollinidae [PALEON] An extinct family of ostracodes in the superfamily Hollinacea distinguished by having a bulbous third lobe on the valve. { hə'lin·ə,dē }

holmquistite [MINERAL] $(Na,K,Ca)Li(Mg,Fe)_3Al_2Si_8O_{22}(OH)_2$ A bluish-black, orthorhombic mineral composed of alkali and silicate of iron, magnesium, lithium, and aluminum. { 'hōm,kwi,stīt }

Holocene [GEOL] An epoch of the Quaternary Period from the end of the Pleistocene, around 10,000 years ago, to the present. Also known as Postglacial; Recent. { 'hō·lə,sēn }

holoclastic [PETR] Being or belonging to ordinary (sedimentary) clastic rock. { ¦häl·ō¦klas·tik }

holocrystalline [PETR] Pertaining to igneous rocks that are entirely crystallized minerals, without glass. { ¦häl·ō'krist·əl·ən }

holohyaline [PETR] Pertaining to an entirely glassy rock. { ¦häl·ō'hī·ə·lən }

Holoptychidae [PALEON] A family of extinct lobefin fishes in the order Osteolepiformes. { ,häl·əp·tə'kī·ə,dē }

holostratotype [GEOL] The originally defined stratotype. { ¦häl·ō'strad·ə,tīp }

Holuridae [PALEON] A group of extinct chondrostean fishes in the suborder Palaeoniscoidei distinguished in having lepidotrichia of all fins articulated but not bifurcated, fins without fulcra, and the tail not cleft. { hə'lùr·ə,dē }

Homacodontidae [PALEON] A family of extinct palaeodont mammals in the superfamily Dichobunoidea. { ,häm·ə·kō'dänt·ə,dē }

homeoblastic [PETR] Of a metamorphic crystalloblastic texture, having constituent minerals of approximately the same size. { ¦hō·mē·ō¦blas·tik }

homilite [MINERAL] $Ca_2(Fe,Mg)B_2Si_2O_{10}$ A black or blackish brown mineral composed of iron calcium borosilicate. { 'hä·mə,līt }

homocline [GEOL] Any rock unit in which the strata exhibit the same dip. { 'hä·mə,klīn }

Homo erectus [PALEON] A type of fossil human from the Pleistocene of Java and China representing a specialized side branch in human evolution. { 'hō·mō ə'rek·təs }

153

Homoistela

Homoistela [PALEON] A class of extinct echinoderms in the subphylum Homalozoa. { hō'mói·stə·lə }

homologous [GEOL] **1.** Referring to strata, in separated areas, that are correlatable (contemporaneous) and are of the same general character or facies, or occupy analogous structural positions along the strike. **2.** Pertaining to faults, in separated areas, that have the same relative position or structure. { hə'mäl·ə·gəs }

honeycomb coral [PALEON] The common name for members of the extinct order Tabulata; has prismatic sections arranged like the cells of a honeycomb. { 'hən·ē,kōm ¦kär·əl }

honeycomb formation [GEOL] A rock stratum containing large cavities or caverns. { 'hən·ē,kōm fȯr,mā·shən }

hopeite [MINERAL] $Zn_3(PO_4)_2 \cdot 4H_2O$ A gray, orthorhombic mineral composed of hydrous phosphate of zinc; specific gravity is 2.76–2.85; dimorphous with parahopeite. { 'hō,pīt }

horizon [GEOL] **1.** The surface separating two beds. **2.** One of the layers, each of which is a few inches to a foot thick, that make up a soil. { hə'rīz·ən }

horizontal displacement See strike slip. { ,här·ə'zänt·əl dis'plās·mənt }

horizontal fold See nonplunging fold. { ,här·ə'zänt·əl 'fōld }

horizontal intensity [GEOPHYS] The strength of the horizontal component of the earth's magnetic field. { ,här·ə'zänt·əl in'ten·səd·ē }

horizontal pressure force [GEOPHYS] The horizontal pressure gradient per unit mass, $-\alpha\nabla_H p$, where α is the specific volume, p the pressure, and ∇_H the horizontal component of the del operator; this force acts normal to the horizontal isobars toward lower pressure; it is one of the three important forces appearing in the horizontal equations of motion, the others being the Coriolis force and friction. { ,här·ə'zänt·əl 'presh·ər ,fȯrs }

horizontal separation See strike slip. { ,här·ə'zänt·əl ,sep·ə'rā·shən }

horn [GEOL] A topographically high, sharp, pyramid-shaped mountain peak produced by the headward erosion of mountain glaciers; the Matterhorn is the classic example. { hȯrn }

hornblende [MINERAL] A general name given to the monoclinic calcium amphiboles that form an extensive solid-solution series between the various metals in the generalized formula $(Ca,Na)_2(Mg,Fe,Al)_5(Al,Si)_8O_{22}(OH,F)_2$. { 'hȯrn,blend }

hornblendite [PETR] A plutonic rock consisting mainly of hornblende. { 'hȯrn·blen,dīt }

horned dinosaur [PALEON] Common name for extinct reptiles of the suborder Ceratopsia. { ¦hȯrnd 'dīn·ə,sȯr }

horned-toad dinosaur [PALEON] The common name for extinct reptiles composing the suborder Ankylosauria. { ¦hȯrnd ,tōd 'dīn·ə,sȯr }

hornfels [PETR] A common name for a class of metamorphic rocks produced by contact metamorphism and characterized by equidimensional grains without preferred orientation. { 'hȯr,felz }

hornfels facies [PETR] Rock formed at depths in the earth's crust not exceeding 6.2 miles (10 kilometers) at temperatures of 250–800°C; includes albite-epidote hornfels facies, pyroxene-hornfels facies, and hornblende-hornfels facies. { 'hȯr,felz ¦fā·shēz }

horn lead See phosgenite. { 'hȯrn ,led }

horn quicksilver See calomel. { 'hȯrn 'kwik,sil·vər }

horn silver See cerargyrite. { 'hȯrn ,sil·vər }

hornstone See chert. { 'hȯrn,stōn }

horse [GEOL] A large rock caught along a fault. { hȯrs }

horseback [GEOL] A low and sharp ridge of sand, gravel, or rock. { 'hȯrs,bak }

horsetail ore [GEOL] An ore occurring in fractures which diverge from a larger fracture. { 'hȯrs,tāl ,ȯr }

horsfordite [MINERAL] Cu_5Sb A silver-white mineral composed of copper-antimony alloy. { 'hȯrs·fər,dīt }

horst [GEOL] **1.** A block of the earth's crust uplifted along faults relative to the rocks on either side. **2.** A mass of the earth's crust limited by faults and standing in relief. **3.** One of the older mountain masses limiting the Alps on the west and north. **4.** A knobby ledge of limestone beneath a thin soil mantle. { hôrst }

hortonolite [MINERAL] $(Fe,Mg,Mn)_2SiO_4$ A dark mineral composed of silicate of iron, magnesium, and manganese; a member of the olivine series. { hôr'tän·əl,īt }

host rock [GEOL] Rock which serves as a host for other rocks or for mineral deposits. { 'hōst ,räk }

howardite [GEOL] An achondritic stony meteorite composed chiefly of calcic plagioclase and orthopyroxene. { 'haú·ər,dīt }

howlite [MINERAL] $Ca_2Bi_5SiO_9(OH)_5$ A white mineral occurring in nodular or earthy form. { 'haú,līt }

huangho deposit [GEOL] A coastal-plain deposit comprising alluvium spread over a level surface (such as a floodplain) but extending into marine beds of equivalent age. { ¦hwän¦hō di,päz·ət }

hudsonite See cortlandite. { 'həd·sə,nīt }

huebnerite [MINERAL] $MnWO_4$ A brownish-red to black manganese member of the wolframite series, occurring in short, monoclinic, prismatic crystals; isomorphous with ferberite. { 'hēb·nə,rīt }

hühnerkobelite [MINERAL] $(Na,Ca)(Fe,Mn)_2(PO_4)_2$ A mineral composed of phosphate of sodium, calcium, iron, and manganese; it is isomorphous with varulite. { ¦hyü·nər¦kō·bə,līt }

hulsite [MINERAL] $(Fe^{2+},Mg)_2(Fe^{3+},Sn)(BO_3)O_2$ A black mineral composed of iron calcium magnesium tin borate. { 'həl,sīt }

humboldtine [MINERAL] $FeC_2O_4·2H_2O$ A mineral composed of hydrous ferrous oxalate. Also known as humboldtite; oxalite. { 'həm,bōl,tēn }

humboldtite See humboldtine. { 'həm,bōl,tīt }

humic [GEOL] Pertaining to or derived from humus. { 'hyü·mik }

humic-cannel coal See pseudocannel coal. { 'hyü·mik ¦kan·əl 'kōl }

humic coal [GEOL] A coal whose attritus is composed mainly of transparent humic degradation material. { 'hyü·mik 'kōl }

humification [GEOL] Formation of humus. { ,hyü·mə·fə'kā·shən }

humin See ulmin. { 'hyü·mən }

humite [MINERAL] **1.** A humic coal mineral. **2.** A series of magnesium neosilicate minerals closely related in crystal structure and chemical composition. { 'hyü,mīt }

hummock [GEOL] A rounded or conical knoll, mound, hillock, or other small elevation, generally of equal dimensions and not ridgelike. Also known as hammock. { 'həm·ək }

hummocky [GEOL] Any topographic surface characterized by rounded or conical mounds. { 'həm·ə·kē }

Humod [GEOL] A suborder of the soil order Spodosol having an accumulation of humus, and of aluminum but not iron. { 'hyü,mäd }

humodurite See translucent attritus. { ¦hyü·mə¦dú,rīt }

humogelite See ulmin. { hyü'mäj·əl,īt }

Humox [GEOL] A suborder of the soil order Oxisol that is high in organic matter, well drained but moist all or nearly all year, and restricted to relatively cool climates and high altitudes for Oxisols. { 'hyü,mäks }

Humult [GEOL] A suborder of the soil order Ultisol, well drained with a moderately thick surface horizon; formed under conditions of high rainfall distributed evenly over the year; common in southeastern Brazil. { 'hyü·məlt }

humus [GEOL] The amorphous, ordinarily dark-colored, colloidal matter in soil; a complex of the fractions of organic matter of plant, animal, and microbial origin that are most resistant to decomposition. { 'hyü·məs }

huntite [MINERAL] $CaMg_3(CO_3)_4$ A white mineral consisting of calcium magnesium carbonate. { 'hən,tīt }

hureaulite [MINERAL] $Mn_5H_2(PO_4)_4·4H_2O$ A monoclinic mineral of varying colors consisting of a hydrated acid phosphate of manganese. { 'hyü·rō,līt }

Huronian [GEOL] The lower system of the restricted Proterozoic. { hyü'rō·nē·ən }

hutchinsonite [MINERAL] $(Pb,Tl)_2(Cu,Ag)As_5S_{10}$ Red mineral composed of sulfide of lead, copper, and arsenic, with varying amounts of thallium and silver, occurring in small orthorhombic crystals. { 'hәch·ən·sә,nīt }

huttonite [MINERAL] $ThSiO_4$ A colorless to pale-green monoclinic mineral composed of silicate of thorium; it is dimorphous with thorite. { 'hәt·ən,īt }

hyacinth See zircon. { 'hī·ә,sinth }

Hyaenodontidae [PALEON] A family of extinct carnivorous mammals in the order Delta-theridia. { hī,ē·nә'dänt·ә,dē }

hyaline [GEOL] Transparent and resembling glass. { 'hī·ә·lәn }

hyalinocrystalline [PETR] Of porphyritic rock texture, having the phenocrysts lying in a glassy ground mass. { hī'al·ә·nō'krist·әl·ən }

hyalite [MINERAL] A colorless, clear or translucent variety of opal occurring as globular concretions or botryoidal crusts in cavities or cracks of rocks. Also known as Müller's glass; water opal. { 'hī·ә,līt }

hyalobasalt See tachylite. { ¦hī·ә·lō·bә'sȯlt }

hyaloclastite [GEOL] A tufflike deposit formed by the flowing of basalt under water and ice and its consequent fragmentation. Also known as aquagene tuff. { ¦hī·ә·lō'kla,stīt }

hyaloophitic [PETR] Of the texture of igneous rocks, being composed principally of a glassy ground mass with little interstitial texture. { ¦hī·ә·lō,ō'fid·ik }

hyalophane [MINERAL] $BaAl_2Si_2O_8$ A colorless feldspar mineral crystallizing in the monoclinic system; isomorphous with adularia. Also known as baryta feldspar. { hī'al·ә,fān }

hyalopsite See obsidian. { ,hī·ә'läp,sīt }

Hyalospongia [PALEON] A class of extinct glass sponges, equivalent to the living Hexac-tinellida, having siliceous spicules made of opaline silica. { ,hi·ә·lō'spәn·jē·ә }

hyalotekite [MINERAL] $(Pb,Ca,Ba)_4BSi_6O_{17}(OH,F)$ A white gray mineral composed of borosilicate and fluoride of lead, barium, and calcium, occurring in crystalline masses. { ¦hī·ә·lō¦tek,tīt }

Hybodontoidea [PALEON] An ancient suborder of extinct fossil sharks in the order Selachii. { ,hī·bә,dän'tȯid·ē·ә }

hybrid [PETR] Pertaining to a rock formed by the assimilation of two magmas. { 'hī·brәd }

hydatogenesis [GEOL] Crystallization and deposition of minerals from aqueous solutions. { ¦hīd·ә·tō'jen·ә·sәs }

hydrargillite See gibbsite. { hī'drär·jә,līt }

hydrated halloysite See endellite. { 'hī,drād·әd hә'lȯi,sīt }

hydraulic ratio [GEOL] The weight of a heavy mineral multiplied by 100 and divided by the weight of a hydraulically equivalent light mineral. { hi'drȯ·lik 'rā·shō }

hydrobasaluminite [MINERAL] $Al_4(SO_4)(OH)_{10}·36H_2O$ Mineral composed of a hydrous sulfate and hydroxide of aluminum. { ¦hī·drō¦bas·ә'lüm·ә,nīt }

hydrobiotite [MINERAL] A light-green, trioctahedral clay mineral of mixed layers of biotite and vermiculite. { ,hī·drō'bī·ә,tīt }

hydroboracite [MINERAL] $CaMgB_6O_{11}·6H_2O$ A white mineral composed of hydrous calcium magnesium borate, occurring in fibrous and foliated masses. { ,hī·drō'bȯr·ә,sīt }

hydrocalumite [MINERAL] $Ca_2Al(OH)_7·3H_2O$ A colorless to light-green mineral composed of a hydrous hydroxide of calcium and aluminum. { ,hī·drō'kal·yә,mīt }

hydrocerussite [MINERAL] $Pb_3(OH)_2(CO_3)_2$ A colorless mineral composed of basic lead carbonate, occurring as crystals in thin hexagonal plates. { ,hī·drō·sә'rә,sīt }

hydrocyanite See chalcocyanite. { ,hī·drә'sī·ә,nīt }

hydrogarnet [MINERAL] One of a group of minerals having the general formula A_3B_2-$(SiO_4)_{3-x}(OH)_{4x}$; isomorphous with certain garnets. { ¦hī·drō'gär·nәt }

hydrogenic rock See aqueous rock. { 'hī·drә,jen·ik 'räk }

hydrogeochemistry [GEOCHEM] The study of the chemical characteristics of ground and surface waters as related to areal and regional geology. { ¦hī·drō,jē·ō'kem·ә·strē }

156

hydrohalite |MINERAL| $Na_2Cl \cdot 2H_2O$ A mineral composed of hydrated sodium chloride, formed only from salty water cooled below $0°C$. { ¦hī·drə'ha,līt }

hydrohalloysite *See* endellite. { ¦hī·drō·hə'lȯi,zīt }

hydrohetaerolite |MINERAL| $Zn_2Mn_4O_8 \cdot H_2O$ A dark brown to brownish-black mineral consisting of a hydrated oxide of zinc and manganese; occurs in massive form. { ¦hī·drō·hə'tir·ə,līt }

hydrokaolin *See* endellite. { ,hī·drə'kā·ə·lən }

hydrolaccolith |GEOL| A frost mound, 0.3–20 feet (0.1–6 meters) in height, having a core of ice and resembling a laccolith in section. Also known as cryolaccolith. { ,hī·drə'lak·ə,lith }

hydrolith |PETR| **1.** A chemically precipitated aqueous rock, such as rock salt. **2.** A rock that is free of organic material. { 'hī·drə,lith }

hydrologic sequence |GEOL| A series of soil sections from differentiated parent material that shows increasing lack of drainage downslope. { ¦hī·drə¦läj·ik 'sē·kwəns }

hydrolyzate |GEOL| A sediment characterized by elements such as aluminum, potassium, or sodium which are readily hydrolyzed. { hī'dräl·ə,zāt }

hydromagnesite |MINERAL| $Mg_4(OH)_2(CO_3)_3 \cdot 3H_2O$ A white, earthy mineral crystallizing in the monoclinic system and found in small crystals, amorphous masses, or chalky crusts. { ,hī·drō'mag·nə,zīt }

hydrometamorphism |GEOL| Alteration of rocks by material carried in solution by water without the influence of high temperature or pressure. { ,hī·drə,med·ə'mȯr·fiz·əm }

hydromica |GEOL| Any of several varieties of muscovite, especially illite, which are less elastic than mica, have a pearly luster, and sometimes contain less potash and more water than muscovite. Also known as hydrous mica. { ¦hī·drō'mī·kə }

hydromorphic |GEOL| Referring to an intrazonal soil with characteristics that were developed in the presence of excess water all or part of the time. { ¦hī·drə'mȯr·fik }

hydrophilite *See* chlorocalcite. { hī'dräf·ə,līt }

hydrostatic assumption |GEOPHYS| **1.** The assumption that the pressure of seawater increases by 1 atmosphere (101,325 pascals) over approximately 33 feet (10 meters) of depth, the exact value depending on the water density and the local acceleration of gravity. **2.** Specifically, the assumption that fluid is not undergoing vertical accelerations, hence the vertical component of the passive gradient force per unit mass is equal to g, the local acceleration due to gravity. { ,hī·drə'stad·ik ə'səm·shən }

hydrotalcite |MINERAL| $Mg_6Al_2(OH)_{16}(CO_3) \cdot 4H_2O$ Pearly-white mineral composed of hydrous aluminum and magnesium hydroxide and carbonate. { ,hī·drə'tal,sīt }

hydrothermal |GEOL| Of or pertaining to heated water, to its action, or to the products of such action. { ,hī·drə'thər·məl }

hydrothermal alteration |GEOL| Rock or mineral phase changes that are caused by the interaction of hydrothermal liquids and wall rock. { ,hī·drə'thər·məl ,ȯl·tə'rā·shən }

hydrothermal deposit |GEOL| A mineral deposit precipitated from a hot, aqueous solution. { ,hī·drə'thər·məl di'päz·ət }

hydrothermal solution |GEOL| Hot, residual watery fluids derived from magmas during the later stages of their crystallization and commonly containing large amounts of dissolved metals which are deposited as ore veins in fissures along which the solutions often move. { ,hī·drə'thər·məl sə'lü·shən }

hydrothermal synthesis |GEOL| Mineral synthesis in the presence of heated water. { ,hī·drə'thər·məl 'sin·thə·səs }

hydrotroilite |MINERAL| $FeS \cdot nH_2O$ A black, finely divided colloidal material reported in many muds and clays; thought to be formed by bacteria on bottoms of marine basins. { ,hī·drō'trȯi,līt }

hydrotungstite |MINERAL| $H_2WO_4 \cdot H_2O$ A mineral composed of hydrous tungstic acid. { ¦hī·drō'təŋz,tīt }

hydrous |MINERAL| Indicating a definite proportion of combined water. { 'hī·drəs }

hydrous mica *See* hydromica. { 'hī·drəs 'mī·kə }

hydroxylapatite |MINERAL| $Ca_5(PO_4)_3OH$ A rare form of the apatite group that crystallizes in the hexagonal system. { hī¦dräk·səl'ap·ə,tīt }

hydroxylherderite [MINERAL] $CaBe(PO_4)(OH)$ A monoclinic mineral composed of a phosphate and hydroxide of calcium and beryllium; isomorphous with herderite. { hīˌdräk·səl'hər·də,rīt }

hydrozincite [MINERAL] $Zn_5(OH)_5(CO_3)_2$ A white, grayish, or yellowish mineral composed of basic zinc carbonate, occurring as masses or crusts. { ˌhī·drō'ziŋ,kīt }

Hyeniales [PALEOBOT] An order of Devonian plants characterized by small, dichotomously forked leaves borne in whorls. { ˌhī·ə'nā·lēz }

Hyeniatae See Hyeniopsida. { ˌhī·ə'nī·ə,tē }

Hyeniopsida [PALEOBOT] An extinct class of the division Equisetophyta. { ˌhī·ə·nē'äp·sə·də }

Hyopssodontidae [PALEON] A family of extinct mammalian herbivores in the order Condylarthra. { ˌhī·äp·sə'dänt·ə,dē }

hypabyssal rock [PETR] Those igneous rocks that rose from great depths as magmas but solidified as minor intrusions before reaching the surface. { ˌhip·ə'bis·əl 'räk }

hypautomorphic See hypidiomorphic. { hiˌpòd·ə'mòr·fik }

hypergene See supergene. { 'hī·pər,jēn }

hypersaline [GEOL] Geologic material with high salinity. { ˌhī·pər'sā,lēn }

hypersthene [MINERAL] $(Mg,Fe)SiO_3$ A grayish, greenish, black, or dark-brown rockforming mineral of the orthopyroxene group, with bronzelike luster on the cleavage surface. { 'hī·pər,sthēn }

hypersthenfels See norite. { ˌhī·pər'sthēn,felz }

Hypertragulidae [PALEON] A family of extinct chevrotainlike pecoran ruminants in the superfamily Traguloidea. { ˌhī·pər·trə'gyül·ə,dē }

hypidiomorphic [PETR] Of the texture of igneous rocks, having the crystals bounded partly by the crystal faces characteristic of the mineral species. Also known as hypautomorphic; subidiomorphic. { hī,pid·ē·ō'mòr·fik }

hypocenter [GEOPHYS] The point along a fault where an earthquake is initiated. { 'hī·pə,sent·ər }

hypocrystalline [PETR] Pertaining to the texture of igneous rock characterized by crystalline components in an amorphous groundmass. Also known as hemicrystalline; hypohyaline; merocrystalline; miocrystalline; semicrystalline. { ˌhī·pō'krist·əl·ən }

hypogene [GEOL] **1.** Of minerals or ores, formed by ascending waters. **2.** Of geologic processes, originating within or below the crust of the earth. { 'hī·pə,jēn }

hypohyaline See hypocrystalline. { ˌhī·pō'hī·ə·lən }

hypomagma [GEOL] Relatively immobile, viscous lava that forms at depth beneath a shield volcano, is undersaturated with gases, and initiates volcanic activity. { ˌhī·pō,mag·mə }

hypothermal [GEOL] Referring to the high-temperature (300–500°C) environment of hypothermal deposits. { ˌhī·pō'thər,məl }

hypothermal deposit [MINERAL] Mineral deposit formed at great depths and high (300–500°C) temperatures. { ˌhī·pō'thər·məl di'päz·ət }

Hypsithermal See Altithermal. { ˌhip·sə'thərm·əl }

hypsometric formula [GEOPHYS] A formula, based on the hydrostatic equation, for either determining the geopotential difference or thickness between any two pressure levels, or for reducing the pressure observed at a given level to that at some other level. { ˌhip·sə'me·trik 'fòr·myə·lə }

Hyracodontidae [PALEON] The running rhinoceroses, an extinct family of perissodactyl mammals in the superfamily Rhinoceratoidea. { ˌhī·rə·kō'dänt·ə,dē }

Hystrichospherida [PALEON] A group of protistan microfossils. { ˌhis·trə·kō'sfer·ə·də }

ianthinite [MINERAL] $2UO_2 \cdot 7H_2O$ A violet mineral composed of hydrous uranium dioxide, occurring as orthorhombic crystals. { ē'an·thə,nīt }

ice age [GEOL] A major interval of geologic time during which extensive ice sheets (continental glaciers) formed over many parts of the world. { 'īs ,āj }

Ice Age See Pleistocene. { 'īs ,āj }

ice calving See calving. { 'īs ,kav·iŋ }

ice cave [GEOL] A cave that is cool enough to hold ice through all or most of the warm season. { 'īs ,kāv }

ice-contact delta [GEOL] A delta formed by a stream flowing between a valley slope and the margin of glacial ice. Also known as delta moraine; morainal delta. { 'īs ¦kän,tak ,del·tə }

ice erosion [GEOL] **1.** Erosion due to freezing of water in rock fractures. **2.** See glacial erosion. { 'īs i'rō·zhən }

ice-laid drift See till. { 'īs ¦lād ,drift }

Iceland agate See obsidian. { 'īs·lənd 'ag·ət }

Iceland crystal See Iceland spar. { 'īs·lənd 'krist·əl }

Iceland spar [MINERAL] A pure, transparent form of calcite found particularly in Iceland; easily cleaved to form rhombohedral crystals that are doubly refracting. Also known as Iceland crystal. { 'īs·lənd 'spär }

ice mound See ground ice mound. { 'īs ,maùnd }

ice push [GEOL] Lateral pressure that is caused by expansion of shoreward-moving ice on a lake or a bay of the sea and that follows a rise in temperature. Also known as ice shove; ice thrust. { 'īs ,pùsh }

ice-rafting [GEOL] The transporting of rock and other minerals, of all sizes, on or within icebergs, ice floes, river drift, or other forms of floating ice. { 'īs ,raf·tiŋ }

ice shove See ice push. { 'īs ,shəv }

ice spar See sanidine. { 'īs ,spär }

ice stone See cryolite. { 'īs ,stōn }

ice thrust See ice push. { 'īs ,thrəst }

ichnite [PALEON] An ichnofossil of the footprint or track of an organism. Also known as ichnolite. { 'ik,nīt }

ichnofacies [GEOL] A recurrent assemblage of ichnofossils that represent certain environmental conditions. { ¦ik·nō¦fā,shēz }

ichnofossil See trace fossil. { 'ik·nə,fäs·əl }

ichnolite [PALEON] **1.** A rock containing a fossilized track or footprint. **2.** See ichnite. { 'ik·nə,līt }

ichnology [PALEON] The study of ichnofossils, especially fossil footprints. { ik'näl·ə·jē }

ichor [GEOL] A fluid rich in mineralizers. { 'ī,kòr }

Ichthyodectidae [PALEON] A family of Cretaceous marine osteoglossiform fishes. { ,ik·thē·ə'dek·tə,dē }

ichthyolith [PALEON] Fossil fish remains. { 'ik·thē·ə,lith }

Ichthyopterygia [PALEON] A subclass of extinct Mesozoic reptiles composed of predatory fish-finned and sea-swimming forms with short necks and a porpoiselike body. { ,ik·thē,äp·tə'rij·ē·ə }

Ichthyornis

Ichthyornis [PALEON] The type genus of Ichthyornithidae. { ik·thē'òr·nəs }

Ichthyornithes [PALEON] A superorder of fossil birds of the order Ichthyornithiformes according to some systems of classification. { ˌik·thē'òr·nəˌthēz }

Ichthyornithidae [PALEON] A family of extinct birds in the order Ichthyornithiformes. { ˌik·thē,òr'nith·ə,dē }

Ichthyornithiformes [PALEON] An order of ancient fossil birds including strong flying species from the Upper Cretaceous that possessed all skeletal characteristics of modern birds. { ˌik·thē,òr·nə·thə'fòr,mēz }

Ichthyosauria [PALEON] The only order of the reptilian subclass Ichthyopterygia, comprising the extinct predacious fish-lizards; all were adapted to a sea life in having tail flukes, paddles, and dorsal fins. { ˌik·thē·ə'sòr·ē·ə }

Ichthyostega [PALEON] Four-legged vertebrates that evolved from their lobe-finned fish ancestors during the later Devonian Period (400–350 million years ago). { ˌik·thē·ə'steg·ə }

Ichthyostegalia [PALEON] An extinct Devonian order of labyrinthodont amphibians, the oldest known representatives of the class. { ˌik·thē·ō·stə'gal·ē·ə }

Ictidosauria [PALEON] An extinct order of mammallike reptiles in the subclass Synapsida including small carnivorous and herbivorous terrestrial forms. { ik'tid·ə'sòr·ē·ə }

iddingsite [MINERAL] A reddish-brown mixture of silicates, forming patches in basic igneous rocks. { 'id·iŋ,zīt }

ideogenous See syngenetic. { ˌid·ē'äj·ə·nəs }

idioblast [GEOL] A mineral constituent of a metamorphic rock formed by recrystallization which is bounded by its own crystal faces. { 'īd·ē·ō,blast }

idiochromatic [MINERAL] Having characteristic color, usually applied to minerals. { ˈid·ē·ō·krō'mad·ik }

idiomorphic See automorphic. { ˈid·ē·ō|mòr·fik }

idocrase See vesuvianite. { 'ī·dō,krās }

idrialite [MINERAL] A mineral composed of crystalline hydrocarbon, $C_{22}H_{14}$. { 'id·rē·ə,līt }

igneous [PETR] Pertaining to rocks which have congealed from a molten mass. { 'ig·nē·əs }

igneous complex [PETR] An assemblage of igneous rocks that are intimately associated and roughly contemporaneous. { 'ig·nē·əs ˈkäm,pleks }

igneous facies [PETR] A part of an igneous rock differing in structure, texture, or composition from the main mass. { 'ig·nē·əs ˈfā·shēz }

igneous mineral [MINERAL] Mineral material forming igneous rock. { 'ig·nē·əs ˈmin·rəl }

igneous petrology [PETR] The study of igneous rocks, their occurrence, composition, and origin. { 'ig·nē·əs pi'träl·ə·jē }

igneous province See petrographic province. { 'ig·nē·əs präv·əns }

Iguanodon [PALEON] A herbivorous ornithopod dinosaur, 30 feet (9 meters) long and weighing 5 tons, that appeared during the Early Cretaceous Period. { i'gwän·ə,dän }

ignimbrite [PETR] A rock deposit (welded or not) resulting from one or more ground-hugging flows of hot volcanic fragments and particles commonly produced during explosive eruptions (pyroclastic flows and tephra fall). Most ignimbrites have a sheet-like shape, cover many thousands of square kilometers, and have chemical compositions that span the range commonly exhibited by igneous rocks (basaltic to rhyolitic). Also known as ash-flow tuff; pyroclastic-flow deposit; welded tuff. { 'ig·nəm,brīt }

IGY See International Geophysical Year.

ihleite See copiapite. { 'ē·lə,īt }

ilesite [MINERAL] (Mn,Zn,Fe)SO$_4$·4H$_2$O A green mineral composed of hydrous manganese zinc iron sulfate. { 'īl,zīt }

ijolite [PETR] A plutonic rock of nepheline and 30–60% mafic materials, generally sodic pyroxene, with accessory apatite, sphene, calcite, and titaniferous garnet. { 'ē·ə,līt }

Illinoian |GEOL| The third glaciation of the Pleistocene in North America, between the Yarmouth and Sangamon interglacial stages. { ¦il·ə¦nȯi·ən }

illite |MINERAL| A group of gray, green, or yellowish-brown micalike clay minerals found in argillaceous sediments; intermediate in composition and structure between montmorillonite and muscovite. { 'i,līt }

illuvial |GEOL| Pertaining to a region or material characterized by the accumulation of soil by the illuviation of another zone or material. { i'lü·vē·əl }

illuvial horizon See B horizon. { i'lü·vē·əl hə'rīz·ən }

illuviation |GEOL| The deposition of colloids, soluble salts, and small mineral particles in an underlying layer of soil. { i,lü·vē'ā·shən }

illuvium |GEOL| Material leached by chemical or other processes from one soil horizon and deposited in another. { i'lü·vē·əm }

ilmenite |MINERAL| FeTiO$_3$ An iron-black, opaque, rhombohedral mineral that is the principal ore of titanium. Also known as mohsite; titanic iron ore. { il·mə,nīt }

ilsemannite |MINERAL| A black, blue-black, or blue mineral composed of hydrous molybdenum oxide or perhaps sulfate, occurring in earthy massive form. { 'il·sə·mə,nīt }

imbricate structure |GEOL| **1.** A sedimentary structure characterized by shingling of pebbles all inclined in the same direction with the upper edge of each leaning downstream or toward the sea. Also known as shingle structure. **2.** Tabular masses that overlap one another and are inclined in the same direction. Also known as schuppen structure; shingle-block structure. { 'im·brə·kət ‚strək·chər }

imbrication |GEOL| Formation of an imbricate structure. Also known as shingling. { ‚im·brə'kā·shən }

imerinite |MINERAL| Na$_2$(Mg,Fe)$_6$Si$_8$O$_{22}$(O,OH)$_2$ A colorless to blue mineral composed of a basic silicate of sodium, iron, and magnesium, occurring as acicular crystals. { ‚im·ə'rē,nīt }

immature soil See azonal soil. { ¦im·ə'chür 'sȯil }

impact cast See prod cast. { 'im,pakt ‚kast }

impact crater |GEOL| A crater formed on a planetary surface by the impact of a projectile. { 'im,pakt ‚krād·ər }

impactite |GEOL| Glassy fused rock or meteor fragments resulting from heat of impact of a meteor on the earth. { 'im,pak,tīt }

impact mark See prod mark. { 'im,pakt ‚märk }

impression |GEOL| A form left on a soft soil surface by plant parts; the soil hardens and usually the imprint is a concave feature. { im'presh·ən }

imprint See overprint. { 'im,print }

impsonite |GEOL| A black, asphaltic pyrobitumen with a high fixed-carbon content derived from the metamorphosis of petroleum. { 'im·sə,nīt }

Inadunata |PALEON| An extinct subclass of stalked Paleozoic Crinozoa characterized by branched or simple arms that were free and in no way incorporated into the calyx. { i¦nä·jə¦näd·ə }

incandescent tuff flow See ash flow. { ‚in·kən'des·ənt 'təf ‚flō }

incarbonization See coalification. { in,kär·bə·nə'zā·shən }

Inceptisol |GEOL| A soil order characterized by soils that are usually moist, with pedogenic horizons of alteration of parent materials but not of illuviation. { in'sep·tə,sȯl }

incised meander |GEOL| A deep, tortuous valley cut by a meandering stream that was rejuvenated. { in'sīzd mē'an·dər }

inclination |GEOL| The angle at which a geological body or surface deviates from the horizontal or vertical; often used synonymously with dip. |GEOPHYS| In magnetic inclination, the dip angle of the earth's magnetic field. Also known as magnetic dip. { ‚iŋ·klə'nā·shən }

inclined bedding |GEOL| A type of bedding in which the strata dip in the direction of current flow. { in'klīnd 'bed·iŋ }

inclined contact |GEOL| A contact plane of gas or oil with water underlying, in which the plane slopes or is inclined. { in'klīnd 'kän,takt }

inclusion |PETR| A fragment of older rock enclosed in an igneous rock. { in'klü·zhən }

incoalation

incoalation *See* coalification. { ‚in·kō'lā·shən }

incoherent [GEOL] Pertaining to a rock or deposit that is loose or unconsolidated, or that is unable to hold together firmly or solidly. { ‚in·kō'hir·ənt }

incompetent bed [GEOL] A bed not combining sufficient firmness and flexibility to transmit a thrust and to lift a load by bending. { in'käm·pəd·ənt 'bed }

incongruous [GEOL] Of a drag fold, having an axis and axial surface not parallel to the axis and axial surface of the main fold to which it is related. { in'käŋ·grü·əs }

incumbent [GEOL] Lying above, said of a stratum that is superimposed or overlies another stratum. { in'kəm·bənt }

inderborite [MINERAL] $CaMgB_6O_{11} \cdot 11H_2O$ A monoclinic mineral composed of hydrous calcium and magnesium borate. { ‚in·dər'bȯ‚rīt }

inderite [MINERAL] $Mg_2B_6O_{11} \cdot 15H_2O$ A hydrated borate mineral. { 'in·də‚rīt }

index bed *See* key bed. { 'in‚deks ‚bed }

index fossil [PALEON] The ancient remains and traces of an organism that lived during a particular geologic time period and that geologically date the containing rocks. { 'in‚deks ‚fäs·əl }

index mineral [PETR] A mineral whose first appearance in passing from low to higher grades of metamorphism indicates the outer limit of a zone. { 'in‚deks ‚min·rəl }

index plane [GEOL] A surface used as a reference point in determining geological structure. { 'in‚deks ‚plān }

indialite [MINERAL] $Mg_2Al_4Si_5O_{18}$ A hexagonal cordierite mineral; it is isotypic with beryl. { 'in·dē·ə‚līt }

indianaite [MINERAL] A white porcelainlike clay mineral; a variety of halloysite found in Indiana. { ‚in·dē'a·nə‚īt }

Indiana limestone *See* spergenite. { ‚in·dē'a·nə 'līm‚stōn }

indicolite [MINERAL] An indigo-blue variety of tourmaline that is used as a gemstone. Also known as indigolite. { in'dik·ə‚līt }

indigenous coal *See* autochthonous coal. { in'dij·ə·nəs kōl }

indigenous limonite [MINERAL] Sulfide-derived limonite that remains fixed at the site of the parent sulfide. { in'dij·ə·nəs 'lī·mə‚nīt }

indigo copper *See* covellite. { 'in·də·gō 'käp·ər }

indigolite *See* indicolite. { 'in·də‚gō‚līt }

indirect stratification *See* secondary stratification. { ‚in·də'rekt ‚strad·ə·fə'kā·shən }

induced magnetization [GEOPHYS] That component of a rock's magnetization which is proportional to, and has the same direction as, the ambient magnetic field. { in'düst ‚mag·nə·tə'zā·shən }

induration [GEOL] **1.** The hardening of a rock material by the application of heat or pressure or by the introduction of a cementing material. **2.** A hardened mass formed by such processes. **3.** The hardening of a soil horizon by chemical action to form a hardpan. { ‚in·də'rā·shən }

industrial diamond [MINERAL] Diamond that is too hard or too radial-grained to be used for jewel cutting. { in'dəs·trē·əl 'dī·mənd }

industrial jewel [MINERAL] A hard stone, such as ruby or sapphire, used for bearings and impulse pins in instruments and for recording needles. { in'dəs·trē·əl ⌐jül }

inertial flow [GEOPHYS] Frictionless flow in a geopotential surface in which there is no pressure gradient; the centrifugal and Coriolis accelerations must therefore be equal and opposite, and the constant inertial wind speed V_i is given by $V_i = fR$, where f is the Coriolis parameter and R the radius of curvature of the path. { i'nər·shəl 'flō }

inertinite [GEOL] A carbon-rich maceral group, which includes micrinite, sclerotinite, fusinite, and semifusinite. { i'nərt·ən‚īt }

inesite [MINERAL] $Ca_2Mn_7Si_{10}O_{28}(OH)_2 \cdot 5H_2O$ A pale-red mineral composed of hydrous manganese calcium silicate, occurring in small prismatic crystals or massive. { 'in·ə‚sīt }

inface *See* scarp slope. { 'in‚fās }

infancy [GEOL] The initial (youthful) or very early stage of the cycle of erosion characterized by smooth, nearly level erosional surfaces dissected by narrow stream gorges, numerous depressions filled by marshy lakes and ponds, and shallow streams.   Also known as topographic infancy.   { 'in·fən·sē }

infiltration |GEOL| Deposition of mineral matter among the pores or grains of a rock by permeation of water carrying the matter in solution.   { ‚in·fil'trā·shən }

infiltration vein |GEOL| Vein deposited in rock by percolating water.   { ‚in·fil'trā·shən vān }

Infracambrian *See* Eocambrian.   { ¦in·frə'kam·brē·ən }

infusorial earth |GEOL| Formerly, and incorrectly, a soft rock or an earthy substance composed of siliceous remains of diatoms.   { ‚in·fyə'sór·ē·əl 'ərth }

ingrown meander |GEOL| A meander of a stream with an undercut bank on one side and a gentle slope on the other.   { 'in‚grōn mē'an·dər }

initial dip *See* primary dip.   { i'nish·əl 'dip }

initial landform |GEOL| A landform that is produced directly by epeirogenic, orogenic, or volcanic activity, and whose original features are only slightly modified by erosion.   { i'nish·əl 'land‚fŏrm }

injected |PETR| Pertaining to intrusive igneous rock or other mobile rock that has erupted through rock walls to neighboring older rocks.   { in'jek·təd }

injection |GEOL| Also known as intrusion; sedimentary injection.   **1.** A process by which sedimentary material is forced under abnormal pressure into a preexisting rock or deposit.   **2.** A structure formed by an injection process.   { in'jek·shən }

injection gneiss |PETR| A composite rock with banding entirely or partly caused by layer-by-layer injection of granitic magma into rock layers.   { in'jek·shən ¦nīs }

inlier |GEOL| A circular or elliptical area of older rocks surrounded by strata that are younger.   { 'in‚lī·ər }

inner core |GEOL| The central part of the earth's core, extending from a depth of 3160 miles (5100 kilometers) to the center of the earth.   Also known as siderosphere.   { ¦in·ər 'kòr }

inner mantle *See* lower mantle.   { ¦in·ər 'mant·əl }

inorganic chert |PETR| Chert derived from siliceous colloids precipitated from silica-saturated waters.   { ¦in·ȯr¦gan·ik 'chərt }

inosilicate |GEOL| A class or structural type of silicate in which the SiO_4 tetrahedrons are linked together by the sharing of oxygens to form linear chains of indefinite length.   { ¦in·ō'sil·ə‚kāt }

in-place stress field *See* ambient stress field.   { ‚in ‚plās 'stres ‚fēld }

inselberg |GEOL| A large, steep-sided residual hill, knob, or mountain, generally rocky and bare, rising abruptly from an extensive, nearly level lowland erosion surface in arid or semiarid regions.   Also known as island mountain.   { 'in·səl‚bərg }

inshore zone |GEOL| The zone of variable width extending from the shoreline at low tide through the breaker zone.   { 'in‚shòr 'zōn }

insoluble residue |GEOL| Material remaining after a geological specimen is dissolved in hydrochloric or acetic acid.   { in'säl·yə·bəl 'rez·ə‚dü }

inspissation |GEOCHEM| Thickening of an oil deposit by evaporation or oxidation, resulting, for example, after long exposure in pitch or gum formation.   { ‚in·spi'sā·shən }

interbedded |GEOL| Having beds lying between other beds with different characteristics.   { ¦in·tər¦bed·əd }

intercalation |GEOL| A layer located between layers of different character.   { in‚tər·kə'lā·shən }

interface *See* seismic discontinuity.   { 'in·tər‚fās }

interference ripple mark |GEOL| A pattern resulting from two sets of symmetrical ripples formed by waves crossing at right angles.   { ‚in·tər'fir·əns 'rip·əl ‚märk }

interfluve |GEOL| The area of land between two rivers, usually an upland or ridge between two adjacent valleys that contain streams flowing in approximately the same direction.   { 'in·tər‚flüv }

intergelisol *See* pereletok.   { ¦in·tər¦jel·ə‚sòl }

interglacial |GEOL| Pertaining to or formed during a period of geologic time between two successive glacial epochs or between two glacial stages.   { ¦in·tər'glā·shəl }

163

intergranular pressure See effective stress. { ¦in·tər'gran·yə·lər 'presh·ər }

intergrowth [MINERAL] A state of interlocking of different mineral crystals because of simultaneous crystallization. { 'in·tər‚grōth }

interlobate moraine See intermediate moraine. { ¦in·tər'lō‚bāt mə'rān }

intermediate layer See sima. { ‚in·tər'mēd·ē·ət ¦lā·ər }

intermediate moraine [GEOL] A type of lateral moraine formed at the junction of two adjacent glacial lobes. Also known as interlobate moraine. { ‚in·tər'mēd·ē·ət mə'rān }

intermontane [GEOL] Located between or surrounded by mountains. { ¦in·tər¦män‚tān }

intermontane glacier [GEOL] A glacier that is formed by the confluence of several valley glaciers and occupies a trough between separate ranges of mountains. { ¦in·tər¦män‚tān 'glā·shər }

intermontane trough [GEOL] **1.** A subsiding area in an island arc of the ocean, lying between the stable elements of a region. **2.** A basinlike area between mountains. { ¦in·tər¦män‚tān 'tròf }

internal cast See steinkern. { in'tərn·əl 'kast }

internal erosion [GEOL] Erosion effected within a compacting sediment by movement of water through the larger pores. { in'tərn·əl i'rō·zhən }

internal sedimentation [GEOL] Accumulation of clastic or chemical sediments derived from the surface of, or within, a more or less consolidated carbonate sediment (mud or silt); deposited in secondary cavities formed in the host rock (after its deposition) by bending of laminae or by internal erosion or solution. { in'tərn·əl ‚sed·ə·mən'tā·shən }

International Geophysical Year [GEOPHYS] An internationally accepted period, extending from July 1957 through December 1958, for concentrated and coordinated geophysical exploration, primarily of the solar and terrestrial atmospheres. Abbreviated IGY. { ¦in·tər¦nash·ən·əl ‚jē·ō'fiz·ə·kəl ‚yir }

internides [GEOL] The internal part of an orogenic belt, farthest away from the craton, which is commonly the site of a eugeosyncline during its early phases and is later subjected to plastic folding and plutonism. Also known as primary arc. { in'tər·nə‚dēz }

interpluvial [GEOL] Pertaining to an episode or period of geologic time that was dryer than the pluvial period occurring before or after it. { ‚in·tər'plü·vē·əl }

intersertal [PETR] Referring to the texture of a porphyritic igneous rock in which the groundmass forms a small proportion of the rock, filling the interstices between unoriented feldspar laths. { ¦in·tər¦sərd·əl }

interstadial [GEOL] Pertaining to a period during a glacial stage in which the ice retreated temporarily. { ‚in·tər'stād·ē·əl }

interstice [GEOL] A pore space within a rock or soil. { in'tərs·təs }

intraclast [GEOL] A fragment of limestone formed by erosion within a basin of deposition and redeposited there to form a new sediment. { 'in·trə‚klast }

intracratonic basin See autogeosyncline. { ¦in·trə·krə'tän·ik 'bā·sən }

intraformational breccia [PETR] A rock resulting from cracking and desiccation-shrinking of a mud after withdrawal of water followed by almost contemporaneous sedimentation. { ¦in·trə‚fòr'māsh·ən·əl 'brech·ə }

intraformational conglomerate [GEOL] **1.** A conglomerate in which clasts and the matrix are contemporaneous in origin. **2.** A conglomerate formed in the midst of a geologic formation. { ¦in·trə‚fòr'māsh·ən·əl kən'gläm·ə·rət }

intraformational fold [GEOL] A minor fold confined to a sedimentary layer lying between undeformed beds. { in·trə‚fòr'māsh·ən·əl 'fōld }

intrastratal solution [GEOCHEM] A chemical attrition of the constituents of a rock after deposition. { ¦in·trə'strad·əl sə'lü·shən }

intratelluric [GEOL] **1.** Pertaining to a phenocryst that is formed earlier than its matrix. **2.** Pertaining to a period in which igneous rocks crystallized prior to their eruption. **3.** Located, formed, or originating at great depths within the earth. { ¦in·trə·tə'lyùr·ik }

intrazonal soil [GEOL] A group of soils with well-developed characteristics that reflect the dominant influence of some local factor of relief, parent material, or age over the usual effect of vegetation and climate. { ,in·trə'zōn·əl 'sȯil }

intrusion [GEOL] **1.** The process of emplacement of magma in preexisting rock. Also known as injection; invasion; irruption. **2.** A large-scale sedimentary injection. Also known as sedimentary intrusion. **3.** Any rock mass formed by an intrusive process. { in'trü·zhən }

intrusive [PETR] Pertaining to material forced while still in a fluid state into cracks or between layers of rock. { in'trü·siv }

invasion [GEOL] **1.** The movement of one material into a porous reservoir area that has been occupied by another material. **2.** See intrusion; transgression. { in'vā·zhən }

inversion [GEOL] **1.** Development of inverted relief through which anticlines are transformed into valleys and synclines are changed into mountains. **2.** The occupancy by a lava flow of a ravine or valley that occurred in the side of a volcano. **3.** A diagenetic process in which unstable minerals are converted to a more stable form without a change in chemical composition. { in'vər·zhən }

inverted See overturned. { in'vərd·əd }

inverted plunge [GEOL] A plunge of a fold whose inclination has been carried past the vertical, so that the plunge is less than 90° in the direction opposite from the original attitude; younger rocks plunge beneath the older rocks. { in'vərd·əd 'plənj }

inverted relief [GEOL] A topographic configuration that is opposite to that of the geologic structure, for example, where a valley occupies the site of an anticline. { in'vərd·əd ri'lēf }

inyoite [MINERAL] $Ca_2B_6O_{11} \cdot 13H_2O$ A colorless, monoclinic mineral consisting of a hydrous calcium borate; hardness is 2 on Mohs scale, and specific gravity is 2. { 'in·yō,īt }

iodargyrite [MINERAL] AgI A yellowish or greenish hexagonal mineral composed of native silver iodide, usually occurring in thin plates. Also known as iodyrite. { ,ī·ə'där·jə,rīt }

iodobromite [MINERAL] Ag(Br,Cl,I) An isometric mineral composed of chloride, iodide, and bromide of silver; it is isomorphous with cerargyrite and bromyrite. { ī,ō·də'brō,mīt }

iodyrite See iodargyrite. { ī'äd·ə,rīt }

ionite See anauxite. { ī·ə,nīt }

Iowan glaciation [GEOL] The earliest substage of the Wisconsin glacial stage; occurred more than 30,000 years ago. { 'ī·ə·wən ,glā·sē'ā·shən }

ipsonite [GEOL] The final stage of weathered asphalt; a black, infusible substance, only slightly soluble in carbon disulfide, containing 50–80% fixed carbon and very little oxygen. { 'ip·sə,nīt }

iron alum See halotrichite. { 'ī·ərn 'al·əm }

iron cordierite See sekaninaite. { 'ī·ərn 'kȯr·dē·ə,rīt }

iron formation [GEOL] Sedimentary, low-grade iron ore bodies consisting mainly of chert or fine-grained quartz and ferric oxide segregated in bands or sheets irregularly mingled. { 'ī·ərn fȯr'mā·shən }

iron glance See specularite. { 'ī·ərn 'glans }

iron hat See gossan. { 'ī·ərn 'hat }

iron mica See lepidomelane. { 'ī·ərn 'mī·kə }

iron ore [GEOL] Rocks or deposits containing compounds from which iron can be extracted. { 'ī·ərn 'ȯr }

iron pyrites See pyrite. { 'ī·ərn 'pī,rīts }

ironshot [MINERAL] Pertaining to a mineral with streaks or spots of iron or iron ore. { 'ī·ərn,shät }

iron spar See siderite. { 'ī·ərn 'spär }

iron spinel See hercynite. { 'ī·ərn spə'nel }

ironstone [PETR] An iron-rich sedimentary rock, either deposited directly as a ferruginous sediment or resulting from chemical replacement. { 'ī·ərn,stōn }

iron-stony meteorite See stony-iron meteorite. { 'ī·ərn ,stō·nē 'mēd·ē·ə,rīt }

165

irrotational strain [GEOL] Strain in which the orientation of the axes of strain does not change. Also known as nonrotational strain. { ¦ir·ə'tā·shən·əl 'strān }

irruption See intrusion. { i'rəp·shən }

Irvingtonian [GEOL] A stage of geologic time in southern California, in the lower Pleistocene, below the Rancholabrean. { ¸ər·viŋ'tō·nē·ən }

Ischnacanthidae [PALEON] The single family of the acanthodian order Ischnacanthiformes. { ¸isk·nə'kan·thə¸dē }

Ischnacanthiformes [PALEON] A monofamilial order of extinct fishes of the order Acanthodii; members were slender, lightly armored predators with sharp teeth, deeply inserted fin spines, and two dorsal fins. { ¸isk·nə¸kan·thə'fōr¸mēz }

Isectolophidae [PALEON] A family of extinct ceratomorph mammals in the superfamily Tapiroidea. { ī¸sek·tə'läf·ə¸dē }

isentropic map [GEOL] A map indicating constant entropy function for facies. { ¦īs·ən'träp·ik 'map }

ishikawaite [MINERAL] A black, orthorhombic mineral consisting essentially of uranium, iron, rare earth, and columbium oxide. { ¸ish·ē'kä·wə¸īt }

ishkyldite [MINERAL] $Mg_{15}Si_{11}O_{27}(OH)_{20}$ A mineral composed of a basic silicate of magnesium. { 'ish·kəl¸dīt }

isinglass [MINERAL] Sheet mica, usually in the form of single cleavage plates; used in furnace and stove doors. { 'īz·ən¸glas }

island mountain See inselberg. { 'ī·lənd ¦maúnt·ən }

isocarb [GEOCHEM] A line on a map that connects points of equal content of fixed carbon in coal. { 'ī·sə¸kärb }

isochemical metamorphism [PETR] Theoretically, a metamorphism involving no great change in its chemical composition. Also known as treptomorphism. { ¦ī·sō'kem·ə·kəl ¸med·ə'mȯr·fiz·əm }

isochemical series [PETR] A series of rocks with identical chemical compositions. { ¦ī·sō'kem·ə·kəl 'sir·ēz }

isochron [GEOCHEM] A line on a graph defined by data for rocks of the same age with the same initial lead isotopic composition, the slope of which is proportional to the age. Also known as geochron. { 'ī·sə¸krän }

isoclasite [MINERAL] $Ca_2(PO_4)(OH) \cdot 2H_2O$ A white mineral composed of a basic hydrous calcium phosphate; occurring in small crystals or columnar forms. { ¸ī·sə'klā¸sīt }

isoclinal See isoclinic line. { ¦ī·sə'klīn·əl }

isoclinal chart [GEOPHYS] A chart showing isoclinic lines. Also known as isoclinic chart. { ¦ī·sə'klīn·əl 'chärt }

isocline [GEOL] A fold of strata so tightly compressed that parts on each side dip in the same direction. { 'ī·sə¸klīn }

isoclinic chart See isoclinal chart. { ¦ī·sə¦klin·ik 'chärt }

isoclinic line [GEOPHYS] A line connecting points on the earth's surface which have the same magnetic dip. Also known as isoclinal. { ¦ī·sə¦klin·ik 'līn }

isodynamic line [GEOPHYS] One of the lines on a map of a magnetic field that connect points having equal strengths of the earth's field. { ¦ī·sō·dī'nam·ik 'līn }

isofacies map [GEOL] A stratigraphic map showing the distribution of one or more facies within a particular stratigraphic unit. { ¦ī·sə¦fā·shēz ¸map }

isogal [GEOPHYS] A contour line on a map connecting points of equal gravity values on the earth's surface. { 'ī·sə¸gal }

isogam [GEOPHYS] A line joining points on the earth's surface having the same value of the acceleration of gravity. { 'ī·sə¸gam }

isogeotherm See geoisotherm. { ¦ī·sō'jē·ə¸thərm }

isogonic line [GEOPHYS] **1.** Any of the lines on a chart or map showing the same direction of the wind vector. **2.** Any of the lines on a chart or map connecting points of equal magnetic variation. { 'ī·sə¦gän·ik ¸līn }

isograd [GEOL] A line on a map joining those rocks comprising the same metamorphic grade. { 'ī·sə¸grad }

isohume [GEOL] A line of a map or chart connecting points of equal moisture content in a coal bed. { 'ī·sə¸hyüm }

isolith [GEOL] A line on a contour-type map that denotes the aggregate thickness of a single lithology in a stratigraphic succession composed of one or more lithologies. { 'ī·sə‚lith }

isolith map [GEOL] A contour-line map depicting the thickness of an exclusive lithology. { 'ī·sə‚lith ‚map }

isomagnetic [GEOPHYS] Of or pertaining to lines connecting points of equality in some magnetic element. { ‚ī·sō·mag¦ned·ik }

isomorph See isomorphic mineral. { 'ī·sə‚mȯrf }

isomorphic mineral [MINERAL] Any two or more crystalline mineral compounds having different chemical composition but identical structure, such as the garnet series or the feldspar group. Also known as isomorph. { ‚ī·sə¦mȯr·fik 'min·rəl }

isopach map [GEOL] Map of the areal extent and thickness variation of a stratigraphic unit; used in geological exploration for oil and for underground structural analysis. { 'ī·sə‚pak ‚map }

isopachous line [GEOL] One of the lines drawn on a map to indicate equal thickness. { ‚ī·sō¦pak·əs ‚līn }

isopor [GEOPHYS] An imaginary line connecting points on the earth's surface having the same annual change in a magnetic element. { 'ī·sə‚pȯr }

isoseismal [GEOPHYS] Pertaining to points having equal intensity of earthquake shock, or to a line on a map of the earth's surface connecting such points. { ‚ī·sə'sīz·məl }

isostasy [GEOPHYS] A theory of the condition of approximate equilibrium in the outer part of the earth, such that the gravitational effect of masses extending above the surface of the geoid in continental areas is approximately counterbalanced by a deficiency of density in the material beneath those masses, while deficiency of density in ocean waters is counterbalanced by an excess in density of the material under the oceans. { ī'säs·tə·sē }

isostatic adjustment See isostatic compensation. { ‚ī·sə'stad·ik ə'jəs·mənt }

isostatic anomaly [GEOPHYS] A gravity anomaly based on a generalized hypothesis that the gravitational effect of masses above sea level is approximately compensated by a density deficiency of the subsurface materials. { ‚ī·sə'stad·ik ə'näm·ə·lē }

isostatic compensation [GEOL] The process in which lateral transport at the surface of the earth by erosion or deposition is compensated by lateral movements in a subcrustal layer. Also known as isostatic adjustment; isostatic correction. { ‚ī·sə'stad·ik ‚käm·pən'sā·shən }

isostatic correction See isostatic compensation. { ‚ī·sə'stad·ik kə'rek·shən }

isotherm [GEOPHYS] A line on a chart connecting all points of equal or constant temperature. { 'ī·sə‚thərm }

isothermal See isotherm. { ‚ī·sə¦thər·məl }

isothermal chart [GEOPHYS] A map showing the distribution of air temperature (or sometimes sea-surface or soil temperature) over a portion of the earth or at some level in the atmosphere; places of equal temperature are connected by lines called isotherms. { ‚ī·sə¦thər·məl 'chärt }

isothermal remanent magnetization [GEOPHYS] A spurious magnetization induced by lightning strikes that produce large surface electrical currents. Abbreviated IRM. { ‚ī·sə¦thər·məl ¦rem·ə·nənt ‚mag·nə·tə'zā·shən }

isotropic fabric [PETR] A random orientation in space of the elements that compose a rock. { ‚ī·sə¦trä·pik 'fab·rik }

itabirite [GEOL] A laminated, metamorphosed, oxide-facies iron formation in which the original chert or jasper bands have been recrystallized into megascopically distinguished grains of quartz and in which the iron is present as thin layers of hematite, magnetite, or martite. { ‚ēd·ə'bi‚rīt }

itacolumite [PETR] A fine-grained, thin-bedded sandstone or a schistose quartzite that contains mica, chlorite, and talc and that exhibits flexibility when split into slabs. Also known as articulite. { ¦id·ə'käl·ə‚mīt }

I-type magma [GEOL] Magma formed from igneous source materials. { 'ī ‚tīp ¦mag·mə }

J

jacinth *See* zircon. { 'jas·ənth }

jack *See* sphalerite. { jak }

jacobsite [MINERAL] $MnFe_2O_4$ A black magnetic mineral composed of an oxide of manganese and iron; a member of the magnetite series. { 'jā·kəb,zīt }

jacupirangite [PETR] An ultramafic plutonic rock that is part of the ijolite series; composed chiefly of titanaugite and magnetite, with a smaller amount of nepheline. { jə'kü·pə·rən,jīt }

jade [MINERAL] A hard, compact, dark-green or greenish-white gemstone composed of either jadeite or nephrite. Also known as jadestone. { jād }

jadeite [MINERAL] $NaAl(SiO_3)_2$ A clinopyroxene mineral occurring as green, fibrous monoclinic crystals; the most valuable variety of jade. { 'jā,dīt }

jadeitite [PETR] A type of metamorphic rock composed of jadeite associated with small amounts of feldspar or feldspathoids. { 'jād·ə,tīt }

jadestone *See* jade. { 'jād,stōn }

jamesonite [MINERAL] $Pb_4FeSb_6S_{14}$ A lead-gray to gray-black mineral that crystallizes in the orthorhombic system, occurs in acicular crystals with fibrous or featherlike forms, and has a metallic luster. Also known as feather ore; gray antimony. { 'jām·sə,nīt }

jarlite [MINERAL] $NaSr_3Al_3F_{16}$ A colorless to brownish mineral composed of aluminofluoride of sodium and strontium. { 'yär,līt }

jarosite [MINERAL] $KFe_3(SO_4)_2(OH)_6$ An ocher-yellow or brown alunite mineral having rhombohedral crystal structure. Also known as utahite. { jə'rō,sīt }

jaspagate *See* agate jasper. { 'jas·pə·gət }

jasper [PETR] A dense, opaque to slightly translucent cryptocrystalline quartz containing iron oxide impurities; characteristically red. Also known as jasperite; jasperoid; jaspis. { 'jas·pər }

jasperite *See* jasper. { 'jas·pə,rīt }

jasperoid *See* jasper. { 'jas·pə,ròid }

jaspilite [PETR] A compact siliceous rock resembling jasper and containing iron oxides in bands. { 'jas·pə,līt }

jaspis *See* jasper. { 'jas·pəs }

jaspoid *See* tachylite. { 'jas,pòid }

Java man [PALEON] An overspecialized, apelike form of *Homo sapiens* from the middle Pleistocene having a small brain capacity, low cranial vault, and massive browridges. { ¦jäv·ə ¦man }

jaw [GEOL] The side of a narrow passage such as a gorge. { jò }

jeffersonite [MINERAL] $Ca(Mn,Zn,Fe)Si_2O_6$ A dark-green or greenish-black mineral composed of pyroxene. { 'jef·ər·sə,nīt }

jelly *See* ulmin. { 'jel·ē }

jeremejevite [MINERAL] $AlBO_3$ A colorless or yellowish mineral composed of aluminum borate that occurs in hexagonal crystals. { ,yer·ə'mā·ə,vīt }

jet coal [GEOL] A hard, lustrous, pure black variety of lignite, occurring in isolated masses in bituminous shale; thought to be derived from waterlogged driftwood. Also known as black amber. { 'jet ¦kōl }

jezekite *See* morinite. { 'jez·ə,kīt }

J function [GEOPHYS] A dimensionless mathematical relationship to correlate capillary pressure data of similar geologic formations. { 'jā ,faŋk·shən }

joaquinite [MINERAL] $NaBa_2Ce_2Fe(Ti,Nb)_2Si_8O_{26}(OH,F)_2$ A honey-yellow mineral composed of sodium iron titanium silicate, occurring in orthorhombic crystals. { wä'kē,nīt }

johannite [MINERAL] $Cu(UO_2)_2(SO_4)_2·6H_2O$ An emerald green to apple green, triclinic mineral consisting of a hydrated basic copper and uranium sulfate. { jō'ha,nīt }

johannsenite [MINERAL] $CaMnSi_2O_6$ A clove-brown, grayish, or greenish clinopyroxene mineral composed of a silicate of calcium and manganese; a member of the pyroxene group. { jō'han·sə,nīt }

johnstrupite [MINERAL] A mineral that is composed of a complex silicate of cerium and other metals, approximately $(Ca,Na)_3(Ce,Ti,Zr)(SiO_4)_2F$; occurs in prismatic crystals. { 'jän·strə,pīt }

joint [GEOL] A fracture that traverses a rock and does not show any discernible displacement of one side of the fracture relative to the other. { jȯint }

joint block [GEOL] A body of rock that is bounded by joints. { 'jȯint ,bläk }

joint drag See kink band. { 'jȯint ,drag }

jointing [GEOL] A condition of rock characterized by joints. { 'jȯint·iŋ }

joint plane [GEOL] The surface of fracturing or potential fracture of a joint. { 'jȯint ,plān }

joint set [GEOL] A group of parallel joints in a geologic formation. { 'jȯint ,set }

joint system [GEOL] Two or more joint sets. { 'jȯint ,sis·təm }

joint vein [GEOL] A small vein in a joint. { 'jȯint ,vān }

jordanite [MINERAL] $(Pb,Tl)_{13}As_7S_{23}$ A lead-gray mineral composed of lead arsenic sulfide, occurring as monoclinic crystals. { 'jȯrd·ən,īt }

joseite [MINERAL] $Bi_3Te(Si,S)$ A mineral composed of telluride of bismuth containing sulfur and selenium. { zhə'zā,īt }

josen See hartite. { 'jō·sən }

josephinite [MINERAL] A mineral consisting of an alloy of iron and nickel; occurs naturally in stream gravel. { |jō·zə|fē,nīt }

julienite [MINERAL] $Na_2Co(SCN)_4·8H_2O$ A blue, tetragonal mineral consisting of a hydrated sodium cobalt thiocyanate. { 'jül·yə,nīt }

Jura See Jurassic. { 'jùr·ə }

Jurassic [GEOL] Also known as Jura. **1.** The second period of the Mesozoic era of geologic time. **2.** The corresponding system of rocks. { jə'ras·ik }

jurupaite [MINERAL] $(Ca,Mg)_2(Si_2O_5)(OH)_2$ A mineral composed of hydrous calcium magnesium silicate. { hə'rüp·ə,īt }

juvenile rift [GEOL] A stage of continental breakup before the onset of actual spreading which precedes the generation of new oceanic lithosphere. { 'jü·və·nəl 'rift }

juvite [PETR] A light-colored nepheline syenite in which the feldspar is exclusively or predominantly orthoclase and the potassium oxide content is higher than the sodium oxide content. { 'jü,vīt }

K

K-A age [GEOL] The radioactive age of a rock determined from the ratio of potassium-40 (^{40}K) to argon-40 (^{40}A) present in the rock. { 'kā'ā' ,āj }

kainite [MINERAL] $MgSO_4 \cdot KCl \cdot 3H_2O$ A white, gray, pink, or black monoclinic mineral, occurring in irregular granular masses; used as a fertilizer and as a source of potassium and magnesium compounds. { 'kī,nīt }

kainosite [MINERAL] $Ca_2(Ce,Y)_2(SiO_4)_3CO_3 \cdot H_2O$ A yellowish-brown mineral composed of a hydrous silicate and carbonate of calcium, cerium, and yttrium. { 'kī·nə,sīt }

kaliborite [MINERAL] $HKMg_2B_{12}O_{21} \cdot 9H_2O$ A colorless to white mineral composed of a hydrous borate of potassium and magnesium. Also known as paternoite. { ,kal·ə'bò,rīt }

kalicinite [MINERAL] $KHCO_3$ A colorless to white or yellowish, monoclinic mineral consisting of potassium bicarbonate; occurs in crystalline aggregates. { kə'lis·ən,īt }

kalinite [MINERAL] $KAl(SO_4)_2 \cdot 11H_2O$ A birefringent mineral of the alum group composed of a hydrous sulfate of potassium and aluminum, occurring in fibrous form. Also known as potash alum. { 'kal·ə,nīt }

kaliophilite [MINERAL] $KAlSiO_4$ A rare hexagonal tectosilicate mineral found in volcanic rocks; high in potassium and low in silica, it is dimorphous with kalsilite. Also known as facellite; phacellite. { ,kal·ē'äf·ə,līt }

kalkowskite [MINERAL] $Fe_2Ti_3O_9$ A rare, brownish or black mineral composed of an oxide of iron and titanium, usually with small amounts of rare-earth elements, niobium, and tantalum. { kal'kòf,sīt }

kalsilite [MINERAL] $KAlSiO_4$ A rare mineral from volcanic rocks in southwestern Uganda; the crystal system is hexagonal; kalsilite is dimorphous with kaliophilite and sometimes contains sodium. { 'kal·sə,līt }

kalunite [MINERAL] The naturally occurring form of alum. { 'kal·ə,nīt }

kamacite [MINERAL] A mineral composed of a nickel-iron alloy and comprising with taenite the bulk of most iron meteorites. { 'kam·ə,sīt }

kame [GEOL] A low, long, steep-sided mound of glacial drift, commonly stratified sand and gravel, deposited as an alluvial fan or delta at the terminal margin of a melting glacier. { kām }

kame terrace [GEOL] A terracelike ridge deposited along the margins of glaciers by meltwater streams flowing adjacent to the valley walls. { kām ,ter·əs }

Kansan glaciation [GEOL] The second glaciation of the Pleistocene epoch in North America; began about 400,000 years ago, after the Aftonian and before the Yarmouth interglacials. { 'kan·zən ,glā·sē'ā·shən }

kansite See mackinawite. { 'kan,zīt }

kaolin [MINERAL] Any of a group of clay minerals, including kaolinite, nacrite, dickite, and anauxite, with a two-layer crystal in which silicon-oxygen and aluminum-hydroxyl sheets alternate; approximate composition is $Al_2O_3 \cdot 2SiO_2 \cdot 2H_2O$. [PETR] A soft, non-plastic white rock composed principally of kaolin minerals. Also known as bolus alba; white clay. { 'kā·ə·lən }

kaolinite [MINERAL] $Al_2Si_2O_5(OH)_4$ A common hydrous aluminum silicate mineral found in sediments, soils, hydrothermal deposits, and sedimentary rocks. It is a member of the kaolin group of minerals, which include dickite, halloysite, nacrite, ordered kaolinite, and disordered kaolinite. { 'kā·ə·lə,nīt }

kaolinization

kaolinization [GEOL] The forming of kaolin by the weathering of aluminum silicate minerals or other clay minerals. { ,kā·ə·lə·nə'zā·shən }

Karnian *See* Carnian. { 'kär·nē·ən }

karren [GEOL] Furrows or channels formed on the surface of soluble bedrock by dissolution of a portion of the rock. Also known as lapies. { kar·ən }

Karroo System [GEOL] Glaciated strata formed in Permian times in southern Africa. { kə'rü ,sis·təm }

karst [GEOL] A topography formed over limestone, dolomite, or gypsum and characterized by sinkholes, caves, and underground drainage. { kärst }

karst base level [GEOL] The level below which karstification ceases in an area of karst topography. { ¦kärst ¦bās ,lev·əl }

karst fenster *See* karst window. { 'kärst ,fen·stər }

karstification [GEOL] Formation of the features of karst topography by the chemical, and sometimes mechanical, action of water in a region of limestone, dolomite, or gypsum bedrock. { ,kär·stə·fə'kā·shən }

karst plain [GEOL] A plain on which karst features are developed. { 'kärst ,plān }

karst window [GEOL] An area over a subterranean stream that is open to the surface and appears as a depression at whose bottom the stream is visible. Also known as karst fenster. { 'kärst ,win·dō }

kasolite [MINERAL] $Pb(UO_2)SiO_4·H_2O$ Yellow-ocher mineral composed of a hydrous lead uranium silicate, occurring in monoclinic crystals. { 'kas·ə,līt }

katazone [GEOL] The lowest depth zone of metamorphism; features include high temperatures (500–700°C), strong hydrostatic pressure, and little or no shearing stress. { 'kad·ə,zōn }

katoptrite *See* catoptrite. { kə'täp,trīt }

kay *See* key. { kā }

Kazanian [GEOL] A European stage of geologic time: Upper Permian (above Kungurian, below Tatarian). { kə'zä·nē·ən }

K bentonite *See* potassium bentonite. { 'kā 'ben·tə,nīt }

Keewatin [GEOL] A division of the Archeozoic rocks of the Canadian Shield. { kē¦wät·ən }

Kegel karst *See* cone karst. { 'kā·gəl ,kärst }

kehoeite [MINERAL] An amorphous mineral composed of a basic hydrous calcium aluminum zinc phosphate, occurring in massive form. { 'kē·ō,īt }

Keilor skull [PALEON] An Australian fossil type specimen of *Homo sapiens* from the Pleistocene. { 'kē·lər ,skəl }

kelyphite *See* corona. { 'kē·lə,fīt }

kelyphytic border *See* kelyphytic rime. { ¦kē·lə¦fid·ik 'bȯr·dər }

kelyphytic rime [PETR] A peripheral zone of pyroxene or amphibole developed around olivine in some igneous rocks. Also known as kelyphytic border. { ¦kē·lə¦fid·ik 'rīm }

kempite [MINERAL] $Mn_2(OH)_3Cl$ An emerald-green orthorhombic mineral composed of a basic manganese oxychloride, occurring in small crystals. { 'kem,pīt }

Kenoran orogeny *See* Algoman orogeny. { kə'nȯr·ən ȯ'räj·ə·nē }

kentrolite [MINERAL] $Pb_2Mn_2Si_2O_9$ A dark reddish-brown mineral composed of a lead manganese silicate. { 'ken·trə,līt }

Kenyapithecus [PALEON] An early member of Hominidae from the Miocene. { ,ken·yə'pith·ə·kəs }

kenyte [MINERAL] A variety of phonolite containing olivine in addition to anorthoclase feldspar, nepheline, acmite-augite, sodic amphibole, apatite, and opaque oxides. { 'ke,nīt }

kerabitumen *See* kerogen. { ¦ker·ə·bə'tü·mən }

keratophyre [PETR] Any dike rock or salic lava that is characterized by the presence of albite or albite oligoclase, chlorite, epidote, and calcite. { 'ker·əd·ō,fī·ər }

kermesite [MINERAL] Sb_2S_2O A cherry-red mineral occurring as tufts of capillary crystals, and formed from an alteration of stibnite. Also known as antimony blende; purple blende; pyrostibite; red antimony. { 'kər·mə,zīt }

kernite |MINERAL| $Na_2B_4O_7 \cdot 4H_2O$ A colorless to white hydrous borate mineral crystallizing in the monoclinic system and having vitreous luster; an important source of boron. Also known as rasorite. { 'kər,nīt }

kerogen |GEOL| The complex, fossilized organic material present in sedimentary rocks, especially in shales; converted to petroleum products by distillation. Also known as kerabitumen; petrologen. { 'ker·ə·jən }

kerogen shale See oil shale. { 'ker·ə·jən ,shāl }

kerosine shale See torbanite. { 'ker·ə,sēn ,shāl }

kersantite |PETR| Dark dike rocks consisting mostly of biotite, plagioclase, and augite. { kər'zan,tīt }

kettle |GEOL| **1.** A bowl-shaped depression with steep sides in glacial drift deposits that is formed by the melting of glacier ice left behind by the retreating glacier and buried in the drift. Also known as kettle basin; kettle hole. **2.** See pothole. { 'ked·əl }

kettle basin See kettle. { 'ked·əl ,bās·ən }

kettle hole See kettle. { 'ked·əl ,hōl }

Keuper |GEOL| A European stage of geologic time, especially in Germany; Upper Triassic. { 'kȯip·ər }

Keweenawan |GEOL| The younger of two Precambrian time systems that constitute the Proterozoic period in Michigan and Wisconsin. { ¦kē·wē¦nȯ·ən }

key |GEOL| A cay, especially one of the islets off the south of Florida. Also spelled kay. { kē }

key bed |GEOL| Also known as index bed; key horizon; marker bed. **1.** A stratum or body of strata that has distinctive characteristics so that it can be easily identified. **2.** A bed whose top or bottom is employed as a datum in the drawing of structure contour maps. { 'kē ,bed }

key horizon See key bed. { 'kē hə,rīz·ən }

K feldspar See potassium feldspar. { 'kā 'fel,spär }

khibinite See mosandrite. { 'kib·ə,nīt }

kidney ore |MINERAL| A form of hematite found in compact masses, concretions, or nodules that are kidney-shaped. { 'kid·nē ,ȯr }

kidney stone See nephrite. { 'kid·nē ,stōn }

kieselguhr See diatomaceous earth. { 'kē·zəl,gu̇r }

kieserite |MINERAL| $MgSO_4 \cdot H_2O$ A white mineral that crystallizes in the monoclinic system, is composed of hydrous magnesium sulfate, and occurs in saline residues. { 'kē·zə,rīt }

Kilkenny coal See anthracite. { kil'ken·ē 'kōl }

Kimberley reefs |GEOL| Gold-bearing reefs in southern Africa that lie above the Main reef and Bird reef groups. Also known as battery reefs. { 'kim·bər·lē ,rēfs }

kimberlite |PETR| A form of mica periodite that is formed mainly of phenocrysts, olivine, phlogopite, and subordinate melilite with minor amounts of pyroxene, apatite, perovskite, and opaque oxides. { 'kim·bər,līt }

Kimmeridgian |GEOL| A European stage of geologic time; middle Upper Jurassic, above Oxfordian, below Portlandian. { ,kim·ə'rij·ē·ən }

kimzeyite |MINERAL| $Ca_3(Zr,Ti)_2(Al,Si)_3O_{12}$ A mineral of the garnet group. { 'kim·zē,īt }

Kinderhookian |GEOL| Lower Mississippian geologic time, above the Chautauquan of Devonian, below Osagian. { ¦kin·dər¦hu̇k·ē·ən }

kink band |GEOL| A deformation band in a single crystal or in foliated rocks in which the orientation is changed due to slipping on several parallel slip planes. Also known as joint drag; knick band; knick zone. { 'kiŋk ,band }

kinzigite |PETR| A coarse-grained metamorphic rock that is formed principally of garnet and biotite, with K feldspar, quartz, mica, cordierite, and sillimanite. { 'kin·zə,gīt }

Kirkbyacea |PALEON| A monomorphic superfamily of extinct ostracods in the suborder Beyrichicopina, all of which are reticulate. { ,kərk·bē'ās·ē·ə }

Kirkbyidae |PALEON| A family of extinct ostracods in the superfamily Kirkbyacea in which the pit is reduced and lies below the middle of the valve. { kərk'bē·ə,dē }

kirovite

kirovite [MINERAL] $(Fe,Mg)SO_4 \cdot 7H_2O$ A mineral composed of a hydrous sulfate of iron and magnesium; it is isomorphous with malanterite and pisanite. { 'kir·ə,vīt }

kirwanite [MINERAL] A type of anthracite coal with a metallic luster. { 'kər·wə,nīt }

klapperstein *See* rattle stone. { 'kläp·ər,shtīn }

klaprothite [MINERAL] $Cu_6Bi_4S_9$ A gray mineral composed of copper bismuth sulfide. { 'klap·rə,thīt }

klebelsbergite [MINERAL] A mineral composed of basic antimony sulfate, occurring between crystals of stibnite. { 'klā·bəlz,bər,gīt }

kleinite [MINERAL] A yellow to orange mineral composed of a basic oxide, sulfate, and chloride of mercury and ammonium. { 'klī,nīt }

klint [GEOL] An exhumed coral reef or bioherm that is more resistant to the processes of erosion than the rocks that enclose it so that the core remains in relief as hills and ridges. { klint }

klintite [GEOL] The dense, hard dolomite composing a klint; gives to the core a strength and resistance to erosion. { 'klin,tīt }

klippe [GEOL] A block of rock that is separated from underlying rocks by a fault that usually has a gentle dip. { klip }

klockmannite [MINERAL] CuSe A slate gray mineral consisting of copper selenide; occurs in granular aggregates. { 'kläk·mə,nīt }

Kloedenellacea [PALEON] A dimorphic superfamily of extinct ostracods in the suborder Kloedenellocopina having the posterior part of one dimorph longer and more inflated than the other dimorph. { ,klēd·ən·ə'lās·ē·ə }

Kloedenellocopina [PALEON] A suborder of extinct ostracods in the order Paleocopa characterized by a relatively straight dorsal border with a gently curved or nearly straight ventral border. { ,klēd·ən,el·ə'käp·ə·nə }

knebelite [MINERAL] $(Fe,Mn)_2SiO_4$ A mineral composed of an iron manganese silicate. { 'nā·bəlīt }

knick *See* knickpoint. { nik }

knick band *See* kink band. { 'nik ,band }

knickpoint [GEOL] A point of sharp change of slope, especially in the longitudinal profile of a stream or of its valley. Also known as break; knick; nick; nickpoint; rejuvenation head; rock step. { 'nik ,pȯint }

knick zone *See* kink band. { 'nik ,zōn }

knob [GEOL] **1.** A rounded eminence, such as a knoll, hillock, or small hill or mountain, and especially a prominent or isolated hill with steep sides. **2.** A peak or other projection at the top of a hill or mountain. { näb }

knoll [GEOL] A mound rising less than 3300 feet (1000 meters) from the sea floor. Also known as sea knoll. { ,nōl }

knopite [MINERAL] A cerium-bearing variety of perovskite. { 'nä,pīt }

knoxvillite *See* copiapite. { 'näks·vi,līt }

kobellite [MINERAL] $Pb_2(Bi,Sb)_2S_5$ A blackish-gray mineral composed of antimony bismuth lead sulfide. { 'kō·bəlīt }

koechlinite [MINERAL] Bi_2MoO_6 A greenish-yellow orthorhombic mineral composed of a bismuth molybdate. { 'kek·lə,nīt }

koenenite [MINERAL] $Mg_5Al_2(OH)_{12}Cl_4$ A very soft mineral composed of a basic magnesium aluminum chloride. { 'kō·nə,nit }

koettigite [MINERAL] $Zn_3(AsO_4)_2 \cdot 8H_2O$ A carmine mineral composed of a hydrated zinc arsenate. { 'ked·i,gīt }

koktaite [MINERAL] $(NH_4)_2Ca(SO_4)_2 \cdot H_2O$ A mineral composed of a hydrous calcium ammonium sulfate. { 'käk·tə,īt }

kolbeckite [MINERAL] A blue to gray mineral composed of a hydrous beryllium aluminum calcium silicate and phosphate. Also known as sterrettite. { 'kōl,be,kīt }

komatiite [PETR] A mantle-derived igneous rock with a high content of magnesium, particularly magnesium oxide. { kō'mäd·ē,īt }

kongsbergite [MINERAL] A silver-rich variety of a native amalgam composed of silver (95) and mercury (5). { 'kaŋz,bər,gīt }

174

kyanite

koninckite [MINERAL] $FePO_4 \cdot 3H_2O$ A yellow mineral composed of a hydrous ferric phosphate. { 'kȯ·niŋ,kīt }

koppite [MINERAL] Mineral composed of a form of pyrochlore containing cerium, iron, and potassium. { 'kä,pīt }

kornelite [MINERAL] $Fe_2(SO_4)_3 \cdot 7H_2O$ A colorless to brown mineral composed of hydrous ferric sulfate. { 'kȯrn·əl,īt }

kornerupine [MINERAL] $(Mg,Fe,Al)_{20}(Si,B)_9O_{43}$ A colorless, yellow, brown, or sea-green mineral composed of magnesium iron borosilicate. { ,kȯr·nə'rü,pēn }

kosmochlor See ureyite. { 'käz·mə,klȯr }

kotoite [MINERAL] $Mg_3(BO_3)_2$ An orthorhombic borate mineral; it is isostructural with jimboite. { 'kōd·ə,wīt }

krausite [MINERAL] $KFe(SO_4)_2 \cdot H_2O$ A yellowish-green mineral composed of hydrous potassium iron sulfate. { 'kraȯ,sīt }

kremersite [MINERAL] $[(NH_4),K]_2FeCl_5 \cdot H_2O$ A red mineral composed of hydrous potassium ammonium iron chloride, occurring in octahedral crystals. { 'krem·ər,zīt }

krennerite [MINERAL] $AuTe_2$ A silver-white to pale-yellow mineral composed of gold telluride and often containing silver. Also known as white tellurium. { 'kren·ə,rīt }

kribergite [MINERAL] $Al_5(PO_4)_3(SO_4)(OH)_4 \cdot 2H_2O$ White, chalklike mineral composed of hydrous basic aluminum sulfate and phosphate. { 'krib·ər,gīt }

krohnkite [MINERAL] $Na_2Cu(SO_4)_2 \cdot 2H_2O$ An azure-blue monoclinic mineral composed of hydrous copper sodium sulfate, occurring in massive form. { 'kreŋ,kīt }

krokidolite See crocidolite. { krə'kid·əl,īt }

kryolithionite [MINERAL] $Na_3Li_3(AlF_6)_2$ Variety of spodumene found in Greenland; has a crystal structure resembling that of garnet. { ,krī·ə'lith·ē·ə,nīt }

Kuehneosauridae [PALEON] The gliding lizards, a family of Upper Triassic reptiles in the order Squamata including the earliest known aerial vertebrates. { ¦kyün,nē·ō'sȯr·ə,dē }

kukersite [GEOL] An organic sediment rich in remains of the alga *Gloexapsamorpha prisca*; found in the Ordovician of Estonia. { 'kü·kər,sīt }

Kungurian [GEOL] A European stage of geologic time; Middle Permian, above Artinskian, below Kazanian. { kuŋ'gür·ē·ən }

kunzite [MINERAL] A pinkish gem variety of spodumene. { 'kunt,sīt }

kurnakovite [MINERAL] $Mg_2B_6O_{11} \cdot 13H_2O$ A white mineral composed of hydrous magnesium borate. { kür'näk·ə,vīt }

kutnahorite [MINERAL] $Ca(Mn,Mg,Fe)(CO_3)_2$ A rare carbonate of calcium and manganese, found with some magnesium and iron substituting for manganese; forms rhombohedral crystals and is isomorphous with dolomite. { ,kət·nə'hȯr,īt }

Kutorginida [PALEON] An order of extinct brachiopod mollusks that is unplaced taxonomically. { ,küd·ər'jīn·ə·də }

kyanite [MINERAL] Al_2SiO_5 A blue or light-green neosilicate mineral; crystallizes in the triclinic system, and luster is vitreous to pearly; occurs in long, thin bladed crystals and crystalline aggregates. Also known as cyanite; disthene; sappare. { 'kī·ə,nīt }

L

labite [MINERAL] $MgSi_3O_6(OH)_2 \cdot H_2O$ A mineral composed of hydrous basic silicate of magnesium. { 'lā,bīt }

labradorite [MINERAL] A gray, blue, green, or brown plagioclase feldspar with composition ranging from $Ab_{50}An_{50}$ to $Ab_{30}An_{70}$, where $Ab = NaAlSi_3O_8$ and $An = CaAl_2Si_2O_8$; in the course of formation when the natural material cools, the feldspar sometimes exhibits a variously colored luster. Also known as Labrador spar. { 'lab·rə,dȯ,rīt }

Labrador spar *See* labradorite. { 'lab·rə,dȯr ¦spär }

Labyrinthodontia [PALEON] A subclass of fossil amphibians descended from crossopterygian fishes, ancestral to reptiles, and antecedent to at least part of other amphibian types. { ,lab·ə,rin·thə'dän·chə }

laccolith [GEOL] A body of igneous rock intruding into sedimentary rocks so that the overlying strata have been notably lifted by the force of intrusion. { 'lak·ə,lith }

lacroixite [MINERAL] A pale yellowish-green mineral composed of basic phosphate of aluminum, calcium, manganese, and sodium (often with fluorine), occurring as crystals. { lə'krwä,zīt }

lacustrine [GEOL] Belonging to or produced by lakes. { lə'kəs·trən }

lacustrine sediments [GEOL] Sediments that are deposited in lakes. { lə'kəs·trən 'sed·ə·məns }

lacustrine soil [GEOL] Soil that is uniform in texture but variable in chemical composition and that has been formed by deposits in lakes which have become extinct. { lə'kəs·trən 'sȯil }

Ladinian [GEOL] A European stage of geologic time: upper Middle Triassic (above Anisian, below Carnian). { lə'din·ē·ən }

lag deposit [GEOL] Residual accumulation of coarse, unconsolidated rock and mineral debris left behind by the winnowing of finer material. { 'lag di,päz·ət }

lag fault [GEOL] A minor low-angle thrust fault occurring within an overthrust; it develops when one part of the mass is thrust farther than an adjacent higher or lower part. { 'lag ,fȯlt }

lag gravel [GEOL] Residual accumulations of particles that are coarser than the material that has blown away. { 'lag ,grav·əl }

lahar [GEOL] **1.** A mudflow or landslide of pyroclastic material occurring on the flank of a volcano. **2.** The deposit of mud or land so formed. { 'lä,här }

lake ore *See* bog iron ore. { 'lāk ,ȯr }

lake peat [GEOL] A sedimentary peat formed near lakes. { 'lāk ,pēt }

lake plain [GEOL] One of the surfaces of the earth that represent former lake bottoms; these featureless surfaces are formed by deposition of sediments carried into the lake by streams. { 'lāk ,plān }

lamina [GEOL] A thin, clearly differentiated layer of sedimentary rock or sediment, usually less than 1 centimeter thick. { 'lam·ə·nə }

laminite [GEOL] Any sedimentary rock composed of millimeter- or finer-scale layers. { 'lam·ə,nīt }

lampadite [MINERAL] A mineral composed chiefly of hydrous manganese oxide with as much as 18% copper oxide and often cobalt oxide. { 'lam·pə,dīt }

lamprobolite *See* basaltic hornblende. { ,lam·prə'bō,līt }

lamprophyllite

lamprophyllite |MINERAL| $Na_2SrTiSi_2O_8$ A mineral composed of titanium strontium sodium silicate. { ,lam·prə'fi,līt }

lamprophyre |PETR| Any of a group of igneous rocks characterized by a porphyritic texture in which abundant, large crystals of dark-colored minerals appear set in a not visibly crystalline matrix. { 'lam·prə,fī·ər }

Lanarkian |GEOL| A European stage of geologic time forming part of the lower Upper Carboniferous, above Lancastrian and below Yorkian, equivalent to lowermost Westphalian. { lə'när·kē·ən }

lanarkite |MINERAL| Pb_2OSO_4 A white, greenish, or gray monoclinic mineral consisting of basic lead sulfate, with specific gravity of 6.92; formed by action of heat and air on galena. { 'lan·ər,kīt }

Lancastrian |GEOL| A European stage of geologic time forming part of the lower Upper Carboniferous, above Viséan and below Lanarkian. { laŋ'kas·trē·ən }

Landenian |GEOL| A European stage of geologic time: upper Paleocene (above Montian, below Ypresian of Eocene). { lan'den·ē·ən }

landesite |MINERAL| A brown mineral consisting of a hydrated phosphate of iron and manganese. { 'lan·də,sīt }

landform map See physiographic diagram. { 'lan,form ,map }

land pebble See land pebble phosphate. { 'land ,peb·əl }

land pebble phosphate |GEOL| A pebble phosphate in a clay or sand bed below the ground surface; a small amount of uranium is often present and is recovered as a by-product; used as a source of phosphate fertilizer. Also known as land pebble; land rock; matrix rock. { 'land ¦peb·əl 'fäs,fāt }

land rock See land pebble phosphate. { 'land ,räk }

landscape agate |MINERAL| A type of chalcedony that is translucent and contains inclusions which give it an appearance reminiscent of familiar natural scenes. Also known as fortification agate. { 'lan,skāp ,ag·ət }

landslide |GEOL| The perceptible downward sliding or falling of a relatively dry mass of earth, rock, or combination of the two under the influence of gravity. Also known as landslip. { 'lan,slīd }

landslide track |GEOL| An exposed path in rock or earth created as the result of a landslide. { 'lan,slīd ,trak }

landslip See landslide. { 'lan,slip }

langbanite |MINERAL| An iron-black hexagonal mineral composed of silicate and oxide of manganese, iron, and antimony, occurring in prismatic crystals. { 'läŋ·bə,nīt }

langbeinite |MINERAL| $K_2Mg_2(SO_4)_3$ Colorless, yellowish, reddish, or greenish hexagonal mineral with vitreous luster, found in salt deposits; used in the fertilizer industry as a source of potassium sulfate. { 'läŋ,bī,nīt }

langite |MINERAL| A blue to green mineral composed of basic hydrous copper sulfate. { 'laŋ,īt }

lansfordite |MINERAL| $MgCO_3 \cdot 5H_2O$ A mineral composed of hydrous basic carbonate of magnesium when extracted from the earth, changing to nesquehovite after exposure to the air. { 'lanz·fər,dīt }

lanthanite |MINERAL| $(La,Ce)_2(CO_3)_3 \cdot 8H_2O$ A colorless, white, pink, or yellow mineral composed of hydrous lanthanum carbonate, occurring in crystals or in earthy form. { 'lan·thə,nīt }

lapies See karren. { lə'pēz }

lapilli |GEOL| Pyroclasts that range from 0.04 to 2.6 inches (1 to 64 millimeters) in diameter. { lə'pi,lī }

lapilli-tuff |GEOL| A pyroclastic deposit that is indurated and consists of lapilli in a fine tuff matrix. { lə'pi,lī ¦təf }

lapis lazuli |PETR| An azure-blue, violet-blue, or greenish-blue, translucent to opaque crystalline rock used as a semiprecious stone; composed chiefly of lazurite and calcite with some haüyne, sodalite, and other minerals. Also known as lazuli. { ¦lap·is 'laz·ə·lē }

Laramic orogeny See Laramidian orogeny. { 'lar·ə·mik o'räj·ə·nē }

Laramide orogeny See Laramidian orogeny. { 'lar·ə·məd o'räj·ə·nē }

Laramide revolution *See* Laramidian orogeny. { 'lar·ə·məd ˌrev·ə'lü·shən }

Laramidian orogeny |GEOL| An orogenic era typically developed in the eastern Rocky Mountains; phases extended from Late Cretaceous until the end of the Paleocene. Also known as Laramic orogeny; Laramide orogeny; Laramide revolution. { ˌlar·ə'mid·ē·ən ö'räj·ə·nē }

larderillite |MINERAL| $(NH_4)B_5O_8·2H_2O$ A white mineral composed of hydrous ammonium borate, occurring as a crystalline powder. { ˌlär·də're‚līt }

lardite *See* agalmatolite. { 'lär‚dīt }

larnite |MINERAL| β-Ca_2SiO_4 A gray mineral that is a metastable monoclinic phase of calcium orthosilicate, stable from 520 to 670°C. Also known as belite. { 'lär‚nīt }

larsenite |MINERAL| $PbZnSiO_4$ A colorless or white mineral composed of lead zinc silicate, occurring in orthorhombic crystals. { 'lars·ən‚īt }

larvikite |PETR| An alkali syenite consisting of cryptoperthite or anorthoclase in rhombic crystals; used as an ornamental building material. { 'lär·vi‚kīt }

lateral accretion |GEOL| The digging away of material at the outer bank of a meandering stream and the simultaneous building up to the water level by deposition of material brought there by pushing and rolling along the stream bottom. { 'lad·ə·rəl ə'krē·shən }

lateral cone *See* adventive cone. { 'lad·ə·rəl kōn }

lateral erosion |GEOL| The action of a stream in undermining a bank on one side of its channel so that material falls into the stream and disintegrates; simultaneously, the stream shifts toward the bank that is being undercut. { 'lad·ə·rəl i'rō·zhən }

lateral fault |GEOL| A fault along which there has been strike separation. Also known as strike-separation fault. { 'lad·ə·rəl 'fölt }

lateral moraine |GEOL| Drift material, usually thin, that was deposited by a glacier in a valley after the glacier melted. { 'lad·ə·rəl mə'rān }

lateral planation |GEOL| Reduction in land in interstream areas in a plane parallel to the stream profile; the reduction is caused by lateral movement of the stream against its banks. { 'lad·ə·rəl plā'nā·shən }

lateral secretion |GEOL| A supposed phenomenon whereby a lode's or vein's mineral content is derived from the adjacent wall rock. { 'lad·ə·rəl si'krē·shən }

laterite |GEOL| Weathered material composed principally of the oxides of iron, aluminum, titanium, and manganese; laterite ranges from soft, earthy, porous soil to hard, dense rock. { 'lad·ə‚rīt }

lateritic soil |GEOL| **1.** Soil containing laterite. **2.** Any reddish soil developed from weathering. Also known as latosol. { ‚lad·ə‚rid·ik 'söil }

laterization |GEOL| Those conditions of weathering that lead to removal of silica and alkalies, resulting in a soil or rock with high concentrations of iron and aluminum oxides (laterite). { ˌlad·ə·rə'zā·shən }

latite |PETR| A not visibly crystalline rock of volcanic origin composed chiefly of sodic plagioclase and alkali feldspar with subordinate quantities of dark-colored minerals in a finely crystalline to glassy groundmass. { 'lā‚tīt }

latitude variation |GEOPHYS| A periodic change in the latitude of any position on the earth's surface, caused by the polar variation. { 'lad·ə‚tüd ˌver·ə'ā·shən }

latosol *See* lateritic soil. { 'lad·ə‚sól }

latrappite |MINERAL| $(Ca,Na)(Nb,Ti,Fe)O_3$ A variety of the mineral perovskite. { 'la·trə‚pīt }

lattice drainage pattern *See* rectangular drainage pattern. { 'lad·əs 'drān·ij ˌpad·ərn }

Lattorfian *See* Tongrian. { lə'tòr·fē·ən }

laubmannite |MINERAL| $Fe_3Fe_6(PO_4)_4(OH)_2$ Mineral composed of basic ferrous iron phosphate and ferric iron phosphate. { 'laùb·mə‚nīt }

Laugiidae |PALEON| A family of Mesozoic fishes in the order Coelacanthiformes. { laü'jī·ə‚dē }

laumonite *See* laumontite. { lō'mä‚nīt }

laumontite |MINERAL| $CaAl_2Si_4O_{12}·4H_2O$ A white zeolite mineral crystallizing in the monoclinic system; loses water on exposure to air, eventually becoming opaque and crumbling. Also known as laumonite; lomonite; lomontite. { lō'män‚tīt }

lauoho o pele *See* Pele's hair.   { ˌlä·ü'ō͝ˌhō ō 'peˌlē }

Laurasia   [GEOL] A continent theorized to have existed in the Northern Hemisphere; supposedly it broke up to form the present northern continents about the end of the Pennsylvanian period.   { lȯ'rā·zhə }

Laurentian Plateau *See* Laurentian Shield.   { lȯ'ren·chən pla'tō }

Laurentian Shield   [GEOL] A Precambrian plateau extending over half of Canada from Labrador southwest along Hudson Bay and northwest to the Arctic Ocean.   Also known as Canadian Shield; Laurentian Plateau.   { lȯ'ren·chən 'shēld }

laurionite   [MINERAL] Pb(OH)Cl A colorless mineral composed of basic lead chloride, occurring in prismatic crystals; it is dimorphous with paralaurionite.   { 'lȯr·ē·əˌnīt }

laurite   [MINERAL] RuS_2 A black mineral composed of ruthenium sulfide (often with osmium), occurring as small crystals or grains.   { 'lȯˌrīt }

lausenite   [MINERAL] $Fe_2(SO_4)_3 \cdot 6H_2O$ A white, monoclinic mineral consisting of hydrated ferric sulfate; occurs in lumpy aggregates of fibers.   { 'lȯs·ənˌīt }

lautarite   [MINERAL] $Ca(IO_3)_2$ A monoclinic mineral composed of calcium iodate that occurs in prismatic crystals.   { 'laȯd·əˌrīt }

lautite   [MINERAL] CuAsS A mineral composed of copper sulfide and copper arsenide.   { 'laȯˌtīt }

lava   [GEOL] **1.** Molten extrusive material that reaches the earth's surface through volcanic vents and fissures.   **2.** The rock mass formed by consolidation of molten rock issuing from volcanic vents and fissures, consisting chiefly of magnesium silicate; used for insulators.   { 'lä·və }

lava blisters   [GEOL] Small, steep-sided swellings that are hollow and raised on the surfaces of some basaltic lava flows; formed by gas bubbles pushing up the lava's viscous surface.   { 'lä·və ˌblis·tərz }

lava cone   [GEOL] A volcanic cone that was formed of lava flows.   { 'lä·və ˌkōn }

lava dome *See* shield volcano.   { 'lä·və ˌdōm }

lava field   [GEOL] A wide area of lava flow; it is commonly several square kilometers in area and forms along the base of a large compound volcano or on the flanks of shield volcanoes.   { 'lä·və ˌfēld }

lava flow   [GEOL] **1.** A lateral, surficial stream of molten lava issuing from a volcanic cone or from a fissure.   **2.** The solidified mass of rock formed when a lava stream congeals.   { 'lä·və ˌflō }

lava fountain   [GEOL] A jetlike eruption of lava that issues vertically from a volcanic vent or fissure.   Also known as fire fountain.   { 'lä·və ˌfaȯnt·ən }

lava lake   [GEOL] A lake of lava that is molten and fluid; usually contained within a summit volcanic crater or in a pit crater on the flanks of a shield volcano.   { 'lä·və ˌlāk }

lava plateau   [GEOL] An elevated tableland or flat-topped highland that is several hundreds to several thousands of square kilometers in area; underlain by a thick succession of lava flows.   { 'lä·və pla'tō }

lava tube   [GEOL] A long, tubular opening under the crust of solidified lava.   { 'lä·və ˌtüb }

lavenite   [MINERAL] $(Na,Ca)_3Zr(Si_2O_7)(O,OH,F)_2$ A mineral composed of complex silicate, occurring in prismatic crystals.   { 'lä·vəˌnīt }

law of superposition   [GEOL] The law that strata underlying other strata must be the older if there has been neither overthrust nor inversion.   { 'lȯ əv ˌsü·pər·pə'zish·ən }

lawrencite   [MINERAL] $(Fe,Ni)Cl_2$ A brown or green mineral composed of ferrous chloride and found as an abundant accessory mineral in iron meteorites.   { 'lär·ənˌsīt }

lawsonite   [MINERAL] $CaAl_2(Si_2O_7)(OH)_2 \cdot H_2O$ A colorless or grayish-blue mineral crystallizing in the orthorhombic system; found in gneisses and schists.   { 'lȯs·ənˌīt }

layer   [GEOL] A tabular body of rock, ice, sediment, or soil lying parallel to the supporting surface and distinctly limited above and below.   [GEOPHYS] One of several strata of ionized air, some of which exist only during the daytime, occurring at altitudes between 30 and 250 miles (50 and 400 kilometers); the layers reflect radio waves at certain frequencies and partially absorb others.   { 'lā·ər }

layer depth effect   [GEOPHYS] The weakening of a sound beam or seismic pulse because of abnormal spreading in passing from a positive gradient layer to an underlying negative layer. { 'lā·ər ¦depth i,fekt }

layered complex [GEOL] An igneous rock body of large dimensions, 5–300 miles (8–480 kilometers) across and as much as 23,000 feet (7000 meters) thick, within which distinct subhorizontal stratification, or layering, is apparent and may be continuous over great distances, in some cases more than 60 miles (100 kilometers). { ¦lā·ərd 'käm,pleks }

layer silicate See phyllosilicate. { 'lā·ər ,sil·ə·kət }

lazuli See lapis lazuli. { 'laz·ə·lē }

lazulite [MINERAL] $(Mg,Fe)Al_2(OH)_2(PO_4)_2$ A violet-blue or azure-blue mineral with vitreous luster; composed of basic aluminum phosphate and occurring in small masses or monoclinic crystals; hardness is 5–6 on Mohs scale, and specific gravity is 3.06–3.12. Also known as berkeyite; blue spar; false lapis. { 'laz·ə,līt }

lazurite [MINERAL] $(Na,Ca)_8(Al,Si)O_{24}(S,SO_4)$ A blue or violet-blue feldspathoid mineral crystallizing in the isometric system; the chief mineral constituent of lapis lazuli. { 'laz·ə,rīt }

leachate [GEOCHEM] A liquid that has percolated through soil and dissolved some soil materials in the process. { 'lē,chāt }

leaching [GEOCHEM] The separation or dissolving out of soluble constituents from a rock or ore body by percolation of water. { 'lēch·iŋ }

lead [GEOL] A small, narrow passage in a cave. { led }

lead glance See galena. { 'led 'glans }

leadhillite [MINERAL] $Pb_4(SO_4)(CO_3)_2(OH)_2$ A yellowish or greenish- or grayish-white monoclinic mineral consisting of basic sulfate and carbonate of lead; dimorphous with susanite. { 'led,hi,līt }

leading stone See lodestone. { 'lēd·iŋ ,stōn }

lead marcasite See sphalerite. { 'led 'mär·kə,zīt }

lead ocher See massicot. { 'led 'ō·kər }

lead spar See anglesite. { 'led 'spär }

lead vitriol See anglesite. { 'led 'vit·rē,ōl }

leaf mold [GEOL] A soil layer or compost consisting principally of decayed vegetable matter. { 'lēf ,mōld }

leakage halo [GEOCHEM] The dispersion of elements along channels and paths followed by mineralizing solutions leading into and away from the central focus of mineralization. { 'lēk·ij ,hā·lō }

leaking mode [GEOPHYS] A surface seismic wave which is imperfectly trapped, so that its energy leaks or escapes across a layer boundary, causing some attenuation. Also known as leaky wave. { 'lēk·iŋ ,mōd }

leaky wave See leaking mode. { 'lēk·ē 'wāv }

lechatelierite [MINERAL] A natural silica glass, occurring in fulgurites and impact craters and formed by the melting of quartz sand at high temperatures generated by lightning or by the impact of a meteorite. { le,shäd·əl'ī,rīt }

lecontite [MINERAL] $Na(NH_4,K)SO_4·2H_2O$ A colorless mineral composed of a hydrous sodium potassium ammonium sulfate; found in bat guano. { lə'kän,tīt }

ledge [GEOL] **1.** A narrow, shelflike ridge or rock protrusion, much longer than high, and usually horizontal, formed in a rock wall or on a cliff face. **2.** A ridge of rocks found underwater, especially one near a shore or connected with and bordering a shore. { lej }

Ledian [GEOL] Lower upper Eocene geologic time. Also known as Auversian. { 'lēd·ē·ən }

lee dune [GEOL] A dune formed to the leeward of a source of sand or of an obstacle. { 'lē ,dün }

left lateral fault [GEOL] A fault in which movement is such that an observer walking toward the fault along an index plane (a bed, vein, or dike) would turn to the left to find the other part of the displaced index plane. Also known as sinistral fault. { 'left ¦lad·ə·rəl 'fólt }

leg [GEOPHYS] A single cycle of more or less periodic motion in a wave train on a seismogram. { leg }

legrandite [MINERAL] $Zn_{14}(OH)(AsO_4)_9 \cdot 12H_2O$ A yellow to nearly colorless mineral composed of basic hydrous zinc arsenate. { lə'gran,dīt }

lehiite [MINERAL] $(Na,K)_2Ca_5Al_8(PO_4)_8(OH)_{12} \cdot 6H_2O$ White mineral composed of hydrous basic calcium aluminum phosphate. { 'lē,hīt }

leifite [MINERAL] $Na_2AlSi_4O_{10}F$ A colorless mineral composed of fluoride and silicate of sodium and aluminum. { 'lē,fīt }

leightonite [MINERAL] $K_2Ca_2Cu(SO_4)_4 \cdot 2H_2O$ A pale-blue mineral composed of hydrous sulfate of copper, calcium, and potassium. { 'lāt·ən,īt }

lengenbachite [MINERAL] $Pb_6(Ag,Cu)_2As_4S_{13}$ A steel gray mineral consisting of lead, silver, and copper arsenic sulfide. { 'leŋ·ən,bä,kīt }

lens [GEOL] **1.** A geologic deposit that is thick in the middle and converges toward the edges, resembling a convex lens. **2.** An irregularly shaped formation consisting of a porous, permeable sedimentary deposit surrounded by impermeable rock. { lenz }

lenticle [GEOL] A bed or rock stratum or body that is lens-shaped. { 'len·tə·kəl }

lentil [GEOL] **1.** A rock body that is lens-shaped and enclosed in a stratum of different material. **2.** A rock stratigraphic unit that is a subdivision of a formation and has limited geographic extent; it thins out in all directions. { 'lent·əl }

Leonardian [GEOL] A North American provincial series: Lower Permian (above Wolfcampian, below Guadalupian). { ¦lā·ə¦när·dē·ən }

leonite [MINERAL] $K_2Mg(SO_4)_2 \cdot 4H_2O$ A colorless, white, or yellowish mineral composed of hydrous magnesium potassium sulfate, occurring in monoclinic crystals. { 'lē·ə,nīt }

leopoldite See sylvite. { 'lē·ə,pōl,dīt }

Leperditicopida [PALEON] An order of extinct ostracods characterized by very thick, straight-backed valves which show unique muscle scars and other markings. { ,le·pər,did·ə'käp·ə·də }

Leperditillacea [PALEON] A superfamily of extinct paleocopan ostracods in the suborder Kloedenellocopina including the unisulcate, nondimorphic forms. { ,le·pər,did·ə'lās·ē·ə }

lepidoblastic [PETR] Of the texture of a metamorphic rock, having a fabric of minerals characterized as flaky or scaly, such as mica. { ¦lep·ə·dō¦blas·tik }

lepidocrocite [MINERAL] α-FeO(OH) A ruby- or blood-red mineral crystallizing in the orthorhombic system; it is associated with limonite in iron ores and is a component of meteorites. { ,lep·ə·dō'krō,sīt }

Lepidodendrales [PALEOBOT] The giant club mosses, an order of extinct lycopods (Lycopodiopsida) consisting primarily of arborescent forms characterized by dichotomous branching, small amounts of secondary vascular tissue, and heterospory. { ,lep·ə·dō,den'drā·lēz }

lepidolite [MINERAL] $K(Li,Al)_3(Si,Al)_4O_{10}(F,OH)_2$ A rose-colored mineral of the mica group crystallizing in the monoclinic system. Also known as lithionite; lithium mica. { lə'pid·əl,īt }

lepidomelane [MINERAL] A black variety of biotite that is characterized by the presence of large amounts of ferric iron. Also known as iron mica. { ,lep·ə·dō'me,lān }

lepisphere [PETR] A microspherical aggregate of platy, blade-shaped crystals of opal-CT. { 'lep·ə,sfir }

Lepospondyli [PALEON] A subclass of extinct amphibians including all forms in which the vertebral centra are formed by ossification directly around the notochord. { ,lep·ə'spänd·əl,ī }

Leptictidae [PALEON] A family of extinct North American insectivoran mammals belonging to the Proteutheria which ranged from the Cretaceous to middle Oligocene. { ,lep'tik·tə,dē }

leptite [PETR] A quartz-feldspathic metamorphic rock that is fine-grained with little or no foliation; formed by regional metamorphism of the highest grade. { 'lep,tīt }

Leptochoeridae [PALEON] An extinct family of palaeodont artiodactyl mammals in the superfamily Dichobunoidea. { ,lcp·tə'kir·ə,dē }

leptogeosyncline [GEOL] A deep oceanic trough that has not been filled with sedimentation and is associated with volcanism. { ,lep·tə,jē·ō'sin,klīn }

Leptolepidae [PALEON] An extinct family of fishes in the order Leptolepiformes representing the first teleosts as defined on the basis of the advanced structure of the caudal skeleton. { ,lep·tə'lep·ə,dē }

Leptolepiformes [PALEON] An extinct order of small, ray-finned teleost fishes characterized by a relatively strong, ossified axial skeleton, thin cycloid scales, and a preopercle with an elongated dorsal portion. { ,lep·tə,lep·ə'fōr,mēz }

letovicite [MINERAL] $(NH_3)_3H(SO_4)_2$ A mineral composed of acid ammonium sulfate. { ,led·ə'vi,sīt }

leucite [MINERAL] $KAlSi_2O_6$ A white or gray rock-forming mineral belonging to the feldspathoid group; at ordinary temperatures the mineral exists as aggregates of trapezohedral crystals with glassy fracture; hardness is 5.5–6.0 on Mohs scale, and specific gravity is 2.45–2.50. Also known as amphigene; grenatite; vesuvian; Vesuvian garnet; white garnet. { 'lü,sīt }

leucite phonolite [PETR] An extrusive rock composed of alkali feldspar, mafic minerals, and leucite. { 'lü,sīt 'fän·əl,īt }

leucitite [PETR] A fine-grained or porphyritic extrusive rock or hypabyssal igneous rock composed mostly of pyroxene and leucite. { 'lü·sə,tīt }

leucochalcite See olivenite. { ,lü·kō'kal,sīt }

leucocratic [PETR] Light-colored as applied to igneous rock containing 0–50% dark-colored minerals. { ¦lü·kə¦krad·ik }

leucophanite [MINERAL] $(Na,Ca)_2BeSi_7(O,F,OH)_7$ Greenish mineral composed of beryllium sodium calcium silicate containing fluorine and occurring in glassy, tabular crystals. { ¦lü·kō'fa,nīt }

leucophosphite [MINERAL] $K_2Fe_4(PO_4)_4(OH)_2·9H_2O$ White mineral composed of hydrous basic phosphate of potassium and iron. { ,lü·kə'fäs,fīt }

leucopyrite See loellingite. { ,lü·kə'pī,rīt }

leucosphenite [MINERAL] $Na_4BaTi_2Si_{10}O_{27}$ A white mineral composed of sodium barium silicotitanate and occurring as wedge-shaped crystals. { ,lü·kə'sfē,nīt }

leucoxene [MINERAL] A mineral composed of rutile with some anatase or sphene; occurs in igneous rocks, usually as an alteration product of ilmenite. { lü'käk,sēn }

leuneburgite [MINERAL] $Mg_3B_2(PO_4)_2(OH)_6·5H_2O$ A colorless mineral consisting of a hydrous basic phosphate of magnesium and boron. { 'lü·nən,bər,gīt }

levee [GEOL] **1.** An embankment bordering one or both sides of a sea channel or the low-gradient seaward part of a canyon or valley. **2.** A low ridge sometimes deposited by a stream on its sides. { 'lev·ē }

level fold See nonplunging fold. { 'lev·əl 'fōld }

levyine See levynite. { lā'vē,īn }

levyite See levynite. { lā'vē,īt }

levyne See levynite. { lā'vēn }

levynite [MINERAL] $NaCa_3Al_7Si_{11}O_{36}·15H_2O$ A white or light-colored mineral of the zeolite group, composed of hydrous silicate of aluminum, sodium, and calcium, and occurring in rhombohedral crystals. Also known as levyine; levyite; levyne. { lā'vē,nīt }

lewisite [MINERAL] $(Ca,Fe,Na)_2$ A titanian romeite mineral. { 'lü·ə,sīt }

lewistonite [MINERAL] $(Ca,K,Na)_5(PO_4)_3(OH)$ White mineral composed of basic calcium potassium sodium phosphate. { 'lü·ə·stə,nīt }

lherzolite [PETR] Peridotite composed principally of olivine with orthopyroxene and clinopyroxene. { 'lərt·sə,līt }

Lias See Liassic. { 'lī·as }

Liassic [GEOL] The Lower Jurassic period of geologic time. Also known as Lias. { lī'as·ik }

Libby effect [GEOCHEM] The increase, since about 1950, in the carbon-14 content

183

libethenite

of the atmosphere, produced by the detonation of thermonuclear devices. { 'lib·ē i,fekt }

libethenite [MINERAL] $Cu_2(PO_4)OH$ An olive-green mineral composed of basic copper sulfate, occurring as small prismatic crystals or in masses. { lə'beth·ə,nīt }

lichenometry [GEOL] Measurement of the diameter of lichens growing on exposed rock surfaces; used for dating geomorphic features, particularly of glacial origin. { ,lī·kə'näm·ə·trē }

liebigite [MINERAL] $Ca_2U(CO_3)_4 \cdot 10H_2O$ An apple- or yellow-green mineral composed of hydrous uranium calcium carbonate; occurs as a coating or concretion in rock. { 'lē·bi,gīt }

Liesegang banding [GEOL] Colored or compositional rings or bands in a fluid-saturated rock due to rhythmic precipitation. Also known as Liesegang rings. { 'lēz·ə,gäŋ ,band·iŋ }

Liesegang rings See Liesegang banding. { 'lēz·ə,gäŋ ,riŋz }

light mineral [MINERAL] **1.** A rock with minerals that have a specific gravity lower than a standard, usually 2.85. **2.** A light-colored mineral. { 'līt ¦min·rəl }

light-red silver ore See proustite. { 'līt ,red 'sil·vər ,ȯr }

light-ruby silver See proustite. { 'līt ,rü·bē 'sil·vər }

lignite [GEOL] Coal of relatively recent origin, intermediate between peat and bituminous coal; often contains patterns from the wood from which it formed. Also known as brown coal; earth coal. { 'lig,nīt }

lignite A See black lignite. { 'lig,nīt 'ā }

lignite B See brown lignite. { 'lig,nīt 'bē }

lignitious coal [MINERAL] A type of coal containing 75–84% elemental carbon. { lig'-nish·əs 'kōl }

lillianite [MINERAL] $Pb_3Bi_2S_6$ A steel-gray mineral composed of lead bismuth sulfide. { 'lil·ē·ə,nīt }

limb [GEOL] One of the two sections of an anticline or syncline on either side of the axis. Also known as flank. { limb }

limburgite [PETR] A dark, glass-rich igneous rock with abundant large crystals of olivine and pyroxene and with little or no feldspar. { 'lim·bər,gīt }

lime-pan playa [GEOL] A playa with a smooth, hard surface composed of calcium carbonate. { 'līm ¦pan 'plī·ə }

limestone [PETR] **1.** A sedimentary rock composed dominantly (more than 95) of calcium carbonate, principally in the form of calcite; examples include chalk and travertine. **2.** Any rock containing 80% or more of calcium carbonate or magnesium carbonate. { 'līm,stōn }

limestone pebble conglomerate [GEOL] A well-sorted conglomerate composed of limestone pebbles resulting from special conditions involving rapid mechanical erosion and short transport distances. { 'līm,stōn ¦peb·əl kən'gläm·ə·rət }

limnite See bog iron ore. { 'lim,nīt }

limonite [MINERAL] A group of brown or yellowish-brown, amorphous, naturally occurring ferric oxides of variable composition; commonly formed secondary material by oxidation of iron-bearing minerals; a minor ore of iron. Also known as brown hematite; brown iron ore. { 'lī·mə,nīt }

linarite [MINERAL] $PbCu(SO_4)(OH)_2$ A deep-blue mineral composed of basic lead copper sulfate and occurring as monoclinic crystals. { 'lī·nə,rīt }

lindackerite [MINERAL] $Cu_6Ni_3(AsO_4)_4(SO_4)(OH)_4$ A light-green or apple-green mineral composed of hydrous basic sulfate and arsenate of nickel and copper; occurs in tabular crystals or massive. { lin'dak·ə,rīt }

lindgrenite [MINERAL] $Cu_3(MoO_4)_2(OH)_2$ A green mineral composed of basic copper molybdate. { 'lin·grə,nīt }

lindstromite [MINERAL] $PbCuBi_3S_6$ A lead-gray to tin-white mineral composed of bismuth copper lead sulfide. { 'linz·trə,mīt }

lineament [GEOL] A straight or gently curved, lengthy topographic feature expressed as depressions or lines of depressions. Also known as linear. { 'lin·ē·ə·mənt }

linear See lineament. { 'lin·ē·ər }

linear cleavage [GEOL] The property of metamorphic rocks of breaking into long planar fragments. { 'lin·ē·ər 'klē·vij }

linear flow structure See platy flow structure. { 'lin·ē·ər 'flō ,strək·chər }

linear parallel texture [PETR] The parallel texture of a rock in which the constituents are parallel to a line, not just to a plane as in plane parallel texture. { 'lin·ē·ər ¦par·ə,lel 'teks·chər }

lineation [GEOL] Any linear structure on or within a rock; examples are ripple marks and flow lines. { ,lin·ē'ā·shən }

line of strike See strike. { 'līn əv 'strīk }

linguloid ripple mark See linguoid ripple mark. { 'liŋ·gyə,lóid 'rip·əl ,märk }

linguoid current ripple See linguoid ripple mark. { 'liŋ·gwóid ¦kə·rənt ,rip·əl }

linguoid ripple mark [GEOL] An aqueous current ripple mark with tonguelike projections which are formed by action of a current of water and which point into the current. Also known as cuspate ripple mark; linguloid ripple mark; linguoid current ripple. { 'liŋ·gwóid 'rip·əl ,märk }

linnaeite [MINERAL] (Co,Ni)₃S₄ A steel-gray mineral with a coppery-red tarnish, occurring in isometric crystals; an ore of cobalt. Also known as cobalt pyrites; linneite. { lə'nē,īt }

linneite See linnaeite. { lə'nē,īt }

Lipalian [GEOL] A hypothetical geologic period that supposedly antedated the Cambrian. { lə'pal·yən }

Lipostraca [PALEON] An order of the subclass Branchiopoda erected to include the single fossil species Lepidocaris rhyniensis. { li'päs·trə·kə }

liptinite See exinite. { 'lip·tə,nīt }

liquid-dominated hydrothermal reservoir [GEOL] Any geothermal system mainly producing superheated water (often termed brines); hot springs, fumaroles, and geysers are the surface expressions of hydrothermal reservoirs; an example is the hot-brine region in the Imperial Valley-Salton Sea area of southern California. { 'lik·wəd ¦däm·ə,nād·əd ,hī·drə¦thər·məl 'rez·əv,wär }

liquid-filled porosity [GEOL] The condition in porous rock or sand formations in which pore spaces contain fresh or salt water, liquid petroleum, pressure-liquefied butane or propane, or tar. { 'lik·wəd ¦fild pə'räs·əd·ē }

liquid limit [GEOL] The moisture content boundary that exists between the plastic and semiliquid states of a sediment. { 'lik·wəd 'lim·ət }

liroconite [MINERAL] Cu₂Al(AsO₄)(OH)₄·4H₂O A light-blue or yellowish-green mineral composed of basic hydrous aluminum copper arsenate, occurring in monoclinic crystals. { lī'räk·ə,nīt }

liskeardite [MINERAL] (Al,Fe)₃(AsO₄)(OH)₆·5H₂O A soft, white mineral composed of basic hydrous aluminum iron arsenate. { li'skär,dīt }

litharenite [PETR] A sandstone that contains more than 25% detrital rock fragments, and more rock fragments than feldspar grains. { li'thər·ə,nīt }

lithian muscovite [MINERAL] A form of the mineral lepidolite containing 3–4% lithium oxide and having a modified two-layer monoclinic muscovite structure. { 'lith·ē·ən 'məs·kə,vīt }

lithic [PETR] Pertaining to stone. { 'lith·ik }

lithic graywacke [PETR] A low-grade graywacke, that is, containing an abundance of unstable materials, especially a sandstone containing less than 75% quartz and chert, 15–75% detrital clay matrix, and more rock fragments than feldspar grains. { 'lith·ik 'grā,wak·ə }

lithic sandstone [PETR] A sandstone that contains more rock fragments than feldspar grains. { 'lith·ik 'san,stōn }

lithic tuff [GEOL] **1.** A tuff that is mostly crystalline rock fragments. **2.** An indurated volcanic ash deposit whose fragments are composed of previously formed rocks that first solidified in the volcanic vent and were then blown out. { 'lith·ik 'təf }

lithifaction See lithification. { ,lith·ə'fak·shən }

lithification [GEOL] **1.** Conversion of a newly deposited sediment into an indurated rock. Also known as lithifaction. **2.** Compositional change of coal to bituminous shale or other rock. { ˌlith·ə·fəˈkā·shən }

lithionite *See* lepidolite. { ˈlith·ē·əˌnīt }

lithiophilite |MINERAL| Li(Mn,Fe)PO₄ A salmon-pink or clove-brown mineral crystallizing in the orthorhombic system; isomorphous with triphylite. { ˌlith·ēˈäf·əˌlīt }

lithiophorite |MINERAL| (Al,Li)MnO₂(OH)₂ A mineral composed of basic manganese aluminum lithium oxide. { ˌlith·ēˈäf·əˌrīt }

Lithistida |PALEON| An order of fossil sponges in the class Demospongia having a reticulate skeleton composed of irregular and knobby siliceous spicules. { ləˈthis·tə·də }

lithium mica *See* lepidolite. { ˈlith·ē·əm ˈmī·kə }

lithoclase |GEOL| A naturally produced rock fracture. { ˈlith·əˌklās }

lithofacies |GEOL| A subdivision of a specified stratigraphic unit distinguished on the basis of lithologic features. { ˌlith·əˈfā·shēz }

lithofacies map |GEOL| The facies map of an area based on lithologic characters; shows areal variation in all aspects of the lithology of a stratigraphic unit. { ˌlith·əˈfā·shēz ˌmap }

lithogenesis |PETR| The branch of science dealing with the formation of rocks, especially the formation of sedimentary rocks. { ˌlith·əˈjen·ə·səs }

lithogeochemical survey |GEOCHEM| A geochemical survey that involves the sampling of rocks. { ˌlith·ō,jē·əˈkem·ə·kəl ˈsər,vā }

lithographic limestone |GEOL| A dense, compact, fine-grained crystalline limestone having a pale creamy-yellow or grayish color. Also known as lithographic stone; litho stone. { ˌlith·əˈgraf·ik ˈlīmˌstōn }

lithographic stone *See* lithographic limestone. { ˌlith·əˈgraf·ik ˈstōn }

lithographic texture |GEOL| The texture of certain calcareous sedimentary rocks characterized by grain size of less than 1/256 millimeter and having a smooth appearance. { ˌlith·əˈgraf·ik ˈteks·chər }

lithologic map |GEOL| A kind of geologic map showing the rock types of a particular area. { ˈlith·əˌläj·ik ˈmap }

lithologic unit *See* rock-stratigraphic unit. { ˈlith·əˌläj·ik ˈyü·nət }

lithology |GEOL| The description of the physical character of a rock as determined by eye or with a low-power magnifier, and based on color, structures, mineralogic components, and grain size. { ləˈthäl·ə·jē }

lithomorphic |GEOL| Referring to a soil whose characteristics are derived from events or conditions of a former period. { ˈlith·əˌmȯr·fik }

lithophile |GEOCHEM| **1.** Pertaining to elements that have become concentrated in the silicate phase of meteorites or the slag crust of the earth. **2.** Pertaining to elements that have a greater free energy of oxidation per gram of oxygen than iron. Also known as oxyphile. { ˈlith·əˌfīl }

lithophysa |GEOL| A large spherulitic hollow or bubble in glassy basalts and certain rhyolites. Also known as stone bubble. { ˌlith·əˈfīs·ə }

lithosiderite *See* stony-iron meteorite. { ˌlith·əˈsīd·əˌrīt }

lithosol |GEOL| A group of shallow soils lacking well-defined horizons and composed of imperfectly weathered fragments of rock. { ˈlith·əˌsȯl }

lithospar |MINERAL| A combination of spodumene and feldspar which occurs naturally. { ˈlith·əˌspär }

lithosphere |GEOL| **1.** The rigid outer crust of rock on the earth about 50 miles (80 kilometers) thick, above the asthenosphere. Also known as oxysphere. **2.** Since the development of plate tectonics theory, a term referring to the rigid, upper 60 miles (100 kilometers) of the crust and upper mantle, above the asthenosphere. { ˈlith·əˌsfir }

lithostatic pressure *See* ground pressure. { ˈlith·əˌstad·ik ˈpresh·ər }

litho stone *See* lithographic limestone. { ˈlith·ō ˌstōn }

lithostratic unit *See* rock-stratigraphic unit. { ˈlith·əˌstrad·ik ˈyü·nət }

lithostratigraphic unit *See* rock-stratigraphic unit. { ˌlith·əˌstrad·əˈgraf·ik ˈyü·nət }

lithostratigraphy |GEOL| A branch of stratigraphy concerned with the description and interpretation of sedimentary successions in terms of their lithic character. { ¦lith·ō·strə'Lig·rə·fē }

lithotope [GEOL] **1.** The environment under which a sediment is deposited. **2.** An area of uniform sedimentation. { 'lith·ə,tōp }

lithotype [GEOL] A macroscopic band in humic coals, analyzed on the basis of physical characteristics rather than botanical origin. { 'lith·ə,tīp }

Litopterna [PALEON] An order of hoofed, herbivorous mammals confined to the Cenozoic of South America; characterized by a skull without expansion of the temporal or squamosal sinuses, a postorbital bar, primitive dentition, and feet that were three-toed or reduced to a single digit. { ,lid·əp'tər·nə }

lit-par-lit [GEOL] Pertaining to the penetration of bedded, schistose, or other foliate rocks by innumerable narrow sheets and tongues of granitic rock. { 'lē,pär'lē }

Little Ice Age [GEOL] A period of expansion of mountain glaciers, marked by climatic deterioration, that began about 5500 years ago and extended to as late as A.D. 1550–1850 in some regions, as the Alps, Norway, Iceland, and Alaska. { 'lid·əl 'īs ,āj }

littoral drift [GEOL] Materials moved by waves and currents of the littoral zone. Also known as longshore drift. { 'lit·ə·rəl 'drift }

littoral sediments [GEOL] Deposits of littoral drift. { 'lit·ə·rəl 'sed·ə·məns }

littoral transport [GEOL] The movement of littoral drift. { 'lit·ə·rəl 'tranz,pórt }

Littorinacea [PALEON] An extinct superfamily of gastropod mollusks in the order Prosobranchia. { ,lid·ə·rə'nās·ē·ə }

livingstonite [MINERAL] $HgSb_4S_7$ A lead-gray mineral with red streak and metallic luster; a source of mercury. { 'liv·iŋ·stə,nīt }

lizard-hipped dinosaur [PALEON] The name applied to members of the Saurichia because of the comparatively unspecialized three-pronged pelvis. { 'liz·ərd ¦hipt 'dī·nə,sòr }

L joint See primary flat joint. { 'el ,jóint }

Llandellian [GEOL] Upper Middle Ordovician geologic time. { lan'del·yən }

Llandoverian [GEOL] Lower Silurian geologic time. { ¦lan·də¦vir·ē·ən }

Llanvirnian [GEOL] Lower Middle Ordovician geologic time. { lan'vir·nē·ən }

load cast [GEOL] An irregularity at the base of an overlying stratum, usually sandstone, that projects into an underlying stratum, usually shale or clay. { lōd ,kast }

load metamorphism See static metamorphism. { 'lōd ,med·ə'mór,fiz·əm }

loadstone See lodestone. { 'lōd,stōn }

loam [GEOL] Soil mixture of sand, silt, clay, and humus. { lōm }

loaming [GEOCHEM] In geochemical prospecting, a method in which samples of material from the surface are tested for traces of a sought-after metal; its presence on the surface presumably indicates a near-surface ore body. { ,lōm·iŋ }

lobate rill mark [GEOL] A flute cast formed by current action. { 'lō,bāt 'ril ,märk }

local attraction See local magnetic disturbance. { 'lō·kəl ə'trak·shən }

local base level See temporary base level. { 'lō·kəl ¦bās ¦lev·əl }

local magnetic disturbance [GEOPHYS] An anomaly of the magnetic field of the earth, extending over a relatively small area, due to local magnetic influences. Also known as local attraction. { 'lō·kəl mag'ned·ik di'stər·bəns }

local peat [GEOL] Peat formed by groundwater. Also known as basin peat. { 'lō·kəl ,pēt }

local relief [GEOL] The vertical difference in elevation between the highest and lowest points of a land surface within a specified horizontal distance or in a limited area. Also known as relative relief. { 'lō·kəl ri'lēf }

lode [GEOL] A fissure in consolidated rock filled with mineral; usually applied to metalliferous deposits. { lōd }

lodestone [MINERAL] The naturally occurring magnetic iron oxide, or magnetite, possessing polarity, and attracting iron objects to itself. Also known as Hercules stone; leading stone; loadstone. { 'lōd,stōn }

Iodranite [GEOL] A stony iron meteorite composed of bronzite and olivine within a fine network of nickel-iron. { 'lō·drə,nīt }

loellingite [MINERAL] $FeAs_2$ A silver-white to steel-gray mineral composed of iron arsenide with some cobalt, nickel, antimony, and sulfur; isomorphous with arsenopyrite; a source of arsenic. Also known as leucopyrite; lauollingite. { 'lel·iŋ‚īt }

loess [GEOL] An essentially unconsolidated, unstratified calcareous silt; commonly it is homogeneous, permeable, and buff to gray in color, and contains calcareous concretions and fossils. { les }

loess kindchen [GEOL] An irregular or spheroidal nodule of calcium carbonate that is found in loess. { 'les ‚kint·chən }

loeweite [MINERAL] $Na_4Mg_2(SO_4)_4 \cdot 5H_2O$ A white to pale-yellow mineral composed of hydrous sulfate of sodium and magnesium. { 'lā·və‚īt }

löllingite See loellingite. { 'lel·iŋ‚īt }

lomonite See laumontite. { lō'mä‚nīt }

lomontite See laumontite. { lō'män‚tīt }

longitudinal dune [GEOL] A type of linear dune ridge that extends parallel to the direction of the dominant dune-building winds. { ‚län·jə'tüd·ən·əl 'dün }

longitudinal fault [GEOL] A fault parallel to the trend of the surrounding structure. { ‚län·jə'tüd·ən·əl 'fólt }

longshore bar [GEOL] A ridge of sand, gravel, or mud built on the seashore by waves and currents, generally parallel to the shore and submerged by high tides. Also known as offshore bar. { 'lóŋ‚shór ‚bär }

longshore drift See littoral drift. { 'lóŋ‚shór ‚drift }

longshore trough [GEOL] A long, wide, shallow depression of the sea floor parallel to the shore. { 'lóŋ‚shór ‚tróf }

lonsdaleite [MINERAL] A mineral composed of a form of carbon; found in meteorites. { 'länz‚dā‚līt }

loparite [MINERAL] $(Ce,Na,Ca)_2(Ti,Nb)_2O_6$ A brown to black mineral; a variety of perovskite containing alkalies and cerium. { 'lō·pə‚rīt }

lopezite [MINERAL] $K_2Cr_2O_7$ An orange-red mineral composed of potassium dichromate. { 'lä·pə‚zīt }

Lophialetidae [PALEON] A family of extinct perissodactyl mammals in the superfamily Tapiroidea. { ‚lä·fē·ə'led·ə‚dē }

Lophiodontidae [PALEON] An extinct family of perissodactyl mammals in the superfamily Tapiroidea. { ‚lä·fē·ə'dänt·ə‚dē }

lopolith [GEOL] A large, floored intrusive body that is sunken centrally into the shape of a basin due to sagging of the underlying country rock. { 'läp·ə‚lith }

lorandite [MINERAL] $TlAsS_2$ A cochineal- to carmine-red or dark lead-gray mineral composed of thallium sulfarsenide, occurring in monoclinic form. { 'lä·rən‚dīt }

loranskite [MINERAL] $(Y,Ce,Ca,Zr)TaO_4$ A black mineral composed of an oxide of yttrium, cerium, calcium, tantalum, and zirconium. { lə'ran‚skīt }

lorettoite [MINERAL] $Pb_7O_6Cl_2$ A honey-yellow to brownish-yellow mineral composed of lead oxychloride. { lə'red·ə‚wīt }

loseyite [MINERAL] $(Mn,Zn)_7(CO_3)_2(OH)_{10}$ A bluish-white or brownish, monoclinic mineral consisting of a basic carbonate of manganese and zinc. { 'lō·zē‚īt }

lotrite See pumpellyite. { 'lō‚trīt }

loughlinite [MINERAL] $Na_2Mg_3Si_6O_{16} \cdot 8H_2O$ A pearly-white mineral that resembles asbestos, consisting of a hydrous silicate of sodium and magnesium. { 'lóf·lə‚nīt }

lovchorrite See mosandrite. { 'ləv·kó‚rīt }

Love wave [GEOPHYS] A horizontal dispersive surface wave, multireflected between internal boundaries of an elastic body, applied chiefly in the study of seismic waves in the earth's crust. { 'ləv ‚wāv }

lovozerite [MINERAL] $(Na,K)_2(Mn,Ca)ZrSi_6O_{16} \cdot 3H_2O$ Mineral composed of hydrous silicate of sodium, potassium, manganese, calcium, and zirconium. { lō'vä·zə‚rīt }

low-angle fault [GEOL] A fault that dips at an angle less than 45°. { 'lō ‚aŋ·gəl ‚fólt }

low-angle thrust See overthrust. { 'lō ‚aŋ·gəl ‚thrəst }

low-energy environment [GEOL] An aqueous sedimentary environment in which there is standing water with a general lack of wave or current action, permitting accumulation of very fine-grained sediments. { 'lō ‚en·ər·jē in'vī·ərn·mənt }

Lower Cambrian [GEOL] The earliest epoch of the Cambrian period of geologic time, ending about 540,000,000 years ago. { 'lō·ər 'kam brē·ən }

Lower Cretaceous [GEOL] The earliest epoch of the Cretaceous period of geologic time, extending from about 140- to 120,000,000 years ago. { 'lō·ər krə'tā·shəs }

Lower Devonian [GEOL] The earliest epoch of the Devonian period of geologic time, extending from about 400- to 385,000,000 years ago. { 'lō·ər də'vō·nē·ən }

Lower Jurassic [GEOL] The earliest epoch of the Jurassic period of geologic time, extending from about 185- to 170,000,000 years ago. { 'lō·ər ju'ras·ik }

lower mantle [GEOL] The portion of the mantle below a depth of about 600 miles (1000 kilometers). Also known as inner mantle; mesosphere; pallasite shell. { 'lō·ər mant·əl }

Lower Mississippian [GEOL] The earliest epoch of the Mississippian period of geologic time, beginning about 350,000,000 years ago. { 'lō·ər ,mis·ə'sip·ē·ən }

Lower Ordovician [GEOL] The earliest epoch of the Ordovician period of geologic time, extending from about 490- to 460,000,000 years ago. { 'lō·ər ,òr·də'vish·ən }

Lower Pennsylvanian [GEOL] The earliest epoch of the Pennsylvanian period of geologic time, beginning about 310,000,000 years ago. { 'lō·ər ,pen·səl'vā·nyən }

Lower Permian [GEOL] The earliest epoch of the Permian period of geologic time, extending from about 275- to 260,000,000 years ago. { 'lō·ər 'pər·mē·ən }

lower plate See footwall. { 'lō·ər ,plāt }

Lower Silurian [GEOL] The earliest epoch of the Silurian period of geologic time, beginning about 420,000,000 years ago. { 'lō·ər sə'lùr·ē·ən }

Lower Triassic [GEOL] The earliest epoch of the Triassic period of geologic time, extending from about 230- to 215,000,000 years ago. { 'lō·ər trī'as·ik }

low-moor bog [GEOL] A bog that is at or slightly below the ground water table. { 'lō ,mür 'bäg }

low-moor peat [GEOL] Peat found in low-moor bogs or swamps and containing little or no sphagnum. Also known as fen peat. { 'lō ¦mür 'pēt }

low quartz [MINERAL] Quartz that has been formed below 573°C; the tetrahedral crystal structure is less symmetrically arranged than a quartz formed at a higher temperature. { 'lō 'kwòrtz }

low-rank graywacke [PETR] A graywacke that is nonfeldspathic. { 'lō ,raŋk 'grā,wak·ə }

low-rank metamorphism [GEOL] A metamorphic process that occurs under conditions of low to moderate pressure and temperature. { 'lō ,raŋk ,med·ə'mór·fiz·əm }

low-tide terrace [GEOL] A flat area of a beach adjacent to the low-water line. { 'lō ¦tīd 'ter·əs }

low-velocity layer [GEOPHYS] A layer in the solid earth in which seismic wave velocity is lower than the layers immediately below or above. { 'lō və¦läs·əd·ē ,lā·ər }

low-volatile coal [GEOL] A coal that is nonagglomerating, has 78% to less than 86% fixed carbon, and 14% to less than 22% volatile matter. { 'lō ¦väl·ət·əl 'kōl }

Loxonematacea [PALEON] An extinct superfamily of gastropod mollusks in the order Prosobranchia. { ,läk·sə,ne·mə'tās·ē·ə }

Ludian [GEOL] A European stage of geologic time in the uppermost Eocene, above the Bartonian and below the Tongrian of the Oligocene. { 'lü·dē·ən }

ludlamite [MINERAL] (Fe,Mg,Mn)$_3$(PO$_4$)$_2$·4H$_2$O A green mineral crystallizing in the monoclinic system and occurring in small, transparent crystals. { 'ləd·lə,mīt }

Ludlovian [GEOL] A European stage of geologic time; Upper Silurian, below Gedinnian of Devonian, above Wenlockian. { ləd'lō·vē·ən }

ludwigite [MINERAL] (Mg,Fe)$_2$FeBO$_5$ A blackish-green mineral that crystallizes in the monoclinic system and occurs in fibrous masses; isomorphous with ronsenite. { 'ləd,wi,gīt }

lueneburgite [MINERAL] Mg$_3$B$_2$(OH)$_6$(PO$_4$)$_2$·6H$_2$O A colorless mineral composed of hydrous basic phosphate of magnesium and boron. { 'lü·nə·bər,gīt }

lueshite [MINERAL] NaNbO$_3$ An orthorhombic mineral having perovskite-type structure; it is dimorphous with natroniobite. { 'lü·əs,hīt }

Luisian [GEOL] A North American stage of geologic time: Miocene (above Relizian, below Mohnian). { lü'ē·shən }

lum *See* trolley. { ləm }

lunar inequality [GEOPHYS] A minute fluctuation of a magnetic needle from its mean position, caused by the moon. { 'lü·nər ,in·i'kwäl·əd·ē }

lunate bar [GEOL] A crescent-shaped bar of sand that is frequently found off the entrance to a harbor. { 'lü,nāt 'bär }

lunette [GEOL] A broad, low crescentic mound of windblown fine silt and clay. { lü'net }

Lusitanian [GEOL] Lower Jurassic geologic time. { ,lü·sə'tan·ē·ən }

luster mottlings [GEOL] The spotted, shimmering appearance of certain rocks caused by reflection of light from cleavage faces of crystals that contain small inclusions of other minerals. { 'ləs·tər ,mät·liŋz }

lutaceous [GEOL] Claylike. { lü'tā·shəs }

lutecite [GEOL] A fibrous, chalcedony-like quartz with optical anomalies that have led to its being considered a distinct species. { 'lüd·ə,sīt }

lutite [GEOL] A consolidated rock or sediment formed principally of clay or clay-sized particles. { 'lü,tīt }

luzonite *See* enargite. { 'lü·zə,nīt }

L wave [GEOPHYS] A phase designation for an earthquake wave that is a surface wave, without respect to type. { 'el ,wāv }

Lydian stone *See* basanite. { 'lid·ē·ən 'stōn }

lydite *See* basanite. { 'li,dīt }

Lyginopteridaceae [PALEOBOT] An extinct family of the Lyginopteridales including monostelic pteridosperms having one or two vascular traces entering the base of the petiole. { ,lī·jə·näp,ter·ə'dās·ē,ē }

Lyginopteridales [PALEOBOT] An order of the Pteridospermae. { ,lī·jə·näp,ter·ə'dā·lēz }

Lyginopteridatae [PALEOBOT] The equivalent name for Pteridospermae. { ,lī·jə·näp·fə'rid·əd,ē }

M

maar [GEOL] A volcanic crater that was created by violent explosion but not accompanied by igneous extrusion; frequently, it is filled by a small circular lake. { mär }

macaluba *See* mud volcano. { ˌmä·kə'lü·bə }

macedonite [MINERAL] $PbTiO_3$ A mineral composed of an oxide of lead and titanium. [PETR] A basaltic rock that contains orthoclase, sodic plagioclase, biotite, olivine, and rare pyriboles. { ¦mas·ə¦dä,nit }

maceral [GEOL] The microscopic organic constituents found in coal. { ¦mas·ə¦ral }

macgovernite [MINERAL] $Mn_5(AsO_3)SiO_3(OH)_2$ A mineral composed of basic manganese arsenite and silicate. Also spelled mcgovernite. { mə'gəv·ər,nīt }

mackayite [MINERAL] $FeTe_2O_5(OH)$ A green mineral composed of basic iron tellurite. { 'mak·ē,īt }

mackinawite [MINERAL] $(Fe,Ni)S$ A tetragonal mineral occurring as a corrosion product in iron pipes. Also known as kansite. { mə'kin·ə,wīt }

macle [MINERAL] **1.** A dark or discolored spot in a mineral specimen. **2.** *See* chiastolite. { 'mak·əl }

Macraucheniidae [PALEON] A family of extinct herbivorous mammals in the order Litopterna; members were proportioned much as camels are, and eventually lost the vertebral arterial canal of the cervical vertebrae. { ˌma,krò·kə'nī·ə,dē }

macroclastic [PETR] Rock that is composed of fragments that are visible without magnification. { ¦mak·rə'klas·tik }

macrocrystalline [PETR] **1.** Pertaining to the texture of holocrystalline rock in which the constituents are visible without magnification. **2.** Pertaining to the texture of a rock with grains or crystals greater than 0.75 millimeter in diameter in recrystallized sediment. { ¦mak·rō'krist·əl·ən }

macrofacies [GEOL] A collection of sedimentary facies that are related genetically. { ¦mak·rō¦fā·shēz }

macrofossil [PALEON] A fossil large enough to be observed with the naked eye. { ¦mak·rō'fäs·əl }

macropore [GEOL] A pore in soil of a large enough size so that water is not held in it by capillary attraction. { 'mak·rə,pòr }

maculose [GEOL] Of a group of contact-metamorphosed rocks or their structures, having spotted or knotted character. { 'mak·yə,lōs }

Maestrichtian [GEOL] A European stage of geologic time: Upper Cretaceous (above Menevian, below Fastiniogian). { ma'strik·tē·ən }

mafic mineral [MINERAL] **1.** A mineral that is composed predominantly of the ferromagnesian rock-forming silicates. **2.** In general, any dark mineral. { 'maf·ik 'min·rəl }

maghemite [MINERAL] $\gamma\text{-}Fe_2O_3$ A mineral form of iron oxide that is strongly magnetic and a member of the magnetite series. { mag'he,mīt }

magma [GEOL] The molten rock material from which igneous rocks are formed. { 'mag·mə }

magma chamber [GEOL] A larger reservoir in the crust of the earth that is occupied by a body of magma. { 'mag·mə ,chäm·bər }

magma geothermal system [GEOL] A geothermal system in which the dominant source of heat is a large reservoir of igneous magma within an intrusive chamber or lava pool; an example is the Yellowstone Park area of Wyoming. { 'mag·mə ¦jē·ō'thər· məl ˌsis·təm }

magma province *See* petrographic province. { 'mag·mə ˌpräv·əns }

magmatic differentiation [PETR] **1.** The process by which the different types of igneous rocks are derived from a single parent magma. **2.** The process by which ores are formed by solidification from magma. Also known as magmatic segregation. { mag'mad·ik ˌdif·ə,ren·chē'ā·shən }

magmatic rock [PETR] A rock derived from magma. { mag'mad·ik 'räk }

magmatic segregation *See* magmatic differentiation. { mag'mad·ik ˌseg·rə'gā·shən }

magmatic stoping [GEOL] A process of igneous intrusion in which magma gradually works its way upward by breaking off and engulfing blocks of the country rock. Also known as stoping. { mag'mad·ik 'stōp·iŋ }

magmatism [PETR] The formation of igneous rock from magma. { 'mag·mə,tiz·əm }

magmosphere *See* pyrosphere. { 'mag·mə,sfir }

magnafacies [GEOL] A major, continuous belt of deposits that is homogeneous in lithologic and paleontologic characteristics and that extends obliquely across time planes or through several time-stratigraphic units. { ¦mag·nə'fā·shēz }

magnesia mica *See* biotite. { mag'nē·zhə 'mī·kə }

magnesian calcite [MINERAL] (Ca,Mg)CO₃ A variety of calcite consisting of randomly substituted magnesium carbonate in a disordered calcite lattice. Also known as magnesium calcite. { mag'nē·zhən 'kal,sīt }

magnesian limestone [PETR] Limestone with at least 90% calcite, a maximum of 10% dolomite, an approximate magnesium oxide equivalent of 1.1–2.1, and an approximate magnesium carbonate equivalent of 2.3–4.4. { mag'nē·zhən 'līm,stōn }

magnesian marble [PETR] A type of magnesian limestone that has been metamorphosed; contains some dolomite. Also known as dolomitic marble. { mag'nē·zhən 'mär·bəl }

magnesiochromite [MINERAL] MgCr₂O₄ A mineral of the spinel group composed of magnesium chromium oxide; it is isomorphous with chromite. Also known as magnochromite. { mag,nē·zhō·'krō,mīt }

magnesiocopiapite [MINERAL] MgFe₄(SO₄)₆(OH)₂·20H₂O A mineral of the copiapite group composed of hydrous basic magnesium and iron sulfate; it is isomorphous with copiapite and cuprocopiapite. { mag,nē·zhō·'kō·pē·ə,pīt }

magnesioferrite [MINERAL] (Mg,Fe)Fe₂O₄ A black, strongly magnetic mineral of the magnetite series in the spinel group. Also known as magnoferrite. { mag,nē·zhō'fe,rīt }

magnesite [MINERAL] MgCO₃ The mineral form of magnesium carbonate, usually massive and white, with hexagonal symmetry; specific gravity is 3, and hardness is 4 on Mohs scale. Also known as giobertite. { 'mag·nə,sīt }

magnesium calcite *See* magnesian calcite. { mag'nē·zē·əm 'kal,sīt }

magnesium-iron mica *See* biotite. { mag'nē·zē·əm 'ī·ərn 'mī·ka }

magnetic annual change [GEOPHYS] The amount of secular change in the earth's magnetic field which occurs in 1 year. Also known as annual magnetic change. { mag'ned·ik 'an·yə·wəl 'chānj }

magnetic annual variation [GEOPHYS] The small, systematic temporal variation in the earth's magnetic field which occurs after the trend for secular change has been removed from the average monthly values. Also known as annual magnetic variation. { mag'ned·ik 'an·yə·wəl ,ver·ē'ā·shən }

magnetic bay [GEOPHYS] A small magnetic disturbance whose magnetograph resembles an indentation of a coastline; on earth, magnetic bays occur mainly in the polar regions and have a duration of a few hours. { mag'ned·ik 'bā }

magnetic character figure *See* C index. { mag'ned·ik ¦kar·ik·tər ,fig·yər }

magnetic daily variation *See* magnetic diurnal variation. { mag'ned·ik ¦dā·lē ver·ē'ā·shən }

magnetic declination *See* declination. { mag'ned·ik ,dek·lə'nā·shən }

magnetic dip *See* inclination. { mag'ned·ik 'dip }

magnetic diurnal variation [GEOPHYS] Oscillations of the earth's magnetic field which have a periodicity of about a day and which depend to a close approximation only on local time and geographic latitude. Also known as magnetic daily variation. { mag'ned·ik dī'ərn·əl ,ver·ē'ā·shən }

magnetic element [GEOPHYS] Magnetic declination, dip, or intensity at any location on the surface of the earth. { mag'ned·ik 'el·ə·mənt }

magnetic equator [GEOPHYS] That line on the surface of the earth connecting all points at which the magnetic dip is zero. Also known as aclinic line. { mag'ned·ik i'kwād·ər }

magnetic iron ore See magnetite. { mag'ned·ik 'ī·ərn 'ȯr }

magnetic latitude [GEOPHYS] Angular distance north or south of the magnetic equator. { mag'ned·ik 'lad·ə,tüd }

magnetic local anomaly [GEOPHYS] A localized departure of the geomagnetic field from its average over the surrounding area. { mag'ned·ik ,lō·kəl ə'näm·ə·lē }

magnetic meridian [GEOPHYS] A line which is at any point in the direction of horizontal magnetic force of the earth; a compass needle without deviation lies in the magnetic meridian. { mag'ned·ik mə'rid·ē·ən }

magnetic north [GEOPHYS] At any point on the earth's surface, the horizontal direction of the earth's magnetic lines of force (direction of a magnetic meridian) toward the north magnetic pole; a particular direction indicated by the needle of a magnetic compass. { mag'ned·ik 'nȯrth }

magnetic observatory [GEOPHYS] A geophysical measuring station employing some form of magnetometer to measure the intensity of the earth's magnetic field. { mag'ned·ik əb'zər·və,tȯr·ē }

magnetic pole [GEOPHYS] In geomagnetism, either of the two points on the earth's surface where the magnetic meridians converge, that is, where the magnetic field is vertical. Also known as dip pole. { mag'ned·ik 'pōl }

magnetic prime vertical [GEOPHYS] The vertical circle through the magnetic east and west points of the horizon. { mag'ned·ik 'prīm 'vərd·ə·kəl }

magnetic profile [GEOPHYS] A profile of a geologic structure showing magnetic anomalies. { mag'ned·ik 'prō,fīl }

magnetic reversal [GEOPHYS] A reversal of the polarity of the earth's magnetic field that has occurred at about one-million-year intervals. { mag'ned·ik ri'vər·səl }

magnetic secular change [GEOPHYS] The gradual variation in the value of a magnetic element which occurs over a period of years. { mag'ned·ik ¦sek·yə·lər 'chānj }

magnetic station [GEOPHYS] A facility equipped with instruments for measuring local variations in the earth's magnetic field. { mag'ned·ik 'stā·shən }

magnetic storm [GEOPHYS] A worldwide disturbance of the earth's magnetic field; frequently characterized by a sudden onset, in which the magnetic field undergoes marked changes in the course of an hour or less, followed by a very gradual return to normalcy, which may take several days. Also known as geomagnetic storm. { mag'ned·ik 'stȯrm }

magnetic stratigraphy See paleomagnetic stratigraphy. { mag'ned·ik strə'tig·rə·fē }

magnetic survey [GEOPHYS] **1.** Magnetometer map of variations in the earth's total magnetic field; used in petroleum exploration to determine basement-rock depths and geologic anomalies. **2.** Measurement of a component of the geomagnetic field at different locations. { mag'ned·ik 'sər,vā }

magnetic temporal variation [GEOPHYS] Any change in the earth's magnetic field which is a function of time. { mag'ned·ik ¦tem·pə·rəl ,ver·ē'ā·shən }

magnetic variation [GEOPHYS] Small changes in the earth's magnetic field in time and space. { mag'ned·ik ,ver·ē'ā·shən }

magnetite [MINERAL] An opaque iron-black and streak-black isometric mineral and member of the spinel structure type, usually occurring in octahedrals or in granular to massive form; hardness is 6 on Mohs scale, and specific gravity is 5.20. Also known as magnetic iron ore; octahedral iron ore. { 'mag·nə,tīt }

magnetoionic duct [GEOPHYS] Duct along the geomagnetic lines of force which exhibits waveguide characteristics for radio-wave propagation between conjugate points on the earth's surface. { mag¦nēd·ō·ī'än·ik 'dəkt }

magnetoioinic theory [GEOPHYS] The theory of the combined effect of the earth's magnetic field and atmospheric ionization on the propagation of electromagnetic waves. { mag¦nēd·ō·ī'än·ik 'thē·ə·rē }

magnetoioinic wave component [GEOPHYS] Either of the two elliptically polarized wave components into which a linearly polarized wave incident on the ionosphere is separated because of the earth's magnetic field. { mag¦nēd·ō·ī'än·ik 'wāv kəm,pō·nənt }

magnetoplumbite [MINERAL] $(Pb,Mn)_2Fe_6O_{11}$ Black mineral consisting of a ferric oxide of plumbite and manganese, and occurring in acute metallic hexagonal crystals. { mag,nēd·ə'pləm,bīt }

magnetostratigraphy [GEOL] A branch of stratigraphy in which sedimentary successions are described and interpreted in terms of remanent magnetization. { mag¦nēd·ō·strə'tig·rə·fē }

magnetotellurics [GEOPHYS] A geophysical exploration technique that measures natural electromagnetic fields to image subsurface electrical resistivity, providing information about the earth's interior composition and structure since naturally occurring rocks and minerals exhibit a broad range of electrical resistivities. { mag,ned·ō·tə'lür·iks }

magnitude [GEOPHYS] A measure of the amount of energy released by an earthquake. { 'mag·nə,tüd }

magnochromite See magnesiochromite. { ,mag·nə'krō,mīt }

magnoferrite See magnesioferrite. { ,mag·nə'fe,rīt }

magnophorite [MINERAL] $NaKCaMg_5Si_8O_{23}OH$ A monoclinic mineral composed of a basic silicate of sodium, potassium, calcium, and magnesium; member of the amphibole group. { ,mag·nə'fór,īt }

main joint See master joint. { 'mān 'jȯint }

major fold [GEOL] A large-scale fold with which minor folds are usually associated. { 'mā·jər 'fōld }

majorite [MINERAL] $Mg_3(Fe,Al,Si)_2(SiO_4)_3$ A garnet mineral that forms in the deep upper mantle in response to the gradual dissolution of pyroxene due to increasing pressure, first identified as an inclusion in diamond. { 'mā·jə,rīt }

major joint See master joint. { 'mā·jər 'jȯint }

malachite [MINERAL] $Cu_2CO_3(OH)_2$ A bright-green monoclinic mineral consisting of a basic carbonate of copper and usually occurring in massive forms or in bundles of radiating fibers; specific gravity is 4.05, and hardness is 3.5–4 on Mohs scale. { 'mal·ə,kīt }

malacolite See diopside. { 'mal·ə·kə,līt }

malchite [PETR] A fine-grained lamprophyre with small, rare phenocrysts or hornblende, labradorite, and sometimes biotite embedded in a matrix of hornblende, andesine, and some quartz. { 'mal,kīt }

maldonite [MINERAL] Au_2Bi A pinkish silver-white mineral consisting of gold and bismuth; occurs in massive granular form. { 'mal·də,nīt }

malladrite [MINERAL] Na_2SiF_6 A hexagonal mineral composed of sodium fluosilicate, occurring as small crystals in volcanic holes in Vesuvius. { mə'lä,drīt }

mallardite [MINERAL] $MnSO_4·7H_2O$ A pale-rose, monoclinic mineral composed of hydrous manganese sulfate. { mə'lär,dīt }

malloseismic [GEOPHYS] Referring to an area that is likely to experience destructive earthquakes several times in a century. { ,mal·ə'sīz·mik }

malm See marl. { mäm }

Malm [GEOL] The Upper Jurassic geologic series, above Dogger and below Cretaceous. { mäm }

malysite [MINERAL] $FeCl_3$ A halogen mineral deposited by sublimation; found most commonly at Mount Vesuvius, Italy. { 'mal·ə,sīt }

mamelon [GEOL] A small, rounded volcano which forms over a vent as a result of the slow extrusion of viscous, silicic lava. { 'mam·ə·lən }

mammillary [MINERAL] Of or pertaining to an aggregate of crystals in the form of a rounded mass. { 'ma·mə,ler·ē }

mammillary structure See pillow structure. { 'ma·mə,ler·ē 'strək·chər }

mammoth [PALEON] Any of various large Pleistocene elephants having long, upcurved tusks and a heavy coat of hair. { 'mam·əth }

Mammutinae [PALEON] A subfamily of extinct proboscidean mammals in the family Mastodontidae. { mə'myüt·ən,ē }

manandonite [MINERAL] $Li_4Al_{14}B_4Si_6O_{29}(OH)_{24}$ A white mineral composed of basic borosilicate of lithium and aluminum. { mə'nan·də,nīt }

manasseite [MINERAL] $Mg_6Al_2(OH)_{16}(CO_3)·4H_2O$ A hexagonal mineral composed of basic hydrous magnesium and aluminum carbonate; it is dimorphous with hydrotalcite. { mə'nas·ē,īt }

manganese epidote See piemontite. { 'maŋ·gə,nēs 'ep·ə,dōt }

manganese nodule [GEOL] Small, irregular black to brown concretions consisting chiefly of manganese salts and manganese oxide minerals; formed in oceans as a result of pelagic sedimentation or precipitation. { 'maŋ·gə,nēs 'naj·ül }

manganite [MINERAL] MnO(OH) A brilliant steel-gray or black polymorphous mineral; crystallizes in the orthorhombic system. Also known as gray manganese ore. { 'maŋ·gə,nīt }

manganolangbeinite [MINERAL] $K_2Mn_2(SO_4)_3$ A rose-red, isometric mineral composed of potassium manganese sulfate; occurs in lava on Vesuvius. { ¦maŋ·gə·nō'laŋ,bī,nīt }

manganosite [MINERAL] MnO An emerald-green isometric mineral occurring in small octahedrons that blacken on exposure; hardness is 5–6 on Mohs scale, and specific gravity is 5.18. { ,maŋ·gə'nō,sīt }

mankato stone [PETR] A variety of limestone containing more than 49% calcium carbonate, with about 4.5% alumina and some silica. { man'kād·ō ,stōn }

mansfieldite [MINERAL] $Al(AsO_4)·2H_2O$ A white to pale-gray orthorhombic mineral composed of hydrous aluminum arsenate; it is isomorphous with scorodite. { 'manz,fēl,dīt }

mantle [GEOL] The intermediate shell zone of the earth below the crust and above the core (to a depth of 2160 miles or 3480 kilometers). { 'mant·əl }

mantled gneiss dome [GEOL] A dome in metamorphic terrains that has a remobilized core of gneiss surrounded by a concordant sheath of the basal part of the overlying metamorphic sequence. { 'mant·əld ¦nīs ,dōm }

mantle rock See regolith. { 'mant·əl ,räk }

manto [GEOL] A sedimentary or igneous ore body occurring in flat-lying depositional layers. { 'man,tō }

marble [PETR] **1.** Metamorphic rock composed of recrystallized calcite or dolomite. **2.** Commercially, any limestone or dolomite taking polish. { 'mär·bəl }

marcasite [MINERAL] FeS_2 A pale bronze-yellow to nearly white mineral, crystallizing in the orthorhombic system; hardness is 6–6.5 on Mohs scale, and specific gravity is 4.89. { 'mär·kə,sīt }

marekanite [GEOL] Rounded to subangular obsidian bodies that occur in masses of perlite. { ¦mär·ə¦ka,nīt }

margarite [GEOL] A string of beadlike globulites; commonly found in glassy igneous rocks. [MINERAL] $CaAl_2(Al_2Si_2)O_{10}(OH)_2$ A pink, reddish, or yellow, brittle mica mineral. { 'mär·gə,rīt }

margarosanite [MINERAL] $PbCa_2(SiO_3)_3$ A colorless or snow-white triclinic mineral composed of lead calcium silicate, occurring in lamellar masses. { ,mär·gə'rōs·ən,īt }

marginal escarpment [GEOL] A seaward slope of a marginal plateau with a gradient of 1:10 or more. { 'mär·jən·əl e'skärp·mənt }

marginal fissure [GEOL] A magma-filled fracture bordering an igneous intrusion. { 'mär·jən·əl 'fish·ər }

marginal moraine See terminal moraine. { 'mär·jən·əl mə'rān }

marginal plain See outwash plain. { 'mär·jən·əl 'plān }

marginal plateau [GEOL] A relatively flat shelf adjacent to a continent and similar topographically to, but deeper than, a continental shelf. { 'mär·jən·əl pla'tō }

marginal salt pan [GEOL] A natural, coastal salt pan. { 'mär·jən·əl 'sólt ,pan }

marginal thrust

marginal thrust [GEOL] One of a series of faults bordering an igneous intrusion and crossing both the intrusion and the wall rock. Also known as marginal upthrust. { 'mär·jən·əl 'thrəst }

marginal upthrust *See* marginal thrust. { 'mär·jən·əl 'əp,thrəst }

marialite [MINERAL] 3NaAlSi$_3$O$_8$·NaCl A scapolite mineral that is isomorphous with meronite. { mə'rē·ə,līt }

marine abrasion [GEOL] Erosion of the ocean floor by sediment moved by ocean waves. Also known as wave erosion. { mə'rēn ə'brā·zhən }

marine arch *See* sea arch. { mə'rēn 'ärch }

marine bridge *See* sea arch. { mə'rēn 'brij }

marine cave *See* sea cave. { mə'rēn 'kāv }

marine-cut terrace [GEOL] A terrace or platform cut by wave erosion of marine origin. Also known as wave-cut terrace. { mə'rēn ¦kət 'ter·əs }

marine geology *See* geological oceanography. { mə'rēn jē'äl·ə·jē }

Marinesian *See* Bartonian. { mar·ə'nē·zhē·ən }

marine stack *See* stack. { mə'rēn 'stak }

marine terrace [GEOL] A seacoast terrace formed by the merging of a wave-built terrace and a wave-cut platform. Also known as sea terrace; shore terrace. { mə'rēn 'ter·əs }

marine transgression *See* transgression. { mə'rēn tranz'gresh·ən }

marker bed [GEOL] **1.** A stratified unit with distinctive characteristics making it an easily recognized geologic horizon. **2.** A rock layer which accounts for a characteristic portion of a seismic refraction time-distance curve. **3.** *See* key bed. { 'märk·ər ,bed }

marl [GEOL] A deposit of crumbling earthy material composed principally of clay with magnesium and calcium carbonate; used as a ertilizer for lime-deficient soils. Also known as malm. { märl }

marlite *See* marlstone. { 'mär,līt }

marlstone [PETR] **1.** A consolidated rock that has about the same composition as marl; considered to be an earthy or impure argillaceous limestone. Also known as marlite. **2.** A hard ferruginous rock of the Middle Lias in England. { 'märl,stōn }

marly [GEOL] Pertaining to, containing, or resembling marl. { 'mär·lē }

marmatite [MINERAL] A dark-brown to black mineral composed of iron-bearing sphalerite. Also known as christophite. { 'mär·mə,tīt }

marmolite [MINERAL] A pale-green serpentine mineral, occurring in thin laminations; a variety of chrysotile. { 'mär·mə,līt }

Marmor [GEOL] A North American stage of Middle Ordovician geologic time, forming the lower subdivision of Chazyan, above Whiterock and below Ashby. { 'mär,mòr }

marrite [MINERAL] PbAgAsS$_3$ A monoclinic mineral, occurring as small crystals in Valais, Switzerland. { 'mä,rīt }

marsh gas [GEOCHEM] Combustible gas, consisting chiefly of methane, produced as a result of decay of vegetation in stagnant water. { 'märsh ,gas }

marshite [MINERAL] CuI A reddish, oil-brown isometric mineral composed of cuprous iodide and occurring as crystals; hardness is 2.5 on Mohs scale, and specific gravity is 5.6. { 'mär,shīt }

marsh ore *See* bog iron ore. { 'märsh ,òr }

martite [MINERAL] Hematite occurring in iron-black octahedral crystals pseudomorphous after magnetite. { 'mär,tīt }

mascagnite [MINERAL] (NH$_4$)$_2$SO$_4$ A yellowish-gray mineral found in guano, near burning coal beds, or as lava incrustation; specific gravity is 1.77; hardness is 2–2.5 on Mohs scale. { ma'skan,yīt }

mascon [GEOL] A large, high-density mass concentration below a ringed mare on the surface of the moon. { 'mas,kän }

mass attraction vertical [GEOPHYS] The vertical which is a function only of the distribution of mass and is unaffected by forces resulting from the motions of the earth. { 'mas ə¦trak·shən ,verd·ə·kəl }

mass erosion [GEOL] A process in which the direct application of gravitational body

196

stresses causes earth and rocks to fall and be carried downslope. Also known as gravity erosion. { 'mas i'rō·zhən }

mass heaving [GEOL] A comprehensive expansion of the ground due to freezing. { 'mas 'hēv·iŋ }

massicot [MINERAL] PbO A yellow, orthorhombic mineral consisting of lead monoxide; found in the western and southern United States. Also known as lead ocher. { 'mas·ə,kät }

massif [GEOL] A massive block of rock within an erogenic belt, generally more rigid than the surrounding rocks, and commonly composed of crystalline basement or younger plutons. { ma'sēf }

massive [GEOL] Of a mineral deposit, having a large concentration of ore in one place. [MINERAL] Of a mineral, lacking an internal structure. [PALEON] Of corallum, composed of closely packed corallites. [PETR] **1.** Of a competent rock, being homogeneous, isotropic, and elastically perfect. **2.** Of a metamorphic rock, having constituents which do not show parallel orientation and are not arranged in layers. **3.** Of igneous rocks, being homogeneous over wide areas and lacking layering, foliation, cleavage, or similar features. { 'mas·iv }

mass movement [GEOL] Movement of a portion of the land surface as a unit. { 'mas 'müv·mənt }

mass wasting [GEOL] Dislodgement and downslope transport of loose rock and soil material under the direct influence of gravitational body stresses. { 'mas ,wāst·iŋ }

master joint [GEOL] A persistent joint plane of greater than average extent, generally constituting the dominant jointing of an area. Also known as main joint; major joint. { 'mas·tər 'jȯint }

mastodon [PALEON] A member of the Mastodontidae, especially the genus *Mammut*. { 'mas·tə,dän }

Mastodontidae [PALEON] An extinct family of elephantoid proboscideans that had low-crowned teeth with simple ridges and without cement. { ,mas·tə'dän·tə,dē }

matched terrace See paired terrace. { 'macht 'ter·əs }

material unit [GEOL] A stratigraphic unit based on rocks and their fossil content without time implication. { mə'tir·ē·əl ,yü·nət }

mathematical geology [GEOL] The branch of geology concerned with the study of probability distributions of values of random variables involved in geologic processes. { ¦math·ə¦mad·ə·kəl jē'äl·ə·jē }

matildite [MINERAL] AgBiS$_2$ An iron black to gray, orthorhombic mineral consisting of silver bismuth sulfide; occurrence is massive or granular. { mə'til,dīt }

matlockite [MINERAL] PbFCl A mineral consisting of lead chloride and fluoride. { 'mat·lə,kīt }

matric forces [GEOL] Forces acting on soil water that are independent of gravity but exist due to the attraction of solid surfaces for water, the attraction of water molecules for each other, and a force in the air-water interface due to the polar nature of water. { 'mā·trik ,fȯrs·əz }

matrix [PETR] The continuous, fine-grained material in which large grains of a sediment or sedimentary rock are embedded. Also known as groundmass. { 'mā·triks }

matrix porosity [GEOL] Core-sample porosity determined from a small sample of the core, in contrast to total porosity, where the whole core is used. { 'mā·triks pə'räs·əd·ē }

matrix rock See land pebble phosphate. { 'mā·triks ,räk }

matrix velocity [GEOPHYS] The velocity of sound through a formation's rock matrix during an acoustic-velocity log. { 'mā·triks və'läs·əd·ē }

mature [GEOL] **1.** Pertaining to a topography or region, and to its landforms, having undergone maximum development and accentuation of form. **2.** Pertaining to the third stage of textural maturity of a clastic sediment. { mə'chur }

matureland [GEOL] The land surface which is characteristic of the mature stage in the erosion cycle. { mə'chur,land }

mature soil See zonal soil. { mə'chur 'sȯil }

197

maturity

maturity [GEOL] **1.** The second stage of the erosion cycle in the topographic development of a landscape or region characterized by numerous and closely spaced mature streams, reduction of level surfaces to slopes, large well-defined drainage systems, and the absence of swamps or lakes on the uplands. Also known as topographic maturity. **2.** A stage in the development of a shore or coast that begins with the attainment of a profile of equilibrium. **3.** The extent to which the texture and composition of a clastic sediment approach the ultimate end product. **4.** The stage of stream development at which maximum vigor and efficiency has been reached. { mə'chúr·əd·ē }

maturity index [GEOL] A measure of the progress of a clastic sediment in the direction of chemical or mineralogic stability; for example, a high ratio of quartz + cherts to feldspar + rock fragments indicates a highly mature sediment. { mə'chúr·əd·ē ,in,deks }

maucherite [MINERAL] $Ni_{11}As_8$ A reddish silver-white mineral composed of nickel arsenide. { 'maú·chə,rīt }

maximum subsidence [GEOL] The maximum amount of subsidence in a basin. { 'mak·sə·məm səb'sīd·əns }

mcgovernite See macgovernite. { mə'gəv·ər,nīt }

meadow ore See bog iron ore. { 'med·ō ,ór }

meander bar See point bar. { mē'an·dər ,bär }

meander belt [GEOL] The zone along the floor of a valley across which a meandering stream periodically shifts its channel. { mē'an·dər ,belt }

meander core [GEOL] A hill encircled by a stream meander. Also known as rock island. { mē'an·dər ,kòr }

meander niche [GEOL] A conical or crescentic opening in the wall of a cave formed by downward and lateral stream erosion. { mē'an·dər ,nich }

meander plain [GEOL] A plain built by the meandering process, or a plain of lateral accretion. { mē'an·dər ,plān }

meander scar [GEOL] A crescentic, concave mark on the face of a bluff or valley wall formed by a meandering stream. { mē'an·dər ,skär }

meander spur [GEOL] An undercut projection of high land that extends into the concave part of, and is enclosed by, a meander. { mē'an·dər ,spər }

mechanical erosion See corrasion. { mi'kan·ə·kəl i'rō·zhən }

mechanical sediment See clastic sediment. { mi'kan·ə·kəl 'sed·ə·mənt }

mechanical weathering [GEOL] The process of weathering by which physical forces break down or reduce a rock to smaller and smaller fragments, involving no chemical change. Also known as physical weathering. { mi'kan·ə·kəl 'weth·ə·riŋ }

medial moraine [GEOL] **1.** An elongate moraine carried in or upon the middle of a glacier and parallel to its sides. **2.** A moraine formed by glacial abrasion of a rocky protuberance near the middle of a glacier. { 'mē·dē·əl mə'rān }

median mass [GEOL] A less disturbed structural block in the middle of an orogenic belt, bordered on both sides by orogenic structure, thrust away from it. Also known as betwixt mountains; Zwischengebirge. { 'mē·dē·ən 'mas }

median particle diameter [GEOL] The middlemost particle diameter of a rock or sediment, larger than 50% of the diameter in the distribution and smaller than the other 50%. { 'mē·dē·ən 'pärd·ə·kəl dī,am·əd·ər }

medina quartzite [MINERAL] A variety of quartz containing 97.8% silica; melting point is about 1700°C. { mə'dē·nə 'kwórt,sīt }

mediterranean See mesogeosyncline. { ,med·ə·tə'rā·nē·ən }

medium-volatile bituminous coal [GEOL] Bituminous coal consisting of 23–31% volatile matter. { 'mē·dē·əm ¦väl·ə·təl bə'tü·mə·nəs 'kōl }

Medullosaceae [PALEOBOT] A family of seed ferns; these extinct plants all have large spirally arranged petioles with numerous vascular bundles. { mə'dəl·ō'sās·ē,ē }

meerschaum See sepiolite. { 'mir,shòm }

megacryst [PETR] Any crystal or grain in an igneous or metamorphic rock that is significantly larger than the surrounding matrix. { 'meg·ə,krist }

198

megacyclothem [GEOL] A cycle of or combination of related cyclothems. { ˌmeg·ə'sī·klə,them }

megaripple [GEOL] A large sand wave. { 'meg·ə,rĭp·əl }

megatectonics [GEOL] The tectonics of the very large structural features of the earth. { ˌmeg·ə,tek'tän·iks }

meionite [MINERAL] $3CaAl_2Si_2O_8·CaCO_3$ A scapolite mineral composed of calcium aluminosilicate and calcium carbonate; it is isomorphous with marialite. { 'mī·ə,nīt }

mélange [GEOL] A heterogeneous medley or mixture of rock materials; specifically, a mappable body of deformed rocks consisting of a pervasively sheared, fine-grained, commonly pelitic matrix, thoroughly mixed with angular and poorly sorted inclusions of native and exotic tectonic fragments, blocks, or slabs, of diverse origins and geologic ages, that may be as much as several kilometers in length. Also known as block clay. { mā'länzh }

melanic See melanocratic. { me'lan·ik }

melanocerite [MINERAL] $(Ca,Ce,Y)_8(BO_3)(SiO_4)_4(F,OH)_4$ A brown or black rhombohedral mineral composed of complex silicate, borate, fluoride, tantalate, or other anion of cerium, yttrium, calcium, and other metals; occurs as crystals. { ¦mel·ə·nō'se,rīt }

melanocratic [GEOL] Dark-colored, referring to igneous rock containing at least 50–60% mafic minerals. Also known as chromocratic; melanic. { ¦mel·ə·nō¦krad·ik }

melanophlogite [MINERAL] A mineral composed chiefly of silicon dioxide and containing some carbon and sulfur. { ¦mel·ə·nō'flō,jīt }

melanostibian [MINERAL] $Mn(Sb,Fe)O_3$ A black mineral consisting of iron and manganese antimonite; occurs as foliated masses and as striated crystals. { ˌmel·ə·nō'stib·ē·ən }

melanotekite [MINERAL] $Pb_2Fe_2Si_2O_9$ A black or dark-gray mineral composed of lead iron silicate. { ˌmel·ə·nō'tek,īt }

melanovanadite [MINERAL] $Ca_2V_{10}O_{25}$ A black mineral composed of a complex oxide of calcium and vanadium. { ¦mel·ə·nō'van·ə,dīt }

melanterite [MINERAL] $FeSO_4·7H_2O$ A green mineral occurring mainly in fibrous or concretionary masses, or in short, monoclinic, prismatic crystals; hardness is 2 on Mohs scale, and specific gravity is 1.90. { mə'lan·tə,rīt }

melaphyre [PETR] Altered basalt, especially of Carboniferous and Permian age. { 'mel·ə,fīr }

melilite [MINERAL] A sorosilicate mineral group of complex composition $[(Na,Ca)_2 (Mg,Al)(Si,Al)_2O_7]$ crystallizing in the tetragonal system; luster is vitreous to resinous, and color is white, yellow, greenish, reddish, or brown; hardness is 5 on Mohs scale, and specific gravity varies from 2.95 to 3.04. { 'mel·ə,līt }

melilitite [PETR] An extrusive rock that is generally olivine-free and composed of more than 90% mafic mineral such as melilite and augite, with minor amounts of feldspathoids and sometimes plagioclase. { mə'lil·ə,tīt }

meliphane See meliphanite. { 'mel·ə,fān }

meliphanite [MINERAL] $(Ca,Na)_2Be(Si,Al)_2(O,OH,F)_7$ A yellow, red, or black mineral composed of sodium calcium beryllium fluosilicate. Also known as meliphane. { mə'lif·ə,nīt }

melissopalynology [PALEOBOT] A branch of palynology that deals with the analysis of bee pollen loads (pollen collected from flowers and then carried back to the hive on the bee's hindlegs) and the pollen component within honeys. { mə,lis·ə,pal·ə'näl·ə·jē }

mellite [MINERAL] $Al_2[C_6(COO)_6]·18H_2O$ A honey-colored mineral with resinous luster composed of the hydrous aluminum salt of mellitic acid, occurring as nodules in brown coal; it is in part a product of vegetable decomposition. { 'me,līt }

melonite [MINERAL] $NiTe_2$ A reddish-white mineral composed of nickel telluride. { 'mel·ə,nīt }

member [GEOL] A rock stratigraphic unit of subordinate rank comprising a specially developed part of a varied formation. { 'mem·bər }

mendip [GEOL] **1.** A buried hill that is exposed as an inlier. **2.** A coastal-plain hill that was originally an offshore island. { 'men,dip }

mendipite [MINERAL] $Pb_3Cl_2O_2$ A white orthorhombic mineral consisting of an oxide and chloride of lead. { 'men·də,pīt }

mendozite [MINERAL] $NaAl(SO_4)_2 \cdot 11H_2O$ A monoclinic mineral of the alum group composed of hydrous sodium aluminum sulfate. { 'men·də,zīt }

meneghinite [MINERAL] $CuPb_{13}Sb_7S_{24}$ A blackish lead gray mineral consisting of lead antimony sulfide. { 'men·ə'gē,nīt }

Meniscotheriidae [PALEON] A family of extinct mammals of the order Condylarthra possessing selenodont teeth and molarized premolars. { mə,nis·kō·thə'rī·ə,dē }

Meramecian [GEOL] A North American provincial series of geologic time: Upper Mississippian (above Osagian, below Chesterian). { ,mer·ə'mē·shən }

meraspis [PALEON] Advanced larva of a trilobite; stage in which the pygidium begins to form. { mə'rap·səs }

Mercalli scale [GEOPHYS] A 12-point scale for classifying the magnitude of an earthquake. { mer'käl·ē ,skāl }

mercallite [MINERAL] $KHSO_4$ A colorless or sky blue, orthorhombic mineral consisting of potassium acid sulfate; occurs as stalactites composed of minute crystals. { mər'kal,īt }

mercurial horn ore *See* calomel. { mər'kyür·ē·əl 'hȯrn ,ȯr }

meridional [GEOL] Pertaining to longitudinal movements or directions, that is, northerly or southerly. { mə'rid·ē·ən·əl }

merismite [PETR] A type of chorismite in which penetration of the diverse units is irregular. { mə'riz,mīt }

merocrystalline *See* hypocrystalline. { ¦mer·ə'krist·əl·ən }

merrihueite [MINERAL] $(K,Na)_2(Fe,Mg)_5Si_{12}O_{30}$ A silicate mineral found only in meteorites. { ,mer·ə'hwā,īt }

merrillite [MINERAL] $Ca_3(PO_4)_2$ Colorless phosphate mineral found only in meteorites. { 'mer·ə,līt }

Mersey yellow coal *See* tasmanite. { 'mər·zē 'yel·ō 'kōl }

merwinite [MINERAL] $Ca_3MgSi_2O_8$ A rare colorless or pale-green neosilicate mineral crystallizing in the monoclinic system; occurs in granular aggregates showing polysynthetic twinning; hardness is 6 on Mohs scale, and specific gravity is 3.15 { 'mər·wə,nīt }

Merycoidodontidae [PALEON] A family of extinct tylopod ruminants in the superfamily Merycoidodontoidea. { ,mer·ə,kȯid·ə'dän·tə,dē }

Merycoidodontoidea [PALEON] A superfamily of extinct ruminant mammals in the infraorder Tylopoda which were exceptionally successful in North America. { 'mer·ə,kȯid·ə,dän'tȯid·ē·ə }

merzlota *See* frozen ground. { ,merz'lō·tə }

Mesacanthidae [PALEON] An extinct family of primitive acanthodian fishes in the order Acanthodiformes distinguished by a pair of small intermediate spines, large scales, superficially placed fin spines, and a short branchial region. { ,mes·ə'kan·thə,dē }

mesh texture *See* reticulate. { 'mesh ,teks·chər }

mesocratic [PETR] Of igneous rock, being intermediate in color between leucocratic and melanocratic due to equal amounts of light and dark constituents. { ,mez·ə'krad·ik }

mesocrystalline [PETR] Of a crystalline rock, containing crystals whose diameters are intermediate between microcrystalline and macrocrystalline rock. { ,mez·ə'krist·əl·ən }

mesogeosyncline [GEOL] A geosyncline between two continents. Also known as mediterranean. { ¦me·zō,jē·ō'sin,klīn }

Mesohippus [PALEON] An early ancestor of the modern horse; occurred during the Oligocene. { ¦me·zō'hip·əs }

mesolite [MINERAL] $Na_2Ca_2Al_6Si_9O_{30} \cdot 8H_2O$ Zeolite mineral composed of hydrous sodium calcium aluminosilicate, usually found in white or colorless tufts of acicular crystals; used as cation exchangers or molecular sieves. { 'mez·ə,līt }

Mesonychidae [PALEON] A family of extinct mammals of the order Condylarthra. { ,me,zän'kid·ə,dē }

200

mesopore [PALEON] A tube paralleling the autopore or chamber in fossil bryozoans. { 'mez·ə,pȯr }

Mesosauria [PALEON] An order of extinct aquatic reptiles which is known from a single genus, *Mesosaurus*, characterized by a long snout, numerous slender teeth, small forelimbs, and webbed hindfeet. { ,me·zō'sȯr·ē·ə }

mesosiderite [GEOL] A stony-iron meteorite containing about equal amounts of silicates and nickel-iron, with considerable troilite. Also known as grahamite. { 'me·zō'sīd·ə,rīt }

mesosphere *See* lower mantle. { 'mez·ə,sfir }

mesostasis [GEOL] The last-formed interstitial material, either glassy or aphanitic, of an igneous rock. { 'me·zō'stā·səs }

Mesosuchia [PALEON] A suborder of extinct crocodiles of the Late Jurassic and Early Cretaceous. { ,me·zō'sü·kē·ə }

mesothermal [MINERAL] Of a hydrothermal mineral deposit, formed at great depth at temperatures of 200–300°C. { 'mez·ə'thər·məl }

mesotil [GEOL] A semiplastic or semifriable derivative of chemically weathered till; forms beneath a partially drained area. { 'mez·ə,til }

Mesozoic [GEOL] A geologic era from the end of the Paleozoic to the beginning of the Cenozoic; commonly referred to as the Age of Reptiles. { 'mez·ə'zō·ik }

mesozone [PETR] The intermediate depth zone of metamorphism in metamorphic rock characterized by moderate temperatures (300–500°C), hydrostatic pressure, and shearing stress. { 'mez·ə,zōn }

metaanthracite [GEOL] Anthracite coal containing at least 98% fixed carbon. { 'med·ə'an·thrə,sīt }

metabentonite [GEOL] Altered bentonite, formed by compaction or metamorphism; it swells very little and lacks the usual high colloidal properties of bentonite. { 'med·ə'bent·ən,īt }

metacinnabar [MINERAL] HgS A black isometric mineral that represents an ore of mercury. Also known as metacinnabarite. { ,med·ə'sin·ə,bär }

metacinnabarite *See* metacinnabar. { ,med·ə'sin·ə·bə,rīt }

Metacopina [PALEON] An extinct suborder of ostracods in the order Podocopida. { ,med·ə'käp·ə·nə }

metacryst [PETR] A large crystal, such as garnet, formed in metamorphic rock by recrystallization. Also known as metacrystal. { 'med·ə,krist }

metacrystal *See* metacryst. { 'med·ə'krist·əl }

metahalloysite [GEOL] A term used in Europe for the less hydrous form of halloysite. Also known as halloysite in the United States. { 'med·ə·hə'lȯi,sīt }

metaharmosis *See* metharmosis. { 'med·ə·här'mō·səs }

metahewettite [MINERAL] CaV$_6$O$_{16}$·9H$_2$O A deep red, probably orthorhombic mineral consisting of hydrated calcium vanadate; occurs as pulverulent masses. { 'med·ə'hyü·ə,tīt }

metahohmannite [MINERAL] Fe$_2$(SO$_4$)$_2$(OH)$_2$·3H$_2$O An orange mineral consisting of a hydrated basic iron sulfate; occurs as pulverulent masses. { 'med·ə'hō·mə,nīt }

metaigneous [PETR] Pertaining to metamorphic rock formed from igneous rock. { 'med·ə'ig·nē·əs }

metalimnion *See* thermocline. { ,med·ə'lim·nē,än }

metalliferous [MINERAL] Pertaining to mineral deposits from which metals can be extracted. { ,med·əl'if·ə·rəs }

metallogenic province [GEOL] A region characterized by a particular mineral assemblage, or by one or more specific types of mineralization. Also known as metallographic province. { mə'tal·ə'jen·ik 'präv·əns }

metallographic province *See* metallogenic province. { mə'tal·ə,graf·ik 'präv·əns }

metamict [MINERAL] Of a radioactive mineral, exhibiting lattice disruption due to radiation damage while the original external morphology is retained. { 'med·ə,mikt }

metamorphic aureole *See* aureole. { 'med·ə'mȯr·fik 'ȯr·ē,ōl }

metamorphic breccia [PETR] Breccia formed by metamorphism. { 'med·ə'mȯr·fik 'brech·ə }

metamorphic differentiation

metamorphic differentiation [PETR] Processes by which different mineral assemblages develop in some sequence from an initially uniform parent rock. { 'med·ə¦mȯr·fik ¸dif·ə¸ren·chē'ā·shen }

metamorphic facies [PETR] All rocks of any composition that have reached chemical equilibrium with respect to certain ranges of pressure and temperature during metamorphism, characterized by the stability of specific index minerals. Also known as densofacies. { ¦med·ə¦mȯr·fik 'fā·shēz }

metamorphic facies series [PETR] A group of metamorphic facies characteristic of an individual area, represented in a pressure-temperature diagram by a curve or group of curves illustrating the range of the different types of metamorphism and metamorphic facies. { ¦med·ə¦mȯr·fik 'fā·shēz ¸sir·ēz }

metamorphic overprint *See* overprint. { ¦med·ə¦mȯr·fik 'ō·vər¸print }

metamorphic rock [PETR] A rock formed from preexisting solid rocks by mineralogical, structural, and chemical changes, in response to extreme changes in temperature, pressure, and shearing stress. { ¦med·ə¦mȯr·fik 'räk }

metamorphic rock reservoir [GEOL] Uncommon type of formation for oil reservoir; developed when secondary porosity results from fracturing or weathering. { ¦med·ə¦mȯr·fik ¦räk 'rez·əv¸wär }

metamorphic zone *See* aureole. { ¦med·ə¦mȯr·fik 'zōn }

metamorphism [PETR] The mineralogical and structural changes of solid rock in response to environmental conditions at depth in the earth's crust. { ¦med·ə¦mȯr ¸fiz·əm }

metaquartzite [MINERAL] A quartzite formed by metamorphic recrystallization. { ¦med·ə'kwȯrt¸zīt }

metaripple [GEOL] An asymmetrical sand ripple. { 'med·ə¸rip·əl }

metarossite [MINERAL] CaV$_2$O$_6$·2H$_2$O A light yellow mineral consisting of hydrated calcium vanadate; occurs as masses and veinlets. { ¦med·ə'rȯ¸sīt }

metasediment [GEOL] A sediment or sedimentary rock which shows evidence of metamorphism. [PETR] Metamorphic rock formed from sedimentary rock. { ¦med·ə'sed· ə·mənt }

metasideronatrite [MINERAL] Na$_4$Fe$_2$(SO$_4$)$_4$(OH)$_2$·3H$_2$O A yellow mineral composed of basic hydrous iron sodium sulfate. { ¦med·ə¸sid·ə·rə'nā¸trīt }

metasilicate [MINERAL] A salt of the hypothetical metasilicic acid H$_2$SiO$_3$. Also known as bisilicate. { ¦med·ə'sil·ə¸kāt }

metasomatic [PETR] Pertaining to the process or the result of metasomatism. { ¦med· ə·sō'mad·ik }

metasomatism [PETR] A variety of metamorphism in which one mineral or a mineral assemblage is replaced by another of different composition without melting. { ¸med· ə'sō·mə¸ tiz·əm }

metatorbernite [MINERAL] Cu(UO$_2$)$_2$(PO$_4$)$_2$·8H$_2$O A green secondary mineral composed of hydrous copper uranium phosphate; similar to torbernite, but with less water content. { ¦med·ə'tȯr·bər¸nīt }

metavariscite [MINERAL] AlPO$_6$·2H$_2$O A green monoclinic mineral composed of hydrous aluminum phosphate; it is isomorphous with phosphosiderite. { ¦med· ə'var·ə¸sīt }

metavauxite [MINERAL] FeAl$_2$(PO$_4$)$_2$(OH)$_2$·8H$_2$O A colorless mineral composed of hydrous basic phosphate of iron and aluminum; similar to vauxite, but with more water. { ¦med·ə'vȯk¸sīt }

metavoltine [MINERAL] A yellowish-brown or orange-brown to greenish-brown, hexagonal mineral consisting of a hydrated basic sulfate of iron and potassium; occurs in tabular form or as aggregates. { ¦med·ə'vōl¸tēn }

metazeunerite [MINERAL] Cu(UO$_2$)$_2$(AsO$_4$)$_2$·8H$_2$O A grass to emerald green, tetragonal mineral consisting of a hydrated arsenate of copper and uranium; occurs in tabular form. { ¦med·ə'zȯi·nə¸rīt }

meteoric stone *See* stony meteorite. { ¸mēd·ē'ȯr·ik 'stōn }

meteorite [GEOL] Any meteoroid that has fallen to the earth's surface. { 'mēd·ē·ə¸rīt }

meteorite crater [GEOL] An impact crater on the surface of the earth or of a celestial body caused by a meteorite; a characteristic feature on the earth is the upturned rim, which formed as the rocks rebounded following the impact. { 'mēd·ē·ə,rīt ,krād·ər }

meteorolite See stony meteorite. { ,med·ē'ór·ə,līt }

metharmosis |GEOL| Changes that occur in a buried sediment after uplift or consolidation but before the onset of weathering. Also spelled metaharmosis. { mə'thär·mə'səs }

Mexican onyx See onyx marble. { 'mek·si·kən 'än·iks }

meyerhofferite |MINERAL| $Ca_2B_6O_{11}·7H_2O$ A colorless, hydrated borate mineral that crystallizes in the triclinic system. { 'mī·ər,häf·ə,rīt }

Miacidae |PALEON| The single, extinct family of the carnivoran superfamily Miacoidea. { mī'as·ə,dē }

Miacoidea |PALEON| A monofamilial superfamily of extinct carnivoran mammals; a stem group thought to represent the progenitors of the earliest member of modern carnivoran families. { ,mī·ə'kòid·ē·ə }

miagite See corsite. { 'mī·ə,jīt }

miargyrite |MINERAL| $AgSbS_2$ An iron-black to steel-gray mineral that crystallizes in the monoclinic system. { mī'är·jə,rīt }

miarolithite |PETR| A chorismite type of igneous rock having miarolitic cavities or vestiges thereof. { ,mē·ə'rō·lə,thīt }

miarolitic |PETR| Of igneous rock, characterized by small irregular cavities into which well-formed crystals of the rock-forming mineral protrude. { ¦mē·ə·rō¦lid·ik }

mica |MINERAL| A group of phyllosilicate minerals (with sheetlike structures) of general formula $(K,Na,Ca)(Mg,Fe,Li,Al)_{2-3}(Al,Si)_4O_{10}(OH,F)_2$ characterized by low hardness $(2-2^1/_2)$ and perfect basal cleavage. { 'mī·kə }

mica book |MINERAL| A crystal of mica, usually large and irregular, whose cleavage plates resemble the leaves of a book. Also known as book. { 'mī·kə ,bùk }

micaceous |GEOL| Pertaining to or resembling mica. { mī'kā·shəs }

micaceous arkose |PETR| A sandstone containing 25–90% feldspars and feldspathic crystalline rock fragments, 10–50% micas and micaceous metamorphic rock fragments, and 0–65% quartz, chert, and metamorphic quartzite. { mī'kā·shəs 'är,kōs }

mica schist |PETR| A schist which is composed essentially of mica and quartz and whose characteristic foliation is mainly due to the parallel orientation of the mica flakes. { 'mī·kə ,shist }

michenerite |MINERAL| A silver-white mineral (PdBiTe) that is a major source of palladium. { 'mich·ə·nə,rīt or 'mich·nə,rīt }

micrinite |PETR| An opaque granular variety of inertinite of medium hardness showing no plant-cell structure. { 'mī·krə,nīt }

micrite |PETR| A semiopaque crystalline limestone matrix that consists of chemically precipitated calcite mud, whose crystals are generally 1–4 micrometers in diameter. { 'mī,krīt }

microbarm |GEOPHYS| That portion of the record of a microbarograph between any two or a specified small number of the successive crossings of the average pressure level in the same direction; analogous to microseism. { 'mī·krə,bärm }

microbreccia |GEOL| A poorly sorted sandstone containing large, angular sand particles in a fine silty or clayey matrix. { ¦mī·krō'brech·ə }

microcline |MINERAL| $KAlSi_3O_8$ A triclinic potassium-rich feldspar, usually containing minor amounts of sodium; may be clear, white, pale-yellow, brick-red, or green, and is generally characterized by crosshatch twinning. { 'mī·krə,klīn }

microcoquina |PETR| A clastic limestone composed wholly or partially of cemented sand-size particles of shell detritus. { ¦mī·krō·kə'kē·nə }

Microdomatacea |PALEON| An extinct superfamily of gastropod mollusks in the order Aspidobranchia. { ,mī·krə,dō·mə'tās·ē·ə }

microearthquake |GEOPHYS| An earthquake with a low intensity, usually less than 3 on the Richter scale. Also known as microquake. { ,mī·krō'ərth,kwāk }

microfacies |PETR| The composition, features, or appearance of a rock or mineral in thin section under the microscope. { ¦mī·krō'fā·shēz }

203

microfossil [PALEON] A small fossil which is studied and identified by means of the microscope. { ¦mī·krō'fäs·əl }

microlite [MINERAL] (Na,Ca)$_2$(Ta,Nb)$_2$O$_6$(O,OH,F) A pale-yellow, reddish, brown, or black isometric mineral composed of sodium calcium tantalum oxide with a small amount of fluorine; it is isomorphous with pyrochlore. Also known as djalmaite. { 'mī·krə,līt }

microlithology [PETR] Microscopic study of the characteristics of rocks. { ¦mī·krō·li'thäl·ə·jē }

microlitic [PETR] Of the texture of a porphyritic igneous rock, having a groundmass composed of an aggregate of microlites in a generally glassy base. { ¦mī·krə¦lid·ik }

micropaleontology [PALEON] A branch of paleontology that deals with the study of microfossils. { ¦mī·krō,pā·lē·ən'täl·ə·jē }

micropegmatite [PETR] Microcrystalline graphic granite. { ¦mī·krō'peg·mə,tīt }

microperthite [MINERAL] Perthite in which the lamellae are visible only under the microscope. { ¦mī·krō'pər,thīt }

microphyric [PETR] Of the texture of an igneous rock, containing microscopic phenocrysts (longest dimension 0.2 millimeter). Also known as microporphyritic. { ¦mī·krō¦fir·ik }

micropoikilitic [PETR] Of the texture of an igneous rock, having poikilitic character visible only under the microscope. { ¦mī·krō,pói·kə'lid·ik }

micropore [GEOL] A pore small enough to hold water against the pull of gravity and to retard water flow. { 'mī·krə,pór }

microporphyritic See microphyric. { ¦mī·krō,pór·fə'rid·ik }

micropulsation [GEOPHYS] A short-period geomagnetic variation in the range of about 0.2–600 seconds, typically exhibiting an oscillatory waveform. { ¦mī·krō·pəl'sā·shən }

microquake See microearthquake.

Microsauria [PALEON] An order of Carboniferous and early Permian lepospondylous amphibians. { ,mī·krō'sór·ē·ə }

microseism [GEOPHYS] A weak, continuous, oscillatory motion in the earth having a period of 1–9 seconds and caused by a variety of agents, especially atmospheric agents; not related to an earthquake. { 'mī·krə,sīz·əm }

microspherulitic [PETR] Of the texture of an igneous rock, having spherulitic character visible only under the microscope. { ¦mī·krō,sfer·ə'lüd·ik }

microstylolite [PETR] A stylolite in which the surface relief is less than 1 millimeter. { ¦mī·krə'stīl·ə,līt }

microtectonics See structural petrology. { ¦mī·krō,tek'tän·iks }

microtektite [GEOL] An extremely small tektite, 1 millimeter or less in diameter. { ,mī·krə'tek,tīt }

Microtragulidae [PALEON] A group of saltatorial caenolistoid marsupials that appeared late in the Cenozoic and paralleled the small kangaroos of Australia. { ,mī·krō·trə'gyül·ə,dē }

microvitrain [GEOL] A coal lithotype; fine vitrain-like lenses or laminae in clarain. { ¦mī·krō'vi,trān }

mid-Atlantic ridge [GEOL] The mid-oceanic ridge in the Atlantic. { ,mid·ət'lan·tik 'rij }

Middle Cambrian [GEOL] The geologic epoch occurring between Upper and Lower Cambrian, beginning approximately 540,000,000 years ago. { 'mid·əl 'kam·brē·ən }

Middle Cretaceous [GEOL] The geologic epoch between the Upper and Lower Cretaceous, beginning approximately 120,000,000 years ago. { 'mid·əl krə'tā·shəs }

Middle Devonian [GEOL] The geologic epoch occurring between the Upper and Lower Devonian, beginning approximately 385,000,000 years ago. { 'mid·əl di'vō·nē·ən }

Middle Jurassic [GEOL] The geologic epoch occurring between the Upper and Lower Jurassic, beginning approximately 170,000,000 years ago. { 'mid·əl jə'ras·ik }

Middle Mississippian [GEOL] The geologic epoch between the Upper and Lower Mississippian. { 'mid·əl ,mis·ə'sip·ē·ən }

Middle Ordovician [GEOL] The geologic epoch occurring between the Upper and Lower

Ordovician, beginning approximately 460,000,000 years ago. { 'mid·əl ‚ȯr·də'vish· ən }

Middle Pennsylvanian [GEOL] The geologic epoch between the Upper and Lower Pennsylvanian. { 'mid·əl ‚pen·səl'vā·nyə }

Middle Permian [GEOL] The geologic epoch occurring between the Upper and Lower Permian, beginning approximately 260,000,000 years ago. { 'mid·əl 'pər·mē·ən }

Middle Silurian [GEOL] The geologic epoch between the Upper and Lower Silurian. { 'mid·əl si'lür·ē·ən }

Middle Triassic [GEOL] The geologic epoch occurring between the Upper and Lower Triassic, beginning approximately 215,000,000 years ago. { 'mid·əl trī'as·ik }

midfan [GEOL] The portion of an alluvial fan between the fanhead and the outer, lower margins. { 'mid‚fan }

mid-ocean canyon See deep-sea channel. { 'mid¦ō·shən 'kan·yən }

mid-oceanic ridge [GEOL] A continuous, median, seismic mountain range on the floor of the ocean, extending through the North and South Atlantic oceans, the Indian Ocean, and the South Pacific Ocean; the topography is rugged, elevation is 0.6–1.8 miles (1–3 kilometers), width is about 900 miles (1500 kilometers), and length is over 52,000 miles (84,000 kilometers). Also known as mid-ocean ridge; mid-ocean rise; oceanic ridge. { 'mid‚ō·shē¦an·ik 'rij }

mid-ocean ridge See mid-oceanic ridge. { 'mid¦ō·shən 'rij }

mid-ocean rift See rift valley. { 'mid¦ō·shən 'rift }

mid-ocean rise See mid-oceanic ridge. { 'mid¦ō·shən 'rīs }

miersite [MINERAL] (Cu,Ag)I A canary yellow, isometric mineral consisting of copper and silver iodide. { 'mir‚zīt }

migma [GEOL] A mixture of solid rock materials and rock melt with mobility or potential mobility. { 'mig·mə }

migmatite [PETR] A mixed rock exhibiting crystalline textures in which a truly metamorphic component is streaked and mixed with obviously once-molten material of a more or less granitic character. { 'mig·mə‚tīt }

migmatization [PETR] Formation of migmatite; involves either injection or in-place melting. { ‚mig·mə·də'zā·shən }

migration [GEOL] **1.** Movement of a topographic feature from one place to another, especially movement of a dune by wind action. **2.** Movement of liquid or gaseous hydrocarbons from their source into reservoir rocks. { mī'grā·shən }

migratory dune See wandering dune. { 'mī·grə‚tȯr·ē 'dün }

Milankovitch cycles [GEOPHYS] Periodic variations in the earth's position relative to the sun as the earth orbits, affecting the distribution of the solar radiation reaching the earth and causing climatic changes that have profound impacts on the abundance and distribution of organisms, best seen in the fossil record of the Quaternary Period (the last 1.6 million years). { mē·lən'kō·vich ‚sīk·əlz }

milarite [MINERAL] $K_2Ca_4Be_4Al_2Si_{24}O_{62} \cdot H_2O$ A colorless to greenish, glassy, hexagonal mineral composed of a hydrous silicate of potassium, calcium, beryllium, and aluminum, occurring in crystals. { 'mē‚lä‚rīt }

milky quartz [MINERAL] An opaque, milk-white variety of crystalline quartz, often with a greasy luster; milkiness is due to the presence of air-filled cavities. Also known as greasy quartz. { 'mil·kē 'kwȯrts }

millerite [MINERAL] NiS A brass to bronze-yellow mineral that crystallizes in the hexagonal system and usually contains trace amounts of cobalt, copper, and iron; hardness is 3–3.5 on Mohs scale, and specific gravity is 5.5; it generally occurs in fine crystals, chiefly as nodules in clay ironstone. Also known as capillary pyrites; hair pyrites; nickel pyrites. { 'mil·ə‚rīt }

millisite [MINERAL] $(Na,K)CaAl_6(PO_4)_4(OH)_9 \cdot 3H_2O$ White mineral composed of a basic hydrous phosphate of sodium, potassium, calcium, and aluminum. { 'mil·ə‚sīt }

millstone See buhrstone. { 'mil‚stōn }

mimetene See mimetite. { 'mim·ə‚tēn }

mimetesite See mimetite. { mə'med·ə‚zīt }

mimetic

mimetic [PETR] Of a tectonite, having a deformation fabric, formed by mimetic crystallization, that reflects and is influenced by preexisting anisotropic structure. { mə'med·ik }

mimetic crystallization [PETR] Recrystallization or neomineralization in metamorphism which reproduces preexistent structures. { mə'med·ik ‚krist·əl·ə'zā·shən }

mimetite [MINERAL] $Pb_5(AsO_4)_3Cl$ A yellow to yellowish-brown mineral of the apatite group, commonly containing calcium or phosphate; a minor ore of lead. Also known as mimetene; mimetesite. { 'mim·ə‚tīt }

minasragrite [MINERAL] $(VO)_2H_2(SO_4)_3 \cdot 15H_2O$ A blue, monoclinic mineral consisting of hydrated acid vanadyl sulfate; occurs in efflorescences and as aggregates or masses. { ‚mē·näs'rä‚grīt }

Mindel glaciation [GEOL] The second glacial stage of the Pleistocene in the Alps. { 'min·dəl ‚glā·sē'ā·shən }

Mindel-Riss interglacial [GEOL] The second interglacial stage of the Pleistocene in the Alps; follows the Mindel glaciation. { 'min·dəl 'ris ‚in·tər'glā·shəl }

mineragraphy See ore microscopy. { ‚min·ə'räg·rə·fē }

mineral [GEOL] A naturally occurring substance with a characteristic chemical composition expressed by a chemical formula; may occur as individual crystals or may be disseminated in some other mineral or rock; most mineralogists include the requirements of inorganic origin and internal crystalline structure. { 'min·rəl }

mineral caoutchouc See elaterite. { 'min·rəl 'kaü‚chùk }

mineral charcoal See fusain. { 'min·rəl 'chär‚kōl }

mineral deposit [GEOL] A mass of naturally occurring mineral material, usually of economic value. { 'min·rəl di‚päz·ət }

mineral facies [PETR] Rocks of any origin whose components have been formed within certain temperature-pressure limits characterized by the stability of certain index minerals. { 'min·rəl 'fā‚shēz }

mineralization [GEOL] **1.** The process of fossilization whereby inorganic materials replace the organic constituents of an organism. **2.** The introduction of minerals into a rock, resulting in a mineral deposit. { ‚min·rə·lə'zā·shən }

mineralize [GEOL] To convert to, or impregnate with, mineral material; applied to processes of ore vein deposition and of fossilization. { 'min·rə‚līz }

mineralizer [GEOL] A gas or fluid dissolved in a magma that aids in the concentration and crystallization of ore minerals. { 'min·rə‚līz·ər }

mineralogenetic epoch [GEOL] A geologic time period during which mineral deposits formed. { ¦min·rə·lō·jə'ned·ik 'ep·ək }

mineralogenetic province [GEOL] Geographic region where conditions were favorable for the concentration of useful minerals. { ¦min·rə·lō·jə'ned·ik 'prä·vəns }

mineralogical phase rule [MINERAL] Any of several variations of the Gibbs phase rule, taking into account the number of degrees of freedom consumed by the fixing of physical-chemical variables in the natural environment; it assumes that temperature and pressure are fixed externally and that consequently the number of phases (minerals) in a system (rock) will not usually exceed the number of components. { ‚min·rə'läj·ə·kəl 'fāz ‚rül }

mineralogist [MINERAL] A person who studies the occurrence, description, mode of formation, and uses of minerals. { ‚min·ə'räl·ə·jəst }

mineralography See ore microscopy. { ‚min·rə'läg·rə·fē }

mineraloid [MINERAL] A naturally occurring, inorganic material that is amorphous and is therefore not considered to be a mineral. Also known as gel mineral. { 'min·rə‚lòid }

mineral resources [GEOL] Valuable mineral deposits of an area that are presently recoverable and may be so in the future; includes known ore bodies and potential ore. { 'min·rəl ri'sórs·əz }

mineral sequence See paragenesis. { 'min·rəl 'sē·kwəns }

mineral soil [GEOL] Soil composed of mineral or rock derivatives with little organic matter. { 'min·rəl ‚sòil }

mineral suite [MINERAL] **1.** A group of associated minerals in one deposit. **2.** A representative group of minerals from a certain locality. **3.** A group of specimens showing variations, as in color or form, in a single mineral species. { 'min·rəl 'swēt }

mineral tallow *See* hatchettite. { 'min·rəl 'tal·ō }

mineral wax *See* ozocerite. { 'min·rəl ¦waks }

minette [PETR] A syenitic variety of lamprophyre composed principally of biotite phenocrysts in a matrix of orthoclase and biotite. { mə'net }

minium [MINERAL] Pb_3O_4 A scarlet or orange-red mineral consisting of an oxide of lead; found in Wisconsin and the western United States. Also known as red lead. { 'min·ē·əm }

minus-cement porosity [GEOL] The porosity that would characterize a sedimentary material if it contained no chemical cement. { 'mī·nəs si¦ment pə'räs·əd·ē }

minyulite [MINERAL] $KAl_2(PO_4)_2(OH,F)\cdot 4H_2O$ A white mineral composed of hydrous basic potassium aluminum phosphate. { 'min·yə,līt }

Miocene [GEOL] A geologic epoch of the Tertiary period, extending from the end of the Oligocene to the beginning of the Pliocene. { 'mī·ə,sēn }

miocrystalline *See* hypocrystalline. { ¦mī·ō'krist·əl·ən }

miogeocline [GEOL] A nonvolcanic (nonmagmatic) continental margin, characterized by carbonate, shale, and sandstone sediments. { ,mī·ō'jē·ə,klīn }

miogeosyncline [GEOL] The nonvolcanic portion of an orthogeosyncline, located adjacent to the craton. { ¦mī·ō¦jē·ō'sin,klīn }

Miosireninae [PALEON] A subfamily of extinct sirenian mammals in the family Dugongidae. { 'mī·ō·sə'ren·ə,nē }

mirabilite [MINERAL] $Na_2SO_4\cdot 10H_2O$ A yellow or white monoclinic mineral consisting of hydrous sodium sulfate, occurring as a deposit from saline lakes, playas, and springs, and as an efflorescence; the pure crystals are known as Glauber's salt. { mə'rab·ə,līt }

mire [GEOL] Wet spongy earth, as of a marsh, swamp, or bog. { mīr }

mirror glance *See* wehrlite. { 'mir·ər ,glans }

mirror stone *See* muscovite. { 'mir·ər ,stōn }

misenite [MINERAL] $K_8H_6(SO_4)_7$ A white mineral composed of native acid potassium sulfate. { mə'ze,nīt }

mispickel *See* arsenopyrite. { 'mi,spik·əl }

Mississippian [GEOL] The fifth period of the Paleozoic Era beginning about 350 million years ago and ending about 320 million years ago. The Mississippian System (referring to rocks) or Period (referring to the time during which these rocks were deposited) is employed in North America as the lower (or older) subdivision of the Carboniferous, as used in Europe and on other continents. { ¦mis·ə¦sip·ē·ən }

Missourian [GEOL] A North American provincial series of geologic time: lower Upper Pennsylvanian (above Desmoinesian, below Virgilian). { mə'zùr·ē·ən }

mitscherlichite [MINERAL] $K_2CuCl_4\cdot 2H_2O$ A greenish-blue, tetragonal mineral consisting of potassium copper chloride dihydrate. { 'mich·ər·lə,kīt }

mixed-layer mineral [MINERAL] A mineral having an interstratified structure consisting of alternating layers of two different clays or of a clay and some other mineral. { 'mikst ¦lā·ər 'min·rəl }

mixed ore [GEOL] Any ore with both oxidized and unoxidized minerals. { 'mikst 'òr }

mixite [MINERAL] $Cu_{11}Bi(AsO_4)_5(OH)_{10}\cdot 6H_2O$ A green to whitish mineral composed of a hydrous basic arsenate of copper and bismuth. { 'mik,sīt }

Mixodectidae [PALEON] A family of extinct insectivores assigned to the Proteutheria; a superficially rodentlike group confined to the Paleocene of North America. { ,mik·sə'dek·tə,dē }

mixtite *See* diamictite. { 'miks,tīt }

mizzonite [MINERAL] A mineral of the scapolite group, composed of 54 to 57% silica. Also known as dipyre. { 'miz·ə,nīt }

moat [GEOL] **1.** A ringlike depression around the base of a seamount. **2.** A valleylike depression around the inner side of a volcanic cone, between the rim and the lava dome. { mōt }

mobile belt [GEOL] A long, relatively narrow crustal region of tectonic acitivity. { 'mō·bəl ¦belt }

mobilization [GEOL] Any process by which solid rock becomes sufficiently soft and plastic to permit it to flow or to permit geochemical migration of the mobile components. { ,mō·bə·lə'zā·shən }

mock lead See sphalerite. { 'mäk 'led }

mock ore See sphalerite. { 'mäk 'ór }

mode [PETR] The mineral composition of a rock, usually expressed as percentages of total weight or volume. { mōd }

moder [GEOL] Humus consisting of plant material that is undergoing alteration from the living to the decayed state and is intermediate in acidity between mor and mull. { 'mōd·ər }

Moeritheriidae [PALEON] The single family of the extinct order Moeritherioidea. { ,mir·ə·thə'rī·ə,dē }

Moeritherioidea [PALEON] A suborder of extinct sirenian mammals considered as primitive proboscideans by some authorities and as a sirenian offshoot by others. { ,mir·ə,thir·ē'óid·ē·ə }

mofette [GEOL] A small opening emitting carbon dioxide in an area of late-stage volcanic activity. { mō'fet }

mohavite See tincalconite. { mō'hä,vīt }

Mohawkian [GEOL] A North American stage of middle Ordovician geologic time, above Chazyan and below Edenian. { mō'hók·ē·ən }

Mohnian [GEOL] A North American stage of geologic time: Miocene (above Luisian, below Delmontian). { 'mō·nē·ən }

Moho See Mohorovičić discontinuity. { 'mō·hō }

Mohole drilling [GEOL] Drilling aimed at penetration of the earth's crust, through the Mohorovičić discontinuity, to sample the mantle. { 'mō,hōl ,dril·iŋ }

Mohorovičić discontinuity [GEOPHYS] A seismic discontinuity that separates the earth's crust from the subjacent mantle, inferred from travel time curves indicating that seismic waves undergo a sudden increase in velocity. Also known as Moho. { ,mō·hō'rō·və,chich dis,känt·ən'ü·əd·ē }

mohsite See ilmenite. { 'mō,sīt }

Mohs scale [MINERAL] An empirical scale consisting of 10 minerals with reference to which the hardness of all other minerals is measured; it includes, from softest (designated 1) to hardest (10): talc, gypsum, calcite, fluorite, apatite, orthoclase, quartz, topaz, corundum, and diamond. { 'mōz ,skāl }

moissanite [MINERAL] SiC A carbide mineral found in meteorites; identical with artificial carborundum. { 'móis·ən,īt }

moisture film cohesion See apparent cohesion. { 'móis·chər ¦film kō,hē·zhən }

molasse [GEOL] A paralic sedimentary facies consisting mainly of shale, subgraywacke sandstone, and conglomerate; it is more clastic and less rhythmic than the preceding flysch and is generally postorogenic. { mə'läs }

mold [GEOL] Soft, crumbling friable earth. [PALEON] An impression made in rock or earth material by an inner or outer surface of a fossil shell or other organic structure; a complete mold would be the hollow space. { mōld }

moldauite See moldavite. { mōl'daú,īt }

moldavite [GEOL] A translucent, olive-to brownish-green or pale-green tektite from western Czechoslovakia, characterized by surface sculpturing due to solution etching. Also known as moldauite; pseudochrysolite; vitavite. [MINERAL] A variety of ozocerite from Moldavia. { mōl'dä,vīt }

molecular fossils See biomarkers. { mə¦lek·yə·lər 'fäs·əlz }

Mollisol [GEOL] An order of soils having dark or very dark, friable, thick A horizons high in humus and bases such as calcium and magnesium; most have lighter-colored or browner B horizons that are less friable and about as thick as the A horizons; all but a few have paler C horizons, many of which are calcareous. { 'mal·ə,säl }

molybdenite |MINERAL| MoS_2 A metallic, lead-gray mineral that crystallizes in the hexagonal system and is commonly found in scales or foliated masses; hardness is 1.5 on Mohs scale, and specific gravity is 4.7; it is chief ore of molybdenum. { mə'lib·də,nīt }

molybdic ocher See molybdite. { mə'lib·dik 'ō·kər }

molybdine See molybdite. { mə'lib,dēn }

molybdite |MINERAL| MoO_3 A mineral, much of which is actually ferrimolybdite. Also known as molybdic ocher; molybdine. { mə'lib,dīt }

molybdophyllite |MINERAL| $(Pb,Mg)_2SiO_4 \cdot H_2O$ A colorless, white, or pale-green mineral composed of a silicate of lead and magnesium. { mə¦lib·dō'fi,līt }

molysite |MINERAL| $FeCl_3$ A brownish-red or yellow mineral composed of native ferric chloride, occurring in lava at Vesuvius. { 'mäl·ə,sīt }

monadnock |GEOL| A remnant hill of resistant rock rising abruptly from the level of a peneplain; commonly represents an outcrop of rock that has withstood erosion. Also known as torso mountain. { mə'nad,näk }

monalbite |MINERAL| A modification of albite with monoclinic symmetry that is stable under equilibrium conditions at temperatures (about 1000°C) near the melting point. { ,mō'nal,bīt }

monazite |MINERAL| A yellow or brown rare-earth phosphate monoclinic mineral with appreciable substitution of thorium for rare-earths and silicon for phosphorus; the principal ore of the rare earths and of thorium. Also known as cryptolite. { 'män·ə,zīt }

monchiquite |PETR| A lamprophyre composed of olivine, pyroxene, and usually mica or amphibole phenocrysts embedded in a glass or analcime groundmass. { 'man·chə,kwīt }

monetite |MINERAL| $CaHPO_4$ A yellowish-white mineral consisting of an acid calcium hydrogen phosphate, occurring in crystals. { 'män·ə,tīt }

monimolite |MINERAL| $(Pb,Ca)_3Sb_2O_8$ Yellowish to brownish or greenish mineral composed of lead calcium antimony oxide; it may contain ferrous iron. { mə'nim·ə,līt }

Monobothrida |PALEON| An extinct order of monocyclic camerate crinoids. { ,män·ō'bäth·rə·də }

monocline |GEOL| A stratigraphic unit that dips from the horizontal in one direction only, not as part of an anticline or syncline. { 'män·ə,klīn }

Monocyathea |PALEON| A class of extinct parazoans in the phylum Archaeocyatha containing single-walled forms. { ,män·ō·sī'ā·thē·ə }

monogeosyncline |GEOL| A primary geosyncline that is long, narrow, and deeply subsided; composed of the sediments of shallow water and situated along the inner margin of the borderlands. { ¦män·ō,jē·ō'sin,klīn }

monomineralic |PETR| Of a rock, composed entirely or principally of a single mineral. { ¦män·ō,min·ə¦ral·ik }

monopyroxene clinoaugite See clinopyroxene. { ¦män·ō·pə'räk,sēn ¦klī·nō¦ȯ,gāt }

montanite |MINERAL| $Bi_2O_3 \cdot TeO_3 \cdot 2H_2O$ A yellowish mineral consisting of a hydrated tellurate of bismuth; occurs in soft and earthy to compact form. { män'ta,nīt }

montebrasite |MINERAL| $LiAlPO_4(OH)$ A mineral composed of basic lithium aluminum phosphate; it is isomorphous with amblygonite and natromontebrasite. { ,män·tə'brä,zīt }

montgomeryite |MINERAL| $Ca_2Al_2(PO_4)_3(OH) \cdot 7H_2O$ A green to colorless mineral composed of hydrous basic calcium aluminum phosphate. { mənt'gəm·rē,īt }

Montian |GEOL| A European stage of geologic time: Paleocene (above Danian, below Thanetian). { 'män·chən }

monticellite |MINERAL| $CaMgSiO_4$ A colorless or gray mineral of the olivine structure type; isomorphous with kirsch steinite. { ,män·tə'se,līt }

montmorillonite |MINERAL| **1.** A group name for all clay minerals with an expanding structure, except vermiculite. **2.** The high-alumina end member of the montmorillonite group; it is grayish, pale red, or blue and has some replacement of aluminum ion by magnesium ion. **3.** Any mineral of the montmorillonite group. { ,mänt·mə'ril·ə,nīt }

209

montroydite

montroydite [MINERAL] HgO Natural mercury oxide mineral from Texas. { 'män ˌtrói,dīt }

monzonite [PETR] A phaneritic (visibly crystalline) plutonic rock composed chiefly of sodic plagioclase and alkali feldspar, with subordinate amounts of dark-colored minerals, intermediate between syenite and dorite. { 'män·zə,nīt }

moonstone [MINERAL] An alkali feldspar or cryptoperthite that is semitransparent to translucent and exhibits a bluish to milky-white, pearly, or opaline luster; used as a gemstone if flawless. Also known as hecatolite. { 'mün,stōn }

moor coal [GEOL] A friable lignite or brown coal. { 'múr ˌkōl }

mooreite [MINERAL] (Mg,Zn,Mn)$_8$(SO$_4$)$_4$(OH)$_{14}$·4H$_2$O A glassy white mineral composed of hydrous basic magnesium zinc manganese sulfate. { 'múr,īt }

mor See ectohumus. { mòr }

morainal apron See outwash plain. { mə'rān·əl 'ā·prən }

morainal delta See ice-contact delta. { mə'rān·əl 'del·tə }

morainal plain See outwash plain. { mə'rān·əl 'plān }

moraine [GEOL] An accumulation of glacial drift deposited chiefly by direct glacial action and possessing initial constructional form independent of the floor beneath it. { mə'rān }

moraine bar [GEOL] A terminal moraine serving as a bar, rising out of deep water at some distance from the shore. { mə'rān ˌbär }

moraine kame [GEOL] One of a group of kames characterized by the same topography, constitution, and position as a terminal moraine. { mə'rān ˌkām }

moraine plateau [GEOL] A relatively flat area within a hummocky moraine, generally at the same elevation as, or a little higher than, the summits of surrounding knobs. { mə'rān plə'tō }

morass ore See bog iron ore. { mə'ras ˌòr }

moravite [MINERAL] Fe$_2$(N,Fe)$_4$Si$_7$O$_{20}$(OH)$_4$ A black mineral of the chlorite group, composed of basic iron aluminum silicate, occurring as fine scales. { mə'rā,vīt }

mordenite [MINERAL] (Ca,Na$_2$,K$_2$)$_4$Al$_8$Si$_{40}$O$_{96}$·28H$_2$O A zeolite mineral crystallizing in the orthorhombic system and found in minute crystals or fibrous concretions. Also known as arduinite; ashtonite; flokite; ptilolite. { 'mórd·ən,īt }

morencite See nontronite. { mə'ren,sīt }

morenosite [MINERAL] NiSO$_4$·7H$_2$O An apple-green or light-green mineral composed of hydrous nickel sulfate, occurring in crystals or fibrous crusts. Also known as nickel vitriol. { mə'ren·ə,sīt }

morganite See vorobyevite. { 'mór·gə,nīt }

morinite [MINERAL] Na$_2$Ca$_3$Al$_3$H(PO$_4$)$_4$F$_6$·8H$_2$O A mineral composed of hydrous acid phosphate of sodium, calcium, and aluminum. Also known as jezekite. { 'mór·ə,nīt }

morphogenetic region [GEOL] A region in which, under certain climatic conditions, the predominant geomorphic processes will contribute regional characteristics to the landscape that contrast with those of other regions formed under different climatic conditions. { ˌmór·fə·jə¦ned·ik 'rē·jən }

morphographic map See physiographic diagram. { ¦mór·fə¦graf·ik 'map }

mortar structure [PETR] A cataclastic structure produced by dynamic metamorphism of crystalline rocks and characterized by a mica-free aggregate of finely crushed grains of quartz and feldspar filling the interstices between or forming borders on the edges of larger, rounded relicts. Also known as cataclastic structure; murbruk structure; porphyroclastic structure. { 'mórd·ər ˌstrək·chər }

morvan [GEOL] The area where two peneplains intersect. Also known as skiou. { 'mór·vən }

mosaic [PETR] **1.** Pertaining to a granoblastic texture in a rock formed by dynamic metamorphism in which the boundaries between individual grains are straight or slightly curved. Also known as cyclopean. **2.** Pertaining to a texture in a crystalline sedimentary rock in which contacts at grain boundaries are more or less regular. { mō'zā·ik }

mosandrite [MINERAL] A reddish-brown or yellowish-brown mineral composed of a

210

silicate of sodium, calcium, titanium, zirconium, and cerium.   Also known as khibinite; lovchorrite; rinkite, rinkolite.   { mō'san,drīt }

mosasaur   [PALEON] Any reptile of the genus *Mosasaurus*; large, aquatic, fish-eating lizards from the Cretaceous which are related to the monitors but had paddle-shaped limbs.   { 'mō·sə,sȯr }

moschellandsbergite   [MINERAL] Ag_2Hg_3 A silver-white mineral consisting of a silver and mercury compound; occurs in dodecahedral crystals and in massive and granular forms.   { ‚mō·shə'lanz·bər‚gīt }

moscovite *See* muscovite.   { 'mäs·kə‚vīt }

mosesite   [MINERAL] $Hg_2N(SO_4,MoO_4)\cdot H_2O$ A mineral composed of a hydrous nitride of mercury and various anions.   { 'mō·zə‚zīt }

moss agate   [MINERAL] A milky or almost transparent chalcedony containing dark inclusions in a dendritic pattern.   { 'mȯs 'ag·ət }

mossite   [MINERAL] $Fe(Nb,Ta)_2O_6$ A mineral composed of an iron tantalum oxide; it is isomorphous with tapiolite.   { 'mȯ‚sīt }

mother lode   [GEOL] A main unit of mineralized matter that may not have economic value but to which workable veins are related.   { 'məth·ər ‚lōd }

mother-of-coal *See* fusain.   { ¦məth·ər əv 'kōl }

mother-of-emerald *See* prase.   { 'məth·ər əv 'em·rəld }

mother rock *See* source rock.   { 'məth·ər ‚räk }

mottled   [GEOL] Of a soil, irregularly marked with spots of different colors.   [PETR] Of a sedimentary rock, marked with spots of various colors.   { 'mäd·əld }

mottramite   [MINERAL] $(Cu,Zn)Pb(VO_4)(OH)$ A mineral composed of a basic lead copper zinc vanadate; it is isomorphous with descloizite.   Also known as cuprodescloizite; psittacinite.   { 'mä·trə‚mīt }

moulin pothole *See* giant's kettle.   { mü'lan 'pät‚hōl }

mound   [GEOL] **1.** A low, isolated, rounded natural hill, usually of earth.   Also known as tuft.   **2.** A structure built by fossil colonial organisms.   { maùnd }

mountain brown ore   [GEOL] Name used in Virginia for limonite or brown iron ore.   { 'maùnt·ən 'braùn 'ȯr }

mountain butter *See* halotrichite.   { 'maùnt·ən 'bəd·ər }

mountain cork   [MINERAL] **1.** A white or gray variety of asbestos composed of thick, interwoven fibers and having a corklike weight and texture.   Also known as rock cork.   **2.** A fibrous clay mineral, such as sepiolite.   { 'maùnt·ən 'kȯrk }

mountain crystal *See* rock crystal.   { 'maùnt·ən 'krist·əl }

mountain mahogany *See* obsidian.   { 'maùnt·ən mə'häg·ə·nē }

mountain pediment   [GEOL] A plain of combined erosion and transportation at the base of and surrounding a desert mountain range; at a distance it has the appearance of a broad triangular mass.   { 'maùnt·ən 'ped·ə·mənt }

mountain soap *See* saponite.   { 'maùnt·ən ‚sōp }

mountain tallow *See* hatchettite.   { 'maùnt·ən ‚tal·ō }

mountain wood   [GEOL] **1.** A compact, fibrous, gray to brown type of asbestos which has an appearance similar to dry wood.   Also known as rock wood.   **2.** A fibrous clay mineral; for example, sepiolite or palygorskite.   { 'maùnt·ən ‚wùd }

muck   [GEOL] Dark, finely divided, well-decomposed, organic matter intermixed with a high percentage of mineral matter, usually silt, forming a surface deposit in some poorly drained areas.   { mək }

mud   [GEOL] An unindurated mixture of clay and silt with water; it is slimy with a consistency varying from that of a semifluid to that of a soft and plastic sediment.   [PETR] The silt plus clay portion of a sedimentary rock.   { məd }

mud ball   [GEOL] A rounded mass of mud or mudstone up to 8 inches (20 centimeters) in diameter in a sedimentary rock.   Also known as chalazoidite; tuff ball.   { 'məd ‚bȯl }

mud cone   [GEOL] A cone of sulfurous mud built around the opening of a mud volcano or mud geyser, with slopes as steep as 40° and diameters ranging upward to several hundred yards.   Also known as puff cone.   { 'məd ‚kōn }

mud crack   [GEOL] An irregular fracture formed by shrinkage of clay, silt, or mud under

211

the drying effects of atmospheric conditions at the surface. Also known as desiccation crack; sun crack. { 'məd ,krak }

mud crack polygon See mud polygon. { 'məd ,krak 'päl·ə,gän }

mud flat [GEOL] A relatively level, sandy or muddy coastal strip along a shore or around an island; may be alternately covered and uncovered by the tide or may be covered by shallow water. Also known as flat. { 'məd ,flat }

mudflow [GEOL] A flowing mass of fine-grained earth material having a high degree of fluidity during movement. { 'məd,flō }

mudlump [GEOL] A diapiric sedimentary structure consisting of clay or silt and forming an island in deltaic areas; produced by the loading action of rapidly deposited delta front sands upon lighter-weight prodelta clays. { 'məd,ləmp }

mud polygon [GEOL] A nonsorted polygon whose center lacks vegetation but whose peripheral fissures contain peat and plants. Also known as mud crack polygon. { 'məd 'päl·ə,gän }

mud pot [GEOL] A type of hot spring which contains boiling mud, typically sulfurous and often multicolored; tends to be associated with geysers and other hot springs in volcanic zones. Also known as painted pot; sulfur-mud pool. { 'məd ,pät }

mudslide [GEOL] A slow-moving mudflow in which movement is mainly by sliding upon a discrete boundary shear surface. { 'məd,slīd }

mudstone [GEOL] An indurated equivalent of mud in the form of a blocky or massive, fine-grained sedimentary rock containing approximately equal proportions of silt and clay; lacks the fine lamination or fissility of shale. { 'məd,stōn }

mud volcano [GEOL] A conical accumulation of variable admixtures of sand and rock fragments, the whole resulting from eruption of wet mud and impelled upward by fluid or gas pressure. Also known as hervidero; macaluba. { 'məd väl'kā·nō }

mugearite [PETR] A dark-colored, fine-grained igneous rock in which the chief feldspar is oligoclase, plus orthoclase and olivine with some apatite and opaque oxides; originates by differentiation and volcanic crystallization of the primary magma. { myü'jē·ə,rīt }

mull [GEOL] Granular forest humus that is incorporated with mineral matter. { məl }

Müller's glass See hyalite. { 'mil·ərz ,glas }

mullion [GEOL] In folded sedimentary and metamorphic rocks, a columnar structure in which the rock columns seem to intersect. { 'məl·yən }

mullite [MINERAL] $Al_6Si_2O_{13}$ An orthorhombic mineral consisting of an aluminum silicate that is resistant to corrosion and heat; used as a refractory. Also known as porcelainite. { 'mə,līt }

multicycle [GEOL] Pertaining to a landscape or landform produced by more than one cycle of erosion. { 'məl·tə,sī·kəl }

multiple fault See step fault. { 'məl·tə·pəl 'fólt }

multiple reflection [GEOPHYS] A seismic wave which has more than one reflection. Also known as repeated reflection; secondary reflection. { 'məl·tə·pəl ri'flek·shən }

Multituberculata [PALEON] The single order of the nominally mammalian suborder Allotheria; multituberculates had enlarged incisors, the coracoid bones were fused to the scapula, and the lower jaw consisted of the dentary bone alone. { ,məl·tē·tə,bər·kyə'läd·ə }

mundic See pyrite. { 'mən,dik }

murbruk structure See mortar structure. { 'mər,brük ,strək·chər }

Murchisoniacea [PALEON] An extinct superfamily of gastropod mollusks in the order Prosobranchia. { ¦mər·chə,sən·ē'ā·shē·ə }

muromontite [MINERAL] $Be_2FeY_2(SiO_4)_3$ A mineral composed of yttrium iron beryllium silicate. { ,myür·ə'män,tīt }

Muschelkalk [GEOL] A European stage of geologic time equivalent to the Middle Triassic, above Bunter and below Keuper. { 'müsh·əl,kälk }

muscovite [MINERAL] $KAl_2(AlSi_3)O_{10}(OH)_2$ One of the mica group of minerals, occurring in some granites and abundant in pegmatites; it is colorless, whitish, or pale brown, and the crystals are tabular sheets with prominent base and hexagonal or rhomboid outline; hardness is 2–2.5 on Mohs scale, and specific gravity is 2.7–3.1. Also known as common mica; mirror stone; moscovite; Muscovy glass; potash mica; white mica. { 'məs·kə,vīt }

Muscovy glass *See* muscovite. { 'məs·kə·vē 'glas }

mustard-seed coal [GEOL] Anthracite that will pass through circular holes in a screen which measure 3/64 inch (1.2 millimeter) in diameter. { 'məs·tərd,sēd ,kōl }

muthmannite [MINERAL] (Ag,Au)Te A bright brass yellow mineral consisting of silver-gold telluride; occurs as tabular crystals. { 'mut·mə,nīt }

mylonite [PETR] A hard, coherent, often glassy-looking rock that has suffered extreme mechanical deformation and granulation but has remained chemically unaltered; appearance is flinty, banded, or streaked, but the nature of the parent rock is easily recognized. { 'mī·lə,nīt }

mylonite gneiss [PETR] A metamorphic rock intermediate in character between mylonite and schist. { 'mī·lə,nīt 'nīs }

mylonitic structure [PETR] A structure characteristic of mylonites, produced by extreme microbrecciation and shearing which gives the appearance of a flow structure. { ¦mī·lə¦nid·ik 'strək·chər }

mylonitization [GEOL] Rock deformation produced by intense microbrecciation without appreciable chemical alteration of granulated materials. { mī,län·ə·tə'zā·shən }

myrmekite [PETR] Intergrowth of plagioclase feldspar and vermicular quartz in an igneous rock. { 'mər·mə,kīt }

myrmekitic [PETR] **1.** Pertaining to the texture of an igneous rock marked by intergrowths of feldspar and vermicular quartz. **2.** Having characteristic properties of myrmekite. { ¦mər·mə¦kid·ik }

N

nacrite [MINERAL] $Al_2Si_2O_5(OH)_4$ A crystallized clay mineral of the kaolinite group; structurally distinct in being the most closely stacked in the c-axis direction. { 'nā,krīt }

nadorite [MINERAL] $PbSbO_2Cl$ A smoky brown or brownish-yellow to yellow, orthorhombic mineral consisting of an oxychloride of lead and antimony. { 'nad·ə,rīt }

nagatelite [MINERAL] Black mineral composed of phosphosilicate of an aluminum, rare-earth elements, calcium, and iron; occurs in tabular masses. { ˌnag·ə'te,līt }

nagyagite [MINERAL] $Pb_5Au(Te,Sb)_4S_{5-8}$ A lead-gray mineral consisting of a sulfide of lead, gold, tellurium, and antimony. Also known as black tellurium; tellurium glance. { 'nag·yə,jīt }

nahcolite [MINERAL] $NaHCO_3$ A white, monoclinic mineral consisting of natural sodium bicarbonate. { 'nä·kə,līt }

naif [MINERAL] Of a gemstone, having a true or natural luster when uncut. Also spelled naife. { nä'ēf }

naife See naif. { nä'ēf }

nailhead striation [GEOL] A glacial striation with a definite or blunt head or point of origin, generally narrowing or tapering in the direction of ice movement and coming to an indefinite end. { 'nāl,hed strī'ā·shən }

naked karst [GEOL] Karst that is developed in a region without soil cover, so that its topographic features are well exposed. { 'nā·kəd 'kärst }

nakhlite [GEOL] An achondritic stony meteorite composed of an aggregate of diopside and olivine. { 'näk,līt }

Namurian [GEOL] A European stage of geologic time; divided into a lower stage (Lower Carboniferous or Upper Mississippian) and an upper stage (Upper Carboniferous or Lower Pennsylvanian). { nə'myúr·ē·ən }

naphthine See hatchettite. { 'naf,thēn }

napoleonite See corsite. { nə'pōl·yə,nīt }

Napoleonville [GEOL] A North American (Gulf Coast) stage of geologic time; a subdivision of the Miocene, above Anahauc and below Duck Lake. { nə'pōl·ē·ən,vil }

nappe [GEOL] A sheetlike, allochthonous rock unit that is formed by thrust faulting or recumbent folding or both. { nap }

nari See caliche. { 'när·ē }

Narizian [GEOL] A North American stage of geologic time; a subdivision of the upper Eocene, above Ulatisian and below Fresnian. { nə'rizh·ən }

narsarsukite [MINERAL] $Na_2(Ti,Fe)Si_4(O,F)$ Mineral composed of sodium titanium iron fluoride and silicate. { ˌnär·sə'sə,kīt }

nasonite [MINERAL] $Ca_4Pb_6Si_6O_{21}Cl_2$ A white mineral composed of silicate and chloride of calcium and lead and occurring in granular masses. { 'nās·ən,īt }

nasturan See pitchblende. { ¦nas·tə¦ran }

native [GEOCHEM] Pertaining to an element found in nature in a nongaseous state. { 'nād·iv }

native asphalt [GEOL] Exudations or seepages of asphalt occurring in nature in a liquid or semiliquid state. Also known as natural asphalt. { 'nād·iv 'as,fólt }

native coal See natural coke. { 'nād·iv 'kōl }

native element [GEOL] Any of 20 elements, such as copper, gold, and silver, which occur naturally uncombined in a nongaseous state; there are three groups—metals, semimetals, and nonmetals. { 'nād·iv 'el·ə·mənt }

native metal [GEOCHEM] A metallic native element; includes silver, gold, copper, iron, mercury, iridium, lead, palladium, and platinum. { 'nād·iv 'med·əl }

native paraffin See ozocerite. { 'nād·iv 'par·ə·fən }

native uranium [GEOCHEM] Uranium as found in nature; a mixture of the fertile uranium-238 isotope (99.3%), the fissionable uranium-235 isotope (0.7%), and a minute percentage of other uranium isotopes. Also known as natural uranium; normal uranium. { 'nād·iv yə'rā·nē·əm }

natric horizon [GEOL] A soil horizon that has the properties of an argillic horizon, but also displays a blocky, columnar, or prismatic structure and has a subhorizon with an exchangeable-sodium saturation of over 15%. { 'nā·trik hə'rīz·ən }

natroalunite [MINERAL] $NaAl_3(SO_4)_2(OH)_6$ Mineral composed of basic sodium aluminum sulfate. Also known as almerite. { ¦nā·trō'al·ə,nīt }

natrochalcite [MINERAL] $NaCu_2(SO_4)(OH)·H_2O$ An emerald-green mineral composed of hydrous basic sulfate of sodium and copper. { ¦nā·trō'kal,sīt }

natrolite [MINERAL] $Na_2Al_2Si_3O_{10}·2H_2O$ A zeolite mineral composed of hydrous silicate of sodium and aluminum; usually occurs in slender acicular or prismatic crystals. { 'nā·trə,līt }

natromontebrasite [MINERAL] $(Na,Li)Al(PO_4)(OH,F)$ A mineral composed of hydrous basic phosphate of sodium, lithium, and aluminum; it is isomorphous with montebrasite and amblygonite. Also known as fremontite. { ¦nā·trō,män·tē'brä,zīt }

natron [MINERAL] $Na_2CO_3·10H_2O$ A white, yellow, or gray mineral that crystallizes in the monoclinic system, is soluble in water, and generally occurs in solution or in saline residues. { 'nā·trən }

natrophilite [MINERAL] $NaMn(PO_4)$ A mineral composed of sodium manganese phosphate. { nə'trä·fə,līt }

natural arch [GEOL] **1.** A landform similar to a natural bridge but not formed by erosive agencies. **2.** See natural bridge. { 'nach·rəl 'ärch }

natural asphalt See native asphalt. { 'nach·rəl 'as,fȯlt }

natural bitumen [GEOL] Native mineral pitch, tar, or asphalt. { 'nach·rəl bə'tü·mən }

natural bridge [GEOL] An archlike rock formation spanning a ravine or valley and formed by erosion. Also known as natural arch. { 'nach·rəl 'brij }

natural coke [GEOL] Coal that has been naturally carbonized by contact with an igneous intrusion, or by natural combustion. Also known as black coal; blind coal; carbonite; cinder coal; coke coal; cokeite; finger coal; native coal. { 'nach·rəl 'kōk }

natural glass [GEOL] An amorphous, vitreous inorganic material that has solidified from magma too quickly to crystallize. { 'nach·rəl 'glas }

natural levee [GEOL] An elongate embankment compounded of sand and silt and deposited along both banks of a river channel during times of flood. { 'nach·rəl 'lev·ē }

natural remanent magnetization [GEOPHYS] The magnetization of rock which exists in the absence of a magnetic field and has been acquired from the influence of the earth's magnetic field at the time of their formation or, in certain cases, at later times. Abbreviated NRM. { 'nach·rəl 'rem·ə·nənt ,mag·nə·tə'zā·shən }

natural tunnel [GEOL] A cave that is nearly horizontal and is open at both ends. Also known as tunnel cave. { 'nach·rəl 'tən·əl }

natural uranium See native uranium. { 'nach·rəl yu'rā·nē·əm }

natural well [GEOL] A sinkhole or other natural opening which resembles a well extending below the water table and from which groundwater can be withdrawn. { 'nach·rəl 'wel }

naujaite [PETR] A coarse hypidiomorphic-granular sodalite-rich nepheline syenite that contains microcline and small amounts of albite, analcime, acmite, and sodium amphiboles and is characterized by a poikilitic texture. { 'naù·jə,īt }

naumannite [MINERAL] Ag_2Se An iron-black mineral that crystallizes in the isometric system; consists of silver selenide, and occurs massive or in crystals; specific gravity is 8. { 'naủ·mə,nı̄t }

Navajo sandstone |GEOL| A fossil dune formation of Jurassic age found in the Colorado Plateau of the United States. { 'nä·və,hō 'san,stŏn }

Navarroan |PALEON| A North American (Gulf Coast) stage of Upper Cretaceous geologic time, above the Tayloran and below the Midwayan of the Tertiary. { ,nav·ə'rō·ən }

navite |MINERAL| A porphyritic basalt containing phenocrysts of altered olivine, augite, and basic plagioclase in a groundmass of labradorite and augite. { 'nä,vı̄t }

Neanderthal man |PALEON| A type of fossil human that is a subspecies of *Homo sapiens* and is distinguished by a low broad braincase, continuous arched browridges, projecting occipital region, short limbs, and large joints. { nē'an·dər,täl 'man }

Nebraskan drift |GEOL| Rock material transported during the Nebraskan glaciation; it is buried below the Kansan drift in Iowa. { nə'bras·kən 'drift }

Nebraskan glaciation |GEOL| The first glacial stage of the Pleistocene epoch in North America, beginning about 1,000,000 years ago, and preceding the Aftonian interglacial stage. { nə'bras·kən glā·sē·ā·shən }

nebulite |PETR| A chorismite in which one of the textural elements occurs in nebulitic lenticular masses. { 'neb·yə,lı̄t }

nebulitic |PETR| **1.** Having indistinct boundaries between textural elements. **2.** Of or pertaining to a nebulite. { ,neb·yə'lid·ik }

neck |GEOL| *See* pipe. { nek }

Necrolestidae |PALEON| An extinct family of insectivorous marsupials. { ¦ne·krō 'les·tə,dē }

Nectridea |PALEON| An order of extinct lepospondylous amphibians characterized by vertebrae in which large fan-shaped hemal arches grow directly downward from the middle of each caudal centrum. { nek'trid·ē·ə }

needle |GEOL| A pointed, elevated, and detached mass of rock formed by erosion, such as an aiguille. |MINERAL| A needle-shaped or acicular mineral crystal. { 'nēd·əl }

needle coal |MINERAL| Lignite containing fibrous needle-shaped masses formed from the vascular bundles of palm stems. { 'nēd·əl ,kōl }

needle ore |MINERAL| **1.** Iron ore of very high metallic luster, found in small quantities, which may be separated into long, slender filaments resembling needles. **2.** *See* aikinite. { 'nēd·əl ,ȯr }

negative area |GEOL| *See* negative element. { 'neg·əd·iv 'er·ē·ə }

negative element |GEOL| A large structural feature or part of the earth's crust, characterized through a long geologic time period by frequent and conspicuous downward movement (subsidence) or by extensive erosion, or by an uplift that is considerably less rapid or less frequent than that of adjacent positive elements. Also known as negative area. { 'neg·əd·iv 'el·ə·mənt }

negative landform |GEOL| **1.** A relatively depressed or low-lying topographic form, such as a valley, basin, or plain. **2.** A volcanic feature formed by a lack of material (such as a caldera). { 'neg·əd·iv 'land,fȯrm }

negative movement |GEOL| **1.** A downward movement of the earth's crust relative to an adjacent part of the crust, such as produced by subsidence. **2.** A relative lowering of the sea level with respect to the land, such as produced by a positive movement of the earth's crust or by a retreat of the sea. { 'neg·əd·iv 'müv·mənt }

negative shoreline *See* shoreline of emergence. { 'neg·əd·iv 'shȯr,lı̄n }

nelsonite |PETR| A group of hypabyssal rocks composed mainly of ilmenite and apatite. { 'nel·sə,nı̄t }

nemalite |MINERAL| A fibrous brucite that contains ferrous oxide. { 'nem·ə,lı̄t }

nematath |GEOL| A submarine ridge across an Atlantic-type ocean basin which is not an orogenic structure, but which is composed of otherwise undeformed continental crust that has been stretched across a sphenochasm or rhombochasm. { 'nem·ə,tath }

nematoblastic |PETR| Pertaining to a metamorphic rock with a homeoblastic texture due to development during recrystallization of slender prismatic crystals. { ¦nem·ə·də¦blas·tik }

Nematophytales [PALEOBOT] A group of fossil plants from the Silurian and Devonian periods that bear some resemblance to the brown seaweeds (Phaeophyta). { ¦nem·əd·ō·fī'tā·lēz }

Neoanthropinae [PALEON] A subfamily of the Hominidae in some systems of classification, set up to include *Homo sapiens* and direct ancestors of *H. sapiens.* { ¦nē·ō·an'thräp·ə,nē }

neoautochthon [GEOL] A stable basement or autochthon formed where a nappe has ceased movement and has become defunct. { ¦nē·ō·ȯ'täk·thən }

Neocathartidae [PALEON] An extinct family of vulturelike diurnal birds of prey (Falconiformes) from the Upper Eocene. { ¦nē·ō·kə'thärd·ə,dē }

Neocomian [GEOL] A European stage of Lower Cretaceous geologic time; includes Berriasian, Valanginian, Hauterivian, and Barremian. { ¦nē·ə¦kō·mē·ən }

neocryst [GEOL] An individual crystal of a secondary mineral in an evaporite. { 'nē·ə,krist }

neoformation *See* neogenesis. { ¦nē·ō·fȯr'mā·shən }

Neogene [GEOL] An interval of geologic time incorporating the Miocene and Pliocene of the Tertiary period; the Upper Tertiary. { 'nē·ə,jēn }

neogenesis [GEOL] The formation of new minerals, as by diagenesis or metamorphism. Also known as neoformation. { ¦nē·ō'jen·ə·səs }

neoglaciation [GEOL] The removal of glacier ice growth in certain mountain areas during the Little Ice Age, following its shrinkage or disappearance during the Altithermal interval. { ¦nē·ō·glā·sē'ā·shən }

neomagma [GEOL] Magma formed by partial or complete refusion of preexisting rocks under the conditions of plutonic metamorphism. { ¦nē·ō'mag·mə }

neomineralization [GEOCHEM] Chemical interchange within a rock whereby its mineral constituents are converted into entirely new mineral species. { ¦nē·ō,min·rə·lə'zā·shən }

neosilicate [MINERAL] A structural type of silicate mineral characterized by linkage of isolated SiO_4 tetrahedra by ionic bonding only; an example is olivine. { ¦nē·ō'sil·ə·kāt }

neosome [GEOL] A geometric element of a composite rock or mineral deposit, appearing to be younger than the main rock mass. { 'nē·ə,sōm }

neostratotype [GEOL] A stratotype established after the holostratotype has been destroyed or is otherwise not usable. { ¦nē·ō'strad·ə,tīp }

neotectonic map [GEOL] A map depicting neotectonic structures. { ¦nē·ō·tek'tän·ik 'map }

neotectonics [GEOL] The study of the most recent structures and structural history of the earth's crust, after the Miocene. { ¦nē·ō·tek'tän·iks }

neovolcanic [PETR] Referring to extrusive rocks that are of Tertiary or younger age. { ¦nē·ō·väl'kan·ik }

nepheline [MINERAL] A mineral of variable composition, with its purest state represented by the formula $NaAlSiO_4$; calcium, magnesium, iron (Fe^{2+} and Fe^{3+}), and titanium are usually present in only minor or trace amounts. { 'nef·ə,lēn }

nepheline basalt *See* olivine nephelinite. { 'nef·ə,lēn bə'sȯlt }

nepheline monzonite [PETR] A nepheline syenite in which sodic plagioclase exceeds the quantity of alkali feldspar. { 'nef·ə,lēn 'män·zə,nīt }

nepheline phonolite [PETR] The fine-grained equivalent of nepheline syenite. { 'nef·ə,lēn fän·ə,līt }

nepheline syenite [PETR] A phaneritic plutonic rock with granular texture, composed largely of alkali feldspar, nepheline, and dark-colored materials. { 'nef·ə,lēn 'sī·ə,nīt }

nephelinite [PETR] A dark-colored, aphanitic rock of volcanic origin, composed essentially of nepheline and pyroxene; texture is usually porphyritic with large crystals of augite and nepheline in a very-fine-grained matrix. { ne'fel·ə,nīt }

nephelite *See* nepheline. { 'nef·ə,līt }

nephrite [MINERAL] An exceptionally tough, compact, fine-grained, greenish or bluish amphibole constituting the less valuable type of jade; formerly worn as a remedy for kidney diseases. Also known as greenstone; kidney stone. { ne'frīt }

neptunian dike [GEOL] A sedimentary dike formed by infilling of sediment, generally sand, in an undersea fissure or hollow. { nep'tü·nē·ən 'dīk }

neptunianism See neptunism. { nep'tü·nē·ə,niz·əm }

neptunian theory See neptunism. { nep'tü·nē·ən 'thē·ə·rē }

neptunic rock [GEOL] **1.** A rock that is formed in the sea. **2.** See sedimentary rock. { nep'tün·ik 'räk }

neptunism [GEOL] The obsolete theory that all rocks of the earth's crust were deposited from or crystallized out of water. Also known as neptunianism; neptunian theory. { 'nep·tə,niz·əm }

neptunite [MINERAL] $(Na,K)_2(Fe,Mn)TiSi_4O_{12}$ Black mineral composed of silicate of sodium, potassium, iron, manganese, and titanium. { 'nep·tə,nīt }

nepuite See garnierite. { 'nep·yə,wīt }

Nesophontidae [PALEON] An extinct family of large, shrewlike lipotyphlans from the Cenozoic found in the West Indies. { ,nes·ə'fän·tə,dē }

nesosilicate [MINERAL] A mineral (such as olivine) composed of independent silicon-oxygen tetrahedra bonded by ionic bonds, without sharing of oxygens. { ¦nes·ō'sil·ə·kət }

nesquehonite [MINERAL] $MgCo_3·3H_2O$ A colorless to white, orthorhombic mineral consisting of hydrated magnesium carbonate. { ,nes·kwə'hō,nīt }

nest [GEOL] A concentration of some relatively conspicuous element of a geologic feature, such as pebbles or inclusions, within a sand layer or igneous rock. { nest }

nested [GEOL] **1.** Pertaining to volcanic cones, craters, or calderas that occur one within another. **2.** Pertaining to two or more calderas that intersect, having been formed at different times or by different explosions. { 'nes·təd }

net [GEOL] **1.** In structural petrology, coordinate network of meridians and parallels, projected from a sphere at intervals of 2°; used to plot points whose spherical coordinates are known and to study the distribution and orientation of planes and points. Also known as projection net; stereographic net. **2.** A form of horizontal patterned ground whose mesh is intermediate between a circle and a polygon. { net }

net slip [GEOL] On a fault, the distance between two formerly adjacent points on either side of the fault; defines direction and relative amount of displacement. Also known as total slip. { 'net 'slip }

Neurodontiformes [PALEON] A suborder of Conodontophoridia having a lamellar internal structure. { ¦nür·ō,dänt·ə'fōr·mēz }

neutral shoreline [GEOL] A shoreline whose essential features are independent of either the submergence of a former land surface or the emergence of a former underwater surface. { 'nü·trəl 'shȯr,līn }

Nevadan orogeny [GEOL] Orogenic episode during Jurassic and Early Cretaceous geologic time in the western part of the North American Cordillera. Also known as Nevadian orogeny; Nevadic orogeny. { nə'vad·ən ȯ'raj·ə·nē }

Nevadian orogeny See Nevadan orogeny. { nə'vad·ē·ən ȯ'raj·ə·nē }

Nevadic orogeny See Nevadan orogeny. { nə'vad·ik ȯ'raj·ə·nē }

newberyite [MINERAL] $MgH(PO_4)·3H_2O$ A white, orthorhombic member of the brushite mineral group; it is isostructural with gypsum. { 'nüb·rē,īt }

new global tectonics [GEOL] Comprehensive theory relating the formation of mountain belts, island arcs, and ocean trenches to the relative movement of regionally extensive lithospheric plates which are delineated by the major seismic belts of the earth. { 'nü 'glō·bəl tek'tän·iks }

New Red Sandstone [GEOL] The red sandstone facies of the Permian and Triassic systems exposed in the British Isles. { 'nü 'red 'san,stōn }

Niagaran [GEOL] A North American provincial geologic series, in the Middle Silurian. { nī'ag·rən }

niccolite [MINERAL] $NiAs$ A pale-copper-red, hexagonal mineral with metallic luster;

an important ore of nickel; hardness is 5–5.5 on Mohs scale.   Also known as arsenical nickel; copper nickel; nickeline.   { 'nik·ə,līt }

niche |GEOL| A shallow cave or reentrant produced by weathering and erosion near the base of a rock face or cliff or beneath a waterfall.   { nich }

nick |GEOL| See knickpoint.   { nik }

nickel-antimony glance See ullmannite.   { 'nik·əl 'ant·ə,mō·nē 'glans }

nickel bloom See annabergite.   { 'nik·əl ,blüm }

nickel glance See gersdorffite.   { 'nik·əl 'glans }

nickeline See niccolite.   { ¦nik·ə¦lēn }

nickel ocher See annabergite.   { 'nik·əl 'ō·kər }

nickel pyrites See millerite.   { 'nik·əl 'pī,rīts }

nickel vitriol See morenosite.   { 'nik·əl 'vit·rē,ól }

nickpoint See knickpoint.   { 'nik,póint }

nieve penitente |GEOL| A jagged pinnacle or spike of snow or firn, up to several meters in height.   Also known as penitent.   { nē'ā·vā ,pen·ə'ten,tā }

niggliite |MINERAL| PtSn or PtTe A silver-white mineral consisting of a platinum telluride compound.   { 'nig·lē,īt }

niklesite |PETR| A pyroxenite containing the three pyroxenes: diopside, enstatite, and diallage.   { 'nik·lə,sīt }

niningerite |MINERAL| (Mg,Fe,Mn)S A mineral found only in meteorites.   { nə'nin·jə,rīt }

niobite See columbite.   { 'nī·ə,bīt }

nip |GEOL| **1.** A small, low cliff or break in slope which is produced by wavelets at the high-water mark.   **2.** The point on the bank of a meander lake where erosion takes place due to crowding of the stream current toward the lake.   **3.** Thinning of a coal seam, particularly if caused by tectonic movements.   Also known as want.   { nip }

nitrate mineral |MINERAL| Any of several generally rare minerals characterized by a fundamental ionic structure of NO_3^-; examples are soda niter, niter, and nitrocalcite.   { 'nī,trāt ,min·rəl }

nitratine See soda niter.   { 'nī·trə,tēn }

nitrogen balance |GEOCHEM| The net loss or gain of nitrogen in a soil.   { 'nī·trə·jən ,bal·əns }

nitromagnesite |MINERAL| $Mg(NO_3)_2 \cdot 6H_2O$ Mineral consisting of magnesium nitrate, occurring as an efflorescence in limestone caverns.   { ¦nī·trō'mag·nə,sīt }

nival gradient |GEOL| The angle between a nival surface and the horizon.   { 'nī·vəl ,grād·ē·ənt }

nival surface |GEOL| The hypothetical planar surface containing all of the different snowlines of the same geologic time period.   { 'nī·vəl ,sər·fəs }

nivation |GEOL| Rock or soil erosion beneath a snowbank or snow patch, due mainly to frost action but also involving chemical weathering, solifluction, and meltwater transport of weathering products.   Also known as snow patch erosion.   { nī'vā·shən }

nivation cirque See nivation hollow.   { nī'vā·shən ,sərk }

nivation hollow |GEOL| A small, shallow depression formed, and occupied during part of the year, by a snow patch or snowbank that, through nivation, is thought to initiate glaciation.   Also known as nivation cirque; snow niche.   { nī'vā·shən,häl·ō }

nivation ridge See winter-talus ridge.   { nī'vā·shən ,rij }

niveal |GEOL| Property of features and effects resulting from the action of snow and ice.   { 'niv·ē·əl }

nivenite |MINERAL| UO_2 A velvet-black member of the uranite group; contains rare-earth metals cerium and yttrium; a source of uranium.   { 'niv·ə,nīt }

niveoglacial |GEOL| Pertaining to the combined action of snow and ice.   { ¦niv·ē·ō'glā·shəl }

niveolian |GEOL| Pertaining to simultaneous accumulation and intermixing of snow and airborne sand at the side of a gentle slope.   { ¦niv·ē¦ō·lē·ən }

nocerite See fluoborite.   { 'nō·sə,rīt }

nocturnal radiation See effective terrestrial radiation.   { näk'tərn·əl ,rād·ē'ā·shən }

node |GEOL| That point along a fault at which the direction of apparent displacement changes. { nōd }

nodular chert |GEOL| Chert occurring as nodular or concretionary segregations (chert nodules). { 'näj·ə·lər 'chərt }

nodule |GEOL| A small, hard mass or lump of a mineral or mineral aggregate characterized by a contrasting composition from and a greater hardness than the surrounding sediment or rock matrix in which it is embedded. { 'näj·ül }

Noeggerathiales |PALEOBOT| A poorly defined group of fossil plants whose geologic range extends from Upper Carboniferous to Triassic. { ‚neg·ə‚rath·ē'ā·lēz }

nominal diameter |GEOL| The diameter computed for a hypothetical sphere which would have the same volume as the calculated volume for a specific sedimentary particle. Also known as equivalent diameter. { 'näm·ə·nəl dī'am·əd·ər }

nonbanded coal |GEOL| Coal without lustrous bands, composed mainly of clarain or durain without nitrain. { 'nän‚ban·dəd 'kōl }

noncaking coal |GEOL| Hard or dull coal that does not cake when heated. Also known as free-burning coal. { 'nän‚kāk·iŋ 'kōl }

Noncalcic Brown soil |GEOL| A great soil group having a slightly acidic, light-pink or reddish-brown A horizon and a light-brown or dull-red B horizon, and developed under a mixture of grass and forest vegetation in a subhumid climate. Also known as Shantung soil. { ¦nän‚kal·sik 'braún ‚sȯil }

noncapillary porosity |GEOL| The property of a volume of large interstices in a rock or soil that do not hold water by capillarity. { ¦nän‚kap·ə‚ler·ē pə'räs·əd·ē }

nonclastic |GEOL| Of the texture of a sediment or sedimentary rock, formed chemically or organically and showing no evidence of a derivation from preexisting rock or mechanical deposition. Also known as nonmechanical. { ¦nän'kla‚stik }

noncohesive See cohesionless. { ‚nän·kō'hē·siv }

nonconformity |GEOL| A type of unconformity in which rocks below the surface of unconformity are either igneous or metamorphic. { ¦nän·kən'fȯr·məd·ē }

noncyclic terrace |GEOL| One of a series of terraces representing previous valley floors formed during periods when continued valley deepening accompanied lateral erosion. { ¦nän'sī·klik 'ter·əs }

nondepositional unconformity See paraconformity. { ¦nän‚dep·ə'zish·ən·əl ‚ən·kən'fȯr·məd·ē }

nonesite |PETR| A porphyritic basalt composed of enstatite, labradorite, and augite phenocrysts in a groundmass of plagiocase and augite. { 'nän·ə‚sīt }

nongraded |GEOL| Pertaining to a soil or an unconsolidated sediment consisting of particles of essentially the same size. { ¦nän'grād·əd }

nonmechanical See nonclastic. { ¦nän·mi'kan·ə·kəl }

nonpenetrative |GEOL| Of a type of deformation, affecting only part of a rock, such as kink bands. { ¦nän'pen·ə‚trād·iv }

nonplunging fold |GEOL| A fold with a horizontal axial surface. Also known as horizontal fold; level fold. { ¦nän¦plən·jiŋ 'fōld }

nonrotational strain See irrotational strain. { ‚nän·rō'tā·shən·əl 'strān }

nonsorted polygon |GEOL| A form of patterned ground which has a dominantly polygonal mesh and an unsorted appearance due to the absence of border stones, and whose borders are generally marked by wedge-shaped fissures narrowing downward. { 'nän‚sȯrd·əd 'päl·i‚gän }

nonsystematic joint |GEOL| A joint that is not part of a set. { ¦nän‚sis·tə'mad·ik 'jȯint }

nontectonite |PETR| Any rock whose fabric shows no influence of movement of adjacent grains; for example, a rock formed by mechanical settling. { ¦nän'tek·tə‚nīt }

nontronite |MINERAL| $Na(Al,Fe,Si)O_{10}(OH)_2$ An iron-rich clay mineral of the montmorillonite group that represents the end member in which the replacement of aluminum by ferric ion is essentially complete. Also known as chloropal; gramenite; morencite; pinguite. { 'nän·trə‚nīt }

nonwetting sand |GEOL| Sand that resists infiltration of water; consists of angular particles of various sizes and occurs as a tightly packed lens. { 'nän‚wed·iŋ 'sand }

221

norbergite [MINERAL] $Mg_3SiO_4(F,OH)_2$ A yellow or pink orthorhombic mineral composed of magnesium silicate with fluoride and hydroxyl; it is a member of the humite group. { 'nȯr,bər,gīt }

nordenskioldine [MINERAL] $CaSn(BO_3)_2$ A colorless or sulfur-, lemon-, or wine-yellow, hexagonal mineral consisting of a borate of calcium and tin; occurs in tabular form and as lenslike crystals. { 'nȯrd·ən,shēl·dən }

nordmarkite [PETR] A quartz-bearing alkalic syenite that has microperthite as its main component with smaller amounts of oligocase, quartz, and biotite and is characterized by granitic or trachytoid texture. { 'nȯrd,mär,kīt }

Norian [GEOL] A European stage of Upper Triassic geologic time that lies above the Carnian and below the Rhaetian. { 'nȯr·ē·ən }

norite [PETR] A coarse-grained plutonic rock composed principally of basic plagioclase with orthopyroxene (hypersthene) as the dominant mafic material. Also known as hypersthenfels. { 'nȯ,rīt }

norm [PETR] The theoretical mineral composition of a rock expressed in terms of standard mineral molecules as determined by means of chemical analyses. { nȯrm }

normal aeration [GEOL] The complete renewal of soil air to a depth of 8 inches (20 centimeters) about once each hour. { 'nȯr·məl e'rā·shən }

normal anticlinorium [GEOL] An anticlinorium in which axial surfaces of the subsidiary folds converge downward. { 'nȯr·məl,ant·i·klə'nȯr·ē·əm }

normal consolidation [GEOL] Consolidation of a sedimentary material in equilibrium with overburden pressure. { 'nȯr·məl kən,säl·ə'dā·shən }

normal cycle [GEOL] A cycle of erosion whereby a region is reduced to base level by running water, especially by the action of rivers. Also known as fluvial cycle of erosion. { 'nȯr·məl 'sī·kəl }

normal dip See regional dip. { 'nȯr·məl 'dip }

normal dispersion [GEOPHYS] The dispersion of seismic waves in which the recorded wave period increases with time. { 'nȯr·məl di'spər·zhən }

normal displacement See dip slip. { 'nȯr·məl di'splās·mənt }

normal erosion [GEOL] Erosion effected by prevailing agencies of the natural environment, including running water, rain, wind, waves, and organic weathering. Also known as geologic erosion. { 'nȯr·məl i'rō·zhən }

normal fault [GEOL] A fault, usually of 45–90°, in which the hanging wall appears to have shifted downward in relation to the footwall. Also known as gravity fault; normal slip fault; slump fault. { 'nȯr·məl 'fȯlt }

normal fold See symmetrical fold. { 'nȯr·məl 'fōld }

normal horizontal separation See offset. { 'nȯr·məl ,här·ə'zänt·əl ,sep·ə'rā·shən }

normal moveout [GEOPHYS] In seismic prospecting, the increase in stepout time that results from an increase in distance from source to detector when there is no dip. { 'nȯr·məl 'müv,au̇t }

normal polarity [GEOPHYS] Natural remanent magnetism nearly identical to the present ambient field. { 'nȯr·məl pə'lar·əd·ē }

normal ripple mark [GEOL] An aqueous current ripple mark consisting of a simple asymmetrical ridge that may have various configurations. { 'nȯr·məl 'rip·əl ,märk }

normal slip fault See normal fault. { 'nȯr·məl 'slip ,fȯlt }

normal soil [GEOL] A soil having a profile that is more or less in equilibrium with the environment. { 'nȯr·məl 'sȯil }

normal synclinorium [GEOL] A synclinorium in which the axial surfaces of the subsidiary folds converge upward. { 'nȯr·məl ,sin·klə'nȯr·ē·əm }

normal uranium See native uranium. { 'nȯr·məl yü'rā·nē·əm }

normative mineral See standard mineral. { 'nȯr·məd·iv 'min·rəl }

north geomagnetic pole See north pole. { 'nȯrth ǀjē·ō,magǀned·ik 'pōl }

north magnetic pole See north pole. { 'nȯrth magǀned·ik 'pōl }

north pole [GEOPHYS] The geomagnetic pole in the Northern Hemisphere, at approximately latitude 78.5°N, longitude 69°W. Also known as north geomagnetic pole; north magnetic pole. { 'nȯrth 'pōl }

northupite [MINERAL] $Na_3MgCl(CO_3)_2$ A white, yellow, gray, or colorless isometric mineral composed of magnesium sodium carbonate; occurs in octahedral crystals. { 'nor·thə,pīt }

nose [GEOL] **1.** A plunging anticline that is short and without closure. **2.** A projecting and generally overhanging buttress of rock. **3.** The projecting end of a hill, spur, ridge, or mountain. **4.** The central forward part of a parabolic dune. { nōz }

nosean See noselite. { 'nō·zē·ən }

noselite [MINERAL] $Na_4Al_3Si_3O_{12}·SO_4$ A gray, blue or brown mineral of the sodalite group; similar to haüynite; hardness is 5.5 on Mohs scale. Also known as nosean. { 'nōz·ə,līt }

notch [GEOL] A deep, narrow cut near the high-water mark at the base of a sea cliff. { näch }

Nothosauria [PALEON] A suborder of chiefly marine Triassic reptiles in the order Sauropterygia. { ,näth·ə'sor·ē·ə }

Notiomastodontinae [PALEON] A subfamily of extinct elephantoid proboscidean mammals in the family Gomphotheriidae. { ,nōd·ē·ō,mas·tə'dän·tə,nē }

Notioprogonia [PALEON] A suborder of extinct mammals comprising a diversified archaic stock of Notoungulata. { ,nōd·ē·ō·prə'gō·nē·ə }

Notoryctidae [PALEON] An extinct family of Australian insectivorous mammals in the order Marsupialia. { ,nōd·ə'rik·tə,dē }

Notoungulata [PALEON] An extinct order of hoofed herbivorous mammals, characterized by a skull with an expanded temporal region, primitive dentition, and primitive feet with five toes, the weight borne mainly by the third digit. { ,nōd·ō,əŋ·gyə'läd·ə }

noumeite See garnierite. { 'nü·mē,īt }

nourishment [GEOL] The replenishment of a beach, either naturally (such as by littoral transport) or artificially (such as by deposition of dredged materials). { 'nər·ish·mənt }

novaculite [GEOL] A siliceous sedimentary rock that is dense, hard, even-textured, light-colored, and characterized by dominance of microcrystalline quartz over chalcedony. Also known as razor stone. { nə'vak·yə,līt }

novaculitic chert [GEOL] A gray chert that fragments into slightly rough, splintery pieces. { nə¦vak·yə¦lid·ik 'chərt }

NRM See natural remanent magnetization.

nubbin [GEOL] **1.** One of the isolated bedrock knobs or small hills forming the last remnants of a mountain crest or mountain range that has succumbed to desert erosion. **2.** A residual boulder, commonly granitic, occurring on a desert dome or broad pediment. { 'nəb·ən }

nuée ardente [GEOL] A turbulent, rapidly flowing, and sometimes incandescent gaseous cloud erupted from a volcano and containing ash and other pyroclastics in its lower part. Also known as glowing cloud; Pelean cloud. { ¦nü¦ā är'dänt }

nugget [GEOL] A small mass of metal found free in nature. { 'nəg·ət }

nunatak [GEOL] An isolated hill, knob, ridge, or peak of bedrock projecting prominently above the surface of a glacier and completely surrounded by glacial ice. { 'nən·ə,tak }

O

oblique fault *See* diagonal fault. { ə'blēk 'fȯlt }

oblique joint *See* diagonal joint. { ə'blēk 'jȯint }

oblique slip fault [GEOL] A fault which has slippage along both the strike and dip of the fault plane. { ə'blēk 'slip ˌfȯlt }

Obolellida [PALEON] A small order of Early and Middle Cambrian inarticulate brachiopods, distinguished by a shell of calcium carbonate. { ˌäb·ə'lel·ə·də }

obsequent [GEOL] Of a stream, valley, or drainage system, being in a direction opposite to that of the original consequent drainage. { 'äb·sə·kwənt }

obsequent fault-line scarp [GEOL] A fault-line scarp which faces in the direction opposite to that of the original fault scarp or in which the structurally upthrown block is topographically lower than the downthrown block. { 'äb·sə·kwənt 'fȯlt ˌlīn ˌskärp }

obsidian [GEOL] A jet-black volcanic glass, usually of rhyolitic composition, formed by rapid cooling of viscous lava; generally forms the upper parts of lava flows. Also known as hyalopsite; Iceland agate; mountain mahogany. { äb'sid·ē·ən }

obsidianite *See* tektite. { äb'sid·ē·ə,nīt }

obstructed stream [GEOL] A stream whose valley has been blocked by a landslide, glacial moraine, sand dune, or lava flow; it frequently consists of a series of ponds or small lakes. { əb'strək·təd 'strēm }

obstruction moraine [GEOL] A moraine formed where the movement of ice is obstructed, for example, by a ridge of bedrock. { əb'strək·shən mə'rān }

occult mineral [MINERAL] A mineral component of rock which cannot be seen through a microscope, but whose presence can be detected by chemical analyses. { ə'kəlt 'min·rəl }

ocean basin [GEOL] The great depression occupied by the ocean on the surface of the lithosphere. { 'ō·shən 'bā·sən }

ocean floor [GEOL] The near-horizontal surface of the ocean basin. { 'ō·shən 'flȯr }

ocean-floor spreading *See* sea-floor spreading. { 'ō·shən ˌflȯr ˌspred·iŋ }

oceanic basalt [PETR] Rocks of the oceanic island volcanoes. { ˌō·shē'an·ik bə'sȯlt }

oceanic crust [GEOL] A thick mass of igneous rock which lies under the ocean floor. { ˌō·shē'an·ik 'krəst }

oceanic heat flow [GEOPHYS] The amount of thermal energy escaping from the earth through the ocean floor per unit area and unit time. { ˌō·shē'an·ik 'hēt ˌflō }

oceanic island [GEOL] Any island which rises from the deep-sea floor rather than from shallow continental shelves. { ˌō·shē'an·ik 'ī·lənd }

oceanic ridge *See* mid-oceanic ridge. { ˌō·shē'an·ik 'rij }

oceanic rise [GEOL] A long, broad elevation of the bottom of the ocean. { ˌō·shē'an·ik 'rīz }

oceanite [PETR] A picritic basalt in which olivine is a great deal more abundant than plagioclase. { 'ō·shə,nīt }

oceanization [GEOL] Process by which continental crust (sial) is converted into oceanic crust (sima). { ˌō·shə·nə'zā·shən }

oceanology *See* oceanography. { ˌō·shə'näl·ə·jē }

ocellar [PETR] Of the texture of an igneous rock, having crystalline aggregates of phenocrysts arranged radially or tangentially around larger euhedral crystals or which form rounded branching forms. { ō'sel·ər }

ocellus

ocellus [PETR] A phenocryst in an ocellar rock. { ō'sel·əs }

ocher [MINERAL] A yellow, brown, or red earthy iron oxide, or any similar earthy, pulverulent metallic oxides used as pigments. { 'ō·kər }

Ochoan [GEOL] A North American provincial series that is uppermost in the Permian, lying above the Guadalupian and below the lower Triassic. { ō'chō·ən }

Ochrept [GEOL] A suborder of the soil order Inceptisol, with horizon below the surface, lacking clay, sesquioxides, or humus; widely distributed, occurring from the margins of the tundra region through the temperate zone, but not into the tropics. { 'ō·krept }

octahedral borax See tincalconite. { ¦äk·tə¦hē·drəl 'bȯr‚aks }

octahedral coordination [MINERAL] An atomic structure where six cations surround every anion, and vice versa. { ¦äk·tə¦hē·drəl kō'ȯrd·ən‚ā·shən }

octahedral copper ore See cuprite. { ¦äk·tə¦hē·drəl 'käp·ər ‚ȯr }

octahedral iron ore See magnetite. { ¦äk·tə¦hē·drəl 'ī·ərn ‚ȯr }

octahedrite [GEOL] The most common iron meteorite, containing 6–18% nickel in the metal phase and having intimate intergrowths lying parallel to the octahedral planes. [MINERAL] See anatase. { ‚äk·tə'hē‚drīt }

octaphyllite [MINERAL] **1.** A group of mica minerals that contain eight cations per ten oxygen and two hydroxyl ions. **2.** Any mineral of this group, such as biotite. { ¦äk·tə'fi‚līt }

odinite [PETR] A grayish-green lamprophyre composed of labradorite and augite or diallage; sometimes containing hornblende, phenocrysts in a groundmass of fine lath-shaped or equigranular feldspar, and a felty mesh of acicular hornblende crystals. { 'ōd·ən‚īt }

Odontognathae [PALEON] An extinct superorder of the avian subclass Neornithes, including all large, flightless aquatic forms and other members of the single order Hesperornithiformes. { ō‚dän'täg·nə‚thē }

Oepikellacea [PALEON] A dimorphic superfamily of extinct ostracods in the order Paleocopa, distinguished by convex valves and the absence of any trace of a major sulcus in the external configuration. { ‚ē‚pik·ə'läs·ē·ə }

offlap [GEOL] The successive lateral contraction extent of strata (in an upward sequence) due to their deposition in a shrinking sea or on the margin of a rising landmass. Also known as regressive overlap. { 'ȯf‚lap }

off-reef facies [GEOL] Facies of the inclined strata made up of reef detritus deposited along the seaward margin of a reef. { 'ȯf¦rēf 'fā·shēz }

offset [GEOL] **1.** The movement of an upcurrent part of a shore to a more seaward position than a downcurrent part. **2.** A spur from a mountain range. **3.** A level terrace on the side of a hill. **4.** The horizontal displacement component in a fault, measured parallel to the strike of the fault. Also known as normal horizontal separation. { 'ȯf‚set }

offset deposit [GEOL] A mineral deposit, especially of sulfides, formed partly by magmatic segregation and partly by hydrothermal solution and located near the source rock. { 'ȯf‚set di'päz·ət }

offset ridge [GEOL] A ridge consisting of resistant sedimentary rock that has been made discontinuous as a result of faulting. { 'ȯf‚set 'rij }

offset stream [GEOL] A stream displaced laterally or vertically by faulting. { 'ȯf‚set 'strēm }

offshore [GEOL] The comparatively flat zone of variable width extending from the outer margin of the shoreface to the edge of the continental shelf. { 'ȯf¦shȯr }

offshore bar See longshore bar. { 'ȯf¦shȯr 'bär }

offshore beach See barrier beach. { 'ȯf¦shȯr 'bēch }

offshore slope [GEOL] The frontal slope below the outer edge of an offshore terrace. { 'ȯf¦shȯr 'slōp }

offshore terrace [GEOL] A wave-built terrace in the offshore zone composed of gravel and coarse sand. { 'ȯf¦shȯr 'ter·əs }

ogive [GEOL] One of a periodically repeated series of dark, curved structures occurring down a glacier that resemble a pointed arch. { 'ō‚jīv }

oikocryst [PETR] One of the enclosing crystals in a poikilitic fabric. { 'ȯik·ə‚krist }

oil *See* petroleum. { ȯil }

oil accumulation *See* oil pool. { 'ȯil ə,kyü·myə,lā·shən }

oil column [GEOL] The difference in elevation between the highest and lowest portions of various producing zones of an oil-producing formation. { 'ȯil ‚käl·əm }

oil floor [GEOL] In a sedimentary basin, the depth below which there is no economic oil accumulation. { 'ȯil ‚flȯr }

oil pool [GEOL] An accumulation of petroleum locally confined by subsurface geologic features. Also known as oil accumulation; oil reservoir. { 'ȯil ‚pül }

oil reservoir *See* oil pool. { 'ȯil 'rez·əv,wär }

oil rock [GEOL] A rock stratum containing oil. { 'ȯil ‚räk }

oil sand [GEOL] An unconsolidated, porous sand formation or sandstone containing or impregnated with petroleum or hydrocarbons. { 'ȯil ‚sand }

oil seep [GEOL] The emergence of liquid petroleum at the land surface as a result of slow migration from its buried source through minute pores or fissure networks. Also known as petroleum seep. { 'ȯil ‚sēp }

oil shale [GEOL] A finely layered brown or black shale that contains kerogen and from which liquid or gaseous hydrocarbons can be distilled. Also known as kerogen shale. { 'ȯil ‚shāl }

oil trap [GEOL] An accumulation of petroleum which, by a combination of physical conditions, is prevented from escaping laterally or vertically. Also known as trap. { 'ȯil ‚trap }

Oiluvium *See* Pleistocene. { ȯi'lü·vē·əm }

oil-water contact *See* oil-water surface. { 'ȯil ¦wȯd·ər 'kän,takt }

oil-water interface *See* oil-water surface. { 'ȯil ¦wȯd·ər 'in·tər,fās }

oil-water surface [GEOL] The datum of a two-dimensional oil-water interface. Also known as oil-water contact; oil-water interface. { 'ȯil ¦wȯd·ər 'sər·fəs }

oil zone [GEOL] The formation or horizon from which oil is produced, usually immediately under the gas zone and above the water zone if all three fluids are present and segregated. { 'ȯil ‚zōn }

okaite [PETR] An ultramafic igneous rock composed chiefly of melilite and haüyne, with accessory biotite, perovskite, apatite, calcite, and opaque oxides. { ō'kā,īt }

okenite [MINERAL] $CaSi_2O_4(OH)_2 \cdot H_2O$ A whitish mineral consisting of calcium silicate and occurring in fibrous masses. { 'ō·kə,nīt }

old age [GEOL] The last stage of the erosion cycle in the development of the topography of a region in which erosion has reduced the surface almost to base level and the land forms are marked by simplicity of form and subdued relief. Also known as topographic old age. { 'ōld 'āj }

Oldhaminidina [PALEON] A suborder of extinct articulate brachiopods in the order Strophomenida distinguished by a highly lobate brachial valve seated within an irregular convex pedicle valve. { ‚ōl·də·mə'nī·də·nə }

oldhamite [MINERAL] CaS A pale-brown mineral known only from meteorites; unstable under earth conditions; member of the galena group with face-centered isometric structure. { 'ōl·də,mīt }

old lake [GEOL] **1.** A lake in an advanced stage of filling by sediments. **2.** A eutrophic or dystrophic lake. **3.** A lake whose shoreline exhibits an advanced stage of development. { 'ōld ¦lāk }

oldland [GEOL] **1.** An extensive area (as the Canadian Shield) of ancient crystalline rocks reduced to low relief by long, continuous erosion from which the materials of later sedimentary rocks were derived. **2.** A region of older land, projected above sea level behind a coastal plain, that supplied the material of which the coastal-plain strata were formed. { 'ōld,land }

old mountain [GEOL] A mountain that was formed before the beginning of the Tertiary Period. { 'ōld ¦maunt·ən }

Old Red Sandstone [GEOL] A Devonian formation in Great Britain and northwestern Europe, of nonmarine, predominantly red sedimentary rocks, consisting principally of sandstone, conglomerates, and shales. { 'ōld 'red 'san,stōn }

227

Olenellidae

Olenellidae [PALEON] A family of extinct arthropods in the class Trilobita. { ˌō·lə'nel·ə,dē }

Oligocene [GEOL] The third oldest of the seven geological epochs of the Cenozoic Era, beginning 34 million years ago and ending 24 million years ago. It corresponds to an interval of geological time (and rocks deposited during that time) from the close of the Eocene Epoch to the beginning of the Miocene Epoch. { ə'lig·ə,sēn }

oligoclase [MINERAL] A plagioclase feldspar mineral with a composition ranging from $Ab_{90}An_{10}$ to $Ab_{70}An_{30}$, where Ab = $NaAlSi_3O_8$ and An = $CaAl_2O_8$. { 'äl·ə·gō,klās }

oligoclasite [PETR] A granular plutonic rock composed almost entirely of oligoclase. Also known as oligosite. { ¦äl·ə·gō'kla,sīt }

oligomictic [PETR] Of a clastic sedimentary rock, composed of a single rock type. { ə,lig·ə'mik·tik }

oligopelic [GEOL] Property of a lake bottom deposit which contains very little clay. { ə,lig·ə'pel·ik }

oligophyre [PETR] A light-colored diorite containing oligoclase phenocrysts in a groundmass of the same minerals. { ə'lig·ə,fīr }

Oligopygidae [PALEON] An extinct family of exocyclic Euechinoidia in the order Holectypoida which were small ovoid forms of the Early Tertiary. { ¦äl·ə·gō¦pij·ə,dē }

oligosite See oligoclasite. { ə'lig·ə,sīt }

olistolith [GEOL] An exotic block or other rock mass that has been transported by submarine gravity sliding or slumping and is included in the binder of an olistostrome. { ə'lis·tə,lith }

olistostrome [GEOL] A sedimentary deposit composed of a chaotic mass of heterogeneous material that is intimately mixed; accumulated in the form of a semifluid body by submarine gravity sliding or slumping of unconsolidated sediments. { ə'lis·tə,strōm }

oliveiraite [MINERAL] $Zr_3Ti_2O_{10}·2H_2O$ An isotropic mineral consisting of an oxide of titanium and zirconium. { ˌäl·ə·və'rā,īt }

olivenite [MINERAL] $Cu_2(AsO_4)(OH)$ An olive-green, dull-brown, gray, or yellow mineral crystallizing in the orthorhombic system and consisting of a basic arsenate of copper. Also known as leucochalcite; wood copper. { ō'liv·ə,nīt }

olivine [MINERAL] $(Mg,Fe_2)SiO_4$ A neosilicate group of olive-green magnesium-iron silicate minerals crystallizing in the orthorhombic system and having a vitreous luster; hardness is $6\frac{1}{2}$–7 on Mohs scale; specific gravity is 3.27–3.37. { 'äl·ə,vēn }

olivine basalt [PETR] Any of a group of olivine-bearing basalts. { 'äl·ə,vēn bə'sólt }

olivine-bronzite chondrite [GEOL] A type of chondritic meteorite that contains about equal amounts of olivine and bronzite. { 'äl·ə,vēn 'brän,zīt 'kän,drīt }

olivine diabase [PETR] An igneous rock composed principally of olivine and formed from tholeiitic magmas by differentiation in thick sills. { 'äl·ə,vēn 'dī·ə,bās }

olivine-hypersthene chondrite [GEOL] A type of chondritic meteorite generally containing more olivine than hypersthene; the hypersthene contains 12–20% iron, giving the meteorite a relatively dark color, and the metal grains usually contain 7–12% nickel. { 'äl·ə,vēn 'hī·pər,sthēn 'kän,drīt }

olivine nephelinite [PETR] An extrusive igneous rock differing in composition from nephelinite only by the presence of olivine. Also known as ankaratrite; nepheline basalt. { 'äl·ə,vēn nə'fel·ə,nīt }

olivine-pigeonite chondrite [GEOL] A type of chondritic meteorite in which olivine is the predominant mineral and pigeonite is secondary, and metal inclusions are usually rich in nickel. { 'äl·ə,vēn 'pij·ə,nīt 'kän,drīt }

ollenite [PETR] A type of hornblende schist characterized by abundant epidote, sphene, and rutile. { 'äl·ə,nīt }

omission [GEOL] The elimination or nonexposure of certain stratigraphic beds at the surface of any specified section because of disruption and displacement of the beds by faulting. { ō'mish·ən }

omphacite [MINERAL] A grassy- to pale-green, granular or foliated, high-temperature aluminous clinopyroxene mineral with a vitreous luster that commonly occurs in the rock eclogite; a variety of augite. { 'äm·fə,sīt }

oncolite |GEOL| A small, variously shaped (often spheroidal), concentrically laminated, calcareous sedimentary structure resembling an oolith; formed by accretion of successive, layered masses of gelatinous sheaths of blue-green algae. { 'äŋ·kō,līt }

Onesquethawan |GEOL| A North American stage in the Lower and Middle Devonian, lying above the Deerparkian and below the Cazenovian. { ,än·ə'skweth·ə,wän }

onionskin weathering |GEOL| A type of spheroidal weathering in which successive shells of decayed rock resembling the layers of an onion are produced. { 'ən,yən,skin 'weth·ə·riŋ }

onlap |GEOL| A type of overlap characterized by regular and progressive pinching out of the strata toward the margins of a depositional basin; each unit transgresses and extends beyond the point of reference of the underlying unit. Also known as transgressive overlap. { 'òn,lap }

Onychodontidae |PALEON| A family of Lower Devonian lobefin fishes in the order Osteolepiformes. { ¦än·ə·kō'dänt·ə,dē }

onyx |MINERAL| **1.** Banded chalcedonic quartz, in which the bands are straight and parallel; natural colors are usually red or brown with white, although black is occasionally encountered. **2.** See onyx marble. { 'än·iks }

onyx agate |MINERAL| A banded agate with straight, parallel, alternating bands of white and different tones of gray. { 'än·iks 'ag·ət }

onyx marble |MINERAL| A hard, compact, dense, generally translucent variety of calcite resembling true onyx and usually banded. Also known as alabaster; Algerian onyx; Gibraltar stone; Mexican onyx; onyx; oriental alabaster. { 'än·iks 'mär·bəl }

onyx opal |MINERAL| Common opal with straight, parallel markings. { 'än·iks 'ō·pəl }

oolicast |PETR| A small, nearly spherical feature occurring in an oolith as a result of a selective dissolution that did not destroy the matrix but left an opening that was subsequently filled. { ō'äl·ə,kast }

oolicastic porosity |PETR| The porosity produced in an oolitic rock by removal of the ooids and formation of oolicasts. { ō¦äl·ə¦kas·tik pə'räs·əd·ē }

oolite |PETR| A sedimentary rock, usually a limestone, composed principally of cemented ooliths. Also known as eggstone; roestone. { 'ō·ə,līt }

oolith |PETR| A small (0.25–2.0 millimeters), rounded accretionary body in a sedimentary rock; generally formed of calcium carbonate by inorganic precipitation or by replacement; ooliths generally exhibit concentric or radial internal structure. { 'ō·ə,lith }

oolitic chert |PETR| Chert composed chiefly of ooliths. { ,ō·ə'lid·ik 'chərt }

oolitic limestone |PETR| An even-textured limestone made up almost entirely of calcareous ooliths with essentially no matrix. { ,ō·ə'lid·ik 'līm,stōn }

oomicrite |PETR| A limestone containing at least 25% ooliths and no more than 25% intraclasts in which the carbonate-mud matrix (micrite) is more abundant than the sparry-calcite cement. { ,ō·ə'mī,krīt }

oomicrudite |PETR| An oomicrite containing ooliths that are more than 1 millimeter in diameter. { ,ō·ə'mī·krə,dīt }

oospararenite |PETR| An oosparite containing medium sand or coarse sand-sized ooliths. { ,ō·ə·spə'rar·ə,nīt }

oosparite |PETR| A limestone containing at least 25% ooliths and no more than 25% intraclasts in which the sparry-calcite cement is more abundant that the carbonate-mud matrix. { ,ō·ə'spa,rīt }

oosparrudite |PETR| An oosparite containing ooliths that are more than 1 millimeter in diameter. { ,ō·ə'spar·ə,dīt }

oovoid |PETR| A void in the center of an incompletely replaced oolith. { 'ō·ə,vòid }

ooze |GEOL| **1.** A soft, muddy piece of ground, such as a bog, usually resulting from the flow of a spring or brook. **2.** A marine pelagic sediment composed of at least 30% skeletal remains of pelagic organisms, the rest being clay minerals. **3.** Soft mud or slime, typically covering the bottom of a lake or river. { üz }

opacite |PETR| Masses of opaque, microscopic grains in rocks, particularly in the groundmass of an igneous rock. { 'äp·ə,sīt }

opal |MINERAL| A natural hydrated form of silica; it is amorphous, usually occurs in botryoidal or stalactic masses, has a hardness of 5–6 on Mohs scale, and specific gravity is 1.9–2.2. { 'ō·pəl }

opal agate [PETR] A variety of banded opal that displays different shades of color, is agatelike in structure, and consists of alternating layers of opal and chalcedony. { 'ō·pəl 'ag·ət }

opal-CT [PETR] A poorly ordered crystalline form of silica thought to be the intermediate phase in quartz chert formation. { 'ō·pəl 'sē'tē }

opaline [MINERAL] **1.** Any of several minerals related to or resembling opal. **2.** An earthy form of gypsum. { 'ō·pə,lēn }

opalized wood See silicified wood. { 'ō·pə,līzd 'wúd }

opaque attritus [GEOL] Attritus that does not contain large quantities of transparent humic degradation matter. { ō'pāk ə'trīd·əs }

open fault [GEOL] A fault, or section of a fault, whose two walls have become separated along the fault surface. { 'ō·pən 'fólt }

open fold [GEOL] A fold having only moderately compressed limbs. { 'ō·pən ,fōld }

open rock [GEOL] Any stratum sufficiently open or porous to contain a significant amount of water or to convey water along its bed. { 'ō·pən ¦räk }

open sand [GEOL] A formation of sandstone that has porosity and permeability sufficient to provide good storage for oil. { 'ō·pən ¦sand }

open-space structure [GEOL] A structure in a carbonate sedimentary rock formed by a partial or complete occupation by internal sediments or cement. { 'ō·pən ,spās 'strək·chər }

operational unit [GEOL] An arbitrary stratigraphic unit that is distinguished by objective criteria for some practical purpose. Also known as parastratigraphic unit. { ,äp·ə'rā·shən·əl 'yü·nət }

ophicalcite [PETR] A recrystallized limestone composed of calcite and serpentine and formed by dedolomitization of a siliceous dolomite. { ¦äf·ə'kal,sīt }

Ophiocistioidea [PALEON] A small class of extinct Echinozoa in which the domed aboral surface of the test was roofed by polygonal plates and carried an anal pyramid. { ,äf·ē·ō,sis·tē'óid·ē·ə }

ophiolite [PETR] A distinctive assemblage of mafic plus ultramafic rocks, generally considered to be fragments of oceanic lithosphere that have been tectonically emplaced onto continental margins and island arcs. { 'äf·ē·ə,līt }

ophiolitic eclogite [PETR] Any of the eclogites which are products of early orogenic volcanism and which by later metamorphism transformed into rocks of the high-pressure facies series. { ¦äf·ē·ə¦lid·ik 'ek·lə,jīt }

ophite [PETR] A diabase in which the ophitic structure is retained even though the pyroxene is altered to uralite. { 'ä,fīt }

ophitic [PETR] Of the holocrystalline, hypidiomorphic-granular texture of an igneous rock, exhibiting lath-shaped plagioclase crystals partly or wholly included within pyroxene crystals. { ä'fid·ik }

opoka [PETR] A porous, flinty, and calcareous sedimentary rock, with conchoidal or irregular fracture, consisting of fine-grained opaline silica (up to 90%), and hardened by the presence of silica of organic origin. { ō'päk·ə }

optical calcite [MINERAL] The type of calcite used to make Nicol prisms. { 'äp·tə·kəl 'kal,sīt }

optimum moisture content [GEOL] The water content at which a specified compactive force can compact a soil mass to its maximum dry unit weight. { 'äp·tə·məm 'móis·chər ,kän,tent }

orange sapphire [MINERAL] An orange variety of gem corundum (sapphire). Also known as padparadsha. { 'är·inj 'sa,fīr }

orbicular [PETR] Of the structure of a rock, containing large quantities of orbicules. { ór'bik·yə·lər }

orbicule [GEOL] A nearly spherical body, up to 2 centimeters (0.8 inch) or more in diameter, in which the components are arranged in concentric layers. { 'ór·bə,kyül }

orbite [PETR] An igneous rock containing large phenocrysts of hornblende, or plagioclase and hornblende, in a groundmass with the composition of malachite. { 'ór,bīt }

Ordovician |GEOL| The second period of the Paleozoic era, above the Cambrian and below the Silurian, from approximately 500 million to 440 million years ago. { ˌȯrd·ə'vish·ən }

ore |GEOL| **1.** The naturally occurring material from which economically valuable minerals can be extracted. **2.** Specifically, a natural mineral compound of the elements, of which one element at least is a metal. **3.** More loosely, all metalliferous rock, though it contains the metal in a free state. **4.** Occasionally, a compound of nonmetallic substances, as sulfur ore. { ȯr }

ore bed |GEOL| An economic aggregation of minerals occurring between or in rocks of sedimentary origin. { 'ȯr ˌbed }

orebody |GEOL| Generally, a solid and fairly continuous mass of ore, which may include low-grade ore and waste as well as pay ore, but is individualized by form or character from adjoining country rock. { 'ȯr,bäd·ē }

ore chimney See pipe. { 'ȯr ˌchim·nē }

ore cluster |GEOL| A group of interconnected ore bodies. { 'ȯr ˌkləs·tər }

ore control |GEOL| A geologic feature that has influenced the ore deposition. { 'ȯr ˌkən'trōl }

ore deposit |GEOL| Rocks containing minerals of economic value in such amount that they can be profitably exploited. { 'ȯr di,päz·ət }

ore district |GEOL| A combination of several ore deposits into one common whole or system. { 'ȯr ˌdis,trikt }

ore-lead age |GEOL| An estimate of the age of the earth made by comparing the relative progress of the two radioactive decay schemes ^{235}U-^{207}Pb and ^{238}U-^{206}Pb. { 'ȯr 'led ˌāj }

ore microscopy |MINERAL| The use of a reflecting microscope to study polished sections of ore minerals. Also known as mineragraphy; mineralography. { 'ȯr mī'kräs·kə·pē }

orendite |PETR| A porphyritic extrusive rock containing phlogopite phenocrysts in a nepheline-free reddish-gray groundmass of leucite, sanidine, phlogopite, amphibole, and diopside. { 'ȯr·ən,dīt }

oreodont |PALEON| Any member of the family Merycoidodontidae. { 'ȯr·ē·ō,dänt }

ore of sedimentation See placer. { 'ȯr əv ˌsed·ə·mən'tā·shən }

ore pipe See pipe. { 'ȯr ˌpīp }

ore shoot |GEOL| **1.** A large, generally vertical, pipelike ore body that is economically valuable. Also known as shoot. **2.** A large and usually rich aggregation of mineral in a vein. { 'ȯr ˌshüt }

organic geochemistry |GEOCHEM| A branch of geochemistry which deals with naturally occurring carbonaceous and biologically derived substances which are of geological interest. { ȯr'gan·ik ˌjē·ō'kem·ə·strē }

organic lattice See growth lattice. { ȯr'gan·ik 'lad·əs }

organic mound See bioherm. { ȯr'gan·ik 'maúnd }

organic reef |GEOL| A sedimentary rock structure of significant dimensions erected by, and composed almost exclusively of the remains of, corals, algae, bryozoans, sponges, and other sedentary or colonial organisms. { ȯr'gan·ik 'rēf }

organic rock |PETR| A sedimentary rock composed principally of the remains of plants and animals. { ȯr'gan·ik 'räk }

organic soil |GEOL| Any soil or soil horizon consisting chiefly of, or containing at least 30% of, organic matter; examples are peat soils and muck soils. { ȯr'gan·ik 'sȯil }

organic texture |GEOL| A sedimentary texture resulting from the activity of organisms such as the secretion of skeletal material. { ȯr'gan·ik 'teks·chər }

organic weathering |GEOL| Biological processes and changes that contribute to the breakdown of rocks. Also known as biological weathering. { ȯr'gan·ik 'weth·ə·riŋ }

organogenic |GEOL| Property of a rock or sediment derived from organic substances. { ȯrˌgan·əˌjen·ik }

organolite |GEOL| Any rock consisting mainly of organic material. { ȯr'gan·ə,līt }

oriental alabaster See onyx marble. { ˌȯr·ē'ent·əl 'al·ə,bas·tər }

oriental amethyst [MINERAL] A violet to purple variety of sapphire. { ˌȯr·ē'ent·əl 'am·ə,thist }

oriental jasper *See* bloodstone. { ˌȯr·ē'ent·əl 'jas·pər }

oriental topaz [MINERAL] A yellow variety of corundum, used as a gem. { ˌȯr·ē'ent·əl 'tō,paz }

orientation diagram [GEOL] Any point or contour diagram used in structural petrology. { ˌȯr·ē·ən'tā·shən ˌdī·ə,gram }

oriented [GEOL] Pertaining to a specimen that is so marked as to show its exact, original position in space. { 'ȯr·ē,ent·əd }

original dip *See* primary dip. { ə'rij·ən·əl 'dip }

original interstice [PETR] An interstice that formed contemporaneously with the enclosing rock. Also known as primary interstice. { ə'rij·ən·əl 'in·tər,stīs }

original valley [GEOL] A valley formed by hypogene action or by epigene action other than that of running water. { ə'rij·ən·əl 'val·ē }

Ornithischia [PALEON] An order of extinct terrestrial reptiles, popularly known as dinosaurs; distinguished by a four-pronged pelvis, and a median, toothless predentary bone at the front of the lower jaw. { ˌȯr·nə'this·kē·ə }

Ornithomimus [PALEON] A 13-foot-long (4-meter) omnivorous theropod dinosaur from the Late Cretaceous Period that had large hips, a long tail, and strong hindlimbs, and closely resembled ostriches. { ˌȯr·nə·thō'mīm·əs }

Ornithopoda [PALEON] A suborder of extinct reptiles in the order Ornithischia including all bipedal forms in the order. { ˌȯr·nə'thäp·ə·də }

orocline [GEOL] An orogenic belt with a change in horizontal direction, either a horizontal curvature or a sharp bend. Also known as geoflex. { 'ȯr·ə,klīn }

orocratic [GEOL] Pertaining to a period of time in which there is much diastrophism. { ¦ȯr·ə¦krad·ik }

orogen *See* orogenic belt. { 'ȯr·ə·jən }

orogene *See* orogenic belt. { 'ȯr·ə,jēn }

orogenesis *See* orogeny. { ˌȯr·ə'jen·ə·səs }

orogenic belt [GEOL] A linear region that has undergone folding or other deformation during the orogenic cycle. Also known as fold belt; orogen; orogene. { ¦ȯr·ə¦jen·ik 'belt }

orogenic cycle [GEOL] A time interval during which a mobile belt evolved into an orogenic belt, passing through preorogenic, orogenic, and postorogenic stages. Also known as geotectonic cycle. { ¦ȯr·ə¦jen·ik 'sī·kəl }

orogenic sediment [GEOL] Any sediment that is produced as the result of an orogeny or that is directly attributable to the orogenic region in which it is later found. { ¦ȯr·ə¦jen·ik 'sed·ə·mənt }

orogenic unconformity [GEOL] An angular unconformity produced locally in a region affected by mountain-building movements. { ¦ȯr·ə¦jen·ik ˌən·kən'fȯr·məd·ē }

orogeny [GEOL] The process or processes of mountain formation, especially the intense deformation of rocks by folding and faulting which, in many mountainous regions, has been accompanied by metamorphism, invasion of molten rock, and volcanic eruption; in modern usage, orogeny produces the internal structure of mountains, and epeirogeny produces the mountainous topography. Also known as orogenesis; tectogenesis. { ȯ'räj·ə·nē }

orogeosyncline [GEOL] A geosyncline that later became an area of orogeny. { ¦ȯr·ō¦jē·ō'sin,klīn }

orographic [GEOL] Pertaining to mountains, especially in regard to their location and distribution. { ¦ȯr·ə¦graf·ik }

orotath [GEOL] An orogenic belt that has been stretched substantially in a lengthwise direction. { 'ȯr·ə,tath }

orpiment [MINERAL] As$_2$S$_3$ A lemon-yellow mineral, crystallizing in the monoclinic system, and generally occurring in foliated or columnar masses; luster is resinous and pearly on the cleavage surface, hardness is 1.5–2 on Mohs scale, and specific gravity is 3.49. Also known as yellow arsenic. { 'ȯr·pə·mənt }

Orthacea [PALEON] An extinct group of articulate brachiopods in the suborder Orthidina in which the delthyrium is open. { ȯr'thās·ē·ə }

Orthent [GEOL] A suborder of the soil order Entisol, well drained and of medium or fine texture, usually shallow to bedrock and lacking evidence of horizonation; occurs mostly on strong slopes. { 'ȯr·thənt }

Orthid [GEOL] A suborder of the soil order Aridisol, mostly well drained, gray or brownish-gray with little change from top to bottom of the soil profile; occupies younger, but not the youngest, land surfaces in deserts. { 'ȯr·thəd }

Orthida [PALEON] An order of extinct articulate brachiopods which includes the oldest known representatives of the class. { 'ȯr·thə·də }

Orthidina [PALEON] The principal suborder of the extinct Orthida, including those articulate brachiopods characterized by biconvex, finely ribbed shells with a straight hinge line and well-developed interareas on both valves. { ȯr'thid·ən·ə }

orthite [MINERAL] Allanite in the form of slender prismatic or acicular crystals. { 'ȯr,thīt }

orthobituminous coal [GEOL] Bituminous coal that contains 87–89% carbon, analyzed on a dry, ash-free basis. { ¦ȯr·thō·bə'tü·mən·əs 'kōl }

orthochem [GEOCHEM] A precipitate formed within a depositional basin or within the sediment itself by direct chemical action. { 'ȯr·thə,kem }

orthochronology [GEOL] Geochronology based on a standard succession of biostratigraphically significant faunas or floras, or based on irreversible evolutionary processes. { ,ȯr·thə·krə'näl·ə·jē }

orthoclase [MINERAL] $KAlSi_3O_8$ A colorless, white, cream-yellow, flesh-reddish, or gray potassium feldspar that usually contains some sodium feldspar, either as albite or analbite or in some intermediate state; it is or appears to be monoclinic. Also known as common feldspar; orthose; pegmatolite. { 'ȯr·thə,klās }

orthoconglomerate [GEOL] A conglomerate with an intact gravel framework held together by mineral cement and deposited by ordinary water currents. { ,ȯr·thə·kən'gläm·ə·rət }

orthocumulate [PETR] A cumulate composed chiefly of one or more cumulus minerals plus the crystallization products of the intercumulus liquid. { ,ȯr·thə'kyü·myə·lət }

Orthod [GEOL] A suborder of the soil order Spodosol having accumulations of humus, aluminum, and iron; widespread in Canada and the former Soviet Union. { 'ȯr,thäd }

orthodolomite [PETR] **1.** A primary dolomite, or one formed by sedimentation. **2.** A dolomite rock so well cemented that the particles interlock. { ,ȯr·thə'dō·lə,mīt }

orthoferrosilite [MINERAL] An orthopyroxene consisting of the orthorhombic silicate $FeSiO_3$. { ¦ȯr·thō,fer·ə'sil,īt }

orthogeosyncline [GEOL] A linear geosynclinal belt lying between continental and oceanic cratons, and having internal volcanic belts (eugeosynclinal) and external nonvolcanic belts (miogeosynclinal). Also known as geosynclinal couple; primary geosyncline. { ¦ȯr·thō,jē·ə'sin,klīn }

orthogneiss [GEOL] Gneiss originating from igneous rock. { 'ȯr·thə,nīs }

orthohydrous coal [GEOL] Coal that contains 5–6% hydrogen, analyzed on a dry, ash-free basis. { ¦ȯr·thə¦hī·drəs 'kōl }

ortholignitous coal [GEOL] Coal that contains 75–80% carbon, analyzed on a dry, ash-free basis. { ¦ȯr·thō·lig'nīd·əs 'kōl }

orthomagmatic stage [GEOL] The principal stage in the crystallization of silicates from a typical magma; up to 90% of the magma may crystallize during this stage. Also known as orthotectic stage. { ¦ȯr·thō,mag'mad·ik 'stāj }

orthomimic feldspars [MINERAL] A group of feldspars that by repeated twinning simulate a higher degree of symmetry with rectangular cleavages. { ¦ȯr·thə¦mim·ik 'fel,spärz }

orthophotograph [GEOL] A photographic copy, prepared from a photograph formed by a perspective projection, in which the displacements due to tilt and relief have been removed. { ,ȯr·thə'fōd·ə,graf }

orthophyric [PETR] Of the texture of the matrix of certain igneous rocks, having feldspar

233

crystals with quadratic or short and stumpy rectangular cross sections. { ‚ȯr‧thə'fir‧ik }

Orthopsidae [PALEON] A family of extinct echinoderms in the order Hemicidaroida distinguished by a camarodont lantern. { ȯr'thäp‧sə‚dē }

orthopyroxene [MINERAL] A series of pyroxene minerals crystallizing in the orthorhombic system; members include enstatite, bronzite, hypersthene, ferrohypersthene, eulite, and orthoferrosilite. { ‚ȯr‧thə‧pə'räk‚sēn }

orthoquartzite [PETR] A clastic sedimentary rock composed almost entirely of detrital quartz grains; a quartzite of sedimentary origin. Also known as orthoquartzitic sandstone; sedimentary quartzite. { ¦ȯr‧thə'kwȯrt‚sīt }

orthoquartzitic conglomerate [GEOL] A lithologically homogeneous, light-colored orthoconglomerate composed of quartzose residues that is commonly interbedded with pure quartz sandstone. Also known as quartz-pebble conglomerate. { ¦ȯr‧thə‧kwȯrt¦sid‧ik kən'gläm‧ə‧rət }

orthoquartzitic sandstone See orthoquartzite. { ¦ȯr‧thə‧kwȯrt¦sid‧ik 'san‚stōn }

orthorhombic pyroxene [MINERAL] A member of the mineral series enstatite-orthoferrosilite, crystallizing in the orthorhombic system, space group *Pbca*. { ¦ȯr‧thə¦räm‧bik pə'räk‚sēn }

orthoschist [PETR] A schist derived from igneous rocks. { 'ȯr‧thə‚shist }

orthose See orthoclase. { 'ȯr‚thōs }

orthosite [PETR] A light-colored coarse-grained igneous rock composed almost entirely of orthoclase. { 'ȯr‧thə‚sīt }

orthostratigraphy [GEOL] Standard stratigraphy based on fossils which identify recognized biostratigraphic zones. { ¦ȯr‧thō‧strə'tig‧rə‧fē }

orthotectic stage See orthomagmatic stage. { ¦ȯr‧thə'tek‧tik ‚stāj }

orthotill [GEOL] A till formed by immediate release of material from transported ice, such as by ablation and melting. { 'ȯr‧thə‚til }

Orthox [GEOL] A suborder of the soil order Oxisol that is moderate to low in organic matter, well drained, and moist all or nearly all year; believed to be extensive at low altitudes in the heart of the humid tropics. { 'ȯr‚thäks }

orvietite [PETR] An extrusive rock composed of approximately equal amounts of plagioclase and sanidine; includes leucite, augite, minor biotite, and olivine, and accessory apatite and opaque oxides. { 'ȯr‧vē‧ə‚tīt }

oryctocoenosis [PALEON] The part of a thanatocoenosis that has been preserved as a fossil. { ə‚rik‧tə‧sə'nō‧səs }

Osagean [GEOL] A provincial series of geologic time in North America; Lower Mississippian (above Kinderhookian, below Meramecian). { ō'sā‧jē‧ən }

osar See esker. { 'ō‚sär }

oscillation ripple See oscillation ripple mark. { ‚äs‧ə'lā‧shən ‚rip‧əl }

oscillation ripple mark [GEOL] A symmetric ripple mark having a sharp, narrow, and relatively straight crest between broadly rounded troughs, formed by the motion of water agitated by oscillatory waves on a sandy base at a depth shallower than wave base. Also known as oscillation ripple; oscillatory ripple mark; wave ripple mark. { ‚äs‧ə'lā‧shən 'rip‧əl ‚märk }

oscillatory ripple mark See oscillation ripple mark. { 'äs‧ə‧lə‚tȯr‧ē 'rip‧əl ‚märk }

ossipite [PETR] A coarse-grained variety of troctolite containing labradorite, olivine, magnetite, and a small amount of diallage. { 'äs‧ə‚pīt }

Osteolepidae [PALEON] A family of extinct fishes in the order Osteolepiformes. { ‚äs‧tē‧ō'lep‧ə‚dē }

Osteolepiformes [PALEON] A primitive order of fusiform lobefin fishes, subclass Crossopterygii, generally characterized by rhombic bony scales, two dorsal fins placed well back on the body, and a well-ossified head covered with large dermal plating bones. { ‚äs‧tē‧ō‚lep‧ə'fȯr‧mēz }

osteolith [PALEON] A fossil bone. { 'äs‧tē‧ə‚lith }

Osteostraci [PALEON] An order of extinct jawless vertebrates; they were mostly small, with the head and part of the body encased in a solid armor of bone, and the posterior part of the body and the tail covered with thick scales. { ‚äs‧tē'äs‧trə‚sī }

ostracoderm [PALEON] Any of various extinct jawless vertebrates covered with an external skeleton of bone which together with the Cyclostomata make up the class Agnatha. { 'ä·strə·kō,dərm }

osumilite [MINERAL] $(K,Na)(Mg,Fe^{2+})_2(Al,Fe^{3+})_3(Si,Al)_{12}O_{30}·H_2O$ A mineral that crystallizes in the hexagonal system and is commonly mistaken for cordierite. { ä'sü·mə,līt }

otavite [MINERAL] $CdCO_3$ A mineral that crystallizes in the hexagonal system and is isostructural with calcite. { 'ōd·ə,vīt }

ottrelite [MINERAL] A gray to black variety of chloritoid containing manganese. { 'ä·trə,līt }

ouachitite [PETR] A biotite monchiquite with no olivine and a glassy or analcime groundmass. { 'wä·chə,tīt }

outcrop [GEOL] Exposed stratum or body of ore at the surface of the earth. Also known as cropout. { 'aüt,kräp }

outcrop curvature See settling. { 'aüt,kräp 'kər·və·chər }

outcrop map [GEOL] A type of geologic map that shows the distribution and shape of actual outcrops, leaving those areas without outcrops blank. { 'aüt,kräp ,map }

outer bar [GEOL] A bar formed at the mouth of an ebb channel of an estuary. { 'aüd·ər 'bär }

outer beach [GEOL] The part of a beach that is ordinarily dry and reached only by the waves generated by a violent storm. { 'aüd·ər 'bēch }

outer core [GEOL] The outer or upper zone of the earth's core, extending to a depth of 3160 miles (5100 kilometers), and including the transition zone. { 'aüd·ər 'kȯr }

outer mantle See upper mantle. { 'aüd·ər 'mant·əl }

outface See dip slope. { 'aüt,fās }

outflow cave [GEOL] A cave from which a stream issues or is known to have issued. { 'aüt,flō ,kāv }

outlier [GEOL] A group of rocks separated from the main mass and surrounded by outcrops of older rocks. { 'aüt,lī·ər }

outwash [GEOL] **1.** Sand and gravel transported away from a glacier by streams of meltwater and either deposited as a floodplain along a preexisting valley bottom or broadcast over a preexisting plain in a form similar to an alluvial fan. Also known as glacial outwash; outwash drift; overwash. **2.** Soil material washed down a hillside by rainwater and deposited on more gently sloping land. { 'aüt,wäsh }

outwash apron See outwash plain. { 'aüt,wäsh ,ā·prən }

outwash cone [GEOL] A cone-shaped deposit consisting chiefly of sand and gravel found at the edge of shrinking glaciers and ice sheets. { 'aüt,wäsh ,kōn }

outwash drift See outwash. { 'aüt,wäsh ,drift }

outwash fan [GEOL] A fan-shaped accumulation of outwash deposited by meltwater streams in front of the terminal moraine of a glacier. { 'aüt,wäsh ,fan }

outwash plain [GEOL] A broad, outspread flat or gently sloping alluvial deposit of outwash in front of or beyond the terminal moraine of a glacier. Also known as apron; frontal apron; frontal plain; marginal plain; morainal apron; morainal plain; outwash apron; overwash plain; sandur; wash plain. { 'aüt,wäsh ,plān }

outwash terrace [GEOL] A dissected and incised valley train or benchlike deposit extending along a valley downstream from an outwash plain or terminal moraine. { 'aüt,wäsh ,ter·əs }

outwash train See valley train. { 'aüt,wäsh ,trān }

ouvarovite See uvarovite. { ü'vär·ə,vīt }

oven [GEOL] **1.** A rounded, saclike, chemically weathered pit or hollow in a rock (especially a granitic rock) which has an arched roof and resembles an oven. **2.** See spouting horn. { 'əv·ən }

overbank deposit [GEOL] Fine-grained sediment (silt and clay) deposited from suspension on a floodplain by floodwaters from a stream channel. { ¦ō·vər¦baŋk di,päz·ət }

overburden [GEOL] **1.** Rock material overlying a mineral deposit or coal seam. Also known as baring; top. **2.** Material of any nature, consolidated or unconsolidated, that overlies a deposit of useful materials, ores, or coal, especially those deposits

235

overconsolidation

that are mined from the surface by open cuts.   **3.** Loose soil, sand, or gravel that lies above the bedrock.   { 'ō·vər‚bərd·ən }

overconsolidation   [GEOL] Consolidation of sedimentary material exceeding that which is normal for the existing overburden.   { ¦ō·vər·kən‚säl·ə'dā·shən }

overdeepening   [GEOL] The erosive process by which a glacier deepens and widens an inherited preglacial valley to below the level of the subglacial surface.   { ¦ō·vər'dēp·ə·niŋ }

overflow channel   [GEOL] A channel or notch cut by the overflow waters of a lake, especially the channel draining meltwater from a glacially dammed lake.   { 'ō·vər‚flō ‚chan·əl }

overfold   [GEOL] A fold that is overturned.   { 'ō·vər‚fōld }

overgrowth   [MINERAL] A mineral deposited on and growing in oriented, crystallographic directions on the surface of another mineral.   { 'ō·vər‚grōth }

overhang   [GEOL] The part of a salt plug that projects from the top.   { 'ō·vər‚haŋ }

overite   [MINERAL] Ca₃Al₈(PO₄)₈(OH)₆·15H₂O A mineral composed of hydrous basic calcium aluminum phosphate.   { 'ō·və‚rīt }

overlap   [GEOL] **1.** Movement of an upcurrent part of a shore to a position extending seaward beyond a downcurrent part.   **2.** Extension of strata over or beyond older underlying rocks.   **3.** The horizontal component of separation measured parallel to the strike of a fault.   { 'ō·vər‚lap }

overlap fault   [GEOL] A fault structure in which the displaced strata are doubled back upon themselves.   { 'ō·vər‚lap ‚fȯlt }

overload   [GEOL] The amount of sediment that exceeds the ability of a stream to transport it and is therefore deposited.   { 'ō·vər‚lōd }

overprint   [GEOCHEM] A complete or partial disturbance of an isolated radioactive system by thermal, igneous, or tectonic activities which results in loss or gain of radioactive or radiogenic isotopes and, hence, a change in the radiometric age that will be given the disturbed system.   [GEOL] The development or superposition of metamorphic structures on original structures.   Also known as imprint; metamorphic overprint; superprint.   { 'ō·vər‚print }

oversaturated   See silicic.   { ¦ō·vər'sach·ə‚rād·əd }

oversteepening   [GEOL] The process by which an eroding alpine glacier steepens the sides of an inherited preglacial valley.   { ¦ō·vər'stēp·ə·niŋ }

overstep   [GEOL] **1.** An overlap characterized by the regular truncation of older units of a complete sedimentary sequence by one or more later units of the sequence.   **2.** A stratum deposited on the upturned edges of underlying strata.   { 'ō·vər‚step }

overthrust   [GEOL] **1.** A thrust fault that has a low dip or a net slip that is large.   Also known as low-angle thrust; overthrust fault.   **2.** A thrust fault with the active element being the hanging wall.   { 'ō·vər‚thrəst }

overthrust black   See overthrust nappe.   { 'ō·vər‚thrəst ‚blak }

overthrust fault   See overthrust.   { 'ō·vər‚thrəst ‚fȯlt }

overthrust nappe   [GEOL] The body of rock making up the hanging wall of a large-scale overthrust.   Also known as overthrust block; overthrust sheet; overthrust slice.   { 'ō·vər‚thrəst ‚nap }

overthrust sheet   See overthrust nappe.   { 'ō·vər‚thrəst ‚shēt }

overthrust slice   See overthrust nappe.   { 'ō·vər‚thrəst ‚slīs }

overturned   [GEOL] Of a fold or the side of a fold, tilted beyond the perpendicular.   Also known as inverted; reversed.   { 'ō·vər‚tərnd }

overwash   [GEOL] **1.** A mass of water representing the part of the wave advancing up a beach that runs over the highest part of the berm (or other structure) and that does not flow directly back to the sea or lake.   **2.** See outwash.   { 'ō·vər‚wäsh }

overwash mark   [GEOL] A narrow, tonguelike ridge of sand formed by overwash on the landward side of a berm.   { 'ō·vər‚wäsh ‚märk }

overwash plain   See outwash plain.   { 'ō·vər‚wäsh ‚plān }

oxalite   See humboldtine.   { 'äk·sə‚līt }

oxammite [MINERAL] $(NH_4)_2C_2O_4 \cdot H_2O$ A yellowish-white, orthorhombic mineral consisting of ammonium oxalate monohydrate; occurs as lamellar masses. { 'äk·sə,mīt }

oxbow [GEOL] The abandoned, horseshoe-shaped channel of a former stream meander after the stream formed a neck cutoff. Also known as abandoned channel. { 'äks,bō }

Oxfordian [GEOL] A European stage of geologic time, in the Upper Jurassic (above Callovian, below Kimmeridgean). Also known as Divesian. { äks'fȯr·dē·ən }

oxidate [GEOL] A sediment made up of iron and manganese oxides and hydroxides crystallized from aqueous solution. { 'äk·sə,dāt }

oxide mineral [MINERAL] A naturally occurring material in oxide form such as silicon dioxide, SiO_2, magnetite, Fe_3O_4, or lime, CaO. { 'äk,sīd 'min·rəl }

oxidite See shale ball. { 'äk·sə,dīt }

oxidized zone [GEOL] A region of mineral deposits which has been altered by oxidizing surface waters. { 'äk·sə,dīzd ,zōn }

Oxisol [GEOL] A soil order characterized by residual accumulations of inactive clays, free oxides, kaolin, and quartz; mostly tropical. { 'äk·sə,sȯl }

oxoferrite [GEOL] A variety of naturally occurring iron with some ferrous oxide in solid solution. { ¦äk·sō'fe,rīt }

Oxyaenidae [PALEON] An extinct family of mammals in the order Deltatheridea; members were short-faced carnivores with powerful jaws. { ,äk·sē'en·ə,dē }

oxybiotite [MINERAL] Phenocrystic biotite with increased amounts of Fe(III). { ¦äk·sē'bī·ə,tīt }

oxygen deficit [GEOCHEM] The difference between the actual amount of dissolved oxygen in lake or sea water and the saturation concentration at the temperature of the water mass sampled. { 'äk·sə·jən ,def·ə·sət }

oxygen isotope fractionation [GEOCHEM] The use of temperature-dependent variations of the oxygen-18/oxygen-16 ratio in the carbonate shells of marine organisms, to measure water temperature at the time of deposition. { 'äk·sə·jən 'īs·ə,tōp ,frak·shə'nā·shən }

oxygen ratio See acidity coefficient. { 'äk·sə·jən ,rā·shō }

oxyheeite [MINERAL] $Pb_5Ag_2Sb_6S_{15}$ A light steel gray to silver white mineral consisting of lead and silver antimony sulfide; occurs as acicular needles or in massive form. { ,äk·sē'hē,īt }

oxyhornblende See basaltic hornblende. { ¦äk·sē'hȯrn,blend }

oxyphile See lithophile. { 'äk·sə,fīl }

oxysphere See lithosphere. { 'äk·sə,sfir }

Ozawainellidae [PALEON] A family of extinct protozoans in the superfamily Fusulinacea. { ō¦zä·wə·i'nel·ə,dē }

ozocerite [GEOL] A natural, brown to jet black paraffin wax occurring in irregular veins; consists principally of hydrocarbons, is soluble in water, and has a variable melting point. Also known as ader wax; earth wax; fossil wax; mineral wax; native paraffin; ozokerite. { ō'zäs·ə,rīt }

ozokerite See ozocerite. { ō'zäk·ə,rīt }

P

paar |GEOL| A depression produced by the moving apart of crustal blocks rather than by subsidence within a crustal block. { pär }

pachnolite |MINERAL| $NaCaAlF_6 \cdot H_2O$ Colorless to white mineral composed of hydrous sodium calcium aluminum fluoride, occurring in monoclinic crystals. { 'pak·nə,līt }

pachoidal structure See flaser structure. { pə'kȯid·əl ,strək·chər }

pachycephalosaur |PALEON| A bone-headed dinosaur, composing the family Pachy-cephalosauridae. { ,pak·ə'sef·ə·lə,sȯr }

Pachycephalosauridae |PALEON| A family of ornithischian dinosaurs characterized by a skull with a solid rounded mass of bone 4 inches (10 centimeters) thick above the minute brain cavity. { ,pak·ə,sef·ə·lə'sȯr·ə,dē }

Pacific suite |PETR| A large group of igneous rocks characterized by calcic and calc-alkalic rocks, especially in the region of the circum-Pacific orogenic belt. Also known as anapeirean; circum-Pacific province. { pə'sif·ik 'swēt }

Pacific-type continental margin |GEOL| A continental margin typified by that of the western Pacific where oceanic lithosphere descends beneath an adjacent continent and produces an intervening island arc system. { pə'sif·ik ,tīp ,känt·ən'ent·əl 'mär·jən }

packing |GEOL| The arrangement of solid particles in a sediment or in sedimentary rock. { 'pak·iŋ }

packing density |GEOL| A measure of the extent to which the grains of a sedimentary rock occupy the gross volume of the rock in contrast to the spaces between the grains; equal to the cumulative grain-intercept length along a traverse in a thin section. { 'pak·iŋ ,den·səd·ē }

packing proximity |GEOL| In a sedimentary rock, an estimate of the number of grains that are in contact with adjacent grains; equal to the total percentage of grain-to-grain contacts along a traverse measured on a thin section. { 'pak·iŋ präk,sim·əd·ē }

packsand |PETR| A very fine-grained sandstone that is so loosely consolidated by a slight calcareous cement that it can be readily cut by a spade. { 'pak,sand }

packstone |PETR| A sedimentary carbonate rock whose granular material is arranged in a self-supporting framework, yet also contains some matrix of calcareous mud. { 'pak,stōn }

padparadsha See orange sapphire. { pad'par·əd,shä }

pagoda stone |GEOL| **1.** A Chinese limestone showing in section fossil orthoceratites arranged in pagodalike designs. **2.** An agate whose markings resemble pagodas. { pə'gōd·ə,stōn }

pagodite See agalmatolite. { 'pag·ə,dīt }

paha |GEOL| A low, elongated, rounded glacial ridge or hill which consists mainly of drift, rock, or windblown sand, silt, or clay but is capped with a thick cover of loess. { pä'hä }

pahoehoe |GEOL| A type of lava flow whose surface is glassy, smooth, and undulating; the lava is basaltic, glassy, and porous. Also known as ropy lava. { pə'hō·ē,hō·ē }

paigeite |MINERAL| $(Fe,Mg)FeBO_5$ A black mineral composed of iron magnesium borate, occurring as fibrous aggregates. { 'pā,jīt }

painted pot See mud pot. { ¦pānt·əd 'pät }

paint pot |GEOL| A mud pot containing multicolored mud. { 'pānt ,pät }

paired terrace [GEOL] One of two stream terraces that face each other at the same elevation from opposite sides of the stream valley and represent the remnants of the same floodplain or valley floor. Also known as matched terrace. { ¦perd 'ter·əs }

paisanite *See* ailsyte. { 'pīs·ən,īt }

Palaeacanthaspidoidei [PALEON] A suborder of extinct, placoderm fishes in the order Rhenanida; members were primitive, arthrodire-like species. { ¦pāl·ē·ə,kan·thə·spi'dóid·ē,ī }

Palaeechinoida [PALEON] An extinct order of echinoderms in the subclass Perischoechinoidea with a rigid test in which the ambulacra bevel over the adjoining interambulacra. { ¦pāl·ē·ə·kī'nóid·ē·ə }

Palaeoconcha [PALEON] An extinct order of simple, smooth-hinged bivalve mollusks. { ¦pāl·ē·ō'kaŋ·kə }

Palaeocopida [PALEON] An extinct order of crustaceans in the subclass Ostracoda characterized by a straight hinge and by the anterior location for greatest height of the valve. { ¦pāl·ē·ō'käp·ə·də }

Palaeoisopus [PALEON] A singular, monospecific, extinct arthropod genus related to the pycnogonida, but distinguished by flattened anterior appendages. { ¦pāl·ē·ō'ī·sə·pəs }

Palaeomastodontinae [PALEON] An extinct subfamily of elaphantoid proboscidean mammals in the family Mastodontidae. { ¦pāl·ē·ō,mas·tə'dänt·ən,ē }

Palaeomerycidae [PALEON] An extinct family of pecoran ruminants in the superfamily Cervoidea. { ¦pāl·ē·ō·mə'ris·ə,dē }

Palaeoniciformes [PALEON] A large extinct order of chondrostean fishes including the earliest known and most primitive ray-finned forms. { ¦pāl·ē·ō,nis·ə'fór·mēz }

Palaeoniscoidei [PALEON] A suborder of extinct fusiform fishes in the order Palaeonisciformes with a heavily ossified exoskeleton and thick rhombic scales on the body surface. { ¦pāl·ē·ō·nis'kóid·ē,ī }

Palaeopantopoda [PALEON] A monogeneric order of extinct marine arthropods in the subphylum Pycnogonida. { ¦pāl·ē·ō·pan'täp·ə·də }

Palaeoryctidae [PALEON] A family of extinct insectivorous mammals in the order Deltatheridia. { ¦pāl·ē·ō'rik·tə,dē }

Palaeospondyloidea [PALEON] An ordinal name assigned to the single, tiny fish *Palaeospondylus*, known only from Middle Devonian shales in Carthness, Scotland. { ¦pāl·ē·ō,spän·də'lóid·ē·ə }

Palaeotheriidae [PALEON] An extinct family of perissodactylous mammals in the superfamily Equoidea. { ¦pāl·ē·ō·thə'rī·ə,dē }

palagonite [GEOL] A brown to yellow altered basaltic glass found as interstitial material or amygdules in pillow lavas. { pə'lag·ə,nīt }

palagonite tuff [PETR] A pyroclastic rock composed of angular fragments of palagonite. { pə'lag·ə,nīt ¦təf }

palasite [GEOL] The most abundant of the intermediate types of meteorites, consisting of olivine enclosed in a nickel-iron matrix. { 'pal·ə,sīt }

paleic surface [GEOL] A smooth, preglacial erosion surface. { pə'lē·ik 'sər·fəs }

paleoagrostology [PALEOBOT] The study of fossil grasses. { ¦pāl·ē·ō,ag·rə'stäl·ə·jē }

paleoalgology [PALEOBOT] The study of fossil algae. Also known as paleophycology. { ¦pāl·ē·ō·al'gäl·ə·jē }

paleobiochemistry [PALEON] The study of chemical processes used by organisms that lived in the geologic past. { ¦pāl·ē·ō,bī·ō'kem·ə·strē }

paleobioclimatology [PALEON] The study of climatological events affecting living organisms for millennia or longer. { ¦pāl·ē·ō,bī·ō,klī·mə'täl·ə·jē }

paleobiocoenosis [PALEON] An assemblage of organisms that lived together in the geologic past as an interrelated community. Also known as paleocoenosis. { ¦pāl·ē·ō,bī·ō·sə'nō·səs }

paleobiology [PALEON] The branch of paleontology concerned with the biologic aspects of the history of life. { ,pā·lē·ō·bī'äl·ə·jē }

paleobotanic province [GEOL] A large region defined by similar fossil floras. { ¦pāl·ē·ō·bə'tan·ik 'präv·əns }

paleobotany [PALEON] The branch of paleontology concerned with the study of ancient and fossil plants and vegetation of the geologic past. { ¦pāl·ē·ō'bät·ən·ē }

Paleocene [GEOL] The oldest of the seven geological epochs of the Cenozoic Era, spanning 65 million to 55 million years ago. Comprising the Tertiary and Quaternary periods in modern usage, it is also the oldest of the five epochs constituting the Tertiary Period. It represents an interval of geological time (and rocks deposited during that time) extending from the termination of the Cretaceous Period of the Mesozoic Era to the dawn of the Eocene Epoch. { 'pāl·ē·ə,sēn }

paleochannel [GEOL] A remnant of a stream channel cut in older rock and filled by the sediments of younger overlying rock. { ¦pāl·ē·ō¦chan·əl }

Paleocharaceae [PALEOBOT] An extinct group of fossil plants belonging to the Charophyta distinguished by sinistrally spiraled gyrogonites. { ¦pāl·ē·ō·kə'rās·ē,ē }

paleoclimate [GEOL] The climate of a given period of geologic time. Also known as geologic climate. { ¦pāl·ē·ō¦klī·mət }

paleoclimatic sequence [GEOL] The sequence of climatic changes in geologic time; it shows a succession of oscillations between warm periods and ice ages, but superimposed on this are numerous shorter oscillations. { ¦pāl·ē·ō·klə'mad·ik 'sē·kwəns }

paleoclimatology [GEOL] The study of climates in the geologic past, involving the interpretation of glacial deposits, fossils, and paleogeographic, isotopic, and sedimentologic data. { ¦pāl·ē·ō,klī·mə'täl·ə·jē }

paleocoenosis See paleobiocoenosis. { ¦pāl·ē·ō·sə'nō·səs }

Paleocopa [PALEON] An order of extinct ostracodes distinguished by a long, straight hinge. { ¸pāl·ē'äk·ə·pə }

paleocurrent [GEOL] Ancient fluid current flow whose orientation can be inferred by primary sedimentary structures and textures. { ¦pāl·ē·ō'kə·rənt }

paleodepth [PALEON] The water level at which an ancient organism or group of organisms flourished. { 'pāl·ē·ō,depth }

paleoecology [PALEON] The ecology of the geologic past. { ¦pāl·ē·ō·i'käl·ə·jē }

paleoequator [GEOL] The position of the earth's equator in the geologic past as defined for a specific geologic period and based on geologic evidence. { ¦pāl·ē·ō·i'kwäd·ər }

paleofluminology [GEOL] The study of ancient stream systems. { ¦pāl·ē·ō,flü·mə'näl·ə·jē }

Paleogene [GEOL] A geologic time interval comprising the Oligocene, Eocene, and Paleocene of the lower Tertiary period. Also known as Eogene. { 'pāl·ē·ō,jēn }

paleogeographic event See palevent. { ¦pāl·ē·ō,jē·ə'graf·ik i'vent }

paleogeographic stage See palstage. { ¦pāl·ē·ō,jē·ə'graf·ik 'stāj }

paleogeography [GEOL] The geography of the geologic past; concerns all physical aspects of an area that can be determined from the study of the rocks. Paleogeography is used to describe the changing positions of the continents and the ancient extent of land, mountains, shallow sea, and deep ocean basins. { ¦pāl·ē·ō·jē'äg·rə·fē }

paleogeologic map [GEOL] An areal map of the geology of an ancient surface immediately below a buried unconformity, showing the geology as it appeared at some time in the geologic past at the time the surface of unconformity was completed and before the overlapping strata were deposited. { ¦pāl·ē·ō,jē·ə'läj·ik 'map }

paleogeology [GEOL] The geology of the past, applied particularly to the interpretation of the rocks at a surface of unconformity. { ¦pāl·ē·ō·jē'äl·ə·jē }

paleogeomorphology [GEOL] A branch of geomorphology concerned with the recognition of ancient erosion surfaces and the study of ancient topographies and topographic features that are now concealed beneath the surface and have been removed by erosion. Also known as paleophysiography. { ¦pāl·ē·ō,jē·ō·mȯr'fäl·ə·jē }

paleoherpetology [PALEON] The study of fossil reptiles. { ¦pāl·ē·ō,hər·pə'täl·ə·jē }

paleohydrology [GEOL] The study of ancient hydrologic features preserved in rock. { ¦pāl·ē·ō·hī'dräl·ə·jē }

paleoichnology [PALEON] The study of trace fossils in the fossil state. Also spelled palichnology. { ¦pāl·ē·ō·ik'näl·ə·jē }

paleoisotherm [GEOL] The locus of points of equal temperature for some former period of geologic time. { ¦pāl·ē·ō'ī·sə,thərm }

paleokarst [GEOL] A rock or area that has undergone the karst process and subsequently been buried under sediments. { ¦pāl·ē·ō,kärst }

paleolatitude [GEOL] The latitude of a specific area on the earth's surface in the geologic past. { ¦pāl·ē·ō'lad·ə,tüd }

paleolimnology [GEOL] 1. The study of the past conditions and processes of ancient lakes. 2. The study of the sediments and history of existing lakes. { ¦pāl·ē·ō·lim'näl·ə·jē }

paleolithologic map [GEOL] A paleogeologic map indicating lithologic variations at a buried horizon or within a restricted zone at a specific time in the geologic past. { ¦pāl·ē·ō,lith·ə'läj·ik 'map }

paleomagnetics [GEOPHYS] The study of the direction and intensity of the earth's magnetic field throughout geologic time. { ¦pāl·ē·ō·mag'ned·iks }

paleomagnetic stratigraphy [GEOPHYS] The use of natural remanent magnetization in the identification of stratigraphic units. Also known as magnetic stratigraphy. { ¦pāl·ē·ō·mag¦ned·ik strə'tig·rə·fē }

paleomalacology [PALEON] A branch of paleontology concerned with the study of mollusks. { ¦pāl·ē·ō,mal·ə'käl·ə·jē }

paleometeoritics [GEOL] The study of variation of extraterrestrial debris as a function of time over extended parts of the geologic record, especially in deep-sea sediments and possibly in sedimentary rocks, and, for more recent periods, in ice. { ¦pāl·ē·ō,mēd·ē'òr·iks }

paleomorphology [PALEON] The study of the form and structure of fossil remains in order to describe the original anatomy of an organism. { ¦pāl·ē·ō·mòr'fäl·ə·jē }

paleomycology [PALEOBOT] The study of fossil fungi. { ¦pāl·ē·ō·mī'käl·ə·jē }

Paleonthropinae [PALEON] A former subfamily of fossil man in the family Hominidae; set up to include the Neanderthalers together with Rhodesian man. { ,pāl·ē·ən'thräp·ə,nē }

paleopalynology [PALEON] A field of palynology concerned with fossils of microorganisms and of dissociated microscopic parts of megaorganisms. { ¦pāl·ē·ō,pal·ə'näl·ə·jē }

Paleoparadoxidae [PALEON] A family of extinct hippopotamuslike animals in the order Desmostylia. { ¦pāl·ē·ō,par·ə'däk·sə,dē }

paleopedology [GEOL] The study of soils of past geologic ages, including determination of their ages. { ¦pāl·ē·ō·pə'däl·ə·jē }

paleophycology See paleoalgology. { ¦pāl·ē·ō·fī'käl·ə·jē }

paleophysiography See paleogeomorphology. { ¦pāl·ē·ō,fiz·ē'äg·rə·fē }

Paleophytic [PALEOBOT] A paleobotanic division of geologic time, signifying that period during which the pteridophytes flourished, sometime between the evolution of the algae and the appearance of the first gymnosperms. Also known as Pteridophytic. { ¦pāl·ē·ə¦fid·ik }

paleoplain [GEOL] An ancient degradational plain that is buried beneath later deposits. { 'pāl·ē·ə,plān }

paleopole [GEOL] A pole of the earth, either magnetic or geographic, in past geologic time. { 'pāl·ē·ə,pōl }

paleosalinity [GEOL] The salinity of a body of water in the geologic past, as evaluated on the basis of chemical analyses of sediment or formation water. { ¦pāl·ē·ō·sə'lin·əd·ē }

paleoseismology [PALEON] The study of geological evidence for past earthquakes. { ,pā·lē·ō·sīz'mäl·ə·jē }

paleoslope [GEOL] The direction of initial dip of a former land surface, such as an ancient continental slope. { 'pāl·ē·ə,slōp }

paleosol [GEOL] A soil horizon that formed on the surface during the geologic past, that is, an ancient soil. Also known as buried soil; fossil soil. { 'pāl·ē·ə,sòl }

paleosome [GEOL] A geometric element of a composite rock or mineral deposit which appears to be older than an associated younger rock element. { 'pāl·ē·ə,sōm }

paleospecies [PALEON] The species that are given ancestor and descendant status in a phyletic lineage, depending on the geological strata in which they are found. { 'pē·lē·ō,spē,shēz }

paleostructure [GEOL] The geologic structure of a region or sequence of rocks in the geologic past. { ¦pāl·ē·ō'strək·chər }

paleotectonic map [GEOL] Regional map that shows the structural patterns that existed during a particular period of geologic time, for example, the Lower Cretaceous in western Canada. { ¦pāl·ē·ō·tek¦tän·ik 'map }

paleotemperature [GEOL] **1.** The temperature at which a geologic process took place in ancient past. **2.** The mean climatic temperature at a given time or place in the geologic past. { ¦pāl·ē·ō'tem·prə·chər }

paleothermal [GEOL] Pertaining to warm climates of the geologic past. { ¦pāl·ē· ō'thər·məl }

paleothermometry [GEOL] Measurement or estimation of past temperatures. { ¦pāl· ē·ō·thər'mäm·ə·trē }

paleotopography [GEOL] The topography of a given area in the geologic past. { ¦pāl· ē·ō·tə'päg·rə·fē }

Paleozoic [GEOL] The era of geologic time from the end of the Precambrian (600 million years before present) until the beginning of the Mesozoic era (225 million years before present). { ¦pāl·ē·ə¦zō·ik }

paleozoology [PALEON] The branch of paleontology concerned with the study of ancient animals as recorded by fossil remains. { ¦pāl·ē·ō·zō'äl·ə·jē }

palette [GEOL] A broad sheet of calcite representing a solutional remnant in a cave. Also known as shield. { 'pal·ət }

palevent [GEOL] A relatively sudden and short-lived paleogeographic happening, such as the short, static existence of a particular depositional environment, or a rapid geographic change separating two palstages. Also known as paleogeographic event. { 'pal·ə·vənt }

palichnology See paleoichnology. { ,pal·ik'näl·ə·jē }

palimpsest [GEOL] **1.** Referring to a kind of drainage in which a modern, anomalous drainage pattern is superimposed upon an older one, clearly indicating different topographic and possibly structural conditions at the time of development. **2.** In sedimentology, autochthonous sediment deposits which exhibit some of the attributes of the source sediment. [PETR] Of a metamorphic rock, having remnants of the original structure or texture preserved. { pə'lim·səst }

palinspastic map [GEOL] A paleogeographic or paleotectonic map showing restoration of the features to their original geographic positions, before thrusting or folding of the crustal rocks. { ¦pal·ən¦spas·tik 'map }

Palisade disturbance [GEOL] Appalachian orogenic episode occurring during Triassic time which produced a series of faultlike basins. { ,pal·ə'sād di'stər·bəns }

palisades [GEOL] A series of sharp cliffs. { ,pal·ə'sādz }

palladium amalgam See potarite. { pə'lād·ē·əm ə'mal·gəm }

palladium gold See porpezite. { pə'lād·ē·əm 'gōld }

pallasite [GEOL] **1.** A stony-iron meteorite composed essentially of large single glassy crystals of olivine embedded in a network of nickel-iron. **2.** An ultramafic rock, of either meteoric or terrestrial origin, which contains more than 60% iron in the former, or more iron oxides than silica in the latter. { 'pal·ə,sīt }

pallasite shell See lower mantle. { 'pal·ə,sīt ,shel }

palmierite [MINERAL] $(K,Na)_2Pb(SO_4)_2$ A white hexagonal mineral that is composed of potassium sodium lead sulfate. { pä'mi,rīt }

palstage [GEOL] A period of time when paleogeographic conditions were relatively static or were changing gradually and progressively with relation to such factors as sea level, surface relief, or the distance of the shoreline from the region in question. Also known as paleogeographic stage. { 'pal ,stāj }

palygorskite [MINERAL] **1.** A chain-structure type of clay mineral. **2.** A group of lightweight, tough, fibrous clay minerals showing extensive substitution of aluminum for magnesium. { ,pal·ə'gȯr,skīt }

243

palynofacies [PALEON] An assemblage of palynomorphs in a portion of a sediment, representing local environmental conditions, but not representing the regional palynoflora. { ¦pal·ə·nō'fā·shēz }

palynology [PALEON] The study of spores, pollen, microorganisms, and microscopic fragments of megaorganisms that occur in sediments. { ¦pal·ə'näl·ə·jē }

palynomorph [PALEON] A microscopic feature such as a spore or pollen that is of interest in palynological studies. { pə'lin·ə,mórf }

palynostratigraphy [PALEON] The stratigraphic application of palynologic methods. { ¦pal·ə·nō·strə'tig·rə·fē }

pan [GEOL] **1.** A shallow, natural depression or basin containing a body of standing water. **2.** A hard, cementlike layer, crust, or horizon of soil within or just beneath the surface; may be compacted, indurated, or very high in clay content. { pan }

panabase See tetrahedrite. { 'pan·ə,bās }

panautomorphic rock See panidiomorphic rock. { ¦pan,ód·ə'mór·fik 'räk }

pandermite See priceite. { 'pan·dər,mīt }

panethite [MINERAL] A phosphate mineral known only in meteorites; contains sodium, potassium, magnesium, calcium, iron, and manganese. { 'pan·ə,thīt }

panfan See pediplain. { 'pan,fan }

Pangaea [GEOL] A postulated former supercontinent supposedly composed of all the continental crust of the earth, and later fragmented by drift into Laurasia and Gondwana. Also spelled Pangea. { pan'jē·ə }

Pangea See Pangaea. { pan'jē·ə }

panidiomorphic rock [GEOL] An igneous rock that is completely or predominantly idiomorphic. Also known as panautomorphic rock. { ¦pan¦id·ē·ō¦mór·fik 'räk }

Pannonian [GEOL] A European stage of geologic time comprising the lower Pliocene. { pə'nō·nē·ən }

panplain [GEOL] A broad, level plain formed by coalescence of several adjacent flood plains. Also spelled panplane. { 'pan,plān }

panplanation [GEOL] The action or process of formation or development of a panplain. { ,pan·plə'nā·shən }

panplane See panplain. { 'pan,plān }

pantellerite [PETR] A green to black extrusive rock characterized by acmite-augite or diopside, anorthoclase, and cossyrite phenocrysts in an acmite or feldspar matrix that is either pumiceous, partly glassy, fine-grained holocrystalline trachytic, or microlitic. { pan'tel·ə,rīt }

Panthalassa [GEOL] The hypothetical proto-ocean surrounding Pangea, supposed by some geologists to have combined all the oceans or areas of oceanic crust of the earth at an early time in the geologic past. { ,pan·thə'las·ə }

Pantodonta [PALEON] An extinct order of mammals which included the first large land animals of the Tertiary. { ,pan·tə'dän·tə }

Pantolambdidae [PALEON] A family of middle to late Paleocene mammals of North America in the superfamily Pantolambdoidea. { ,pan·tə'lam·də,dē }

Pantolambdodontidae [PALEON] A family of late Eocene mammals of Asia in the superfamily Pantolambdoidea. { ,pan·tə,lam·də'dän·tə,dē }

Pantolambdoidea [PALEON] A superfamily of extinct mammals in the order Pantodonta. { ,pan·tə·lam'dóid·ē·ə }

Pantolestidae [PALEON] An extinct family of large aquatic insectivores referred to the Proteutheria. { ,pan·tə'les·tə,dē }

Pantotheria [PALEON] An infraclass of carnivorous and insectivorous Jurassic mammals; early members retained many reptilian features of the jaws. { ,pan·tə'thir·ē·ə }

paper shale [GEOL] A shale that easily separates on weathering into very thin, tough, uniform, and somewhat flexible layers or laminae suggesting sheets of paper. { 'pā·pər ¦shāl }

paper spar [GEOL] A crystallized variety of calcite occurring in thin lamellae or paperlike plates. { 'pā·pər ¦spär }

Pappotheriidae [PALEON] A family of primitive, tenreclike Cretaceous insectivores assigned to the Proteutheria. { ,pap·ə·thə'rī·ə,dē }

parabituminous coal [GEOL] Bituminous coal that contains 84–87% carbon, analyzed on a dry, ash-free basis. { ¦par·ə·bə'tüm·ə·nəs 'kōl }

parabolic dune [GEOL] A long, scoop-shaped sand dune having a ground plan approximating the form of a parabola, with the horns pointing windward (upwind). Also known as blowout dune. { ¦par·ə¦bäl·ik 'dün }

parabutlerite See butlerite. { ¸par·ə'bət·lə¸rīt }

parachronology [GEOL] **1.** Practical dating and correlation of stratigraphic units. **2.** Geochronology based on fossils that supplement, or replace, biostratigraphically significant fossils. { ¸par·ə·krə'näl·ə·jē }

paraclinal [GEOL] Referring to a stream or valley that is oriented in a direction parallel to the fold axes of a region. { ¦par·ə¦klīn·əl }

paraconformity [GEOL] A type of unconformity in which strata are parallel; there is little apparent erosion and the unconformity surface resembles a simple bedding plane. Also known as nondepositional unconformity; pseudoconformity. { ¦par·ə·kən'for·məd·ē }

paraconglomerate [GEOL] A conglomerate that is not a product of normal aqueous flow but is deposited by such modes of mass transport as subaqueous turbidity currents and glacier ice; characterized by a disrupted gravel framework, often unstratified, and notable for a matrix of greater than gravel-sized fragments. { ¦par·ə·kən'gläm·ə·rət }

paracoquimbite [MINERAL] $Fe_2(SO_4)_3 \cdot 9H_2O$ A pale-violet rhombohedral mineral composed of hydrous ferric iron sulfate; it is dimorphous with coquimbite. { ¦par·ə·kə'kim¸bīt }

Paracrinoidea [PALEON] A class of extinct Crinozoa characterized by the numerous, irregularly arranged plates, uniserial armlike appendages, and no clear distinction between adoral and aboral surfaces. { ¦par·ə·krə'noid·ē·ə }

paraffin coal [GEOL] A type of light-colored bituminous coal from which oil and paraffin are produced. { 'par·ə·fən ¸kōl }

paraffin dirt [GEOL] A clay soil appearing rubbery or curdy and occurring in the upper several inches of a soil profile near gas seeps; probably formed by biodegradation of natural gas. { 'par·ə·fən ¸dərt }

paragenesis [MINERAL] **1.** The association and order of crystallization of minerals in a rock or vein. **2.** The effect of one mineral on the development of another. Also known as mineral sequence; paragenetic sequence. { ¸par·ə'jen·ə·səs }

paragenetic mineralogy [MINERAL] The study of mineral paragenesis, usually accompanying the analysis of the general geologic structures within and around the ore body. { ¦par·ə·jə'ned·ik ¸min·ə'räl·ə·jē }

paragenetic sequence See paragenesis. { ¦par·ə·jə'ned·ik 'sē·kwəns }

parageosyncline [GEOL] An epeirogenic geosynclinal basin located within a craton or stable area. { ¦par·ə¸jē·ō'sin¸klīn }

paraglomerate [GEOL] A conglomerate which contains more matrix than gravel-sized fragments and was deposited by subaqueous turbidity flows and glacier ice rather than normal aqueous flow. Also known as conglomeratic mudstone. { ¸par·ə'gläm·ə·rət }

paragneiss [GEOL] A gneiss showing a sedimentary parentage. { 'par·ə¸nīs }

paragonite [MINERAL] $NaAl_2(AlSi_3)O_{10}(OH)_2$ A yellowish or greenish monoclinic mica species that contains sodium and usually occurs in metamorphic rock. Also known as soda mica. { pə'rag·ə¸nīt }

parahilgardite [MINERAL] $Ca_8(B_6O_{11})_3Cl \cdot 4H_2O$ A triclinic mineral composed of hydrous borate and chloride of calcium; it is dimorphous with hilgardite. { ¦par·ə'hil·gär¸dīt }

parahopeite [MINERAL] $Zn_3(PO_4)_2 \cdot 4H_2O$ A colorless mineral composed of hydrous phosphate of zinc, occurring in tabular triclinic crystals; it is dimorphous with hopeite. { ¦par·ə'hō¸pīt }

paralaurionite [MINERAL] PbCl(OH) A white mineral composed of basic lead chloride; it is dimorphous with laurionite. { ¦par·ə'lor·ē·ə¸nīt }

paraliageosyncline [GEOL] A geosyncline developing along a present-day continental margin, such as the Gulf Coast geosyncline. { pə¸ral·yə¸jē·ō'sin¸klīn }

245

paralic [GEOL] Pertaining to deposits laid down on the landward side of a coast. { pə'ral·ik }

paralic coal basin [GEOL] Coal deposits formed along the margin of the sea. { pə'ral·ik 'kōl ,bas·ən }

parallel fold *See* concentric fold. { 'par·ə,lel 'fōld }

parallel ripple mark [GEOL] A ripple mark characterized by a relatively straight crest and an asymmetric profile. { 'par·ə,lel 'rip·əl ,märk }

parallel roads [GEOL] A series of horizontal beaches or wave-cut terraces occurring parallel to each other at different levels on each side of a glacial valley. { 'par·ə,lel 'rōdz }

parallel texture [PETR] A rock texture characterized by tabular-to-prismatic crystals oriented parallel to a plane or line. { 'par·ə,lel 'teks·chər }

parallochthon [GEOL] Rocks that were brought from intermediate distances and deposited near an allochthonous mass during transit. { ,par·ə'läk,thän }

paramelaconite [MINERAL] A black tetragonal mineral composed of cupric and cuprous oxides, occurring in pyramidal crystals. { ¦par·ə·mə'lak·ə,nīt }

paramorph [MINERAL] A mineral exhibiting paramorphism. { 'par·ə,mórf }

paramorphism [MINERAL] The property of a mineral whose internal structure has changed without change in composition or external form. Formerly known as allomorphism. { ¦par·ə'mór,fiz·əm }

Paranyrocidae [PALEON] An extinct family of birds in the order Anseriformes, restricted to the Miocene of South Dakota. { pə,ran·ə'räs·ə,dē }

Paraparchitacea [PALEON] A superfamily of extinct ostracods in the suborder Kloedenellocopina including nonsulcate, nondimorphic forms. { ¦par·ə,pär·kə'tās·ē·ə }

pararammelsbergite [MINERAL] NiAs₂ A tin white, orthorhombic or pseudoorthorhombic mineral consisting of nickel diarsenide; occurrence is usually in massive form. { ¦par·ə'ram·əlz,bər,gīt }

pararipple [GEOL] A large, symmetric ripple whose surface slopes gently and which shows no assortment of grains. { 'par·ə,rip·əl }

paraschist [PETR] A schist derived from sedimentary rocks. { 'par·ə,shist }

Paraseminotidae [PALEON] A family of Lower Triassic fishes in the order Palaeonisciformes. { ,par·ə,sem·ə'näd·ə,dē }

parasitic cone *See* adventive cone. { ¦par·ə¦sik·ik 'kōn }

parastratigraphic unit *See* operational unit. { ¦par·ə,strad·ə'graf·ik 'yü·nət }

parastratigraphy [GEOL] **1.** Supplemental stratigraphy based on fossils other than those governing the prevalent orthostratigraphy. **2.** Stratigraphy based on operational units. { ¦par·ə·strə'tig·rə·fē }

parastratotype [GEOL] Another section in the original locality where a stratotype was defined. { ¦par·ə'strad·ə,tīp }

Parasuchia [PALEON] The equivalent name for Phytosauria. { ,par·ə'sü·kē·ə }

paratacamite [MINERAL] Cu₂(OH)₃Cl Rhombohedral mineral composed of basic copper chloride; it is dimorphous with tacamite. { par,ad·ə'ka,mīt }

Parathuramminacea [PALEON] An extinct superfamily of foraminiferans in the suborder Fusulinina, with a test having a globular or tubular chamber and a simple, undifferentiated wall. { ,par·ə·thə,ram·ə'nās·ē·ə }

paratill [GEOL] A till formed by ice-rafting in a marine or lacustrine environment; includes deposits from ice floes and icebergs. { 'par·ə,til }

parautochthonous [GEOL] Pertaining to a mobilized part of an autochthonous granite moved higher in the crust or into a tectonic area of lower pressure and characterized by variable and diffuse contacts with country rocks. [PETR] Pertaining to a rock that is intermediate in tectonic character between autochthonous and allochthonous. { ¦par·ə·ó'täk·thə·nəs }

paravauxite [MINERAL] FeAl₂(PO₄)₂(OH)₂·8H₂O A colorless mineral composed of hydrous basic iron aluminum phosphate; contains more water than vauxite. { ¦par·ə'vók,sīt }

parawollastonite [MINERAL] CaSiO₃ A monoclinic mineral composed of silicate of calcium; it is dimorphous with wollastonite. { ,par·ə'wól·ə·stə,nīt }

246

Pareiasauridae [PALEON] A family of large, heavy-boned terrestrial reptiles of the late Permian, assigned to the order Cotylosauria. { pə'rī·ə'sòr·ə'dē }

parental magma [GEOL] The naturally occurring mobile rock material from which a particular igneous rock solidified or from which another magma was derived. { pə'rent·əl 'mag·mə }

parent material [GEOL] The unconsolidated mineral or organic material from which the true soil develops. { 'per·ənt mə,tir·ē·əl }

parent rock [GEOL] **1.** The rock mass from which parent material is derived. **2.** See source rock. { 'per·ənt ,räk }

parisite [MINERAL] $(Ce,La)_2Ca(CO_3)_3F_2$ A brownish-yellow secondary mineral composed of a carbonate and a fluoride of calcium, cerium, and lanthanum. { 'par·ə,sīt }

parkerite [MINERAL] $Ni_3(Bi,Pb)_2S_2$ A bright-bronze mineral composed of nickel bismuth lead sulfide. { 'pär·kə,rīt }

parogenetic [GEOL] Formed previous to the enclosing rock; especially said of a concretion formed in a different (older) rock from its present (younger) host. { ,par·ə'jen·ik }

paroxysmal eruption See Vulcanian eruption. { ¦par·ək¦siz·məl i'rəp·shən }

parsettensite [MINERAL] $Mn_5Si_6O_{13}(OH)_8$ A copper-red mineral composed of hydrous silicate of manganese. { pär'set·ən,zīt }

parsonsite [MINERAL] $Pb_2(UO_2)(PO_4)_2·2H_2O$ A pale-yellow to brownish mineral composed of hydrous lead uranyl phosphate, occurring as a powder. { 'pär·sən,zīt }

partial pediment [GEOL] **1.** A broadly planate, gravel-capped, interstream bench or terrace. **2.** A broad, planate erosion surface which is formed by the coalescence of contemporaneous, valley-restricted benches developed at the same elevation in proximate valleys, and which would produce a pediment if uninterrupted planation were to continue at this level. { 'pär·shəl 'ped·ə·mənt }

partial pluton [GEOL] That part of a composite intrusion representing a single intrusive episode. { 'pär·shəl 'plü,tän }

partial thermoremanent magnetization [GEOPHYS] The thermoremanent magnetization acquired by cooling in an ambient field over only a restricted temperature interval, as opposed to the entire temperature range from Curie point to room temperature. Abbreviated PTRM. { 'pär·shəl ¦thər·mō'rem·ə·nənt ,mag·nə·tə'zā·shən }

particle diameter [GEOL] The diameter of a sedimentary particle considered as a sphere. { 'pärd·ə·kəl dī,am·əd·ər }

particle size [GEOL] The general dimensions of the particles or mineral grains in a rock or sediment based on the premise that the particles are spheres; commonly measured by sieving, by calculating setting velocities, or by determining areas of microscopic images. { 'pärd·ə·kəl ,sīz }

particle-size analysis [GEOL] A determination of the distribution of particles in a series of size classes of a soil, sediment, or rock. Also known as size analysis; size-frequency analysis. { 'pärd·ə·kəl ¦sīz ə,nal·ə·səs }

parting [GEOL] **1.** A bed or bank of waste material dividing mineral veins or beds. **2.** A soft, thin sedimentary layer following a surface of separation between thicker strata of different lithology. **3.** A surface along which a hard rock can be readily separated or is naturally divided into layers. [MINERAL] Fracturing a mineral along planes weakened by deformation or twinning. { 'pärd·iŋ }

parting cast [GEOL] A sand-filled tension crack produced by creep along the sea floor. { 'pärd·iŋ ,kast }

parting lineation [GEOL] A small-scale primary sedimentary structure made up of a series of parallel ridges and grooves formed parallel to the current. Also known as current lineation. { 'pärd·iŋ ,lin·ē'ā·shən }

parting plane lineation [GEOL] A parting lineation on a laminated surface, consisting of subparallel, linear, shallow grooves and ridges of low relief, generally less than 1 millimeter. { 'pärd·iŋ ¦plān ,lin·ē,ā·shən }

parting-step lineation [GEOL] A parting lineation characterized by subparallel, steplike ridges where the parting surface cuts across several adjacent laminae. { 'pärd·iŋ ¦step 'lin·ē,ā·shən }

partiversal |GEOL| Pertaining to formations that dip in different directions roughly as far as a semicircle. { ¦pard·ə¦vər·səl }

partridgeite *See* bixbyite. { 'pär·trə,jīt }

parvafacies |GEOL| A body of rock constituting the part of any magnafacies that occurs between designated time-stratigraphic planes or key beds traced across the magnafacies. { ¦pär·və'fā·shēz }

pascoite |MINERAL| $Ca_2V_6O_{17} \cdot 11H_2O$ A dark-red-orange to yellow-orange mineral composed of hydrous vanadate of calcium. { 'pas·kə,wīt }

passage bed |GEOL| A stratum marking a transition from rocks of one geological system to those of another. { 'pas·ij ,bed }

passive fold |GEOL| A fold in which the mechanism of folding, either flow or slip, crosses the boundaries of the strata at random. { 'pas·iv 'fōld }

passive margin |GEOL| A continental margin formed by rifting during continental breakup. { 'pas·iv 'mär·jən }

passive permafrost |GEOL| Permafrost that will not refreeze under present climatic conditions after being disturbed or destroyed. Also known as fossil permafrost. { 'pas·iv 'pər·mə,fròst }

patch reef |GEOL| **1.** A small, irregular organic reef with a flat top forming a part of a reef complex. **2.** A small, thick, isolated lens of limestone or dolomite surrounded by rocks of different facies. **3.** *See* reef patch. { 'pach ,rēf }

Patellacea |PALEON| An extinct superfamily of gastropod mollusks in the order Aspidobranchia which developed a cap-shaped shell and were specialized for clinging to rock. { ,pad·əl'ās·ē·ə }

Paterinida |PALEON| A small extinct order of inarticulated brachiopods, characterized by a thin shell of calcium phosphate and convex valves. { ,pad·ə'rin·əd·ə }

paternoite *See* kaliborite. { ,päd·ər'nō,īt }

patina |GEOL| A thin, colored film produced on a rock surface by weathering. { 'pat· ən·ə *or* pə'tē·nə }

patronite |MINERAL| A black vanadium sulfide mineral; mined as a vanadium ore in Minasragra, Peru. { 'pa·trə,nīt }

patterned ground |GEOL| Any of several well-defined, generally symmetrical forms, such as circles, polygons, and steps, that are characteristic of surficial material subject to intensive frost action. { 'pad·ərnd 'graünd }

paulingite |MINERAL| An isometric zeolite mineral consisting of an aluminosilicate of potassium, calcium, and sodium. { 'pòl·iŋ,īt }

paulopost *See* deuteric. { 'pòl·ə,pōst }

pavement |GEOL| A bare rock surface that suggests a paved road surface or other pavement in smoothness, hardness, horizontality, surface extent, or close packing of units. { 'pāv·mənt }

pavonite |MINERAL| $AgBi_3S_5$ A mineral composed of silver bismuth sulfide. { 'pa· və,nīt }

pawdite |PETR| A dark-colored, fine-grained, granular hypabyssal rock composed of magnetite, titanite, biotite, hornblende, calcic plagioclase, and traces of quartz. { 'pò,dīt }

PDB *See* PeeDee belemnite.

peachblossom ore *See* erythrite. { 'pēch,bläs·əm ,òr }

pea coal |GEOL| A size of anthracite that will pass through a 13/16-inch (20.6-millimeter) round mesh but not through a 9/16-inch (14.3-millimeter) round mesh. { 'pē ,kōl }

peacock copper *See* peacock ore. { 'pē,käk 'käp·ər }

peacock ore |MINERAL| A copper mineral, such as bornite, having an iridescent tarnished surface upon exposure to air. Also known as peacock copper. { 'pē,käk 'òr }

pea gravel |GEOL| A type of gravel whose individual particles are about the size of peas. { 'pē ,grav·əl }

peak |GEOL| **1.** The conical or pointed top of a hill or mountain. **2.** An individual mountain or hill taken as a whole, used especially when it is isolated or has a pointed, conspicuous summit. { pēk }

peak plain [GEOL] A high-level plain formed by a series of summits of approximately the same elevation, often described as an uplifted and fully dissected peneplain. Also known as summit plain. { 'pēk ,plān }

peak zone [PALEON] An informal biostratigraphic zone consisting of a body of strata characterized by the exceptional abundance of some taxon (or taxa) or representing the maximum development of some taxon. { 'pēk ,zōn }

pea ore [MINERAL] A variety of pisolitic limonite or bean ore occurring in small, rounded grains or masses about the size of a pea. { 'pē ,ȯr }

pearceite [MINERAL] $Ag_{16}As_2S_{11}$ A black mineral composed of sulfide of arsenic and silver. { 'pir,sīt }

pearlite See perlite. { 'pər,līt }

pearl sinter See siliceous sinter. { 'pərl ¦sin·tər }

pearl spar [MINERAL] A crystalline carbonate having a pearly luster; an example is ankerite. { 'pərl ,spär }

pearlstone See perlite. { 'pərl,stōn }

peat [GEOL] A dark-brown or black residuum produced by the partial decomposition and disintegration of mosses, sedges, trees, and other plants that grow in marshes and other wet places. { pēt }

peat bed See peat bog. { 'pēt ,bed }

peat bog [GEOL] A bog in which peat has formed under conditions of acidity. Also known as peat bed; peat moor. { 'pēt ,bäg }

peat breccia [GEOL] Peat that has been broken up and then redeposited in water. Also known as peat slime. { 'pēt,brech·ə }

peat coal [GEOL] A coal transitional between peat and lignite. { 'pēt ,kōl }

peat formation [GEOCHEM] Decomposition of vegetation in stagnant water with small amounts of oxygen, under conditions intermediate between those of putrefaction and those of moldering. { 'pēt fȯr'mā·shən }

peat moor See peat bog. { 'pēt ,mu̇r }

peat-sapropel [GEOL] A product of the degradation of organic matter that is transitional between peat and sapropel. Also known as sapropel-peat. { 'pēt 'sap·rə,pel }

peat slime See peat breccia. { 'pēt ,slīm }

peat soil [GEOL] Soil containing a large amount of peat; it is rich in humus and gives an acid reaction. { 'pēt ,sȯil }

pebble [GEOL] A clast, larger than a granule and smaller than a cobble, having a diameter in the range of 0.16–2.6 inches (4–64 millimeters). Also known as pebblestone. [MINERAL] See rock crystal. { 'peb·əl }

pebble armor [GEOL] A desert armor made up of rounded pebbles. { 'peb·əl ,är·mər }

pebble bed [GEOL] Any pebble conglomerate, especially one in which the pebbles weather conspicuously and become loose. Also known as popple rock. { 'peb·əl ,bed }

pebble coal [GEOL] Coal that is transitional between peat and brown coal. { 'peb·əl ,kōl }

pebble conglomerate [PETR] A consolidated rock consisting mainly of pebbles. { 'peb·əl kən'gläm·ə·rət }

pebble dike [GEOL] **1.** A clastic dike composed largely of pebbles. **2.** A tabular body containing sedimentary fragments in an igneous matrix. { 'peb·əl ,dīk }

pebble peat [GEOL] Peat that is formed in a semiarid climate by the accumulation of moss and algae, no more than 0.25 inch (6 millimeters) in thickness, under the surface pebbles of well-drained soils. { 'peb·əl ,pēt }

pebble phosphate [GEOL] A secondary phosphorite of either residual or transported origin, consisting of pebbles or concretions of phosphatic material. { 'peb·əl 'fäs,fāt }

pebblestone See pebble. { 'peb·əl,stōn }

pebbly mudstone |GEOL| A delicately laminated till-like conglomeratic mudstone. { 'peb·lē 'məd,stōn }

pebbly sand |GEOL| An unconsolidated sedimentary deposit containing at least 75% sand and up to a maximum of 25% pebbles. { 'peb·lē 'sand }

pebbly sandstone |GEOL| A sandstone that contains 10–20% pebbles. { 'peb·lē 'san,stōn }

pectolite |MINERAL| NaCa$_2$Si$_3$O$_8$(OH) A colorless, white, or gray inosilicate, crystallizing in the monoclinic system and having a vitreous to silky luster; hardness is 5 on Mohs scale, and specific gravity is 2.75. { 'pek·tə,līt }

ped |GEOL| A naturally formed unit of soil structure. { ped }

pedalfer |GEOL| A soil in which there is an accumulation of sesquioxides; it is characteristic of a humid region. { pə'dal·fər }

pedality |GEOL| The physical nature of a soil as expressed by the features of its constituent peds. { pe'dal·əd·ē }

pedestal |GEOL| A relatively slender column of rock supporting a wider rock mass and formed by undercutting as a result of wind abrasion or differential weathering. Also known as rock pedestal. { 'ped·əst·əl }

pedestal boulder |GEOL| A rock mass supported on a rock pedestal. Also known as pedestal rock. { 'ped·əst·əl ,bōl·dər }

pedestal rock See pedestal boulder. { 'ped·əst·əl ,räk }

pediment |GEOL| A piedmont slope formed from a combination of processes which are mainly erosional; the surface is chiefly bare rock but may have a covering veneer of alluvium or gravel. Also known as conoplain; piedmont interstream flat. { 'ped·ə·mənt }

pedimentation |GEOL| The actions or processes by which pediments are formed. { ,ped·ə·mən'tā·shən }

pediment gap |GEOL| A broad opening formed by the enlargement of a pediment pass. { 'ped·ə·mənt ,gap }

pediment pass |GEOL| A flat, narrow tongue that extends from a pediment on one side of a mountain to join a pediment on the other side. { 'ped·ə·mənt ,pas }

pediocratic |GEOL| Pertaining to a period of time in which there is little diastrophism. { ,ped·ē·ə'krad·ik }

pediplain |GEOL| A rock-cut erosion surface formed in a desert by the coalescence of two or more pediments. Also known as desert peneplain; desert plain; panfan. { 'ped·ə,plān }

pediplanation |GEOL| The actions or processes by which pediplanes are formed. { ,ped·ə·plə'nā·shən }

pediplane |GEOL| Any planate erosion surface formed in the piedmont area of a desert, either bare or covered with a veneer of alluvium. { 'ped·ə,plān }

pedocal |GEOL| A soil containing a concentration of carbonates, usually calcium carbonate; it is characteristic of arid or semiarid regions. { 'ped·ə,kal }

pedogenesis See soil genesis. { ¦ped·ō¦jen·ə·səs }

pedogenics |GEOL| The study of the origin and development of soil. { ¦ped·ō¦jen·iks }

pedogeochemical survey |GEOCHEM| A geochemical prospecting survey in which the materials sampled are soil and till. { ¦ped·ō,jē·ō'kem·ə·kəl 'sər,vā }

pedogeography |GEOL| The study of the geographic distribution of soils. { ¦ped·ō·jē'äg·rə·fē }

pedography |GEOL| The systematic description of soils; an aspect of soil science. { pə'däg·rə·fē }

pedolith |GEOL| A surface formation that has undergone one or more pedogenic processes. { 'ped·ə,lith }

pedologic age |GEOL| The relative maturity of a soil profile. { ¦ped·ō¦läj·ik 'āj }

pedologic unit |GEOL| A soil considered without regard to its stratigraphic relations. { ¦ped·ō¦läj·ik 'yü·nət }

pedology See soil science. { pe'däl·ə·jē }

pedon |GEOL| The smallest unit or volume of soil that represents or exemplifies all the horizons of a soil profile; it is usually a horizontal, hexagonal area of about 1 square meter, or possibly larger. { 'pə,dän }

pedorelic |GEOL| Referring to a soil feature that is derived from a preexisting soil horizon. { ¦ped·ō¦rel·ik }

pedosphere |GEOL| That shell or layer of the earth in which soil-forming processes occur. { 'ped·ə,sfir }

pedotubule |GEOL| A soil feature consisting of skeleton grains, or skeleton grains plus plasma, and having a tubular external form (either single tubes or branching systems of tubes) characterized by relatively sharp boundaries and relatively uniform cross-sectional size and shape (circular or elliptical). { ¦ped·ō'tüb·yül }

PeeDee belemnite |GEOCHEM| Limestone from the PeeDee Formation in South Carolina (derived from the Cretaceous marine fossil *Belemnitella americana*), the carbon and oxygen isotope ratios of which are used as an international reference standard. Abbreviated PDB. { ,pē,dē bə'lem,nīt }

peel thrust |GEOL| A sedimentary sheet peeled off a sedimentary sequence, usually along a bedding plane. { 'pēl ,thrəst }

pegmatite |PETR| Any extremely coarse-grained, igneous rock with interlocking crystals; pegmatites are relatively small, are relatively light colored, and range widely in composition, but most are of granitic composition; they are principal sources for feldspar, mica, gemstones, and rare elements.   Also known as giant granite; granite pegmatite. { 'peg·mə,tīt }

pegmatitic stage |GEOL| A stage in the normal sequence of crystallization of magma containing volatiles when the residual fluid is sufficiently enriched in volatile materials to permit the formation of coarse-grained rocks, that is pegmatites. { ¦peg·mə¦tid·ik ,stāj }

pegmatitization |GEOL| Formation of or replacement by a pegmatite. { ,peg·mə,tīd·ə'za·shən }

pegmatoid |PETR| An igneous rock that has the coarse-grained texture of a pegmatite but that lacks graphic intergrowths or typically granitic composition. { 'peg·mə,tóid }

pegmatolite *See* orthoclase. { peg'mad·əl,īt }

peg model |GEOL| Three-dimensional model used to illustrate and study stratigraphic and structural conditions of subsurface geology; consists of a flat platform onto which vertical pegs of varying heights are mounted to represent the contours of various strata. { 'peg ,mäd·əl }

Peking man |PALEON| *Sinanthropus pekinensis*. An extinct human type; the braincase was thick, with a massive basal and occipital torus structure and heavy browridges. { 'pē,kiŋ 'man }

pelagic |GEOL| Pertaining to regions of a lake at depths of 33–66 feet (10–20 meters) or more, characterized by deposits of mud or ooze and by the absence of vegetation. Also known as eupelagic. { pə'laj·ik }

pelagic limestone |GEOL| A fine-textured limestone formed in relatively deep water by the concentration of calcareous tests of pelagic Foraminifera. { pə'laj·ik 'līm,stōn }

pelagochthonous |GEOL| Referring to coal derived from a submerged forest or from driftwood. { ,pel·ə'gäk·thə·nəs }

pelagosite |GEOL| A superficial calcareous crust a few millimeters thick, generally white, gray, or brownish with a pearly luster, formed in the intertidal zone by ocean spray and evaporation, and composed of calcium carbonate with higher contents of magnesium carbonate, strontium carbonate, calcium sulfate, and silica than are found in normal limy sediments. { pə'lag·ə,sīt }

peldon |PETR| A very hard, smooth, compact sandstone with conchoidal fracture, occurring in coal measures. { 'pel·dən }

Pelean cloud *See* nuée ardente. { pə'lē·ən 'klaud }

pelelith |GEOL| Vesicular or pumiceous lava in the throat of a volcano. { pə'lā,lith }

Pele's hair |GEOL| A spun volcanic glass formed naturally by blowing out during quiet fountaining of fluid lava.   Also known as capillary ejecta; filiform lapilli; lauoho o pele. { ,pā,läz 'her }

Pele's tears [GEOL] Volcanic glass in the form of small, solidified drops which precede pendants of Pele's hair. { 'pā,lāz 'tirz }

pelite [GEOL] A sediment or sedimentary rock, such as mudstone, composed of fine, clay- or mud-size particles. Also spelled pelyte. { 'pē,līt }

pelitic [GEOL] Pertaining to, characteristic of, or derived from pelite. { pə'lid·ik }

pelitic hornfels [PETR] A fine-grained metamorphic rock derived from pelite. { pə'lid·ik 'hȯrn,felz }

pelitic schist [PETR] A foliated crystalline metamorphic rock derived from pelite. { pə'lid·ik 'shist }

pellet [GEOL] A fine-grained, sand-size, spherical to elliptical aggregate of clay-sized calcareous material, devoid of internal structure, and contained in the body of a well-sorted carbonate rock. { 'pel·ət }

pell-mell structure [GEOL] A sedimentary structure characterized by absence of bedding in a coarse deposit of waterworn material; it may occur where deposition is too rapid for sorting or where slumping has destroyed the layered arrangement. { 'pel¦mel 'strək·chər }

pellodite See pelodite. { 'pel·ə,dīt }

pelmicrite [GEOL] A limestone containing less than 25% each of intraclasts and ooliths, having a volume ratio of pellets to fossils greater than 3 to 1, and with the micrite matrix more abundant than the sparry-calcite cement. { 'pel·mə,krīt }

pelodite [GEOL] A lithified glacial rock flour which is composed of glacial pebbles in a silt or clay matrix and which was formed by redeposition of the fine fraction of a till. Also spelled pellodite. { 'pel·ə,dīt }

pelogloea [GEOL] Marine detrital slime from settled plankton. { ¦pel·ə¦glē·ə }

pelphyte [GEOL] A lake-bottom deposit consisting mainly of fine, nonfibrous plant remains. { 'pel,fīt }

pelsparite [PETR] A limestone containing less than 25% each of intraclasts and ooliths, having a volume ratio of pellets to fossils greater than 3 to 1, and with the sparry-calcite cement more abundant than the micrite matrix. { 'pel,spä,rīt }

Pelycosauria [PALEON] An extinct order of primitive, mammallike reptiles of the subclass Synapsida, characterized by a temporal fossa that lies low on the side of the skull. { ,pel·ə·kə'sȯr·ē·ə }

pelyte See pelite. { 'pe,līt }

pencatite [PETR] A recrystallized limestone containing periclase or brucite and calcite in approximately equal molecular proportions. { 'peŋ·kə,tīt }

pencil cleavage [GEOL] Cleavage in which fracture produces long, slender pieces of rock. { 'pen·səl ,klē·vij }

pencil gneiss [GEOL] A gneiss that splits into thin, rodlike quartz-feldspar crystal aggregates. { 'pen·səl ,nīs }

pencil ore [GEOL] Hard, fibrous masses of hematite that can be broken up into splinters. { 'pen·səl ,ȯr }

pencil stone See pyrophyllite. { 'pen·səl ,stōn }

pendant See roof pendant. { 'pen·dənt }

pendent terrace [GEOL] A connecting ribbon of sand that joins an isolated point of rock with a neighboring coast. { 'pen·dənt ,ter·əs }

penecontemporaneous [GEOL] Of a geologic process or the structure or mineral that is formed by the process, occurring immediately following deposition but before consolidation of the enclosing rock. { ¦pēn·ē·kən,tem·pə'rā·nē·əs }

peneplain See base-leveled plain. { 'pēn·ə,plān }

peneplanation [GEOL] The actions or processes by which peneplains are formed. { ,pēn·ə·plə'nā·shən }

penetration frequency See critical frequency. { ,pen·ə'trā·shən ,frē·kwən·sē }

penetration funnel [GEOL] An impact crater, generally funnel-shaped, formed by a small meteorite striking the earth at a relatively low velocity and containing nearly all the impacting mass within it. { ,pen·ə'trā·shən ,fən·əl }

penetrative [GEOL] Referring to a texture of deformation that is uniformly distributed in a rock, without notable discontinuities; for example, slaty cleavage. { 'pen·ə,trā·div }

penfieldite [MINERAL] $Pb_2(OH)Cl_3$ A white hexagonal mineral composed of basic chloride of lead, occurring in hexagonal prisms. { 'pen‚fēl‚dīt }

penikkavaarite [PETR] An intrusive rock composed chiefly of augite, barkevikite, and green hornblende in a feldspathic groundmass. { ‚pen·ə'ka·və‚rīt }

penitent See nieve penitente. { 'pen·ə·tənt }

pennantite [MINERAL] $Mn_9Al_6Si_5O_{20}(OH)_{16}$ Orange mineral composed of basic manganese aluminum silicate; member of the chlorite group; it is isomorphous with thuringite. { 'pen·ən‚tīt }

penninite [MINERAL] $(Mg,Fe,Al)_6(Si,Al)_4O_{11}(OH)_8$ An emerald-green, olive-green, pale-green, or bluish mineral of the chlorite group crystallizing in the monoclinic system, with a hardness of 2–2.5 on Mohs scale, and specific gravity of 2.6–2.85. { 'pen·ə‚nīt }

Pennsylvanian [GEOL] A division of late Paleozoic geologic time, extending from 320 to 280 million years ago, varyingly considered to rank as an independent period or as an epoch of the Carboniferous period; named for outcrops of coal-bearing rock formations in Pennsylvania. { ¦pen·sal¦vā·nyən }

Penokean See Animikean. { pə'nō·kē·ən }

penroseite [MINERAL] $(Ni,Co,Cu)Se_2$ A lead gray, isometric mineral consisting of a selenide of nickel, copper, and cobalt; occurs in reniform masses. { 'pen‚rō‚zīt }

pentahydrite [MINERAL] $MgSO_4 \cdot 5H_2O$ A triclinic mineral composed of hydrous magnesium sulfate; it is isostructural with chalcanthite. { ‚pen·tə'hī‚drīt }

Pentamerida [PALEON] An extinct order of articulate brachiopods. { ‚pen·tə'mer·ə·də }

Pentameridina [PALEON] A suborder of extinct brachiopods in the order Pentamerida; dental plates associated with the brachiophores were well developed, and their bases enclosed the dorsal adductor muscle field. { ‚pen·tə·mə'rid·ən·ə }

pentlandite [MINERAL] $(Fe,Ni)_9S_8$ A yellowish-bronze mineral having a metallic luster and crystallizing in the isometric system; hardness is 3.5–4 on Mohs scale, and specific gravity is 4.6–5.0; the major ore of nickel. { 'pent·lən‚dīt }

Penutian [GEOL] A North American stage of geologic time: lower Eocene (above Bulitian, below Ulatasian). { pə'nü·shən }

peperite [GEOL] A breccialike material in marine sedimentary rock, considered to be either a mixture of lava with sediment, or shallow intrusions of magma into wet sediment. { 'pep·ə‚rīt }

peralkaline [PETR] Of igneous rock, having a molecular proportion of aluminum lower than that of sodium oxide and potassium oxide combined. { pər'al·kə‚līn }

peraluminous [PETR] Of igneous rock, having a molecular proportion of aluminum oxide greater than that of sodium oxide and potassium oxide combined. { ‚pər·ə'lü·mə·nəs }

perbituminous [GEOL] Referring to bituminous coal containing more than 5.8% hydrogen, analyzed on a dry, ash-free basis. { ¦pər·bə'tü·mə·nəs }

perched block [GEOL] A large, detached rock fragment presumed to have been transported and deposited by a glacier, and perched in a conspicuous and precarious position on the side of a hill. Also known as balanced rock; perched boulder; perched rock. { 'pərcht 'bläk }

perched boulder See perched block. { 'pərcht 'bōl·dər }

perched rock See perched block. { 'pərcht 'räk }

perching bed [GEOL] A body of rock, generally stratiform, that supports a body of perched water. { 'pərch·iŋ ‚bed }

percussion mark [GEOL] A small, crescent-shaped scar produced on a hard, dense pebble by a blow. { pər'kəsh·ən ‚märk }

percylite [MINERAL] $PbCuCl_2(OH)_2$ Mineral made up of a basic chloride of copper and lead and occurring as cubic blue crystals, with a hardness of 2.5. { 'pər·sē·līt }

pereletok [GEOL] A frozen layer of ground, at the base of the active layer, which may persist for one or several years. Also known as intergelisol. { ‚per·ə·lə'täk }

perezone [GEOL] A zone in which sediments accumulate along coastal lowlands; includes lagoons and brackish-water bays. { 'per·ə‚zōn }

perfemic rock [GEOL] An igneous rock in which the ratio of salicalic to femic minerals is less than 1:7. { pər'fem·ik 'räk }

perforation deposit [GEOL] An isolated kame consisting of material that accumulated in a vertical shaft which pierced a glacier and afforded no outlet for water at the bottom. { ‚pər·fə'rā·shən di‚päz·ət }

pergelic [GEOL] Referring to a soil temperature regime in which the mean annual temperature is less than 0°C and there is permafrost. { pər'jel·ik }

pergelisol table See permafrost table. { ¦pər¦jel·ə‚sól }

perhydrous coal [GEOL] Coal that contains more than 6% hydrogen, analyzed on a dry, ash-free basis. { ¦pər'hī·drəs 'kōl }

periblinite [GEOL] A variety of provitrinite consisting of cortical tissue. { pə'rib·lə‚nīt }

periclase [MINERAL] MgO Native magnesia; a mineral occurring in granular forms or isometric crystals, with hardness of 6 on Mohs scale, and specific gravity of 3.67–3.90. Also known as periclasite. { 'per·ə‚klās }

periclasite See periclase. { ‚per·ə'klā‚sīt }

periclinal [GEOL] Referring to strata and structures that dip radially outward from, or inward toward, a center, forming a dome or a basin. { ¦per·ə¦klīn·əl }

pericline [GEOL] A fold characterized by central orientation of the dip of the beds. [MINERAL] A variety of albite elongated, and often twinned, along the *b* axis. { 'per·ə‚klīn }

pericline ripple mark [GEOL] A ripple mark arranged in an orthogonal pattern either parallel to or transverse to the current direction and having a wavelength up to 80 centimeters and amplitude up to 30 centimeters. { 'per·ə‚klīn 'ripəl ‚märk }

peridot [MINERAL] **1.** A gem variety of olivine that is transparent to translucent and pale-, clear-, or yellowish-green in color. **2.** A variety of tourmaline approaching olivine in color. { 'per·ə‚dät }

peridotite [PETR] A dark-colored, ultrabasic phaneritic igneous rock composed largely of olivine, with smaller amounts of pyroxene or hornblende. { pə'rid·ə‚tīt }

peridotite shell See upper mantle. { pə'rid·ə‚tīt ‚shel }

perigenic [GEOL] Referring to a rock constituent or mineral formed at the same time as the rock it is part of, but not formed at the specific location it now occupies in the rock. { ¦per·ə¦jen·ik }

periglacial [GEOL] Of or pertaining to the outer perimeter of a glacier, particularly to the fringe areas immediately surrounding the great continental glaciers of the geologic ice ages, with respect to environment, topography, areas, processes, and conditions influenced by the low temperature of the ice. { ¦per·ə'glā·shəl }

perimagmatic [GEOL] Referring to a hydrothermal mineral deposit located near its magmatic source. { ¦per·ə·mag'ned·ik }

period [GEOL] A unit of geologic time constituting a subdivision of an era; the fundamental unit of the standard geologic time scale. { 'pir·ē·əd }

peripediment [GEOL] The segment of a pediplane extending across the younger rocks or alluvium of a basin which is always beyond but adjacent to the segment developed on the older upland rocks. { ¦per·ə'ped·ə·mənt }

peripheral depression See ring depression. { pə'rif·ə·rəl di'presh·ən }

peripheral faults [GEOL] Arcuate faults bounding an elevated or depressed area such as a diapir. { pə'rif·ə·rəl 'fóls }

peripheral sink See rim syncline. { pə'rif·ə·rəl 'siŋk }

Periptychidae [PALEON] A family of extinct herbivorous mammals in the order Condylartha distinguished by specialized, fluted teeth. { ‚per·əp'tik·ə‚dē }

peristerite [MINERAL] A gem variety of albite (An_2-An_{24}) that resembles moonstone and has a blue or bluish-white luster characterized by sharp internal reflections of blue, green, and yellow. { pə'ris·tə‚rīt }

perlite [GEOL] A rhyolitic glass with abundant spherical or convolute cracks that cause it to break into small pearllike masses or pebbles, usually less than a centimeter across; it is commonly gray or green with a pearly luster and has the composition of rhyolite. Also known as pearlite; pearlstone. { 'pər‚līt }

permafrost [GEOL] Perennially frozen ground, occurring wherever the temperature remains below 0°C for several years, whether the ground is actually consolidated by ice or not and regardless of the nature of the rock and soil particles of which the earth is composed. { 'pər·mə,frȯst }

perlitic |PETR| **1.** Of the texture of a glassy igneous rock, exhibiting small spheruloids formed from cracks due to contraction during cooling. **2.** Pertaining to or characteristic of perlite. { ¦pər¦lid·ik }

permafrost island |GEOL| A small, shallow, isolated patch of permafrost surrounded by unfrozen ground. { 'pər·mə,frȯst 'ī·lənd } .

permafrost line |GEOL| A line on a map representing the border of the arctic permafrost. { 'pər·mə,frȯst ,līn }

permafrost table |GEOL| The upper limit of permafrost. Also known as pergelisol table. { 'pər·mə,frȯst ,tā·bəl }

permanent extinction |GEOL| The extinction of a lake by destruction of the lake basin, because of such processes as deposition of sediments, erosion of the basin rim, filling with vegetation, or catastrophic events. { 'pər·mə·nənt ik'stiŋk·shən }

permeability |GEOL| The capacity of a porous rock, soil, or sediment for transmitting a fluid without damage to the structure of the medium. Also known as conductivity; perviousness. { ,pər·mē·ə'bil·əd·ē }

permeability trap |GEOL| An oil trap formed by lateral variation within a reservoir bed which seals the contained hydrocarbons through a change of permeability. { ,pər·mē·ə'bil·əd·ē ,trap }

permeable bed |GEOL| A porous reservoir formation through which hydrocarbon fluids (oil or gas) or water (waterflood or interstitial) can flow. { 'pər·mē·ə·bəl 'bed }

permeation gneiss |PETR| A gneiss formed as a result of or modified by the passage of geochemically mobile materials through or into solid rock. { ,pər·mē'ā·shən ¦nīs }

Permian |GEOL| The last period of geologic time in the Paleozoic era, from 280 to 225 million years ago. { 'pər·mē·ən }

permineralization |GEOL| A fossilization process whereby additional minerals are deposited in the pore spaces of originally hard animal parts. { pər,min·rə·lə'zā·shən }

Permo-Carboniferous |GEOL| **1.** The Permian and Carboniferous periods considered as one unit. **2.** The Permian and Pennsylvanian periods considered as a single unit. **3.** The rock unit, or the period of geologic time, transitional between the Upper Pennsylvanian and the Lower Permian periods. { ¦pər·mō,kär·bə'nif·ə·rəs }

perovskite |MINERAL| $CaTiO_3$ A natural, yellow, brownish-yellow, reddish, brown, or black mineral and a structure type which includes no less than 150 synthetic compounds; the crystal structure is ideally cubic, it occurs as rounded cubes modified by the octahedral and dodecahedral forms, luster is subadamantine to submetallic, hardness is 5.5 on Mohs scale, and specific gravity is 4.0. { pə'rävz,kīt }

perpendicular slip |GEOL| The component of a fault slip measured at right angles to the trace of the fault on any intersecting surface. { ¦pər·pən¦dik·yə·lər 'slip }

perpendicular slope |GEOL| A very steep slope or precipitous face, as on a mountain. { ¦pər·pən¦dik·yə·lər 'slōp }

perpendicular throw |GEOL| The distance between two points which were formerly adjacent in a faulted bed, vein, or other surface, measured at right angles to the surface. { ¦pər·pən¦dik·yə·lər 'thrō }

Perret phase |GEOL| That stage of a volcanic eruption that is characterized by the emission of much high-energy gas that may significantly enlarge the volcanic conduit. { 'per·ət ,fāz }

perryite |MINERAL| $(Ni,Fe)_5(Si,P)_2$ A mineral found only in meteorites. { 'per·ē,īt }

persalic rock |GEOL| An igneous rock in which the ratio of salic to femic minerals is greater than 7:1. { pər'sal·ik 'räk }

persilicic See silicic. { ¦pər·sə'lis·ik }

perthite |GEOL| A parallel to subparallel intergrowth of potassium and sodium feldspar; the potassium-rich phase is usually the host from which the sodium-rich phase evolves. { 'pər,thīt }

perthitic |GEOL| Of a texture produced by perthite, exhibiting sodium feldspar as small

strings, blebs, films, or irregular veinlets in a host of potassium feldspar.
{ pər'thid·ik }

perthosite [PETR] A light-colored syenite composed almost entirely of perthite, with less than 3% mafic minerals. { 'pər·thə,sīt }

Peru saltpeter See soda niter. { pə'rü 'sȯlt'pēd·ər }

perviousness See permeability. { 'pər·vē·əs·nəs }

Petalichthyida [PALEON] A small order of extinct dorsoventrally flattened fishes belonging to the class Placodermi; the external armor is in two shields of large plates. { ,ped·ə·lik'thē·ə·də }

petalite [MINERAL] LiAlSi$_4$O$_{10}$ A white, gray, or colorless monoclinic mineral composed of silicate of lithium and aluminum, occurring in foliated masses or as crystals. { 'ped·əl,īt }

Petalodontidae [PALEON] A family of extinct cartilaginous fishes in the order Bradyodonti distinguished by teeth with deep roots and flattened diamond-shaped crowns. { ,ped·əl·ə'dänt·ə,dē }

petrifaction [GEOL] A fossilization process whereby inorganic matter dissolved in water replaces the original organic materials, converting them to a stony substance. { ,pe·trə'fak·shən }

petrified wood See silicified wood. { 'pe·trə,fīd 'wu̇d }

petrochemistry [GEOCHEM] An aspect of geochemistry that deals with the study of the chemical composition of rocks. { ¦pe·trō'kem·ə·strē }

petrofabric See fabric. { ¦pe·trō'fab·rik }

petrofabric analysis See structural petrology. { ¦pe·trō'fab·rik ə'nal·ə·səs }

petrofabric diagram See fabric diagram. { ¦pe·trō'fab·rik 'dī·ə,gram }

petrofabrics See structural petrology. { ¦pe·trō'fab·riks }

petrofacies See petrographic facies. { ¦pe·trō'fā·shēz }

petrogenesis [PETR] That branch of petrology dealing with the origin of rocks, particularly igneous rocks. Also known as petrogeny. { ¦pe·trō'jen·ə·səs }

petrogenic grid [PETR] A diagram whose coordinates are parameters of the rock-forming environment on which equilibrium curves are plotted indicating the limits of the stability fields of specific minerals and mineral assemblages. { ¦pe·trō¦jen·ik 'grid }

petrogeny See petrogenesis. { pə'träj·ə·nē }

petrogeometry See structural petrology. { ¦pe·trō·jē'äm·ə·trē }

petrographer [GEOL] An individual who does petrography. { pə'träg·rə·fər }

petrographic facies [GEOL] Facies distinguished principally by composition and appearance. Also known as petrofacies. { ¦pe·trə¦graf·ik 'fā·shēz }

petrographic period [GEOL] The extension in time of a rock association. { ¦pe·trə¦graf·ik 'pir·ē·əd }

petrographic province [GEOL] A broad area in which similar igneous rocks are formed during the same period of igneous activity. Also known as comagmatic region; igneous province; magma province. { ¦pe·trə¦graf·ik 'präv·əns }

petrography [GEOL] The branch of geology that deals with the description and systematic classification of rocks, especially by means of microscopic examination. { pə'träg·rə·fē }

petroleum [GEOL] A naturally occurring complex liquid hydrocarbon which after distillation yields combustible fuels, petrochemicals, and lubricants; can be gaseous (natural gas), liquid (crude oil, crude petroleum), solid (asphalt, tar, bitumen), or a combination of states. { pə'trō·lē·əm }

petroleum geology [GEOL] The branch of economic geology dealing with the origin, occurrence, movement, accumulation, and exploration of hydrocarbon fuels. { pə'trō·lē·əm jē'äl·ə·jē }

petroleum seep See oil seep. { pə'trō·lē·əm ,sēp }

petroleum trap [GEOL] Stable underground formation (geological or physical) of such nature as to trap and hold liquid or gaseous hydrocarbons; usually consists of sand or porous rock surrounded by impervious rock or clay formations. { pə'trō·lē·əm ,trap }

petroliferous [GEOL] Containing petroleum. { ,pe·trə'lif·ə·rəs }

petrologen See kerogen. { pə'träl·ə·jən }

petrologist [GEOL] An individual who studies petrology. { pə'träl·ə·jəst }

petrology [GEOL] The branch of geology concerned with the origin, occurrence, structure, and history of rocks, principally igneous and metamorphic rock. { pə'träl·ə·jē }

petromict [GEOL] Of a sediment, composed of metastable rock fragments. { 'pe·trə,mikt }

petromorph [GEOL] A speleothem or cave formation that is exposed to the surface by erosion of the limestone in which the cave was formed. { 'pe·trə,mòrf }

petromorphology See structural petrology. { |pe·trō·mór'fäl·ə·jē }

petrophysics [GEOL] Study of the physical properties of reservoir rocks. { |pe·trō|fiz·iks }

petrotectonics [GEOL] Extension of the field of structural petrology to include analysis of the movements that produced the rock's fabric. Also known as tectonic analysis. { |pe·trō·tek'tän·iks }

petzite [MINERAL] Ag_3AuTe_2 A steel-gray to iron-black mineral consisting of a silver gold telluride; hardness is 2.5–3 on Moh's scale, and specific gravity is 8.7–9.0. { 'pet,sīt }

peuroseite [MINERAL] $(Ni,Cu,Pb)Se_2$ A gray mineral composed of nickel copper lead selenide, occurring in columnar masses. { pyù'rō,zīt }

pezograph See regmaglypt. { 'pez·ə,graf }

phacellite See kaliophilite. { 'fas·əl,īt }

phacolith [GEOL] A minor, concordant, lens-shaped, and usually granitic intrusion into folded sedimentary strata. { 'fak·ə,lith }

phanerite [PETR] An igneous rock having phaneritic texture. { 'fan·ə,rīt }

phaneritic [PETR] Of the texture of an igneous rock, being visibly crystalline. Also known as coarse-grained; phanerocrystalline; phenocrystalline. { ,fan·ə'rid·ik }

phanerocryst See phenocryst. { 'fan·ə·rō,krist }

phanerocrystalline See phaneritic. { |fan·ə·rō'krist·əl·ən }

Phanerorhynchidae [PALEON] A family of extinct chondrostean fishes in the order Palaeonisciformes having vertical jaw suspension. { |fan·ə·rō'riŋ·kə,dē }

Phanerozoic [GEOL] The part of geologic time for which there is abundant evidence of life, especially higher forms, in the corresponding rock, essentially post-Precambrian. { |fan·ə·rō|zō·ik }

phantom [GEOL] A bed or member that is absent from a specific stratigraphic section but is usually present in a characteristic position in a sequence of similar geologic age. [PETR] See ghost. { 'fan·təm }

phantom horizon [GEOL] In seismic reflection prospecting, a line constructed so that it is parallel to the nearest actual dip segment at all points along a profile. { 'fan·təm hə'rīz·ən }

pharmacolite [MINERAL] $CaH(AsO_4)·2H_2O$ A white to grayish monoclinic mineral composed of hydrous acid arsenate of calcium, occurring in fibrous form. { fär'mak·ə,līt }

pharmacosiderite [MINERAL] $Fe_3(AsO_4)_2(OH)_3·5H_2O$ Green or yellowish-green mineral composed of a hydrous basic iron arsenate and commonly found in cubic crystals. Also known as cube ore. { |fär·mə·kō'sīd·ə,rīt }

phenacite See phenakite. { 'fen·ə,sīt }

Phenacodontidae [PALEON] An extinct family of large herbivorous mammals in the order Condylarthra. { fə,näk·ə'dänt·ə,dē }

phenakite [MINERAL] Be_2SiO_4 A colorless, white, wine-yellow, pink, blue, or brown glassy mineral that crystallizes in the rhombohedral system; used as a minor gemstone. Also spelled phenacite. { 'fen·ə,kīt }

phenicochroite See phoenicochroite. { ,fēn·ə'käk·rə,wīt }

phenoclastic rock [PETR] A nonuniformly sized clastic rock containing phenoclasts. { |fēn·ə|klas·tik 'räk }

phenoclasts [PETR] The larger, conspicuous fragments in a sediment or sedimentary rock, such as cobbles in a conglomerate. { 'fen·ə,klasts }

phenocryst [PETR] A large, conspicuous crystal in a porphyritic rock. Also known as phanerocryst. { 'fēn·ə,krist }

phenocrystalline See phaneritic. { |fēn·ə'krist·əl·ən }

phenoplast

phenoplast [PETR] A large rock fragment in a rudaceous rock that was plastic at the time of its incorporation into the matrix. { 'fē·nə,plast }

phi grade scale [GEOL] A logarithmic transformation of the Wentworth grade scale in which the diameter value of the particle is replaced by the negative logarithm to the base 2 of the particle diameter (in millimeters). { 'fī 'grād ,skāl }

philipstadite [MINERAL] $Ca_2(Fe,Mg)_5(Si,Al)_8O_{22}(OH)_2$ Monoclinic mineral composed of basic silicate of calcium, iron, magnesium, and aluminum; member of the amphibole group. { 'fil·əp,stä,dīt }

phillipsite [MINERAL] $(K_2,Na_2CA)Al_2Si_4O_{12}·H_2O$ A white or reddish zeolite mineral crystallizing in the orthorhombic system; occurs in complex fibrous crystals, which make up a large part of the red-clay sediments in the Pacific Ocean. { 'fil·əp,sīt }

phlebite [PETR] Roughly banded or veined metamorphite or migmatite. { 'fle,bīt }

phlogopite [MINERAL] $K_2[Mg,Fe(II)]_6(Si_6,Al_2)O_{20}(OH)_4$ A yellow-brown to copper mineral of the mica group occurring in disseminated flakes, foliated masses, or large crystals; hardness is 2.5–3.0 on Mohs scale, and specific gravity is 2.8–3.0. Also known as bronze mica; brown mica. { 'fläg·ə,pīt }

phoenicite See phoenicochroite. { 'fē·nə,sīt }

phoenicochroite [MINERAL] Pb_2CrO_5 A red mineral composed of basic chromate of lead, occurring in crystals and masses. Also known as beresovite; phenicochroite; phoenicite. { ,fēn·ə'kä·krə,wīt }

Pholidophoridae [PALEON] A generalized family of extinct fishes belonging to the Pholidophoriformes. { fə,lid·ə'fȯr·ə,dē }

Pholidophoriformes [PALEON] An extinct actinopterygian group composed of mostly small fusiform marine and fresh-water fishes of an advanced holostean level. { fə,lid·ə,fȯr·ə'fȯr,mēz }

phonolite [PETR] A light-colored, aphanitic rock of volcanic origin, composed largely of alkali feldspar, feldspathoids, and smaller amounts of mafic minerals. { 'fō·nə,līt }

phorogenesis [GEOL] The shifting or slipping of the earth's crust relative to the mantle. { ¦fȯr·ə'jen·ə·səs }

phosgenite [MINERAL] $Pb_2Cl_2(CO_3)$ A white, yellow, or grayish mineral that crystallizes in the tetragonal system, has adamantine luster, hardness of 3 on Mohs scale, and specific gravity of 6–6.3. Also known as cromfordite; horn lead. { 'fäz·jə,nīt }

phosphate [MINERAL] A mineral compound characterized by a tetrahedral ionic group of phosphate and oxygen, PO_4^{3-}. { 'fä,sfāt }

phosphatic nodule [GEOL] A dark, usually black, earthy mass or pebble of variable size and shape, having a hard shiny surface and occurring in marine strata. { fä'sfad·ik 'naj·yül }

phosphatization [GEOCHEM] Conversion to a phosphate or phosphates; for example, the diagenetic replacement of limestone, mudstone, or shale by phosphate-bearing solutions, producing phosphates of calcium, aluminum, or iron. { ,fäs·fad·ə'zā·shən }

phosphoferrite [MINERAL] $(Fe,Mn)_3(PO_4)_2·3H_2O$ A white or greenish orthorhombic mineral composed of hydrous phosphate of ferrous iron manganese phosphate; exhibits micalike cleavage. { ¦fäs·fō'fe,rīt }

phosphophyllite [MINERAL] $Zn_2(FeMn)(PO_4)_2·4H_2O$ Colorless to pale-blue mineral composed of hydrous zinc ferrous iron manganese phosphate; exhibits micalike cleavage. { ,fäs·fō'fi,līt }

phosphorite [PETR] A sedimentary rock composed chiefly of phosphate minerals. { 'fäs·fə,rīt }

phosphorization [GEOCHEM] Impregnation or combination with phosphorus or a compound of phosphorus; for example, the diagenetic process of phosphatization. { ,fäs·fə·rə'zā·shən }

phosphorroesslerite [MINERAL] $MgH(PO_4)·7H_2O$ A yellowish, monoclinic mineral consisting of a hydrated acid magnesium phosphate. { ¦fäs·fə'res·lə,rīt }

phosphosiderite [MINERAL] $FePO_4·2H_2O$ A pinkish-red mineral crystallizing in the monoclinic system, dimorphous with strengite and isomorphous with metavariscite. { ¦fä·sfō'sīd·ə,rīt }

258

phosphuranylite [MINERAL] $(UO_2)(PO_4)_2 \cdot 6H_2O$ A yellow secondary mineral composed of hydrous uranyl phosphate, occurring in powder form, it is phosphorescent when exposed to radium emanations. { ‚fäs·fyə'ran·əl‚īt }

photoclinometry [GEOL] A technique for ascertaining slope information from an image brightness distribution, used especially for studying the amount of slope to a lunar crater wall or ridge by measuring the density of its shadow. { ¦fōd·ō·klə'näm·ə·trē }

photogeologic anomaly [GEOL] Any systematic deviation of a photogeologic factor from the expected norm in a given area. { ¦fōd·ō‚jē·ə'läj·ik ə'näm·ə·lē }

photogeologic map [GEOL] A compilation of interpretations of a series of aerial photographs, including annotations of geologic features. { ¦fōd·ō‚jē·ə'läj·ik 'map }

photogeology [GEOL] The geologic interpretation of landforms by means of aerial photographs. { ¦fōd·ō‚jē'äl·ə·jē }

photogeomorphology [GEOL] The study of landforms by means of aerial photographs. { ¦fōd·ō‚jē·ō·mȯr'fäl·ə·jē }

phreatic [GEOL] Of a volcanic explosion of material such as steam or mud, not being incandescent. { frē'ad·ik }

phreatic gas [GEOL] A gas formed by the contact of atmospheric or surface water with ascending magma. { frē'ad·ik 'gas }

phreatomagmatic [GEOL] Pertaining to a volcanic explosion that extrudes both magmatic gases and steam; it is caused by the contact of the magma with groundwater or ocean water. { frē¦ad·ō·mag'mad·ik }

phthanite See chert. { 'tha‚nīt }

phyllite [PETR] A metamorphic rock intermediate in grade between slate and schist, and derived from argillaceous sediments; has a silky sheen on the cleavage surface. { 'fi‚līt }

phyllofacies [GEOL] A facies differentiated on the basis of stratification characteristics, especially the stratification index. { ‚fil·ō'fā·shēz }

Phyllolepida [PALEON] A monogeneric order of placoderms from the late Upper Devonian in which the armor is broad and low with a characteristic ornament of concentric and transverse ridges on the component plates. { ‚fil·ə'lep·ə·də }

phyllomorphic stage [GEOL] The most advanced geochemical stage of diagenesis, characterized by authigenic development of micas, feldspars, and chlorites at the expense of clays. { ¦fil·ə¦mȯr·fik ‚stāj }

phyllonite [PETR] A metamorphic rock occupying an intermediate position between phyllite and mylonite. { 'fil·ə‚nīt }

phyllosilicate [MINERAL] A structural type of silicate mineral in which flat sheets are formed by the sharing of three of the four oxygen atoms in each tetrahedron with neighboring tetrahedrons. Also known as layer silicate; sheet mineral; sheet silicate. { ¦fil·ō'sil·ə·kət }

physical exfoliation [GEOL] A type of exfoliation caused by physical forces; for example, by the freezing of water that has penetrated fine cracks in rock or by the removal of overburden concealing deeply buried rocks. { 'fiz·ə·kəl eks‚fō·lē'ā·shən }

physical geology [GEOL] That branch of geology concerned with understanding the composition of the earth and the physical changes occurring in it, based on the study of rocks, minerals, and sediments, their structures and formations, and their processes of origin and alteration. { 'fiz·ə·kəl jē'äl·ə·jē }

physical residue [GEOL] A residue which results from physical, as opposed to chemical, weathering processes. { 'fiz·ə·kəl 'rez·ə‚dü }

physical stratigraphy [GEOL] Stratigraphy based on the physical aspects of rocks, especially the sedimentologic aspects. { 'fiz·ə·kəl strə'tig·rə·fē }

physical time [GEOL] Geologic time as measured by some physical process, such as the radioactive decay of elements. { 'fiz·ə·kəl 'tīm }

physical weathering See mechanical weathering. { 'fiz·ə·kəl 'weth·ə·riŋ }

physiographic diagram [GEOL] A small-scale map showing landforms by the systematic application of a standardized set of simplified pictorial symbols that represent the appearance such forms would have if viewed obliquely from the air at an angle

259

of about 45°. Also known as landform map; morphographic map. { ¦fiz·ē·ə¦graf·ik 'dī·ə,gram }

physiographic feature [GEOL] A prominent or conspicuous physiographic form or noticeable part thereof. { ¦fiz·ē·ə¦graf·ik 'fē·chər }

physiographic form [GEOL] A landform considered with regard to its origin, cause, or history. { ¦fiz·ē·ə¦graf·ik 'fórm }

physiographic province [GEOL] A region having a pattern of relief features or landforms that differs significantly from that of adjacent regions. { ¦fiz·ē·ə¦graf·ik 'präv·əns }

phyteral [GEOL] Morphologically recognizable forms of vegetal matter in coal. { 'fīd·ə·rəl }

phytocollite [GEOL] A black, gelatinous, nitrogenous humic body occurring beneath or within peat deposits. { fī'täk·ə,līt }

phytolith [PALEON] A fossilized part of a living plant that secreted mineral matter. { 'fīd·ə,lith }

Phytosauria [PALEON] A suborder of Late Triassic long-snouted aquatic thecodonts resembling crocodiles but with posteriorly located external nostrils, absence of a secondary palate, and a different structure of the pelvic and pectoral girdles. { ,fīd·ə'sór·ē·ə }

Piacention See Plaisancian. { ,pē·ə'sen·chən }

pickeringite [MINERAL] $MgAl_2(SO_4)_4·22H_2O$ A white or faintly colored mineral composed of hydrous sulfate of magnesium and aluminum, occurring in fibrous masses. { 'pik·riŋ,īt }

picotite [MINERAL] A dark-brown variety of hercynite that contains chromium and is commonly found in dunites. Also known as chrome spinel. { 'pik·ə,tīt }

picrite [PETR] A medium- to fine-grained igneous rock composed chiefly of olivine, with smaller amounts of pyroxene, hornblende, and plagioclase felspar. { 'pi,krīt }

picrolite See antigorite. { 'pik·rə,līt }

picromerite [MINERAL] $K_2Mg(SO_4)_2·6H_2O$ A white mineral composed of hydrous sulfate of magnesium and potassium, occurring as crystalline encrustations. { pi'kräm·ə,rīt }

picropharmacolite [MINERAL] $(Ca,Mg)_3(AsO_4)_2·6H_2O$ Mineral composed of hydrous calcium magnesium arsenate. { ¦pik·rō·fär'mak·ə,līt }

piecemeal stoping [GEOL] Magmatic stoping in which only isolated blocks of roof rock are assimilated. { 'pēs,mēl ,stōp·iŋ }

piedmont [GEOL] Lying or formed at the base of a mountain or mountain range, as a piedmont terrace or a piedmont pediment. { 'pēd,mänt }

piedmont alluvial plain See bajada. { 'pēd,mänt ə'lüv·ē·əl 'plān }

piedmont angle [GEOL] The sharp break of slope between a hill and a plain, such as the angle at the junction of a mountain front and the pediment at its base. { 'pēd ,mänt ¦aŋ·gəl }

piedmont bench See piedmont step. { 'pēd,mänt ¦bench }

piedmont benchland [GEOL] One of several successions or systems of piedmont steps. Also known as piedmont stairway; piedmont treppe. { 'pēd,mänt 'bench,land }

piedmont flat See piedmont step. { 'pēd,mänt ,flat }

piedmont gravel [GEOL] Coarse gravel derived from high ground by mountain torrents and spread out on relatively flat ground where the velocity of the water is decreased. { 'pēd,mänt ¦grav·əl }

piedmont interstream flat See pediment. { 'pēd,mänt 'in·tər,strēm ,flat }

piedmontite See piemontite. { 'pēd,män,tīt }

piedmont plain See bajada. { 'pēd,mänt ¦plān }

piedmont plateau [GEOL] A plateau lying between the mountains and the plains or the ocean. { 'pēd,mänt pla'tō }

piedmont scarp [GEOL] A small, low cliff formed in alluvium on a piedmont slope at the foot of a steep mountain range; due to dislocation of the surface, especially by faulting. Also known as scarplet. { 'pēd,mänt ¦skärp }

piedmont slope See bajada. { 'pēd,mänt ¦slōp }

piedmont stairway *See* piedmont benchland. { 'pēd,mänt 'ster,wā }

piedmont step [GEOL] A terracelike or benchlike piedmont feature that slopes outward or downvalley. Also known as piedmont bench; piedmont flat. { 'pēd,mänt ¦step }

piedmont treppe *See* piedmont benchland. { 'pēd,mänt 'trep·ə }

piemontite [MINERAL] Ca$_2$(Al,Mn^{3-},Fe)$_3$Si$_3$O$_{12}$(OH) Reddish-brown epidote mineral that contains manganese. Also known as manganese epidote; piedmontite. { 'pē,män,tīt }

piercement *See* diapir. { 'pirs·mənt }

piercement dome *See* diapir. { 'pirs·mənt ,dōm }

piercing fold *See* diapir. { 'pirs·iŋ ,fōld }

piezocrystallization [GEOL] Crystallization of a magma under pressure, such as the pressure associated with orogeny. { pē¦ā·zō,krist·əl·ə'zā·shən }

piezogene [GEOL] Pertaining to the formation of minerals primarily under the influence of pressure. { pē'ā·zō,jēn }

piezoglypt *See* regmaglypt. { pē'ā·zō,glipt }

pigeonite [MINERAL] (Mg,Fe^{2+},Ca)(MgFe^{2+})Si$_2$O$_6$ Clinopyroxene mineral species intermediate in composition between clinoenstatite and diopside, found in basic igneous rocks. { 'pij·ə,nīt }

pike [GEOL] A mountain or hill which has a peaked summit. { pīk }

pillar [GEOL] **1.** A natural formation shaped like a pillar. **2.** A joint block produced by columnar jointing. **3.** *See* stalacto-stalagmite. { 'pil·ər }

pillow breccia [PETR] A deposit of pillow structures and fragments of lava in a matrix of tuff. { 'pil·ō ,brech·ə }

pillow lava [GEOL] Any lava characterized by pillow structure and presumed to have formed in a subaqueous environment. Also known as ellipsoidal lava. { 'pil·ō ,läv·ə }

pillow structure [GEOL] A primary sedimentary structure that resembles a pillow in size and shape. Also known as mammillary structure. [PETR] A pillow-shaped structure visible in some extrusive lavas attributed to the congealment of lava under water. { 'pil·ō ,strək·chər }

pilotaxitic [GEOL] Pertaining to the texture of the groundmass of a holocrystalline igneous rock in which lath-shaped microlites (usually of plagioclase) are arranged in a glass-free felty mesh, often aligned along the flow lines. { ¦pī·lō·tak'sid·ik }

Piltdown man [PALEON] An alleged fossil man based on fragments of a skull and mandible that were eventually discovered to constitute a skillful hoax. { 'pilt ,daún ,man }

pimple mound [GEOL] A low, flattened, roughly circular or elliptical dome consisting of sandy loam that is entirely distinct from the surrounding soil; peculiar to the Gulf coast of eastern Texas and southwestern Louisiana. { 'pim·pəl ,maúnd }

pimple plain [GEOL] A plain distinguished by the presence of numerous, conspicuous pimple mounds. { 'pim·pəl ,plān }

pinakiolite [MINERAL] Mg$_3$Mn$_3$B$_2$O$_{10}$ A black mineral composed of borate of magnesium and manganese; it is polymorphous with orthopinakiolite. { pə'näk·ē·ə,līt }

pinch [GEOL] Thinning of a rock layer, as where a vein narrows. { pinch }

pinch-and-swell structure [GEOL] A structural condition common in pegmatites and veins of quartz in metamorphosed rocks; the vein is pinched at frequent intervals, leaving expanded parts between. { ¦pinch ən 'swel ,strək·chər }

pingo remnant [GEOL] A rimmed depression formed by the rupturing of a pingo summit which results in the exposure of the ice core to melting followed by partial or total collapse. Also known as pseudokettle. { 'piŋ·gō ¦rem·nənt }

pinguite *See* nontronite. { 'piŋ,gwīt }

pinhole chert [PETR] Chert containing weathered pebbles which are pierced by minute holes or pores. { 'pin,hōl 'chərt }

pinite [MINERAL] A compact gray, green, or brown mica, chiefly muscovite derived from other minerals such as cordierite. { 'pē,nīt }

pinnacle [GEOL] **1.** A sharp-pointed rock rising from the bottom, which may extend above the surface of the water, and may be a hazard to surface navigation; due to the sheer rise from the sea floor, no warning is given by sounding. **2.** Any high tower or spire-shaped pillar of rock, alone or cresting a summit. { 'pin·ə·kəl }

pinnate joint *See* feather joint. { 'pi͵nāt ͵jóint }

pinnoite [MINERAL] $Mg(BO_2)_2 \cdot 3H_2O$ A yellow mineral composed of hydrous borate of magnesium, occurring in nodular masses. { 'pin·ə͵wīt }

pinolite [PETR] A metamorphic rock containing magnesite (breunnerite) as crystals and as granular aggregates in a schistose matrix (phyllite or talc schist). { 'pin·əl͵īt }

pintadoite [MINERAL] $Ca_2V_2O_7 \cdot 9H_2O$ A green mineral consisting of a hydrated calcium vanadate; occurs as an efflorescence. { ͵pin·tə'dō͵īt }

piotine *See* saponite. { 'pī·ə͵tēn }

pipe [GEOL] **1.** A vertical, cylindrical ore body. Also known as chimney; neck; ore chimney; ore pipe; stock. **2.** A tubular cavity of varying depth in calcareous rocks, often filled with sand and gravel. **3.** A vertical conduit through the crust of the earth below a volcano, through which magmatic materials have passed. Also known as breccia pipe. { pīp }

pipe amygdule [GEOL] An elongate amygdule occurring toward the base of a lava flow, probably formed by the generation of gases or vapor from the underlying material. { 'pīp ə'mig͵dyül }

pipe clay [GEOL] A mass of fine clay, usually lens-shaped, which forms the surface of bedrock and upon which often rests the gravel of old river beds. { 'pīp ͵klā }

pipernoid texture [GEOL] The eutaxitic texture of certain extrusive igneous rocks in which dark patches and stringers occur in a light-colored groundmass. { 'pī·pər͵nóid ͵teks·chər }

pipe rock [PETR] A marine sandstone containing abundant scolites. { 'pīp ͵räk }

pipestone [PETR] A pink or mottled argillaceous stone; carved by the Indians into tobacco pipes. { 'pīp͵stōn }

pipe vesicle [GEOL] A slender vertical cavity, a few centimeters or tens of centimeters in length, extending upward from the base of a lava flow. { 'pīp ͵ves·ə·kəl }

pirssonite [MINERAL] $Na_2Ca(CO_3)_2 \cdot 2H_2O$ A colorless or white orthorhombic mineral composed of hydrous carbonate of sodium and calcium. { 'pirs·ən͵īt }

pisanite [MINERAL] $(Fe,Cu)SO_4 \cdot 7H_2O$ A blue mineral composed of hydrous sulfate of copper and iron; it is isomorphous with kirovite and melanterite. { pə'zä͵nīt }

pisolite [PETR] A sedimentary rock composed principally of pisoliths. { 'pī·zə͵līt }

pisolith [GEOL] Small, more or less spherical particles found in limestones and dolomites, having a diameter of 2–10 millimeters and often formed of calcium carbonate. { 'pī·zə͵lith }

pisolitic [PETR] Pertaining to pisolite or to the characteristic texture of such a rock. { ͵pī·zə͵lid·ik }

pisolitic tuff [GEOL] Of a tuff, composed of accretionary lapilli or pisolites. { ͵pī·zə͵lid·ik 'təf }

pisoparite [PETR] A limestone which contains at least 25% pisoliths and no more than 25% intraclasts and in which the sparry-calcite cement is more abundant than the carbonate-mud matrix (micrite). { pī'zäp·ə͵rīt }

pitch *See* plunge. { pich }

pitchblende [MINERAL] A massive, brown to black, and fine-grained, amorphous, or microcrystalline variety of uraninite which has a pitchy to dull luster and contains small quantities of uranium. Also known as nasturan; pitch ore. { 'pich͵blend }

pitch coal *See* bituminous lignite. { 'pich ͵kōl }

pitching fold *See* plunging fold. { 'pich·iŋ ͵fōld }

pitch opal [MINERAL] A yellowish to brownish inferior quality of common opal displaying a luster resembling that of pitch. { 'pich ͵ō·pəl }

pitch ore *See* pitchblende. { 'pich ͵òr }

pitchstone [GEOL] A type of volcanic glass distinguished by a waxy, dull, resinous, pitchy luster. Also known as fluolite. { 'pich͵stōn }

pit-run gravel [GEOL] A natural deposit of a mixture of gravel, sand, and foreign materials. { 'pit ͵rən ͵grav·əl }

262

pitted outwash plain [GEOL] An outwash plain characterized by numerous depressions such as kettles, shallow pits, and potholes. { 'pid·əd 'aút,wäsh ,plān }

pitted pebble [GEOL] A pebble having marked concavities not related to the texture of the rock in which it appears or to differential weathering. { 'pid·əd 'peb·əl }

pitticite [MINERAL] A mineral of varying color composed of a hydrous sulfate-arsenate of iron. { 'pid·ə,sīt }

Pityaceae [PALEOBOT] A family of fossil plants in the order Cordaitales known only as petrifactions of branches and wood. { ,pid·ē'ās·ē,ē }

pivotal fault See rotary fault. { 'piv·əd·əl 'fólt }

placanticline [GEOL] A gentle, anticlinallike uplift of the continental platform, usually asymmetric and without a typical outline. { plak'ant·i,klīn }

placer [GEOL] A mineral deposit at or near the surface of the earth, formed by mechanical concentration of mineral particles from weathered debris. Also known as ore of sedimentation. { 'plās·ər }

placic horizon [GEOL] A black to dark red soil horizon that is usually cemented with iron and is not very permeable. { 'plā·sik hə'rīz·ən }

Placodermi [PALEON] A large and varied class of Paleozoic fishes characterized by a complex bony armor covering the head and the front portion of the trunk. { ,plak·ə'dər·mē }

Placodontia [PALEON] A small order of Triassic marine reptiles of the subclass Euryapsida characterized by flat-crowned teeth in both the upper and lower jaws and on the palate. { ,plā·kə'dän·chə }

Plaggept [GEOL] A suborder of the soil order Inceptisol, with very thick surface horizons of mixed mineral and organic materials resulting from manure or human wastes added over long periods of time. { 'plä·gept }

plagiaplite [PETR] An aplite composed chiefly of plagioclase (oligoclase to andesine), possibly green hornblende, and accessory quartz, biotite, and muscovite. { 'plā·jē·ə,plīt }

Plagiaulacida [PALEON] A primitive, monofamilial suborder of multituberculate mammals distinguished by their dentition (dental formula I 3/0 C 0/0 Pm 5/4 M 2/2), having cutting premolars and two rows of cusps on the upper molars. { ¦plā·jē·ə·yü'läs·ə·də }

Plagiaulacidae [PALEON] The single family of the extinct mammalian suborder Plagiaulacida. { ¦plā·jē·ə·yü'läs·ə,dē }

plagioclase [MINERAL] **1.** A type of triclinic feldspars having the general formula $(Na,Ca)Al(Si,Al)Si_2O_8$; they are common rock-forming minerals. **2.** A series in the plagioclase group which can be divided into a number of varieties based on the relative proportion of the solid solution end members, albite and anorthite (An): albite (An 0–10) oligoclase (An 10–30), andesine (An 30–50), labradorite (An 50–70), bytownite (An 70–90), and anorthite (An 90–100). Also known as sodium-calcium feldspar. { 'plā·jē·ə,klās }

plagionite [MINERAL] $Pb_5Sb_8S_{17}$ A lead-gray mineral with metallic appearance, composed of sulfide of lead and antimony. { 'plā·jē·ə,nīt }

Plagiosauria [PALEON] An aberrant Triassic group of labyrinthodont amphibians. { ¦plā·jē·ə'sòr·ē·ə }

plain [GEOL] A flat, gently sloping region of the sea floor. Also known as submarine plain. { plān }

plain of denudation [GEOL] A surface that has been reduced to sea level or to just above sea level by the agents of erosion (usually considered to be of subaerial origin). { 'plān əv ,dē·nü'dā·shən }

plain of lateral planation [GEOL] An extensive, smooth, apronlike surface developed at the base of a mountain or escarpment by the widening of valleys and the coalescence of floodplains as a result of lateral planation. { 'plān əv ¦lad·ə·rəl plā'nā·shən }

plain of marine denudation [GEOL] A plane or nearly plane surface worn down by the gradual encroachment of ocean waves upon the land; or a plane or nearly plane imaginary surface representing such a plain after uplift and partial subaerial erosion. Also known as plain of submarine denudation. { 'plān əv mə¦rēn ,dē·nü'dā·shən }

plain of marine erosion [GEOL] A theoretical platform representing a plane surface of unlimited width produced below sea level by the complete cutting away of the land by marine processes acting over a very long period of stillstand. { 'plān əv məˌrēn i'rō·zhən }

plain of submarine denudation See plain of marine denudation. { 'plān əv ˌsəb·məˌrēn ˌdē·nü'dā·shən }

plains-type fold [GEOL] An anticlinal or domelike structure of the continental platform which has no typical outline and for which there is no corresponding synclinal structure. { 'plānz ˌtīp ˌfōld }

plain tract [GEOL] The lower part of a stream, characterized by a low gradient and a wide floodplain. { 'plān ˌtrakt }

Plaisancian [GEOL] A European stage of geologic time: lower Pliocene (above Pontian of Miocene, below Astian). Also known as Piacention; Plaisanzian. { plā'zän·chən }

Plaisanzian See Plaisancian. { plā'zän·zhən }

plaiting [GEOL] A texture in some schists that results from the intersection of relict bedding planes with well-developed cleavage planes. Also known as gaufrage. { 'plād·iŋ }

planar cross-bedding [GEOL] Cross-bedding characterized by planar surfaces of erosion in the lower bounding surface. { 'plā·nər 'kròs ˌbed·iŋ }

planar flow structure See platy flow structure. { 'plā·nər 'flō ˌstrək·chər }

planate [GEOL] Referring to a surface that has been flattened or leveled by planation. { 'plā,nāt }

planation [GEOL] Erosion resulting in flat surfaces, caused by meandering streams, waves, ocean currents, wind, or glaciers. { plā'nā·shən }

plane bed [GEOL] A sedimentary bed without elevations or depressions larger than the maximum size of the bed material. { 'plān ˌbed }

plane parallel texture [PETR] The parallel texture of a rock in which the constituents are parallel to a plane, but not to a line as in linear parallel texture. { 'plān ˌpar·ə,lel 'teks·chər }

planetary geology [GEOL] A science that applies geologic principles and techniques to the study of planets and their natural satellites. Also know as planetary geoscience. { 'plan·ə,ter·ē jē'äl·ə·jē }

planetary geoscience See planetary geology. { 'plan·ə,ter·ē ˌjē·ō'sī·əns }

planetary vorticity effect [GEOPHYS] The effect of the variation of the earth's vorticity with latitude in altering the relative vorticity of a flow with a meridional component; a fluid with a free surface in a rotating cylinder exhibits a corresponding effect, owing to the shrinking or stretching of radially displaced columns. { 'plan·ə,ter·ē vòr'tis·əd·ē i,fekt }

planoclastic rock [PETR] An even-grained or uniformly sized clastic rock. { ˌplan·ə,kla·stik 'räk }

planoconformity [GEOL] The relation between conformable strata that are approximately uniform in thickness and sensibly parallel throughout. { ˌplā·nō·kən'fòr·məd·ē }

Planosol [GEOL] An intrazonal, hydromorphic soil having a clay pan or hardpan covered with a leached surface layer; developed in a humid to subhumid climate. { 'plan·ə,sòl }

plasma [GEOL] The part of a soil material that can be, or has been, moved, reorganized, or concentrated by soil-forming processes. [MINERAL] A faintly translucent or semitranslucent and bright green, leek green, or nearly emerald green variety of chalcedony, sometimes having white or yellowish spots. { 'plaz·mə }

plasma mantle [GEOPHYS] A thick layer of plasma just inside the magnetopause characterized by a tailward bulk flow with a speed of 60 to 120 miles (100 to 200 kilometers) per second and by a gradual decrease of density, temperature, and speed as the depth inside the magnetosphere increases. { 'plaz·mə 'mant·əl }

plasmapause [GEOPHYS] The sharp outer boundary of the plasmasphere, at which the plasma density decreases by a factor of 100 or more. { 'plaz·mə,pòz }

plasma sheet [GEOPHYS] A region of relatively hot plasma outside the plasmasphere, which reaches, during quiet times, from an altitude of about 30,000 miles (50,000 kilometers) to at least past the moon's orbit in a long tail extending away from the sun; composed of particles with typical thermal energies of 2 to 4 kiloelectronvolts. { 'plaz·mə ˌshēt }

plaster conglomerate [GEOL] A conglomerate composed entirely of boulders derived from a partially exhumed monadnock forming a wedgelike mass of its flank. { 'plas·tər kən'gläm·ə·rət }

plastic equilibrium [GEOL] State of stress within a soil mass or a portion thereof that has been deformed to such an extent that its ultimate shearing resistance is mobilized. { 'plas·tik ˌē·kwə'lib·rē·əm }

plasticity index [GEOL] The percent difference between moisture content of soil at the liquid and plastic limits. { plas'tis·əd·ē ˌin,deks }

plasticlast [GEOL] An intraclast consisting of calcareous mud that has been torn up while still soft. { 'plas·tə,klast }

plastic limit [GEOL] The water content of a sediment, such as a soil, at the point of transition between the plastic and semisolid states. { 'plas·tik 'lim·ət }

plastic zone [GEOL] A region located adjacent to the rupture zone of an explosion crater and at an increased distance from the shot site, differing from the rupture zone by having less fracturing and only small permanent deformations. { 'plas·tik ˌzōn }

plate [GEOL] **1.** A smooth, thin, flat fragment of rock, such as a flagstone. **2.** A large rigid, but mobile, block involved in plate tectonics; thickness ranges from 30 to 150 miles (50 to 250 kilometers) and includes both crust and a portion of the upper mantle. { plāt }

plateau [GEOL] A broad, comparatively flat and poorly defined elevation of the sea floor, commonly over 60 meters (200 feet) in elevation. { pla'tō }

plateau basalt [GEOL] One or a succession of high temperature basaltic lava flows from fissure eruptions which accumulate to form a plateau. Also known as flood basalt. { pla'tō bə'sòlt }

plateau gravel [GEOL] A sheet, spread, or patch of surficial gravel, often compacted, occupying a flat area on a hilltop, plateau, or other high region at a height above that normally occupied by a stream terrace gravel. { pla'tō ¦grav·əl }

plateau mountain [GEOL] A pseudomountain produced by the dissection of a plateau. { pla'tō ¦maùnt·ən }

plateau plain [GEOL] An extensive plain surmounted by a sublevel summit area and bordered by escarpments. { pla'tō ˌplān }

plate tectonics [GEOL] Global tectonics based on a model of the earth characterized by a small number (10–25) of semirigid plates which float on some viscous underlayer in the mantle; each plate moves more or less independently and grinds against the others, concentrating most deformation, volcanism, and seismic activity along the periphery. Also known as raft tectonics. { 'plāt tek'tän·iks }

platform [GEOL] **1.** Any level or almost level surface; a small plateau. **2.** A continental area covered by relatively flat or gently tilted, mainly sedimentary strata which overlay a basement of rocks consolidated during earlier deformations; platforms and shields together constitute cratons. { 'plat,fòrm }

platform beach [GEOL] A looped bar or ridge of sand and gravel formed on a wave-cut platform. { 'plat,fòrm ˌbēch }

platform facies See shelf facies. { 'plat,fòrm ˌfā·shēz }

platform reef [GEOL] An organic reef, generally small but more extensive than a patch reef, with a flat upper surface. { 'plat,fòrm ˌrēf }

platiniridium [MINERAL] A silver-white cubic mineral composed of platinum, iridium, and related metals, occurring in grains. { ¦plat·ən·ə'rid·ē·əm }

platinite See platynite. { 'plat·ən,īt }

platte [GEOL] A resistant knob of rock in a glacial valley or rising in the midst of an existing glacier, often causing a glacier to split near its snout. { 'plad·ə }

plattnerite [MINERAL] PbO_2 An iron-black mineral consisting of lead dioxide, occurring in masses with submetallic luster. { 'plat·nə,rīt }

platy [GEOL] **1.** Referring to a sedimentary particle whose length is more than three times its thickness. **2.** Referring to a sandstone or limestone that splits into laminae having thicknesses in the range of 2 to 10 millimeters. { 'plad·ē }

Platybelondoninae [PALEON] A subfamily of extinct elephantoid mammals in the family Gomphotheriidae consisting of species with digging specializations of the lower tusks. { ˌplad·ē,bel·ən'dän·ə,nē }

Platyceratacea [PALEON] A specialized superfamily of extinct gastropod mollusks which adapted to a coprophagous life on crinoid calices. { ˌplad·ē,ser·ə'tās·ē·ə }

platy flow structure [PETR] Structure of an igneous rock characterized by tabular sheets which suggest stratification, and formation by contraction during cooling. Also known as linear flow structure; planar flow structure. { 'plad·ē 'flō ,strək·chər }

platynite [MINERAL] PbBi$_2$(Se,S)$_3$ An iron-black mineral composed of selenide and sulfide of lead and bismuth; occurs in thin metallic plates resembling graphite. Also spelled platinite. { 'plad·ə,nīt }

Platysomidae [PALEON] A family of extinct palaeonisciform fishes in the suborder Platysomoidei; typically, the body is laterally compressed and rhombic-shaped, with long dorsal and anal fins. { ˌplad·ē'säm·ə,dē }

Platysomoidei [PALEON] A suborder of extinct deep-bodied marine and fresh-water fishes in the order Palaeonisciformes. { ˌplad·ē·sə'mȯid·ē,ī }

playa [GEOL] **1.** A low, essentially flat part of a basin or other undrained area in an arid region. **2.** A small, generally sandy land area at the mouth of a stream or along the shore of a bay. **3.** A flat, alluvial coastland, as distinguished from a beach. { 'plī·ə }

Playfair's law [GEOL] The law that each stream cuts its own valley, the valley being proportional in size to its stream, and the stream junctions in the valley are accordant in level. { 'plā,ferz ,lȯ }

Pleistocene [GEOL] The older of the two epochs of the Quaternary Period, spanning about 1.8 million to 10,000 years ago. It represents the interval of geological time (and rocks accumulated during that time) extending from the end of the Pliocene Epoch (and the end of Tertiary Period) to the start of the Holocene Epoch. It is commonly characterized as an epoch when the earth entered its most recent phase of widespread glaciation. Also known as Ice Age; Oiluvium. { 'plī·stə,sēn }

pleonaste See ceylonite. { 'plē·ə,nast }

Plesiocidaroida [PALEON] An extinct order of echinoderms assigned to the Euechinoidea. { ˌplē·sē·ō,sik·ə'rȯid·ə }

Plesiosauria [PALEON] A group of extinct reptiles in the order Sauropterygia constituting a highly specialized offshoot of the nothosaurs. { ˌplē·sē·ə'sȯr·ē·ə }

Pleuracanthodii [PALEON] An order of Paleozoic sharklike fishes distinguished by two-pronged teeth, a long spine projecting from the posterior braincase, and direct backward extension of the tail. { plu̇,rak·ən'thō·dē,ī }

Pleuromeiaceae [PALEOBOT] A family of plants in the order Pleuromiales, but often included in the Isoetales due to a phylogenetic link. { ˌplu̇r·ō·mē'ās·ē,ē }

Pleuromeiales [PALEOBOT] An order of Early Triassic lycopods consisting of the genus Pleuromeia; the upright branched stem had grasslike leaves and a single terminal strobilus. { ˌplu̇r·ō·mē'ā·lēz }

Pleurotomariacea [PALEON] An extinct superfamily of gastropod mollusks in the order Aspidobranchia. { ˌplu̇r·əd·ə,mar·ē'ās·ē·ə }

plexus [GEOL] An area on a subglacial deposit that encloses a giant's kettle. { 'plek·səs }

plication [GEOL] Intense, small-scale folding. { plī'kā·shən }

Pliensbachian [GEOL] A European stage of geologic time: Lower Jurassic (above Sinemurian, below Toarcian). { plēnz'bäk·ē·ən }

Plinian eruption See Vulcanian eruption. { 'plin·ē·ən i'rəp·shən }

plinth [GEOL] The lower and outer part of a seif dune, beyond the slip-face boundaries, that has never been subjected to sand avalanches. { plinth }

plinthite [GEOL] In a soil, a material consisting of a mixture of clay and quartz with other diluents, that is rich in sesquioxides, poor in humus, and highly weathered. { 'plin,thīt }

Pliocene [GEOL] The youngest of the five geological epochs of the Tertiary Period. The Pliocene represents the interval of geological time (and rocks deposited during that time) extending from the end of the Miocene Epoch to the beginning of the Pleistocene Epoch of the Quaternary Period. Modern time scales assign the duration of 5.0 million to 1.8 million years ago to the Pliocene. { 'plī·ə,sēn }

Pliohyracinae [PALEON] An extinct subfamily of ungulate mammals in the family Procaviidae. { ,plī·ō·hī'ras·ə,nē }

pliothermic [GEOL] Pertaining to a period in geologic history characterized by more than average climatic warmth. { |plī·ō|thər·mik }

plow sole [GEOL] A pressure pan representing a layer of soil compacted by repeated plowing to the same depth. { 'plaů ,sōl }

plucking [GEOL] A process of glacial erosion which involves the penetration of ice or rock wedges into subglacial niches, crevices, and joints in the bedrock; as the glacier moves, it plucks off pieces of jointed rock and incorporates them. Also known as glacial plucking; quarrying. { 'plək·iŋ }

plug [GEOL] **1.** A vertical pipelike magmatic body representing the conduit to a former volcanic vent. **2.** A crater filling of lava, the surrounding material of which has been removed by erosion. **3.** A mass of clay, sand, or other sediment filling the part of a stream channel abandoned by the formation of a cutoff. { pləg }

plug dome [GEOL] A volcanic dome characterized by an upheaved, consolidated conduit filling. { 'pləg ,dōm }

plug reef [GEOL] A small, triangular reef that grows with its apex pointing seaward through openings between linear shelf-edge reefs. { 'pləg ,rēf }

plum [GEOL] A clast embedded in a matrix of a different kind, especially a pebble in a conglomerate. { pləm }

plumbago See graphite. { ,pləm'bā·gō }

plumb line [GEOPHYS] A continuous curve to which the direction of gravity is everywhere tangential. { 'pləm ,līn }

plumboferrite [MINERAL] PbFe$_4$O$_7$ A dark hexagonal mineral composed of lead iron oxide. { |pləm·bō'fe,rīt }

plumbogummite [MINERAL] **1.** PbAl$_3$(PO$_4$)$_2$(OH)$_5$·H$_2$O A mineral composed of hydrous basic lead aluminum phosphate. **2.** A group of isostructural minerals, that includes gorceixite, goyazite, crandallite, deltaite, florencite, and dussertite, as well as plumbogummite. { |pləm·bō'gə,mīt }

plumbojarosite [MINERAL] PbFe$_6$(SO$_4$)$_4$(OH)$_{12}$ A mineral composed of basic lead iron sulfate; it is isostructural with jarosite. { |pləm·bō·jə'rō,sīt }

plume structure [GEOL] On the surface of a master joint, a ridgelike tracing in a plumelike pattern, usually oriented parallel to the upper and lower surfaces of the constituent rock unit. Also known as plumose structure. { 'plüm ,strək·chər }

plumose structure See plume structure. { 'plü,mōs ,strək·chər }

plunge [GEOL] The inclination of a geologic structure, especially a fold axis, measured by its departure from the horizontal. Also known as pitch; rake. { plənj }

plunge basin [GEOL] A deep, large hollow or cavity scoured in the bed of a stream at the foot of a waterfall or cataract by the force and eddying effect of the falling water. { 'plənj ,bās·ən }

plunging cliff [GEOL] A sea cliff bordering directly on deep water, having a base that lies well below water level. { 'plənj·iŋ 'klif }

plunging fold [GEOL] A fold having a relatively steep plunge. Also known as pitching fold. { 'plənj·iŋ 'fōld }

plush copper ore See chalcotrichite. { 'pləsh 'käp·ər 'ȯr }

plutology [GEOL] The study of the interior of the earth. { plü'täl·ə·jē }

pluton [GEOL] **1.** An igneous intrusion. **2.** A body of rock formed by metasomatic replacement. { 'plü,tän }

plutonian See plutonic. { plü'tō·nē·ən }

plutonic

plutonic [GEOL] Pertaining to rocks formed at a great depth. Also known as abyssal; deep-seated; plutonian. { plü'tän·ik }

plutonic breccia [GEOL] Breccia consisting of older annular rock fragments enclosed in younger plutonic rock. { plü'tän·ik 'brech·ə }

plutonic metamorphism [GEOL] Deep-seated regional metamorphism at high temperatures and pressures, often accompanied by strong deformation. { plü'tän·ik ‚med· ə'mòr‚fiz·əm }

plutonic rock [GEOL] A rock formed at considerable depth by crystallization of magma or by chemical alteration. { plü'tän·ik 'räk }

plutonism [GEOL] **1.** Pertaining to the processes associated with pluton formation. **2.** The theory that the earth formed by solidification of a molten mass. { 'plüt· ən‚iz·əm }

pluvial [GEOL] Of a geologic process or feature, effected by rain action. { 'plü·vē·əl }

pluvial lake [GEOL] A lake formed during a period of exceptionally heavy rainfall; specifically, a Pleistocene lake formed during a period of glacial advance and now either extinct or only a remnant. { 'plü·vē·əl 'lāk }

pluviofluvial [GEOL] Pertaining to the combined action of rainwater and streams. { ‚plü·vē·ō‚flü·vē·əl }

pneumatogenic [GEOL] Referring to a rock or mineral deposit formed by a gaseous agent. { ‚nü·məd·ō‚jen·ik }

pneumatolysis [GEOL] Rock alteration or mineral crystallization effected by gaseous emanations from solidifying magma. { ‚nü·mə'täl·ə·səs }

pneumatolytic [GEOL] Formed by gaseous agents. { ‚nü·məd·ō‚lid·ik }

pneumatolytic metamorphism [PETR] Contact metamorphism by the chemical action of magmatic gases. { ‚nü·məd·ō‚lid·ik ‚med·ə'mòr‚fiz·əm }

pneumatolytic stage [GEOL] The stage in the cooling of a magma in which the solid and gaseous phases are in equilibrium. { ‚nü·məd·ō‚lid·ik 'stāj }

pneumotectic [GEOL] Referring to processes and products of magmatic consolidation affected to some degree by the gaseous constituents of the magma. { ‚nü· mō‚tek·tik }

pocket [GEOL] **1.** A cavity that contains a deposit such as a gas or an ore. **2.** An enclosed or sheltered place along a coast, such as a reentrant between rocky, cliffed headlands or a bight on a lee shore. { 'päk·ət }

pocket beach [GEOL] A small, narrow beach formed in a pocket, commonly crescentic in plan, with the concave edge toward the sea, and displaying well-sorted sands. { 'päk·ət ‚bēch }

pocket valley [GEOL] A valley whose head is enclosed by steep walls at the base of which underground water emerges as a spring. { 'päk·ət ‚val·ē }

pod [GEOL] An orebody of elongate, lenticular shape. Also known as podiform orebody. { päd }

podiform orebody See pod. { 'päd·ə‚fòrm 'òr‚bäd·ē }

Podzol [GEOL] A soil group characterized by mats of organic matter in the surface layer and thin horizons of organic minerals overlying gray, leached horizons and dark-brown illuvial horizons; found in coal forests to temperate coniferous or mixed forests. { 'päd‚zòl }

podzolic soil See red-yellow podzolic soil. { päd‚zäl·ik 'sòil }

podzolization [GEOL] The process by which a soil becomes more acid because of the depletion of bases, and develops surface layers that have been leached of clay. { ‚päd·zə·lə'zā·shən }

poikilitic [PETR] Of the texture of an igneous rock, having small crystals of one mineral randomly scattered without common orientation in larger crystals of another mineral. { ‚pòi·kə'lid·ik }

poikiloblast [GEOL] A large crystal (xenoblast) formed by recrystallization during metamorphism and containing numerous inclusions of small idioblasts. { pòi'kil· ə‚blast }

poikiloblastic [PETR] Of a metamorphic texture, simulating the poikilitic texture of

268

igneous rocks in having small idioblasts of one constituent lying within larger xenoblasts. Also known as sieve texture. { pói¦kil·ə¦blas·tik }

poikilocrystallic See poikilotopic. { pói¦kil·ə·kri¦stal·ik }

poikilophitic |GEOL| Referring to ophitic texture characterized by lath-shaped feldspar crystals completely included in large, anhedral pyroxene crystals. { pói¦kil·ə¦fid·ik }

poikilotope |GEOL| A large crystal enclosing smaller crystals of another mineral in a sedimentary rock showing poikilotopic fabric. { pói'kil·ə‚tōp }

poikilotopic |GEOL| Referring to the fabric of a crystalline sedimentary rock in which the constituent crystals are multisized and larger crystals enclose smaller crystals of another mineral. Also known as poikilocrystallic. { pói¦kil·ə¦täp·ik }

point bar |GEOL| One of a series of low, arcuate sand and gravel ridges formed on the inside of a growing meander by the gradual accretions. Also known as meander bar. { 'póint ‚bär }

point diagram |PETR| A fabric diagram in which a point represents the preferred orientation of each individual fabric element. Also known as scatter diagram. { 'póint ‚dī·ə‚gram }

Poisson relation |GEOPHYS| A model of elastic behavior used in experimental structural geology that takes the Poisson ratio as equal to 0.25. { pwä'sōn ri‚lā·shən }

polarity epoch |GEOPHYS| A period of time during which the earth's magnetic field was predominantly of a single polarity. { pə'lar·əd·ē ‚ep·ək }

polarity event |GEOPHYS| A period of no more than about 100,000 years when the earth's magnetic polarity was opposite to the predominant polarity of that polarity epoch. { pə'lar·əd·ē i‚vent }

polarity zone |GEOL| In stratigraphy, a material unit that is defined in terms of magnetic polarity, that is, reversals of the earth's magnetic field. { pə'lar·əd·ē ‚zōn }

polar migration See polar wandering. { 'pō·lər mī'grā·shən }

polar variation |GEOPHYS| A small movement of the earth's axis of rotation relative to the geoid, the resultant of the Chandler wobble and other smaller movements. { 'pō·lər ‚ver·ē'ā·shən }

polar wandering |GEOL| Migration during geologic time of the earth's poles of rotation and magnetic poles. Also known as Chandler motion; polar migration. { 'pō·lər 'wan·də·riŋ }

pollucite |MINERAL| $(Cs,Na)_2Al_2Si_4O_{12}·H_2O$ A colorless, transparent zoolite mineral composed of hydrous silicate of cesium, sodium, and aluminum, occurring massive or in cubes; used as a gemstone. Also known as pollux. { pə'lü‚sīt }

pollux See pollucite. { 'päl·əks }

polyargyrite |MINERAL| $Ag_{24}Sb_2S_{15}$ A gray to black mineral composed of antimony silver sulfide. { ‚päl·ē'är·jə‚rīt }

polybasite |MINERAL| $(Ag,Cu)_{16}Sb_2S_{11}$ An iron-black to steel-gray metallic-looking mineral; an ore of silver. { ‚päl·i'bā‚sīt }

polyclinal fold |GEOL| One of a group of adjacent folds, the axial surfaces of which are oriented randomly, but which have similar surface axes. { ‚päl·i¦klīn·əl 'fōld }

polycrase |MINERAL| $(Y,Ca,Ce,U,Th)(Ti,Cb,Ta)_2O_6$ Black mineral composed of titanate, columbate, and tantalate of yttrium-group metals; it is isomorphous with euxenite and occurs in granite pegmatites. { 'päl·i‚krās }

Polydolopidae |PALEON| A Cenozoic family of rodentlike marsupial mammals. { ‚päl·i·də'läp·ə‚dē }

polydymite |MINERAL| Ni_3S_4 A mineral of the linnaeite group consisting of nickel sulfide. { pə'lid·ə‚mīt }

polygene |GEOL| An igneous rock composed of two or more minerals. Also known as polymere. { 'päl·i‚jēn }

polygenetic |GEOL| **1.** Resulting from more than one process of formation or derived from more than one source, or originating or developing at various places and times. **2.** Consisting of more than one type of material, or having a heterogeneous composition. Also known as polygenic. { ¦päl·i·jə'ned·ik }

polygenic See polygenetic. { ¦päl·i¦jen·ik }

polygeosyncline [GEOL] A geosynclinal-geoanticlinal belt that lies along the continental margin and receives sediments from a borderland on its oceanic side. { ¦päl·i,jē-ō'sin,klīn }

Polygnathidae [PALEON] A family of Middle Silurian to Cretaceous conodonts in the suborder Conodontiformes, having platforms with small pitlike attachment scars. { ,pal·ig'nath-ə,dē }

polygonal ground [GEOL] A ground surface consisting of polygonal arrangements of rock, soil, and vegetation formed on a level or gently sloping surface by frost action. Also known as cellular soil. { pə'lig·ən·əl 'graund }

polygonal karst [GEOL] A karst pattern that is characteristic of tropical types such as cone karsts, with the surface completely divided into a polygonal network. { pə'lig·ən·əl 'kärst }

polyhalite [MINERAL] $K_2MgCa_2(SO_4)_4 \cdot 2H_2O$ A sulfate mineral usually found in fibrous brick-red masses due to iron. { ,päl·i'ha,līt }

polymere See polygene. { 'päl·ə,mir }

polymetamorphic diaphthoresis [GEOL] Retrograde changes during a second phase of metamorphism that is clearly separated from a previous, higher-grade metamorphic period. { ¦päl·i,med-ə'mòr·fik dī,af·thə'rē·səs }

polymetamorphism [GEOL] Polyphase or multiple metamorphism whereby two or more successive metamorphic events have left their imprint upon the same rocks. { ¦päl·i,med-ə'mòr,fiz·əm }

polymictic [PETR] Of a clastic sedimentary rock, being made up of many rock types or of more than one mineral species. { ¦päl·i¦mik·tik }

polymignite See polymignyte. { ,päl·i'mig,nīt }

polymignyte [MINERAL] $(Ca,Fe,Y,Zr,Th)(Nb,Ti,Ta)O_4$ A black mineral composed of niobate, titanate, and tantalate of cerium-group metals, with calcium and iron. Also spelled polymignite. { ,päl·i'mig,nīt }

polzenite [PETR] **1.** A group of lamprophyres characterized by the presence of olivine and melilite. **2.** Any rock in this group. { 'päl·zə,nīt }

pondage land [GEOL] Land on which water is stored as dead water during flooding, and which does not contribute to the downstream passage of flow. Also known as flood fringe. { 'pän·dij ,land }

Pontian [GEOL] A European stage of geologic time in the uppermost Miocene, above the Sarmatian and below the Plaisancian of the Pliocene; it has also been regarded as the lowermost Pliocene. { 'pän·chən }

pontic [GEOL] Pertaining to sediments or facies deposited in comparatively deep and motionless water, such as an association of black shales and dark limestones deposited in a stagnant basin. { 'pän·tik }

pool [GEOL] Underground accumulation of petroleum. { pül }

popple rock See pebble bed. { 'päp·əl ,räk }

porcelainite See mullite. { 'pòr·slə,nīt }

porcelain jasper [GEOL] A hard, naturally baked, impure clay (or porcellanite) which because of its red color had long been considered a variety of jasper. { 'pòr·slən 'jas·pər }

porcelaneous [GEOL] Resembling unglazed porcelain. { ¦pòr·sə¦lā·nē·əs }

porcelaneous chert [PETR] A hard, opaque to subtranslucent smooth chert, having a smooth fracture surface and a typically china-white appearance resembling chinaware or glazed porcelain. { ¦pòr·sə¦lā·nē·əs 'chərt }

porcellanite [PETR] A hard, dense siliceous rock, such as impure chert or indurated clay or shale. { pòr'sel·ə,nīt }

pore [GEOL] An opening or channelway in rock or soil. { pòr }

pore compressibility [GEOL] The fractional change in reservoir-rock pore volume with a unit change in pressure upon that rock. { 'pòr kəm,pres-ə'bil·əd·ē }

pore-size distribution [GEOL] Variations in pore sizes in reservoir formations; each type of rock has its own typical pore size and related permeability. { 'pòr¦sīz ,dis·trə'byü·shən }

pore space [GEOL] The pores in a rock or soil considered collectively. Also known as pore volume. { 'pȯr ‚spās }

pore volume See pore space. { 'pȯr ‚väl·yəm }

porosity trap See stratigraphic trap. { pə'räs·əd·ē ‚trap }

Poroxylaceae [PALEOBOT] A monogeneric family of extinct plants included in the Cordaitales. { pə‚räk·sə'lās·ē‚ē }

porpezite [MINERAL] A mineral consisting of a native alloy of palladium (5–10%) and gold. Also known as palladium gold. { 'pȯr·pə‚zīt }

porphrite See porphyry. { 'por‚frīt }

porphyritic [PETR] Pertaining to or resembling porphyry. { ‚pȯr·fə¦rid·ik }

porphyroblast [PETR] A relatively large crystal formed in a metamorphic rock. { pȯr'fir·ə‚blast }

porphyroblastic [PETR] Pertaining to the texture of recrystallized metamorphic rock having large idioblasts of minerals possessing high form energy in a finer-grained crystalloblastic matrix. { pȯr¦fir·ə¦blas·tik }

porphyrocrystallic See porphyrotopic. { pȯr¦fir·ō·kri'stal·ik }

porphyroclastic structure See mortar structure. { pȯr¦fir·ō¦klas·tik 'strək·chər }

porphyrogranulitic [PETR] Referring to ophitic texture characterized by large phenocrysts of feldspar and augite or olivine in a groundmass of smaller lath-shaped feldspar crystals and irregular augite grains; a combination of porphyritic and intergranular textures. { pȯr¦fir·ō‚gran·yə'lid·ik }

porphyroid [PETR] **1.** A blastoporphyritic, or sometimes porphyroblastic, metamorphic rock of igneous origin. **2.** A feldspathic metasedimentary rock having the appearance of a porphyry. { 'pȯr·fə‚rȯid }

porphyroskelic [GEOL] Pertaining to an arrangement in a soil fabric whereby the plasma occurs as a dense matrix in which skeleton grains are set like phenocrysts in a porphyritic rock. { pȯr¦fir·ə¦skel·ik }

porphyrotope [GEOL] A large crystal enclosed in a finer-grained matrix in a sedimentary rock showing porphyrotopic fabric. { pȯr'fir·ə‚tōp }

porphyrotopic [GEOL] Referring to the fabric of a crystalline sedimentary rock in which the constituent crystals are of more than one size and in which larger crystals are enclosed in a finer-grained matrix. Also known as porphyrocrystallic. { pȯr¦fir·ə¦täp·ik }

porphyry [PETR] An igneous rock in which large phenocrysts are enclosed in a very-fine-grained to aphanitic matrix. Formerly known as porphrite. { 'pȯr·fə·rē }

Porterfield [GEOL] A North American geologic stage of the Middle Ordovician, forming the lower division of the Mohawkian, and lying above Ashby and below Wilderness. { 'pȯrd·ər‚fēld }

Portlandian [GEOL] A European geologic stage of the Upper Jurassic, above Kimmeridgian, below Berriasian of Cretaceous. { pȯrt'land·ē·ən }

portlandite [MINERAL] Ca(OH)$_2$ A colorless, hexagonal mineral consisting of calcium hydroxide; occurs as minute plates. { 'pȯrt·lən‚dīt }

positive landform [GEOL] An upstanding topographic form, such as a mountain, hill, plateau, or cinder cone. { 'päz·əd·iv 'land‚fȯrm }

positive movement [GEOL] **1.** Uplift or emergence of the earth's crust relative to an adjacent area of the crust. **2.** A relative rise in sea level with respect to land level. { 'päz·əd·iv 'müv·mənt }

positive shoreline See shoreline of submergence. { 'päz·əd·iv 'shȯr‚līn }

Postglacial See Holocene. { pōst'glā·shəl }

posthumous structure [GEOL] Folds, faults, and other structural features in covering strata which revive or mimic the structure of older underlying rocks that are generally more deformed. { 'päs·chə·məs 'strək·chər }

postmagmatic [GEOL] Pertaining to geologic reactions or events occurring after the bulk of the magma has crystallized. { ‚pōst·mag'mad·ik }

postorogenic [GEOL] Of a geologic process or event, occurring after a period of orogeny. { ‚pōst‚ȯr·ə'jen·ik }

potamogenic rock

potamogenic rock [PETR] A sedimentary rock formed by precipitation from river water. { ¦päd·ə·mō¦jen·ik räk }

potarite [MINERAL] PdHg A silver-white isometric mineral composed of palladium and mercury alloy. Also known as palladium amalgam. { pə'tä,rīt }

potash alum *See* kalinite. { 'päd,ash 'al·əm }

potash bentonite *See* potassium bentonite. { 'päd,ash 'bent·ən,īt }

potash feldspar *See* potassium feldspar. { 'päd,ash 'fel,spär }

potash kettle *See* giant's kettle. { 'päd,ash 'ked·əl }

potash mica *See* muscovite. { 'päd,ash 'mī·kə }

potassic [PETR] Referring to a rock which contains a significant amount of potassium. { pə'tas·ik }

potassium-argon dating [GEOL] Dating of archeological, geological, or organic specimens by measuring the amount of argon accumulated in the matrix rock through decay of radioactive potassium. { pə'tas·ē·əm 'är,gän 'dād·iŋ }

potassium bentonite [GEOL] A clay of the illite group that contains potassium and is formed by alteration of volcanic ash. Also known as K bentonite; potash bentonite. { pə'tas·ē·əm 'bent·ən,īt }

potassium feldspar [MINERAL] Any alkali feldspar (orthoclase, microcline, sonidine, adularia) containing the molecule $KAlSi_3O_8$. Incorrectly known as K feldspar; potash feldspar. { pə'tas·ē·əm 'fel,spär }

potato stone [GEOL] A potato-shaped geode, especially one consisting of hard, silicified limestone with an internal lining of quartz crystals. { pə'tā·dō ,stōn }

pothole [GEOL] **1.** A shaftlike cave opening upward to the surface. **2.** Any bowl-shaped, cylindrical, or circular hole formed by the grinding action of a stone in the rocky bed of a river or stream. Also known as churn hole; colk; eddy mill; evorsion hollow; kettle; pot. **3.** A vertical, or nearly vertical shaft in limestone. Also known as aven; cenote. **4.** A small depression with steep sides in a coastal marsh; contains water at or below low-tide level. Also known as rotten spot. { 'pät,hōl }

potrero [GEOL] An elongate, islandlike beach ridge, surrounded by mud flats and separated from the coast by a lagoon and barrier island, made up of a series of accretionary dune ridges. { pə'trer·ō }

Poulter seismic method [GEOPHYS] A type of air shooting in which the explosive is set on poles above the ground. { 'pōl·tər 'sīz·mik ,meth·əd }

powder avalanche [GEOL] Loose powder snow rapidly descending a mountainside. { ¦paúd·ər ¦av·ə,lanch }

powellite [MINERAL] $Ca(WMo)O_4$ A commercially important tungsten mineral, crystallizing in the tetragonal system; isomorphous with scheelite ($CaWO_4$). { ,paú·ə,līt }

pozzolan [GEOL] A finely ground burnt clay or shale resembling volcanic dust, found near Pozzuoli, Italy; used in cement because it hardens underwater. { 'pät·sə·lən }

prairie soil [GEOL] A group of zonal soils having a surface horizon that is dark or grayish brown, which grades through brown soil into lighter-colored parent material; it is 2–5 feet (0.6–1.5 meters) thick and develops under tall grass in a temperate and humid climate. { 'prer·ē ,sóil }

prase [MINERAL] **1.** A translucent and dull leek green or light-grayish yellow-green variety of chalcedony. **2.** Crystalline quartz containing a multitude of green hairlike crystals of actinolite. Also known as mother-of-emerald. { präz }

prase opal *See* prasopal. { 'präz 'ō·pəl }

prasinite [PETR] A greenschist in which the proportions of the hornblende-chlorite-epidote assemblage are more or less equal. { 'präz·ən,īt }

prasopal [MINERAL] A green variety of common opal containing chromium. Also spelled prase opal. { 'präz,ō·pəl }

prealpine facies [GEOL] A geosynclinal facies characteristic of neritic areas, displaying thick limestone deposits and coarse terrigenous material and resembling epicontinental platform sediments. { prē'al,pīn 'fā·shēz }

Precambrian [GEOL] All geologic time prior to the beginning of the Paleozoic era (before 600,000,000 years ago); equivalent to about 90% of all geologic time. { prē 'kam·brē·ən }

precious stone [MINERAL] **1.** Any genuine gemstone. **2.** A gemstone of high commercial value because of its beauty, rarity, durability, and hardness; examples are diamond, ruby, sapphire, and emerald. { 'presh·əs 'stōn }

precipice [GEOL] A very steeply inclined, vertical, or overhanging wall or surface of rock. { 'pres·ə·pəs }

precipitation facies [GEOL] Facies characteristics that provide evidence of depositional conditions; revealed mainly by sedimentary structures (such as cross-bedding and ripple marks) and by primary constituents (especially fossils). { prə,sip·ə'tā·shən ,fā·shēz }

preconsolidation pressure [GEOL] The greatest effective stress exerted on a soil; result of this pressure from overlying materials is compaction. Also known as prestress. { ¦prē·kən,säl·ə'dā·shən ,presh·ər }

predazzite [PETR] A recrystallized limestone that resembles pencatite, but contains less brucite than calcite. { 'pred·ə,zīt }

predozzite [PETR] Limestone rich in periclase and brucite. { 'pred·ə,zīt }

preferred orientation [PETR] The nonrandom orientation of planar or linear fabric elements in structural petrology. { pri'fərd ,ȯr·ē·ən'tā·shən }

preglacial [GEOL] **1.** Pertaining to the geologic time immediately preceding the Pleistocene epoch. **2.** Of material, underlying glacial deposits. { prē'glā·shəl }

prehnite [MINERAL] $Ca_2Al_2Si_3O_{10}(OH)_2$ A light-green to white mineral sorosilicate crystallizing in the orthorhombic system and generally found in reniform and stalactitic aggregates with crystalline surface; it has a vitreous luster, hardness is 6–6.5 on Mohs scale, and specific gravity is 2.8–2.9. { 'prā,nīt }

preliminary waves [GEOPHYS] The body of waves of an earthquake, including both P waves and S waves. { pri'lim·ə,ner·ē ¦wāvz }

preorogenic [GEOL] The initial phase of an orogenic cycle during which geosynclines form. { ¦prē,ȯr·ə'jen·ik }

pressolution See pressure solution. { 'pres·ə,lü·shən }

pressolved [GEOL] Referring to a sedimentary bed or rock in which the grains have undergone pressure solution. { pri'zälvd }

pressure breccia See tectonic breccia. { 'presh·ər ,brech·ə }

pressure fringe See pressure shadow. { 'presh·ər ,frinj }

pressure pan [GEOL] An induced soil pan which has a higher bulk density and a lower total porosity than the soil directly above or below it and is produced as a result of pressure applied by normal tillage operations or by other artificial means. { 'presh·ər ,pan }

pressure penitente [GEOL] A nieve penitente composed of brilliantly white ice which is shaped into a slender ridge by lateral pressure of converging morainal streams and by melting of the adjacent debris-covered ice. { 'presh·ər ,pen·ə'ten·tā }

pressure plateau [GEOL] An uplifted area of a thick lava flow, measuring up to 10 or 13 feet (3 or 4 meters), the uplift of which is due to the intrusion of new lava from below that does not reach the surface. { 'presh·ər pla,tō }

pressure release [GEOPHYS] The outward-expanding force of pressure which is released within rock masses by unloading, as by erosion of superincumbent rocks or by removal of glacial ice. { 'presh·ər ri,lēs }

pressure-release jointing [GEOL] Exfoliation that occurs in once deeply buried rock that erosion has brought nearer the surface, thus releasing its confining pressure. { 'presh·ər ri¦lēs ,jȯint·iŋ }

pressure ridge [GEOL] **1.** A seismic feature resulting from transverse pressure and shortening of the land surface. **2.** An elongate upward movement of the congealing crust of a lava flow. **3.** A ridge of glacier ice. { 'presh·ər ,rij }

pressure shadow [PETR] In structural petrology, an area adjoining a porphyroblast, characterized by a growth fabric rather than a deformation fabric, as seen in a section perpendicular to the b axis of the fabric. Also known as pressure fringe; strain shadow. { 'presh·ər ,shad·ō }

pressure solution [PETR] In a sedimentary rock, solution occurring preferentially at the grain boundary surfaces. Also known as pressolution. { 'presh·ər sə,lü·shən }

prestress [GEOL] *See* preconsolidation pressure. { ¦prē'stres }

presuppression [GEOPHYS] In seismic prospecting, the suppression of the early events on a seismic record for control of noise and reflections on that portion of the record. { ¦prē·sə'presh·ən }

previtrain [GEOL] The woody lenses in lignite that are equivalent to vitrain in coal of higher rank. { prē'vi‚trān }

Priabonian [GEOL] A European stage of geologic time in the upper Eocene, believed to consist of Auversian and Bartonian. { ‚prē·ə'bō·nē·ən }

priceite [MINERAL] $Ca_4B_{10}O_{19} \cdot 7H_2O$ A snow-white earthy mineral composed of hydrous calcium borate, occurring as a massive. Also known as pandermite. { 'prī‚sīt }

primary [GEOL] **1.** A young shoreline whose features are produced chiefly by nonmarine agencies. **2.** Of a mineral deposit, unaffected by supergene enrichment. { 'prī‚mer·ē }

primary arc [GEOL] **1.** A curved segment of elongated mountain zones that are the areas of the earth's major and most recent tectonic activity. **2.** *See* internides. { 'prī‚mer·ē 'ärk }

primary basalt [PETR] Theoretically, the original magma from which all other rock types are supposedly obtained by various processes. { 'prī‚mer·ē bə'sólt }

primary clay *See* residual clay. { 'prī‚mer·ē 'klā }

primary crater [GEOL] **1.** An impact crater produced directly by the high-velocity impact of a meteorite or other projectile. **2.** *See* true crater. { 'prī‚mer·ē 'krād·ər }

primary dip [GEOL] The slight dip assumed by a bedded deposit at its moment of deposition. Also known as depositional dip; initial dip; original dip. { 'prī‚mer·ē 'dip }

primary fabric *See* apposition fabric. { 'prī‚mer·ē 'fab·rik }

primary flat joint [GEOL] An approximately horizontal joint plane in igneous rocks. Also known as L joint. { 'prī‚mer·ē 'flat ‚jóint }

primary geosyncline *See* orthogeosyncline. { 'prī‚mer·ē ¦jē·ō'sin‚klīn }

primary gneiss [PETR] A rock that exhibits planar or linear structures characteristic of metamorphic rocks but lacks observable granulation or recrystallization and is therefore considered to be of igneous origin. { 'prī‚mer·ē 'nīs }

primary gneissic banding [PETR] A kind of banding developed in certain igneous (plutonic) rocks of heterogeneous composition, produced by the admixture of two magmas only partly miscible or by magma intimately admixed with country rock into which it has been injected along planes of bedding or foliation. { 'prī‚mer·ē ¦nī‚sik 'band·iŋ }

primary interstice *See* original interstice. { 'prī‚mer·ē in'tər‚stəs }

primary magma [GEOL] A magma that originates below the earth's crust. { 'prī‚mer·ē 'mag·mə }

primary mineral [MINERAL] A mineral that is formed at the same time as the rock in which it is contained, and that retains its original form and composition. { 'prī‚mer·ē 'min·rəl }

primary orogeny [GEOL] Orogeny that is characteristic of the internides and that involves deformation, regional metamorphism, and granitization. { 'prī‚mer·ē ó'räj·ə·nē }

primary porosity [GEOL] Natural porosity in petroleum reservoir sands or rocks. { 'prī‚mer·ē pə'räs·əd·ē }

primary rocks [PETR] Rocks whose constituents are newly formed particles that have never been constituents of previously formed rocks and that are not the products of alteration or replacement, such as limestones formed by precipitation from solution. { 'prī‚mer·ē 'räks }

primary sedimentary structure [GEOL] A sedimentary structure produced during deposition, such as ripple marks and graded bedding. { 'prī‚mer·ē ‚sed·ə'men·trē ‚strək·chər }

primary stratification [GEOL] Stratification which develops when sediments are first deposited. Also known as direct stratification. { 'prī‚mer·ē ‚strad·ə·fə'kā·shən }

primary stratigraphic trap [GEOL] A stratigraphic trap formed by the deposition of clastic materials (such as shoestring sands, lenses, sand patches, bars, or cocinas) or through chemical deposition (such as organic reefs or biostromes). { 'prī,mer·ē ¦strad·ə¦graf·ik 'trap }

primary stress field *See* ambient stress field. { 'prī,mer·ē 'stres ,fēld }

primary structure [GEOL] A structure, in an igneous rock, that formed at the same time as the rock, but before its final consolidation. { 'prī,mer·ē 'strək·chər }

primary tectonite [PETR] A tectonite with depositional fabric. { 'prī,mer·ē 'tek·tə,nīt }

primary wave [GEOPHYS] The first seismic wave that reaches a station from an earthquake. { 'prī,mer·ē 'wāv }

Primitiopsacea [PALEON] A small dimorphic superfamily of extinct ostracodes in the suborder Beyrichicopina; the velum of the male was narrow and uniform, but that of the female was greatly expanded posteriorly. { prī,mid·ē·äp'sā·shə }

principle of uniformity *See* uniformitarianism. { 'prin·sə·pəl əv ,yü·nə'fȯr·məd·ē }

Prioniodidae [PALEON] A family of conodonts in the suborder Conodontiformes having denticulated bars with a large denticle at one end. { ,prī·ə,nī'äd·ə,dē }

Prioniodinidae [PALEON] A family of conodonts in the suborder Conodontiformes characterized by denticulated bars or blades with a large denticle in the middle third of the specimen. { ,prī·ə,nī·ə'din·ə,dē }

priorite [MINERAL] (Y,Ce,Th)(Ti,Nb)$_2$O$_6$ A mineral composed of titanoniobate of rare-earth metals; it is isomorphous with eschynite. Also known as blomstrandine. { 'prī·ə,rīt }

prism [GEOL] A long, narrow, wedge-shaped sedimentary body with a width-thickness ratio greater than 5 to 1 but less than 50 to 1. { 'priz·əm }

prismatic jointing *See* columnar jointing. { priz'mad·ik 'jȯint·iŋ }

prismatic structure *See* columnar jointing. { priz'mad·ik 'strək·chər }

prism crack [GEOL] A mud crack that develops in regular or irregular polygonal patterns on the surface of drying mud puddles and that breaks the sediment into prisms standing normal to the bedding. { 'priz·əm ,krak }

Proanura [PALEON] Triassic forerunners of the Anura. { prō'an·yə·rə }

probertite [MINERAL] NaCaB$_5$O$_9$·5H$_2$O A colorless mineral crystallizing in the monoclinic system, consisting of hydrous sodium calcium borate. { 'präb·ər,tīt }

Procellarian [GEOL] Pertaining to lunar lithologic map units and topographic forms constituting, or closely associated with, the maria. { ,prō·sə'lar·ē·ən }

Procolophonia [PALEON] A subclass of extinct cotylosaurian reptiles. { ,präk·ə·lə'fō·nē·ə }

prod cast [GEOL] The cast of a prod mark. Also known as impact cast. { 'präd ,kast }

prodelta [GEOL] The part of a delta lying beyond the delta front, and sloping gently down to the basin floor of the delta; it is entirely below the water level. { 'prō,del·tə }

prodelta clay [GEOL] Fine sand, silt, and clay transported by the river and deposited on the floor of a sea or lake beyond the main body of a delta. { 'prō,del·tə ,klā }

Prodinoceratinae [PALEON] A subfamily of extinct herbivorous mammals in the family Untatheriidae; animals possessed a carnivorelike body of moderate size. { ¦präd·ən·ō·sə'rat·ən,ē }

prod mark [GEOL] A short tool mark oriented parallel to the current and gradually deepening downcurrent. Also known as impact mark. { 'präd ,märk }

Productinida [PALEON] A suborder of extinct articulate brachiopods in the order Strophomenida characterized by the development of spines. { ,prä·dək'tin·ə·də }

profile [GEOL] **1.** The outline formed by the intersection of the plane of a vertical section and the ground surface. Also known as topographic profile. **2.** Data recorded by a single line of receivers from one shot point in seismic prospecting. [GEOPHYS] A graphic representation of the variation of one property, such as gravity, usually as ordinate, with respect to another property, usually linear, such as distance. [PETR] In structural petrology, a cross section of a homoaxial structure. { 'prō,fīl }

profile line [GEOL] The top line of a profile section, representing the intersection of a vertical plane with the surface of the ground. { ¦prō,fīl ,līn }

profile of equilibrium [GEOL] **1.** The slope of the floor of a sea, ocean, or lake, taken in a vertical plane, when deposition of sediment is balanced by erosion. **2.** The longitudinal profile of a graded stream. Also known as equilibrium profile; graded profile. { ¦prō‚fīl əv ‚ē·kwə'lib·rē·əm }

profile section |GEOL| A diagram or drawing that shows along a given line the configuration or slope of the surface of the ground as it would appear if it were intersected by a vertical plane. { ¦prō‚fīl ‚sek·shən }

Proganosauria |PALEON| The equivalent name for Mesosauria. { prō‚gan·ə'sȯr·ē·ə }

proglacial |GEOL| Of streams, deposits, and other features, being immediately in front of or just beyond the outer limits of a glacier or ice sheet, and formed by or derived from glacier ice. { prō'glā·shəl }

progradation |GEOL| Seaward buildup of a beach, delta, or fan by nearshore deposition of sediments transported by a river, by accumulation of material thrown up by waves, or by material moved by longshore drifting. { ¦prō·grə'dā·shən }

prograde metamorphism |GEOL| Metamorphic changes in response to a higher pressure or temperature than that to which the rock was last adjusted. { ¦prō'grād ‚med·ə'mȯr‚fiz·əm }

prograding shoreline |GEOL| A shoreline that is being built seaward by accumulation or deposition. { ¦prō'grād·iŋ 'shȯr‚līn }

progressive metamorphism |GEOL| Systematic change in metamorphic grade from lower to higher in any metamorphic terrain. { prə'gres·iv ‚med·ə'mȯr‚fiz·əm }

progressive sand wave |GEOL| A sand wave characterized by downcurrent migration. { prə'gres·iv 'sand ‚wāv }

progressive sorting |GEOL| Sorting of sedimentary particles in the downcurrent direction, resulting in a systematic downcurrent decrease in the mean grain size of the sediment. { prə'gres·iv 'sȯrd·iŋ }

Progymnospermopsida |PALEON| A class of plants intermediate between ferns and gymnosperms; comprises the Devonian genus *Archaeopteris*. { prō¦jim·nō‚spər'mäp·səd·ə }

projection net *See* net. { prə'jek·shən ‚net }

Prolacertiformes |PALEON| A suborder of extinct terrestrial reptiles in the order Eosuchia distinguished by reduction of the lower temporal arcade. { prō¦las·ər·də'fȯr‚mēz }

prolapsed bedding |GEOL| Bedding characterized by a series of flat folds with near-horizontal axial planes contained entirely within a bed which has undisturbed boundaries. { 'prō‚lapst 'bed·iŋ }

proluvium |GEOL| A complex, friable, deltaic sediment accumulated at the foot of a slope as a result of an occasional torrential washing of fragmental material. { prō'lü·vē·əm }

promontory |GEOL| **1.** A high, prominent projection or point of land, or a rock cliff, jutting out boldly into a body of water. **2.** A cape, either low-lying or of considerable height, with a bold termination. **3.** A bluff or prominent hill overlooking or projecting into a lowland. { 'präm·ən‚tȯr·ē }

prong reef |GEOL| A wall reef that has developed irregular buttresses normal to its axis in both leeward and (to a smaller degree) seaward directions. { präŋ ‚rēf }

propaedeutic stratigraphy *See* prostratigraphy. { ‚prō·pi'düd·ik strə'tig·rə·fē }

propylite |PETR| A modified andesite, altered by hydrothermal processes, resembling a greenstone and consisting of calcite, epidote, serpentine, quartz, pyrite, and iron ore. { 'prō·pə‚līt }

propylization |PETR| A hydrothermal process by which propylite is formed from andesite by the introduction of or replacement by an assemblage of minerals. { ‚prō·pəl·ə'zā·shən }

Prorastominae |PALEON| A subfamily of extinct dugongs (Dugongidae) which occur in the Eocene of Jamaica. { ‚prȯr·ə'stäm·ə‚nē }

Prosauropoda |PALEON| A division of the extinct reptilian suborder Sauropodomorpha; they possessed blunt teeth, long forelimbs, and extremely large claws on the first finger of the forefoot. { ‚prä·sȯ'räp·əd·ə }

prosopite |MINERAL| $CaAl_2(F,OH)_8$ A colorless mineral composed of basic calcium aluminum fluoride. { 'präs·ə‚pīt }

prostratigraphy [GEOL] Preliminary stratigraphy, including lithologic and paleonto-logic studies, without consideration of the time factor. Also known as propaedeutic stratigraphy; protostratigraphy. { ‚prä·strə'tig·rə·fē }

protactinium-ionium age method [GEOL] A method of calculating the ages of deep-sea sediments formed during the last 150,000 years from measurements of the ratio of protactinium-231 to ionium (thorium-230), based on the gradual change of this ratio over time because of the difference in half-lives. { ¦prōd‚ak'tin·ē·əm ī'ō·nē·əm 'āj ‚meth·əd }

protalus rampart [GEOL] An arcuate ridge consisting of boulders and other coarse debris marking the downslope edge of an existing or melted snowbank. { prō'tal·əs 'ram‚pärt }

protectite [PETR] A rock formed by the crystallization of a primary magma. { prə'tek‚tīt }

Proterosuchia [PALEON] A suborder of moderate-sized thecodont reptiles with lightly built triangular skulls, downturned snouts, and palatal teeth. { ‚präd·ə·rō'sü·kē·ə }

Proterotheriidae [PALEON] A group of extinct herbivorous mammals in the order Litopterna which displayed an evolutionary convergence with the horses in their dentition and in reduction of the lateral digits of their feet. { ‚präd·ə·rō·thə'rī·ə‚dē }

Proterozoic [GEOL] Geologic time between the Archean and Paleozoic eras, that is, from 2500 million to 550 million years ago. Also known as Algonkian. { ¦präd·ə·rə¦zō·ik }

Protoceratidae [PALEON] An extinct family of pecoran ruminants in the superfamily Traguloidea. { ¦prōd·ō·sə'rad·ə‚dē }

protoclastic [PETR] Of igneous rocks, characterized by granulation and deformation of the earlier-formed minerals due to differential flow of the magma before solidification. { ¦prōd·ō¦klas·tik }

protodolomite [MINERAL] A crystalline calcium-magnesium carbonate with a disordered lattice in which the metallic ions occur in the same crystallographic layers instead of in alternate layers as in the dolomite mineral. { ¦prōd·ō'dō·lə‚mīt }

Protodonata [PALEON] An extinct order of huge dragonflylike insects found in Permian rocks. { ¦prōd·ō·də'näd·ə }

protoenstatite [MINERAL] An artificial, unstable, altered form of MgSiO$_3$ produced by thermal decomposition of talc; convertible to enstatite by grinding or heating to a high temperature. { ¦prōd·ō'en·stə‚tīt }

Protoeumalacostraca [PALEON] The stem group of the crustacean series Eumalacostraca. { ‚prōd·ō‚yü·mə·lə'käs·trə·kə }

protointraclast [GEOL] A limestone component that resulted from a premature attempt at resedimentation while it was still in an unconsolidated and viscous or plastic state, and that never existed as a free clastic entity. { ‚prōd·ō'in·trə‚klast }

protolith [PETR] The original, unmetamorphosed rock from which a given metamorphic rock is formed. { 'prōd·ə‚lith }

protomylonite [PETR] A mylonitic rock that develops from contact-metamorphosed rock; granulation and flowage are caused by overthrusts following the contact surfaces between the intrusion and the country rock. { ‚prōd·ō'mī·lə‚nīt }

Protopteridales [PALEOBOT] An extinct order of ferns, class Polypodiatae. { ‚prōd·ō‚ter·ə'dā·lēz }

protoquartzite [PETR] A well-sorted sandstone that is intermediate in composition between subgraywacke and orthoquartzite, consisting of 75–95% quartz and chert, with less than 15% detrital clay matrix and 5–25% unstable materials in which there is a greater abundance of rock fragments than feldspar grains. Also known as quartzose subgraywacke. { ¦prōd·ō'kwȯrt‚sīt }

Protosireninae [PALEON] An extinct superfamily of sirenian mammals in the family Dugongidae found in the middle Eocene of Egypt. { ‚prōd·ō·sə'ren·ə‚nē }

protostratigraphy See prostratigraphy. { ¦prōd·ō·strə'tig·rə·fē }

Protosuchia [PALEON] A suborder of extinct crocodilians from the Late Triassic and Early Jurassic. { ¦prōd·ō'sü·kē·ə }

proustite [MINERAL] Ag_3AsS_3 A cochineal-red mineral that crystallizes in the rhombohedral system, consists of silver arsenic sulfide, is isomorphous with pyrargyrite, and occurs massively and in crystals. Also known as light-red silver ore; light-ruby silver. { 'prü,sīt }

provenance [GEOL] The location, topography, and composition of the source area for any sedimentary rock. Also known as source area; sourceland. { 'präv·ə·nəns }

provincial series [GEOL] A time-stratigraphic series recognized only in a particular region and involving a major division of time within a period. { prə'vin·chəl 'sir·ēz }

provitrain [GEOL] Vitrain in which some plant structure can be discerned by microscope. Also known as telain. { prō'vi,trān }

provitrinite [GEOL] A variety of vitrinite characteristic of provitrain and including the varieties periblinite, suberinite, and xylinite. { prō'vi·trə,nīt }

proximal [GEOL] Of a sedimentary deposit, composed of coarse clastics and formed near the source. { 'präk·sə·məl }

Psamment [GEOL] A suborder of the soil order Entisol, characterized by a texture of loamy fine sand or coarser sand, and by a coarse fragment content of less than 35. { 'sa,ment }

psammite See arenite. { 'sa,mīt }

psammitic See arenaceous. { sə'mid·ik }

Psammodontidae [PALEON] A family of extinct cartilaginous fishes in the order Bradyodonti in which the upper and lower dentitions consisted of a few large quadrilateral plates arranged in two rows meeting in the midline. { ,sam·ə'dänt·ə,dē }

psephicity [GEOL] A coefficient of roundability of a pebble- or sand-size mineral fragment, expressed as the ratio of specific gravity to hardness (as measured in the air) or the quotient of specific gravity minus one divided by hardness (as measured in water). { sə'fis·əd·ē }

psephite [GEOL] A sediment or sedimentary rock composed of fragments that are coarser than sand and which are set in a qualitatively and quantitatively varying matrix; equivalent to a rudite or, generally, a conglomerate. { sē,fīt }

psephyte [GEOL] A lake-bottom deposit consisting mainly of coarse, fibrous plant remains. { sē,fīt }

pseudoallochem [GEOL] An object resembling an allochem but produced in place within a calcareous sediment by a secondary process such as recrystallization. { ,sü·dō'al·ə,kem }

Pseudoborniales [PALEOBOT] An order of fossil plants found in Middle and Upper Devonian rocks. { ¦sü·dō,bȯr·nē'ā·lēz }

pseudobreccia [PETR] Limestone that is partially and irregularly dolomitized and is characterized by a mottled, breccialike appearance. Also known as recrystallization breccia. { ¦sü·dō'brech·ə }

pseudobrookite [MINERAL] Fe_2TiO_5 A brown or black mineral consisting of iron titanium oxide and occurring in orthorhombic crystals; specific gravity is 4.4–4.98. { ¦sü·dō'brü,kīt }

pseudocannel coal [GEOL] Cannel coal that contains much humic matter. Also known as humic-cannel coal. { ¦sü·dō¦kan·əl 'kōl }

pseudochrysolite See moldavite. { ¦sü·dō'kris·ə,līt }

pseudocol [GEOL] A landform represented by a constriction of a stream valley diverted by a glacial ponding, formed by the cutting through of a cover of drift and subsequent exposure of a former col. { 'süd·ə,kȯl }

pseudoconcretion [GEOL] A subspherical, secondary sedimentary structure resembling a true concretion but not formed by orderly precipitation of mineral matter in the pores of a sediment. { ¦sü·dō·kän'krē·shən }

pseudoconformity See paraconformity. { ¦sü·dō·kən'fȯr·məd·ē }

pseudoconglomerate [GEOL] A rock that resembles, or may easily be mistaken for, a true or normal (sedimentary) conglomerate. { ¦sü·dō·kən'gläm·ə·rət }

pseudocotunnite [MINERAL] K_2PbCl_4 A yellow or yellowish-green, orthorhombic mineral consisting of a potassium lead chloride. { ¦sü·dō·kə'tə,nīt }

pseudo cross-bedding [GEOL] **1.** An inclined bedding produced by deposition in response to ripple-mark migration and characterized by foreset beds that appear to dip into the current.   **2.** A structure resembling cross bedding, caused by distortion-free slumping and sliding of a semiconsolidated mass of sediments (such as sandy shales).   { 'sü·dō 'krós ,bed·iŋ }

pseudodiffusion   [GEOL]   Mixing of thin superpositioned layers of slowly accumulated marine sediments by the action of water motion or subsurface organisms.   { ¦sü·dō·di'fyü·zhən }

pseudofault   [GEOL]   A faultlike feature resulting from weathering along joint, shrinkage, or bedding planes.   { 'süd·ə,fólt }

pseudofibrous peat   [GEOL]   Peat that is fibrous in texture but is plastic and incoherent.   { ¦sü·dō'fī·brəs 'pēt }

pseudogalena   See sphalerite.   { ¦sü·do·gə'lē·nə }

pseudogley   [GEOL]   A densely packed, silty soil that is alternately waterlogged and rapidly dried out.   { 'sü·dō,glā }

pseudogradational bedding   [GEOL]   A structure in metamorphosed sedimentary rock in which the original textural graduation (coarse at the base, finer at the top) appears to be reversed because of the formation of porphyroblasts in the finer-grained part of the rock.   { ¦sü·dō·grā'dā·shən·əl 'bed·iŋ }

pseudokarst   [GEOL]   A topography that resembles karst but that is not formed by the dissolution of limestone; usually a rough-surfaced lava field in which ceilings of lava tubes have collapsed.   { 'süd·ə,kärst }

pseudokettle   See pingo remnant.   { ¦sü·dō'ked·əl }

pseudoleucite   [MINERAL]   A pseudomorph after leucite consisting of a mixture of nepheline, orthoclase, and analcime.   { ¦sü·dō'lü,sīt }

pseudomalachite   [MINERAL] $Cu_5(PO_4)_2(OH)_4 \cdot H_2O$ An emerald green to dark green and blackish-green, monoclinic mineral consisting of a hydrated basic copper phosphate. Also known as tagilite.   { ¦sü·dō'mal·ə,kīt }

pseudomicroseism   [GEOPHYS]   A microseism due to instrumental effects.   { ,sü·dō'mī·krə,sīz·əm }

pseudomorph   [MINERAL]   An altered mineral whose crystal form has the outward appearance of another mineral species.   Also known as false form.   { 'süd·ə,mórf }

pseudomountain   [GEOL]   A mountain formed by differential erosion, in contrast to one produced by uplift.   { ¦sü·dō¦maúnt·ən }

pseudonodule   [GEOL]   A primary sedimentary structure consisting of a ball-like mass of sandstone enclosed in shale or mudstone; characterized by a rounded base with upturned or inrolled edges and resulting from the settling of sand into underlying clay or mud which has welled up between isolated sand masses.   Also known as sand roll.   { ¦sü·dō'näj,ül }

pseudo-oolith   [GEOL]   A spherical or roundish pellet or particle (generally less than l millimeter in diameter) in a sedimentary rock, externally resembling an oolith in size or shape but of secondary origin and amorphous or crypto- or microcrystalline, and lacking the radial or concentric internal structure of an oolith.   Also known as false oolith.   { ¦sü·dō'ō,ō,lith }

pseudoporphyritic   [PETR]   Pertaining to a rock that is not a true porphyry, but resembles one because of rapid growth of some of the crystals.   { ¦sü·dō,pòr·fə'rid·ik }

pseudo ripple mark   [GEOL]   A bedding-plane feature that resembles a ripple mark but is formed by lateral pressure caused by slumping or by local, small-scale tectonic deformation.   { ¦sü·dō 'rip·əl ,märk }

pseudospharolith   [MINERAL]   A spherulite consisting of two minerals, one with parallel and one with inclined extinction, growing from the same center.   { ¦sü·dō'sfar·ə,lith }

pseudostratification   See sheeting structure.   { ¦sü·dō,strad·ə·fə'kā·shən }

Pseudosuchia   [PALEON]   A suborder of extinct reptiles of the order Thecodontia comprising bipedal, unarmored or feebly armored forms which resemble dinosaurs in many skull features but retain a primitive pelvis.   { ,sü·dō'sü·kē·ə }

pseudotachylite   [PETR]   A black rock that resembles tachylite; carries fragmental enclosures and shows evidence of having been at high temperature.   { ¦sü·dō'tak·ə,līt }

pseudotillite   [GEOL]   A nonglacial tillite-like rock, such as a pebbly mudstone, formed

279

on land by the flow of nonglacial mud or deposited by a subaqueous turbidity flow. { ¦sü·dō'däd·əl‚īt }

pseudounconformity |GEOL| A stratigraphic relationship that appears unconformable but is characterized by a superabundance or an excess accumulation of sediment, due to factors like submarine slumping which occurs penecontemporaneously with sedimentation off the sides of a rising anticline or dome. { ¦sü·dō·kən'fȯr·məd·ē }

pseudovitrinite |GEOL| A maceral of coal that is superficially similar to vitrinite but that is higher in reflectance from polished surfaces in oil immersion and has slitted structure, remnant cellular structures, uncommon fracture patterns, higher relief, and paucity or absence of pyrite inclusions. { ¦sü·dō'vi·trə‚nīt }

pseudovitrinoid |GEOL| Pseudovitrinite occurring in bituminous coal. { ¦sü·dō'vi·trə‚nȯid }

pseudovolcano |GEOL| A large crater or circular hollow believed not to be associated with recent volcanic activity, such as a crater which is the result of cauldron subsidence or of a phreatic explosion in the distant past. { ¦sü·dō·väl'kā·nō }

psilomelane |MINERAL| BaMn₉O₁₆(OH)₄ A massive, hard, black, botryoidal manganese oxide mineral mixture with a specific gravity ranging from 3.7 to 4.7. { ‚sī·lō'me‚lān }

Psilophytales |PALEOBOT| A group formerly recognized as an order of fossil plants. { ‚sī·lō‚fī'tā·lēz }

Psilophytineae |PALEON| The equivalent name for Rhyniopsida. { ‚sī·lō‚fī'tin·ē‚ē }

psittacinite See mottramite. { sə'tas·ə‚nīt }

Pteridophytic See Paleophytic. { tə¦rid·ə¦fid·ik }

Pteridospermae |PALEOBOT| Seed ferns, a class of the Cycadicae comprising extinct plants characterized by naked seeds borne on large fernlike fronds. { ‚ter·ə·dō'spər‚mē }

Pteridospermophyta |PALEOBOT| The equivalent name for Pteridospermae. { ‚ter·ə·dō·spər'mäf·əd·ē }

pterodactyl |PALEON| The common name for members of the extinct reptilian order Pterosauria. { ¦ter·ə¦dak·təl }

Pterodactyloidea |PALEON| A suborder of Late Jurassic and Cretaceous reptiles in the order Pterosauria distinguished by lacking tails and having increased functional wing length due to elongation of the metacarpals. { ¦ter·ə¦dak·tə'lȯid·ē·ə }

pteropod ooze |GEOL| A pelagic sediment containing at least 45% calcium carbonate in the form of tests of marine animals, particularly pteropods. { 'ter·ə‚päd 'üz }

Pterosauria |PALEON| An extinct order of flying reptiles of the Mesozoic era belonging to the subclass Archosauria; the wing resembled that of a bat, and a large heeled sternum supported strong wing muscles. { ‚ter·ə'sȯr·ē·ə }

Ptilodontoidea |PALEON| A suborder of extinct mammals in the order Multituberculata. { ¦til·ō·dän'tȯid·ē·ə }

ptilolite See mordenite. { 'til·ə‚līt }

PTRM See partial thermoremanent magnetization.

Ptyctodontida |PALEON| An order of Middle and Upper Devonian fishes of the class Placodermi in which both the head and trunk shields are present, and the joint between them is a well-differentiated and variable structure. { ‚tik·tə'dänt·əd·ə }

ptygma |GEOL| Pegmatitic material with migmatite or gneiss, resembling disharmonic folds. Also known as ptygmatic fold. { 'tig·mə }

ptygmatic fold See ptygma. { tig'mad·ik 'fōld }

pucherite |MINERAL| BiVO₄ A reddish-brown orthorhombic mineral composed of bismuth vanadate, occurring as small crystals. { 'pü·kə‚rīt }

pudding ball See armored mud ball. { 'pud·iŋ ‚bȯl }

puddingstone |GEOL| In the United Kingdom, a conglomerate consisting of rounded pebbles whose colors are in marked contrast with the matrix, giving a section of the rock the appearance of a raisin pudding. { 'pud·iŋ‚stōn }

puff cone See mud cone. { 'pəf ‚kōn }

pulaskite |PETR| A light-colored, feldspathoid-bearing, granular or trachytoid alkali syenite composed chiefly of orthoclase, soda pyroxene, arfvedsonite, and nepheline. { pü'las‚kīt }

pull-apart [GEOL] A precompaction sedimentary structure having the appearance of boudinage and consisting of beds that have been stretched and pulled apart into relatively short slabs. { 'púl ə,pärt }

pulverite [PETR] A sedimentary rock composed of silt- or clay-sized aggregates of non-clastic origin with a texture simulating a lutite of clastic origin. { 'pəl·və,rīt }

pumice [GEOL] A rock froth, formed by the extreme puffing up of liquid lava by expanding gases liberated from solution in the lava prior to and during solidification. Also known as foam; pumice stone; pumicite; volcanic foam. { 'pəm·əs }

pumice fall [GEOL] Pumice falling from a volcano eruption cloud. { 'pəm·əs ,fól }

pumiceous [GEOL] Pertaining to the texture of a pyroclastic rock, such as pumice, characterized by numerous small cavities presenting a spongy, frothy appearance. { pyü'mish·əs }

pumice stone *See* pumice. { 'pəm·əs ,stōn }

pumicite *See* pumice. { 'pəm·ə,sīt }

pumilith [GEOL] A lithified deposit of volcanic ash. { 'pəm·ə,lith }

pumpellyite [MINERAL] $Ca_2Al_3Si_3O_{12}(OH)$ A greenish epidote-like mineral that is probably related to clinozoisite. Also known as lotrite; zonochlorite. { ,pəm'pel·ē,īt }

pumpellyite-prehnite-quartz facies [PETR] A variety of low-temperature, moderate-pressure metamorphism. { ¦pəm 'pel·ē,īt ¦prā,nīt ¦kwórts 'fā·shēz }

Purbeckian [GEOL] A stage of geologic time in the United Kingdom: uppermost Jurassic (above Bononian, below Cretaceous). { pər'bek·ē·ən }

pure coal *See* vitrain. { 'pyúr ¦kōl }

purple blende *See* kermesite. { 'pər·pəl ¦blend }

purpurite [MINERAL] $(Mn,Fe)PO_4$ A dark-red or purple mineral composed of ferric-manganic phosphate; it is isomorphous with heterosite. { 'pər·pyə,rīt }

push moraine [GEOL] A broad, smooth, arc-shaped ridge consisting of material mechanically pushed or shoved along by an advancing glacier. Also known as push-ridge moraine; shoved moraine; thrust moraine; upsetted moraine. { 'púsh mə,rān }

push-ridge moraine *See* push moraine. { 'púsh ¦rij mə,rān }

Pustulosa [PALEON] An extinct suborder of echinoderms in the order Phanerozonida found in the Paleozoic. { ,pəs·chə'lō·sə }

puy [GEOL] A small, remnant volcanic cone. { pwē }

P wave [GEOPHYS] A body wave that can pass through all layers of the earth. It is fastest of all seismic waves, traveling at a velocity of 3–4 miles (5–7 kilometers) per second in the crust and 5–6 miles (8–9 kilometers) per second in the upper mantle. Also known as compressional wave; longitudinal wave; primary wave. { 'pē ,wāv }

pycnite [MINERAL] A variety of topaz occurring in massive columnar aggregations. { 'pik,nīt }

pycnocline [GEOPHYS] A change in density of ocean or lake water or rock with displacement in some direction, especially a rapid change in density with vertical displacement. { 'pik·nə,klīn }

Pycnodontiformes [PALEON] An extinct order of specialized fishes characterized by a laterally compressed, disk-shaped body, long dorsal and anal fins, and an externally symmetrical tail. { ,pik·nə,dänt·ə'fór,mēz }

Pygasteridae [PALEON] The single family of the extinct order Pygasteroida. { ,pī·gə'ster·ə,dē }

Pygasteroida [PALEON] An order of extinct echinoderms in the superorder Diadematacea having four genital pores, noncrenulate tubercles, and simple ambulacral plates. { ,pī·gə·stə'róid·ə }

pyrargyrite [MINERAL] Ag_3SbS_3 A deep ruby-red to black mineral, crystallizing in the hexagonal system, occurring in massive form and in disseminated grains, and having an adamantine luster; hardness is 2.5 on Mohs scale, and specific gravity is 5.85; an important silver ore. Also known as dark-red silver ore; dark ruby silver. { pī'rär·jə,rīt }

Pyrenean orogeny [GEOL] A short-lived orogeny that occurred during the late Eocene, between the Bartonian and Ludian stages. { ,pir·ə'nē·ən ó'räj·ə·nē }

pyrite [MINERAL] FeS_2 A hard, brittle, brass-yellow mineral with metallic luster, crystallizing in the isometric system; hardness is 6–6.5 on Mohs scale, and specific gravity is 5.02. Also known as common pyrite; fool's gold; iron pyrites; mundic. { 'pī͵rīt }

pyritization [GEOL] A common process of hydrothermal alteration involving introduction of or replacement by pyrite. { ͵pī͵rīd·ə'zā·shən }

pyritobitumen [GEOL] Any of various dark-colored, relatively hard, nonvolatile hydrocarbon substances often associated with mineral matter, which decompose upon heating to yield bitumens. Also known as pyrobitumen. { pə¦rīd·ō·bə'tü·mən }

pyroaurite [MINERAL] $Mg_6Fe_2(OH)_{16}·CO_3·4H_2O$ A goldlike or brownish rhombohedral mineral composed of hydrous basic magnesium iron carbonate. { ¦pī·rō'ȯ͵rīt }

pyrobelonite [MINERAL] $PbMn(VO_4)(OH)$ A fire-red to deep brilliant-red mineral composed of basic vanadate of manganese and lead, occurring as crystal needles. { ¦pī·rō'bel·ə͵nīt }

pyrobiolite [PETR] An organic rock containing organic remains that have been altered by volcanic action. { ¦pī·rō'bī·ə͵līt }

pyrobitumen See pyritobitumen. { ¦pī·rō·bə'tü·mən }

pyroborate See borax. { ¦pī·rō'bȯ͵rāt }

pyrochlore [MINERAL] $(Na,Ca)_2(Nb,Ta)_2O_6(OH,F)$ Pale-yellow, reddish, brown, or black mineral, crystallizing in the isometric system, and occurring in pegmatites derived from alkalic igneous rocks. Also known as pyrrhite. { 'pī·rə͵klȯr }

pyrochroite [MINERAL] $Mn(OH)_2$ A hexagonal mineral composed of naturally occurring manganese hydroxide; it is white when fresh, but darkens upon exposure. { ͵pī·rə'krō͵īt }

pyroclast [GEOL] An individual pyroclastic fragment or clast. { 'pī·rə͵klast }

pyroclastic flow [GEOL] Ash flow not involving high-temperature conditions. { ¦pī·rə¦klas·tik 'flō }

pyroclastic-flow deposit See ignimbrite. { ͵pī·rə͵klas·tik 'flō dī͵päz·ət }

pyroclastic ground surge [GEOL] The relatively thin mantle of rock found around a volcanic vent; the thickness is not uniform, the internal stratification is not parallel to the top and bottom of the layer, and the extent is a few kilometers from the source. { ¦pī·rə¦klas·tik 'graünd ͵sərj }

pyroclastic rock [PETR] A rock that is composed of fragmented volcanic products ejected from volcanoes in explosive events. { ¦pī·rə¦klas·tik 'räk }

pyrogenesis [GEOL] The intrusion and extrusion of magma and its derivatives. { ͵pī·rō'jen·ə·səs }

pyrogenetic mineral [MINERAL] An anhydrous mineral of an igneous rock, usually crystallized at high temperature in a magma containing relatively few volatile components. { ¦pī·rō·jə'ned·ik 'min·rəl }

pyrolusite [MINERAL] MnO_2 An iron-black mineral that crystallizes in the tetragonal system and is the most important ore of manganese; hardness is 1–2 on Mohs scale, and specific gravity is 4.75. { ͵pī·rə'lü͵sīt }

pyromagma [GEOL] A highly mobile lava, oversaturated with gases, that exists at shallower depths than hypomagma. { ͵pī·rō'mag·mə }

pyromelane See brookite. { ¦pī·rō'me͵lān }

pyrometamorphism [PETR] Contact metamorphism at temperatures near the melting points of the component minerals. { ¦pī·rō͵med·ə'mȯr͵fiz·əm }

pyrometasomatism [PETR] Forming of contact-metamorphic mineral deposits at high temperatures by emanations from the intrusive rock, involving replacement of the enclosing rock with the addition of materials. { ¦pī·rō͵med·ə'sō·mə͵tiz·əm }

pyromorphite [MINERAL] $Pb_5(PO_4)_3Cl$ A green, yellow, brown, gray, or white mineral of the apatite group, crystallizing in the hexagonal system; a minor ore of lead. Also known as green lead ore. { ¦pī·rō'mȯr͵fīt }

pyrope [MINERAL] $Mg_3Al_2(SiO_4)_3$ A mineral species of the garnet group characterized by a deep fiery-red color and occurring in basic and ultrabasic igneous rocks. { 'pī͵rōp }

pyrophane See fire opal. { 'pī·rə͵fān }

pyrophanite [MINERAL] $MnTiO_3$ A blood-red rhombohedral mineral consisting of manganese titanate; it is isomorphous with ilmenite. { pī'räf·ə͵nīt }

pyrophyllite [MINERAL] $AlSi_2O_5(OH)$ A white, greenish, gray, or brown phyllosilicate mineral that resembles talc and occurs in a foliated form or in compact masses in quartz veins, granites, and metamorphic rocks. Also known as pencil stone. { ,pī·rō'fi,līt }

pyroretinite [MINERAL] A type of retinite found in the brown coals of Aussig (Usti and Labem), in the Czech Republic. { ,pī·rō'ret·ən,īt }

pyroschist [PETR] A schist or shale that has a sufficiently high carbon content to burn with a bright flame or to yield volatile hydrocarbons when heated. { 'pī·rə,shist }

pyrosmalite [MINERAL] $(Mn,Fe)_4Si_3O_7(OH,Cl)_6$ A colorless, pale-brown, gray, or gray-green mineral composed mainly of basic iron manganese silicate with chlorine. { pī'räz·mə,līt }

pyrosphere [GEOL] The zone of the earth below the lithosphere, consisting of magma. Also known as magmosphere. { 'pī·rə,sfir }

pyrostibite See kermesite. { ,pī·rə'sti,bīt }

pyrostilpnite [MINERAL] Ag_3SbS_3 A hyacinth-red mineral composed of silver antimony sulfide, occurring in monoclinic crystal tufts; it is polymorphous with pyrargerite. { ,pī·rə'stilp,nīt }

Pyrotheria [PALEON] An extinct monofamilial order of primitive, mastodonlike, herbivorous, hoofed mammals restricted to the Eocene and Oligocene deposits of South America. { ,pī·rō'thir·ē·ə }

Pyrotheriidae [PALEON] The single family of the Pyrotheria. { ,pī·rō·thə'rī·ə,dē }

pyroxene [MINERAL] A family of diverse and important rock-forming minerals having infinite (Si_2O_6) single inosilicate chains as their principal motif; colors range from white through yellow and green to brown and greenish black; hardness is 5.5–6 on Mohs scale, and specific gravity is 3.2–4.0. { pə'räk,sēn }

pyroxene alkali syenite [GEOL] A quartz-poor (less than 20%) member of the charnockite series, characterized by the presence of microperthite. { pə'räk,sēn 'al·kə,lī 'sī·ə,nīt }

pyroxene monzonite [GEOL] A quartz-poor (less than 20%) member of the charnockite series, containing approximately equal amounts of microperthite and plagioclase. { pə'räk,sēn 'män·zə,nīt }

pyroxene syenite [GEOL] A quartz-poor (less than 20%) member of the charnockite series, containing more microperthite than plagioclase. { pə'räk,sēn 'sī·ə,nīt }

pyroxenite [PETR] A heavy, dark-colored, phaneritic igneous rock composed largely of pyroxene with smaller amounts of olivine and hornblende, and formed by crystallization of gabbroic magma. { pə'räk·sə,nīt }

pyroxenoids [MINERAL] A mineral group (including wollastonite and rhodonite) compositionally similar to pyroxene, but SiO_4 tetrahedrons are connected in rings rather than chains. { pə'räk·sə,nóidz }

pyrrhite See pyrochlore. { 'pi,rīt }

pyrrhotite [MINERAL] $Fe_{1-x}S$ A common reddish-brown to brownish-bronze mineral that occurs as rounded grains to large masses, more rarely as tabular pseudohexagonal crystals and rosettes; hardness is 4 on Mohs scale, and specific gravity is 4.6 (for the composition Fe_7S_8). { 'pir·ə,tīt }

Q

Quadrijugatoridae [PALEON] A monomorphic family of extinct ostracods in the super-family Hollinacea. { ¦kwä·drə¸jü·gə'tȯr·ə¸dē }

quake sheet [GEOL] A well-defined bed resembling a slump sheet but produced by an earthquake and resulting in the formation of a load cast without horizontal slip. { kwāk ¸shēt }

quaking bog [GEOL] A peat bog floating or growing over water-saturated land which shakes or trembles when walked on. { 'kwāk·iŋ 'bäg }

quantitative geomorphology [GEOL] The assignment of dimensions of mass, length, and time to all descriptive parameters of landform geometry and geomorphic processes, followed by the derivation of empirical mathematical relationships and formulation of rational mathematical models relating these parameters. { 'kwän·ə·tād·iv ¸jē·ō·mȯr'fäl·ə·jē }

quantum mineralogy [MINERAL] A branch of mineralogy concerned with the application of quantum mechanics to mineralogical systems. { 'kwän·təm ¸min·ə'räl·ə·jē }

quaquaversal [GEOL] Of strata and geologic structures, dipping outward in all directions away from a central point. { ¦kwä·kwə¦vər·səl }

quarrying [GEOL] See plucking. { 'kwär·ē·iŋ }

quartz [MINERAL] SiO₂ A colorless, transparent rock-forming mineral with vitreous luster, crystallizing in the trigonal trapezohedral class of the rhombohedral subsystem; hardness is 7 on Mohs scale, and specific gravity is 2.65; the most abundant and widespread of all minerals. { kwȯrts }

quartzarenite [PETR] A quartz-rich sandstone with framework grains separated predominantly by cement rather than matrix; essentially an orthoquartzite. { kwȯrt'sar·ə¸nīt }

quartz basalt [PETR] An igneous rock with more than 5% quartz. { 'kwȯrts bə'sȯlt }

quartz-bearing diorite See quartz diorite. { 'kwȯrts ¸ber·iŋ 'dī·ə¸rīt }

quartz crystal [MINERAL] See rock crystal. { 'kwȯrts ¦krist·əl }

quartz diorite [PETR] A group of plutonic rocks having the composition of diorite but with large amounts of quartz (greater than 20%). Also known as quartz-bearing diorite; tonalite. { 'kwȯrts 'dī·ə¸rīt }

quartz-flooded limestone [PETR] A limestone characterized by an abundance of quartz particles that had been imported suddenly from a nearby source by wind or water currents, but that gradually become sparser in an upward direction and completely disappear within a few centimeters. { 'kwȯrts ¸fləd·əd 'līm¸stōn }

quartz graywacke [PETR] A graywacke containing abundant grains of quartz and chert and less than 10% each of feldspars and rock fragments. { 'kwȯrts 'grā¸wak·ə }

quartzite [PETR] A granoblastic metamorphic rock consisting largely or entirely of quartz; most quartzites are formed by metamorphism of sandstone. { 'kwȯrt¸sīt }

quartzitic sandstone [PETR] Sandstone consisting of 100% quartz grains cemented with silica. { kwȯrt'sid·ik 'san¸stōn }

quartz lattice See rhyodacite. { 'kwȯrts 'lad·əs }

quartz monzonite [PETR] Granitic rock in which 10–50% of the felsic constituents are quartz, and in which the ratio of alkali feldspar to total feldspar is between 35% and 65%. Also known as adamellite. { 'kwȯrts 'män·zə¸nīt }

quartzose [GEOL] Referring to a substance which contains quartz as a principal constituent. { 'kwȯrt͵sōs }

quartzose arkose [PETR] A sandstone containing 50–85% quartz, chert, and metamorphic quartzite, 15–25% feldspars and feldspathic crystalline rock fragments, and 0–25% micas and micaceous metamorphic rock fragments. { 'kwȯrt͵sōs 'är͵kōs }

quartzose chert [PETR] A vitreous, sparkly, shiny chert, which under high magnification shows a heterogeneous mixture of pyramids, prisms, and faces of quartz, but also includes chert in which the secondary quartz is largely anhedral. { 'kwȯrt͵sōs 'chərt }

quartzose graywacke [PETR] **1.** A sandstone containing 50–85% quartz, chert, and metamorphic quartzite, 15–25% micas and micaceous metamorphic rock fragments, and 0–25% feldspars and feldspathic crystalline rock fragments. **2.** A graywacke that has lost its micaceous constituents through abrasion and thus tends to approach an orthoquartzite. { 'kwȯrt͵sōs 'grā͵wak·ə }

quartzose sandstone [PETR] Sandstone consisting of more than 95% clear quartz grains and less than 5% matrix. Also known as quartz sandstone. { 'kwȯrt͵sōs 'san͵stōn }

quartzose shale [PETR] A green or gray shale composed predominantly of rounded quartz grains of silt size, commonly associated with highly mature sandstones (orthoquartzites), representing the reworking of residual clays as transgressive seas encroached on old land areas. { 'kwȯrt͵sōs 'shāl }

quartzose subgraywacke See protoquartzite. { 'kwȯrt͵sōs 'səb͵grā͵wak·ə }

quartz-pebble conglomerate See orthoquartzitic conglomerate. { 'kwȯrts ͵peb·əl kən'gläm·ə·rət }

quartz porphyry [PETR] A porphyritic extrusive or hypabyssal rock containing quartz and alkali feldspar phenocrysts embedded in a microcrystalline or cryptocrystalline matrix. Also known as granite porphyry. { 'kwȯrts 'pȯr·fə·rē }

quartz sandstone See quartzose sandstone. { 'kwȯrts 'san͵stōn }

quartz schist [PETR] A schist whose foliation is due mainly to streaks and lenticles of nongranular quartz. { 'kwȯrts 'shist }

quartz syenite [PETR] A group of plutonic rocks having the characteristics of syenite but with a greater amount of quartz (5–20%). { 'kwȯrts 'sī·ə͵nīt }

quartz topaz See citrine. { 'kwȯrts 'tō͵paz }

quasi-cratonic [GEOL] Pertaining to a part of oceanic crust marginal to the continent which is considered to be former continental material that stretched and foundered during expansion. Also known as semicratonic. { ¦kwä·zē krə'tän·ik }

quasi-equilibrium [GEOL] The state of balance or grade in a stream cross section, whereby conditions of approximate equilibrium tend to be established in a reach of the stream as soon as a rather smooth longitudinal profile has been established in that reach, even though downcutting may go on. { ¦kwä·zē ͵ē·kwə'lib·rē·əm }

Quaternary [GEOL] The second period of the Cenozoic geologic era, following the Tertiary, and including the last 2–3 million years. { 'kwät·ən͵er·ē }

queenslandite See Darwin glass. { 'kwēnz·lən͵dīt }

Queenston shale [GEOL] A red bed series from the Ordovician found in Niagara Gorge; it is composed of deltaic red shale. { 'kwēnz·tən 'shāl }

queenstownite See Darwin glass. { 'kwēn·stə͵nīt }

quenite [PETR] A fine-grained, dark-colored hypabyssal rock composed of anorthite, chrome diopside, with less olivine and a small amount of bronzite. { 'kwe͵nīt }

quenselite [MINERAL] $PbMnO_2(OH)$ A pitch black mineral consisting of an oxide of lead and manganese; occurs in tabular form. { 'kwens·əl͵īt }

quenstedtite [MINERAL] $Fe_2(SO_4)_3 \cdot 10H_2O$ A pale violet to reddish-violet, triclinic mineral consisting of hydrated ferric sulfate; occurs in aggregates of crystals. { 'kwen͵ste͵tīt }

quick [GEOL] **1.** Referring to a sediment that, when mixed with or absorbing water, becomes extremely soft, incoherent, or loose, and is capable of flowing easily under load or by force of gravity. **2.** Referring to a soil in which a decrease in effective stress allows water to flow upward with sufficient velocity to reduce significantly the soil's bearing capacity.   **3.** Referring to a highly porous soil that readily absorbs heat.   { kwĭk }

quick clay   [GEOL] Clay that loses its shear strength after being disturbed.   { 'kwik ‚klā }

quicksand   [GEOL] A highly mobile mass of fine sand consisting of smooth, rounded grains with little tendency to mutual adherence, usually thoroughly saturated with upward-flowing water; tends to yield under pressure and to readily swallow heavy objects on the surface.   Also known as running sand.   { 'kwik‚sand }

quickstone   [PETR] A consolidated rock that flowed under the influence of gravity before lithification.   { 'kwik‚stōn }

quilted surface   [GEOL] A land surface characterized by broad, rounded, uniformly convex hills separating valleys that are comparatively narrow.   { 'kwil·təd 'sər·fəs }

R

radar interferometry   [GEOPHYS] A microwave remote sensing method for combining imagery collected over time by radar systems on board airplane or satellite platforms to map the elevations, movements, and changes of the earth's surface. Such detectable changes include earthquakes, volcanoes, glaciers, landslides, and underground explosions, as well as fires, floods, forestry operations, moisture changes, and vegetation growth.   { ¦rā,där ˌin·tər·fə'räm·ə·trē }

radial drainage pattern   [GEOL] A drainage pattern characterized by radiating streams diverging from a high central area.   Also known as centrifugal drainage pattern.   { 'rād·ē·əl 'drān·ij ˌpad·ərn }

radial faults   [GEOL] Faults arranged like the spokes of a wheel, radiating from a central point.   { 'rād·ē·əl 'fóls }

radiation budget   [GEOPHYS] A quantitative statement of the amounts of radiation entering and leaving a given region of the earth.   { ˌrād·ē'ā·shən ˌbəj·ət }

radiation chart   [GEOPHYS] Any chart or diagram which permits graphical solution of the (generally unintegrable) flux integrals arising in problems of atmospheric infrared radiation transfer.   { ˌrād·ē'ā·shən ˌchärt }

radioactive mineral   [MINERAL] Any mineral species that contains uranium or thorium as an essential part of the chemical composition; examples are uraninite, pitchblende, carnotite, coffinite, and autunite.   { ¦rād·ē·ō'ak·tiv 'min·rəl }

radiochronology   [GEOL] An absolute-age dating method based on the existing ratio between radioactive parent elements (such as uranium-238) and their radiogenic daughter isotopes (such as lead-206).   { ˌrad·ē·ō·krə'näl·ə·jē }

radiogeology   [GEOCHEM] The study of the distribution patterns of radioactive elements in the earth's crust and the role of radioactive processes in geologic phenomena.   { ¦rād·ē·ō·jē'äl·ə·jē }

radioglaciology   [GEOPHYS] The study of glacier ice by means of radar, especially the sounding of ice depth.   { ¦rād·ē·ō,glā·sē'äl·ə·jē }

radiolarian chert   [GEOL] A homogeneous cryptocrystalline radiolarite with a well-developed matrix.   { ¦rād·ē·ō¦lar·ē·ən 'chərt }

radiolarian earth   [GEOL] A porous, unconsolidated siliceous sediment formed from the opaline silica skeletal remains of Radiolaria; formed from radiolarian ooze.   { ¦rād·ē·ō¦lar·ē·ən 'ərth }

radiolarian ooze   [GEOL] A siliceous ooze containing the skeletal remains of the Radiolaria.   { ¦rād·ē·ō¦lar·ē·ən 'üz }

radiolarite   [GEOL] **1.** A whitish, hard, consolidated equivalent of radiolarian earth. **2.** Radiolarian ooze that has been indurated.   { ˌrād·ē·ō'la,rīt }

radiolitic   [PETR] **1.** Pertaining to the texture of an igneous rock, characterized by radial, fanlike groupings of acicular crystals, resembling sectors of spherulites. **2.** Referring to limestones in which the components radiate from central points, with the cement making up less than 50% of the total rock.   { ¦rād·ē·ō¦lid·ik }

radiometric age   [GEOL] Geologic age expressed in years determined by quantitatively measuring radioactive elements and their decay products.   { ¦rād·ē·ō¦me·trik 'āj }

Radstockian   [GEOL] A European stage of geologic time forming the upper Upper Carboniferous, above Staffordian and below Stephanian, equivalent to uppermost Westphalian.   { rad'stäk·ē·ən }

rafaelite [PETR] A nepheline-free orthoclase-bearing hypabyssal rock that also contains analcime and calcic plagioclase. { 'raf·ē·ə,līt }

raft [GEOL] **1.** A rock fragment caught up in a magma and drifting freely, more or less vertically. **2.** *See* float coal. { raft }

rafting [GEOL] Transporting of rock by floating ice or floating organic materials (such as logs) to places not reached by water currents. { 'raft·iŋ }

raft tectonics *See* plate tectonics. { 'raft tek,tän·iks }

rag [PETR] Any of various hard, coarse, rubbly, or shell rocks that weather with a rough, irregular surface, such as a flaggy sandstone or limestone used as a building stone. Also known as ragstone. { rag }

raglanite [PETR] A nepheline syenite composed of oligoclase, nepheline, and corundum with minor amounts of mica, calcite, magnetite, and apatite. { 'rag·lə,nīt }

ragstone *See* rag. { 'rag,stōn }

rainbow granite [PETR] A type of granite having either a black or dark-green background with pink, yellowish, or reddish mottling, or a pink background with dark mottling. { 'rān,bō 'gran·ət }

raindrop impressions *See* rain prints. { 'rān,dräp im'presh·ənz }

raindrop imprints *See* rain prints. { 'rān,dräp 'im,prins }

rain pillar [GEOL] A minor landform consisting of a column of soil or soft rock capped and protected by pebbles or concretions, produced by the differential erosion from the impact of falling rain. { 'rān ,pil·ər }

rain prints [GEOL] Small, shallow depressions formed in soft sediment or mud by the impact of falling raindrops. Also known as raindrop impressions; raindrop imprints. { 'rān ,prins }

rainwash [GEOL] **1.** The washing away of loose surface material by rainwater after it has reached the ground but before it has been concentrated into definite streams. **2.** Material transported and accumulated, or washed away, by rainwater. { 'rān,wäsh }

raised beach [GEOL] An ancient beach raised to a level above the present shoreline by uplift or by lowering of the sea level; often bounded by inland cliffs. { 'rāzd 'bēch }

rake *See* plunge. { rāk }

ralstonite [MINERAL] NaMgAl$_5$F$_{12}$(OH)$_6$·3H$_2$O A colorless, white, or yellowish mineral composed of hydrous basic sodium magnesium aluminum fluoride, occurring in octahedral crystals. { 'ròl·stə,nīt }

Ramapithecinae [PALEON] A subfamily of Hominidae including the protohominids of the Miocene and Pliocene. { ,räm·ə·pə'thes·ən,ē }

Ramapithecus [PALEON] The genus name given to a fossilized upper jaw fragment found in the Siwalik hills, India; closely related to the human family. { ,räm·ə'pith·ə·kəs }

rambla [GEOL] A dry ravine, or the dry bed of an ephemeral stream. { 'ram·blə }

ramdohrite [MINERAL] Pb$_3$Ag$_2$Sb$_6$S$_{13}$ A dark-gray mineral composed of a lead silver antimony sulfur compound. { 'räm,dō,rīt }

rammelsbergite [MINERAL] NiAs$_2$ A gray mineral composed of nickel diarsenide; it is dimorphous with pararammelsbergite. Also known as white nickel. { 'ram·əlz,bər,gīt }

rampart [GEOL] **1.** A narrow, wall-like ridge, 3–7 feet (1–2 meters) high, built up by waves along the seaward edge of a reef flat, and consisting of boulders, shingle, gravel, or reef rubble, commonly capped by dune sand. **2.** A wall-like ridge of unconsolidated material formed along a beach by the action of strong waves and current. **3.** A crescentic or ringlike deposit of pyroclastics around the top of a volcano. { 'ram,pärt }

rampart wall [GEOL] A rimming wall formed along the outer or seaward margin of a terrace, as on various high limestone Pacific islands. { 'ram,pärt ¦wòl }

ramp valley [GEOL] A trough between faults, forced downward by lateral pressure. { 'ramp ,val·ē }

ramsdellite [MINERAL] MnO$_2$ An orthorhombic mineral composed of manganese dioxide; it is dimorphous with pyrolusite. { 'ramz·de,līt }

Rancholabrean [GEOL] A stage of geologic time in southern California, in the upper Pleistocene, above the Irvingtonian. { ͵ran·chō·lə'brā·ən }

randannite [MINERAL] An earthy form of opal. { ran'da͵nīt }

rang [PETR] A unit of subdivision in the C.I.P.W. (Cross-Iddings-Pirsson-Washington) classification of igneous rocks. { räŋ }

range zone [GEOL] Formal biostratigraphic zone made up of a body of strata comprising the total horizontal (geographic) and vertical (stratigraphic) range of occurrence of a specified taxon of a group of taxa. { 'rānj ͵zōn }

rank [GEOL] **1.** A coal classification based on degree of metamorphism. **2.** *See* stack. { raŋk }

rankinite [MINERAL] $Ca_3Si_2O_7$ A monoclinic mineral composed of calcium silicate. { 'raŋ·kə͵nīt }

ransomite [MINERAL] $Cu(Fe,Al)_2(SO_4)_4 \cdot 7H_2O$ A sky-blue mineral composed of hydrous copper iron aluminum sulfate. { 'ran·sə͵mīt }

rapakivi [PETR] Granite or quartz monzonite characterized by orthoclase phenocrysts mantled with plagioclase. Also known as wiborgite. { ͵rä·pə'kē·vē }

rapakivi texture [PETR] An igneous and metamorphic rock texture in which spherical potassium feldspar crystals are surrounded by a rim of sodium feldspar, both within a finer-grained matrix. { ͵rä·pə'kē·vē 'teks·chər }

rare-earth mineral [MINERAL] A mineral containing lanthanides and yttrium as essential constituents. The total atomic ratio of lanthanides and yttrium is greater than any other element within at least one crystallographic site. Examples are monazite, xenotime, and bastnaesite. { 'rer ͵ərth 'min·rəl }

rasorite *See* kernite. { 'rā·zə͵rīt }

raspite [MINERAL] $PbWO_4$ A yellow or brownish-yellow mineral composed of lead tungstate, occurring as monoclinic crystals. { 'ra͵spīt }

rate-of-change map [GEOL] A derived stratigraphic map that shows the rate of change of structure, thickness, or composition of a given stratigraphic unit. { 'rāt əv 'chānj ͵map }

rate of sedimentation [GEOL] The amount of sediment accumulated in an aquatic environment over a given period of time, usually expressed as thickness of accumulation per unit time. Also known as sedimentation rate. { 'rāt əv ͵sed·ə·mən'tā·shən }

rathite [MINERAL] $Pb_{13}As_{18}S_{40}$ A dark-gray mineral with metallic luster composed of sulfide of lead and arsenic; occurs as orthorhombic crystals. { 'rä͵tīt }

ratio map [GEOL] A facies map that depicts the ratio of thicknesses between rock types in a given stratigraphic unit. { 'rā·shō ͵map }

rattlesnake ore [GEOL] A gray, black, and yellow mottled ore of carnotite and vanoxite; its spotted appearance resembles that of a rattlesnake. { 'rad·əl͵snāk ͵ȯr }

rattle stone [GEOL] A concretion composed of concentric laminae of different compositions, in which the more soluble layers have been removed by solution, leaving the central part detached from the outer part, such as a concretion of iron oxide filled with loose sand that rattles on shaking. Also known as klapperstein. { 'rad·əl ͵stōn }

rauhaugite [PETR] A carbonatite that contains ankerite. { raü'haü͵gīt }

Rauracian [GEOL] A substage of Upper Jurassic geologic time in Great Britain forming the middle Lusitanian, above the Argovian and below the Sequanian. { raü'rā·shən }

ravelly ground [GEOL] Rock that breaks into small pieces when drilled and tends to cave or slough into the hole when the drill string is pulled, or binds the drill string by becoming wedged or locked between the drill rod and the borehole wall. { 'rav·lē 'graünd }

ravinement [GEOL] **1.** The formation of a ravine or ravines. **2.** An irregular junction which marks a break in sedimentation, such as an erosion line occurring where shallow-water marine deposits have cut down into slightly eroded underlying beds. { rə'vēn·mənt }

raw humus *See* ectohumus. { 'rȯ 'hyü·məs }

raw map [GEOPHYS] A seismic map in which the z coordinate is time. { 'rȯ 'map }

291

Rayleigh wave [GEOPHYS] In seismology, a surface wave with a retrograde, elliptical motion at the free surface. Also known as R wave. { 'rā·lē ,wāv }

ray parameter [GEOPHYS] A function p that is constant along a given seismic ray, given by $p = rv^{-1} \sin i$, where r is the distance from the center O of the earth, v is the velocity, and i is the angle that the ray at a point P makes with the radius OP. { 'rā pə,ram·əd·ər }

razorback [GEOL] A sharp, narrow ridge. { 'rā·zər,bak }

razor stone See novaculite. { 'rā·zər ,stōn }

reaction border See reaction rim. { rē'ak·shən ,bȯrd·ər }

reaction pair [MINERAL] Any two minerals, one of which is formed at the expense of the other by reaction with liquid. { rē'ak·shən ,per }

reaction principle [MINERAL] The concept of a reaction series for the principal rock-forming minerals. { rē'ak·shən ,prin·sə·pəl }

reaction rim [PETR] A surficial rim around one mineral produced by the reaction of the core mineral with the surrounding magma. Also known as reaction border. { rē'ak·shən ,rim }

reaction series [MINERAL] Any series of minerals in which early formed varieties react with the melt to yield new minerals; two different types of reaction series exist, continuous and discontinuous. { rē'ak·shən ,sir·ēz }

realgar [MINERAL] AsS A red to orange mineral crystallizing in the monoclinic system, having a resinous luster and found in short, vertical striated crystals; specific gravity is 3.48, and hardness is 1.5–2 on Mohs scale. Also known as red arsenic; red orpiment; sandarac. { rē'al,gär }

rebound [GEOL] The isostatic readjustment upward of a landmass depressed by glacial loading. { 'rē,baúnd }

Recent See Holocene. { 'rē·sənt }

recess [GEOL] **1.** An indentation occurring in a surface, bounded by a straight line. **2.** An area having the axial traces of folds concave toward the outer edge of the folded belt. { 'rē,ses }

recession [GEOL] **1.** The backward movement, or retreat, of an eroded escarpment. **2.** A continuing landward movement of a shoreline or beach undergoing erosion. Also known as retrogression. **3.** The withdrawal of a body of water (as a sea or lake), thereby exposing formerly submerged areas. { ri'sesh·ən }

recessional moraine [GEOL] **1.** An end moraine formed during a temporary halt in the final retreat of a glacier. **2.** A moraine formed during a minor readvance of the ice front during a period of glacial recession. Also known as stadial moraine. { ri'sesh·ən·əl mə'rān }

reclined fold See recumbent fold. { ,ri'klīnd 'fōld }

recomposed granite [PETR] An arkose composed of consolidated feldspathic residue that has been reworked and decomposed so slightly that upon cementation the rock resembles granite except that its grain is less even and it contains a greater percentage of quartz. Also known as reconstructed granite. { ,rē·kəm'pōzd 'gran·ət }

recomposed rock [PETR] A rock produced in place by the cementation of the fragmental products of surface weathering; for example, a recomposed granite. { ,rē·kəm'pōzd 'räk }

reconstitution [GEOL] The formation of new chemicals, minerals, or structures under the influence of metamorphism. { rē,kän·stə'tü·shən }

reconstructed granite See recomposed granite. { ,rē·kən'strək·təd 'gran·ət }

recrystallization [PETR] The formation of new mineral grains in crystalline form in a rock under the influence of metamorphic processes. { rē,krist·əl·ə'zā·shən }

recrystallization breccia See pseudobreccia. { rē,krist·əl·ə'zā·shən ,brech·ə }

recrystallization flow [GEOL] Flow in which there is molecular rearrangement by solution and redeposition, solid diffusion, or local melting. { rē,krist·əl·ə'zā·shən ,flō }

rectangular cross ripple mark [GEOL] An oscillation cross ripple mark consisting of two sets of ripples which intersect at right angles, enclosing a rectangular pit. { rek'taŋ·gyə·lər ¦krȯs 'rip·əl ,märk }

red oxide of zinc

rectangular drainage pattern |GEOL| A drainage pattern characterized by many right-angle bends in both the main streams and their tributaries. Also known as lattice drainage pattern. { rek'taŋ·gyə·lər 'drān·ij ,pad·ərn }

rectification |GEOL| The simplification and straightening of the outline of an initially irregular and crenulate shoreline through the cutting back of headlands and offshore islands by marine erosion, and through deposition of waste from erosion or of sediment brought down by neighboring rivers. { ,rek·tə·fə'kā·shən }

rectilinear shoreline |GEOL| A long, relatively straight shoreline. { ¦rek·tə'lin·ē·ər 'shȯr,līn }

rectorite |GEOL| A white clay-mineral mixture with a regular interstratification of two mica layers (pyrophyllite and vermiculite) and one or more water layers. Also known as allevardite. { 'rek·tə,rīt }

recumbent fold |GEOL| An overturned fold with a nearly horizontal axial surface. Also known as reclined fold. { ri'kəm·bənt 'fōld }

recurrent folding |GEOL| A type of folding due to periodic deformation or subsidence and characterized by thinning or possible disappearance of formations at the crest. Also known as revived folding. { ri'kər·ənt 'fōld·iŋ }

red antimony See kermesite. { 'red 'an·tə,mō·nē }

red arsenic See realgar. { 'red 'ärs·ən·ik }

redbed |GEOL| Continentally deposited sediment composed principally of sandstone, siltsone, and shale; red in color due to the presence of ferric oxide (hematite). Also known as red rock. { 'red,bed }

red clay |GEOL| A fine-grained, reddish-brown pelagic deposit consisting of relatively large proportions of windblown particles, meteoric and volcanic dust, pumice, shark teeth, manganese nodules, and debris transported by ice. Also known as brown clay. { 'red 'klā }

red cobalt See erythrite. { 'red 'kō,bȯlt }

red copper ore See cuprite. { 'red 'käp·ər ,ȯr }

Reddish-Brown Lateritic soil |GEOL| One of a zonal, lateritic group of soils developed from a mottled red parent material and characterized by a reddish-brown surface horizon and underlying red clay. { 'red·ish ¦braȯn ,lad·ə'rid·ik 'sȯil }

Reddish-Brown soil |GEOL| A group of zonal soils having a reddish, light brown surface horizon overlying a warmer, more reddish horizon and a light-colored lime horizon. { 'red·ish ¦braȯn 'sȯil }

red earth |GEOL| Leached, red, deep, clayey soil that is characteristic of a tropical climate. Also known as red loam. { 'red ¦ərth }

redeposition |GEOL| Formation into a new accumulation, such as the deposition of sedimentary material that has been picked up and moved (reworked) from the place of its original deposition, or the solution and reprecipitation of mineral matter. { rē,dep·ə'zish·ən }

red hematite See hematite. { 'red 'hē·mə,tīt }

redingtonite |MINERAL| (Fe,Mg,Ni)(Cr,Al)$_2$(SO$_4$)$_4$·22H$_2$O A pale-purple mineral composed of a hydrous sulfate of iron, magnesium, nickel, chromium, and aluminum. { 'red·iŋ·tə,nīt }

red iron ore See hematite. { 'red 'ī·ərn ,ȯr }

red lead ore See crocoite. { 'red 'led ,ȯr }

red loam See red earth. { 'red 'lōm }

red magnetism |GEOPHYS| The magnetism of the north-seeking end of a freely suspended magnet; this is the magnetism of the earth's south magnetic pole. { 'red 'mag·nə,tiz·əm }

red mud |GEOL| A reddish terrigenous mud composed of up to 25% calcium carbonate and deriving its color from the presence of ferric oxide; found on the sea floor near deserts and near the mouths of large rivers. { 'red 'məd }

red ocher See hematite. { 'red 'ō·kər }

red orpiment See realgar. { 'red 'ȯr·pə·mənt }

red oxide of zinc See zincite. { 'red 'äk,sīd əv 'ziŋk }

293

redoxomorphic stage [GEOCHEM] The earliest geochemical stage of diagenesis characterized by mineral changes primarily due to oxidation and reduction reactions. { ri¦däk·sə¦mȯr·fik ‚stāj }

red rock See redbed. { 'red ‚räk }

redruthite See chalcocite. { red'rü‚thīt }

redstone [PETR] **1.** Any reddish sedimentary rock, such as red-colored sandstone. **2.** A deep-red, clayey sandstone or silt-stone representing a floodplain micaceous arkose. { 'red‚stōn }

reduction [GEOL] The lowering of a land surface by erosion. { ri'dək·shən }

reduction index [GEOL] The rate of wear of a sedimentary particle subject to abrasion, expressed as the difference between the mean weight of the particle before and after transport divided by the product of mean weight before transport and the distance traveled. { ri'dək·shən ‚in‚deks }

reduction sphere [GEOL] A white, leached, spheroidal mass produced in a reddish or brownish sandstone by a localized reducing environment, commonly surrounding an organic nucleus or a pebble and ranging in size from a poorly defined speck to a large, perfect sphere more than 10 inches (25 centimeters) in diameter. { ri'dək·shən ‚sfir }

reduzate [GEOL] A sediment accumulated under reducing conditions and consequently rich in organic carbon and in iron sulfide minerals; examples are coal and black shale. { 'rej·yə‚zāt }

Red-Yellow Podzolic soil [GEOL] Any of a group of acidic, zonal soils having a leached, light-colored surface layer and a subsoil containing clay and oxides of aluminum and iron, varying in color from red to yellowish red to a bright yellowish brown. { 'red 'yel·ō päd'zäl·ik 'sȯil }

red zinc ore See zincite. { 'red 'ziŋk ‚ȯr }

reedmergnerite [MINERAL] NaBSi$_3$O$_8$ A colorless, triclinic borate mineral that represents the boron analog of albite. { rēd'mər·nya‚rīt }

reef [GEOL] **1.** A ridge- or moundlike layered sedimentary rock structure built almost exclusively by organisms. **2.** An offshore chain or range of rock or sand at or near the surface of the water. { rēf }

reef breccia [PETR] A rock formed by the consolidation of limestone fragments broken off from a reef by the action of waves and tides. { 'rēf 'brech·ə }

reef cap [GEOL] A deposit of fossil-reef material overlying or covering an island or mountain. { 'rēf ‚kap }

reef cluster [GEOL] A group of reefs of wholly or partly contemporaneous growth, found within a circumscribed area or geologic province. { 'rēf ‚kləs·tər }

reef complex [GEOL] The solid reef core and the heterogeneous and contiguous fragmentary material derived from it by abrasion. { 'rēf ‚käm‚pleks }

reef conglomerate See reef talus. { 'rēf kən‚gläm·ə·rət }

reef core [GEOL] The rock mass constructed in place, and within the rigid growth lattice formed by reef-building organisms. { 'rēf ‚kȯr }

reef debris See reef detritus. { 'rēf də‚brē }

reef detritus [GEOL] Fragmental material derived from the erosion of an organic reef. Also known as reef debris. { 'rēf di‚trīd·əs }

reef edge [GEOL] The seaward margin of the reef flat, commonly marked by surge channels. { 'rēf ‚ej }

reef flank [GEOL] The part of the reef that surrounds, interfingers with, and locally overlies the reef core, often indicated by massive or medium beds of reef talus dipping steeply away from the reef core. { 'rēf ‚flaŋk }

reef flat [GEOL] A flat expanse of dead reef rock which is partly or entirely dry at low tide; shallow pools, potholes, gullies, and patches of coral debris and sand are features of the reef flat. { 'rēf ‚flat }

reef front [GEOL] The upper part of the outer or seaward slope of a reef, extending to the reef edge from above the dwindle point of abundant living coral and coralline algae. { 'rēf ‚frənt }

reef-front terrace [GEOL] A shelflike or benchlike eroded surface, sometimes veneered

with organic growth, sloping seaward to a depth of 8–15 fathoms (15–27 meters).
{ 'rēf ¦frȯnt ˌter· əs }

reef knoll |GEOL| **1.** A bioherm or fossil coral reef represented by a small, prominent, rounded hill, up to 330 feet (100 meters) high, consisting of resistant reef material, being either a local exhumation of an original reef feature or a feature produced by later erosion. **2.** A present-day reef in the form of a knoll; a small reef patch developed locally and built upward rather than outward. { 'rēf ˌnōl }

reef limestone |PETR| Limestone composed of the remains of sedentary organisms such as sponges, and of sediment-binding organic constituents such as calcareous algae. Also known as coral rock. { 'rēf 'līm ˌstōn }

reef milk |GEOL| A very-fine-grained matrix material of the back-reef facies, consisting of white, opaque microcrystalline calcite derived from abrasion of the reef core and reef flank. { 'rēf ˌmilk }

reef patch |GEOL| A single large colony of coral formed independently on a shelf at depths less than 220 feet (70 meters) in the lagoon of a barrier reef or of an atoll. Also known as patch reef. { 'rēf ˌpach }

reef pinnacle |GEOL| A small, isolated spire of rock or coral, especially a small reef patch. { 'rēf ˌpin·ə·kəl }

reef rock |PETR| A hard, unstratified rock composed of sand, shale, and the calcareous remains of sedentary organisms, cemented by calcium carbonate. { 'rēf ˌräk }

reef segment |GEOL| A part of an organic reef lying between passes, gaps, or channels. { 'rēf ˌseg·mənt }

reef slope |GEOL| The face of a reef rising from the sea floor. { 'rēf ˌslōp }

reef talus |GEOL| Massive inclined strata composed of reef detritus deposited along the seaward margin of an organic reef. Also known as reef conglomerate. { 'rēf ˌtā·ləs }

reef tufa |GEOL| Drusy, prismatic, fibrous calcite deposited directly from supersaturated water upon the void-filling internal sediment of the calcite mudstone of a reef knoll. { 'rēf ˌtüf·ə }

reef wall |GEOL| A wall-like upgrowth of living coral and the skeletal remains of dead coral and other reef-building organisms, which reaches an intertidal level and acts as a partial barrier between adjacent environments. { 'rēf ˌwȯl }

reentrant |GEOL| A prominent, generally angular indentation into a coastline. { rē'en·trənt }

reevesite |MINERAL| $Na_6Fe_2(OH)_{16}(CO_3)\cdot4H_2O$ Hydrous oxide mineral known only in meteorites. { 'rēv ˌzīt }

reference locality |GEOL| A locality containing a reference section, established to supplement the type locality. { 'ref·rəns lō'kal·əd·ē }

reference section |GEOL| A rock section, or group of sections, designated to supplement the type section, or sometimes to supplant it (as where the type section is no longer exposed), and to afford a standard for correlation for a certain part of the geologic column. { 'ref·rəns ˌsek·shən }

reflected buried structure |GEOL| The distortion of surface beds that reflect a similar structural distortion of underlying formations. { ri'flek·təd 'ber·ēd 'strək·chər }

reflector |GEOPHYS| A layer or horizon that reflects seismic waves. { ri'flek·tər }

refolding |GEOL| A process by which folds of one generation are subjected to and stressed by a force of different orientation. { rē'fōld·iŋ }

refoliation |GEOL| A foliation that is subsequent to and oriented differently from an earlier foliation. { ri ˌfō·lē'ā·shən }

Refugian |GEOL| A North American stage of geologic time in the Eocene and Oligocene, above the Fresnian and below the Zemorrian. { rə'fyü·jē·ən }

reg |GEOL| An extensive, nearly level, low desert plain from which fine sand has been removed by wind, leaving a sheet of coarse, smoothly angular, wind-polished gravel and small stones lying on an alluvial soil, strongly cemented by mineralized solutions to form a broad desert pavement. Also known as gravel desert. { reg }

regime |GEOL| The existence in a stream channel of a balance between erosion and deposition over a period of years. { rə'zhēm }

295

regional dip [GEOL] The nearly uniform and generally low-angle inclination of strata over a wide area. Also known as normal dip. { 'rēj·ən·əl 'dip }

regional geology [GEOL] The geology of a large region, treated from the viewpoint of the spatial distribution and position of stratigraphic units, structural features, and surface forms. { 'rēj·ən·əl jē'äl·ə·jē }

regional metamorphism [GEOL] Geological metamorphism affecting an extensive area. { 'rēj·ən·əl ,med·ə'mȯr,fiz·əm }

regional metasomatism [GEOL] Metasomatic processes affecting extensive areas whereby the introduced material may be derived from partial fusion of the rocks involved from deep-seated magmatic sources. { 'rēj·ən·əl ¦med·ə'sō·mə,tiz·əm }

regional slope [GEOL] The generally uniform dip of rock strata or land surface over a wide area. { 'rēj·ən·əl 'slōp }

regional slope deposit [GEOL] A sedimentary deposit widely distributed as a thin sheet over a regional slope. { 'rēj·ən·əl ¦slōp di,päz·ət }

regional unconformity [GEOL] A continuous unconformity extending throughout a wide region that may be nearly continentwide, and usually represents a long period of time. { 'rēj·ən·əl ,ən·kən'fȯr·məd·ē }

regmagenesis [GEOL] Diastrophic production of regional strike-slip displacements. { ¦reg·mə'jen·ə·səs }

regmaglypt [GEOL] Any of various small, well-defined, characteristic indentations or pits on the surface of meteorites, frequently resembling the imprints of fingertips in soft clay. Also known as pezograph; piezoglypt. { 'reg·mə,glipt }

regolith [GEOL] The layer rock or blanket of unconsolidated rocky debris of any thickness that overlies bedrock and forms the surface of the land. Also known as mantle rock. { 'reg·ə,lith }

Regosol [GEOL] In early United States soil classification systems, one of an azonal group of soils that form from deep, unconsolidated deposits and have no definite genetic horizons. { 'reg·ə,säl }

regradation [GEOL] The formation by a stream of a new profile of equilibrium, as when the former profile, after gradation, became deformed by crustal movements. { ,rē·grā'dā·shən }

regression [GEOL] The theory that some rivers have sources on the rainier sides of mountain ranges and gradually erode backward until the ranges are cut through. { ri'gresh·ən }

regression conglomerate [GEOL] A coarse sedimentary deposit formed during a retreat (recession) of the sea. { ri'gresh·ən kən,gläm·ə·rət }

regressive metamorphism See retrograde metamorphism. { ri'gress·iv ¦med·ə·mȯr,fiz·əm }

regressive overlap See offlap. { ri'gres·iv 'o·vər,lap }

regressive reef [GEOL] One of a series of nearshore reefs or bioherms superimposed on basinal deposits during the rising of a landmass or the lowering of the sea level, and developed more or less parallel to the shore. { ri'gres·iv 'rēf }

regressive ripple [GEOL] An asymmetric ripple mark formed by a current but oriented in a direction opposite to the general movement of current flow (steep side facing upcurrent). { ri'gres·iv 'rip·əl }

regressive sediment [GEOL] A sediment deposited during the retreat or withdrawal of water from a land area or during the emergence of the land, and characterized by an offlap arrangement. { ri'gres·iv 'sed·ə·mənt }

regur [GEOL] One of a group of calcareous, intrazonal soils characterized by dark color and a high clay content. Also known as black cotton soil. { 'reg·ər }

Reichenbach's lamellae [GEOL] Thin, platy inclusions of foreign minerals (usually troilite, schreibersite, or chromite) occurring in iron meteorites. { 'rī·kən,bäks lə'mel·ē }

rejuvenate [GEOL] The act of stimulating a stream to renewed erosive activity either by tectonic uplift or a drop in sea level. { ri'jü·və,nāt }

rejuvenated fault scarp [GEOL] A fault scarp revived by renewed movement along an old fault line after partial dissection or erosion of the initial scarp.   Also known as revived fault scarp.   { ri'jü·və‚näd·əd 'fȯlt ‚skärp }

rejuvenation   [GEOL] The restoration of youthful features to fluvial landscapes; the renewal of youthful vigor to low-gradient streams is usually caused by regional upwarping of broad areas formerly at or near base level.   { ri‚jü·və'nā·shən }

rejuvenation head See knickpoint.   { ri‚jü·və'nā·shən ‚hed }

relative age   [GEOL] The geologic age of a fossil organism, rock, or geologic feature or event defined relative to other organisms, rocks, or features or events rather than in terms of years.   { 'rel·əd·iv 'āj }

relative chronology   [GEOL] Geochronology in which the time order is based on superposition or fossil content rather than on an age expressed in years.   { 'rel·əd·iv krə'näl·ə·jē }

relative dating   [GEOL] The proper chronological placement of a feature, object, or happening in the geologic time scale without reference to its absolute age.   { 'rel·əd·iv 'dād·iŋ }

relative geologic time   [GEOL] Nonabsolute geological time in which events may be placed relatively to one another.   { 'rel·əd·iv ‚jē·ə‚läj·ik 'tīm }

relative permeability   [GEOL] Specific permeability of a porous rock formation to a particular phase (oil, water, gas) at a particular saturation and a particular saturation distribution; for example, ratio of effective permeability to a specified phase to the rock's absolute permeability.   { 'rel·əd·iv ‚pər·mē·ə'bil·əd·ē }

relative relief See local relief.   { 'rel·əd·iv ri'lēf }

relative time   [GEOL] Geologic time determined by the placing of events in a chronologic order of occurrence, especially time as determined by organic evolution or superposition.   { 'rel·əd·iv 'tīm }

relaxation   [GEOL] In experimental structural geology, the diminution of applied stress with time, as the result of any of various creep processes.   { ‚rē‚lak'sā·shən }

released mineral   [MINERAL] A mineral formed during the crystallization of a magma due to failure of an earlier phase to react with the liquid portion of the magma.   { ri'lēst 'min·rəl }

release fracture   [GEOL] A fracture formed as a result of a decrease in the maximum principal stress.   { ri'lēs ‚frak·chər }

release joint See sheeting structure.   { ri'lēs ‚jȯint }

relic   [GEOL] **1.** A landform that remains intact after decay or disintegration or that remains after the disappearance of the major portion of its substance.   **2.** A vestige of a particle in a sedimentary rock, such as a trace of a fossil fragment.   { 'rel·ik }

relict   [GEOL] **1.** Referring to a topographic feature that remains after other parts of the feature have been removed or have disappeared.   **2.** Pertaining to a mineral, structure, or feature of a rock which represents features of an earlier rock and which persists in spite of processes tending to destroy it, such as metamorphism.   { 'rel·ikt }

relict dike   [GEOL] In a granitized mass, a tabular, crystalloblastic body that represents a dike which was emplaced prior to, and which was relatively resistant to, the granitization process.   { 'rel·ikt 'dīk }

relict mineral   [MINERAL] A mineral of a rock that persists from an earlier rock.   { 'rel·ikt 'min·rəl }

relict permafrost   [GEOL] Permafrost formed in the past which persists in areas where it would not form today.   { 'rel·ikt 'pər·mə‚frȯst }

relict sediment   [GEOL] A sediment which was in equilibrium with its environment when first deposited but which is unrelated to its present environment even though it is not buried by later sediments, such as a shallow-marine sediment on the deep ocean floor.   { 'rel·ikt 'sed·ə·mənt }

relict soil   [GEOL] A soil formed on a preexisting landscape but not subsequently buried under younger sediments.   { 'rel·ikt 'sȯil }

relict texture   [GEOL] In mineral deposits, an original texture that persists after partial replacement.   { 'rel·ikt 'teks·chər }

relief limonite [MINERAL] Indigenous limonite that is porous and cavernous in texture. { ri'lēf 'līm·ə,nīt }

Relizean stage [GEOL] A subdivision of the Miocene in the California-Oregon-Washington area. { rə'lē·zē·ən ,stāj }

remanent magnetization [GEOPHYS] That component of a rock's magnetization whose direction is fixed relative to the rock and which is independent of moderate, applied magnetic fields. { 'rem·ə·nənt ,mag·nə·tə'zā·shən }

remolded soil [GEOL] Soil that has had its natural internal structure modified or disturbed by manipulation so that it lacks shear strength and gains compressibility. { rē'mōl·dəd 'sȯil }

remolding index [GEOL] The ratio of the modulus of deformation of a soil in the undisturbed state to that of a soil in the remolded state. { rē'mōld·iŋ ,in,deks }

renardite [MINERAL] $Pb(UO_2)_4(PO_4)_2(OH)_4 \cdot 7H_2O$ A yellow mineral composed of hydrous basic lead uranyl phosphate. { rə'när,dīt }

Rendoll [GEOL] A suborder of the soil order Mollisol, formed in highly calcareous parent materials, mostly restricted to humid, temperate regions; the soil profile consists of a dark upper horizon grading to a pale lower horizon. { 'ren,däl }

Rendzina [GEOL] One of an intrazonal, calcimorphic group of soils characterized by a brown to black, friable surface horizon and a light-gray or yellow, soft underlying horizon; found under grasses or forests in humid to semiarid climates. { rent'sin·ə }

rensselaerite [PETR] A soft, compact, fibrous talc pseudomorphous after pyroxene and found in Canada and northern New York. { 'ren·sə·lə,rīt }

repeated reflection *See* multiple reflection. { ri'pēd·əd ri'flek·shən }

repetition [GEOL] The duplication of certain stratigraphic beds at the surface or in any specified section owing to disruption and displacement of the beds by faulting or intense folding. { ,rep·ə'tish·ən }

Repettian [GEOL] A North American stage of lower Pliocene geologic time, above the Delmontian and below the Venturian. { rə'pesh·ən }

replacement [GEOL] Growth of a new or chemically different mineral in the body of an old mineral by simultaneous capillary solution and deposition. [PALEON] Substitution of inorganic matter for the original organic constituents of an organism during fossilization. { ri'plās·mənt }

replacement deposit [MINERAL] A mineral deposit formed by the in-position replacement of one mineral for another. { ri'plās·mənt di,päz·ət }

replacement dike [GEOL] A dike which is made by gradual transformation of wall rock by solutions along fractures or permeable zones. { ri'plās·mənt ,dīk }

replacement texture [GEOL] The texture exhibited where one mineral has replaced another. { ri'plās·mənt ,teks·chər }

replacement vein [GEOL] A mineral vein formed by the gradual transformation of an original vein by secondary fluids. { ri'plās·mənt vān }

replenishment [GEOL] The stage in development of a cavern in which the presence of air in the passages allows the deposition of speleothems. { ri'plen·ish·mənt }

resedimentation [GEOL] **1.** Sedimentation of material derived from a preexisting sedimentary rock, that is, redeposition of sedimentary material. **2.** Mechanical deposition of material in cavities of postdepositional age, such as the deposition of carbonate muds and silts by internal mechanical erosion or solution of a limestone. **3.** The general process of subaqueous, downslope movement of sediment under the influence of gravity, such as the formation of a turbidity-current deposit. { rē,sed·ə·mən'tā·shən }

resequent [GEOL] Referring to a geologic or topographic feature that resembles or agrees with a consequent feature but that developed from the feature at a later date. { rē'sē·kwənt }

resequent fault-line scarp [GEOL] A fault-line scarp which faces in the same direction as the original fault scarp or in which the downthrown block is topographically lower than the upthrown block. { rē'sē·kwənt 'fȯlt ,līn ,skärp }

reservoir |GEOL| **1.** A subsurface accumulation of crude oil or natural gas under adequate trap conditions. **2.** An area covered by névé where snow collects to form a glacier. **3.** A space within the earth that is occupied by magma. { 'rez·əv,wär }

reservoir fluid |GEOL| The subterranean fluid trapped by a reservoir formation; can include natural gas, liquid and vapor petroleum hydrocarbons, and interstitial water. { 'rez·əv,wär ,flü·əd }

reservoir pressure |GEOL| **1.** The pressure on fluids (water, oil, gas) in a subsurface formation. Also known as formation pressure. **2.** The pressure under which fluids are confined in rocks. { 'rez·əv,wär ,presh·ər }

reservoir rock |GEOL| Friable, porous sandstone containing deposits of oil or gas. { 'rez·əv,wär ,räk }

residual |GEOL| **1.** Of a mineral deposit, formed by either mechanical or chemical concentration. **2.** Pertaining to a residue left in place after weathering of rock. **3.** Of a topographic feature, representing the remains of a formerly great mass or area and rising above the surrounding surface. { rə'zij·ə·wəl }

residual anticline |GEOL| In salt tectonics, a relative structural high resulting from the depression of two adjacent rim synclines. Also known as residual dome. { rə'zij·ə·wəl 'ant·i,klīn }

residual clay |GEOL| Very finely divided clay material formed in place by weathering of rock. Also known as primary clay. { rə'zij·ə·wəl 'klā }

residual compaction |GEOL| The difference between the amount of compaction that will ultimately occur for a given increase in applied stress, and that which has occurred at a specified time. { rə'zij·ə·wəl kəm'pak·shən }

residual dome *See* residual anticline. { rə'zij·ə·wəl ¦dōm }

residual kame |GEOL| A ridge or mound of sand or gravel formed by the denudation of glaciofluvial material that had been deposited in glacial lakes or on the flanks of hills of till. { rə'zij·ə·wəl 'kām }

residual liquid |GEOL| The volatile components of a magma that remain in the magma chamber after much crystallization has taken place. { rə'zij·ə·wəl 'lik·wəd }

residual liquor *See* rest magma. { rə'zij·ə·wəl 'lik·ər }

residual map |GEOL| A stratigraphic map that displays the small-scale variations (such as local features in the sedimentary environment) of a given stratigraphic unit. { rə'zij·ə·wəl ¦map }

residual material |GEOL| Unconsolidated or partly weathered parent material of a soil, presumed to have developed in place (by weathering) from the consolidated rock on which it lies. { rə'zij·ə·wəl mə'tir·ē·əl }

residual mineral |GEOL| A mineral that has been concentrated in place by weathering and leaching of rock. { rə'zij·ə·wəl ¦min·rəl }

residual ochre |GEOL| An earthy, red, yellow, or brownish iron oxide powder of iron oxide (usually the mineral limonite) produced during chemical weathering. { rə'zij·ə·wəl 'ō·kər }

residual sediment *See* resistate. { rə'zij·ə·wəl 'sed·ə·mənt }

residual stress field *See* ambient stress field. { rə'zij·ə·wəl 'stres ,fēld }

residual swelling |GEOL| The difference between the original prefreezing level of the ground and the level reached by the settling after the ground is completely thawed. { rə'zij·ə·wəl 'swel·iŋ }

residual valley |GEOL| An intervening trough between uplifted mountains. { rə'zij·ə·wəl 'val·ē }

residue |GEOL| The in-place accumulation of rock debris which remains after weathering has removed all but the least soluble constituent. { 'rez·ə,dü }

resinite |GEOL| A variety of exinite composed of resinous compounds, often in elliptical or spindle-shaped bodies. { 'rez·ən,īt }

resin opal |MINERAL| A wax-, honey-, or ocher-yellow variety of common opal with a resinous luster or appearance. { 'rez·ən 'ō·pəl }

resinous coal |GEOL| Coal in which large proportions of resinous material are contained in the attritus. { 'rez·ən·əs 'kōl }

resinous luster |GEOL| The luster on the fractured surfaces of certain minerals (such as opal, sulfur, amber, and sphalerite) and rocks (such as pitchstone) that resemble the appearance of resin. { 'rez·ən·əs 'ləs·tər }

resin tin See rosin tin. { 'rez·ən ,tin }

resistate [GEOL] A sediment consisting of minerals that are chemically resistant and are enriched in the residues of weathering processes. Also known as residual sediment. { ri'zis,tāt }

resistivity factor See formation factor. { ,rē,zis'tiv·əd·ē ,fak·tər }

resorbed reef [GEOL] A reef characterized by embayed margins and by the numerous isolated patches of reef that are closely distributed about the main mass. { rē'sȯrbd 'rēf }

resorption [PETR] The process by which a magma redissolves previously crystallized minerals. { rē'sȯrp·shən }

rest hardening [GEOL] The increase of strength, with time, of a clay subsequent to its deposition, remolding, or modification by the application of shear stress. { 'rest ,härd·ən·iŋ }

rest magma [GEOL] The part of magma that remains after many minerals have crystallized from it during a long series of differentiations. Also known as residual liquor. { 'rest ,mag·mə }

restricted [GEOL] Referring to tectonic transport or movement in which elongation of particles is transverse to the direction of movement. { ri'strik·təd }

restricted basin [GEOL] A depression in the floor of the ocean in which the water circulation is topographically restricted and therefore generally is oxygen-depleted. Also known as barred basin; silled basin. { ri'strik·təd 'bās·ən }

resurgent [GEOL] Referring to magmatic water or gases that were derived from sources on the earth's surface, from its atmosphere, or from country rock of the magma. { ri'sər·jənt }

resurgent cauldron [GEOL] A cauldron in which the cauldron block has been uplifted following subsidence, usually in the form of a structural dome. { ri'sər·jənt 'kȯl·drən }

resurrected [GEOL] Pertaining to a surface, landscape, or feature (such as a mountain, peneplain, or fault scarp) that has been restored by exhumation to its previous status in the existing relief. Also known as exhumed. { ¦rez·ə¦rek·təd }

retgersite [MINERAL] $NiSO_4·6H_2O$ A deep emerald green, tetragonal mineral consisting of a hydrated nickel sulfate. { 'ret·gər,sīt }

reticular See reticulate. { re'tik·yə·lər }

reticulate [GEOL] **1.** Referring to a vein or lode with netlike texture. **2.** Referring to rock texture in which crystals are partly altered to a secondary material, forming a network that encloses the remnants of the original mineral. Also known as mesh texture; reticular; reticulated. { rə'tik·yə·lət }

reticulated See reticulate. { rə'tik·yə,lād·əd }

reticulated bar [GEOL] One of a group of slightly submerged sandbars in two sets, both of which are diagonal to the shoreline, forming a crisscross pattern. { rə'tik·yə,lād·əd 'bär }

Reticulosa [PALEON] An order of Paleozoic hexactinellid sponges with a branching form in the subclass Hexasterophora. { rə,tik·yə'lō·sə }

retinalite [MINERAL] A massive, honey-yellow or greenish serpentine mineral with a waxy or resinous luster; a variety of chrysolite. { 'ret·ən·əl,īt }

retinasphalt [MINERAL] A light-brown variety of retinite usually found with lignite. { ,ret·ən'a,sfȯlt }

retinite [MINERAL] A fossil resin, such as glessite, krantzite, muckite, and ambrite, composed of 6–15% oxygen, lacking succinic acid, and found in brown coals and peat. { 'ret·ən,īt }

retrograde metamorphism [PETR] Formation of metamorphic minerals of a lower grade of metamorphism at the expense of minerals which are characteristic of a higher grade. Also known as diaphthoresis; retrogressive metamorphism. { 're·trə,grād ,med·ə'mȯr,fiz·əm }

retrograde reservoir [GEOL] Hydrocarbon reservoir in which hydrocarbons are initially in the vapor phase; as pressure is reduced, the bubble-point line is passed and liquids are formed; upon further pressure reduction, a vapor phase is again formed. { 're·trə,grād 'rez·əv,wär }

retrograding shoreline [GEOL] A shoreline that is being moved landward by wave erosion. { 're·trə,grād·iŋ 'shȯr,līn }

retrogression [GEOL] *See* recession. { ,re·trə'gresh·ən }

retrogressive metamorphism *See* retrograde metamorphism. { ¦re·trə¦gres·iv ,med·ə'mȯr,fiz·əm }

return [GEOPHYS] Any of those surface waves on the record of a large earthquake which have traveled around the earth's surface by the long (greater than 180°) arc between epicenter and station, or which have passed the station and returned after traveling the entire circumference of the earth. { ri'tərn }

return stroke *See* return streamer. { ri'tərn ,strōk }

retzian [MINERAL] $Mn_2Y(AsO_4)(OH)_4$ A chocolate brown to chestnut brown, orthorhombic mineral consisting of a basic arsenate of calcium, rare earths, and manganese. { 'ret·sē·ən }

reversal of dip [GEOL] Change in the dip direction of bedding near a fault such that the beds curve toward the fault surface in a direction exactly opposite that of the drag folds. Also known as dip reversal. { ri'vər·səl əv 'dip }

reversed *See* overturned. { ri'vərst }

reversed arc [GEOL] A curved belt of islands which is concave toward the open ocean, the opposite of most island arcs. { ri'vərst 'ärk }

reversed polarity [GEOPHYS] Natural remanent magnetism opposite that of the present geomagnetic field. { ri'vərst pə'lar·əd·ē }

reverse fault *See* thrust fault. { ri'vərs 'fȯlt }

reverse-flowage fold [GEOL] A fold in which flow from deformation has thickened the anticlinal crests and thinned the synclinal troughs, contrary to the normal flow pattern of a flow fold. { ri'vərs ¦flō·ij 'fōld }

reverse saddle [GEOL] A mineral deposit associated with the trough of a synclinal fold and following the bedding plane. Also known as trough reef. { ri'vərs ¦sad·əl }

reverse similar fold [GEOL] A fold whose strata are thickened on the limbs and thinned on the axes, contrary to the pattern of a similar fold. { ri'vərs 'sim·ə·lər 'fōld }

reverse slip fault *See* thrust fault. { ri'vərs 'slip ,fȯlt }

reverse slope [GEOL] A hill descending away from a ridge. { ri'vərs ¦slōp }

reversing dune [GEOL] A dune that tends to develop unusual height but migrates only a limited distance because seasonal shifts in dominant wind direction cause it to move alternately in nearly opposite directions. { ri'vərs·iŋ ,dün }

revet-crag [GEOL] One of a series of narrow, pointed outliers or ridges of eroded strata inclined like a revetment against a mountain spur. { rə'vet ,krag }

revived fault scarp *See* rejuvenated fault scarp. { ri'vīvd 'fȯlt ,skärp }

revived folding *See* recurrent folding. { ri'vīvd 'fōld·iŋ }

revolution [GEOL] A little-used term to describe a time of profound crustal movements, on a continentwide or worldwide scale, which led to abrupt geographic, climatic, and environmental changes that were related to changes in forms of life. { ,rev·ə'lü·shən }

rework [GEOL] Any geologic material that has been removed or displaced by natural agents from its origin and incorporated in a younger formation. { 'rē,wərk }

rezbanyite [MINERAL] $Pb_3Cu_2Bi_{10}S_{19}$ A metallic-gray mineral composed of sulfide of lead, copper, and bismuth. { rez'ban,yīt }

rhabdite [MINERAL] *See* schreibersite. { 'rab,dīt }

rhabdoglyph [PALEON] A trace fossil consisting of a presumable worm trail appearing on the undersurface of flysch beds (sandstones) as a nearly straight bulge with little or no branching. { 'rab·də,glif }

rhabdophane [MINERAL] $(Ce,Y,La,Di)(PO_4)\cdot H_2O$ A brown, pinkish, or yellowish-white mineral consisting of a hydrated phosphate of cerium, yttrium, and rare earths. { 'rab·də,fān }

Rhachitomi

Rhachitomi [PALEON] A group of extinct amphibians in the order Temnospondyli in which pleurocentra were retained. { rə'kid·ə,mī }

Rhaetian [GEOL] A European stage of geologic time; the uppermost Triassic (above Norian, below Hettangian of Jurassic). Also known as Rhaetic. { 'rē·shən }

Rhaetic See Rhaetian. { 'rēd·ik }

Rhamphorhynchoidea [PALEON] A Jurassic suborder of the Pterosauria characterized by long, slender tails with an expanded tip. { ,ram·fə·riŋ'kȯid·ē·ə }

rhegmagenesis [GEOL] Orogeny characterized by the development of large-scale strike-slip faults. { ¦reg·mə'jen·ə·səs }

rheid [GEOL] A substance (below its melting point) which deforms by viscous flow during applied stress at an order of magnitude at least three times that of elastic deformation under similar circumstances. { 'rē·əd }

rheid fold [GEOL] A fold whose strata deform by viscous flow as if they were fluid. { 'rē·əd ,fōld }

rheidity [GEOL] Relaxation time of a substance, divided by 1000. { rē'id·əd·ē }

Rhenanida [PALEON] An order of extinct marine fishes in the class Placodermi distinguished by mosaics of small bones between the large plates in the head shield. { re'nan·ə·də }

rheoignimbrite [GEOL] An ignimbrite, on the slope of a volcanic crater, that has developed secondary flowage due to high temperatures. { ¦rē·ō'ig·nim,brīt }

rheomorphic intrusion [PETR] The injection of country rock that has become mobilized into the igneous intrusion that caused the rheomorphism. { ¦rē·ə¦mȯr·fik in'trü·zhən }

rheomorphism [PETR] Mobilization of a rock by at least partial fusion accompanied by, and sometimes promoted by, addition of new material by diffusion. { ¦rē·ə¦mȯr,fiz·əm }

rhexistasy [GEOL] The mechanical breaking up and transport of old soils or other surface residual materials. { rek'sis·tə·sē }

rhizoconcretion See root cast. { ¦rī·zō·kän'krē·shən }

Rhizodontidae [PALEON] An extinct family of lobefin fishes in the order Osteolepiformes. { ,rī·zō'dänt·ə,dē }

rhizosphere [GEOL] The soil region subject to the influence of plant roots and characterized by a zone of increased microbiological activity. { 'rī·zə,sfir }

Rhodanian orogeny [GEOL] A short-lived orogeny that occurred at the end of the Miocene Period. { rō'dān·ē·ən ȯ'räj·ə·nē }

Rhodesian man [PALEON] A type of fossil man inhabiting southern and central Africa during the late Pleistocene; the skull was large and low, marked by massive brow-ridges, with a cranial capacity of 1300 cubic centimeters or less. { rō'dē·zhən 'man }

rhodite [MINERAL] A mineral consisting of a native alloy of rhodium (about 40) and gold. { 'rō,dīt }

rhodizite [MINERAL] $CsAl_4Be_4B_{11}O_{25}(OH)_4$ A white mineral composed of a basic borate of cesium, aluminum, and beryllium, occurring as isometric crystals. { 'rōd·ə,zīt }

rhodochrosite [MINERAL] $MnCO_3$ A rose-red to pink or gray mineral form of manganese carbonate with hexagonal symmetry but occurring in massive or columnar form; isomorphous with calcite and siderite, has a hardness of 3.5–4 on Mohs scale, and a specific gravity of 3.7; a minor ore of manganese. { ,rōd·ə'krō,sīt }

rhodolite [MINERAL] A violet-red garnet species composed of a mixture of almandite and pyrope in about a 3:1 ratio. { 'rōd·əl,īt }

rhodonite [MINERAL] $MnSiO_3$ A pink or brown mineral inosilicate crystallizing in the triclinic system and commonly found in cleavable to compact masses or in embedded grains; luster is vitreous, hardness is 5.5–6 on Mohs scale, and specific gravity is 3.4–3.7. { 'rōd·ən,īt }

Rhombifera [PALEON] An extinct order of Cystoidea in which the thecal canals crossed the sutures at the edges of the plates, so that one-half of any canal lay in one plate and the other half on an adjoining plate. { räm'bif·ə·rə }

rhombochasm [GEOL] A parallel-sided gap in the sialic crust occupied by simatic crust, probably caused by spreading and separation. { 'räm·bə,kaz·əm }

rhomboclase [MINERAL] $HFe^{3+}(SO_4)_2 \cdot 4H_2O$ A colorless mineral composed of hydrous acid ferric sulfate, occurring in rhombic plates. { 'räm·bə,klās }

rhombohedral iron ore *See* hematite; siderite. { ¦räm·bō¦hē-drəl 'ī-ərn ,òr }

rhomboid ripple mark [GEOL] An aqueous current ripple mark characterized by a reticular arrangement of diamond-shaped tongues of sand, with each tongue having two acute angles, one pointing upcurrent and the other pointing downcurrent. { 'räm ,bòid 'rip·əl ,märk }

rhomboporoid cryptostome [PALEON] Any of a group of extinct bryozoans in the order Cryptostomata that built twiglike colonies with zooecia opening out in all directions from the central axis of each branch. { ¦räm·bō¦pòr,òid 'krip·tə,sōm }

rhomb-porphyry [PETR] A porphyritic alkaline syenite composed of an alkali feldspar groundmass with augites having rhombohedral cross sections as the principal phenocryst minerals. { 'räm'pòr·fə·rē }

rhourd [GEOL] A pyramid-shaped sand dune, formed by the intersection of other dunes. { ròrd }

rhyacolite *See* sanidine. { rī'ak·ə,līt }

Rhynchosauridae [PALEON] An extinct family of generally large, stout, herbivorous lepidosaurian reptiles in the order Rhynchocephalea. { ,riŋ·kə'sòr·ə,dē }

Rhynchotheriinae [PALEON] A subfamily of extinct elephantoid mammals in the family Gomphotheriidae comprising the beak-jawed mastodons. { ,riŋ·kə·thə'rī·ə,nē }

Rhyniatae *See* Rhyniopsida. { rī'nī·ə,dē }

Rhyniophyta [PALEOBOT] A subkingdom of the Embryobionta including the relatively simple, uppermost Silurian-Devonian vascular plants. { ,rī·nē'äf·əd·ə }

Rhyniopsida [PALEOBOT] A class of extinct plants in the subkingdom Rhyniophyta characterized by leafless, usually dichotomously branched stems that bore terminal sporangia. { ,rī·nē'äp·səd·ə }

rhyodacite [PETR] A group of extrusive porphyritic igneous rocks containing quartz, plagioclase, and biotite phenocrysts in a fine-grained to glassy groundmass composed of alkali feldspar and silica minerals. Also known as dellenite; quartz lattice. { rī'äd·ə,sīt }

rhyolite [PETR] A light-colored, aphanitic volcanic rock composed largely of alkali feldspar and free silica with minor amounts of mafic minerals; the extrusive equivalent of granite. { 'rī·ə,līt }

rhyolitic glass [GEOL] Volcanic glass that is chemically equivalent to rhyolite. { ¦rī·ə¦lid·ik 'glas }

rhyolitic lava [GEOL] A highly viscous, silica-rich lava. { ¦rī·ə¦lid·ik 'lä·və }

rhyolitic magma [PETR] A type of magma formed by differentiation from basaltic magma in combination with assimilation of siliceous material, or by melting of portions of the earth's sialic layer. { ¦rī·ə¦lid·ik 'mag·mə }

rhyolitic tuff [GEOL] A tuff composed of fragments of rhyolitic lava. { ¦rī·ə¦lid·ik 'təf }

rhythmic accumulations [GEOL] Regular patterns of ripples and cusps in sediment on the beach or the sea floor, formed by currents and waves. { 'rith·mik ə,kyü·mə'lā·shənz }

rhythmic crystallization [PETR] In igneous rocks, a phenomenon in which different minerals crystallize in concentric layers, giving rise to orbicular texture. { 'rith·mik ,krist·əl'ā·shən }

rhythmic layering [GEOL] A type of layering in an igneous intrusion which is easily observable and in which there is repetition of zones of varying composition. { 'rith·mik 'lā·ər·iŋ }

rhythmic sedimentation [GEOL] A repetitious, regular sequence of rock units formed by sedimentary succession and indicating a frequent, predictable recurrence of the same sequence of conditions. { 'rith·mik ,sed·ə·men'tā·shən }

rhythmic stratification [GEOL] The occurrence of sediment layers in repetitive patterns, such as a regular alternation of layers of lime and clay. { 'rith·mik ,strad·ə·fə'kā·shən }

rhythmic succession [GEOL] A succession of rock units showing continual and repeated changes of lithology. { 'rith·mik sək'sesh·ən }

303

rhythmite |GEOL| An independent unit of a rhythmic succession or of beds that were developed by rhythmic sedimentation. { 'rith·mīt }

rib |GEOL| A layer or dike of rock forming a small ridge on a steep mountainside. { rib }

rib-and-furrow |GEOL| The bedding-plane expression for micro-cross-bedding, consisting of sets of small, transverse arcuate markings confined to long, narrow, parallel grooves oriented parallel to the current flow and separated by narrow ridges. { 'rib ən 'fər·ō }

riband jasper See ribbon jasper. { 'rib·ənd 'jas·pər }

ribbed moraine |GEOL| One of a group of irregularly subparallel, locally branching, generally smoothly rounded and arcuate ridges that are convex in the downstream direction of a glacier but that curve upstream adjacent to eskers. { 'ribd mə'rān }

ribble See ripple till. { 'rib·əl }

ribbon |PETR| One of a set of parallel bands in a rock or mineral. { 'rib·ən }

ribbon banding |PETR| A banding produced in the bedding of a sedimentary rock by thin strata of contrasting colors, giving the rock an appearance which suggests bands of ribbons. { 'rib·ən ,band·iŋ }

ribbon bomb |GEOL| An elongate and flattened volcanic bomb derived from ropes of lava. { 'rib·ən ,bäm }

ribbon diagram |GEOL| A continuous geologic cross section that is drawn in perspective along a curved or sinuous line. { 'rib·ən ¦dī·ə,gram }

ribbon jasper |GEOL| Banded jasper with parallel, ribbonlike stripes of alternating colors or shades of color. Also known as riband jasper. { 'rib·ən ,jas·pər }

ribbon reef |GEOL| A linear reef within the Great Barrier Reef off the northeast coast of Australia, having inwardly curved extremities, and forming a festoon along the precipitous edge of the continental shelf. { 'rib·ən ,rēf }

ribbon rock |PETR| A rock showing a succession of thin layers of differing composition or appearance. { 'rib·ən ,räk }

ribbon slate |PETR| Slate produced by incomplete metamorphism of clearly visible residual bedding planes that cut across the cleavage surface. { 'rib·ən ,slāt }

ribbon structure |GEOL| A succession of thin layers of different mineralogy and texture often contorted and deformed. { 'rib·ən ,strək·chər }

ribbon vein See banded vein. { 'rib·ən ,vān }

rice coal |GEOL| Anthracite that will pass through circular holes in a screen, the holes measuring 5/16 inch (7.9 millimeters), but not 3/16 inch (4.8 millimeters), in diameter. { 'rīs ,kōl }

richellite |MINERAL| $Ca_3Fe_{10}(PO_4)_8(OH,F)_{12}\cdot nH_2O$ A yellow mineral composed of hydrous basic iron calcium fluophosphate; occurs in masses. { rə'she,līt }

Richmondian |GEOL| A North American stage of geologic time: Upper Ordovician (above Maysvillian, below Lower Silurian). { rich'mən·dē·ən }

richterite |MINERAL| $(Na,K)_2(Mg,Mn,Ca)_6Si_8O_{22}(OH)_2$ A brown, yellow, or rose-red monoclinic mineral composed of basic silicate of sodium, potassium, magnesium, manganese, and calcium; a member of the amphibole group. { 'rik·tə,rīt }

Richter scale |GEOPHYS| A scale of numerical values of earthquake magnitude ranging from 1 to 9. { 'rik·tər ,skāl }

rickardite |MINERAL| Cu_4Te_3 A deep-purple mineral composed of copper telluride, occurring in masses. { 'rik·ər,dīt }

ricolettaite |PETR| A dark-colored syenite-gabbro containing anorthite as the plagioclase, along with olivine and augite. { ,rik·ə'led·ə,īt }

rideau |GEOL| A small ridge or mound of earth, or a slightly elevated piece of ground. { ri'dō }

ridge |GEOL| An elongate, narrow, steep-sided elevation of the earth's surface or the ocean floor. { rij }

ridge fault |GEOL| A fault structure that is a set of two faults bounding a horst. { 'rij ,fólt }

ridge-top trench |GEOL| A trench, occasionally found at or near the crest of high, steep-sided mountain ridges, formed by the creep displacement of a large slab of rock along shear surfaces more or less parallel with the side slope of the ridge. { 'rij ˌtäp ˌtrench }

riebeckite [MINERAL] $Na_2(Fe,Mg)_5Si_8O_{22}(OH)_2$ A blue or black monoclinic amphibole occurring as a primary constituent in some acid- or sodium-rich igneous rocks. { 'rē,be,kīt }

riebungsbreccia [GEOL] A breccia developed during folding. { 'rē·bəŋz,brech·ə }

Riecke's principle [MINERAL] The principle that solution of a mineral occurs most readily at points of greatest external pressure, and crystallization occurs most readily at points of least external pressure; applied to recrystallization in metamorphic rock. { 'rē·kəz ˌprin·sə·pəl }

riedenite [PETR] An igneous rock composed of large tabular biotite crystals in a granular groundmass of nosean, biotite, pyroxene, and small amounts of sphene and apatite. { 'rēd·ən,īt }

riegel [GEOL] A low, traverse ridge of bedrock on the floor of a glacial valley. Also known as rock bar; threshold; verrou. { 'rē·gəl }

rift [GEOL] **1.** A narrow opening in a rock caused by cracking or splitting. **2.** A high, narrow passage in a cave. { rift }

rift-block mountain [GEOL] A mountain range which is a horst block bounded by normal faults. { 'rift ¦bläk 'maunt·ən }

rift-block valley [GEOL] A valley which occupies a graben. { 'rift ¦bläk 'val·ē }

rift lake *See* sag pond. { 'rift ˌlāk }

rift valley [GEOL] A deep, central cleft with a mountainous floor in the crest of a midoceanic ridge. Also known as central valley; midocean rift. { 'rift ˌval·ē }

rift-valley lake *See* sag pond. { 'rift ¦val·ē 'lāk }

right-lateral fault *See* dextral fault. { 'rīt ¦lad·ə·rəl 'fȯlt }

right-lateral slip fault *See* dextral fault. { 'rīt ¦lad·ə·rəl 'slip ˌfȯlt }

right side up *See* right way up. { 'rīt 'sīd 'əp }

right-slip fault *See* dextral fault. { 'rīt ˌslip ˌfȯlt }

right way up [GEOL] The state of strata where the present upward succession of layers is the original (normal) order of deposition. Also known as right side up. { 'rīt 'wā 'əp }

rill [GEOL] A small, transient runnel. { ril }

rillenstein [GEOL] A pattern of tiny solution grooves of about 1 millimeter or less in width, formed on the limestone surface of a karstic region. { 'ril·ən,stīn }

rill erosion [GEOL] The formation of numerous, closely spaced rills due to the uneven removal of surface soil by streamlets of running water. Also known as rilling; rill wash; rillwork. { 'ril i'rō·zhən }

rilling *See* rill erosion. { 'ril·iŋ }

rill mark [GEOL] A small, dendritic channel formed on beach mud or sand by a rill, especially if on the lee side of a partially buried obstruction. { 'ril ˌmärk }

rillstone *See* ventifact. { 'ril,stōn }

rill wash *See* rill erosion. { 'ril ˌwäsh }

rillwork *See* rill erosion. { 'ril,wərk }

rima [GEOL] A long, narrow aperture, cleft, or fissure. { 'rī·mə }

rim cement [GEOL] A thin layer of calcium carbonate, hematite, or silica developed on the surface of detrital grains during diagenesis. { 'rim si,mənt }

rim gypsum [GEOCHEM] Gypsum in thin films between anhydrite crystals, believed to have been introduced in solution rather than produced by replacement. { 'rim ,jip·səm }

rimmed kettle [GEOL] A morainal depression with raised edges. { 'rimd 'ked·əl }

rimmed solution pool [GEOL] A pool in rock with a hardened rim resulting from deposition of lime during evaporation at low tide. { 'rimd sə'lü·shən ,pül }

rimming wall [GEOL] A steep, ridgelike erosional remnant of continuous layers of porous, permeable, poorly cemented, detrital limestones, believed to form under tropical or subtropical conditions by surface-controlled secondary cementation of an original steep slope and followed by differential erosion that brings the cemented zone into relief. { 'rim·iŋ ,wȯl }

rimpylite [MINERAL] A group name for several green and brown hornblendes with high contents of $(Al,Fe)_2O_3$. { 'rim·pə,līt }

rim ridge [GEOL] A minor ridge of till defining the edge of a moraine plateau. { 'rim ,rij }

rimrock [GEOL] A top layer of resistant rock on a plateau outcropping with vertical or near vertical walls. { 'rim,räk }

rimstone [GEOL] A calcium-containing deposit ringing an overflowing basin such as a hot spring. { 'rim,stōn }

rim syncline [GEOL] In salt tectonics, a local depression that develops as a border around a salt dome, as the salt in the underlying strata is displaced toward the dome. Also known as peripheral sink. { 'rim 'sin,klīn }

rincon [GEOL] **1.** A small, secluded valley. **2.** A bend in a stream. { riŋ'kōn }

ring complex [GEOL] An association of two ring-shaped igneous intrusive forms, ring dikes and cone sheets. { 'riŋ ,käm,pleks }

ring current [GEOPHYS] A westward electric current which is believed to circle the earth at an altitude of several earth radii during the main phase of geomagnetic storms, resulting in a large worldwide decrease in the geomagnetic field horizontal component at low latitudes. { 'riŋ ,kə·rənt }

ring depression [GEOL] The annular, structurally depressed area surrounding the central uplift of a cryptoexplosion structure; faulting and folding may be involved in its formation. Also known as peripheral depression; ring syncline. { 'riŋ di,presh·ən }

ring dike [GEOL] A roughly circular dike that is vertical or inclined away from the center of the arc. Also known as ring-fracture intrusion. { 'riŋ ,dīk }

ring fault [GEOL] **1.** A fault that bounds a rift valley. **2.** A steep-sided fault pattern that is cylindrical in outline and associated with cauldron subsidence. Also known as ring fracture. { 'riŋ ,fȯlt }

ring fissure [GEOL] A roughly circular desiccation crack formed on a playa around a point source (generally a phreatophyte). { 'riŋ ,fish·ər }

ring fracture See ring fault. { 'riŋ ,frak·chər }

ring-fracture intrusion See ring dike. { 'riŋ ¦frak·chər in,trü·zhən }

ring-fracture stoping [GEOL] Large-scale magmatic stoping that is associated with cauldron subsidence. { 'riŋ ¦frak·chər ,stōp·iŋ }

ringite [GEOL] An igneous rock formed by the mixing of silicate and carbonatite magmas. { 'riŋ,īt }

ring silicate See cyclosilicate. { 'riŋ 'sil·ə·kət }

ring structure [GEOL] A formation on the surface of the earth, moon, or a planet, having a ring-shaped trace in plan. { 'riŋ ,strək·chər }

ring syncline See ring depression. { 'riŋ 'sin,klīn }

rinkite See mosandrite. { 'riŋ,kīt }

rinkolite See mosandrite. { 'riŋ·kə,līt }

rinneite [MINERAL] NaK_3FeCl_6 A colorless, pink, violet, or yellow mineral composed of sodium potassium iron chloride, occurring in granular masses. { 'rin·ē,īt }

rip channel [GEOL] A channel, often more than 2 meters (6.6 feet) deep, carved on the shore by a rip current. { 'rip ,chan·əl }

ripe [GEOL] Referring to peat, in an advanced state of decay. { rīp }

ripidolite [MINERAL] $(Mg,Fe^{2+})_9Al_6Si_5O_{20}(OH)_{16}$ A mineral of the chlorite group; consists of basic magnesium iron aluminum silicate. Also known as aphrosiderite. { rə'pid·əl,īt }

ripple [GEOL] A very small ridge of sand resembling or suggesting a ripple of water and formed on the bedding surface of a sediment. { 'rip·əl }

ripple bedding [GEOL] A bedding surface characterized by ripple marks. { 'rip·əl ,bed·iŋ }

ripple biscuit [GEOL] A bedding structure produced by lenticular lamination of sand in a bay or lagoon. { 'rip·əl ,bis·kət }

ripple drift [GEOL] A pattern of cross-lamination formed by sedimentary deposits on both sides of a migrating ripple. { 'rip·əl ,drift }

ripple index |GEOL| On a rippled surface, the ratio of the crest-to-crest distance to the crest-to-trough distance. { 'rip·əl ,in,deks }

ripple lamina |GEOL| An internal sedimentary structure formed in sand or silt by currents or waves, as opposed to a ripple mark formed externally on a surface. { 'rip·əl ,lam· ə·nə }

ripple load cast |GEOL| A load cast of a ripple mark showing evidence of penecontemporaneous deformation in the accumulation of its trough and crest and in the oversteepening of the component laminae. { 'rip·əl 'lōd ,kast }

ripple mark |GEOL| **1.** A surface pattern on incoherent sedimentary material, especially loose sand, consisting of alternating ridges and hollows formed by wind or water action. **2.** One of the ridges on a ripple-marked surface. { 'rip·əl ,märk }

ripple scour |GEOL| A shallow, linear trough with transverse ripple marks. { 'rip·əl ,skaûr }

ripple symmetry index |GEOL| A measure of the degree of symmetry of a ripple mark, equal to the ratio of the length of the gentle (upcurrent) side to the steep (downcurrent) side. { 'rip·əl 'sim·ə·trē ,in,deks }

ripple till |GEOL| A till sheet containing low, winding smooth-topped ridges lying at right angles to the direction of ice movement, and grouped into narrow belts up to 48 miles (80 kilometers) long that are generally parallel to the direction of ice movement. Also known as ribble. { 'rip·əl ¦til }

rise |GEOL| A long, broad elevation which rises gently from its surroundings, such as the sea floor. { rīz }

rise pit |GEOL| A pit through which an underground stream rises to the surface with a calm and steady flow. { 'rīz ,pit }

riser |GEOL| A steplike topographic feature, such as a steep slope between terraces. { 'rīz·ər }

Riss |GEOL| **1.** A European stage of geologic time: Pleistocene (above Mindel, below Würm). **2.** The third stage of glaciation of the Pleistocene in the Alps. { ris }

Rissoacea |PALEON| An extinct superfamily of gastropod mollusks. { ,ris·ə'wās·ē·ə }

Riss-Würm |GEOL| The third interglacial stage of the Pleistocene in the Alps, following the Riss glaciation and preceding the Würm glaciation. { 'ris'virm }

river bar |GEOL| A ridgelike accumulation of alluvium in the channel, along the banks, or at the mouth of a river. { 'riv·ər ,bär }

river basin |GEOL| The area drained by a river and all of its tributaries. { 'riv·ər ,bās·ən }

riverbed |GEOL| The channel which contains, or formerly contained, a river. { 'riv·ər,bed }

river bottom |GEOL| The low-lying alluvial land along a river. Also known as river flat. { 'riv·ər ,bäd·əm }

river-deposition coast |GEOL| A deltaic coast characterized by lobate seaward bulges crossed by river distributaries and bordered by lowlands. { 'riv·ər ,dep·ə'zish·ən ,kōst }

river drift |GEOL| Rock material deposited by a river in one place after having been moved from another. { 'riv·ər ,drift }

river flat See river bottom. { 'riv·ər ,flat }

river morphology |GEOL| The study of the channel pattern and the channel geometry at several points along a river channel, including the network of tributaries within the drainage basin. Also known as channel morphology; fluviomorphology; stream morphology. { 'riv·ər mȯr'fäl·ə·jē }

river plain See alluvial plain. { 'riv·ər ,plān }

river run gravel |GEOL| Natural gravel as found in deposits that have been subjected to the action of running water. { 'riv·ər ¦rən ,grav·əl }

river terrace See stream terrace. { 'riv·ər ,ter·əs }

riverwash |GEOL| **1.** Soil material that has been transported and deposited by rivers. **2.** An alluvial deposit in a river bed or flood channel, subject to erosion and deposition during recurring flood periods. { 'riv·ər,wäsh }

riving [GEOL] The splitting off, cracking, or fracturing of rock, especially by frost action. { 'rīv·iŋ }

road [GEOL] One of a series of erosional terraces in a glacial valley, formed as the water level dropped in an ice-dammed lake. { rōd }

roaring sand [GEOL] A sounding sand, found on a desert dune, that sets up a low roaring sound that sometimes can be heard for a distance of 1200 feet (400 meters). { 'ror·iŋ 'sand }

robinsonite [MINERAL] $Pb_7Sb_{12}S_{25}$ A mineral composed of lead antimony sulfide. { 'räb·ən·sə,nīt }

rocdrumlin See rock drumlin. { 'räk¦drəm·lən }

roche moutonnée [GEOL] A small, elongate hillock of bedrock sculptured by a large glacier so that its long axis is oriented in the direction of ice movement; the upstream side is gently inclined, smoothly rounded, but striated, and the downstream side is steep, rough, and hackly. { 'rōch ¦müt·ən¦ā }

rock [PETR] **1.** A consolidated or unconsolidated aggregate of mineral grains consisting of one or more mineral species and having some degree of chemical and mineralogic constancy. **2.** In the popular sense, a hard, compact material with some coherence, derived from the earth. { räk }

rock asphalt See asphalt rock. { 'räk 'as,fólt }

rock association [PETR] A group of igneous rocks within a petrographic province that are related chemically and petrographically, generally in a systematic manner such that chemical data for the rocks plot as smooth curves on variation diagrams. Also known as rock kindred. { 'räk ə,sō·shē¦ā·shən }

rock bar See riegel. { 'räk ,bär }

rock bench See structural bench. { 'räk ,bench }

rockbridgeite [MINERAL] $Fe^{2+}Fe_6^{3+}(PO_4)_4(OH)_8$ A basic phosphate mineral containing iron; isomorphous with frondelite. { 'räk,bri,jīt }

rock-bulk compressibility [GEOL] One of three types of rock compressibility (matrix, bulk, and pore); the fractional change in volume of the bulk volume of the rock with a unit change in pressure. { 'räk ¦bəlk kəm,pres·ə'bil·əd·ē }

rock cave See shelter cave. { 'räk ,kāv }

rock cleavage [PETR] The capacity of a rock to split along certain parallel surfaces more easily than along others. { 'räk ,klē·vij }

rock control [GEOL] The influences of differences in earth materials on development of landforms. { 'räk kən,trōl }

rock cork See mountain cork. { 'räk ,kórk }

rock creep [GEOL] A form of slow flowage in rock materials evident in the downhill bending of layers of bedded or foliated rock and in the slow downslope migration of large blocks of rock away from their parent outcrop. { 'räk ,krēp }

rock crystal [MINERAL] A transparent, colorless form of quartz with low brilliance; used for lenses, wedges, and prisms in optical instruments. Also known as berg crystal; crystal; mountain crystal; pebble; quartz crystal. { 'räk ,krist·əl }

rock cycle [GEOL] The interrelated sequence of events by which rocks are initially formed, altered, destroyed, and reformed as a result of magmatism, erosion, sedimentation, and metamorphism. { 'räk ,sī·kəl }

rock-defended terrace [GEOL] **1.** A river terrace having a ledge or outcrop of resistant rock at its base which serves as protection against undermining. **2.** A marine terrace having a mass of resistant rock at the base of the cliff which protects against wave erosion. { 'räk di¦fen·dəd 'ter·əs }

rock desert [GEOL] An upland desert in which bedrock is either exposed or is covered with a thin veneer of coarse rock fragments. { 'räk ,dez·ərt }

rock drum See rock drumlin. { 'räk ¦drəm }

rock drumlin [GEOL] A smooth, streamlined hill modeled by glacial erosion, which has a core of bedrock usually veneered with a layer of glacial till and which resembles a true drumlin in outline and form but is generally less symmetrical and less regularly shaped. Also known as drumlinoid; false drumlin; rocdrumlin; rock drum. { 'räk ¦drəm·lən }

rock element [PETR] The coherent, intact piece of rock that is the basic constituent of the rock system and which has physical, mechanical, and petrographic properties that can be described or measured by laboratory tests. { 'räk ,el·ə·mənt }

rocket lightning [GEOPHYS] A rare form of lightning whose luminous channel seems to advance through the air with only the speed of a skyrocket. { 'räk·ət ,līt·niŋ }

rock failure [GEOL] Fracture of a rock that has been stressed beyond its ultimate strength. { 'räk ,fāl·yər }

rockfall [GEOL] **1.** The fastest-moving landslide; free fall of newly detached bedrock segments from a cliff or other steep slope; usually occurs during spring thaw. **2.** The rock material moving in or moved by a rockfall. { 'räk,fȯl }

rock fan [GEOL] A fan-shaped bedrock surface whose apex is where a mountain stream debouches upon a piedmont slope, and which occupies an area where a pediment meets the mountain slope. { 'räk ,fan }

rock-floor robbing [GEOL] A form of sheetflood erosion in which sheetfloods remove crumbling debris from rock surfaces in desert mountains. { 'räk ¦flȯr ,räb·iŋ }

rock flour [GEOL] A fine, chemically unweathered powder of rock-forming minerals produced by pulverization of rock fragments during natural transport or crushing. Also known as glacial flour. { 'räk ,flau̇·ər }

rockforming [GEOL] Referring to any minerals which commonly occur in important proportions in common rocks. { 'räk,fȯrm·iŋ }

rock fragment [PETR] A component of a sedimentary rock consisting of polymineralic or polygranular sand grains that are abraded particles of igneous, sedimentary, or metamorphic rocks. { 'räk ,frag·mənt }

rock glacier [GEOL] Boulders and fine material cemented by ice about a meter below the surface. Also known as talus glacier. { 'räk ,glā·shər }

rock-glacier creep [GEOL] A rapid talus creep of tongues of debris in a cold region, caused by the expansive force of the alternate freeze and thaw of ice in the interstices of the debris. { 'räk ¦glā·shər ,krēp }

rock gypsum [MINERAL] Massive, coarsely crystalline to earthy, finely granular type of gypsum found in gyp rock. { 'räk ,jip·səm }

rocking stone [GEOL] A stone or boulder, often of great size, so finely poised upon its foundation (as on the side of a hill or cliff) that it can be moved slightly backward and forward with little force (as with the hand) and still retain its original position. Also known as roggan. { 'räk·iŋ ,stōn }

rock island See meander core. { 'räk ,ī·lənd }

rock kindred See rock association. { 'räk ,kin·drəd }

rock magnetism [GEOPHYS] The natural remanent magnetization of igneous, metamorphic, and sedimentary rocks resulting from the presence of iron oxide minerals. { 'räk 'mag·nə,tiz·əm }

rock matrix compressibility [GEOL] One of three types of rock compressibility (matrix, bulk, and pore); the fractional change in volume of the solid rock material (grains) with a unit change in pressure. { 'räk ¦mā·triks kəm,pres·ə'bil·əd·ē }

rock meal See rock milk. { 'räk ,mēl }

rock mechanics [GEOPHYS] Application of the principles of mechanics and geology to quantify the response of rock when it is acted upon by environmental forces, particularly when human-induced factors alter the original ambient forces. { 'räk mi,kan·iks }

rock milk [MINERAL] A soft, white, earthy or powdery variety of calcite. Also known as agaric mineral; bergmehl; forril farina; rock meal. { 'räk ,milk }

rock pedestal See pedestal. { 'räk ,ped·ə·stəl }

rock pediment [GEOL] A pediment formed on the surface of bedrock. { 'räk ,ped·ə·mənt }

rock permeability [GEOL] The ability of a rock to receive, hold, or pass fluid materials (oil, water, and gas) by nature of the interconnections of its internal porosity. { 'räk ,pər·mē·ə'bil·əd·ē }

rock phosphate See phosphorite. { 'räk 'fä,sfāt }

rock pillar [GEOL] **1.** A column of rock produced by differential weathering or erosion, as along a joint plane. **2.** In a cave, a pillar-type structure that is residual bedrock rather than a stalactostalagmite. { 'räk ˌpil·ər }

rock pool |GEOL| A tidal pool formed along a rocky shoreline. { 'räk ˌpül }

rock pressure |GEOPHYS| **1.** Stress in underground geologic material due to weight of overlying material, residual stresses, and pressures resulting from swelling clays. **2.** *See* ground pressure. { 'räk ˌpresh·ər }

rock river |GEOL| A very long and narrow rock stream. { 'räk ˌriv·ər }

rock salt *See* halite. { 'räk ˌsȯlt }

rock shelter |GEOL| A cave that is formed by a ledge of overhanging rock. { 'räk ˌshel·tər }

rock silk |MINERAL| A silky variety of asbestos. { 'räk ˌsilk }

rockslide |GEOL| The sudden, rapid downward movement of newly detached bedrock segments over a surface of weakness, such as of bedding, jointing, or faulting. Also known as rock slip. { 'räkˌslīd }

rock slip *See* rockslide. { 'räk ˌslip }

rock stack |GEOL| A rocky crag that has been uplifted from an old sea floor. { 'räk ˌstak }

rock step *See* knickpoint. { 'räk ˌstep }

rock-stratigraphic unit |GEOL| A lithologically homogeneous body of strata characterized by certain observable physical features, or by the dominance of a certain rock type or combination of rock types; rock-stratigraphic units include groups, formations, members, and beds. Also known as geolith; lithologic unit; lithostratic unit; lithostratigraphic unit; rock unit. { 'räk ¦strad·əˈgraf·ik 'yü·nət }

rock stream |GEOL| Rocks moving (or already moved) in a mass down a slope under the influence of their own weight. { 'räk ˌstrēm }

rock system |GEOPHYS| In rock mechanics, all natural environmental factors that can influence the behavior of that portion of the earth's crust that will become part of an engineering structure. { 'räk ˌsis·təm }

rock terrace |GEOL| A stream terrace on the side of a valley composed of resistant bedrock which remains during erosion of weaker overlying and underlying beds. { 'räk ˌter·əs }

rock type |PETR| **1.** One of the three major rock groups: igneous, sedimentary, metamorphic. **2.** A rock having a unique, identifiable set of characters, such as basalt. { 'räk ˌtīp }

rock unit *See* rock-stratigraphic unit. { 'räk ˌyü·nət }

rock varnish |GEOL| A dark coating on rock surfaces exposed to the atmosphere. It is composed of about 30% manganese and iron oxides, up to 70% clay minerals, and over a dozen trace and rare-earth minerals. Although found in all terrestrial environments, it is mostly developed and best preserved in arid regions. Also know as desert varnish. { 'räk ˌvär·nəsh }

rock wood *See* mountain wood. { 'räk ˌwu̇d }

rod |GEOL| A rodlike sedimentary particle characterized by a width-length ratio less than 2/3 and a thickness-width ratio more than 2/3. Also known as roller. { räd }

rodding |PETR| In metamorphic rocks, a linear structure in which the stronger parts, such as vein quartz or quartz pebbles, have been shaped into parallel rods. { 'räd·iŋ }

rodingite |PETR| A medium- to coarse-grained, commonly calcium-enriched gabbroic rock containing grossular and diallage as essential minerals. { 'rōd·iŋˌgīt }

rodite *See* diogenite. { 'rōˌdīt }

roedderite |MINERAL| $(Na,K)_2(Mg,Fe)_5Si_{12}O_{30}$ A silicate meteorite mineral. { 'rädˌəˌrīt }

roemerite |MINERAL| $FeFe_2(SO_4)_4 \cdot 14H_2O$ A rust-brown to yellow mineral composed of hydrous ferric and ferrous iron sulfate. { 'rām·əˌrīt }

roesslerite |MINERAL| $MgH(AsO_4) \cdot 7H_2O$ A monoclinic mineral composed of hydrous acid magnesium arsenate; it is isomorphous with phosphorroesslerite. { 'res·ləˌrīt }

roestone *See* oolite. { 're̱ˌstōn }

rofla |GEOL| An extremely narrow, tortuous gorge, frequently formed by meltwater streams flowing from a glacier. { 'rō·flə }

rogenstein [GEOL] An oolite in which the ooliths are united by argillaceous cement. { 'rō·gən,stīn }

roggan See rocking stone. { 'räg·ən }

roll [GEOL] A primary sedimentary structure produced by deformation involving subaqueous slump or vertical foundering. { rōl }

roller See rod. { 'rō·lər }

rolling beach [GEOL] At the base of a sea cliff, the upper part of an accumulation of boulder sand pebbles which is being ground to sand and finer particles. { 'rōl· iŋ 'bēch }

Romanche trench [GEOL] A 24,320-foot-deep (7370-meter) trench in the Mid-Atlantic Ridge near the equator. { rō'mänsh 'trench }

romeite [MINERAL] $(Ca,Fe,Mn,Na)_2(Sb,Ti)_2O_6(O,OH,F)$ A honey-yellow to yellowish-brown mineral composed of oxide of calcium, iron, manganese, sodium, antimony, and titanium, occurring in minute octahedrons. { 'rō·mē,īt }

rongstockite [GEOL] A medium- to fine-grained plutonic rock composed of zoned plagioclase, orthoclase, some cancrinite, augite, mica, hornblende, magnetite, sphene, and apatite. { raŋ'stä,kīt }

roof [GEOL] **1.** The rock above an orebody. **2.** The country rock bordering the upper surface of an igneous intrusion. { rüf }

roofed dike [GEOL] A dike that has an upward termination. { "rüft 'dīk }

roof foundering [GEOL] Collapse of overlying rock into a magma chamber following excavation of a large quantity of magma. { 'rüf ,faün·driŋ }

roof pendant [GEOL] Downward projection or sag into an igneous intrusion of the country rock of the roof. Also known as pendant. { 'rüf ,pen·dənt }

room [GEOL] An open area in a cave. { rüm }

rooseveltite [MINERAL] $BiAsO_4$ A gray mineral consisting of bismuth arsenate; occurs as thin botryoidal crusts. { 'rōz·vəl,tīt }

root [GEOL] **1.** The lower limit of an ore body. Also known as bottom. **2.** The part of a fold nappe that was originally linked to its root zone. { rüt }

root cast [GEOL] A slender, tubular, near-vertical, and commonly downward-branching sedimentary structure formed by the filling of a tubular opening left by a root. Also known as rhizoconcretion. { 'rüt ,kast }

root clay See underclay. { 'rüt ,klā }

rootless vent [GEOL] A source of lava that is not directly connected to a volcanic vent or magma source. { 'rüt·ləs 'vent }

root sheath [GEOL] A hollow root cast. { 'rüt ,shēth }

root zone [GEOL] **1.** The area where a low-angle thrust fault steepens and descends into the crust. **2.** The source of the root of a fold nappe. { 'rüt ,zōn }

ropy lava See pahoehoe. { 'rō·pē 'lä·və }

rosasite [MINERAL] $(Cu,Zn)_2(OH)_2(CO_3)$ A green to bluish-green and sky blue mineral consisting of a carbonate-hydroxide of copper and zinc. { 'rō·zə,sīt }

roscherite [MINERAL] $(Ca,Mn,Fe)_2Al(PO_4)(OH)·2H_2O$ A dark-brown mineral composed of hydrous basic phosphate of aluminum, calcium, manganese, and iron, occurring as monoclinic crystals. { räsh·ə,rīt }

roscoelite [MINERAL] $K(V,Al,Mg)_3Si_3O_{10}(OH)_2$ Tan, grayish-brown, or greenish-brown vanadium-bearing mica mineral occurring in minute scales or flakes. { 'rä,skō,īt }

rose diagram [GEOL] A circular graph indicating values in several classes of vector properties of rocks such as cross-bedding direction. { 'rōz 'dī·ə,gram }

roselite [MINERAL] $(Ca,Co)_2(Co,Mg)(AsO_4)_2·2H_2O$ A pink or rose-colored, monoclinic mineral consisting of a hydrated arsenate of calcium, cobalt, and magnesium. { 'rōz·ə,līt }

rose opal [MINERAL] An opaque variety of common opal having a fine red color. { 'rōz 'ō·pəl }

rose quartz [MINERAL] A pink variety of crystalline quartz; commonly massive and used as a gemstone. Also known as Bohemian ruby. { 'rōz 'kwòrts }

rosette [MINERAL] Rose-shaped, crystalline aggregates of barite, marcasite, or pyrite formed in sedimentary rock. { rō'zet }

rosieresite [MINERAL] A yellow to brown mineral composed of hydrous aluminum phosphate containing lead and copper, occurring in stalactitic masses. { ,rō,zē'er·ə,sīt }

rosin tin [MINERAL] A red or yellow variety of cassiterite. Also known as resin tin. { 'räz·ən ¦tin }

Rosiwal analysis [PETR] A quantitive method of estimating the volume percentages of the minerals in a rock, in which thin sections of a rock are examined under a microscope which has a micrometer to measure the linear intercepts of each mineral along a particular set of lines. { 'räz·ə,wól ə,nal·ə·səs }

rossite [MINERAL] CaV$_2$O$_6$·4H$_2$O A yellow, triclinic mineral consisting of a hydrated calcium vanadate. { 'ró,sīt }

rosterite See vorobyevite. { 'rä·stə,rīt }

rosthornite [MINERAL] A brown to garnet-red variety of retinite with a low (4.5) oxygen content, found in lenticular masses in coal. { 'räs·thər,nīt }

rotary fault [GEOL] A fault in which displacement is downward at one point and upward at another point. Also known as pivotal fault; rotational fault. { 'rōd·ə·rē 'fólt }

rotational bomb [GEOL] A bomb whose shape is formed by spiral motion or rotation during flight. { rō'tā·shən·əl 'bäm }

rotational fault See rotary fault. { rō'tā·shən·əl 'fólt }

rotational landslide [GEOL] A landslide in which shearing takes place on a well-defined, curved shear surface, concave upward in cross section, producing a backward rotation in the displaced mass. { rō'tā·shən·əl 'lan,slīd }

rotational movement [GEOL] Apparent fault-block displacement in which the blocks have rotated relative to one another, so that alignment of formerly parallel features is disturbed. { rō'tā·shən·əl 'müv·mənt }

rotational wave See shear wave; S wave. { rō'tā·shən·əl 'wāv }

Rotliegende [GEOL] A European series of geologic time: Lower and Middle Permian. { 'rōt,lē·gən·də }

rotten spot See pothole. { 'rät·ən ,spät }

rougemontite [GEOL] A coarse-grained igneous rock composed of anorthite, titanaugite, and small amounts of olivine and iron ore. { 'rüzh,män,tīt }

roundness [GEOL] The degree of abrasion of sedimentary particles; expressed as the radius of the average radius of curvature of the edges or corners to the radius of curvature of the maximum inscribed sphere. { 'raúnd·nəs }

roundstone [GEOL] Any naturally rounded rock fragment of any size larger than a sand grain (diameter greater than 2 millimeters), such as a boulder, cobble, pebble, or granule. { 'raúnd,stōn }

routivarite [GEOL] A fine-grained igneous rock containing orthoclase, plagioclase, quartz, and garnet. { ¦rüd·ə¦va,rīt }

rouvillite [GEOL] A light-colored theralite composed predominantly of labradorite and nepheline, with small amounts of titanaugite, hornblende, pyrite, and apatite. { 'rüv·ə,līt }

rouvite [MINERAL] CaU$_2$V$_{12}$O$_{36}$·20H$_2$O A purplish- to bluish-black mineral consisting of a hydrated vanadate of calcium and uranium; occurs as dense masses, crusts, and coatings. { 'rü,vīt }

roweite [MINERAL] (Mn,Mg,Zn)Ca(BO$_2$)$_2$(OH)$_2$ A light-brown mineral composed of basic borate of calcium, manganese, magnesium, and zinc. { 'rō,īt }

R tectonite [PETR] A tectonite in which the fabric is believed to have resulted from rotation. { 'är 'tek·tə,nīt }

rubble [GEOL] 1. A loose mass of rough, angular rock fragments, coarser than sand. 2. See talus. { 'rəb·əl }

rubble drift [GEOL] 1. A rubbly deposit (or congeliturbate) formed by solifluction under periglacial conditions. 2. A coarse mass of angular debris and large blocks set in an earthy matrix of glacial origin. { 'rəb·əl ,drift }

rubble tract [GEOL] The part of the reef flat immediately behind and on the lagoon side of the reef front, paved with cobbles, pebbles, blocks, and other coarse reef fragments. { 'rəb·əl ,trakt }

rubellite |MINERAL| The red to red-violet variety of the gem mineral tourmaline; hardness is 7–7.5 on Mohs scale, and specific gravity is near 3.04 { 'rü·bə,līt }

rubicelle |MINERAL| A yellow or orange-red gem variety of spinel. { ¦rü·bə¦sel }

rubidium-strontium dating |GEOL| A method for determining the age of a mineral or rock based on the decay rate of rubidium-87 to strontium-87. { rü'bid·ē·əm 'strän·chəm 'dād·iŋ }

ruby |MINERAL| The red variety of the mineral corundum; in its finest quality, the most valuable of gemstones. { 'rü·bē }

ruby copper ore See cuprite. { 'rü·bē 'käp·ər ¦ȯr }

ruby mica |MINERAL| The finest grade of Indian mica; used for electrical capacitors. { 'rü·bē 'mī·kə }

ruby silver |MINERAL| Either of two red silver sulfide minerals: pyrorgyrite (dark-ruby silver) and proustite (light-ruby silver). { 'rü·bē 'sil·vər }

ruby spinel |MINERAL| A clear-red gem variety of spinel, containing small amounts of chromium and having the color but none of the other attributes of true ruby. { 'rü·bē spə'nel }

ruby zinc See zincite. { 'rü·bē 'ziŋk }

rudaceous |PETR| Of or pertaining to a sedimentary rock composed of a large quantity of fragments that are larger than sand grains (diameter greater than 2 millimeters). { rü'dā·shəs }

rudistids |PALEON| Fossil sessile bivalves that formed reefs during the Cretaceous in the southern Mediterranean or the Tethyan belt. { rü'dis·tədz }

rudite |GEOL| A sedimentary rock composed of fragments coarser than sand grains. { 'rü,dīt }

Rudzki anomaly |GEOPHYS| A gravity anomaly calculated by replacing the surface topography by its mirror image within the geoid. { 'rüd·skē a,näm·ə·lē }

ruffle |GEOL| A ripple mark produced by an eddy. { 'rəf·əl }

ruffled groove cast |GEOL| A groove cast with a feather pattern, consisting of a groove with lateral wrinkles that join the main cast in the downcurrent direction at an acute angle. { 'rəf·əld ¦grüv ,kast }

ruggedness number |GEOL| A dimensionless number that expresses the geometric characteristics of a drainage system; derived from the product of maximum basin relief and drainage density within the drainage basin. { 'rəg·əd·nəs ,nəm·bər }

Rugosa |PALEON| An order of extinct corals having either simple or compound skeletons with internal skeletal structures consisting mainly of three elements, the septa, tabulae, and dissepiments. { ,rü'gō·sə }

ruin agate |MINERAL| A brown variety of agate displaying, on a polished surface, markings that resemble or suggest the outlines of ruins or ruined buildings. { 'rü·ən 'ag·ət }

ruin marble |PETR| A brecciated limestone that, when cut and polished, gives a mosaic effect suggesting the appearance of ruins or ruined buildings. { 'rü·ən 'mär·bəl }

rule of V's |GEOL| The outcrop of a formation that crosses a valley forms an acute angle (a V) that points in the direction in which the formation lies underneath the stream. { 'rül əv 'vēz }

run |GEOL| **1.** A ribbonlike, flat-lying, irregular orebody following the stratification of the host rock. **2.** A branching or fingerlike extension of the feeder of an igneous intrusion. { rən }

runite See graphic granite. { 'rü,nīt }

runnel |GEOL| A troughlike hollow on a tidal sand beach which carries water drainage off the beach as the tide retreats. { 'rən·əl }

running sand See quicksand. { 'rən·iŋ ,sand }

run-up See swash. { 'rən,əp }

runway |GEOL| The channel of a stream. { 'rən,wā }

Rupelian |GEOL| A European stage of middle Oligocene geologic time, above the Tongrian and below the Chattian. Also known as Stampian. { rü'pel·yən }

rupture See fracture. { 'rəp·chər }

313

rupture zone [GEOL] The region immediately adjacent to the boundary of an explosion crater, characterized by excessive in-place crushing and fracturing where the stresses produced by the explosion have exceeded the ultimate strength of the medium. { 'rəp·chər ,zōn }

russellite [MINERAL] Bi_2WO_6 A pale yellow to greenish, tetragonal mineral consisting of an oxide of bismuth and tungsten; occurs as fine-grained compact masses. { 'rəs·ə,līt }

rusting [GEOL] The formation of red, yellow, or brown iron oxide minerals by oxidation of mineral deposits. { 'rəst·iŋ }

rutherfordine [MINERAL] $(UO_2)(CO_3)$ A yellow mineral composed of uranyl carbonate, occurring as masses of fibers. { 'rəth·ər·fər,dēn }

rutilated quartz [MINERAL] Sagenitic quartz characterized by the presence of enclosed needlelike crystals of rutile. Also known as Venus hairstone. { 'rüd·əl,ād·əd 'kwȯrts }

rutile [MINERAL] TiO_2 A reddish-brown tetragonal mineral common in acid igneous rocks, in metamorphic rocks, and as residual grain in beach sand. { 'rü,tēl }

rutterite [PETR] A medium-grained, equigranular, dark-pink plutonic rock composed chiefly of microperthite, microcline, and albite, with small amounts of nepheline, biotite, amphibole, graphite, and magnetite. { 'rəd·ə,rīt }

R wave *See* Rayleigh wave. { 'är ,wāv }

S

Saalic orogeny [GEOL] A short-lived orogeny that occurred early in the Permian period, between the Autunian and Saxonian stages. { 'sä·lik ȯ'räj·ə·nē }

sabach *See* caliche. { ˌsä‚bäk }

Sabinas [GEOL] A North American (Gulf Coast) provincial series in Upper Jurassic geologic time, below the Coahuilan. { sə'bēn·əs }

sabkha *See* sebkha. { 'sab·kə }

sabulous *See* arenaceous. { 'sab·yə·ləs }

saccharoidal [PETR] The texture of a rock that is crystalline or granular. Also known as sucrosic; sugary. { ¦sak·ə¦rȯid·əl }

saccus *See* vesicle. { 'sak·əs }

sackungen [GEOL] Deep-seated rock creep which has produced a ridge-top trench by gradual settlement of a slablike mass into an adjacent valley. { 'saˌku̇ŋ·ən }

saddle [GEOL] **1.** A gap that is broad and gently sloping on both sides. **2.** A relatively flat ridge that connects the peaks of two higher elevations. **3.** That part along the surface axis or axial trend of an anticline that is a low point or depression. { 'sad·əl }

saddleback [GEOL] A hill or ridge with a concave outline along its crest. { 'sad·əlˌbak }

saddle fold [GEOL] A flexural fold perpendicular to the parent fold and having an additional flexure at its crest. { 'sad·əl ˌfōld }

saddle point *See* col. { 'sad·əl ˌpȯint }

saddle reef [GEOL] A mineral deposit associated with the crest of an anticlinal fold and following the bedding plane, usually found in vertical succession. Also known as saddle vein. { 'sad·əl ˌrēf }

saddle vein *See* saddle reef. { 'sad·əl ˌvān }

safflorite [MINERAL] $CoAs_2$ A cobalt arsenide mineral that occurs in tin-white masses, and is dimorphous with smaltite; found in Canada, Morocco, and the United States. { 'saf·ləˌrīt }

sag [GEOL] **1.** A pass or gap in a ridge or mountain range shaped like a saddle. **2.** A shallow depression in a relatively flat land surface. **3.** A regional basin with gently sloping sides. { sag }

sagenite [MINERAL] A variety of rutile that is acicular and occurs in reticulated twin groups of crystals crossing at 60°. { 'saj·əˌnīt }

sagenitic [GEOL] Containing acicular minerals. { ˌsaj·ə'nid·ik }

Sagenocrinida [PALEON] A large order of extinct, flexible crinoids that occurred from the Silurian to the Permian. { ¦saj·ə·nō'krī·nə·də }

Saghathiinae [PALEON] An extinct subfamily of hyracoids in the family Procaviidae. { ˌsag·ə'thī·ə‚nē }

sag pond [GEOL] A small body of water occupying an enclosed depression or sag formed where active or recent fault movement has impounded drainage. Also known as fault-trough lake; rift lake; rift-valley lake. { 'sag ˌpänd }

sahlinite [MINERAL] $Pb_{14}(AsO_4)_2O_9Cl_4$ A pale sulfur-yellow, monoclinic mineral consisting of a basic chloride-arsenate of lead; occurs in aggregates of small scales. { 'sä·ləˌnīt }

sahlite *See* salite. { 'säˌlīt }

Saint Peter sandstone [GEOL] An artesian aquifer of early Lower Paleozoic age which underlies part of Minnesota, Wisconsin, Iowa, Illinois, and Indiana. { 'sänt 'pēd·ər 'san,stōn }

Sakmarian [GEOL] A European stage of geologic time; the lowermost Permian, above Stephanian of Carboniferous and below Artinskian. { säk'mär·ē·ən }

sal *See* sial. { sal }

Salado formation [GEOL] A red-bed formation from the Permian found in southeast New Mexico; contains rock salt and potash salts. { sə'lä·dō fȯr,mā·shən }

salammoniac [MINERAL] NH_4Cl A white, isometric, crystalline mineral composed of native ammonium chloride. { ,sal·ə'mō·nē,ak }

salband *See* selvage. { 'sal,band }

salcrete [GEOL] A thin, hard crust of salt-cemented sand grains, occurring on a marine beach that is occasionally or periodically saturated by saline water. { 'sal,krēt }

saléeite [MINERAL] $Mg(UO_2)_2(PO_4)_2 \cdot 10H_2O$ A lemon-yellow mineral composed of hydrous phosphate of magnesium and uranium. { sə'lā,īt }

salesite [MINERAL] $Cu(IO_3)(OH)$ A bluish-green mineral composed of basic iodate of copper. { 'sāl,zīt }

salfemic rock [GEOL] An igneous rock in which the ratio of salic to femic minerals is greater than 3:5 and less than 5:3. { sal'fē·mik 'räk }

salic [GEOL] A soil horizon enriched with secondary salts, at least 2 percent, and measuring at least 6 inches (15 centimeters) in thickness. [MINERAL] Pertaining to certain light-colored minerals, such as quartz and feldspars, that are rich in silica or magnesium and commonly occur in igneous rock. { 'sal·ik }

salient [GEOL] **1.** A landform that projects or extends outward or upward from its surroundings. **2.** An area in which the axial traces of folds are convex toward the outer edge of the folded belt. { 'sāl·yənt }

saliferous stratum [GEOL] A stratum that contains, produces, or is impregnated with salt. Also known as saliniferous stratum. { sə'lif·ə·rəs 'strad·əm }

salina [GEOL] An area, such as a salt flat, in which deposits of crystalline salts are formed or found. { sə'lē·nə }

salinastone [GEOL] A sedimentary rock composed mostly of saline minerals which are usually precipitated but may be fragmental. { sə'lē·nə,stōn }

saline-alkali soil [GEOL] A salt-affected soil with a content of exchangeable sodium greater than 15, with much soluble salts, and with a pH value usually less than 9.5. { 'sā,lēn 'al·kə,lī ,sȯil }

salinelle [GEOL] A mud volcano erupting saline mud. { ,sa·lə'nel }

saline soil [GEOL] A nonalkali, salt-affected soil with a high content of soluble salts, with exchangeable sodium of less than 15, and with a pH value less than 8.5. { 'sā,lēn ,sȯil }

saliniferous stratum *See* saliferous stratum. { ,sal·ə'nif·ə·rəs 'strad·əm }

salinization [GEOL] In a soil of an arid, poorly drained region, the accumulation of soluble salts by the evaporation of the waters that bore them to the soil zone. { ,sal·ən·ə'zā·shən }

salite [MINERAL] $(Mg,Fe)_2Si_2O_6$ A grayish-green to black mineral variety of diopside containing more magnesium than iron; member of the clinopyroxene group. Also spelled sahlite. { 'sa,līt }

salitrite [PETR] A lamprophyre composed chiefly of titanite and diopside with acmite, accessory apatite, microcline, and occasionally anorthoclase and baddeleyite. { 'sal·ə,trīt }

salmonsite [MINERAL] A buff-colored mineral composed of hydrous phosphate of manganese and iron occurring in cleavable masses. { 'sam·ən,zīt }

salt-affected soil [GEOL] A general term for a soil that is not suitable for the growth of crops because of an excess of salts, exchangeable sodium, or both. { 'sȯlt iˌfek·təd 'sȯil }

salt-and-pepper sand [GEOL] A sand composed of a mixture of light- and dark-colored grains. { ˌsȯlt ən ˌpep·ər 'sand }

salt anticline [GEOL] A structure like a salt dome but with a linear salt core. Also known as salt wall. { 'sȯlt 'ant·i,klīn }

saltation |GEOL| Transport of a sediment in which the particles are moved forward in a series of short intermittent bounces from a bottom surface. { sȯl'tā·shən }

saltation load |GEOL| The part of the bed load that is bouncing along the stream bed or is moved, directly or indirectly, by the impact of bouncing particles. { sȯl'tā·shən ,lōd }

salt bottom |GEOL| A flat piece of relatively low-lying ground encrusted with salt. { 'sȯlt ,bäd·əm }

salt burst |GEOL| Rock destruction caused by crystallization of soluble salts that enter the pores. { 'sȯlt ,bərst }

salt dome |GEOL| A diapiric or piercement structure in which there is a central, equidimensional salt plug. { 'sȯlt ,dōm }

salt-dome breccia |GEOL| A breccia found in deep shale sequences and occurring as a dome-shaped mass in a broad zone surrounding a salt plug. { 'sȯlt ¦dōm 'brech·ə }

salt field |GEOL| An area overlying a usually workable salt deposit of economic value. { 'sȯlt ,fēld }

salt flat |GEOL| The level, salt-encrusted bottom of a lake or pond that is temporarily or permanently dried up. { 'sȯlt ,flat }

salt glacier |GEOL| A gravitational flow of salt down the slopes of a salt plug, following the preexisting structure. { 'sȯlt ,glā·shər }

salt hill |GEOL| An abrupt hill of salt, with sinkholes and pinnacles at its summit. { 'sȯlt ,hil }

saltierra |GEOL| A deposit of salt left by evaporation of a shallow salt lake. { ,sal·tē'er·ə }

salt pan |GEOL| **1.** An undrained, usually small and shallow, natural depression or hollow in which water accumulates and evaporates, leaving a salt deposit. **2.** A shallow lake of brackish water occupying such a depression. { 'sȯlt ,pan }

saltpeter cave |GEOL| A cave in which there are deposits of saltpeter earth. { sȯlt'pēd·ər ,kāv }

saltpeter earth |GEOL| A deposit containing calcium nitrate and found in caves. { sȯlt 'pēd·ər ,ərth }

salt pillow |GEOL| An embryonic salt dome rising from its source bed, still at depth. { 'sȯlt ,pil·ō }

salt pit |GEOL| A pit in which sea water is received and evaporated and from which salt is obtained. { 'sȯlt ,pit }

salt plug |GEOL| The salt core of a salt dome. { 'sȯlt ,pləg }

salt polygon |GEOL| A surface of salt on a playa, having three to eight sides marked by ridges of material formed as a result of the expansive forces of crystallizing salt, and ranging in width from an inch or so to 100 feet (30 meters). { 'sȯlt 'päl·i,gän }

salt stock |GEOL| An immature salt dome comprising a pluglike salt diapir that has pierced the overlying strata. { 'sȯlt ,stäk }

salt tectonics |GEOL| The study of the structure and mechanism of emplacement of salt domes. Also known as halokinesis. { 'sȯlt tek'tän·iks }

salt wall *See* salt anticline. { 'sȯlt ,wȯl }

salt weathering |GEOL| The granular disintegration or fragmentation of rock material produced by saline solutions or by salt-crystal growth. { 'sȯlt ,weth·ə·riŋ }

samarskite |MINERAL| $(Y,Ce,U,Ca,Fe,Pb,Th)(Nb,Ta,Ti,Sn)_2O_6$ A velvet-black to brown metamict orthorhombic mineral with splendent vitreous to resinous luster occurring in granite pegmatites. Also known as ampangabeite; uranotantalite. { sə'mär ,skīt }

sampleite |MINERAL| $NaCaCu_5(PO_4)_4Cl·5H_2O$ A blue mineral composed of hydrous phosphate and chloride of sodium, calcium, and copper. { 'sam·pə,līt }

samsonite |MINERAL| $Ag_4MnSb_2S_6$ A black mineral composed of sulfide of silver, manganese, and antimony occurring in monoclinic prismatic crystals. { 'sam·sə,nīt }

sanbornite |MINERAL| $BaSi_2O_5$ A white triclinic mineral composed of barium silicate. { 'san,bȯr,nīt }

sanakite |PETR| A glassy andesite composed of bronzite, augite, magnetite, and a few large plagioclase and garnet crystals. { 'san·ə,kīt }

sand |GEOL| Unconsolidated granular material consisting of mineral, rock, or biological fragments between 63 micrometers and 2 millimeters in diameter, usually produced primarily by the chemical or mechanical breakdown of older source rocks, but may also be formed by the direct chemical precipitation of mineral grains or by biological processes. { sand }

sand apron |GEOL| A deposit of sand along the shore of a lagoon of a reef. { 'sand ‚ā·prən }

sandarac *See* realgar. { 'san·də‚rak }

sand avalanche |GEOL| Movement of large masses of sand down a dune face when the angle of repose is exceeded or when the dune is disturbed. { 'sand ‚av·ə‚lanch }

sandbag |GEOL| In the roof of a coal seam, a deposit of glacial debris formed by scour and fill subsequent to coal formation. { 'san‚bag }

sandbank |GEOL| A deposit of sand forming a mound, hillside, bar, or shoal. { 'san‚baŋk }

sandbar |GEOL| A bar or low ridge of sand bordering the shore and built up, or near, to the surface of the water by currents or wave action. Also known as sand reef. { 'san‚bär }

sandblasting |GEOL| Abrasion affected by the action of hard, windblown mineral grains. { 'san‚blast·iŋ }

sand cay *See* sandkey. { 'san ‚kē }

sand cone |GEOL| **1.** A cone-shaped deposit of sand, produced especially in an alluvial cone. **2.** A low debris cone whose protective veneer consists of sand. { 'san ‚kōn }

sand crystal |GEOL| A large crystal loaded up to 60% with detrital sand inclusions formed in a sandstone during or as a result of cementation. { 'san ‚krist·əl }

sand dike |GEOL| A sedimentary dike consisting of sand that has been squeezed or injected upward into a fissure. { 'san ‚dīk }

sand drift |GEOL| **1.** Movement of windblown sand along the surface of a desert or shore. **2.** An accumulation of sand against the leeward side of a fixed obstruction. { 'san ‚drift }

sand drip |GEOL| A rounded or crescentic surface form on a beach sand, resulting from the sudden absorption of overwash. { 'san ‚drip }

sand dune |GEOL| A mound of loose windblown sand commonly found along low-lying seashores above high-tide level. { 'san ‚dün }

sandfall *See* slip face. { 'san‚fól }

sand flat |GEOL| A sandy tidal flat barren of vegetation. { 'san ‚flat }

sand flood |GEOL| A vast body of sand moving or borne along a desert, as in the Arabian deserts. { 'san ‚fləd }

sand gall *See* sand pipe. { 'san ‚gól }

sand glacier |GEOL| **1.** An accumulation of sand that is blown up the side of a hill or mountain and through a pass or saddle, and then spread out on the opposite side to form a wide, fan-shaped plain. **2.** A horizontal plateau of sand terminated by a steep talus slope. { 'san ‚glā·shər }

sand hill |GEOL| A ridge of sand, especially a sand dune in a desert region. { 'san ‚hil }

sand hole |GEOL| A small pit (7–8 millimeters in depth and a little less wide than deep) with a raised margin, formed on a beach by waves expelling air from a formerly saturated mass of sand. { 'san ‚hōl }

sand horn |GEOL| A pointed sand deposit extending from the shore into shallow water. { 'san ‚hórn }

sandkey |GEOL| A small sandy island parallel with the shore. Also known as sand cay. { 'san‚kē }

sand levee *See* whaleback dune. { 'san ‚lev·ē }

sand lobe |GEOL| A rounded sand deposit extending from the shore into shallow water. { 'san ‚lōb }

sand pavement |GEOL| A sandy surface derived from coarse-grained sand ripples, developed on the lower, windward slope of a dune or rolling sand area during a period of intermittent light, variable winds. { 'san ‚pāv·mənt }

318

sand pipe [GEOL] A pipe formed in sedimentary rocks, filled with considerable sand and some gravel. Also known as sand gall. { 'san ,pīp }

sand plain [GEOL] A small outwash plain formed by deposition of sand transported by meltwater streams flowing from a glacier. { 'san ,plān }

sand reef See sandbar. { 'san ,rēf }

sand ridge [GEOL] **1.** Any low ridge of sand formed at some distance from the shore, and either submerged or emergent, such as a longshore bar or a barrier beach. **2.** One of a series of long, wide, extremely low, parallel ridges believed to represent the eroded stumps of former longitudinal sand dunes. **3.** A crescent-shaped landform found on a sandy beach, such as a beach cusp. **4.** See sand wave. { 'san ,rij }

sand river [GEOL] A river that deposits much of its sand load along its middle course, to be subsequently removed by the wind. { 'san ,riv·ər }

sandrock [GEOL] A field term for a sandstone that is not firmly cemented. { 'san,räk }

sand roll See pseudonodule. { 'san ,rōl }

sand run [GEOL] **1.** A fluidlike motion of dry sand. **2.** A mass of dry sand in motion. { 'san ,rən }

sand sea [GEOL] **1.** An extensive assemblage of sand dunes of several types in an area where a great supply of sand is present; characterized by an absence of travel lines, or directional indicators, and by a wavelike appearance of dunes separated by troughs. **2.** The flat, rain-smoothed plain of volcanic ash and other pyroclastics on the floor of a caldera. { 'san 'sē }

sand shadow [GEOL] A lee-side accumulation of sand, as a small turret-shaped dune, formed in the shelter of, and immediately behind, a fixed obstruction, such as clumps of vegetation. { 'san ,shad·ō }

sandshale [GEOL] A sedimentary deposit consisting of thin alternating beds of sandstone and shale. { 'san,shāl }

sand-shale ratio [GEOL] The ratio between the thickness or percentage of sandstone and that of shale in a geologic section. { 'san 'shāl 'rā·shō }

sand sheet [GEOL] A thin accumulation of coarse sand or fine gravel having a flat surface. { 'san ,shēt }

sandspit [GEOL] A spit consisting principally of sand. { 'san,spit }

sand splay [GEOL] A floodplain splay consisting of coarse sand particles. { 'san ,splā }

sandstone [PETR] A detrital sedimentary rock consisting of individual grains of sand-size particles 0.06 to 2 millimeters in diameter either set in a fine-grained matrix (silt or clay) or bonded by chemical cement. { 'san,stōn }

sandstone dike [GEOL] A dike made of sandstone or lithified sand. { 'san,stōn 'dīk }

sandstone sill [GEOL] A tabular mass of sandstone that has been emplaced by sedimentary injection parallel to the structure or by bedding of preexisting rock in the manner of an igneous sill. { 'san,stōn 'sil }

sand streak [GEOL] A low, linear ridge formed at the interface of sand and air or water, oriented parallel to the direction of flow, and having a symmetric cross section. { 'san ,strēk }

sand stream [GEOL] A small sand delta spread out at the mouth of a gully, or a deposit of sand along the bed of a small creek, formed by a torrential rain. { 'san ,strēm }

sand strip [GEOL] A long, narrow ridge of sand extending for a long distance downwind from each horn of a dune. { 'san ,strip }

sandur See outwash plain. { 'san·dər }

sandwash [GEOL] A sandy or gravel stream bed, devoid of vegetation, containing water only during a sudden and heavy rainstorm. { 'san,wäsh }

sand wave [GEOL] A large, ridgelike primary structure resembling a water wave on the upper surface of a sedimentary bed that is formed by high-velocity air or water currents. Also known as sand ridge. { 'san ,wāv }

sand wedge [GEOL] A wedge-shaped accumulation of sand with the apex downward formed by the filling in of winter contraction cracks. { 'san ,wej }

sandy bentonite See arkosic bentonite. { 'san·dē 'bent·ən,īt }

sandy chert [PETR] Chert formed in sandy beds by replacement of cement, or the filling of pore spaces, with silica. { 'san·dē 'chərt }

Sangamon |GEOL| The third interglacial stage of the Pleistocene epoch in North America, following the Illinoian glacial and preceding the Wisconsin. { 'saŋ·gə,mən }

sanidal |GEOL| Pertaining to the continental shelf. { 'san·əd·əl }

sanidine |MINERAL| $KAlSi_3O_8$ An alkali feldspar mineral occurring in clear, glassy crystals embedded in unaltered acid volcanic rocks; a high-temperature, disordered form. Also known as glassy feldspar; ice spar; rhyacolite. { 'san·ə,dēn }

sanidinite |PETR| A type of igneous rock composed chiefly of sanidine. { sə'nid·ən,īt }

sanmartinite |MINERAL| $ZnWO_4$ A mineral composed of zinc tungstate. { san'mart·ən,īt }

sannaite |PETR| An extrusive rock containing phenocrysts of barkevikite, pyroxene, and biotite (in order of decreasing abundance) in a fine-grained to dense groundmass of alkali feldspar, acmite, chlorite, calcite, and pseudomorphs of mica after nepheline. { 'san·ə,īt }

sansicl |GEOL| An unconsolidated sediment, consisting of a mixture of sand, silt, and clay, in which no component forms 50% or more of the whole aggregate. { 'san ,sik·əl }

Santonian |GEOL| A European stage of geologic time in the Upper Cretaceous, above the Coniacian and below the Campanian. { san'tō·nē·ən }

santorinite |PETR| **1.** A light-colored extrusive rock containing approximately 60–65% silica and calcic plagioclase (labradorite to anorthite) as the only feldspar. **2.** A hypersthene andesite containing plagioclase crystals that have labradorite cores and sodic rims and a groundmass with microlites of sodic oligoclase. { san'tȯr·ə,nīt }

sanukite |PETR| An andesite characterized by orthopyroxene as the mafic mineral, andesine as the plagioclase, and a glassy groundmass. { 'san·ə,kīt }

saponite |MINERAL| A soft, soapy, white or light-buff to bluish or reddish trioctahedral montmorillonitic clay mineral consisting of hydrous magnesium aluminosilicate and occurring in masses in serpentine and basaltic rocks. Also known as bowlingite; mountain soap; piotine; soapstone. { 'sap·ə,nīt }

sappare See kyanite. { 'sa,per }

sapphire |MINERAL| Any of the gem varieties of the mineral corundum, especially the blue variety, except those that have medium to dark tones of red that characterize ruby; hardness is 9 on Mohs scale, and specific gravity is near 4.00. { 'sa,fīr }

sapphire quartz |MINERAL| An indigo-blue opaque variety of quartz. { 'sa,fīr 'kwȯrts }

sapphirine |MINERAL| $(MgFe)_{15}(Al,Fe)_{34}Si_7O_{80}$ A green or pale-blue mineral composed of silicate and oxide of magnesium, iron, and aluminum; usually occurs in granular form. { 'saf·ə,rēn }

sapping |GEOL| Erosion along the base of a cliff by the wearing away of softer layers, thus removing the support for the upper mass which breaks off into large blocks and falls from the cliff face. Also known as undermining. { 'sap·iŋ }

Saprist |GEOL| A suborder of the soil order Histosol consisting of residues in which plant structures have been largely obliterated by decay; saturated with water most of the time. { 'sa,prist }

saprogenous ooze |GEOL| Ooze formed of putrefying organic matter. { sə'präj·ə·nəs 'üz }

saprolite |GEOL| A soft, earthy red or brown, decomposed igneous or metamorphic rock that is rich in clay and formed in place by chemical weathering. Also known as saprolith; sathrolith. { 'sap·rə,līt }

saprolith See saprolite. { 'sap·rə,lith }

sapropel |GEOL| A mud, slime, or ooze deposited in more or less open water. { 'sap·rə,pel }

sapropel-clay |GEOL| A sedimentary deposit in which the amount of clay is greater than that of sapropel. { 'sap·rə,pel ,klā }

sapropelic coal |GEOL| Coal formed by putrefaction of organic matter under anaerobic conditions in stagnant or standing bodies of water. Also known as sapropelite. { ¦sap·rə¦pel·ik 'kōl }

sapropelite See sapropelic coal. { 'sap·rə,pe,līt }

sapropel-peat See peat-sapropel. { 'sap·rə,pel ,pēt }

sarcopside [MINERAL] $(Fe,Mn,Mg)_3(PO_4)_2$ A mineral composed of a phosphate of manganese, magnesium, and iron. { sär'käp·səd }

sard [MINERAL] A translucent brown, reddish-brown, or deep orange-red variety of chalcedony. Also known as sardine; sardius. { särd }

Sardic orogeny [GEOL] A short-lived orogeny that occurred near the end of the Cambrian period. { 'sär·dik ȯ'räj·ə·nē }

sardine *See* sard. { sär'dēn }

sardius *See* sard. { 'sär·dē·əs }

sardonyx [MINERAL] An onyx characterized by parallel layers of sard, a deep orange-red variety of chalcedony, and a mineral of different color. { sär'dän·iks }

sarkinite [MINERAL] $Mn_2(AsO_4)(OH)$ A flesh-red monoclinic mineral composed of hydrous manganese arsenate, occurring in crystals. { 'sär·kə,nīt }

Sarmatian [GEOL] A European stage of geologic time: the upper Miocene, above Tortonian, below Pontian. { sär'mā·shən }

sarmientite [MINERAL] $Fe_2(AsO_4)(SO_4)(OH)·5H_2O$ A yellow mineral composed of basic hydrous arsenate and sulfate of iron; it is isomorphous with diadochite. { ,sär·mē'en,tīt }

sarnaite [GEOL] A feldspathoid-bearing syenite composed of cancrinite and acmite. { 'sär·nə,īt }

sarospatakite [GEOL] A micaceous clay mineral composed of mixed layers of illite and montmorillonite. { ,sar·ə'späd·ə,kīt }

sartorite [MINERAL] $PbAs_2S_4$ A dark-gray monoclinic mineral, occurring in crystalline form. { 'sär·də,rīt }

sassoline *See* sassolite. { 'sas·ə,lēn }

sassolite [MINERAL] H_3BO_3 A white or gray mineral consisting of native boric acid usually occurring in small pearly scales as an incrustation or as tabular triclinic crystals. Also known as sassoline. { 'sas·ə,līt }

satellitic crater *See* secondary crater. { ¦sad·ə¦lid·ik 'krād·ər }

sathrolith *See* saprolite. { 'sath·rə,lith }

satin spar [MINERAL] A white, translucent, fine fibrous variety of gypsum having a silky luster. Also known as satin stone. { 'sat·ən 'spär }

satin stone *See* satin spar. { 'sat·ən 'stōn }

saturated mineral [MINERAL] A mineral that forms in the presence of free silica. { 'sach·ə,rād·əd 'min·rəl }

saturated permafrost [GEOL] Permafrost that contains no more ice than the ground could hold if the water were in the liquid state. { 'sach·ə,rād·əd 'pər·mə,frȯst }

saturated rock [PETR] An igneous rock composed principally of saturated minerals. { 'sach·ə,rād·əd 'räk }

saturation curve [GEOL] A curve showing the weight of solids per unit volume of a saturated soil mass as a function of water content. { ,sach·ə'rā·shən ¦kərv }

saturation line [PETR] The line, on a variation diagram of an igneous rock series, that represents saturation with respect to silica; rocks to the right of the line are oversaturated and those to the left, undersaturated. { ,sach·ə'rā·shən ,līn }

Saucesian [GEOL] A North American stage of geologic time in the Oligocene and Miocene, above the Zemorrian and below the Relizian. { sȯ'sē·zhən }

sauconite [MINERAL] The zinc-bearing end member of the montmorillonite group; a trioctahedral clay mineral. { 'sȯ·kə,nīt }

saucyite [PETR] A glassy rhyolitic rock composed of large sanidine phenocrysts in a groundmass of orthoclase microlites and minute crystals of biotite, augite, sphene, zircon, and magnetite. { 'sȯ·sē,īt }

Saurichthyidae [PALEON] A family of extinct chondrostean fishes bearing a superficial resemblance to the Aspidorhynchiformes. { ,sȯr·ək'thī·ə,dē }

Saurischia [PALEON] The lizard-hipped dinosaurs, an order of extinct reptiles in the subclass Archosauria characterized by an unspecialized, three-pronged pelvis. { sȯ'ris·kē·ə }

Sauropoda [PALEON] A group of fully quadrupedal, seemingly herbivorous dinosaurs from the Jurassic and Cretaceous periods in the suborder Sauropodomorpha; members had small heads, spoon-shaped teeth, long necks and tails, and columnar legs. { so'räp·əd·ə }

Sauropodomorpha [PALEON] A suborder of extinct reptiles in the order Saurischia, including large, solid-limbed forms. { so¦räp·əd·ə'mȯr·fə }

Sauropterygia [PALEON] An order of Mesozoic marine reptiles in the subclass Euryapsida. { sȯ,räp·tə'rij·ē·ə }

saussurite [MINERAL] A white or grayish, tough, compact mineral aggregate composed chiefly of a mixture of albite or oligoclase and zoisite or epidote. { 'sȯ·sə,rīt }

saussuritization [GEOL] A metamorphic process involving replacement of plagioclase in basalts and gabbros by a fine-grained aggregate of zoisite, epidote, albite, calcite, sericite, and zeolites. { sȯ'sür·əd·ə'zā·shən }

savic orogeny [GEOL] A short-lived orogeny that occurred in late Oligocene geologic time, between the Chattian and Aquitanian stages. { 'sav·ik ȯ'räj·ə·nē }

saw-cut [GEOL] A large canyon that cuts abruptly across a terrace, so that it is visible only from locations near its edge. { 'sȯ ,kət }

Saxonian [GEOL] A European stage of geologic time in the Middle Permian, above the Autonian and below the Thuringian. { sak'sō·nē·ən }

saxonite [PETR] A peridotite composed chiefly of olivine and orthopyroxene. { 'sak·sə,nīt }

scabland [GEOL] Elevated land that is essentially flat-lying and covered with basalt and has only a thin soil cover, sparse vegetation, and usually deep, dry channels. { 'skab,land }

scabrock [GEOL] **1.** An outcropping of scabland. **2.** Weathered material of a scabland surface. { 'skab,räk }

scacchite [MINERAL] MnCl$_2$ A mineral composed of native manganese chloride, found in volcanic regions. { ska,kīt }

scaglia [GEOL] A dark, very-fine-grained, somewhat calcareous shale usually developed in the Upper Cretaceous and Lower Tertiary periods of the northern Apennines. { 'skal·yə }

scallop See scalloping. { 'skäl·əp }

scalloped upland [GEOL] The region near or at the divide of an upland into which glacial cirques have cut from opposite sides. { 'skäl·əpt 'əp·lənd }

scalloping [GEOL] A sedimentary structure superficially resembling an oscillation ripple mark, and having a concave side that is always oriented toward the top of the bed. Also known as scallop. { 'skäl·ə·piŋ }

scalped anticline See breached anticline. { 'skalpt 'ant·i,klīn }

scapolite [MINERAL] A white, gray, or pale-green complex aluminosilicate of sodium and calcium belonging to the tectosilicate group of silicate minerals; crystallizes in the tetragonal system and is vitreous; hardness is 5–6 on Mohs scale, and specific gravity is 2.65–2.74. Also known as wernerite. { 'skap·ə,līt }

scapolitization [GEOL] Introduction of or replacement by scapolite. { skap·ə,lid·ə'zā·shən }

scar [GEOL] **1.** A steep, rocky eminence, such as a cliff or precipice, where bare rock is well exposed. Also known as scaur; scaw. **2.** See shore platform. { skär }

scarp See escarpment. { skärp }

scarped plain [GEOL] A terrain characterized by a succession of faintly inclined or gently folded strata. { 'skärpt 'plān }

scarp face See scarp slope. { 'skärp ,fās }

scarplet See piedmont scarp. { 'skärp·lət }

scarpline [GEOL] A relatively straight line of cliffs of considerable extent, produced by faulting or erosion along a fault. { 'skärp,līn }

scarp slope [GEOL] The steep face of a cuesta, or asymmetric ridge, facing in an opposite direction to the dip of the strata. Also known as front slope; inface; scarp face. { 'skärp ,slōp }

scatter diagram [PETR] See point diagram. { 'skad·ər ,dī·ə,gram }

scaur See scar. { skär }

322

scaw *See* scar. { skȯ }

schafarzikite [MINERAL] $Fe_5Sb_4O_{11}$ A red to brown mineral composed of iron antimony oxide. { 'shä·fər,zi,kīt }

schairerite [MINERAL] $Na_3(SO_4)(F,Cl)$ A colorless rhombohedral mineral composed of sodium sulfate with fluorine and chlorine, occurring in crystals. { 'shī·rə,rīt }

schalstein [PETR] A slaty rock formed by shearing basaltic or andesitic tuff or lava. { 'shäl,stīn }

scheelite [MINERAL] $CaWO_4$ A yellowish-white mineral crystallizing in the tetragonal system and occurring in tabular or massive form in pneumatolytic veins associated with quartz; an ore of tungsten. { 'shā,līt }

schefflerite [MINERAL] $(Ca,Mn)(Mg,Fe,Mn)Si_2O_6$ Brown to black variety of pyroxene that crystallizes in the monoclinic system and contains manganese and frequently iron. { 'shef·lə,rīt }

scheteligite [MINERAL] $(Ca,Y,Sb,Mn)_2(Ti,Ta,Nb,W)_2O_6(O,OH)$ A mineral composed of oxide of calcium, rare-earth metals, antimony, manganese, titanium, columbium, and tantalum. { shə'tel·ə,gīt }

schirmerite [MINERAL] $PbAg_4Bi_4S_9$ A mineral composed of lead, silver, and bismuth sulfide. { 'shər·mə,rīt }

schist [GEOL] A large group of coarse-grained metamorphic rocks which readily split into thin plates or slabs as a result of the alignment of lamellar or prismatic minerals. { shist }

schist-arenite [PETR] A light-colored sandstone containing more than 20% rock fragments derived from an area of regionally metamorphosed rocks. { 'shist 'a·rə,nīt }

schistose [GEOL] Pertaining to rocks exhibiting schistosity. { 'shis,tōs }

schistosity [GEOL] A type of cleavage characteristic of metamorphic rocks, notably schists and phyllites, in which the rocks tend to split along parallel planes defined by the distribution and parallel arrangement of platy mineral crystals. { shis'täs·əd·ē }

schizolite [MINERAL] A light-red variety of pectolite containing manganese. { 'skiz·ə,līt }

schlieren [PETR] Irregular streaks with shaded borders in some igneous rocks, representing the segregation of light and dark minerals or altered inclusions, elongated by flow. { 'shlir·ən }

schlieren arch [GEOL] An intrusive igneous body with flow layers which occur along its borders but which are poorly developed or absent in its interior. { 'shlir·ən ,ärch }

schlieren dome [GEOL] An intrusive body more or less completely outlined by flow layers which culminate in one central area. { 'shlir·ən ,dōm }

Schmidt net [GEOL] A coordinate or reference system used to plot a Schmidt projection. { 'shmit ,net }

Schmidt projection [GEOL] A Lambert azimuthal equal-area projection of the lower hemisphere of a sphere onto the plane of a meridian; used in structural geology. { 'shmit prə,jek·shən }

schoepite [MINERAL] $UO_3·2H_2O$ A yellow secondary mineral composed of hydrous uranium oxide. { 'ske,pīt }

schönfelsite [PETR] A form of basalt containing embedded crystals of olivine and augite in a complex, dense fine-grained groundmass. { 'shən,fel,zīt }

schorl *See* schorlite. { shȯrl }

schorlite [MINERAL] The black, iron-rich, opaque variety of tourmaline. Also known as schorl. { 'shȯr,lit }

schorlomite [MINERAL] $Ca_3(Fe,Ti)_2(Si,Ti)_3O_{12}$ A black mineral of the garnet group that has a vitreous luster and usually occurs in masses; hardness is 7–7.5 on Mohs scale, and specific gravity is 3.81–3.88. { 'shȯr·lə,mīt }

schreibersite [MINERAL] $(Fe,Ni)_3P$ A silver-white to tin-white magnetic meteorite mineral crystallizing in the tetragonal system and occurring in tables or plates as oriented inclusions in iron meteorites. Also known as rhabdite. { 'shrī·bər,sīt }

schriesheimite [PETR] An amphibole peridotite that contains diopside. { 'shrē·shē·ə,mīt }

323

schroeckingerite

schroeckingerite [MINERAL] $NaCa_3(UO_2)(CO_3)(SO_4)F \cdot 10H_2O$ A yellowish secondary mineral composed of hydrous sodium calcium uranyl carbonate, sulfate, and fluoride. { 'shrek·iŋ·ə,rīt }

schrötterite [MINERAL] An opaline variety of allophane that is rich in aluminum. { 'shrād·ə,rīt }

schrund line [GEOL] The base of the bergschrund, or deep crevasse, at a late stage in the excavation of a cirque; the schrund line separates the steep slope of the cirque wall from the gentler slope below. { 'shrünt ,līn }

Schubertellidae [PALEON] An extinct family of marine protozoans in the superfamily Fusulinacea. { ,shü·bər'tel·ə,dē }

schultenite [MINERAL] $PbHAsO_4$ A colorless mineral composed of lead hydrogen arsenate occurring in tabular orthorhombic crystals. { 'shült·ən,īt }

schungite [GEOL] Amorphous carbon-rich material occurring in Precambrian schists. { 'shúŋ,gīt }

schuppen structure See imbricate structure. { 'shúp·ən ,strək·chər }

Schwagerinidae [PALEON] A family of fusulinacean protozoans that flourished during the Early and Middle Pennsylvanian and became extinct during the Late Permian. { ,shwäg·ə'rin·ə,dē }

schwartzembergite [MINERAL] $Pb_5(IO_3)Cl_3O_3$ A mineral composed of lead iodate, chloride, and oxide. { 'shwȯrt·səm,bər,gīt }

scissors fault [GEOL] A fault on which the offset or separation along the strike increases in one direction from an initial point and decreases in the other direction. Also known as differential fault. { 'siz·ərz ,fȯlt }

sclerotinite [GEOL] A variety of inertinite composed of fungal sclerotia. { 'skler·ə·tə,nīt }

scolecite [MINERAL] $CaAl_2Si_3O_{10}$ A zeolite mineral that occurs in delicate, radiating groups of white fibrous or acicular crystals; sometimes shows wormlike motion upon heating. { 'skäl·ə,sīt }

scolecodont [PALEON] Any of the paired, pincerlike jaws occurring as fossils of annelid worms. { skō'lē·kə,dänt }

scolite [GEOL] Any of the small tubes in rock believed to be the fossilized burrows of worms. { 'skō,līt }

scopulite [GEOL] A crystallite in the form of a rod with terminal brush or plume. { 'skäp·yə,līt }

score See scoring. { skȯr }

scoria [GEOL] Vesicular, cindery, dark lava formed by the escape and expansion of gases in basaltic or andesitic magma; generally denser and darker than pumice. { 'skȯr·ē·ə }

scoria cone [GEOL] A volcanic cone composed of a vesicular, cindery crust on the surface of lava that is basaltic or andesitic in nature. { 'skȯr·ē·ə ,kōn }

scoria mound [GEOL] A volcanic knoll composed of vesicular, cindery crust on the surface of lava that is basaltic or andesitic in nature. { 'skȯr·ē·ə ,maúnd }

scoria tuff [GEOL] A deposit of fragmented scoria in a fine-grained tuff matrix. { 'skȯr·ē·ə ,təf }

scoring [GEOL] **1.** The formation of parallel scratches, lines, or grooves in a bedrock surface by the abrasive action of rock fragments transported by a moving glacier. **2.** A scratch, line, or groove produced by this process. Also known as score. { 'skȯr·iŋ }

scorodite [MINERAL] $FeAsO_4 \cdot 2H_2O$ A pale leek-green or liver-brown orthorhombic mineral consisting of ferric arsenate; isomorphous with mansfieldite and represents a minor ore of arsenic. { 'skȯr·ə,dīt }

scorzalite [MINERAL] $FeAl_2(PO_4)_2(OH)_2$ A blue mineral composed of basic iron aluminum phosphate; it is isomorphous with lazulite. { 'skȯr·zə,līt }

Scotch-type volcano [GEOL] A volcanic form characterized by concentric cuestas and produced by cauldron subsidence. { 'skäch ¦tīp väl'kā·nō }

scour See tidal scour. { 'skaú·ər }

scour and fill [GEOL] The process of first digging out and then refilling a channel

324

instigated by the action of a stream or tide; refers particularly to the process that occurs during a period of flood. { 'skaủ·ər ən 'fil }

scour channel [GEOL] A large, groovelike erosional feature produced in sediments by scour. { 'skaủ·ər ˌchan·əl }

scour depression [GEOL] A crescent-shaped hollow in the stream bed near the outside of the stream's bend, caused by water that scours below the grade of the stream. { 'skaủ·ər di,presh·ən }

scouring [GEOL] **1.** An erosion process resulting from the action of the flow of air, ice, or water. **2.** *See* glacial scour. { 'skaủr·iŋ }

scouring velocity [GEOL] The velocity of water which is necessary to dislodge stranded solids from the stream bed. { 'skaủr·iŋ və,läs·əd·ē }

scour lineation [GEOL] A smooth, low, narrow (2–5 centimeters or 1–2 inches wide) ridge formed on a sedimentary surface and believed to result from the scouring action of a current of water. { 'skaủ·ər ,lin·ē,ā·shən }

scour mark [GEOL] A mark produced by the cutting or scouring action of a current flowing over the bottom of a river or body of water. { 'skaủ·ər ,mark }

scourway [GEOL] A channel created by a powerful water current, particularly the temporary channels formed by streams on the edge of a Pleistocene ice sheet. { 'skaủ·ər,wā }

scratch hardness test [MINERAL] A determination of the resistance of a mineral to scratching by testing it with minerals on the Mohs scale. { 'skrach ¦härd·nəs ,test }

scree [GEOL] **1.** A mound of loose, angular material, less than 4 inches (10 centimeters). **2.** *See* talus. { skrē }

scroll [GEOL] One of a series of crescent-shaped sediments on the inner bank of a moving channel, deposited there by the stream. { skrōl }

scroll meander [GEOL] A type of forced-cut meander, in which the scrolls built on the inner bank cause erosion of the outer bank. { 'skrōl mē'an·dər }

scyelite [PETR] A coarse-grained ultramafic igneous rock characterized by poikilitic texture resulting from the inclusion of olivine crystals in crystals of other minerals, especially amphiboles. { 'sī·ə,līt }

Scythian stage [GEOL] A stage in the lesser Triassic series of the alpine facies. Also known as Werfenian stage. { 'sith·ē·ən ,stāj }

sea arch [GEOL] An opening through a headland, formed by wave erosion or solution (as by the enlargement of a sea cave, or by the meeting of two sea caves from opposite sides), which leaves a bridge of rock over the water. Also known as marine arch; marine bridge; sea bridge. { 'sē ,ärch }

seabeach [GEOL] A beach along the margin of the sea. { 'sē,bēch }

seabed *See* sea floor. { 'sē,bed }

sea bottom *See* sea floor. { 'sē ,bäd·əm }

sea bridge *See* sea arch. { 'sē ,brij }

sea cave [GEOL] A split or hollow opening, usually at sea level, in the base of a sea cliff, formed by waves acting on weak parts of the weathered rock. Also known as marine cave; sea chasm. { 'sē ,kāv }

sea channel [GEOL] A long, narrow, U-shaped or V-shaped shallow depression of the sea floor, usually occurring on a gently sloping plain or fan. { 'sē ,chan·əl }

sea chasm *See* sea cave. { 'sē ,kaz·əm }

sea cliff [GEOL] An erosional landform, produced by wave action, which is either at the seaward edge of the coast or at the landward side of a wave-cut platform and which denotes the inner limit of the beach erosion. { 'sē ,klif }

sea fan *See* submarine fan. { 'sē ,fan }

sea floor [GEOL] The bottom of the ocean. Also known as seabed; sea bottom. { 'sē ,flȯr }

sea-floor spreading [GEOL] The hypothesis that the ocean floor is spreading away from the midoceanic ridges and is being conveyed landward by convective cells in the earth's mantle, carrying the continental blocks as passive passengers; the ocean floor moves away from the midoceanic ridge at the rate of 0.4 to 4 inches (1 to 10 centimeters) per year and provides the source of power in the hypothesis of plate tectonics. Also known as ocean-floor spreading; spreading concept; spreading floor hypothesis. { 'sē ¦flȯr ‚spred·iŋ }

sea-foam *See* sepiolite. { 'sē ‚fōm }

sea gully *See* slope gully. { 'sē ‚gəl·ē }

sea knoll *See* knoll. { 'sē ‚nōl }

sea level [GEOL] The level of the surface of the ocean; especially, the mean level halfway between high and low tide, used as a standard in reckoning land elevation or sea depths. { 'sē ‚lev·əl }

sealing-wax structure [GEOL] A primary sedimentary flow structure produced by slumping, characterized by the lack of a sharply defined slip plane at the base or a contemporaneous erosion plane at the top, and occupying a zone of highly fluid contortion in an otherwise normal sedimentary succession. { 'sēl·iŋ ¦waks ‚strək·chər }

seam [GEOL] **1.** A stratum or bed of coal or other mineral. **2.** A thin layer or stratum of rock. **3.** A very narrow coal vein. { sēm }

seamanite [MINERAL] $Mn_3(PO_4)(BO_3)·3H_2O$ A pale- to wine-yellow orthorhombic mineral that is a phosphate and borate of manganese; occurs in crystals. { 'sē·mə‚nīt }

seamount [GEOL] A mountain rising from the ocean floor as a result of submarine volcanism. { 'sē‚maůnt }

seamount chain [GEOL] Several seamounts in a line with bases separated by a relatively flat sea floor. { 'sē‚maůnt ‚chān }

seamount group [GEOL] Several closely spaced seamounts not in a line. { 'sē ‚maůnt ‚grüp }

seamount range [GEOL] Three or more seamounts having connected bases and aligned along a ridge or rise. { 'sē‚maůnt ‚rānj }

sea mud [GEOL] A rich, slimy deposit in a salt marsh or along a seashore, sometimes used as a manure. Also known as sea ooze. { 'sē ‚məd }

sea ooze *See* sea mud. { 'sē ‚üz }

sea peak [GEOL] A peaked elevation of the sea floor, rising 3300 feet (1000 meters) or more from the floor. { 'sē ‚pēk }

seaquake [GEOPHYS] An earth tremor whose epicenter is beneath the ocean and can be felt only by ships in the vicinity of the epicenter. Also known as submarine earthquake. { 'sē‚kwāk }

searlesite [MINERAL] $NaB(SiO_3)_2·H_2O$ A white mineral composed of hydrous sodium borosilicate occurring as spherulites. { 'sərl‚zīt }

seascarp [GEOL] A submarine cliff that is relatively long, high, and straight. { 'sē‚skärp }

seashore [GEOL] **1.** The strip of land that borders a sea or ocean. Also known as seaside; shore. **2.** The ground between the usual tide levels. Also known as seastrand. { 'sē‚shȯr }

seaside *See* seashore. { 'sē‚sīd }

sea slope [GEOL] The slope of land toward the sea. { 'sē ‚slōp }

seasonally frozen ground [GEOL] Ground that is frozen during low temperatures and remains so only during the winter season. Also known as frost zone. { 'sēz·ən·lē ¦frō·zən 'graůnd }

seastrand *See* seashore. { 'sē‚strand }

seat clay *See* underclay. { 'sēt ‚klā }

seat earth *See* underclay. { 'sēt ‚ərth }

sea terrace *See* marine terrace. { 'sē ‚ter·əs }

sea valley [GEOL] A relatively shallow, wide depression with gentle slopes in the sea floor, the bottom of which grades continuously downward. { 'sē ‚val·ē }

seawall [GEOL] A steep-faced, long embankment situated by powerful storm waves along a seacoast at high-water mark. { 'sē‚wȯl }

sebastianite [PETR] A plutonic rock composed of euhedral anorthite, biotite, and some augite and apatite, but without feldspathoids and quartz. { si'bas·chə‚nīt }

sebcha *See* sebkha. { 'seb·kə }

sebka *See* sebkha. { 'seb·kə }

sebkha [GEOL] A geologic feature, in North Africa, which is a smooth, flat, plain usually high in salt; after a rain the plain may become a marsh or a shallow lake until the water evaporates. Also known as sabkha; sebcha; sebka; sibjet. { 'seb·kə }

secondary [GEOL] A term with meanings that changed from early to late in the 19th century, when the term was confined to the entire Mesozoic era; it was finally replaced by Mesozoic era. { 'sek·ən,der·ē }

secondary clay [GEOL] A clay that has been transported from its place of formation and redeposited elsewhere. { 'sek·ən,der·ē 'klā }

secondary coast [GEOL] A relatively stable seacoast or shoreline whose features are the result of present-day marine processes. { 'sek·ən,der·ē 'kōst }

secondary consolidation [GEOL] Consolidation of sedimentary material, at essentially constant pressure, resulting from internal processes such as recrystallization. { 'sek·ən,der·ē kən,säl·ə'dā·shən }

secondary crater [GEOL] An impact crater produced by the relatively low-velocity impact of fragments ejected from a large primary crater. Also known as satellitic crater. { 'sek·ən,der·ē 'krād·ər }

secondary enlargement [MINERAL] Overgrowth by chemical deposition on a mineral grain of additional material of identical composition in optical and crystallographic continuity with the original grain; crystal faces characteristic of the original mineral often result. Also known as secondary growth. { 'sek·ən,der·ē in'lärj·mənt }

secondary enrichment [GEOL] The addition to a vein or ore body of material that originated later in time from the oxidation of decomposed ore masses that overlie the vein. { 'sek·ən,der·ē in'rich·mənt }

secondary geosyncline [GEOL] A geosyncline appearing at the culmination of or after geosynclinal orogeny. { 'sek·ən,der·ē ,jē·ō'sin,klīn }

secondary growth See secondary enlargement. { 'sek·ən,der·ē 'grōth }

secondary interstices [GEOL] Openings in a rock that formed after the enclosing rock was formed. { 'sek·ən,der·ē in'tər·stə,sēz }

secondary limestone [PETR] Limestone deposited from solution in cracks and cavities of other rocks. { 'sek·ən,der·ē 'līm,stōn }

secondary mineral [MINERAL] A mineral produced in an enclosing rock after the rock was formed as a result of weathering or metamorphic or solution activity, and usually at the expense of a primary material that came into existence earlier. { 'sek·ən,der·ē 'min·rəl }

secondary porosity [GEOL] The interstices that appear in a rock formation after it has formed, because of dissolution or stress distortion taking place naturally or artificially as a result of the effect of acid treatment or the injection of coarse sand. { 'sek·ən,der·ē pə'räs·əd·ē }

secondary reflection See multiple reflection; shoot. { 'sek·ən,der·ē ri'flek·shən }

secondary stratification [GEOL] The layering that occurs when sediments that were at one time deposited are resuspended and redeposited. Also known as indirect stratification. { 'sek·ən,der·ē ,strad·ə·fə'kā·shən }

secondary stratigraphic trap See stratigraphic trap. { 'sek·ən,der·ē ¦strad·ə¦graf·ik 'trap }

secondary structure [GEOL] A structure such as a fault, fold, or joint resulting from tectonic movement that started after the rock in which it is found was emplaced. [PALEON] A coarse structure usually between the thin sheets in the protective wall of a tintinnid. { 'sek·ən,der·ē 'strək·chər }

secondary tectonite [GEOL] A tectonite having a deformation fabric. { 'sek·ən,der·ē 'tek·tə,nīt }

secondary wave See S wave. { 'sek·ən,der·ē 'wāv }

second bottom [GEOL] The first terrace rising over a floodplain. { 'sek·ənd 'bäd·əm }

second-derivative map [GEOPHYS] A map of the second vertical derivative of a potential field such as the earth's gravity or magnetic field. { 'sek·ənd də¦riv·əd·iv 'map }

secretion [GEOL] A secondary structure formed of material deposited (from solution) within an empty cavity in any rock, especially a deposit formed on or parallel to the walls of the cavity, the first layer being the outer one. { si'krē·shən }

327

sectile [MINERAL] Pertaining to a mineral whose texture is tenacious enough to be cut with a knife.   { 'sek·təl }

section [GEOL] **1.** An inclined or vertical surface that is uncovered either naturally (as a sea cliff or stream bank) or artificially (as a strip mine or road cut) through a part of the earth's crust.   **2.** A description or scale drawing of the successive rock units or geologic structures shown by the exposed surface, or their appearance if cut through by any intersecting plane.   **3.** See columnar section; geologic section; type section; thin section.   { 'sek·shən }

secular variation [GEOPHYS] The changes, measured in hundreds of years, in the magnetic field of the earth.   Also known as geomagnetic secular variation.   { 'sek·yə·lər ,ver·ē'ā·shən }

secundine dike [GEOL] A dike which has been intruded into hot country rock.   { 'sek·ən,dīn 'dīk }

sedentary soil [GEOL] Soil that still lies on the rock from which it was formed.   { 'sed·ən,ter·ē 'sȯil }

sedifluction [GEOL] The subaquatic or subaerial movement of material in unconsolidated sediments, occurring in the primary stages of diagenesis.   { ,sed·ə'flək·shən }

sediment [GEOL] **1.** A mass of organic or inorganic solid fragmented material, or the solid fragment itself, that comes from weathering of rock and is carried by, suspended in, or dropped by air, water, or ice; or a mass that is accumulated by any other natural agent and that forms in layers on the earth's surface such as sand, gravel, silt, mud, fill, or loess.   **2.** A solid material that is not in solution and either is distributed through the liquid or has settled out of the liquid.   { 'sed·ə·mənt }

sedimentary breccia [PETR] A rock composed of fragments that are larger than 2 millimeters in diameter and are the result of sedimentary processes; characterized by imperfect mechanical sorting of its materials and by a higher concentration of fragments from one local source or by a wide variety of materials mixed together in no particular pattern.   Also known as sharpstone conglomerate.   { ¦sed·ə¦men·trē 'brech·ə }

sedimentary cycle See cycle of sedimentation.   { ¦sed·ə¦men·trē 'sī·kəl }

sedimentary differentiation [GEOL] The progressive separation (by erosion and transportation) of a well-defined rock mass into physically and chemically unlike products that are resorted and deposited as sediments in more or less separate areas.   { ¦sed·ə¦men·trē ,dif·ə,ren·chē'ā·shən }

sedimentary dike [GEOL] A tabular mass of sedimentary material that cuts across the structure or bedding of preexisting rock in the manner of an igneous dike and that is formed by the filling of a crack or fissure by forcible injection or intrusion of sediments under abnormal pressure, or by simple infilling of sediments.   { ¦sed·ə¦men·trē 'dīk }

sedimentary facies [GEOL] A stratigraphic facies differing from another part or parts of the same unit in both lithologic and paleontologic characters.   { ¦sed·ə¦men·trē 'fā·shēz }

sedimentary injection See injection.   { ¦sed·ə¦men·trē in'jek·shən }

sedimentary insertion [GEOL] The emplacement of sedimentary material among deposits or rocks already formed, such as by infilling, injection, or intrusion, or through localized subsidence due to solution of underlying rock.   { ¦sed·ə¦men·trē in'sər·shən }

sedimentary intrusion See intrusion.   { ¦sed·ə¦men·trē in'trü·zhən }

sedimentary laccolith [GEOL] An intrusion of plastic sedimentary material (such as clayey salt breccia) forced up under high pressure and penetrating parallel or nearly parallel to the bedding planes of the invaded formation; characterized by a very irregular thickness.   { ¦sed·ə¦men·trē 'lak·ə,lith }

sedimentary lag [GEOL] Delay between the formation of potential sediment by weathering and its removal and deposition.   { ¦sed·ə¦men·trē 'lag }

sedimentary petrography [PETR] The description and classification of sedimentary rocks.   Also known as sedimentography.   { ¦sed·ə¦men·trē pə'träg·rə·fē }

sedimentary petrology |PETR| The study of the composition, characteristics, and origin of sediments and sedimentary rocks.   { ¦sed·ə¦men·trē pə'träl·ə·jē }

sedimentary quartzite *See* orthoquartzite.   { ¦sed·ə¦men·trē 'kwòrt‚sīt }

sedimentary rock |PETR| A rock formed by consolidated sediment deposited in layers. Also known as derivative rock; neptunic rock, stratified rock.   { ¦sed·ə¦men·trē 'räk }

sedimentary structure |GEOL| A structure in sedimentary rocks, such as cross-bedding, ripple marks, and sandstone dikes, produced either contemporaneously with deposition (primary sedimentary structures) or shortly after deposition (secondary sedimentary structures).   { ¦sed·ə¦men·trē 'strək·chər }

sedimentary tectonics |GEOL| Folding and deformation in geosynclinal basins caused by subsidence and buckling of strata.   { ¦sed·ə¦men·trē tek'tän·iks }

sedimentary trap |GEOL| An area in which sedimentary material accumulates instead of being transported farther, as in an area between high-energy and low-energy environments.   { ¦sed·ə¦men·trē 'trap }

sedimentary tuff |GEOL| A tuff containing a small amount of nonvolcanic detrital material.   { ¦sed·ə¦men·trē 'təf }

sedimentary volcanism |GEOL| The expelling, extruding, or breaking through of overlying formations by a mixture of sediment, water, and gas, driven by the gas under pressure.   { ¦sed·ə¦men·trē 'väl·kə‚niz·əm }

sedimentation |GEOL| **1.** The act or process of accumulating sediment in layers. **2.** The process of deposition of sediment.   { ‚sed·ə·mən'tā·shən }

sedimentation basin |GEOL| A depression in the ocean floor with a wide, flat bottom in which sediment accumulates.   { ‚sed·ə·mən'tā·shən ‚bās·ən }

sedimentation curve |GEOL| A curve showing cumulatively, and in successive units of time, the amount of sediment accumulated or removed from an originally uniform suspension.   { ‚sed·ə·mən'tā·shən ‚kərv }

sedimentation diameter |GEOL| The diameter of a sedimentary particle, determined from the measurement of a hypothetical sphere of the same gravity and settling velocity as those of a given sedimentary particle in the same fluid.   { ‚sed·ə·mən'tā·shən dī‚am·əd·ər }

sedimentation radius |GEOL| One-half of the sedimentation diameter.   { ‚sed·ə·mən'tā·shən ‚rād·ē·əs }

sedimentation rate *See* rate of sedimentation.   { ‚sed·ə·mən'tā·shən ‚rāt }

sedimentation trend |GEOL| The direction in which sediments were laid down.   { ‚sed·ə·mən'tā·shən ‚trend }

sedimentation trough |GEOL| A depression in the ocean floor with a narrow U- or V-shaped bottom in which sediment accumulates.   { ‚sed·ə·mən'tā·shən ‚tròf }

sedimentation unit |GEOL| A sedimentary deposit formed during one distinct act of sedimentation.   { ‚sed·ə·mən'tā·shən ‚yü·nət }

sediment-delivery ratio |GEOL| The ratio of sediment yield of a drainage basin to the total amount of sediment moved by sheet erosion and channel erosion.   { 'sed·ə·mənt di¦liv·ə·rē ‚rā·shō }

sedimentography *See* sedimentary petrography.   { ‚sed·ə·mən'täg·rə·fē }

sedimentology |GEOL| The science concerned with the description, classification, origin, and interpretation of sediments and sedimentary rock.   { ‚sed·ə·mən'täl·ə·jē }

sediment-production rate |GEOL| Sediment yield per unit of drainage area, derived by dividing the annual sediment yield by the area of the drainage basin.   { 'sed·ə·mənt prə'dək·shən ‚rāt }

sediment vein |GEOL| A sedimentary dike formed by the filling of a fissure from above with sedimentary material.   { 'sed·ə·mənt ‚vān }

sediment yield |GEOL| The amount of material eroded from the land surface by runoff and delivered to a stream system.   { 'sed·ə·mənt ‚yēld }

seed fern |PALEOBOT| The common name for the extinct plants classified as Pteridospermae, characterized by naked seeds borne on large, fernlike fronds.   { 'sēd ‚fərn }

Seelandian |GEOL| A European stage of geologic time in the lowermost Paleocene.   { zā'län·dē·ən }

329

seep [GEOL] An area, generally small, where water, or another liquid such as oil, percolates slowly to the land surface. { sēp }

seepage face [GEOL] A belt on a slope, such as the bank of a stream, along which water emerges at atmospheric pressure and flows down the slope. { 'sēp·ij ,fās }

segregated vein [GEOL] A fissure filled with mineral matter derived from country rock by the action of percolating water. Also known as exudation vein. { 'seg·rə,gād·əd 'vān }

segregation [GEOL] The formation of a secondary feature within a sediment after deposition due to chemical rearrangement of minor constituents. { ,seg·rə'gā·shən }

segregation banding [PETR] A compositional band in gneisses that is the result of segregation of material from an originally homogeneous rock. { ,seg·rə'gā·shən ,band·iŋ }

seif dune [GEOL] A large, tapering, longitudinal dune or chain of sand dunes with a sharp crest that in profile consists of a succession of peaks and cols. { 'sāf ,dün }

seismic activity *See* seismicity. { 'sīz·mik ak,tiv·əd·ē }

seismic anisotropy [GEOPHYS] The dependence of seismic velocity on the direction of propagation. { 'sīz·mik ,an·ə'sä·trə·pē }

seismic area *See* earthquake zone. { 'sīz·mik ,er·ē·ə }

seismic belt [GEOPHYS] An elongate seismic zone, such as that in the Circum-Pacific. { 'sīz·mik ,belt }

seismic discontinuity [GEOPHYS] **1.** A surface at which velocities of seismic waves change abruptly. **2.** A boundary between seismic layers of the earth. Also known as interface; velocity discontinuity. { 'sīz·mik ,dis·känt·ən'ü·əd·ē }

seismic efficiency [GEOPHYS] The proportion of the total available strain energy which is radiated as seismic waves. { 'sīz·mik i'fish·ən,sē }

seismic-electric effect [GEOPHYS] The variation of resistivity with elastic deformation of rocks. { 'sīz·mik i'lek·trik i,fekt }

seismic event [GEOPHYS] An earthquake or a somewhat similar transient earth motion caused by an explosion. { 'sīz·mik i,vent }

seismic gradient [GEOPHYS] The variation of seismic velocity with distance in a specified direction. Also known as velocity gradient. { 'sīz·mik 'grād·ē·ənt }

seismic hazard [GEOPHYS] Any physical phenomenon, such as ground shaking or ground failure, that is associated with an earthquake and that may produce adverse effects on human activities. { 'sīz·mik 'haz·ərd }

seismic intensity [GEOPHYS] The average rate of flow of seismic-wave energy through a unit section perpendicular to the direction of propagation. { 'sīz·mik in'ten·səd·ē }

seismicity [GEOPHYS] The phenomena of earth movements. Also known as seismic activity. { sīz'mis·əd·ē }

seismic map [GEOPHYS] A contour map constructed from seismic data, the z coordinate of which could be either time or depth. { 'sīz·mik 'map }

seismic prospecting [GEOPHYS] Geophysical prospecting based on the analysis of elastic waves generated in the earth by artificial means. { 'sīz·mik 'präs,pek·tiŋ }

seismic ray [GEOL] The path along which seismic energy travels. { 'sīz·mik 'rā }

seismic reflector [GEOPHYS] A subsurface profile that is generated by seismic data and indicates a distinctive type of sediment geometry produced by sea-level changes; used to correlate stratigraphic sequences. { ¦sīz·mik ri'flek·tər }

seismic risk [GEOPHYS] **1.** An assortment of earthquake effects that range from ground shaking, surface faulting, and landsliding to economic loss and casualties. **2.** The probability that social or economic consequences of earthquakes will equal or exceed specified values at a site, at several sites, or in an area, during a specified exposure time. { 'sīz·mik 'risk }

seismic stratigraphy [GEOL] A branch of stratigraphy in which sediments and sedimentary rocks are interpreted in a geometrical context from seismic reflectors. { 'sīz·mik strə'tig·rə·fē }

seismic tomography [GEOPHYS] The estimation of seismic wave velocities throughout a region of interest from the travel times of either transmitted or reflected waves, generally through numerical models and iterative procedures. { 'sīz,mik tō'mäg·rə·fē }

seismic velocity |GEOPHYS| The rate of propagation of an elastic wave, usually measured in kilometers per second. { 'sīz·mik və'läs·əd·ē }

seismic vertical |GEOL| 1. The point on the earth's surface directly over the point within the earth from which an earthquake impulse originates. 2. The vertical line between the surface point and the point of origin. { 'sīz·mik 'vərd·ə·kəl }

seismology |GEOPHYS| 1. The study of earthquakes. 2. The science of strain-wave propagation in the earth. { sīz'mäl·ə·jē }

sekaninaite |MINERAL| A violet variety of cordierite in which magnesium is largely replaced by ferrous iron. Also known as iron cordierite. { sə'kän·ən·ə,īt }

selagite |PETR| A mica trachyte characterized by abundant tabular biotite crystals in a holocrystalline groundmass of orthoclase and diopside, and possibly quartz and olivine. { 'sel·ə,jīt }

selective fusion |GEOL| The fusion of only a portion of a mixture, such as a rock. { si'lek·tiv 'fyü·zhən }

selective replacement |GEOL| The replacement of one mineral by another, preferentially within an altered rock mass. { si'lek·tiv ri'plās·mənt }

selenite |MINERAL| The clear, colorless variety of gypsum crystallizing in the monoclinic system and occurring in crystals or in crystal mass. Also known as spectacle stone. { 'sel·ə,nīt }

selenite butte |GEOL| A small tabular mound, rising 3.3–10 feet (1–3 meters) above a playa, composed of lake sediments capped with a veneer of selenite formed by deflation of the playa or by the effects of rising groundwater. { 'sel·ə,nīt 'byüt }

self-reversal |GEOPHYS| Acquisition by a rock of a natural remanent magnetization opposite to the ambient magnetic field direction at the time of rock formation. { ¦self ri¦vər·səl }

self-rising ground |GEOL| The puffy, irregular, surface or near-surface zone of certain playas, formed by the effects of capillary rise of groundwater. { 'self ¦rīz·iŋ 'graùnd }

seligmannite |MINERAL| $PbCuAsS_3$ A metallic gray orthorhombic mineral, occurring in crystals. { 'sel·əg·mə,nīt }

sellaite |MINERAL| MgF_2 A colorless mineral composed of magnesium fluoride occurring in tetragonal prismatic crystals. { 'sel·ə,īt }

selvage |PETR| The marginal zone of an igneous mass, generally characterized by a fine-grain, or sometimes glassy, texture. Also known as salband. { 'sel·vij }

semianthracite |GEOL| Coal which is between bituminous coal and anthracite in metamorphic rank, and which has a fixed-carbon content of 86–92%. { ,sem·ē'an·thrə,sīt }

semibituminous coal |GEOL| Coal that is harder and more brittle than bituminous coal, has a high fuel ratio, contains 10–20% volatile matter, and burns without smoke; ranks between bituminous and semianthracite coals. { ¦sem·i·bə'tü·mə·nəs 'kōl }

semibolson |GEOL| A wide desert basin or valley whose central playa is absent or poorly developed, and which is drained by an intermittent stream that flows through canyons at each end and reaches a surface outlet. { ¦sem·i'bōls·ən }

semibright coal |GEOL| A type of banded coal defined microscopically as consisting of between 80 and 61% bright ingredients such as vitrain, clarain, and fusain, with clarodurain and durain composing the remainder. { 'sem·i,brīt 'kōl }

semicratonic See quasi-cratonic. { ¦sem·i·krə'tän·ik }

semicrystalline See hypocrystalline. { ¦sem·i'krist·əl·ən }

semidull coal |GEOL| A type of banded coal consisting mainly of clarodurain and durain, with from 40 to 21% bright ingredients such as vitrain, clarain, and fusain. { 'sem·i,dəl 'kōl }

semifusinite |GEOL| A coal maceral with a well-defined woody structure and optical properties intermediate between those of nitrinite and those of fusinite. { ¦sem·i'fyüz·ən,īt }

semischist |PETR| A partly metamorphosed sedimentary rock, exhibiting some foliation. { 'sem·i,shist }

semisplint coal |GEOL| Banded coal that is intermediate between bright-banded and splint coal, and has 20–30% opaque attritus and more than 5% anthraxylon. { ¦sem·i'splint 'kōl }

semseyite |MINERAL| $Pb_9Sb_8S_{21}$ A gray to black mineral composed of lead antimony sulfide. { 'sem·sē,īt }

senaite |MINERAL| $(Fe,Mn,Pb)TiO_3$ A black mineral consisting of a lead- and manganese-bearing ilmenite; occurs as rough crystals and rounded fragments. { 'sen·ə,īt }

senarmontite |MINERAL| Sb_2O_3 A colorless or grayish mineral composed of native antimony trioxide occurring in masses or as octahedral crystals. { ,sen·ər'män,tīt }

Senecan |GEOL| A North American provincial series of geologic time, forming the lower part of the Upper Devonian, above the Erian and below the Chautauquan. { 'sen·i·kən }

senescence |GEOL| The part of the erosion cycle at which the stage of old age begins. { si'nes·əns }

senesland |GEOL| A land surface intermediate between a matureland and a peneplain. { 'sen·əs,land }

sengierite |MINERAL| $Cu(UO_2)_2(VO_4)_2·8-10H_2O$ A yellowish-green mineral composed of hydrous copper uranyl vanadate. { 'seŋ·ē·ə,rīt }

senile |GEOL| Pertaining to the stage of senility of the cycle or erosion. { 'sē,nīl }

senility |GEOL| The stage of the cycle of erosion in which erosion of a land surface has reached a minimum, most of the hills have disappeared, and base level has been approached. { si'nil·əd·ē }

Senonian |GEOL| A European stage of geologic time, forming the Upper Cretaceous, above the Turonian and below the Danian. { sə'nō·nē·ən }

sensitive clay |GEOL| A clay whose shear strength is reduced to a very small fraction of its former value on remolding at constant moisture content. { 'sen·səd·iv 'klā }

sensitivity |GEOL| The effect of remolding on the consistency of a clay or cohesive soil, regardless of the physical nature of the causes of the change. { ,sen·sə'tiv·əd·ē }

separate *See* soil separate. { 'sep·rət }

separation |GEOL| The apparent relative displacement on a fault, measured in any given direction. { ,sep·ə'rā·shən }

sepiolite |MINERAL| $Mg_4(Si_2O_5)_3(OH)_2·6H_2O$ A soft, lightweight, absorbent, white to light-gray or light-yellow clay mineral, found principally in Asia Minor; used for tobacco pipe bowls and ornamental carvings. Also known as meerschaum; seafoam. { 'sē·pē·ə,līt }

septarian |GEOL| Pertaining to the irregular polygonal pattern of internal cracks developed in septaria. { sep'tar·ē·ən }

septarian boulder *See* septarium. { sep'tar·ē·ən 'bōl·dər }

septarian nodule *See* septarium. { sep'tar·ē·ən 'näj·ül }

septarium |GEOL| A large (32–36 inches or 80–90 centimeters in diameter), spheroidal concretion, usually composed of argillaceous carbonate, characterized by internal cracking into irregular polygonal blocks that become cemented together by crystalline minerals. Also known as beetle stone; septarian boulder; septarian nodule; turtle stone. { sep'tar·ē·əm }

Sequanian |GEOL| Upper Lower Jurassic (Upper Lusitanian) geologic time. Also known as Astartian. { sə'kwä·nē·ən }

sequence |GEOL| **1.** A sequence of geologic events, processes, or rocks, arranged in chronological order. **2.** A geographically discrete, major informal rock-stratigraphic unit of greater than group or supergroup rank. Also known as stratigraphic sequence. **3.** A body of rock deposited during a complete cycle of sea-level change. { 'sē·kwəns }

sequence stratigraphy |GEOL| A branch of stratigraphy that subdivides the sedimentary record along continental margins and in interior basins into a succession of depositional sequences as regional and interregional correlative units. { 'sē·kwəns strə'tig·rə·fē }

sequential landform |GEOL| One of an orderly succession of smaller landforms that are developed by the erosion, weathering, and mass wasting of larger initial landforms. { si'kwen·chəl 'land,fȯrm }

serandite [MINERAL] $Na(Mn,Ca)_2Si_3O_8(OH)$ A rose-red mineral composed of a basic silicate of manganese, lime, potash, and soda occurring in monoclinic crystals. { 'ser·ən,dīt }

seriate [GEOL] Having crystals that vary gradually in size. { 'sir·ē,āt }

sericite [MINERAL] A white, fine-grained potassium mica, usually muscovite in composition, having a silky luster and found as small flakes in various metamorphic rocks. { 'ser·ə,sīt }

sericitic sandstone [PETR] A sandstone in which sericite (derived by decomposition of feldspar) intermingles with finely divided quartz and fills the voids between quartz grains. { ¦ser·ə¦sīd·ik 'san,stōn }

sericitization [GEOL] A hydrothermal or metamorphic process involving the introduction of or replacement by sericite. { ‚ser·ə‚sīd·ə'zā·shən }

series [GEOL] **1.** A number of rocks, minerals, or fossils that can be arranged in a natural sequence due to certain characteristics, such as succession, composition, or occurrence. **2.** A time-stratigraphic unit, below system and above stage, composed of rocks formed during an epoch of geologic time. { 'sir·ēz }

serpentine [MINERAL] $(Mg,Fe)_3Si_2O_5(OH)_4$ A group of green, greenish-yellow, or greenish-gray ferromagnesian hydrous silicate rock-forming minerals having greasy or silky luster and a slightly soapy feel; translucent varieties are used for gemstones as substitutes for jade. { 'sər·pən,tēn }

serpentine jade [MINERAL] A variety of the mineral serpentine resembling jade in appearance and used as an ornamental stone. { 'sər·pən,tēn 'jād }

serpentine rock See serpentinite. { 'sər·pən,tēn 'räk }

serpentinite [PETR] A rock composed almost entirely of serpentine minerals. Also known as serpentine rock. { 'sər·pən,tē,nīt }

serpentinization [GEOL] A hydrothermal process by which magnesium-rich silicate minerals are converted into or replaced by serpentine minerals. { ‚sər·pən,tē·nə'zā·shən }

serpent kame See esker. { 'sər·pənt 'kām }

serpierite [MINERAL] $(Cu,Zn,Ca)_5(SO_4)_2(OH)_6 \cdot 3H_2O$ A bluish-green mineral composed of hydrous basic sulfate of copper, zinc, and calcium; occurs in tabular crystals and tufts. { 'sər·pē·ə,rīt }

serrate [GEOL] Pertaining to topographic features having a notched or toothed edge, or a saw-edge profile. { 'se,rāt }

serrate ridge See arête. { 'se,rāt 'rij }

Serridentinae [PALEON] An extinct subfamily of elephantoids in the family Gomphotheriidae. { ‚ser·ə'dent·ən,ē }

set [GEOL] A group of essentially conformable strata or cross-strata, separated from other sedimentary units by surfaces of erosion, nondeposition, or abrupt change in character. { set }

settlement [GEOL] The subsidence of surficial material (such as coastal sediments) due to compaction. { 'sed·əl·mənt }

settling [GEOL] The sag in outcrops of layered strata, caused by rock creep. Also known as outcrop curvature. { 'set·liŋ }

Sevier orogeny [GEOL] The deformation that occurred along the eastern edge of the Great Basin in Utah (eastern edge of the Cordilleran miogeosyncline) during times intermediate between the Nevadan orogeny to the west and the Laramide orogeny to the east, culminating early in the Late Cretaceous. { se'vyā ȯ'räj·ə·nē }

seybertite See clintonite. { 'sī·bər,dīt }

Seymouriamorpha [PALEON] An extinct group of labyrinthodont Amphibia of the Upper Carboniferous and Permian in which the intercentra were reduced. { sē,mȯr·ē·ə'mȯr·fə }

shadow zone [GEOPHYS] The zone, between 103 and 143° from the epicenter of an earthquake, in which direct seismic waves do not arrive because of refraction and absorption by the earth's core. { 'shad·ō ‚zōn }

shaft [GEOL] A passage in a cave that is vertical or nearly vertical. { shaft }

shake wave See S wave. { 'shāk ‚wāv }

shale [PETR] A fine-grained laminated or fissile sedimentary rock made up of silt- or clay-size particles; generally consists of about one-third quartz, one-third clay materials, and one-third miscellaneous minerals, including carbonates, iron oxides, feldspars, and organic matter. { shāl }

shale ball [GEOL] A meteorite partly or wholly converted to iron oxides by weathering. Also known as oxidite. { 'shāl ‚bȯl }

shale break [GEOL] A thin layer or parting of shale between harder strata or within a bed of sandstone or limestone. { 'shāl ‚brāk }

shale crescent [GEOL] A crescent formed by the filling of a ripple-mark trough by shale. { 'shāl ‚kres·ənt }

shale reservoir [GEOL] Underground hydrocarbon reservoir in which the reservoir rock is a brittle, siliceous, fractured shale. { 'shāl 'rez·əv‚wär }

shalification [GEOL] The formation of shale. { ‚shāl·ə·fə'kā·shən }

shallow-focus earthquake [GEOPHYS] An earthquake whose focus is located within 70 kilometers of the earth's surface. { 'shal·ō ¦fō·kəs 'ərth‚kwāk }

shallow inland seas [GEOL] Epeiric seas which periodically cover cratonic areas as a result of continental subsidence or eustatic rises in sea level. { 'shal·ō 'in·lənd 'sēz }

shallow marginal seas [GEOL] Epeiric seas along the cratonic margins. { 'shal·ō 'märj·ən·əl 'sēz }

shaly [GEOL] Pertaining to, composed of, containing, or having the properties of shale, especially readily split along close-spaced bedding planes. { 'shāl·ē }

shaly bedding [GEOL] Laminated bedding varying between 2 and 10 millimeters in thickness. { 'shāl·ē 'bed·iŋ }

shandite [MINERAL] Ni$_3$Pb$_2$S$_2$ A rhombohedral mineral composed of nickel lead sulfide, occurring in crystals. { 'shan‚dīt }

shantung [GEOL] A monadnock in the process of burial by huangho deposits. { shan'təŋ }

Shantung soil See Noncalcic Brown soil. { shan'təŋ ‚sȯil }

shard [GEOL] A vitric fragment in pyroclastics, having a characteristically curved surface of fracture. { shärd }

sharkskin pahoehoe [GEOL] A type of pahoehoe displaying numerous tiny spines or spicules on the surface. { 'shärk‚skin pə'hō·ē‚hō·ē }

shark-tooth projection [GEOL] Sharp pointed projections several centimeters in length, formed by the pulling apart of plastic lava. { 'shärk ¦tüth prə'jek·shən }

sharpite [MINERAL] (UO$_2$)(CO$_3$)·H$_2$O A greenish-yellow mineral composed of hydrous basic uranyl carbonate. { 'shär‚pīt }

sharp sand [GEOL] An angular-grain sand free of clay, loam, and other foreign particles. { 'shärp 'sand }

sharpstone [GEOL] Any rock fragment having angular edges and corners and being more than 2 millimeters in diameter. { 'shärp‚stōn }

sharpstone conglomerate See sedimentary breccia. { 'shärp‚stōn kən'gläm·ə·rət }

shatter breccia [PETR] A tectonic breccia composed of angular fragments that show little rotation. { 'shad·ər 'brech·ə }

shatter cone [GEOL] A striated conical rock fragment along which fracturing has occurred. { 'shad·ər ‚kōn }

shatter zone [GEOL] An area of randomly fissured or cracked rock that may be filled by mineral deposits, forming a network pattern of veins. { 'shad·ər ‚zōn }

shattuckite [MINERAL] Cu$_5$(SiO$_3$)$_4$H$_2$O A blue mineral composed of basic copper silicate, occurring in fibrous masses. { 'shad·ə‚kīt }

sheaf structure [GEOL] A bundled arrangement of crystals that is characteristic of certain fibrous minerals, such as stibnite. { 'shēf ‚strək·chər }

shear cleavage See slip cleavage. { 'shir ‚klē·vij }

shear fold [GEOL] A similar fold whose mechanism is shearing or slipping along closely spaced planes that are parallel to the fold's axial surface. Also known as glide fold; slip fold. { 'shir ‚fōld }

shear-gravity wave [GEOPHYS] A combination of gravity waves and a Helmholtz wave on a surface of discontinuity of density and velocity. { 'shir 'grav·əd·ē ,wāv }

shear joint [GEOL] A joint that is a shear fracture; it is a potential plane of shear. Also known as slip joint. { 'shir ,jȯint }

shear moraine [GEOL] A debris-laden surface or zone found along the margin of any ice sheet or ice cap, dipping in toward the center of the ice sheet but becoming parallel to the bed at the base. { 'shir mə'rān }

shear plane See shear surface. { 'shir ,plān }

shear slide [GEOL] A landslide, especially a slump, produced by shear failure usually along a plane of weakness such as a bedding or cleavage plane. { 'shir ,slīd }

shear sorting [GEOL] Sorting of sediments in which the smaller grains tend to move toward the zone of greatest shear strain, and the larger grains toward the zone of least shear. { 'shir ,sȯrd·iŋ }

shear structure [GEOL] A local structure in which earth stresses have been relieved by many small, closely spaced fractures. { 'shir ,strək·chər }

shear surface [GEOL] A surface along which differential movement has taken place parallel to the surface. Also known as shear plane. { 'shir ,sər·fəs }

shear wave See S wave. { 'shir ,wāv }

shear zone [GEOL] A tabular area of rock that has been crushed and brecciated by many parallel fractures resulting from shear strain; often becomes a channel for underground solutions and the seat of ore deposition. { 'shir ,zōn }

sheer [GEOL] A steep face of a cliff. { shir }

sheet [GEOL] **1.** A thin flowstone coating of calcite in a cave. **2.** A tabular igneous intrusion, especially when concordant or only slightly discordant. { shēt }

sheet crack [GEOL] A planar crack attributed to shrinkage of sediment due to dewatering. { 'shēt ,krak }

sheet deposit [GEOL] A stratiform mineral deposit that is more or less horizontal and extensive relative to its thickness. { 'shēt di,päz·ət }

sheet drift [GEOL] An evenly spread deposit of glacial drift that did not significantly alter the form of the underlying rock surface. { 'shēt ,drift }

sheeted fissure [GEOL] A closely spaced fissure. { 'shēd·əd 'fish·ər }

sheeted vein [GEOL] A vein filling a shear zone. { 'shēd·əd 'vān }

sheeted zone [GEOL] An area of mineral deposits consisting of sheeted veins. { 'shēd·əd 'zōn }

sheet erosion [GEOL] Erosion of thin layers of surface materials by continuous sheets of running water. Also known as sheetflood erosion; sheetwash; surface wash; unconcentrated wash. { 'shēt i,rō·zhən }

sheetflood erosion See sheet erosion. { 'shēd,fləd i,rō·zhən }

sheeting [GEOL] The process by which thin sheets, slabs, scales, plates, or flakes of rock are successively broken loose or stripped from the outer surface of a large rock mass in response to release of load. Also known as exfoliation. { 'shēd·iŋ }

sheeting plane [PETR] In igneous rocks, the primary cleavage plane or parting. { 'shēd·iŋ ,plān }

sheeting structure [GEOL] A fracture or joint formed by pressure-release jointing or exfoliation. Also known as exfoliation joint; expansion joint; pseudostratification; release joint; sheet joint; sheet structure. { 'shēd·iŋ ,strək·chər }

sheet joint See sheeting structure. { 'shēt ,jȯint }

sheet mineral See phyllosilicate. { 'shēt ,min·rəl }

sheet sand See blanket sand. { 'shēt ,sand }

sheet sandstone [GEOL] A thin, blanket-shaped deposit of sandstone of regional extent. { 'shēt 'san,stōn }

sheet silicate See phyllosilicate. { 'shēt 'sil·ə·kət }

sheet spar [GEOL] A sheet crack filled with spar. { 'shēt ,spär }

sheet structure See sheeting structure. { 'shēt ,strək·chər }

sheetwash [GEOL] **1.** The detritus deposited by a sheetflood. **2.** See sheet erosion. { 'shēt, wäsh }

shelf |GEOL| **1.** Solid rock beneath alluvial deposits. **2.** A flat, projecting ledge of rock. **3.** *See* continental shelf. { shelf }

shelf break |GEOL| An obvious steepening of the gradient between the continental shelf and the continental slope. { 'shelf ,brāk }

shelf channel |GEOL| A valley formed in a shelf by erosion. { 'shelf ,chan·əl }

shelf edge |GEOL| The demarcation, without dramatic change in gradient, between continental shelf and continental slope. { 'shelf ,ej }

shelf facies |GEOL| A sedimentary facies characterized by carbonate rocks and fossil shells and produced in the neritic environments of marginal shelf seas. Also known as foreland facies; platform facies. { 'shelf ,fā·shēz }

shelfstone |GEOL| A speleothem formed at the water's edge as a horizontally projecting ledge. { 'shelf,stōn }

shell |GEOL| **1.** The crust of the earth. **2.** A thin hard layer of rock. { shel }

shell marl |GEOL| A light-colored calcareous deposit formed on the bottoms of small fresh-water lakes, composed largely of uncemented mollusk shells and precipitated calcium carbonate, along with the hard parts of minute organisms. { 'shel ,märl }

shell sand |GEOL| A loose aggregate that is largely composed of shell fragments of sand size. { 'shel ,sand }

shelly |GEOL| **1.** Pertaining to a sediment or sedimentary rock containing the shells of animals. **2.** Pertaining to land abounding in or covered with shells. { 'shel·ē }

shelly facies |GEOL| A nongeosynclinal sedimentary facies that is commonly character-ized by abundant calcareous fossil shells, dominant carbonate rocks (limestones and dolomites), mature orthoquartzitic sandstones, and a paucity of shales. { 'shel·ē 'fā·shēz }

shelly pahoehoe |GEOL| A type of pahoehoe characterized by open tubes and blisters on the surface. { 'shel·ē pə'hō·ē,hō·ē }

shelter cave |GEOL| A cave which extends only a short way underground, and whose roof of overlying rock usually extends beyond its sides. Also known as rock cave. { 'shel·tər ,kāv }

shelter porosity |GEOL| A type of primary interparticle porosity created by the shelter-ing effect of relatively large sedimentary particles which prevent the infilling of pore space by finer clastic particles. { 'shel·tər pə'räs·əd·ē }

shergottite |GEOL| An achondritic stony meteorite that is composed chiefly of pigeon-ite and maskelynite. { 'shər·gə,tīt }

sheridanite |MINERAL| $(Mg,Al)_6(Al,Si)_4O_{10}(OH)_8$ Pale-green to colorless talclike mineral composed of basic magnesium aluminum silicate. { 'shər·ə·də,nīt }

sherry topaz |MINERAL| A brownish-yellow to yellow-brown variety of topaz resembling sherry wine in color. { 'sher·ē 'tō,paz }

shield |GEOL| **1.** The very old, rigid core of relatively stable rocks within a continent around which younger sedimentary rocks have been deposited. Also known as continental shield. **2.** *See* palette. { shēld }

shield basalt |GEOL| A basaltic lava flow from a group of small, close-spaced shield-volcano vents that coalesced to form a single unit. { shēld bə'sólt }

shield cone |GEOL| A cone or dome-shaped volcano built up by successive outpourings of lava. { shēld ,kōn }

shielding factor |GEOPHYS| The ratio of the strength of the magnetic field at a direc-tional compass to its strength if there were no disturbing material; usually expressed as a decimal. { 'shēld·iŋ ,fak·tər }

shield volcano |GEOL| A broad, low volcano shaped like a flattened dome and built of basaltic lava. Also known as basaltic dome; lava dome. { shēld väl,kā·nō }

shift |GEOL| The relative displacement of the units affected by a fault but outside the fault zone itself. { shift }

shifting |GEOL| The movement of the crest of a divide away from a more actively eroding stream (as on the steeper slope of an asymmetric ridge) toward a weaker stream on the gentler slope. { 'shift·iŋ }

shingle |GEOL| Pebbles, cobble, and other beach material, coarser than ordinary gravel but roughly the same size and occurring typically on the higher parts of a beach. { 'shiŋ·gəl }

shingle barchan |GEOL| A dunelike ridge formed of shingle perpendicular to the beach in shallow water. { 'shiŋ·gəl bär'kän }

shingle beach |GEOL| A narrow beach composed of shingle and commonly having a steep slope on both its landward and seaward sides. Also known as cobble beach. { 'shiŋ·gəl ‚bēch }

shingle-block structure See imbricate structure. { 'shiŋ·gəl ‚bläk 'strək·chər }

shingle rampart |GEOL| A rampart of shingle built along a reef on the seaward edge. { 'shiŋ·gəl 'ram‚pärt }

shingle ridge |GEOL| A steeply sloping bank of shingle heaped upon and parallel with the shore. { 'shiŋ·gəl ‚rij }

shingle structure See imbricate structure. { 'shin·gəl ‚strək·chər }

shingling See imbrication. { 'shiŋ·gliŋ }

shoal |GEOL| A submerged elevation that rises from the bed of a shallow body of water and consists of, or is covered by, unconsolidated material, and may be exposed at low water. { 'shōl }

shoal breccia |PETR| A breccia formed by the action of waves and tides on a shoal, and resulting from diastrophism or aggradation. { 'shōl ‚brech·ə }

shoal reef |GEOL| A reef formed in irregular masses amid submerged shoals of calcareous reef detritus. { 'shōl ‚rēf }

shock breccia |PETR| A fragmental rock formed by the action of shock waves, such as suevite formed by meteorite impact. { 'shäk ‚brech·ə }

shock lithification |GEOL| The conversion of originally loose fragmental materials into coherent aggregates by the action of shock waves, such as those generated by explosions or meteorite impacts. { 'shäk ‚lith·ə·fə‚kā·shən }

shock loading |GEOPHYS| The process of subjecting material to the action of high-pressure shock waves generated by artificial explosions or by meteorite impact. { 'shäk ‚lōd·iŋ }

shock melting |GEOPHYS| Fusion of material as a result of the high temperatures produced by the action of high-pressure shock waves. { 'shäk ‚melt·iŋ }

shock metamorphism |PETR| The complete permanent changes (physical, chemical, mineralogic, morphologic) in rocks caused by transient high-pressure shock waves that act over short-time intervals, ranging from a few microseconds to a fraction of a minute. { 'shäk ‚med·ə'mȯr‚fiz·əm }

shock zone |GEOL| A volume of rock in or around an impact or explosion crater in which a distinctive shock-metamorphic deformation or transformation effect is present. { 'shäk ‚zōn }

shoestring |GEOL| A long, relatively straight and narrow sedimentary body having a width/thickness ratio of less than 5:1, usually 1:1. { shü‚striŋ }

shoestring rill |GEOL| One of several long, narrow, uniform channels, closely spaced and roughly parallel with one another, that merely score the homogeneous surface of a relatively steep slope of bare soil or weak, clay-rich bedrock, and that develop wherever overland flow is intense. { 'shü‚striŋ ‚ril }

shoestring sand |GEOL| A shoestring composed of sand and usually buried in mud or shale, usually a sandbar or channel fill. { 'shü‚striŋ ‚sand }

shonkinite |PETR| A dark-colored syenite composed principally of augite and orthoclase with some olivine, hornblende, biotite, and nepheline. { 'shäŋ·kə‚nīt }

shoot |GEOL| See ore shoot. |GEOPHYS| The energy that goes up through the strata from a seismic profiling shot and is reflected downward at the surface or at the base of the weathering; appears either as a single wave or unites with a wave train that is traveling downward. Also known as secondary reflection. { shüt }

shore |GEOL| 1. The narrow strip of land immediately bordering a body of water. 2. See seashore. { shȯr }

shore drift See littoral drift. { 'shȯr ‚drift }

shoreface |GEOL| The narrow, steeply sloping zone between the seaward limit of the shore at low water and the nearly horizontal offshore zone. { 'shȯr‚fās }

shoreface terrace |GEOL| A wave-built terrace in the shoreface region, composed of gravel and coarse sand swept from the wave-cut bench into deeper water. { 'shȯr,fās ,ter·əs }

shoreline |GEOL| The intersection of a specified plane of water, especially mean high water, with the shore; a limit which changes with the tide or water level. Also known as strandline; waterline. { 'shȯr,līn }

shoreline cycle |GEOL| The cycle of changes through which sequential forms of coastal features pass during shoreline development, from the establishment of a water level to the time when the water can do no more work. { 'shȯr,līn ,sī·kəl }

shoreline-development ratio |GEOL| A ratio indicating the degree of irregularity of a lake shoreline, given as the length of the shoreline to the circumference of a circle whose area is equal to that of the lake. { 'shȯr,līn di'vel·əp·mənt ,rā·shō }

shoreline of depression |GEOL| A shoreline of submergence that implies an absolute subsidence of the land. { 'shȯr,līn əv di'presh·ən }

shoreline of elevation |GEOL| A shoreline of emergence that implies an absolute rise of the land. { 'shȯr,līn əv ,el·ə'vā·shən }

shoreline of emergence |GEOL| A straight or gently curving shoreline formed by the dominant relative emergence of the floor of an ocean or a lake. Also known as emerged shoreline; negative shoreline. { 'shȯr,līn əv i'mər·jəns }

shoreline of submergence |GEOL| A shoreline, characterized by bays, promontories, and other minor features, formed by the dominant relative submergence of a landmass. Also known as positive shoreline; submerged shoreline. { 'shȯr,līn əv səb'mər·jəns }

shore platform |GEOL| The horizontal or gently sloping surface produced along a shore by wave erosion. Also known as scar. { 'shȯr ,plat,fȯrm }

shore terrace |GEOL| **1.** A terrace produced along the shore by wave and current action. **2.** See marine terrace. { 'shȯr ,ter·əs }

shortite |MINERAL| $Na_2Ca_2(CO_3)_3$ A mineral composed of sodium and calcium carbonate. { 'shȯr,tīt }

shoshonite |PETR| A basaltic rock composed of olivine and augite phenocrysts in a groundmass of labradorite with orthoclase rims, olivine, augite, a small amount of leucite, and some dark-colored glass. { shə'shō,nīt }

shot copper |GEOL| Small, rounded particles of native copper, molded by the shape of vesicles in basaltic host rock, and resembling shot in size and shape. { 'shät ,käp·ər }

shoulder |GEOL| **1.** A short, rounded spur protruding laterally from the slope of a mountain or hill. **2.** The sloping segment below the summit of a mountain or hill. **3.** A bench on the flanks of a glaciated valley, located at the sharp change of slope where the steep sides of the inner glaciated valley meet the more gradual slope above the level of glaciation. **4.** A joint structure on a joint face produced by the intersection of plume-structure ridges with fringe joints. { 'shōl·dər }

shoved moraine See push moraine. { 'shəvd mə'rān }

shrinkage |GEOL| The decrease in volume of soil, sediment, fill, or excavated earth due to the reduction of voids by mechanical compaction, superimposed loads, natural consolidation, or drying. { 'shriŋ·kij }

shrinkage crack |GEOL| A small crack produced in fine-grained sediment or rock by the loss of contained water during drying or dehydration. { 'shriŋ·kij ,krak }

shrinkage index |GEOL| The numerical difference between the plastic limit of a material and its shrikage limit. { 'shriŋ·kij ,in,deks }

shrinkage limit |GEOL| That moisture content of a soil below which a decrease in moisture content will not cause a decrease in volume, but above which an increase in moisture will cause an increase in volume. { 'shriŋ·kij ,lim·ət }

shrinkage pore |GEOL| An irregular pore formed in muddy sediment by shrinkage. { 'shriŋ·kij ,pȯr }

shrinkage ratio |GEOL| The ratio of a volume change to the moisture-content change above the shrinkage limit. { 'shriŋ·kij ,rā·shō }

shrub-coppice dune [GEOL] A small dune formed on the leeward side of bush-and-clump vegetation. { 'shrəb ¦käp əs 'dün }

shungite [GEOL] A hard, black, amorphous, coallike material composed of more than 98% carbon. { 'shəŋ,īt }

shutterridge [GEOL] A ridge formed by vertical, lateral, or oblique displacement of a fault traversing a ridge-and-valley topography with the displaced part of a ridge shutting in the adjacent ravine or canyon. { 'shəd·ə,rij }

sial [PETR] A petrologic term for the silica- and alumina-rich upper rock layers of the earth's crust; gives rise to granite magma; the bulk of the continental blocks is sialic. Also known as granitic layer; sal. { 'sī,al }

siberite [MINERAL] A violet-red or purplish lithian variety of tourmaline. { sī'bi,rīt }

sibjet See sebkha. { 'sib·jət }

sicklerite [MINERAL] (Li,Mn)(PO₄) A dark-brown mineral composed of hydrous lithium manganese phosphate occurring in cleavable masses. { 'sik·lə,rīt }

side canyon [GEOL] A ravine or other valley smaller than a canyon, through which a tributary flows into the main stream. { 'sīd ,kan·yən }

sideraerolite See stony-iron meteorite. { ¦sid·ə·ra¦er·ə,līt }

siderite [MINERAL] FeCO₃ A brownish, gray, or greenish rhombohedral mineral composed of ferrous carbonate; hardness is 4 on Mohs scale, and specific gravity is 3.9. Also known as chalybite; iron spar; rhombohedral iron ore; siderose; sparry iron; spathic iron; white iron ore. { 'sid·ə,rīt }

sideroferrite [GEOL] A variety of native iron occurring as grains in petrified wood. { ,sid·ə·rə'fe,rīt }

siderogel [MINERAL] A mineral consisting of truly amorphous FeO(OH) and occurring in some bog iron ores. { 'sid·ə·rə,jel }

siderolite See stony-iron meteorite. { 'sid·ə·rə,līt }

sideromelane [MINERAL] Any iron-rich mafic mineral. { ,sid·ə·rə'me,lān }

sideronatrite [MINERAL] Na₂Fe(SO₄)(OH)·3H₂O A yellow mineral composed of basic hydrous sodium iron sulfate occurring in fibrous masses. { ,sid·ə·rə'nā,trīt }

sideronitic texture [GEOL] In mineral deposits, a mesh of silicate minerals so shattered and pressed as to force out solutions and other volatiles. { ¦sid·ə·rə¦nid·ik 'teks·chər }

siderophyllite [MINERAL] An iron-rich variety of biotite. { ,sid·ə·rə'fil,īt }

siderophyre [GEOL] A stony-iron meteorite containing bronzite and tridymite crystals in a nickel-iron network. Also known as siderophyry. { 'sid·ə·rə,fīr }

siderophyry See siderophyre. { ,sid·ə'räf·ə·rē }

siderose See siderite. { 'sid·ə,rōs }

siderosphere See inner core. { 'sid·ə·rə,sfir }

siderotil [MINERAL] (Cu,Fe)SO₄·5H₂O A white to yellowish or pale greenish-white mineral consisting of ferrous sulfate pentahydrate; occurs as fibrous crusts and groups of needlelike crystals. { 'sid·ə·rə,til }

sideswipe [GEOPHYS] **1.** A phenomenon wherein two cross reflections come from a single seismograph, due to the almost simultaneous arrival of reflection energy from both limbs of a syncline or from two nearby, steeply dipping fault scarps. **2.** In refraction shooting, the lateral deflection of a minimum-time path to include a nearby, steeply dipping, high-velocity boundary such as a flank of a salt dome. { 'sīd,swīp }

siegenite [MINERAL] (Co,Ni)₃S₄ A mineral composed of nickel cobalt sulfide. { 'sē·gə,nīt }

sierozem [GEOL] A soil found in cool to temperate arid regions, characterized by a brownish-gray surface on a lighter layer based on a carbonate or hardpan layer. { 'sir·ə,zem }

sieve deposition [GEOL] The formation of coarse-grained lobate masses on an alluvial fan whose material is sufficiently coarse and permeable to permit complete infiltration of water before it reaches the toe of the fan. { 'siv ,dep·ə,zish·ən }

sieve lobe [GEOL] A coarse-grained lobate mass produced by sieve deposition on an alluvial fan. { 'siv ,lōb }

sieve texture See poikiloblastic. { 'siv ,teks·chər }

sigmoidal dune [GEOL] A dune with an S-shaped ridge crest formed by the merger of crescentic dunes. { sig'mȯid·əl 'dün }

sigmoidal fold [GEOL] A recumbent fold having an axial surface which resembles the Greek letter sigma. { sig'mȯid·əl 'fōld }

silcrete [GEOL] A conglomerate of sand and gravel cemented by silica. { 'sil‚krēt }

silex [MINERAL] A pure or finely ground quartz. { 'sī‚leks }

silexite [GEOL] Chert occurring in calcareous beds. [PETR] Igneous rock composed mainly of primary quartz. { sī'lek‚sīt }

silica [MINERAL] SiO_2 Naturally occurring silicon dioxide; occurs in five crystalline polymorphs (quartz, tridymite, cristobalite, coesite, and stishovite), in cryptocrystalline form (as chalcedony), in amorphous and hydrated forms (as opal), and combined in silicates. { 'sil·ə·kə }

silica sand [GEOL] Sand having a very high percentage of silicon dioxide; a source of silicon. { 'sil·ə·kə ‚sand }

silica stone [PETR] A sedimentary rock composed of siliceous minerals. { 'sil·ə·kə ‚stōn }

silicate [MINERAL] Any of a large group of minerals whose crystal lattice contains SiO_4 tetrahedra, either isolated or joined through one or more of the oxygen atoms. { 'sil·ə·kət }

silication [GEOL] The conversion to or the replacement by silicates. { ‚sil·ə'kā·shən }

siliceous [PETR] Describing a rock containing abundant silica, especially free silica. { sə'lish·əs }

siliceous earth [GEOL] A loose, friable, soft, porous, lightweight, fine-grained, and usually white siliceous sediment, usually derived from the remains of organisms. { sə'lish·əs 'ərth }

siliceous limestone [PETR] **1.** A dense, dark, commonly thin-bedded limestone representing an intimate admixture of calcium carbonate and chemically precipitated silica that are believed to have accumulated simultaneously. **2.** A silicified limestone, bearing evidence of replacement of calcite by silica. { sə'lish·əs 'līm‚stōn }

siliceous ooze [GEOL] An ooze composed of siliceous skeletal remains of organisms, such as radiolarians. { sə'lish·əs 'üz }

siliceous sediment [GEOL] Fine-grained sediment and sedimentary rock mainly composed of the microscopic remains of the unicellular, silica-secreting plankton diatoms and radiolarians. Minor constituents include extremely small shards of sponge spicules and other microorganisms such as silicoflagellates. Siliceous sedimentary rock sequences are often highly porous and can form excellent petroleum source and reservoir rocks. { sə'lish·əs 'sed·ə·mənt }

siliceous shale [PETR] A hard, fine-grained rock with the texture of shale and with as much as 85% silica. { sə'lish·əs 'shāl }

siliceous sinter [MINERAL] A white, lightweight, porous, opaline variety of silica, deposited by a geyser or hot spring. Also known as fiorite; geyserite; pearl sinter; sinter. { sə'lish·əs 'sin·tər }

silicic [PETR] Describing magma or igneous rock rich in silica (usually at least 65); granite is a silicic rock. Also known as oversaturated; persilicic. { sə'lis·ik }

siliciclastic See siliclastic. { ‚sil·ə·si'klas·tik }

silicification [GEOL] Introduction of or replacement by silica. Also known as silification. { sə‚lis·ə·fə'kā·shən }

silicified wood [GEOL] A material formed by the silicification of wood, generally in the form of opal or chalcedony, in such a manner as to preserve the original form and structure of the wood. Also known as agatized wood; opalized wood; petrified wood; woodstone. { sə'lis·ə‚fīd 'wud }

silicinate [GEOL] Pertaining to the silica cement of a sedimentary rock. { sə'lis·ən‚āt }

siliclastic [PETR] Pertaining to clastic noncarbonate rocks which are almost exclusively silicon-bearing, either as forms of quartz or as silicates. Also known as siliciclastic. { ‚sil·ə'klas·tik }

silicomagnesiofluorite [MINERAL] $Ca_4Mg_3Si_2O_5(OH)_2F_{10}$ A mineral composed of basic calcium magnesium fluoride and silicate. { ‚sil·ə·kō·mag‚nē·zē·ō'flur‚īt }

340

silification *See* silicification. { ˌsil·ə·fə'kā·shən }

silk [GEOL] Microscopic needle-shaped crystalline inclusions of rutile in a natural gem from which subsurface reflections produce a whitish sheen resembling that of a silk fabric. { silk }

sill [GEOL] **1.** Submarine ridge in relatively shallow water that separates a partly closed basin from another basin or from an adjacent sea. **2.** A tabular igneous intrusion that is oriented parallel to the planar structure of surrounding rock. { sil }

silled basin *See* restricted basin. { 'sild 'bās·ən }

sillenite [MINERAL] Bi_2O_3 A mineral composed of native bismuth oxide, is polymorphous with bismite, and occurs as earthy masses. { 'sil·ə,nīt }

sillimanite [MINERAL] Al_2SiO_5 A brown, pale-green, or white neosilicate mineral with vitreous luster crystallizing in the orthorhombic system; commonly occurs in slender crystals, often in fibrous aggregates; hardness is 6–7 on Mohs scale, and specific gravity is 3.23. Also known as fibrolite. { 'sil·ə·mə,nīt }

silt [GEOL] **1.** A rock fragment or a mineral or detrital particle in the soil having a diameter of 0.002–0.05 millimeter that is, smaller than fine sand and larger than coarse clay. **2.** Sediment carried or deposited by water. **3.** Soil containing at least 80% silt and less than 12% clay. { silt }

silting [GEOL] The deposition or accumulation of stream-deposited silt that is suspended in a body of standing water. { 'silt·iŋ }

siltite *See* siltstone. { 'sil,tīt }

silt loam [GEOL] A soil containing 50–88% silt, 0–27% clay, and 0–50% sand. { 'silt ˌlōm }

silt shale [PETR] A consolidated sediment consisting of no more than 10% sand and having a silt/clay ratio greater than 2:1. { 'silt ˌshāl }

silt soil [GEOL] A soil containing 80% or more of silt, and not more than 12% of clay and 20% of sand. { 'silt ˌsȯil }

siltstone [GEOL] Indurated silt having a shalelike texture and composition. Also known as siltite. { 'silt,stōn }

silttil [GEOL] A chemically decomposed and eluviated till consisting of a friable, brownish, open-textured silt that contains a few small siliceous pebbles. { 'sil,til }

Silurian [GEOL] **1.** A period of geologic time of the Paleozoic era, covering a time span of between 430–440 and 395 million years ago. **2.** The rock system of this period. { si'lùr·ē·ən }

silver glance *See* argentite. { 'sil·vər 'glans }

sima [PETR] A petrologic term for the lower layer of the earth's crust, composed of silica- and magnesia-rich rocks; source of basaltic magma; sima is equivalent to the lower part of the continental crust and the bulk of the oceanic crust. Also known as intermediate layer. { 'sī·mə }

similar fold [GEOL] A fold in deformed beds in which the successive folds resemble each other. { 'sim·ə·lər 'fōld }

simple crater [GEOL] A meteorite impact crater of relatively small diameter, characterized by a uniformly concave-upward shape and a maximum depth in the center, and lacking a central uplift. { 'sim·pəl 'krād·ər }

simple cross-bedding [GEOL] Cross-bedding in which the lower bounding surfaces are nonerosional surfaces. { 'sim·pəl 'krȯs ˌbed·iŋ }

simple dike [PETR] An igneous dike emplaced in a single episode. { 'sim·pəl 'dīk }

simple ore [GEOL] An ore of a single metal. { 'sim·pəl 'ȯr }

simple shear [GEOPHYS] Strain caused by differential movements on one set of parallel planes which results in internal rotation of fabric elements. { 'sim·pəl 'shir }

simple valley [GEOL] A valley that maintains a constant relation to the general structure of the underlying strata. { 'sim·pəl 'val·ē }

simpsonite [MINERAL] $AlTaO_4$ A hexagonal mineral composed of aluminum tantalum oxide and occurring in short crystals. { 'sim·sə,nīt }

sincosite [MINERAL] $Ca(VO)_2(PO_4)_2 \cdot 5H_2O$ A leek-green mineral composed of hydrous calcium vanadyl phosphate and occurring in tetragonal scales or plates. { 'siŋ·kə,sīt }

Sinemurian [GEOL] A European stage of geologic time; Lower Jurassic, above Hattangian and below Pliensbachian. { sin·ə'myür·ē·ən }

singing sand See sounding sand. { 'siŋ·iŋ ¦sand }

single-cycle mountain [GEOL] A fold mountain that has been destroyed without reelevation of any of its important parts. { 'siŋ·gəl ¦sī·kəl 'maunt·ən }

sinhalite [MINERAL] MgAl(BO₄) A mineral composed of magnesium aluminum borate; sometimes used as a gem. { 'sin·ə,līt }

sinistral fault See left lateral fault. { 'sin·əs·trəl 'folt }

sinistral fold [GEOL] An asymmetric fold whose long limb, when viewed along its dip, appears to have a leftward offset. { 'sin·əs·trəl 'fōld }

sink [GEOL] **1.** A circular or ellipsoidal depression formed by collapse on the flank of or near to a volcano. **2.** A slight, low-lying desert depression containing a central playa or saline lake with no outlet, as where a desert stream comes to an end or disappears by evaporation. { siŋk }

sinkhole [GEOL] Closed surface depressions in regions of karst topography produced by solution of surface limestone or the collapse of cavern roofs. { 'siŋk,hōl }

sinkhole plain [GEOL] A regionally extensive plain or plateau characterized by well-developed karst features. { 'siŋk,hōl ,plān }

sinoite [MINERAL] Si₂N₂O A nitride mineral known only in meteorites. { 'sīn·ə,wīt }

sinople [MINERAL] A blood-red or brownish-red (with a tinge of yellow) variety of quartz containing inclusions of hematite. { 'sin·ə·pəl }

sinter [MINERAL] See siliceous sinter. [PETR] A chemical sedimentary rock deposited by precipitation from mineral waters, especially siliceous sinter and calcareous sinter. { 'sin·tər }

siphon [GEOL] A passage in a cave system that connects with a water trap. { 'sī·fən }

Siphonotretacea [PALEON] A superfamily of extinct, inarticulate brachiopods in the suborder Acrotretidina of the order Acrotretida having an enlarged, tear-shaped, apical pedicle valve. { ¦sī·fə·nō·trə'tās·ē·ə }

siserskite [MINERAL] A light steel gray mineral consisting of an alloy of osmium and iridium; occurs in tabular form. { 'sis·ər,kīt }

sitaparite See bixbyite. { sə'tap·ə,rīt }

size analysis See particle-size analysis. { 'sīz ə,nal·ə·səs }

size-frequency analysis See particle-size analysis. { 'sīz 'frē·kwən·sē ə,nal·ə·səs }

sjogrenite [MINERAL] Mg₆Fe₂(OH)₁₆(CO₃)·4H₂O A hexagonal mineral composed of hydrous basic magnesium iron carbonate. { 'shō·grə,nīt }

skarn [GEOL] A lime-bearing silicate derived from nearly pure limestone and dolomite with the introduction of large amounts of silicon, aluminum, iron, and magnesium. { skärn }

skeleton grain [GEOL] A relatively stable and not readily translocated grain of soil material, concentrated or reorganized by soil-forming processes. { 'skel·ət·ən ,grān }

skeleton texture [PETR] Descriptive of the texture of limestone that consists of an in-place accumulation of skeletal material, that is, the hard parts secreted by organisms. { 'skel·ət·ən ,teks·chər }

skerry [GEOL] A low, small, rugged and rocky island or reef. { 'sker·ē }

skialith [PETR] A vague remnant of country rock assimilated in granite. { 'skī·ə,lith }

skid boulder [GEOL] An isolated angular block of stone resting on the floor of a playa, derived from an outcrop near the playa margin, and associated with a trail or mark indicating that the boulder has recently slid across the mud surface. { 'skid ,bōl·dər }

Skiddavin See Arenigian. { skə'dav·ən }

skiou See morvan. { skyō }

skip cast [GEOL] The cast of a skip mark. { 'skip ,kast }

skip mark [GEOL] A crescent-shaped mark that is one of a linear pattern of regularly spaced marks made by an object that skipped along the bottom of a stream. { 'skip ,märk }

skleropelite [PETR] An argillaceous or allied rock which has been indurated by low-grade metamorphism, is more massive and dense than shale, and differs from slate by the absence of cleavage. { sklə'räp·ə,līt }

342

skolite [MINERAL] A scaly, dark-green variety of glauconite rich in aluminum and calcium and deficient in ferric iron. { 'skō‚līt }

skomerite [PETR] A fine-grained, compact extrusive rock containing microscopic grains and crystals of augite, olivine, and phenocrysts of decomposed plagioclase (probably albite) in a groundmass of plagioclase, thought to be more calcic than the phenocrysts. { 'skäm·ə‚rīt }

skutterudite [MINERAL] (Co,Ni)As₃ A tin-white mineral with metallic luster composed of cobalt and nickel arsenides; crystallizes in the isometric system but commonly is massive; hardness is 5.5–6 on Mohs scale, and specific gravity is 6.6; it is a minor ore of cobalt and nickel. { 'skəd·ə‚rə‚dīt }

slab [GEOL] A cleaved or finely parallel jointed rock, which splits into tabular plates from 1 to 4 inches (2.5 to 10 centimeters) thick. Also known as slabstone. { slab }

slab jointing [GEOL] Jointing produced in rock by the formation of numerous cleaved or closely spaced parallel fissures dividing the rock into thin slabs. { 'slab ‚jȯint·iŋ }

slab pahoehoe [GEOL] A pahoehoe whose surface consists of a jumbled arrangement of slabs of flow crust. { 'slab pə'hō·ē‚hō·ē }

slabstone See slab. { 'slab‚stōn }

slack [GEOL] A hollow or depression between lines of shore dunes or in a sandbank or mudbank on a shore. { slak }

slaking [GEOL] **1.** Crumbling and disintegration of earth materials when exposed to air or moisture. **2.** The breaking up of dried clay when saturated with water. { 'slāk·iŋ }

slate [PETR] A group name for various very-fine-grained rocks derived from mudstone, siltstone, and other clayey sediment as a result of low-degree regional metamorphism; characterized by perfect fissility or slaty cleavage which is a regular or perfect planar schistosity. { slāt }

slate ribbon [GEOL] A relict ribbon sructure on the cleavage surface of slate, in which varicolored and straight, wavy, or crumpled stripes cross the cleavage surface. { 'slāt ‚rib·ən }

slaty cleavage See flow cleavage. { 'slād·ē 'klē·vij }

slavikite [MINERAL] MgFe₃³⁺(SO₄)₄(OH)₃·18H₂O A greenish-yellow mineral composed of hydrous basic magnesium ferric sulfate and occurring as rhombohedral crystals. { 'slav·ə‚kīt }

slice [GEOL] An arbitrary section of some uniform standard, such as thickness of a stratigraphic unit that is otherwise indivisible for purposes of analytic study. { slīs }

slickens [GEOL] A layer of fine silt deposited by a flooding stream. { 'slik·ənz }

slickenside [GEOL] A surface that is polished and smoothly striated and results from slippage along a fault plane. { 'slik·ən‚sīd }

slickolite [GEOL] A vertically discontinuous slip-scratch surface made by slippage and shearing and developed on sharply dipping bedding planes of limestone that shapes the wall of a solution cavity. { 'slik·ə‚līt }

slide [GEOL] **1.** A vein of clay intersecting and dislocating a vein vertically, or the vertical dislocation itself. **2.** A rotational or planar mass movement of earth, snow, or rock resulting from failure under shear stress along one or more surfaces. { slīd }

sliding See gravitational sliding. { 'slīd·iŋ }

slip [GEOL] The actual relative displacement along a fault plane of two points which were formerly adjacent on either side of the fault. Also known as actual relative movement; total displacement. { slip }

slip bedding [GEOL] Convolute bedding formed as the result of subaqueous sliding. { 'slip ‚bed·iŋ }

slip block [GEOL] A separate rock mass that has slid away from its original position and come to rest down the slope without undergoing much deformation. { 'slip ‚bläk }

slip cleavage [GEOL] Cleavage that is superposed on slaty cleavage or schistosity, characterized by spaced cleavage with thin tabular bodies of rock between the cleavage planes. Also known as close-joints cleavage; crenulation cleavage; shear cleavage; strain-slip cleavage. { 'slip ‚klē·vij }

slip face [GEOL] The steeply sloping leeward surface of a sand dune. Also known as sandfall. { 'slip ‚fās }

slip fold See shear fold. { 'slip ,fōld }

slip joint |GEOL| See shear joint. { 'slip ,jȯint }

slip-off slope |GEOL| The long, low, gentle slope on the inside of the downstream face of a stream meander. { 'slip ,ȯf ,slōp }

slip plane |GEOL| A planar slip surface. { 'slip ,plān }

slip sheet |GEOL| A stratum or rock on the limb of an anticline that has slid down and away from the anticline; a gravity collapse structure. { 'slip ,shēt }

slip surface |GEOL| The displacement surface of a landslide. { 'slip ,sər·fəs }

slope |GEOL| The inclined surface of any part of the earth's surface. { slōp }

slope correction |GEOL| A tape correction applied to a distance measured on a slope in order to reduce it to a horizontal distance, between the vertical lines through its end points. Also known as grade correction. { 'slōp kə,rek·shən }

slope failure |GEOL| The downward and outward movement of a mass of soil beneath a natural slope or other inclined surface; four types of slope failure are rockfall, rock flow, plane shear, and rotational shear. { 'slōp ,fāl·yər }

slope gully |GEOL| A small, discontinuous submarine valley, usually formed by slumping along a fault scarp or the slope of a river delta. Also known as sea gully. { 'slōp ,gəl·ē }

slope stability |GEOL| The resistance of an inclined surface to failure by sliding or collapsing. { 'slōp stə,bil·əd·ē }

slope wash |GEOL| **1.** The mass-wasting process, assisted by nonchanneled running water, by which rock and soil is transported down a slope, specifically, sheet erosion. **2.** The material that is or has been transported. { 'slōp ,wäsh }

slud |GEOL| **1.** Muddy material which has moved downslope by solifluction. **2.** Ground that behaves as a viscous fluid, including material moved by solifluction and by mechanisms not limited to gravitational flow. { sləd }

sludge |GEOL| A soft or muddy bottom deposit as on tideland or in a stream bed. { sləj }

sludging See solifluction. { 'sləj·iŋ }

slump |GEOL| A type of landslide characterized by the downward slipping of a mass of rock or unconsolidated debris, moving as a unit or several subsidiary units, characteristically with backward rotation on a horizontal axis parallel to the slope; common on natural cliffs and banks and on the sides of artificial cuts and fills. { sləmp }

slump ball |GEOL| A relatively flattened mass of sandstone resembling a large concretion, measuring from 0.8 inch to 10 feet (2 centimeters to 3 meters) across, commonly thinly laminated with internal contortions and a smooth or lumpy external form, and formed by subaqueous slumping. { 'sləmp ,bȯl }

slump basin |GEOL| A shallow basin near the base of a canyon wall and on a shale hill or ridge, formed by small, irregular slumps. { 'sləmp ,bās·ən }

slump bedding |GEOL| Also known as slurry bedding. **1.** Any disturbed bedding. **2.** Convolute bedding produced by subaqueous slumping or lateral movement of newly deposited sediment. { 'sləmp ,bed·iŋ }

slump fault See normal fault. { 'sləmp ,fȯlt }

slump fold |GEOL| An intraformational fold produced by slumping of soft sediments, as at the edge of the continental shelf. { 'sləmp ,fōld }

slump overfold |GEOL| A fold consisting of hook-shaped masses of sandstone produced during slumping. { 'sləmp 'ō·vər,fōld }

slump scarp |GEOL| A low cliff or rim of thin solidified lava occurring along the margins of a lava flow and against the valley walls or around steptoes after the central part of the lava crust collapsed due to outflow of still-molten underlying layers. { 'sləmp ,skärp }

slump sheet |GEOL| A well-defined bed of limited thickness and wide horizontal extent, containing slump structures. { 'sləmp ,shēt }

slump structure |GEOL| Any sedimentary structure produced by subaqueous slumping. { 'sləmp ,strək·chər }

slurry bedding See slump bedding. { 'slər·ē ,bed·iŋ }

slurry slump |GEOL| A slump in which the incoherent sliding mass is mixed with water and disintegrates into a quasiliquid slurry { 'slər·ē ,sləmp }

slush avalanche |GEOL| A rapid and far-reaching downslope transport of rock debris released by snow supersaturated with meltwater and marking the catastrophic opening of ice- and snow-dammed brooks to the spring flood. { 'sləsh 'av·ə,lanch }

smaltite |MINERAL| (Co,Ni)As$_{3-x}$ A metallic-gray isometric mineral composed of nickel cobalt arsenide. { 'smȯl,tīt }

smaragd See emerald. { 'sma,ragd }

smaragdite |MINERAL| A green amphibole mineral that is pseudomorphous after pyroxene in rocks such as eclogite. { smə'rag,dīt }

smectite |MINERAL| Dioctahedral (montmorillonite) and trioctahedral (saponite) clay minerals, and their chemical varieties characterized by swelling properties and high cation-exchange capacities. { 'smek,tīt }

smithite |MINERAL| AgAsS$_2$ A red monoclinic mineral composed of silver arsenic sulfide and occurring as small crystals. { 'smi,thīt }

smithsonite |MINERAL| ZnCO$_3$ White, yellow, gray, brown, or green secondary carbonate mineral associated with sphalerite and commonly reniform, botryoidal, stalactitic, or granular; hardness is 5 on Mohs scale, and specific gravity is 4.30–4.45; it is an ore of zinc. Also known as calamine; dry-bone ore; szaskaite; zinc spar. { 'smith·sə,nīt }

smokestone See smoky quartz. { 'smōk,stōn }

smoky quartz |MINERAL| A smoky-yellow, smoky-brown, or brownish-gray, often transparent variety of crystalline quartz containing inclusions of carbon dioxide; may be used as a semiprecious stone. Also known as cairngorm; smokestone. { 'smōk·ē 'kwȯrts }

smooth chert |GEOL| A hard, dense, homogeneous chert (insoluble residue) characterized by a conchoidal-to-even fracture surface that is devoid of roughness and by a lack of crystallinity, granularity, or other distinctive structure. { 'smüth 'chərt }

smooth phase |GEOL| The part of stream traction whereby a mass of sediment travels as a sheet with gradually increasing density from the surface downward. { 'smüth ,fāz }

smothered bottom |GEOL| A sedimentary surface on which complete, well-preserved, and commonly very fragile and delicate fossils were saved by an influx of mud that buried them instantly. { 'sməth·ərd 'bäd·əm }

SMOW See standard mean ocean water. { smaůw or ¦es¦em¦ō'dəb·əl,yü }

SNC group |GEOL| A group of meteorites comprising the shergottites, nakhlites, and chassignites, which are all believed to have originated from Mars. { ¦es¦en'sē ,grüp }

snowflake obsidian |PETR| An obsidian that contains white, gray, or reddish spherulites ranging in size from microscopic to a meter or more in diameter. { 'snō,flāk äb'sid·ē·ən }

snowflush |GEOL| An accumulation of drifted snow, windblown soil, and wind-transported seeds on a lee slope, characteristically marked during the winter by a dark patch of soil. { 'snō,fləsh }

snow niche See nivation hollow. { 'snō ,nich }

snow patch erosion See nivation. { 'snō ¦pach i,rō·zhən }

soaprock See soapstone. { 'sōp,räk }

soapstone |MINERAL| 1. A mineral name applied to steatite or to massive talc. Also known as soaprock. 2. See saponite. |PETR| A metamorphic rock characterized by massive, schistose, or interlaced fibrous texture and a soft unctuous feel. { 'sōp,stōn }

sodaclase See albite. { 'sōd·ə,klās }

soda-granite |PETR| 1. A granite in which soda is more abundant than potash. 2. A granite that contains soda-plagioclase instead of the orthoclase found in normal granite. { 'sōd·ə ,gran·ət }

sodalite |MINERAL| Na$_2$Al$_3$Si$_3$O$_{12}$Cl A blue or sometimes white, gray, or green mineral tectosilicate of the feldspathoid group, crystallizing in the isometric system, with vitreous luster, hardness of 5 on Mohs scale, and specific gravity of 2.2–2.4; used as an ornamental stone. { 'sōd·əl,īt }

soda mica

soda mica *See* paragonite. { 'sōd·ə 'mī·kə }

soda microcline *See* anorthoclase. { 'sōd·ə 'mī·krə,klīn }

soda niter [MINERAL] NaNO₃ A colorless to white mineral composed of sodium nitrate, crystallizing in the rhombohedral division of the hexagonal system; hardness is $1\frac{1}{2}$ to 2 on Mohs scale and specific gravity is 2.266. Also known as nitratine; Peru saltpeter. { 'sōd·ə 'nīd·ər }

soddyite [MINERAL] $(UO_2)_{12}Si_5O_{22} \cdot 14H_2O$ A pale-yellow orthorhombic mineral composed of hydrous uranium silicate and occurring in fine-grained aggregates or crystals. { 'säd·ē,īt }

sodium-calcium feldspar *See* plagioclase. { 'sōd·ē·əm 'kal·sē·əm 'fel,spär }

sodium feldspar *See* albite. { 'sōd·ē·əm 'fel,spär }

sodium illite *See* brammalite. { 'sōd·ē·əm 'i,līt }

soffione [GEOL] A jet of steam and other vapors issuing from the ground in a volcanic area. { ,sä·fē'ō·nē }

soffosian knob *See* frost mound. { sə'fō·zhən 'näb }

soft coal *See* bituminous coal. { 'sóft 'kōl }

soft rock [PETR] **1.** A broad designation for sedimentary rock. **2.** A rock that is relatively nonresistant to erosion. { 'sóft 'räk }

Sohm Abyssal Plain [GEOL] A basin in the North Atlantic, about 2400 fathoms (4390 meters) deep, between Newfoundland and the Mid-Atlantic Ridge. { 'sōm ə'bis·əl 'plān }

soil [GEOL] **1.** Unconsolidated rock material over bedrock. **2.** Freely divided rock-derived material containing an admixture of organic matter and capable of supporting vegetation. { sȯil }

soil air [GEOL] The air and other gases in spaces in the soil; specifically, that which is found within the zone of aeration. Also known as soil atmosphere. { 'sȯil 'er }

soil atmosphere *See* soil air. { 'sȯil ¦at·mə,sfir }

soil blister *See* frost mound. { 'sȯil ,blis·tər }

soil chemistry [GEOCHEM] The study and analysis of the inorganic and organic components and the life cycles within soils. { 'sȯil ¦kem·ə·strē }

soil colloid [GEOL] Colloidal complex of soils composed principally of clay and humus. { 'sȯil ¦kä,lȯid }

soil complex [GEOL] A mapping unit used in detailed soil surveys; consists of two or more recognized classifications. { 'sȯil ¦käm,pleks }

soil creep [GEOL] The slow, steady downhill movement of soil and loose rock on a slope. Also known as surficial creep. { 'sȯil ,krēp }

soil element [GEOL] A unit that represents an arbitrarily small volume of soil within a soil mass. { 'sȯil ,el·ə·mənt }

soil erosion [GEOL] The detachment and movement of topsoil by the action of wind and flowing water. { 'sȯil i,rōzh·ən }

soil flow *See* solifluction. { 'sȯil ,flō }

soil fluction *See* solifluction. { 'sȯil ,flək·shən }

soil formation *See* soil genesis. { 'sȯil ,fȯr·mā·shən }

soil genesis [GEOL] The mode by which soil originates, with particular reference to processes of soil-forming factors responsible for the development of true soil from unconsolidated parent material. Also known as pedogenesis; soil formation. { 'sȯil ,jen·ə·səs }

soil physics [GEOPHYS] The study of the physical characteristics of soils; concerned also with the methods and instruments used to determine these characteristics. { ¦sȯil ¦fiz·iks }

soil profile [GEOL] A vertical section of a soil, showing horizons and parent material. { ¦sȯil ¦prō,fīl }

soil science [GEOL] The study of the formation, properties, and classification of soil; includes mapping. Also known as pedology. { 'sȯil ,sī·əns }

soil separate [GEOL] Any of a group of rock or mineral particles, separated from a soil sample, having diameters less than 0.8 inch (2 millimeters) and ranging within the limits of one of the standard classifications of soil particle size.   Also known as separate.   { 'sȯil ¦sep·rət }

soil series |GEOL| A family of soils having similar profiles, and developing from similar original materials under the influence of similar climate and vegetation.   { 'sȯil ¸sir·ēz }

soil shear strength |GEOL| The maximum resistance of a soil to shearing stresses.   { 'sȯil 'shir ¸streŋkth }

soil stripes |GEOL| Alternating bands of fine and coarse material in a soil structure.   { 'sȯil ¸strīps }

soil structure |GEOL| Arrangement of soil into various aggregates, each differing in the characteristics of its particles.   { 'sȯil ¸strək·chər }

soil survey |GEOL| The systematic examination of soils, their description and classification, mapping of soil types, and the assessment of soils for various agricultural and engineering uses.   { 'sȯil 'sər¸vā }

soil-water belt See belt of soil water.   { 'sȯil ¦wȯd·ər ¸belt }

soil-water zone See belt of soil water.   { 'sȯil ¦wȯd·ər ¸zōn }

sole |GEOL| **1.** The bottom of a sedimentary stratum.   **2.** The middle and lower portion of the shear surface of a landslide.   **3.** The underlying fault plane of a thrust nappe.   Also known as sole plane.   { sōl }

sole injection |GEOL| An igneous intrusion that was put in place along a thrust plane.   { 'sōl in¸jek·shən }

sole mark |GEOL| An irregularity or penetration on the undersurface of a sedimentary stratum.   { 'sōl ¸märk }

Solenopora |PALEOBOT| A genus of extinct calcareous red algae in the family Solenoporaceae that appeared in the Late Cambrian and lasted until the Early Tertiary.   { ¸säl·ə'näp·rə }

Solenoporaceae |PALEOBOT| A family of extinct red algae having compact tissue and the ability to deposit calcium carbonate within and between the cell walls.   { sō¸lē·nə·pə'rās·ē¸ē }

sole plane See sole.   { 'sōl ¸plān }

solfatara |GEOL| A fumarole from which sulfurous gases are emitted.   { ¸säl·fə'tär·ə }

solifluction |GEOL| A rapid soil creep, especially referring to downslope soil movement in periglacial areas.   Also known as sludging; soil flow; soil fluction.   { ¦säl·ə'flək·shən }

solifluction lobe |GEOL| An isolated, tongue-shaped feature of the land surface with a steep front and a smooth upper surface formed by more rapid solifluction on certain sections of the slope.   Also known as solifluction tongue.   { ¦säl·ə'flək·shən ¦lōb }

solifluction mantle |GEOL| The locally derived, unsorted material moved downslope by solifluction.   Also known as flow earth.   { ¦säl·ə'flək·shən ¦mant·əl }

solifluction sheet |GEOL| A broad deposit of a solifluction mantle.   { ¦säl·ə'flək·shən ¦shēt }

solifluction stream |GEOL| A narrow, streamlike deposit of a solifluction mantle.   { ¦säl·ə'flək·shən ¦strēm }

solifluction tongue See solifluction lobe.   { ¦säl·ə'flək·shən ¦təŋ }

solodize |GEOL| To improve a soil by removing alkalies from it.   { 'sō·lə¸dīz }

Solod soil See Soloth soil.   { 'sō·ləd ¸sȯil }

Solo man |PALEON| A relatively late but primitive form of fossil man from Java; this form had a small brain, heavy horizontal browridges, and a massive cranial base.   { 'sō·lō 'man }

Solonchak soil |GEOL| One of an intrazonal, balamorphic group of light-colored soils rich in soluble salts.   { ¦säl·ən¦chäk ¸sȯil }

Solonetz soil |GEOL| One of an intrazonal group of black alkali soils having a columnar structure.   { ¦säl·ə¦nets ¸sȯil }

Soloth soil |GEOL| One of an intrazonal halomorphic group of soils formed from saline material; the surface layer is soft and friable, and overlies a light-colored leached horizon which, in turn, overlies a dark horizon.   Also known as Solod soil.   { 'sō·lət ¸sȯil }

solum |GEOL| The upper part of a soil profile, composed of A and B horizons in mature soil. Also known as true soil. { 'sō·ləm }

solution groove |GEOL| One of a series of continuous, subparallel furrows developed on an inclined or vertical surface of a soluble and homogeneous rock (such as the limestone walls of a cave) by the slow corroding action of trickling water. { sə'lü·shən ‚grüv }

solution pool |GEOL| A pool in a rock that is formed by the dissolution of the rock in ocean water. { sə'lü·shən ‚pül }

solution potholes |GEOL| Potholes produced in carbonate rocks by dissolution. { sə'lü·shən ‚pät‚hōlz }

solution transfer |GEOL| A process whereby pressure solution of detrital mineral grains at contact areas is followed by recrystallization on the less strained parts of the grain surfaces. { sə'lü·shən ‚tranz·fər }

somma |GEOL| The rim of a volcano. { 'säm·ə }

sordawalite See tachylite. { sór'dä·wə‚līt }

sorosilicate |MINERAL| A structural type of silicate whose crystal lattice has two SiO_4 tetrahedra sharing one oxygen atom. { ‚sór·ō'sil·ə·kət }

sorotiite |GEOL| A type of meteorite similar to the pallasites, with troilite substituting for olivine. { sə'räd·ē‚īt }

sorted |GEOL| **1.** Pertaining to a nongenetic group of patterned-ground features displaying a border of stones, including boulders, commonly alternating with very small particles, including silt, sand, and clay. **2.** Pertaining to an unconsolidated sediment or a cemented detrital rock consisting of particles of essentially uniform size or of particles lying within the limits of a single grade. { 'sórd·əd }

sorted polygon |GEOL| A patterned ground having a sorted appearance due to a border of stones and characterized by a polygonal mesh. Also known as stone polygon. { 'sórd·əd 'päl·i‚gän }

sorting |GEOL| The process by which similar in size, shape, or specific gravity sedimentary particles are selected and separated from associated but dissimilar particles by the agent of transportation. { 'sórd·iŋ }

sorting coefficient |GEOL| A sorting index equal to the square root of the ratio of the larger quartile (the diameter having 25% of the cumulative size-frequency distribution larger than itself) to the smaller quartile (the diameter having 75% of the cumulative size-frequency distribution larger than itself). { 'sórd·iŋ ‚kō·i‚fish·ənt }

sorting index |GEOL| A measure of the degree of sorting in a sediment based on the statistical spread of the frequency curve of particle sizes. { 'sórd·iŋ ‚in‚deks }

sounding sand |GEOL| Sand that emits musical, humming, or crunching sounds when disturbed. Also known as singing sand. { 'saúnd·iŋ ‚sand }

source area See provenance. { 'sórs ‚er·ē·ə }

source bed |GEOL| The original stratigraphic horizon from which secondary sulfide minerals were derived. { 'sórs ‚bed }

sourceland See provenance. { 'sórs‚land }

source rock |GEOL| **1.** Rock from which fragments have been derived which form a later, usually sedimentary rock. Also known as mother rock; parent rock. **2.** Sedimentary rock, usually shale and limestone, deposited together with organic matter which was subsequently transformed to liquid or gaseous hydrocarbons. { 'sórs ‚räk }

South African jade See Transvaal jade. { 'saúth 'af·ri·kən 'jād }

south geomagnetic pole |GEOPHYS| The geomagnetic pole in the Southern Hemisphere at approximately 78.5°S, longitude 111°E, 180° from the north geomagnetic pole. Also known as south pole. { 'saúth ¦jē·ō·mag'ned·ik ‚pōl }

south pole |GEOPHYS| See south geomagnetic pole. { 'saúth 'pōl }

souzalite |MINERAL| $(Mg,Fe)_3(Al,Fe)_4(PO_4)_4(OH)_6·2H_2O$ A green mineral composed of hydrous basic phosphate of magnesium, iron, and aluminum. { 'sō·zə‚līt }

spall |GEOL| **1.** A fragment removed from the surface of a rock by weathering. **2.** A relatively thin, sharp-edged fragment produced by exfoliation. **3.** A rock fragment produced by chipping with a hammer. { spól }

348

spalling [GEOL] The chipping or fracturing with an upward heaving, of rock caused by a compressional wave at a free surface. { 'spȯl·iŋ }

spangolite [MINERAL] $Cu_6Al(SO_4)(OH)_{12}Cl \cdot 3H_2O$ A dark-green hexagonal mineral composed of hydrous basic sulfate and chloride of aluminum and copper and occurring as crystals. { 'spaŋ·gə,līt }

spar [MINERAL] Any transparent or translucent, nonmetallic, light-colored, readily cleavable, crystalline mineral; examples are calespar and fluorspar. { spär }

sparagmite [GEOL] Late Precambrian fragmental rocks of Scandinavia, characterized by high proportions of microcline. { spə'rag,mīt }

sparite *See* sparry calcite. { 'spä,rīt }

Sparnacean [GEOL] A European stage of geologic time; upper upper Paleocene, above Thanetian, below Ypresian of Eocene. { spär'nāsh·ən }

sparry calcite [MINERAL] A clean, coarse-grained calcite crystal. Also known as calcsparite; sparite. { 'spär·ē 'kal,sīt }

sparry cement [GEOL] Clear, relatively coarse-grained calcite in the interstices of any sedimentary rock. { 'spär·ē si'ment }

sparry iron *See* siderite. { 'spär·ē 'ī·ərn }

spartalite *See* zincite. { 'spärd·əl,īt }

spathic iron *See* siderite. { 'spath·ik 'ī·ərn }

spatter cone [GEOL] A low, steep-sided cone of small pyroclastic fragments built up on a fissure or vent. Also known as agglutinate cone; volcanello. { 'spad·ər ,kōn }

spatter rampart [GEOL] A low, circular ridge of pyroclastics built up around the margins of small volcanoes. { 'spad·ər ,ram,pärt }

specific retention [GEOL] The ratio of the volume of water that a given body of rock or soil will retain after saturation, and the pull of gravity to the volume of the body itself. { spə¦sif·ik ri'ten·chən }

spectacle stone *See* selenite. { 'spek·tə·kəl ,stōn }

specular hematite [MINERAL] A variety of hematite with a blue-gray color and bright metallic luster. { 'spek·yə·lər 'hē·mə,tīt }

specular iron *See* specularite. { 'spek·yə·lər 'ī·ərn }

specularite [MINERAL] A black or gray variety of hematite with brilliant metallic luster, occurring in micaceous or foliated masses, or in tabular or disklike crystals. Also known as gray hematite; iron glance; specular iron. { 'spek·yə·lə,rīt }

spelean [GEOL] Of or pertaining to a feature in a cave. { spə'lē·ən }

speleology [GEOL] The study and exploration of caves. { ,spē·lē'äl·ə·jē }

speleothem [GEOL] A secondary mineral deposited in a cave by the action of water. Also known as cave formation. { 'spē·lē·ə,them }

spencerite [MINERAL] $Zn_4(PO_4)_2(OH)_2 \cdot 3H_2O$ A pearly white monoclinic mineral composed of hydrous basic zinc phosphate and occurring in scaly masses and small crystals. { 'spen·sə,rīt }

spending beach [GEOL] In a wave basin, the beach on which the entering waves spend themselves, except for the small remainder entering the inner harbor. { 'spend·iŋ ,bēch }

spergenite [GEOL] A biocalcarenite containing ooliths and fossil debris and having a maximum quartz content of 10%. Also known as Bedford limestone; Indiana limestone. { 'spər·jə,nīt }

sperrylite [MINERAL] $PtAs_2$ A tin-white isometric mineral composed of platinum arsenide; the only platinum compound known to occur in nature; hardness is 6–7 on Mohs scale, and specific gravity is 10.60. { 'sper·ē,līt }

spessartite [MINERAL] $Mn_3Al_2(SiO_4)_3$ A mineral composed of manganese aluminum silicate with small amounts of iron, magnesium, or other elements. [PETR] A lamprophyre composed of a sodic plagioclase groundmass in which green hornblende phenocrysts are embedded; also contains accessory olivine, biotite, apatite, and opaque oxides. { 'spes·ər,tīt }

Sphaeractinoidea [PALEON] An extinct group of fossil marine hydrozoans distinguished in part by the relative prominence of either vertical or horizontal trabeculae and by the presence of long, tabulate tubes called autotubes. { sfir,ak·tə'nȯid·ē·ə }

sphaerite [MINERAL] Light-gray or bluish mineral composed of hydrous aluminum phosphate and occurring in global concretions. { 'sfir,īt }

sphaerolitic *See* spherulitic. { ¦sfir·ə¦lid·ik }

sphalerite [MINERAL] (Zn,Fe)S The low-temperature form and common polymorph of zinc sulfide; a usually brown or black mineral that crystallizes in the hextetrahedral class of the isometric system, occurs most commonly in coarse to fine, granular, cleanable masses, has resinous luster, hardness of 3.5 on Mohs scale, and specific gravity of 4.1. Also known as blende; false galena; jack; lead marcasite; mock lead; mock ore; pseudogalena; steel jack. { 'sfal·ə,rīt }

Sphenacodontia [PALEON] A suborder of extinct reptiles in the order Pelycosauria which were advanced, active carnivores. { sfə,näk·ə'dän·chə }

sphene [MINERAL] CaTiSiO$_5$ A brown, green, yellow, gray, or black neosilicate mineral common as an accessory mineral in igneous rocks; it is monoclinic and has resinous luster; hardness is 5–5.5 on Mohs scale; specific gravity is 3.4–3.5. Also known as grothite; titanite. { sfēn }

sphenochasm [GEOL] A triangular gap of oceanic crust separating two continental blocks and converging to a point. { ,sfē·nə'kaz·əm }

sphenolith [GEOL] A wedgelike igneous intrusion that is partly concordant and partly discordant. { 'sfēn·əl,ith }

Sphenyllopsida [PALEOBOT] An extinct class of embryophytes in the division Equisetophyta. { ,sfēn·əl'äp·səd·ə }

spherical weathering *See* spheroidal weathering. { 'sfir·ə·kəl 'weth·ə·riŋ }

spheroidal recovery [GEOPHYS] The hypothetical return of the earth to spheroid form after it has been distorted. { sfir'ȯid·əl ri'kəv·ə·rē }

spheroidal weathering [GEOL] Chemical weathering in which concentric or spherical shells of decayed rock are successively separated from a block of rock; commonly results in the formation of a rounded boulder of decomposition. Also known as concentric weathering; spherical weathering. { sfir'ȯid·əl 'weth·ə·riŋ }

spherulite [GEOL] A spherical body or coarsely crystalline aggregate having a radial internal structure arranged about one or more centers. { 'sfir·ə,līt }

spherulitic [PETR] Relating to the texture of a rock composed of numerous spherulites. Also known as globular; sphaerolitic. { ¦sfir·ə¦lid·ik }

Sphinctozoa [PALEON] A group of fossil sponges in the class Calcarea which have a skeleton of massive calcium carbonate organized in the form of hollow chambers. { ,sfiŋk·tə'zō·ə }

spiculite [PETR] A spindle-shaped belonite thought to have formed by the coalescence of globulites. { 'spik·yə,līt }

spilite [PETR] An altered basalt containing albitized feldspar accompanied by low-temperature, hydrous crystallization products such as chlorite, calcite, and epidote. { 'spī,līt }

spinel [MINERAL] **1.** MgAl$_2$O$_4$ A colorless, purplish-red, greenish, yellow, or black mineral, usually forming octahedral crystals, and characterized by great hardness; used as a gemstone. **2.** A group of minerals of general formula AB$_2$O$_4$, where A is magnesium, ferrous iron, zinc, or manganese, or a combination of them, and B is aluminum, ferric iron, or chromium. { spə'nel }

spinodal decomposition [MINERAL] An unmixing process in which crystals with bulk composition in the central region of the phase diagram undergo exsolution. { spī'nōd·əl dē,käm·pə'zish·ən }

Spiriferida [PALEON] An order of fossil articulate brachiopods distinguished by the spiralium, a pair of spirally coiled ribbons of calcite supported by the crura. { ,spī·rə'fer·əd·ə }

Spiriferidina [PALEON] A suborder of the extinct brachiopod order Spiriferida including mainly ribbed forms having laterally or ventrally directed spires, well-developed interareas, and a straight hinge line. { spī,rif·ə·rə'dī·nə }

splash erosion [GEOL] Erosion resulting from the impact of falling raindrops. { 'splash i,rōzh·ən }

splent coal *See* splint coal. { 'splent ¦kōl }

spliced |GEOL| Relating to veins that pinch out and are overlapped at that point by another parallel vein. { splīst }

splint See splint coal. { splint }

splint coal |GEOL| A hard, dull, blocky, grayish-black, banded bituminous coal characterized by an uneven fracture and a granular texture; burns with intense heat. Also known as splent coal; splint. { 'splint ‚kōl }

split |GEOL| A coal seam that cannot be mined as a single unit because it is separated by a parting of other sedimentary rock. Also known as coal split; split coal. { split }

split coal See split. { 'split ‚kōl }

spodic horizon |GEOL| A soil horizon characterized by illuviation of amorphous substances. { 'späd·ik hə'rīz·ən }

Spodosol |GEOL| A soil order characterized by accumulations of amorphous materials in subsurface horizons. { 'späd·ə‚sól }

spodumene |MINERAL| LiAlSi$_2$O$_6$ A white to yellowish-, purplish-, or emerald-green clinopyroxene mineral occurring in prismatic crystals; hardness is 6.5–7 on Mohs scale, and specific gravity 3.13–3.20; an ore of lithium. Also known as triphane. { 'spä·jə‚mēn }

spongework |GEOL| A pattern of small irregular interconnecting cavities on walls of limestone caves. { 'spənj‚wərk }

Spongiomorphida |PALEON| A small, extinct Mesozoic order of fossil colonial Hydrozoa in which the skeleton is a reticulum composed of perforate lamellae parallel to the upper surface and of regularly spaced vertical elements in the form of pillars. { ‚spən·jē-ō'mór·fə·də }

Spongiomorphidae |PALEON| The single family of extinct hydrozoans comprising the order Spongiomorphida. { ‚spən·jē-ō'mór·fə‚dē }

spongolite |GEOL| A rock or sediment composed chiefly of the remains of sponges. Also known as spongolith. { 'späŋ·gə‚līt }

spongolith See spongolite. { 'spaŋ·gə‚lith }

sporinite |GEOL| A variety of exinite composed of spore exines which have been compressed parallel to the stratification. { 'spór·ə‚nīt }

spotted phyllite |PETR| A phyllite rock containing dark spots that represent the beginning of porphyroblast development. { 'späd·əd 'fī‚līt }

spotted slate |PETR| A type of slate containing dark spots that represent the beginning of porphyroblast development. { 'späd·əd 'slāt }

spouting horn |GEOL| A sea cave with a rearward or upward opening through which water spurts or sprays after waves enter the cave. Also known as chimney; oven. { 'spaúd·iŋ ‚hórn }

spreading concept See sea-floor spreading. { 'spred·iŋ ‚kän‚sept }

spreading-floor hypothesis See sea-floor spreading. { 'spred·iŋ ‚flór hī‚päth·ə·səs }

spur |GEOL| A ridge or rise projecting from a larger elevational feature. { spər }

spurrite |MINERAL| Ca$_5$(SiO$_4$)$_2$(CO$_3$) A light-gray mineral occurring in granular masses. { 'spər‚īt }

stability |GEOL| **1.** The resistance of a structure, spoil heap, or clay bank to sliding, overturning, or collapsing. **2.** Chemical durability, resistance to weathering. { stə'bil·əd·ē }

stack |GEOL| An erosional, coastal landform that is a steep-sided, pillarlike rocky island or mass that has been detached by wave action from a shore made up of cliffs; applies particularly to a stack that is columnar in structure and has horizontal stratifications. Also known as marine stack; rank. { stak }

stade |GEOL| A substage of a glacial stage marked by a secondary advance of glaciers. { stād }

stadial moraine See recessional moraine. { 'stād·ē·əl mə'rān }

Staffellidae |PALEON| An extinct family of marine protozoans (superfamily Fusulinacea) that persisted during the Pennsylvanian and Early Permian. { sta'fel·ə‚dē }

Staffordian |GEOL| A European stage of geologic time forming the middle Upper Carboniferous, above Yorkian and below Radstockian, equivalent to part of the upper Westphalian. { sta'fórd·ē·ən }

stage [GEOL] **1.** A developmental phase of an erosion cycle in which landscape features have distinctive characteristic forms. **2.** A phase in the historical development of a geologic feature. **3.** A major subdivision of a glacial epoch. **4.** A time-stratigraphic unit ranking below series and above chronozone, composed of rocks formed during an age of geologic time. { stāj }

stainierite *See* heterogenite. { 'stī·nē·ə,rīt }

stalactite [GEOL] A conical or roughly cylindrical speleothem formed by dripping water and hanging from the roof of a cave; usually composed of calcium carbonate. { stə'lak,tīt }

stalacto-stalagmite [GEOL] A columnar deposit formed by the union of a stalactite with its complementary stalagmite. Also known as column; pillar. { stə¦lak·tō stə'lag,mīt }

stalagmite [GEOL] A conical speleothem formed upward from the floor of a cave by the action of dripping water; usually composed of calcium carbonate. { stə'lag,mīt }

Stampian *See* Rupelian. { 'stam·pē·ən }

standard mean ocean water [GEOL] An international reference standard used to determine oxygen and hydrogen isotopic content. Abbreviated SMOW. { ¦stan·dərd ¦mēn 'ō·shən ,wȯd·ər }

standard mineral [MINERAL] A mineral that, on the basis of chemical analyses, is theoretically capable of being present in a rock. Also known as normative mineral. { 'stan·dərd 'min·rəl }

stanfieldite [MINERAL] $Ca_4(Mg,Fe,Mn)_5(PO_4)_6$ A phosphate mineral found only in meteorites. { 'stan,fēl,dīt }

stannite [MINERAL] Cu_2FeSnS_4 A steel-gray or iron-black mineral crystallizing in the tetragonal system and occurring in granular masses; luster is metallic, hardness is 4 on Mohs scale, and specific gravity is 4.3–4.53. Also known as bell-metal ore; tin pyrites. { 'sta,nīt }

star ruby [MINERAL] An asteriated variety of ruby with normally six chatoyant rays. { 'stär 'rü·bē }

star sapphire [MINERAL] A variety of sapphire exhibiting a six-pointed star resulting from the presence of microscopic crystals in various orientations within the gemstone. { 'stär 'sa,fīr }

starved basin [GEOL] A sedimentary basin in which rate of subsidence exceeds rate of sedimentation. { 'stärvd 'bās·ən }

static granitization [PETR] The formation of a granitic rock by a metasomatic process in the absence of compressive forces or strains. { 'stad·ik ,gran·əd·ə'zā·shən }

static metamorphism [GEOL] Regional metamorphism caused by heat and solvents at high lithostatic pressures. Also known as load metamorphism. { 'stad·ik ,med·ə'mȯr,fiz·əm }

staurolite [MINERAL] $FeAl_4(SiO_4)_2(OH)_2$ A reddish-brown to black neosilicate mineral that crystallizes in the orthorhombic system, has resinous to vitreous luster, hardness is 7–7.5 on Mohs scale, and specific gravity is 3.7. Also known as cross-stone; fairy stone; grenatite; staurotide. { 'stȯr·ə,līt }

staurotide *See* staurolite. { 'stȯr·ə,tīd }

steatite [PETR] A compact, massive, fine-ground rock composed principally of talc, but with much other material. { 'stē·ə,tīt }

steatization [GEOL] Introduction of or replacement by talc or steatite. { stē,ad·ə'zā·shən }

S tectonite [PETR] A tectonite whose fabric is dominated by planar surfaces of formation or deformation, such as slate. { 'es 'tek·tə,nīt }

steel jack *See* sphalerite. { 'stēl 'jak }

Stegodontinae [PALEON] An extinct subfamily of elephantoid proboscideans in the family Elephantidae. { ,steg·ə'dänt·ə,nē }

Stegosauria [PALEON] A suborder of extinct reptiles of the order Ornithischia comprising the plated dinosaurs of the Jurassic which had tiny heads, great triangular plates arranged on the back in two alternating rows, and long spikes near the end of the tail. { ,steg·ə'sȯr·ē·ə }

steigerite [MINERAL] $4AlVO_4 \cdot 13H_2O$ A canary-yellow mineral composed of hydrous aluminum vanadate and occurring in masses. { 'stī·gə,rīt }

Steinheim man [PALEON] A prehistoric man represented by a skull, without mandible, found near Stuttgart, Germany; the browridges are massive, the face is relatively small, and the braincase is similar in shape to that of *Homo sapiens*. { 'shtīn,hīm ,man }

steinkern [GEOL] **1.** Rock material formed from consolidated mud or sediment that filled a hollow organic structure, such as a fossil shell. **2.** The fossil formed after dissolution of the mold. Also known as endocast; internal cast. { 'shtīn,kərn }

Stenomasteridae [PALEON] An extinct family of Euechinoidea, order Holasteroida, comprising oval and heart-shaped forms with fully developed pore pairs. { ,sten·ə·mas'ter·ə,dē }

Stensioellidae [PALEON] A family of Lower Devonian placoderms of the order Petalichthyida having large pectoral fins and a broad subterminal mouth. { ,sten·shō'el·ə,dē }

Stenurida [PALEON] An order of Ophiuroidea, comprising the most primitive brittlestars, known only from Paleozoic sediments. { stə'nūr·əd·ə }

step [GEOL] A hitch or dislocation of the strata. { step }

step fault [GEOL] One of a set of closely spaced, parallel faults. Also known as distributive fault; multiple fault. { 'step ,fȯlt }

Stephanian [GEOL] A European stage of Upper Carboniferous geologic time, forming the Upper Pennsylvanian, above the Westphalian and below the Sakmarian of the Permian. { stə'fān·ē·ən }

stephanite [MINERAL] Ag_5SbS_4 An iron-black mineral crystallizing in the orthorhombic system and having a metallic luster; an ore of silver. Also known as black silver; brittle silver ore; goldschmidtine. { 'stef·ə,nīt }

step-out time [GEOPHYS] In seismic prospecting, the time differentials in arrivals of a given peak or trough of a reflected or refracted event for successive detector positions on the earth's surface. { 'step ¦aút ,tīm }

steptoe [GEOL] An isolated protrusion of bedrock, such as the summit of a hill or mountain, in a lava flow. { 'step,tō }

stercorite [MINERAL] $Na(NH_4)H(PO_4) \cdot 4H_2O$ A white to yellowish and brown, triclinic mineral consisting of a hydrated acid phosphate of sodium and ammonium. { 'stər·kə,rīt }

stereographic net See net. { ¦ster·ē·ə¦graf·ik 'net }

Stereospondyli [PALEON] A group of labyrinthodont amphibians from the Triassic characterized by a flat body without pleurocentra and with highly developed intercentra. { ,ster·ē·ə'spän·də,lī }

sternbergite [MINERAL] $AgFe_2S_3$ A dark-brown or black mineral composed of silver iron sulfide and occurring as tabular crystals or flexible laminae. { 'stərn,bər,gīt }

sterrettite See kolbeckite. { 'ster·ə,tīt }

stewartite [GEOL] A steel-gray, iron-containing variety of bort that has magnetic properties. [MINERAL] $Mn_3(DO)_2 \cdot 4H_2O$ A brownish-yellow mineral composed of hydrous manganese phosphate occurring in minute crystals or fibrous tufts in pegmatites. { 'stü·ər,tīt }

Sthenurinae [PALEON] An extinct subfamily of marsupials of the family Diprotodontidae, including the giant kangaroos. { sthə'nūr·ə,nē }

stibiconite [MINERAL] $Sb_3O_6(OH)$ A pale yellow to yellowish- or reddish-white mineral consisting of a basic or hydrated oxide of antimony; occurs in massive form, as a powder, and in crusts. { 'stib·ə·kə,nīt }

stibiocolumbite [MINERAL] $Sb(Nb,Ta,Cb)O_4$ A dark brown to light yellowish- or reddish-brown, orthorhombic mineral consisting of an oxide of antimony and tantalum-columbium. { ¦stib·ē·ō'käl·əm,bīt }

stibium See antimonite. { 'stib·ē·əm }

stibnite See antimonite. { 'stib,nīt }

stichtite [MINERAL] $Mg_6Cr_2(CO_3)(OH)_{16} \cdot 4H_2O$ A lilac-colored rhombohedral mineral composed of hydrous basic carbonate of magnesium and chromium. { 'sti,kīt }

stilbite [MINERAL] $Ca(Al_2Si_7O_{18}) \cdot 7H_2O$ A white, brown, or yellow mineral belonging to the zeolite family of silicates; crystallizes in the monoclinic system, occurs in sheaflike aggregates of tabular crystals, and has pearly luster; hardness is 3.5–4 on Mohs scale, and specific gravity is 2.1–2.2. Also known as desmine. { 'stil‚bīt }

stillstand |GEOL| A period during which a land area, a continent, or an island remains stationary with respect to the interior of the earth or to sea level. { 'stil‚stand }

stilpnomelane [MINERAL] K(Fe,Mg,Al)$_3$Si$_4$O$_{10}$(OH)$_2$·H$_2$O A black or greenish-black mineral composed of basic hydrous potassium iron magnesium aluminum silicate; occurs as fibers, incrustations, and foliated plates. { ‚stilp·nō'me‚lān }

stinkstone |GEOL| A stone containing decomposing organic matter that gives off an offensive odor when rubbed or struck. { 'stiŋk‚stōn }

stipoverite See stishovite. { stə'päv·ə‚rīt }

stishovite [MINERAL] SiO$_2$ A polymorph of quartz, a dense, fine-grained mineral formed under very high pressure (about 1 × 10^6 pounds per square inch or 7 × 10^9 pascals); it is the only mineral in which the silicon atom has a coordination number of six; specific gravity is 4.28. Also known as stipoverite. { 'stish·ə‚vīt }

stock |GEOL| See pipe. |PETR| A usually discordant, batholithlike body of intrusive igneous rock not exceeding 40 square miles (103.6 square kilometers) in surface exposure and usually discordant. { stäk }

stockwork |GEOL| A mineral deposit in the form of a network of veinlets diffused in the country rock. { 'stäk‚wərk }

stokesite [MINERAL] CaSnSi$_3$O$_9$·2H$_2$O A colorless orthorhombic mineral composed of hydrous calcium tin silicate occurring in crystals. { 'stōk‚sīt }

stolzite [MINERAL] PbWO$_4$ A tetragonal mineral composed of native lead tungstate; it is isomorphous with wulfenite and dimorphous with raspite. { 'stōl‚zīt }

stone |GEOL| **1.** A small fragment of rock or mineral. **2.** See stony meteorite. { stōn }

stone bubble See lithophysa. { 'stōn ‚bəb·əl }

stone coal See anthracite. { 'stōn ‚kōl }

stone polygon See sorted polygon. { 'stōn 'päl·i‚gän }

stone ring |GEOL| A ring of stones surrounding a central area of finer material; characteristic of sorted circle and sorted polygon. { 'stōn 'riŋ }

stony-iron meteorite |GEOL| Any of the rare meteorites containing at least 25% of both nickel-iron and heavy basic silicates. Also known as iron-stony meteorite; lithosiderite; sideraerolite; siderolite; syssiderite. { 'stō·nē ‖ī·ərn 'mēd·ē·ə‚rīt }

stony meteorite |GEOL| Any meteorite composed principally of silicate minerals, especially olivine, pyroxene, and plagioclase. Also known as aerolite; asiderite; meteoric stone; meteorolite; stone. { 'stō·nē 'mēd·ē·ə‚rīt }

storm beach |GEOL| A ridge composed of gravel or shingle built up by storm waves at the inner margin of a beach. { 'stȯrm ‚bēch }

storm delta See washover. { 'stȯrm ‚del·tə }

storm microseism |GEOPHYS| A microseism lasting 25 or more seconds, caused by ocean waves. { 'stȯrm 'mī·krə‚sīz·əm }

stoss |GEOL| Of the side of a hill, knob, or prominent rock, facing the upstream side of a glacier. { stäs }

stoss-and-lee topography |GEOL| A type of glaciated landscape in which small hills or other landforms exhibit gentle eroded slopes on the up-glacier or upstream side and less eroded, steeper slopes on the lee side. { ‖stäs ənd ‖lē tə'päg·rə·fē }

strain shadow See pressure shadow. { 'strān ‚shad·ō }

strain-slip |GEOL| A rock fracture resulting in a slight displacement. { 'strān ‚slip }

strain-slip cleavage See slip cleavage. { 'strān ‖slip ‚klē·vij }

strand |GEOL| A beach bordering a sea or an arm of an ocean. { strand }

strand flat See wave-cut platform. { 'strand ‚flat }

strandline |GEOL| **1.** A beach raised above the present sea level. **2.** The level at which a body of standing water meets the land. **3.** See shoreline. { 'strand‚līn }

strath |GEOL| **1.** A broad, elongate depression with steep sides on the continental shelf. **2.** An extensive remnant of a broad, flat valley floor that has undergone degradation following uplift. { strath }

strath terrace [GEOL] An extensive remnant of a strath from a former erosion cycle. { 'strath ‚ter·əs }

stratification [GEOL] An arrangement or deposition of sedimentary material in layers, or of sedimentary rock in strata. { ‚strad·ə·fə'kā·shən }

stratification index [GEOL] A measure of the beddedness of a stratigraphic unit, expressed as the number of beds in the unit per 100 feet (30 meters) of section. { ‚strad·ə·fə'kā·shən ‚in‚deks }

stratification plane [GEOL] A demarcation between two layers of sedimentary rock, often signifying that the layers were deposited under different conditions. { ‚strad·ə·fə'kā·shən ‚plān }

stratified drift [GEOL] Fluvioglacial drift composed of material deposited by a meltwater stream or settled from suspension. { 'strad·ə‚fīd 'drift }

stratified rock See sedimentary rock. { 'strad·ə‚fīd 'räk }

stratiform [GEOL] **1.** Descriptive of a layered mineral deposit of either igneous or sedimentary origin. **2.** Consisting of parallel bands, layers, or sheets. { 'strad·ə‚fȯrm }

stratigrapher [GEOL] A geologist who deals with stratified rocks, for example, the classification, nomenclature, correlation, and interpretation of rocks. { strə'tig·rə·fər }

stratigraphic geology See stratigraphy. { ¦strad·ə¦graf·ik jē'äl·ə·jē }

stratigraphic map [GEOL] A map showing the areal distribution, configuration, or aspect of a stratigraphic unit or surface, such as an isopach map or a lithofacies map. { ¦strad·ə¦graf·ik 'map }

stratigraphic oil fields [GEOL] Hydrocarbon reserves in stratigraphic (sedimentary) traps formed by the positioning of clastic materials through chemical deposition. { ¦strad·ə¦graf·ik 'ȯil ‚fēlz }

stratigraphic separation See stratigraphic throw. { ¦strad·ə¦graf·ik ‚sep·ə'rā·shən }

stratigraphic sequence See sequence. { ¦strad·ə¦graf·ik 'sē·kwəns }

stratigraphic throw [GEOL] The thickness of the strata which originally separated two beds brought into contact at a fault. Also known as stratigraphic separation. { ¦strad·ə¦graf·ik 'thrō }

stratigraphic trap [GEOL] Sealing of a reservoir bed due to lithologic changes rather than geologic structure. Also known as porosity trap; secondary stratigraphic trap. { ¦strad·ə¦graf·ik 'trap }

stratigraphic unit [GEOL] A stratum of rock or a body of strata classified as a unit on the basis of character, property, or attribute. { ¦strad·ə¦graf·ik 'yü·nət }

stratigraphy [GEOL] A branch of geology concerned with the form, arrangement, geographic distribution, chronologic succession, classification, correlation, and mutual relationships of rock strata, especially sedimentary. Also known as stratigraphic geology. { strə'tig·rə·fē }

stratotype [GEOL] A specifically bounded type section of rock strata to which a time-stratigraphic unit is ascribed, ideally consisting of a complete and continuously exposed and deposited sequence of correlatable strata, and extending from a readily identifiable basal boundary to a readily identifiable top boundary. { 'strad·ə‚tīp }

stratovolcano [GEOL] A volcano constructed of lava and pyroclastics, deposited in alternating layers. Also known as composite volcano. { ¦strad·ō·väl'kā·nō }

stratum [GEOL] A mass of homogeneous or gradational sedimentary material, either consolidated rock or unconsolidated soil, occurring in a distinct layer and visually separable from other layers above and below. { 'strad·əm }

stray [GEOL] A lenticular rock formation encountered unexpectedly in drilling an oil or a gas well; it differs from an adjacent persistent formation in lithology and hardness. { strā }

stray sand [GEOL] A stray composed of sandstone. { 'strā 'sand }

streak [MINERAL] The color of a powdered mineral, obtained by rubbing the mineral on a streak plate. { strēk }

stream-built terrace See alluvial terrace. { 'strēm ¦bilt 'ter·əs }

stream capacity [GEOL] The ability of a stream to carry detritus, measured at a given point per unit of time. { 'strēm kə,pas·əd·ē }

stream channel [GEOL] A long, narrow, sloping troughlike depression where a natural stream flows or may flow. Also known as streamway. { 'strēm ,chan·əl }

stream-channel form ratio [GEOL] The mathematical relationship between a stream channel width, depth, and channel perimeter. { 'strēm ,chan·əl 'fȯrm ,rā·shō }

stream erosion [GEOL] The progressive removal of exposed matter from the surface of a stream channel by a stream. { 'strēm i,rō·zhən }

stream frequency [GEOL] A measure of topographic texture expressed as the ratio of the number of streams in a drainage basin to the area of the basin. Also known as channel frequency. { 'strēm ,frē·kwən·sē }

stream gradient [GEOL] The angle, measured in the direction of flow, between the water surface (for large streams) or the channel flow (for small streams) and the horizontal. Also known as stream slope. { 'strēm ,grād·ē·ənt }

stream-gradient ratio [GEOL] Ratio of the stream gradient of a stream channel of one order to the stream gradient of the next higher order channel in the same drainage basin. Also known as channel gradient ratio. { 'strēm ,grād·ē·ənt ,rā·shō }

stream load [GEOL] Solid material transported by a stream. { 'strēm ,lōd }

stream morphology See river morphology. { 'strēm ,mȯr'fäl·ə·jē }

streamsink [GEOL] An opening in the surface of the earth down which a stream disappears underground. { 'strēm,siŋk }

stream slope See stream gradient. { 'strēm ,slōp }

stream terrace [GEOL] One of a series of level surfaces on a stream valley flanking and parallel to a stream channel and above the stream level, representing the uneroded remnant of an abandoned floodplain or stream bed. Also known as river terrace. { 'strēm ,ter·əs }

stream tin [GEOL] The mineral cassiterite occurring as pebbles in alluvial deposits. { 'strēm ,tin }

stream transport [GEOL] Movement of rock material in and by a stream. { 'strēm 'tranz,pȯrt }

streamway See stream channel. { 'strēm,wā }

strengite [MINERAL] $FePO_4 \cdot 2H_2O$ A pale-red mineral crystallizing in the orthorhombic system, isomorphous with variscite and dimorphous with phosphosiderite, and specific gravity 2.87. { 'streŋ,īt }

stress mineral [MINERAL] Any mineral whose formation in metamorphosed rock is favored by shearing stress. { 'stres ,min·rəl }

stretched pebbles [GEOL] Pebbles in a sedimentary rock which have been elongated from their original shape by deformation. { 'strecht 'peb·əlz }

stretch fault See stretch thrust. { 'strech ,fȯlt }

stretch thrust [GEOL] A reverse fault developed as a result of shear in the middle limb of an overturned fold. Also known as stretch fault. { 'strech ,thrəst }

striated ground See striped ground. { 'strī,ād·əd 'graúnd }

striation [GEOL] One of a series of parallel or subparallel scratches, small furrows, or lines on the surface of a rock or rock fragment; usually inscribed by rock fragments embedded at the base of a moving glacier. [MINERAL] One of a series of parallel, shallow depressions or narrow bands on the cleavage face of a mineral caused either by growth twinning or oscillatory growth of different crystal faces. { strī'ā·shən }

strigovite [MINERAL] $Fe_3(Al,Fe)_3Si_3O_{11}(OH)_7$ A dark-green mineral of the chlorite group, composed of basic aluminum iron silicate; occurs as crystalline incrustations. { 'strig·ə,vīt }

strike [GEOL] The direction taken by a structural surface, such as a fault plane, as it intersects the horizontal. Also known as line of strike. { strīk }

strike fault [GEOL] A fault whose strike is parallel with that of the strata involved. { 'strīk ,fȯlt }

strike joint [GEOL] A joint that strikes parallel to the bedding or cleavage of the constituent rock. { 'strīk ,jȯint }

strike separation |GEOL| The distance of separation on either side of a fault surface of two formerly adjacent beds. { 'strīk ‚sep·ə‚rā·shən }

strike-separation fault See lateral fault. { 'strīk ‚sep·ə‚rā·shən ‚fȯlt }

strike-shift fault See strike-slip fault. { 'strīk ‚shift ‚fȯlt }

strike slip |GEOL| The component of the slip of a fault that is parallel to the strike of the fault. Also known as horizontal displacement; horizontal separation. { 'strīk ‚slip }

strike-slip fault |GEOL| A fault whose direction of movement is parallel to the strike of the fault. Also known as strike-shift fault. { 'strīk ‚slip ‚fȯlt }

string |GEOL| A very small vein, either independent or occurring as a branch of a larger vein. Also known as stringer. { striŋ }

stringer |GEOL| See string. { 'striŋ·ər }

stringer lode |GEOL| A lode that consists of many narrow veins in a mass of country rock. { 'striŋ·ər ‚lōd }

striped ground |GEOL| A pattern of alternating stripes formed by frost action on a sloping surface. Also known as striated ground; striped soil. { 'strīpt 'graůnd }

striped soil See striped ground. { 'strīpt 'sȯil }

stripped illite See degraded illite. { 'stript 'il‚īt }

stripped plain |GEOL| The upper, exposed surface of a resistant stratum that forms a stripped structural surface when extended over a considerable area. { 'stript 'plān }

stripped structural surface |GEOL| An erosion surface formed in an area underlain by horizontal or gently sloping strata of unequal resistance where the overlying softer beds have been removed by erosion. Also known as stripped surface. { 'stript ‚strək·chə·rəl 'sər·fəs }

stripped surface See stripped structural surface. { 'stript 'sər·fəs }

stromatite |GEOL| Chorismite having flat or folded parallel layers of two or more textural elements. Also known as stromatolith. { 'strō·mə‚tīt }

stromatolite |GEOL| A structure in calcareous rocks consisting of concentrically laminated masses of calcium carbonate and calcium-magnesium carbonate which are believed to be of calcareous algal origin; these structures are irregular to columnar and hemispheroidal in shape, and range from 1 millimeter to many meters in thickness. Also known as callenia. { strə'mad·əl‚īt }

stromatolith |GEOL| **1.** A complex sill-like igneous intrusion interfingered with sedimentary strata. **2.** See stromatite. { strə'mad·əl‚ith }

Stromatoporoidea |PALEON| An extinct order of fossil colonial organisms thought to belong to the class Hydrozoa; the skeleton is a coenosteum. { strə‚mad·ə·pə‚rȯid·ē·ə }

Strombacea |PALEON| An extinct superfamily of gastropod mollusks in the order Prosobranchia. { sträm'bās·ē·ə }

strombolian |GEOL| A type of volcanic eruption characterized by fire fountains of lava from a central crater. { sträm'bō·lē·ən }

stromeyerite |MINERAL| CuAgS A metallic-gray orthorhombic mineral with a blue tarnish composed of silver copper sulfide occurring in compact masses. { 'strō‚mī·ə‚rīt }

strontianite |MINERAL| $SrCO_3$ A pale-green, white, gray, or yellowish mineral of the aragonite group having orthorhombic symmetry and occurring in veins or as masses; hardness is 3.5 on Mohs scale, and specific gravity is 3.76. { 'strän·chē·ə‚nīt }

Strophomenida |PALEON| A large diverse order of articulate brachiopods which first appeared in Lower Ordovician times and became extinct in the Late Triassic. { ‚strä·fə'men·əd·ə }

Strophomenidina |PALEON| A suborder of extinct, articulate brachiopods in the order Strophomenida characterized by a concavo-convex shell, the pseudodeltidium and socket plates disposed subparallel to the hinge. { ‚strä·fə‚men·ə'dī·nə }

structural analysis |PETR| See structural petrology. { 'strək·chə·rəl ə'nal·ə·səs }

structural bench |GEOL| A bench typifying the resistant edge of a terrace that is being reduced by erosion. Also known as rock bench. { 'strək·chə·rəl 'bench }

structural contour map |GEOL| A map representation of a subsurface stratigraphic unit; depicts the configuration of a rock surface by means of elevation contour lines. { 'strək·chə·rəl 'kän,tür ,map }

structural fabric *See* fabric. { 'strək·chə·rəl ¦fab·rik }

structural geology [GEOL] A branch of geology concerned with the form, arrangement, and internal structure of the rocks. { 'strək·chə·rəl jē'äl·ə·jē }

structural high [GEOL] Any of various structural features such as a crest, culmination, anticline, or dome. { 'strək·chə·rəl 'hī }

structural low [GEOL] Any of various structural features such as a basin, a syncline, a saddle, or a sag. { 'strək·chə·rəl 'lō }

structural petrology [PETR] The study of the internal structure of a rock to determine its deformational history. Also known as fabric analysis; microtectonics; petrofabric analysis; petrofabrics; petrogeometry; petromorphology; structural analysis. { 'strək·chə·rəl pi'träl·ə·jē }

structural terrace [GEOL] A terracelike landform developed where generally steeply inclined and otherwise uniformly dipping strata locally flatten. { 'strək·chə·rəl 'ter·əs }

structural trap [GEOL] Containment in a reservoir bed of oil or gas due to flexure or fracture of the bed. { 'strək·chə·rəl 'trap }

structural valley [GEOL] A valley whose form and origin is attributable to the underlying geologic structure. { 'strək·chə·rəl 'val·ē }

structure [GEOL] **1.** An assemblage of rocks upon which erosive agents have been or are acting. **2.** The sum total of the structural features of an area. [MINERAL] The form taken by a mineral, such as tabular or fibrous. [PETR] A macroscopic feature of a rock mass or rock unit, best seen in an outcrop. { 'strək·chər }

structure contour [GEOL] A contour that portrays a structural surface, such as a fault. Also known as subsurface contour. { 'strək·chər ,kän,tür }

structure-contour map [GEOL] A map that uses structure contour lines to portray subsurface configuration. Also known as structure map. { 'strək·chər ¦kän,tür ,map }

structure map *See* structure-contour map. { 'strək·chər ,map }

structure section [GEOL] A vertical section showing the observed or inferred geologic structure on a vertical surface or plane. { 'strək·chər ,sek·shən }

struvite [MINERAL] $Mg(NH_4)PO_4 \cdot 6H_2O$ A colorless to yellow or pale-brown mineral consisting of a hydrous ammonium magnesium phosphate, and occurring in orthorhombic crystals; hardness is 2 on Mohs scale, and specific gravity is 1.7. { 'strü,vīt }

stuffed mineral [MINERAL] A mineral having extra ions of a foreign element within its larger interstices. { 'stəft 'min·rəl }

sturtite [MINERAL] A black mineral composed of hydrous silicate of iron, manganese, calcium, and magnesium; occurs in compact masses. { 'stərd,īt }

stylolite [GEOL] An irregular surface, generally parallel to a bedding plane, in which small toothlike projections on one side of the surface fit into cavities of complementary shape on the other surface; interpreted to result diagenetically by pressure solution. { 'stī·lə,līt }

stylotypite *See* tetrahedrite. { 'stī·lə,tī,pīt }

S-type magma [GEOL] Magma formed from sedimentary source material. { 'es ¦tīp 'mag·mə }

subaerial [GEOL] Pertaining to conditions and processes occurring beneath the atmosphere or in the open air, that is, on or adjacent to the land surface. { ¦səb'er·ē·əl }

subage [GEOL] A subdivision of a geologic age. { 'səb,āj }

subalkaline [GEOCHEM] Pertaining to a soil in which the pH is 8.0 to 8.5, usually in a limestone or salt-marsh region. { ¦səb'al·kə,līn }

subaqueous dune [GEOL] A dune resulting from entrainment of grains by the flow of moving water. { ¦səb'ā·kwē·əs 'dün }

subarkose [GEOL] Sandstone that is intermediate in composition between arkose and pure quartz sandstone; it contains less feldspar than arkose. { səb'är,kōs }

subbituminous coal [GEOL] Black coal intermediate in rank between lignite and bituminous coal; has more carbon and less moisture than lignite. { ¦səb·bə'tü·mə·nəs 'kōl }

subbottom reflection [GEOPHYS] The return of sound energy from a discontinuity in material below the surface of the sea bottom. { ¦səb'bäd·əm ri'flek·shən }

subcapillary interstice [GEOL] An interstice in which the molecular attraction of its walls extends across the entire opening; it is smaller than a capillary interstice. { ¦səb'kap·ə,ler·ē in'tər·stəs }

subconchoidal [GEOL] Pertaining to a fracture that is partly or vaguely conchoidal in shape. { ¦səb·kən'kȯid·əl }

subcrop [GEOL] An occurrence of strata beneath the subsurface of an inclusive stratigraphic unit that succeeds an unconformity on which there is marked overstep. { 'səb,kräp }

subduction [GEOL] The process by which one crustal block descends beneath another, such as the descent of the Pacific plate beneath the Andean plate along the Andean Trench. { səb'dək·shən }

subduction zones [GEOL] Regions where portions of the earth's tectonic plates are diving beneath other plates, into the earth's interior. They are defined by deep oceanic trenches, lines of volcanoes parallel to the trenches, and zones of large earthquakes that extend from the trenches landward. { səb'dək·shən ,zōnz }

suberinite [GEOL] A variety of provitrinite composed of corky tissue. { sü'ber·ə,nīt }

subfeldspathic [GEOL] Referring to mature lithic wacke or arenite containing an abundance of quartz grains with less than 10% feldspar grains. { ¦səb·fel'spath·ik }

subgelisol [GEOL] Unfrozen ground beneath permafrost. { ¦səb'jel·ə,sȯl }

subglacial [GEOL] Pertaining to the area in or at the bottom of, or immediately beneath, a glacier. { ¦səb'glā·shəl }

subglacial moraine See ground moraine. { ¦səb'glā·shəl mə'rān }

subgraywacke [PETR] An argillaceous sandstone with a composition intermediate between graywacke and orthoquartzite; a clay matrix is usually present but it amounts to less than 15%. { ¦səb'grā,wak·ə }

subhedral [MINERAL] **1.** Pertaining to an individual mineral crystal that is partly bounded by its own crystal faces and partly bounded by surfaces formed against preexisting crystals. **2.** Descriptive of a crystal having partially developed crystal faces. { ¦səb¦hē·drəl }

subidiomorphic See hypidiomorphic. { ¦səb,id·ē·ə'mȯr·fik }

subjacent [GEOL] Being lower than but not directly underneath. { ,səb'jās·ənt }

subjacent igneous body [GEOL] An igneous intrusion without a known floor, and which presumably enlarges downward. { ,səb'jās·ənt 'ig·nē·əs 'bäd·ē }

sublacustrine [GEOL] Existing or formed on the bottom of a lake. { ¦səb·lə'kəs·trən }

sublacustrine channel [GEOL] A channel eroded in a lake bed either before the lake existed or by a strong current in the lake. { ¦səb·lə'kəs·trən ,chan·əl }

sublimation vein [GEOL] A vein of mineral that has condensed from a vapor. { ,səb·lə'mā·shən ¦vān }

sublitharenite [PETR] A sandstone which contains between 5 and 25% rock fragments and in which the rock fragments are more abundant than feldspar grains. { ¦səb·li'thar·ə,nīt }

submarine canyon [GEOL] Steep-sided valleys winding across the continental shelf or continental slope, probably originally produced by Pleistocene stream erosion, but presently the site of turbidity flows. { ¦səb·mə'rēn 'kan·yən }

submarine cave See submarine fan. { ¦səb·mə'rēn 'kāv }

submarine delta See submarine fan. { ¦səb·mə'rēn 'del·tə }

submarine earthquake See seaquake. { ¦səb·mə'rēn 'ərth,kwāk }

submarine fan [GEOL] A shallow marine sediment that is fan- or cone-shaped and lies off the seaward opening of large rivers and submarine canyons. Also known as abyssal cave; abyssal fan; sea fan; submarine cave; submarine delta; subsea apron. { ¦səb·mə'rēn 'fan }

submarine geology See geological oceanography. { ¦səb·mə'rēn jē'äl·ə·jē }

submarine isthmus [GEOL] A submarine elevation joining two land areas and separating two basins or depressions by a depth less than that of the basins. { ¦səb·mə'rēn 'is·məs }

submarine peninsula [GEOL] An elevated portion of the submarine relief resembling a peninsula. { ¦səb·mə'rēn pə'nin·sə·lə }

submarine pit [GEOL] A cavity on the bottom of the sea. Also known as submarine well. { ¦səb·mə'rēn 'pit }

submarine relief [GEOL] Relative elevations of the ocean bed, or the representation of them on a chart. { ¦səb·mə'rēn ri'lēf }

submarine topography [GEOL] Configuration of a surface such as the sea bottom or of a surface of given characteristics within the water mass. { ¦səb·mə'rēn tə'päg·rə·fē }

submarine trough See trough. { ¦səb·mə'rēn 'trȯf }

submarine valley See valley. { ¦səb·mə'rēn 'val·ē }

submarine weathering [GEOL] A slow alteration of the form, texture, and composition of the sea floor from chemical, thermal, and biological causes. { ¦səb·mə'rēn 'weth·ə·riŋ }

submarine well See submarine pit. { ¦səb·mə'rēn 'wel }

submerged coastal plain [GEOL] The continental shelf as the seaward extension of a coastal plain on the land. Also known as coast shelf. { səb'mərjd 'kōst·əl 'plān }

submerged lands [GEOL] Lands covered by water at any stage of the tide, as distinguished from tidelands which are attached to the mainland or an island and are covered or uncovered with the tide; tidelands presuppose a high-water line as the upper boundary, submerged lands do not. { səb'mərjd 'lanz }

submerged shoreline See shoreline of submergence. { səb'mərjd 'shȯr,līn }

submergence [GEOL] A change in the relative levels of water and land either from a sinking of the land or a rise of the water level. { səb'mər·jəns }

subsea apron See submarine fan. { 'səb,sē 'ā·prən }

subsequent [GEOL] Referring to a geologic feature that followed in time the development of a consequent feature of which it is a part. { 'səb·sə·kwənt }

subsequent fold See cross fold. { 'səb·sə·kwənt 'fōld }

subsequent valley [GEOL] A valley eroded by a stream developed subsequent to the system of which it is a part. { 'səb·sə·kwənt 'val·ē }

subsidiary fracture See tension fracture. { səb'sid·ē,er·ē 'frak·chər }

subsoil [GEOL] **1.** Soil underlying surface soil. **2.** See B horizon. { 'səb,sȯil }

substratum [GEOL] Any layer underlying the true soil. { ¦səb'strad·əm }

subsurface contour See structure contour. { ¦səb'sər·fəs 'kän,túr }

subsurface geology [GEOL] The study of geologic features beneath the land or sea-floor surface. Also known as underground geology. { ¦səb'sər·fəs jē'äl·ə·jē }

Subulitacea [PALEON] An extinct superfamily of gastropod mollusks in the order Prosobranchia which possessed a basal fold but lacked an apertural sinus. { ,səb·yə·lə'tās·ē·ə }

succession [GEOL] A group of rock units or strata that succeed one another in chronological order. { sək'sesh·ən }

succinite [MINERAL] An amber-colored variety of grossularite. { 'sək·sə,nīt }

sucrosic See saccharoidal. { sü'krō·sik }

sudburite [GEOL] A basic basalt composed of hypersthene, augite, and magnetite, among other minerals. { 'səd·bə,rīt }

sudden commencement [GEOPHYS] Magnetic storms which start suddenly (within a few seconds) and simultaneously all over the earth. { 'səd·ən kə'mens·mənt }

suevite [GEOL] A grayish or yellowish fragmental rock associated with meteorite impact craters; resembles tuff breccia or pumiceous tuff but is of nonvolcanic origin. { 'swā,vīt }

sugary See saccharoidal. { 'shúg·ə·rē }

sulfate mineral [MINERAL] A mineral compound characterized by the sulfate radical SO_4. { 'səl,fāt 'min·rəl }

sulfide mineral [MINERAL] A mineral compound characterized by the linkage of sulfur with a metal or semimetal. { 'səl,fīd ,min·rəl }

sulfoborite [MINERAL] $Mg_6H_4(BO_3)_4(SO_4)_2 \cdot 7H_2O$ A mineral composed of hydrous acid sulfate and borate of magnesium. { ,səl·fə'bȯr,īt }

sulfofication [GEOCHEM] Oxidation of sulfur and sulfur compounds into sulfates, occurring in soils by the agency of bacteria { ‚səl·fə·fə'kā·shən }

sulfohalite [MINERAL] $Na_6(SO_4)_2FCl$ A mineral composed of sulfate, chloride, and fluoride of sodium. { ¦səl·fō'ha‚līt }

sulfophile element [GEOCHEM] An element occurring preferentially in an oxygen-free mineral. Also known as thiophile element. { ¦səl·fə'fīl ‚el·ə·mənt }

sulfur [MINERAL] A yellow orthorhombic mineral occurring in crystals, masses, or layers, and existing in several allotropic forms; the native form of the element. { 'səl·fər }

sulfur ball [GEOL] A bubble of hot volcanic gas encased in a sulfurous mud skin that solidified on contact with air. { 'səl·fər ‚bȯl }

sulfur-mud pool See mud pot. { 'səl·fər ¦məd ‚pül }

sullage [GEOL] Mud, silt, or other sediments carried and deposited by flowing water. { 'səl·ij }

sulvanite [MINERAL] Cu_3VS_4 A bronze-yellow mineral composed of copper vanadium sulfide occurring in masses. { 'səl·və‚nīt }

summit plain See peak plain. { 'səm·ət ‚plān }

sun crack See mud crack. { 'sən ‚krak }

sundtite See andorite. { 'sən‚tīt }

sun opal See fire opal. { 'sən ‚ō·pəl }

sunstone [MINERAL] An aventurine feldspar containing minute flakes of hematite; usually brilliant and translucent, it emits reddish or golden billowy reflection. Also known as heliolite. { 'sən ‚stōn }

supercapillary interstice [GEOL] An interstice that is too large to hold water above the free water surface by surface tension; it is larger than a capillary interstice. { ¦sü·pər'kap·ə‚ler·ē in'tər·stəs }

supercontinent [GEOL] A large continental mass, such as Pangea, that existed early in geologic time and from which smaller continents formed and separated by fragmentation and drifting. { 'sü·pər‚känt·ən·ənt }

superficial deposit See surficial deposit. { ¦sü·pər¦fish·əl di'päz·ət }

supergene [MINERAL] Referring to mineral deposits or enrichments formed by descending solutions. Also known as hypergene. { 'sü·pər‚jēn }

supergroup [GEOL] A lithostratigraphic material unit of the highest order. { 'sü·pər‚grüp }

superimposed [GEOL] Pertaining to layered or stratified rocks. { ¦sü·pər·im'pōzd }

superimposed fan [GEOL] An alluvial fan developed on, and having a steeper gradient than, an older fan. { ¦sü·pər·im'pōzd 'fan }

superimposed fold See cross fold. { ¦sü·pər·im'pōzd 'fōld }

superimposed glacier [GEOL] A glacier whose course is maintained despite different preexisting structures and lithologies as the glacier erodes downward. { ¦sü·pər·im'pōzd 'glā·shər }

superimposed valley [GEOL] A valley eroded by or containing a superimposed stream. { ¦sü·pər·im'pōzd 'val·ē }

superincumbent [GEOL] Pertaining to a superjacent layer, especially one that is situated so as to exert pressure. { ¦sü·pər·in'kəm·bənt }

superjacent [GEOL] Pertaining to a stratum situated immediately upon or over a particular lower stratum or above an unconformity. { ¦sü·pər'jā·sənt }

supermature [GEOL] Pertaining to a texturally mature clastic sediment whose grains have become rounded. { ¦sü·pər·mə'chür }

superposed stream See consequent stream. { ¦sü·pər'pōzd 'strēm }

superposition [GEOL] **1.** The order in which sedimentary layers are deposited, the highest being the youngest. **2.** The process by which the layering occurs. { ‚sü·pər·pə'zish·ən }

superprint See overprint. { 'sü·pər‚print }

supracrustal rocks [GEOL] Rocks that overlie basement rock. { ¦sü·prə'krəst·əl 'räks }

supratidal sediment [GEOL] The sediment deposited immediately above the high-tide level. { ¦sü·prə'tīd·əl 'sed·ə·mənt }

supratidal zone [GEOL] Pertaining to the shore area immediately marginal to and above the high-tide level. { ¦sü·prə'tīd·əl zōn }

surface creep [GEOL] A stage of the wind erosion process in which grains of sand move each other along the surface. { 'sər·fəs ,krēp }

surface deposit *See* surficial deposit. { 'sər·fəs di,päz··ət }

surface geology [GEOL] The scientific study of the features at the surface of the earth. { 'sər·fəs jē,äl·ə·jē }

surface phase [GEOCHEM] A thin rock layer differing in geochemical properties from those of the volume phases on either side. Also known as volume phase. { 'sər·fəs ,fāz }

surface soil [GEOL] The soil extending 5 to 8 inches (13 to 20 centimeters) below the surface. { 'sər·fəs ,sȯil }

surface wash *See* sheet erosion. { 'sər·fəs ,wäsh }

surficial creep *See* soil creep. { sər'fish·əl 'krēp }

surficial deposit [GEOL] Unconsolidated alluvial, residual, or glacial deposits overlying bedrock or occurring on or near the surface of the earth. Also known as superficial deposit; surface deposit. { sər'fish·əl di'päz·ət }

surficial geology [GEOL] The scientific study of surficial deposits, including soils. { sər'fish·əl jē'äl·ə·jē }

surf ripple [GEOL] A ripple mark formed on a sandy beach by wave-generated currents. { 'sərf ,rip·əl }

sursassite [MINERAL] Mn$_5$Al$_4$Si$_5$O$_{21}$·3H$_2$O A mineral which is composed of hydrous manganese aluminum silicate. { ,sər'sa,sīt }

susannite [MINERAL] Pb$_4$(SO$_4$)$_2$(CO$_3$)$_2$(OH)$_2$ A greenish or yellowish, rhombohedral mineral that is dimorphous with leadhillite. { sü'za,nīt }

suspended load [GEOL] The part of the stream load that is carried for a long time in suspension. Also known as suspension load. { sə'spen·dəd 'lōd }

suspension load *See* suspended load. { sə'spen·shən ¦lōd }

sussexite [MINERAL] MnBO$_2$OH A white mineral composed of basic manganese borate occurring in fibrous veins. { 'səs·iks,īt }

sutured [PETR] Referring to rock texture in which mineral grains or irregularly shaped crystals interfere with their neighbors, producing interlocking, irregular contacts without interstitial spaces. { 'sü·chərd }

svabite [MINERAL] Ca$_5$(AsO$_4$)$_3$F A colorless, yellow, rose, or reddish-brown mineral composed of fluoride-arsenate of calcium. { 'sfä,bīt }

svanbergite [MINERAL] SrAl$_3$(PO$_4$)(SO$_4$)(OH)$_6$ A colorless to yellow mineral composed of basic phosphate and sulfate of strontium and aluminum; it is isomorphous with corkite, hinsdalite, and woodhouseite. { 'sfän,bər,gīt }

swale [GEOL] **1.** A slight depression, sometimes swampy, in the midst of generally level land. **2.** A shallow depression in an undulating ground moraine due to uneven glacial deposition. **3.** A long, narrow, generally shallow, troughlike depression which lies between two beach ridges and is aligned roughly parallel to the coastline. { swāl }

swallow hole [GEOL] An opening that occurs occasionally at the bottom of a sinkhole which permits direct drainage from the surface into an underground channel. { 'swäl·ō ,hōl }

Swanscombe man [PALEON] A partial skull recovered in Swanscombe, Kent, England, which represents an early stage of *Homo sapiens* but differing in having a vertical temporal region and a rounded occipital profile. { 'swanz·kəm ¦man }

swartzite [MINERAL] CaMg(UO$_2$)(CO$_3$)$_3$·12H$_2$O A green monoclinic mineral composed of hydrous carbonate of calcium, magnesium, and uranium. { 'swȯrt,sīt }

swash [GEOL] **1.** A narrow channel or ground within a sand bank, or between a sand bank and the shore. **2.** A bar over which the sea washes. { swäsh }

swash mark [GEOL] A fine, wavy or arcuate line or minute ridge consisting of fine sand, seaweed, and other debris on a beach; marks the farthest advance of wave uprush. Also known as debris line; wave line; wavemark. { 'swäsh ,märk }

S wave [GEOPHYS] A seismic body wave propagated in the crust or mantle of the earth by a shearing motion of material; speed is 1.9–2.5 miles (3–4 kilometers) per second in the crust and 2.7–2.9 miles (4.4–4.6 kilometers) in the mantle. Also known as distortional wave; equivoluminal wave; rotational wave; secondary wave; shake wave; shear wave; tangential wave; transverse wave. { 'es ‚wāv }

swedenborgite [MINERAL] $NaBe_4SbO_7$ A colorless to wine-yellow mineral composed of sodium beryllium antimony oxide. { 'swēd·ən‚bȯr‚gīt }

swell [GEOL] **1.** The volumetric increase of soils on being removed from their compacted beds due to an increase in void ratio. **2.** A local enlargement or thickening in a vein or ore deposit. **3.** A low dome or quaquaversal anticline of considerable areal extent; long and generally symmetrical waves contribute to the mixing processes in the surface layer and thus to its sound transmission properties. **4.** Gently rising ground, or a rounded hill above the surrounding ground or ocean floor. { swel }

swelled ground [GEOL] A soil or rock that expands when wetted. { 'sweld 'graùnd }

swelling clay [GEOL] Clay that can absorb large amounts of water, such as bentonite. { 'swel·iŋ ‚klā }

swinestone [PETR] Limestone containing black bituminous matter, which gives off an objectionable odor when rubbed. { 'swīn‚stōn }

Sycidales [PALEON] A group of fossil aquatic plants assigned to the Charophyta, characterized by vertically ribbed gyrogonites. { ‚sis·ə'dā·lēz }

syenite [PETR] A visibly crystalline plutonic rock with granular texture composed largely of alkali feldspar, with subordinate plagioclose and mafic minerals; the intrusive equivalent of trachyte. { 'sī·ə‚nīt }

syenodiorite [PETR] Plutonic rock consisting of acid plagioclase, orthoclase, and a ferromagnesian mineral. { ‚sī·ə·nō'dī·ə‚rīt }

syenogabbro [PETR] Plutonic rock consisting of basic plagioclase, orthoclase, and a dark mineral such as augite. { ‚sī·ə·nō'ga‚brō }

sylvanite [MINERAL] $(Au,Ag)Te_2$ A steel-gray, silver-white, or brass-yellow mineral that crystallizes in the monoclinic system and often occurs in implanted crystals. Also known as goldschmidtite; graphic tellurium; white tellurium; yellow tellurium. { 'sil·və‚nīt }

sylvine See sylvite. { 'sil‚vīn }

sylvite [MINERAL] KCl A salty-tasting, white or colorless isometric mineral, occurring in cubes or crystalline masses or as a saline residue; the chief ore of potassium. Also known as leopoldite; sylvine. { 'sil‚vīt }

symmetrical fold [GEOL] A fold whose limbs have approximately the same angle of dip relative to the axial surface. Also known as normal fold. { sə'me·trə·kəl 'fōld }

symmetric ripple mark [GEOL] A ripple mark whose cross-section profile is symmetric. { sə'me·trik 'rip·əl ‚märk }

Symmetrodonta [PALEON] An order of the extinct mammalian infraclass Pantotheria distinguished by the central high cusp, flanked by two smaller cusps and several low minor cusps, on the upper and lower molars. { ‚sim·ə·trə'dänt·ə }

symmict [GEOL] Referring to a sedimentation unit that is structureless and in which coarse- and fine-grained particles are mixed more extensively in the lower part. { 'sim·ikt }

symmictite [PETR] An eruptive breccia that is homogenized and is made up of a mixture of country rock and intrusive rock. { sə'mik‚tīt }

symmicton See diamicton. { sə'mikt·ən }

symplectite See symplektite. { sim'plek‚tīt }

symplektite [MINERAL] An intimate intergrowth of two different minerals. Also spelled symplectite. { sim'plek‚tīt }

symplesite [MINERAL] $Fe_2(AsO_4)_3 \cdot 8H_2O$ A blue to bluish-green triclinic mineral composed of hydrous iron arsenate. { 'sim·plə‚sīt }

synadelphite [MINERAL] $(Mn,Mg,Ca,Pb)(AsO_4)(OH)_5$ A black mineral composed of basic arsenate of manganese, often with magnesium, calcium, lead, or other metals. { ‚sin·ə'del‚fīt }

synantectic [MINERAL] Refers to a mineral that was formed by the reaction of two other minerals. { ‚sin·ən‚tek·tik }

synantexis [GEOL] Deuteric alteration. { ¦sin·ən¦tek·səs }

synchisite See synchysite. { 'siŋ·kə‚sīt }

synchronous [GEOL] Geological rock units or features formed at the same time. { 'siŋ· krə·nəs }

synchronous pluton [GEOL] Any pluton whose time of emplacement coincides with a major orogeny. { 'siŋ·krə·nəs 'plü‚tän }

synchysite [MINERAL] (Ce,La)Ca(CO₃)₂F A mineral composed of fluoride and carbonate of calcium, cerium, and lanthanum. Also spelled synchisite. { 'siŋ·kə‚sīt }

synclinal axis See trough surface. { sin'klīn·əl 'ak·səs }

synclinal valley [GEOL] Pertaining to a topographic valley whose sides coincide with a synclinal fold. { sin'klīn·əl 'val·ē }

syncline [GEOL] A fold having stratigraphically younger rock material in its core; it is concave upward. { 'sin‚klīn }

synclinorium [GEOL] A composite synclinal structure in a region of lesser folds. { ‚sin· klə'nȯr·ē·əm }

syngenesis [GEOL] In place formation of unconsolidated sediments. { sin'jen·ə·səs }

syngenetic [GEOL] **1.** Pertaining to a primary sedimentary structure formed contemporaneously with sediment deposition. **2.** Pertaining to a mineral deposit formed contemporaneously with the enclosing rock. Also known as ideogenous. { ¦sin· jə¦ned·ik }

syngenite [MINERAL] K₂Ca(SO₄)₂·H₂O A colorless or white mineral composed of hydrous potassium calcium sulfate occurring in tabular crystals. { 'sin·jə‚nīt }

synkinematic See syntectonic. { ¦sin‚kin·ə'mad·ik }

synorogenic [GEOL] Referring to a geologic process occurring at the same time as orogenic activity. { ¦sin‚ȯr·ə'jen·ik }

syntaxial overgrowth [MINERAL] A crystallographically oriented overgrowth of two alternating, chemically identical substances. { sin'tak·sē·əl 'ō·vər‚grōth }

syntectic [GEOL] See syntexis. { sin'tek·tik }

syntectonic [GEOL] Refers to a geologic process or event occurring during tectonic activity. Also known as synkinematic. { ¦sin·tek¦tän·ik }

syntexis [GEOL] Magma made by the melting of two or more rock types and the assimilation of country rock. Also known as syntectic. { sin'tek·səs }

synthem [GEOL] A chronostratigraphic unit that defines an unconformity-bounded regional body of sediments and represents a cycle of sedimentation in response to changes in relative sea level or tectonics. { 'sin‚them }

Syntrophiidina [PALEON] A suborder of extinct articulate brachiopods of the order Pentamerida characterized by a strong dorsal median fold. { sin‚träf·ē·ə'dī·nə }

Synxiphosura [PALEON] An extinct heteorgeneous order of arthropods in the subclass Merostomata possibly representing an explosive proliferation of aberrant, terminal, and apparently blind forms. { ¦sin‚zif·ə'sùr·ə }

Syringophyllidae [PALEON] A family of extinct corals in the order Tabulata. { sə‚riŋ· gō'fil·ə‚dē }

syrinx [PALEON] A tube surrounding the pedicle in certain fossil brachiopods. { 'sir·iŋks }

syserskite [MINERAL] Mineral composed of an alloy of osmium (50–80%) and iridium (20–50%). { 'sis·ər‚skīt }

syssiderite See stony-iron meteorite. { sə'sid·ə‚rīt }

system [GEOL] **1.** A major time-stratigraphic unit of worldwide significance, representing the basic unit of Phanerozic rocks. **2.** A group of related structures, such as joints. **3.** A chronostratigraphic unit, below erathem and above series. { 'sis·təm }

systematic joints [GEOL] Joints occurring in patterns or sets and oriented perpendicular to the boundaries of the constituent rock unit. { ‚sis·tə'mad·ik 'jȯins }

systems tract [GEOL] A discrete package of distinctive sediment types (facies) that are laid down during different phases of a cycle of sea-level change. { 'sis·təmz ‚trakt }

szaibelyite [MINERAL] (Mn,Mg)(BO₂)(OH) A white to buff or straw yellow, orthorhombic mineral consisting of a basic borate of manganese and magnesium; occurs as veinlets, masses, or embedded nodules. { sā'bel‚yīt }

szaskaite *See* smithsonite. { sə'skā,īt }

szmikite [MINERAL] $MnSO_4 \cdot H_2O$ A monoclinic mineral composed of hydrous manganese sulfate. { 'smi,kīt }

szomolnokite [MINERAL] $FeSO_4 \cdot H_2O$ A yellow or brown monoclinic mineral composed of hydrous ferrous sulfate. { sə'mäl·nə,kīt }

T

tabbyite [MINERAL] A variety of solid asphalt found in the western United States; used as rubber filler and with roofing materials. { 'ta·bē,īt }

tabetisol *See* talik. { tə'bed·ə,sȯl }

tablemount *See* guyot. { 'tā·bəl,maůnt }

table reef [GEOL] A small, isolated organic reef which has a flat top and does not enclose a lagoon. { 'tā·bəl ,rēf }

tabula [PALEON] A transverse septum that closes off the lower part of the polyp cavity in certain extinct corals and hydroids. { 'tab·yə·lə }

tabular [GEOL] Referring to a sedimentary particle whose length is two to three times its thickness. { 'tab·yə·lər }

tabular spar *See* wollastonite. { 'tab·yə·lər 'spär }

Tabulata [PALEON] An extinct Paleozoic order of corals of the subclass Zoantharia characterized by an exclusively colonial mode of growth and by secretion of a calcareous exoskeleton of slender tubes. { ,tab·yə'läd·ə }

tachyhydrite [MINERAL] $CaMg_2Cl_6·12H_2O$ A honey yellow, hexagonal mineral consisting of a hydrated chloride of calcium and magnesium; occurs in massive form. { ,tak·ə'hī,drīt }

tachylite [GEOL] A black, green, or brown volcanic glass formed from basaltic magma. Also known as basalt glass; basalt obsidian; hyalobasalt; jaspoid; sordawalite; wichtisite. { 'tak·ə,līt }

Taconian orogeny [GEOL] A process of formation of mountains in the latter part of the Ordovician period, particularly in the northern Appalachians. Also known as Taconic orogeny. { tə'kō·nē·ən ȯ'räj·ə·nē }

Taconic orogeny *See* Taconian orogeny. { tə'kän·ik ȯ'räj·ə·nē }

taconite [GEOL] The siliceous iron formation from which high-grade iron ores of the Lake Superior district have been derived; consists chiefly of fine-grained silica mixed with magnetite and hematite. { 'tak·ə,nīt }

tactite [PETR] A rock with a complex mineralogical composition, formed by contact metamorphism and metasomatism of carbonate rocks. { 'tak,tīt }

taele *See* frozen ground. { 'tā·lə }

Taeniodonta [PALEON] An order of extinct quadrupedal land mammals, known from early Cenozoic deposits in North America. { ,tē·nē·ə'dänt·ə }

Taeniolabidoidea [PALEON] An advanced suborder of the extinct mammalian order Multituberculata having incisors that were self-sharpening in a limited way. { ,tē·nē·ō,lab·ə'dȯid·ē·ə }

taeniolite [MINERAL] $KLiMg_2Si_4O_{10}F_2$ A white or colorless mica mineral. { 'tē·nē·ə,līt }

taenite [MINERAL] A meteoritic mineral consisting of a nickel-iron alloy, with a nickel content varying from about 27 to 65%. { 'tē,nīt }

tagilite *See* pseudomalachite. { 'tag·ə,līt }

Tahuian [GEOL] A local Eocene time subdivision in Australia whose identification is based on foraminiferans. { tə'wī·ən }

talc [MINERAL] $Mg_3Si_4O_{10}(OH)_2$ A whitish, greenish, or grayish hydrated magnesium silicate mineral crystallizing in the monoclinic system; it is extremely soft (hardness is 1 on Mohs scale) and has a characteristic soapy or greasy feel. { talk }

talcose rock

talcose rock |PETR| A rock having a soft and soapy feel, that is, resembling talc. { 'tal,kōs ,räk }

talc schist |PETR| A schist in which talc is the dominant schistose material. { 'talk ¦shist }

talik |GEOL| A Russian term applied to permanently unfrozen ground in regions of permafrost; usually applies to a layer which lies above the permafrost but below the active layer, that is, when the permafrost table is deeper than the depth reached by winter freezing from the surface. Also known as tabetisol. { 'tä·lik }

talus |GEOL| Also known as rubble; scree. **1.** Coarse and angular rock fragments derived from and accumulated at the base of a cliff or steep, rocky slope. **2.** The accumulated heap of such fragments. { 'tal·əs }

talus creep |GEOL| The slow, downslope movement of talus. { 'tal·əs ,krēp }

talus glacier See rock glacier. { 'tal·əs ,glā·shər }

talus slope |GEOL| A steep, concave slope consisting of an accumulation of talus. Also known as debris slope. { 'tal·əs ,slōp }

tamarugite |MINERAL| NaAl(SO₄)₂·6H₂O A colorless, monoclinic mineral consisting of a hydrated sulfate of sodium and aluminum; occurs as crystals and masses. { ,tam·ə'rü,gīt }

tangeite See calciovolborthite. { 'tan·jē,īt }

tangential wave See S wave. { tan'jen·chəl 'wāv }

tantalite |MINERAL| (Fe,Mn)Ta₂O₆ An iron-black mineral that crystallizes in the orthorhombic system and commonly occurs in short prismatic crystals; luster is submetallic, hardness is 6 on Mohs scale, and specific gravity is 7.95; principal ore of tantalum. { 'tant·əl,īt }

tanteuxenite |MINERAL| (Y,Ce,Ca)(Ta,Nb,O)₂(O,OH)₆ A brown or black variety of euxenite with tantalum substituting for niobium. Also known as delorenzite; eschwegeite. { tan'tyük·sə,nīt }

taphocoenosis See thanatocoenosis. { ¦taf·ō·sē'nō·səs }

taphonomy |PALEON| The study of fossil preservation, including all events during the transition of organisms from the biosphere to the lithosphere. { tə'fän·ə·mē }

taphrogenesis See taphrogeny. { ,taf·rə'jen·ə·səs }

taphrogeny |GEOL| The formation of rift or trench phenomena, characterized by block faulting and associated subsidence. Also known as taphrogenesis. { tə'fräj·ə·nē }

taphrogeosyncline |GEOL| A geosyncline formed as a rift basin between faults. { ¦taf·rō,jē·ō'sin,klīn }

tapiolite |MINERAL| Fe(Ta,Nb)₂O₆ A mineral that is isomorphous with mossite; occurs in pegmatites or detrital deposits; an ore of tantalum. { 'tap·ē·ə,līt }

taranakite |MINERAL| KAl₃(PO₄)₃(OH)·9H₂O A white, gray, or yellowish-white mineral consisting of a hydrated basic phosphate of potassium and aluminum. { ,tar·ə'nä,kīt }

tarapacaite |MINERAL| K₂CrO₄ A bright canary yellow, orthorhombic mineral consisting of potassium chromate; occurs in tabular form. { ,tar·ə·pə'kä,īt }

tarbuttite |MINERAL| Zn₂(PO₄)(OH) A triclinic mineral of varying color, consisting of basic zinc phosphate. { 'tär·bə,tīt }

tarnish |MINERAL| The altered color and luster of a mineral surface; characteristic of copper-bearing minerals. { 'tär·nish }

tar sand |GEOL| A type of oil sand; a sand whose interstices are filled with asphalt that remained after the escape of the lighter fractions of crude oil. { 'tär ,sand }

tar seep |GEOL| Natural tar that, because of its close proximity to the ground surface, seeps from cracks in the earth or from between rocks, often forming pits or pools. { 'tär ,sēp }

tasmanite |GEOL| An impure coal, transitional between cannel coal and oil shale. Also known as combustible shale; Mersey yellow coal; white coal; yellow coal. { 'taz·mə,nīt }

tavistockite |MINERAL| Ca₃Al₂(PO₄)₂(OH)₆ A white, orthorhombic mineral consisting of a basic phosphate of calcium and aluminum. { 'tav·ə,stä,kīt }

Taxocrinida [PALEON] An order of flexible crinoids distributed from Ordovician to Mississippian. { ‚tak·sə'krı·nəd·ə }

taylorite See bentonite. { 'tā·lə‚rīt }

Tchernozem See Chernozem. { 'chər·nə‚zem }

teallite [MINERAL] $PbSnS_2$ A grayish-black, orthorhombic mineral consisting of lead tin sulfide. { 'tē‚līt }

tear fault [GEOL] A very steep to vertical fault associated with and perpendicular to the strike of an overthrust fault. { 'tar ‚fȯlt }

tectite See tektite. { 'tek‚tīt }

tectofacies [GEOL] A lithofacies that is interpreted tectonically. { ‚tek·tə'fā·shēz }

tectogene [GEOL] A long, relatively narrow downward fold of sialic crust considered to be an early phase in mountain-building processes. Also known as geotectogene. { 'tek·tə‚jēn }

tectogenesis See orogeny. { ‚tek·tə'jen·ə·səs }

tectonic analysis See petrotectonics. { tek'tän·ik ə'nal·ə·səs }

tectonic breccia [PETR] A breccia developed from brittle rocks, formed as a result of crustal movements and produced by lateral or vertical pressure. Also known as dynamic breccia; pressure breccia. { tek'tän·ik 'brech·ə }

tectonic conglomerate See crush conglomerate. { tek'tän·ik kən'gläm·ə·rət }

tectonic cycle [GEOL] The orogenic cycle which relates larger crustal features, such as mountain belts, to a series of stages of development. Also known as geosynclinal cycle. { tek'tän·ik 'sī·kəl }

tectonic framework [GEOL] The relationship in space and time of subsiding, stable, and rising tectonic elements in a sedimentary source area. { tek'tän·ik 'frām‚wərk }

tectonic land [GEOL] Linear fold ridges and volcanic islands which existed for a short time in the interior sections of an orogenic belt during the geosynclinal phase. { tek'tän·ik 'land }

tectonic lens [GEOL] An elongate, sausage-shaped body of rock formed by distortion of a continuous incompetent layer enclosed between competent layers, similar to a boudin, but genetically distinct. { tek'tän·ik 'lenz }

tectonic map [GEOL] A map which shows the architecture of the upper portion of the earth's crust. { tek'tän·ik 'map }

tectonic moraine [GEOL] An aggregation of boulders incorporated in the base of an overthrust mass. { tek'tän·ik mə'rān }

tectonic patterns [GEOL] The arrangement of the large structural units of the earth's crust, such as mountain systems, shields or stable areas, basins, arches, and volcanic archipelagoes. { tek'tän·ik 'pad·ərnz }

tectonic plate [GEOL] Any one of the internally rigid crustal blocks of the lithosphere which move horizontally across the earth's surface relative to one another. Also known as crustal plate. { tek'tän·ik 'plāt }

tectonic rotation [GEOL] Internal rotation of a tectonite in the direction of transport. { tek'tän·ik rō'tā·shən }

tectonics [GEOL] A branch of geology that deals with regional structural and deformational features of the earth's crust, including the mutual relations, origin, and historical evolution of the features. Also known as geotectonics. { tek'tän·iks }

tectonite [PETR] A rock in which the history of its deformation is reflected in its fabric. { 'tek·tə‚nīt }

tectonomagnetism [GEOPHYS] Study of magnetic anomalies due to tectonic stress. { ‚tek·tə·nō'mag·nə‚tiz·əm }

tectonophysicist [GEOPHYS] One who studies elastic deformation of flow and rupture of constituent materials of the earth's crust and makes deductions concerning the forces that cause these deformations. { ‚tek·tə·nō'fiz·ə‚sist }

tectonophysics [GEOPHYS] A branch of geophysics dealing with the physical processes involved in forming geological structures. { ‚tek·tə·nō'fiz·iks }

tectosilicate [MINERAL] A structural type of silicate in which all four oxygen atoms of the silicate tetrahedra are shared with neighboring tetrahedra; tectosilicates include quartz, the feldspars, the feldspathoids, and zeolites. Also known as framework silicate. { ¦tek·tō'sil·ə,kāt }

tectosome [GEOL] A body of strata representing a tectotope. { 'tek·tə,sōm }

tectosphere [GEOL] The region of the earth's crust occupied by the tectonic plates. { 'tek·tə,sfir }

teepleite [MINERAL] $Na_2BO_2Cl \cdot 2H_2O$ A mineral composed of hydrous chloride and borate of sodium. { 'tē·pə,līt }

teineite [MINERAL] $CuTeO_3 \cdot 2H_2$ A greenish to yellowish, probably triclinic mineral consisting of a hydrated sulfate-tellurate of copper; occurs as crystals. { 'tā,nīt }

tektite [GEOL] A collective term applied to certain objects of natural glass of debatable origin that are widely strewn over the land and in sediments under the oceans; composition and size vary, and overall shapes resemble splash forms; most tektites are believed to be of extraterrestrial origin. Also known as obsidianite; tectite. { 'tek,tīt }

telain See provitrain. { 'te,lān }

telemagmatic [GEOL] Pertaining to a hydrothermal mineral deposit that is distant from its magmatic source. { ¦tel·ə·mag'mad·ik }

Teleosauridae [PALEON] A family of Jurassic reptiles in the order Crocodilia characterized by a long snout and heavy armor. { ¦tel·ē·ə'sȯr·ə,dē }

telescope structure [GEOL] An alluvial fan structure characterized by younger fans with flatter gradients spreading out between older fans with steeper gradients. { 'tel·ə,skōp ,strək·chər }

teleseism [GEOPHYS] An earthquake that is far from the recording station. { 'tel·ə,sīz·əm }

teleseismology [GEOPHYS] The aspect of seismology dealing with records made at a distance from the source of the impulse. { ,tel·ə·sīz'mäl·ə·jē }

telethermal [GEOL] Pertaining to a hydrothermal mineral deposit precipitated at a shallow depth and at a mild temperature. { ¦tel·ə'thər·məl }

telinite [GEOL] A variety of provitrinite composed of plant cell-wall material. { 'tē·lə,nīt }

telluric current See earth current. { tə'lůr·ik ,kə·rənt }

tellurics [GEOPHYS] A geophysical exploration technique that measures variations in the conductivity (or resistivity) of rocks; often used for metallic mineral prospecting. { tə'lůr·iks }

tellurite [MINERAL] TeO_2 A white or yellowish orthorhombic mineral consisting of tellurium dioxide, and occurring in crystals; it is dimorphous with paratellurite. { 'tel·yə,rīt }

tellurium glance See nagyagite. { tə'lůr·ē·əm 'glans }

tellurobismuthite [MINERAL] Bi_2Te_3 A pale lead gray, hexagonal mineral consisting of a bismuth and tellurium compound; occurs as irregular plates or foliated masses. { ¦tel·yə·rō'biz·mə,thīt }

Temnospondyli [PALEON] An order of extinct amphibians in the subclass Labyrinthodontia having vertebrae with reduced pleurocentra and large intercentra. { ,temnō'spän·də,lī }

temporal unit [GEOL] A stratigraphic unit defined in terms of time-related characteristics. { 'tem·prəl ,yü·nət }

temporary base level [GEOL] Any base level, other than sea level, below which a land area temporarily cannot be reduced by erosion. Also known as local base level. { 'tem·pə,rer·ē 'bās ,lev·əl }

tennantite [MINERAL] $(Cu,Fe)_{12}As_4S_{13}$ A lead-gray mineral crystallizing in the isometric system; it is isomorphous with tetrahedrite; an important ore of copper. { 'ten·ən,tīt }

tenorite [MINERAL] CuO A triclinic mineral that occurs in small, shining, steel-gray scales, in black powder, or in black earthy masses; an ore of copper. { 'ten·ə,rīt }

tension crack [GEOL] An extension fracture caused by tensile stress. { 'ten·chən ,krak }

tension fault [GEOL] A fault in which crustal tension is a factor, such as a normal fault. Also known as extensional fault. { 'ten·chən ,folt }

tension fracture [GEOL] A minor rock fracture developed at right angles to the direction of maximum tension. Also known as subsidiary fracture. { 'ten·chən ,frak·chər }

tension joint [GEOL] A joint that is a tension fracture. { 'ten·chən ,joint }

tepee structure [GEOL] A disharmonic sedimentary structure consisting of a fold that resembles an inverted depressed V in cross section. { 'te·pe ,strək·chər }

tepetate See caliche. { ,tep·ə'täd·e }

tephra [GEOL] All pyroclastics of a volcano. { 'tef·rə }

tephrite [PETR] A group of basaltic extrusive rocks composed chiefly of calcic plagioclase, augite, and nepheline or leucite, with some sodic sanidine. { 'te,frīt }

tephrochronology [GEOL] The dating of different layers of volcanic ash for the establishment of a sequence of geologic and archeologic occurrences. { ,tef·rō·krə'näl·ə·jē }

tephroite [MINERAL] Mn_2SiO_4 An olivine mineral that occurs with zinc and manganese minerals. { 'tef·rō,īt }

tephrostratigraphy [GEOL] The use of pyroclastic layers, in particular volcanic ash, as a correlational tool in the study of stratigraphic sequences. { ,tef·rō·strə'tig·rə·fē }

Teratornithidae [PALEON] An extinct family of vulturelike birds of the Pleistocene of western North America included in the order Falconiformes. { ,ter·ə·tòr'nith·ə,dē }

Terebratellidina [PALEON] An extinct suborder of articulate brachiopods in the order Terebratulida in which the loop is long and offers substantial support to the side arms of the lophophore. { ,ter·ə·brə,tel·ə'dīn·ə }

terlinguaite [MINERAL] Hg_2OCl A sulfur yellow to greenish-yellow, monoclinic mineral consisting of an oxychloride of mercury. { tər'liŋ·gwə,īt }

terminal moraine [GEOL] An end moraine that extends as an arcuate or crescentic ridge across a glacial valley; marks the farthest advance of a glacier. Also known as marginal moraine. { 'tər·mən·əl mə'rān }

ternary diagram [PETR] A triangular diagram that graphically depicts the composition of a three-component mixture or ternary system. { 'tər·nə·rē 'dī·ə,gram }

Ternifine man [PALEON] The name for a fossil human type, represented by three lower jaws and a parietal bone discovered in France and thought to be from the upper part of the middle Pleistocene. { 'tər·nə,fēn 'man }

terrace [GEOL] **1.** A horizontal or gently sloping embankment of earth along the contours of a slope to reduce erosion, control runoff, or conserve moisture. **2.** A narrow coastal strip sloping gently toward the water. **3.** A long, narrow, nearly level surface bounded by a steeper descending slope on one side and by a steeper ascending slope on the other side. **4.** A benchlike structure bordering an undersea feature. { 'ter·əs }

terracette [GEOL] A small steplike form developed on the surface of a slumped soil mass along a steep grassy incline. { 'ter·ə,set }

terra miraculosa See bole. { 'ter·ə mi,rak·yə'lō·sə }

terrane [GEOL] A rock formation, a cluster of rock formations, or the general area of outcrops. { tə'rān }

terra rossa [GEOL] A reddish-brown soil overlying limestone bedrock. { 'ter·ə 'ròs·ə }

terrestrial electricity [GEOPHYS] Electric phenomena and properties of the earth; used in a broad sense to include atmospheric electricity. Also known as geoelectricity. { tə'res·trē·əl ,i,lek'tris·əd·ē }

terrestrial gravitation [GEOPHYS] The effect of gravitational attraction of the earth. { tə'res·trē·əl ,grav·ə'tā·shən }

terrestrial magnetism See geomagnetism. { tə'res·trē·əl 'mag·nə,tiz·əm }

terrestrial radiation [GEOPHYS] Electromagnetic radiation originating from the earth and its atmosphere at wavelengths determined by their temperature. Also known as earth radiation; eradiation. { tə'res·trē·əl ,rād·ē'ā·shən }

terrestrial sediment [GEOL] A sedimentary deposit on land above tidal reach. { tə'res·trē·əl 'sed·ə·mənt }

371

terrigenous sediment

terrigenous sediment [GEOL] Shallow marine sedimentary deposits composed of eroded terrestrial material. { tə'rij·ə·nəs 'sed·ə·mənt }

Tertiary [GEOL] The older major subdivision (period) of the Cenozoic era, extending from the end of the Cretaceous to the beginning of the Quaternary, from 70,000,000 to 2,000,000 years ago. { 'tər·shē,er·ē }

teschemacherite [MINERAL] $(NH_4)HCO_3$ A colorless to white or yellowish, orthorhombic mineral consisting of ammonium bicarbonate; occurs as compact, crystalline masses. { 'tesh·ə,mäk·ə,rīt }

teschenite [PETR] A granular hypabyssal rock composed principally of calcic plagioclase, augite, and sometimes hornblende, with some brotite and analcime. { 'tesh·ə,nīt }

Tethys [GEOL] **1.** A sea which existed for extensive periods of geologic time between the northern and southern continents of the Eastern Hemisphere. **2.** A composite geosyncline from which many structures of the present Alpine-Himalayan orogenic belt were formed. { 'tē·thəs }

Tetracorallia [PALEON] The equivalent name for Rugosa. { ,te·trə·kə'ral·yə }

tetradymite [MINERAL] Bi_2Te_2S A pale steel-gray mineral that usually occurs in foliated masses in auriferous veins; has metallic luster, hardness of 1.5–2 on Mohs scale, and specific gravity of 7.2–7.6. { tə'trad·ə,mīt }

tetrahedrite [MINERAL] $(Cu,Fe,Zn,Ag)_{12}Sb_4S_{13}$ A grayish-black mineral crystallizing in the isometric system as tetrahedrons and occurring in massive or granular form; luster is metallic, hardness is 3.5–4 on Mohs scale, and specific gravity is 4.6–5.1; an important ore of copper. Also known as fahlore; gray copper ore; panabase; stylotypite. { ,te·trə'hē,drīt }

Tetralophodontinae [PALEON] An extinct subfamily of proboscidean mammals in the family Gomphotheridae. { ¦te·trə,läf·ə'dänt·ə,dē }

texture [GEOL] The physical nature of the soil according to composition and particle size. [PETR] The physical appearance or character of a rock; applied to the megascopic or microscopic surface features of a homogeneous rock or mineral aggregate, such as grain size, shape, and arrangement. { 'teks·chər }

thalassocratic [GEOL] **1.** Pertaining to a thalassocraton. **2.** Referring to a period of high sea level in the geologic past. { thə¦las·ə¦krad·ik }

thalassocraton [GEOL] A craton that is part of the oceanic crust. { ,thal·ə'säk·rə,tän }

thalassophile element [GEOCHEM] An element that is relatively more abundant in sea water than in normal continental waters, such as sodium and chlorine. { thə'las·ə,fīl ,el·ə·mənt }

Thalattosauria [PALEON] A suborder of extinct reptiles in the order Eosuchia from the Middle Triassic. { thə,lad·ə'sȯr·ē·ə }

thalweg [GEOL] **1.** A line connecting the lowest points along a stream bed or a valley. Also known as valley line. **2.** A line crossing all contour lines on a land surface perpendicularly. { 'täl,veg }

thanatocoenosis [PALEON] The assemblage of dead organisms or fossils that occurred together in a given area at a given moment of geologic time. Also known as death assemblage; taphocoenosis. { ,than·ə·tō·sə'nō·səs }

Thanetian [GEOL] A European stage of geologic time; uppermost Paleocene, above Montian, below Ypresian of Eocene. { thə'nē·shən }

Thecideidina [PALEON] An extinct suborder of articulate brachiopods doubtfully included in the order Terebratulida. { thə,sid·ē·ə'dīn·ə }

Thecodontia [PALEON] An order of archosaurian reptiles, confined to the Triassic and distinguished by the absence of a supratemporal bone, parietal foramen, and palatal teeth, and by the presence of an antorbital fenestra. { ,thek·ə'dän·chə }

thenardite [MINERAL] Na_2SO_4 A colorless, grayish-white, yellowish, yellow-brown, or reddish, orthorhombic mineral consisting of sodium sulfate. { thə'när,dīt }

theralite [PETR] A dark-colored, visibly crystalline rock composed chiefly of pyroxene with smaller amounts of calcic plagioclase and nepheline. { 'ther·ə,līt }

Therapsida [PALEON] An order of mammallike reptiles of the subclass Synapsida which

372

tholeiite

first appeared in mid-Permian times and persisted until the end of the Triassic. { thə'rap·səd·ə }

thermal aureole *See* aureole. { 'thər·məl 'ȯr·ē,ōl }

thermal gradient [GEOPHYS] The rate of temperature change with distance; for example, its increase with depth below the surface of the earth. { 'thər·məl 'grād·ē·ənt }

thermal metamorphism [PETR] Metamorphism that results from temperature-controlled and induced chemical reconstitution of preexisting rocks, with little influence of pressure. Also known as thermometamorphism. { 'thər·məl ¦med·ə¦mȯr,fiz·əm }

thermal structure [PETR] A distinct structural pattern, such as a dome or anticline, defined by the arrangement of metamorphic zones of increasing grade. { 'thər·məl 'strək·chər }

thermocline [GEOPHYS] **1.** A temperature gradient as in a layer of sea water, in which the temperature decrease with depth is greater than that of the overlying and underlying water. Also known as metalimnion. **2.** A layer in a thermally stratified body of water in which such a gradient occurs. { 'thər·mə,klīn }

thermokarst topography [GEOL] An irregular land surface formed in a permafrost region by melting ground ice. { 'thər·mə,kärst tə'päg·rə·fē }

thermometamorphism *See* thermal metamorphism. { ¦thər·mō¦med·ə¦mȯr,fiz·əm }

thermonatrite [MINERAL] $Na_2CO_3 \cdot H_2O$ A colorless to white, grayish, or yellowish, orthorhombic mineral consisting of sodium carbonate monohydrate; occurs as a crust or efflorescence. { ,thər·mə'nā,trīt }

thermoremanent magnetization [GEOPHYS] The permanent magnetization of igneous rocks, acquired at the time of cooling from the molten state. { ¦thər·mō'rem·ə·nənt ,mag·nəd·ə'zā·shən }

Theropoda [PALEON] A suborder of carnivorous bipedal saurischian reptiles which first appeared in the Upper Triassic and culminated in the uppermost Cretaceous. { thi'räp·əd·ə }

Theropsida [PALEON] An order of extinct mammallike reptiles in the subclass Synapsida. { thi'räp·səd·ə }

thick-bedded [GEOL] Pertaining to a sedimentary bed that ranges in thickness from 60 to 120 centimeters (2–4 feet). { 'thik ,bed·əd }

thick-skinned structure [GEOL] Any large-scale structure, such as a fold or fault, believed to have originated as a result of basement movement beneath overlying rocks. { 'thik ¦skind 'strək·chər }

thill *See* underclay. { thil }

thin-bedded [GEOL] Pertaining to a sedimentary bed that ranges in thickness from 2 inches to 2 feet (5 to 60 centimeters). { 'thin ,bed·əd }

thin-out [GEOL] Gradual thinning of a stratum, vein, or other body of rock until the upper and lower surfaces meet and the rock disappears. { 'thin'aút }

thin section [GEOL] A piece of rock or mineral specifically prepared to study its optical properties; the sample is ground to 0.03-millimeter thickness, then polished and placed between two microscope slides. Also known as section. { 'thin 'sek·shən }

thin-skinned structure [GEOL] Any large-scale structure, such as a fold or fault, confined to and originating within a thin layer of rocks above a surface of décollement. { 'thin ¦skind 'strək·chər }

thiophile element *See* sulfophile element. { 'thī·ə,fīl ,el·ə·mənt }

thiospinel [MINERAL] Any mineral with the spinel structure having the general formula AR_2S_4. { ¦thī·ō·spə'nel }

thixotropic clay [GEOL] A clay that weakens when disturbed and increases in strength upon standing. { ¦thik·sə¦träp·ik 'klā }

Thlipsuridae [PALEON] A Paleozoic family of ostracod crustaceans in the suborder Platycopa. { thlip'sùr·ə,dē }

tholeiite [PETR] **1.** A group of basalts composed principally of plagioclase, pyroxene, and iron oxide minerals as phenocrysts in a glassy groundmass. **2.** Any rock in the group. { thō'lē·ə,īt }

373

thomsenolite [MINERAL] $NaCaAlF_6 \cdot H_2O$ A colorless to white, monoclinic mineral consisting of a hydrated aluminofluoride of sodium and calcium; it is dimorphous with pachnolite. { 'täm·sə·nə,līt }

thomsonite [MINERAL] $NaCa_2Al_5Si_5O_{20} \cdot 6H_2O$ Snow-white zeolite mineral forming orthorhombic crystals and occurring in masses of radiating crystals; hardness is 5–5.5 on Mohs scale. { 'täm·sə,nīt }

thoreaulite [MINERAL] $SnTa_2O_7$ A brown, monoclinic mineral consisting of an oxide of tin and tantalum; occurs as rough, prismatic crystals. { 'thȯ·rō,līt }

thorianite [MINERAL] ThO_2 A radioactive mineral that crystallizes in the isometric system, occurs in worn cubic crystals, is brownish black to reddish brown in color, and has resinous luster; hardness is 7 on the Mohs scale, and specific gravity is 9.7–9.8. { 'thȯr·ē·ə,nīt }

thorite [MINERAL] $ThSiO_4$ A brownish-yellow to brownish-black and black radioactive mineral that is tetragonal in crystallization; hardness is about 4.5 on Mohs scale, and specific gravity is 4.3–5.4. { 'thȯr,īt }

thorogummite [MINERAL] A silicate mineral and chemical variant of thorium silicate, with similar properties; isostructural with thorite and zircon; it is deficient in silica and contains small amounts of OH in substitution for oxygen. { ,thȯr·ə'gə,mīt }

thortveitite [MINERAL] $(Sc,Y)_2Si_2O_7$ A grayish-green mineral occurring in orthorhombic crystals; a source of scandium. { tȯrt'vī,tīt }

thread [GEOL] An extremely small vein, even thinner than a stringer. { thred }

thread-lace scoria [GEOL] Scoria whose vesicle walls have collapsed and are represented only by a network of threads. { 'thred ¦lās 'skȯr·ē·ə }

three-point method [GEOL] A method used to determine the dip and strike of a structural surface from three points of varying elevation along the surface. { 'thrē ¦pȯint 'meth·əd }

threshold [GEOL] See riegel. { 'thresh,hōld }

threshold velocity [GEOPHYS] The minimum velocity at which wind or water begins to move particles of soil, sand, or other material at a given place under specified conditions. { 'thresh,hōld və,läs·əd·ē }

through valley [GEOL] **1.** A depression eroded across a divide by glacier ice or meltwater streams. **2.** A valley excavated by a through glacier. { 'thrü ,val·ē }

throw [GEOL] The vertical component of dip separation on a fault, or generally the amount of vertical displacement on any fault. { thrō }

thrust [GEOL] Overriding movement of one crystal unit over another. Also known as mountain thrust. { thrəst }

thrust block See thrust nappe. { 'thrəst ,bläk }

thrust fault [GEOL] A low-angle (less than a 45° dip) fault along which the hanging wall has moved up relative to the footwall. Also known as reverse fault; reverse slip fault; thrust slip fault. { 'thrəst ,fȯlt }

thrust moraine See push moraine. { 'thrəst mə·rān }

thrust nappe [GEOL] The body of rock that makes up the hanging wall of a thrust fault. Also known as thrust block; thrust plate; thrust sheet; thrust slice. { 'thrəst ,nap }

thrust plate See thrust nappe. { 'thrəst ,plāt }

thrust sheet See thrust nappe. { 'thrəst ,shēt }

thrust slice See thrust nappe. { 'thrəst ,slīs }

thrust slip fault See thrust fault. { 'thrəst 'slip ,fȯlt }

thucolite [GEOL] Concentrations of carbonaceous matter in ancient sedimentary rocks. { 'thü·kə,līt }

thulite [MINERAL] A pink, rose-red, or purplish-red variety of epidote that contains manganese; used as an ornamental stone. { 'thü,līt }

Thuringian [GEOL] A European stage of Upper Permian geologic time, above the Saxonian and below the Triassic. { thə'rin·jē·ən }

Thylacoleonidae [PALEON] An extinct family of carnivorous marsupials in the superfamily Phalangeroidea. { ,thī·lə,kō·lē'än·ə,dē }

tidal delta [GEOL] A sand bar or shoal formed in the entrance of an inlet by the action of reversing tidal currents. { 'tīd·əl 'del·tə }

---

tin pyrites

tidal flat [GEOL] A marshy, sandy, or muddy nearly horizontal coastal flatland which is alternately covered and exposed as the tide rises and falls. { 'tīd·əl 'flat }

tidal inlet [GEOL] A natural inlet maintained by tidal currents. { 'tīd·əl 'in·lət }

tidalite [GEOL] Any sediment transported and deposited by tidal currents. { 'tīd·əl‚īt }

tidal scour [GEOL] Sea-floor erosion caused by strong tidal currents, resulting in removal of inshore sediments and formation of deep holes and channels. Also known as scour. { 'tīd·əl 'skaúr }

tie bar [GEOL] See tombolo. { 'tī ‚bär }

tiemannite [MINERAL] HgSe A steel gray to blackish-lead gray mineral consisting of mercuric selenide; commonly occurs in massive form. { 'tē·mə‚nīt }

tiger's-eye [MINERAL] A yellowish-brown crystalline variety of quartz; a translucent, fibrous, broadly chatoyant gemstone that may be dyed other colors. { 'tī·gərz ‚ī }

tight fold See closed fold. { 'tīt 'fōld }

tight sand [GEOL] A sand whose interstices are filled with finer grains of the matrix material, thus effectively destroying porosity and permeability. Also known as close sand. { 'tīt 'sand }

tilasite [MINERAL] CaMg(AsO$_4$)F A gray, gray-violet, olive green, or apple green, monoclinic mineral consisting of a fluorarsenate of calcium and magnesium. { 'til·ə‚sīt }

till [GEOL] Unsorted and unstratified drift consisting of a heterogeneous mixture of clay, sand, gravel, and boulders which is deposited by and underneath a glacier. Also known as boulder clay; glacial till; ice-laid drift. { til }

till billow [GEOL] An undulating mass of glacial drift that is disposed in an irregular pattern with regard to the direction of movement of the ice. { 'til ‚bil·ō }

tilleyite [MINERAL] Ca$_5$(Si$_2$O$_7$)(CO$_3$)$_2$ A white mineral consisting of a carbonate and silicate of calcium. { 'til·ē‚īt }

tillite [PETR] A sedimentary rock formed by lithification of till, especially pre-Pleistocene till. { 'ti‚līt }

Tillodontia [PALEON] An order of extinct quadrupedal land mammals known from early Cenozoic deposits in the Northern Hemisphere and distinguished by large, rodentlike incisors, blunt-cuspid cheek teeth, and five clawed toes. { ‚til·ə'dän·chə }

tilloid [GEOL] A nonglacial till-like deposit. [PETR] A rock of uncertain origin which resembles tillite. { 'ti‚lóid }

till plain [GEOL] An extensive, relatively flat area overlying a till. { 'til ‚plān }

till sheet [GEOL] A sheet, layer, or bed of till. { 'til ‚shēt }

tilt block [GEOL] A tilted fault block. { 'tilt ‚bläk }

tilted interface [GEOL] Oil-water interface in which water moves in a generally linear direction under an oil accumulation which is, for instance, in an anticline. { 'til·təd 'in·tər‚fās }

tilth [GEOL] The physical condition of a soil as expressed in terms of fitness for growth of specified plants or crops. { tilth }

time correlation [GEOL] A correlation of age or mutual time relations between stratigraphic units in separated areas. { 'tīm ‚kär·ə'lā·shən }

time line [GEOL] **1.** A line that indicates equal geologic age in a correlation diagram. **2.** A rock unit represented by a time line. { 'tīm ‚līn }

time-rock unit See time-stratigraphic unit. { 'tīm 'räk ‚yü·nət }

time-stratigraphic facies [GEOL] A stratigraphic facies based on the amount of geologic time during which deposition and nondeposition of sediment occurred. { 'tīm ¦strad·ə¦graf·ik ‚fā·shēz }

time-stratigraphic unit [GEOL] A stratigraphic unit based on geologic age or time of origin. Also known as chronolith; chronolithologic unit; chronostratic unit; chronostratigraphic unit; time-rock unit. { 'tīm ¦strad·ə¦graf·ik ‚yü·nət }

time-transgressive See diachronous. { 'tīm tranz‚gres·iv }

tincal See borax. { 'tin‚kal }

tincalconite [MINERAL] Na$_2$B$_4$O$_7$·5H$_2$O A colorless to dull-white mineral, crystallizing in the rhombohedral system; one of the principal ores of borax and boron compounds. Also known as mohavite; octahedral borax. { tin'kal·kə‚nīt }

tin pyrites See stannite. { 'tin 'pī‚rīts }

375

tin stone See cassiterite. { 'tin ,stōn }

tinticite [MINERAL] $Fe_3(PO_4)_2(OH)_3 \cdot 3^1/_2 H_2O$ A creamy white mineral with a yellowish-green tint, consisting of a hydrated basic iron phosphate. { 'tin·ti,kīt }

titanaugite [MINERAL] $Ca(Mg,Fe,Ti)(Si,Al)_2O_6$ A variety of augite rich in titanium and occurring in basaltic rocks. { ¦tīt·ən'ȯ,gīt }

titanic iron ore See ilmenite. { tī'tan·ik 'ī·ərn ,ȯr }

titanite See sphene. { 'tīt·ən,īt }

Titanoideidae [PALEON] A family of extinct land mammals in the order Pantodonta. { ,tīt·ən·ȯi'dē·ə,dē }

titanothere [PALEON] Any member of the family Brontotheriidae. { tī'tan·ə,thir }

Tithonian [GEOL] Southern European equivalent of the Portlandian stage (uppermost Jurassic) of geologic time. { ti'thō·nē·ən }

tjaele See frozen ground. { 'chā·lē }

Toarcian [GEOL] A European stage of geologic time; Lower Jurassic (above Pliensbachian, below Bajocian). { tō'är·shən }

tobacco jack See wolframite. { tə'bak·ō ,jak }

todorokite [GEOL] A hydrated manganese oxide mineral containing calcium, barium, potassium, sodium, and sometimes magnesium; a major constituent of manganese nodules, which occur in large quantities ($>10^{12}$ tons) on the ocean floors. { tə'dȯr·ə,kīt }

toe [GEOL] The leading edge of a thrust nappe. { tō }

tombolo [GEOL] A sand or gravel bar or spit that connects an island with another island or an island with the mainland. Also known as connecting bar; tie bar; tying bar. { 'täm·bə,lō }

tombolo cluster See complex tombolo. { 'täm·bə,lō ,kləs·tər }

tombolo series See complex tombolo. { 'täm·bə,lō ,sir·ēz }

tonalite See quartz diorite. { 'tōn·əl,īt }

Tongrian [GEOL] A European stage of geologic time; lower Oligocene (above Ludian of Eocene, below Rupelian). Also known as Lattorfian. { 'täŋ·grē·ən }

tongue [GEOL] **1.** A minor rock-stratigraphic unit of limited geographic extent; it disappears laterally in one direction. **2.** A lava flow branching from a larger flow. { təŋ }

tonstein [GEOL] Kaolinitic bands in certain coalfields which have characteristic fossil fauna from short-lived but widespread marine invasions. { 'tän,shtīn }

tool mark [GEOL] Any of the wide variety of current marks, such as groove marks, prod marks, and skip marks, produced by the continuous contact or intermittent impact of solid, current-borne objects against a muddy bottom. { 'tül ,märk }

top See overburden. { täp }

topaz [MINERAL] $Al_2SiO_4(F,OH)$ A red, yellow, green, blue, or brown neosilicate mineral that crystallizes in the orthorhombic system and commonly occurs in prismatic crystals with pyramidal terminations; hardness is 8 on Mohs scale, and specific gravity is 3.4–3.6; used as a gemstone. { 'tō,paz }

topaz quartz See citrine. { 'tō,paz 'kwȯrtz }

topographic infancy See infancy. { ¦täp·ə¦graf·ik 'in·fən·sē }

topographic maturity See maturity. { ¦täp·ə¦graf·ik mə'chùr·əd·ē }

topographic old age See old age. { ¦täp·ə¦graf·ik 'ōld 'āj }

topographic profile See profile. { ¦täp·ə¦graf·ik 'prō,fīl }

topographic youth See youth. { ¦täp·ə¦graf·ik 'yüth }

topset bed [GEOL] One of the nearly horizontal sedimentary layers deposited on the top surface of an advancing delta. { 'täp,set ¦bed }

topsoil [GEOL] **1.** Soil presumed to be fertile and used to cover areas of special planting. **2.** Surface soil, usually corresponding with the A horizon, as distinguished from subsoil. { 'täp,sȯil }

torbanite [GEOL] A variety of coal that resembles a carbonaceous shale in outward appearance; it is fine-grained, black to brown, and tough. Also known as bituminite; kerosine shale. { 'tȯr·bə,nīt }

torbernite [MINERAL] $Cu(UO_2)_2(PO_4)_2 \cdot 8\text{-}12H_2O$ A green radioactive mineral crystallizing in the tetragonal system and occurring in tabular crystals or in foliated form. Also known as chalcolite; copper uranite; cuprouranite, uran mica. { 'tȯr·bər‚nīt }

torose load cast [GEOL] One of a group of elongate load casts with alternate contractions and swellings, which may terminate down current in bulbous, teardrop, or spiral forms. { 'tȯ‚rōs 'lōd ‚kast }

Torrert [GEOL] A suborder of the soil order Vertisol; it is the driest soil of the order and forms cracks that tend to remain open; occurs in arid regions. { 'tȯr·ərt }

torreyite [MINERAL] $(Mg,Mn,Zn)_7(SO_4)(OH)_{12}\cdot4H_2O$ A bluish-white mineral consisting of a hydrated basic sulfate of magnesium, manganese, and zinc; occurs in massive form. { 'tȯr·ē‚īt }

Torrox [GEOL] A suborder of the soil order Oxisol that is low in organic matter, well drained, and dry most of the year; believed to have been formed under rainier climates of past eras. { 'tȯr‚äks }

torsion fault See wrench fault. { 'tȯr·shən ‚fȯlt }

torso mountain See monadnock. { 'tȯr·sō ‚maùnt·ən }

Tortonian [GEOL] A European stage of geologic time: Miocene (above Helvetian, below Sarmatian). { tȯr'tō·nē·ən }

total displacement See slip. { 'tōd·əl di'splās·mənt }

total porosity [GEOL] The ratio of total void space in porous oil-reservoir rock to the bulk volume of the rock itself. { 'tōd·əl pə'räs·əd·ē }

total slip See net slip. { 'tōd·əl 'slip }

tourmaline [MINERAL] $(Na,Ca)(Al,Fe,Li,Mg)_3Al_6(BO_3)_3Si_6O_{18}(OH)_4$ Any of a group of cyclosilicate minerals with a complex chemical composition, vitreous to resinous luster, and variable color; crystallizes in the ditrigonal-pyramidal class of the hexagonal system, has piezoelectric properties, and is used as a gemstone. { 'tùr·mə‚lēn }

Tournaisian [GEOL] European stage of lowermost Carboniferous time. { tùr'nā·zhən }

Toxasteridae [PALEON] A family of Cretaceous echinoderms in the order Spatangoida which lacked fascioles and petals. { ‚täk·sə'ster·ə‚dē }

Toxodontia [PALEON] An extinct suborder of mammals representing a central stock of the order Notoungulata. { ‚täk·sə'dän·chə }

trace [GEOL] The intersection of two geological surfaces. { trās }

trace element [GEOCHEM] An element found in small quantities (usually less than 1.0%) in a mineral. Also known as accessory element; guest element. { 'trās ‚el·ə·mənt }

trace fossil [GEOL] A trail, track, or burrow made by an animal and found in ancient sediments such as sandstone, shale, or limestone. Also known as ichnofossil. { 'trās ‚fäs·əl }

trace slip [GEOL] That component of the net slip in a fault which is parallel to the trace of an index plane on a fault plane. { 'trās ‚slip }

trace-slip fault [GEOL] A fault whose net slip is trace slip. { 'trās ‚slip ‚fȯlt }

trachybasalt [PETR] An extrusive rock characterized by calcic plagioclase and sanidine, with augite, olivine, and possibly minor analcime or leucite. { ¦tra·kē·bə'sȯlt }

trachyte [PETR] The light-colored, aphanitic rock (the volcanic equivalent of syenite), composed largely of alkali feldspar with minor amounts of mafic minerals. { 'tra‚kīt }

trachytoid texture [GEOL] The texture of a phaneritic extrusive igneous rock in which the microlites of a mineral, not necessarily feldspar, in the groundmass have a subparallel or randomly divergent alignment. { 'trak·ə‚tȯid ‚teks·chər }

traction [GEOL] Transport of sedimentary particles along and parallel to a bottom surface of a stream channel by rolling, sliding, dragging, pushing, or saltation. { 'trak·shən }

trail [GEOL] A line of rock fragments that were picked up by glacial ice at a localized outcropping and left scattered along a fairly well-defined tract during the movement of a glacier. { trāl }

trajectory [GEOPHYS] The path followed by a seismic wave. { trə'jek·trē }

transcurrent fault [GEOL] A strike-slip fault characterized by a steeply inclined surface. Also known as transverse thrust. { ¦tranz¦kə·rənt 'fȯlt }

transform fault |GEOL| A strike-slip fault with offset ridges characteristic of a midoceanic ridge. { 'tranz,fȯrm ,fȯlt }

transgression |GEOL| Geologic evidence of landward extension of the sea. Also known as invasion; marine transgression. { tranz'gresh·ən }

transgressive deposit |GEOL| Sediment deposited during transgression of the sea or during subsidence of the land. { tranz'gres·iv di'päz·ət }

transgressive overlap *See* onlap.

transition zone |GEOL| **1.** A region within the upper mantle bordering the lower mantle, at a depth of 246–600 miles (410–1000 kilometers), characterized by a rapid increase in density of about 20% and an increase in seismic wave velocities. **2.** A region within the outer core, transitional to the inner core. { tran'zish·ən ,zōn }

translational fault |GEOL| A fault in which there has been uniform movement in one direction and no rotational component of movement. Also known as translatory fault. { tran'slā·shən·əl 'fȯlt }

translational movement |GEOL| Movement, as of fault blocks, that is uniform, without rotation, so that parallel features maintain their orientation. { tran'slā·shən·əl 'müv·mənt }

translatory fault *See* translational fault. { 'tran·slə,tȯr·ē 'fȯlt }

translucent attritus |GEOL| Attritus composed principally of transparent humic degradation matter. Also known as humodurite. { tran'slüs·əns ə'trīd·əs }

transportation |GEOL| A phase of sedimentation concerned with movement by natural agents of sediment or any loose or weathered material from one place to another. { ,tranz·pər'tā·shən }

Transvaal jade |MINERAL| A mineral that is not a true jade but a green grossularite garnet. Also known as South African jade. { trans'väl 'jād }

transverse bar |GEOL| A slightly submerged sand bar extending perpendicular to the shoreline. { trans¦vərs 'bär }

transverse basin *See* exogeosyncline. { trans¦vərs 'bās·ən }

transverse dune |GEOL| A sand dune with a nearly straight ridge crest formed by the merger of crescentic dunes; elongated at right angles to the direction of prevailing winds, with a gentle windward slope and a steep leeward slope. { trans¦vərs 'dün }

transverse fault |GEOL| A fault whose strike is more or less perpendicular to the general structural trend of the region. { trans¦vərs 'fȯlt }

transverse fold *See* cross fold. { trans¦vərs 'fōld }

transverse joint *See* cross joint. { trans¦vərs 'jȯint }

transverse ripple mark |GEOL| A ripple mark formed nearly perpendicular to the direction of the current. { trans¦vərs 'rip·əl ,märk }

transverse thrust *See* transcurrent fault. { trans¦vərs 'thrəst }

transverse valley |GEOL| **1.** A valley perpendicular to the general strike of the underlying strata. **2.** A valley cutting perpendicularly across a ridge, range, or chain of mountains. Also known as cross valley. { trans¦vərs 'val·ē }

transverse wave *See* S wave. { trans¦vərs 'wāv }

trap |GEOL| *See* oil trap. |PETR| Any dark-colored, fine-grained, nongranitic, hypabyssal or extrusive rock. Also known as trappide; trap rock. { trap }

trapdoor fault |GEOL| A circular fault that is hinged at one end. { 'trap¦dȯr ,fȯlt }

trappide *See* trap. { 'tra,pīd }

trap rock *See* trap. { 'trap ¦räk }

traveling dune *See* wandering dune. { 'trav·əl·iŋ 'dün }

travel-time curve |GEOPHYS| A plot of P-, S-, and L-wave travel times used by seismologists to locate earthquakes. { 'trav·əl ,tīm ,kərv }

traverse |GEOL| A line of survey or sampling across a thin section of geological region. { tra'vərs }

travertine |GEOL| Concretionary limestone deposited at the mouth of a hot spring. { 'trav·ər,tēn }

treanorite *See* allanite. { 'trā·nə,rīt }

Trematosauria |PALEON| A group of Triassic amphibians in the order Temnospondyli. { ,trem·əd·ə'sȯr·ē·ə }

tremolite [MINERAL] $Ca_2Mg_5Si_8O_{22}(OH)_2$ Magnesium-rich monoclinic calcium amphibole that forms one end member of a group of solid-solution series with iron, sodium, and aluminum; occurs in long blade-shaped or short stout prismatic crystals and also in masses or compound aggregates. { 'trem·ə,līt }

tremor [GEOPHYS] A minor earthquake. Also known as earthquake tremor; earth tremor. { 'trem·ər }

trench [GEOL] A long, narrow, deep depression of the sea floor, with relatively steep sides. Also known as submarine trench. { trench }

trend [GEOL] The direction of an outcrop of a layer, vein, fold, or other kind of geologic feature. Also known as direction. { trend }

Trentonian [GEOL] A North American stage of geologic time; Middle Ordovician (above Wilderness, below Edenian); equivalent to the upper Mohawkian. { tren'tō·nē·ən }

Trepostomata [PALEON] An extinct order of ectoproct bryozoans in the class Stenolaemata characterized by delicate to massive colonies composed of tightly packed zooecia with solid calcareous zooecial walls. { ,trep·ə'stō·məd·ə }

treptomorphism See isochemical metamorphism. { ¦trep·tə'mór,fiz·əm }

triangular facet [GEOL] A triangular-shaped steep-sloped hill or cliff formed usually by the erosion of a fault-truncated hill. { trī'aŋ·gyə·lər 'fas·ət }

Triassic [GEOL] The first period of the Mesozoic era, lying above Permian and below Jurassic, 180–225 million years ago. { trī'a,sik }

tributary glacier [GEOL] A glacier that flows into a larger glacier. { 'trib·yə,ter·ē 'glā·shər }

Triceratops [PALEON] Herbivorous dinosaur, 30 feet (9 meters) long and weighing 6 tons, from the Late Cretaceous Period that had long sharp horns over each eye and a short horn on its nose. { trī'ser·ə,täps }

trichalcite [MINERAL] $Cu_5Ca(AsO_4)_2(CO_3)(OH)_4·6H_2O$ A verdigris green to blue-green, orthorhombic mineral consisting of hydrated copper arsenate. Also known as tyrolite. { trī'kal,sīt }

trichite [PETR] A black, straight or curved, hairlike crystallite. { 'tri,kīt }

Triconodonta [PALEON] An extinct mammalian order of small flesh-eating creatures of the Mesozoic era having no angle or a pseudoangle on the lower jaw and triconodont molars. { trī,kän·ə'dänt·ə }

tridymite [MINERAL] SiO_2 A white or colorless crystal occurring in minute, thin, tabular crystals or scales; a high-temperature polymorph of quartz. { 'trid·ə,mīt }

trigonite [MINERAL] $MnPb_3H(AsO_3)_3$ A sulfur yellow to yellowish-brown or dark brown, monoclinic mineral consisting of an acid arsenite of lead and manganese; occurs in domatic form. { 'trī·gə,nīt }

Trigonostylopoidea [PALEON] A suborder of Paleocene-Eocene ungulate mammals in the order Astrapotheria. { ,trig·ə·nō,stil·ə'póid·ē·ə }

Trilobita [PALEON] The trilobites, a class of extinct Cambrian-Permian arthropods characterized by an exoskeleton covering the dorsal surface, delicate biramous appendages, body segments divided by furrows on the dorsal surface, and a pygidium composed of fused segments. { ,trī·lə'bīd·ə }

Trilobitoidea [PALEON] A class of Cambrian arthropods that are closely related to the Trilobita. { ,trī·lō·bə'tóid·ē·ə }

Trimerellacea [PALEON] A superfamily of extinct inarticulate brachiopods in the order Lingulida; they have valves, usually consisting of calcium carbonate. { trə,mer·ə'lās·ē·ə }

Trimerophytatae See Trimerophytopsida. { trə,mer·ə'fīd·ə,tē }

Trimerophytopsida [PALEOBOT] A group of extinct land vascular plants with leafless, dichotomously branched stems that bear terminal sporangia. { trə,mer·ə·fə'täp·səd·ə }

triphane See spodumene. { 'trī,fān }

triphylite [MINERAL] $Li(Fe^{2+},Mn^{2+})PO_4$ A grayish-green or bluish-gray mineral crystallizing in the orthorhombic system; it is isomorphous with lithiophilite. { 'trif·ə,līt }

379

triplite [MINERAL] $(Mn,Fe,Mg,Ca)_2(PO)_4(F,OH)$ A dark brown, chestnut brown, reddish-brown, or salmon pink, monoclinic mineral consisting of a fluophosphate of iron, manganese, magnesium, and calcium; occurs in massive form. { 'trip,līt }

tripoli [GEOL] A lightweight, porous, friable, siliceous sedimentary rock that may have a white, gray, pink, red, or yellow color; used for polishing metals and stones. { 'trip·ə·lē }

tripolite See diatomaceous earth. { 'trip·ə,līt }

trippkeite [MINERAL] $CuAs_2O_4$ A greenish-blue, tetragonal mineral consisting of copper arsenite. { 'trip·kē,īt }

tripuhyite [MINERAL] $FeSb_2O_6$ A greenish-yellow to dark brown mineral consisting of iron antimonate; occurs as microcrystalline aggregates. { ,trip·ə'wē,īt }

Trochacea [PALEON] A recent subfamily of primitive gastropod mollusks in the order Aspidobranchia. { trō'kāsh·ē·ə }

Trochiliscales [PALEOBOT] A group of extinct plants belonging to the Charophyta in which the gyrogonites are dextrally spiraled. { trə,kil·ə'skā·lēz }

troctolite [PETR] A gabbro composed principally of calcic plagioclase and olivine. Also known as forellenstein. { 'träk·tə,līt }

troegerite [MINERAL] $(UO_2)_3(AsO_4)_2 \cdot 12H_2O$ A lemon yellow, tetragonal mineral consisting of a hydrated uranium arsenate. { 'treg·ə,rīt }

troilite [MINERAL] FeS A meteorite mineral crystallizing in the hexagonal system; a variety of pyrrhotite. { 'trôi,līt }

trolley [GEOL] A basin-shaped depression in strata. Also known as lum. { 'träl·ē }

trona [MINERAL] $Na_2(CO_3) \cdot Na(HCO_3) \cdot 2H_2O$ A gray-white or yellowish-white mineral that crystallizes in the monoclinic system and occurs in fibrous or columnar layers or masses. Also known as urao. { 'trō·nə }

Tropept [GEOL] A suborder of the order Inceptisol, characterized by moderately dark A horizons with modest additions of organic matter, B horizons with brown or reddish colors, and slightly pale C horizons; restricted to tropical regions with moderate or high rainfall. { 'trä,pept }

tropospheric duct See duct. { ¦trōp·ə¦sfir·ik 'dəkt }

trough [GEOL] **1.** A small, straight depression formed just offshore on the bottom of a sea or lake and on the landward side of a longshore bar. **2.** Any narrow, elongate depression in the surface of the earth. **3.** An elongate depression on the sea floor that is wider and shallower than a trench. Also known as submarine trench. **4.** The line connecting the lowest points of a fold. { trôf }

trough crossbedding [GEOL] A variety of crossbedding in which the lower crossbedding surfaces are smoothly curved, rather than planar. { 'trôf 'krôs,bed·iŋ }

trough fault [GEOL] One of a set of two faults bounding a graben. { 'trôft ,fôlt }

trough plane See trough surface. { 'trôf ,plān }

trough reef See reverse saddle. { 'trôf ,rēf }

trough surface [GEOL] A surface or plane connecting the troughs of the bed of a syncline. Also known as synclinal axis; trough plane. { 'trôf ,sər·fəs }

trough valley See U-shaped valley. { 'trôf ,val·ē }

Trucherognathidae [PALEON] A family of conodonts in the order Conodontophorida in which the attachment scar permits the conodont to rest on the jaw ramus. { ,trü·chə·räg'nath·ə,dē }

trudellite [MINERAL] $Al_{10}(SO_4)_3Cl_{12}(OH)_{12} \cdot 30H_2O$ An amber yellow, hexagonal mineral consisting of a hydrated basic sulfate-chloride of aluminum; occurs as compact masses. { trü'de,līt }

true crater [GEOL] The primary depression formed by impact or explosion before modification by slumping or by deposition of ejected material. Also known as primary crater. { 'trü 'krād·ər }

true dip See dip. { 'trü 'dip }

true formation resistivity [GEOPHYS] Electrical resistivity of a clean (nonshaly) porous reservoir formation containing hydrocarbons and formation water; value is greater than the resistivity when there is added water incursion. { 'trü fôr'mā·shən ,rē·zis'tiv·əd·ē }

true soil *See* solum. { 'trü 'sȯil }

Tryblidiidae [PALEON] An extinct family of Paleozoic mollusks. { ‚trib·lə'dī·ə‚dē }

tschermakite [MINERAL] $Ca_2Mg_3(Al,Fe^{3+})_2(Al_2Si_6)O_{22}(OH,F)_2$ An amphibole mineral. { 'chər·mə‚kīt }

tsumebite [MINERAL] $Pb_2Cu(PO_4)(SO_4)(OH)$ An emerald green, monoclinic mineral consisting of a hydrated basic phosphate and sulfate of lead and copper. { 'tsü·mə‚bīt }

tsunamiite [GEOL] **1.** A sedimentary deposit resulting from a tsunami generated by an asteroid or comet impact. **2.** Rock deposited by a tsunami. Also known as tsunamite. { tsü'näm·ē‚īt }

tsunamite *See* tsunamiite. { 'tsü·nə‚mīt }

tube [GEOL] A passage in a cave having smooth sides and an elliptical to nearly circular cross section. { 'tüb }

tufa [GEOL] A spongy, porous limestone formed by precipitation from evaporating spring and river waters, often onto leaves and stems of neighboring plants. Also known as calcareous sinter; calcareous tufa. { 'tü·fə }

tufaceous [GEOL] Pertaining to or similar to tufa. { tü'fā·shəs }

tuff [GEOL] Consolidated volcanic ash, composed largely of fragments (less than 4 millimeters) produced directly by volcanic eruption; much of the fragmented material represents finely comminuted crystals and rocks. { təf }

tuffaceous [GEOL] Pertaining to sediments which contain up to 50% tuff. { tə'fā·shəs }

tuff ball *See* mud ball. { 'təf ‚bȯl }

tuff lava *See* welded tuff. { 'təf ‚läv·ə }

tuft *See* mound. { təft }

tumuli lava [GEOL] A type of lava flow forming ovoid mounds, a few feet high and a few tens of feet long, caused by buckling up of the crust. { 'tü·myə‚lī 'lä·və }

tungstate [MINERAL] Any species of mineral containing the radical WO_4, such as wolframite. { 'təŋ‚stāt }

tungstenite [MINERAL] WS_2 A dark lead gray mineral consisting of tungsten disulfide; occurs in massive form, in scaly or feathery aggregates. { 'təŋ·stə‚nīt }

tungstite [MINERAL] $WO_3·H_2O$ A bright yellow, golden yellow, or yellowish-green mineral thought to consist of hydrated tungsten oxide; occurs in massive form and as platy crystals. { 'təŋ‚stīt }

tunnel cave *See* natural tunnel. { 'tən·əl ‚kāv }

turanite [MINERAL] $Cu_5(VO_4)_2(OH)_4$ An olive green, orthorhombic mineral consisting of basic copper vanadate; occurs as reniform crusts and spherical concretions. { 'tür·ə‚nīt }

turbidite [GEOL] Any sediment or rock transported and deposited by a turbidity current, generally characterized by graded bedding, large amounts of matrix, and commonly exhibiting a Bouma sequence. { 'tər·bə‚dīt }

turbidity factor [GEOPHYS] A measure of the atmospheric transmission of incident solar radiation; if I_0 is the flux density of the solar beam just outside the earth's atmosphere, I the flux density measured at the earth's surface with the sun at a zenith distance which implies an optical air mass m, and $I_{m,w}$ the intensity which would be observed at the earth's surface for a pure atmosphere containing 1 centimeter of precipitable water viewed through the given optical air mass, then turbidity factor θ is given by $\theta = (\ln I_0 - \ln I)/(\ln I_0 - \ln I_{m,w})$. { tər'bid·əd·ē ‚fak·tər }

Turkey stone *See* turquoise. { 'tər‚kē ‚stōn }

Turonian [GEOL] A European stage of geologic time: Upper or Middle Cretaceous (above Cenomanian, below Coniacian). { tü'rō·nē·ən }

turquoise [MINERAL] $CuAl_6(PO_4)_4(OH)_8·4H_2O$ A semitranslucent sky-blue, bluish-green, apple-green, or greenish-gray mineral that crystallizes in the triclinic system and occurs in veinlets or as crusts of massive, concretionary, and stalactite shapes; an important gem mineral. Also known as calaite; Turkey stone. { 'tər‚kwȯiz }

turtle stone *See* septarium. { 'tərd·əl ‚stōn }

tychite [MINERAL] $Na_6Mg_2(SO_4)(CO_3)_4$ A white, isometric mineral consisting of a sulfate-carbonate of sodium and magnesium. { 'tī‚kīt }

tying bar *See* tombolo. { 'tī·iŋ ‚bär }

type C1 carbonaceous chondrite

type C1 carbonaceous chondrite [GEOL] A type of carbonaceous chondrite that is strongly magnetic, has a lower density than the other two types, contains sulfates, and has a carbon content of about 3.5%. { 'tīp ¦sē¦wən ‚kär·bə'nā·shəs 'kän‚drīt }

type C2 carbonaceous chondrite [GEOL] A type of carbonaceous chondrite that is weakly magnetic or nonmagnetic, has most of its sulfur present as free sulfur, and contains about 2.5% carbon. { 'tīp ¦sē¦tü ‚kär·bə'nā·shəs 'kän‚drīt }

type C3 carbonaceous chondrite [GEOL] A type of carbonaceous chondrite that has a lower percentage of water and a higher density than the other two types, and usually consists largely of olivine. { 'tīp ¦sē¦thrē ‚kär·bə'nā·shəs 'kän‚drīt }

type locality [GEOL] **1.** The place at which a stratigraphic unit is typically displayed and from which it derives its name. **2.** The place where a geologic feature was first recognized and described. { 'tīp lō‚kal·əd·ē }

type section [GEOL] That sequence of strata identified as the original sequence for a location or area; the standard against which other stratigraphy of parts of the area are compared. Also known as section. { 'tīp ‚sek·shən }

Typotheria [PALEON] A suborder of extinct rodentlike herbivores in the order Notoungulata. { ‚tī·pə'thir·ē·ə }

Tyrannosaur [PALEON] A large carnivorous therapod dinosaur 40 feet (12 meters) long and weighing 6 tons, from the Late Cretaceous Period that had powerful hindlimbs, short forelimbs, a large skull (4 feet long), and very powerful jaws. { tə'ran·ə‚sȯr }

tyrolite See trichalcite. { 'tir·ə‚līt }

tyuyamunite [MINERAL] $Ca(UO_2)_2(VO_4)_2 \cdot 5\text{-}8H_2O$ A yellow orthorhombic mineral occurring in incrustations as a secondary mineral; an ore of uranium. Also known as calciocarnotite. { ‚tyü·ə'mü‚nīt }

U

Udalf |GEOL| A suborder of the soil order Alfisol; brown soil formed in a udic moisture regime and in a mesic or warmer temperature regime. { 'ü,dälf }

Udert |GEOL| A suborder of the soil order Vertisol; formed in a humid region so that surface cracks remain open only for 2–3 months. { 'üd,ərt }

Udoll |GEOL| A suborder of the Mollisol soil order; found in humid, temperate, and warm regions where maximum rainfall comes during growing season; has thick, very dark A horizons, brown B horizons, and paler C horizons. { 'üd,ól }

Udult |GEOL| A suborder of the soil order Ultisol; organic-carbon content is low, argillic horizons are reddish or yellowish; formed in a udic moisture regime. { 'üd,əlt }

U figure See U index. { 'yü ,fig·yər }

uhligite |MINERAL| A black, pseudoisometric mineral consisting of an oxide of titanium and calcium, with zirconium and aluminum replacing titanium. { ü·lə,gīt }

U index |GEOPHYS| The difference between consecutive daily mean values of the horizontal component of the geomagnetic field. Also known as U figure. { 'yü ,in,deks }

Uintatheriidae |PALEON| The single family of the extinct mammalian order Dinocerata. { yủ,win·tə·thə'rī·ə,dē }

Uintatheriinae |PALEON| A subfamily of extinct herbivores in the family Uintatheriidae including all horned forms. { yủ,win·tə·thə'rī·ə,nē }

Ulatisian |GEOL| A mammalian age in a local stage classification of the Eocene in use on the Pacific Coast based on foraminifers. { ,yü·lə'tē·zhən }

ulexite |MINERAL| $NaCaB_5O_9·8H_2O$ A white mineral that crystallizes in the triclinic system and forms rounded reniform masses of extremely fine acicular crystals. Also known as cotton ball. { 'ü·lek,sīt }

ullmannite |MINERAL| NiSbS A steel-gray to black mineral consisting of nickel antimonide and sulfide, usually with a little arsenic, occurring massive, and having a metallic luster. Also known as nickel-antimony glance. { 'əl·mə,nīt }

ulmic acid See ulmin. { 'əl·mik 'as·əd }

ulmin |GEOL| Alkali-soluble organic substances derived from decaying vegetable matter; occurs as amorphous brown to black gel material. Also known as carbohumin; fundamental jelly; fundamental substance; gelose; humin; humogelite; jelly; ulmic acid; vegetable jelly. { 'əl·mən }

ulrichite See uraninite. { 'əl·rə,kīt }

Ultisol |GEOL| A soil order characterized by typically moist soils, with horizons of clay accumulation and a low supply of bases. { 'əl·tə,sól }

ultrabasic |PETR| Of igneous rock, having a low silica content, as opposed to the higher silica contents of acidic, basic, and intermediate rocks. { ¦əl·trə'bā·sik }

ultrabasite See diaphorite. { ¦əl·trə'bā,sīt }

ultralow-velocity zone |GEOPHYS| Thin, mushy layer detected in some places along the earth's core-mantle boundary where seismic waves slow down. { ,əl·trə·lō·və'läs·əd·ē ,zōn }

ultramafic |PETR| Referring to igneous rock composed principally of mafic minerals, such as olivine and pyroxene. { ¦əl·trə'maf·ik }

ultravulcanian |GEOL| A type of volcanic eruption characterized by periodic violent gaseous explosions of lithic dust and solid blocks, with little if any fiery scoria. { ¦əl·trə·vəl'kā·nē·ən }

umangite

umangite [MINERAL] Cu_3Se_2 A dark cherry red mineral consisting of copper selenide; occurs in massive form, in small grains or fine granular aggregates. { ü'maŋ,gīt }

Umbrept [GEOL] A suborder of the Inceptisol soil order; has dark A horizon more than 10 inches (25 centimeters) thick, brown B horizons, and slightly paler C horizons; soil is strongly acid, and clay minerals are crystalline; occurs in cool or temperate climates. { 'əm,brept }

unaka [GEOL] A large residual mass rising above a peneplain that is less well developed than one having a monadnock. { ü'näk·ə }

unakite [PETR] An altered igneous rock composed principally of epidote, pink ortho-clase, and quartz. { 'ü·nə,kīt }

unconcentrated wash See sheet erosion. { ¦ən'käns·ən,trād·əd 'wäsh }

unconformable [GEOL] Pertaining to strata that do not conform in position, dip, or strike to the older underlying rocks. { ¦ən·kən'for·mə·bəl }

unconformity [GEOL] The relation between adjacent rock strata whose time of deposi-tion was separated by a period of nondeposition or of erosion; a break in a strati-graphic sequence. { ¦ən·kən'for·məd·ē }

unconsolidated material [GEOL] Loosely arranged or unstratified sediment whose par-ticles are not cemented together. { ¦ən·kən'säl·ə,dād·əd mə'tir·ē·əl }

underclay [GEOL] A layer of clay or other fine-grained detrital material underlying a coal bed or comprising the floor of a coal seam. Also known as coal clay; root clay; seat clay; seat earth; thill; underearth; warrant. { 'ən·dər,klā }

underclay limestone [GEOL] A thin, fresh-water limestone that is relatively free of fossils and is dense and nodular; found in underlying coal deposits. { 'ən·dər,klā 'līm,stōn }

undercliff [GEOL] A subordinate cliff or terrace formed by material which has fallen or slid from above. { 'ən·dər,klif }

underconsolidation [GEOL] Less than normal consolidation of sedimentary material for the existing overburden. { ¦ən·dər·kən,säl·ə'dā·shən }

undercutting [GEOL] Erosion of material at the base of a steep slope, cliff, or other exposed rock. { ¦ən·dər¦kəd·iŋ }

underearth See underclay. { 'ən·dər,ərth }

underflow conduit [GEOL] A permeable deposit underlying a surface stream channel. { 'ən·dər,flō 'kän,dü·ət }

underground geology See subsurface geology. { ¦ən·dər¦graúnd jē'äl·ə·jē }

underlie [GEOL] To lie or be situated under; to occupy a lower position, or to pass beneath. { 'ən·dər,lī }

undermining See sapping. { 'ən·dər,mīn·iŋ }

undersaturated [PETR] Pertaining to igneous rock composed of unsaturated minerals, that is, without free silica. { ¦ən·dər¦sach·ə,rād·əd }

underthrust [GEOL] A thrust fault in which the lower, active rock mass has been moved under the upper, passive rock mass. { ¦ən·dər¦thrəst }

unfreezing [GEOL] The upward movement of stones to the surface as a result of repeated freezing and thawing of the containing soil. { ¦ən'frēz·iŋ }

ungemachite [MINERAL] $K_3Na_9Fe(SO_4)_6(OH)_3 \cdot 9H_2O$ A colorless to pale yellow, hexago-nal mineral consisting of a hydrated basic sulfate of potassium, sodium, and iron; occurs in tabular form. { 'əŋ·gə,mä,kīt }

uniformitarianism [GEOL] Classically, the concept that the present is the key to the past; the principle that contemporary geologic processes have occurred in the same regular manner and with essentially the same intensity throughout geologic time, and that events of the geologic past can be explained by phenomena observable today. Also known as actualism; principle of uniformity. { ,yü·nə,for·mə'ter·ē·ə,niz·əm }

unsaturated [MINERAL] Referring to a mineral that will not form in the presence of free silica. { ¦ən'sach·ə,rād·əd }

unsaturated zone See zone of aeration. { ¦ən'sach·ə,rād·əd 'zōn }

uphole time [GEOPHYS] The time that a seismic pulse requires to travel from an explo-sion at some depth in a shot hole to the surface of the earth. { 'əp,hōl ,tīm }

384

upper |GEOL| Pertaining to rocks or strata that normally overlie those of earlier formations of the same subdivision of rocks. { 'əp·ər }

Upper Cambrian |GEOL| The latest epoch of the Cambrian period of geologic time, beginning approximately 510 million years ago. { 'əp·ər 'kam·brē·ən }

Upper Carboniferous |GEOL| The European epoch of geologic time equivalent to the Pennsylvanian of North America. { 'əp·ər ‚kär·bə'nif·ə·rəs }

Upper Cretaceous |GEOL| The late epoch of the Cretaceous period of geologic time, beginning about 90 million years ago. { 'əp·ər kri'tā·shəs }

Upper Devonian |GEOL| The latest epoch of the Devonian period of geologic time, beginning about 365 million years ago. { 'əp·ər də'vō·nē·ən }

Upper Huronian See Animikean. { 'əp·ər hyü'rō·nē·ən }

Upper Jurassic |GEOL| The latest epoch of the Jurassic period of geologic time, beginning approximately 155 million years ago. { 'əp·ər jü'ras·ik }

upper mantle |GEOL| The portion of the mantle lying above a depth of about 600 miles (1000 kilometers). Also known as outer mantle; peridotite shell. { 'əp·ər 'mant·əl }

Upper Mississippian |GEOL| The latest epoch of the Mississippian period of geologic time. { 'əp·ər ‚mis·ə'sip·ē·ən }

Upper Ordovician |GEOL| The latest epoch of the Ordovician period of geologic time, beginning approximately 440 million years ago. { 'əp·ər ‚ȯr·də'vish·ən }

Upper Pennsylvanian |GEOL| The latest epoch of the Pennsylvanian period of geologic time. { 'əp·ər ‚pen·səl'vā·nyən }

Upper Permian |GEOL| The latest epoch of the Permian period of geologic time, beginning about 245 million years ago. { 'əp·ər 'pər·mē·ən }

Upper Silurian |GEOL| The latest epoch of the Silurian period of geologic time. { 'əp·ər sə'lür·ē·ən }

Upper Triassic |GEOL| The latest epoch of the Triassic period of geologic time, beginning about 200 million years ago. { 'əp·ər trī'as·ik }

upsetted moraine See push moraine. { ‚əp‚sed·əd mə'rān }

upthrow |GEOL| **1.** The fault side that has been thrown upward. **2.** The amount of vertical fault displacement. { 'əp‚thrō }

upwarp |GEOL| A broad anticline with gently sloping limbs formed as a result of differential uplift. { 'əp‚wȯrp }

Uralean |GEOL| A stage of geologic time in Russia: uppermost Carboniferous (above Gzhelian, below Sakmarian of Permian). { yü'rāl·ē·ən }

uralite |MINERAL| A green variety of secondary amphibole; it is usually fibrous or acicular and is formed by alteration of pyroxene. { 'yür·ə‚līt }

uralitization |GEOL| **1.** A process of replacement whereby pyroxene undergoes alteration resulting in uralite. **2.** Development of amphibole from pyroxene. { yə‚ral·əd·ə'zā·shən }

uraninite |MINERAL| UO_2 A black, brownish-black, or dark-brown radioactive mineral that is isometric in crystallization; often contains impurities such as thorium, radium, cerium, and yttrium metals, and lead; the chief ore of uranium; hardness is 5.5–6 on Mohs scale, and specific gravity of pure UO_2 is 10.9, but that of most natural material is 9.7–7.5. Also known as coracite; ulrichite. { 'yür·ə·nə‚nīt }

uranium age |GEOL| The age of a mineral as calculated from the numbers of ionium atoms present originally, now, and when equilibrium is established with uranium. { yə'rā·nē·əm ‚āj }

uranium-lead dating |GEOL| A method for calculating the geologic age of a material in years based on the radioactive decay rate of uranium-238 to lead-206 and of uranium-235 to lead-207. { yə'rā·nē·əm 'led 'dād·iŋ }

uranium ocher See gummite. { yə'rā·nē·əm 'ō·kər }

uran-mica See torbernite. { 'yür‚an 'mī·kə }

uranocircite |MINERAL| $Ba(UO_2)_2(PO_4)_2·8H_2O$ A yellow-green, tetragonal mineral consisting of a hydrated phosphate of barium and uranium; occurs as crystals. { ‚yür·ə·nō'sər‚sīt }

uranophane [MINERAL] $Ca(UO_2)_2Si_2O_7 \cdot 6H_2O$ A yellow or orange-yellow radioactive secondary mineral; it is dimorphous with β-uranophane. Also known as uranotile. { yə'ran·ə,fān }

uranopilite [MINERAL] $(UO_2)_6(SO_4)(OH)_{10} \cdot 12H_2O$ A bright yellow, lemon yellow, or golden yellow, monoclinic mineral consisting of a hydrated basic sulfate of uranium; occurs as encrustations and masses. { ,yùr·ə·nō'pī,līt }

uranosphaerite [MINERAL] $Bi_2O_3 \cdot 2UO_3 \cdot 3H_2O$ An orange-yellow or brick red, orthorhombic mineral consisting of a hydrated oxide of bismuth and uranium. { ,yùr·ə·nō'sfi,rīt }

uranospinite [MINERAL] $Ca(UO_2)_2(AsO_4)_2 \cdot 8H_2O$ A lemon yellow to siskin green, tetragonal mineral consisting of a hydrated arsenate of calcium and uranium; occurs in tabular form. { ,yùr·ə'näs·pə,nīt }

uranotantalite See samarskite. { ,yùr·ə·nō'tant·əl,īt }

uranothorite [MINERAL] A uranium-bearing variety of thorite. { ¦yùr·ə·nō'thòr,īt }

uranotile See uranophane. { yə'ran·ə,tīl }

urao See trona. { 'yü·raù }

urban geology [GEOL] The study of geological aspects of planning and managing high-density population centers and their surroundings. { ¦ər·bən jē'äl·ə·jē }

ureilite [GEOL] An achondritic stony meteorite consisting principally of olivine and clinobronzite, with some nickel-iron, troilite, diamond, and graphite. { yə'rē·ə,līt }

ureyite [MINERAL] $NaCrSi_2O_6$ A meteoritic mineral of the pyroxene group. Also known as cosmochlore; kosmochlor. { 'yùr·ē,īt }

urstromthal [GEOL] A large channel cut by a stream of water from melting ice, flowing along the edge of an ice sheet. { 'ùr,strōm,täl }

U-shaped valley [GEOL] A type of valley with a broad floor and steep walls produced by glacial erosion. Also known as trough valley; U valley. { 'yü ¦shāpt 'val·ē }

Ustalf [GEOL] A suborder of the soil order Alfisol; red or brown soil formed in a ustic moisture regime and in a mesic or warmer temperature regime. { 'üst,älf }

Ustert [GEOL] A suborder of the Vertisol soil order; has a faint horizon and is dry for an appreciable period or more than one period of the year. { 'üst,ərt }

Ustoll [GEOL] A suborder of the soil order Mollisol; formed in a ustic moisture regime and in a mesic or warmer temperature regime; may have a calcic, petrocalcic, or gypsic horizon. { 'üst,ól }

Ustox [GEOL] A suborder of the soil order Oxisol that is low to moderate in organic matter, well drained, and dry for at least 90 cumulative days each year. { 'üst,äks }

Ustult [GEOL] A suborder of the soil order Ultisol; brownish or reddish, with low to moderate organic-carbon content; a well-drained soil of warm-temperate and tropical climates with moderate or low rainfall. { 'üst,əlt }

utahite See jarosite. { 'yü·tò,īt }

U valley See U-shaped valley. { 'yü ,val·ē }

uvanite [MINERAL] $U_2V_6O_{21} \cdot 15H_2O$ A brownish-yellow, orthorhombic mineral consisting of a hydrated uranium vanadate; occurs as crystalline masses and coatings. { 'yü·və,nīt }

uvarovite [MINERAL] $Ca_3Cr_2(SiO_4)_3$ The emerald-green, calcium-chromium end member of the garnet group. Also known as ouvarovite; uwarowite. { ü'var·ə,vīt }

uwarowite See uvarovite. { ü'var·ə,vīt }

V

vacuole *See* vesicle. { 'vak·yə,wōl }

vadose zone *See* zone of aeration. { 'vā,dōs ,zōn }

vaesite [MINERAL] NiS_2 An isometric mineral with pyrite structure composed of sulfide of nickel. { 'vä,sīt }

valencianite [MINERAL] A variety of potassium feldspar from Mexico. { və'len·chə,nīt }

valentinite [MINERAL] Sb_2O_3 A colorless to snow white mineral consisting of antimony trioxide. { 'val·ən,tē,nīt }

vallerite [MINERAL] $CuFeS_2$ A sulfide mineral found in meteorites. { və'lir,īt }

valley [GEOL] A relatively shallow, wide depression of the sea floor with gentle slopes. Also known as submarine valley. { 'val·ē }

valley bottom *See* valley floor. { 'val·ē ,bäd·əm }

valley fill [GEOL] Unconsolidated sedimentary deposit which fills or partly fills a valley. { 'val·ē ,fil }

valley flat [GEOL] The small plain at the bottom of a narrow valley with steep sides. { 'val·ē ,flat }

valley floor [GEOL] The broad, flat bottom of a valley. Also known as valley bottom; valley plain. { 'val·ē ,flȯr }

valley plain *See* valley floor. { 'val·ē ,plān }

valley train [GEOL] A long, narrow body of outwash, deposited by meltwater far beyond the margin of an active glacier and extending along the floor of a valley. Also known as outwash train. { 'val·ē ,trān }

Valvatacea [PALEON] A superfamily of extinct gastropod mollusks in the order Prosobranchia. { ,val·və'tā·shə }

vanadate [MINERAL] Any of several mineral compounds characterized by pentavalent vanadium and oxygen in the anion; an example is vanadinite. { 'van·ə,dāt }

vanadinite [MINERAL] $Pb_5(VO_4)_3Cl$ A red, yellow, or brown opatite mineral often occurring as globular masses encrusting other minerals in lead mines; an ore of vanadium and lead hardness is 2.75–3 on Mohs scale, and specific gravity is 6.66–7.10. { və'nād·ən,īt }

vandenbrandite [MINERAL] $CuO \cdot UO_3 \cdot 2H_2O$ A dark green to black mineral consisting of a hydrated oxide of copper and uranium; occurs in small crystals and massive form. { ,van·dən'bran,dīt }

vanoxite [MINERAL] $(V_4)^{4+}(V_2)^{5+}O_{13} \cdot 8H_2O$ A black mineral consisting of a hydrous oxide of vanadium; occurs as microscopic crystals and in massive form. { va'näk,sīt }

vanthoffite [MINERAL] $Na_6Mg(SO_4)_4$ A colorless mineral consisting of a sulfate of sodium and magnesium; occurs in massive form. { van'tȯ,fīt }

vapor-dominated hydrothermal reservoir [GEOL] Any geothermal system mainly producing dry steam; the Geysers area of northern California and the Larderelle region of Italy are two examples. { 'vā·pər ¦dom·ə,nād·əd ¦hī·drə¦thər·məl 'rez·əv,wär }

variation *See* declination. { ,ver·ē'ā·shən }

variation diagram [PETR] A diagram constructed by plotting the chemical compositions of rocks in an igneous rock series in order to show the genetic relationships and the nature of the processes that have affected the series. Also known as Harker diagram. { ,ver·ē'ā·shən ,dī·ə,gram }

variation of latitude [GEOPHYS] Change of the latitude of a place on earth because of the irregular movement of the north and south poles; the movement is caused by the earth's shifting on its axis. { ‚ver·ē'ā·shən əv 'lad·ə‚tüd }

variation per day |GEOPHYS| The change in the value of any geophysical quantity during 1 day. { ‚ver·ē'ā·shən pər 'dā }

variation per hour |GEOPHYS| The change in the value of any geophysical quantity during 1 hour. { ‚ver·ē'ā·shən pər 'aůr }

variation per minute |GEOPHYS| The change in the value of any geophysical quantity during 1 minute. { ‚ver·ē'ā·shən pər 'min·ət }

variole |GEOL| A spherule the size of a pea, usually consisting of radiating plagioclase or pyroxene crystals. { 'ver·ē‚ōl }

variolitic |PETR| Referring to the texture of basic igneous rock composed of varioles in a finer-grained matrix. { ¦ver·ē·ə¦lid·ik }

Variscan orogeny |GEOL| The late Paleozoic orogenic era in Europe, extending through the Carboniferous and Permian. Also known as Hercynian orogeny. { va'ris·kən ò'räj·ə·nē }

varulite |MINERAL| $(Na,Ca)(Mn,Fe)_2(PO_4)_2$ An olive green, orthorhombic mineral consisting of a phosphate of sodium, calcium, manganese, and iron; occurs in massive form. { 'vär·ə‚līt }

varve |GEOL| A sedimentary bed, layer, or sequence of layers deposited in a body of still water within a year's time, and usually during a season. Also known as glacial varve. { 'värv }

varve clay *See* varved clay. { 'värv ‚klā }

varved clay |GEOL| A lacustrine sediment of distinct layers consisting of varves. Also known as varve clay. { 'värvd ‚klā }

vashegyite |MINERAL| $2Al_4(PO_4)_3(OH)_3 \cdot 27H_2O$ A white or pale green to yellow and brownish mineral consisting of a hydrous basic aluminum phosphate; occurs in massive and microcrystalline forms. { 'väsh‚he‚jīt }

vaterite |MINERAL| $CaCO_3$ A rare hexagonal mineral consisting of unstable calcium carbonate; it is trimorphous with calcite and aragonite. { 'väd·ə‚rīt }

vauquelinite |MINERAL| $Pb_2Cu(CrO_4)PO_4(OH)$ A monoclinic mineral of varying color, consisting of a basic chromate-phosphate of lead and copper. { 'vōk·lə‚nīt }

vauxite |MINERAL| $FeAl_2(PO_4)_2(OH)_2 \cdot 7H_2O$ A sky blue to Venetian blue, triclinic mineral consisting of a hydrated basic phosphate of iron and aluminum. { 'vòk‚sīt }

veatchite |MINERAL| $Sr_2B_{11}O_{16}(OH)_5 \cdot H_2O$ A white mineral consisting of hydrous strontium borate. { 'vē‚chīt }

Vectian *See* Aptian. { 'vek·chən }

vectorial structure *See* directional structure. { vek'tòr·ē·əl 'strək·chər }

vegetable jelly *See* ulmin. { 'vej·tə·bəl ‚jel·ē }

vein |GEOL| A mineral deposit in tabular or shell-like form filling a fracture in a host rock. { vān }

veined gneiss |PETR| A composite gneiss with irregular layering. { 'vānd 'nīs }

veinite |GEOL| A genetic type of veined gneiss in which the vein material was secreted from the rock itself. { 'vā‚nīt }

vein quartz |PETR| A rock composed chiefly of sutured quartz crystals of pegmatitic or hydrothermal origin of variable size. { 'vān ‚kwòrtz }

Velociraptor |PALEON| A carnivorous theropod dinosaur, 7 feet (2 meters) long, with birdlike features from the Late Cretaceous that had strong grasping hands with claws, powerful hindlimbs, and jaws containing sharp teeth. { və'läs·ə‚rap·tər }

velocity discontinuity *See* seismic discontinuity. { və'läs·əd·ē dis‚känt·ən'ü·əd·ē }

velocity gradient |GEOPHYS| *See* seismic gradient. { və'läs·əd·ē ‚grād·ē·ənt }

venite |PETR| Migmatite having mobile portions which were formed by exudation from the rock itself. { 'vē‚nīt }

vent |GEOL| The opening of a volcano on the surface of the earth. { vent }

ventifact |GEOL| A stone or pebble whose shape, wear, faceting, cut, or polish is the result of sandblasting. Also known as glyptolith; rillstone; wind-cut stone; wind-grooved stone; wind-polished stone; wind-scoured stone; wind-shaped stone. { 'ven·tə‚fakt }

Venturian [GEOL] A North American stage of middle Pliocene geologic time, above Repeltian and below Wheelerian. { ven'chür·ē·ən }

Venus hairstone *See* rutilated quartz. { 'vē·nəs 'her,stōn }

Verbeekinidae [PALEON] A family of extinct marine protozoans in the superfamily Fusulinacea. { ,ver,bā'kin·ə,dē }

vergence [GEOL] The direction of overturning or of inclination of a fold. { 'vər·jəns }

vermiculite [MINERAL] (Mg,Fe,Al)$_3$(Al,Si)$_4$O$_{10}$(OH)$_2$·4H$_2$O A clay mineral constituent similar to chlorite and montmorillonite, and consisting of trioctahedral mica sheets separated by double water layers; sometimes used as a textural material in painting, or as an aggregate in certain plaster formulations used in sculpture. { vər'mik·yə,līt }

vernadskite *See* antlerite. { vər'nadz,kīt }

vernal [GEOPHYS] Pertaining to spring. { 'vərn·əl }

verrou *See* riegel. { və'rü }

vertical dip slip *See* vertical slip. { 'vərd·ə·kəl 'dip ,slip }

vertical intensity [GEOPHYS] The magnetic intensity of the vertical component of the earth's magnetic field, reckoned positive if downward, negative if upward. { 'vərd·ə·kəl in'ten·səd·ē }

vertical separation [GEOL] The vertical component of the dip slip in a fault. { 'vərd·ə·kəl ,sep·ə'rā·shən }

vertical slip [GEOL] The vertical component of the net slip in a fault. Also known as vertical dip slip. { 'vərd·ə·kəl 'slip }

Vertisol [GEOL] A soil order formed in regoliths high in clay; subject to marked shrinking and swelling with changes in water content; low in organic content and high in bases. { 'vərd·ə,sȯl }

vesicle [GEOL] A cavity in lava formed by entrapment of a gas bubble during solidification. Also known as air sac; bladder; saccus; vacuole; wing. { 'ves·ə·kəl }

vesicular structure [PETR] A structure that is common in many volcanic rocks and which forms when magma is brought to or near the earth's surface; may form a structure with small cavities, or produce a pumiceous structure or a scoriaceous structure. { və'sik·yə·lər 'strək·chər }

vesuvian *See* leucite; vesuvianite. { və'sü·vē·ən }

Vesuvian eruption *See* Vulcanian eruption. { və'sü·vē·ən i'rəp·shən }

Vesuvian garnet *See* leucite. { və'sü·vē·ən 'gär·nət }

vesuvianite [MINERAL] Ca$_{10}$Mg$_2$Al$_4$(SiO$_4$)$_5$(Si$_2$O$_7$)$_2$(OH)$_4$ A brown, yellow, or green mineral found in contact-metamorphosed limestones. Also known as idocrase; vesuvian. { və'sü·vē·ə,nīt }

veszelyite [MINERAL] (Cu,Zn)$_3$(PO$_4$)(OH)$_3$·2H$_2$O A greenish-blue to dark blue, monoclinic mineral consisting of a hydrated basic phosphate of copper and zinc. { 'ves·əl,yīt }

villiaumite [MINERAL] NaF A carmine, isometric mineral consisting of sodium fluoride; occurs in masssive form. { vē'yō,mīt }

Vindobonian [GEOL] A European stage of geologic time, middle Miocene. { ,vin·də'bō·nē·ən }

violarite [MINERAL] Ni$_2$FeS$_4$ A violet-gray mineral of the linnaeite group consisting of a sulfide of nickel and iron; found in meteorites. { vī'ō·lə,rīt }

viscous magnetization *See* viscous remanent magnetization. { 'vis·kəs ,mag·nəd·ə'zā·shən }

viscous remanent magnetization [GEOPHYS] A process in which grains of magnetic minerals, which are either too small or too finely divided by undergrowths of different chemical composition to retain a permanent magnetization indefinitely, acquire a new direction of magnetization when the direction of the earth's magnetic field changes. Abbreviated VRM. Also known as viscous magnetization. { 'vis·kəs 'rem·ə·nənt ,mag·nəd·ə'zā·shən }

Viséan [GEOL] A European stage of lower Carboniferous geologic time forming the lowermost Upper Mississippian, above Tournaisian and below lower Namurian. { vi'sā·ən }

visor tin [MINERAL] Twin crystals of cassiterite characterized by a notch. { 'vī·zər 'tin }

389

vitavite *See* moldavite. { 'vīd·ə,vīt }

vitrain |GEOL| A brilliant black coal lithotype with vitreous luster and cubical cleavage. Also known as pure coal. { 'vi,trān }

vitreous copper *See* chalcocite. { 'vi·trē·əs 'käp·ər }

vitreous silver *See* argentite. { 'vi·trē·əs 'sil·vər }

vitric |GEOL| Referring to a pyroclastic material which is characteristically glassy, that is, contains more than 75% glass. { 'vi·trik }

vitric tuff |GEOL| Tuff composed principally of volcanic glass fragments. { 'vi·trik 'təf }

vitrification |GEOL| Formation of a glassy or noncrystalline material. { ,vi·tra·fə'kā·shən }

vitrinite |GEOL| A maceral group that is rich in oxygen and composed of humic material associated with peat formation; characteristic of vitrain. { 'vi·trə,nīt }

vitrinoid |GEOL| Vitrinite occurring in bituminous coking coals; characterized by a reflectance of 0.5–2.0%. { 'vi·trə,nȯid }

vitriol stone |MINERAL| A hard, crystalline material, mainly a mixture of ferric sulfate and aluminum sulfate, that is extracted from weathered pyritic schist and used in the manufacture of sulfuric acid. { 'vi·trē·əl ,stōn }

vitrophyre |PETR| Any porphyritic igneous rock whose groundmass is glassy. Also known as glass porphyry. { 'vi·trə,fīr }

vivianite |MINERAL| $Fe_3(PO_4)_2 \cdot 8H_2O$ A colorless, blue, or green mineral in the unaltered state (darkens upon oxidation); crystallizes in the monoclinic system and occurs in earth form and as globular and encrusting fibrous masses. Also known as blue iron earth; blue ocher. { 'vi·vē·ə,nīt }

vogesite |PETR| A syenitic lamprophyre composed of phenocrysts of hornblende in a groundmass of orthoclase and hornblende. { 'vō·gə,sīt }

voglite |MINERAL| An emerald green to grass green, triclinic mineral consisting of a hydrated carbonate of calcium, copper, and uranium; occurs as coatings of scales. { 'vō,glīt }

volatile component |GEOL| A component of magma whose vapor pressures are high enough to allow them to be concentrated in any gaseous phase. Also known as volatile flux. { 'väl·əd·əl kəm'pō·nənt }

volatile flux *See* volatile component. { 'väl·əd·əl 'fləks }

volborthite |MINERAL| $Cu_3(UO_4)_2 \cdot 3H_2O$ An olive green to green and yellowish-green, monoclinic mineral consisting of hydrated copper vanadate. { 'väl,bȯr,thīt }

volcanello *See* spatter cone. { ,väl·kə'nel·ō }

volcanic ash |GEOL| Fine pyroclastic material; particle diameter is less than 4 millimeters. { väl'kan·ik 'ash }

volcanic bombs |GEOL| Pyroclastic ejecta; the lava fragments, liquid or plastic at the time of ejection, acquire rounded forms, markings, or internal structure during flight or upon landing. { väl'kan·ik 'bämz }

volcanic breccia |PETR| A pyroclastic rock that is composed of angular volcanic fragments having a diameter larger than 2 millimeters and that may or may not have a matrix. { väl'kan·ik 'brech·ə }

volcanic foam *See* pumice. { väl'kan·ik 'fōm }

volcanic gases |GEOL| Volatile matter composed principally of about 90% water vapor, and carbon dioxide, sulfur dioxide, hydrogen, carbon monoxide, and nitrogen, released during an eruption of a volcano. { väl'kan·ik 'gas·əz }

volcanic glass |GEOL| Natural glass formed by the cooling of molten lava, or one of its liquid fractions, too rapidly to allow crystallization. { väl'kan·ik 'glas }

volcanicity *See* volcanism. { ,väl·kə'nis·əd·ē }

volcaniclastic rock |PETR| Clastic rock containing volcanic material in any proportion. { väl,kan·ə,klas·tik 'räk }

volcanic mud |GEOL| Sediment containing large quantities of ash from a volcanic eruption, mixed with water. { väl'kan·ik 'məd }

volcanic mudflow |GEOL| The flow of volcanic mud down the slope of a volcano. { väl'kan·ik 'məd,flō }

volcanic neck |GEOL| A residual remnant of the pipe or throat of a volcano that was filled with solidified lava after its final eruption. { väl'kan·ik 'nek }

volcanic rift zone |GEOL| A zone comprising volcanic fissures with underlying dike assemblages; occurs in Hawaii. { väl'kan·ik 'rift ‚zōn }

volcanic rock |GEOL| Finely crystalline or glassy igneous rock resulting from volcanic activity at or near the surface of the earth. Also known as extrusive rock. { väl'kan·ik 'räk }

volcanics |PETR| Igneous rocks that solidified after reaching or nearing the earth's surface. { väl'kan·iks }

volcanic vent |GEOL| The channelway or opening of a volcano through which magma ascends to the surface; two general types are fissure and pipelike vents. { väl'kan·ik 'vent }

volcanism |GEOL| The movement of magma and its associated gases from the interior into the crust and to the surface of the earth. Also known as volcanicity. { 'väl·kə‚niz·əm }

volcano [GEOL| **1.** A mountain or hill, generally with steep sides, formed by the accumulation of magma erupted through openings or volcanic vents. **2.** The vent itself. { väl'kā·nō }

volcanology |GEOL| The branch of geology that deals with volcanism. { ‚väl·kə'näl·ə·jē }

voltaite |MINERAL| A greenish-black to black, isometric mineral consisting of a hydrated potassium iron sulfate. { 'väl·tə‚īt }

voltzite |MINERAL| Zn_5S_4O A rose red, yellowish, or brownish mineral consisting of an oxysulfide of zinc; occurs in implanted spherical globules and as a crust. { 'vält‚sīt }

volume phase See surface phase. { 'väl·yəm ‚fāz }

vorobievite See vorobyevite. { və'rō‚bē·ə‚vīt }

vorobyevite |MINERAL| A rose-red, purplish-red, or pinkish cesium-containing variety of beryl; used as a gem. Also known as morganite; rosterite; vorobievite; worobieffite. { və'rō‚bē·ə‚vīt }

vougesite |PETR| A lamprophyre having an orthoclase and hornblende groundmass in which are embedded hornblende phenocrysts. { 'vüzh‚sīt }

vrbaite |MINERAL| $Tl_4Hg_3Sb_2As_8S_{20}$ A dark gray-black, orthorhombic mineral that occurs in small crystals. { 'vər·bə‚īt }

VRM See viscous remanent magnetization.

V-shaped valley |GEOL| A valley having a cross-sectional profile in the form of the letter V, commonly produced by stream erosion. Also known as V valley. { 'vē ¦shāpt 'val·ē }

vug |PETR| A small cavity in a vein or rock usually lined with minerals differing in composition from those of the enclosing rock. Also known as bughole. { vəg }

Vulcanian eruption |GEOL| A volcanic eruption characterized by periodic explosive events. Also known as paroxysmal eruption; Plinian eruption; Vesuvian eruption. { ¦vəl¦kā·nē·ən i'rəp·shən }

V valley See V-shaped valley. { 'vē ‚val·ē }

W

wacke |PETR| Sandstone composed of a mixture of angular and unsorted or poorly sorted fragments of minerals and rocks and an abundant matrix of clay and fine silt. { 'wak·ə }

wackestone |PETR| A limestone composed of mud (micrite) containing more than 10% particles (grains) with diameters greater than 20 micrometers scattered throughout. { 'wak·ə,stōn }

wad |MINERAL| A massive, generally soft, amorphous, earthy, dark-brown or black mineral composed principally of manganese oxides with some other minerals, and formed by decomposition of manganese minerals. Also known as black ocher; bog manganese; earthy manganese. { wäd }

Wadati-Benioff zone See Benioff zone. { ¦wä¦dä·tē 'ben·ē,óf ,zōn }

wadi |GEOL| In the desert regions of southwestern Asia and northern Africa, a stream bed or channel, or a steep-sided ravine, gulley, or valley, which carries water only during the rainy season. Also spelled wady. { 'wäd·ē }

wady See wadi. { 'wäd·ē }

wagnerite |MINERAL| $Mg_2(PO_4)F$ A yellow, grayish, flash-red, or greenish, monoclinic mineral consisting of magnesium fluophosphate. { 'väg·nə,rīt }

wairakite |MINERAL| $CaAl_2Si_4O_{12}·2H_2O$ A zeolite mineral that is isostructural with analcime. { 'wī·rə,kīt }

wall |GEOL| The side of a cave passage. { wól }

wall reef |GEOL| A linear, steep-sided coral reef constructed on a reef wall. { 'wól ¦rēf }

wall rock |GEOL| Rock that encloses a vein. { 'wól ¦räk }

wall-rock alteration |GEOL| Alteration of wall rock adjacent to hydrothermal veins by the fluid responsible for formation of the mineral deposit. { 'wól ¦räk ,ól·tə'rā·shən }

walpurgite |MINERAL| $Bi_4(UO_2)(AsO_4)_2O_4·3H_2O$ A wax yellow to straw yellow, triclinic mineral consisting of a hydrated arsenate of bismuth and uranium. Also known as waltherite. { wäl'pər,jīt }

waltherite See walpurgite. { 'väl·tə,rīt }

wander See apparent wander. { 'wän·dər }

wandering dune |GEOL| A sand dune that has moved as a unit in the leeward direction of the prevailing winds, and that is characterized by the lack of vegetation to anchor it. Also known as migratory dune; traveling dune. { 'wän·də·riŋ 'dün }

want See nip. { wänt }

wardite |MINERAL| $Na_4CaAl_{12}(PO_4)_8(OH)_{18}·6H_2O$ A blue-green to pale green, tetragonal mineral consisting of a hydrated basic phosphate of sodium, calcium, and aluminum. { 'wór,dīt }

warp |GEOL| **1.** An upward or downward flexure of the earth's crust. **2.** A layer of sediment deposited by water. { wórp }

warrant See underclay. { 'wär·ənt }

warringtonite See brochantite. { 'wär·iŋ·tə,nīt }

warwickite |MINERAL| $(Mg,Fe)_3Ti(BO_4)_2$ A dark brown to dull black, orthorhombic mineral consisting of a titanoborate of magnesium and iron; occurs as prismatic crystals. { 'wór·i,kīt }

wash |GEOL| **1.** An alluvial placer. **2.** A piece of land washed by a sea or river. **3.** See alluvial cone. { wäsh }

wash-built terrace

wash-built terrace *See* alluvial terrace. { 'wäsh ¦bilt 'ter·əs }

wash load [GEOL] The finer part of the total sediment load of a stream which is supplied from bank erosion or an external upstream source, and which can be carried in large quantities. { 'wäsh ,lōd }

washover [GEOL] Material deposited by overwash, especially a small delta produced by storm waves and built on the landward side of a bar or barrier. Also known as storm delta; wave delta. { 'wäsh,ō·vər }

wash plain *See* outwash plain. { 'wäsh ,plān }

wash slope [GEOL] The gentle slope on a hillside occurring below the gravity slope and lying at the foot of an escarpment or steep rock face; usually covered by an accumulation of talus. Also known as haldenhang. { 'wäsh ,slōp }

waste plain *See* alluvial plain. { 'wāst ,plān }

water-bearing strata [GEOL] Ground layers below the standing water level. { 'wȯd·ər ¦ber·iŋ 'strad·ə }

water gap [GEOL] A deep and narrow pass that cuts to the base of a mountain ridge, and through which a stream flows; the Delaware Water Gap is an example. { 'wȯd·ər ,gap }

waterline *See* shoreline. { 'wȯd·ər,līn }

water opal *See* hyalite. { 'wȯd·ər ,ō·pəl }

water trap [GEOL] A chamber or part of a cave system that is filled with water, due to the dipping of the roof or ceiling below the water level. { 'wȯd·ər ,trap }

wattevilleite [MINERAL] $Na_2Ca(SO_4)_2 \cdot 4H_2O$ A snow white mineral consisting of a hydrated sulfate of sodium and calcium; occurs as aggregates of acicular or hairlike crystals. { 'wät·vi₁līt }

wave-built platform *See* alluvial terrace. { 'wāv ¦bilt 'plat,fȯrm }

wave-built terrace *See* alluvial terrace. { 'wāv ¦bilt 'ter·əs }

wave-cut bench [GEOL] A level or nearly level narrow platform produced by wave erosion and extending outward from the base of a wave-cut cliff. Also known as beach platform; high-water platform. { 'wāv ¦kət 'bench }

wave-cut cliff [GEOL] A cliff formed by the erosive action of waves on rock. { 'wāv ¦kət 'klif }

wave-cut notch [GEOL] An indentation cut into a sea cliff at water level by wave action. { 'wāv ¦kət 'näch }

wave-cut plain *See* wave-cut platform. { 'wāv ¦kət 'plān }

wave-cut platform [GEOL] A gently sloping surface which is produced by wave erosion and which extends into the sea for a considerable distance from the base of the wave-cut cliff. Also known as cut platform; erosion platform; strand flat; wave-cut plain; wave-cut terrace; wave platform. { 'wāv ¦kət 'plat,fȯrm }

wave-cut terrace *See* wave-cut platform. { 'wāv ¦kət 'ter·əs }

wave delta *See* washover. { 'wāv ,del·tə }

wave erosion *See* marine abrasion. { 'wāv i,rō·zhən }

wave line *See* swash mark. { 'wāv ,līn }

wavellite [MINERAL] $Al_3(PO_4)_2(OH)_3 \cdot 5H_2O$ A white to yellow, green, or black mineral crystallizing in the orthorhombic system and occurring in small hemispherical aggregates. { 'wā·və,līt }

wavemark *See* swash mark. { 'wāv,märk }

wave platform *See* wave-cut platform. { 'wāv 'plat,fȯrm }

wave ripple mark *See* oscillation ripple mark. { 'wāv 'rip·əl ,märk }

waxy [MINERAL] A type of mineral luster that is soft like that of wax. { 'wak·sē }

weathered layer [GEOPHYS] The zone of the earth which lies immediately below the surface and is characterized by low wave velocities. { 'weth·ərd 'lā·ər }

weathering [GEOL] Physical disintegration and chemical decomposition of earthy and rocky materials on exposure to atmospheric agents, producing an in-place mantle of waste. Also known as clastation; demorphism. { 'weth·ə,riŋ }

weathering correction [GEOPHYS] A velocity correction which is applied to seismic data, necessitated by the diminished velocity of seismic wave propagation in weathered rock. { 'weth·ə,riŋ kə,rek·shən }

weathering-potential index |GEOL| A measure of the susceptibility of a rock or mineral to weathering. { 'weṯẖ·ə,riŋ pəˈten·chəl ,in,dėks }

weathering rind |GEOL| The outer layer of a pebble, boulder, or other rock fragment that was formed as a result of chemical weathering. { 'weṯẖ·ər·iŋ ,rīnd }

weathering velocity |GEOPHYS| The velocity of propagation of seismic waves through weathered rock. { 'weṯẖ·ə,riŋ və,läs·əd·ē }

weather pit |GEOL| A shallow depression (depth up to 6 inches or 15 centimeters) on the flat or gently sloping summit of large exposures of granite or granitic rocks, attributed to strongly localized solvent action of impounded water. { 'weṯẖ·ər ,pit }

weberite |MINERAL| Na$_2$MgAlF$_7$ A light gray, orthorhombic mineral consisting of an aluminofluoride of sodium and magnesium; occurs as grains and masses. { 'vā·bə,rīt }

websterite *See* aluminite. { 'web·stə,rīt }

weddellite |MINERAL| CaC$_2$O$_4$·2H$_2$O A colorless to white or yellowish-brown to brown, tetragonal mineral consisting of calcium oxalate dihydrate. { wəˈde,līt }

wehrlite |MINERAL| BiTe A mineral that is a native alloy of bismuth and tellurium. Also known as mirror glance. |PETR| A peridotite composed principally of olivine and clinopyroxene with accessory opaque oxides. { 'wer,līt }

weibullite |MINERAL| Pb$_4$Bi$_6$S$_9$Se$_4$ A steel gray mineral consisting of lead bismuth sulfide with selenium replacing the sulfide; occurs in indistinct prismatic crystals in massive form. { 'wī,bù,līt }

weinschenkite |MINERAL| **1.** YPO$_4$·2H$_2$O A white mineral consisting of a hydrous yttrium phosphate. Also known as churchite. **2.** A dark-brown variety of hornblende high in ferric iron, aluminum, and water. { 'vīn,sheŋ,kīt }

weissite |MINERAL| Cu$_5$Te$_3$ A dark bluish-black mineral consisting of copper telluride; occurs in massive form. { 'wī,sīt }

welded tuff |PETR| A pyroclastic deposit hardened by the action of heat, pressure from overlying material, and hot gases. Also known as tuff lava. { 'wel·dəd 'təf }

welding |GEOL| Consolidation of sediments by pressure; water is squeezed out and cohering particles are brought within the limits of mutual molecular attraction. { 'weld·iŋ }

well-sorted |GEOL| Referring to a sorted sediment that consists of particles of approximately the same size and has a sorting coefficient of less than 2.5. { 'wel ¦sȯrd·əd }

Wenlockian |GEOL| A European stage of geologic time: Middle Silurian (above Tarannon, below Ludlovian). { wenˈläk·ē·ən }

Wentworth classification |GEOL| A logarithmic grade for size classification of sediment particles starting at 1 millimeter and using the ratio of 1/2 in one direction (and 2 in the other), providing diameter limits to the size classes of 1, 1/2, 1/4, etc. and 1, 2, 4, etc. { 'went,wərth ,klas·ə·fəˈkā·shən }

Wentworth scale |GEOL| A geometric grade scale for sedimentary particles ranging from clay particles (diameter less than 1/250 millimeter) to boulders (diameters greater than 256 millimeters), in which the size classes are related to one another by a constant ratio of 1/2 (4, 2, 1, 1/2, etc.). { 'went,wərth ,skāl }

Werfenian stage *See* Scythian stage. { verˈfē·nē·ən ,stäj }

wernerite *See* scapolite. { 'ver·nə,rīt }

Westphalian |GEOL| A European stage of Upper Carboniferous geologic time, forming the Middle Pennsylvanian, above upper Namurian and below Stephanian. { west 'fäl·yən }

wetted perimeter |GEOL| The portion of the perimeter of a steam channel cross section which is in contact with the water. { 'wed·əd pəˈrim·əd·ər }

whaleback dune |GEOL| A smooth, elongated mound or hill of desert sand shaped generally like a whale's back; formed by passage of a succession of longitudinal dunes along the same path. Also known as sand levee. { 'wāl,bak ,dün }

Wheelerian |GEOL| A North American stage of upper Pliocene geologic time, above the Venturian and below the Hallian. { wēˈlir·ē·ən }

wherryite |MINERAL| A light green mineral consisting of a basic carbonate-sulfate of lead and copper; occurs in massive form. { 'wer·ē,īt }

whewellite |MINERAL| $Ca(C_2O_4) \cdot H_2O$ A colorless or yellowish or brownish, monoclinic mineral consisting of calcium oxalate monohydrate; occurs as crystals. { 'hyü·ə,līt }

white clay See kaolin. { 'wīt 'klā }

white coal See tasmanite. { 'wīt 'kōl }

white cobalt See cobaltite. { 'wīt 'kō,bȯlt }

white feldspar See albite. { 'wīt 'fel,spär }

white garnet See leucite. { 'wīt 'gär·nət }

white iron ore See siderite. { 'wīt 'ī·ərn ,ȯr }

white mica See muscovite. { 'wīt 'mī·kə }

white nickel See rammelsbergite. { 'wīt 'nik·əl }

white olivine See forsterite. { 'wīt 'äl·ə,vēn }

Whiterock |GEOL| A North American stage of lowermost Middle Ordovician geologic time, above lower Ordovician and below Marmor. { 'wīt,räk }

white schorl See albite. { 'wīt 'shȯrl }

white tellurium See sylvanite. { 'wīt tə'lu̇r·ē·əm }

whitleyite |GEOL| An achondritic stony meteorite consisting essentially of enstatite with fragments of black chondrite. { 'wit·lē,īt }

whitlockite |MINERAL| $Ca_9(Mg,Fe)H(PO_4)_7$ A rare mineral that forms hexagonal crystals. { 'wit,lä,kīt }

wiborgite See rapakivi. { 'wī,bȯr,gīt }

wichtisite See tachylite. { 'wik·tə,sīt }

Widmanstatten patterns |GEOL| Characteristic figures that appear on the surface of an iron meteorite when the meteorite is cut, polished, and etched with acid. { 'vit·mən,shtät·ən ,pad·ərnz }

Wiik classification |GEOL| A classification of carbonaceous chondrites into three types, C_1, C_2, and C_3. { 'wik ,klas·ə·fə'kā·shən }

Wilderness |GEOL| A North American stage of Middle Ordovician geologic time, above Porterfield and below Trentonian. { 'wil·dər·nəs }

wildflysch |GEOL| A type of flysch facies that represents a stratigraphic unit with irregularly sorted boulders resulting from fragmentation, and twisted, confused beds resulting from slumping or sliding due to the influence of gravity. { 'vilt,flish }

wilkeite |MINERAL| $Ca_5(SiO_4,PO_4,SO_4)_3(O,OH,F)$ A rose red or yellow, hexagonal mineral consisting of a basic sulfate-silicate-phosphate of calcium. { 'wil·kē,īt }

willemite |MINERAL| Zn_2SiO_4 A white, greenish-yellow, green, reddish, or brown mineral that forms rhombohedral crystals and exhibits intense bright-yellow fluorescence in ultraviolet light; a minor ore of zinc. { 'wil·ə,mīt }

Williamsoniaceae |PALEOBOT| A family of extinct plants in the order Cycadeoidales distinguished by profuse branching. { ,wil·yəm,sō·nē'ās·ē,ē }

wind-cut stone See ventifact. { 'win ¦kət 'stōn }

wind erosion |GEOL| Detachment, transportation, and deposition of loose topsoil or sand by the action of wind. { 'wind i,rō·zhən }

wind gap |GEOL| A shallow, relatively high-level notch in the upper part of a mountain ridge, usually an abandoned water gap. Also known as air gap; wind valley. { 'win ,gap }

wind-grooved stone See ventifact. { 'win ¦grüvd 'stōn }

window |GEOL| A break caused by erosion of a thrust sheet or a large recumbent anticline that exposes the rocks beneath the thrust sheet. Also known as fenster. |GEOPHYS| Any range of wavelengths in the electromagnetic spectrum to which the atmosphere is transparent. { 'win·dō }

wind-polished stone See ventifact. { 'win ¦päl·əsht 'stōn }

windrow |GEOL| Any accumulation of material formed by wind or tide action. { 'win,drō }

wind-scoured stone See ventifact. { 'win ¦skau̇rd 'stōn }

wind-shaped stone See ventifact. { 'win ¦shāpt 'stōn }

wind valley See wind gap. { 'win ,val·ē }

wing See vesicle. { wiŋ }

winter-talus ridge |GEOL| A wall-like arcuate ridge on the floor of a cirque formed by freezing activity that dislodged boulders from a cirque wall covered with a snowbank. Also known as nivation ridge.   { 'win·tər 'lā·ləs ,rij }

Wisconsin   [GEOL] Pertaining to the fourth, and last, glacial stage of the Pleistocene epoch in North America; followed the Sangamon interglacial, beginning about 85,000 ± 15,000 years ago and ending 7000 years ago.   { wi'skän·sən }

witherite   [MINERAL] $BaCO_3$ A yellowish- or grayish-white mineral of the aragonite group that has orthorhombic symmetry, hardness of 31/4 on Mohs scale, and specific gravity 4.3.   { 'with·ə,rīt }

wittichenite   [MINERAL] Cu_3BiS_3 A steel gray to tin white, orthorhombic mineral consisting of copper bismuth sulfide; occurs in tabular and massive form.   { 'wid·ə·kə,nīt }

wittite   [MINERAL] $Pb_5Bi_6(S,Se)_{14}$ A light lead gray, orthorhombic or monoclinic mineral consisting of a sulfide of lead and bismuth.   { 'wi,tīt }

wolfachite   [MINERAL] $Ni(As,Sb)S$ A silver white to tin white mineral consisting of nickel, arsenic, and antimony sulfide; occurs in small crystals and in aggregates.   { 'vōl,fäk,īt }

Wolfcampian   [GEOL] A North American provincial series of geologic time; lowermost Permian (below Leonardian, above Virgilian of Pennsylvania).   { wúlf'kam·pē·ən }

wolfeite   [MINERAL] $(Fe,Mn)_2(PO_4)(OH)$ A pinkish, wine yellow to yellowish-brown or reddish-brown, monoclinic mineral consisting of a basic phosphate of iron and manganese.   { 'wúl,fīt }

wolfram See wolframite.   { 'wúl·frəm }

wolframine See wolframite.   { 'wúl·frə,mēn }

wolframite   [MINERAL] $(Fe,Mn)WO_4$ A brownish- or grayish-black mineral occurring in short monoclinic, prismatic, bladed crystals; the most important ore of tungsten. Also known as tobacco jack; wolfram; wolframine.   { 'wúl·frə,mīt }

wollastonite   [MINERAL] $CaSiO_3$ A white to gray inosilicate mineral (a pyroxenoid) that crystallizes in the triclinic system in tabular crystals and has a pearly or silky luster on the cleavages; hardness is 5–5.5 on Mohs scale, and specific gravity is 2.85.   Also known as tabular spar.   { 'wúl·ə·stə,nīt }

wood coal See bituminous wood.   { 'wúd 'kōl }

wood copper See olivenite.   { 'wúd 'käp·ər }

woodhouseite   [MINERAL] $CaAl_3(PO_4)(SO_4)(OH)_6$ A colorless to flesh-colored or white, hexagonal mineral consisting of a basic sulfate-phosphate of calcium and aluminum; occurs in small crystals and tabular form.   { 'wúd,haú,sīt }

woodstone See silicified wood.   { 'wúd,stōn }

wood tin   [MINERAL] A riniform, brownish variety of cassiterite with fibers radiating concentrically and resembling dry wood.   Also known as dneprovskite.   { 'wúd 'tin }

woodwardite   [MINERAL] $Cu_4Al_2(SO_4)(OH)_{12}·2-4H_2O$ A greenish-blue to turquoise blue mineral consisting of a hydrated basic sulfate of copper and aluminum; occurs as botryoidal concretions and in spherulitic form.   { 'wúd·wər,dīt }

woody lignite See bituminous wood.   { 'wúd·ē 'lig,nīt }

world rift system   [GEOL] The system of interconnected midocean ridges which is the locus of tensional splitting and magma upwelling believed responsible for sea-floor spreading.   { 'wərld 'rift ,sis·təm }

worobieffite See vorobyevite.   { wə'rō·bē·ə,fīt }

wrench fault   [GEOL] A lateral fault with a more or less vertical fault surface.   Also known as basculating fault; torsion fault.   { 'rench ,fólt }

wulfenite   [MINERAL] $PbMoO_4$ A yellow, orange, orange-yellow, or orange-red tetragonal mineral occurring in tabular crystals or granular masses; an ore of molybdenum. Also known as yellow lead ore.   { 'wúl·fə,nīt }

Würm   [GEOL] **1.** A European stage of geologic time: uppermost Pleistocene (above Riss, below Holocene).   **2.** Pertaining to the fourth glaciation of the Pleistocene epoch in the Alps, equivalent to the Wisconsin glaciation in North America, following the Riss-Würm interglacial.   { vúrm }

wurtzilite   [GEOL] A black, massive, sectile, infusible, asphaltic pyrobitumen derived from the metamorphosis of petroleum.   { 'wərt·sə,līt }

wurtzite [MINERAL] (Zn,Fe)S A brownish-black hexagonal mineral consisting of zinc sulfide and occurring in hemimorphic pyramidal crystals, or in radiating needles and bundles. { 'wərt,sīt }

wustite [MINERAL] FeO Iron oxide. { 'wùs,tīt }

Wynyardiidae [PALEON] An extinct family of herbivorous marsupial mammals in the order Diprotodonta. { ‚win·yər'dī·ə,dē }

X

xanthochroite *See* greenockite. { ¦zan·thrə'krō,īt }

xanthoconite |MINERAL| Ag_3AsS_3 A dark red to dull orange to clove brown mineral consisting of silver arsenic sulfide. { zan'thäk·ə,nīt }

xanthophyllite *See* clintonite. { ,zan·thə'fi,līt }

xanthosiderite *See* goethite. { ¦zan·thō'sī·də,rīt }

xanthoxenite |MINERAL| $Ca_2Fe(PO_4)_2(OH)·1^1/_2H_2O$ A pale yellow to brownish-yellow, monoclinic or triclinic mineral consisting of a hydrated basic phosphate of calcium and iron; occurs as masses and crusts. { zan,thäk·sə,nīt }

xenoblast |MINERAL| A mineral which has grown during metamorphism without development of its characteristic crystal faces. Also known as allotrioblast. { 'zēn·ə,blast }

xenolith |PETR| An inclusion in an igneous rock which is not genetically related, such as an unmelted fragment of country rock. Also known as accidental inclusion; exogenous inclusion. { 'zēn·ə,lith }

xenomorphic *See* allotriomorphic. { ¦zēn·ə¦mȯr·fik }

xenothermal |MINERAL| Pertaining to a mineral deposit formed at high temperature but at shallow to moderate depth. { ¦zēn·ə¦thər·məl }

xenotime |MINERAL| $Y(PO_4)$ A tetragonal mineral of varying color, consisting of yttrium phosphate. { 'zēn·ə,tīm }

Xenungulata |PALEON| An order of large, digitigrade, extinct, tapirlike mammals with relatively short, slender limbs and five-toed feet with broad, flat phalanges; restricted to the Paleocene deposits of Brazil and Argentina. { zə,nún¦·gyə 'läd·ə }

Xeralf |GEOL| A suborder of the soil order Alfisol, having good drainage, and found in regions with rainy winters and dry summers in mediterranean climates; the surface horizons tend to become massive and hard during the dry seasons, with some soils having duripans that interfere with root growth. { 'zir,älf }

Xerert |GEOL| A suborder of the soil order Vertisol, formed in a Mediterranean climate; wide surface cracks open and close once a year. { 'zir,ərt }

Xeroll |GEOL| A suborder of the soil order Mollisol, formed in a xeric moisture regime; may have a calcic, petrocalcic, or gypsic horizon, or a duripan. { 'zir,ȯl }

xerothermal period *See* xerothermic period. { ¦zir·ə¦thər·məl 'pir·ē·əd }

xerothermic period |GEOL| A postglacial interval of a warmer, drier climate. Also known as xerothermal period. { ¦zir·ə¦thər·mik 'pir·ē·əd }

Xerult |GEOL| A suborder of the soil order Ultisol, formed in a xeric moisture regime; brownish or reddish soil with a low to moderate organic-carbon content. { 'zir,əlt }

Xiphodontidae |PALEON| A family of primitive tylopod ruminants in the superfamily Anaplotherioidea from the late Eocene to the middle Oligocene of Europe. { ,zif·ə'dänt·ə,dē }

xylinite |GEOL| A variety of provitrinite consisting of xylem or lignified tissue. { 'zī·lə,nīt }

xyloid coal *See* bituminous wood. { 'zī,lȯid 'kōl }

xyloid lignite *See* bituminous wood. { 'zī,lȯid 'lig,nīt }

yardang |GEOL| A long, irregular ridge with a sharp crest sited between two round-bottomed troughs that have been carved by wind erosion in a desert region. { 'yär₁daŋ }

yardang trough |GEOL| A long, shallow, round-bottomed groove, furrow, or trough cut into a desert floor by wind erosion and separated by a yardang from the neighboring trough. { 'yär₁daŋ 'tròf }

Yarmouth interglacial |GEOL| The second interglacial stage of the Pleistocene epoch in North America, following the Kansan glacial stage and before the Illinoian. { 'yär·məth ¦in·tər'glā·shəl }

yellow arsenic See orpiment. { yel·ō 'ärs·ən·ik }

yellow coal See tasmanite. { 'yel·ō 'kōl }

yellow copperas See copiapite. { 'yel·ō 'käp·rəs }

yellow lead ore See wulfenite. { 'yel·ō 'led ‚òr }

yellow mud |GEOL| Mud containing sediment having a characteristic yellow color, resulting from certain iron compounds. { 'yel·ō 'məd }

yellow pyrite See chalcopyrite. { 'yel·ō 'pī₁rīt }

yellow quartz See citrine. { 'yel·ō 'kwòrts }

yellow tellurium See sylvanite. { 'yel·ō tə'lùr·ē·əm }

yoked basin See zeugogeosyncline. { 'yōkt 'bās·ən }

Yorkian |GEOL| A European stage of geologic time forming part of the lower Upper Carboniferous, above Lanarkian and below Staffordian, equivalent to part of the lower Westphalian. { 'yòr·kē·ən }

Younginiformes |PALEON| A suborder of extinct small lizardlike reptiles in the order Eosuchia, ranging from the Middle Permian to the Lower Triassic in South Africa. { ‚yəŋ·gə·nə'fòr₁mēz }

youth |GEOL| The first stage of the cycle of erosion in which the original surface or structure is the dominant topographic feature; characterized by broad, flat-topped interstream divides, numerous swamps and shallow lakes, and progressive increase of local relief. Also known as topographic youth. { yüth }

yttrocrasite |MINERAL| (Y,Th,U,Ca)₂Ti₄O₁₁ A black, orthorhombic mineral consisting of an oxide of rare earths and titanium. { ‚i·trə'krā₁sīt }

yttrotantalite |MINERAL| (Y,U,Fe)(Ta,Nb)O₄ A black or brown, orthorhombic mineral consisting of an oxide of iron, yttrium, uranium, columbium, and tantalum; occurs in prismatic and tabular form. { ‚i·trə'tant·əl₁īt }

yugawaralite |MINERAL| CaAl₂Si₆O₁₆·4H₂O A zeolite mineral consisting of hydrous calcium aluminum silicate. { ‚yü·gə'wär·ə₁līt }

Z

Zalambdalestidae |PALEON| A family of extinct insectivorous mammals belonging to the group Proteutherea; they occur in the Late Cretaceous of Mongolia. { zə‚lam·də'les·tə‚dē }

zaratite |MINERAL| $Ni_3(CO_3)(OH)_4 \cdot 4H_2O$ An emerald-green mineral consisting of a hydrous basic nickel carbonate and occurring in incrustations or compact masses. { 'zär·ə‚tīt }

Zechstein |GEOL| A European series of geologic time, especially in Germany: Upper Permian (above Rothliegende). { 'zek‚shtīn }

Zemorrian |GEOL| A North American stage of Oligocene and Miocene geologic time, above Refugian and below Saucesian. { zə'mȯr·ē·ən }

zeolite |MINERAL| **1.** A group of white or colorless, sometimes red or yellow, hydrous tectosilicate minerals characterized by an aluminosilicate tetrahedral framework, ion-exchangeable large cations, and loosely held water molecules permitting reversible dehydration. **2.** Any mineral of the zeolite group, such as analcime, chabazite, natrolite, and stilbite. { 'zē·ə‚līt }

zeolite facies |PETR| Metamorphic rocks formed in the transitional period from diagenesis to metamorphism, at pressures of about 2000–3000 bars and temperatures of 200–300°C. { 'zē·ə‚līt 'fā·shēz }

zeolitization |GEOL| Introduction of or replacement by a zeolite mineral. { zē‚äl·əd·ə'zā·shən }

zero curtain |GEOL| The layer of ground between the active layer and permafrost where the temperature remains nearly constant at 0°C. { 'zir·ō ‚kərt·ən }

zeugogeosyncline |GEOL| A geosyncline in a craton or stable area, within which is also an uplifted area, receiving clastic sediments. Also known as yoked basin. { ‚zü·gō‚jē·ō'sin‚klīn }

zeunerite |MINERAL| $Cu(UO_2)_2(AsO_4)_2 \cdot 10\text{-}16H_2O$ A green secondary mineral of the autunite group consisting of a hydrous copper uranium arsenate; it is isomorphous with uranospinite. { 'zȯi·nə‚rīt }

zeylanite *See* ceylonite. { 'zā·lə‚nīt }

zincaluminite |MINERAL| $Zn_6Al_6(SO_4)_2(OH)_{26} \cdot 5H_2O$ A white to bluish-white and pale blue mineral consisting of a basic hydrated sulfate of zinc and aluminum; occurs in tufts and crusts. { ‚zink·ə'lü·mə‚nīt }

zincite |MINERAL| (Zn,Mn)O A deep-red to orange-yellow brittle mineral; an ore of zinc. Also known as red oxide of zinc; red zinc ore; ruby zinc; spartalite. { 'zin‚kīt }

zinckenite *See* zinkenite. { 'zin·kə‚nīt }

zinc spar *See* smithsonite. { 'zink 'spär }

zinc spinel *See* gahnite. { 'zink spə'nel }

zinkenite |MINERAL| $Pb_6Sb_{14}S_{27}$ A steel-gray orthorhombic mineral consisting of a lead antimony sulfide and occurring in crystals and in masses; has metallic luster, hardness of 3–3.5 on Mohs scale, and specific gravity of 5.30–5.35. Also spelled zinckenite. { 'zin·kə‚nīt }

zinnwaldite |MINERAL| $K_2(Li,Fe,Al)_6(Si,Al)_8O_{20}(OH,F)_4$ A pale-violet, yellowish, brown, or dark-gray mica mineral; an iron-bearing variety of lepidolite; the characteristic mica of greisens. { 'tsin‚väl‚dīt }

zippeite [MINERAL] $(UO_2)_2(SO_4)(OH)_2 \cdot nH_2O$ An orange-yellow, orthorhombic mineral consisting of a hydrated basic sulfate of uranium. { 'tsip·ə,īt }

zircon [MINERAL] $ZrSiO_4$ A brown, green, pale-blue, red, orange, golden-yellow, grayish, or colorless neosilicate mineral occurring in tetragonal prisms; it is the chief source of zirconium; the colorless varieties provide brilliant gemstones. Also known as hyacinth; jacinth; zirconite. { 'zər,kän }

zirconite See zircon. { 'zər·kə,nīt }

zirkelite [MINERAL] A black mineral consisting of an oxide of zirconium, titanium, calcium, ferrous iron, thorium, uranium, and rare earths. { 'zər·kə,līt }

zodiacal pyramid [GEOPHYS] The pattern formed by the zodiacal light. Also known as zodiacal cone. { zō'dī·ə·kəl 'pir·ə·mid }

zoisite [MINERAL] $Ca_2Al_3Si_3O_{12}(OH)$ A white, gray, brown, green, or rose-red orthorhombic mineral of the epidote group consisting of a basic calcium aluminum silicate and occurring massive or in prismatic crystals. { 'zȯi,sīt }

zonal soil [GEOL] In early classification systems in the United States, a soil order including soils with well-developed characteristics that reflect the influence of agents of soil genesis. Also known as mature soil. { 'zōn·əl 'sȯil }

zonal theory [GEOL] A theory of the formation of mineral deposition and sequence patterns, based on the changes in a mineral-bearing fluid as it passes upward from a magmatic source. { 'zōn·əl 'thē·ə·rē }

zonation [GEOL] The condition of being arranged in zones. { zō'nā·shən }

zone [GEOL] A belt, layer, band, or strip of earth material such as rock or soil. { zōn }

zone of accumulation See B horizon. { 'zōn əv ə,kyü·mə'lā·shən }

zone of aeration [GEOL] The subsurface sediment above the water table containing air and water. Also known as unsaturated zone; vadose zone; zone of suspended water. { 'zōn əv e'rā·shən }

zone of cementation [GEOL] The layer of the earth's crust in which unconsolidated deposits are cemented by percolating water containing dissolved minerals from the overlying zone of weathering. Also known as belt of cementation. { 'zōn əv ,sē,men'tā·shən }

zone of illuviation See B horizon. { 'zōn əv i,lü·vē'ā·shən }

zone of soil water See belt of soil water. { 'zōn əv 'sȯil ,wȯd·ər }

zone of suspended water See zone of aeration. { 'zōn əv sə¦spen·dəd ,wȯd·ər }

zonochlorite See pumpellyite. { ,zō·nō'klȯr,īt }

zorsite [MINERAL] $Ca_2Al_3Si_3O_{12}(OH)$ White, gray, brown, green, or rose-red orthorhombic mineral of the epidote group; an essential constituent of saussurite. { 'zȯr,sīt }

Zosterophyllatae See Zosterophyllopsida. { ¦zäs·tə·rō'fil·ə,tē }

Zosterophyllopsida [PALEOBOT] A group of early land vascular plants ranging from the Lower to the Upper Devonian; individuals were leafless and rootless. { ¦zäs·tə·rō·fə'läp·səd·ə }

Zwischengebirge See median mass. { 'tsfish·ən,gə'bir·gə }

Appendix

Appendix

Equivalents of commonly used units for the U.S. Customary System and the metric system

1 inch = 2.5 centimeters (25 millimeters)	1 centimeter = 0.4 inch	1 inch = 0.083 foot
1 foot = 0.3 meter (30 centimeters)	1 meter = 3.3 feet	1 foot = 0.33 yard (12 inches)
1 yard = 0.9 meter	1 meter = 1.1 yards	1 yard = 3 feet (36 inches)
1 mile = 1.6 kilometers	1 kilometer = 0.62 mile	1 mile = 5280 feet (1760 yards)
1 acre = 0.4 hectare	1 hectare = 2.47 acres	
1 acre = 4047 square meters	1 square meter = 0.00025 acre	
1 gallon = 3.8 liters	1 liter = 1.06 quarts = 0.26 gallon	1 quart = 0.25 gallon (32 ounces; 2 pints)
1 fluid ounce = 29.6 milliliters	1 milliliter = 0.034 fluid ounce	1 pint = 0.125 gallon (16 ounces)
32 fluid ounces = 946.4 milliliters		1 gallon = 4 quarts (8 pints)
1 quart = 0.95 liter	1 gram = 0.035 ounce	1 ounce = 0.0625 pound
1 ounce = 28.35 grams	1 kilogram = 2.2 pounds	1 pound = 16 ounces
1 pound = 0.45 kilogram	1 kilogram = 1.1×10^{-3} ton	1 ton = 2000 pounds
1 ton = 907.18 kilograms		

$$°F = (1.8 \times °C) + 32$$

$$°C = (°F - 32) \div 1.8$$

Appendix

Conversion factors for the U.S. Customary System, metric system, and International System

A. Units of length

Units	cm	m	in.	ft	yd	mi
1 cm =	1	0.01	0.3937008	0.03280840	0.01093613	6.213712×10^{-6}
1 m =	100.	1	39.37008	3.280840	1.093613	6.213712×10^{-4}
1 in. =	2.54	0.0254	1	0.08333333...	0.02777777...	1.578283×10^{-5}
1 ft =	30.48	0.3048	12.	1	0.3333333...	$1.893939... \times 10^{-4}$
1 yd =	91.44	0.9144	36.	3.	1	$5.681818... \times 10^{-4}$
1 mi =	1.609344×10^{5}	1.609344×10^{3}	6.336×10^{4}	5280.	1760.	1

B. Units of area

Units	cm^2	m^2	$in.^2$	ft^2	yd^2	mi^2
1 cm^2 =	1	10^{-4}	0.1550003	1.076391×10^{-3}	1.195990×10^{-4}	3.861022×10^{-11}
1 m^2 =	10^{4}	1	1550.003	10.76391	1.195990	3.861022×10^{-7}
1 $in.^2$ =	6.4516	6.4516×10^{-4}	1	$6.944444... \times 10^{-3}$	7.716049×10^{-4}	2.490977×10^{-10}
1 ft^2 =	929.0304	0.09290304	144.	1	0.1111111...	3.587007×10^{-8}
1 yd^2 =	8361.273	0.8361273	1296.	9.	1	3.228306×10^{-7}
1 mi^2 =	2.589988×10^{10}	2.589988×10^{6}	4.014490×10^{9}	2.78784×10^{7}	3.0976×10^{6}	1

C. Units of volume

Units	m^3	cm^3	liter	$in.^3$	ft^3	qt	gal
1 m³ =	1	10^6	10^3	6.102374×10^4	35.31467×10^{-3}	1.056688	264.1721
1 cm³ =	10^{-6}	1	10^{-3}	0.06102374	3.531467×10^{-5}	1.056688×10^{-3}	2.641721×10^{-4}
1 liter =	10^{-3}	1000.	1	61.02374	0.03531467	1.056688	0.2641721
1 in.³ =	1.638706×10^{-5}	16.38706	0.01638706	1	5.787037×10^{-4}	0.01731602	4.329004×10^{-3}
1 ft³ =	2.831685×10^{-2}	28316.85	28.31685	1728.	1	2.992208	7.480520
1 qt =	9.463529×10^{-4}	946.3529	0.9463529	57.75	0.03342014	1	0.25
1 gal (U.S.) =	3.785412×10^{-3}	3785.412	3.785412	231.	0.1336806	4.	1

D. Units of mass

Units	g	kg	oz	lb	metric ton	ton
1 g =	1	10^{-3}	0.03527396	2.204623×10^{-3}	10^{-6}	1.102311×10^{-6}
1 kg =	1000.	1	35.27396	2.204623	10^{-3}	1.102311×10^{-3}
1 oz (avdp) =	28.34952	0.02834952	1	0.0625	2.834952×10^{-5}	3.125×10^{-5}
1 lb (avdp) =	453.5924	0.4535924	16.	1	4.535924×10^{-4}	$5. \times 10^{-4}$
1 metric ton =	10^8	1000.	35273.96	2204.623	1	1.102311
1 ton =	907184.7	907.1847	32000.	2000.	0.9071847	1

Appendix

Conversion factors for the U.S. Customary System, metric system, and International System (cont.)

E. Units of density

Units	$g \cdot cm^{-3}$	$g \cdot L^{-1}, kg \cdot m^{-3}$	$oz \cdot in.^{-3}$	$lb \cdot in.^{-3}$	$lb \cdot ft^{-3}$	$lb \cdot gal^{-1}$
$1\ g \cdot cm^{-3}$	$= 1$	$1000.$	0.5780365	0.03612728	62.42795	8.345403
$1\ g \cdot L^{-1}, kg \cdot m^{-3}$	$= 10^{-3}$	1	5.780365×10^{-4}	3.612728×10^{-5}	0.06242795	8.345403×10^{-3}
$1\ oz \cdot in.^{-3}$	$= 1.729994$	1729.994	1	0.0625	$108.$	14.4375
$1\ lb \cdot in.^{-3}$	$= 27.67991$	27679.91	$16.$	1	$1728.$	$231.$
$1\ lb \cdot ft^{-3}$	$= 0.01601847$	16.01847	9.259259×10^{-3}	5.787037×10^{-4}	1	0.1336806
$1\ lb \cdot gal^{-1}$	$= 0.1198264$	119.8264	4.749536×10^{-3}	4.329004×10^{-3}	7.480519	1

F. Units of pressure

Units	$Pa, N \cdot m^{-2}$	$dyn \cdot cm^{-2}$	bar	atm	$kgf \cdot cm^{-2}$	$mmHg\ (torr)$	$in.\ Hg$	$lbf \cdot in.^{-2}$
$1\ Pa, 1\ N \cdot m^{-2}$	$= 1$	10	10^{-5}	9.869233×10^{-6}	1.019716×10^{-5}	7.500617×10^{-3}	2.952999×10^{-4}	1.450377×10^{-4}
$1\ dyn \cdot cm^{-2}$	$= 0.1$	1	10^{-6}	9.869233×10^{-7}	1.019716×10^{-6}	7.500617×10^{-4}	2.952999×10^{-5}	1.450377×10^{-5}
$1\ bar$	$= 10^{5}$	10^{6}	1	0.9869233	1.019716	750.0617	29.52999	14.50377
$1\ atm$	$= 101325$	1013250	1.01325	1	1.033227	$760.$	29.92126	14.69595
$1\ kgf \cdot cm^{-2}$	$= 98066.5$	980665	0.980665	0.9678411	1	735.5592	28.95903	14.22334
$1\ mmHg\ (torr)$	$= 133.3224$	1333.224	1.333224×10^{3}	1.315789×10^{-3}	1.359510×10^{-3}	1	$0.0393700 8$	0.0193678
$1\ in.\ Hg$	$= 3386.388$	33863.88	0.03386388	0.03342105	0.03453155	25.4	1	0.4911541
$1\ lbf \cdot in.^{-2}$	$= 6894.757$	68947.57	0.06894757	0.06804596	0.07030696	51.71493	2.036021	1

G. Units of energy

Units	g mass (energy equiv)	J	eV	cal	cal_{IT}	Btu_{IT}	kWh	hp-h	ft-lbf	$ft^3 \cdot lbf \cdot in.^{-2}$	liter-atm
1 g mass (energy equiv)	= 1	8.987752×10^{13}	5.609589×10^{32}	2.148076×10^{3}	2.146640×10^{13}	8.518555×10^{10}	2.496542×10^{7}	3.347918×10^{7}	6.628878×10^{13}	4.603388×10^{11}	8.870024×10^{11}
1 J	= 1.112650×10^{-14}	1	6.241510×10^{18}	0.2390057	0.2388459	9.478172×10^{-4}	$2.777777... \times 10^{-7}$	3.725062×10^{-7}	0.7375622	5.121960×10^{-3}	9.869233×10^{-3}
1 eV	= 1.782662×10^{-33}	1.602176×10^{-19}	1	3.829293×10^{-20}	3.826733×10^{-20}	1.518570×10^{-22}	4.450490×10^{-26}	5.968206×10^{-26}	1.181705×10^{-19}	8.206283×10^{-22}	1.581225×10^{-21}
1 cal	= 4.655328×10^{-14}	4.184	2.611448×10^{19}	1	0.9993312	3.965667×10^{-3}	$1.162222... \times 10^{-6}$	1.558562×10^{-6}	3.085960	2.143028×10^{-2}	0.04129287
1 cal_{IT}	= 4.658443×10^{-14}	4.1868	2.613195×10^{19}	1.000669	1	3.968321×10^{-3}	1.163×10^{-6}	1.559609×10^{-6}	3.088025	2.144462×10^{-2}	0.04132050
1 Btu_{IT}	= 1.173908×10^{-11}	1055.056	6.585141×10^{21}	252.1644	251.9958	1	2.930711×10^{-4}	3.930148×10^{-4}	778.1693	5.403953	10.41259
1 kWh	= 4.005540×10^{-8}	3600000.	2.246944×10^{25}	860420.7	859845.2	3412.142	1	1.341022	2655224.	18349.06	35529.24
1 hp-h	= 2.986931×10^{-8}	2384519.	1.675545×10^{25}	641615.6	641186.5	2544.33	0.7456998	1	1980000.	13750.	26494.15
1 ft-lbf	= 1.508551×10^{-14}	1.355818	8.462351×10^{18}	0.3240483	0.3238315	1.285067×10^{-3}	3.766161×10^{-7}	$5.050505... \times 10^{-7}$	1	$6.944444... \times 10^{-3}$	0.01338088
1 ft^3 lbf $\cdot$ in.$^{-2}$	= 2.172313×10^{-12}	195.2378	1.218579×10^{21}	46.66295.	46.63174	0.1850497	5.423272×10^{-5}	$7.272727... \times 10^{-5}$	144.	1	1.926847
1 liter-atm	= 1.127393×10^{-12}	101.325	6.324210×10^{20}	24.21726	24.20106	0.09603757	2.814583×10^{-5}	3.774419×10^{-5}	74.73349	0.5189825	1

Appendix

Periodic table

(The atomic numbers are listed above the symbols identifying the elements. The heavy line separates metals from nonmetals.)

s																		p
1																		18
1 H Hydrogen	2												13	14	15	16	17	2 He Helium
3 Li Lithium	4 Be Beryllium												5 B Boron	6 C Carbon	7 N Nitrogen	8 O Oxygen	9 F Fluorine	10 Ne Neon
11 Na Sodium	12 Mg Magnesium	3	4	5	6	7	8	9	10	11	12		13 Al Aluminum	14 Si Silicon	15 P Phosphorus	16 S Sulfur	17 Cl Chlorine	18 Ar Argon
19 K Potassium	20 Ca Calcium	21 Sc Scandium	22 Ti Titanium	23 V Vanadium	24 Cr Chromium	25 Mn Manganese	26 Fe Iron	27 Co Cobalt	28 Ni Nickel	29 Cu Copper	30 Zn Zinc		31 Ga Gallium	32 Ge Germanium	33 As Arsenic	34 Se Selenium	35 Br Bromine	36 Kr Krypton
37 Rb Rubidium	38 Sr Strontium	39 Y Yttrium	40 Zr Zirconium	41 Nb Niobium	42 Mo Molybdenum	43 Tc Technetium	44 Ru Ruthenium	45 Rh Rhodium	46 Pd Palladium	47 Ag Silver	48 Cd Cadmium		49 In Indium	50 Sn Tin	51 Sb Antimony	52 Te Tellurium	53 I Iodine	54 Xe Xenon
55 Cs Cesium	56 Ba Barium	71 Lu Lutetium	72 Hf Hafnium	73 Ta Tantalum	74 W Tungsten	75 Re Rhenium	76 Os Osmium	77 Ir Iridium	78 Pt Platinum	79 Au Gold	80 Hg Mercury		81 Tl Thallium	82 Pb Lead	83 Bi Bismuth	84 Po Polonium	85 At Astatine	86 Rn Radon
87 Fr Francium	88 Ra Radium	103 Lr Lawrencium	104 Rf Rutherfordium	105 Db Dubnium	106 Sg Seaborgium	107 Bh Bohrium	108 Hs Hassium	109 Mt Meitnerium	110	111	112		113	114	115	116	117	118

f

57 La Lanthanum	58 Ce Cerium	59 Pr Praseodymium	60 Nd Neodymium	61 Pm Promethium	62 Sm Samarium	63 Eu Europium	64 Gd Gadolinium	65 Tb Terbium	66 Dy Dysprosium	67 Ho Holmium	68 Er Erbium	69 Tm Thulium	70 Yb Ytterbium
89 Ac Actinium	90 Th Thorium	91 Pa Protactinium	92 U Uranium	93 Np Neptunium	94 Pu Plutonium	95 Am Americium	96 Cm Curium	97 Bk Berkelium	98 Cf Californium	99 Es Einsteinium	100 Fm Fermium	101 Md Mendelevium	102 No Nobelium

Appendix

Principal regions of a standard earth model

Layer	Approximate depth range, mi (km)
Ocean layer	0–1.8 (0–3)
Upper and lower crust	1.8–15 (3–24)
Lithosphere below the crust	15–50 (24–80)
Asthenosphere	50–140 (80–220)
Upper mantle above phase or compositional changes near 240 mi (400 km)	140–240 (220–400)
Transition region between phase or compositional changes near 240 and 416 mi (400 and 670 km)	240–416 (400–670)
Lower mantle above core-mantle boundary layer	416–1703 (670–2741)
Core-mantle boundary layer	1703–1796 (2741–2891)
Outer core	1796–3200 (2891–5150)
Inner core	3200–3959 (5150–6371)

Physical properties of some common rocks

Rock	Specific gravity	Porosity, %	Compressive strength, psi*	Tensile strength, psi*
Igneous				
Granite	2.67	1	30,000–50,000	500–1000
Basalt	2.75	1	25,000–30,000	
Sedimentary				
Sandstone	2.1–2.5	5–30	5,000–15,000	100–200
Shale	1.9–2.4	7–25	5,000–10,000	
Limestone	2.2–2.5	2–20	2,000–20,000	400–850
Metamorphic				
Marble	2.5–2.8	0.5–2	10,000–30,000	700–1000
Quartzite	2.5–2.6	1–2	15,000–40,000	
Slate	2.6–2.8	0.5–5	15,000–30,000	

*1 psi = 6.9 kPa.

Appendix

Approximate concentration of ore elements in earth's crust and In ores

Element	In average igneous rocks, %	In ores, %
Iron	5.0	50
Copper	0.007	0.5–5
Zinc	0.013	1.3–13
Lead	0.0016	1.6–16
Tin	0.004	0.01*–1
Silver	0.00001	0.05
Gold	0.0000005	0.0000015*–0.01
Uranium	0.0002	0.2
Tungsten	0.003	0.5
Molybdenum	0.001	0.6

*Placer deposits.

Elemental composition of earth's crust based on igneous and sedimentary rock

Element	Weight %	Atomic %	Volume %
Oxygen	46.71	60.5	94.24
Silicon	27.69	20.5	0.51
Titanium	0.62	0.3	0.03
Aluminum	8.07	6.2	0.44
Iron	5.05	1.9	0.37
Magnesium	2.08	1.8	0.28
Calcium	3.65	1.9	1.04
Sodium	2.75	2.5	1.21
Potassium	2.58	1.4	1.88
Hydrogen	0.14	3.0	

Some historical volcanic eruptions

Volcano	Year	Estimated casualties	Principal causes of death
Merapi (Indonesia)	1006	>1,000	Explosions
Kelut (Indonesia)	1586	10,000	Lahars (mudflows)
Vesuvius (Italy)	1631	18,000	Lava flows, mudflows
Etna (Italy)	1669	10,000	Lava flows, explosions
Merapi (Indonesia)	1672	>300	Nuées ardentes, lahars
Awu (Indonesia)	1711	3,200	Lahars
Papandayan (Indonesia)	1772	2,957	Explosions
Laki (Iceland)	1783	10,000	Lava flows, volcanic gas, starvation*
Asama (Japan)	1783	1,151	Lava flows, lahars
Unzen (Japan)	1792	15,000	Lahars, tsunami
Mayon (Philippines)	1814	1,200	Nuées ardentes, lava flows
Tambora (Indonesia)	1815	92,000	Starvation*
Galunggung (Indonesia)	1822	4,000	Lahars
Awu (Indonesia)	1856	2,800	Lahars
Krakatau (Indonesia)	1883	36,000	Tsunami
Awu (Indonesia)	1892	1,500	Nuées ardentes, lahars
Mont Pelée, Martinique (West Indies)	1902	36,000	Nuées ardentes
Soufrière, St. Vincent (West Indies)	1902	1,565	Nuées ardentes
Taal (Philippines)	1911	1,332	Explosions
Kelut (Indonesia)	1919	5,000	Lahars
Lamington (Papua New Guinea)	1951	3,000	Nuées ardentes, explosions
Merapi (Indonesia)	1951	1,300	Lahars
Agung (Indonesia)	1963	3,800	Nuées ardentes, lahars
Taal (Philippines)	1965	350	Explosions
Mount St. Helens (United States)	1980	57	Lateral blast, mudflows
El Chichón (Mexico)	1982	>2,000	Explosions, nuées ardentes
Nevado del Ruiz (Colombia)	1985	>25,000	Mudflows
Unzen (Japan)	1991	41	Nuées ardentes
Pinatubo (Philippines)	1991	>300	Nuées ardentes, mudflows, ash fall (roof collapse)
Merapi (Indonesia)	1994	>41	Nuées ardentes from dome collapse
Soufrière Hills, Montserrat (West Indies)	1997	19	Nuées ardentes

*Deaths directly attributable to the destruction or reduction of food crops, livestock, agricultural lands, pasturage, and other disruptions of food chain.

Appendix

Compositions of important rock types in the earth's crust and the average continental crust

Composition	Anorthosite	Peridotite	Oceanic basalt	Andesite	Dacite	Granodiorite	Granite	Graywacke	Sandy shale	Continental crust upper 9 mi (15 km)
Chemical					*Weight, %*					
SiO_2	54.0	44.0	50.0	60.0	65.5	66.0	70.5	64.0	65.5	66.0
TiO_2	.8	.2	1.5	.8	.3	0.5	.3	.5	.5	0.5
Al_2O_3	24.0	2.5	15.5	17.5	15.0	15.5	14.6	14.5	14.0	15.5
Fe_2O_3	.8	1.0	1.5	3.0	.8	2.0	1.6	1.5	3.5	3.0
FeO	2.5	8.0	8.0	3.2	2.5	2.6	1.8	3.5	2.0	3.0
MgO	1.5	40.0	7.0	2.8	2.0	2.0	.8	2.2	1.7	2.0
CaO	10.0	2.5	10.5	6.0	3.7	4.0	2.0	2.6	2.5	4.2
Na_2O	4.5	.1	2.9	3.5	3.8	3.6	3.5	3.2	1.5	3.5
K_2O	.1	.02	.25	3.0	2.4	2.8	4.3	2.0	4.0	3.0
Mineralogical					*Approximate volume*					
Olivine	—	*	†	—	—	—	—	—	—	—
Fe, T, Mg oxides	†	†	†	*	—	—	—	—	—	†
Pyroxene	†	*	*	*	—	—	—	—	—	—
Amphibole	—	—	—	*	†	†	—	*	†	†
Plagioclase	*	—	*	*	*	*	†	†	*	*
K-feldspar	—	—	—	†	†	*	*	*	*	*
Micas	—	—	—	—	—	*	*	*	*	*
Quartz	—	—	—	—	*	*	*	*	*	*
Chlorites	—	—	—	—	—	—	—	*	—	—
Clay minerals	—	—	—	—	—	—	—	†	*	—

*Major constituent.
†Subordinate mineral.

Dental formulas of some mammals

Animal	Teeth				Total
	I	C	Pm	M	
Human	2/2	1/1	2/2	3/3	32
Cony	3/3	1/1	4/4	4/4	48
Beaver	1/1	0/0	1/1	3/3	20
Cat	3/3	1/1	3/2	1/1	30
Dog	3/3	1/1	4/4	2/3	42
Sheep	0/3	0/1	3/3	3/3	32
Lynx	3/3	1/1	2/2	1/1	28
Rat	1/1	0/0	0/0	3/3	16
Horse	3/3	1/1	4/4	3/3	44
Mole	3/3	1/1	4/4	3/3	44
Squirrel	1/1	0/0	2/1	3/3	22
Reindeer	0/3	0/1	3/3	3/3	32
Pig	3/3	1/1	4/4	3/3	44
Common seal	3/2	1/1	4/4	1/1	34
Skunk	3/3	1/1	3/3	1/2	34
Raccoon	3/3	1/1	4/4	2/2	40
Bear	3/3	1/1	4/4	2/3	42

Appendix

Geologic column and scale of time

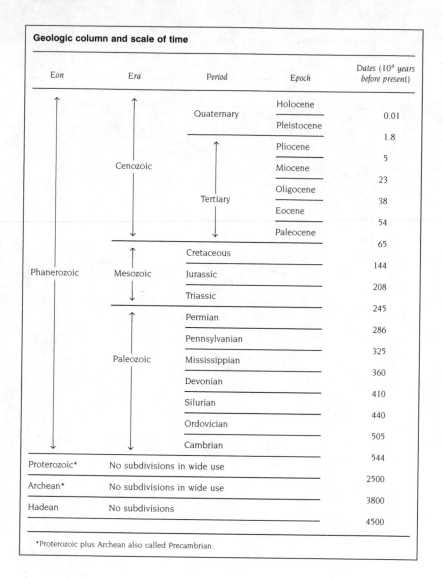

Eon	Era	Period	Epoch	Dates (10^4 years before present)
Phanerozoic	Cenozoic	Quaternary	Holocene	
				0.01
			Pleistocene	
				1.8
		Tertiary	Pliocene	
				5
			Miocene	
				23
			Oligocene	
				38
			Eocene	
				54
			Paleocene	
				65
	Mesozoic		Cretaceous	
				144
			Jurassic	
				208
			Triassic	
				245
	Paleozoic		Permian	
				286
			Pennsylvanian	
				325
			Mississippian	
				360
			Devonian	
				410
			Silurian	
				440
			Ordovician	
				505
			Cambrian	
				544
Proterozoic*	No subdivisions in wide use			
				2500
Archean*	No subdivisions in wide use			
				3800
Hadean	No subdivisions			
				4500

*Proterozoic plus Archean also called Precambrian.

Types of volcanic structure

Name	Characteristics
Shield	Low height, broad area; formed by successive fluid flows accumulating around a single, central vent
Cinder cone	Cone of moderate size with apex truncated; circular in plan, gently sloping sides; composed of pyroclastic particles, usually poorly consolidated
Spatter cone	Small steep-sided cone with well-defined crater composed of pyroclastic particles, well consolidated (agglomerate)
Composite cone	Composed of interlayered flows and pyroclastics; flows from sides (flank flows) common, as are radial dike swarms; slightly concave in profile, with central crater
Caldera	Basins of great size but relatively shallow; formed by explosive decapitation of stratocones, by collapse into underlying magma chamber, or both
Plug dome	Domal piles of viscous (usually rhyolitic) lava, growing by subsurface accretion and accompanied by outer fragmentation
Cryptovolcanic structures	Circular areas of highly fractured rocks in regions generally free of other structural disturbances; believed to have formed either by subsurface explosions or by sinking of cylindrical rock masses over magma chambers

Mohs scale*

Hardness	Mineral	Hardness	Mineral
1	Talc	6	Orthoclase
2	Gypsum	7	Quartz
3	Calcite	8	Topaz
4	Fluorite	9	Corundum
5	Apatite	10	Diamond

*Hardness or resistance to scratching is defined by comparison with 10 selected minerals, which are numbered in order of increasing hardness. Minerals lower in the scale are scratched by those with higher numbers.

Appendix

Hardness, specific gravity, and refractive indices of gem materials

Gem material	Hardness (Mohs scale)	Specific gravity	Refractive index
Amber	2–2 1/2	1.05	1.54
Beryl	7 1/2–8	2.67–2.85	1.57–1.58
Synthetic emerald	7 1/2–8	2.66–2.7	1.56–1.563 to 1.57–1.58
Chrysoberyl and synthetic	8 1/2	3.73	1.746–1.755
Corundum and synthetic	9	4.0	1.76–1.77
Diamond			
Synthetic cubic	10	3.52	2.42
Zirconia	8 1/2	5.80	2.15
Feldspar	6–6 1/2	2.55–2.75	1.5–1.57
Garnet			
Almandite	7 1/2	4.05	1.79
Pyrope	7–7 1/2	3.78	1.745
Rhodolite	7–7 1/2	3.84	1.76
Andradite	6 1/2–7	3.84	1.875
Grossularite	7	3.61	1.74
Spessartite	7–7 1/2	4.15	1.80
Hematite	5 1/2–6 1/2	5.20	
Jade			
Jadeite	6 1/2–7	3.34	1.66–1.68
Nephrite	6–6 1/2	2.95	1.61–1.63
Lapis lazuli	5–6	2.4–3.05	1.50
Malachite	3 1/2–4	3.34–3.95	1.66–1.91
Opal	5–6 1/2	2.15	1.45
Pearl	3–4	2.7	
Peridot	6 1/2–7	3.34	1.654–1.690
Quartz			
Crystalline and synthetic	7	2.66	1.54–1.55
Chalcedony	6 1/2–7	2.60	1.535–1.539
Spinel and flux synthetic	8	3.60	1.718
Synthetic spinel, flame	8	3.64	1.73
Spodumene	6–7	3.18	1.66–1.676
Topaz	8	3.53	1.61–1.63
Tourmaline	7–7 1/2	3.06	1.62–1.64
Turquois	5–6	2.76	1.61–1.65
Zircon	7 1/2	4.70	1.925–1.98
Metamict	7	4.00	1.81
Zoisite (tanzanite)	6–7	3.35	1.691–1.70

Praise for
A Darker Domain

"McDermid pulls us deeply into the lives of the victims and their families. . . . Pirie is a complicated heroine made all the more appealing by her everywoman demeanor."   —*USA Today*

"[McDermid] writes with great sympathy and understanding about the labor strike and with a sharp eye for class differences. This rich novel keeps the reader asking 'And what happens next?' right up to an abrupt ending that, though it explains the conspiracy with which the novel began, leaves more than a few unanswered questions and issues to ponder."   —*Boston Globe*

"Though Scottish crime writer Val McDermid is much better known in the UK than in the U.S., I'm hoping that her new novel, *A Darker Domain*, will attract a huge readership on this side of the Atlantic. . . . The complexities of [the] cases . . . will keep the reader turning the pages of this nicely written, complex story. This is a novel that reinforces my belief that you learn something from every book you read. It was fascinating to me to read about the terribly difficult lives of the miners (McDermid's knowledge of the miners' experiences come from stories that her miner grandfather told her as a child) and how, in their strike of 1984, the miners were betrayed by both their union leadership and the British government."   —NPR's "Morning Edition"

"Tough-minded, richly described. . . . combines a thrilling story with heartbreaking questions of social justice and history."
—*Seattle Times*

"Complex and layered plotlines come together, and McDermid does an excellent job creating tension around a cold case. Sure to be a hit with McDermid's large fan base, it should also appeal to those who read other Scottish police mysteries, such as Stuart MacBride. Those who enjoyed the cold-case aspect may also enjoy Theorin's *Echoes of the Dead*."   —*Booklist*

"This renowned Scottish writer's fascinating new book is dark, less violent than some previous titles and heartbreakingly vivid. Its greatest strength is its depiction of the horrendous miners' strike of the mid-1980s. . . . This is a wonderfully written book you can't put down." —*Romantic Times*

"This richly plotted and superbly written tale involves misdirected and conflicting loyalties, the abuse of power, and an examination of what family really means. . . . Val McDermid has done it again—*A Darker Domain* is a masterpiece." —*Out*

"Tension propels the plot forward. . . . Geography, pathology, archaeology, anthropology mingled with common sense, intuition and the brains of the two women investigating two arms of the monster cases afford readers a fascinating history of what the powers that were did to an entire industry and the people they crushed. Murders are uncovered and disappearances put to rest. Fans of McDermid and newcomers alike should find *A Darker Domain* a masterful novel that will challenge their armchair detective skills." —Bookreporter.com

"A top flight mystery. . . . When you crack open the spine of *A Darker Domain*, McDermid's skillful writing and sense of place will ensnare you amid centuries-old castles, cunning crooks, and an unflappable lady detective. McDermid seamlessly shuffles the reader through time and place between 1985 and 2007, from Scotland to Italy and in-between. In her first adventure, DI Pirie quickly becomes one of the most engaging characters in crime fiction; a bulldog who doesn't back down to her clueless boss or to the rich and powerful. *A Darker Domain* holds the reader glued until the subtle finale, uncovering corruption and lies with every turn of the page. The truth not only stings, it kills. Having Val McDermid as your Scottish emissary is both an honor and pleasure. Book your flight today and travel alongside the grand mistress of crime." —*Madison County Herald*

"With *A Darker Domain*, Scottish author Val McDermid boldly steps away from her famous Tony Hill series to deliver a brand-new standalone psychological thriller/mystery filled with mystery, betrayal and darkness. . . . McDermid's writing skills shine through in every sentence and every word of this beautifully written novel. Readers will feel like they are experiencing the turmoil and confusion of these two stories firsthand. Written against a fascinating backdrop and spanning over two decades, McDermid paints a perfect, realistic picture of the characters' lives and makes the miners' strike accessible to readers who are both familiar and unfamiliar with the event. . . . A thoroughly thought-out story. . . . An enjoyable, wonderfully written read."   —Blogcritics.org

"Much of this novel conveys a visceral sense of place, time and custom—especially life in villages in Fife. McDermid grew up near these mining towns, and she describes them in vivid, melancholy strokes. . . . Her characters are drawn well enough to elicit sympathy or shivers."   —Scripps Howard News Service

"When discussing Britain's best mystery writers, you must include Val McDermid, who, at fifty-three, has cranked out more than twenty-five books, nearly all of them highly acclaimed. Her latest novel, which is not part of any of her several popular series, may be her finest. . . . McDermid's skills as a writer have never been more clearly in evidence than in this superb psycho-thriller."   —*Tampa Tribune*

"With this new novel, McDermid takes the complexity of her storytelling to a new level. . . . The narrative moves from present to past and from viewpoint to viewpoint with barely a break to cue the reader."   —*The Writer*

Charles Bush

About the Author

Scottish crime writer Val McDermid is the author of twenty-three novels. Her books have won the Gold Dagger Award for Best Crime Novel of the Year, a *Los Angeles Times* Book Prize, have been selected for the *New York Times'* 100 Notable Books of the Year list, and have been nominated for the Edgar Award. She lives in the north of England.

A DARKER DOMAIN

ALSO BY VAL MCDERMID

The Grave Tattoo
The Distant Echo
Killing the Shadows
A Place of Execution

TONY HILL NOVELS
Fever of the Bone
Beneath the Bleeding
The Torment of Others
The Last Temptation
The Wire in the Blood
The Mermaids Singing

KATE BRANNIGAN NOVELS
Star Struck
Blue Genes
Clean Break
Crack Down
Kick Back
Dead Beat

LINDSAY GORDON NOVELS
Hostage to Murder
Booked for Murder
Conferences Are Murder
Deadline for Murder
Common Murder
Report for Murder

SHORT STORIES
Stranded

NONFICTION
A Suitable Job for a Woman

A DARKER DOMAIN

A NOVEL

VAL McDERMID

HARPER

NEW YORK • LONDON • TORONTO • SYDNEY

THIS BOOK IS DEDICATED TO THE MEMORY *of Meg and Tom McCall, my maternal grandparents. They showed me love, they taught me about community, and they never forgot the shame of standing in line at a soup kitchen to feed their bairns. Thanks to them, I grew up loving the sea, the woods, and the work of Agatha Christie. No small debt.*

HARPER

FIRST HARPER PAPERBACKS EDITION PUBLISHED 2010.

Designed by Emily Cavett Taff

Library of Congress Cataloging-in-Publication Data

McDermid, Val.
A darker domain : a novel / by Val McDermid.—1st ed.
p. cm.
ISBN 978-0-06-168898-0

1. Police—Scotland—Fife—Fiction. 2. Murder—Inves-tigation—Scotland—Fife—Fiction. 3. Kidnapping—Scotland—Fife—Fiction. 4. Cold cases (Criminal investigation)—Fiction. 5. Fife (Scotland)—Fiction. I. Title.

PR6063.C37 D37   2009
823'.914—dc22   2008033707

ISBN 978-0-06-168899-7 (pbk.)

HB 12.12.2019

ACKNOWLEDGMENTS

It started when Kari "Mrs. Shapiro" Furre made a bizarre discovery in the *casa rovina* down the hill. The Giorgi family of the Chiocciola contrada offered their suggestions; the wonderful Mamma Rosa fed us like kings and taught us like little children; Marino Garaffi continues to breed the finest pigs, even the ones that get stuck. Their friendship, their kindness, and their generosity brightens my summers.

In Fife, I owe thanks to my mother for her memories; to the many miners and musicians whose songs and stories weave in and out of my childhood memories; to the fellow Raith Rovers supporter who suggested it was about time I wrote another book set in the Kingdom; and to the communities I grew up among who were shattered by the 1984 strike and its aftermath.

Professor Sue Black was generous as ever with her expertise and reminds me that the mistakes are mine.

Some of the people who made this book possible are beyond thanks. My father Jim McDermid, my miner grandfathers Tom McCall and Donald McDermid, and my brevet uncle Doddy Arnold all opened doors into the world of working men, a world whose demands cut short their lives.

And finally, a nod of appreciation to the team who always push me to make the book the best I can manage—my editor, Julia Wisdom; my copy editor, Anne O'Brien; and my agent, Jane Gregory. Not forgetting Kelly and Cameron, whose patience is entirely remarkable.

Wednesday, 23rd January 1985; Newton of Wemyss

The voice is soft, like the darkness that encloses them. "You ready?"

"As ready as I'll ever be."

"You've told her what to do?" Words tumbling now, tripping over each other, a single stumble of sounds.

"Don't worry. She knows what's what. She's under no illusions about who's going to carry the can if this goes wrong." Sharp words, sharp tone. "She's not the one I'm worrying about."

"What's that supposed to mean?"

"Nothing. It means nothing, all right? We've no choices. Not here. Not now. We just do what has to be done." The words have the hollow ring of bravado. It's anybody's guess what they're hiding. "Come on, let's get it done with."

This is how it begins.

Wednesday, 27th June 2007; Glenrothes

The young woman strode across the foyer, low heels striking a rhythmic tattoo on vinyl flooring dulled by the passage of thousands of feet. She looked like someone on a mission, the civilian clerk thought as she approached his desk. But then, most of them did. The crime prevention and public information posters that lined the walls were invariably wasted on them as they approached, lost in the slipstream of their determination.

She bore down on him, her mouth set in a firm line. Not bad looking, he thought. But like a lot of the women who showed up here, she wasn't exactly looking her best. She could have done with a bit more make-up, to make the most of those sparkly blue eyes.

And something more flattering than jeans and a hoodie. Dave Cruickshank assumed his fixed professional smile. "How can I help you?" he said.

The woman tilted her head back slightly, as if readying herself for defence. "I want to report a missing person."

Dave tried not to show his weary irritation. If it wasn't neighbours from hell, it was so-called missing persons. This one was too calm for it to be a missing toddler, too young for it to be a runaway teenager. A row with the boyfriend, that's what it would be. Or a senile granddad on the lam. The usual bloody waste of time. He dragged a pad of forms across the counter, squaring it in front of him and reaching for a pen. He kept the cap on; there was one key question he needed answered before he'd be taking down any details. "And how long has this person been missing?"

"Twenty-two and a half years. Since Friday the fourteenth of December 1984, to be precise." Her chin came down and truculence clouded her features. "Is that long enough for you to take it seriously?"

Detective Sergeant Phil Parhatka watched the end of the video clip then closed the window. "I tell you," he said, "if ever there was a great time to be in cold cases, this is it."

Detective Inspector Karen Pirie barely raised her eyes from the file she was updating. "How?"

"Stands to reason. We're in the middle of the war on terror. And I've just watched my local MP taking possession of 10 Downing Street with his missus." He jumped up and crossed to the minifridge perched on top of a filing cabinet. "What would you rather be doing? Solving cold cases and getting good publicity for it, or trying to make sure the muzzers dinnae blow a hole in the middle of our patch?"

"You think Gordon Brown becoming prime minister makes Fife a target?" Karen marked her place in the document with her index finger and gave Phil her full attention. It dawned on her that

for too long she'd had her head too far in the past to weigh up present possibilities. "They never bothered with Tony Blair's constituency when he was in charge."

"Very true." Phil peered into the fridge, deliberating between an Irn Bru and a Vimto. Thirty-four years old and still he couldn't wean himself off the soft drinks that had been treats in childhood. "But these guys call themselves Islamic jihadists and Gordon's a son of the manse. I wouldn't want to be in the chief constable's shoes if they decide to make a point by blowing up his dad's old kirk." He chose the Vimto. Karen shuddered.

"I don't know how you can drink that stuff," she said. "Have you never noticed it's an anagram of vomit?"

Phil took a long pull on his way back to his desk. "Puts hairs on your chest," he said.

"Better make it two cans, then." There was an edge of envy in Karen's voice. Phil seemed to live on sugary drinks and saturated fats but he was still as compact and wiry as he'd been when they were rookies together. She just had to look at a fully leaded Coke to feel herself gaining inches. It definitely wasn't fair.

Phil narrowed his dark eyes and curled his lip in a good-natured sneer. "Whatever. The silver lining is that maybe the boss can screw some more money out of the government if he can persuade them there's an increased threat."

Karen shook her head, on solid ground now. "You think that famous moral compass would let Gordon steer his way towards anything that looked that self-serving?" As she spoke, she reached for the phone that had just begun to ring. There were other, more junior officers in the big squad room that housed the Cold Case Review Team, but promotion hadn't altered Karen's ways. She'd never got out of the habit of answering any phone that rang in her vicinity. "CCRT, DI Pirie speaking," she said absently, still turning over what Phil had said, wondering if, deep down, he had a hankering to be where the live action was.

"Dave Cruickshank on the front counter, Inspector. I've got somebody here, I think she needs to talk to you." Cruickshank

sounded unsure of himself. That was unusual enough to grab Karen's attention.

"What's it about?"

"It's a missing person," he said.

"Is it one of ours?"

"No, she wants to report a missing person."

Karen suppressed an irritated exhalation. Cruickshank really should know better by now. He'd been on the front desk long enough. "So she needs to talk to CID, Dave."

"Well, yeah. Normally, that would be my first port of call. But see, this is a bit out of the usual run of things. Which is why I thought it would be better to run it past you, see?"

Get to the point. "We're cold cases, Dave. We don't process fresh inquiries." Karen rolled her eyes at Phil, smirking at her obvious frustration.

"It's not exactly fresh, Inspector. This guy went missing twenty-two years ago."

Karen straightened up in her chair. "Twenty-two years ago? And they've only just got round to reporting it?"

"That's right. So does that make it cold, or what?"

Technically, Karen knew Cruickshank should refer the woman to CID. But she'd always been a sucker for anything that made people shake their heads in bemused disbelief. Long shots were what got her juices flowing. Following that instinct had brought her two promotions in three years, leap-frogging peers and making colleagues uneasy. "Send her up, Dave. I'll have a word with her."

She replaced the phone and pushed back from the desk. "Why the fuck would you wait twenty-two years to report a missing person?" she said, more to herself than to Phil as she raided her desk for a fresh notebook and a pen.

Phil pushed his lips out like an expensive carp. "Maybe she's been out of the country. Maybe she only just came back and found out this person isn't where she thought they were."

"And maybe she needs us so she can get a declaration of death. Money, Phil. What it usually comes down to." Karen's smile was

wry. It seemed to hang in the air in her wake as if she were the Cheshire Cat. She bustled out of the squad room and headed for the lifts.

Her practised eye catalogued and classified the woman who emerged from the lift without a shred of diffidence visible. Jeans and fake-athletic hoodie from Gap. This season's cut and colours. The shoes were leather, clean and free from scuffs, the same colour as the bag that swung from her shoulder over one hip. Her mid-brown hair was well cut in a long bob just starting to get a bit ragged along the edges. Not on welfare, then. Probably not a schemie. A nice, middle-class woman with something on her mind. Mid to late twenties, blue eyes with the pale sparkle of topaz. The barest skim of make-up. Either she wasn't trying or she already had a husband. The skin round her eyes tightened as she caught Karen's appraisal.

"I'm Detective Inspector Pirie," she said, cutting through the potential stand-off of two women weighing each other up. "Karen Pirie." She wondered what the other woman made of her—a wee fat woman crammed into a Marks and Spencer suit, mid-brown hair needing a visit to the hairdresser, might be pretty if you could see the definition of her bones under the flesh. When Karen described herself thus to her mates, they would laugh, tell her she was gorgeous, make out she was suffering from low self-esteem. She didn't think so. She had a reasonably good opinion of herself. But when she looked in the mirror, she couldn't deny what she saw. Nice eyes, though. Blue with streaks of hazel. Unusual.

Whether it was what she saw or what she heard, the woman seemed reassured. "Thank goodness for that," she said. The Fife accent was clear, though the edges had been ground down either by education or absence.

"I'm sorry?"

The woman smiled, revealing small, regular teeth like a child's first set. "It means you're taking me seriously. Not fobbing me off with the junior officer who makes the tea."

"I don't let my junior officers waste their time making tea,"

Karen said drily. "I just happened to be the one who answered the phone." She half-turned, looked back, and said, "If you'll come with me?"

Karen led the way down a side corridor to a small room. A long window gave on to the parking lot and, in the distance, the artificially uniform green of the golf course. Four chairs upholstered in institutional grey tweed were drawn up to a round table, its cheerful cherry wood polished to a dull sheen. The only indicator of its function was the gallery of framed photographs on the wall, all shots of police officers in action. Every time she used this room, Karen wondered why the brass had chosen the sort of photos that generally appeared in the media after something very bad had happened.

The woman looked around her uncertainly as Karen pulled out a chair and gestured for her to sit down. "It's not like this on the telly," she said.

"Not much about Fife Constabulary is," Karen said, sitting down so that she was at ninety degrees to the woman rather than directly opposite her. The less confrontational position was usually the most productive for a witness interview.

"Where's the tape recorders?" The woman sat down, not pulling her chair any closer to the table and hugging her bag in her lap.

Karen smiled. "You're confusing a witness interview with a suspect interview. You're here to report something, not to be questioned about a crime. So you get to sit on a comfy chair and look out the window." She flipped open her pad. "I believe you're here to report a missing person?"

"That's right. His name's—"

"Just a minute. I need you to back up a wee bit. For starters, what's your name?"

"Michelle Gibson. That's my married name. Prentice, that's my own name. Everybody calls me Misha, though."

"Right you are, Misha. I also need your address and phone number."

Misha rattled out details. "That's my mum's address. I'm sort of acting on her behalf, if you see what I mean?"

Karen recognized the village, though not the street. Started out as one of the hamlets built by the local laird for his coal miners when the workers were as much his as the mines themselves. Ended up as commuterville for strangers with no links to the place or the past. "All the same," she said, "I need your details too."

Misha's brows lowered momentarily, then she gave an address in Edinburgh. It meant nothing to Karen, whose knowledge of the social geography of the capital, a mere thirty miles away, was parochially scant. "And you want to report a missing person," she said.

Misha gave a sharp sniff and nodded. "My dad. Mick Prentice. Well, Michael, really, if you want to be precise."

"And when did your dad go missing?" This, thought Karen, was where it would get interesting. If it was ever going to get interesting.

"Like I told the guy downstairs, twenty-two and a half years ago. Friday, 14th December 1984 was the last time we saw him." Misha Gibson's brows drew down in a defiant scowl.

"It's kind of a long time to wait to report someone missing," Karen said.

Misha sighed and turned her head so she could look out of the window. "We didn't think he was missing. Not as such."

"I'm not with you. What do you mean, 'not as such'?"

Misha turned back and met Karen's steady gaze. "You sound like you're from round here."

Wondering where this was going, Karen said, "I grew up in Methil."

"Right. So, no disrespect, but you're old enough to remember what was going on in 1984."

"The miners' strike?"

Misha nodded. Her chin stayed high, her stare defiant. "I grew up in Newton of Wemyss. My dad was a miner. Before the strike, he worked down the Lady Charlotte. You'll mind what folk

used to say round here—that nobody was more militant than the Lady Charlotte pitmen. Even so, there was one night in December, nine months into the strike, when half a dozen of them disappeared. Well, I say 'disappeared,' but everybody knew the truth. That they'd gone to Nottingham to join the blacklegs." Her face bunched in a tight frown, as if she was struggling with some physical pain. "Five of them, nobody was too surprised that they went scabbing. But according to my mum, everybody was stunned that my dad had joined them. Including her." She gave Karen a look of pleading. "I was too wee to remember. But everybody says he was a union man through and through. The last guy you'd expect to turn blackleg." She shook her head. "Still, what else was she supposed to think?"

Karen understood only too well what such a defection must have meant to Misha and her mother. In the radical Fife coalfield, sympathy was reserved for those who toughed it out. Mick Prentice's action would have granted his family instant pariah status. "It can't have been easy for your mum," she said.

"In one sense, it was dead easy," Misha said bitterly. "As far as she was concerned, that was it. He was dead to her. She wanted nothing more to do with him. He sent money, but she donated it to the hardship fund. Later, when the strike was over, she handed it over to the Miners' Welfare. I grew up in a house where my father's name was never spoken."

Karen felt a lump in her chest, somewhere between sympathy and pity. "He never got in touch?"

"Just the money. Always in used notes. Always with a Nottingham postmark."

"Misha, I don't want to come across like a bitch here, but it doesn't sound to me like your dad's a missing person." Karen tried to make her voice as gentle as possible.

"I didn't think so either. Till I went looking for him. Take it from me, Inspector. He's not where he's supposed to be. He never was. And I need him found."

The naked desperation in Misha's voice caught Karen by sur-

prise. To her, that was more interesting than Mick Prentice's whereabouts. "How come?" she said.

Tuesday, 19th June 2007; Edinburgh

It had never occurred to Misha Gibson to count the number of times she'd emerged from the Sick Kids' with a sense of outrage that the world continued on its way in spite of what was happening inside the hospital behind her. She'd never thought to count because she'd never allowed herself to believe it might be for the last time. Ever since the doctors had explained the reason for Luke's misshapen thumbs and the scatter of café-au-lait spots across his narrow back, she had nailed herself to the conviction that somehow she would help her son dodge the bullet his genes had aimed at his life expectancy. Now it looked as if that conviction had finally been tested to destruction.

Misha stood uncertain for a moment, resenting the sunshine, wanting weather as bleak as her mood. She wasn't ready to go home yet. She wanted to scream and throw things, and an empty flat would tempt her to lose control and do just that. John wouldn't be home to hold her or to hold her back; he'd known about her meeting with the consultant so of course work would have thrown up something insurmountable that only he could deal with.

Instead of heading up through Marchmont to their sandstone tenement, Misha cut across the busy road to the Meadows, the green lung of the southern city centre where she loved to walk with Luke. Once, when she'd looked at their street on Google Earth, she'd checked out the Meadows too. From space, it looked like a rugby ball fringed with trees, the criss-cross paths like laces holding the ball together. She'd smiled at the thought of her and Luke scrambling over the surface like ants. Today, there were no smiles to console Misha. Today, she had to face the fact that she might never walk here with Luke again.

She shook her head, trying to dislodge the maudlin thoughts.

Coffee, that's what she needed to gather her thoughts and get things into proportion. A brisk walk across the Meadows, then down to George IV Bridge, where every shop front was a bar, a café, or a restaurant these days.

Ten minutes later, Misha was tucked into a corner booth, a comforting mug of latte in front of her. It wasn't the end of the line. It couldn't be the end of the line. She wouldn't let it be the end of the line. There had to be some way to give Luke another chance.

She'd known something was wrong from the first moment she'd held him. Even dazed by drugs and drained by labour, she'd known. John had been in denial, refusing to set any store by their son's low birth weight and those stumpy little thumbs. But fear had clamped its cold certainty on Misha's heart. Luke was different. The only question in her mind had been how different.

The sole aspect of the situation that felt remotely like luck was that they were living in Edinburgh, a ten-minute walk from the Royal Hospital for Sick Children, an institution that regularly appeared in the "miracle" stories beloved of the tabloids. It didn't take long for the specialists at the Sick Kids' to identify the problem. Nor to explain that there would be no miracles here.

Fanconi Anaemia. If you said it fast, it sounded like an Italian tenor or a Tuscan hill town. But the charming musicality of the words disguised their lethal message. Lurking in the DNA of both Luke's parents were recessive genes that had combined to create a rare condition that would condemn their son to a short and painful life. At some point between the ages of three and twelve, he was almost certain to develop aplastic anaemia, a breakdown of the bone marrow that would ultimately kill him unless a suitable donor could be found. The stark verdict was that without a successful bone marrow transplant, Luke would be lucky to make it into his twenties.

That information had given her a mission. She soon learned that, without siblings, Luke's best chance of a viable bone marrow transplant would come from a family member—what the doctors

called a mismatched related transplant. At first, this had confused Misha. She'd read about bone marrow transplant registers and assumed their best hope was to find a perfect match there. But according to the consultant, a donation from a mismatched family member who shared some of Luke's genes had a lower risk of complications than a perfect match from a donor who wasn't part of their extended kith and kin.

Since then, Misha had been wading through the gene pool on both sides of the family, using persuasion, emotional blackmail, and even the offer of reward on distant cousins and elderly aunts. It had taken time, since it had been a solo mission. John had walled himself up behind a barrier of unrealistic optimism. There would be a medical breakthrough in stem cell research. Some doctor somewhere would discover a treatment whose success didn't rely on shared genes. A perfectly matched donor would turn up on a register somewhere. John collected good stories and happy endings. He trawled the Internet for cases that had proved the doctors wrong. He came up with medical miracles and apparently inexplicable cures on a weekly basis. And he drew his hope from this. He couldn't see the point of Misha's constant pursuit. He knew somehow it would be all right. His capacity for denial was Olympic.

It made her want to kill him.

Instead, she'd continued to clamber through the branches of their family trees in search of the perfect candidate. She'd come to her final dead end only a week or so before today's terrible judgement. There was only one possibility left. And it was the one possibility she had prayed she wouldn't have to consider.

Before her thoughts could go any further down that particular path, a shadow fell over her. She looked up, ready to be sharp with whoever wanted to intrude on her. "John," she said wearily.

"I thought I'd find you hereabouts. This is the third place I tried," he said, sliding into the booth, awkwardly shunting himself round till he was at right angles to her, close enough to touch if either of them had a mind to.

"I wasn't ready to face an empty flat."

"No, I can see that. What did they have to say?" His craggy face screwed up in anxiety. Not, she thought, over the consultant's verdict. He still believed his precious son was somehow invincible. What made John anxious was her reaction.

She reached for his hand, wanting contact as much as consolation. "It's time. Six months tops without the transplant." Her voice sounded cold even to her. But she couldn't afford warmth. Warmth would melt her frozen state, and this wasn't the place for an outpouring of grief or love.

John clasped her fingers tight inside his. "It's maybe not too late," he said. "Maybe they'll—"

"Please, John. Not now."

His shoulders squared inside his suit jacket, his body tensing as he held his dissent close. "So," he said, an outbreath that was more sigh than anything else. "I suppose that means you're going looking for the bastard?"

Wednesday, 27th June 2007; Glenrothes

Karen scratched her head with her pen. *Why do I get all the good ones?* "Why did you leave it so long to try to trace your father?"

She caught a fleeting expression of irritation round Misha's mouth and eyes. "Because I'd been brought up thinking my father was a selfish blackleg bastard. What he did cast my mother adrift from her own community. It got me bullied in the play park and later on at school. I didn't think a man who dumped his family in the shit like that would be bothered about his grandson."

"He sent money," Karen said.

"A few quid here, a few quid there. Blood money," Misha said. "Like I said, my mum wouldn't touch it. She gave it away. I never saw the benefit of it."

"Maybe he tried to make it up to your mum. Parents don't always tell us the uncomfortable truths."

Misha shook her head. "You don't know my mum. Even with

Luke's life at stake, she wasn't comfortable with me trying to track down my dad."

To Karen, it seemed a thin reason for avoiding a man who might provide the key to a boy's future. But she knew how deep feelings ran in the old mining communities, so she let it lie. "You say he wasn't where he was supposed to be. What happened when you went looking for him?"

Thursday, 21st June 2007; Newton of Wemyss

Jenny Prentice pulled a bag of potatoes out of the vegetable rack and set about peeling them, her body bowed over the sink, her back turned to her daughter. Misha's question hung unanswered between them, reminding them both of the barrier her father's absence had put between them from the beginning. Misha tried again. "I said—"

"I heard you fine. There's nothing the matter with my hearing," Jenny said. "And the answer is, I've no bloody idea. How would I know where to start looking for that selfish scabbing sack of shite? We've managed fine without him the last twenty-two years. There's been no cause to go looking for him."

"Well, there's cause now." Misha stared at her mother's rounded shoulders. The weak light that spilled in through the small kitchen window accentuated the silver in her undyed hair. She was barely fifty, but she seemed to have bypassed middle age, heading straight for the vulnerable stoop of the little old lady. It was as if she knew this attack would arrive one day and had chosen to defend herself by becoming piteous.

"He'll not help," Jenny scoffed. "He showed what he thought of us when he left us to face the music. He was always out for number one."

"Maybe so. But I've still got to try for Luke's sake," Misha said. "Was there never a return address on the envelopes the money came in?"

Jenny cut a peeled potato in half and dropped it in a pan of salted water. "No. He couldn't even be bothered to put a wee letter in the envelope. Just a bundle of dirty notes, that's all."

"What about the guys he went with?"

Jenny cast a quick contemptuous glance at Misha. "What about them? They don't show their faces round here."

"But some of them have still got family here or in East Wemyss. Brothers, cousins. They might know something about my dad."

Jenny shook her head firmly. "I've never heard tell of him since the day he walked out. Not a whisper, good or bad. The other men he went with, they were no friends of his. The only reason he took a lift with them was he had no money to make his own way south. He'll have used them like he used us and then he'll have gone his own sweet way once he got where he wanted to be." She dropped another potato in the pan and said without enthusiasm, "Are you staying for your dinner?"

"No, I've got things to see to," Misha said, impatient at her mother's refusal to take her quest seriously. "There must be somebody he's kept in touch with. Who would he have talked to? Who would he have told what he was planning?"

Jenny straightened up and put the pan on the old-fashioned gas cooker. Misha and John offered to replace the chipped and battered stove every time they sat down to the production number that was Sunday dinner, but Jenny always refused with the air of frustrating martyrdom she brought to every offer of kindness. "You're out of luck there too." She eased herself on to one of the two chairs that flanked the tiny table in the cramped kitchen. "He only had one real pal. Andy Kerr. He was a red-hot Commie, was Andy. I tell you, by 1984, there weren't many still keeping the red flag flying, but Andy was one of them. He'd been a union official well before the strike. Him and your father, they'd been best pals since school." Her face softened for a moment and Misha could almost make out the young woman she'd been. "They were always up to something, those two."

"So where do I find this Andy Kerr?" Misha sat down opposite her mother, her desire to be gone temporarily abandoned.

Her mother's face twisted into a wry grimace. "Poor soul. If you can find Andy, you'll be quite the detective." She leaned across and patted Misha's hand. "He's another one of your father's victims."

"How do you mean?"

"Andy adored your father. He thought the sun shone out of his backside. Poor Andy. The strike put him under terrible pressure. He believed in the strike, he believed in the struggle. But it broke his heart to see the hardship his men were going through. He was on the edge of a nervous breakdown, and the local executive forced him to go on the sick not long before your father shot the craw. Nobody saw him after that. He lived out in the middle of nowhere, so nobody noticed he was away." She gave a long, weary sigh. "He sent a postcard to your dad from some place up north. But of course, he was blacklegging by then, so he never got it. Later, when Andy came back, he left a note for his sister, saying he couldn't take any more. Killed himself, the poor soul."

"What's that got to do with my dad?" Misha demanded.

"I always thought your dad going scabbing was the straw that broke the camel's back." Jenny's expression was pious shading into smug. "That was what drove Andy over the edge."

"You can't know that." Misha pulled away in disgust.

"I'm not the only one around here that thinks the same thing. If your father had confided in anybody, it would have been Andy. And that would have been one burden too many for that fragile wee soul. He took his own life, knowing that his one real friend had betrayed everything he stood for." On that melodramatic note, Jenny got to her feet and lifted a bag of carrots from the vegetable rack. It was clear she had shot her bolt on the subject of Mick Prentice.

Karen sneaked a look at her watch. Whatever fine qualities Misha Gibson might possess, brevity was not one of them. "So Andy Kerr turned out to be literally a dead end?"

"My mother thinks so. But apparently they never found his body. Maybe he didn't kill himself after all," Misha said.

"They don't always turn up," Karen said. "Sometimes the sea claims them. Or else the wilderness. There's still a lot of empty space in this country." Resignation took possession of Misha's face. She was, Karen thought, a woman inclined to believe what she was told. If anyone knew that, it would be her mother. Perhaps things weren't quite as clear cut as Jenny Prentice wanted her daughter to think.

"That's true," Misha said. "And my mother did say that he left a note. Will the police still have the note?"

Karen shook her head. "I doubt it. If we ever had it, it will have been given back to his family."

"Would there not have been an inquest? Would they not have needed it for that?"

"You mean a Fatal Accident Inquiry," Karen said. "Not without a body, no. If there's a file at all, it'll be a missing-person case."

"But he's not missing. His sister had him declared dead. Their parents both died in the Zeebrugge ferry disaster, but apparently their dad had always refused to believe Andy was dead so he hadn't changed his will to leave the house to the sister. She had to go to court to get Andy pronounced dead so she would inherit. That's what my mother said, anyway." Not a flicker of doubt disturbed Misha's expression.

Karen made a note, *Andy Kerr's sister*, and added a little asterisk to it. "So if Andy killed himself, we're back with scabbing as the only reasonable explanation of your dad's disappearance. Have you made any attempts to contact the guys he's supposed to have gone away with?"

Monday, 25th June 2007; Edinburgh

Ten past nine on a Monday morning, and already Misha felt exhausted. She should be at the Sick Kids by now, focusing on Luke. Playing with him, reading to him, cajoling therapists into expanding their regimes, discussing treatment plans with medical staff, using all her energy to fill them with her conviction that her son could be saved. And if he could be saved, they all owed it to him to shovel every scrap of therapeutic intervention his way.

But instead, she was sitting on the floor, back to the wall, knees bent, phone cradled in her lap, notepad at her side. She told herself she was summoning the courage to make a phone call, but she knew in a corner of her mind exhaustion was the real reason for her inactivity.

Other families used the weekends to relax, to recharge their batteries. But not the Gibsons. For a start, fewer staff were on duty at the hospital, so Misha and John felt obliged to pile even more energy than usual into Luke. There was no respite when they came home either. Misha's acceptance that the last best hope for their son lay in finding her father had simply escalated the conflict between her missionary ardour and John's passive optimism.

This weekend had been harder going than usual. Having a time limit put on Luke's life imbued each moment they shared with more value and more poignancy. It was hard to avoid a kind of melodramatic sentimentality. As soon as they'd left the hospital on Saturday Misha had picked up the refrain she'd been delivering since she'd seen her mother. "I need to go to Nottingham, John. You know I do."

He shoved his hands into the pockets of his rain jacket, thrusting his head forward as if he was butting against a high wind. "Just phone the guy," he said. "If he's got anything to tell you, he'll tell you on the phone."

"Maybe not." She took a couple of steps at a trot to keep pace

with him. "People always tell you more face to face. He could maybe put me on to the other guys that went down with him. They might know something."

John snorted. "And how come your mother can only remember one guy's name? How come she can't put you on to the other guys?"

"I told you. She's put everything out of her mind about that time. I really had to push her before she came up with Logan Laidlaw's name."

"And you don't think it's amazing that the only guy whose name she can remember has no family in the area? No obvious way to track him down?"

Misha pushed her arm through his, partly to make him slow down. "But I did track him down, didn't I? You're too suspicious."

"No, I'm not. Your mother doesn't understand the power of the Internet. She doesn't know about things like online electoral rolls or 192.com. She thinks if there's no human being to ask, you're screwed. She didn't think she was giving you anything you could use. She doesn't want you poking about in this, she's not going to help you."

"That makes two of you then." Misha pulled her arm free and strode out ahead of him.

John caught up with her on the corner of their street. "That's not fair," he said. "I just don't want you getting hurt unnecessarily."

"You think watching my boy die and not doing anything that might save him isn't hurting me?" Misha felt the heat of anger in her cheeks, knew the hot tears of rage were lurking close to the surface. She turned her face away from him, blinking desperately at the tall sandstone tenements.

"We'll find a donor. Or they'll find a treatment. All this stem cell research, it's moving really fast."

"Not fast enough for Luke," Misha said, the familiar sensation of weight in her stomach slowing her steps. "John, please. I need to go to Nottingham. I need you to take a couple of days off work, cover for me with Luke."

"You don't need to go. You can talk to the guy on the phone."

"It's not the same. You know that. When you're dealing with clients, you don't do it over the phone. Not for anything important. You go out and see them. You want to see the whites of their eyes. All I'm asking is for you to take a couple of days off, to spend time with your son."

His eyes flashed dangerously and she knew she'd gone too far. John shook his head stubbornly. "Just make the phone call, Misha."

And that was that. Long experience with her husband had taught her that when John took a position he believed was right, going over the same ground only gave him the opportunity to build stronger fortifications. She had no fresh arguments that could challenge his decision. So here she was, sitting on the floor, trying to shape sentences in her head that would persuade Logan Laidlaw to tell her what had happened to her father since he'd walked out on her more than twenty-two years earlier.

Her mother hadn't given her much to base a strategy on. Laidlaw was a waster, a womanizer, a man who, at thirty, had still acted like a teenager. He'd been married and divorced by twenty-five, building the sour reputation of a man who was too handy with his fists around women. Misha's picture of her father was patchy and partial, but even with the bias imposed by her mother, Mick Prentice didn't sound like the sort of man who would have had much time for Logan Laidlaw. Still, hard times made for strange company.

At last, Misha picked up the phone and keyed in the number she'd tracked down via Internet searches and directory enquiries. He'd probably be out at work, she thought on the fourth ring. Or asleep.

The sixth ring cut off abruptly. A deep voice grunted an approximate hello.

"Is that Logan Laidlaw?" Misha said, working to keep her voice level.

"I've got a kitchen and I don't want any insurance." The Fife

accent was still strong, the words bumping into each other with the familiar rise and fall.

"I'm not trying to sell you anything, Mr. Laidlaw. I just want to talk to you."

"Aye, right. And I'm the prime minister."

She could sense he was on the point of ending the call. "I'm Mick Prentice's daughter," she blurted out, strategy hopelessly holed beneath the waterline. Across the distance, she could hear the liquid wheeze of his breathing. "Mick Prentice from Newton of Wemyss," she tried.

"I know where Mick Prentice is from. What I don't know is what Mick Prentice has to do with me."

"Look, I realize the two of you might not see much of each other these days, but I'd really appreciate anything you could tell me. I really need to find him." Misha's own accent slipped a few gears till she was matching his own broad tongue.

A pause. Then, with a baffled note, "Why are you talking to me? I haven't seen Mick Prentice since I left Newton of Wemyss way back in 1984."

"OK, but even if you split up as soon as you got to Notting-ham, you must have some idea of where he ended up, where he was heading for?"

"Listen, hen, I don't have a clue what you're on about. What do you mean, split up as soon as we got to Nottingham?" He sounded irritated, what little patience he had evaporating in the heat of her demands.

Misha gulped a deep breath, then spoke slowly. "I just want to know what happened to my dad after you got to Nottingham. I need to find him."

"Are you wrong in the head or something, lassie? I've no idea what happened to your dad after I came to Nottingham, and here's for why. I was in Nottingham and he was in Newton of Wemyss. And even when we were both in the same place, we weren't what you would call pals."

The words hit like a splash of cold water. Was there something

wrong with Logan Laidlaw's memory? Was he losing his grip on the past? "No, that's not right," she said. "He came to Nottingham with you."

A bark of laughter, then a gravel cough. "Somebody's been winding you up, lassie," he wheezed. "Trotsky would have crossed a picket line before the Mick Prentice I knew. What makes you think he came to Nottingham?"

"It's not just me. Everybody thinks he went to Nottingham with you and the other men."

"That's mental. Why would anybody think that? Do you not know your own family history?"

"What do you mean?"

"Christ, lassie, your great-grandfather. Your father's granddad. Do you not know about him?"

Misha had no idea where this was going, but at least he hadn't hung up on her as she'd earlier feared he would. "He was dead before I was even born. I don't know anything about him, except that he was a miner too."

"Jackie Prentice," Laidlaw said with something approaching relish. "He was a strike breaker back in 1926. After it was settled, he had to be moved to a job on the surface. When your life depends on the men in your team, you don't want to be a scab underground. Not unless everybody else is in the same boat, like with us. Christ knows why Jackie stayed in the village. He had to take the bus to Dysart to get a drink. There wasn't a bar in any of the Wemyss villages that would serve him. So your dad and your granddad had to work twice as hard as anybody else to be accepted down the pit. No way would Mick Prentice throw that respect away. He'd sooner starve. Aye, and see you starve with him. Wherever you got your info, they don't know what the hell they're talking about."

"My mother told me. It's what everybody says in the Newton." The impact of his words left her feeling as if all the air had been sucked from her.

"Well, they're wrong. Why would anybody think that?"

"Because the night you went to Nottingham was the last night anyone in the Newton saw him or heard from him. And because my mother occasionally gets money in the post with a Nottingham postmark."

Laidlaw breathed heavily, a concertina wheeze in her ear. "By Christ, that's wild. Well, sweetheart, I'm sorry to disappoint you. There was five of us left Newton of Wemyss that December night. But your dad wasn't among us."

Wednesday, 27th June 2007; Glenrothes

Karen stopped at the canteen for a chicken salad sandwich on the way back to her desk. Criminals and witnesses could seldom fool Karen, but when it came to food, she could fool herself seventeen ways before breakfast. The sandwich, for example. Wholegrain bread, a swatch of wilted lettuce, a couple of slices of tomato and cucumber, and it became a health food. Never mind the butter and the mayo. In her head, the calories were cancelled by the benefits. She tucked her notebook under her arm and ripped open the plastic sandwich box as she walked.

Phil Parhatka looked up as she flopped into her chair. Not for the first time, the angle of his head reminded her that he looked like a darker, skinnier version of Matt Damon. There was the same jut of nose and jaw, the straight brows, *The Bourne Identity* haircut, the expression that could swing from open to guarded in a heartbeat. Just the colouring was different. Phil's Polish ancestry was responsible for his dark hair, brown eyes, and thick pale skin; his personality had contributed the tiny hole in his left earlobe, a piercing that generally accommodated a diamond stud when he was off duty. "How was it for you?" he said.

"More interesting than I expected," she admitted, getting up again to fetch herself a Diet Coke. Between bites and swallows, she gave him a concise précis of Misha Gibson's story.

"And she believes what this old geezer in Nottingham told

her?" he said, leaning back in his chair and linking his fingers behind his head.

"I think she's the sort of woman who generally believes what people tell her," Karen said.

"She'd make a lousy copper, then. So, I take it you'll be passing it across to Central Division to get on with?"

Karen took a chunk out of her sandwich and chewed vigorously, the muscles of her jaw and temple bulging and contracting like a stress ball under pressure. She swallowed before she'd finished chewing properly, then washed the mouthful down with a swig of Diet Coke. "Not sure," she said. "It's kind of interesting."

Phil gave her a wary look. "Karen, it's not a cold case. It's not ours to play with."

"If I pass it over to Central, it'll wither on the vine. Nobody over there's going to bother with a case where the trail went cold twenty-two years ago." She refused to meet his disapproving eye. "You know that as well as I do. And according to Misha Gibson, her kid's drinking in the last-chance saloon."

"That still doesn't make it a cold case."

"Just because it wasn't opened in 1984 doesn't mean it's not cold now." Karen waved the remains of her sandwich at the files on her desk. "And none of this lot are going anywhere any time soon. Darren Anderson—nothing I can do till the cops in the Canaries get their fingers out and find which bar his ex-girlfriend's working in. Ishbel Mackindoe—waiting for the lab to tell me if they can get any viable DNA from the anonymous letters. Patsy Millar—can't get any further with that till the Met finish digging up the garden in Haringey and do the forensics."

"There's witnesses in the Patsy Millar case that we could talk to again."

Karen shrugged. She knew she could pull rank on Phil and shut him up that way, but she needed the ease between them too much. "They'll keep. Or else you can take one of the DCs and give them some on-the-job training."

"If you think they need on-the-job training, you should give

them this stone-cold missing person case. You're a DI now, Karen. You're not supposed to be chasing about on stuff like this." He waved a hand towards the two DCs sitting at their computers. "That's for the likes of them. What this is about is that you're bored." Karen tried to protest but Phil carried on regardless. "I said when you took this promotion that flying a desk would drive you mental. And now look at you. Sneaking cases out from under the woolly suits at Central. Next thing, you'll be going off to do your own interviews."

"So?" Karen screwed up the sandwich container with more force than was strictly necessary and tossed it in the bin. "It's good to keep my hand in. And I'll make sure it's all above board. I'll take DC Murray with me."

"The Mint?" The tone in Phil's voice was incredulous, the look on his face offended. "You'd take the Mint over me?"

Karen smiled sweetly. "You're a sergeant now, Phil. A sergeant with ambitions. Staying in the office and keeping my seat warm will help your aspirations become a reality. Besides, the Mint's not as bad as you make out. He does what he's told."

"So does a collie dog. But a dog would show more initiative."

"There's a kid's life at stake, Phil. I've got more than enough initiative for both of us. This needs to be done right and I'm going to make sure it is." She turned to her computer with an air of having finished with the conversation.

Phil opened his mouth to say more, then thought better of it when he saw the repressive glance Karen flashed in his direction. They'd been drawn to each other from the start of their careers, each recognizing nonconformist tendencies in the other. Having come up the ranks together had left the pair of them with a friendship that had survived the challenge of altered status. But he knew there were limits to how far he could push Karen, and he had a feeling he'd just butted up against them. "I'll cover for you here, then," he said.

"Works for me," Karen said, her fingers flying over the keys.

"Book me out for tomorrow morning. I've a feeling Jenny Prentice might be a wee bit more forthcoming to a pair of polis than she was to her daughter."

Thursday, 28th June 2007; Edinburgh

Learning to wait was one of the lessons that courses in journalism didn't teach. When Bel Richmond had had a full-time job on a Sunday paper, she had always maintained that she was paid not for a forty-hour week but for the five minutes when she talked her way across a doorstep that nobody else had managed to cross. That left a lot of time for waiting. Waiting for someone to return a call. Waiting for the next stage of the story to break. Waiting for a contact to turn into a source. Bel had done a lot of waiting and, while she'd become skilled at it, she had never learned to love it.

She had to admit she'd passed the time in surroundings that were a lot less salubrious than this. Here, she had the physical comforts of coffee, cookies, and newspapers. And the room she'd been left in commanded the panoramic view that had graced a million shortbread tins. Running the length of Princes Street, it featured a clutch of keynote tourist sights—the castle, the Scott Monument, the National Gallery, and Princes Street Gardens. Bel spotted other significant architectural eye candy, but she didn't know enough about the city to identify it. She'd visited the Scottish capital only a few times, and conducting this meeting here hadn't been her choice. She'd wanted it in London, but her reluctance to show her hand in advance had turfed her out of the driving seat and into the role of supplicant.

Unusually for a freelance journalist, she had a temporary research assistant. Jonathan was a journalism student at City University, and he'd asked his tutor to assign him to Bel for his work experience assignment. Apparently he liked her style. She'd been

mildly gratified by the compliment and delighted at the prospect
of having eight weeks free from drudgery. And so it was Jonathan
who had made the first contact with Maclennan Grant Enter-
prises. The message he'd returned with was simple. If Ms. Richmond was not prepared to state her reason for wanting a meeting
with Sir Broderick Maclennan Grant, Sir Broderick was not prepared to meet her. Sir Broderick did not give interviews. Further
arm's-length negotiations had led to this compromise.

And now Bel was, she thought, being put in her place. Being
forced to cool her heels in a hotel meeting room. Being made to
understand that someone as important as the personal assistant
to the chairman and principal shareholder of the country's twelfth
most valuable company had more pressing calls on her time than
dancing attendance on some London hack.

She wanted to get up and pace, but she didn't want to reveal
any lack of composure. Giving up the high ground was not something that had ever come naturally to her. Instead she straightened
her jacket, made sure her shirt was tucked in properly, and picked
a stray piece of grit from her emerald suede shoes.

At last, precisely fifteen minutes after the agreed time, the door
opened. The woman who entered in a flurry of tweed and cashmere resembled a school mistress of indeterminate age but one
accustomed to exerting discipline over her pupils. For one crazy
moment, Bel nearly jumped to her feet in a Pavlovian response to
her own teenage memories of terrorist nuns. But she managed to
restrain herself and stood up in a more leisurely manner.

"Susan Charleson," the woman said, extending a hand. "Sorry
to keep you waiting. As Harold Macmillan once said, 'Events,
dear boy. Events.'"

Bel decided not to point out that Harold Macmillan had been
referring to the job of prime minister, not wet nurse to a captain of
industry. She took the warm dry fingers in her own. A moment's
sharp grip, then she was released. "Annabel Richmond."

Susan Charleson ignored the armchair opposite Bel and headed
instead for the table by the window. Wrong-footed, Bel scooped

up her bag and the leather portfolio beside it and followed. The women sat down opposite each other and Susan smiled, her teeth like a line of chalky toothpaste between the dark pink lipstick. "You wanted to see Sir Broderick," she said. No preamble, no small talk about the view. Just straight to the chase. It was a technique Bel had used herself on occasion, but that didn't mean she enjoyed the tables being turned.

"That's right."

Susan shook her head. "Sir Broderick does not speak to the press. I fear you've had a wasted journey. I did explain all that to your assistant, but he wouldn't take no for an answer."

It was Bel's turn to produce a smile without warmth. "Good for him. I've obviously got him well trained. But there seems to be a misunderstanding. I'm not here to beg for an interview. I'm here because I think I have something Sir Broderick will be interested in." She lifted the portfolio on to the table and unzipped it. From inside, she took a single A3 sheet of heavyweight paper, face down. It was smeared with dirt and gave off a faint smell, a curious blend of dust, urine, and lavender. Bel couldn't resist a quick teasing look at Susan Charleson. "Would you like to see?" she said, flipping the paper.

Susan took a leather case from the pocket of her skirt and extracted a pair of tortoiseshell glasses. She perched them on her nose, taking her time about it, but her eyes never left the stark black-and-white images before her. The silence between the women seemed to expand, and Bel felt almost breathless as she waited for a response. "Where did you come by this?" Susan said, her tone as prim as a Latin mistress.

Monday, 18th June 2007; Campora, Tuscany, Italy

At seven in the morning, it was almost possible to believe that the baking heat of the previous ten days might not show up for work. Pearly daylight shimmered through the canopy of oak and chest-

nut leaves, making visible the motes of dust that spiralled upwards from Bel's feet. She was moving slowly enough to notice because the unpaved track that wound down through the woods was rutted and pitted, the jagged stones scattered over it enough to make any jogger conscious of the fragility of ankles.

Only two more of these cherished early-morning runs before she'd have to head back to the suffocating streets of London. The thought provoked a tiny tug of regret. Bel loved slipping out of the villa while everyone else was still asleep. She could walk barefoot over cool marble floors, pretending she was chatelaine of the whole place, not just another holiday tenant carving off a slice of borrowed Tuscan elegance.

She'd been coming on holiday with the same group of five friends since they'd shared a house in their final year at Durham. That first time, they'd all been cramming for their finals. One set of parents had a cottage in Cornwall that they'd colonized for a week. They'd called it a study break, but in truth, it had been more of a holiday that had refreshed and relaxed them, leaving them better placed to sit exams than if they'd huddled over books and articles. And although they were modern young women not given to superstition, they'd all felt that their week together had somehow been responsible for their good degrees. Since then, they'd gathered together every June for a reunion, committed to pleasure.

Over the years, their drinking had grown more discerning, their eating more epicurean, and their conversation more outrageous. The locations had become progressively more luxurious. Lovers were never invited to share the girls' week. Occasionally, one of their number had a little wobble, claiming pressure of work or family obligations, but they were generally whipped back into line without too much effort.

For Bel, it was a significant component of her life. These women were all successful, all private sources she could count on to smooth her path from time to time. But still, that wasn't the main reason this holiday was so important to her. Partners had

come and gone, but these friends had been constant. In a world where you were measured by your last headline, it felt good to have a refuge where none of that mattered. Where she was appreciated simply because the group enjoyed themselves more with her than without her. They'd all known each other long enough to forgive each other's faults, to accept each other's politics, and to say what would be unsayable in any other company. This holiday formed part of the bulwark she constantly shored up against her own insecurities. Besides, it was the only holiday she took these days that was about what she wanted. For the past half dozen years, she'd been bound to her widowed sister Vivianne and her son Harry. The sudden death of Vivianne's husband from a heart attack had left her emotionally stranded and practically struggling. Bel had barely hesitated before throwing her lot in with her sister and her nephew. On balance it had been a good decision, but even so she still treasured this annual work-free break from a family life she hadn't expected to be living. Especially now that Harry was teetering on the edge of teenage existential angst. So this year, even more than in the past, the holiday had to be special, to outdo what had gone before.

It was hard to imagine how they could improve on this, she thought as she emerged from the trees and turned into a field of sunflowers preparing to burst into bloom. She speeded up a little as she made her way along the margin, her nose twitching at the aromatic perfume of the greenery. There was nothing she'd change about the villa, no fault she could find with the informal gardens and fruit trees that surrounded the loggia and the pool. The view across the Val d'Elsa was stunning, with Volterra and San Gimignano on the distant skyline.

And there was the added bonus of Grazia's cooking. When they'd discovered that the "local chef" trumpeted on the website was the wife of the pig farmer down the hill, they'd been wary of taking up the option of having her come to the villa and prepare a typically Tuscan meal. But on the third afternoon, they'd all been too stunned by the heat to be bothered with cooking so

they'd summoned Grazia. Her husband Maurizio had delivered her to the villa in a battered Fiat Panda that appeared to be held together with string and faith. He'd also unloaded boxes of food covered in muslin cloths. In fractured English, Grazia had thrown them out of the kitchen and told them to relax with a drink on the loggia.

The meal had been a revelation—nutty salamis and prosciutto from the rare Cinta di Siena pigs Maurizio bred, coupled with fragrant black figs from their own tree; spaghetti with pesto made from tarragon and basil; quails roasted with Maurizio's vegetables, and long fingers of potatoes flavoured with rosemary and garlic; cheeses from local farms, and finally, a rich cake heavy with limoncello and almonds.

The women never cooked dinner again.

Grazia's cooking made Bel's morning runs all the more necessary. As forty approached, she struggled harder to maintain what she thought of as her fighting weight. This morning, her stomach still felt like a tight round ball after the meltingly delicious *melanzane alla parmigiana* that had provoked her into an excessive second helping. She'd go a little farther than usual, she decided. Instead of making a circuit of the sunflower field and climbing back up to their villa, she would take a track that ran from the far corner through the overgrown grounds of a ruined *casa colonica* she'd noticed from the car. Ever since she'd spotted it on their first morning, she'd indulged a fantasy of buying the ruin and transforming it into the ultimate Tuscan retreat, complete with swimming pool and olive grove. And of course, Grazia on hand to cook. Bel had few qualms about poaching, neither in fantasy nor reality.

But she knew herself well enough to understand it would never be more than a pipe dream. Having a retreat implied a willingness that was alien to her, to step away from the world of work. Maybe when she was ready to retire she could contemplate devoting herself to such a restoration project. Except that she recognized that

as another daydream. Journalists never really retired. There was always another story on the horizon, another target to pursue. Not to mention the terror of being forgotten. All reasons why past relationships had failed to stay the course, all reasons why the future probably held the same imperfections. Still, it would be fun to take a closer look at the old house, to see just how bad a state it was in. When she'd mentioned it to Grazia, she'd pulled a face and called it *rovina*. Bel, whose Italian was fluent, had translated it for the others: "ruin." Time to find out whether Grazia was telling the truth or just trying to divert the interest of the rich English women.

The path through the long grass was still surprisingly clear, bare soil packed hard by years of foot traffic. Bel took the opportunity to pick up speed, then slowed as she reached the edge of the gated courtyard in front of the old farmhouse. The gates were dilapidated, hanging drunkenly from hinges that were barely attached to the tall stone posts. A heavy chain and padlock held them fastened. Beyond, the courtyard's broken paving was demarcated with tufts of creeping thyme, chamomile, and coarse weeds. Bel shook the gates without much expectation. But that was enough to reveal that the bottom corner of the right-hand gate had parted company completely with its support. It could readily be pulled clear enough to allow an adult through the gap. Bel slipped through and let go. The gate creaked faintly as it settled back into place, returning to apparent closure.

Close up, she could understand Grazia's description. Anyone taking this project on would be in thrall to the builders for a very long time. The house surrounded the courtyard on three sides, a central wing flanked by a matching pair of arms. There were two storeys, with a loggia running round the whole of the upper floor, doors and windows giving on to it, providing the bedrooms with easy access to fresh air and common space. But the loggia floor sagged, what doors remained were skewed, and the lintels above the windows were cracked and oddly angled. The window panes

on both floors were filthy, cracked, or missing. But still the solid lines of the attractive vernacular architecture were obvious and the rough stones glowed warm in the morning sun.

Bel couldn't have explained why, but the house drew her closer. It had the raddled charm of a former beauty sufficiently self-assured to let herself go without a fight. Unpruned bougainvillea straggled up the peeling ochre stucco and over the low wall of the loggia. If nobody chose to fall in love with this place soon, it was going to be overwhelmed by vegetation. In a couple of generations, it would be nothing more than an inexplicable mound on the hill-side. But for now, it still had the power to bewitch.

She picked her way across the crumbling courtyard, passing cracked terracotta pots lying askew, the herbs they'd contained sprawling and springing free, spicing the air with their fragrances. She pushed against a heavy door made of wooden planks hanging from a single hinge. The wood screeched against an uneven floor of herringbone brick, but it opened wide enough for Bel to enter a large room without squeezing. Her first impression was of grime and neglect. Cobwebs were strung in a maze from wall to wall. The windows were mottled with dirt. A distant scurrying had Bel peering around in panic. She had no fear of news editors, but four-legged rats filled her with revulsion.

As she grew accustomed to the gloom, Bel realized the room wasn't completely empty. A long table stood against one wall. Opposite was a sagging sofa. Judging by the rest of the place, it should have been rotten and filthy, but the dark red upholstery was still relatively clean. She filed the oddity for further consideration.

Bel hesitated for a moment. None of her friends, she was sure, would be urging her to penetrate deeper into this strange deserted house. But she had built her career on a reputation for fearless-ness. Only she knew how often the image had concealed levels of anxiety and uncertainty that had reduced her to throwing up in gutters and strange toilets. Given what she'd faced down in her determination to secure a story, how scary could an abandoned ruin be?

A doorway in the far corner led to a cramped hallway with a worn stone staircase climbing up to the loggia. Beyond, she could see another dark and grubby room. She peered in, surprised to see a thin cord strung across one corner from which half a dozen metal coat hangers dangled. A knitted scarf was slung round the neck of one of the hangers. Beneath it, she could see a crumpled pile of camouflage material. It looked like one of the shooting jackets on sale at the van that occupied the turnout opposite the café on the main Colle Val d'Elsa road. The women had been laughing about it just the other day, wondering when exactly it had become fashionable for Italian men of all ages to look as if they'd just walked in from a tour of duty in the Balkans. *Weird,* she thought. Bel cautiously climbed the stairs to the loggia, expecting the same sense of long-abandoned habitation.

But as soon as she emerged from the stairwell, she realized she'd stepped into something very different. When she turned to her left and glanced in the first door, she understood this house was not what it seemed. The rancid mustiness of the lower floor was only a faint note here, the air almost as fresh as it was outside. The room had obviously been a bedroom, and fairly recently at that. A mattress lay on the floor, a bedspread flung back casually across the bottom third. It was dusty but had none of the ingrained grime the lower floor had led Bel to expect. Again, a cord was strung across the corner. There were a dozen empty hangers, but the final three held slightly crumpled shirts. Even from a distance, she could see they were past their best, fade lines across the sleeves and collars.

A pair of tomato crates acted as bedside tables. One held a stump of candle in a saucer. A yellowed copy of the *Frankfurter Allgemeine Zeitung* lay on the floor next to the bed. Bel picked it up, noting that the date was less than four months ago. So that gave her an idea of when this place had been last abandoned. She lifted one of the shirtsleeves and pressed it to her nose. Rosemary and marijuana. Faint but unmistakable.

She went back to the loggia and checked out the other rooms. The pattern was similar. Three more bedrooms containing a hand-

ful of leftovers—a couple of T-shirts, paperbacks and magazines
in English, Italian, and German, half a bottle of wine, the stub of
a lipstick, a leather sandal whose sole had parted company with
its upper—the sort of things you would leave behind if you were
moving out with no thought of who might come after. In one, a
bunch of flowers stuck in an olive jar had dried to fragility.

The final room on the west side was the biggest so far. Its windows had been cleaned more recently than any of the others, its
shutters renovated and its walls whitewashed. Standing in the
middle of the floor was a silk-screen printing frame. Trestle tables
set against one wall contained plastic cups stained inside with
dried pigments, and brushes stiff with neglect. A scatter of spots
and blots marked the floor. Bel was intrigued, her curiosity overcoming any lingering nervousness at being alone in this peculiar
place. Whoever had been here must have cleared out in a hurry.
Leaving a substantial silk-screen frame behind wasn't what you
would do if your departure was planned.

She backed out of the studio and made her way along the loggia
to the wing opposite. She was careful to stay close to the wall, not
trusting the undulating brick floor with her weight. She passed the
bedroom doors, feeling like a trespasser on the *Mary Celeste*. A
silence unbroken even by birdsong accentuated the impression.
The last room before the corner was a bathroom whose nauseat-
ing mix of odours still hung in the air. A coil of hosepipe lay on
the floor, its tail end disappearing through a hole in the masonry
near the window. So they had improvised some sort of running
water, though not enough to make the toilet anything less than
disgusting. She wrinkled her nose and backed away.

Bel rounded the corner just as the sun cleared the corner of
the woods, flooding her in sudden warmth. It made her entry into
the final room all the more chilling. Shivering at the dank air, she
ventured inside. The shutters were pulled tight, making the interior almost too dim to discern anything. But as her eyes adjusted,
she gained a sense of the room. It was the twin of the studio in
scale, but its function was quite different. She crossed to the near-

est window and struggled with the shutter, finally managing to haul it halfway open. It was enough to confirm her first impression; this had been the heart of the occupation of the *casa rovina*. A battered old cooking range connected to a gas cylinder stood by a stone sink. The dining table was scarred and stripped to the bare wood, but it was solid and had beautifully carved legs. Seven unmatched chairs sat around it, an eighth overturned a few feet away. A rocking chair and a couple of sofas lined the walls. Odd bits of crockery and cutlery lay scattered around, as if the inhabitants couldn't be bothered collecting them when they'd left.

As Bel walked back from the window, a rickety table caught her eye. Standing behind the door, it was easy to miss. An untidy scatter of what appeared to be posters lay across it. Fascinated, she moved towards it. Two strides and she stopped short, her sharp gasp echoing in the dusty air.

Before her on the limestone flags was an irregular stain, perhaps three feet by eighteen inches. Rusty brown, its edges were rounded and smooth, as if it had flowed and pooled rather than spilled. It was thick enough to obscure the flags beneath. One section on the farthest edge looked smudged and thinned, as if someone had tried to scrub it clean and soon given up. Bel had covered enough stories of domestic violence and sexual homicide to recognize a serious bloodstain when she saw it.

Startled, she stepped back, head swivelling from side to side, heart thudding so hard she thought it might choke her. What the hell had happened here? She looked around wildly, noticing other dark stains marking the floor beyond the table. *Time to get out of here*, the sensible part of her mind was screaming. But the devil of curiosity muttered in her ear. *There's been nobody here for months. Look at the dust. They're long gone. They're not going to be back any time soon. Whatever happened here was good reason for them to clear out. Check out the posters . . .*

Bel skirted the stain, giving it as wide a berth as she could without touching any of the furniture. All at once, she felt a taint in the air. Knew it was imagination, but still it seemed real. Back to the

room, face to the door, she crab-walked to the table and looked down at the posters strewn across it.

The second shock was almost more powerful than the first.

Bel knew she was pushing too hard up the hill, but she couldn't pace herself. She could feel the sweat from her hand coating the good-quality paper of the rolled-up poster. At last the track emerged from the trees and became less treacherous as it approached their holiday villa. The road sloped down almost imperceptibly, but gravity was enough to give her tired legs an extra boost, and she was still moving fast when she rounded the corner of the house to find Lisa Martyn stretched out on the shady terrace in a pool chair with Friday's *Guardian* for company. Bel felt relief. She needed to talk to someone and, of all her companions, Lisa was least likely to turn her revelations into dinner party gossip. A human rights lawyer whose compassion and feminism seemed as ineluctable as every breath she took, Lisa would understand the potential of the discovery Bel thought she had made. And her right to handle it as she saw fit.

Lisa dragged her eyes away from the newspaper, distracted by the unfamiliar heave of Bel's breath. "My God," she said. "You look like you're about to stroke out."

Bel put the poster down on a chair and leaned over, hands on knees, dragging breath into her lungs, regretting those secret, stolen cigarettes. "I'll be—OK in—a minute."

Lisa struggled ungainly out of the chair and hurried into the kitchen, returning with a towel and bottle of water. Bel stood straight, took the water, and poured half over her head, snorting as she breathed it in by accident. Then she rubbed her head with the towel and slumped into a chair. She swallowed a long draught of water while Lisa returned to her pool chair. "What was all that about?" Lisa said. "You're the most dignified jogger I know. Never seen an out-of-breath Bel before. What's got you into such a state?"

"I found something," Bel said. Her chest was still struggling, but she could manage short bursts of speech. "At least, I think I found something. And if I'm right, it's the story of my career." She reached for the poster. "I was kind of hoping you might be able to tell me whether I've completely lost the plot."

Intrigued, Lisa tossed the paper to the ground and sat up. "So, what is it, this thing that might be something?"

Bel unrolled the heavy paper, weighing it down at the corners with a pepper grinder, a coffee mug, and a couple of dirty ashtrays. The image on the A3 sheet was striking. It had been designed to look like a stark black-and-white woodcut in the German Expressionist style. At the top of the page, a bearded man with an angular shock of hair leaned over a screen, his hands holding wooden crosses from which three marionettes dangled. But these were no ordinary marionettes. One was a skeleton, the second a goat, and the third a representation of Death with his hooded robe and scythe. There was something indisputably sinister about the image. Across the bottom, enclosed by a funereal black border, was a blank area about three inches deep. It was the sort of space where a small bill might be posted announcing a performance.

"Fuck me," Lisa said. At last, she looked up. "Catriona Maclennan Grant," she said. There was wonder in her voice. "Bel . . . where the hell did you find this?"

Thursday, 28th June 2007; Edinburgh

Bel smiled. "Before I answer that, I want to clarify a few things."

Susan Charleson rolled her eyes. "You can't imagine you're the first person who's walked through the door with a faked-up copy of the ransom poster. I'll tell you what I've told them. The reward is contingent on finding Sir Broderick's grandson alive or demonstrating conclusively that he is dead. Not to mention bringing Catriona Maclennan Grant's killers to justice."

"You misunderstand me," Bel said, smile mischievous but not giving an inch. "Ms. Charleson, I'm really not interested in Sir Broderick's money. But I do have one condition."

"You're making a mistake here." Susan Charleson's voice had acquired an edge. "This is a police matter. You're in no position to be imposing conditions."

Bel placed a hand firmly on the poster. "I can walk out the door now with this poster and forget I ever saw it. I'd have little difficulty in lying to the police. I'm a journalist, after all." She was beginning to enjoy herself far more than she'd anticipated. "Your word against mine, Ms. Charleson. And I know you don't want me to walk out on you. One of the skills a successful journalist has to learn is how to read people. And I saw the way you reacted when you looked at this. You know this is the real thing, not some faked-up copy."

"You've a very aggressive attitude." Susan Charleson sounded almost nonchalant.

"I like to think of it as assertive. I didn't come here to fall out with you, Ms. Charleson. I want to help. But not for free. In my experience, the rich don't appreciate anything they don't have to pay for."

"You said you weren't interested in money."

"That's true. And I'm not. I am, however, interested in reputation. And my reputation is built on being not just first with the story but with getting to the story behind the story. I think there are areas where I can help unravel this story more effectively than official channels. I'm sure you'll agree once I've explained where this poster came from. All I'm asking is that you don't obstruct me looking into the case. And beyond that, that you and your boss cooperate when it comes to sharing information about what was going on around the time Catriona was kidnapped."

"That's quite a significant request. Sir Broderick is not a man who compromises his privacy readily. You'll appreciate I don't have the authority to grant what you are asking."

Bel shrugged one shoulder delicately. "Then we can meet again

when you have an answer." She slid the poster across the table, opening the portfolio to replace it there.

Susan Charleson stood up. "If you can spare me a few more minutes, I might be able to give you an answer now."

Bel knew at that point that she had won. Susan Charleson wanted this too badly. She would persuade her boss to accept the deal. Bel hadn't been this excited in years. This wasn't just a slew of news stories and features, though there wasn't a paper in the world that wouldn't be interested. Especially after the Madeleine McCann case. With access to the mysterious Brodie Grant plus the chance of discovering the fate of his grandson, this was potentially a bestseller. *In Cold Blood* for the new millennium. It would be her ticket for the gravy train.

Bel gave a little snort of laughter. Maybe she could use the proceeds to buy the *casa rovina* and bring things full circle. It was hard to imagine what could be neater.

Thursday, 28th June 2007; Newton of Wemyss

It had been a few years since Karen had last taken the single-track road to Newton of Wemyss. But it was obvious that the hamlet had undergone the same transformation as its sister villages on the main road. Commuters had fallen ravenous upon all four of the Wemyss villages, seeing rustic possibilities in what had been grim little miners' rows. One-bedroom hovels had been knocked through to make lavish cottages, back yards transformed by conservatories that poured light into gloomy living-kitchens. Villages that had shrivelled and died following the Michael pit disaster in '67 and the closures that followed the 1984 strike had found a new incarnation as dormitories whose entire idea of community was a pub quiz night. In the village shops you could buy a scented candle but not a pint of milk. The only way you could tell there had ever been a mining community was the scale model of pit winding gear that straddled the point where the private steam railway had once

crossed the main road laden with open trucks of coal bound for the railhead at Thornton Junction. Now, the whitewashed miners' rows looked like an architect's deliberate choice of what a vernacular village ought to look like. Their history had been overwhelmed by a designer present.

Since her last visit, Newton of Wemyss had spruced itself up. The modest war memorial stood on a triangle of shaven grass in the centre. Wooden troughs of flowers stood around it at perfect intervals. Immaculate single-storey cottages lined the village green, the only break in the low skyline the imposing bulk of the local pub, the Laird o' Wemyss. It had once been owned collectively by the local community under the Gothenburg system, but the hard times of the eighties had forced it to close. Now it was a destination restaurant, its "Scottish Fusion" cuisine drawing visitors from as far afield as Dundee and Edinburgh and its prices lifting it well out of her budget. Karen wondered how far Mick Prentice would have had to travel for a simple pint of heavy if he'd stayed put in Newton.

She consulted the Mapquest directions she'd printed out and pointed to a road at the apex of the triangle to her driver, DC Jason "the Mint" Murray. "You want to go down the lane there," she said. "Towards the sea. Where the pit used to be."

They left the village centre behind immediately. Shaggy hedgerows fringed a field of lush green wheat on the right. "All this rain, it's making everything grow like the clappers," the Mint said. It had taken him the full twenty-five-minute journey from the office to summon up a comment.

Karen couldn't be bothered with a conversation about the weather. What was there to say? It had rained all bloody summer so far. Just because it wasn't raining right this minute didn't mean it wouldn't be wet by the end of the day. She looked over to her left where the colliery buildings had once stood. She had a vague memory of offices, pithead baths, a canteen. Now it had been razed to its concrete foundation, weeds forcing through jagged cracks as they reclaimed it. Marooned beyond it was a single un-

touched miners' row; eight raddled houses stranded in the middle of nowhere by the demolition of the buildings that had provided the reason for their existence. Beyond them was a thick stand of tall sycamores and beeches, a dense windbreak between the houses and the edge of the cliff that plunged down thirty feet to the coastal path below. "That's where the Lady Charlotte used to be," she said.

"Eh?" The Mint sounded startled.

"The pit, Jason."

"Oh. Right. Aye. Before my time." He peered through the windscreen, making her wonder uneasily if he needed glasses. "Which house is it, guv?"

She pointed to the one second from the end. The Mint eased the car round the potholes as carefully as if it had been his own and came to a halt at the end of Jenny Prentice's path.

In spite of Karen's phone call setting up the meeting, Jenny took her time answering the door, which gave them plenty of opportunity to examine the cracked concrete flags and the depressing patch of weedy gravel in front of the house. "If this was mine . . . ," the Mint began, then trailed off, as if it was all too much to contemplate.

The woman who answered the door had the air of someone who had spent her days lying down so life could more easily trample over her. Her lank greying hair was tied back haphazardly, strands escaping at both sides. Her skin was lined and puckered, with broken veins mapping her cheeks. She wore a nylon overall that came to mid-thigh over cheap black trousers whose material had gone bobbly. The overall was a shade of lavender found nowhere in nature. Karen's parents still lived in a street populated by ex-miners and their kin in unfashionable Methil, but even the most dysfunctional of their neighbours would have taken more trouble with their appearance when they knew they were in for any kind of official visit. Karen didn't even bother trying to avoid judging Jenny Prentice on her appearance. "Good morning, Mrs. Prentice," she said briskly. "I'm DI Pirie. We spoke on the phone. And this is DC Murray."

Jenny nodded and sniffed. "You'd better come in."

The living room was cramped but clean. The furniture, like the carpet, was unfashionable but not at all shabby. A room for special occasions, Karen thought, and a life where there were few of those.

Jenny waved them towards the sofa and perched on the edge of an armchair opposite. She was clearly not going to offer them any sort of refreshment. "So. You're here because of our Misha. I thought you lot would have something better to do, all the awful things I keep reading about in the newspapers."

"A missing husband and father is a pretty awful thing, wouldn't you say?" Karen said.

Jenny's lips tightened, as if she'd felt the burn of indigestion. "Depends on the man, Inspector. The kind of guy you run into doing your job, I don't imagine too many of their wives and kids are that bothered when they get taken away."

"You'd be surprised. A lot of their families are pretty devastated. And at least they know where their man is. They don't have to live with uncertainty."

"I didn't think I was living with uncertainty. I thought I knew damn fine where Mick was until our Misha started raking about trying to find him."

Karen nodded. "You thought he was in Nottingham."

"Aye. I thought he'd went scabbing. To be honest, I wasn't that sorry to see the back of him. But I was bloody livid that he put that label round our necks. I'd rather he was dead than a blackleg, if you really want to know." She pointed at Karen. "You sound like you're from round here. You must know what it's like to get tarred with that brush."

Karen tipped her head in acknowledgement. "All the more galling now that it looks like he didn't go scabbing after all."

Jenny looked away. "I don't know that. All I know is that he didn't go to Nottingham that night with that particular bunch of scabs."

"Well, we're here to try to establish what really happened. My colleague here is going to take some notes, just to make sure I don't misremember anything you tell me." The Mint hastily took out his notebook and flipped it open in a nervous flurry of pages. Maybe Phil had been right about his deficiencies, Karen thought. "Now, I need his full name and date of birth."

"Michael James Prentice. Born 20th January 1955."

"And you were all living here at the time? You and Mick and Misha?"

"Aye. I've lived here all my married life. Never really had a choice in the matter."

"Have you got a photo of Mick you could let us have? I know it's a long time ago, but it could be helpful."

"You can put it on the computer and make it older, can't you?" Jenny went to the sideboard and opened a drawer.

"Sometimes it's possible." But too expensive unless there's a more pressing reason than your grandson's leukaemia.

Jenny took out an immaculate black leather album and brought it back to the chair. When she opened it, the covers creaked. Even upside down and from the other side of the room, Karen could see it was a wedding album. Jenny quickly turned past the formal wedding shots to a pocket at the back, thickly stuffed with snaps. She pulled out a bundle and flicked through them. She paused at a couple, then finally settled on one. She handed Karen a rectangular picture. It showed the heads and shoulders of two young men grinning at the camera, corners of the beer glasses in shot as they toasted the photographer. "That's Mick on the left," Jenny said. "The good-looking one."

She wasn't lying. Mick Prentice had tousled dark blond hair, cut in the approximation of a mullet that George Michael had boasted in his Wham! period. Mick had blue eyes, ridiculously long eyelashes, and a dangerous smile. The sickle crescent of a coal tattoo sliced through his right eyebrow, saving him from being too pretty. Karen could see exactly why Jenny Prentice had fallen for her hus-

band. "Thanks," she said. "Who's the other guy?" A raggedy mop of brown hair, long, bony face, a few faint acne scars pitting the sunken cheeks, lively eyes, a triangular grin like the Joker in the Batman comics. Not a looker like his pal, but something engaging about him all the same.

"His best pal. Andy Kerr."

The best pal who killed himself, according to Misha. "Misha told me your husband went missing on Friday the fourteenth of December 1984. Is that your recollection?"

"That's right. He went out in the morning with his bloody paints and said he'd be back for his tea. That was the last I saw him."

"Paints? He was doing a bit of work on the side?"

Jenny made a sound of disdain. "As if. Not that we couldn't have used the money. No, Mick painted watercolours. Can you credit it? Can you imagine anything more bloody useless in the 1984 strike than a miner painting watercolours."

"Could he not have sold them?" the Mint chipped in, leaning forward and looking keen.

"Who to? Everybody round here was skint and there was no money for him to go someplace else on the off chance." Jenny gestured at the wall behind them. "He'd have been lucky to get a couple of pounds apiece."

Karen swivelled round and looked at the three cheaply framed paintings on the wall. West Wemyss, Macduff Castle, and the Lady's Rock. To her untutored eye, they looked vivid and lively. She'd have happily given them house room, though she didn't know how much she'd have been willing to pay for the privilege back in 1984. "So, how did he get into that?" Karen asked, turning back to face Jenny.

"He did a class at the Miners' Welfare the year Misha was born. The teacher said he had a gift for it. Me, I think she said the same to every one of them that was halfway good looking."

"But he kept it up?"

"It got him out of the house. Away from the dirty nappies and

the noise." Bitterness seemed to come off Jenny Prentice in waves. Curious but heartening that it didn't seem to have infected her daughter. Maybe that had something to do with the stepfather she'd spoken about. Karen reminded herself to ask about the other man in Jenny's life, another who seemed notable for his absence.

"Did he paint much during the strike?"

"Every day it was fair he was out with his kitbag and his easel. And if it was raining, he was down the caves with his pals from the Preservation Society."

"The Wemyss caves, do you mean?" Karen knew the caves that ran back from the shore deep into the sandstone cliffs between East Wemyss and Buckhaven. She'd played in them a few times as a child, oblivious to their historical significance as a major Pictish site. The local kids had treated them as indoor play areas, which was one of the reasons why the Preservation Society had been set up. Now there were railings closing off the deeper and more dangerous sections of the cave network, and amateur historians and archaeologists had preserved them as a playground for adults. "Mick was involved with the caves?"

"Mick was involved in everything. He played football, he painted his pictures, he messed about in the caves, he was up to his eyes in the union. Anything and everything was more important than spending time with his family." Jenny crossed one leg over the other and folded her arms across her chest. "He said it kept him sane during the strike. I think it just kept him out the road of his responsibilities."

Karen knew this was fertile soil for her inquiries, but she could afford to leave it for later. Jenny's suppressed anger had stayed put for twenty-two years. It wasn't about to go anywhere now. Something much more immediate interested her. "So, during the strike, where did Mick get the money for paints? I don't know much about art, but I know it costs a few bob for proper paper and paint." She couldn't imagine any striking miner spending money on art supplies when there was no money for food or heating.

"I don't want to get anybody into trouble," Jenny said.

Yeah, right. "It was twenty-three years ago," Karen said flatly. "I'm really not interested in small-scale contra from the time of the miners' strike."

"One of the art teachers from the high school lived up at Coaltown. He was a wee cripple guy. One leg shorter than the other and a humphy back. Mick used to do his garden for him. The guy paid him in paints." She gave a little snort. "I said could he not pay him in money or food. But apparently the guy was paying out all his wages to the ex-wife. The paints he could nick from the school." She refolded her arms. "He's dead now anyway."

Karen tried to tamp down her dislike of this woman, so different from the daughter who had beguiled her into this case. "So what was it like between you, before Mick disappeared?"

"I blame the strike. OK, we had our ups and downs. But it was the strike that drove a wedge between us. And I'm not the only woman in this part of the world who could say the same thing."

Karen knew the truth of that. The terrible privations of the strike had scarred just about every couple she had known back then. Domestic violence had erupted in improbable places; suicide rates had risen; marriages had shattered in the face of implacable poverty. She hadn't understood it at the time, but she did now. "Maybe so. But everybody's story's different. I'd like to hear yours."

Friday, 14th December 1984; Newton of Wemyss

"I'll be back for my tea," Mick Prentice said, slinging the big canvas bag across his body and grabbing the slender package of his folded easel.

"Tea? What tea? There's nothing in the house to eat. You need to be out there finding food for your family, not messing about painting the bloody sea for the umpteenth time," Jenny shouted, trying to force him to halt on his way out the door.

He turned back, his gaunt face twisted in shame and pain. "You

think I don't know that? You think we're the only ones? You think if I had any idea how to make this better I wouldn't be doing it? Nobody has any fucking food. Nobody has any fucking money." His voice caught in his throat like a sob. He closed his eyes and took a deep breath. "Down the Welfare last night, Sam Thomson said there was talk of a food delivery from the Women Against Pit Closures. If you get yourself down there, they're supposed to be here about two o'clock." It was so cold in the kitchen that his words formed a cloud in front of his lips.

"More handouts. I can't remember the last time I actually chose what I was going to cook for the tea." Jenny suddenly sat down on one of the kitchen chairs. She looked up at him. "Are we ever going to get to the other side of this?"

"We've just got to hold out a bit longer. We've come this far. We can win this." He sounded as if he was trying to convince himself as much as her.

"They're going back, Mick. All the time, they're going back. It was on the news the other night. More than a quarter of the pits are back working. Whatever Arthur Scargill and the rest of the union executives might say, there's no way we can win. It's just a question of how bloody that bitch Thatcher will make the losing."

He shook his head vehemently. "Don't say that, Jenny. Just because there are a few pockets down south where they've caved in. Up here, we're rock solid. So's Yorkshire. And South Wales. And we're the ones that matter." His words sounded hollow, and there was no conviction in his face. They were, she thought, all beaten. They just didn't know when to lie down.

"If you say so," she muttered, turning away. She waited till she heard the door close behind him, then slowly got up and put her coat on. She picked up a heavy-duty plastic sack and left the freezing chill of the kitchen for the damp cold of the morning. This was her routine these days. Get up and walk Misha to school. At the school gate, the bairn would be given an apple or an orange, a bag of chips, and a chocolate cookie by the Friends of the Lady Char-

lotte, a ragtag and bobtail bunch of students and public sector workers from Kirkcaldy who made sure none of the kids started the day on an empty stomach. At least, not on school mornings.

Then back to the house. They'd given up taking milk in their tea, when they could get tea. Some mornings, a cup of hot water was all Jenny and Mick had to start the day. That hadn't happened often, but once was enough to remind you how easy it would be just to fall off the edge.

After a hot drink, Jenny would take her sack into the woods and try to collect enough firewood to give them a few hours of heat in the evening. Between the union executives always calling them "comrade" and the wood gathering, she felt like a Siberian peasant. At least they were lucky to live right by a source of fuel. It was, she knew, a lot harder for other folk. It was their good fortune that they'd kept their open fireplace. The miners' perk of cheap coal had seen to that.

She went about her task mechanically, paying little attention to her surroundings, turning over the latest spat between her and Mick. It sometimes seemed it was only the hardship that kept them together, only the need for warmth that kept them in the same bed. The strike had brought some couples closer together, but plenty had split like a log under an axe after those first few months, once their reserves had been bled dry.

It hadn't been so bad at the start. Since the last wave of strikes in the seventies, the miners had earned good money. They were the kings of the trade union movement—well paid, well organized, and well confident. After all, they'd brought down Ted Heath's government back then. They were untouchable. And they had the cash to prove it.

Some spent up to the hilt—foreign holidays where they could expose their milk-white skin and coal tattoos to the sun, flash cars with expensive stereos, new houses that looked great when they moved in but started to scuff round the edges almost at once. But most of them, made cautious by history, had a bit put by. Enough to cover the rent or the mortgage, enough to feed the family and

pay the fuel bills for a couple of months. What had been horrifying was how quickly those scant savings had disappeared. Early
on, the union had paid decent money to the men who piled into
cars and vans and minibuses to join flying pickets to working pits,
power stations, and coking plants. But the police had grown in-
creasingly heavy-handed in making sure the flyers never made it to
their destinations, and there was little enthusiasm for paying men
for failing to reach their objectives. Besides, these days the union
bosses were too busy trying to hide their millions from the government's sequestrators to be bothered wasting money in a fight they
had to know in their hearts was doomed. So even that trickle of
cash had run dry, and the only thing left for the mining communities to swallow had been their pride.

Jenny had swallowed plenty of that over the past nine months.
It had started right at the beginning when she'd heard the Scottish miners would support the Yorkshire coalfield in the call for a
national strike not from Mick but from Arthur Scargill, president
of the National Union of Mineworkers. Not personally, of course.
Just his yapping harangue on the TV news. Instead of coming
straight back from the Miners' Welfare meeting to tell her, Mick
had been hanging out with Andy and his other union pals, drinking at the bar like money was never going to be a problem. Celebrating King Arthur's battle-cry in the time-honoured way. *The
miners united will never be defeated.*

The wives knew the hopelessness of it all, right from the
start. You go into a coal strike at the beginning of winter, when
the demand from the power stations is at its highest. Not in the
spring, when everybody's looking to turn off their heating. And
when you go for major industrial action against a bitch like Margaret Thatcher, you cover your back. You follow the labour laws.
You follow your own rules. You stage a national ballot. You don't
rely on a dubious interpretation of a resolution passed three years
before for a different purpose. Oh yes, the wives had known it was
futile. But they'd kept their mouths shut and, for the first time ever,
they'd built their own organization to support their men. Loyalty,

that was what counted in the pit villages and mining communities.

And so Mick and Jenny were still hanging together. Jenny sometimes wondered if the only reason Mick was still with her and Misha was because he had nowhere else to go. Parents dead, no brothers or sisters, there was no obvious refuge. She'd asked him once and he'd frozen like a statue for a long moment. Then he'd scoffed at her, denying he wanted to be gone, reminding her that Andy would always put him up in his cottage if he wanted to be away. So, no reason why she should have imagined that Friday was different from any other.

Thursday, 28th June 2007; Newton of Wemyss

"So this wasn't the first time he'd gone off with his paints for the day?" Karen said. Whatever was going on in Jenny Prentice's head, it was clearly a lot more than the bare bones she was giving up.

"Four or five times a week, by the end."

"What about you? What did you do for the rest of the day?"

"I went up the woods for some kindling, then I came back and watched the news on the telly. It was quite the day, that Friday. King Arthur was in court for police obstruction at the Battle of Orgreave. And Band Aid got to number one. I tell you, I could have spat in their faces. All that effort for bairns thousands of miles away when there were hungry kids on their own doorsteps. Where was Bono and Bob Geldof when our kids were waking up on Christmas morning with bugger all in their stockings?"

"It must have been hard to take," Karen said.

"It felt like a slap in the face. Nothing glamorous about helping the miners, was there?" A bitter little smile lit up her face. "Could have been worse, though. We could have had to put up with that sanctimonious shite Sting. Not to mention his bloody lute."

"Right enough." Karen couldn't hide her amusement. Gallows humour was never far from the surface in these mining communities. "So, what did you do after the TV news?"

"I went down the Welfare. Mick had said something about a food handout. I got in the queue and came home with a packet of pasta, a tin of tomatoes, and two onions. And a pack of dried Scotch broth mix. I mind I felt pretty pleased with myself. I collected Misha from the school and I thought it might cheer us up if we put up the Christmas decorations, so that's what we did."

"When did you realize it was late for Mick to be back?"

Jenny paused, one hand fiddling with a button on her overall. "That time of year, it's early dark. Usually, he'd be back not long after me and Misha. But with us doing the decorations, I didn't really notice the time passing."

She was lying, Karen thought. But why? And about what?

Friday, 14th December 1984; Newton of Wemyss

Jenny had been one of the first in the queue at the Miners' Welfare, and she'd hurried home with her pitiful bounty, determined to get a pot of soup going so there would be something tasty for the tea. She rounded the pithead baths building, noticing all her neighbours' houses were in darkness. These days, nobody left a welcoming light on when they went out. Every penny counted when the fuel bills came in.

When she turned in at her gate, she nearly jumped out of her skin. A shadowy figure rose from the darkness, looming huge in her imagination. She made a noise halfway between a gasp and a moan.

"Jenny, Jenny, calm down. It's me. Tom. Tom Campbell. I'm sorry, I didn't mean to scare you." The shape took form and she recognized the big man standing by her front door.

"Christ, Tom, you gave me the fright of my life," she complained, moving past him and opening the front door. Conscious of the breathtaking chill of the house, she led the way into the kitchen. Without hesitation, she filled her soup pan with water and put it on the stove, the gas ring giving out a tiny wedge of

heat. Then she turned to face him in the dimness of the afternoon light. "How are you doing?"

Tom Campbell shrugged his big shoulders and gave a half-hearted smile. "Up and down," he said. "It's ironic. The one time in my life I really needed my pals and this strike happens."

"At least you've got me and Mick," Jenny said, waving him to a chair.

"Well, I've got you, anyway. I don't think I'd be on Mick's Christmas card list, always supposing anybody was sending any this year. Not after October. He's not spoken to me since then."

"He'll get over it," she said without a shred of conviction. Mick had always had his reservations about the wider ripples of the schoolgirl friendship between Jenny and Tom's wife Moira. The women had been best pals forever, Moira standing chief bridesmaid at Jenny and Mick's wedding. When it came time to return the favour, Jenny had been pregnant with Misha. Mick had pointed out that her increasing size was the perfect excuse to turn Moira down, what with having to buy the bridesmaid's dress in advance. It wasn't a suggestion, more an injunction. For although Tom Campbell was by all accounts a decent man and a handsome man and an honest man, he was not a miner. True, he worked at the Lady Charlotte. He went underground in the stomach-juddering cage. He sometimes even got his hands dirty. But he was not a miner. He was a pit deputy. A member of a different union. A management man there to see that the health and safety rules were followed and that the lads did what they were supposed to. The miners had a term for the easiest part of any task—"the deputy's end." It sounded innocuous enough, but in an environment where every member of a gang knew his life depended on his colleagues, it expressed a world of contempt. And so Mick Prentice had always held something in reserve when it came to his dealings with Tom Campbell.

Mick had resented the invitations to dinner at their detached house in West Wemyss. He'd mistrusted Tom's invitations to join

him at the football. He'd even begrudged the hours Jenny spent at Moira's bedside during her undignified but swift death from cancer a couple of years before. And when Tom's union had dithered and swithered over joining the strike a couple of months before, Mick had raged like a toddler when they'd finally come down on the side of the bosses.

Jenny suspected part of the reason for his anger was the kindness Tom had shown them since the strike had started to bite. He'd taken to stopping by with little gifts—a bag of apples, a sack of potatoes, a soft toy for Misha. They'd always come with plausible excuses—a neighbour's tree with a glut, more potatoes in his allotment than he could possibly need, a raffle prize from the bowling club. Mick had always grumbled afterwards. "Patronizing shite," he'd said.

"He's trying to help us without shaming us," Jenny said. It didn't hurt that Tom's presence reminded her of happier times. Somehow, when he was there, she felt a sense of possibility again. She saw herself reflected in his eyes, and it was as a younger woman, a woman who had ambitions for her life to be different. So although she knew it would annoy Mick, Jenny was happy for Tom to sit at her kitchen table and talk.

He drew a limp but heavy parcel from his pocket. "Can you use a couple of pounds of bacon?" he said, his brow creasing in anxiety. "My sister-in-law, she brought it over from her family's farm in Ireland. But it's smoked, see, and I can't be doing with smoked bacon. It gives me the scunners. So I thought, rather than it go to waste . . . He held it out to her.

Jenny took the package without a second's hesitation. She gave a little snort of self-deprecation. "Look at me. My heart's all a flutter over a couple of pounds of bacon. That's what Margaret Thatcher and Arthur Scargill have done between them." She shook her head. "Thank you, Tom. You're a good man."

He looked away, unsure what to say or do. His eyes fixed on the clock. "Do you not need to pick up the bairn? I'm sorry, I wasn't

thinking about the time when I was waiting, I just wanted to . . ." He got to his feet, his face pink. "I'll come again."

She heard the stumble of his boots in the hall then the click of the latch. She tossed the bacon on to the counter and turned off the pan of water. It would be a different soup now.

Moira had always been the lucky one.

Thursday, 28th June 2007; Newton of Wemyss

Jenny's eyes came off the middle distance and focused on Karen. "I suppose it was about seven o'clock when it dawned on me that Mick hadn't come home. I was angry, because I'd actually got a half-decent tea to put on the table. So I got the bairn to her bed, then I got her next door to sit in so I could run down the Welfare and see if Mick was there." She shook her head, still surprised after all these years. "And of course, he wasn't."

"Had anybody seen him?"

"Apparently not."

"You must have been worried," Karen said.

Jenny shrugged one shoulder. "Not really. Like I said, we hadn't exactly parted on the best of terms. I just thought he'd taken the huff and gone over to Andy's."

"The guy in the photo?"

"Aye. Andy Kerr. He was a union official. But he was on the sick from his work. Stress, they said. And they were right. He'd killed himself within the month. I often thought Mick going scabbing was the last straw for Andy. He worshipped Mick. It would have broken his heart."

"So that's where you assumed he was?" Karen prompted her.

"That's right. He had a cottage out in the woods, in the middle of nowhere. He said he liked the peace and quiet. Mick took me out there one time. It gave me the heebie jeebies. It was like the witch's house in one of Misha's fairy stories—there was no sign

of it till suddenly you were there, right in front of it. You wouldn't catch me living there."

"Could you not have phoned to check?" the Mint butted in. Both women stared at him with a mixture of amusement and indulgence.

"Our phone had been cut off months before, son," Jenny said, exchanging a look with Karen. "And this was long before mobiles."

By now, Karen was gagging for a cup of tea, but she was damned if she was going to put herself in Jenny Prentice's debt. She cleared her throat and continued. "When did you start to worry?"

"When the bairn woke me up in the morning and he still wasn't home. He'd never done that before. It wasn't as if we'd had a proper row on the Friday. Just a few cross words. We'd had worse, believe me. When he wasn't there in the morning, I really started to think there was something badly wrong."

"What did you do?"

"I got Misha fed and dressed and took her down to her pal Lauren's house. Then walked out through the woods to Andy's place. But there was nobody there. And then I remembered Mick had said that now he was on the sick, Andy was maybe going to go off up to the Highlands for a few days. Get away from it all. Get his head straight. So of course he wasn't there. And by then, I was really starting to get scared. What if there had been an accident? What if he'd been taken ill?" The memory still had the power to disturb Jenny. Her fingers picked endlessly at the hem of her overall.

"I went up to the Welfare to see the union reps. I figured if anybody knew where Mick was, it would be them. Or at least they'd know where to start looking." She stared down at the floor, her hands clasped tight in her lap. "That's when the wheels really started to come off my life."

Saturday, 15th December 1984; Newton of Wemyss

Even in the morning, without the press of bodies to raise the temperature, the Miners' Welfare Institute was warmer than her house, Jenny noticed as she walked in. Not by much, but enough to be perceptible. It wasn't the sort of thing that usually caught her attention, but today she was trying to think of anything except the absence of her husband. She stood hesitant for a moment in the entrance hall, trying to decide where to go. The NUM strike offices were upstairs, she vaguely remembered, so she made for the ornately carved staircase. On the first-floor landing, it all became much easier. All she had to do was to follow the low mutter of voices and the high thin layer of cigarette smoke.

A few yards down the hall, a door was cracked ajar, the source of the sound and the smell. Jenny tapped nervously and the room went quiet. At last, a cautious voice said, "Come in."

She slid round the door like a church mouse. The room was dominated by a U-shaped table covered in tartan oilcoth. Half a dozen men were slouched around it in varying states of despondency. Jenny faltered when she realized the man at the top corner was someone she recognized but did not know. Mick McGahey, former Communist, leader of the Scottish miners. The only man, it was said, who could stand up to King Arthur and make his voice heard. The man who had been deliberately kept from the top spot by his predecessor. If Jenny had a pound for every time she'd heard someone say how different it would have been if McGahey had been in charge, her family would have been the best-fed and best-dressed in Newton of Wemyss. "I'm sorry," she stuttered. "I just wanted a word . . ." Her eyes flickered round the room, wondering which of the men she knew would be best to focus on.

"It's all right, Jenny," Ben Reekie said. "We were just having a wee meeting. We're pretty much done here, eh, lads?" There was a discontented murmur of agreement. But Reekie, the local secretary, was good at taking the temperature of a meeting and moving things along. "So, Jenny, how can we help you?"

She wished they were alone but didn't have the nerve to ask for it. The women had learned a lot in the process of supporting their men, but face to face their assertiveness still tended to melt away. But it would be all right, she told herself. She'd lived in this co-cooned world all her adult life, a world that centred on the pit and the Welfare, where there were no secrets and the union was your mother and your father. "I'm worried about Mick," she said. No point in beating about the bush. "He went out yesterday morning and never came back. I was wondering if maybe . . . ?"

Reekie rested his forehead on his fingers, rubbing it so hard he left alternating patches of white and red across the centre. "Jesus Christ," he hissed from between clenched teeth.

"And you expect us to believe you don't know where he is?" The accusation came from Ezra Macafferty, the village's last survivor of the lock-outs and strikes of the 1920s.

"Of course I don't know where he is." Jenny's voice was plaintive, but a dark fear had begun to spread its chill across her chest. "I thought maybe he'd been in here. I thought somebody might know."

"That makes six," McGahey said. She recognized the rough deep rumble of his voice from TV interviews and open-air rallies. It felt strange to be in the same room with it.

"I don't understand," she said. "Six what? What's going on?" Their eyes were all on her, boring into her. She could feel their contempt but didn't understand what it was for. "Has something happened to Mick? Has there been an accident?"

"Something's happened, all right," McGahey said. "It looks like your man's away scabbing to Nottingham."

His words seemed to suck the air from her lungs. She stopped breathing, letting a bubble form round her so the words would bounce off. It couldn't be right. Not Mick. Dumb, she shook her head hard. The words started to seep back in but they still made no sense. "Knew about the five . . . thought there might be more . . . always a traitor in the ranks . . . disappointed . . . always a union man."

"No," she said. "He wouldn't do that."

"How else do you explain him not being here?" Reekie said. "You're the one that came to us looking for him. We know a van load went down last night. And at least one of them is a pal of your Mick. Where the hell else is he going to be?"

Thursday, 28th June 2007; Newton of Wemyss

"I couldn't have felt worse if they'd accused me of being a whore," Jenny said. "I suppose, in their eyes, that's exactly what I was. My man away scabbing, it would be no time at all before I'd be living on immoral earnings."

"You never doubted that they were right?"

Jenny pushed her hair back from her face, momentarily stripping away some of the years and the docility. "Not really. Mick was pals with Iain Maclean, one of the ones that went to Nottingham. I couldn't argue with that. And don't forget what it was like back then. The men ran the game and the union ran the men. When the women wanted to take part in the strike, the first battle we had to fight was against the union. We had to beg them to let us join in. They wanted us where we'd always been—in the back room, keeping the home fires burning. Not standing by the braziers on the picket lines. But even though we got Women Against Pit Closures off the ground, we still knew our place. You'd have to be bloody strong or bloody stupid to try and blow against the wind round here."

It wasn't the first time Karen had heard a version of this truth. She wondered whether she'd have done any better in the same position. It felt good to think she'd stand by her man a bit more sturdily. But in the face of the community hostility Jenny Prentice must have faced, Karen reckoned she'd probably have caved in too. "Fair enough," she said. "But now that it looks like Mick might not have gone scabbing after all, have you got any idea what might have happened to him?"

Jenny shook her head. "Not a scooby. Even though I couldn't believe it, the scabbing kind of made sense. So I never thought about any other possibility."

"Do you think he'd just had enough? Just upped and left?"

She frowned. "See, that wouldn't be like Mick. To leave without the last word? I don't think so. He'd have made sure I knew it was all my fault." She gave a bitter laugh.

"You don't think he might have gone without a word as a way of making you suffer even more?"

Jenny's head reared back. "That's sick," she protested. "You make him sound like some kind of a sadist. He wasn't a cruel man, Inspector. Just thoughtless and selfish like the rest of them."

Karen paused for a moment. This was always the hardest part when interviewing the relatives of the missing. "Had he fallen out with anybody? Did he have any enemies, Jenny?"

Jenny looked as if Karen had suddenly switched into Urdu. "Enemies? You mean, like somebody that would kill him?"

"Maybe not mean to kill him. Maybe just fight him?"

This time, Jenny's laugh had genuine warmth. "By Christ, that's funny coming from you." She shook her head. "The only physical fights Mick ever got into in all the years we were married were with your lot. On the picket lines. At the demonstrations. Did he have enemies? Aye, the thin blue line. But this isn't South America, and I don't recall any talk about the disappeared of the miners' strike. So the answer to your question is no, he didn't have the kind of enemies that he'd get into a fight with."

Karen studied the carpet for a long moment. The gung-ho violence of the police against the strikers had poisoned community relations for a generation or more. Never mind that the worst offenders came from outside forces, bused in to make up the numbers and paid obscene amounts of overtime to oppress their fellow citizens in ways most people chose to avoid knowing about. The fallout from their ignorance and arrogance affected every officer in every coalfield force. Still did, Karen reckoned. She took a deep breath and looked up. "I'm sorry," she said. "The

way they treated the miners, it was inexcusable. I like to think we wouldn't act like that now, but I'm probably wrong. Are you sure there wasn't anybody he'd had a run-in with?"

Jenny didn't even pause for thought. "Not that I knew about. He wasn't a troublemaker. He had his principles, but he didn't use them as excuses to pick fights. He stood up for what he believed in, but he was a talker, not a fighter."

"What if the talking didn't work? Would he back down?"

"I'm not sure I follow you."

Karen spoke slowly, feeling her way into the idea. "I'm wondering if he bumped into this Iain Maclean that day and tried to talk him out of going to Nottingham. And if Iain wouldn't change his mind, and maybe had his pals there to back him up . . . Would Mick have got into a fight with them, maybe?"

Jenny shook her head firmly. "No way. He'd have said his piece and, if that didn't work, he'd have walked away."

Karen felt frustrated. Even after the passage of so much time, cold cases usually provided one or two loose ends to pick away at. But so far, there seemed to be nothing to reach for here. One last question, then she was out of this place. "Do you have any idea at all where Mick might have gone painting that day?"

"He never said. The only thing I can tell you is that in the winter he often went along the shore to East Wemyss. That way, if it came on rain, he could go down to the caves and shelter there. The preservation group, they had a wee bothy at the back of one of the caves with a camping stove where they could brew up. He had keys, he could make himself right at home," she added, the acid back in her voice. "But I've no idea whether he was there that day or not. He could have been anywhere between Dysart and Buckhaven." She looked at her watch. "That's all I know."

Karen got to her feet. "I appreciate your time, Mrs. Prentice. We will be continuing our inquiries and I'll keep you informed." The Mint scrambled to his feet and followed her and Jenny to the front door.

"I'm not bothered for myself, you understand," Jenny said

when they were halfway down the path. "But see if you can find him for the bairn's sake."

It was, Karen thought, the first sign of emotion she'd shown all morning. "Get your notebook out," she said to the Mint as they got into the car. "Follow-ups. Talk to the neighbour. See if she remembers anything about the day Mick Prentice disappeared. Talk to somebody from the cave group, see who's still there from 1984. Get another picture of what Mick Prentice was really like. Check in the files for anything about this Andy Kerr, NUM official, supposedly committed suicide around the time Mick disappeared. What's the story there? And we need to track down these five scabs and get Nottingham to have a chat with them." She opened the passenger door again as the Mint finished scribbling. "And since we're here already, let's have a crack at the neighbour."

She was barely two steps from the car when her phone rang. "Phil," she said.

No pleasantries, just straight to the point. "You need to get back here right now."

"Why?"

"The Macaroon is on the warpath. Wants to know why the hell you're not at your desk."

Simon Lees, Assistant Chief Constable (Crime), was temperamentally different from Karen. She was convinced his bedtime reading consisted of the Police, Public Order and Criminal Justice (Scotland) Act 2006. She knew he was married with two teenage children, but she had no idea how that could have happened to a man so obsessively organized. It was sod's law that on the first morning in months when she was doing something off the books the Macaroon should come looking for her. He seemed to believe that it was his divine right to know the whereabouts of any of the officers under his command, whether on or off duty. Karen wondered how close he'd come to stroking out on discovering she was not occupying the desk where he expected to find her. Not close enough, by the sounds of it. "What did you tell him?"

"I said you were having a meeting with the evidence store team

to discuss streamlining their cataloguing procedures," Phil said. "He liked the idea, but not the fact that it wasn't listed in your electronic appointments list."

"I'm on my way," Karen said, confusing the Mint by getting back into the car. "Did he say why he was looking for me?"

"To me? A mere sergeant? Gimme a break, Karen. He just said it was 'of the first importance.' Somebody probably stole his digestive biscuits."

Karen gestured impatiently at the Mint. "Home, James, and don't spare the horses." He looked at her as if she was mad, but he did start the car and drive off. "I'm coming in," she said. "Get the kettle on."

Glenrothes

The double helix of frustration and irritation twisted in Simon Lees's gut. He shifted in his chair and rearranged the family photos on his desk. What was wrong with these people? When he'd gone looking for DI Pirie and failed to find her where she should be, DS Parhatka had acted as if that were perfectly fine. There was something fundamentally lackadaisical about the detectives in Fife. He'd realized that within days of arriving from Glasgow. It amazed him that they'd ever managed to put anyone behind bars before he'd arrived with his analytical methods, streamlined investigations, sophisticated crime linkage, and the inevitable rise in the detection rate.

What riled him even more was the fact that they seemed to have no gratitude for the modern methods he'd brought to the job. He even had the suspicion that they were laughing at him. Take his nickname. Everybody in the building seemed to have a nickname, most of which could be construed as mildly affectionate. But not him. He'd discovered early on that he'd been dubbed the Macaroon because he shared the surname of a confectionary firm whose most famous product had become notorious because

of an ancient advertising jingle; its cheerful racism would provoke rioting in the streets if it were to be aired in twenty-first-century Scotland. He blamed Karen Pirie; it was no coincidence that the nickname had surfaced after his first run-in with her. It had been typical of most of their encounters. He wasn't quite sure how it happened, but she always seemed to wrong-foot him.

Lees still smarted at that early memory. He'd barely got his feet under the table but he'd started as he meant to go on, instigating a series of training days. Not the usual macho posturing or tedious revision of the rules of engagement, but fresh approaches to issues of modern policing. The first tranche of officers had assembled in the training suite, and Lees had started his preamble, explaining how they would spend the day developing strategies for policing a multicultural society. His audience had looked mutinous, and Karen Pirie had led the charge. "Sir, can I make a point?"

"Of course, Detective Inspector Pirie." His smile had been genial, hiding his annoyance at being interrupted before he'd even revealed the agenda.

"Well, sir, Fife's not really what you'd call multicultural. We don't have many people here who are not indigenous Brits. Apart from the Italians and the Poles, that is, and they've been here so long we've forgotten they're not from here."

"So racism's all right by you, is it, Inspector?" Maybe not the best reply, but he'd been driven to it by the apparently Neanderthal attitude she'd expressed. Not to mention that bland pudding face she presented whenever she said anything that might be construed as inflammatory.

"Not at all, sir." She'd smiled, almost pityingly. "What I would say is that, given we have a limited training budget, it might make more sense to deal first with the sort of situations we're more likely to encounter day to day."

"Such as? How hard to hit people when we arrest them?"

"I was thinking more of strategies to deal with domestic violence. It's a common call-out and it can easily escalate. Too many people are still dying every year because a domestic has got out

of hand. And we don't always know how to deal with it without inflaming the situation. I'd say that was my number one priority right now, sir."

And with that short speech, she'd cut the ground from under his feet. There was no way back for him. He could carry on with the planned training, knowing that everyone in the room was laughing at him. Or he could postpone till he could put together a programme to deal with DI Pirie's suggestion and lose face completely. In the end, he'd told them to spend the rest of the day researching the subject of domestic violence in preparation for another training day.

Two days later, he'd overheard himself referred to as the Macaroon. Oh yes, he knew who to blame. But as with everything she did to undermine him, there was nothing he could pin directly on her. She'd stand there, looking as shaggy, stolid, and inscrutable as a Highland cow, never saying or doing anything that he could complain about. And she set the style for the rest of them, even though she was stranded on the fringes in the Cold Case Review Team where she should be able to wield no influence whatsoever. But somehow, thanks to Pirie, dealing with the detectives of all three divisions was like herding cats.

He tried to avoid her, tried to sideline her via his operational directives. Until today, he'd thought it was working. Then the phone had rung. "Assistant Chief Constable Lees," he'd announced as he picked up the phone. "How may I be of assistance?"

"Good morning, ACC Lees. My name is Susan Charleson. I'm personal assistant to Sir Broderick Maclennan Grant. My boss would like to talk to you. Is this a good time?"

Lees straightened up in his chair, squaring his shoulders. Sir Broderick Maclennan Grant was notorious for three things—his wealth, his misanthropic reclusiveness, and the kidnap and murder of his daughter Catriona twenty-odd years before. Unlikely though it seemed, his PA calling the ACC Crime could mean only that there had been some sort of development in the case. "Yes, of course, perfect time, couldn't be better." He dredged his memory for de-

tails, only half listening to the woman on the phone. Daughter and grandson kidnapped, that was it. Daughter killed in a botched ransom handover, grandson never seen again. And now it looked as if he was going to be the one to have the chance finally to solve the case. He tuned in to the woman's voice again.

"If you'll bear with me, I'll put you through now," she said.

The hollow sound of dead air, then a dark, heavy voice said, "This is Brodie Maclennan Grant. And you're the assistant chief constable?"

"That's right, Sir Broderick. ACC Lees. Simon Lees."

"Are you aware of the unsolved murder of my daughter Catriona? And the kidnapping of my grandson Adam?"

"Of course, naturally, there's not an officer in the land who—"

"We believe some new evidence has come to light. I'd be obliged if you'd arrange for Detective Inspector Pirie to come to the house tomorrow morning to discuss it with me."

Lees actually held the phone away from his face and stared at it. Was this some kind of elaborate practical joke? "DI Pirie? I don't quite . . . I could come," he gabbled.

"You're a desk man. I don't need a desk man." Brodie Grant's voice was dismissive. "DI Pirie is a detective. I liked the way she handled that Lawson business."

"But . . . but it should be a more senior officer who deals with this," Lees protested.

"Isn't DI Pirie in charge of your Cold Case Review Team?" Grant was beginning to sound impatient. "That's senior enough for me. I don't care about rank, I care about effectiveness. That's why I want DI Pirie at my house at ten tomorrow morning. That should give her enough time to acquaint herself with the basic facts of the case. Good day, Mr. Lees." The line went dead, and Simon Lees was left alone with his rising blood pressure and his bad mood.

Much as it grieved him, he had no choice but to find DI Pirie and brief her. At least he could make it sound as though sending her was his idea. But in spite of there being no appointment

in the electronic diary system he had instituted for his senior detectives, she hadn't been at her desk. It was all very well, officers doing things on their own initiative, but they had to learn to leave a record of their movements.

He was on the point of marching back down to the CCRT squad room to find out why DI Pirie hadn't appeared yet when a sharp rap on the door was followed without any interval by the entrance of DI Pirie. "Did I invite you to come in?" Lees said, glowering across the room at her.

"I thought it was urgent, sir." She kept walking and sat down in the visitor's chair across the desk from him. "DS Parhatka gave me the impression that whatever it is you wanted me for, it couldn't wait."

What an advert for the service, he thought crossly. Shaggy brown hair flopping into her eyes, the merest smudge of make-up, teeth that really could have done with some serious orthodontics. He supposed she was probably a lesbian, given her penchant for trouser suits that really were a mistake given the breadth of her hips. Not that he had anything against lesbians, his internal governor reminded him. He just thought it gave people the wrong impression about today's police service. "Sir Broderick Maclennan Grant called me earlier this morning," he said. The only sign of interest was a slight parting of her lips. "You know who Sir Broderick Maclennan Grant is, I take it?"

Karen looked puzzled by the question. She leaned back in her seat and recited, "Third richest man in Scotland, owns half of the profitable parts of the Highlands. Made his money building roads and houses and running the transport systems that serve them. Owns a Hebridean island but lives mostly in Rotheswell Castle near Falkland. Most of the land between there and the sea belongs either to him or to the Wemyss estate. His daughter Cat and her baby son Adam were kidnapped by an anarchist group in 1985. Cat was shot dead when the ransom handover went wrong. Nobody knows what happened to Adam. Grant's wife committed suicide a couple of years later. He remarried about ten years ago.

He has a wee boy who must be about five or six." She grinned. "How did I do?"

"It's not a contest, Inspector." Lees felt his hands closing into fists and lowered them below the desk. "It appears that there may be some fresh evidence. And since you are in charge of cold cases, I thought you should deal with it."

"What sort of evidence?" She leaned on the arm of her chair. It was almost a slouch.

"I thought it best that you confer directly with Sir Broderick. That way there can be no possibility of confusion."

"So he didn't actually tell you?"

Lees could have sworn she was enjoying this. "I've arranged for you to meet him at Rotheswell Castle tomorrow morning at ten. I need hardly remind you how important it is that we are seen to be taking this seriously. I want Sir Broderick to understand this matter will have our full attention."

Karen stood up abruptly, her eyes suddenly cold. "He'll get exactly the same attention as every other bereaved parent I deal with. I don't make distinctions among the dead, sir. Now, if that's all, I've got a case file to assimilate before morning." She didn't wait for a dismissal. She just turned on her heel and walked out, leaving Lees feeling that she didn't make many distinctions among the living either.

Yet again, Karen Pirie had left him feeling like an idiot.

Rotheswell Castle

Bel Richmond took a last quick look through her file on Catriona Maclennan Grant, double-checking that her list of questions covered all the angles. Broderick Maclennan Grant's inability to suffer fools was as notorious as his dislike of publicity. Bel suspected that he would pounce on the first sign of unpreparedness on her part and use that as an excuse to break the deal she had brokered with Susan Charleson.

Truth to tell, she was still amazed that she had pulled it off. She stood up, closing her laptop and pausing to check her look in the mirror. *Tits and teeth. You don't get a second chance to make a first impression.* Country house weekend, that was the look she'd gone for. She'd always been good at camouflage. Another of the many reasons she was so good at what she did. Blending in, becoming "one of us," whoever the "us" happened to be, was a necessary evil. So if she was sleeping under Brodie Grant's baronial roof, she needed to look the part. She straightened the Black Watch tartan dress she'd borrowed from Vivianne, checked her kitten heels for scuffs, pushed her crow-black hair behind one ear, and parted her scarlet lips in a smile. A glance at her watch confirmed it was time to head downstairs and discover what the formidable Susan Charleson had lined up.

As she turned the corner of the wide staircase, she had to jink to one side to avoid a small boy careering up. He brought his flailing limbs under control on the half-landing, gasped, "Sorry," then hurtled on upwards. Bel blinked and raised her eyebrows. It had been a couple of years since she'd last had a similar small boy encounter, and she hadn't missed it a bit. She carried on down, but before she reached the bottom, a woman wearing cords the colour of butter and a dark red shirt swung round the newel post then stopped dead, taken by surprise. "Oh, sorry, I didn't mean to startle you," she said. "You haven't seen a small boy go past, have you?"

Bel gestured over her shoulder with her thumb. "He went that-away."

The woman nodded. Now she was nearer, Bel could see she was a good ten years older than she'd first thought; late thirties, at least. Good skin, thick chestnut hair, and a trim build gave the illusion a helping hand. "Monster," the woman said. They met a couple of steps from the bottom. "You must be Annabel Richmond," she said, extending a slender hand that was chilly in spite of the comfortable warmth trapped inside the thick walls of the castle. "I'm Judith. Brodie's wife."

Of course she was. How could Bel have imagined a nanny so perfectly groomed? "Lady Grant," she said, wincing inside.

"Judith, please. Even after all these years married to Brodie, I still want to look over my shoulder when someone calls me Lady Grant." She sounded as though she wasn't just saying it out of fake humility.

"And I'm Bel, apart from my by-line."

Lady Grant smiled, her eyes already scanning the stairs above. "Bel it is. Look, I can't stop now, I have to capture the monster. I'll see you at dinner." And she was off, taking the stairs two at a time.

Feeling overdressed in comparison with the chatelaine of Rotheswell, Bel made her way back down the stone-flagged hallways to Susan Charleson's office. The door was open and Susan, who was talking on the phone, beckoned her in. "Fine. Thank you for organizing that, Mr. Lees." She replaced the phone and came round the desk, ushering Bel back towards the door. "Perfect timing," she said. "He likes punctuality. Is your room to your liking? Do you have everything you need? Is the wireless access working?"

"It's all perfect," Bel said. "Lovely view too." Feeling as if she'd wandered into a BBC2 drama scripted by Stephen Poliakoff, she allowed herself to be led back through the maze of corridors whose walls were lined with poster-sized photographs of the Scottish landscape printed on canvas to resemble paintings. She was surprised by how cosy it felt. But then, this wasn't quite her idea of a castle. She'd expected something like Windsor or Alnwick. Instead, Rotheswell was more like a fortified manor with turrets. The interior resembled a country house rather than a medieval banqueting hall. Substantial but not as intimidating as she'd feared.

By the time they stopped in front of a pair of tall arched mahogany doors, she was beginning to regret not having thought of breadcrumbs.

"Here we are," Susan said, opening one of the doors and lead-

ing Bel into a billiard room panelled in dark wood with shutters over the windows. The only light came from an array of lamps above the full-size table. As they walked in, Sir Broderick Maclennan Grant looked up from sighting down his cue. A thick shock of startling silver hair falling boyishly over a broad forehead, eyebrows a pair of silver bulwarks over eyes so deep set their colour was guesswork, a parrot's bill of a nose, and a long thin mouth over a square chin made him instantly recognizable; the lighting made him a dramatic figure.

Bel knew what to expect from photographs, but she was startled by the crackle of electricity she felt in his presence. She'd been in the company of powerful men and women before, but she'd felt this instant charisma only a handful of times. She understood at once how Brodie Grant had built his empire from the ground up.

He straightened up and leaned on his cue. "Miss Richmond, I take it?" His voice was deep and almost grudging, as if he hadn't used it enough.

"That's right, Sir Broderick." Bel wasn't sure whether to advance or stay put.

"Thank you, Susan," Grant said. As the door closed behind her, he waved towards a pair of well-worn leather armchairs flanking a carved marble fireplace. "Sit yourself down. I can play and talk at the same time." He returned to study his shot while Bel shifted one of the chairs so she could watch him more directly.

She waited while he played a couple of shots, the silence rising between them like a drowning tide. "This is a beautiful house," she said finally.

He grunted. "I don't do small talk, Miss Richmond." He cued swiftly and two balls collided with a crack like a gunshot. He chalked his cue and studied her for a long moment. "You're probably wondering how on earth you managed this. Direct access to a man notorious for his loathing of the media spotlight. Quite an achievement, eh? Well, I'm sorry to disappoint you, but you just got lucky." He walked round the table, frowning at the position of the balls, moving like a man twenty years younger.

"That's how I've got some of my best stories," Bel said calmly. "It's a big part of what successful journalism is about, the knack of being in the right place at the right time. I don't have a problem with luck."

"Just as well." He studied the balls, cocking his head for a different angle. "So, are you not wondering why I've chosen to break my silence after all these years?"

"Yes, of course I am. But to be honest, I don't think your reasons for talking now will have much to do with what I end up writing. So it's more personal curiosity than professional."

He stopped halfway through his preparation for a shot and straightened up, staring at her with an expression she couldn't read. He was either furious or curious. "You're not what I expected," he said. "You're tougher. That's good."

Bel was accustomed to being underestimated by the men in her world. She was less used to them admitting their mistake. "Damn right, I'm tough. I don't rely on anybody else to fight my battles."

He turned to face her, leaning on the table and folding his arms over his cue. "I don't like being in the public eye," he said. "But I'm a realist. Back in 1985, it was possible for someone like me to exert a degree of influence over the media. When Catriona and Adam were kidnapped, to a large extent we controlled what was printed and broadcast. The police cooperated with us too." He sighed and shook his head. "For all the good it did us." He leaned the cue on the table and came to sit opposite Bel.

He sat in the classic alpha male pose: knees spread wide, hands on his thighs, shoulders back. "The world is a different place now," he said. "I've seen what you people do to parents who have lost children. Mohamed Al Fayed, made to look like a paranoid buffoon. Kate McCann, turned into a modern-day Medea. Put one foot wrong and they bury you. Well, I'm not about to let that happen. I'm a very successful man, Miss Richmond. And I got that way by accepting that there are things I don't know, and understanding that the way to overcome that is to employ experts and listen to them. As far as this business goes, you are my hired

gun. Once the word gets out that there is new evidence, the media will go wild. But I will not be talking to anyone but you. Everything goes through you. So whatever image reaches the public will be the one you generate. This place was built to withstand a siege, and my security is state of the art. None of the reptiles gets near me or Judith or Alec."

Bel felt a smile tugging at the corners of her mouth. Exclusive access was every hack's wet dream. Usually she had to work her arse off to get it. But here it was, on a plate and for free. Still, let him keep on thinking that she was the one doing him a favour. "And what's in it for me? Apart from becoming the journalist that all the others love to hate?"

The thin line of Grant's lips compressed further, and his chest rose as he breathed deeply. "I will talk to you." The words came out as if they'd been ground between a pair of millstones. It was clearly meant to be a moment reminiscent of Moses descending from Mount Sinai.

Bel was determined not to be impressed. "Excellent. Shall we make a start then?" She reached into her bag and produced a digital recorder. "I know this is not going to be easy for you, but I need you to tell me about Catriona. We'll get to the kidnapping and its consequences, but we're going to have to go back before that. I want to have a sense of what she was like and what her life was like."

He stared into the middle distance, and for the first time Bel saw a man who looked his seventy-two years. "I'm not sure I'm the best person for that," he said. "We were too alike. It was always head to head with me and Catriona." He pushed himself out of the armchair and went back to the billiard table. "She was always volatile, even when she was wee. She had toddler tantrums that could shake the walls of this place. She grew out of the tantrums but not out of the tempers. Still, she could always charm her way right back into your good graces. When she put her mind to it." He glanced up at Bel and smiled. "She knew her own mind. And you couldn't shift her once she was set on something."

Grant moved round the table, studying the balls, lining up his next shot. "And she had talent. When she was a child, you never saw her without a pencil or a paintbrush in her hand. Drawing, painting, modelling with clay. She never stopped. She didn't grow out of it like most kids do. She just got better at it. And then she discovered glass." He bent over the table and stunned the cue ball into the red, slotting it into the middle pocket. He respotted the red and studied the angles.

"You said you were always head to head with each other. What were the flashpoints?" Bel said when he showed no sign of continuing his reminiscences.

Grant gave a little snort of laughter. "Anything and everything. Politics. Religion. Whether Italian food was better than Indian. Whether Mozart was better than Beethoven. Whether abstract art had any meaning. Whether we should plant beech or birch or Scots pine in the Check Bar wood." He straightened up slowly. "Why she didn't want to take over the company. That was a big one. I didn't have a son then. And I've never had a problem with women in business. I saw no reason why she shouldn't take over MGE once she'd learned how it all works. She said she'd rather stick needles in her eyes."

"She didn't approve of MGE?" Bel asked.

"No, it wasn't anything to do with the company or its policies. What she wanted was to be an artist in glass. Sculpting, blowing, casting—anything you could do with glass, she wanted to be the best. And that didn't leave any room for building roads or houses."

"That must have been a disappointment."

"Broke my heart." Grant cleared his throat. "I did everything I could to talk her out of it. But she wouldn't be talked out of it. She went behind my back, applied for a place at Goldsmiths in London. And she got it." He shook his head. "I was all for cutting her adrift without a penny, but Mary—my wife, Cat's mother—she shamed me into agreeing to support her. She pointed out that, for somebody who hated being in the public eye, I'd be throwing a

hell of a bone to the tabloids. So I let myself be talked into it." He gave a wry smile. "Almost reconciled myself to it too. And then I found out what was really going on."

Wednesday, 13th December 1978; Rotheswell Castle

Brodie Grant swung the Land Rover into a gravel-scattering turn and ground to a halt yards from the kitchen door of Rotheswell Castle. He stamped into the house, a chocolate Lab at his heels. He strode through the kitchen, leaving a swirl of freezing air in his wake, barking at the dog to stay. He moved through the house with the speed and certainty of a man who knows precisely where he is going.

At last he burst into the prettily decorated room where his wife indulged her passion for quilting. "Did you know about this?" he said. Mary looked up, startled. She could hear the rush of his breathing from across the room.

"About what, Brodie?" she said. She'd been married to a force of nature long enough not to be ruffled by a grand entrance.

"You talked me into this." He threw himself into a low armchair, struggling to untangle his legs. " 'It's what she wants, Brodie. She'll never forgive you if you stand in her way, Brodie. You followed your dreams, Brodie. Let her follow hers.' That's what you said. So I did. Against my better judgement, I said I would back her up. Finance her bloody degree. Keep my mouth shut about what a bloody waste of time it is. Stop reminding her how few artists ever make any kind of a living from their self-indulgent bloody carry-on. Not till they're dead, anyway." He banged his fist on the arm of the chair.

Mary continued piecing her fabric and smiled. "You did, Brodie. And I'm very proud of you for it."

"And now look where it's got us. Look what's really going on."

"Brodie, I've no idea what you're talking about. Do you think you could explain? And with due consideration for your blood

pressure?" She'd always had the gift of gently teasing him out of his extreme positions. But today, it wasn't working well. Brodie's dander was up, and it was going to take more than an application of sweet reason to restore him to his normal humour.

"I've been out with Sinclair. Checking the drives for the shoot on Friday."

"And how were the drives?"

"Perfectly fine. They're always fine. He's a good keeper. But that's not the point, Mary." His voice rose again, incongruous in the cosy room with its stacked riot of fabrics on the shelves.

"No, Brodie. I realize that. What is the point, exactly?"

"Fergus bloody Sinclair, that's what. I told Sinclair. Back in the summer, when his bloody son was sniffing round Cat. I told him to keep the boy away from my daughter, and I thought he'd listened to me. But now this." He waved his hands as if he was throwing a pile of hay in the air.

Mary finally put down her work. "What's the matter, Brodie? What's happened?"

"It's what's *going* to happen. You know how we breathed a sigh of relief when he signed up for his bloody estate management degree at Edinburgh? Well, it turns out that wasn't the only iron in his bloody fire. He's only gone and accepted a place at London University. He's going to be in the same bloody city as our daughter. He'll be all over her like a rash. Bloody gold-digging peasant." He scowled and smacked his fist down on the chair again. "I'm going to settle his hash, you see if I don't."

To his astonishment, Mary was laughing, rocking back and forward at her piecing table, tears glistening at the corners of her eyes. "Oh, Brodie," she gasped. "I can't tell you how funny this is."

"Funny?" he howled. "That bloody boy's going to ruin Cat and you think it's funny?"

Mary jumped to her feet and crossed the room to her husband. Ignoring his protests, she sat on his lap and ran her fingers through his thick hair. "It's all right, Brodie. Everything's going to be fine."

"I don't see how." He jerked away from her hand.

"Me and Cat, we've been trying to figure out how to tell you for the past week."

"Tell me what, woman?"

"She's not going to London, Brodie."

He straightened up, almost toppling Mary on to the floor. "What do you mean, not going to London? Is she giving up this daftness? Is she coming to work with me?"

Mary sighed. "Don't be silly. You know in your heart she's doing what she should be doing. No, she's been offered a scholarship. It's a combination of academic study and working in a designer glass factory. Brodie, it's absolutely the best training in the world. And they want our Catriona."

For a long moment, he allowed himself to be torn between pride and fear. "Where about?" he said at last.

"It's not so far, Brodie." Mary ran the back of her hand down his cheek. "It's only Sweden."

"Sweden? Bloody Sweden? Jesus Christ, Mary. Sweden?"

"You make it sound like the ends of the earth. You can fly there from Edinburgh, you know. It takes less than two hours. Honestly, Brodie. Listen to yourself. This is wonderful. It's the best possible start for her. And you won't have to worry about Fergus being in the same place. He's not likely to turn up in a small town between Stockholm and Uppsala, is he?"

Grant put his arms round his wife and rested his chin on her head. "Trust you to find the silver lining." His mouth curled in a cruel smile. "It's certainly going to put Fergus bloody Sinclair's gas at a peep."

Thursday, 28th June 2007; Rotheswell Castle

"So you argued with Cat about boyfriends as well?" Bel said. "Was it all of them, or just Fergus Sinclair in particular?"

"She didn't have that many boyfriends. She was too focused on

her work. She went out for a few months with one of the sculp-
tors at the glass factory. I met him a couple of times. Swedish,
but a sensible enough lad all the same. I could see she wasn't serious, though, so there was no need to argue about him. But Fergus
Sinclair was a different kettle of fish." He paced the perimeter
of the table, the anger obvious. "The police never took him seriously as a suspect, but I wondered at the time whether he might
have been behind what happened to Cat and Adam. He certainly
couldn't accept it when she finally cut the ties between them. And
he couldn't accept that she wouldn't acknowledge him as Adam's
father. At the time, I thought it was possible he took the law into
his own hands. Though it's hard to see him having the wit to put
something that complicated together."

"But Cat continued her relationship with Fergus after she went
to Sweden?"

Tiredness seemed suddenly to hit Grant and he dropped back
into the chair opposite Bel. "They were very close. They'd run
about together when they were kids. I should have put a stop to it
but it never crossed my mind that it would ever come to anything.
They were so different. Cat with her art and Sinclair with no
more ambition than to follow his father into keepering. Different
class, different aspirations. The only thing that I could see pulling
them together was that life had landed them in the same place. So
yes, when she came back in the holidays and he was around, they
got back together again. She made no secret of it, even though she
knew how I felt about Sinclair. I kept hoping she'd meet someone she deserved but it never happened. She kept going back to
Sinclair."

"And yet you didn't sack his father? Move him off the estate?"

Grant looked shocked. "Good God, no. Have you any idea
how hard it is to find a keeper as good as Willie Sinclair? You
could interview a hundred men before you'd find one with his in-
stincts for the birds and the land. A decent man, too. He knew his
son wasn't in Cat's league. He was ashamed that he couldn't stop
Fergus chasing Cat. He wanted to bar him from the family home,

but his wife wouldn't have it." He shrugged. "I can't say I blame her. Women are always soft with their sons."

Bel tried to hide her surprise. She'd assumed Grant would stop at nothing to have his own way where his daughter was concerned. He was apparently more complex than she'd given him credit for. "What happened after she came back from Sweden?"

Grant rubbed his face with his hands. "It wasn't pretty. She wanted to move out. Set up a studio where she could work and sell things from, somewhere with living quarters attached. She had her eye on a couple of properties on the estate. I said the price of my support was that she stop seeing Sinclair." For the first time, Bel saw sadness seeping round the edges of the simmering anger. "It was stupid of me. Mary said so at the time, and she was right. They were both furious with me, but I wouldn't give in. So Cat went her own way. She spoke to the Wemyss estate and rented a property from them. An old gatehouse with what had been a logging shed, set back from the main road. Perfect for attracting customers. A parking area in front of the old gates, studio and display space, and living quarters tucked away behind the walls. All the privacy you could want. And everybody knew. Catriona Maclennan Grant had gone to the Wemyss estate to spite her old man."

"If she needed your support, how did she pay for it all?" Bel asked.

"Her mother equipped the studio, paid the first year's rent, and stocked the kitchen till Cat started selling pieces." He couldn't suppress a smile. "Which didn't take long. She was good, you know. Very good. And her mother saw to it that all her friends went there for wedding presents and birthday gifts. I was never angrier with Mary than I was then. I was outraged. I felt thwarted and disrespected and it really did not help when bloody Sinclair came back from university and picked up where he'd left off."

"Were they living together?"

"No. Cat had more sense than that. I look back at it now and I sometimes think she only went on seeing him to spite me. It didn't

last that long after she'd set the studio up. It was pretty much over about eighteen months before . . . before she died."

Bel did her mental arithmetic and came up with the wrong answer. "But Adam was only six months old when they were kidnapped. So how could Fergus Sinclair be his father if he split up with Cat eighteen months earlier?"

Grant sighed. "According to Mary, it wasn't a clean break. Cat kept telling Sinclair it was over but he wouldn't take no for an answer. These days, you'd call it harassment. Apparently he kept turning up with a pathetic puppy face and Cat didn't always have the strength to send him away. And then she got pregnant." He stared at the floor. "I'd always imagined what it would be like to be a grandfather. To see the family line continue. But when Cat told us, all I felt was anger. That bastard Sinclair had wrecked her future. Saddled her with his child, ruined her chances of the career she'd dreamed of. The one good thing she did was refuse to have anything more to do with him. She wouldn't acknowledge him as the father, she wouldn't see him or talk to him. She made it plain that, this time, it really was over and done."

"How did he take that?"

"Again, I got it second-hand. This time from Willie Sinclair. He said the boy was devastated. But all I cared about was that he'd finally got the message that he was never going to be part of this family. Willie advised the boy to put some distance between himself and Cat, and for once, he listened. Within a few weeks, he'd got a job in Austria, working on some hunting estate near Salzburg. And he's worked in Europe ever since."

"And now? You still think he might have been responsible for what happened?"

Grant made a face. "If I'm honest, no. Not really. I don't think he had the brains to come up with such a complicated plot. I'm sure he'd have loved to get his hands on his son and take his revenge on Cat at the same time, but it's much more likely that it was some politically motivated bastards who thought it would be

clever to get me to fund their revolution." Wearily, he got to his feet. "I'm tired now. The police are coming tomorrow morning and we'll be going through all the other stuff then. We'll see you at dinner, Miss Richmond." He walked out of the room, leaving Bel with plenty to ponder. And to transcribe. When Brodie Grant had said he would talk to her, she hadn't imagined for a moment he would provide her with this rich seam of information. She was going to have to consider very carefully how to present him to the world's media. One foot wrong and she knew the mine would be closed down. And now she'd had a taste of what lay within, that was definitely the last thing she wanted.

Glenrothes

The Mint was staring at the computer screen as if it was an artefact from outer space when Karen got back to her office. "What have you got for me?" she asked. "Have you tracked down the five scabs yet?"

"None of them's got a criminal record," he said.

"And?"

"I wasn't sure where else to look."

Karen rolled her eyes. Her conviction that the Mint had been dumped on her as a form of sabotage by the Macaroon intensified daily. "Google. Electoral rolls. 192.com. Vehicle licensing. Make a start there, Jason. And then fix me up a site meeting with the cave preservation person. Better leave tomorrow clear, see if you can get him to meet me on Saturday morning."

"We don't work Saturdays usually," the Mint said.

"Speak for yourself," Karen muttered, making a note to herself to ask Phil to come with her. Scots law's insistence on corroboration for all evidence made it hard to be a complete maverick.

She woke her computer from hibernation and tracked down the contact details of her opposite number in Nottingham. To her relief, DCI Des Mottram was at his desk, receptive to her re-

quest. "I think it's probably a dead end, but it's one that needs to be checked out," she said.

"And you don't fancy a trip down to the Costa del Trent," he said, amused resignation in his voice.

"It's not that. I've just had a major case reopen today and there's no way I can spare a couple of bodies on something that probably won't take us any further forward except in a negative way."

"Don't worry about it. I know how it goes. It's your lucky day, though, Karen. We got two new CID aides on Monday and this is exactly the kind of thing I can use to break them in. Nothing too complicated, nothing too dodgy."

Karen gave him the names of the men. "I've got one of my lads looking for last known addresses. Soon as he's got anything, I'll get him to e-mail you." A few more details, and she was done. Right on cue, Phil Parhatka walked back into the room, a bacon roll transmitting a message straight to the pleasure centres of Karen's brain. "Mmm," she groaned. "Christ, that smells glorious."

"If I'd known you were back, I'd have got you one. Here, we'll go halves." He took a knife out of his drawer and cut the roll in half, tomato sauce squirting over his fingers. He handed over her share, then licked his fingers. What more, Karen wondered, could a woman ask for in a man?

"What did the Macaroon want?" Phil said.

Karen bit into the roll and spoke through a mouthful of soft sweet dough and salty bacon. "New development in the Catriona Maclennan Grant case."

"Really? What's happened?"

Karen grinned. "I don't know. King Brodie didn't bother to tell the Macaroon. He just told him to send me round tomorrow morning. So I need to get myself up to speed smartish. I've already sent for the records, but I'm going to check it out online first. Listen . . ." She drew him to one side. "The Mick Prentice business. I need to talk to somebody on Saturday and obviously the Mint doesn't do Saturdays. Any chance I can talk you into coming along with me?"

"Coming along where?"

"The Wemyss caves."

"Really?" Phil perked up. "We get to go behind the railings?"

"I expect so," Karen said. "I didn't know you were into the caves."

"Karen, I used to be a wee boy."

She rolled her eyes. "Right enough."

"Besides, the caves have got really cool stuff. Pictish inscriptions and drawings. Iron Age carvings. I like the idea of being a secret squirrel and taking a look at the things you don't usually get to see. Sure, I'll come with you. Have you logged the case yet?"

Karen looked embarrassed. "I want to see where it goes. It was a hard time round here. If something bad happened to Mick Prentice, I want to get to the bottom of it. And you know how the media are always poking around in what we're doing in CCRT. I've a feeling this is one where we've got a better chance of finding out what happened if we can keep the lid on it a bit."

Phil finished his roll and wiped his mouth with the back of his hand. "Fair enough. You're the boss. Just make sure the Macaroon can't use it as a stick to beat you with."

"I'll watch my back. Listen, are you busy right now?"

He tossed the empty paper bag in the bin with an overhead action, preening himself when it landed right in the middle. "Nothing I can't put to one side."

"See what you can dig up on a guy called Andy Kerr. He was an NUM official during the strike. Lived in a cottage in the middle of Wemyss woods. He was on the sick with depression around the time Mick went missing. Supposedly topped himself, but the body was never recovered."

Phil nodded. "I'll see what I can find."

As he returned to his own desk, Karen was Googling Catriona Maclennan Grant. The first hit took her to a two-year-old broadsheet newspaper feature published to mark the twentieth anniversary of the young sculptor's death. Three paragraphs in, Karen felt a physical jolt in the middle of her chest. "It's amazing how

few people are available to talk about this case," she read. "Cat Grant's father has never spoken to the press about what happened. Her mother killed herself two years after the death of her daughter. Her ex-boyfriend, Fergus Sinclair, refuses to be interviewed. And the officer in charge of the case is also beyond our reach—he is himself serving life for murder."

"Oh Christ," she groaned. She hadn't even seen the case file but already this was turning into the assignment from hell.

Kirkcaldy

It was after ten when Karen walked through her front door with a bundle of files and a fish supper. The notion that she was playing at keeping house had never deserted her. Maybe it was something to do with the house itself, an identikit box on a 1960s warren development to the north of Kirkcaldy. The sort of place people started out in, clinging to the hope it wasn't going to be where they ended up. Low-crime suburbia, a place where you could let kids play out in the street so long as you didn't live on one of the through roads. Traffic accidents, not abductions, were what parents feared here. Karen could never quite remember why she'd bought it, though it had seemed like a good idea at the time. She suspected the appeal had been that it came completely refurbished, probably by somebody who'd got the idea from a TV property development programme. She'd bought the furniture with the house, right down to the pictures on the walls. She didn't care that she hadn't chosen the stuff she lived among. It was the kind of thing she'd probably have picked anyway and it had saved her the hassle of a Sunday in IKEA. And nobody could deny that it was a million times nicer than the faded floral clutter her parents inhabited. Her mother kept waiting for her to revert to type, but it wasn't going to happen. When she had a weekend off, Karen wanted nothing more than a curry with her pals and a significant amount of time on the sofa watching football and old films. Not homemaking.

She dumped everything on the dining table and went in search of a plate and cutlery. She still had some standards, for God's sake. She tossed her coat over a chair and sat down to her meal, flipping a folder open and reading as she ate. She'd worked her way through the Grant case files earlier and made a note of the questions she wanted answers to. Now finally she had the chance to look over the material Phil had gathered for her.

As she'd expected, the original missing persons report could hardly have been more sketchy. Back then, the disappearance of an unmarried, childless adult male with a history of clinical depression barely dented the police consciousness. It was nothing to do with the fact that the miners' strike had stretched the force's staffing levels almost to breaking point and everything to do with the fact that, back then, missing persons were not a priority. Not unless they were small children or attractive young women. Even these days, only the fact of Andy Kerr's medical problems would have guaranteed him mild interest.

He'd been reported missing by his sister Angie on Christmas Eve. He'd failed to show up at their parents' home for the traditional family celebration. Angie, home from teacher training college for the holidays, had left a couple of messages on his answering machine in the previous week, trying to arrange meeting up for a drink. Andy hadn't responded, but that wasn't unusual. He'd always been dedicated to his job, but since the strike had begun, he'd become workaholic.

Then on the afternoon of Christmas Eve, Mrs. Kerr had admitted that Andy was on sick leave for depression. Angie had persuaded her father to drive her over to Andy's cottage in the Wemyss woods. The place had been cold and deserted, the fridge empty of fresh food. A note was propped up against the sugar bowl on the kitchen table. Amazingly, it had been bagged and included in the file. *If you're reading this, it's probably because you're worried about me. Don't be. I've had enough. It's just one thing after another and I can't take it any more. I've gone away to try and get my head straight. Andy.*

It wasn't exactly a suicide note, but if you found a body near a message like that, you wouldn't be expecting a murder victim. And the sister had said Andy liked to go mountain walking. She could see why the uniform who'd checked out the cottage and the surrounding woodland had recommended no further action aside from circulating the information to other forces in Scotland. A comment on the file in a different hand noted that Angie Kerr had applied to have her brother declared dead in 1992 and the application had been granted.

The last page was in Phil's familiar writing. "The Kerr parents died in the Zeebrugge ferry disaster in 1987. Angie couldn't claim their estate till she could have Andy declared dead. When she finally got probate in 1993, she sold up and emigrated to New Zealand. She teaches piano in Nelson on the South Island, works from home." Angie Kerr's full address and phone number followed.

She'd had a rough time of it, Karen thought. Losing her brother and both parents in the space of a couple of years was tough enough, without having to go through the process of having Andy formally declared dead. No wonder she'd wanted to move to the other side of the world. Where, she noted, it would now be half past eleven in the morning. A perfectly civilized time to call someone.

One of the few things Karen had bought for her home was an answering machine that allowed her to make digital recordings of her phone calls, recordings which she could then transfer via a USB connection to her computer. She'd tried to persuade the Macaroon to acquire some for the office, but he'd seemed unimpressed. Probably because it hadn't been his idea. Karen wouldn't have minded betting something similar would turn up in the main CID office before long, the brainchild of ACC Lees himself. Never mind. At least she could use the system at home and reclaim the cost of the calls.

A woman answered on the third ring, the Scots accent obvious even in the two syllables of "Hello?"

Karen introduced herself then said, "Is this Angie Kerr?"

"Kerr as was. Mackenzie as is. Is this about my brother? Have you found him?" She sounded excited, pleased almost.

"I'm afraid not, no."

"He didn't kill himself, you know. I've always thought he had an accident. Came off a mountain somewhere. No matter how depressed he was, Andy would never have killed himself. He wasn't a coward." Defiance travelled well.

"I'm sorry," Karen said. "I really have no answers for you. But we are looking again at events around the time he went missing. We're investigating the disappearance of Mick Prentice, and your brother's name came up."

"Mick Prentice." Angie sounded disgusted. "Some friend he turned out to be."

"What do you mean?"

"I don't think it's any coincidence that he went scabbing just before Andy took off."

"Why do you say that?"

A short pause, then Angie said, "Because it would feel like the worst betrayal. Those guys had been friends since the first day at school. Mick becoming a scab would have broken Andy's heart. And I think he saw it coming."

"What makes you say that?"

"The last time I saw him, he knew there was something going on with Mick."

Sunday, 2nd December 1984; Wemyss Woods

A visit home was never complete for Angie without time spent with her brother. She tried to get back at least once a term, but although the bus ride from Edinburgh was only an hour, it sometimes seemed too big an undertaking. She knew the problem was the different kind of distance that was growing between her and her parents as she moved more freely through a world that was alien to theirs: lectures, student societies, parties where drugs

were as common as drink, and a conversational range that out-
stripped anything she'd ever encountered back in Fife. Not that
there weren't opportunities for broadening one's intellectual horizons there. But the reading rooms and WEA courses and Burns
Clubs were for the men. Women had never had the access or the
time. The men did their shifts underground, then their time was
their own. But the women's work truly was never done, especially
for those whose landlords were the old coal companies or the na-
tionalized coal board. Angie's own grandmother hadn't had running hot water or a bath in her home until she'd been in her sixties.
So the men didn't easily take to women with an education.

Andy was one of the exceptions. His move from the coal face
to working for the union had exposed him to the wider equality policies pursued by the trade union movement. There might
not be women working in the pits, but contact with other unions
had persuaded Andy that the world would not end if you treated
women as fellow members of the human race. And so brother and
sister had grown closer, replacing their childhood squabbling with
genuine debate. Now Angie looked forward to Sunday afternoons
spent with her brother, tramping through the woods or nursing
mugs of hot chocolate by the fire.

That afternoon, Andy had met her off the bus at the end of the
track that led deep into the woods to his cottage. They'd planned
to skirt the woods and walk down to the shore, but the sky threatened rain so they opted to head back for the cottage. "I've got the
fire on for you coming," Andy had said as they set off. "I feel guilty
about having the money for the coal, so I don't usually bother. I
just put another sweater on."

"That's daft. Nobody blames you for still getting a wage."

Andy shook his head. "That's where you're wrong. There's
plenty think we should be kicking back our wages into the union
pot."

"And who does that help? You're doing a job. You're supporting the men on strike. You deserve to be paid." She linked her arm
with his, understanding how embattled he felt.

"Aye, and a lot of the strikers think they should be getting something from the union too. I've heard a few of them down the Welfare saying that if the union had been paying strike pay, they wouldn't be having to work so hard to keep the funds out of the hands of the sequestrators. They wonder what the union funds are for, if not to support the members when there's a strike on." He sighed, head down as if he was walking into a high wind. "And they've got a point, you know?"

"I suppose so. But if you've willingly handed over the decision-making to your leaders, which they've done by agreeing to strike without a national ballot, then you can't really start to complain when they make decisions you're not so keen on." Angie looked closely at her brother, seeing how the lines of strain round his eyes had deepened since she'd last seen him. His skin looked waxy and unhealthy, like that of a man who has spent too long indoors without vitamin supplements. "And it doesn't help anybody if you let them wind you up about it."

"I don't feel like I'm much help to anybody right now," he said, so quietly it was almost lost in the scuffle of dead leaves beneath their feet.

"That's just silly," Angie protested, knowing it wasn't enough but not knowing what else to say.

"No, it's the truth. The men I represent, their lives are falling apart. They're losing their homes because they can't pay the mortgage. Their wives have sold their wedding rings. Their kids go to school hungry. They've got holes in their shoes. It's like a bloody Third World country here, only we don't have charities raising money to help us with our disaster. And I can't do anything about it. How do you think that makes me feel?"

"Pretty shitty," Angie said, hugging his arm tighter to her. There was no resistance; it was like embracing the stuffed draught excluder their mother used to keep the living room as stifling as she could manage. "But you can only do the best you can. Nobody expects you to solve all the problems of the strike."

"I know." He sighed. "But I used to feel part of this commu-

nity. I've belonged here all my life. These days, it feels like the guys on strike are on one side of the fence and everybody else is on the other side. Union officials, pit deputies, managers, fucking Tory government—we're all the enemy."

"Now you're really talking rubbish. There's no way you're on the same side as the Tories. Everybody knows that." They walked on in silence, quickening their pace as the promise of rain became a reality. It sheeted down in cold hard drops. The bare branches above their heads offered little protection against the penetrating downpour. Angie let go his arm and began to run. "Come on, I'll race you," she said, exhilarated somehow by the drenching cold. She didn't check to see whether he was following her. She just hurtled pell-mell through the trees, jinking and swerving to stay with the winding path. As always, emerging into the clearing where the cottage hunkered down seemed impossibly sudden. It sat there like something out of the Brothers Grimm, a low squat building with no charm except its isolation. The slate roof, grey harling, black door and window frames would easily have qualified it as the home of the wicked witch in the eyes of any passing child. A wooden lean-to sheltered a coal box, a wood pile, and Andy's motorbike and sidecar.

Angie ran to the porch and turned round, panting. There was no sign of Andy. A couple of minutes passed before he trudged out of the trees, light brown hair plastered dark to his head. Angie felt deflated at the failure of her attempt to lighten his spirits. He said nothing as he led the way into the cottage, as neat and spartan as a barracks. The only decoration was a series of wildlife posters that had been given away free with one of the Scottish Sunday papers. One set of shelves was crammed with books on natural history and politics; another with LPs. It couldn't have been less like the rooms she frequented in Edinburgh, but Angie liked it better than any of them. She shook her head like a dog to shed the raindrops from her dark blond hair, tossed her coat over a chair, and curled up in one of the secondhand armchairs that flanked the fire. Andy went straight through to the scullery to make the hot chocolate.

As she waited for him to come through, Angie fretted over how she might lift his mood. Usually she made him laugh with tales of her fellow students and their antics, but she sensed that wasn't going to work today. It would feel too much like insensitive tales of the over-privileged. Maybe the answer was to remind him of the people who still believed in him.

He came back with two steaming mugs on a tray. Usually they had cookies, but clearly anything that smacked of luxury was off the menu today. "I've been giving most of my wages to the hardship fund," he said, noticing her noticing. "Just keeping enough for the rent and the basics."

They sat facing each other, nursing their hot drinks to let the warmth seep back into their cold hands. Angie spoke first. "You shouldn't pay attention to them. The people who really know you don't think you're one of the enemy. You should listen to people like Mick who know who you are. What you are."

"You think?" His mouth twisted in a bitter expression. "How can the likes of Mick know who I am when I don't know who they are any more?"

"What do you mean, you don't know who Mick is any more? The two of you have been best pals for twenty-odd years. I don't believe the strike has changed either of you that much."

"You'd think so, wouldn't you?" Andy stared into the fire, his eyes dull and his shoulders sagging. "Men round here, we're not supposed to talk about our feelings. We live in this atmosphere of comradeship and loyalty and mutual dependence, but we never talk about what's going on inside us. But me and Mick, we weren't like that. We used to tell each other everything. There was nothing we couldn't talk about." He pushed his damp hair back from his high narrow forehead. "But lately something's changed. I feel like he's holding back. Like there's something really important that he can't bring himself to talk about."

"But that could be anything," Angie said. "Something between him and Jenny, maybe. Something it wouldn't be right to talk to you about."

Andy snorted. "You think he doesn't talk about Jenny? I know all about that marriage, trust me. I could draw you a map of the fault lines between that pair. No, it's not Jenny. The only thing I can think is that he agrees with the rest of them. That I'm neither use nor ornament to them right now."

"You sure you're not imagining things? It doesn't sound like Mick."

"I wish I was. But I'm not. Even my best pal thinks I'm not fit to be trusted any more. I just don't know how long I can go on doing my work, feeling like this."

Now Angie was starting to feel genuinely worried. Andy's despair was clearly far beyond anything she knew how to deal with. "Andy, don't take me wrong, but you need to go and see the doctor."

He made a noise like a laugh strangled at birth. "What? Aspirin and Disprin, the painkilling twins? You think I'm losing my marbles? You think that pair would know what to do about it if I was? You think I need temazepam like half the bloody women round here? Happy pills to make it not matter?"

"I want to help you, Andy. And I don't have the skills. You need to talk to somebody that knows what they're doing, and the doctor's a good place to start. Even Aspirin and Disprin know more than I do about depression. I think you're depressed, Andy. Like, clinically depressed, not just miserable."

He looked as if he was going to cry. "You know the worst thing about what you just said? I think you might be right."

Thursday, 28th June 2007; Kirkcaldy

It sounded plausible. Andy Kerr had sensed Mick Prentice was keeping something from him. When it appeared Mick had joined the scabs and gone to Nottingham, it might have been enough to push someone in a fragile state over the edge. But it looked as if Mick Prentice hadn't gone to Nottingham at all. The question, Karen

thought, was whether Andy Kerr knew what had really happened to his best friend. And whether he was involved in his disappearance. "And you never spoke to Andy after that Sunday?" she asked.

"No. I tried to ring him a couple of times, but I just got the answering machine. And I didn't have a phone where I was living so he couldn't call me back. Mum told me the doctor had signed him off his work with depression, but that was all I knew."

"Do you think it's possible he and Mick went off somewhere together?"

"What? You mean, just turned their back on everybody and waltzed off into the sunset like Butch Cassidy and the Sundance Kid?"

Karen winced. "Not that, exactly. More like they'd both had enough and couldn't see any other way out. No question that Andy was having his problems. And you suggested Mick and Jenny weren't getting along too well. Maybe they just decided on a clean break?"

She could hear Angie breathing on the other side of the world. "Andy wouldn't do that to us. He would never have hurt us like that."

"Could Mick have talked him into it? You said they'd been pals since school. Who was the leader? Who was the follower? There's always one who leads and one who follows. You know that, Angie. Was Mick the leader?" No one pushed more gently but firmly than Karen on a roll.

"I suppose so. Mick was the extrovert, Andy was much quieter. But they were a team. They were always in trouble, but not in a bad way. Not with the police. Just always in trouble at school. They'd booby-trap chemistry experiments with fireworks. Glue teachers' desks shut. Andy was good with words and Mick was artistic, so they'd print up posters with fake school announcements. Or Mick would forge notes from teachers letting the pair of them off classes they didn't like. Or they'd mess about in the library, swapping the dust jackets on the books. I'd have had a breakdown if I'd ever had pupils like them. But they grew out of it. By the time

of the strike, they'd both settled down into their lives." There was more than a hint of regret in her voice. "So yes, theoretically Mick might have talked Andy into doing a runner. But it wouldn't have lasted. They'd have come back. They couldn't stay away. Their roots were too deep."

"You tore yours up," Karen observed.

"I fell in love with a New Zealander, and all my family were dead," Angie said flatly. "I wasn't leaving anybody behind to grieve."

"Fair enough. Can we go back to Mick? You said Andy had implied there were problems in his marriage?"

"She trapped him into that marriage, you know. Andy always thought she got pregnant on purpose. She was supposed to be on the pill, but amazingly it didn't work and the next thing was Misha was on the way. She knew Mick came from a decent family, the kind of people who don't run away from their responsibilities. So of course he married her." There was a bitter edge in her tone that made Karen wonder whether Angie had carried a torch for Mick Prentice before her New Zealander came along.

"Not the best of starts, then."

"They seemed happy enough to begin with." Angie's grudging admission came out slowly. "Mick treated her like a little princess and she lapped it up. But she didn't like it one little bit when the hard times hit. I thought at the time that she'd pushed him into scabbing because she'd had enough of being skint."

"But she really suffered after he went," Karen said. "It was a terrible stigma, being the wife of a scab. She wouldn't have let him leave her behind to face that on her own."

Angie made a dismissive noise in the back of her throat. "She had no idea what it would be like until it hit her. She didn't get it. She wasn't one of us, you know. People talk about the working class as if it's just one big lump, but the demarcation lines are just as well defined as they are among any other class. She was born and bred in East Wemyss, but she wasn't one of us. Her dad didn't get his hands dirty. He worked in the Co-operative. He served

behind the counter in the store. He wore a collar and tie to his work. I bet he never voted Labour in his life. So I'm not sure how clearly she understood what would happen to her if Mick went on the black."

It made sense. Karen understood viscerally what Angie was saying. She knew people like that from her own community. People who didn't fit anywhere, who had a deep groove across their backsides from a lifetime of sitting on the fence. It lent weight to the idea that Mick Prentice might have gone scabbing. Except that he hadn't. "The thing is, Angie, it looks like Mick didn't go scabbing that night. Our preliminary inquiries indicate that he didn't join the five men who went to Nottingham."

A shocked silence. Then Angie said, "He could have gone somewhere else on his own."

"He had no money. No means of transport. He didn't take anything with him when he went out that morning except his painting gear. Whatever happened to him, I don't think he went scabbing."

"So what did happen to him?"

"I don't know that yet," Karen said. "But I plan to find out. And here's the question I have to start asking. Let's assume Mick didn't go scabbing. Who might have had a reason for wanting him out of the way?"

Friday, 29th June 2007; Nottingham

Femi Otitoju entered the fourth address into Google Earth and studied the result. "Come on, Fem," Mark Hall muttered. "The DCI's got his eye on us. He's wondering what the hell you're doing, playing around on the computer after he's given us an assignment."

"I'm working out the most efficient order to do the interviews in, so we don't waste half the day backtracking." She looked at the four names and addresses supplied by some DC in Fife and numbered them according to her logic. "And I've told you. Don't

call me Fem." She printed the list and folded it neatly into her un-scuffed handbag. "My name is Femi."

Mark rolled his eyes and followed her out of the Cold Case Review office, flashing a nervous smile at DCI Mottram as they went. He'd been gagging for his temporary transfer to CID, but if he'd been warned that it would mean working with Femi Otitoju, he might have had second thoughts. The word round the station when they were both still in uniform was that, in Otitoju's case, PC stood for Personal Computer. Her uniform had always been immaculate, her shoes polished to a military sheen. Her plain clothes followed the same pattern. Neatly pressed anonymous grey suit, blinding white shirt, impeccable hair. And shoes still polished like mirrors. Everything she did was by the book; everything was precise. Not that Mark had anything against doing things prop-erly. But he'd always believed there was a place for spontaneity, especially in an interview. If the person you were talking to veered off at a tangent, it didn't hurt to follow for a while. Sometimes it was among the tangents that the truth was hiding. "So these four were all miners from Fife who broke the strike to go down the pits here?" he said.

"That's right. There were originally five of them, but one of them, Stuart McAdam, died two years ago of lung cancer."

How did she remember that stuff? And why did she bother? "And who are we going to see first?"

"William John Fraser. Known as Billy. Fifty-three years old, married with two grown-up children, one at Leeds University, the other at Loughborough. He's a self-employed electrician now." She hitched her bag higher on her shoulder. "I'll drive, I know where we're going."

They emerged in the windy parking lot behind the station and headed for an unmarked CID pool car. It would, Mark knew, be full of someone else's rubbish. CID and cars were like dogs and lampposts, he'd discovered. "Won't he be at work now?" He opened the passenger door to find the footwell held plastic sand-wich containers, empty Coke cans, and five Snickers bar wrap-

pers. Something white snapped at the corner of his peripheral vision. Otitoju was waving an empty carrier bag at him. "There you go," she said. "Stick the rubbish in there and I'll take it to the bin."

Mark reminded himself that she did have her uses after all. They hit the main ring road, still busy even after the worst of the morning rush, and headed west. The road was flanked with dirty red-brick houses and the sort of businesses that managed to hang on by a fingernail in the teeth of classier opposition elsewhere. Convenience stores, nail studios, hardware shops, launderettes, fast-food outlets, and hairdressers. It was depressing driving past it. Mark was grateful for his city-centre flat in a converted lace mill. It might be small, but he didn't have to deal with this crap in his personal life. And there was a great Chinese just round the corner that delivered.

Fifteen minutes round the ring road and they turned off into a pleasant enclave of semi-detached brick cottages. They looked as if they'd been built in the 1930s; solid, unpretentious, and nicely proportioned. Billy Fraser's house was on a corner plot, with a substantial, well-established garden. "I've lived in this city all my life and I didn't even know this place existed," Mark said.

He followed Otitoju up the path. The door was answered by a woman who couldn't have been much over five feet. She had the look of someone just past her best; silver strands in her light brown bob, jawline starting to soften, a few more pounds than was comfortable. Mark thought she was in pretty good nick for her age. He dived straight in before Otitoju could scare her. "Mrs. Fraser?"

The woman nodded, looking anxious. "Yes, that's me." Local accent, Mark noted. So he hadn't brought a wife from Fife. "And you are . . . ?"

"I'm Mark Hall and this is my colleague Femi Otitoju. We're police officers and we need to have a word with Billy. It's nothing to worry about," he added hastily, seeing the look of panic on

Mrs. Fraser's face. "Someone he used to know back in Fife has been reported missing and we need to ask Billy a few questions."

The woman shook her head. "You'll be wasting your time, duck. Billy's not kept in touch with anyone from Fife except the lads he came down here with. And that was more than twenty years ago."

"The man we're interested in went missing more than twenty years ago," Otitoju said bluntly. "So we do need to speak to your husband. Is he at home?" Mark felt like kicking her as he watched Mrs. Fraser's face close down on them. Otitoju had definitely been behind the door when sisterhood got handed out.

"He's at work."

"Can you tell us where he's working, flower?" Mark said, trying to get back on a conversational keel.

He could practically see the mental debate on the woman's face. "Wait a minute," she said at last. She returned with a large-format diary open at that day's date. She turned it to face him. "There."

Otitoju was already scribbling the address down on her precious sheet of paper. Mrs. Fraser caught sight of the names. "You're in luck," she said. "Johnny Ferguson's working with him today. You'll be able to kill two birds with one stone." From the expression on her face, she wasn't convinced that was a metaphor.

The two ex-miners were working a scant five-minute drive away, refitting a shop on the main drag. "From kebab shop to picture framing in one easy move," Mark said, reading the clues. Fraser and Ferguson were hard at work, Fraser chiselling out a channel for cables, Ferguson demolishing the bench seat running along one wall for the takeaway's customers. They both stopped what they were doing when the two police officers entered, eyeing them warily. It was funny, Mark thought, how some people always recognized a cop instantly, while others seemed oblivious to whatever signals he and his kind gave off. It was nothing to do with guilt or innocence, as he'd naïvely thought at first. Just an instinct for the hunter.

Otitoju introduced them and explained why they were there. Fraser and Ferguson both looked bemused. "Why would anybody think he'd have come with us?" Ferguson said.

"More to the point, why would anybody think we'd have taken him?" Billy Fraser wiped the back of his hand across his mouth in a gesture of disgust. "Mick Prentice thought the likes of us were beneath him. Even before we went scabbing, he looked down on other folk. Thought he was better than us."

"Why would he think that?" Mark asked.

Fraser pulled a packet of Bensons out of his overalls. Before he could get the cigarette out of the packet, Otitoju had placed her smooth hand over the rough one. "That's against the law now, Mr. Fraser. This is a place of work. You can't smoke in here."

"Aw, for fuck's sake," Fraser complained, turning away as he shoved his smokes back in his pocket.

"Why would Mick Prentice think he was better than you?" Mark said again.

Ferguson took up the challenge. "Some men went on strike because the union told them to. And some went on strike because they were convinced they were right and they knew what was best for the rest of us. Mick Prentice was one of the ones who thought they knew best."

"Aye," Fraser said bitterly. "And he had his pals in the union taking care of him." He rubbed fingers and thumb together in the universal representation of money.

"I don't understand," Mark said. "I'm sorry, mate, I'm too young to remember the strike. But I thought one of the big problems was that you didn't get strike pay?"

"You're right, son," Fraser said. "But for a while, the lads that went on the flying pickets got cash in hand. So when there was any picketing duty available, it was always the same ones that got the nod. And if your face didn't fit, there was nothing for you. But Mick's face fit better than most. His best pal was an NUM official, see?"

"It was harder for some of us than others," Ferguson added. "I expect Prentice's pal slipped him the odd fiver or bag of food when the picketing money ran out. Most of us weren't that lucky. So no, Mick Prentice didn't come with us. And Billy's right. We wouldn't have had him if he'd asked."

Otitoju was prowling round the room, scrutinizing their work as if she were a building inspector. "The day you left. Did you see Mick Prentice at all?"

The two men exchanged a look that seemed furtive to Mark. Ferguson quickly shook his head. "Not really," he said.

"How can you 'not really' see somebody?" Otitoju demanded, turning back towards them.

Friday, 14th December 1984; Newton of Wemyss

Johnny Ferguson stood in the dark at the bedroom window where he could see the main road through the village. The room wasn't cold, but he was shivering slightly, the hand cupping his rollie trembling, interrupting the smooth rise of the smoke. "Come on, Stuart," he muttered under his breath. He took another drag off his cigarette and looked again at the cheap watch on his wrist. Ten minutes late. His right foot began tapping involuntarily.

Nothing was stirring. It was barely nine o'clock, but there was hardly a light showing. People couldn't afford the electricity. They went down the Welfare for a bit of light and heat or they went to bed, hoping they might sleep long enough for the nightmare to be over when they woke. For once, though, the quiet of the streets didn't bother Ferguson. The fewer people the better to witness what was happening tonight. He knew exactly what he was about to do and it scared the living shit out of him.

Suddenly, a pair of headlights swung into sight round the corner of Main Street. Against the dim street lights, Ferguson could make out the shape of a transit van. The old shape, not the

new one that the police used as troop carriers in their operations against the miners. As the van drew closer, he could see it was dark in colour. Finally, Stuart was here.

Ferguson pinched out his cigarette. He took a last look round the bedroom where he'd slept for the last three years, ever since he'd taken the tenancy on the tiny house. It was too gloomy to see much, but then there wasn't much to be seen. What couldn't be sold had been broken up for firewood. Now there was just the mattress on the floor with an ashtray and a tattered Sven Hassel paperback beside it. Nothing left to regret. Helen was long gone, so he might as well turn his back on the fucking lot of them.

He clattered downstairs and opened the door just as Stuart was about to knock. "Ready?" Stuart said.

A deep breath. "As ready as I'll ever be." He pushed a hold-all towards Stuart with a foot and grabbed another holdall and a black trash bag. Ten fucking years at the coal face, and that was all he had to show for it.

They took two steps of the four that would bring them to the van and suddenly they weren't alone. A figure came hustling round the corner like a man on a mission. A couple of yards closer and the shape resolved itself into Mick Prentice. Ferguson felt a cold hand clutch his chest. Christ, that was all they needed. Prentice ripping into them, shouting the odds and doors opening all the way down the street.

Stuart threw his holdall into the back of the van, where Billy Fraser was already settled on a pile of bags. He turned to face Prentice, ready to make something of it if he had to.

But the rage they expected wasn't raining down on them. Instead, Prentice just stood there, looking like he was going to burst into tears. He looked at them and shook his head. "No, lads. No. Dinnae do it," he said. He kept on saying it. Ferguson could hardly believe this was the same man who'd chivvied them and rallied them and goaded them into staying loyal to the union. It was, he thought, a measure of how this strike had broken them.

Ferguson pushed past Prentice, stowed his bags, and climbed

in beside Fraser, who pulled the doors closed behind him. "Fucking amazing," Fraser said.

"He looked like he just took a punch to the gut," Ferguson said. "The guy's lost it."

"Just be grateful," Fraser said. "Last thing we needed was him going off like a fucking rocket, bringing the place down about us." He raised his voice as the engine roared into life. "Let's go, Stu. The new life starts here."

Friday, 29th June 2007

"Were there any witnesses to this encounter?" Otitoju said.

"Stuart's dead now, so I'm the only witness left," Fraser said. "I was in the van. The back door was open and I saw the whole thing. Johnny's right. Prentice looked gutted. Like it was a personal affront, what we were doing."

"It might have been a different story if it had been Iain in the van and not you," Ferguson said.

"Why might that have made a difference?" Mark said.

"Iain and him were pals. Prentice might have felt the need to try and talk him out of it. But Iain was the last pick-up, so I guess we were off the hook. And that was the last time we saw Prentice," Ferguson said. "I've still got family up there. I heard he'd taken off, but I just assumed he'd gone off with that pal of his, the union guy. I can't remember his name—"

"Andy something," Fraser said. "Aye, when you told me they were both on the missing list, I thought they'd decided to bugger off and make a fresh start somewhere else. You have to understand, people's lives were falling apart by then. Men did things you'd never have thought they were capable of." He turned away and walked to the door, stepping outside and taking out his cigarettes.

"He's right," Ferguson said. "And mostly we didn't want to think too much about it. Come to that, we still don't want to. So

unless there's anything else, we'll say good day to you." He picked up his crowbar and returned to his task.

Unable to think of anything else to ask, Mark started for the door. Otitoju hesitated briefly before following him to the car. They sat in silence for a moment, then Mark said, "It must have been bloody awful."

"It doesn't excuse their lawlessness," Otitoju said. "The miners' strike drove a wedge between us and the people we serve. They made us look brutal even though we were provoked. They say even the Queen was shocked by the battle of Orgreave, but what did people expect? We're supposed to keep her peace. If people don't consent to be policed, what else can we do?"

Mark stared at her. "You scare me," he said.

She looked surprised. "I sometimes wonder if you're in the right job," she said.

Mark looked away. "You and me both, flower."

Rotheswell Castle

In spite of her determination to deal with Sir Broderick Maclennan Grant on precisely the same terms as she would anyone else, Karen had to admit her stomach was off message. Anxiety always affected her digestive tract, putting her off her food and precipitating urgent dashes to the toilet. "If I had more interviews like this, I wouldn't need to think about going on a diet," she said as she and Phil set off for Rotheswell Castle.

"Ach, dieting's overrated," said Phil from the comfortable vantage point of a man whose weight hadn't wavered since he'd turned eighteen, no matter what he ate or drank. "You're fine just the way you are."

Karen wanted to believe him, but she couldn't. Nobody could find her chunky figure appealing, not unless they were a lot more hard up for female company than Phil need be. "Aye, right." She opened her briefcase and ran through the key points of the case

file for Phil's benefit. She'd barely reached the end of her summary when they turned into the gateway of Rotheswell. They could see the castle in the distance beyond the bare branches of a stand of trees, but before they could approach their identities had to be verified. They both had to get out of the car and hold their warrant cards up to the CCTV camera. Eventually, the solid wooden gates swung open, allowing the car access to a sort of security airlock. Phil drove forward, Karen walking beside the car. The wooden gates swung shut behind them, leaving them contained as if in a giant cattle pen. Two security men emerged from a guardhouse and inspected the exterior and interior of the car, Karen's briefcase, and the pockets of Phil's duffel coat.

"He's got better security than the prime minister," Karen said as they finally drove up the drive.

"Easier to get a new prime minister than a new Brodie Grant," Phil said.

"I bet that's what he thinks, anyway."

As they approached the house, an elderly man in a waterproof jacket and a tweed cap rounded the nearest turret and waved them towards the far side of the gravel apron in front of the house. By the time they'd parked, he'd vanished, leaving them no option but to approach the massive studded wooden doors in the middle of the frontage. "Where's Mel Gibson when you need him?" Karen muttered, raising a hefty iron door knocker and letting it fall with a satisfying bang. "It's like a very bad film."

"And we still don't know why we're here." Phil looked glum. "Hard to see what could live up to this build-up."

Before Karen could reply, the door swung open on silent hinges. A woman who reminded her of her primary school teacher said, "Welcome to Rotheswell. I'm Susan Charleson, Sir Broderick's personal assistant. Come on in."

They filed into an entrance hall that, provided the grand staircase had been removed, could comfortably have accommodated Karen's house. She had no chance to take much in other than a general atmosphere of rich colour and warmth before they were

hustled a short distance down a wide corridor. "You're DI Pirie, I presume," Susan Charleson said. "But I'm not familiar with your colleague's name and rank."

"Detective Sergeant Phil Parhatka," he said with as much pomp as he could muster in the teeth of her formality.

"Good, now I can introduce you," she said, stepping to one side and opening a door. She waved them into a drawing room where CID could comfortably have staged their annual Burns' Supper. They'd have had to push some of the furniture back to the walls to make room for the country dancing, but still, it wouldn't have been much of a squeeze.

There were three people in the room, but Karen was instantly focused on the one who radiated the charisma. Brodie Grant might be knocking at the wrong side of seventy, but he was still more glamorous than either of the women who flanked him. He stood to one side of the substantial carved stone mantel, left hand cupping his right elbow, right hand casually holding a slim cigar, face as still and striking as the magazine cover shot she'd found on Google Images. He wore a grey-and-white tweed jacket whose weight suggested cashmere and silk rather than Harris or Donegal, a black turtleneck, matching trousers, and the sort of shoes Karen had only ever seen on the feet of rich Americans. She thought they were called tasselled oxfords or something. They looked like something a kiltie doll would wear rather than a captain of industry. She was so busy studying his weird footwear that she almost missed the introductions.

She looked up in time to catch the faintest twitch of a smile on the mouth of Lady Grant, elegant in a heather mixture suit with the classic velveteen collar that somehow always spoke of money and class to Karen. But the smile felt strangely complicit.

Susan Charleson introduced the other woman. "This is Annabel Richmond, a freelance journalist." Wary now, Karen nodded an acknowledgement. What the hell was a journalist doing here? If she knew one thing about Brodie Grant, it was that he was so

allergic to the media he should be going into anaphylactic shock any moment now.

Brodie Grant stepped forward and indicated with a wave of his cigar that they should sit on a sofa within loudhailing distance of the fireplace. Karen perched on the edge, conscious that it was the type of seat that would swallow her, making anything other than a clumsy exit impossible. "Miss Richmond is here at my request for two reasons," Grant said. "One, I'll come to in a moment. The other is that she'll be acting as liaison between the media and the family. I will not be giving press conferences or making sentimental televised appeals. So she is your first port of call if you're looking for something to feed the reptiles."

Karen inclined her head. "That's your prerogative," she said, trying to sound as if she was making a concession out of the goodness of her heart. Anything to claw back some control. "I understand from Mr. Lees that you believe some new evidence has emerged relating to the kidnap of your daughter and grandson?"

"It's new evidence all right. No doubt of that. Susan?" He glanced expectantly at her. Smart enough to anticipate her boss's demands, she was already advancing towards them with a plastic-covered sheet of plywood. As she drew near, she turned it round to face Karen and Phil.

Karen felt a shimmer of disappointment. "This isn't the first time we've seen something like this," she said, studying the monochrome print of the puppeteer and his sinister marionettes. "I came across three or four instances in the files."

"Five, actually," Grant said. "But none like this. The previous ones were all dismissed because they diverged in some way from the originals. The reproductions DCI Lawson distributed to the media at the time were subtly altered so we could weed out any copycats. All the ones that have turned up since were copies of the altered versions."

"And this one's different?" Karen said.

Grant nodded his approval. "Spot on, Inspector. It's identical

in every respect. I'm well aware that the reward I've offered is a temptation to some people. I kept my own copy of the original so I could compare anything that was brought directly to me. As this was." He gave a wan smile. "Not that I needed a copy. I'll never forget a single detail. The first time I clapped eyes on it, it burned itself on to my memory."

Saturday, 19th January 1985

Mary Grant poured her husband a second cup of coffee before he noticed he'd finished his first. She'd been doing it for so many years it still surprised him that his cup needed so many refills when he stayed in hotels. He turned the page of his newspaper and grunted. "Some good news at last. Lord Wolfenden's shuffled off this mortal coil."

Mary's expression was one of weary resignation rather than shock. "That's a horrible thing to say, Brodie."

Without lifting his eyes, he said, "The man made the world a more horrible place, Mary. So I'm not sorry he's gone."

Years of marriage had knocked most of the combativeness out of Mary Grant. But even if she'd intended to say anything, she wouldn't have had the chance. To the surprise of both Grants, the door to the breakfast room burst open without a knock and Susan Charleson practically ran in. Brodie dropped his paper on to his scrambled eggs, taking in her pink cheeks and shortness of breath.

"I'm sorry," she gabbled. "But you have to see this." She thrust a large manila envelope at him. On the front was his name and address, with the words "Private" and "Confidential" written in thick black marker above and below.

"What in the name of God is it that can't wait till after breakfast?" he said, poking two fingers into the envelope to reveal a thick piece of paper folded into quarters.

"This," Susan said, pointing to the envelope. "I put it back

in the envelope because I didn't want anybody else to get a look at it."

Making an impatient noise, Grant pulled out the paper and unfolded it. It looked like an advertising poster for a macabre puppet show. In stark black and white, a puppeteer leaned over the set, manipulating a group of marionettes that included a skeleton and a goat. It reminded him of the kind of prints he'd once seen on a TV programme about the art Hitler hated. Even while he was thinking this, his eyes were scanning the bottom section of the poster. Where one would expect to find details of the puppet show performance there was a very different message.

YOUR GREEDY EXPLOITATIVE CAPITALISM IS ABOUT TO BE PUNISHED. WE HAVE YOUR DAUGHTER AND YOUR GRANDSON. DO AS WE TELL YOU IF YOU WANT TO SEE THEM AGAIN. NO POLICE. JUST GO ABOUT YOUR BUSINESS AS USUAL. WE ARE WATCHING YOU. WE WILL CONTACT YOU AGAIN SOON. THE ANARCHIST COVENANT OF SCOTLAND.

"Is this some kind of sick joke?" Grant said, throwing it down on the table and pushing his chair back. As he stood up Mary snatched the poster, then dropped it as if it had burned her fingers.

"Oh my God," she breathed. "Brodie?"

"It's a trick," he said. "Some sick bastard trying to give us a bit of a fright."

"No," Susan said. "There's more." She picked up the envelope where it had fallen to the floor and shook out a Polaroid photo. Silently she handed it to Grant.

He saw his only daughter tied to a chair. A slash of packing tape covered her mouth. Her hair was a mess; a smudge of dirt or a bruise marked her left cheek. Between her and the camera, a gloved hand held enough of the front page of the previous day's *Daily Record* to leave no room for doubt. He felt his legs give way and he collapsed back into his chair, his eyelids fluttering as he tried to regain control of himself. Mary reached for the photo-

graph, but he shook his head and held it tight against his chest. "No," he said. "No, Mary."

There was a long silence, then Susan said, "What do you need me to do?"

Grant couldn't form words. He didn't know what he thought, what he felt, or what he wanted to say. It was an experience as alien and as unlikely as taking recreational drugs. He was always in charge of himself and most of what happened around him. To be powerless was something that hadn't happened for so long he had forgotten how to cope with it.

"Do you want me to call the chief constable?" Susan said.

"It says not to," Mary said. "We can't take risks with Catriona and Adam."

"To hell with that," Grant said in a pale approximation of his normal voice. "I'm not being pushed around by a bunch of bloody anarchists." He forced himself upright, sheer will overcoming the fear that was already eating him from the inside out. "Susan, call the chief constable. Explain the situation. Tell him I want the best officer he has who doesn't look like a policeman. I want him at the office in an hour. And now I'm going to the office. Going about my business as usual, if they really are watching."

"Brodie, how can you?" White-faced, Mary looked stricken. "We have to do what they tell us."

"No, we don't. We just have to look as if we are." His voice was stronger now. Having fixed on the bare beginnings of a plan gave him strength to recover himself. He could deal with the fear if he could make himself believe he was doing something to resolve the situation. "Susan, get things moving." He went to Mary and patted her on the shoulder. "It'll be all right, Mary. I promise you." If he couldn't see her face, he didn't have to deal with her doubt or terror. He had enough to worry about without that extra burden.

Dysart, Fife

Other men might have paced the floor waiting for the police to arrive. Brodie Grant had never been one to waste his energy on pointless activity. He sat in his office chair, swivelled away from the desk so he faced the spectacular view across the Forth estuary to Berwick Law, Edinburgh, and the Pentlands. He stared out over the stippled grey water, ordering his thoughts to avoid any waste of time once the police arrived. He hated to squander anything, even that which could be readily replaced.

Susan, who had followed him to work at the usual time, came through the door that separated her office from his. "The police are here," she said. "Shall I bring them in?"

Grant swung round in his chair. "Yes. Then leave us." He registered the look of surprise on her face. She was accustomed to being privy to all his secrets, to knowing more than Mary ever cared to. But this time, he wanted the circle to be as small as possible. Even Susan was one person too many.

She ushered in two men in painters' overalls, then pointedly closed the door behind her. Grant was pleased with the ruse. "Thank you for coming so promptly. And so discreetly," he said, studying the pair of them. They looked too young for so important a task. The elder, lean and dark, was probably in his early to mid-thirties, the other, fair and ruddy, his late twenties.

The dark one spoke first. To Grant's surprise, his introduction went straight to the heart of his own reservations. "I'm Detective Inspector James Lawson," he said. "And this is Detective Constable Rennie. We've been personally briefed by the chief constable. I know you're probably thinking I'm on the young side to be running an operation like this, but I've been chosen because of my experience. Last year the wife of one of the East Fife players was abducted. We managed to resolve the matter without anybody getting hurt."

"I don't remember hearing anything about that," Grant said.

"We managed to keep the lid on very successfully," Lawson said, the briefest of proud smiles flitting across his face.

"Wasn't there a trial? How could you keep that out of the papers?"

Lawson shrugged. "The kidnapper pleaded guilty. It was over and done with before the press even noticed. We're pretty good at news management here in Fife." His quick smile flashed again. "So you see, sir, I'm the man with the relevant experience."

Grant gave him a long appraising stare. "I'm glad to hear it." He took a pair of tweezers from his drawer and delicately shifted the blank sheet of paper that he'd placed over the ransom poster. "This is what arrived in the post this morning. Accompanied by this—" Carefully lifting it by its edges, he turned over the Polaroid.

Lawson moved closer and studied them intently. "And you're sure this is your daughter?"

For the first time, Grant's grip on his self-control slipped a fraction. "You think I don't know my own daughter?"

"No, sir. But for the record, I have to be sure you're sure."

"I'm sure."

"In that case, there's not much room for doubt," Lawson said. "When did you last see or hear from your daughter?"

Grant made an impatient gesture with his hand. "I don't know. I suppose I saw her last about two weeks ago. She'd brought Adam over to visit. Her mother will have spoken to her or seen her since then. You know how women are." The sudden guilt he felt was not so much a twinge as a slow pulse. He didn't regret anything he'd said or done; he regretted only that it had caused a rift between him and Cat.

"We'll talk to your wife," Lawson said. "It would be helpful for us to get an idea of when this happened."

"Catriona has her own business. Presumably if the gallery was closed, someone will have noticed. There must be hundreds, thousands of people who drive past every day. She was scrupulous about the open and closed sign." He gave a tight, wintry smile. "She had a good head for business." He pulled a pad towards him and

scribbled down the address and directions to Catriona's gallery.

"Of course," Lawson said. "But I thought you didn't want the kidnappers to know you had come to us?"

Grant was taken aback by his own stupidity. "I'm sorry. You're right. I'm not thinking straight. I . . ."

"That's my job, not yours." There was kindness in Lawson's tone. "You can rest assured that we'll make no inquiries that raise suspicion. If we can't find something out in an apparently natural way, we leave it alone. The safety of Catriona and Adam is paramount. I promise you that."

"That's a promise I expect you to keep. Now, what's the next step?" Grant was back in command of himself but unnerved by the emotions that kept throwing him off balance.

"We'll be putting a tap and a trace on your phone lines in case they try to contact you that way. And I'm going to need you to go to Catriona's home. It's what the kidnappers would expect. You have to be my eyes inside her house. Anything out of place, anything unusual, you need to make a note of it. You'll have to carry a briefcase or something, so if for example there's two mugs on the table, you can bring them out to us. We'll also need something of Catriona's so that we can get her prints. A hairbrush would be ideal, then we get her hair too." Lawson sounded eager.

Grant shook his head. "You'll have to get my wife to do that. I'm not very observant." He wasn't about to admit he'd crossed the threshold only once, and that reluctantly. "She'll be happy to have something to do. To feel useful."

"Fine, we'll see to that." Lawson tapped the poster with a pen. "On the surface, this looks like a political act rather than a personal one. And we'll be checking out intelligence on any group that might have the resources and the determination to pull off something like this. I need to ask you, though—have you had any run-ins with any special interest group? Some organization that might have a few hotheads on the fringes who would think this was a good idea?"

Grant had already asked himself the same question while he'd

been waiting for the police. "The only thing I can think of is a problem we had a year or so ago with one of those 'save the whale' outfits. We had a development on the Black Isle that they claimed would adversely affect the habitat of some bunch of dolphins in the Moray Firth. All nonsense, of course. They tried to stop our construction crew—the usual stunt, lying in front of the JCBs. One of them got hurt. Their own stupid fault, which was how the authorities saw it. But that was the end of it. They went off with their tail between their legs and we got on with the development. And the dolphins are perfectly fine, by the way."

Lawson had visibly perked up at Grant's information. "Nevertheless, we'll have to check it out," he said.

"Mrs. Charleson will have all the files. She'll be able to tell you what you need to know."

"Thanks. I also have to ask you if there's anyone you can think of who has a personal grievance against you. Or anyone in your family."

Grant shook his head. "I've tramped on a lot of toes in my time. But I can't think of anything I've done that would provoke someone into doing this. Surely this is about money, not spite? Everybody knows I'm one of the richest men in Scotland. It's not a secret. To me, that's the obvious motive here. Some bastard wants to get their hands on my hard-earned cash. And they think this is the way to do it."

"It's possible," Lawson agreed.

"It's more than possible. It's the most likely scenario. And I'm damned if they're going to get away with it. I want my family back, and I want them back without giving an inch to these bastards." Grant slammed the flat of his hand down on the desk. The two policemen jumped at the sudden crack.

"That's why we're here," Lawson said. "We'll do everything possible to produce the outcome you're looking for."

Back then, Grant's confidence was still intact. "I expect nothing less," he said.

Friday, 29th June 2007; Rotheswell Castle

Listening to Grant's account of that first morning after the world had changed, what struck Karen was everybody's assumption that this had all been about Brodie Grant. Nobody seemed to have considered that the person being punished here was not Grant himself but his daughter. "Did Catriona have any enemies?"

Grant gave her an impatient frown. "Catriona? How could she have enemies? She was a single parent and an artist in glass. She didn't live the sort of life that generated personal animosity." He sighed and pursed his lips.

Karen told herself not to be daunted by his attitude. "Sorry. I expressed myself badly. I should have asked if you knew of anyone she'd upset."

Grant gave her a small nod of satisfaction, as if she'd passed a test she hadn't even known about. "The father of her child. He was upset, all right. But I never really thought he had it in him, and your colleagues could never find any evidence to connect him to the crime."

"Are you talking about Fergus Sinclair?" Karen asked.

"Who else? I thought you'd brought yourself up to speed with the background?" Grant demanded.

Karen was beginning to feel sorry for everyone obliged to put up with Brodie Grant's level of irritation. She suspected it wasn't reserved for her. "There's only one mention of Sinclair in the file," she said. "In the notes of an interview with Lady Grant, Sinclair is mentioned as Adam's putative father."

Grant snorted. "Putative? Of course he was the boy's father. They'd been seeing each other on and off for years. But what do you mean, there's only one reference to Sinclair? There must be more. They went to Austria to interview him."

"Austria?"

"He worked over there. He's a qualified estate manager. He's worked in France and Switzerland since, but he went back to Austria about four years ago. Susan can give you all the details."

"You've kept tabs on him?" It wasn't surprising, Karen thought.

"No, Inspector. I told you: I never thought Sinclair had the gumption to pull this off. So why would I keep tabs on him? The only reason I know where Sinclair's living is that his father is still my head keeper." Grant shook his head. "I can't believe this isn't all in the file."

Karen was thinking the same thing, but she didn't want to admit it. "And as far as you know, was there anybody else Catriona had upset?"

Grant's face was as wintry as his hair. "Only me, Inspector. Look, it's obvious from where this new evidence has turned up that this had nothing to do with Cat personally. It's obviously political. Which makes it about what I stand for, not whose heart Cat had broken."

"So where did this poster turn up?" Phil said. Karen was grateful for the interruption. He was good at jumping in and steering interviews in more productive directions when she was in danger of getting bogged down.

"In a ruined farmhouse in Tuscany. Apparently the place had been squatted." He extended his arm towards the journalist. "This is the other reason Miss Richmond is here. She's the person who found it. Doubtless you'll want to talk to her." He pointed to the poster. "You'll want to take that with you, too. I expect there will be tests you need to do on it. And, Inspector . . . ?"

Karen recovered her breath in the face of his high-handedness. "Yes?"

"I don't want to read about this in the paper tomorrow morning." He glared at her as if defying her to respond.

Karen held her fire for a moment, trying to compose a reply that encompassed what she wanted to say and left out anything that might be misconstrued. Grant's expression changed to a prompt. "Whatever we release to the media and the timing of any release will be an operational decision," she said at last. "It will be

made by me and, where appropriate, my superior officers. I fully understand how painful all this is for you, but I'm sorry, sir. We have to base our decisions on what we think is most likely to produce the best outcome. You might not always agree with that, but I'm afraid you don't get a veto." She waited for the explosion, but none came. She supposed he'd save that for the Macaroon or his bosses.

Instead, Grant nodded mildly to Karen. "I have confidence in you, Inspector. All I ask is that you communicate with Miss Richmond here in advance, so we're forearmed against the mob." He ran a hand through his thick silver hair in a gesture that looked well practised. "I have high hopes that this time the police will get to the truth. With all the advances in forensic science, you should have a head start on Inspector Lawson." He turned away in what was clearly a dismissal.

"I expect I'll have some further questions for you," Karen said, determined not to cede all control of the encounter. "If Catriona didn't have any enemies, maybe you could think of the names of some of her friends who might be able to help us. Sergeant Parhatka will let you know when I want to talk to you again. In the meantime—Miss Richmond?"

The woman inclined her head and smiled. "I'm at your disposal, Inspector."

At least someone round here had a vague notion of how things were supposed to work. "I'd like to see you in my office this afternoon. Shall we make it four o'clock?"

"What's wrong with interviewing Miss Richmond here? And now?" Grant said.

"This is my investigation," Karen said. "I'll conduct my interviews where it suits me. And because of other ongoing inquiries, it suits me to be in my office this afternoon. Now, if you'll excuse us?" She got to her feet, registering Lady Grant's guarded amusement and Susan Charleson's prim disapproval. Grant himself was still as a statue.

"It's all right, Susan, I'll see the officers out," Lady Grant said, jumping to her feet and heading for the door before the other woman gathered her self-possession.

As they followed her down the hallway, Karen said, "This must be hard for you."

Lady Grant half-turned, walking backwards with the assurance of someone who knows every inch of her territory. "Why do you say that?"

"Watching your husband revisit such a terrible time. I wouldn't want to see someone I cared about going through all that."

Lady Grant looked puzzled. "He lives with it every day, Inspector. He may not give that impression, but he dwells on it. Sometimes I catch him looking at our son Alec, and I know he's thinking about what might have been, with Adam. About what he's lost. Having something fresh to focus on is almost a relief for him." She swivelled on her toes and turned her back to them again. As they followed her, Karen caught Phil's eye and was surprised by the anger she saw there.

"Still, you wouldn't be human if a part of you wasn't hoping we won't find Adam alive and well," Phil said, the lightness of his tone a contrast to the darkness of his expression.

Lady Grant stopped in her tracks and whirled round, eyebrows drawn down. A blush of pink spread up her neck. "What the hell do you mean by that?"

"I think you know exactly what I mean, Lady Grant. We find Adam and suddenly your boy Alec isn't Brodie's only heir," Phil said. It took guts, Karen thought, to assume the role of the investigation's lightning rod.

For a moment Lady Grant looked as if she might slap his face. Karen could see her chest rising and falling with the effort of holding herself in check. Finally, she forced herself to assume the familiar pose of civility. "Actually," she said, her words clipped and tight, "you're looking at this from precisely the wrong angle. Brodie's absolute commitment to uncovering his grandson's fate fills me with confidence about Alec's future. A man so bound by

obligation to his own flesh and blood is never going to let our son down. Believe it or not, Sergeant, Brodie's quest for the truth gives me hope. Not fear." She turned on her heel and marched to the front door, where she pointedly held it open for them.

Once the door had closed behind them, Karen said, "Jeez, Phil, why not tell us what you really think? What brought that on?"

"I'm sorry." He opened the passenger door for her, a small courtesy he seldom bothered to extend. "I'd had enough of playing at Miss Marple. All that country house murder bollocks. Bloodless and civilized. I just wanted to see if I could provoke an honest reaction."

Karen grinned. "I think it's safe to say you did that. I just hope we don't get buried in the fallout."

Phil snorted. "You're not exactly behind the door when it comes to being a hardarse. 'This is my investigation,'" he mimicked, not unkindly.

She settled herself in the car. "Aye, well. The illusion of being in charge. It was nice while it lasted."

Nottingham

The beauties of the Nottingham Arboretum were not so much diminished as rendered invisible by the sheets of rain that blinded DC Mark Hall as he followed Femi Otitoju up the path that led to the Chinese Bell Tower. She'd finally shown some emotion, but it wasn't exactly what Mark had been hoping for.

Logan Laidlaw had been even less pleased than Ferguson and Fraser to see them. Not only had he refused to let them across the threshold of his flat, he'd told them he had no intention of repeating what he'd told Mick Prentice's daughter. "Life's too bloody short to waste my energy going over that stuff twice," he'd said, then slammed the door shut in their faces.

Otitoju had turned the deep purple of pickled beet, breathing heavily through her nose. Her hands had bunched into fists and

she'd actually drawn her foot back as if she was going to kick the door. Pretty wild, considering there wasn't that much of her. Mark had put a hand on her arm. "Leave it, Femi. He's within his rights. He doesn't have to talk to us."

Otitoju had swung round, her whole body compressed in anger. "It shouldn't be allowed," she said. "They should have to talk to us. It should be against the law for people to refuse to answer our questions. It should be an offence."

"He's a witness, not a criminal," Mark said, alarmed by his vehemence. "It's what they told us when we were doing our induction. Police by consent, not coercion."

"It's not right," Otitoju said, storming back to the car. "They expect us to solve crimes, but they don't give us the tools to do the job. Who the hell does he think he is?"

"He's somebody whose opinion of the police was set in stone back in 1984. Have you never seen the news reports from back then? Mounted police rode into the pickets, like they were Cossacks or something. If we used our batons like that, we'd be up on a charge. It wasn't our finest hour. So it's not really surprising that Mr. Laidlaw doesn't feel like talking to us."

She shook her head. "It just makes me wonder what he has to hide."

The drive across town from Iain Maclean's house to the Arboretum hadn't done much to improve her temper. Mark caught up with her. "Leave this to me, OK?" he said.

"You think I can't conduct an interview?"

"No, I don't think that. But I know enough about ex-miners to know they're a pretty macho bunch. You saw back there with Ferguson and Fraser—they didn't take kindly to you asking questions."

Otitoju stopped abruptly and threw her head back, letting the rain course down her face like cold tears. She straightened up and sighed. "Fine. Let's pander to their prejudices. You do the chat." Then she set off again, this time at a more measured pace.

They arrived at the Chinese Bell Tower to find two middle-

aged men in council overalls sheltering from the downpour. The narrow pillars that supported the elegant roof offered little protection from the scatterings of rain thrown around by the gusty wind, but it was better than being completely out in the open. "I'm looking for Iain Maclean," Mark said, glancing from one to the other.

"That would be me," the shorter of the two said, bright blue eyes sparkling in a tanned face. "And who are you?"

Mark identified them both. "Is there somewhere we can go and get a cup of tea?"

The two men looked at each other. "We're supposed to be tidying up the borders, but we were just about to give up and go back to the greenhouses," Maclean said. "There's no café here, but you could come back with us and we could brew up there."

Ten minutes later, they were squashed into a corner at the back of a large plastic-covered tunnel, out of the way of the other gardeners, whose curious stares quickly subsided once they realized there was no drama happening. The smell of humus hung heavy in the air, reminding Mark of his granddad's allotment shed. Iain Maclean wrapped his large hands round a mug of tea and waited for them to speak. He'd shown no surprise at their arrival, nor had he asked them why they were there. Mark suspected Fraser or Ferguson had warned him.

"We wanted to talk to you about Mick Prentice," he began.

"What about Mick? I've not seen him since we moved south," Maclean said.

"Neither has anybody else," Mark said. "Everybody assumed he'd gone south with you, but that's not what we've been hearing today."

Maclean scratched the silver bristles that covered his head in a neat crew cut. "Aye well. I'd heard folk thought that back in the Newton. It just shows you how willing they are to think the worst. There's no way Mick would have joined us. I don't see how anybody who knew him could think that."

"You never contradicted them?"

"What would be the point? As far as they're concerned I'm a dirty blackleg miner. Nothing I have to say in anybody's defence would carry any weight in the Newton."

"To be fair, it's not just a matter of jumping to conclusions. His wife's had money sent to her on and off since he left. The postmark was Nottingham. That's one of the main reasons everybody thought he'd done the unthinkable."

"I can't explain that. But I'm telling you this: Mick Prentice could no more go scabbing than fly to the moon."

"That's what everybody keeps telling us," Mark said. "But people do things that seem out of character when they're desperate. And by all accounts, Mick Prentice was desperate."

"Not that desperate."

"You did it."

Maclean stared into his coffee. "I did. And I've never been so ashamed all my days. But my wife was pregnant with our third. I knew there was no way we could bring another bairn into that life. So I did what I did. I talked about it with Mick beforehand." He flashed a swift glance at Mark. "We were pals, me and him. We were at the school together. I wanted to explain to him why I was doing it." He sighed. "He said he understood why I was set on going. That he felt like getting out too. But scabbing wasn't for him. I don't know where he went, but I knew for sure it wasn't down another pit."

"When did you know he'd gone missing?"

He screwed his face up as he thought. "It's hard to say. I think it was maybe when the wife came down to join me. So that would make it round about the February. But it might have been after that. The wife, she's still got family back in the Wemyss. We don't go back there. We wouldn't be welcome. Folk have got long memories, you know? But we stay in touch and sometimes they come here for a visit." A pale apology of a smile crossed his face. "The wife's nephew, he's a student at the university down here. Just finishing his second year. He comes round for his dinner now and

again. So aye, I heard Mick had gone on the missing list, but I couldn't tell you for sure when I knew."

"Where do you think he went? What do you think happened?" In his eagerness, Mark forgot the cardinal rule of asking only one question at a time. Maclean ignored both of them.

"How come you're interested in Mick all of a sudden?" he said. "Nobody's come looking for him all these years. What's the big deal now?"

Mark explained why Misha Gibson had finally reported her father missing. Maclean shifted awkwardly in his seat, his coffee slopping over his fingers. "That's hellish. I mind when Misha was just a wee lassie herself. I wish I could help. But I don't know where he went," he said. "Like I said, I've not seen hide nor hair of him since I left the Newton."

"Have you heard from him?" Otitoju chipped in.

Maclean gave her a flat hard look. On his weatherbeaten face, it rested as impassive as Mount Rushmore. "Don't get smart with me, hen. No, I haven't heard from him. As far as I'm concerned, Mick Prentice fell off Planet Iain the day I came down here. And that's exactly what I expected."

Mark tried to rebuild the rapport, injecting sympathy into his voice. "I understand that," he said. "But what do you think happened to Mick? You were his pal. If anyone can come up with an answer, it would be you."

Maclean shook his head. "I really don't know."

"If you had to hazard a guess?"

Again he scratched his head. "I tell you what. I thought him and Andy had taken off together. I thought they'd both had enough, that they'd buggered off somewhere else to start all over again. Clean sheet, and that."

Mark remembered Prentice's friend's name from the briefing document. But there had been no mention of them leaving together. "Where would they go? How could they just disappear without a trace?"

Maclean tapped the side of his nose. "Andy was a Commie, you know. And that was when Lech Wałesa and Solidarity was a big deal in Poland. I always thought the pair of them had buggered off there. Plenty of pits in Poland, and that wouldn't have felt like scabbing. No way, no how."

"Poland?" Mark felt as if he needed a crash course in twentieth-century political history.

"They were trying to overthrow totalitarian Communism," Otitoju said crisply. "To replace it with a sort of workers' socialism."

Maclean nodded. "That would have been right up Andy's street. I figured he'd talked Mick into going with him. That would explain how nobody heard from them. Stuck digging coal behind the Iron Curtain."

"It's been in mothballs for a while now, the Iron Curtain," Mark said.

"Aye, but who knows what kind of life they made for themselves over there? Could be married with kids, could have put the past right behind them. If Mick had a new family, he wouldn't be wanting the old one emerging from the woodwork, would he?"

Suddenly Mark had one of those revelatory moments where he could see the wood hidden among the trees. "It was you that sent the money, wasn't it? You put cash in an envelope and sent it to Jenny Prentice because you thought Mick wouldn't be sending her any money from Poland."

Maclean seemed to shrink back against the translucent polythene wall. His face screwed up so tight it was hard to see his bright blue eyes. "I was only trying to help. I've done OK since I came down here. I always felt sorry for Jenny. Seemed like she'd ended up with the sticky end because Mick didn't have the courage of his convictions."

It seemed an odd way to put it, Mark thought. He could have left it at that; it wasn't his case, after all, and he could do without the aggravation pulling at a loose thread might bring. But on the other hand, he wanted to make the most of this posting. He

wanted to parlay the position of CID aide into a permanent transfer to the detective division. So going the extra mile was definitely
part of the plan. "Is there something you're not telling us, Iain?"
he said. "Some other reason Mick had for taking off the way he
did, without a word to anyone?"

Maclean drained his coffee and put the mug down. His hands,
disproportionately large from a lifetime of hard manual work,
clasped and unclasped themselves. He looked like a man uncomfortable with the contents of his own head. He took a deep breath
and said, "I suppose it makes no odds now. You can't make somebody pay when they're the wrong side of the grave."

Otitoju was about to break Maclean's silence, but Mark gripped
her arm in warning. She subsided, her mouth a compressed line,
and they waited.

At last, Maclean spoke. "I've never told anybody this. For all
the good keeping schtum's done. You have to understand, Mick
was a big union man. And of course, Andy was a full-time NUM
official. Feet under the table, well in there with the top men. I don't
doubt Andy told Mick a lot of things he maybe shouldn't have."
He gave a wan smile. "He was always trying to impress Mick, to
be his best pal. We were all in the same class at school. The three
of us used to hang about together. But you know how it is with
threesomes. There's always the leader and the other two trying to
keep in with him, trying to edge out the other one. That's how it
was with us. Mick in the middle, trying to keep the peace. He was
good at it too, clever at finding ways to keep the pair of us happy.
Never letting either one of us get the upper hand. Well, not for
long anyway."

Mark could see Maclean relaxing as he remembered the relative ease of those early days. "I know just what you mean," he said
quietly.

"Anyway, we all stayed pals. Me and the wife, we'd go out in a
foursome with Mick and Jenny. Him and Andy would play football together. Like I said, he was good at finding things that made
the both of us feel like we had a wee bit of something special. So

anyway, a couple of weeks before I came away down here, we spent the day together. We walked along to Dysart harbour. He set up his easel and painted, and I fished. I told him what I had planned and he tried to talk me out of it. But I could tell his heart wasn't really in it. So I asked him what was bothering him." He stopped again, his strong fingers working against each other.

"And what was it?" Mark said, leaning forward to close Otito-ju's stiff presence out of the circle, to make it a male environment.

"He said he thought one of the full-time officials had his hand in the till." Then he locked eyes with Mark, who could sense the terrible betrayal that lay behind Maclean's words. "We were all skint and starving, and one of the guys who was supposed to be on our side was lining his own pockets. It might not sound like a big deal now, but back then, it shook me to the very core."

Thursday, 30th November 1984; Dysart

A mackerel was tugging at his line, but Iain Maclean paid it no mind. "You're fucking joking," he said. "Nobody would do that."

Mick Prentice shrugged, never taking his eyes off the cartridge paper pinned to his easel. "You don't have to believe me. But I know what I know."

"You must have misunderstood. No union official would steal from us. Not here. Not now." Maclean looked as if he was going to burst into tears.

"Look, I'll tell you what I know." Mick swept his brush across the paper, leaving a blur of colour along the horizon. "I was in the office last Tuesday. Andy had asked me to come in and help him with the welfare requests, so I was going through the letters we'd had in. I tell you, it would break your heart, reading what folk are going through." He cleaned his brush and mixed a green-ish grey colour on his pocket-sized palette. "So I'm going through this stuff in the wee cubbyhole off the main office, and this official is out front. Anyway, some woman came in from Lundin

Links. Tweed suit, stupid mohair beret. You know the kind—
Lady Bountiful, looking out for the peasants. She said they'd had
a coffee morning at the golf club and they'd raised two hundred
and thirty-two pounds to help the poor families of the striking
miners."

"Good for them," Maclean said. "Better going to us than
Thatcher's bloody crew."

"Right enough. So he thanks her and off she goes. Now, I didn't
actually see where the money went, but I can tell you it didn't go
in the safe."

"Aw, come on, Mick. That proves nothing. Your guy might
have been taking it straight to the branch. Or the bank."

"Aye, right." Mick gave a humourless laugh. "Like we put
money in the bank these days when the sequestrators are breathing down our neck."

"All the same," Maclean said, feeling offended somehow.

"Look, if that was all, I wouldn't be bothered. But there's
more. One of Andy's jobs is to keep a tally of money that comes
in from donations and the like. All that money's supposed to be
passed through to the branch. I don't know what happens to it
then, whether it comes back to us as handouts or whether it ends
up at the court of King Arthur, salted away in some bloody Swiss
bank account. But everybody that collects any money is supposed
to tell Andy and he writes it up in a wee book."

Maclean nodded. "I remember having to tell him what we made
when we were doing the street collections back in the summer."

Mick paused briefly, staring out at the point where the sea met
the land. "I was at Andy's the other night. The book was sitting
on the table. When he went to the toilet, I took a wee look. And
the donation from Lundin Links wasn't there."

Maclean jerked so roughly on his line that he lost the fish.
"Fuck," he said, reeling in furiously. "Maybe Andy wasn't up to
date."

"I wish it was that simple. But that's not it. The last entries in
Andy's book were four days after that money was handed in."

Maclean threw his rod on the stone flags at his feet. He could feel tears pricking his eyes. "That's a fucking disgrace. And you expect me to feel guilty about going to Nottingham? At least that's an honest day's work for an honest day's pay, not stealing. I can't believe that."

"I couldn't believe it either. But how else can you explain it?" Mick shook his head. "And this is a guy who's still on a wage."

"Who is it?"

"I shouldn't tell you. Not till I've decided what I'm going to do about it."

"It's obvious what you've got to do. You've got to tell Andy. If there's an innocent explanation, he'll know what it is."

"I can't tell Andy," Mick protested. "Christ, sometimes I feel like walking away from the whole fucking mess. Drawing a line and starting again someplace else." He shook his head. "I can't tell Andy, Iain. He's already depressed. I tell him this, it could send him over the edge."

"Well, tell somebody else. Somebody from the branch. You've got to nail the bastard. Who is it? Tell me. A couple of weeks, I'm going to be out of here. Who am I going to tell?" Maclean felt the need to know burning inside him. It was one more thing that could help him believe he was doing the right thing. "Tell me, Mick."

The wind whipped Mick's hair into his eyes, saving him from the desperation in Maclean's face. But the need to share his burden was too heavy to ignore. He pushed his hair back and looked his friend in the eye. "Ben Reekie."

Friday, 29th June 2007; Glenrothes

Karen had to admit she was impressed. Not only had the Nottingham team done a great job, but DC Femi Otitoju had typed up her report and e-mailed it in record time. Mind you, Karen thought, she'd probably have done the same in her shoes. Given the quality of the information she and her partner had been able to extract,

any officer on CID trial would be desperate to make the most of it.

And there was something here to make the most of. DC Otitoju and her oppos had found out who had been muddying the waters by sending money to Jenny Prentice from Nottingham. And crucially, she'd also given the first possible answer to the question of who might be happy to see the back of Mick Prentice. Feelings were running high by then, the union growing in unpopularity in many quarters. Violence had erupted more times than anyone could count, and not always between police and strikers. Mick Prentice could have found himself consumed by the fire he was playing with. If he'd confronted Ben Reekie with what he knew; if Ben Reekie was guilty as charged; and if Andy Kerr had been dragged into the affair because of his connection to the other two, then there was motive for getting rid of both of the men who had gone missing around the same time. Maybe Angie Kerr had been right about her brother. Maybe he hadn't killed himself. Maybe Mick Prentice and Andy Kerr were both victims of a killer—or killers—desperate to protect the reputation of a crooked union official.

Karen shuddered. "Too much imagination," she said out loud.

"What's that?" Phil dragged his eyes from the computer screen to frown at her.

"Sorry. Just giving myself a telling off for being melodramatic. I tell you, though, if this Femi Otitoju ever fancies a move north, I'd swap her for the Mint so fast it would make his eyes water."

"Not that that's saying much," Phil said. "By the way, what are you doing here? Shouldn't you be talking to the lovely Miss Richmond?"

"She left a message." Karen glanced at her watch. "She'll be here in a wee while."

"What's the hold-up been?"

"Apparently she had to talk to some newspaper lawyer about an article she wrote."

Phil tutted. "Just like Brodie Grant. Still think we're the servant class, that lot. Maybe you should keep her waiting."

"I can't be bothered getting into stupid game playing. Here, have a look at this. The paragraph I've highlighted." She passed Otitoju's report over to Phil and waited for him to read it. As soon as he lifted his eyes from the page, she spoke. "That's a sighting of Mick Prentice a good twelve hours after he walked out of the house. And it sounds like he wasn't himself."

"It's weird. If he was taking off, why was he still hanging around at that time of night? Where had he been? Where was he going? What was he waiting for?" Phil scratched his chin. "Makes no sense to me."

"Me neither. But we're going to have to try and find out. I'll add it to my list." She sighed. "Somewhere below having a proper conversation with the Italian police."

"I thought you'd spoken to them?"

She nodded. "An officer at their headquarters in Siena, some guy called Di Stefano that Pete Spinks in Child Protection dealt with a couple of years back. He speaks pretty good English, but he needs more info."

"So you'll be looking at Monday now?"

Karen nodded. "Aye. He said not to expect anybody in their office after two o'clock on a Friday."

"Nice work if you can get it," Phil said. "Speaking of which, do you fancy a quick drink after you're finished talking to Annabel Richmond? I've got to go round to my brother's for dinner, but I've time for a swift half."

Karen was torn. The prospect of a drink with Phil was always enticing, but her absence from the office meant that her admin load had gone unattended for too long. And she couldn't catch up tomorrow because they were off to the caves. She toyed with the idea of slipping out for a quick drink, then coming back to the office. But she knew herself well enough to know that once she'd escaped from her desk she would find any excuse to avoid returning to the paperwork. "Sorry," she said. "I need to clear the decks."

"Maybe tomorrow, then? We could treat ourselves to lunch at the Laird o' Wemyss."

Karen laughed. "Have you won the pools? Do you know what that place costs?"

Phil winked. "I know they have a special deal for lunch on the last Saturday of the month. Which would be tomorrow."

"And I thought I was the detective round here. OK, you've got a deal." Karen turned her attention back to her notes, making sure she knew exactly what to ask Annabel Richmond.

Karen's phone rang five minutes before the agreed time. The journalist was in the building. She asked a uniform to show Richmond to the interview room where she'd met Misha Gibson, then gathered her papers together and headed downstairs. She walked in to find her witness leaning on the window sill and staring out at the thin strands of cloud stretched across the sky. "Thanks for coming in, Miss Richmond," Karen said.

She turned, her smile apparently genuine. "It's Bel, please," she said. "I should be thanking you for being so accommodating. I appreciate your flexibility." She crossed to the table and sat down, fingers intertwined, seeming relaxed. "I hope I'm not keeping you late."

Karen wondered when she'd last been home at five on a Friday and couldn't come up with an answer. "I wish," she said.

Bel's laugh was warm and conspiratorial. "Tell me about it. I suspect your work culture is scarily similar to mine. I must say, by the way, that I'm impressed."

Karen knew it was a ploy but rose to the bait anyway. "Impressed by what?"

"Brodie Grant's pulling power. I didn't imagine I'd be dealing with the woman who put Jimmy Lawson behind bars."

Karen felt the blush rising up her neck, knew she'd be looking blotchy and ugly, and wanted to kick the furniture. "I don't talk about that," she said.

Again that mellow, inviting laugh. "I don't imagine it's a popu-

lar topic of conversation between you and your colleagues. It must make them watch their backs, knowing you were responsible for pinning three murders on your boss."

She made it sound as though Karen had fitted Lawson up. In truth, once she'd been invited to think the unthinkable, the evidence had been there for the finding. One twenty-five-year-old rape and murder, two fresh kills to cover that past misdemeanour. Not nailing Lawson would have been the fit-up. It was tempting to tell Bel Richmond just that. But Karen knew that responding would start a conversation that could only go places she didn't want to revisit. "Like I said, I don't talk about it." Bel cocked her head, gave a smile that Karen read as rueful but confident. Not a defeat but a delay. Karen's smile was inward, knowing the journalist was wrong on that score.

"So, how do you want to do this, Inspector Pirie?" Bel said.

Stolidly refusing to be seduced by Bel's charm, Karen kept her voice official. "What I need right now is for you to be my eyes and ears and take me through what happened, step by step. How you found it, where you found it. The whole story. All the details you can remember."

"It started with my morning run," Bel began. Karen listened carefully as she told the story of her discovery again. She took notes, jotting down questions to ask afterwards. Bel appeared to be candid and comprehensive in her account, and Karen knew better than to break the flow of a helpful witness on a roll. The only sounds she made were wordless murmurs of encouragement.

At last Bel came to the end of her story. "To be honest, I'm surprised you recognized the poster right away," Karen said. "I'm not sure I would have."

Bel shrugged. "I'm a hack, Inspector. It was a huge story at the time. I had just reached the age where I thought I might like to be a journalist. I'd started paying proper attention to newspapers and news bulletins. More than the average person. I guess the image lodged in the deep recesses of my brain."

"I can see how that would happen. But, given that you under-stood the significance of it, I'm surprised you didn't bring it straight to us rather than Sir Broderick." Karen let the unspoken accusation lie in the air between them.

Bel's answer came smoothly. "Two reasons, really. First, I had no idea who to contact. I thought if I just walked into my local police station it might not be treated very seriously. And second, the last thing I wanted to do was to waste police time. For all I knew, this was some sick copy. I reckoned Sir Broderick and his people would know at once whether this was something that ought to be taken seriously."

Slick answer, Karen thought. Not that she expected Bel Rich-mond to admit any interest in the substantial reward Brodie Grant still offered. Nor in the prospect of gaining unrivalled access to the ultimate source. "Fair enough," she said. "Now, you said you had the impression that whoever had been living there had cleared out in a hurry. And you told me about what looked like a blood-stain in the kitchen. Did it seem to you that the two things were connected?"

A moment's silence, then Bel said, "I'm not sure how I would be able to make a judgement about that."

"If the stain on the floor was old, or it wasn't blood, it could be part of the landscape. Chairs sitting on it, that sort of thing."

"Oh, right. Yes, I hadn't thought of it in those terms. No, I don't think it was part of the scenery. There was a chair over-turned near it." She spoke slowly, obviously summoning the scene in her mind. "One section looked like someone had tried to clean it up then realized it was pointless. The floor's made of stone slabs, not glazed tiles. So the stone soaked up the blood."

"Were there any other posters or printed material?"

"Not that I saw. But I didn't search the place. To be honest, the poster freaked me out so much I couldn't wait to get out." She gave a little laugh. "Not really the image of the intrepid investiga-tive hack, am I?"

Karen couldn't be bothered bolstering her ego. "The poster freaked you out? Not the blood?"

Again a pause for thought. "You know something? That hadn't occurred to me till now. You're right. It *was* the poster, not the blood. And I don't really know why."

Saturday, 30th June 2007; East Wemyss

The sea wall was new since Karen had last visited East Wemyss. She'd deliberately arrived early so she could take a walk around the lower part of the village. They'd sometimes walked along the foreshore between there and Buckhaven when she'd been a kid. She remembered a run-down fag-end of a place, shabby and forlorn. Now it was spruced up and smart, old houses recently harled white or sandstone red and new ones looking fresh out of the box. The deconsecrated church of St. Mary's-by-the-Sea had been saved from dilapidation and turned into a private home. Thanks to the EU, a sea wall had been built with sturdy blocks of local stone to hold the Firth of Forth at bay. She walked along the Back Dykes, trying to get her bearings. The woodland behind the manse was gone, replaced by new houses. Same with the old factory buildings. And the skyline ahead of her was transformed now the pit winding gear and the bing were gone. If she hadn't known it was the same place, she'd have been hard pressed to recognize it.

She had to admit it was an improvement, though. It was easy to be sentimental about the old days and forget the appalling conditions so many people were forced to live in. They were economic slaves too, trapped by poverty into shopping only at the local establishments. Even the Co-operative, supposedly run for the benefit of its members, was pricey compared to the shops in Kirkcaldy High Street. It had been a hard way of life, the community spirit its only real compensation. The loss of that small offset must have been a killer blow for Jenny Prentice.

Karen turned back towards the parking lot, looking along the

seashore to the striated red sandstone bluff that marked the start of the string of deep caves huddled along the base of the cliff. In her memory, they were quite separate from the village, but now a row of houses butted right up against the outside edge of the Court Cave. And there were information boards for the tourists, telling them about the caves' five-thousand-year history of habitation. The Picts had lived there. The Scots had used them as smithies and glassworks. The back wall of the Doo Cave was pocked with dozens of literal pigeonholes. Down through time, the caves had been used by the locals for purposes as diverse as clandestine political meetings, family picnics on rainy days, and romantic trysts. Karen had never dropped her knickers there, but she knew girls who had and thought none the worse of them for it.

Walking back, she saw Phil's car draw up where tarmac gave way to the coastal path. Time to explore a different conjunction of past and present. By the time she reached the parking lot, Phil had been joined by a tall, stooped man with a gleaming bald head, dressed in the kind of jacket and trousers that the middle classes had to buy before they could attempt any walk more challenging than a stroll to the local pub. All zippers and pockets and high-tech materials. Nobody Karen had grown up with had special clothes or boots for walking. You just went out for a walk in your street clothes, maybe adding an extra layer in the winter. Didn't stop them doing eight or nine miles before dinner.

Karen mentally shook herself as she approached the two men. Sometimes she freaked herself out, thinking like her granny. Phil introduced her to the other man, Arnold Haigh. "I've been secretary of the Wemyss Caves Preservation Society since 1981," he said proudly in an accent that had its roots a few hundred miles south of Fife. He had a long thin face with an incongruous snub nose and teeth that gleamed an unnatural white against weather-beaten skin.

"That's real dedication," Karen said.

"Not really." Haigh chuckled. "No one else has ever wanted the job. What exactly is it you wanted to talk to me about? I mean,

I know it's Mick Prentice, but I haven't even thought of him in years."

"Why don't we take a look at the caves and we can talk as we go?" Karen suggested.

"Surely," Haigh said graciously. "We can stop off in the Court Cave and the Doo Cave, then have a cup of coffee in the Thane's Cave."

"A cup of coffee?" Phil sounded bemused. "They've got a café down here?"

Haigh chuckled again. "Sorry, Sergeant. Nothing so grand. The Thane's Cave was closed to the public after the rock fall of 1985, but the society has keys to the railings. We thought it was appropriate to maintain the tradition of the caves having a useful function, so we set up a little clubhouse area in a safe part of the cave. It's all very ad hoc, but we enjoy it." He strode off towards the first cave, not seeing the look of mock horror Phil gave Karen.

The first sign that the cliffs were less than solid was a hole in the sandstone that had been bricked up years before. Some of the bricks were missing, revealing darkness within. "Now, that opening and the passage behind it is man-made," Haigh said, pointing to the brickwork. "As you can see, the Court Cave juts out further than the others. Back in the nineteenth century, high tide reached the cave mouth, cutting off East Wemyss from Buckhaven. The lasses who gutted the herring couldn't get between the two villages at high tide, so a passage was cut through the west side of the cave, which allowed them to pass along the shore safely. Now, if you'll follow me, we'll go in by the east entrance."

When she'd said "talk as we go," this hadn't been quite what Karen had in mind. Still, since they were doing this in their own time, for once there was no hurry and, if it settled Haigh down, it could work to their advantage. Glad that she'd chosen jeans and sneakers, she followed the men round the front of the cave and up a path by a low fence. Near the cave the fence had been trampled down, and they stepped over the bent wires and made their way

into the cave, where the beaten-earth floor was surprisingly dry, given the amount of rain there had been in recent weeks. The fact that the roof was supported by a brick column with a sign warning DANGER: NO ENTRY was less reassuring.

"Some people believe the cave got its name from King James the Fifth, who liked to go among his people in disguise," Haigh said, switching on a powerful flashlight and shining it up into the roof. "He was said to have held court here among the Gypsies who lived here at the time. But I think it's more likely that this was where the baronial courts were held in the Middle Ages."

Phil was roaming around, his face eager as a schoolboy's on the best-ever day trip. "How far back does it go?"

"After about twenty metres, the floor rises to the roof. There used to be a passage that ran three miles inland to Kennoway, but a roof fall closed the opening at this end, so the Kennoway entrance was sealed up for safety's sake. Makes you wonder, doesn't it? What were they up to here that they needed a secret passage to Kennoway?" Haigh chuckled again. Karen could only imagine how irritated this little tic would make her by the time they'd finished their interview.

She left the two men exploring the cave and walked back into the fresh air. The sky was dappled grey with the promise of rain. The sea reflected the sky and came up with a few more shades of its own. She turned back to the lush green summer growth and the brilliant colours of the sandstone, both still vibrant in spite of the gloominess of the weather. Before long, Phil emerged, Haigh still talking at his back. He gave Karen a rueful grin; she returned a stony face.

Next came the Doo Cave and a lecture on the historical necessity of keeping pigeons for fresh meat in winter. Karen listened with half an ear; then when Haigh paused for a moment she said, "The colours are amazing in here. Did Mick paint inside the caves?"

Haigh looked startled by the question. "Yes, as a matter of fact he did. Some of his watercolours are on display at the cave infor-

mation centre. It's the various mineral salts in the rock that create the vivid colours."

Before he could get into his stride on that subject, Karen asked another question. "Was he here a lot during the strike?"

"Not really. He was helping with the flying pickets to start with, I believe. But we didn't see him any more than usual. Less, if anything, as autumn and winter wore on."

"Did he say why that was?"

Haigh looked blank. "No. Never occurred to me to ask him. We're all volunteers, we all do what we can manage."

"Shall we get that cup of coffee now?" Phil said, his struggle between duty and pleasure obvious to Karen though not, thankfully, to Haigh.

"Good idea," Karen said, leading them back into daylight. Getting to the Thane's Cave was harder work, involving a clamber over the rocks and concrete that acted as a rough breakwater between the sea and the foot of the cliffs. Karen remembered the beach being lower, the sea less close, and she said so.

Haigh agreed, explaining that over the years the level of the beach had risen, partly because of the spoil from the coal mines. "I've heard some of the older residents talk about golden sands along here when they were children. Hard to credit now," he said, waving a hand at the grainy black of the tiny smooth fragments of coal that filled the spaces between the rocks and pebbles.

They emerged on to a grassy semi-circle. Perched on the cliff above them was the sole remaining tower of Macduff Castle— something else Karen remembered from her childhood. There had been more ruins around the tower, but they'd been removed by the council on the grounds of health and safety some years before. She remembered her father complaining about it at the time.

In the base of the cliff were several openings. Haigh headed for a sturdy metal grille protecting a narrow entrance a mere five feet tall. He unlocked the padlock and asked them to wait. He went inside, disappearing round a turn in the narrow passage. He

returned almost immediately with three hard hats. Feeling like
an idiot, Karen put one on and followed him inside. The first few
yards were a tight fit, and she heard Phil cursing behind her as he
banged an elbow against the wall. But soon it opened out into a
wide chamber whose ceiling disappeared into darkness.

Haigh groped in a niche in the wall and suddenly the pale
yellow of battery-operated lights cast a soft glow round the cave.
Half a dozen rickety wooden chairs sat round a Formica-topped
table. On a deep ledge about three feet above the ground sat a
camping stove, half a dozen litre bottles of water, and mugs. The
makings for tea and coffee were enclosed in plastic boxes. Karen
looked around and just knew that the mainstays of the cave preservation group were all men. "Very cosy," she said.

"Supposedly there was a secret passage from this cave to the
castle above," Haigh said. "Legend has it that was how Macduff
escaped when he came home to find his wife and children slain
and Macbeth in possession." He gestured to the chairs. "Take a
seat, please," he said, fiddling with stove and kettle. "So, why the
interest in Mick after all this time?"

"His daughter has only just got round to reporting him missing," Phil said.

Haigh half-turned, puzzled. "But he's not missing, surely? I
thought he'd gone off to Nottingham with another bunch of lads?
Good luck to them, I thought. There was nothing here but misery
back then."

"You didn't disapprove of the blackleg miners, then?" Karen
asked, trying not to make it sound too sharp.

Haigh's chuckle echoed spookily. "Don't get me wrong. I've
got nothing against trade unions. Working people deserve to be
treated decently by their employers. But the miners were betrayed
by that self-serving egomaniac Arthur Scargill. A true case of lions
led by a jackass. I watched this community fall apart. I saw terrible suffering. And all for nothing." He spooned coffee into mugs,
shaking his head. "I felt sorry for the men, and their families. I did
what I could—I was the regional manager for a specialist food im-

porter, and I brought as many samples as I could back to the village. But it was just a drop in the ocean. I totally understood why Mick and his friends did what they did."

"You didn't think there was something selfish about him leaving his wife and child behind? Not knowing what had happened to him?"

Haigh shrugged, his back to them. "To be honest, I didn't know much about his personal circumstances. He didn't discuss his home life."

"What did he talk about?" Karen asked.

Haigh brought over two plastic tubs, one containing sachets of sugar pilfered from motorway service stations and hotel bedrooms, the other little pots of non-dairy creamer from the same sources. "I don't really recall, so it was probably the usual. Football. TV. Projects to raise money for work on the caves. Theories about what the various carvings meant." Again the chuckle. "I suspect we're a bit dull to outsiders, Inspector. Most hobbyists are."

Karen thought about lying but couldn't be bothered. "I'm just trying to get an impression of what Mick Prentice was like."

"I always thought he was a decent, straightforward sort of bloke." Haigh brought the coffees over, taking almost exaggerated care not to spill any. "To be honest, apart from the caves, we didn't have a great deal in common. I thought he was a talented painter, though. We all encouraged him to paint the caves, inside and out. It seemed appropriate to have a creative record, since the main fame of the caves rests on their Pictish carvings. Some of the best are here in the Thane's Cave." He picked up his flashlight and targeted it at a precise spot on the wall. He didn't have to think about it. In the direct line of the beam, they could see the unmistakable shape of a fish, tail down, carved in the rock. In turn, he revealed a running horse and something that could have been a dog or a deer. "We lost some of the cupping designs in the fall of '85, but luckily Mick had done some paintings of them not long before."

"Where was the fall?" Phil said, peering towards the back of the cave.

Haigh led them to the furthest corner, where a jumble of rocks were piled almost to the roof. "There was a small second chamber linked by a short passage." Phil stepped forward to take a closer look, but Haigh grabbed his arm and yanked him back. "Careful," he said. "Where there's been a recent fall, we can never be sure how secure the roof is."

"Is it unusual to have cave-ins?" Karen said.

"Big ones like this? They used to happen quite regularly when the Michael pit was still working. But it closed in 1967 after—"

"I know about the Michael disaster," Karen interrupted. "I grew up in Methil."

"Of course." Haigh looked suitably rebuked. "Well, since they stopped working underground, there hasn't been much movement in the caves. We haven't had a major fall since this one, in fact."

Karen felt the twitch of her copper's instinct. "When exactly was the fall?" she said slowly.

Haigh seemed surprised at her line of questioning, giving Phil a glance of what felt like male complicity. "Well, we can't be precise about it. To be honest, from mid-December to mid-January is pretty much a dead time for us. Christmas and New Year and all that. People are busy, people are away. All we can say with any certainty is that the passage was clear on 7th December. One of our members was here that day, taking detailed measurements for a grant proposal. As far as we know, I was the next person in the cave. It's my wife's birthday on 24th January and we had some friends visiting from England. I brought them along to see the caves and that's when I discovered the fall. It was quite a shock. Of course, I cleared them out at once and called the council when we got back."

"So, some time between 7th December 1984 and 24th January 1985, the roof fell in?" Karen wanted to be sure she had it right. Two and two were coming together in her head, and she was pretty sure they weren't making five.

"That's right. Though I think myself it was earlier rather than later," Haigh said. "The air was clear in the cave. And that takes longer than you might think. You could say the dust had well and truly settled."

Newton of Wemyss

Phil looked at Karen with concern. In front of her was a perfectly presented pithivier of pigeon breast, surrounded by tiny new potatoes and a tower of roasted baby carrots and courgettes. The Laird o' Wemyss was more than living up to its reputation. But the plate had been sitting before Karen for at least a minute and she hadn't even lifted her cutlery. Instead of tucking in, she was staring at her plate, a frown line between her eyebrows. "Are you all right?" he said cautiously. Sometimes women behaved in strange and unpredictable ways around food.

"Pigeons," she said. "Caves. I can't get my mind off that fall."

"What about it? Cave falls happen. That's why they've got signs up warning people. And padlocked railings to keep them out. Health and safety, that's the bosses' mantra these days." He cut off a piece of his crispy fillet of sea bass and loaded it on his fork with the sesame hoisin vegetables.

"But you heard that guy. This is the only significant roof fall in any of the caves since the pit closed back in '67. What if it wasn't an accident?"

Phil shook his head, chewing and swallowing hastily. "You're doing that melodrama thing again. This is not Indiana Jones and the Wemyss Caves, Karen. It's a guy who went on the missing list when his life was shite."

"Not one guy, Phil. Two of them. Mick and Andy. Best pals. Not the kind to go scabbing. Not the kind to leave loved ones behind without a word."

Phil put down his fork and knife. "Did it ever occur to you that they might have been an item? Mick and his best pal Andy with

the isolated cottage deep in the heart of the woods? Being gay in a place like Newton of Wemyss back in the early eighties can't have been the easiest thing in the world."

"Of course it occurred to me," Karen said. "But you can't just run with theories that have absolutely nothing to back them up. Nobody we've spoken to has even hinted at it. And believe me, if Fife has one thing in common with Brokeback Mountain, it's that folk talk. Don't get me wrong. I'm not dismissing it. But until I have something to base it on, I've got to file it right at the back of my mind."

"Fair enough," Phil said, starting in on his food again. "But you've got no more foundation for your notion that there's somebody buried under an unnatural cave fall."

"I never said anybody was buried," Karen said.

He grinned. "I know you, Karen. There's no other reason you'd be interested in a pile of rock."

"Maybe so," she said without a trace of defensiveness. "But I'm not just punting wild ideas. If there's one group of people who know all about shot-firing to bring down rock precisely where they want it, it's miners. And the shot-firers also had access to explosives. If I was looking for someone to blow up a cave, the first person I would go to would be a miner."

Phil blinked. "I think you need to eat. I think you've got low blood sugar."

Karen glowered at him for a moment, then she picked up her knife and fork and attacked the food with her usual gusto. Once she'd demolished a few mouthfuls, she said, "That takes care of the low blood sugar. And I still think I'm on to something. If Mick Prentice didn't go on the missing list of his own free will, he disappeared because somebody wanted him out of the way. Lo and behold, we have somebody that wanted him out of the way. What did Iain Maclean tell us?"

"That Prentice discovered Ben Reekie had his hand in the union's till," Phil said.

"Exactly. Pocketing money that was supposed to go to the

branch. From all we've heard about Mick, he wouldn't have let that pass. And it's hard to see how he could pursue it without Andy being involved, since he was the one keeping the records. I don't think it was in their natures to do nothing about it. And if it had become common knowledge, Reekie would have been lynched, and you know it. That's a very tasty motive, Phil."

"Maybe so. But if it was two against one, how did Reekie kill the pair of them? How did he get the bodies in the cave? How did he get his hands on explosive charges in the middle of a strike?"

Karen's grin had always managed to disarm him. "I don't know yet. But, if I'm right, sooner or later I will know. I promise you that, Phil. And try this for starters: we know when Mick went missing, but we don't have an exact date for Andy's disappearance. It's entirely possible they were killed separately. They could have been killed in the cave. And as for getting hold of explosives—Ben Reekie was a union official. All sorts of people will have owed him favours. Don't pretend you don't know that."

Phil finished his fish and pushed his plate away from him. He raised his hands, palms towards Karen, indicating surrender. "So what do we do now?"

"Clear those rocks and see what's behind them," she said, as if the answer was obvious.

"And how are we going to do that? As far as the Macaroon's concerned, you're not even investigating this. And even if it was official, there's no way he'd stretch his precious budget to cover an archaeological dig for a pair of bodies that probably aren't there."

Karen paused with a forkful of pigeon breast halfway to her mouth. "What did you just say?"

"There's no budget."

"No, no. You said 'an archaeological dig.' Phil, if it wasn't for this pigeon coming between us, I could kiss you. You are a genius."

Phil's heart sank. It was hard to avoid the feeling that this was another fine mess he'd got himself into.

Kirkcaldy

Sometimes it was more sensible to make work calls from home. Until she'd actually got things under way and had her pitch firmly in place, Karen didn't want the Macaroon to get a sniff of what she was up to. Phil's words had set off a chain reaction in her brain. She wanted that rock fall cleared. The dates Arnold Haigh had given her offered the promise of being able to sneak it past the Macaroon under the pretext of a possible connection to the Grant case, but the cheaper she could make it, the less likely he would be to ask too many questions.

She settled herself down at the dining table with phone, note-pad, and contacts book. Comfortable though she was with new technology, Karen still maintained a physical record of names, addresses, and phone numbers. She reasoned that if the world ever went into electronic meltdown she would still be able to find the people she needed. It had naturally occurred to her that, in that event, there would be no functioning telephones and the transport network would also be in meltdown, but nevertheless her contacts book felt like a security blanket. And if it ever came to it, much easier to destroy without trace than any electronic memory.

She flicked it open at the appropriate page and ran her finger down the list till she came to Dr. River Wilde. The forensic anthropologist had been one of the mentors on a course Karen had attended aimed at improving the scientific awareness of detectives with responsibility at crime scenes. On the face of it, it would have been hard to find much common ground between the two women, but they had formed an instant if unlikely bond. Although neither of them would ever have explained it thus, it was something to do with the way they both appeared to play the game while subtly undermining the authority of those who had failed to earn their respect.

Karen liked the way River never tried to blind her audience with science. Whether lecturing to a group of cops whose scientific

education had ended in their teens or sharing an anecdote in the bar, she managed to convey complicated information in terms that a lay person could understand and appreciate. Some of her stories were horrifying; others reduced her listeners to helpless laughter; still others gave them pause.

The other thing that made River a great potential ally was that the man in her life was a cop. Karen hadn't met him, but from everything River had said, he sounded like her kind of cop. No bullshit, just a driven desire to get to the heart of things the straight way. So she'd come away from the forensics course with a greater understanding of her job but also with what felt like a new friendship. And that was rare enough to be worth nurturing. Since then, the women had met up a couple of times in Glasgow, the mid-point between Fife and River's base in the Lake District. They'd enjoyed their nights out, occasions that had cemented what their first encounter had started. Now Karen would find out if River had been serious when she'd offered her students as a cut-price team for exploratory work that couldn't really justify a big-budget spend.

River answered her mobile on the second ring. "Rescue me," she said.

"From what?"

"I'm sitting on the verandah of a wooden hut watching Ewan's terrible cricket team and praying for rain. The things we do for love."

Chance would be a fine thing. "At least you're not making the teas."

River snorted. "No way. I made that clear right from the start. No washing of sports kit, no slaving away in primitive kitchens. I get the hard stare from a lot of the other WAGs, but if they think I'm bothered, they're confusing me with someone who gives a shit. So how's tricks with you?"

"Complicated."

"So, nothing new there, then. We need to get together, have a night out. Uncomplicate yourself."

"Sounds good to me. And we might just manage it sooner than you think."

"Ah-hah. You've got something brewing?"

"You could say that. Listen, you remember you once said that you had a small army of students at your disposal if I ever needed help on the cheap?"

"Sure," River said easily. "You trying to get something done off the books?"

"Sort of." Karen explained the bare bones of the scenario. River made small noises of encouragement as she spoke.

"OK," she said when Karen had finished. "So we need forensic archaeologists first, preferably the big strong ones who can hump rocks. Can't use the final-year students because they're still doing exams. But it's nearly the end of term and I can press-gang the first- and second-years. Plus any of the anthros I can get my hands on. I can call it a field trip, make them think there's Brownie points to be had. When do you need us?"

"How about tomorrow?"

There was a long silence. Then River said, "Morning or afternoon?"

The phone call with River left Karen feeling all revved up with nowhere to go. She used some of her sudden access of energy to arrange accommodation for the students at the campsite on the links at nearby Leven. She tried to watch a DVD of *Sex and the City* but it only irritated her. It was always like this when she was in the middle of a case. No appetite for anything but the hunt. Hating being stalled because it was the weekend, or tests took time, or nothing could be done till the next bit of information fell into place.

She tried to distract herself by cleaning. Trouble was, she never spent long enough in the house to make much mess. After an hour's blitzing, there was nothing left that warranted attention.

"To hell with it," she muttered, grabbing her car keys and

making for the door. Strictly speaking, the laws of evidence required that she shouldn't be flying solo when she was talking to witnesses. But Karen told herself she was only colouring in the background, not actually taking evidence. And if she stumbled across something that might be relevant later in court, she could always send a couple of officers back another day to take a formal statement.

The drive back to Newton of Wemyss took less than twenty minutes. There was no sign of life in the isolated enclave where Jenny Prentice lived. No children played; nobody sat in their garden to enjoy the late afternoon sunshine. The short terrace of houses had assumed a dispirited air that would take more than a bit of summer weather to disperse.

This time, Karen approached the house next door to Jenny Prentice. She was still on a quest to get a sense of what Mick Prentice had really been like. Someone who was close enough to the family to be entrusted with the care of Misha must have had some dealings with her father.

Karen knocked and waited. She was just about to give up and head back to her car when the door cracked open on the chain. A tiny wizened face peered out at her from beneath a mass of heavy grey curls.

"Mrs. McGillivray?"

"I don't know you," the old woman said.

"No." Karen took out her official ID and held it up in front of the smeared lenses of the big glasses that made faded blue eyes swim large behind them. "I'm a police officer."

"I didn't call the police," the woman said, cocking her head and frowning at Karen's warrant card.

"No, I know that. I just wanted to have a wee word with you about the man who used to live next door." Karen gestured with her thumb towards Jenny's house.

"Tom? He's been dead years."

Tom? Who was Tom? Oh shit, she'd forgotten to ask Jenny Prentice about Misha's stepdad. "Not Tom, no. Mick Prentice."

"Mick? You want to talk about Mick? What are the police doing with Mick? Has he done something wrong?" She sounded confused, which filled Karen with foreboding. She'd spent enough time trying to get coherent information out of old people to know that it could be an uphill struggle with dubious results.

"Nothing like that, Mrs. McGillivray," Karen reassured her. "We're just trying to find out what happened to him all those years ago."

"He let us all down, that's what happened," the old woman said primly.

"Right enough. But I just need to clear up some of the details. I wonder if I could come in and have a wee chat with you?"

The woman exhaled heavily. "Are you sure you've got the right house? Jenny's the one you want. There's nothing I can tell you."

"To be honest, Mrs. McGillivray, I'm trying to get an idea of what Mick was really like." Karen switched on her best smile. "Jenny's a wee bit biased, if you get my drift?"

The old woman chuckled. "She's a besom, Jenny. Not a good word to say about him, has she? Well, lassie, you'd better come through." A rattle as the chain came off, then Karen was admitted to a stuffy interior. There was an overwhelming smell of lavender, with bass notes of stale fat and cheap cigarettes. She followed Mrs. McGillivray's bent figure through to the back room, which had been knocked through to make a kitchen diner. It looked like the work had been done in the seventies and nothing had been changed since, including the wallpaper. The various fades and stains bore witness to sunlight, cooking, and smoking. The low sun streamed in, slanting a gold light across the worn furniture.

A caged budgie chattered alarmingly as they walked in. "Quiet now, Jocky. This is a nice police lady come to talk to us." The budgie let out a stream of chirrups which sounded as if it was swearing at them, then subsided. "Sit yourself down. I'll get the kettle on."

Karen didn't really want a cup of tea but knew the conversation would go better if she let the old woman fuss around her. They

ended up facing each other across a surprisingly well-scrubbed table, a pot of tea and a plate of obviously home-made cookies between them. The sun lit Mrs. McGillivray like stage lighting, revealing details of make-up that had clearly been applied without the benefit of her glasses. "He was a lovely lad, Mick. A braw-looking fella, with that blond hair and big shoulders. He always had a smile and a cheery word for me," she confided as she poured the tea into china cups so fine you could see the sunlight in the tea. "I've been a widow thirty-two years now, and never had a better neighbour than young Mick Prentice. He'd always turn his hand to any wee job that I couldn't manage. It was never a trouble to him. A lovely laddie, right enough."

"It must have been hard on them, the strike." Karen helped herself to one of the proffered bourbon creams.

"It was hard on everybody. But that's not why Mick went away scabbing."

"No?" *Keep it casual, don't show you're particularly interested.*

"She drove him to it. Keeping company with that Tom Campbell right under his nose. No man would put up with that, and Mick had his pride."

"Tom Campbell?"

"He was never away from the door. Jenny had been a pal of his wife. She helped nurse the poor soul when she had the cancer. But after she died, it was like he couldn't stay away from Jenny. You had to wonder what had been going on all along." Mrs. McGillivray winked conspiratorially.

"You're saying Jenny was having an affair with Tom Campbell?" Karen bit her tongue on the questions she wanted to ask but knew she'd be better leaving till later. *Who was Tom Campbell? Where is he now? Why did Jenny not mention him?*

"I won't say what I can't swear to. All I know is that there was hardly a day went by when he didn't come calling. And always when Mick was out of the house. He never came empty-handed either. Wee parcels of this, packets of that. During the strike, Mick

used to say his Jenny could make a pound go further than any other woman in the Newton. I never told him the reason why."

"How come Tom Campbell had stuff to hand out? Was he not a miner, then?"

Mrs. McGillivray looked like the tea she'd just drunk had turned to vinegar. "He was a deputy." Karen suspected she'd have accorded more respect to the word "paedophile."

"And you think Mick found out what was going on between them?"

She nodded emphatically. "Everybody else in the Newton knew what was what. It's the usual story. The other half is always the last to know. And if anybody had their doubts, Tom Campbell was in there fast enough after Mick took his leave."

Too late, Karen remembered she hadn't followed up the subject of Misha's stepfather. "He moved in with Jenny?"

"A few months went by before he moved in. Keeping up appearances, for what it was worth. Then he had his feet right under Mick's table."

"Did he not have a house of his own? On a deputy's money, I'd have thought . . ."

"Oh aye, he had a braw house along at West Wemyss. But Jenny wouldn't move. She said it was for the bairn's sake. That Mick going had been upheaval enough for Misha without being uprooted from her own home." Mrs. McGillivray pursed her lips and shook her head. "But you know, I've often wondered. I don't think she ever loved Tom Campbell the way she loved Mick. She liked what he could give her, but I think her heart aye belonged to Mick. For all her carrying on, I never quite believed that Jenny stopped loving Mick. I think she stayed put because deep down she believes Mick'll come back one day. And she wants to be sure he knows where to find her."

It was, Karen thought, a theory based on soap opera sentimentality. But it did have the merit of making sense of what seemed otherwise inexplicable. "So what happened with her and Tom?"

"He rented out his own house and moved in next door. I never

had much to do with him. He didn't have Mick's easy way with folk. And things were never easy between the Lady Charlotte boys and the deputies, especially after the pit was closed in 1987." The old woman shook her head, jiggling the lank grey curls. "But Jenny got her come-uppance." Her smile was gleeful.

"How come?"

"He died. Took a massive heart attack on the golf course at Lundin Links. It must be getting on for ten years ago. And when the will was read, Jenny got a hell of a shock. He'd left everything in trust for Misha. She got the lot when she was twenty-five, and Jenny never saw a penny." Mrs. McGillivray raised her teacup in a toast. "Served her right, if you ask me."

Karen couldn't find it in her heart to disagree. She drained her cup and pushed back her chair. "You've been very helpful," she said.

"He was round here the very day Mick went to Nottingham," Mrs. McGillivray said. It was the verbal equivalent of grabbing someone by the arm to prevent them leaving.

"Tom Campbell?"

"The very same."

"When did he show up?" Karen asked.

"It must have been round about three o'clock. I like to listen to the afternoon play on the radio in the front room. I saw him coming up the path then hanging about waiting for Jenny to get back. I think she'd been down the Welfare—she'd got some packets and tins, one of the hand-outs they picked up there."

"You seem to remember it very clearly."

"I mind it so well because that morning was the last time I ever saw Mick. It stuck in my mind." She poured herself another cup of tea.

"How long did he stay? Tom Campbell, I mean."

Mrs. McGillivray shook her head. "Now there I can't help you. After the play was finished, I went down to the green to catch the bus for Kirkcaldy. I'm not able for it now, but I used to like to go to the big Tesco down by the bus station. I'd get the bus in and a

taxi back. So I don't know how long he stayed." She took a long drink of her tea. "I sometimes wondered, you know."

"Wondered what?"

The old woman looked away. Reached into the pocket of her saggy cardigan and pulled out a packet of Benson & Hedges. Extracted a cigarette and took her time lighting it. "I wondered if he paid Mick off."

"You mean, paid him to leave town?" Karen couldn't hide her incredulity.

"It's not such a daft idea. Like I said, Mick had his pride. He wouldn't have stayed where he thought he wasn't wanted. So if he was set on going anyway, maybe he took Tom Campbell's money."

"Surely he'd have had too much self-respect for that?"

Mrs. McGillivray breathed out a thin stream of smoke. "It would be dirty money either way. Maybe Tom Campbell's money felt a wee bit cleaner than the coal board's? And besides, when he left that morning, it didn't look like he was going any further than the shore, to do his painting. If Tom Campbell paid him, he wouldn't have needed to come back for his clothes or anything, would he?"

"You're sure he didn't come back for his stuff later?"

"I'm sure. Trust me, there's no secrets in this row."

Karen's eyes were on the old woman but her mind was racing. She didn't believe for a minute that Mick Prentice had sold his place in the marital bed to Tom Campbell. But maybe Tom Campbell had wanted to take that place badly enough to come up with a different scenario to dispose of his rival.

So much for picking up a bit of character background. Karen bit back a sigh and said, "I'd like to send a couple of officers round to see you on Monday morning. Maybe you could tell them what you've just told me?"

Mrs. McGillivray perked up. "That would be lovely. I could bake some scones."

Just because she was stuck at Rotheswell like a self-immured Rapunzel didn't mean Bel Richmond could turn her back on the rest of her work. Even if she was deprived of access to Grant, she didn't have to twiddle her thumbs. She'd spent most of the day writing up an interview for a *Guardian* feature. It was almost done, but she needed a bit of distance before the final polish. A visit to the pool house concealed in a nearby stand of pine trees would do the trick, she thought, pulling her swimsuit from her bag. Halfway across the room, the house phone rang.

Susan Charleson's voice was crisp and clear. "Are you busy?"

"I was just going for a swim."

"Sir Broderick has an hour free. He'd like to continue your background briefing."

There was clearly no room for discussion. "Fine." Bel sighed. "Where'll I find him?"

"He'll see you downstairs in the Land Rover. He thought you might like to see where Catriona lived."

She couldn't complain about that. Anything that added colour to the story was well worth her time. "Five minutes," she said.

"Thank you."

Quickly, Bel changed into jeans and her weatherproof jacket, thanking the fashionista gods that stylized construction boots had come into vogue, allowing her to look vaguely as if she was ready for country life. She grabbed her recorder and hurried downstairs. A shiny Land Rover Defender sat outside the front door, engine running. Brodie Grant sat behind the wheel. Even from a distance, she could see his gloved fingers drumming on the steering wheel.

Bel climbed aboard and gave him her best smile. She hadn't seen him since the bizarre interview with the cops the day before. She'd eaten a working lunch alone in her room, and he'd been missing from the dinner table. Judith had said he was at some men-only

charity boxing dinner, and she sounded relieved to be missing it. Their conversation had been anodyne; either Judith herself or the ever-present Susan had steered it sideways whenever it threatened to become in any way revealing. Bel had felt frustrated and exploited.

But now she was alone with him again she could forgive all that. She considered asking him if he really thought he could control Karen Pirie like the lord of the manor in a 1930s crime drama, but thought better of it. Best to use the time to beef up her background on the case. "Thanks for taking me to see Cat's place," she said.

"We won't be able to go inside," he said, releasing the handbrake and setting off round the back of the house and down a track that led through the pine trees. "It's had several sets of tenants since, so you're not really missing anything. So, what did you think of Inspector Pirie?"

There was no clue in his face or voice to what he wanted to hear, so Bel settled on the truth. "I think she's one of those people it's easy to underestimate," she said. "I suspect she's a smart operator."

"She is," Grant said. "I expect you know that she's the reason the former Assistant Chief Constable of this county is serving life in prison. A man who was apparently beyond suspicion. But she was capable of questioning his probity. And once she started, she didn't stop till she'd established beyond doubt that he was a cold-blooded killer. Which is why I want her on this. Back when Catriona died, we were all guilty of thinking along the traditional tramlines. And look where that got us. If we're going to have a second bite at the cherry, I want someone who will think outside the box."

"Makes sense," Bel said.

"So what do you want to talk about next?" he said as they emerged from the trees into a clearing that ended with a high wall and another of the airlock-style gates like the one Bel had entered

by when she first arrived. Clearly nobody got into the grounds of Rotheswell unless they were welcome. Grant slowed enough to let the security guards be sure who was behind the wheel, then accelerated on to the main road.

"What happened next?" she said, switching on her recorder and holding it out between them. "You got the first demand and started working with the police. How did things go after that?"

He stared ahead resolutely, showing no signs of emotion. As they rolled past chequered fields of ripening grain and grazing, the sun slipping in and out of louring grey clouds, his words spilled out in an unsettling flow. It was hard for Bel to keep any kind of professional distance. Living with her nephew Harry had given her insight enough to readily imagine the anguish of a parent in Brodie Grant's situation. That understanding generated enough sympathy to absolve him from almost any criticism. "We waited," he said. "I've never known time to drag like it did then."

Monday, 21st January 1985; Rotheswell Castle

For a man who didn't have the patience to let a pint of Guinness settle, waiting to hear from the Anarchist Covenant of Scotland was exquisite torture. Grant roamed Rotheswell like a pinball, almost literally bouncing off walls and doorways in his efforts to stop himself imploding. There was no sense or logic to his movements, and when he and his wife crossed paths, he could barely find the words to respond to her anxious enquiries.

Mary seemed to be much more in control, and he came close to resenting her for it. She had been to Cat's cottage and reported to both him and Lawson that, apart from an overturned chair in the kitchen, nothing seemed out of place. The sell-by date on the milk had been Sunday, indicating she hadn't been gone more than a few days at most.

The nights were worse than the days. He didn't sleep so much

as collapse when physical exhaustion overcame him. Then he'd wake with a start, disorientated and unrefreshed. As soon as consciousness reasserted itself, he wished he was unconscious again. He knew he was supposed to be behaving normally, but that was beyond him. Susan cancelled all his engagements and he holed up inside the walls at Rotheswell.

By Monday morning, he was as close to a wreck as he'd ever felt. The face he saw in the mirror looked as if it belonged in a prisoner-of-war camp, not a rich man's castle. He didn't even care that those around him could see his vulnerability. All he wanted was for the post to arrive, to bring with it something concrete, something that could liberate him from impotence and give him a task. Even if it was only raising whatever ransom the bastards wanted. If it had been up to him, he would have staked out the sorting office in Kirkcaldy, stopping his postie like an old-fashioned highwayman and demanding his mail. But he accepted the madness of that. Instead he paced to and fro behind the letterbox where the castle's mail would drop to the mat at some point between half past eight and nine o'clock.

Lawson and Rennie were already on the spot. They'd arrived in a plumber's van wearing tradesmen's overalls via the back drive at eight. Now they sat stolidly in the hall, waiting for the post. Mary, stunned with the Valium he'd insisted she take, sat on the bottom step in her pyjamas and dressing gown, arms wrapped around her calves, chin on her knees. Susan moved among them with teas and coffees, her normal composure hiding God alone knew what. Grant certainly had no idea how she had held everything together over the past couple of days.

Lawson's radio crackled an incomprehensible message and moments later there was a rustle and a clatter of the letterbox. The day's bundle of mail cascaded to the floor, Grant falling on it like a starving man on the promise of food. Lawson was almost as fast, grabbing the big manila envelope seconds after Grant's fingers closed on it. "I'll take that," he said.

Grant yanked it from him. "No, you bloody won't. It's addressed to me and you'll see it in good time." He clasped it to his chest and stood up, backing away from Lawson and Rennie.

"OK, OK," Lawson said. "Just take it easy, sir. Why don't you sit down next to your wife?"

To his own surprise Grant did as Lawson suggested, subsiding on the stairs beside Mary. He stared at the envelope, suddenly unwilling to discover what demands were about to be made of him. Then Mary laid her hand on his arm and it felt like an unexpected transfusion of strength. He ripped back the flap and pulled out a thick wad of paper. Unfolding it, he saw that this time there were two copies of the puppeteer poster. Before he could take in the words written inside the box at the foot of each, he spotted the Polaroid. He went to cover it, but Mary was too fast for him, reaching across and grabbing it.

This time, Cat's mouth wasn't taped up. Her expression was angry and defiant. She was bound to a chair with loops of parcel tape, the wall behind her a white blank. A gloved hand held the previous day's *Sunday Mail* in the foreground of the picture.

"Where's Adam?" Mary demanded.

"We have to assume he's there. It's a bit harder to get a baby to pose," Lawson said.

"But there's no proof. He could be dead for all you know." Mary put her hand to her mouth as if trying to push the treacherous words back.

"Don't be silly," Grant said, putting his arm round her and injecting spurious warmth into his voice. "You know what Catriona's like. There's no way she'd be this cooperative if they'd done anything to Adam. She'd be screaming like a banshee and throwing herself to the floor, not sitting there all meek and still." He squeezed her shoulders. "It's going to be all right, Mary."

Lawson waited for a moment, then said, "Can we take a look at the message?"

Grant's eyelids flickered and he nodded. He spread the top

poster open on his knees and read its message, written in the same thick black marker as the previous one.

> WE WANT A MILLION. £200,000 IN USED, NON-SEQUENTIAL £20 NOTES, IN A HOLDALL. THE REST IN UNCUT DIAMONDS. THE HAND-OVER WILL BE ON WEDNESDAY EVENING. WHEN YOU HAND OVER THE RANSOM, YOU WILL GET ONE OF THEM BACK. YOU GET TO CHOOSE WHICH ONE.

"Jesus Christ," Grant said. He passed the poster to Lawson, who had gloved up in anticipation. The second sheet offered no more cheer.

> WHEN WE AUTHENTICATE THE DIAMONDS AND KNOW THE MONEY IS SAFE, WE WILL RELEASE THE OTHER HOSTAGE. REMEMBER, NO POLICE. DO NOT FUCK WITH US. WE KNOW WHAT WE ARE DOING AND WE ARE NOT AFRAID TO SPILL BLOOD FOR THE CAUSE. THE ANARCHIST COV-ENANT OF SCOTLAND.

"What have you done to track these people down?" Grant demanded. "How close are you to finding my family?"

Lawson held a hand up while he studied the second poster. He passed it to Rennie and said, "We're doing everything we can. We've spoken to Special Branch and MI5, but neither of them has any knowledge of an activist group called the Anarchist Covenant of Scotland. We managed to get a fingerprint man and an evidence officer into Catriona's cottage under cover of darkness on Saturday night. So far we've no direct leads from that, but we're working on it. Also, we had an officer posing as a customer asking around to see if anyone knew when Catriona's workshop would be open. We've established that she was definitely working on Wednesday but nobody can confirm they saw any sign of her after that. We've had no reports of anything untoward in the area. No suspicious vehicles or behaviour. We—"

"What you're saying is that you have nothing and you know nothing," Grant interrupted brutally.

Lawson didn't even flinch. "That's often the way in kidnap cases. Unless the snatch happens in a public place, there's little to go on. And where there's a small child involved, it's very easy to control the adult, so you don't even get the sort of struggle that generates forensic evidence. Generally, the handover is the point where we can make real progress."

"But you can't do anything then. Can't you read, man? They're going to hold on to one of them till they're sure we haven't double-crossed them," Grant said.

"Brodie, they're both going to be there at the handover," Mary said. "Look, it says we get to choose one of them."

Grant snorted. "And which one are we going to choose? It's bloody obvious that we'd go for Adam. The most vulnerable one. The one who can't look out for himself. Nobody in their right mind would leave a six-month-old baby with some bunch of anarchist terrorists if they had any choice. They'll bring Adam and leave Catriona behind wherever they're keeping them. That's what I'd do if it was me." He looked to Lawson for confirmation.

The policeman refused to meet his eyes. "That's certainly one possibility," he said. "But whatever they do, we have options. We can try to follow them. We can put a tracking device in the holdall and another among the diamonds."

"And if that doesn't work? What's to stop them coming back for more?" Grant said.

"Nothing. It's entirely possible they'll ask for a second ransom." Lawson looked deeply uncomfortable.

"Then we'll pay," Mary said calmly. "I want my daughter and my grandson back safely. Brodie and I will do whatever it takes to achieve that. Won't we, Brodie?"

Grant felt cornered. He knew what the answer was supposed to be, but he was surprised by his ambivalence. He cleared his throat. "Of course we will, Mary." This time Lawson's eyes locked on his, and Grant understood that he might have given a little too much

away. He needed to remind the cop that he had something at stake too. "And so will Mr. Lawson, Mary. I promise you that."

Lawson folded the posters together and slid them back into the envelope. "We're all a hundred per cent committed to getting Catriona and Adam back safely," he said. "And the first thing on the agenda is that you have to start making arrangements with your bank."

"My bank? You mean, we're giving them the real thing?" Grant felt incredulous. If he'd ever thought about anything like this, he'd assumed the police had a stash of marked counterfeit ready for such contingencies.

"It would be very dangerous at this point to do anything else," Lawson said. He stared at the carpet, the very picture of embarrassment. "I take it you do have the money?"

Saturday, 30th June 2007; Newton of Wemyss

"Cheeky bastard tried to look like he was embarrassed to ask, but I could tell he was actually enjoying putting me on the spot," Grant said, stepping on the gas as they left Coaltown of Wemyss behind them. "Don't get me wrong. Lawson never put a foot out of place in the whole investigation. I've no reason to suspect he was anything other than totally committed to catching the bastards who took Catriona and Adam. But I could tell there was a part of him that was secretly enjoying watching me get my come-uppance."

"Why was that, do you think?"

Grant slowed as a gap appeared in the high wall they were driving alongside. "Envy, pure and simple. Doesn't matter what label you put on it—class warfare, machismo, chip on the shoulder. It comes down to the same thing. There's a lot of people out there who resent what I have." He pulled off the road into a large turnout. The wall angled inwards on both sides, giving way in the middle to tall gates made of a thick lattice of wood painted black, built to resemble a medieval portcullis. Set into the wall on

one side was the frontage of a two-storey house, built from the same blocks of local red sandstone as the wall itself. Net curtains blanked the windows, none of them twitching at the sound of the Land Rover's engine. "And those same people resented Catriona too. It's ironic, isn't it? People assumed Catriona got such a great start in her professional life because of me. They never realized it was in spite of me."

He cut the engine and got out, slamming the door behind him. Bel followed, intrigued by the insights he was giving her, the unwitting as much as the witting. "And you? Is their envy of you ironic too?"

Grant swung round on his heel and glowered at her. "I thought you'd done your research?"

"I have. I know you started out in a miner's row in Kelty. That you built your business from nothing. But a couple of places in the cuttings there's a big hint that your marriage didn't exactly hurt your meteoric rise." Bel knew she was playing with fire here, but if she was going to capitalize on this unique access and parlay it into something career-changing, she needed to get beneath the surface to the material nobody else had even suspected, never mind reached down into.

Grant's heavy brows drew together in a glare, and for a moment she thought she was going to experience the withering blast of his temper. But something shifted in his expression. She could see the effort it took, but he managed a twisted little smile and shrugged. "Yes, Mary's father did have power and influence in areas that were crucial to the building of my business." He spread his arms in a gesture of helplessness. "And yes, marrying her did me nothing but good in a professional sense. But here's the thing, Bel. My Mary was smart enough to know she'd be miserable if she married a man who didn't love her. And that's why she chose me." His smile slowly faded. "I never had a choice in the matter. And I never had a choice when she chose to leave me behind." Abruptly he turned away and strode towards the heavy gates.

Friday, 23rd January 1987; Eilean Dearg

They spent so little time together these days. The thought had plagued Grant at every meal he'd eaten at Rotheswell all week. Breakfast without her. Lunch without her. Dinner without her. There had been guests; business associates, politicians, and of course, Susan. But none of them had been Mary. The time without her had reached critical mass this week. He couldn't go on with this distance between them. He needed her now as much as he ever had. Nothing made Cat's death easier, but Mary made it bearable. And now her absence, today of all days, was entirely unbearable.

She'd left on Monday, saying she needed to be on her own. On the island, she would have the peace she wanted. There were no staff there. It took only twenty minutes to walk round, but a couple of miles out to sea felt a long way from anywhere and anyone. Grant liked to go there for the thinking as much as the fishing. Mary mostly left him to it, joining only him occasionally. He couldn't remember her ever going there alone. But she'd been adamant.

Of course there was no phone line. She had a car phone, but the car would be in the hotel parking lot on Mull, half a mile from the jetty. And besides, there would be no signal for a car phone in the wilderness of the Hebridean chain. He hadn't even heard her voice since she'd said goodbye on Monday.

And now he'd had enough of the silence. Two years to the day since his daughter had died and his grandson had disappeared, Grant did not want to be alone with his pain. He tried not to be too harsh on himself over what had gone wrong, but guilt had still scarred his heart. He sometimes wondered if Mary blamed him too, if that was why she absented herself so often. He had tried to tell her the only people who should carry responsibility for Catriona's death were the men who had kidnapped her, but he could barely convince himself, never mind her.

He'd set off after an early breakfast, phoning ahead to the hotel

to make sure someone would be available to take him over to his island. He'd had to pull off the road a couple of times when the grief clogging his throat threatened to overwhelm him. He'd arrived while there was still a faint smudge of daylight in the sky, but by the time they'd crossed the water, dusk was well advanced. But the path to the lodge was broad and well tended, so he had no fear of straying.

As Grant grew near, he was surprised to see no lights showing. When she was quilting, Mary had an array of lights that would shame a theatre rig. Maybe she wasn't quilting. Maybe she was sitting in the sun room at the back of the house, watching the last threads of light across the western sky. Grant quickened his step, refusing to acknowledge the ragged claws of fear dragging across his chest.

The door wasn't locked. It swung open on oiled hinges. He reached for the light and the hall sprung into sharp relief. "Mary," he called. "It's me." The dead air seemed to absorb his words, preventing them from carrying any distance.

Grant strode down the hall, throwing doors open as he went, calling his wife's name, panic tightening his scalp and making him tearful. Where the hell was she? She wouldn't be outside. Not this late. Not when it was this cold.

He found her in the sun room. But she wasn't watching the sunset. Mary Grant would never watch the sunset again. A scatter of pills and an empty vodka bottle broke the secret of her silence. Her skin was already cool.

Saturday, 30th June 2007; Newton of Wemyss

Bel caught up with Grant by the heavy beams of the gate. Close up, she could see there was a smaller entrance cut into one of the gates, big enough to take a small van or a large car. On the other side was a rutted track leading deep into thick woodland.

"She left a note," he said. "I have it by heart still. 'I'm so sorry,

Brodie. I can't do this any more. You deserve better and I can't get better. I can't bear to see your pain and I can't bear my own. Please, try to love again. I pray you can.'" His face twisted in a bitter smile. "Judith and Alec. That's me doing what she told me. Have you heard of the Iditarod race?"

Startled by the abrupt change in subject, Bel could only stutter, "Yes. In Alaska. Dog sleds."

"One of the biggest hazards they face is something called drum ice. What happens is that the water recedes from under the ice, leaving a thin skin over an air pocket. From above, it looks just like the rest of the ice field. But you put any weight on it and you fall through. And you can't get out because the sides are sheer ice. That's what losing Catriona and Adam and Mary feels like sometimes. I don't know when the ground under my feet is going to stop supporting me." He cleared his throat and pointed at a small wooden barn just visible on the edge of the trees. "That was Catriona's workshop and showroom. It was in better nick back then. When she was open for business, she had a couple of A-boards by the roadside. She'd leave the inner gate ajar, enough for people to walk in and out but not wide enough for cars. There was plenty of room for people to park out here." He waved his hand at the ample space where he'd left the Land Rover. The subject of his first wife was clearly closed. But he'd given her a wonderful gift with the image of the drum ice. Bel knew she could make something remarkable out of that.

She surveyed the scene. "But, theoretically, whoever kidnapped her could have opened the gate wide enough to drive through? Then they'd have been pretty much invisible from the road."

"That was what the police thought initially, but the only tyre tracks they found belonged to Catriona's own car. They must have parked out here, where it's hard standing. Anyone driving past could have seen them. They were taking a hell of a risk."

Bel shrugged. "Yes and no. If they physically had hold of Adam, Cat would have done what she was told."

Grant nodded. "Even a woman as bloody-minded as my daugh-

ter would have put her son first. I've no doubts about that." He turned away. "I still blame myself."

It seemed an extreme reaction, even for a control freak. "How do you mean?" Bel asked.

"I relied too heavily on the police. I should have taken more responsibility for the way things played out. I tried. Just not hard enough."

Wednesday, 23rd January 1985; Rotheswell Castle

"We know what we're doing," Lawson said. He was beginning to sound tetchy, which didn't fill Grant with confidence. "We can end this tonight."

"You should have the area under surveillance," Grant said. "They could already be in place."

"I imagine they know roughly when the mail is delivered," Lawson said. "If they wanted to get the jump on us, they would have dug in before we even got the message with the arrangements. So it makes no odds, really."

Grant stared down at that morning's Polaroid. This time, Cat was lying on her side on a bed, Adam leaning wide-eyed against her. Again, the *Daily Record* provided proof of life. At least, proof of life for the previous day. "Why there?" he said. "It's such an odd place. It's not like you can make a quick getaway."

"Maybe that's why they chose it. If they can't get away quickly, you can't either. They're still going to have one hostage. They can use her as a bargaining chip to make you keep your distance until they can get to their vehicle," Lawson said. He spread out the large-scale map Rennie had brought in. The site for the hand-over was circled in red. "The Lady's Rock. It's about halfway between the old pithead at East Wemyss and the eastern edge of West Wemyss. The nearest points they can drive to are here, at the start of the woods . . ." Lawson tapped the map. "Or here. In the parking lot at West Wemyss. If I was them, I wouldn't choose

West Wemyss. It's further from the main road. It takes a crucial few minutes longer to get on to the grid."

"More options when you do get there, though," Grant pointed out. "Towards Dysart or the Boreland, towards Coaltown, or down the Check Bar Road to the Standing Stone, and then more or less anywhere."

"We'll have all the options covered," Lawson said.

"You can't take any chances," Grant said. "They'll have the ransom. They might sacrifice Cat for the sake of their getaway."

"What do you mean?"

"If I was a kidnapper who had my hands on the ransom and I realized your men were on my tail, I'd throw my hostage out of the car," Grant said, sounding much cooler than he felt. "You'd stop for her because you are civilized. They know that. They can afford to gamble on it."

"We won't take any risks," Lawson said.

Grant threw his hands up in frustration. "That's not the right answer either. You can't play safety first in a situation like this. You have to be willing to take calculated risks. You have to go with the moment. You can't be rigid. You have to be flexible. I didn't get to the top of the tree by not taking any risks."

Lawson gave him a measured stare. "And if I take a risk that I think is necessary, and it backfires? Will you be the one shouting loudest for my head on the block?"

Grant closed his eyes for a moment. "Of course I bloody will," he said. "Now, I've got two lives and a million pounds riding on this. You need to convince me you know what you're doing. Can we run through this again?"

Saturday, 30th June 2007; Newton of Wemyss

"I knew I'd let her down. Right then, I knew it." Grant sighed heavily. "Still, I kept believing that if it all went to hell, someone would come forward. That someone must have seen something."

"It didn't happen." It was a flat statement.

"No. It didn't happen." He turned and looked at Bel. His expression was perplexed. "Nobody ever came forward. Not about the actual kidnap itself. Not about where they were held. Nobody ever gave the police a single piece of credible eye-witness testimony. Oh, there were the usual nutters. And people calling in good faith. But after they were investigated, every single report was dismissed."

"That seems odd," Bel said. "Usually there's something. Even if it's only a falling out among thieves."

"I think so too. The police never seemed to think it was peculiar. But I've always wondered how they managed it without there being a single witness to any of it."

Bel looked pensive. "Maybe there wasn't a falling out among thieves because they weren't thieves."

"What do you mean?"

"I'm not quite sure," she said slowly.

Grant looked frustrated. "That's the trouble with this case." He set off towards the Land Rover. "Nobody's ever been sure about bloody anything. The only thing that's certain is that my daughter is dead."

Sunday, 1st July 2007; East Wemyss

Karen had never had a particularly high opinion of students. It was one reason why she'd opted to join the police straight from school, in spite of her teachers' attempts to persuade her to go to university. She didn't see the point of building up four years of debt when she could be earning good money and doing a proper job. Nothing she'd seen of the lives of her former schoolmates had made her feel she'd made a mistake.

But River Wilde's crew was forcing her to admit that maybe students weren't all self-indulgent slackers. They'd arrived just before eleven; they'd unloaded their gear and set up their tarpau-

lins and floodlights by noon; and they'd organized a pizza run, bolted their food, and begun the difficult but delicate task of shifting tons of rock and rubble by hand. Once they had established a rhythm with picks, trowels, sieves, and brushes, River left them to it and joined Karen where she sat at the cave society's table, feeling pretty much redundant.

"Very impressive," Karen said.

"They don't get out much," River said. "Well, not in a professional sense, anyway. They're raring to go."

"How long do you think it'll take to clear the obstruction?"

River shrugged. "Depends how far back it goes. It's impossible to guess. One of my postgrads has his first degree in earth sciences, and he says that sandstone is notoriously unpredictable when it starts to move. Once we get some clearance up at the top, we can stick a drill probe in. That should give us an idea of how far back it goes. If we hit clear air, we can shove a fibre-optic camera down. Then we'll have a much better sense of what we're dealing with."

"I really appreciate this," Karen said. "I'm taking a bit of a flyer here."

"So I gathered. You want to fill me in? Or is it better if I don't know?"

Karen grinned. "You're doing me the favour. Better you know what the score is." She took River through the key points of her investigation, elaborating where River asked for more detail. "What do you think?" she said at last. "You think I can finesse it?"

River held out a hand, waggling it from side to side to indicate it could go either way. "How smart is your boss?" she asked.

"He's a numpty," Karen said. "All the insight of a shag-pile carpet."

"In that case, you might get lucky."

Before Karen could reply, a familiar shape emerged from the gloom of the cave entry. "Are you lassies not one short?" Phil said, coming into the light and pulling up a chair.

"What're you on about?" Karen said.

"Hubble bubble, toil and trouble," he said. "Trick of the light. Sorry, boss." He thrust out a hand. "You must be Dr. Wilde. I have to say, I thought Karen was a one-off, but apparently I was wrong."

"He means that in a nice way," Karen said, rolling her eyes. "Phil, you have to learn to play nice with strange women. Especially ones who know seventeen different undetectable ways to kill you."

"Excuse me," River said, apparently offended. "I know a hell of a lot more than seventeen ways."

Ice broken, Phil had River explain what her team were hoping to achieve. He listened carefully, and when she had finished, he stared across at the students. They'd already made a visible dent in the top corner where fallen rocks met the roof. "No offence," he said, "but I hope this all turns out to be a waste of time."

"You still hoping Mick Prentice is alive and well and digging holes in Poland, like Iain Maclean suggested?" Karen said, pity withering her tone.

"I'd rather that than find him under those rocks."

"And I'd rather my numbers had come up on the lottery last night," Karen said.

"Nothing wrong with a bit of optimism," River said kindly. She got to her feet. "I'd better do some leading by example. I'll call you if anything comes up."

There was no difficulty in finding two parking places in Jenny Prentice's terraced street. Phil followed Karen up the path, muttering under his breath that the Macaroon was going to throw a fit when he found out about River's big dig.

"It's all under control," Karen said. "Don't worry." The door opened abruptly and Jenny Prentice glared at them. "Good afternoon, Mrs. Prentice. We'd like to have a wee chat with you." Steel in the eyes and the voice.

"Aye well, I don't want to have any kind of chat with you just now. It's not convenient."

"It is for us," Phil said. "Do you want to do it here where the neighbours can tune in? We could come in, if you'd rather do it that way?"

Another figure appeared behind Jenny. Karen couldn't help being pleased when she recognized Misha Gibson. "Who is it, Mum?" she said, then realized. "Inspector Pirie—have you got news?" The hope that sprang into her eyes felt like a reproach.

"Nothing concrete," Karen said. "But you were right. Your dad didn't go to Nottingham with the scabs. Whatever happened to him, it wasn't that."

"So if you've not come with news, why are you here?"

"We've one or two questions we need to ask your mum," Phil said.

"Nothing that can't wait for tomorrow," Jenny said, folding her arms across her chest.

"All the same, no reason not to get them out of the way today," Karen said, smiling at Misha.

"I don't see my daughter that often," Jenny said. "I don't want to waste the time we've got talking to you."

"It won't take long," Karen said. "And it does concern Misha too."

"Come on, Mum. They've come all the way out here, the least we can do is invite them in," Misha said, steering her mother away from her position on the threshold. The look Jenny gave them would have shrivelled smaller souls, but she conceded and swung away from them, back into the front room they'd spoken in last time.

Karen refused the tea Misha offered, barely allowing mother and daughter to settle before she went straight to the point. "When we last spoke, you never mentioned Tom Campbell."

"Why should I?" Jenny couldn't keep the hostility from her voice.

"Because he was here the day your husband disappeared. And not for the first time, either."

"Why shouldn't he be here? He was a friend of the family. He was very generous to us during the strike." Jenny's mouth clamped tight as a mousetrap.

"What are you suggesting, Inspector?" Misha sounded genuinely puzzled.

"I'm not suggesting anything. I'm asking Jenny why she's never mentioned that Campbell was here that day."

"Because it was irrelevant," Jenny said.

"How long was it after Mick disappeared that you and Tom started having a relationship?" The question hung alongside the dust motes that inhabited the air.

"You've got a very nasty mind," Jenny said.

Karen shrugged. "It's a matter of record that he moved in here. That you lived together as a family. That his will left everything to Misha. All I'm asking was how much time elapsed between Mick vanishing and Tom getting his feet under the table."

Jenny flashed an unreadable look at her daughter. "Tom was a good man. You've got no right to come here with your innuendos and slanders. The man wasn't long widowed. His wife was my best pal. He needed friends about him. And he was a deputy, so most of the men didn't want to know."

"I'm not disputing any of that," Karen said. "I'm just trying to get the facts straight. It doesn't help me find Mick, you not telling me the whole story. So how long was it before Tom and you moved from friendship to something more?"

Misha made an impatient noise. "Tell her what she wants to know, Mum. She'll just get it from somebody else otherwise. It's got to be better coming from you than the local sweetie wives."

Jenny stared at her feet, studying battered slippers nearly through at the toes as if the answer was written there and she didn't have the right glasses on. "We were both lonely. We'd both been abandoned, it felt like. And he was good to us, very good to us." There was a long pause, then Misha put out her hand to cover

her mother's clenched fist. "I asked him to my bed six weeks to the day after Mick walked out on us. We'd have starved if it hadn't been for Tom. We were both looking for comfort."

"Nothing wrong with that." The gentle words came, surprisingly, from Phil. "We're not here to make judgements."

Jenny gave the barest of nods. "He moved in with us in May."

"And he was a great stepdad," Misha said. "He couldn't have done a better job if he'd been my real dad. I loved Tom."

"We both did," Jenny said. Karen couldn't help thinking she was trying to convince herself as much as them. She remembered Mrs. McGillivray's contention that Jenny Prentice's heart had only ever belonged to Mick.

"Did you ever wonder whether Tom had anything to do with Mick leaving?"

Jenny's head snapped back, her eyes blazing at Karen. "What the hell is that supposed to mean? You think Tom did something to Mick? You think he did away with Mick?"

"You tell me. Did he?" Karen was as implacable as Jenny was roused.

"You're barking up the wrong tree," Misha said, her voice loud and defiant. "Tom wouldn't hurt a fly."

"I didn't say anything about Campbell causing Mick any harm. I find it extremely interesting that you both leapt to the assumption that that's what I meant," Karen said. Jenny looked baffled, Misha furious. "What I was wondering was whether Mick realized there was some bond between you and Tom. By all accounts, he was a proud man. Maybe he decided it would be best for everybody if he left the field clear for a man you seemed to prefer."

"You're talking pure shite," Jenny hissed. "There was nothing going on between me and Tom back then."

"No? Well, maybe Tom thought there might be if he could take Mick out of the picture. He had plenty of money. Maybe he bought Mick off." It was an outrageous suggestion, she knew. But outrage often precipitated interesting outcomes.

Jenny pulled her hand away from Misha and shifted away from

her daughter. "This is your fault," she shouted at her daughter. "I don't have to listen to this. In my own home, she dares to slander the man that gave you everything. What have you brought down on us, Michelle? What have you done?" Tears spilled down her cheeks as she drew her hand back and slapped Misha hard across the face.

Karen was on her feet and moving. But she wasn't fast enough. Jenny had made it out of the room before anyone could stop her. Stunned, Misha pressed a hand to her scarlet cheek. "Leave her," she shouted. "You've done enough damage for one day." She caught her breath, then collected herself. "I think you should leave," she said.

"I'm sorry things got out of hand," Karen said. "But that's the trouble with taking the lid off the box. You never know what's going to pop out."

Monday, 2nd July 2007; Glenrothes

ACC Simon Lees stared at the piece of paper Karen Pirie had placed in front of him. He'd read it three times and still it made no sense. He knew he was going to have to ask her for an explanation and that somehow he would end up on the back foot. It felt so unfair. First thing on Monday morning, and the sanctuary of his office was already breached. "I'm not entirely clear why we're paying for this—" he checked the paper again, trying to shake the suspicion that Pirie was indulging in a twisted practical joke—'Dr. River Wilde to lead a team of students in a 'forensic dig' in a cave at East Wemyss."

"Because it's going to cost us about a tenth of what the forensic science service would charge us. And I know how you like us to get value for money," Karen said.

Lees thought she knew full well that wasn't what he had meant. "I'm not referring to the budget implications," he said peevishly.

"What I'm trying to understand is why this"—he cast his hands upwards in a gesture of frustration—"this circus is happening at all."

"I thought I was to leave no stone unturned in my investigation of Catriona Maclennan Grant's kidnap," Karen said sweetly.

Was she making fun of him? Or did she really not understand what she'd just said? "I didn't mean that literally, Inspector. What the hell is all this in aid of?" He waved the budget requisition form at her.

"It came to my notice in the course of my inquiries that there had been a somewhat unusual roof fall in one of the Wemyss caves in January 1985. I say unusual, because since the Michael pit closed in 1967, the ground has settled and there have been no other major falls." Karen savoured the look of bafflement on Lees's face. "When I looked into this further, I found out that the fall had been discovered on Thursday, 24th January."

"And?" Lees looked uncomprehending.

"That's the day after Catriona was killed, sir."

"I know that, Inspector. I am familiar with the case. But I still fail to see what a roof fall in an obscure cave has to do with anything." He fiddled with the photograph frame on his desk.

"Well, sir, it's like this." Karen leaned back in her chair. "As far as the locals are concerned, the caves aren't really obscure. Everybody knows about them. Most folk have played in them at least once when they were wee. Now, one of the things that we never found out back then was where Catriona and Adam were being held. We never had any witness reports that tied them in to any particular location. And I got to thinking. That time of year, the caves are pretty well deserted. It's too cold for kids to be playing outside, and there's never enough bright daylight to tempt passers-by beyond the first few feet of any of the caves."

Lees felt himself drawn into her narrative in spite of himself. She didn't deliver reports the way his other officers did. Mostly it drove him slightly crazy, but sometimes, like today, he couldn't

resist the shape of her storytelling. "You're saying the caves could have been a potential hiding place for the kidnappers? Isn't that a bit Enid Blyton?" he said, trying to reassert himself.

"Very popular, Enid Blyton, sir. Maybe she could even be called inspirational. Anyway, the cave in question, the Thane's Cave, has a gated railing along the front to keep people out these days. But back then, there was just a fence across the access passageway. It wasn't meant to be impregnable. The cave society used the Thane's Cave as a kind of club-house. Still do, as a matter of fact. The railing was just there to discourage the casual explorer. So it wouldn't have been difficult for anyone to gain access."

"But they'd have been like rats in a trap if they'd been found," Lees protested.

"Well, that's the thing. We can't be entirely sure about that. There's always been a story that there was a passageway down from Macduff Castle to the cave."

"Oh for heaven's sake, Inspector. Have you been taking drugs? This is insane."

"With respect, sir. It makes a kind of sense. We know the kidnappers escaped from the scene in a boat. Police witnesses said at the time it sounded like a small outboard. But by the time they scrambled the chopper and got the spotlight on the sweep, there was no sign of any wee boats anywhere in range of the Lady's Rock. Now, it was a high tide that night. What if they just shot a couple of miles up the shore and hid the boat in the cave? They'd have got an inflatable in, no bother. They dump it with the rest of their make-shift camp, then get out, bringing the roof down behind them."

Lees shook his head. "It sounds like a cross between *The Dangerous Book for Boys* and *Die Hard*. How exactly do you think they went about"—he paused to do that thing with two fingers that indicated quotation marks and also for some reason irritated his wife out of all proportion to the offence—"bringing the roof down behind them?"

Karen smiled far too brightly for his liking. "I've no idea, sir. Hopefully Dr. Wilde's team will be able to tell us. I'm pretty sure we'll find something behind that rock fall that will justify all this expense."

Lees held his head in his hands. "I think you've lost your mind, Inspector."

"Never mind," she said, getting to her feet. "It's the Brodie Grant case. You can spend pretty much what you like, sir. This is one time when nobody's going to question the budget."

Lees could feel the blood pounding in his ears. "Are you taking the piss?"

Immediately he regretted swearing, not least because she looked as if she thought it was definitely an improvement. "No, sir," Karen said soberly. "I'm taking this case very seriously."

"You've got a funny way of showing it." Lees slammed his palms down on the desk. "I want to see some proper police work here, not a day trip to Kirrin Island. It's time you did some digging into the past. It's time you went to talk to Lawson." That would teach her who was boss.

But somehow she'd already defused his little bomb. "I'm glad you think so, sir. I've arranged an appointment for"—she consulted her watch—"three hours' time. So if you don't mind, I'm away to put the pedal to the metal and head for the Blue Toon."

"Pardon?" Why could these Fifers not speak plain English?

Karen sighed. "I've to drive to Peterhead." She headed for the door. "I keep forgetting you're not from here." She cast a quick look over her shoulder. "You don't really get us, do you, sir?"

But before he could respond, she was gone, the door left wide open. Like a barn door behind a cow, he thought bitterly, getting up to slam it shut. What had he done to deserve this bloody woman? And how the hell was he going to come out of the Brodie Grant case smelling of roses when he was forced to rely on the investigative skills of a woman who thought it might be interesting to dig up a bloody cave?

With a sense of relief, Bel Richmond turned off the SS2, the treacherous divided highway that corkscrews down Tuscany from Florence to Siena. As usual, the Italian drivers had scared the living shit out of her, driving too fast and too close, wing mirrors almost touching as they shot past her in tight bends that seemed to make the narrow lanes even smaller. The fact that she was in a hire car only magnified the unpleasantness. Bel thought of herself as a pretty good driver, but Italy never failed to shred her nerves. And thanks to this latest assignment, she was feeling sufficiently shredded, thank you very much.

On Sunday evening, she'd eaten dinner off a tray in her room. Her choice; she'd been invited to join the Grants in the dining room, but she'd pleaded the demands of work. The reality was more prosaic, but its selfishness made it impossible to admit to. In truth, Bel craved her own company. She wanted to hang out of the window smoking the red Marlboros Vivianne had nagged her into supposedly giving up months before. She wanted to watch some crap TV, and she wanted to gossip on the phone with any of the women friends whose connection made her feel better. She wanted to run away home and play some shoot-'em-up Playstation game with Harry. It was always the same when she found herself living at close quarters with the subjects of her journalism. There was only so much intimacy she could take.

But her pleasure in her own company had been short-lived. She'd barely started watching the first episode of a new U.S. cop show when there was a knock at the door. Bel muted the TV, put down her glass of wine, and got up from the sofa. She opened the door to find Susan Charleson, a thin plastic folder in her hand. "I'm sorry to interrupt," she said. "But I'm afraid this is rather urgent."

Disguising the ill grace she felt, Bel stepped back and waved her in. "Come in," she sighed.

"May I?" Susan gestured to the sofa.

"Make yourself at home." Bel sat down at the opposite end, leaving as much space between them as possible. She hadn't taken to Susan Charleson. Behind the chilly efficiency, there was nothing to latch on to, no glimmer of sisterly warmth to build the conspiracy of friendship on. "How can I help you?"

Susan cocked her head to one side and gave a wry little smile. "You'll have realized that Sir Broderick is given to quick decisions that he expects everyone else to turn into realities."

"That's one way of putting it," Bel said. *Used to getting his own way* might have been a better one. "So what has he decided he needs from me?"

"You're pretty quick off the mark yourself," Susan said. "That's probably why he likes you." She gave Bel a measured look. "He doesn't like many people. When he does, he rewards us very well."

Flattery and bribery, the twisted twins. Thank heavens she'd reached a point in her career where she could feed and clothe herself without having to cave in to their poisoned gifts. "I do things because they interest me. If they don't interest me, I don't do them well, so there's not much point, really."

"Fair enough. He'd like you to go to Italy."

Whatever she'd been expecting, it hadn't been that. "Why?"

"Because he thinks the Italian police have no investment in this case so they won't be working it very hard. If DI Pirie goes out there, or sends one of her team, she'll be hampered by the language and by being an outsider. He thinks you might do better, given that you speak Italian. Not to mention the fact that you're just back from there and presumably have some recent acquaintance with the locals. Not the police, obviously. But the locals who might actually know something of what's been going on at that ruined villa." Susan smiled at her. "If nothing else, you get an expenses-paid trip back to Tuscany."

Bel didn't have to think about it for long. This was probably

the only chance she'd have to get the jump on the police in terms of new information. "How do you know I speak Italian?" She stalled, not wanting to look too much of a pushover.

A wintry smile. "It's not just journalists who know how to do research."

I asked for that. "When does he want me to go?"

Susan held out the folder. "There's a flight to Pisa at six tomorrow morning. You're booked on it, and there's a hire car arranged at the airport. I didn't reserve accommodation—I thought you'd rather sort out your own. You will, of course, be reimbursed."

Bel was taken aback. "Six in the morning?"

"It's the only direct flight. I've checked you in. You'll be driven to the airport. It only takes forty minutes at that time of the morning—"

"Yes, fine," Bel said impatiently. "You were very sure I'd agree."

Susan put the folder on the sofa between them and stood up. "It was a pretty safe bet."

So here she was, bouncing down a dirt road in the Val d'Elsa past fields of sunflowers just bursting into dramatic flower, the hot beat of excitement pulsing in her throat. She didn't know if Brodie Grant's name would open doors in Italy as easily as it did in Scotland, but she had a sneaking suspicion that he'd know exactly how to manipulate the bone-deep corruption that underpinned everything here. There was nothing in Italy these days that couldn't be reduced to a transaction.

Except friendship, of course. And thanks to that, at least she had a roof over her head. The villa, of course, was out of the question. Not because of the cost—she was pretty sure she could have made Brodie Grant spring for it—but because it was high season in Tuscany. But she'd been lucky. Grazia and Maurizio had converted one of their old barns into holiday apartments, and the smallest, a studio with a tiny terrace, had been available. When she'd called from the airport, Grazia had tried to offer it to her for free. It had taken Bel almost ten minutes to explain that someone

was paying her expenses so Grazia should overcharge as much as she liked.

Bel turned off the track on to a narrower rutted lane that wound up through a forest of oaks and chestnuts. After a mile or so, she emerged on a small plateau with an olive grove and a field of maize. At the far end was a tight cluster of houses beyond a hand-painted sign that read BOSCOLATA. Bel negotiated the tight turns and carried on, back into the trees. As she rounded the second bend after Boscolata, she slowed and peered through the undergrowth at the ruined villa where this trail had started. There was nothing to show it was of any interest, other than a piece of red-and-white tape tied half-heartedly to the gate. So much for the Italian police investigation.

Another five minutes' tortuous driving and Bel pulled into Grazia's farmyard. A tan hound with droopy ears and a pink nose danced at the end of his chain, barking with all the bravado of a dog who knows nobody is going to come close enough to bite. Before Bel could open her door, Grazia appeared on the steps leading down from the loggia, wiping her hands on her apron, her face crinkling in a broad smile.

Extravagant greetings and the settling of Bel into the beautifully appointed studio took half an hour and had the advantage of helping Bel recover the rhythms of the language. Then the two women settled down with a cup of coffee in Grazia's dim kitchen, the thick stone walls keeping the summer heat at bay as they had done for hundreds of years. "And now, you have to tell me why you are back so soon," Grazia said. "You said it was something to do with work?"

"Sort of," Bel said, wrestling her Italian back into shape. "Tell me, have you noticed anything going on down at the ruined villa recently?"

Grazia gave her a suspicious look. "How do you know about that? The carabinieri were there on Friday. They took a look around, then they went to talk to the people in Boscolata. But what has this got to do with you?"

"When we were here on holiday, I went exploring in the old villa. I found something there that connects to an unsolved crime back in England. A case from twenty years ago."

"What kind of crime?" Grazia looked anxious. The swollen joints of her hands moved restlessly on the table.

"A woman and her baby son were kidnapped. But something went wrong when the ransom was being handed over. The woman was killed, and they never found out what happened to the child." Bel spread her hands and shrugged. Somehow, such gestures came more naturally when she was speaking Italian.

"And you found something here connected to that?"

"Yes. The kidnappers called themselves anarchists and they delivered their demands in the form of a poster. I found a poster just like it down at the villa."

Grazia shook her head in amazement. "The world is getting smaller and smaller. So when did you go to the carabinieri?"

"I didn't. I didn't think they'd believe me. Or if they did, they wouldn't be interested in something that happened back in the UK twenty-odd years ago. I waited till I got home, then I went to the woman's father. He's a very rich man, a powerful man. The sort of person who makes things happen."

Grazia gave a grim little laugh. "It would take a man like that to make the carabinieri get off their backsides and come all the way out here from Siena. That explains why they were so interested in who had been living in the villa."

"Yes. I thought it looked as if squatters had been living there."

Grazia nodded. "The villa belonged to Paolo Totti. He died, maybe a dozen years ago. A silly man, very vain. He'd spent all his money buying a big house to impress everyone, but he didn't have enough left over to look after the place like it deserved. And then he died without a will. His family have been fighting over the villa ever since. It drags on through the courts and every year the villa falls down a little bit more. Nobody from the family does anything to repair it in case they end up with nothing to show. They stopped coming near it years ago. So sometimes people

move in for a while. They stay for summer then they go. The last
lot, they stayed longer." Grazia finished her coffee and stood up.
"All I know is gossip, but we'll go down to Boscolata and talk to
my friends there. They'll tell you a damn sight more than they told
those bossy carabinieri."

Peterhead, Scotland

Karen studied James Lawson as he approached. No more ramrod
bearing, head high and back straight. His shoulders were slumped,
his steps small and tight. Three years in jail had put ten years on
him. He lowered himself into the chair across the table from her,
fidgeting and fussing till at last he settled. A small attempt at con-
trolling some part of the interview, she thought.

Then he looked up. He still had the flat, hard cop stare, his eyes
burning, his face stony. "Karen," he said, acknowledging her with
a tiny nod. His lips, pale and bluish, compressed in a tight line.

She couldn't see any point in small talk. There was nothing to
be said that wouldn't lead straight to recrimination and bitterness.
"I need your help," she said.

Lawson's mouth relaxed into a sneer. "Who do you think you
are? Clarice Starling? You'd need to lose a few pounds before you
could give Jodie Foster a run for her money."

Karen reminded herself that Lawson had attended the same interrogation courses she had. He knew all about probing for your
opponent's weaknesses. But then, so did she. "It might be worth
going on a diet for Hannibal Lecter," she said. "But not for a dis-
graced cop that's pulled his last trout out of Loch Leven."

Lawson raised his eyebrows. "Did they send you on a smartarse course before you took your inspector's exam? If you're supposed to be buttering me up, you're not exactly going about it the
right way."

Karen gave a resigned shake of the head. "I haven't got the time
or the energy for this. I'm not here to flatter your ego. We both

know how these things play out. You help me, your life inside these four walls gets a wee bit less horrible for a while. You walk away from me, who knows what shitty little stunt is going to make your life that wee bit more miserable? Up to you, Jimmy."

"It's Mr. Lawson to you."

She shook her head. "That would imply more respect than you deserve. And you know it." Her point made, she'd refrain from calling him anything. She could hear him breathing hard through his nose, a faint wheeze at the end of each exhalation.

"You think you could make my life any more miserable?" He glared at her. "You have no bloody idea. They keep me in isolation because I'm an ex-cop. You're the first visitor I've had this year. I'm too old and too ugly to interest anybody else. I don't smoke and I don't need any more phone cards." He gave a faint snort of laughter, phlegm bubbling in his throat. "How much worse do you think you can make it?"

She stared back at him, unflinching. She knew what he'd done and there was no place for pity or compassion in her heart for him. She didn't care if they spat in his food. Or worse. He had betrayed her and everyone else who had worked with him. Most of the cops Karen knew were in the job for decent motives. They made sacrifices for the job, they cared that it was done properly. Discovering that a man whose orders they'd followed without flinching was a triple killer had shattered morale in the CID. The fractures were still healing. Some people still blamed Karen, arguing that it would have been better to let sleeping dogs lie. She didn't know how they could sleep at night.

"They tell me you use the library a lot," she said. His eyes flinched. She knew she had him. "It's important to keep your mind active, isn't it? Otherwise you really do go stir crazy. I hear you can download books and music on a wee MP3 player from the library these days. Listen any time you've a mind to."

He looked away, folding and unfolding his fingers. "You still on cold cases?" The concession of the words seemed to take energy he could ill spare.

"It's my department now. Robin Maclennan retired." Karen kept her voice neutral and her face impassive.

Lawson looked over her shoulder at the blank wall behind her. "I was a good cop. I didn't leave many loose ends for you carrion crows to pick over," he said.

Karen gave him a measured stare. He'd killed three people and tried to frame a vulnerable man for two murders, and yet he still thought of himself as a good cop. The capacity of criminals for self-delusion never ceased to amaze her. She wondered that he could sit there with a straight face after the laws he'd broken, the lies he'd told, and the lives he'd shattered. "You cleared a lot of cases," was the best she could manage. "But I've got what looks like new evidence on one that's still open."

Lawson's expression didn't change, but she sensed a flicker of interest as he shifted slightly in his chair. "Catriona Maclennan Grant," he said, allowing himself a self-satisfied smirk. "For you to come yourself, it has to be murder. And that's the only unsolved murder where I was SIO."

"Nothing wrong with your deductive skills," Karen said.

"So, what? You finally got something to nail the bastard, after all this time?"

"What bastard?"

"The ex-boyfriend, of course . . ." Lawson's grey skin furrowed as he dredged his memory for details. "Fergus Sinclair. Game-keeper. She'd given him the push, wouldn't let him be a father to his kid."

"You think Fergus Sinclair kidnapped her and the baby? Why would he do that?"

"To get his hands on his kid and enough money to keep the pair of them in high style," Lawson said, as if he were instructing a small child in the obvious. "Then he killed her during the hand-over so she couldn't finger him. We all knew he'd done it, we just couldn't prove it."

Karen leaned forward. "There's nothing about that in the file," she said.

"Of course there isn't." Lawson made a derisive noise in his throat. "Christ, Karen, do you think we were stupid back then?"

"You didn't have to disclose everything to the defence in 1985," she pointed out. "No operational reason why you shouldn't have left a wee pointer for anyone coming after you."

"All the same, we didn't put anything on paper that we couldn't back up with solid evidence."

"Fair enough. But there's nothing in the file to suggest you even looked at him. No interview notes or tapes, no statements. The only mention in the file is in a statement from Lady Grant saying she believed Sinclair was the father of Catriona's son but that her daughter had always refused to confirm that."

Lawson looked away. "Brodie Maclennan Grant's a powerful man. We all agreed, right up to chief constable level. Nothing went in the file that we couldn't back up a hundred and ten per cent." He cleared his throat. "Even though we thought Sinclair was the obvious suspect, we didn't want to sign his death warrant."

Karen's mouth opened and closed. Her eyes widened. "You thought Brodie Grant would have Sinclair killed?"

"You didn't see the pain he was in after Cat died. I wouldn't have put it past him." His mouth snapped shut and he glared at her defiantly.

She'd thought Brodie Grant was a harsh, driven man. But it had never crossed Karen's mind to consider him a potential commissioner of death. "You were wrong about that," she said. "Sinclair was always safe. Grant doesn't think he had it in him."

Lawson snorted. "He might be saying that now. But at the time, you could feel the hatred coming off him for that lad."

"And you looked close at Sinclair?"

Lawson nodded. "He seemed promising. He had no alibi. He was working abroad. Austria, I think it was. Estate management, that's his line." He frowned again, scratching his clean-shaven chin. He started speaking slowly, speeding up as memory took shape. "We sent a team over to talk to him. They didn't find anything that let him off the hook. He'd been off work on holiday for

the crucial time—kidnap, ransom notes, handover, getaway. And
the guy we consulted at the art school said the poster was in the
German Expressionist style, which kind of tied in with where he
was living."

He shrugged. "But Sinclair said he'd been on a skiing holiday.
Moving from one resort to the next. Sleeping in his Land Rover to
save money. He had ski-lift passes for all the relevant dates, paid
for with cash. We couldn't prove he hadn't been where he said
he'd been. More to the point, we couldn't prove he was where we
thought he was. It was the only real lead we had, and it took us
nowhere."

Monday, 21st January 1985; Kirkcaldy

Lawson flicked through the folder again, as if he might find some-
thing he'd missed on his previous pass. It was still painfully thin.
Without raising his head, he called across the office to DC Pete
Rennie. "Has nothing come in from the crime scene lads yet?"

"I just spoke to them. They're working as fast as they can, but
they're not optimistic. They said it looks like they're dealing with
people who are smart enough not to leave traces." Rennie sounded
both apologetic and anxious, as if he knew this was somehow
going to be his fault.

"Useless wankers," Lawson muttered. After the initial flash of
excitement provoked by the second note from the kidnappers, it
had been a day of mounting frustration. He'd had to accompany
Grant to the bank, where they'd had a difficult meeting with a
senior official who had mounted his high horse, announcing the
bank had a policy of non-cooperation with kidnappers. And that
was without either of them saying a word about the reason for
Grant's request. They'd ended up having to speak to a director of
the bank before they'd made any headway.

Then Grant had taken him to some fancy gentleman's club in
Edinburgh and sat him down with a large whisky in spite of his

protestations about being on duty. When the waiter put his drink in front of him, he ignored it and waited for Grant to say what was on his mind. This was one investigation where Lawson knew better than to appear to be in the driving seat.

"I've got kidnap insurance, you know," Grant had said without preamble.

Lawson had wanted to ask how that worked, but he didn't want to look like some provincial numpty who didn't know what he was doing. "Have you spoken to them?"

"Not so far." Grant swirled the malt round inside the crystal tumbler. The heavy phenolic smell of the whisky rose in a miasma that made Lawson feel faintly sick.

"Can I ask why not?"

Grant took out a cigar and started the fiddly process of trimming and lighting. "You know how it is. They'll want to come in mob-handed. The price of the ransom will be them running the show."

"Is that a problem?" Lawson was feeling a little out of his depth. He took a sip of whisky and nearly spat it out. It tasted like the kind of cough medicine his grandmother had sworn by. It didn't seem to belong to the same family as the dram of Famous Grouse he enjoyed by his own fireside.

"I'm worried it will get out of hand. They've got two hostages. If they get so much as a sniff that we've set them up, who knows what they're capable of?" He lit the cigar and screwed up his eyes to peer at Lawson through the smoke. "What I need to know is whether you're confident you can bring this to a successful conclusion. Do I need to take a chance on outsiders? Or can you get my daughter and grandson back to me?"

Lawson tasted the sweet, cloying smoke in his throat. "I believe I can," he said, wondering if his own career was about to go the same way as the cigar.

And that was how they'd left it. So here he was now, still at his desk while the evening crept inexorably towards night. Nothing was happening, except that his words seemed more and more fool-

hardy. He glared at Rennie. "Have you managed to track down Fergus Sinclair yet?"

Rennie's shoulders hunched and he squirmed in his seat. "Yes and no," he said. "I found out where he's working and I spoke to his boss. But he's not around. Sinclair, I mean. He's away on holiday. Skiing, apparently. And nobody knows where."

"Skiing?"

"He went off in his Land Rover with his skiing gear," Rennie said defensively, as if he'd personally packed up Sinclair's stuff.

"So he could be anywhere?"

"I suppose so."

"Including here? In Fife?"

"There's no evidence of that." Rennie's mouth seemed to slip sideways, as if his jaw had just realized it was on thin ice.

"Have you been on to the airlines? Airports? Channel ports? Have you made them go through their passenger lists?"

Rennie looked away. "I'll get on to it right away."

Lawson pinched the bridge of his nose between his thumb and forefinger. "And get on to the passport office. I want to know if Fergus Sinclair has ever applied for a passport for his son."

Monday, 2nd July 2007; Peterhead

"I was always convinced that Sinclair was involved somehow. It's not as if there were that many people who knew her routine well enough to do the snatch," Lawson said, a touch of defensiveness in his voice now.

Karen felt perplexed. "But what about the baby? If he did all that to get his hands on his son, where's Adam now?"

Lawson shrugged. "That's your million-dollar question, isn't it? Maybe Adam didn't survive the shootout. Maybe Sinclair had some woman lined up to take care of the kid for him. If I was you, I'd take a look at his life now. See if there's some lad in it the right age." He sat back, folding his hands in his lap. "So you've

not come up with anything significant? This just a fishing expedition?"

She reached for the rolled-up poster she'd propped against her chair and slipped it free from its elastic band. She let it uncurl facing Lawson. He started to reach for it then stopped, giving her an interrogative look. "Go ahead," she said. "It's a copy."

Lawson carefully unfurled the paper. He studied the stark black-and-white artwork, running a finger over the puppeteer and his marionettes; the skeleton, Death and the goat. "That's the poster the kidnappers used to communicate with Brodie Maclennan Grant." He pointed to the blank area at the bottom of the poster. "There, where you'd paste on the details of the show, that's where the messages would be written." He gave her a look of resignation. "But you know all that already. Where did this come from?"

"It turned up in an abandoned house in Tuscany. The place is falling down, been empty for years. According to the locals, it had been squatted on and off. The last lot cleared out overnight. No warning, no goodbyes. They left a lot of gear behind. Half a dozen of these posters included."

Lawson shook his head. "Pretty meaningless. We've had a few posters like this turn up over the years. Because Sinclair faked it up to look like some anarchist group hitting on Brodie Maclennan Grant, every now and again you'd get wankers using the poster to promote some direct action or festival or whatever. We checked them out every time, and there was never a connection to what happened to Catriona." He waved a hand dismissively.

Karen smiled. "You think I didn't know that? At least that much made it into the files. But this is different. None of the copies that turned up before was exact. There were differences in the detail, the way there would be if you were copying it off old newspaper cuttings. But this one's different. It's exactly the same. Forensics say it's identical. That it came off the same silk screen."

Lawson's eyes brightened, the spark of interest suddenly obvious in him. "You're kidding?"

"They've had all weekend to make their minds up. They say

there's no doubt. But why would you keep the screen all these years? It's the one piece of evidence that ties the kidnappers into the crime."

Lawson smirked. "Maybe they didn't keep the screen. Maybe they just hung on to the posters."

Karen shook her head. "Not according to the document examiner. Neither the paper nor the ink had been developed in 1985. This was produced recently. On the original screen."

"It doesn't make sense."

"Like so many other things about this case," Karen muttered. Without realizing, she had slipped into her historic relationship with the man opposite. She was the junior officer, pricking him into making sense of the scraps she laid at his feet.

Unconsciously, Lawson responded, relaxing into the conversation for the first time. "What other things?" he said. "Once we'd fixed on Sinclair, it all came together."

"I don't see it. Why would Fergus Sinclair kill Cat at the handover?"

"Because she could identify him."

The impatience in his voice stung Karen, reminding her of their present roles. "I understand that. But why kill her then? Why not kill her beforehand? With her alive at the handover, he was setting up a really complicated situation. He had to control Cat and the baby, get his hands on the ransom, then shoot Cat and get away with the baby in the resulting confusion. He couldn't even be sure that he'd kill her. Not in the dark, with everybody milling around. It would have made life a lot simpler for him to have killed her before the ransom handover. Why didn't he kill her earlier?"

"Proof of life," Lawson said with the satisfaction of a man trumping an ace. "Brodie required proof of life before he'd go ahead."

"No, that doesn't fly," Karen said. "The kidnapper still had the bairn. He could use Adam for proof of life. You're not telling me Brodie Grant would refuse to pay the ransom if he didn't have proof of life for Cat too."

"No ... He'd have paid whether Cat was alive or dead."
Lawson frowned. "I hadn't thought of it like that. You're right. It
doesn't make sense."

"Of course, if it wasn't Sinclair, she might not have had to
die." Karen's eyes went dreamy as she tried the idea on for size.
"It might have been a stranger. She might not have been able to
identify him. Maybe it was an accident?"

Lawson cocked his head to one side and gave her a specula-
tive look. Karen felt as if her fitness for purpose were being as-
sessed. He did a little drum roll with his fingers on the edge of
the chipped table. "Sinclair could have been the kidnapper, Karen.
But not necessarily the killer. You see, there's something else that
wasn't in the report."

Wednesday, 23rd January 1985; Newton of Wemyss

The tension was excruciating. The bulk of the Lady's Rock took a
bite out of the starry sky, blocking out the shoreline beyond. The
cold nibbled at Lawson's nose and ears and around the narrow
bracelet between his leather gloves and the cuffs of his sweater.
The air held the acrid tang of coal smoke and salt. The nearby sea
was only a faint rumble and whisper on this windless night. The
waning moon gave just enough light for him to see the taut features of Brodie Maclennan Grant a few yards away, just clear of
the trees that sheltered Lawson himself. One hand held the holdall with the cash, the diamonds, and the tracking transmitters, the
other grasped his wife's elbow tightly. Lawson imagined the pain
radiating from that pincer grip and was glad he wasn't on the re-
ceiving end of it. Mary Maclennan Grant's face was in shadow,
her head bowed. Lawson imagined she was shivering inside her
fur coat, and not from the cold.

What he couldn't see were the half-dozen men he had stationed
among the trees. That was just as well. If he couldn't see them,
neither could the kidnappers. He'd hand-picked them, choosing

the ones he believed to be both clever and brave, two qualities
that coincided less often than he liked to admit. A couple of them
were firearms trained, one with a handgun, the other on top of the
Lady's Rock with an assault rifle, complete with night sights. They
were under orders not to shoot except on his direct order. Lawson
sincerely hoped he was over-reacting by having them there.

He'd managed to pry some other uniforms from their routine
duties guarding the pit heads and the power stations. Their bud-
dies had resented their detachment, all the more since Lawson
hadn't been in a position to explain the reason for their temporary
secondment to his command. These extra officers were stationed
at the rough ground at either end of the wood, the nearest points
to the rendezvous where vehicles could be parked. Between them,
they should be able to prevent a getaway if Lawson and his immediate team bungled the take-down at the handover.

Which was more than a possibility. This was a nightmare of a
set-up. He'd tried to persuade Grant to say no, to insist on another
place for the handover. Anything but a bloody beach in the middle
of the night. He might as well have saved his breath. As far as
Grant was concerned, Lawson and his men were there as a sort of
private security force. He acted as if he was doing them enough of
a favour by inviting them along against the express instructions
of whoever had taken his daughter and grandson. In spite of what
he'd said about the kidnap insurers' team, he didn't seem to appreciate how much could go wrong. It really didn't bear thinking
about.

Lawson snatched a look at the luminous dial of his watch. Three
minutes to go. It was so still, he'd have expected to hear their car
engine in the distance. But acoustics were always unpredictable in
the open. He'd noticed when he walked the path during his earlier
reconnaissance how the looming bulk of the Lady's Rock acted as
a baffle, cutting off the sound of the sea as effectively as a set of
ear protectors. God alone knew how the woodland would distort
the sound of an approaching vehicle.

Then without warning a brilliant burst of white light from the

direction of the rock wiped out his night vision. All Lawson could make out was the mesmerizing circle of light. Without conscious thought, he stepped further back into the trees, afraid his cover was blown.

"Jesus Christ," Brodie Grant yelped, letting go of his wife and taking a couple of steps forward.

"Stay where you are." A disembodied shout from beyond the light. Lawson tried to place the accent, but there was nothing distinctive about it other than its Scottishness.

Lawson could make out Grant's profile, all colour stripped from his skin by the bleaching white light. His lips were stretched back over his teeth in a snarl. Unease squirmed in Lawson's stomach like acid indigestion. How the hell had the kidnappers got into position at the side of the rock without him seeing them? The moonlight had been enough to illuminate the path in both directions. He'd expected a vehicle. They had two hostages, after all. They could hardly march them a mile up the beach from West Wemyss or East Wemyss. The steep cliff behind him ruled out Newton of Wemyss.

The kidnapper shouted again. "OK, let's do it. Just like we said. Mrs. Grant, you walk towards us with the money."

"Not without proof of life," Grant bellowed.

The words were barely out of his mouth when a figure stumbled out in front of the light, a stark marionette that reminded Lawson of the posters the kidnappers had used to deliver their demands. As his eyes adjusted, he could see it was Cat. "It's me, Daddy," she called, her voice hoarse. "Mummy, bring me the money."

"What about Adam?" Grant shouted, grabbing his wife by the shoulder as she reached for the holdall. Mary nearly tripped and fell, but her husband had no eyes for her. "Where's my grandson, you bastards?"

"He's all right. As soon as they have the money and the diamonds, they'll hand him over," Cat shouted, desperation obvious in her voice. "Please, Mummy, bring the money like you're supposed to."

"Damn it," Grant said. He thrust the holdall at his wife. "Go on, do what she says."

This was out of control, Lawson knew it. To hell with the radio silence he'd called for. He reached for his radio and spoke as clearly as he dared. "Tango One and Tango Two. This is Tango Lima. Despatch officers to shore side of Lady's Rock. Do it now. Do not reply. Just deploy. Do it now."

As he spoke, he could see Mary walking uncertainly towards her daughter, shoulders hunched. He estimated there was about thirty-five yards between them. It seemed to him that Mary was covering more of the distance than her daughter. As they came within touching distance, he could see Cat reaching for the bag.

To his surprise, that was the moment Mary chose to cast aside the conditioning of thirty years' marriage to Brodie Grant. Instead of doing what she'd been told, first by the kidnappers' note and second by her husband, Mary clung on to the holdall in spite of Cat's efforts to pull it from her. He could hear Cat's exasperation as she said, "For Christ's sake, Mother, give me the bloody thing. You don't know what you're dealing with here."

"Give her the bloody bag, Mary," Grant yelled. Lawson could hear the man's breath rattling in his chest.

Then the kidnapper's voice rang out again. "Hand it over, Mrs. Grant. Or you won't see Adam again."

Lawson registered the horror on Cat's face as she looked desperately over her shoulder into the light. "No, wait," she shouted. "It's all going to be fine." She seemed to wrest the bag from her mother and take a step backwards.

Suddenly Grant sprang forward half a dozen paces, his hand disappearing inside his overcoat. "Damn it," he said. Then his voice rose. "I want my grandson and I want him now." His hand emerged, the dull sheen of an automatic pistol obvious in the glare of the light. "Nobody move. I've got a gun and I'm not afraid to use it. Bring Adam out now."

Later, Lawson would wonder at the collection of bad clichés that was Brodie Maclennan Grant. But at that moment, all he

could feel was the weight of catastrophe as time seemed to slow. He started to run towards Grant as the businessman raised his arms in a two-handed shooter's stance. But before Lawson could take a second step, the light cut out, leaving him blind and helpless. He saw the flash of a muzzle near him, heard a shot, smelled cordite. Then a replay of the same sequence but this time from a distance. He tripped over a fallen branch and fell headlong. Heard a scream. A child crying. A high-pitched voice repeating, "Fuck." Then realized the voice was his.

A third shot rang out, this time from the woods. Lawson tried to stand, but hot spikes of pain spiralled up through his ankle. He rolled on his side, scrabbling for flashlight and radio. "Hold your fire," he yelled into the radio. "Hold your fire, that's an order." As he spoke, he could see flashlight beams criss-crossing the area as his men swarmed round the base of the rock.

"They've got a fucking boat," he heard somebody shout. Then a roar louder than the waves as the engine caught. Lawson closed his eyes momentarily. What a fiasco. He should have tried harder to make Grant refuse this set-up. It had been doomed from the start. He wondered what they'd managed to get away with. The kid, certainly. The money, probably. The daughter, maybe.

But he was wrong about Catriona Maclennan Grant. Terribly, horribly wrong.

Monday, 2nd July 2007; Peterhead

"Brodie Maclennan Grant had a gun?" Karen's voice rose to a squeak. "He fired a gun? And you kept that out of the report?"

"I had no choice. And it seemed like a good idea at the time," Lawson said with the cynical air of a man quoting his superiors.

"A good idea? Cat Grant died that night. In what sense was it a good idea?" Karen couldn't believe what she was hearing. The idea of such cavalier behaviour was completely alien to her.

Lawson sighed. "The world's changed, Karen. We didn't have a

police complaints commissioner. We didn't have the kind of scrutiny you live with."

"Obviously," she said drily, remembering why he was where he was. "But still. You managed to cover up a civilian discharging a weapon in the middle of a police operation? Money talks, right enough."

Lawson shook his head impatiently. "It wasn't just money talking, Karen. The chief constable was thinking PR as well. Grant's only child was dead. His grandchild was missing. As far as the public was concerned, he was a victim. If we'd prosecuted him for the firearms offences, it would have looked like we were being vindictive—we can't catch the real villains, so we'll have you instead—that sort of thing. The view was that nobody's interests were served by revealing that Grant had been armed."

"Could it have been Grant's shot that killed Cat?" Karen demanded, forearms on the table, head thrusting like a rugby forward's.

Lawson shifted in his chair, leaning his weight to one side. "She was shot in the back. Work it out for yourself."

Karen leaned back in her chair, not liking the answer she came up with but knowing there would be nothing better coming from the man opposite her. "You were a right bunch of fucking cowboys in the old days, weren't you?" There was no admiration in her tone.

"We got the job done," Lawson said. "The public got what they wanted."

"The public didn't know the half of it, by all accounts." She sighed. "So we've got three gunshots, not the two that appear in the report?"

He nodded. "For all the difference it makes." He shifted again, angling his body towards the door.

"Is there anything else I should know about that didn't make it to the case file?" Karen said, reasserting herself as the person in control of the interview.

Lawson tilted his head back, eyeing the corner where the walls

and ceiling met. He exhaled noisily, then pushed his lips out. "I think that's it," he finally said. He dragged his eyes back to meet her weary gaze. "We thought it was Fergus Sinclair at the time. And nothing's happened since to change my mind on that one."

Campora, Tuscany

The warmth of the Tuscan sun melted the stiffness in Bel's shoulders. She was sitting in the shade of a chestnut tree tucked away behind the cluster of houses at the tail end of Boscolata. If she craned her neck, she could see one corner of the terracotta tiled roof of Paolo Totti's ruined villa. Her more immediate view was, however, rather more appealing. On a low table in front of her was a jug of red wine, a bottle of water, and a bowl of figs. Around the table, her primary sources. Giulia, a young woman with a tumbled mane of black hair and skin marked with the angry puce scars of old acne, who made hand-painted toys for the tourist market in a converted pig sty; and Renata, a blond Dutch woman with a complexion the colour of Gouda, who worked part-time in the restoration department of the Pinoteca Nationale in nearby Siena. According to Grazia, who was leaning against the tree trunk shelling a sack of peas, the carabinieri had talked to both of the women already.

The social pleasantries had to be observed, and Bel contained herself while they chatted together. Eventually, Grazia moved them on. "Bel is also interested in what happened at the Totti villa," she said.

Renata nodded portentously. "I always thought someone would come asking about that," she said in perfectly enunciated Italian that sounded like computer-generated speech.

"Why?" Bel asked.

"They left so suddenly. One day they were here, the next day they were gone," Renata said.

"They went without a word," Giulia said, looking sulky. "I

couldn't believe it. Dieter was supposed to be my boyfriend, but he didn't even say goodbye. I was the one who discovered they were gone. I went over to have coffee with Dieter that morning, just like I always did when they weren't setting off early for a show. And the place was deserted. As if they'd thrown everything they could grab hold of into the vans and just taken off. I haven't heard from that bastard Dieter since."

"When was this?" Bel asked.

"At the end of April. We had plans for the Mayday holiday, but that all came to nothing." Giulia was clearly still pissed off.

"How many of them were there?" Bel said. Between them, Giulia and Renata counted them off on their fingers. Dieter, Maria, Rado, Sylvia, Matthias, Peter, Luka, Ursula, and Max. A mongrel mix from all over Europe. A motley crew that seemed on the face of it to have nothing to do with Cat Grant. "What were they doing there?" she asked.

Renata grinned. "I suppose you would say they were borrowing the place. They turned up last spring in two battered old camper vans and a flashy Winnebago and just moved in. They were very friendly, very sociable." She shrugged. "We're all a little alternative here in Boscolata. This place was a ruin back in the seventies when a few of us moved in illegally. Gradually we bought the properties and restored them to what you see now. So we were pretty sympathetic to our new neighbours."

"They became our friends," Giulia said. "The carabinieri are crazy, acting like they're criminals or something."

"So they just turned up without warning? How did they know the house was there?"

"Rado had a job at the cement works down in the valley a couple of years ago. He told me he used to go walking in the woods, and he came across the villa. So when they needed a place that was accessible to the main towns in this part of Tuscany, he remembered the villa and they came to stay," Giulia said.

"So what exactly did they do?" Bel asked, searching for some connection to the past at the heart of her inquiries.

Renata said, "They ran a puppet theatre." She seemed surprised that Bel hadn't known this. "Marionettes. Street theatre. During the tourist season, they had regular pitches. Firenze, Siena, Volterra, San Gimignano, Greve, Certaldo Alto. They did festivals too. Every little town in Tuscany has a festival of something—porcini mushrooms, antique salami-slicing machines, vintage tractors. So BurEst performed anywhere there was an audience."

"BurEst? How do I spell that?" Bel said.

Renata obliged. "It's short for Burattinaio Estemporaneo. They did a lot of improvisation."

"The poster from the villa—a black-and-white drawing of a puppeteer with some pretty strange marionettes—was that what they used for advertising?" Bel asked.

Renata shook her head. "Only for special performances. I only ever saw them use it when they did a performance in Colle Val d'Elsa for All Souls Day. Mostly, they used one with bright colours, sort of commedia dell'arte. A modern twist on the more traditional images of puppetry. It reflected their performance better than the monochrome poster."

"Were they popular?" Bel asked.

"I think they did OK," Giulia said. "They'd been in the south of France the summer before they came here. Dieter said Italy was a better place to work. He said the tourists were more open-minded and the locals were more tolerant of them. They didn't make a huge amount of money, but they did OK. They always had food on the table and plenty of wine. And they made everybody welcome."

"She's right," Renata said. "They weren't scroungers. If they ate dinner at your house, next time you ate with them." One corner of her mouth twitched downwards. "That's less usual than you'd think in these circles. They talk a lot about sharing and communality, but mostly they are even more selfish than the people they despise."

"Except for Ursula and Matthias," Giulia said. "They were more private. They didn't really socialize like the others did."

Renata snorted. "That's because Matthias thought he was in charge." She poured more wine for them all and continued. "It was Matthias who started the company, and he still liked everyone to treat him like he was the circus ringmaster. And Ursula, his woman, she bought into that whole thing. Matthias obviously took the lion's share of the income too. They had the best van, their clothes were always expensive hippy style. I think it was partly a generation thing—Matthias must be in his fifties, but most of the others were much younger. Twenties, early thirties at the very most."

It was all fascinating, but Bel was struggling to see what the link might be to Cat Grant's death and the disappearance of her son. This Matthias character sounded like the only one old enough to have had any connection to those distant events. "Does he have a son, Matthias?" she asked.

Both women looked at each other, puzzled. "There was no child with him," Renata said. "I never heard him speak of a son."

Giulia picked up a fig and bit into it, the purple flesh splitting and spilling seeds down her fingers. "He had a friend who came to visit sometimes. A British guy. He had a son."

Like all good reporters, Bel had an unquantifiable instinct for where the story lay. And that instinct told her she'd just hit gold. "How old was the son?"

Giulia licked her fingers clean while she considered. "Twenty? Maybe a bit older, but not much."

There were a dozen questions butting against each other inside Bel's head, but she knew better than to blurt them out in an urgent stream. She took a slow sip of her wine and said, "What else do you remember about him?"

Giulia shrugged. "I saw him a couple of times, but I only ever met him properly once. His name was Gabriel. He spoke perfect Italian. He said he'd grown up in Italy, he didn't remember ever living in England. He was studying, but I don't know where or what." She made an apologetic face. "Sorry, I wasn't that interested in him."

OK, it wasn't decisive. But it felt like a possibility. "What did he look like?"

Giulia looked more uncertain. "I don't know how to describe him. Tall, light brown hair. Quite good looking." She screwed her face up. "I'm not good at this kind of thing. What's so interesting about him anyway?"

Renata saved Bel from having to answer. "Was he at the New Year party?" she asked.

Giulia's face cleared. "Yes. He was there with his father."

"So he might be on a photograph," Renata said. She turned to Bel. "I had my camera with me. I took dozens of photographs that night. Let me get my laptop." She jumped up and headed back to her house.

"What about Gabriel's dad?" Bel asked. "You said he was British?"

"That's right."

"So how did he know Matthias? Was he British too?"

Giulia looked dubious. "I thought he was German. He and Ursula got together years ago in Germany. But he spoke Italian just like his friend. They sounded the same. So maybe he was British too. I don't know."

"What was Gabriel's dad called?"

Giulia sighed. "I'm not much use to you. I don't remember his name. I'm sorry. He was just another man my dad's age, you know? I was with Dieter, I wasn't interested in some old guy in his fifties."

Bel hid her disappointment. "Do you know what he does for a living? Gabriel's dad, I mean."

Giulia brightened, pleased that she knew the answer to something. "He's a painter. He paints landscapes for the tourists. He sells to a couple of galleries—one in San Gimignano and one in Siena. He also goes to the same sort of festivals that BurEst performed at, and sells his work there."

"Is that how he met Matthias?" Bel asked, trying not to feel

disappointed that the mysterious Gabriel's father was not estate manager Fergus Sinclair. After all, an artist would fit right in with Cat's background. Maybe Adam's father was someone she knew from her student days. Or someone she'd met at a gallery or exhibition back in Scotland. There would be time to explore those possibilities later. Right now, she needed to pay attention to Giulia.

"I don't think so. I think they knew each other from way back."

As she spoke, Renata returned with her laptop. "Are you talking about Matthias and Gabriel's father? It's funny. It didn't seem as though they really liked each other that much. I don't know why I think that, but I do. It was more . . . you know how sometimes you stay in touch with someone because she's the only person left who shares the same past? You might not like them so much, but they give a connection back to something that was important. Sometimes it's family, sometimes it's a time of your life when important things happened. And you want to hold on to that link. That's how it seemed to me when I saw them together." As she spoke, her fingers were flying over the keyboard, summoning up a library of pictures. She placed the laptop where Giulia and Bel could see the screen, then came round behind them, leaning in so she could advance the shots.

It looked like half the parties Bel had ever been to. People sitting at tables drinking. People mugging at the camera. People dancing. People getting more red in the face, more blurred around the eyes, and more uncoordinated as the evening wore on. Both of the Boscolata women giggled and exclaimed, but neither identified either Gabriel or his father.

Bel had almost given up hope when Giulia suddenly called out and pointed at the screen. "There. That's Gabriel in the corner." It wasn't the clearest of shots, but Bel didn't think she was seeing things. There were fifty years between them, but it wasn't hard to discern a resemblance between this boy and Brodie Grant. Cat's features had been a feminine translation of her father's striking

looks. Unlikely though it was, a simulacrum of the original was staring out at her from a New Year party in an Italian squat. The same deep-set eyes, parrot nose, strong chin, and the distinctive thick shock of hair, only blond rather than silver. She dug in her handbag and pulled out a memory stick.

"Can I get a copy of that?" she asked.

Renata paused, looking thoughtful. "You didn't answer when Giulia asked why you are interested in this boy. Maybe you should answer now."

East Wemyss, Fife

River stripped off her heavy-duty gloves and straightened her back, trying not to groan. The trouble with working alongside her students was that she couldn't reveal any signs of weakness. Admittedly, they were a dozen or more years younger than her, but River was determined to demonstrate she was at least as fit as they were. So they might complain of aching arms and sore backs from shifting rock and rubble, but she had to maintain her Superwoman act. She suspected the only person she was kidding was herself, but that made no odds. The deception had to be maintained for the sake of her self-image.

She walked across the cave to where three of the students were sifting the dirt freed by the shifting of the rocks. So far, nothing of archaeological or forensic interest had turned up, but their enthusiasm seemed undiminished. River could remember her own earliest investigations; how the very fact of being involved in a real case was excitement enough to overcome the tedium of a repetitive, apparently fruitless task. She saw her own reactions mirrored in these students, and it made her happy to think that she had some responsibility for making sure the next generation of forensic investigators would bring that same commitment to speaking for the dead.

"Anything?" she said as she emerged from the shadows into the brilliance that illuminated their huddle.

Heads were shaken, negatives muttered. One of the archaeology postgrads looked up. "It'll get interesting when the labourers have finished clearing the rocks."

River grinned. "Don't let my anthropologists hear you calling them labourers." She glanced back at them affectionately. "With luck, they should have the bulk of the rocks out of the way by the end of the afternoon." They'd all been surprised by the discovery that the rock fall was only a few feet deep. In River's experience, cave falls tended to extend a long way back. A fault had to grow to a substantial size before it reached critical mass and brought a previously stable roof down. So when it collapsed, it took a lot of rock with it. But this was different. And that made it very interesting indeed.

Already, they'd removed the top seven or eight feet all the way back. A couple of the more intrepid among them had climbed up for a look when River had been off fetching lunchtime pies and sandwiches for everyone. They'd reported that it looked clear beyond the fall itself, apart from a few boulders that had rolled down from the main pile.

River walked outside to make a couple of phone calls, appreciating the salt air as a bonus. She'd barely finished speaking to her department secretary when one of the students burst out of the narrow entrance.

"Dr. Wilde," he shouted. "You need to come and see this."

Campora, Tuscany

Bel had pitched her story to provoke the maximum emotional response. From the stunned silence of Renata and Giulia, it looked as if she'd achieved her goal.

"That's so sad. I'd have been in bits if that had happened in my

family," Giulia finally said, taking ownership of it in the style of a woman raised on soap operas and celebrity magazines. "That poor baby boy."

Renata was more objective. "And you think Gabriel might be that boy?"

Bel shrugged. "I have no idea. But that poster is the first definite lead there has been in over twenty years. And Gabriel looks incredibly like the missing boy's grandfather. It might be wishful thinking, but I wonder if we're on to something here."

Renata nodded. "So we must help in every way we can."

"I'm not talking to the carabinieri again," Giulia said. "Pigs."

"Hey," Grazia complained, rousing herself from her pea-shelling. "Don't you be insulting pigs. Our pigs are wonderful creatures. Intelligent. Useful. Not like the carabinieri."

Renata held her hand out. "Give me the memory stick. There's no point in talking to the carabinieri because they don't care about this case. Not like you do. Not like the family does. That's why we need to share everything with you." Expertly, she copied the picture on to Bel's memory stick. "Now we need to see if there are any more photos of Gabriel and his father."

By the end of the search, they had three shots where Gabriel appeared, though none of them was any clearer than the first. Renata had also found two images of his father—one in profile, one where half of his face was obscured by someone else's head. "Do you think anybody else has photos from that night?" Bel asked.

Both women looked dubious. "I don't remember anyone else taking pictures," Renata said. "But with mobile phones, who knows? I'll ask around."

"Thanks. And it would be helpful if you could ask if anyone else knew Gabriel or his father." Bel took the precious memory stick. As soon as she had the chance, she would send it to a colleague who specialized in enhancing dodgy photos of the great and the good doing things they shouldn't be doing with people they shouldn't be doing them with.

"I have a better idea," Grazia said. "Why don't I bring a pig down tonight? We can put it on the spit and you can meet everyone else. A nice bit of pork and a few glasses of wine and they'll be ready to tell you all they know about this Gabriel and his father."

Renata grinned and raised her glass in a toast. "I'll drink to that. But I'm warning you, Grazia, your pig might roast in vain. I don't think this guy was very sociable. I don't remember him joining in much with the party."

Grazia gathered her peapods together and stuffed them in a plastic bag. "Never mind. It's a good excuse to have some fun with my neighbours. Bel, are you staying down here or do you want a ride back up the hill?"

Now she had the prospect of gossiping with the whole community, Bel felt less urgency. "I'll come back now, and see you girls later," she said, draining her wine.

"Don't you want to know about the blood?" Giulia asked.

Caught halfway out of her chair, Bel nearly fell over. "The blood on the floor, you mean?" she said.

"Oh. You know about it." Giulia sounded disappointed.

"I know there's a bloodstain on the floor of the kitchen," Bel said. "But that's all I know."

"We went and had a look after the carabinieri left on Friday," Giulia said. "And the bloodstain was different from when I first saw it. The day after they left."

"Different how?"

"It's all brown and rusty now and soaked into the stone. But back then, it was still quite red and shiny. Like it was fresh."

"And you didn't call the cops?" Bel tried not to show her disbelief.

"It wasn't up to us," Renata said. "If the BurEst people had thought it was a police matter, they'd have made the call." She shrugged. "I know it seems strange to you, and if it had happened in Holland, I don't know that I would have done nothing. But things are different here. Nobody on the left trusts them. You saw how the Italian police reacted at the G8 in Genoa, the way they

treated the protesters. Giulia asked a few of us whether she should call the police and we all agreed that the only thing that would achieve would be to give the cops an excuse to blame the puppeteers, no matter what happened."

"So you just blanked it?"

Renata shrugged. "It was in the kitchen. Who's to say it wasn't animal blood? It was none of our business."

Kirkcaldy

Karen crawled along the street, checking the house numbers. This was the first time she'd visited Phil Parhatka's new house in the centre of Kirkcaldy. He'd moved in three months before; kept promising a housewarming party but so far he hadn't delivered. Once upon a time, Karen had harboured dreams of them buying a place together some day. But she'd got past that. A guy like Phil was never going to be drawn to a dumpy wee thing like her, especially once her latest promotion set her in authority over him. Some men might like the idea of screwing the boss. Karen knew instinctively that wasn't part of Phil's fantasy life. So she'd chosen the maintenance of their friendship and close working relationship over what she classified as adolescent hankerings. If she was going to have to settle for being a career-driven spinster, she could at least make sure the career was as satisfying as it could be.

Part of the recipe for that job satisfaction was having someone to bounce ideas off. No individual detective was smart enough to see the whole picture of a complex investigation. Everyone needed a sounding board who saw things differently and was smart enough to articulate those differences. It was especially important in cold cases where, instead of leading a substantial team of officers, an SIO might have only one or two bodies at her disposal. And those foot soldiers usually didn't have the experience to make their input as valuable as she wanted. For Karen, Phil ticked all

the boxes. And judging by the number of times he ran his cases past her, it was a two-way street.

Usually, they put their heads together in her office or in a quiet corner of a pub halfway between her house and his. But when she'd rung him on her way back from Peterhead, he'd already had a couple of glasses of wine. "I'm probably legal, but only just," he'd said. "Why don't you come round to my place? You can help me choose my living-room curtains."

Karen spotted the house number she was looking for and parked across Phil's driveway. She sat for a moment, wedded to the cop habit of checking out her surroundings before committing to leaving her vehicle. It was a quiet, unassuming street of stone semi-detached houses, square and solid, apparently as sound as when they'd first been built at the tail end of the nineteenth century. Gravel driveways and neat flowerbeds. Drawn curtains upstairs where children slept, shut off from the persistent daylight by heavy liners. She remembered how hard it had been to fall asleep on light summer evenings as a child. But her bedroom curtains had been thin. And her street had been noisy with music and conversations from the pub on the corner. Not like this. It was hard to believe the town centre was five minutes' walk away. It felt like the distant suburbs.

Alerted by the sound of her car, Phil had the door open before Karen was out of the driver's seat. Silhouetted against the light, he looked bigger. His pose contained the casual threat of the doorkeeper; one arm raised to lean on the door jamb, one leg crossed over the other, head cocked. But there was nothing threatening in his expression. His round, dark eyes twinkled in the light, and his smile crinkled his cheeks into creases. "Come away in," he greeted her, stepping back and gesturing for her to enter.

She stepped on to a perfect replica of a traditional Victorian tiled hallway, terracotta squares broken up with lozenges of white, blue, and claret. "Very nice," she said, noting the dado rail and the Lincrusta beneath it.

"My brother's girlfriend's an architectural historian. She's

been through the place like a dose of salts. It's going to look like a bloody National Trust property before she's finished," he grumbled good-naturedly. "Turn right at the end of the hall."

Karen burst out laughing as she entered the room. "Christ, Phil," she giggled. "It was Colonel Mustard, in the Library, with the Lead Piping. You should be wearing a smoking jacket, not a Raith Rovers shirt."

He gave a rueful shrug. "You've got to see the funny side. Me, a cop, with the perfect body-in-the-library scenario." He waved a hand at the dark wooden bookshelves, the leather-topped desk, and the club chairs that flanked the elaborate fireplace. The room clearly hadn't been big to start with, but now it felt positively overstuffed. "She says this is what the master of the house would have had."

"In a house this size?" Karen said. "I think she's got delusions of grandeur. And somehow, I don't think he'd have gone for the tartan carpet."

The pink of embarrassment flushed his ears. "Apparently that's post-modern irony." He raised his eyebrows sceptically. "It's not all it seems, though," he said, brightening as he fiddled with one of the books. A section of shelving swung open to reveal a plasma screen TV.

"Thank God for that," Karen said. "I was beginning to wonder. Not much like the old place, is it?"

"I think I've outgrown the boy racer style of living," Phil said.

"Time to settle down?"

He shrugged, not meeting her eyes. "Maybe." He pointed to a chair and dropped into the one opposite. "So how was Lawson?"

"A changed man. And not in a good way. I've been thinking about it, driving back. He was always a tough bastard, but right up until we found out what he'd really been up to, I felt his motives were the right ones, you know? But the stuff he told me today . . . I don't know. It almost felt as if he was taking his chance to get his own back."

"What do you mean? What did he tell you?"

Karen held a hand up. "I'll get to that in a minute. I just want to let off steam, I suppose. I felt like he said what he did out of malice. Because he knew it would damage the reputation of the force, not because he wants to help us solve what happened to Cat and Adam Grant."

As she spoke, Phil reached for his pack of cigarillos and lit up. He hardly smoked in her company these days, she realized. There were so few places it was permissible. The familiar bittersweet aroma filled Karen's nostrils, strangely comforting after the day she'd had. "Does it matter what his motives are?" he said. "As long as what he's telling us is true?"

"Maybe not. And as it turns out, he did have something very interesting to tell us. Something that sheds a whole new light on what happened the night Cat Grant died. Apparently it wasn't just the cops and the kidnappers who were armed that night. Our pillar of society, Sir Broderick Maclennan Grant, had a gun with him. And he used it."

Phil's mouth hung open, smoke leaking into the air. "Grant had a shooter? You're kidding. How come we're only hearing about this now?"

"Lawson says the cover-up came from on high. Grant was a victim, nobody would be served by charging him. Bad PR, all that shit. But I think that decision completely altered the outcome." Karen pulled a file folder from her bag. She took out the drawing of the crime scene made by the forensic team at the time and spread it out between them. She pointed out where everyone had been standing. "Got that?" she asked.

Phil nodded.

"So what happened?" Karen said.

"The light went out, our guy fired high and wide, then there was another shot from behind Cat. The shot that killed her."

Karen shook her head. "Not according to Lawson. What he's saying now is that Cat and her mother were wrestling with the bag of money. Cat managed to get the bag and started to turn. Then Grant drew his firearm and demanded to see Adam. The

light went out, Grant fired. There was a second shot, from beyond Cat. Then PC Armstrong fired wide."

Phil frowned, digesting what she'd said. "OK," he said slowly. "I don't quite see how that changes things."

"The bullet that killed Cat hit her in the back and exited through her chest. Into the sand. They never found the bullet. The wound wasn't consistent with Armstrong's weapon, so, given that Grant's gun was never mentioned, there was only one possible public explanation. The kidnappers killed Cat. Which made it a murder hunt."

"Oh, fuck," Phil groaned. "And of course, that's what totally puts the kybosh on any possibility of getting Adam back. These guys know they're going down for life, no question now that Cat's dead. They've got a bag of money and the kid. No way are they going to put themselves up for another confrontation with Grant. They're going to melt away into the night. And Adam's just a liability now. He's worthless to them, alive or dead."

"Exactly. And we both know which side of the scales the weight comes down on. But there's more. The argument's always been that the nature of the wound plus the fact that Cat was shot in the back pointed inevitably to the kidnappers. But according to Lawson, Grant's gun could have inflicted the fatal wound. He says Cat had started to turn back towards the kidnappers when the light went out." She looked bleakly at Phil. "The chances are Grant killed his own daughter."

"And the cover-up cost him his grandson." Phil took a long drag on his cigarillo. "You going to talk to Brodie Grant about this?"

Karen sighed. "I don't see how I can avoid it."

"Maybe you should let the Macaroon deal with it?"

Karen laughed with genuine delight. "What a joy that would be. But we both know he'd throw himself off a tall building to dodge that bullet. No, I'm going to have to front him up myself. I'm just not sure of the best way to handle it. Maybe I'll wait till I see what the Italians have got for me. See if there's anything to sugar

the pill." Before Phil could reply, Karen's phone rang. "Bloody thing," she muttered as she took it out. Then she read the screen and smiled. "Hello, River," she said. "How are you doing?"

"Never better." River's voice crackled and spat in her ear. "Listen, I think you need to get down here."

"What? Have you found something?"

"This is a crap connection, Karen. Better if you just come straight down."

"OK. Twenty minutes." She ended the call. "Get your slippers off, Sherlock. Bugger Brodie Grant. The good doctor has something for us."

Boscolata

Bel had to admit that Grazia knew how to create the perfect ambience for loosening tongues. As the sun slowly sank behind the distant hills and the lights of medieval hill towns scattered their dark slopes like handfuls of glitter, the inhabitants of Boscolata gorged on moist suckling pig accompanied by mounds of slow-roasted potatoes redolent with garlic and rosemary, and bowls of tomato salad pungent with basil and tarragon. Boscolata provided flagons of wine from their own vines, and Maurizio had added bottles of his home-made *vin santo* to the feast.

The knowledge that this unexpected celebration was in honour of Bel inclined them favourably towards her. She moved among them, chatting easily about all manner of things. But always, the conversation moved back to the puppeteers who had squatted Paolo Totti's villa. Gradually, she was able to conjure up a mental dossier of the people who had lived there. Rado and Sylvia, a Kosovan Serb and a Slovenian who had a gift for making puppets. Matthias, who had set up the company in the first place and now designed and built the sets. His woman, Ursula, responsible for organizing their schedule and greasing the wheels to make it possible. Maria and Peter from Austria, the principal puppeteers, and

the three-year-old daughter they were determined to keep out of
the formal school system. Dieter, a Swiss who was responsible for
lighting and sound. Luka and Max, the second-string puppeteers
who put up the posters, did most of the donkey work, and got to
run their own show when a special presentation clashed with one
of their regular pitches.

And then there were the visitors. Apparently, there had been
plenty of those. Gabriel and his father hadn't stood out particularly, except that the father was clearly a friend of Matthias rather
than a friend of the house. He kept himself to himself. Always
polite but never actually open. Opinions varied as to his name.
One thought it was David, another Daniel, a third Darren.

As the evening wore on, Bel began to wonder if there was any
substance to her gut reaction to the photograph Renata had shown
her. Everything else seemed so very insubstantial. Then, as she
helped herself to a glass of *vin santo* and a handful of *cantuccini*,
a teenage boy sidled up to her.

"You're the one who wants to know about BurEst, right?" he
mumbled.

"That's right."

"And that lad, Gabe?"

"What do you know?" Bel said, moving closer to him, letting
him feel they were in a conspiracy of two.

"He was there, the night they legged it."

"Gabriel, you mean?"

"That's right. I didn't say anything before, because I was supposed to be at school, only I wasn't, you know?"

Bel patted his arm. "Believe me, I know all about it. I didn't
really get on with school either. Much more interesting things to
be doing."

"Yeah, well. Anyway, I was in Siena, and I saw Matthias walking up from the station with Gabe. Matthias had been away for a
couple of days. I didn't have anything better to do, so I followed
them. They walked across town to the car park by the Porta
Romana, and they came out in Matthias's van."

"Were they talking? Did they seem to be friendly?"

"They looked pretty fed up. They had their heads down, they weren't saying much. Not unfriendly, as such. Just like they were both pissed off about something."

"Did you see them again? Back here?"

The boy gave a jerky half-shrug. "I never saw them. But when I got back, Matthias's van was there. The others had gone off all the way to Grossetto to do a special performance. That's a good couple of hours' drive, so they'd gone by the time I got back. I just assumed Matthias and Gabe were in the villa." He gave a lairy grin. "Doing who knows what."

Judging by the blood on the floor, Bel thought, it wasn't anything like as much fun as this unimaginative young man was picturing. The real question was whose the blood was. Had BurEst fled because they'd come back to find their leader dead in a pool of his own blood? Or had they scattered because their leader had Gabriel's blood on his hands? "Thanks," she said, turning away and refilling a glass that had somehow become empty. She drifted away from the chattering crowd and walked along the fringe of the vineyard. Her informant had given her plenty to think about. Matthias had been gone for a few days. He came back with Gabriel. The two had been alone in the villa. By the middle of the following morning, the whole troupe had cleared out in a hurry, leaving the same posters once used by the Anarchist Covenant of Scotland and a large bloodstain on the floor.

You didn't have to be much of a detective to figure out that something had gone horribly wrong. But to whom? And maybe more important, why?

East Wemyss

Summer in Scotland, Karen thought bitterly as she scrambled down the path to the Thane's Cave. Still daylight at nine o'clock, a thin drizzle soaking her, and the midges biting like there was no

tomorrow. She could see them in a cloud round Phil's head as she
followed him down to the beach. She was sure they were worse
now than when she was a kid. Bloody global warming. The wee
beasties got more vicious and the weather got worse.

As the path levelled out, she could see a couple of River's students huddled under an overhang, enjoying a fly fag. Maybe if
she stood upwind, their smoke would see the midges off. Beyond
them, River herself was pacing, phone to her ear, head down, long
dark hair pulled back in a ponytail that stuck through the back of
her baseball cap. What chilled Karen more than the rain was the
gleam of the white paper overall River was wearing. The anthropologist turned, caught sight of them, and brought her call to an
abrupt end. "Just telling Ewan not to expect me home for a few
days," she said ruefully.

"So what have you got?" Karen asked, urgency stripping courtesy to the bone.

"Come on in and I'll show you."

They followed her into the cave, the working lights creating
an abstract pattern of darkness and light that took a moment to
adjust to. The clearing crew had stopped work and were sitting
around eating sandwiches and drinking cans of soft drinks. Karen
and Phil were magnets for their interest, and their eyes never left
the cops.

River led the way to where the rock fall had blocked the passage leading back into the rock. Almost all of the boulders and
small stones had been shifted, leaving a narrow opening. She
played a powerful flashlight over the remaining rubble, showing
that the actual fall was only about four feet deep. "We were surprised to find how shallow this fall was. We would have expected
it to go back twenty feet or more. That made me suspicious right
from the word go."

"What do you mean?" Phil asked.

"I'm not a geologist. But as I understand it from my colleagues
in earth sciences, it takes a lot of pressure for a natural cave-in
to happen. When they were mining underground around here,

it produced a lot of stress in the rocks above, so you would get big fractures and falls. It's that scale of geological pressure that causes roof falls in old caves like this. They've been here for eight thousand years. They don't just collapse for no reason at all. But when they do go, it's like pulling the keystone out of a bridge. And you get a big fall." As she spoke, River kept moving the flashlight beam around, showing that the roof was surprisingly sound on either side of the fall. "On the other hand, if you know what you're doing, a small explosive charge will create a controlled fall that affects only a relatively small area." She raised her eyebrows at Karen. "The kind of thing that's done down mines all the time."

"You're saying this fall was created deliberately?" Karen said.

"You'd need an expert to give you a definitive yes or no, but based on what little I do know, I would say it looks that way to me." She swung round and shone the flashlight at a section of the cave wall about five feet above the ground. There was a roughly conical hole in the rock, black streaks staining the red sandstone. "That looks like a shot hole to me," River said.

"Shit," Karen said. "What now?"

"Well, when I saw this, I thought we needed to step very carefully once we'd cleared a path through. So I put on the J-suit and went through by myself. There's maybe three metres of passageway, then it opens out into quite a big chamber. Maybe five metres by four metres." River sighed. "It's going to be a bastard to process."

"And there's a reason to process it?" Phil asked.

"Oh yes. There's a reason." She shone the flashlight at their feet. "You can see the floor's just packed earth. Just inside the chamber, on the left, the earth is loose. It had been tramped down, but I could see it was different in texture from the rest of the floor. I set up some lights and a camera and started moving soil." River's voice had become cool and distant. "I didn't have to go far. About six inches down, I found a skull. I haven't moved it. I wanted you to see it in situ before we do anything further." She waved them back from the fall. "You'll need suits," she said, turning to the stu-

dents. "Jackie, could you bring me over suits and bootees for DI Pirie and DS Parhatka?"

As they suited up, River ran through their options. It boiled down to either letting the students work on under River's close supervision or bringing in the force's own CSI team. "It's your decision," River said. "All I would say is that we're not only the budget option, we're the recently trained specialist option. I don't know what your level of expertise is in archaeology and anthropology, but I'm betting a small force like Fife is not going to have a team of leading-edge specialists on the payroll."

Karen gave her the look that reduced her DCs to childhood. "We've not had a case like this while I've been serving. Anything out of the usual run of things, we use outside experts all the time. The main issue is making sure the evidence will hold up in court. I know you're a qualified expert witness, but your students are not. I'm going to have to run this past the Macaroon, but I think we should continue with your crew. There have to be two video cameras running at all times, though, and you have to be on site whenever they're working." She fastened her suit, glad that Jackie had given her one big enough to accommodate her generous proportions. CSIs weren't always so considerate. She thought they sometimes did it on purpose, to make her feel uncomfortable in what they regarded as their domain. "Let's have a look at it, then."

River handed them each a flashlight. "I haven't taped off an approach route," she said, strapping on a head lamp. "Just stay as far to the left as you can."

They followed her bobbing light into the darkness. Karen gave a last look over her shoulder, but it was hard to see anything beyond Phil's silhouette. The quality of the air changed as they passed the remains of the rock fall, the saltiness replaced by a faint mustiness tinged with the acid of old bird and bat droppings. A dull glow ahead of them indicated the spotlight on the video camera that was still running.

River stopped as the walls fell back and broadened out into the chamber. Her flashlight augmented the camera light, revealing a

small area of the earthen floor where the soil had been scraped back to create a shallow depression. Gleaming dull against the reddish brown earth was the unmistakable outline of a human skull.

"You were right," Phil said softly.

"You have no idea how much that pisses me off," Karen said heavily, taking in all the details. She turned away, gathering her thoughts. "Poor bastard, whoever you are."

Tuesday, 3rd July 2007; Glenrothes

Karen pulled into her parking space at headquarters and turned off the engine. She sat for a long moment, watching the rain reclaim her windscreen. This was not going to be the easiest morning of her career. She had a body, but technically it was the wrong body. She had to stop the Macaroon going off at half-cock and assuming this was one of Catriona Maclennan Grant's kidnappers. And to do that, she would have to admit she'd been working on something he didn't know about. Phil had been right. She shouldn't have indulged her desire for hands-on policing. It was small consolation that she'd made more headway in the case of Mick Prentice than the woolly suits would have done. Getting out of this without a formal reprimand would be a result.

Sighing, she grabbed her files and ran through the driving rain. She pushed the door open, head down, heading straight for the lifts. But Dave Cruickshank's voice made her break stride. "DI Pirie," he called. "There's a lady here to see you."

Karen turned as Jenny Prentice rose hesitantly from a chair in the reception area. She'd obviously made an effort. Her grey hair was neatly braided, and her outfit was clearly the one she kept for best. The dark red wool coat would normally have been insanely warm for July, but not this year. "Mrs. Prentice," Karen said, hoping the sinking of her heart wasn't as obvious on the outside.

"I need to speak to you," Jenny said. "It'll not take long," she added, seeing Karen glance at the wall clock.

"Good. Because I've not got long," Karen said. There was a small interview room off the foyer, and she led the way there. She dumped her folders on a chair in one corner, then sat opposite Jenny across a small table. She wasn't in the mood for coaxing. "I take it you've come to answer the questions I tried to ask you yesterday?"

"No," Jenny said, as mulish as Karen herself could be. "I've come to tell you to call it off."

"Call what off?"

"This so-called missing person hunt for Mick." Her eyes locked defiantly with Karen's. "He's not missing. I know where he is."

It was the last thing Karen had expected to hear. "What do you mean, you know where he is?"

Jenny shrugged. "I don't know how else to put it. I've known for years where he was. And that he wanted nothing more to do with us."

"So why keep it a secret? Why am I only hearing this now? Don't you understand the concept of wasting police time?" Karen knew she was almost shouting, but she didn't care.

"I didn't want to upset Misha. How would you feel if somebody told you your father wanted nothing to do with you? I wanted to spare her."

Karen stared at her uncertainly. Jenny's voice and expression held conviction. But Karen couldn't afford to take her at face value. "What about Luke? Surely you want to do everything you can to save him? Doesn't Misha have the right to ask for his help?"

Jenny looked at her with contempt. "You think I haven't already asked him? I begged him. I sent him photos of wee Luke to try and change his mind. But he just said the boy was nothing to do with him." She looked away. "I think he's got a new family now. We don't matter to him. Men seem to manage that better than women."

"I'm going to need to talk to him," Karen said.

Jenny shook her head. "No way."

"Look, Mrs. Prentice," Karen said through mounting irrita-

tion, "a man has been reported missing. You say he's not but I have only your word for that. I need to confirm what you're telling me. I wouldn't be doing my job right if I didn't."

"And what happens then?" Jenny gripped the edge of the table. "What do you say when Misha asks you how the investigation's going? Do you lie to her? Is that part of your job? Do you lie to her and hope she never finds out the truth from some other polis somewhere down the line? Or do you tell the truth and let Mick break her heart all over again?"

"It's not my job to make those judgements. I'm supposed to find out the truth and then it's out of my hands. You need to tell me where Mick is, Mrs. Prentice." Karen knew she was hard to resist when she brought the full force of her personality to bear. But this defiant little woman was giving as good as she got.

"All I'm telling you is that you're wasting your time looking for a missing person that isn't missing. Call it off, Inspector. Just call it off."

Something about Jenny Prentice was striking a bum note. Karen couldn't identify what it was, but until she could, she wasn't giving an inch. She stood up and pointedly stepped away to pick up her folders. "I don't believe you. And anyway, you're too late, Jenny," she said, turning back to face her. "We've found a body."

She'd read about the colour draining from people's faces, but she'd never seen it before. "That can't be right." Jenny's voice was a whisper.

"It's right enough, Jenny. And the place we found it—thanks to you, we know it's a place where Mick used to hang about." Karen opened the door. "We'll be in touch." She waited pointedly while Jenny came to herself and shuffled out the door, a woman utterly reduced by words. For once, Karen had little sympathy. Whatever Jenny Prentice's motives for that little performance, Karen was certain now that a performance was what it had been. Jenny had no more idea of where Mick Prentice was than Karen herself.

All she had to do now was figure out why it was so important to Jenny that the police give up the hunt. Another encounter, an-

other puzzle. They seemed to be walking hand in hand these days. Some weeks, you couldn't buy a straight answer.

"But that's fantastic news, Inspector." It wasn't often that Karen Pirie's reports brought Simon Lees satisfaction, far less delight. But he couldn't hide the fact that he was doubly pleased at what she had to tell him today. Not only had they uncovered a body that would progress a case dormant for over twenty years, but they'd also achieved it on a shoestring budget.

Then a horrible thought occurred to him. "It is an adult skeleton?" he said, apprehension tightening his chest.

"Yes, sir."

Why was she looking so miserable about it? She'd acted on a hunch and it had come good. In her shoes, he'd be like a dog with two tails. Well, actually, that was pretty much how he felt anyway. This was his operation ultimately; its results reflected credit on him as much as on his officers. For once, she'd brought him sunshine instead of shit. "Well done," he said briskly, pushing his chair back. "I think we should go straight over to Rotheswell and break the good news to Sir Broderick." Her pudding face ran through a series of different expressions, ending in what looked very like consternation. "What's wrong? You haven't told him already?"

"No, I haven't," she said slowly. "And that's because I'm really not convinced this has anything to do with Adam Grant's disappearance."

He understood the words, but it made no sense. She'd organized this whole operation on the basis that the cave fall had been discovered after the ransom disaster. She'd implied that one of the kidnappers could be lying underneath the rubble. He would never have authorized it otherwise. But now she seemed to be suggesting this body had nothing to do with the case she was supposed to be investigating. It was Alice-through-the-looking-glass stuff. "I don't understand," he said plaintively. "You told me you thought there might be a boat. Implied there might be a body. And you

find a body. But instead of celebrating being right, you're telling me it's the wrong body."

"I couldn't have put it better myself," she said, daring to smile.

"But why?" He could hear himself almost howling, and he cleared his throat noisily. "Why?" he repeated, an octave lower.

She twisted in her seat and crossed her legs. "It's a bit difficult to explain."

"I don't care. Start somewhere. Preferably the beginning." Lees couldn't stop his hands clenching and unclenching. He wished he still had the stress ball his kids had given him one Christmas, the stress ball he'd thrown away because he was far too much in control to need something like that.

"We had a very unusual case come in the other day," she began. She sounded hesitant, a version of herself he'd never seen before. If this wasn't so infuriating, he'd almost have been able to enjoy that. "A man reported missing by his daughter."

"That's hardly unusual," he snapped.

"It is when the disappearance happened in 1984. At the height of the miners' strike," Karen shot straight back, all hesitancy gone. "I took a wee look at it, and discovered there were a couple of people who had good reason for wanting this guy out of the way. Both of them worked in the mining industry. Both of them knew about shot-firing rock. Neither of them would have been too hard-pressed to get their hands on explosives. And like I tried to explain to you before, sir, everybody round here knows about the caves." She paused momentarily and glared at him. It was a look that bordered on insubordination. "I knew you would never sanction digging out the rock fall on account of one striking miner on the missing list."

"So you lied?" Lees pounced. He wasn't taking this cavalier rebelliousness any longer.

"No, I didn't lie," she said calmly. "I was just a bit creative with the truth. That cave fall really was discovered after Catriona Mac-lennan Grant died. And the chopper couldn't find the boat the kidnappers escaped in. What I gave you was a reasonable hypoth-

esis. But on the balance of probabilities, I'm saying this is more likely to be the body of Mick Prentice than some unknown kidnapper."

Lees could feel the blood pumping in his head. "Unbelievable."

"Actually, sir, I think you'd have to say we got a result. I mean, it's not like we spent all this money for nothing. At least we've got a body to show for it. OK, it maybe gives us more questions than answers. But you know, sir, we talk about it being our job to speak for the dead, to get justice for people who can't get it for themselves. If you look at it like that, this is an opportunity to serve."

Lees felt something snap inside his head. "An opportunity? What planet are you on? It's a bloody nightmare. You're supposed to be focusing all your resources on finding who killed Catriona Grant and what happened to her son, not farting around on some missing persons case from 1984. What am I supposed to say to Sir Broderick? 'We'll get round to your family once Inspector Pirie can be bothered.' You think you're a law unto yourself," he raged. "You just drive a coach and horses through protocol. You follow your hunches as if they were based on something more than a woman's intuition. You . . . you . . ."

"Careful, sir. You're bordering on sexism there," Karen said sweetly, her eyes wide with assumed innocence. "Men have intuition too. Only, you call it logic. Look on the bright side. If it is Mick Prentice, we've already put together a lot of information about what was going on around the time of his disappearance. We've got a head start on that murder inquiry. And it's not like we're ignoring the Grant case. I'm working closely with the Italian police, but these things take time. Of course, if I was to go out to Italy, it might speed things up . . . ?"

"You're going nowhere. Once this is all over you may not even be—" The phone rang across the end of his threat. He grabbed it. "I thought I said no calls, Emma? . . . Yes, I know who Dr. Wilde is . . ." He sighed harshly. "Fine. Send her in." He replaced the phone carefully and glared at Karen. "We will be revisiting this. But Dr. Wilde is here. Let's see what she has to say."

The woman who walked in was not what he'd expected. For a start, she looked like an adolescent still waiting for her growth spurt. Barely five feet tall, she was lean as a whippet. Dark hair pulled back from a face dominated by large grey eyes and a wide mouth accentuated the comparison. She wore construction boots, jeans, and a denim shirt faded almost white in places under a battered waxed waterproof jacket. Lees had never seen anyone who looked less like an academic. She held out a slim hand, saying, "You must be Simon Lees. It's a pleasure to meet you."

He looked at her hand, imagining the places it had been and the things it had touched. Trying not to shudder, he gripped her cool fingers briefly and gestured towards the other visitor's chair. "Thank you for your help," he said, attempting to put his anger at Karen back in its box for now.

"My pleasure," River said, sounding as if she meant it. "It's a great opportunity for me to work a live case with my students. They get a lot of lab experience, but you can't compare that to the real thing. And they've done a terrific job."

"So it seems. Now, am I to assume you are here because you have something to report?" He knew he sounded stiff as one of her cadavers, but it was the only way he could keep himself under control. River exchanged a quick unreadable look with Karen, and he felt his temper rising again. "Or do you need access to more facilities? Is that it?"

"No. We have access to what we need. I just wanted to bring DI Pirie up to speed, and when DS Parhatka told me she was in a meeting with you, I thought I'd grab the chance to meet you. I hope I haven't interrupted anything?" River leaned forward, giving him the full benefit of a smile that reminded him of Julia Roberts'. It was hard to maintain anger in the teeth of a smile like that.

"Not at all," he said, feeling calmer by the second. "It's always good to put a face to the name."

"Even when it's such a stupid name," River said ruefully. "Hippy parents, before you ask. Now, you'll want to know what I've learned so far." She took out her pocket organizer and hit a

couple of keys. "We worked late into the night to clear the skeleton and remove it from the shallow grave." She turned to Karen. "I've given Phil a copy of the video." Back to her organizer. "I did a preliminary examination early this morning and I can give you some information. Our skeleton is a male. He's over twenty and less than forty. There is some hair, but it's hard to tell what colour it was originally. It's taken up stain from the soil. He's had some dental work, so once you narrow down the possibilities we can follow up on that. And we'll be able to get DNA."

"When was he buried?" Lees asked.

River shrugged. "There are more extensive and expensive and time-consuming tests that we can do. But right now it's hard to be precise about how long he's been in the ground. However, I can say with a high degree of certainty that he was still alive for at least part of 1984."

"That's amazing," Lees exclaimed. "You people in forensics astonish me."

Karen gave him a cool stare. "Loose change in his pockets, was there?" she said to River.

"Actually, no pockets left to speak of," River said. "He was wearing cotton and wool so it's mostly gone. The coins were lying inside the pelvic girdle." She smiled at Lees again. "Sorry, not science this time. Just observation."

Lees cleared his throat, feeling foolish. "Is there anything else you can tell us at this stage?"

"Oh yes," River said. "He absolutely didn't die a natural death."

San Gimignano

As she drove round the parking lot for the third time in search of the elusive space, Bel cast her mind back to her memories of what San Gimignano had been like before it became a UNESCO World Heritage site. No question that it was worth the rating. The

medieval inhabitants had used the soft grey limestone to build a huddled maze of streets around a central piazza with its ancient well. When it threatened to outgrow its massive city walls, they'd simply chosen to build tall rather than sprawl. Dozens of towers speared the skyline, giving a jagged, gap-toothed appearance from the plain below. Definitely unique. Definitely world heritage. And definitely ruined by its status.

Bel had first come to the spectacular Tuscan hill town in the early eighties when the streets were almost empty of tourists. There were proper shops back then—bakers, greengrocers, butchers, cobblers. Shops where you could buy washing powder or underpants or a comb. Locals actually drank coffee in the bars and cafés. Now, it had been transformed. The only opportunity to buy proper food and clothes was at the Thursday market. Apart from that, everything was targeted at tourists. Enotecas selling overpriced *vernaccia* and *chianti* that the locals wouldn't drink if you paid them. Leather stores, all selling identical factory-produced handbags and wallets. Souvenir shops and gelaterie. And of course, art galleries for those with more money than sense. Bel hoped it was the locals who were making the money, because they were the ones paying the highest price.

At least the streets wouldn't be too crowded so early in the day, ahead of the tour buses. Bel finally squeezed into a parking spot and headed for the vast stone portal that guarded the higher entry to the town. She had barely gone a hundred feet when she came to the first art gallery. The owner was just raising his shutters when she arrived. Bel checked him out; probably about her age, smooth skinned and dark haired, stylishly framed glasses that made his eyes look too small, a little too plump for the tight jeans and Ralph Lauren shirt. An appeal to his vanity would probably be the best approach. She waited patiently, then followed him inside. The walls were covered with prints and watercolours filled with the Tuscan clichés—cypress trees, sunflowers, rustic farmhouses, poppies. They were all well executed and pretty, but there wasn't

onc she would have hung on her walls. Production-line paintings for tour-bus punters ticking off the next place on the list. God, she'd become a snob in her old age.

The owner had settled himself behind a leather-topped desk, obviously meant to look antique. Probably about as old as his car, Bel thought. She approached, plastering her least predatory smile on her face. "Good morning," she said. "What a wonderful display of paintings. Anyone would be lucky to have these on their walls."

"We pride ourselves on the quality of our art works," he said without a flicker of irony.

"Amazing. They make the landscape come alive. I wonder if you can help me?"

. He eyed her up from top to toe. She could see him pricing everything from her Harvey Nicks' sundress to her market stall straw bag before deciding how much wattage to put into his own smile. He must have liked what he saw; she got the full benefit of his cosmetic dentistry. "It will be my pleasure," he said. "What is it that you are looking for?" He stood up, adjusting his shirt to hide his extra pounds.

Apologetic smile. "I'm not actually looking for a painting," she said. "I'm looking for a painter. I'm a journalist." Bel took her business card from the pocket in her dress and handed it over, ignoring the wintry look that had replaced the previous warmth. "I'm looking for a British landscape painter who's been living over here, earning a living for the last twenty years or so. The difficult thing is that I don't know his name. It begins with a D—David, Darren, Daniel. Something like that. He has a son in his early twenties, Gabriel." She'd made print-outs of Renata's photos and she took them out of her bag. "This is the son, and this is the painter I want to track down. My editor thinks there's a feature there." She shrugged. "I don't know. I need to talk to him, find out what his story is."

He glanced at the photos. "I don't know him," he said. "All my artists are Italian. Are you sure he's a professional? There are a

lot of amateurs who sell stuff on the pavements. A lot of them are foreigners."

"Oh no, he's a professional all right. He's represented here and in Siena." She spread her hands to take in the stuff on the walls. "Obviously not good enough for you, though." She took the photos back. "Thanks for your time." He had already turned away, heading for his comfy chair surrounded by his soulless paintings. No sale, no more conversation.

There was, she knew, no shortage of galleries. Two more, then she'd have a coffee and a cigarette. Another three, then an ice-cream. Little treats to drag her through the work.

She didn't make it to the ice-cream. At the fifth gallery she tried, she hit gold. It was a light and airy space, paintings and sculptures spread out so they could be appreciated. Bel actually enjoyed walking through to the desk in the back. This time, it was a middle-aged woman behind a modern, functional desk piled with brochures and catalogues. She wore the crumpled linen uniform of the more relaxed class of Italian middle-class womanhood. She looked up from her computer and gave Bel a vague, slightly harassed look. "Can I help?" she said, her words running into each other.

Bel launched into her spiel. A few sentences in, the woman's hand flew to her mouth, her eyes widening in shock. "Oh my God," she said. "Daniel. You mean Daniel?"

Bel pulled the prints out and showed them to the woman. She looked as if she might burst into tears. "That's Daniel," she said. She reached out and touched Gabriel's head with her fingertips. "And Gabe. Poor sweet Gabe."

"I don't understand," Bel said. "Is there a problem?"

The woman took a deep, shuddering breath. "Daniel's dead." She spread her hands in a gesture of sorrow. "He died back in April."

Now it was Bel's turn to feel a jolt. "What happened?"

The woman leaned back in her chair and ran a hand through her curly black hair. "Pancreatic cancer. He was diagnosed just

before Christmas. It was horrible." Tears sparkled in her eyes. "It shouldn't have happened to him. He was ... he was such a lovely man. Very gentle, very reserved. And he loved his boy so much. Gabe's mother, she died giving birth. Daniel brought him up single-handed and he did a great job."

"I'm so sorry," Bel said. At least the blood on the floor of the villa Totti wasn't Daniel's. "I had no idea. I'd just heard about this terrific British artist who'd been making a living out here for years. I wanted to do a feature about him."

"Do you know his work?" The woman got up and beckoned Bel to follow her. They ended up in a small room at the back of the gallery. On the wall were a series of vibrant triptychs, abstract representations of landscape and seascape. "He did watercolours as well," the woman said. "The watercolours were more figurative. He could sell more of them. But these were what he loved."

"They're splendid," said Bel, meaning it. Really wishing she had met the man who had seen the world like this.

"Yes. They are. I hate that there will be no more of them." She reached out and brushed the textured acrylic paint with her fingertips. "I miss him. He was a friend as well as a client."

"I wonder if you can put me in touch with his son?" Bel said, not losing sight of why she was there. "Maybe I could still do that feature. A sort of tribute."

The woman smiled, a sad little curl of the lips. "Daniel always spurned publicity when he was alive. He had no interest in the cult of personality. He wanted his paintings to speak for him. But now ... it would be good to see his work appreciated. Gabe might like it." She nodded slowly.

"Can you give me his phone number? Or address?" Bel said.

The woman looked slightly shocked. "Oh no, I couldn't do that. Daniel always insisted on privacy. Please, give me your card and I will contact Gabe. Ask him if he is willing to talk to you about his father."

"Is he still around, then?"

"Where else would he be? Tuscany is the only home he's ever

known. His friends are all here. We're taking turns to make sure he has at least one decent meal a week."

As they walked back to the desk, it dawned on Bel that she hadn't discovered Daniel's surname. "Have you got a brochure or a catalogue of his work?" she asked.

The woman nodded. "I'll print it out for you."

Ten minutes later, Bel was back out on the street. At last she had something concrete to grab on to. The hunt was on.

Coaltown of Wemyss

The whitewashed cottages that lined the main street were spick and span, their porches supported by rustic tree trunks. They'd always been well maintained because they were what people saw when they travelled through the village. These days, the back streets looked just as smart. But Karen knew it hadn't always been like that. The hovels of Plantation Row had been a notorious slum, ignored by their landlord because what no eye from polite society ever saw was not worth bothering with. But even from the doorstep of this particular cottage, Karen suspected that somehow, if Effie Reekie had found herself in a hellhole, she'd have turned it into a little paradise. The front door looked as if it had been washed down that morning, there wasn't a dead head in the window boxes, and the net curtains hung in perfect pleats. She wondered if Effie and her mother had possibly been twins separated at birth.

"Are you going to knock or what?" Phil said.

"Sorry. I was just having a moment of déjà vu. Or something." Karen pressed the doorbell, feeling guilty for leaving her fingerprint on it.

The door opened almost at once. The sense of being in a time warp continued. Karen hadn't seen a woman with a scarf turbanning her head like that since her grandmother died. With her overall and rolled-up sleeves Effie Reekie resembled a pensioned-off

version of Rosie the Riveter. She looked Karen up and down, as if gauging whether she was clean enough to be allowed across the doorstep. "Aye?" she said. It wasn't a welcome.

Karen introduced herself and Phil. Effie frowned, apparently affronted to have police officers at her door. "I never saw anything or heard anything," she said abruptly. "That's always been my policy."

"We need to talk to you," Karen said gently, sensing the fragility the elderly woman was desperately hiding.

"No, you don't," Effie said.

Phil stepped forward. "Mrs. Reekie," he said, "even if you don't have anything to say to us, I would be your pal for life if you could see your way to making us a cup of tea. I've a throat on me like the Sahara."

She hesitated, looking from one to the other with anxious eyes. Her face scrunched up with the wrestle of hospitality versus vulnerability. "You'd better come in, then," she said at last. "But I've got nothing to tell you."

The kitchen was immaculate. River could have conducted an autopsy on the table without risk of contamination. Karen was pleased to see she'd guessed right. Like her mother, Effie Reekie viewed every available surface as a depository for ornaments and knick-knacks. It was, Karen thought, a desperate waste of the planet's resources. She tried not to think of all the crap she'd brought home from school trips. "You've got a lovely home," she said.

"I've always tried to keep it nice," Effie said as she busied herself with the kettle. "I would never let Ben smoke in the house. That was my man, Ben. He's been dead now five years, but he was somebody round these parts. Everybody knew Ben Reekie. There wouldn't be the bother there is in this street these days if my Ben was still alive. No, siree. There would not."

"It's Ben we need to talk to you about, Mrs. Reekie," Karen said.

She swung round, eyes wide, rabbit in the headlights. "There's

nothing to talk about. He's been dead these five years. Cancer, it was. Lung cancer. Years of smoking. Years of branch committee meetings, all of them smoking like chimneys."

"He was the branch secretary, wasn't he?" Phil asked. He was studying a group of decorative plates mounted on the wall. They represented various milestones in trade union history. "A big job, especially during the strike."

"He loved the men," Effie said vehemently. "He'd have done anything for his men. It broke his heart to see the way that bitch Thatcher brought them down. And Scargill." She brought their tea to the table with a clatter of china. "I never had any time for King Arthur. Into the valley of death, that's where he led them. It would have been a different story if it had been Mick McGahey running the show. A very different story. He had respect for the men. Like my Ben. He had respect for his men." She gave Karen a look that bordered on the desperate.

"I understand that, Mrs. Reekie. But it's time now to set the record straight." Karen knew she was chancing her arm. Mick Prentice could have been mistaken. Ben Reekie might have kept his own counsel. And Effie Reekie might be determined not to think about the way her husband had breached the trust of the men he professed to love.

Effie's whole body seemed to clench. "I don't know what you're talking about." It was a shrill denial, its deceit obvious.

"I think you do, Effie," Phil said, joining the two women at the table. "I think it's been eating away at you for a long time."

Effie covered her face with her hands. "Go away," she said, her words muffled. She was shivering now, like a sheep that had just been sheared.

Karen sighed. "It can't have been easy for you. Seeing how hard everybody else had it, when you were doing all right."

Effie grew still and took her hands away from her face. "What are you talking about?" she said. "You surely don't think he took it for himself?" Affront had given her strength. That or made her careless.

Fuck, fuck, fuck. Karen realized she had completely misjudged the situation. But if she had, so could others. Others like Mick Prentice. Mick Prentice, whose best friend had been a union official. Who might even have been complicit in what Ben Reekie was doing. Thoughts racing, she pulled herself back into the conversation.

"Of course we don't think that," Phil said. "Karen just meant the fact that you still had a wage coming in."

Effie looked uncertainly at them both. "He only did it after they started sequestering the union funds," she said. The words spilled out as if it was a relief to let them loose. "He said, what was the point in passing money through to the branch when they'd just hand it on to Head Office. He said money raised locally should go to support local miners, not be shuffled off to Buffalo." She managed a piteous smile. "That's what he always used to say. 'Not be shuffled off to Buffalo.' He just took some here and there, not enough for the high-ups to notice. And he was very discreet about passing it out. He got Andy Kerr to go through the welfare request letters and he'd hand it out where it was most needed."

"Did anybody find out?" Phil asked. "Anybody catch him at it?"

"What do you think? They'd have strung him up first and asked questions afterwards. The union was sacred round here. He'd never have walked away in one piece if anybody had so much as suspected."

"But Andy knew." Karen wasn't ready to give up yet.

"No, no, he never knew. Ben never said he was giving them money. He just asked Andy to prioritize them, supposedly for branch relief. Except there wasn't any branch relief by then because all the funds were going to national level." Effie rubbed her hands as if they hurt. "He knew he couldn't trust anybody with that. You see, even if they'd believed he was doing it for the men and their families, they'd still have seen it as treason. Everybody was supposed to put the union first, especially officials. What he did, it would have been unforgivable. And he knew it."

San Gimignano

Bel finally found a bar that wasn't crammed with tourists, tucked away in a back street. The only patrons were half a dozen old men playing cards and drinking small glasses of dark purple wine. She ordered an espresso and a water and sat down by the back door, which was open on to a tiny cobbled yard.

She spent a few minutes looking at the catalogue she'd picked up at the gallery. Daniel Porteous had been an artist whose work she'd have happily lived with. But who the hell had he been? What was his background? And had his path truly crossed Cat's, or was Bel making bricks without straw? Just because Daniel Porteous was an artist and he had a loose connection to the place where the posters had been found didn't mean he was involved with the kidnapping. Maybe she was looking at the wrong man. Maybe the link was Matthias, the man who designed the puppets and their stage sets. The man who might either be a killer or a victim.

Still looking at the reproductions of Porteous's work, she called her work experience student Jonathan on her mobile.

"I tried to get hold of you last night," he said. "But your mobile was switched off. So I rang the ice maiden at Rotheswell and she said you were unavailable."

Bel laughed. "She does like to make herself important, doesn't she? Sorry I missed you last night. I was at a party."

"A party? I thought you were supposed to be being Nancy Drew?"

Part of her thought Jonathan's cheeky flirtatiousness was marginally inappropriate. But its absurdity amused her so she let him play. "I am. The party was in Italy."

"In *Italy*? You're in *Italy*?"

Bel quickly brought Jonathan up to speed. "So now you have the inside track," she wound up.

"Wow," Jonathan said. "Who knew this was going to be so exciting? None of my mates are having an internship like this. It's like Woodward and Bernstein, hot on the trail of Watergate."

"It's nothing like that," Bel protested.

"Of course it is. You told me there was blood on the villa floor. People generally don't run from household accidents or suicide, so that rather suggests that somebody was killed. And in a situation that ties in with murder and kidnap going back twenty-two years. Bel, there is at least one very unpleasant person out there and you are definitely hot on his trail."

"At the moment, Jonathan, what I'm on the trail of is a young man who's just lost his father. How scary can that be?" Bel said, her tone light and easy.

Suddenly serious, Jonathan said, "Bel, they're not all as charming and harmless as me. We can be savages. You've done enough stories about rape and murder to have no illusions about that. Stop treating me like a child. This isn't a game. Promise me you'll take it seriously."

Bel sighed. "When I get to something that looks serious, I will take it seriously, Jonathan. I promise. Now, meanwhile, I need you to do something for me."

"Of course, whatever you need. I don't suppose it involves a visit to Tuscany?"

"It involves a visit to the Family Records Centre in Islington to find out what you can about a man called Daniel Porteous. He'd be late forties, early fifties. He died in April in Italy, but I'm not sure where exactly. And besides, Italian death certificates have almost no information on them. So I'm looking for his birth certificate, maybe a marriage certificate. Can you do that for me?"

"I'm on it. I'll get back to you as soon as I've got anything. Thanks, Bel. It's great being involved in something as meaty as this."

"Thanks," Bel said to emptiness. She sipped her espresso and thought. She wasn't convinced the gallery owner would come up trumps as far as Gabriel Porteous was concerned. She was going to have to do some serious digging herself. The records would be in the provincial capital, Siena. There was no point in heading over there now. By the time she made it, everyone would have

disappeared for the day. Afternoons and Italian bureaucracy were unhappy bedfellows.

There was nothing else for it. She was going to have to go back to Campora and lie by Grazia's pool. Maybe call Vivianne, catch up on family life. Sometimes life was just too, too hard.

Edinburgh

Karen reclined the car seat back from bolt upright and settled herself in for the drive to Edinburgh. "I tell you," she said. "My head's nipping with this case. Every time I think I'm making sense of it, something trips me up."

"Which case did you have in mind? The one the Macaroon thinks you're prioritizing or the one you're actually working?" Phil said, turning on to the back road that would bring them to a farm tearoom by the motorway. One thing about cold cases was that you could generally manage to eat at regular times. There wasn't the pressure of the clock ticking before another offence was committed. It was a regime that suited both of them just fine.

"I can't do anything about Cat Grant until I get a proper report from the Italian police. And they're not exactly going hell for leather. No, I'm talking about Mick Prentice. First, everybody thinks he's gone to Nottingham. But now it looks like he never left the Wemyss alive. He never went with the scabs, even though one of them confused the issue by sending money to Jenny. But the one thing we did learn from the scabs is that Mick was alive and well and walking round the Newton a good twelve hours after Jenny claims he walked out."

"Which is odd," Phil said. "If he was leaving her, you'd think he'd be long gone. Unless he was just trying to teach her a lesson. Maybe he'd stayed away for hours to wind her up. Maybe he was on the way back home and something happened to divert him."

"It certainly sounds like something knocked him out of character. The guys going scabbing obviously expected him to lose his

head with them. When they saw him, they thought they were in for a tongue-lashing or a fight. But all they got was him pleading with them, looking like he was ready to burst into tears."

"Maybe that was the night he found out there was something going on between Jenny and Tom Campbell," Phil suggested. "That would have knocked his confidence for six."

"Maybe." She sounded unconvinced. "If you're right, he would have been in a state. He wouldn't have wanted to go home. So maybe he crashed with his pal Andy in the cottage in the woods."

"If he did, why did nobody see him again after that night? You know what it used to be like round here. When people split up, they didn't leave town. They just moved three houses down the street."

Karen sighed. "Fair enough. But he could still have gone to Andy's. It could have played out a different way. We know Andy was on the sick with depression. And we know from his sister that he liked to go up into the Highlands, walking. What if Mick decided to go with him? What if they both had an accident and their bodies are lying in some ravine? You know what it's like up there. Climbers go missing and they're never found. And that's just the ones we know about."

"It's possible." Phil signalled and turned into the parking lot. "But if that's what happened, whose body is it in the cave? I think it's a lot simpler than you're making out, Karen."

They walked into the café in silence. They ordered steak pie, peas, and new potatoes without looking at the menu, then Karen said, "Simpler how?"

"I think you're right, he did go to Andy's. I don't know if he was planning on leaving for good or just putting a bit of space between him and Jenny. But I think he told Andy about Ben Reekie. And I think there was some sort of confrontation. I don't know if Andy lost the place with Mick, or if Ben came round and it all got out of hand. But I think Mick died in that cottage that night."

"What? And they took him down the cave to get rid of

---

him? That seems a bit elaborate. Why not just bury him in the woods?"

"Andy was a country man. He knew bodies don't stay buried in shallow graves in woodland. Putting him in the cave then engineering a rock fall was a much safer place to put him. And a lot more private than trying to dig a grave in the middle of the Wemyss woods. Remember what it was like back then. Every bit of woodland was alive with poachers trying to get a rabbit or even a deer to put on the table."

"You've got a point." Karen smiled an acknowledgement at the waitress who brought their coffees. She added a heaped spoonful of sugar to hers and stirred slowly. "So what happened to Andy? You think he went off and topped himself?"

"Probably. From what you've told me, he sounds like the sensitive kind."

She had to admit it made sense. Phil's distance allowed him to see the case more clearly. Smart though she was, she knew when to step back and let someone else consider the facts. "If you're right, I suppose we'll never know how it panned out. Whether it was between Andy and Mick, or whether Ben Reekie was in the picture too."

Phil smiled, shaking his head. "That's one theory we can't run past Effie Reekie. Not unless we want another body on our hands."

"She'd stroke out on the spot," Karen agreed.

He chuckled. "Of course, this could all be a wild-goose chase if Jenny was telling the truth when she told you to lay off."

Karen snorted. "Fantasy island, that line. I reckon she's trying to shut down the aggravation. She wants us out of her hair so she can get back to her life of martyrdom."

Phil looked surprised. "You think she rates her own peace and quiet above her grandson's life?"

"No. She's incredibly self-absorbed, but I don't think she sees it in those terms. I think deep down she feels some responsibility for

Mick disappearing. And that means she has to carry some of the guilt for his unavailability to be a donor for Luke. So she's trying to offload the guilt by getting us to stop looking for him so she can go back to hiding her head in the sand like before."

Phil scratched his chin. "People are so fucked up." He sighed.

"True enough. At least this jaunt will get us some answers."

"Maybe. But it makes you wonder," Phil said.

"Makes you wonder what, exactly?"

He pulled a face. "We're going all the way to Edinburgh to take a DNA sample so River can compare it to the corpse. But what if Misha's not Mick's kid? What if she's Tom Campbell's bairn?"

Karen gave him an admiring look. "You have a truly evil mind, Phil. I think you're wrong, but it's a beautiful thing all the same."

"You want to take a bet on the DNA showing it's Mick Prentice?"

They both leaned back to let the waitress put the piled plates of food in front of them. The aroma was killer. Karen wanted to pick up the plate and inhale it. But first she had to answer Phil. "No," she said. "And not because I think Misha might be Tom Campbell's kid. There's other possibilities. River says it's the back of the skull that's smashed in, Phil. If Andy Kerr killed Mick Prentice, it was in the heat of the moment. He would never have crept up behind him and caved his head in. Your theory's neat enough, but I'm not convinced." She smiled. "But then, that's why you love me."

He gave her an odd look. "You're always full of surprises."

Karen swallowed a divine mouthful of meat and pastry. "I want some answers, Phil. Real answers, not just the daft notions you and me dream up to fit what we know. I want the truth."

Phil cocked his head, considering her. "Actually," he said, "*that's* why I love you, ma'am."

An hour later, they were standing on the doorstep of the March-mont tenement where Misha Gibson lived. Karen was still won-

dering whether there had been anything more than a tease in Phil's words. She'd thought for a long time that nothing was off limits between them. Apparently she'd been wrong. She certainly wasn't going to ask him what he'd meant. She pressed the buzzer again, but there was no reply.

A voice from behind them said, "Are you looking for Misha?"

"That's right," Phil said.

An elderly man stepped round them, forcing Karen to move away from the door or be trampled on. "You'll not find her in at this time of day. She'll be down at the Sick Kids' with the boy." He looked pointedly at them. "I'm not letting you in and I'm not putting my code in while you're standing there looking."

Karen laughed. "Very commendable, sir. But at the risk of sounding like a cliché, we are the police."

"That's no guarantee of honesty these days," the old man said.

Taken aback, Karen stepped away. What was the world coming to when people thought the police would burgle them? Or worse? She was about to protest when Phil put his hand on her arm. "No point," he said softly. "We've got what we need."

"I tell you," Karen said when they were out of earshot. "They sit watching their American cop shows where every other cop is bent and they think that's what we're like. It makes me mad."

"That's a bit rich, coming from the woman who put the assistant chief constable behind bars. It's not just the Americans," Phil said. "You get people who take the piss everywhere. That's where the scriptwriters get their ideas from."

"Oh, I know. It just offends me. All the years I've been in this job, Lawson's the only truly bad apple I've ever come up against. But that's all it takes for people to lose all respect."

"You know what they say: Trust is like virginity. You can only lose it once. So, you ready for 'good cop, bad cop'?" They paused on the kerb to wait for a break in the traffic and headed on down the hill to the hospital.

"Count me in," Karen said.

Finding Luke Gibson's ward was easy but harrowing. It was

impossible to avoid the presence of ill children, the images of their sickness burning themselves into memory. It was, Karen thought, one of the few upsides of being childless. You didn't have to stand by impotent as your kid suffered.

The door to Luke's room was open and Karen couldn't stop herself watching mother and son together for a few minutes. Luke seemed very small, his face pale and pinched but still hanging on to a young boy's prettiness. Misha was sitting on the bed next to him, reading a Captain Underpants book. She was doing all the voices, making the story come alive for her boy, who laughed out loud at the bad puns and the daft storyline.

Finally, she cleared her throat and stepped inside. "Hi, Misha." She smiled at the boy. "You must be Luke. My name's Karen. I need to have a wee word with your mummy. Is that OK?"

Luke nodded. "Sure. Mum, can I watch my *Dr. Who* DVD if you're going away?"

"I'll be right back," Misha said, scrambling off the bed. "But yeah, you can have the DVD on." She reached for a personal DVD player and set it up for him.

Karen waited patiently, then led her into the corridor, where Phil was waiting. "We need to talk to you," Karen said.

"That's fine," Misha said. "There's a parents' room down the hall." She set off without waiting for a response, and they followed her into a small, brightly decorated room with a coffee vending machine and a trio of sagging couches. "It's where we escape to when it all gets too much." She gestured at the sofas. "It's amazing what you can catnap on after twelve hours sitting by a sick kid's bed."

"We're sorry to intrude—"

"You're not intruding," Misha interrupted. "It's good that you've met Luke. He's a wee doll, isn't he? Now you understand why I'm willing to pursue this even though my mother doesn't like you poking into the past. I told her she was out of order on Sunday. You need to ask these questions if you're going to find my dad."

Karen flashed a quick glance at Phil, who looked as surprised as she felt. "Did you know your mother came to see me this morning?" she said.

Misha frowned. "I had no idea. Did she tell you what you wanted to know?"

"She wanted us to give up looking for your father. She said she didn't think he was missing. That he'd walked out on the pair of you from choice and that he didn't want to come back."

"That makes no sense," Misha said. "Even though he did walk out on us, he wouldn't turn his back on his own grandson if he needed his help. All I've heard about my dad was that he was one of the good guys."

"She says she's trying to protect you," Karen said. "She's scared that if we do find him he'll reject you for a second time."

"Either that or she knows more about his disappearance than she's letting on," Phil said grimly. "What you probably don't know is that we've found a body."

Campora

Bel sat on her tiny terrace, watching the sky and the hills range through the spectrum as the sun set slowly and gloriously. She picked at the cold leftovers of pork and potato that Grazia had left in her fridge, considering her next move. She wasn't relishing the battle with Italian bureaucracy that lay ahead of her, but if she was to find Gabriel Porteous, it would have to be faced. She pulled out Renata's prints again, wondering whether she was imagining the resemblance.

But again, it leapt off the page at her. The deep-set eyes, the curved beak of the nose, the wide mouth. All mimicking Brodie Grant's distinctive features. The mouth was different, it was true. The lips were fuller, more shapely. Definitely more kissable, Bel thought, chiding herself instantly for the thought. The hair was

a different colour too. Both Brodie Grant and his daughter had had hair so dark it was almost black. But this boy's hair was much lighter, even allowing for the bleaching of the Italian sun. His face was broader too. There were points of difference. You wouldn't mistake Gabriel Porteous for the young Brodie Grant, not judging by the photos Bel had seen around Rotheswell. But you might take them for brothers.

Her thoughts were interrupted by the phone. With a sigh, she picked it up. It was a pain that caller ID didn't always work abroad. You could never tell whether the person on the other end was someone you were trying to avoid. And letting calls go to voicemail so you could screen them soon became hideously expensive. Plus, being partly responsible for her nephew meant she could never ignore the mystery callers. "Hello?" she said cautiously.

"Bel? It's Susan Charleson. Is this a good moment?"

"Yes, perfect."

"I got your e-mail. Sir Broderick asked me to tell you he's very pleased with your progress so far. He wanted to know whether you needed anything at this end. We can organize record searches, that sort of thing."

Bel bit back a rueful laugh. She'd spent her working life doing her own dirty work, or else persuading others to do it for her. It hadn't occurred to her that working for Brodie Grant meant she could offload all the boring bits. "It's all in hand," she said. "Where you could give me a hand is on the personal stuff. I can't help thinking there must be a point in her past where Catriona's life intersected either with Daniel Porteous or this Matthias, who might be German or British. I suppose he might even be Swedish, given that's where Catriona studied. I need to find out when and where that happened. I don't know if she kept diaries or an address book? Also, when I get back, I could really do with tracking down her female friends. The sort of women she would confide in."

Susan Charleson gave a well-bred little laugh. "You're going to be disappointed, then. You might think her father plays things

close to his chest, but Catriona made him look like a soul-barer. She was the ultimate cat who walks alone. Her mother was her best friend, really. They were very close. Apart from Mary, the only person who really got inside Cat's head was Fergus." She left the name dangling in the air between them.

"I don't suppose you know where I'll find Fergus?"

"You could talk to his father when you get back. He often visits his family around this time of year," Susan said. "It's not something Willie feels the need to communicate to Sir Broderick. But I'm aware of it."

"Thank you."

"And I'll see what I can do about diaries and address books. Don't hold your breath, though. The trouble with artists is that they let their work do the talking. When will you be back?"

"I'm not sure. It depends how I get on tomorrow. I'll let you know."

There was nothing more to say, no pleasantries. Bel couldn't remember the last time she'd failed so completely to make a connection with another woman. She'd spent her adult life learning how to get people to like her enough to confide things they didn't really want to tell anyone. With Susan Charleson, she had failed. This job that had started out as little more than an off-chance of persuading a famously reclusive man to talk had exposed her to herself in the most unexpected of ways.

What next, she wondered, taking a long sip of her wine. What next?

Wednesday, 4th July 2007; East Wemyss

Some American woman on the radio was belting out a cracking alt.country song about Independence Day. Only this wasn't about the Stars and Stripes, it was about a radical approach to domestic violence. As a police officer, Karen couldn't approve; but as a

woman, she had to admit the song's solution had its appeal. If Phil had been there, she would have bet him a pound to a gold clock that the man she was about to meet wouldn't have had "Independence Day" blasting out of his car radio.

She drove slowly up the narrow street that led to what had been the pithead and offices of the Michael colliery. There was nothing there now apart from a scarred area of hard standing where the canteen and wages office had been. Everything else had been landscaped and transformed. Without the rust-red pylon of the winding gear it was hard to orientate herself. But at the far end of the asphalt, a single car sat pointing out to sea. Her rendezvous.

The car she pulled up beside was an elderly Rover, buffed to within an inch of its life. She felt faintly embarrassed about the collection of dead insects on her number plate. The Rover's door opened in synch with her own and both drivers got out simultaneously, like a choreographed shot from a film. Karen walked to the front of her car and waited for him to join her.

He was shorter than she expected. He must have struggled to make the five-feet-eight minimum for a cop. Maybe his hair had tipped him over the edge. It was steel grey now, but the quiff would have put Elvis to shame. He wouldn't have been allowed the DA and the sideburns when he was a serving officer, but when it came to hairstyle, Brian Beveridge had taken full advantage of retirement.

Like Elvis, he'd piled on the beef since his days of strutting his stuff on the streets of the Wemyss villages. The buttons of his immaculate white shirt strained over a substantial belly, but his legs were incongruously slim and his feet surprisingly dainty. His face had the florid tones and fleshiness of a man heading for a cardiovascular disaster. When he smiled, his cheeks became tight pink balls, as if someone had stuffed them with cotton wool. "DI Pirie?" he enquired cheerfully.

"Karen," she said. "And you must be Brian? Thanks for coming out to meet me." It was like shaking hands with the Pillsbury Dough Boy, all soft, engulfing warmth.

"It's better than dottering about in the garden," he said, his thick Fife accent untempered. "I'm always happy to help. I walked the beat in these villages for thirty years and, if I'm honest, I miss that sense of knowing every pavement and every house. Back then, you could make a career out of being a beat bobby. There was no pressure to go for promotion or the CID." He rolled his eyes. "There I go. I promised my wife I wasn't going to do my Dixon of Dock Green impersonation, but I can't help myself."

Karen laughed. She liked this cheery little man already, even though she was well aware that working alongside him back in the day would have been a different matter. "I bet you remember the Catriona Maclennan Grant case," she said.

Suddenly sombre, he nodded. "I'll never forget that one. I was there that night—of course, you know that, it's why I'm here. But I still dream about it sometimes. The gunshots, the smell of the cordite on the sea air, the screams and cries. All these years later, and what have we got to show for it? Lady Grant in her grave alongside her daughter. Jimmy Lawson in the jail for the rest of his life. And Brodie Grant, master of the bloody universe. New wife, new heir. Funny how things turn out, eh?"

"You can never tell," Karen said, happy to buy into the clichés for the time being. "So, can you talk me through it as we walk down to the Lady's Rock?"

They set off past a row of houses similar to Jenny Prentice's street in Newton of Wemyss, stranded and solitary now the reason for their existence was gone. Soon they entered the woods and the path began to descend, a waist-high stone wall on one side, thick undergrowth flanking it. In the distance, she could see the sparkle of the sea, the sun shining for once as they descended to shore level. "We had teams stationed up at the top here, and the same along at West Wemyss," Beveridge said. "Back then, you couldn't get along the shore towards East Wemyss for the pit bing. But when they made the coastal path, they got money from the EU and trucked all the pit red off the foreshore. You look at it now and you'd never know."

He was right. As they reached the level of the shore, Karen could see right along past East Wemyss to Buckhaven on its high promontory. In 1985, the view would not have existed. She turned towards West Wemyss, surprised that she couldn't actually see the Lady's Rock from where she stood.

Karen followed Beveridge along the path, trying to imagine what it must have been like that night. The file said it had been a new moon. She pictured the sickle sliver in the sky, the pin-prick stars in the freezing night. The Plough like a big saucepan. Orion's belt and dagger, and all the other ones whose names she didn't know. The cops hidden in the woods, breathing with their mouths open so their breath would be chilled before it came out in puffs. She took in the high sycamores, wondering how much smaller they'd been then. Ropes hung from thick branches where kids would swing as they'd done when she was little. To Karen in her heightened state of imagination, they resembled hangman's nooses, motionless in the mild morning air, waiting for tenants. She shivered slightly and hurried to catch up with Beveridge.

He pointed up to the high cliffs where the treetops ended. "Up there, that's the Newton. You can see how sheer the cliffs are. Nobody was coming down there without us knowing. The guys in charge figured the kidnappers had to come along the path one way or the other, so they put most of the team here in the trees." He turned and pointed to what looked like a huge boulder by the side of the path. "And a guy with a rifle up there on top of the Lady's Rock." He gave a derisive snort of laughter. "Facing the wrong way, like."

"It's much smaller than I remember from when I was wee." Looking at it now, Karen found it hard to believe that anybody had bothered to name so insignificant a chunk of sandstone. The side next to the path was a straight cliff about twenty-five feet high, pitted with holes and striated with cracks. Small boy paradise. On the other side, it fell away in a forty-five-degree slope, dotted with tussocks of coarse grass and small shrubs. It had loomed much larger in her imagination.

"It's not just your memory playing tricks on you. I know it doesn't look much now, but twenty years ago, the shore was a lot lower, and the rock was a lot bigger. Come on, I'll show you what I mean."

Beveridge led the way down the side of the rock. The path was little more than grass bent by the passage of feet; a far cry from the EU's well-dressed track. They walked a dozen paces past the rock on to what seemed to be a narrow road made of coarse concrete. A few feet along, a rusted metal ring was set into the concrete. Karen frowned, trying to make sense of it. She let her eye follow the road, which bent at an angle before eventually meeting the sea. "I don't get it," she said.

"It was a quay," Beveridge said. "That's a mooring ring. Twenty years ago, you could bring a decent-sized boat along here. The shore was somewhere between eight and fifteen feet lower than it is now, depending on where you stand. This is how they did it."

"Jesus," Karen said, taking it all in; the sea, the rock, the quay, the splay of the woodland behind them. "Surely we must have heard them coming in?"

Beveridge smiled at her, like a teacher with a favourite pupil. "You'd think so, wouldn't you? But if they were using a small open boat, you could bring it in on the rising tide with just oars. With a good boatman, you wouldn't hear a thing. Besides, when you're up on the path, the rock itself acts as a baffle. You can hardly hear the sea itself. When it came to the getaway, you could give it full throttle, of course. You could be at Dysart or Buckhaven by the time they got the helicopter scrambled."

Karen studied the lie of the land again. "Hard to believe nobody thought about the sea."

"They did." Beveridge spoke abruptly.

"You mean, you did?"

"I did. So did my sergeant." He turned away and stared out to sea.

"Why did nobody listen to you?"

He shrugged. "They listened, I'll give them that. We had a

briefing with DI Lawson and Brodie Grant. The two of them just didn't believe it could be done. A big boat would be too obvious, too easy to identify and chase down. A wee boat would be impossible because you couldn't subdue an adult hostage in an open boat. They said the kidnappers had shown forward planning and intelligence and they wouldn't take stupid risks like that." He turned back to her and sighed. "Maybe we should have pushed harder. Maybe if we had, there would have been a different outcome."

"Maybe," Karen said thoughtfully. So far, everyone had looked at the botched ransom operation from the point of view of the police and Brodie Grant. But there was another angle that deserved consideration. "They did have a point, though, didn't they? How did they manage it in a wee boat? They've got an adult hostage. They've got a baby hostage. They've got to handle the boat and keep the hostages under control, and there can't have been that many of them in a boat small enough to have avoided detection coming in. I wouldn't like to have been running that operation."

"Me neither," Beveridge said. "It would be hard enough getting that lot ashore if everybody was on the same side, never mind at odds with each other."

"Unless they were there a good long while before the actual handover. It would have been dark by four, and the quay itself would hide a wee boat from most lines of sight . . ." She pondered. "When did you guys set up?"

"We'd supposedly had the whole area under surveillance from two. The advance teams were in place by six."

"So theoretically, they could have sneaked in after it got dark and before your boys were on station," she said thoughtfully.

"It's possible," Beveridge said, sounding unconvinced. "But how could they be sure we didn't have the quay staked out? And how could you be sure of keeping a six-month-old kid quiet in the freezing cold for three or four hours?"

Karen walked out along the old quay, marvelling at the movement of the shoreline. The more she found out about the mechanics of this case, the less sense it made. She didn't think she was stupid. But she couldn't get things to add up. There had never been a single verified sighting of Cat or Adam after they'd been snatched. Nobody had witnessed anyone staking out her cottage, or the snatch itself. Nobody had seen them arrive at the ransom handover site. Nobody had seen them escape. If it hadn't been for the very real corpse of Cat Grant, she could almost believe it had never happened.

But it had.

Rotheswell Castle

Brodie Grant handed Bel's report to his wife and started fiddling with the espresso machine in his office. "She's doing surprisingly well," he said. "I wasn't sure about this arrangement of Susan's, but it seems to be paying off. I thought we should use a private investigator, but the journalist seems to be doing just as well."

"She has more at stake than a private eye would, Brodie. I think she's almost as desperate for a result as we are," Susan Charleson said, helping herself to a glass of water and sitting down on the window seat. "With her unparalleled access to you, I suspect she sees a bestseller in this."

"If she helps us get some answers after all this time, she deserves it," Judith said. "You're right, this is an impressive start. What does DI Pirie think?"

Grant and Susan exchanged a quick glance of complicity. "We've not passed it on to her yet," Grant said.

"Why ever not? I imagine she'd find it useful." Judith looked from one to the other, puzzled.

"I think we'll just keep it to ourselves for now," Grant said, pressing the button that forced pressurized hot water through the

coffee to produce an espresso as perfectly as any Italian barista. "My experience with the police last time wasn't exactly a happy one. They cocked things up and my daughter ended up dead. This time, I'd rather leave as little up to them as possible."

"But this is a police matter," Judith protested. "You brought them in. You can't ignore them now."

"Can't I?" His head came up. "Maybe if I'd ignored them last time and done things my way, Cat would still be alive. And it would be Adam . . ." He stopped abruptly, realizing that nothing he said could get him out of the hole he'd just dug.

"Quite," Judith said, her voice sharp as a splinter. She tossed the papers on his desk and walked out.

Grant pulled a face. "Drum ice," he said as the door closed behind his wife. "I didn't handle that as well as I might have. Tricky things, words."

"She'll get over it," Susan said dismissively. "I agree. We should keep this to ourselves for now. The police are notoriously incapable of keeping a lid on information."

"It's not that I'm bothered about. I'm more worried about them cocking it up again. This could be the last chance we have of finding out what happened to my daughter and my grandson, and I don't want to take a chance on it going sour. It matters too much. I should have taken more control last time. I won't make that mistake again."

"We will have to tell the police eventually, if Bel Richmond comes up with a serious suspect," Susan pointed out.

Grant raised his eyebrows. "Not necessarily. Not if he's dead."

"They'll want to clear up the case."

"That's not my problem. Whoever destroyed my family deserves to be dead. Bringing the police into it won't make that happen. If they're already dead, well and good. If they're not—well, we'll cross that bridge when we come to it."

Little shocked Susan Charleson after three decades of working for Brodie Grant. But for once, she felt a tremor pass through her

calm certainties. "I'm going to pretend I never heard that," she said.

"That's probably a good idea," he said, finishing his espresso. "A very good idea."

Glenrothes

Phil was on the phone when Karen got back to the office, handset tucked into his neck, scribbling in his notebook. "And you're sure about that?" she heard him say as she tossed her bag on the desk and headed for the fridge. By the time she returned with a Diet Coke, he was staring glumly at his notes. "That was Dr. Wilde," he said. "She got someone to do a quick and dirty on the DNA. There's no linkage between Misha Gibson and the body in the cave."

"Shit," Karen said. "So that means the body isn't Mick Prentice."

"Or else Mick Prentice wasn't Misha's father."

Karen leaned back in her chair. "It's a good thought, but, if I'm honest, I don't really think that Jenny Prentice was playing away when Mick was still on the scene. We'd have heard about it by now. A place like the Newton, it's a gossip factory. There's always somebody ready to shop their neighbour. I think the chances are that body's not Mick's."

"Plus you said the neighbour was adamant that Jenny was in love with him. That Tom Campbell was a poor second."

"So if we're right about him being Misha's dad, maybe Mick was the one that put the body there. He knew the caves, he probably could have got his hands on explosives. We need to find out if he had any experience of shot-firing. But burying a body in the Thane's Cave would be a pretty good reason for disappearing. And we know somebody else went on the missing list around the same time . . . " Karen reached for her notebook and flicked back

the pages till she found what she was looking for. She glanced at her watch. "You think it's too late to ring somebody at half past eleven?"

Phil looked baffled. "Too late how? It's not even dinner time."

"I mean at night. In New Zealand." She reached for the phone and keyed in Angie Mackenzie's number. "Mind you, it's a murder inquiry now. That always comes ahead of beauty sleep."

A grumpy male voice answered. "Who is this?"

"I'm sorry to bother you, this is Fife Police. I need a word with Angie," Karen said, trying to sound ingratiating.

"Jesus. Do you know what time it is?"

"Yes, I'm sorry. But I do need to talk to her."

"Hang on, I'll get her." Off the phone, she could hear him calling his wife's name.

A full minute passed, then Angie came on the line. "I was in the shower," she said. "Is this DI Pirie?"

"That's right." Karen softened her voice. "I'm really sorry to bother you, but I wanted to let you know we've found human remains behind a rock fall in one of the Wemyss caves."

"And you think it might be Andy?"

"It's possible. The timescale looks like it might fit."

"But what would he have been doing in the caves? He was an outdoor kind of guy. One of the things he liked about being a union official was never having to go underground again."

"We don't know yet that it is your brother," Karen said. "Those are questions for later, Angie. We still have to identify the remains. Do you happen to know who your brother's dentist was?"

"How did he die?"

"We're not sure yet," Karen said. "As you'll appreciate, it's been a long time. It's a bit of a forensic challenge. I'll keep you informed, of course. But in the meantime, we have to treat it as an unexplained death. So, Andy's dentist?"

"He went to Mr. Torrance in Buckhaven. But he died a couple of years before I left Scotland. I don't even know if there's still a

practice there." She sounded slightly panicked. Shock kicking in, Karen thought.

"Don't worry, we'll check it out," she said.

"DNA," Angie blurted out. "Can you get DNA from . . . what you found?"

"Yes, we can. Can we arrange for your local police to take a sample from you?"

"You don't need to. Before I left for New Zealand, I arranged with my lawyer to hold a certified copy of my DNA analysis." There was a crack in her voice. "I thought he'd gone off a mountain. Or maybe walked into a loch with his pockets full of stones. I didn't want him lying unclaimed. My lawyer has instructions to provide my DNA analysis to the police where there's an unidentified body the right age." Karen heard a sob from the other side of the world. "I always hoped . . . "

"I'm sorry," Karen said. "I'll get in touch with your lawyer."

"Alexander Gibb," Angie said. "In Kirkcaldy. I'm sorry, I need to go now." The line went dead abruptly.

"Not too late, then," Phil said.

Karen sighed. Shook her head. "Depends what you mean by 'too late.'"

Hoxton, London

Jonathan speed-dialled Bel's mobile. When she answered, he spoke quickly. "I can't chat, I've got a meeting with my tutor. I've got some stuff to e-mail you, I'll get to it in an hour or so. But here's the headline news—Daniel Porteous is dead."

"I know that," Bel said impatiently.

"What you don't know is that he died in 1959, aged four."

"Oh, shit," said Bel.

"I couldn't have put it better myself. But here's the kicker. In November 1984, Daniel Porteous registered the birth of his son."

Bel felt light-headed, realized she was holding her breath, and released it in a sigh. "No way."

"Trust me, this is the business. Our Daniel Porteous somehow managed to have a son twenty-five years after he died."

"Wild. And who was the mother?"

Jonathan chuckled. "This just gets better, I'm afraid. I'm going to spell it out to you. F-R-E-D-A C-A-L-L-O-W is the name on the birth certificate. Say it out loud, Bel."

"Freda Callow." *Sounds like Frida Kahlo. The cheeky bastard.*

"He has a sense of humour, our Daniel Porteous."

Dundee

Karen found River at the university, sitting at her laptop in a small room lined with shelves of plastic boxes crammed with tiny bones. "What in the name of God is this place?" she said, plonking herself down on the only other chair.

"The professor here is the world's leading expert in the bones of babies and young children. You ever seen a foetus's skull?"

Karen shook her head. "And I don't want to, thank you very much."

River grinned. "OK, I won't make you. Let's just say when you've seen that, you understand where ET came from. So, I take it this isn't a social visit?"

Karen snorted. "Oh, sure. The anatomy department of Dundee University is my number one destination when I want a good day out. No, River, this is not a social visit. I'm here because I need a clear chain of custody on a piece of evidence in a homicide inquiry." She placed a sheet of paper on the desk. Angela Kerr's solicitor had been quick off the mark. "That is the DNA of Andy Kerr's sister Angie. I'm formally requesting that you compare it with the DNA extracted from the human remains discovered in the area known as the Thane's Cave lying between East Wemyss

and Buckhaven. You'll get that in writing as soon as I get back to my desk."

River looked at it with curiosity. "Fast work, Karen. Where did this come from?"

"Angie Mackenzie is a woman of foresight," Karen said. "She lodged it with her lawyer. Just in case a body ever turned up." As she spoke, River was tapping keys on her laptop.

"I'll do you a detailed report in writing," she said slowly, distracted by what she was looking at. "And I'll need to scan this in to be certain . . . But quick and dirty says these two people are closely related." She looked up. "Looks like you might have an ID for your mystery man."

Siena

How, Bel wondered, could Italian investigative journalists cope? She'd thought British bureaucracy wearisome and cumbersome. But compared to Italian red tape, it was open access all areas. First there had been the office-to-office shuttle. Then the form-filling shuffle. Then the blank-stared shut-out from officials who clearly minded their leisure being interrupted by someone who wanted them to do their job. It was a miracle anyone ever managed to find out anything in this country.

Towards the end of the morning, she began to fear that time would run out before she had learned what she needed to know. Then, with minutes to go before the registry office closed for lunch, a bored-looking bottle blonde called her name. Bel rushed to the counter, fully expecting to be fobbed off till the next day. Instead, in exchange for a bundle of unreceipted euros, she was handed two sheets of paper that appeared to have been photocopied on a machine painfully short of toner. One was headed *Certificato di Morte*, the other *Certificato di Residenza*. In the end, she'd got more than she'd bargained for.

The death certificate of Daniel Simeon Porteous stated simply

that he had died on 7th April 2007 at the age of fifty-two at the Policlinico Le Scotte, Siena. His parents were named as Nigel and Rosemary Porteous. And that was it. No cause of death, no address. About as much use as a chocolate teapot, Bel thought bitterly. She considered going to the hospital to see if she could find anything out, but dismissed the idea at once. Breaching the walls of officialdom would be impossible for someone who didn't know the system. And the chances of finding someone bribable who remembered Daniel Porteous after this much time were remote and probably beyond her command of the language.

With a sigh, she turned to the other certificate. It seemed to be a short list of addresses and dates. It didn't take her long to figure out that this was a record of where Daniel had lived since he had come to the Commune di Siena in 1986. And that the last address on the list was where he had been living when he died. Even more surprising was that she knew more or less where it was. Costalpino was the last village she'd driven through on her way from Campora. The main road twisted down through its main street in a series of curves, the road flanked with houses, the occasional shop or bar tucked alongside.

Bel practically ran back to the car in spite of the sweaty heat of the middle of the day. She gasped with gratitude as the air conditioning kicked in and wasted no time getting out of the parking lot and on to the road heading for Costalpino. The man behind the counter of the first bar she came to provided excellent directions, and a mere fifteen minutes after leaving Siena she was parking a few doors down from the house where she expected to find Gabriel Porteous. It was a pleasant street, wider than most in that part of Tuscany. Tall trees shaded the narrow pavements, and waist-high walls topped with iron railings separated small but well-kept villas from each other. Bel felt the pulse of excitement in her throat. If she was right, she could be about to come face to face with Catriona Maclennan Grant's lost son. The police had failed twice, but Bel Richmond was about to show them all how it was done.

So confident was she that she could hardly credit the sign on the front of the yellow stuccoed villa. She checked the numbers again to make sure she was standing before the right house, but there was no mistake. The dark green shutters were pulled tight. The plants in the tall terracotta pots that lined the driveway looked tired and dusty. Occasional weeds were poking through the gravel, and junk mail poked out of the mailbox. All of which reinforced the SE VENDE sign with the name and number of an estate agent in nearby Sovicille. Wherever Gabriel Porteous was, it looked like it wasn't here.

It was a setback. But it wasn't the end of the world. She'd overcome bigger obstacles than this on her way to the stories that had built her reputation as someone who could deliver. All she had to do was formulate a plan of campaign and follow it through. And for once, if she came up against stuff she couldn't do, she could call on Brodie Grant's resources to make it happen. It wasn't exactly a comforting feeling, but it was better than nothing.

Before she headed off for Sovicille, she decided to check out the neighbours. It wouldn't be the first time that somebody who knew they were being looked for went out of their way to make their home look uninhabited. Bel had already noticed a man on the loggia of a villa diagonally opposite the Porteous house. There had been nothing covert about the way he had been watching her walk up the street and study the sign. Time for a little stretching of the truth.

She crossed the road and greeted him with a wave. "Hello," she said.

The man, who could have been anywhere from mid-fifties to mid-seventies, gave her an appraising look, making her wish she'd worn a loose T-shirt rather than the close-fitting spaghetti-strapped top she'd chosen that morning. She loved Italy, but God, she hated the way so many of the men eyed up women as if they were meat on the hoof. This one wasn't even good looking: one eye bigger than the other, a nose like an ill-favoured parsnip, and

hair spurting out of the top of his vest. He smoothed down an eye-brow with his little finger and gave her a crooked smile. "Hello," he said, managing to make it sound freighted with meaning.

"I'm looking for Gabriel," she said. She gestured over her shoulder to the house. "Gabriel Porteous. I'm a friend of the family, from England. I haven't seen Gabriel since Daniel died, and this is the only address I have. But it's up for sale, and it doesn't look like Gabe's still living there."

The man stuck his hands in his pockets and shrugged. "Gabriel hasn't lived here for more than a year now. He's supposed to be studying some place, I don't know where. He was back for a while before his father died, but I haven't seen him for a couple of months now." His smile reappeared, a little wider than before. "If you want to give me your number, I could call you if he shows up?"

Bel smiled. "That's very kind, but I'm only going to be here for a few days. You said Gabe's 'supposed' to be studying." She gave him a look of complicity. "Like you think he's up to his old games?"

It did the trick. "Daniel, he worked hard. He didn't mess about. But Gabe? He's always messing about, hanging out with his friends. I never saw him with a book in his hands. What kind of studying is he going to be doing? If he'd been serious, he'd have signed up at the university in Siena, so he could live at home and only think about his studies. But no, he goes off some place he can have a good time." He tutted. "Daniel was sick for weeks before Gabe showed up."

"Maybe Daniel didn't tell him he was ill. He's always been a very private person," Bel said, making it up as she went along.

"A good son would have visited regularly enough to know," the man said stubbornly.

"And you've no idea where he's studying?"

The man shook his head. "No. I saw him on the train one time. I was coming back from Firenze. So, somewhere up north. Firenze, Bologna, Padova, Perugia. Could be anywhere."

"Oh well. I guess I'll just have to try the estate agent. I really wanted to see him. I feel bad about missing the funeral. Were many of the old crowd here?"

He looked surprised. "It was a private funeral. None of us neighbours knew anything about it till it was all over. I spoke to Gabe afterwards. I wanted to pay my respects, you know? He said his father had wanted it that way. But now, you're talking like there was something to miss." He took out a pack of cigarettes and lit up. "You can't trust kids to tell you the truth."

There was no real reason why she should try to cover her tracks with someone she would never meet again, but she'd always believed in keeping her hand in. "What I was talking about was more of a gathering for some of Daniel's old friends. Not a funeral as such."

He nodded. "The arty crowd. Kept them separate from his friends in the village. I met a couple of them once. They turned up at the villa when a few of us were round there playing cards. Another English guy and a German woman." He hawked and spat over the stone balustrade. "I've got no time for the Germans. That Englishman, though. You'd have thought he was German, the way he acted."

"Matthias?" Bel guessed.

"That's the one. High-handed. Treated Daniel like he was dirt. Like he was the one with the brains and the talent. And very amused to find Daniel playing cards with the locals. The funny thing was, Daniel let him get away with it. We didn't stick around, just finished the hand and left them to it. If that's your arty intellectuals, you can keep them."

"I never had much time for Matthias myself," Bel said. "Anyway, thanks for your help. I'll head over to Sovicille and see if the agents can put me in touch with Gabe."

It was amazing how even the least promising encounter could add to your store of knowledge, Bel thought as she set off again. Now she had a second source who thought Matthias was English,

in spite of his Teutonic name and German partner. A Brit who didn't acknowledge his roots, who had artistic leanings, a connection to the ransom notes and a friendship with the man whose son looked eerily like Cat Grant and her father. It was starting to make a tantalizing shape in her mind.

Two young men, struggling artists, aware of Cat Grant because she moved in the same circles. Aware, too, of her father's wealth. They hatch a plan to feather their nests. Kidnap Cat and her kid, make it look like some political thing. Take off with the ransom and never have to paint for anyone but themselves ever again. Say it fast, it sounds like a great idea. Only it all goes horribly wrong and Cat dies. They're left with the kid and the ransom money, but now they're the focus of a murder manhunt.

Professional criminals would know what to do and be cold-hearted enough to do it. But these are nice, civilized boys who thought they were indulging in something only marginally more serious than an art college prank. They've got a boat, so they just keep going across the North Sea to Europe. Daniel ends up in Italy, Matthias in Germany. And somewhere along the line, they decide not to kill or abandon the child. For whatever reason, they keep him. Daniel brings him up as his son. Cushioned by the ransom money, he sets them up in comfort and then, ironically, becomes a reasonably successful artist. But he can't cash in on his success with media interviews and personality-based marketing because he knows he's a criminal on the run. And he knows his son is not Gabriel Porteous. He's Adam Maclennan Grant, a young man cursed with a distinctive face.

It was an attractive scenario, no doubt about it. It begged questions, true—How did they get their hands on the ransom, given that they were floundering round in the dark trying to find the dead woman who'd been carrying it? How did they outwit the tracking devices the cops had planted on the ransom? How did they get away by boat and not get spotted by the helicopter? How would a couple of art students have got hold of a gun back then? All good questions, but ones she was sure she could finesse one

way or another. She'd have to; this was too good to pass by just for the sake of a few awkward details.

She'd known she was on to a good thing with her unique access to Brodie Grant, but this was infinitely better than she could have hoped for. This was the kind of story that would make her name. Establish her as one of that handful of journalists whose name alone stood for the story. Stanley with the discovery of Dr. Livingstone. Woodward and Bernstein with Watergate. Max Hastings with the liberation of Port Stanley. Now they'd be able to include Annabel Richmond with the unmasking of Adam Maclennan Grant.

There were lots of gaps in the story at this point, but they could be filled in later. What Bel needed now was the young man known as Gabriel Porteous. With or without his cooperation, she needed a sample of his DNA so Brodie Grant could establish whether this really was his missing grandson. And then her fame was assured. Newspaper features, a book, maybe even a movie. It was a thing of beauty.

The estate agent's office was tucked in a side street just off Via Nuova. The window was filled with A4 sheets displaying photographs and a few details of each property. The Porteous villa was there, its rooms and facilities enumerated without comment. Bel pushed open the door and found herself in a small grey office. Grey filing cabinets, grey carpet, pale walls, grey desks. The only inhabitant, a woman in her thirties, was like a bird of paradise by comparison. Her scarlet blouse and turquoise necklace blazed brightly, drawing the eye to her tumble of dark hair and perfectly made-up face. She was definitely making the most of what she had, Bel thought as they found their way through the pleasantries.

"I'm afraid I'm not actually in the market for a property," Bel said, with an apologetic gesture. "I'm trying to contact the owner of the villa you have for sale in Costalpino. I was an old friend of Gabriel Porteous's father, Daniel. Sadly, I was in Australia when Daniel died. I'm back in Italy for a while and I wanted to see Gabriel, to pay my respects. Is it possible for you to put me in touch with him?"

The woman rolled her eyes. "I'm really sorry. I can't do that."

Bel reached for her wallet. "I could pay for your time," she said, using one of the traditional formulae for corruption.

"No, no, it's not that," the woman said, not in the least offended. "When I say I can't, that's what I mean. Not that I won't. I can't." She sounded flustered. "It's very unusual. I don't have an address or a phone number or even an e-mail for Signor Porteous. Not even a mobile phone. I tried to explain this was very unconventional, and he said, so was he. He said now his father was dead, he planned to go travelling and he didn't want to be tied down to his past." She gave a wry little smile. "The sort of thing young men think is very romantic."

"And the rest of us think is impossibly self-indulgent," Bel said. "Gabriel always had a mind of his own. But how are you supposed to sell the house if you can't contact him? How can he agree to a sale?"

The woman spread her hands. "He phones us every Monday. I said to him, 'What if someone comes in on a Tuesday morning with an offer?' He said, 'In the old days people had to wait for letters to go back and forth. It wouldn't kill them to wait till the next Monday if they're serious about buying the house.'"

"And have there been many offers?"

The woman looked glum. "Not at that price. I think he needs to drop at least five thousand before anyone will get serious. But we'll see. It's a nice house, it should find a buyer. He's emptied it, too, which makes the rooms look so much bigger."

Since Bel's next suggestion had been that she take a look round to see if there were any clues to Gabriel's whereabouts, that last revelation came as a disappointment. Instead, she fished a business card from her Filofax. One of the ones that had her name, her mobile number, and her e-mail address. "Never mind," she said. "Perhaps when he rings on Monday you could ask him to get in touch? I knew his father for the best part of twenty years, I'd just like to get together." She handed the card over.

Scarlet fingernails plucked it from her hand. "Sure, I'll pass the

message on. And if you ever want a property around here . . . ?"
She waved at the array of details in the window. "We've got a
great selection. I always say we are on the unfashionable side of
the autostrada so the prices are lower but the properties are just as
beautiful."

Bel walked back to the car, knowing there was nothing else
she could do here. Five days until Gabriel Porteous would get her
message, and then who knew whether he would get in touch? If
he didn't, tracking him down would be a job for a private detective in Italy, someone who knew the ropes and the right hands to
ply with brown envelopes of cash. It would still be her story, but
someone else could do the grunt work. Meanwhile she needed to
get back to Rotheswell to see if she could nail down a chat with
Fergus Sinclair.

Time to exploit the resources Brodie Grant had put at her disposal. She dialled Susan Charleson's number. "Hello, Susan," she
said. "I need a flight back to the UK asap."

Glenrothes

The trouble with cold cases, Karen thought, was that there were
so many brick walls to run into. When there really was nothing
you could do next. No obvious witness to interview. No convenient forensic samples to organize. At times like this, she was at
the mercy of her wits, twisting the Rubik's Cube of what she knew
in the hope that a new pattern would emerge.

She'd interviewed everyone who might have been able to give
her a lead on what had happened to Mick Prentice. In a way, that
should have worked to her advantage when it came to investigating Andy Kerr's death because she'd been talking to them in the
context of a missing-person inquiry. Unless they had something
to hide, people were generally pretty open with the police when it
was a matter of helping to track down those missing and missed.
When it came to murder, they were more reluctant to talk. And

what they did say was hedged with qualifications and anxieties. Theoretically, she knew she should go back to her witnesses and take fresh statements, statements that might lead her to other witnesses who remembered what Andy Kerr was saying and doing leading up to his death. But experience told her it would be a waste of time now suspicious death was on the agenda. Nevertheless, she'd sent the Mint and a bright new CID aide on a fresh round of interviews. Maybe they'd get lucky and pick up on something she'd missed. A girl could always hope.

She turned to the Cat Grant file. She was stalled there, too. Until she had a proper report from the Italian police, it was hard to see where she could make progress. There had been one stroke of luck in that area, however. She'd contacted Fergus Sinclair's parents, hoping to find out where their son was working so she could arrange to interview him. To her surprise, Willie Sinclair had told her his son would be arriving with his wife and children that very evening for their annual Scottish holiday. Tomorrow morning, she would have the chance to talk to Fergus Sinclair. It sounded as if he was the only person left who might be willing to unlock Cat Grant's personality. Her mother was dead, her father was unwilling, and the files offered no clue to any close friendships.

Karen wondered if the lack of friendships was a matter of choice or personality. She knew people so invested in their work that the lack of close human relationships was something they barely noticed. She also knew others who were desperate for intimacy but whose only talent was for driving people away. She counted her blessings; she had friends whose support and laughter filled an important place in the pattern of her days. It might lack a central relationship at its heart, but hers was a life that felt solid and comfortable.

What had Cat Grant's life felt like? Karen had seen women consumed by their children. Witnessing their adoring gaze, she'd felt uneasy. Children were human, not gods to be worshipped. Was Cat's child the centre of her world? Had Adam occupied her entire heart? It looked that way from the outside. Everyone

assumed Fergus was the baby's father, but even if he hadn't been, one thing seemed clear. Adam's father had been banished from his life; it appeared that his mother had wanted him for herself alone.

Or maybe not. Karen wondered if she was looking through the wrong end of the telescope. What if it hadn't been Cat who had cast out Adam's father? What if he'd had his own reasons for refusing to accept a role in his son's life? Maybe he didn't want the responsibility. Maybe he had other responsibilities, another family whose call on him was thrown into relief by the prospect of another child. Maybe he'd only been passing through and had gone before she even knew she was pregnant. There was no denying that there were other possibilities worth considering.

Karen sighed. She'd know more after she'd spoken to Fergus. With luck, he'd help her to narrow down some of her wilder ideas. "Cold cases," she said out loud. They'd break your heart. Like lovers, they tantalized with promises that this time it would be different. It would start out fresh and exciting, you'd try to ignore those little niggles that you felt sure would disappear as you got to understand things better. Then suddenly it would be going nowhere. Wheels spinning in a gravel pit. And before you knew it, it was over. Back to square one.

She glanced up at Phil, who was working computer databases, trying to track down a witness in another case. Probably just as well it had never come to anything between them. Better to have him as a friend than to end up with bitterness and frustration measuring the distance between them.

And then the phone rang. "CCRT, DI Pirie speaking," she said, trying not to sound as pissed off as she felt.

"This is Capitano di Stefano from the carabinieri in Siena," a heavily accented voice said. "You are the officer I have talk-ed to about the Villa Totti near Boscolata?"

"That's right." Karen sat bolt upright, reaching for pen and paper. She remembered di Stefano's style from their previous conversation. His English was surprisingly good as far as vocabulary

and grammar were concerned, but his accent was atrocious. He
pronounced English as if it were an opera libretto, the stresses in
peculiar places and the pronunciation bordering on the bizarre.
None of that mattered. What mattered was the content, and Karen
was prepared to work as hard as necessary to nail that down precisely. "Thanks for calling."

"It is my pleasure," he said, every vowel distinct. "So. We have
visited the villa and talk-ed to the neighbours."

Who knew you could get four syllables out of "neighbours"?
"Thank you. What did you discover?"

"We have found more copies of the poster you e-mailed to us.
Also, we have found the silk screen it was printed from. Now we
are processing fingerprints from the frame and other areas inside
the villa. You understand, many people have been here, and there
are many traces everywhere. As soon as we have process-ed the
prints and the other material, we will transmit our results as well
as copies of prints and DNA sequences. I am sorry, but this aspect
is not a priority for us, you understand?"

"Sure, I understand. Is there any chance that you can send us
some samples so we can run our own tests? Just in the interests
of time, not for any other reason." *Like, everybody in my department thinks you're useless.*

"*Sì.* This is already done. I have sent you samples from the
bloodstain on the floor and other bloodstains in the kitchen and
living area. Also, other evidence where we have multiple samples.
So, I hope this will come to you tomorrow."

"What did the neighbours have to say?"

Di Stefano tutted down the phone. "I think you call these
people lefties. They don't like the carabinieri. They're the kind of
people who go to Genoa for the G8. They are more on the side
of the people living illegally at the Villa Totti. So my men did not
learn a great deal. What we know is that the people living here
ran a travelling puppet show called BurEst. We have some photos
from a local newspaper and my colleague is e-mailing them to you.
We know some names, but these are the kind of people who can

very easily disappear. They live in the world of the black economy. They don't pay taxes. Some of them are probably illegals."

Karen could almost see him spreading his hands in a frustrated shrug. "I appreciate how hard it is. Can you send me a list of the names you do have?"

"I can tell you now. We only have first names for these people. So far, no family names. Dieter, Luka, Maria, Max, Peter, Rado, Sylvia, Matthias, Ursula. Matthias was in charge. I am sending you this list. Some of them, we think we know their nationality, but it's mostly guessing I think."

"Any Brits?"

"It does not look like it, although one of the neighbours thinks that Matthias might have been English because of his accent."

"It's not a very English name."

"Maybe it wasn't always his name," Di Stefano pointed out. "The other thing about people like this, they are always trying to be born again. New name, new history. So, I am sorry. There does not seem to be very much here for you."

"I appreciate what you've been able to do. I know it's hard to justify manpower on something like this."

"Inspector, it looks to me as if there has been a murder in this villa. We are treating this as a possible murder investigation. We try to help you in the course of this, but we are more interested in what we think happened three months ago than what happened twenty-two years ago in your country. We are looking very hard for these people. And tomorrow, we bring in the body dogs and the ground-penetrating radar to see if we can find a burial site. It will be difficult because it is surrounded by woodland. But we must try. So you see, manpower is not the issue here."

"Of course. I didn't mean to suggest you weren't taking it seriously. I know what it's like, believe me."

"There is one more thing we have found out. I don't know if this matters to you, but there has been an English journalist here, asking questions."

Karen was momentarily at a loss. Nothing had been released

to the media. What was a hack doing sniffing around in her case? Then suddenly it dawned on her. "Bel Richmond," she said.

"Annabel," di Stefano said. "She was staying at a farm up the hill. She left this afternoon. She is returning to England tonight. The neighbours, they said that she wanted to know about the BurEst people. A teenager told one of my men that she was also interested in a couple of friends of Matthias. An English painter and his son. But I have no names, no photos, no nothing. Maybe you can speak to her? Maybe the Boscolata neighbours think it's better to talk to a journalist rather than a cop, what do you think?"

"Tragically, I think you might be right," Karen said bitterly. They exchanged pleasantries and empty promises to visit, then the call was over. Karen screwed up a piece of paper and tossed it at Phil. "Can you believe it?"

"What?" He looked up, startled. "Believe what?"

"Fucking Bel Richmond," she said. "Who does she think she is? Brodie Grant's private police force?"

"What's she done?" He stretched his arms above his head, grunting as he unkinked his spine.

"She's only been to Italy." Karen kicked her bin. "Fucking cheeky bitch. Going out there and chatting up the neighbours. The neighbours that won't say much to the police because they're a bunch of unreconstructed lefties. Jesus Christ."

"Wait a minute," Phil said. "Shouldn't we be pleased about that? I mean, that we've got somebody getting the dirt, even if it's not our colleagues in Italy?"

"Can you come over here and look in my e-mail inbox and show me the message from Bel Richmond telling us what she's dug up in bloody Tuscany? Can you maybe let your fingers do the walking through my in-tray and show me the fax she sent with all the information she's gathered out there? Or maybe it's my voice-mail that I've lost the ability to access? Phil, she might have found out all sorts. But we're not the ones she's telling."

Edinburgh Airport to Rotheswell Castle

Bel watched the empty luggage carousel circling, exhaustion rendering her incapable of thought. A drive to Florence airport, mysteriously hidden somewhere in the suburbs, a dismal journey via Charles de Gaulle, an airport surely designed by a latter-day Marquis de Sade, and still miles to go before she could sleep. And not even in her own bed. At last, suitcases and holdalls started to appear. Ominously, hers was absent from the first circuit. She was about to throw a tantrum at the ground services counter when her case finally came limping through, one latch hanging loose from its moorings. In her heart, she knew Susan Charleson had nothing to do with her miseries, but it was nice to have someone to lay irrational blame on. Please God she'd sent someone to pick her up.

Her spirits should have risen when she emerged in the arrivals area to see there was indeed a chauffeur waiting for her. But the fact that it was Brodie Grant himself only emphasized her weariness. She wanted to curl up and sleep or curl up and drink. She did not want to spend the next forty minutes under interrogation. He wasn't even paying her, now she came to think about it. Just fronting her exes and opening doors for her. Which wasn't exactly a bad gig. But in her book, it didn't entitle him to 24/7 service. *Like you're going to tell him that.*

Grant greeted her with a nod and they wrestled momentarily over the suitcase before Bel gave in gracelessly. As they hustled through the terminal, Bel was conscious of eyes on them. Brodie Grant clearly had street recognition. Not many businessmen achieved that. Richard Branson, Alan Sugar. But they were familiar TV faces, on screen for reasons that were nothing to do with business. She didn't think Grant would be noticed in London, but here in Scotland, the punters knew his face in spite of his media shyness. Charisma, or just a big fish in a small pond? Bel wouldn't have liked to hazard a guess.

It wasn't just the punters. Outside the terminal, where signs and PA announcements strictly banned the parking of cars, an

armed police officer was standing next to Grant's Land Rover. He wasn't there to warn Grant or give him a ticket; he was there to make sure nobody messed with the Defender. Grant gave him a patriarchal nod as he loaded the case, then waved graciously as they drove off.

"I'm impressed," Bel said. "I thought it was just royalty that got that sort of treatment."

His face twitched as if he wasn't certain whether she was being critical. "In my country, we respect success."

"What? Three hundred years of English oppression hasn't knocked that out of you?"

Grant started upright, then realized she was teasing him. To her relief, he laughed. "No. You're much keener to knock success than we are. I think you like success too, Annabel. Isn't that why you're up here working with me instead of uncovering some ghastly tale of rape and sex trafficking in London?"

"Partly. And partly because I'm interested in finding out what happened." As soon as the words were out, she could have kicked herself for giving him the perfect opening.

"And what have you found out in Tuscany?" he asked.

As they raced through the night on empty roads, she told him what she had discovered and what she had surmised. "I came back because I don't have the resources to track Gabriel Porteous down," she concluded. "DI Pirie might be able to kick the Italian cops into action—"

"We're not going to be talking to DI Pirie about this," Grant said firmly. "We'll hire a private investigator. He can buy us the information we need."

"You're not going to tell the police what I've found out? You're not sharing the info with them? Or the photos?" She knew she shouldn't be shocked by the antics of the very rich, but she was taken aback by so adamant a response.

"The police are useless. We can wrap it up ourselves. If this boy is Adam, it's a family matter. It's not up to the police to find him."

"I don't understand," Bel said. "When we started this, you were

the one who went to the police. Now you want to shut them out."

There was a long silence. The dashboard lit up his profile against the night, the muscles in his jaw tight and hard. At last, he spoke. "Forgive me, but I don't think you've entirely thought this through, Bel."

"What have I missed?" She felt the old clutch of fear that news editors had always induced with their questioning of her copy.

"You talked about a significant amount of blood on the kitchen floor. You thought someone who lost that much blood would probably be dead. That means there's a body somewhere, and now the police are looking, they'll probably find it. And when they find it, they'll be looking for a killer—"

"And Gabriel was there the night before they all disappeared. You think Gabriel will come under suspicion," Bel said, suddenly getting it. "And if he is your grandson, you want him out of the picture."

"You got there, Bel," he said. "More than that, I don't want the Italian police fitting him up because they can't find the real killer. If he's not around, the temptation is less, especially since there will be other, more attractive suspects on the ground. The Italian private eyes won't just be looking for Gabriel Porteous."

Oh my God, he's going to have someone else fitted up. Just as an insurance policy. Bel felt nauseous. "You mean, you're going to find a scapegoat?"

Grant gave her an odd look. "What an extraordinary suggestion. I'm just going to make sure the Italian police get all the help they deserve." His smile was grim. "We're all citizens of Europe now, Bel."

Thursday, 5th July 2007; Kirkcaldy

Karen had conducted interviews in strange places before, but Ravenscraig Castle would probably have made the top five. When she'd asked Fergus Sinclair to meet her, he'd suggested the venue.

"That way, my wife can take the kids round the castle and down to the shore," he said. "This is our summer holiday. I don't see why we have to be cooped up just because you want to talk to me."

"The weather" would have been as good an answer as any. Karen was sitting on the remains of a wall with her anorak collar turned up against a sharp breeze coming off the sea; Phil sat next to her huddled into his leather jacket. "This better be worth it," he said. "I'm not sure whether it's rheumatism or piles I'm getting here, but I know it's not good for me."

"He's probably used to it. Working on a hunting estate like he does." Karen squinted up at the sky. The cloud was high and thin, but she'd still have put money on rain by lunchtime. "You know that, back in the Middle Ages, this was the St. Clair family seat?"

"That's why this part of Kirkcaldy's called Sinclairtown, Karen." Phil rolled his eyes. "You think he's trying to intimidate us?"

She laughed. "If I can survive Brodie Grant, I can survive a descendant of the St. Clairs of Ravenscraig. Do you think this is him?"

A tall, rangy man walked through the castle gatehouse followed by a woman almost as tall as him and a pair of small sturdy boys, each with a shock of bright blond hair like their mother. The lads looked around them and then they were off, running and jumping, clambering and exploring. The woman turned her face upwards and the man planted a kiss on her forehead, then patted her back as she turned to chase the boys. He looked around and caught sight of the two cops. He raised a hand in greeting and came towards them with quick, long strides.

As he approached, Karen studied a face she'd seen only in twenty-two-year-old photos. He'd aged well, though his face was weathered, the web of fine white lines round his sharp blue eyes a testament to time spent in the sun and the wind. His face was lean, the cheeks hollow, the outline of the bones clear beneath the skin. His light brown hair hung in a fine fringe, making him look almost medieval. He wore a soft plaid shirt tucked into moleskin

trousers, lightweight walking boots on his feet. She stood up and nodded hello. "You must be Fergus Sinclair," she said, extending a hand. "I'm DI Karen Pirie and this is DS Phil Parhatka."

He took her hand in one of those tight grips that always made her want to slap the other's face with her free hand. "I appreciate you meeting me here," he said. "I didn't want my parents subjected to the bad old memories again." His Fife accent was almost completely gone. If she'd been pressed, Karen might even have placed him as a German with exceptionally good English.

"No problem," she lied. "You know why we've reopened the case?"

He sat down on a piece of masonry at right angles to Karen and Phil. "My dad said it was something to do with the ransom poster. Another copy's turned up?"

"That's right. In a ruined villa in Tuscany." Karen waited. He said nothing.

"Not that far from where you live," Phil said.

Sinclair raised his eyebrows. "It's hardly on my doorstep."

"About seven hours' driving, according to the Internet."

"If you say so. I'd have said more like eight or nine. But either way, I'm not sure what you're implying."

"I'm not implying anything, sir. Just setting the location in context for you," Phil said. "The people who were squatting the villa were a group of puppeteers. They called themselves BurEst. The leaders were a couple of Germans called Matthias and Ursula. Ever come across them?"

"Christ," Sinclair said, exasperated. "That's a bit like asking a Scotsman if he ever ran across your auntie from London. I don't think I've ever been to a puppet show. Not even with the kids. And I don't know anyone called Matthias. The only Ursula I know works in my local bank, and I doubt very much she's into puppets in her spare time." He turned to Karen. "I thought you wanted to talk about Cat."

"We do. I'm sorry, I thought you wanted to know why we were reopening the case," she said earnestly, slipping easily into the

Good Cop role. "I suppose you've put it all behind you now. What with having a wife and kids."

He dropped his hands between his knees, interlocking his fingers. "I'll never put it behind me. I still loved her when she died. Even though she'd sent me packing, there wasn't a day went by that I didn't think of her. I wrote so many letters. Sent none of them." He closed his eyes. "But even if I could put Cat behind me, I'll never be able to do the same with Adam." He blinked hard and caught Karen's eye. "He's my son. Cat kept me from him when he was tiny, but the kidnappers have kept him from me for twenty-two and a half years."

"You think he's still alive?" Karen asked gently.

"I know the chances are that he was dead within hours of his mother. But I'm a parent. I can't help hoping that somewhere he's walking around in the world. Having a decent life. That's how I like to think of him."

"You were always sure he was your son," Karen said. "Even though Cat wouldn't acknowledge you as the father, you never wavered."

He twisted his hands together. "Why would I waver? Look, I know my relationship with Cat was on the skids by the time she got pregnant. We'd split up and got back together half a dozen times. We were hardly seeing each other at all. But we did spend the night together almost exactly nine months before Adam was born. When we were having our . . . difficulties, I asked her if there was someone else, but she swore there wasn't. And God knows she had no reason to lie. If anything, she'd have been better off saying she was seeing someone else. I'd have had to accept it was over then. So there wasn't anybody else in the frame." He unclenched his hands and splayed his fingers. "He even had my colouring. I knew he was mine the first time I clapped eyes on him."

"You must have been angry when Cat refused to admit Adam was yours," Karen said.

"I was furious," he said. "I wanted to go to court, to do all the tests."

"So why didn't you?" Phil said.

Sinclair stared down at the ground. "My mum talked me out of it. Brodie Grant hated the idea of me and Cat being together. Considering he came from dirt poverty in Kelty, he had some pretty high and mighty ideas about who was a fit partner for his daughter. And it certainly wasn't a ghillie's son. He was practically dancing a jig when we split up." He sighed. "My mum said if I fought Cat over Adam, Grant would take it out on her and my dad. They live in a tied cottage. Grant once promised my dad they could stay there for the rest of their lives. They've worked all their days for low wages. They've got no other provision for their old age. So I bit the bullet for their sakes. And I took myself off where I didn't have to face Cat or her father every day."

"I know you were asked this at the time, but did you ever consider taking revenge on these people who had wrecked your life?" Karen asked.

Sinclair's face screwed up as if he was in pain. "If I'd had any notion of how to take revenge, I would have. But I didn't have a clue and I didn't have any resources. I was twenty-five years old, I was working as a junior keeper on a hunting estate in Austria. I worked long hours, I spent my spare time learning the language and drinking. Trying to forget what I'd left behind. Believe me, Inspector, the idea of kidnapping Cat and Adam never crossed my mind. I just don't have that kind of mind. Would it have crossed yours?"

Karen shrugged. "I don't know. Happily I've never been put in that position. I do know that if I'd been treated like you were I'd have wanted to get my own back."

Sinclair's sideways nod conceded her point. "Here's what I know. My mum always used to say that living well is the best revenge. And that's what I've tried to do. I'm lucky to have a job I love in a beautiful part of the world. I can shoot and fish and climb and ski. I've got a good marriage and two bright, healthy boys. I don't envy any man, least of all Brodie Grant. That man took everything I valued away from me. Him and his daughter,

they hurt me. No getting away from it. But I've rebuilt my life and it's a good one. I've got history that's left me with scars, but those three"—he pointed to where his wife and sons were scrambling up a grassy bank—"those three make up for a hell of a lot."

It was a pretty speech, but Karen wasn't entirely convinced by it. "I think I'd resent him more, in your shoes."

"Then it's just as well you're not. Resentment isn't a healthy emotion, Inspector. It'll eat you away like cancer." He looked her straight in the eye. "There are those who believe there's a direct connection between the two. Me, I don't want to die of cancer."

"My colleagues interviewed you after Cat died. I expect you remember that quite well?"

His face twisted, and suddenly Karen saw a glimmer of the fires that Fergus Sinclair kept well banked down. "Being treated as a suspect in the death of the woman you love? That's not something you forget very easily," he said, his voice tight with contained anger.

"Asking someone for an alibi isn't necessarily treating them as a suspect," Phil said. She could tell he'd taken a dislike to Sinclair and hoped it wouldn't derail the interview. "We have to exclude people from our inquiries so we don't waste time investigating the innocent. Sometimes alibi evidence is the quickest way to take someone out of the picture."

"Maybe so," Sinclair said, his chin jutting forward defensively. "It didn't feel like that at the time. It felt like your people were putting a hell of a lot of effort into proving I wasn't where I said I was."

Time for the oil on the water, Karen thought. "Is there anything that has occurred to you since then that might be helpful?"

He shook his head. "What could I know that would be helpful? I've never been remotely interested in politics, never mind anarchist splinter groups. The people I mix with don't want a revolution." He gave a self-congratulatory little smile. "Unless maybe it's a revolution in ski design."

"To tell you the truth, we don't think it was an anarchist group," Karen said. "We have pretty good intelligence on the kind

of people who believe in direct action in the furtherance of their political ambitions. And the Anarchist Covenant of Scotland was never heard from before or since."

"Well, they weren't going to draw attention to themselves afterwards, were they? Not with charges of murder and kidnap hanging over their heads."

"Not under that name, no. But they walked away with a million pounds in cash and diamonds. That would be over three million in today's money. If they were dedicated political animals, you'd expect to see chunks of that money turning up in the coffers of radical groups with similar aims. My predecessors on this case asked MI5 for a watching brief. In the five years after Cat's murder, it never happened. None of the groups of fringe nutters suddenly came into money. So we don't think the kidnappers were really a bunch of political activists. We think they were likely closer to home."

Sinclair's expression said it all. "And that's why I'm here." He couldn't keep the sneer off his face.

"Not for the reason you think," Karen said. "You're not here because I suspect you." She held up her hands in a gesture of surrender. "We never managed to put you anywhere near the kidnap or the ransom scene. Your bank accounts never showed any unaccountable funds. Yes, I know, you're pissed off to hear we checked your bank accounts. Don't be. Not if you really care about Cat or Adam. You should be pleased that we've been doing our job the best we can all these years. And that it's pretty much put you in the clear."

"In spite of the poison Brodie Grant has tried to plant about me."

Karen shook her head. "You might be pleasantly surprised on that score. But anyway, here's the point. You're here because you are the only person who really knew Cat. She was too like her father; I suspect they might have ended up best pals, but they were still in the fighting phase. Her mother's dead. She didn't seem to have close female friends. So that leaves you as my only way in to

Cat's life. And I think that's where the secret of her death lies." She pinned Sinclair with the directness of her gaze. "So what's it going to be, Fergus? Are you going to help me?"

Catriona Maclennan Grant spun round on one toe, arms outstretched. "Mine, all mine," she said in mock wicked witch tones. Suddenly she stopped, staggering slightly with dizziness. "What do you think, Fergus? Isn't it just perfect?"

Fergus Sinclair surveyed the dingy room. The gatehouse on the Wemyss estate was nothing like the plain but spotless cottage he'd grown up in. It was even further removed from Rotheswell Castle. It wasn't even as appealing as the student houses he'd lived in. Having stood empty for a couple of years, it held no sense of its previous occupants. But even so, he found it hard to feel enthusiastic about it. It wasn't how he'd imagined them setting up home together. "It'll be fine once we've gone through it with a bucket of paint," he said.

"Of course it will," Cat said. "I want to keep it simple. Bright but simple. Apricot in here, I think." She headed for the door. "Lemon for the hall, stairs, and landing. Sunshine yellow in the kitchen. I'm going to use the other downstairs room as an office, so something neutral." She ran up the stairs, and leaned over the banister, smiling down at him. "Blue for my bedroom. A nice Swedish sort of blue."

Sinclair laughed at her enthusiasm. "Don't I get a say?"

Cat's smile faded. "Why would you get a say, Fergus? It's not your house."

The words slammed into him like a physical blow. "What do you mean? I thought we were going to live together?"

Cat dropped on to the top step and sat there, knees tight together, arms folded round herself. "Why did you think that? I never said anything about that."

The ground beneath his feet seemed not to be stable. Sinclair clutched at the newel post for support. "It's what we always talked about. We'd finish our training and move in together. Me keepering, you doing the glasswork. It's what we planned, Cat." He stared up at her, willing her to admit he was right.

And so she did, but not in a way that made him feel any better. "Fergus, we were hardly more than children then. It's like when you're little and your big cousin says he's going to marry you when you're older. You mean it all when you say it, but then you outgrow the promise."

"No," he protested, starting up the stairs. "No, we weren't children. We knew what we were saying. I still love you as much as I ever did. Every promise I've made to you—I still want to keep them." He pushed himself down next to her, forcing her to shift right up against the wall. He put his arm round her shoulders. But still she kept her arms wrapped round her body.

"Fergus, I want to live by myself," Cat said, staring down to where he'd been a moment before as if she was still speaking directly to him. "This is the first time I've had my own work space and my own living space. My head is bursting with ideas for things I want to make. And for how I want to live."

"I won't interfere with your ideas," Sinclair insisted. "You can have everything just the way you want it."

"But you'll be *here*, Fergus. When I go to bed at night, when I wake up in the morning. I'll have to think about things like what we're going to eat and when we're going to eat it."

"I'll do the cooking," he said. He could feed himself; how hard could it be to feed both of them? "We can do this on your terms."

"I'll still have to think about mealtimes and things happening at set times, not when it feels natural or right for my creative rhythms. I'll have to think about your washing, when you need to be in the bathroom. What you're going to watch on the TV." Cat was rocking to and fro now, the natural anxiety she'd always worked to hide coming to the surface. "I don't want to have to deal with all that."

"But, Cat . . . "

"I'm an artist, Fergus. I'm not saying that like it's some precious state that sets me above everybody else. What I mean is that I'm kind of fucked up. I'm not good at being with people for extended periods of time."

"We seem to do OK together." He could hear the pleading in his voice and he wasn't ashamed. She was worth shaming himself for.

"But we don't actually spend huge chunks of time together, Fergus. Look at the last few years. I've been in Sweden, you've been in London. We've spent the occasional weekend together, but mostly we've seen each other at Rotheswell. We've hardly ever spent more than a couple of nights together. And that suits me fine."

"It doesn't suit me," he said gruffly. "I want to be with you all the time. Like I said, we can do it on your terms."

She slipped out from under his arm and dropped down a couple of steps, turning so she could look at him. "Can't you see how scary that is for me? Just hearing you say it makes me feel claustrophobic. You talk about doing it on my terms, but none of my terms include having someone under the same roof as me. Fergus, you mean so much to me. There isn't anyone else who makes me feel the way you do. Please, please don't spoil that by pushing me or guilt-tripping me into something I can't bear the thought of."

His face felt frozen, as if he was standing on top of Falkland Hill in a gale, the skin whipped hard against his bones, his eyes flayed to tears. "It's what people do when they love each other," he said.

Now she reached out her hand and put it on his knee. "It's one model for loving," she said. "It's the most common one. But part of the reason for that is economic, Fergus. People live together because it's cheaper than living apart. Two can live as cheaply as one. It doesn't mean it's the best way for everybody. Lots of people have relationships that don't conform to that pattern. And those other ways of doing it work just as well. You think me not wanting to live with you means I don't love you. But Fergus, it's the other way round. Living with you would destroy our relationship. I'd go

crazy. I'd want to kill you. It's because I love you that I don't want to live with you."

He pushed her hand away and stood up. "You've spent too fucking long in Sweden," he shouted, feeling his throat tighten. "Listen to yourself. Models for loving. Conforming to patterns. That's not what love is. Love is . . . Love is . . . Cat, where does affection and kindness and helping each other find a place in your world?"

She stood up and leaned against the wall. "Same place they always have. Fergus, we've always been kind to each other. We've always cared for each other. Why do we need to change the shape of our relationship? Why risk all those beautiful things that work so well between us? Even sex. Everybody I know, once they start living together, the sex stops being so exciting. Two, three years down the line, they hardly ever fuck any more. But look at us." She sidestepped up so she was level with him. "We don't take each other for granted. So when we see each other, it's still electric." She stepped forward, one hand flat on his chest, the other cupped under his balls. In spite of himself, he felt the hardening rush of blood. "Come on, Fergus—fuck me," she whispered. "Here. Now."

And so she got her own way. As usual.

Thursday, 5th July 2007

"Like her father, she was very good at getting her own way. She was more subtle than him, but the end result was the same," Sinclair concluded.

For the first time since the Macaroon had briefed her, Karen felt she had a sense of who Catriona Maclennan Grant had been. A woman who knew her own mind. An artist with a vision she was determined to realize. A loner who took pleasure in company when she was in the mood for it. A lover who learned how to accept being pinned down only after she became a mother. A difficult woman but a brave one, Karen suspected. "Can you think

of anybody whose life touched hers that might have wanted to punish her?" she asked.

"Punish her for what?"

"You name it. Her talent. Her privilege. Her father."

He thought about it. "It's hard to imagine. The thing is, she'd just spent four years in Sweden. She just called herself Cat Grant. I don't think anybody over there had the faintest idea who Brodie Maclennan Grant is." He stretched his legs out and crossed them at the ankles. "She did summer school over here the first couple of years she was in Sweden. She hooked up with some of the people she knew from when she was at the Edinburgh College of Art."

Karen sat up straight. "I didn't know she was at the Edinburgh College of Art," she said. "There was nothing in the file about that. All it says is that she studied in Sweden."

Sinclair nodded. "Technically, that's right. But instead of doing the sixth year at her fancy private school in Edinburgh, she did a foundation course at the College of Art. It's probably not on the file because her old man didn't know about it. He absolutely didn't want her to be an artist. So it was a big secret between Cat and her mum. She'd go off every morning on the train and come home at more or less the usual time. But instead of going to school she went to the college. You really didn't know?"

"We really didn't know." Karen looked at Phil. "We need to start looking at the people who were on that foundation course."

"The good news is that there weren't many of them," Sinclair said. "Only ten or a dozen. Of course, she knew other students, but it was the ones on her course that she mainly hung out with."

"Can you remember who her pals were?"

Sinclair nodded. "There were five of them. They liked the same bands, they liked the same artists. They were always going on about modernism and its legacy." He rolled his eyes. "I used to feel like a complete hick from the sticks."

"Names? Details?" Phil putting the pressure on again. He reached for his notebook and flipped it open.

"There was a lassie from Montrose: Diana Macrae. Another

from Peebles, what was her name . . . ? Something Italian . . . Demelza Gardner."

"Demelza's not Italian, it's Cornish," Phil said. Karen silenced
him with a look.

"Whatever. It sounded Italian to me," Sinclair said. "There
were two lads as well. A guy from Crieff or some arsey place in
Perthshire like that: Toby Inglis. And finally, Jack Docherty. He
was a working-class toe-rag from Glasgow. They were all nice
middle-class kids and Jack was their performing monkey. He
didn't seem to mind. He was one of those people who don't care
what kind of attention they get as long as they get some."

"Did she stay in touch with any of them when she went to
Sweden?"

Sinclair stood up, ignoring her, as his boys raced across the
grass towards him. They threw themselves on him in an excited
torrent of what Karen took to be German. Sinclair clung on to
them, struggling forward a couple of steps with them hanging on
like baby chimps. Then he dropped them, said something to them,
ruffled their hair, and sent them off in pursuit of their mother, who
had disappeared towards the steps down to the shore. "Sorry," he
said, coming back and sitting down again. "They always like to be
sure you know what you're missing. To answer your question—
I don't really know. I vaguely remember Cat mentioning one or
other of them a few times, but I didn't pay much attention. I had
nothing in common with them. I never met any of them again
after Cat left the college." He ran a hand over his jaw. "Looking
back at it now, I think that, the older we got, the less Cat and I had
in common. If she'd lived, we would never have got back together
again."

"You might have found some common ground over Adam
eventually," Karen said.

"I'd like to think so." He looked longingly at the gateway his
boys had disappeared through. "Is there anything else? Only, I'd
kind of like to get back to my life."

"Do you think there was anyone from her art college days who

might have harboured ill feeling towards her?" Karen asked.

Sinclair shook his head. "Nothing she ever said would make me think that," he said. "She had a strong personality, but she was a hard person to dislike. I don't remember her ever complaining that she'd been given a hard time by anybody." He stood up again, smoothing down his trousers. "I have to say, I can't believe anyone who knew her would think they could get away with kidnapping her. She was far too good at getting her own way."

Glenrothes

The Mint stabbed the keyboard with his index fingers. He didn't know why they called that fast business "touch typing." Because you couldn't type without touching the keyboard. It was all touch typing, when you got down to it. He also wasn't sure why the boss kept lumbering him with the computer searches unless it was just pure sadism. Everybody thought young guys like him were totally at home in front of a computer, but for the Mint it was like a foreign country where he didn't even know the word for beer.

He'd have been much happier if she'd sent him off with the Hat to the College of Art to talk to real people and pore through yearbooks and physical records. He was better at that. And besides, DS Parhatka was a good laugh. There was nothing funny about trawling through the message boards and membership lists of www.bestdaysofourlives.com searching for the names the boss had dropped on his desk on a tatty page torn from a notebook.

This was so not what he'd joined up for. Where was the action? Where were the dramatic car chases and arrests? Instead of excitement, he got the boss and the Hat acting like they were some ancient comedy partnership, like French and Saunders. Or was it Flanders and Swann? He could never get them straight in his head.

He hadn't even had to monster anybody to get full access to

the website. The woman he'd spoken to had fallen over herself to be helpful. "We've helped the police before, we're always happy to do what we can," she'd gabbled as soon as he'd made his request. Whoever she'd dealt with before had clearly left her in a state of shivering submission. He liked that in a source.

He checked the list of names again. Diana Macrae. Demelza Gardner. Toby Inglis. Jack Docherty. 1977–78 the year he was looking for. After a couple of false clicks, he finally made it to the membership list. Only one of them was there. Diana Macrae was now Diana Waddell, but it wasn't hard to figure that out. He clicked on Diana's profile.

I followed my foundation course at the College of Art with a degree from Glasgow School of Art, specializing in sculpture. After graduating, I started working in the field of art therapy for people with mental illness. I met Desmond, my husband, when we were both working in Dundee. We married in 1990 and we have two children. We live in Glenisla, which we all adore. I have started sculpting in wood again and have a contract with a local garden centre as well as a gallery in Dundee.

A gallery in Dundee, the Mint thought scornfully. Art? In Dundee. About as likely as peace in the Middle East. He skimmed through more rubbish about her husband and kids, then clicked through to her messages and e-mails from fellow ex-students. Why did these people bother? Their lives were as dull as an East Fife home game. After scrolling through a couple of dozen innocuous exchanges, he found a message from someone called Shannon. Do you ever hear from Jack Docherty? she asked.

Darling Jack! We swap Christmas cards. Her smugness penetrated the notoriously nuance-free e-mail. He's out in Western Australia now. He has his own gallery in Perth. He does a lot of work with Aboriginal artists. We have a couple of pieces from him, they're remarkable. He's very happy. He has an Aboriginal boyfriend. Quite

a few years younger than him and very handsome, but he sounds like a sweetheart. Once our two are off to uni, we're planning a trip out to visit.

Two birds with one stone, the Mint thought, scribbling the details down. He continued to the end of Diana's wittering correspondence, then decided he needed a break while he came up with his next move.

A cup of coffee later, he got back to his search. Neither Toby Inglis nor Demelza Gardner showed up anywhere on the College of Art area of the website. But thanks to the way his contact had rolled over, he was able to search the entire website. He typed in the woman's name and to his complete astonishment, he got a hit. He clicked on the result and discovered Gardner described as "totally my favourite teacher." The message was on the site of a high school in Norwich.

At least he had the sense to Google the school. And there was Demelza Gardner. Head of Art. God, this computer stuff was a piece of piss once you got the hang of it. He tried Toby Inglis's name in the search engine and again came up with a hit. The Mint followed the link to a forum where former pupils of a private school in Crieff could rabbit on to their hearts' content about their fabulous bloody lives. It took him a while to unravel the threads of correspondence, but at last he found what he was looking for.

Feeling rather pleased with himself, the Mint tore off the top sheet from his notepad and went off in search of DI Pirie.

It had, Karen thought, gone something like this. She had called Bel Richmond and invited her to come to CCRT for an interview as soon as possible. Preferably within the hour. Bel had refused. Karen had mentioned the small matter of police obstruction.

Then Bel had gone to Brodie Grant and complained that she didn't want to trot off to Glenrothes at Karen Pirie's beck and call. Then Grant had called the Macaroon and explained that Bel

didn't want to be interviewed and DI Pirie had better stop threatening her. Then the Macaroon had summoned her and given her a hard time for upsetting Brodie Grant, and told her to lay off Bel Richmond.

Then Karen had called Bel Richmond again. In her sweetest voice, she had told Bel to present herself at CCRT at two o'clock. "If you're not here," she said, "there will be a squad car at Rotheswell ten minutes after that to arrest you for police obstruction." Then she'd put the phone down.

Now it was a minute to two and Dave Cruickshank had just called her to say Bel Richmond was in the building. "Get a uniform to take her up to Interview One and wait with her till I get there." Karen got herself a Diet Coke out of the fridge and sat down at her desk for five minutes. She took a last swig from her can, then headed down the hall to the interview room.

Bel was sitting at the table in the grey windowless room, looking furious. A red pack of Marlboros sat in front of her, a single cigarette lying next to it. Clearly she'd forgotten the Scots had banned smoking ahead of the English until the uniformed officer had reminded her.

Karen pulled a chair out and dropped into it. The foam cushion had been worn into its shape by other buttocks than hers, and she wriggled to get comfortable. Elbows on the table, she leaned forward. "Don't ever try to fuck with me again," she said, her voice conversational, her eyes like glittering granite.

"Oh, please," Bel said. "Let's not make this a pissing contest. I'm here now, so let it go."

Karen didn't take her eyes off Bel. "We need to talk about Italy."

"Why not? Lovely country. Fabulous food, wine's getting better all the time. And then there's the art—"

"Stop it. I mean it. I will charge you with police obstruction and put you in a cell and leave you there till I can bring you before a sheriff. I am not going to be jerked around by Sir Broderick Maclennan Grant or his minions."

"I'm not a minion of Brodie Grant," Bel said. "I'm an independent investigative journalist."

"Independent? You're living under his roof. Eating his food, drinking his wine. Which I bet is not Italian, by the way. And who paid for that little jaunt to Italy? You're not independent, you're bought and paid for."

"You're wrong."

"No, I'm not. I've got more freedom of action than you have right now, Bel. I can tell my boss to shove it. Come to think of it, I just have. Can you say the same? If it wasn't for the Italian police, I wouldn't even know you'd been talking to people in Tuscany about the Villa Totti. The very fact that you've been reporting to Grant and not talking to us tells me that he owns you."

"That's bullshit. Reporters don't talk to cops about their investigations till their work is finished. That's what's going on here."

Karen shook her head slowly. "I don't think so. And, to tell you the truth, I'm surprised. I didn't think you were that kind of woman."

"You don't know anything about me, Inspector." Bel settled herself more comfortably in the chair, as if she was getting ready for something pleasurable.

"I know you didn't earn your reputation spouting clichés like that." Karen pulled her chair closer to the table, cutting the distance between them to less than a couple of feet. "And I know that you've been a campaigning journalist for almost the whole of your career. You know what people say about you, Bel? They say you're a fighter. They say you're someone who does the right thing even if it's not the easiest. Like the way you took your sister and her boy under your roof when they needed looking after. They say you don't care about the popularity of your position, you drag the truth out kicking and screaming and make people confront it. They say you're a maverick. Somebody who operates to her own set of rules. Somebody who doesn't take orders from the Man." She waited, staring Bel down. The journalist blinked first, but she didn't look away. "You think they'd recognize you now? Taking

your orders from a man like Sir Broderick Maclennan Grant? A man who epitomizes the capitalist system? A man who resisted his daughter's every attempt at self-determination to the point where she ended up putting herself in harm's way? Is that what you've come down to?"

Bel picked up her cigarette and tapped it end to end on the table. "Sometimes you have to find a place inside the enemy's tent so you can find out what he's really like. You of all people should understand that. Cops use undercover all the time when there's no other way of getting a story. Do you have any idea how many press interviews Brodie Grant has given in the last twenty years?"

"Taking a wild guess, I'd say . . . none?"

"Right. When I found a piece of evidence that might just crack this cold case open, I figured there would be a lot of interest in Grant. Publisher-type interest. But only if someone could get alongside Grant and see what he was like for real." She raised one corner of her mouth in a cynical half-smile. "I thought it might as well be me."

"Fair enough. I'm not going to sit here and pick holes in your self-justification. But how does your quest to give the world the definitive book about that miserable family grant you the right to stand above the law?"

"That's not how I see it."

"Of course it's not how you see it. You need to see yourself as the person who's acting on behalf of Cat Grant. The person who's going to bring her son home, dead or alive. The hero. You can't afford to see yourself in a true light. Because that true light shows you up as the person who is standing in the way of all of those things. Well, here's the scoop, Bel. You haven't got the resources to bring this to an end. I don't know what Brodie Grant's promised you, but it's not going to be clean. Not in any sense." Karen could feel her anger coiling inside her, getting ready to spring. She pushed her chair back, putting some space between them.

"The Italian police don't care about what happened to Cat Grant," Bel said.

"You're right. And why should they?" Karen felt her face flush. "But they do care about the person whose blood is all over the kitchen floor of the Villa Totti. So much blood that that person is almost certainly dead. They care about that, and they're doing everything they can to find out what happened there. And in the course of that, there will be information that will help us. That's how we do things. We don't hire private eyes who tailor their reports to what the client wants to hear. We don't construct our own private legal system to serve our own interests. Let me ask you a question, Bel. Just between the two of us." Karen turned to the uniformed constable who was still standing by the door. "Could you give us a minute?"

She waited till he had closed the door behind him. "Under Scots law, I can't use anything you say to me now. There's no corroboration, you see. So here's my question. And I want you to think about it very carefully. You don't need to tell me the answer. I just want to be sure that you've thought about it honestly and sincerely. If you were to find the kidnappers, what do you think Brodie Grant would do with that information?"

The muscles round Bel's mouth tightened. "I think that's a scurrilous implication."

"I didn't imply. You inferred." Karen got up. "I'm not a numpty, Bel. Don't treat me like one." She opened the door. "You can come back in now."

The constable took up his station by the door, and Karen returned to her chair. "You should be ashamed of yourself," she said. "Who the hell do you people think you are, with your private law? Is this what you've spent your career working for? A law for the rich and powerful that can thumb its nose at the rest of us?" *That hit home. About bloody time.*

Bel shook her head. "You misjudge me."

"Prove it. Tell me what you found out in Tuscany."

"Why should I? If you people were any good at your job, you'd have found it out yourself."

"You think I need to defend my ability? The only thing I have

to defend is that our investigations struggle under the weight of rules and regulations and resources. That sometimes means it takes me and my team a while to cover the ground. But you can be sure that, when we do, there's not a blade of grass goes unexamined. If you give a toss about justice, you should tell me." She gave Bel a cold smile. "Otherwise, you might find yourself on the other end of the reporters' notebooks."

"Is that a threat?"

To Karen's ear, it sounded like bluster. Bel was close to spilling, she could sense it. "I don't need to threaten," she said. "Even Brodie Grant knows what a leaky sieve the police are. Stuff just seems to slip out into the public domain. And you know how the press love it when someone camped out on the moral high ground gets caught up in a mudslide." Oh yes, she was right. Bel was definitely growing uneasy.

"Look, Karen—I can call you Karen?" Bel's voice dropped into hot chocolate warmth.

"Call me what you like, it makes no odds to me. I'm not your pal, Bel. I've got six hours to question you without a lawyer and I plan to make the most of every minute. Tell me what you found out in Italy."

"I'm not telling you anything," Bel said. "I want to go outside for a cigarette. I'll just leave my bag here on the table. Careful you don't knock it over, things might spill out." She stood up. "Is that OK with you, Inspector?"

Karen struggled not to smile. "The constable will need to keep you company. But take your time. Make it two. I've got plenty to keep me occupied." Watching Bel leave the room, she couldn't help a momentary flash of admiration for the other woman's style. Give it up without giving in. *Nice one, Bel.*

Her arm brushed against the straw shopping bag, which fell over on its side, fanning a wedge of paper out on the table. Without reading it, Karen scooped it up and hustled down the hall to her office. Into the photocopier, the whole bundle copied inside ten minutes, a set of copies locked in her drawer, the originals in

her hand. And back to the interview room, where she settled down to read.

As she digested Bel's report for Brodie Grant, her mind arranged the bullet points. Mongrel bunch of puppeteers squatting the Villa Totti. Daniel Porteous, British painter, not so much friend of the house as friend of Matthias the boss and his girlfriend. Matthias the set designer and poster maker. Gabriel Porteous, son of Daniel. Seen with Matthias the day before BurEst scattered to the four winds. Blood on the kitchen floor, fresh that morning. Daniel Porteous a fake. Already a fake in November 1984, when he registered his son's fake birth.

She stumbled for a moment on the mother's name, knowing she'd come across it but struggling for context. Then she said it out loud and it clicked. Frida Kahlo. That Mexican artist that Michael Marra wrote the song about. "Frida Kahlo's Visit to the Taybridge Bar." She had a bad time with her man. So, nothing new there, then. But somebody was being a smartarse with the registrar, laughing up his sleeve at some minor civil servant who wouldn't know Frida Kahlo from Michelangelo. Showing off. Thinking he was being clever but not realizing he was saying something about himself in the process. He must have been a skilful forger though, this Daniel Porteous, to turn up with all the necessary documentation to convince the registrar. And bold, to carry it through.

It was all very interesting, but what had convinced Bel that Gabriel Porteous was Adam Maclennan Grant? And, by logical extension, Daniel Porteous his biological father? And by extending the logic further, that Daniel Porteous and Matthias were the kidnappers? Still in touch after all these years, still in possession of the original silk screen. Based on the poster, you could draw the thread through, but it was only circumstantial.

Aware that Bel would be back any moment, Karen flicked forward through the pages, skimming for sense, hunting for something that might anchor the theory to solid fact. The last few pages

were photographs—originals taken at some party, and enlarged sections with captions.

Her stomach flipped and her mind at first refused to accept what she was looking at. Yes, it was true that the boy Gabriel bore a striking resemblance to both Brodie and Cat Grant. But that wasn't what had provoked the turmoil inside. Karen stared at the image of Daniel Porteous, nausea churning her guts. Dear God, what was she to make of this? And then with the suddenness of a light coming on, she realized something that turned everything on its head.

Daniel Porteous had registered the birth of his son three months before the kidnap. He'd assumed a fake identity at least three months ahead of the time that he was going to use it to make his getaway. Fair enough. It demonstrated forethought. But he'd also established the right to take his son with him. "You don't do that if you're planning to ransom him," she said under her breath.

Karen stuffed Bel's papers back in the straw bag and headed for the door. This was insane. She needed to talk to someone who could help her make sense of this. Where the hell was Phil when she needed him?

As she burst out of the interview room, she practically collided with the Mint. He sidestepped, looking startled. "I was looking for you," he said.

Definitely not mutual. "I can't stop now," she said, pushing past him.

"I've got this for you," he said plaintively.

Karen whirled round, grabbed the sheet of paper, and broke into a run. She felt as if an army of messengers were running round inside her head, each with a jigsaw piece. Right now, none of the matching pieces were joining up. But she had a shrewd suspicion that, when they did, the picture would rock everybody back on their heels.

There had been a shift change in the security team since Bel left for her interview with Karen Pirie, so the guard on duty at the gate had to clear her return via taxi with the castle. That knocked on the head any hopes of slipping back quietly. As she paid off the cab, the front door swung open to reveal a grim-faced Grant. Bel assumed a look of pleasure and walked towards him.

No pleasantries today. "What did you tell her?" he demanded.

"Nothing," Bel said. "A good journalist protects her sources and her information. I told her nothing." It was, technically, the truth. She had told Karen Pirie nothing. She hadn't had to. The inspector had come haring out of the building, pausing only to tell Bel she was free to go.

"Something's just broken on another case I'm working, I've got to go to Edinburgh. I'll be in touch. You can go back to Rotheswell as soon as you like," Karen had said. Then she'd given Bel a wink. "And you can put your hand on your heart and tell Brodie you didn't talk."

Secure in the knowledge that she wasn't actually lying, Bel moved into the house, leaving him no choice but to grab her or follow her.

"You're telling me you told her nothing and she just let you go?" He had to extend his stride to its full length to keep up with her as she bustled down the hall to the stairs.

"I made it clear to DI Pirie that I was not going to talk. She recognized there was no point in prolonging the stalemate." Bel glanced over her shoulder. "This isn't the first time in my career I've had to hold information back from the police. I told you there was no need to try and put the frighteners on her."

Grant conceded with a nod. "I'm sorry I didn't take you at your word."

"So you should be," Bel said. "I—" She broke off to reach for her ringing phone. "Bel Richmond," she said, holding a finger up to still Grant.

A torrent of Italian poured into her ear. She made out "Bosco-lata," then recognized the voice of the youth who had seen Gabriel with Matthias the night BurEst had done a runner. "Slowly, just take your time," she protested gently, switching to his language.

"I saw him," the boy said. "Yesterday. I saw Gabe in Siena again. And I knew you wanted to find him, so I followed him."

"You followed him?"

"Yeah, like in the movies. He got on a bus, and I managed to sneak on without him seeing me. We ended up in Greve. You know Greve in Chianti?"

She knew Greve. A perfect little market town stuffed with trendy shops for the rich English, redeemed by a few bars and trattorie where the locals still ate and drank. A meeting place for young people on Fridays and Saturdays. "I know Greve," she said.

"So, we end up in the main piazza and he goes into this bar, sits down with a bunch of other guys about the same age. I stayed outside, but I could see him through the window. He had a couple of beers and a bowl of pasta, then he came out."

"Were you able to follow him?"

"Not really. I thought I could, but he had a Vespa parked a couple of streets back. He went off down the road that heads east out of town."

Near, but not near enough. "You did well," she said.

"I did better. I left it about twenty minutes, then I went into the bar he'd been in. I said I was looking for Gabe, I was supposed to meet him there. His mates said I'd not long missed him. So I went all innocent and said, could they direct me to his place, only I didn't know how to get there."

"Amazing," Bel said, genuinely taken aback by his initiative. Grant started to walk away, but she beckoned him back.

"So they drew me a map," he said. "Pretty cool, huh? Apparently it's, like, one step up from a shepherd's hut."

"What did you do?"

"I got the last bus home," he said, as if it was blindingly obvi-ous. Which she supposed it was if you were a teenage boy.

"And you've got this map?"

"I brought it back with me," he said. "I thought it might be worth something to you. I thought maybe a hundred euros?"

"We'll talk about it. Listen, I'll be back as soon as I can. Don't talk to anybody except Grazia about this, OK?"

"OK."

Bel ended the call and gave Grant the thumbs-up. "Result," she said. "Forget the private eyes. My contact has discovered where Gabriel is living. And now I need to get back to Italy to talk to him."

Grant's face lit up. "That's tremendous news. I'm coming with you. If this boy is my grandson, I want to see him face to face. The sooner the better."

"I don't think so. This needs to be handled carefully," Bel said.

From behind her, a voice chimed in. "She's right, Brodie. We need to know a lot more about this boy before you put your head over the parapet." Judith stepped forward and laid a hand on her husband's arm. "This could all be an elaborate set-up. If these are the people who kidnapped Adam and robbed you twenty-two years ago, we know they're capable of the most cruel behaviour. We don't know anything else for sure. Let Bel handle it." Grant made a protest, but she shushed him. "Bel, do you think you can get a DNA sample without this young man realizing?"

"It's not so hard," Bel said. "One way or another, I'm sure I can manage it."

"I still think I should go," Grant said.

"Of course you do, darling. But the women are right this time. And you will just have to possess your soul in patience. Now, where's the plane?"

Grant sighed. "It's at Edinburgh."

"Perfect. By the time Bel's packed a bag, Susan will have everything arranged." She glanced at her watch. "You said you would take Alec fishing after school, so I can drive Bel over." She smiled at Bel. "Better get cracking. I'll see you downstairs in fifteen?"

Bel nodded, too gobsmacked to argue. If she'd ever wondered how Judith Grant held her own in her marriage, she'd just witnessed a spectacular demonstration. Grant had been totally sandbagged and, short of throwing a temper tantrum, there was no way back for him. She turned and ran up the stairs. *Add another zero to the advance.* This was turning into the story of her career. Everyone who had ever dissed her was going to have to eat their words. It was going to be blissful. OK, there was some tedious legwork to be got through, but there was always tedious legwork. There just wasn't always glory at the end of it.

Kirkcaldy

Karen paced the floor, an unwavering ten steps across the living room, then a swivel and ten steps back. Usually movement helped her get her thoughts lined up in order. But this evening, it wasn't working. The jumble in her head was intractable, like herding cats or wrestling water. She suspected it was because, at some deep level, she was resisting the inevitable conclusion. She needed Phil here to hold her hand while she thought the unthinkable.

Where the hell was he? She'd left a message on his voicemail almost two hours ago, but he hadn't got back to her. It wasn't like him to go off the radar. As that thought circled for the hundredth time, her doorbell pealed out.

She'd never covered the distance to the front door faster. Phil stood on the doorstep, looking sheepish. "I'm sorry," he said. "I went to the National Library in Edinburgh and I had to turn my phone off. I forgot to turn it back on again till a few minutes ago. I thought it would be quicker to just come straight over."

Karen was ushering him into the living room as he spoke. He looked around curiously. "This is nice," he said.

"No, it's not. It's just a machine for living in," she said.

"But it's a good one. It's relaxing. The colours all work with each other. You've got a good eye."

She didn't have the heart to tell him it was someone else's eye. "I didn't ask you round to appreciate the décor," she said. "Do you want a beer? Or a glass of wine?"

"I've got the car," he said.

"Never mind that. You can always get a cab home. Believe me, you are going to need a drink." She thrust the photocopy of Bel's notes at him. "Beer or wine?"

"Have you got some red wine?"

"Read that. I'll be right back." Karen went through to the kitchen, chose the best of the half-dozen reds she had in the rack, unscrewed the cap, and poured two big glasses. The jammy spice of the Australian shiraz tickled her nose as she picked up the drinks. It was the first external thing she'd noticed since she'd left the office.

Phil had made his way through to the dining area and was sitting at the table, intent on the report. She put the glass down by his hand. Absently, he took a swig. Karen couldn't keep still. She sat down, then she stood up. She went through to the kitchen and returned with a plate of cheese crackers. Then she remembered the sheet of paper the Mint had given her. She'd stuffed it in her bag without looking at it.

She tracked her bag down in the kitchen. The Mint's notes weren't exactly the most clear or succinct that she'd ever read, but she got the gist of what he had found out. Three of Cat's friends were clearly of no interest. But the forum message he'd copied about Toby Inglis leapt out at her with all the force of a coiled spring . . . just like Kate Mosse's book. But you'll never guess who we bumped into in a bistro in Perpignan. Only Toby Inglis. You remember how he was going to set the world on fire, be the next Olivier? Well, it obviously hasn't worked out quite the way he planned. He was pretty evasive when it came to the details, but he said he's a theatre director and designer. IMHO he was being a bit economical with the truth. Brian said he looked more like a superannuated hippy. He certainly smelled like one, all patchouli and dope. We asked where we could see one of his productions, but he said he was taking

a summer break. I was dying to do some more digging but then this German woman arrived. I think she thought they were eating there, but he hustled her out of the door as fast as he could. I think he didn't want us talking to her and finding out the truth. Whatever that is. So, after Perpignan . . .

Karen re-read the Mint's scrawl. Could this be Matthias? It certainly sounded like the mysterious Matthias who hadn't been seen since he was spotted in Siena with Gabriel Porteous. Another piece that appeared to belong with the jigsaw but didn't seem to fit.

Karen forced herself to breathe deeply, then joined Phil at the dining table. He'd spread the prints in front of him. He poked one with his finger to align it with the others. "It's him, isn't it?" he said.

"Adam?"

He flapped an impatient hand at her. "Well yes, of course it's Adam. It's got to be Adam. Not just because he looks like his mother and his grandfather. But because the man who's brought him up is Mick Prentice."

Karen experienced a moment of weightlessness. The agitation stilled and she could think straight again. She wasn't losing her mind or letting her imagination run away with her. "Are you sure?"

"He's not actually changed that much," Phil said. "And look, there's the scar—" He traced it with his fingertip. "The coal tattoo through his right eyebrow. The thin blue line. It's Mick Prentice. I'd put money on it."

"Mick Prentice was one of the kidnappers?" Even to her own ears, Karen sounded a bit wobbly.

"I think we both know he was more than that," Phil said.

"The registration," Karen said.

"Exactly. This was all planned even before Mick left Jenny. He'd set up his fake identity so he could start a new life. But there can be only one reason why he needed to set up a fake identity for Adam."

"He wasn't planning to ransom him at all," Karen said. "Because he was Adam's father. Not Fergus Sinclair. Mick Prentice." She took a gulp of red wine. "It was a set-up, wasn't it? There were no anarchists, were there?"

"No." Phil sighed. "It looks like there were two miners. Mick and his pal Andy."

"You think Andy was part of the plot?"

"It looks that way. How else do you account for him ending up buried in the cave at just the right time?"

"But why? Why kill him? He was Mick's best mate," Karen protested. "If he could trust anyone, he could trust Andy. The way you guys operate, he could probably have trusted Andy more than Cat."

"Maybe it was an accident. Maybe he hit his head getting in or out of the boat."

"River said the back of his head was smashed in. That doesn't sound like an accident getting into a boat."

Phil threw his hands in the air in a "whatever" gesture. "He could have tripped, smacked his head on the quay. It was chaos that night. Anything could have happened. I'd put my money on Andy being the co-conspirator."

"And Cat? Was she part of the plan or was she the victim? Were she and Mick still an item or was he trying to get his kid and enough of Brodie Grant's money to set the pair of them up for life?"

Phil scratched his head. "I think she was in on it," he said. "If they'd split up and he'd taken them both, she'd never have let Adam out of her arms. She'd have been too scared of him taking the bairn away from her."

"I can't believe they got away with it," she said.

Phil gathered the prints together and straightened the edges. "Lawson was looking in the wrong direction. And with good reason."

"No, no. I don't mean the kidnap. I mean the affair. Everybody knows everybody else's business in a place like the Newton.

Easier to get away with murder than an extra-marital affair, I'd have said."

"So it looks like we've done what Lawson couldn't do. Solved the kidnapping, tracked down Adam Maclennan Grant."

"Not quite," Karen said. "We don't actually know where he is. And there's the small matter of a lot of blood spilled in Tuscany. Which could be his."

"Or it could have been spilled by him. In which case he's not going to be very keen on being found."

"There's one thing we haven't factored in," Karen said, passing Phil the result of the Mint's searches. "It looks like Matthias the puppeteer might actually be a friend of Cat's from art college. Toby Inglis has a description that you could stretch to cover Matthias, the leader of the motley crew. Where does he fit in the picture?"

Phil looked at the paper. "Interesting. If he was involved in the kidnap, it might be more than embarrassment at his less-than-glittering career that's making him keep a low profile." He finished his glass of wine and tipped it towards Karen. "Any more where that came from?"

She fetched the bottle and refilled his glass. "Any bright ideas?"

Phil took a slow mouthful. "Well, if this Toby is Matthias, he was an old pal of Cat's. Could be that's how he met Mick. It didn't have to be planned, he could just have turned up out of the blue when Mick was there. You know what artists are like."

"I don't, actually. I don't think I've ever met anyone who was at art college."

"My brother's girlfriend was. The one who's doing the make-over at my place."

"And is she prone to being unreliable?" Karen asked.

"No," Phil admitted. "Unpredictable, though. I never know what she's going to inflict on me next. Maybe I should have got you to do the job instead. This is definitely more easy on the eye."

"What I live for," Karen said. "Easy on the eye." There was a charged moment of silence between them, then she hastily cleared

her throat and said, "But here's the thing, Phil. If they'd met when Mick was with Cat, then ran into each other by chance in Italy, how the holy fuck did Mick explain what had happened to Cat and how he'd ended up with the kid?"

"So you're saying he must have been involved in the kidnap too?"

She shrugged. "I don't know. I really don't know. What I do know is that we need to get the Italian police to find the person whose blood isn't on the floor of that villa so we can ask them some pertinent questions."

"Another tall order for the woman who put Jimmy Lawson behind bars." He raised his glass to her.

"I'm never going to live that down, am I?"

"Why would you want to?"

Karen looked away. "Sometimes it feels like a millstone round my neck. Like the man who shot Liberty Valance."

"It's not like that," Phil said. "You nailed Lawson fair and square."

"After somebody else did all the work. Just like this time, with Bel doing the legwork."

"You did the work that mattered, both times. We'd still be back at square one if you hadn't had the cave excavated and the Nottingham guys properly questioned. If you're going to quote the movies, remember how it goes. 'When the legend becomes fact, print the legend.' You are a legend, Karen. And you deserve to be."

"Shut up, you're embarrassing me."

Phil leaned back in his chair and grinned at her. "Do they deliver pizza round here?"

"Why? Are you buying?"

"I'm buying. We deserve a wee celebration, don't you think? We've come a long way towards solving two cold cases. Even if we're landed with Andy Kerr's murder as a sick kind of bonus. You order the pizza, I'll check out your DVDs."

"I should speak to the Italians," Karen said half-heartedly.

"With the time difference, it's nearly eight o'clock there. Do

you really think there's going to be anyone with any seniority around? You might as well wait till morning and talk to the guy you've been dealing with. Relax for once. Switch off. We'll finish the wine, knock off a pizza, and watch a movie. What do you say?"

Yes, yes, yes! "Sounds like a plan," Karen said. "I'll get the menus."

Celadoria, near Greve in Chianti

The sun was heading for the hills, a scarlet ball in her rear-view mirror as Bel drove east out of Greve. Grazia had met her in a bar in the main piazza and handed over the paper directing her to the simple cottage where Gabriel Porteous was living. Just over three kilometres out of town, she found the right turn indicated on the scrawled map. She drove up slowly, keeping an eye out for a pair of stone gateposts on the left. Immediately after them, there was supposed to be a dirt road on the left.

And there it was. A narrow track weaving between rows of vines that followed the contour of the hill; you'd pass it without a second glance if you weren't looking for it. But Bel was looking, and she didn't hesitate. The map had a cross on the left side of the track, but it clearly wasn't drawn to scale. Anxiety began to creep upon her as the distance from the main road grew. Then suddenly, tinted pink by the setting sun, a low stone building appeared in her sights. It looked one step above complete dilapidation. But that wasn't unusual, even somewhere as fashionable as the Chiantishire area of Tuscany.

Bel pulled over and got out, stretching her back after hours of sitting. Before she'd taken a couple of steps, the plank door creaked open and the young man in the photographs appeared in the doorway dressed in a pair of cut-off jeans and a black muscle vest that emphasized evenly bronzed skin. His stance was casual; a hand on the door, the other on the jamb, a look of polite enquiry on his

face. In the flesh, the resemblance to Brodie Grant was striking enough to seem eerie. Only the colouring was different. Where the young Brodie's hair had been as black as Cat's, Gabriel's was caramel coloured, highlighted with sun-streaks of gold. Other than that, they could have been brothers.

"You must be Gabriel," Bel said in English.

He cocked his head to one side, his brows lowering, shading his deep-set eyes even further. "I don't think we've met," he said. He spoke English with the music of Italian underpinning it.

She drew closer and extended a hand. "I'm Bel Richmond. Didn't Andrea from the gallery in San Gimi mention I'd be stopping by?"

"No," he said, folding his hands over his chest. "I don't have any of my father's work for sale. You've wasted your time coming out here."

Bel laughed. It was a light, pretty laugh, one she'd worked on over the years for doorstep moments like this. "You've got me wrong. I'm not trying to rip off you or Andrea. I'm a journalist. I'd heard about your father's work and I wanted to write a feature about him. And then I discovered I was too late." Her face softened and she gave him a small, sympathetic smile. "I am so sorry. To have painted those paintings, he must have been a remarkable man."

"He was," Gabriel said. It sounded as if he begrudged her both syllables. His face remained inscrutable.

"I thought it might still be possible to write something?"

"There's no point, is there? He's gone."

Bel gave him a shrewd look. Reputation or money, that was the question now. She didn't know this lad well enough to know what would get her across the door. And she wanted to be across the door before she dropped the bombshell of what she really knew about him and his father. "It would enhance his reputation," she said. "Make sure his name was established. And that would obviously increase the value of his work too."

"I'm not interested in publicity." He moved backwards, the door starting to inch closed.

Time to throw the dice. "I can see why that might be, Adam." She'd hit home, judging by the swift spasm of shock that passed across his features. "You see, I know a lot more than I told Andrea. Enough to write a story, that's for sure. Do you want to talk about it? Or shall I just go away and write what I know without you having any say in how the world sees you and your dad?"

"I don't know what you're talking about," he said.

Bel had seen enough bluster in her time to recognize it for what it was. "Oh, please," she said. "Don't waste my time." She turned and started to walk back to the car.

"Wait," he shouted after her. "Look, I think you've got hold of the wrong end of the stick. But come in and have a glass of wine anyway." Bel swung round without a second's hesitation and headed back towards him. He shrugged and gave her a puppy-dog grin. "It's the least I can do, seeing as you've come all the way out here."

She followed him into the classic dim Tuscan room that served as living room, dining room, and kitchen. There was even a bed recess beyond the fireplace, but instead of a narrow mattress, it housed a plasma-screen TV and a sound system that Bel would have been happy to have installed in her own home.

A scarred and scrubbed pine table sat off to one side near the cooking range. A pack of Marlboro Lights and a disposable lighter sat next to an overflowing ashtray. Gabriel pulled out a chair for Bel on the far side, then brought over a couple of glasses and an unlabelled bottle of red wine. While his back was turned, she lifted a cigarette butt from the ashtray and slipped it into her pocket. She could leave any time now and she would have what she needed to prove whether this young man really was Adam Maclennan Grant. Gabriel settled down at the head of the table, poured the wine, and raised his glass to her. "Cheers."

Bel clinked her glass against his. "Nice to meet you at last, Adam," she said.

"Why do you keep calling me Adam?" he said, apparently bewildered. He was good, she had to admit. A better dissembler than Harry, who'd never been able to stop his cheeks pinking whenever he lied. "My name's Gabriel." He took a cigarette from the pack and lit it.

"It is now," Bel conceded. "But it's not your real name, any more than Daniel Porteous was your father's real name."

He gave a half-laugh, flipping one hand in the air in a gesture of incomprehension. "See, this is very bizarre to me. You turn up at my house, I've never seen you before, and you start coming out with all this . . . I don't mean to sound rude, but really, there's no other word for it but bullshit. Like I don't know my own name."

"I think you do know your own name. I think you know exactly what I'm talking about. Whoever your father was, Daniel Porteous wasn't his name. And you're not Gabriel Porteous. You're Adam Maclennan Grant." Bel picked up her bag and pulled out a folder. "This is your mother." She extracted a photo of Cat Grant on her father's yacht, head back and laughing. "And this is your grandfather." She added a publicity head shot of Brodie Grant in his early forties. She looked up and saw Gabriel's chest rising and falling in time with his rapid and shallow breathing. "The resemblance is striking, wouldn't you say?"

"So you found a couple of people who look a bit like me. What does that prove?" He drew hard on his cigarette, squinting through the smoke.

"Nothing, in itself. But you turned up in Italy with a man using the identity of a boy who'd died years before. The pair of you showed up not long after Adam Maclennan Grant and his mother were kidnapped. Adam's mother died when the ransom handover went sour, but Adam vanished without trace."

"That's pretty thin," Gabriel said. He wasn't meeting her eyes now. He drained his glass and refilled it. "I don't see any real connection to me and my father."

"The ransom demand was made in a very distinctive format. A poster of a puppeteer. The same poster turned up in a villa near

Siena that was being squatted by a puppet troupe led by a guy called Matthias."

"You've lost me." His eyes might be focused over her shoulder, but his smile was charm itself. Just like his grandfather's.

Bel placed a photo of Gabriel at the Boscolata party on the table. "Wrong answer, Adam. This is you at a party where you and your father were guests of Matthias. It ties the pair of you to a ransom demand that was made for you and your mother twenty-two years ago. Which is more than suggestive, don't you think?"

"I don't know what you're talking about," he said. She recognized the stubborn line of the jaw from her encounters with Brodie Grant. Really, she could leave now and rely on the DNA to do all that was necessary. But she couldn't help herself. The journalist's instinct for running the game and gaining the scoop was too strong.

"Of course you do. This is a great story, Adam. And I am going to write it with or without your help. But there's more, isn't there?"

There was nothing friendly in the look Gabriel gave her. "This is bullshit. You've taken a couple of coincidences and built this fantasy out of them. What are you hoping to get out of it? Money from this Grant guy? Some crappy magazine story? If you've got any reputation at all, you're going to destroy it if you write this."

Bel smiled. His feeble threats told her she had him on the run. Time to go for the throat. "Like I said, there's more. You might think you're safe, Adam, but you're not. There's a witness, you see . . . " She left the sentence dangling.

He crushed out his cigarette and immediately began fiddling with another. "A witness to what?" There was an edge to his voice that made Bel feel she was on the right track.

"You and Matthias were seen together the day before the BurEst troupe disappeared from the Villa Totti. You were at the villa with him that night. The next day, they'd all gone. And so had you."

"So what?" He sounded angry now. "Even if that's true, so

what? I meet up with a friend of my father. My father, who's just
died. Next day, he leaves town with his crew. So fucking what?"

Bel let his words hang in the air. She reached for his cigarettes
and helped herself to one. "So there's a bloodstain the size of a
couple of litres on the kitchen floor. OK, you already know that
bit." She sparked the lighter, the flame's brightness revealing how
much darker it had become in the short time since she'd arrived.
The cigarette lit, she drew smoke into her mouth and let it trickle
out of one corner. "What you probably don't know is that the Italian police have launched a murder hunt." She tapped the cigarette
pointlessly against the edge of the ashtray. "I think it's time you
came clean about what happened back in April."

Thursday, 26th April 2007; Villa Totti, Tuscany

Until the last few days of his father's life, Gabriel Porteous hadn't
understood his closeness to the man who had brought him up
single-handed. The bond between father and son had never been
something he'd thought much about. If he'd been pressed, polite
rather than passionate was how he'd have characterized their relationship, especially when he contrasted it with the dynamic rapport that most of his mates shared with their fathers. He put it
down to Daniel's Britishness. After all, the Brits were supposed to
be uptight and reserved, weren't they? Plus, all his mates had vast
extended families, ranging vertically and horizontally through
time and space. In an environment like that, you had to stake
your claim or sink without trace. But Gabriel and Daniel had only
each other. They didn't have to compete for attention. So being
undemonstrative was OK. Or so he told himself. Pointless to acknowledge a longing for the sort of family he could never have.
Grandparents dead, the only child of only children, he was never
going to be part of a clan like his mates. He'd be stoic, like his dad,
accepting what couldn't be changed. Over the years, he'd shut the
door on his desire for something different, learning to bow to the

inevitable and reminding himself to count the blessings that came with his solitary status.

So when Daniel had told him about the prognosis of his cancer, Gabriel had gone into denial. He couldn't get his head round the thought of life without Daniel. This horrible information didn't make sense in his vision of the world, so he simply went on with his life as if the news hadn't been delivered. No need to come home more often. No need to snatch at every possible opportunity to spend time with Daniel. No need to talk about a future that didn't contain his father. Because it wasn't going to happen. Gabriel wasn't going to be abandoned by the only family he had.

But finally it had been impossible to ignore a reality that was bigger than his capacity for defiance. When Daniel had phoned him from the Policlinico Le Scotte and said in a voice weaker than a whisper that he needed Gabriel to be there, the truth had hit him with the force of a sandbag to the back of the neck. Those final days at his father's bedside had been excruciating for Gabriel, not least because he hadn't allowed himself to prepare for them.

It was too late for the conversation Gabriel finally craved, but in one of his lucid moments, Daniel had told him that Matthias was keeping a letter for him. He could give Gabriel no sense of what the letter contained, only that it was important. It was, Gabriel thought, typical of his father the artist to communicate on paper rather than face to face. He'd given his instructions for his funeral previously in an e-mail. A private service prearranged and paid for in advance in a small but perfect Renaissance church in Florence, Gabriel alone to see him to his grave in an undistinguished cemetery on the western fringes of the city. Daniel had attached an MP3 file of Gesualdo's *Tenebrae Responsories* for his son to upload to his iPod and listen to on the day of his burial. The choice of music puzzled Gabriel; his father always listened to music while he painted, but never anything like this. But there was no explanation for the choice of music. Just another mystery, like the letter left with Matthias.

Gabriel had planned to visit Matthias at the dilapidated villa

near Siena once the first acerbity of his grief had passed. But when he emerged from the graveyard, the puppeteer was waiting for him. Matthias and his partner Ursula had been the nearest to an uncle and aunt that Gabriel had known. They'd always been part of his life, even if they'd never stayed in one place long enough for him to grow familiar with it. They hadn't exactly been emotionally accessible either; Matthias was too wrapped up in himself and Ursula too wrapped up in Matthias. But he'd spent childhood holidays with them while his father went off for a couple of weeks on his own. Gabriel would end the holidays with suntanned skin, wild hair, and skinned knees; Daniel would return with a satchel bursting with new work from further afield: Greece, Yugoslavia, Spain, North Africa. Gabriel was always pleased to see his father, but his delight was tempered by having to say goodbye to the light touch of Ursula and Matthias's childcare.

Now the two men fell into a wordless embrace at the cemetery gates, clinging to each other like the shipwrecked to driftwood, not caring how unstable. At last, they parted, Matthias patting him gently on the shoulder. "Come back with me," he said.

"You've got a letter for me," Gabriel said, falling into step beside him.

"It's at the villa."

A bus to the station, a train to Siena, then Matthias's van back to the Villa Totti, and hardly a word exchanged. Sorrow blanketed them, bowing their heads and slumping their shoulders. By the time they reached the villa, drink was the only solution either of them could face. Thankfully, the rest of the BurEst troupe had set off earlier for a gig in Grossetto, leaving Gabriel and Matthias to bury their dead alone.

Matthias poured the wine and placed a fat envelope in front of Gabriel. "That's the letter," he said, sitting down and rolling a spliff.

Gabriel picked it up and set it down again. He drank most of his glass of wine, then ran a finger round the edge of the envelope. He drank some more, shared the spliff, and continued drinking.

He couldn't imagine anything Daniel had to tell him that would need so much paper. It hinted at revelation, and Gabriel wasn't sure he wanted revelation right now. It was painful enough holding on to the memory of what he had lost.

At some point, Matthias got up and put a CD in a portable player. Gabriel was surprised by the same music he'd listened to earlier, recognizing the strange dissonances. "Dad sent that to me," he said. "He told me to play it today."

Matthias nodded. "Gesualdo. He murdered his wife and her lover, you know. Some say he killed his second son because he wasn't sure if he was really the father. And his father-in-law too, supposedly, because the old man was out for revenge and Gesualdo got his retaliation in first. Then he repented and spent the rest of his life writing church music. It just goes to show. You can do terrible things and still find redemption."

"I don't get it," Gabriel said, uneasy. "Why would he want me to listen to that?" They were already on the second bottle of wine and the third joint. He felt a little fuzzy round the edges, but nothing too serious.

"You really should read the letter," Matthias said.

"You know what's in it," Gabriel said.

"Kind of." Matthias stood up and made for the door. "I'm going out on the loggia for some fresh air. Read the letter, Gabe."

It was hard not to feel there was something portentous about a letter delivered in such circumstances. Hard to avoid the fear that the world would be changed for ever. Gabriel wished he could pass; leave it unopened and let his life move on, unaltered. But he couldn't ignore his father's final message. Hastily, he grabbed it and ripped it open. His eyes watered at the sight of the familiar hand, but he forced himself to read on.

Dear Gabriel,

I always meant to tell you the truth about yourself but it never felt like the right time. Now I'm dying, and you deserve the truth but I'm too scared to tell you in case you walk away

and leave me to face the end by myself. So I'm writing this
letter that you'll get from Matthias after I'm away. Try not
to be too hard on me. I've done some stupid things but I did
them out of love.

The first thing I am going to say is that although I've told
you a lot of lies, the one thing that is the truth, the whole
truth, and nothing but the truth is that I am your father and
I love you more than any other living soul. Hang on to that
when you wish I was alive so that you could kill me.

It's hard to know where to start this story. But here goes.
My name is not Daniel Porteous and I'm not from Glasgow.
My first name is Michael, but everybody called me Mick.
Mick Prentice, that's who I used to be. I was a coal miner,
born and raised in Newton of Wemyss in Fife. I had a wife
and a daughter, Misha. She was four years old when you were
born. But I'm getting ahead of myself here because the two of
you have different mothers and I need to explain that.

The one thing I was any good at, apart from digging
coal, was painting. I was good at art at school but there was
no way somebody like me could do anything about that. I
was headed for the pit and that was that. Then the Miners'
Welfare ran a class in painting and I got the chance to learn
something from a proper artist. It turned out I had a knack
for watercolours. People liked what I painted and I could sell
them for a couple of quid now and again. At least, I could
before the miners' strike in 1984, when folk still had money
for luxuries.

One afternoon in September 1983, I came off the day
shift and the light was amazing, so I took my paints up on
the cliffs on the far side of the village. I was painting a view
of the sea through the tree trunks. The water looked luminous, I can still remember how it looked too beautiful to be
real. Anyway, I was totally into what I was doing, not paying
attention to anything else. And suddenly this voice said,
"You're really good."

And the thing that got me right away was that she didn't
sound surprised. I was used to folk being amazed that a miner

could paint a beautiful landscape. Like it was a monkey doing it
or something. But not her. Not Catriona. Right from that first
moment, she spoke to me like I was on an equal footing to her.

I just about shat myself, mind. I thought I was all alone,
and suddenly somebody right next to me was speaking to me.
She saw how freaked out I was and she laughed and said she
was sorry to disturb me. By then, I'd noticed she was bloody
gorgeous. Hair black as a jackdaw's wing, bone structure
like it had been carved with a flawless chisel. Eyes set deep so
you'd have to get right up close to be sure of the colour (blue
like denim, by the way) and a big smile that could wipe out
the sun. You look so like her sometimes it catches my heart
and makes me want to cry like a bairn.

So there I am in the woods, face to face with this amazing
creature and I can't find a word to say. She stuck her hand out
and said, "I'm Catriona Grant." I practically choked myself
clearing my throat so I could tell her my name. She said she
was an artist too, a sculptor in glass. I was even more amazed
then. The only other artist I'd ever met was the woman who
took the painting classes, and she was no great shakes. But I
just knew Catriona would live up to the job description. She
walked about with this ring of confidence, the sort of thing
you only have when you're the real thing. But I'm running
ahead of myself again.

Anyway, we talked a bit about the kind of work we were
interested in making, and we got along pretty well. Me, I was
just grateful to have anybody to talk to about art. I'd not seen
much art in the flesh, so to speak, just what they had at Kirkcaldy Art Gallery. But it turns out they had some pretty good
stuff there, which maybe helped me a bit in the early days.

Catriona told me she had a studio and a cottage on the
main road and told me to come round and see her set-up.
Then she went on her way and I felt like the light had gone
out of the day.

It took me a couple of weeks to build up to going to see
her studio. It wasn't hard to get to—only a couple of miles
through the woods—but I wasn't sure if she really meant

me to come or if she was just being polite. Shows how little I knew her back then! Catriona never said anything she didn't mean. And by the same token, she never held back when she had something to say.

I went across to see her one day when it was raining and I couldn't get painting. Her cottage was an old gatehouse on the Wemyss estate. It was no bigger than the house I was living in with my wife and kid, but she'd painted it in vibrant colours that made the rooms feel big and sunny even on a miserable grey day. But best of all was the studio and gallery she had out the back. A big glass kiln and plenty of working space, and at the other end, display shelves where people could come and buy. Her work was beautiful. Smooth, rounded lines. Very sensual shapes. And amazing colours. I'd never seen glass like it and even here in Italy you'd be hard pressed to find colours so rich and intense. The glass seemed to be on fire with different colours. You wanted to pick it up and hold it close to you. I wish I had a piece of hers but I never thought I'd need part of her until it was too late. Maybe one day you'll be able to track down something she made and then you'll understand the power of her work.

It was a good afternoon. She made me coffee, proper coffee like you didn't find much in Scotland back then. I had to put extra sugar in, it tasted funny to me at first. And we talked. I couldn't believe the way we talked. Everything under the sun, or so it seemed. It was obvious from the first time she opened her mouth that day in the woods that she was a different class from me, but that afternoon it didn't seem to make a lot of difference.

We arranged to meet again at the studio a few days later. I don't think either of us had any notion that there might be risks in what we were doing. But we were playing with fire. Neither of us had anybody else in our lives that we could talk to the way we could talk to each other. We were young—I was 28 and she was 24, but back then we were a lot more innocent than you and your friends at the same age. And from the very first moment we met, there was electricity between us.

I know you don't want to think about your mum and
dad being in love and all that goes with it, so I won't trouble
you with the details. All I will say is we became lovers soon
enough and I think for both of us it was like coming out into
bright sunlight after you've been used to electric lights. We
were daft about each other.

And of course it was impossible. I learned soon enough
the truth about your mother. She wasn't just any nice middle-
class lassie. She wasn't just plain Catriona Grant. She was the
daughter of a man called Sir Broderick Maclennan Grant. It's
a name that everybody in Scotland knows, like everybody in
Italy knows Silvio Berlusconi. Grant is a builder and devel-
oper. Everywhere you go in Scotland, you see his company's
name on cranes and hoardings. Plus he owns chunks of things
like radio stations and a football club and a whisky distillery
and a haulage company and a chain of leisure centres. He's a
bully as well. He tried to stop Catriona becoming a sculptor.
Everything she did, she did in spite of him. He would never
have stood for her having a relationship with someone as
common as a miner. Never mind a miner who was married to
somebody else.

And yes, I was married to somebody else. I'm not trying
to excuse myself. I never meant to be a cheating bastard, but
Catriona swept me off my feet. I never felt that way about
anybody before or since. You might have noticed I've never
been one for girlfriends. The thing is that nobody could ever
match up to Catriona. The way she made me feel, I don't
think anybody else could do that.

And then she fell pregnant with you. You see, son, you're
not Gabriel Porteous. You're really Adam Maclennan Grant.
Or Adam Prentice, if you prefer that.

When that happened, I would have left my wife for Catriona, no question. I wanted to and I told her so. But she wasn't
long out of a relationship that had been going for years, on
and off. She wasn't ready to live with me and she wasn't ready
for another fight with her father. I don't think anybody even
suspected we knew each other. We were careful. I always

came and went through the woods, and everybody knew I
was a painter, so nobody paid any attention to me wandering
about.

So we agreed to keep things as they were. Most days we
saw each other, even if it was only for twenty minutes or so.
And once you were born, I spent as much time with the two
of you as I could. By then, I was on strike so I didn't have
work to keep me from you.

I'm not going to do your head in by telling you all about
the year-long miners' strike that broke the union and the spirits of the men. There's plenty of books about it. Go and read
David Peace's *GB84* if you want an idea of what it was like.
Or get the DVD of *Billy Elliot*. All you need to know is that
every week that passed made me long for something different,
some life where the three of us could be together.

By the time you were a few months old, Catriona had
changed her mind too. She wanted us to be together. A fresh
start somewhere nobody knew us. The big problem was that
we had no money. Catriona was making a pretty bare living
from her glass work and I wasn't working at all because of the
strike. She could only afford her cottage and studio because
her mother paid the rent. That was a kind of bribe, to get
Catriona to stay near at hand. So we knew her mum wouldn't
be paying for us to set up home anyplace else. We couldn't
stay put either. Me walking out on my wife and daughter at
the height of the strike to go and live with somebody from
the bosses' class would have been seen as worse than being a
scab. They'd have put bricks through our windows. So without a bit of money to get us started, we were screwed.

Then Catriona had this idea. The first time she mentioned
it, I thought she'd lost her mind. But the more she talked
about it, the more she convinced me it would work. The idea
was that we'd fake a kidnap. I'd walk out on my family, make
it look like I'd gone scabbing, and hide at Catriona's. A few
weeks later, you and Catriona would disappear and her father
would get a ransom note. Everybody would think you'd been
kidnapped. We knew her father would pay the ransom, if not

for her then for you. I would take the money, you and
Catriona would go back, then a few weeks later, Catriona
would take you away, saying she was too upset by the kidnapping to carry on living there. And we'd all meet up and start
our life together.

It sounds simple when you say it fast. But it got complicated, and things went to shit. As it turns out, your mother
couldn't have had a worse idea if she'd spent her whole life
working on it.

The first thing we realized when we started making the
detailed plans was that we couldn't do it with just two of us.
We needed an extra pair of hands. Can you imagine trying to
find somebody we could trust to join in with a plan like that?
I didn't know anybody who would be mad enough to join us,
but Catriona did. One of her old pals from the College of Art in
Edinburgh, a guy called Toby Inglis. One of those upper-class
mad bastards who are up for anything. You've always known
him as Matthias, the puppeteer. The man who will have given
you this letter. And he's still a mad bastard, by the way.

He had the bright idea of making the kidnapping look like
a political act. He came up with these posters of a sinister
puppeteer with his marionettes and used them to deliver the
ransom notes as if they were from some anarchist group. It was
a good idea. It would have been a better idea if he'd destroyed
the screen he used to print them, but Toby's always thought
he was one degree smarter than everybody else. So he kept the
screen and he still uses that same poster sometimes for special
performances. Every time I see it, my bowels turn to water. All
it would take is one person to recognize where it comes from
and we'd have found ourselves up to our necks in it.

But I'm getting ahead of myself again. I really wasn't sure
whether I should tell you all of this, and Toby thought maybe
it would be better to let sleeping dogs lie, especially since
you'll be having to cope with me not being around any more.
But the more I thought about it, the more it seemed to me that
you have a right to know the whole truth, even if it's hard for
you to deal with. Just remember the years we've had together.

Remember the good stuff, it's what redeems all the crap I did.
At least, I hope that's how it works.

A very bad thing happened the night I left my wife and
daughter. I walked out in the morning without saying anything about leaving. I'd heard there was a bunch of scabs
going down to Nottingham that night and I figured everybody would think I'd gone with them. I went straight round
to Catriona's and I spent the day looking after you while she
was working. It was bloody cold that day, and we were going
through a lot of wood. After dark, I went out to chop some
more logs.

This is hard for me. I haven't talked about this for twenty-
two years and still it haunts me. When I was growing up, I
had two pals. Like you and Enzo and Sandro. One of them,
Andy Kerr, had become a union official. The strike was hard
on him and he was off work with depression. He lived in a
cottage in the woods about three miles west of Catriona's
place. He loved natural history, and he used to walk the
woods at night so he could watch badgers and owls and that
kind of thing. I loved him like a brother.

I was chopping the wood when he came round the end of
the workshop. I don't know who got the bigger shock. He
asked what the hell I was doing, chopping wood for Catriona
Maclennan Grant. Then he twigged. And he lost it. He came
at me like a madman. I dropped the axe and we fought like
stupid wee boys.

The fight's all a bit of a blur to me. The next thing I
remember is Andy just stopping. Collapsing into me so I had
to put my arms round him to stop him falling. I just stared at
him. I couldn't make sense of it. Then I saw Catriona standing behind him holding the axe. She'd hit him with the blunt
end, but she was strong for a woman and she'd hit him so
hard she'd smashed his skull.

I couldn't believe it. A few hours before, we'd been on top
of the world. And now I was in hell, holding the dead body of
my best pal.

I don't know how I got through the next few hours. My brain seemed to work independent of the rest of me. I knew I had to sort things, to protect Catriona. Andy had a motorbike and sidecar combination. I walked back through the woods to his place and drove the bike back to Catriona's. We put him in the sidecar and I drove down to the Thane's Cave at East Wemyss. There's a set of caves down there that have been used by humans for 5,000 years, and I was involved in the preservation society so I knew what I was doing. I could get the bike right up to the entrance to the Thane's Cave. I carried Andy in the rest of the way and buried him in a shallow grave in the back part of the cave.

I went back a couple of days later and brought the roof down so nobody would find Andy. I knew where to get my hands on some pit explosives—my wife's pal had been married to a pit deputy and I remembered him boasting about having a couple of shots of dynamite in his garden shed.

But back to that night. I wasn't finished. I drove the bike back through East Wemyss and along to the pit bing. I jammed the throttle open and let it pile into the side of the bing. The slag covered it while I stood there.

I walked home in a total daze. Ironically, I ran into the scabs as they were setting off. I've no idea what I said to them, I was deranged.

When I got to Catriona's, she was in a hell of a state. I don't think either of us slept that night. But by the time morning came, we knew we had to go through with her idea. As well as wanting to start a new life, we needed to put some distance between Andy and us. So we started to make our plans.

Ironically, Andy being dead solved one problem we'd had about faking the kidnap—where we could hide you and Catriona without anybody knowing. I hit on the idea of forging a note in Andy's handwriting in case any of his family came by to see why they hadn't heard from him. It wasn't a straight-out suicide note. I didn't want to upset them, so I left it kind of ambiguous. I know that sounds weird, but I'm tell-

ing it like it was, not trying to make myself look like the good guy. Like I said, I've done things I'm ashamed of, but I did all of them out of love.

We let some time pass before we set up the kidnap because we didn't want anybody making a connection between me leaving and the kidnap. Also, we wanted to be sure Andy's family had accepted he'd gone away and wouldn't be coming round on the off-chance. I'm ashamed to say I forged a couple of postcards in his writing and went up north to post them after the New Year so they'd stay away from his cottage and not come looking to see if he was back. We needed to make sure we'd be safe there.

On the day we'd agreed, the three of us went off to Andy's with your toys and your clothes and there we stayed until the night of the ransom handover. Toby wasn't around much—he was sorting out the boats. We'd decided to do the handover in a place where we could escape by boat. We'd told Grant not to tell the police, but we weren't sure if he'd stick to that, so we thought we'd leave the police flat-footed if we got away on the water.

At the time, Toby was living on his father's boat, a four-berth cabin cruiser. He knew about boats, and he'd decided we needed to make our getaway in an inflatable with an outboard engine. He knew somebody who had one up in a boathouse in Johnstown. He reckoned nobody would even notice it was missing until May, so that seemed like a good idea.

Anyway, the night of the handover came and we set off. We'd agreed Catriona was going to get the money, then we'd hand you over to her mother. We'd go off with Catriona, then the next day, she'd turn up by some roadside, supposedly having been dumped once the kidnappers knew the ransom was the real thing. Meanwhile I'd give Toby his third share, he'd go his way, and I'd go mine, finding us somewhere to live and work up in the Highlands.

Nothing went like it was supposed to. The place was crawling with armed police, though we didn't realize it. Toby had a gun too, though I didn't realize that either until we got

out the boat at the rendezvous. And Grant had a gun. It was
a recipe for disaster. And a disaster was what we got.

Even after all this time, thinking about it makes me
choke up. Everything was going to plan, but for some reason,
Catriona's mum made a big performance about handing
the ransom over. Grant lost the place and started waving
his gun around. Then Toby turned off the spotlight and the
shooting started. Catriona got caught in the crossfire. I had
night-vision goggles from the army surplus and I saw her fall
just a few yards away from me. I ran to her. She died in my
arms. It was all over in seconds. She'd dropped the bag with
the ransom when she was shot, and Toby grabbed it. I didn't
know what to do. You were back by the boat, in your carry
cot. We'd planned to leave you there. But I knew I couldn't
leave you, not with your mother dead. I couldn't leave you
behind for Grant to bring up in his image. So we ran for the
boat. I got a hold of your carry cot and threw it back aboard
and we got out of there as fast as we could.

The only thing that went according to plan was what we'd
decided to do to avoid anyone using tracking devices to follow
us. We put the money in another bag that we'd brought with
us and tossed the original over the gunwale. Then I dredged
the bag with the diamonds through the sea. We figured the
water would knock out any transmitter they might have put
in amongst them. It seemed to do the trick, because there was
nobody on our tail as we shot down the coast to Dysart where
Toby's boat had already been moored for a few days. It was
just a few miles, so we got there before the helicopter was in
the air. We could hear it and see it from the boat. After it had
gone, Toby took the inflatable out of the harbour and sank
it off the beach. Then we holed up there till dawn and set off
on the morning tide. I was in a state of shock, to tell you the
truth. A couple of times, I was on the point of walking to the
nearest police station and giving myself up. But Toby held
himself together and saved all of us.

It took us a few weeks to get to Italy. We laundered most
of the money in automatic cash machines and casinos along

the French coast. The lion's share of the ransom was in uncut diamonds, and we hung on to those.

Once we got here, we split up. I left Toby with the boat and I rented a house in the hills outside Lucca for a few months till I decided where I wanted to live. I don't remember much about that time. I was dazed with grief and guilt and the terrible pain of losing Catriona. If it hadn't been for you, I might not have made it through. I still can't believe how it all went so wrong.

I know you probably look at my life and think I had it pretty good. The ransom money bought us the house in Costalpino, and a bit left over that I've got invested. The income from that put the jam on the bread and butter I earned from the painting. I got to spend the rest of my life in a beautiful place, bringing up my son and painting the things I wanted to paint without ever having to worry too much about money.

The only reason you can think I had it pretty good is that you never knew your mother. When she died, she took the light away. You have been the only real light in my life since then, and don't underestimate what a joy it has been for me to spend these years with you. It breaks my heart that I will not live to see what you achieve with the rest of your life. You're a very special person, Adam. I call you that because it is the name we chose together for you.

There's one last thing I want you to do. I want you to make contact with your grandfather. I Googled him last week for the first time: Sir Broderick Maclennan Grant. His friends call him Brodie. He lives in Rotheswell Castle in Fife. His first wife, your grandmother, committed suicide two years after Catriona died. He's got a new wife now, and a son called Alec. So you see, you have a family. You have a grandfather and an uncle who is quite a few years younger than you! Make the most of them, son. You've got a lot of time to make up for, and you're enough of a man now to stand up to a bully like Brodie Grant.

So now you know it all. Blame me or forgive me, it's up

to you. But never doubt that you were conceived and born in
love, and that you have been loved every single day of your
life. Take care of yourself, Adam.

All my love,
Your father, Mick the miner

Gabriel dropped the last sheet on top of the others. He went
back to the first page and read it all again, aware that Matthias
had come back in at some point. It was like reading the synopsis of
a movie. Impossible to connect to his life. Too absurd to be true.
He felt as if the foundations of his life had been removed, leaving
him hanging in the air like a cartoon character holding his breath
for the inevitable catastrophic fall. "Does Ursula know all this?"
he said, knowing it wasn't that important a question, but wanting
to know the answer anyway.

"Some of it." Matthias sat down heavily opposite Gabriel, another bottle of wine in his hand. "She doesn't know who your
mother was, or all of Daniel's story. She knows he set up a fake
kidnap because he wanted to be with you and your mother. But
she doesn't know about the shoot-out at the OK Corral."

The flippancy of Matthias's description of his mother's death
gave Gabriel a jolt. *Toby had a gun too.* He gave a half-hearted
snort of derision. "All these years, I thought I was living among a
bunch of old hippies with a load of outdated leftie ideals. And it
turns out you lot are actually a bunch of criminals on the run after
the worst kind of capitalist crime." He knew there were more important things to talk about, but he had to work his way round to
them, like a dog faced with a hot dinner who starts off nibbling at
the edges because that's all he can cope with. *Toby had a gun too.*

"You're looking at it all wrong, Gabe, my man," Matthias said,
fingers busy with another joint. "Think of us as latter-day Robin
Hoods. Robbing the seriously rich to spread the money round
more fairly."

"You and my dad living the life of Riley, doing exactly what

you want—how exactly does that further the fight against international capitalism?" Gabriel didn't even try to keep the sneer from
his face or his voice. "If my grandfather had been supportive of
my mother's art, none of this would have happened. Don't tell me
you all did this for some higher purpose. You did it because you
wanted your own way and you saw how you could make somebody else pay for it." He waved the joint away impatiently. He
didn't want to lose any of the shreds of clarity left to him.

"Hey now, Gabe, don't be rushing to judgement on us."

"Why not? Isn't that what the Gesualdo is all about? It's like
the last thing he did was invite me to judge him. Should I see him
as a killer or as a man redeemed by his painting? Or redeemed by
loving me and bringing me up the best he could?" Gabriel scrab-
bled through the letter, looking for the last page. "Here it is, in his
own hand: 'Blame me or forgive me, it's up to you.' He wanted me
to make up my own mind about what you did." The heat of anger
was spreading through him, filling him up and making it harder to
be reasonable. *Toby had a gun too.*

"And you should forgive him," Matthias said. "You doubt our
motives, but I tell you, all he wanted was to make a life with you
and Cat. Circumstances were against them. We just tried to re-
dress the balance, that's all, Gabe."

His easy complacency was like a goad to Gabriel. "And when
did that give you the right to make my choices for me?"

"What are you talking about?"

"You and Daniel, you chose what I got to know about who I
am and when I got to know it. You kept me away from my family.
You lied about my history, made me think all I had was Daniel and
you and Ursula. You took away my chance of growing up knowing my grandfather. My grandmother might still be alive if she'd
had me with her."

Matthias blew out a plume of smoke. "Gabe, there was no
going back for us. You think growing up under Brodie Grant's
thumb would have been better than the life you've had?" He

snorted derisively. "You wouldn't say that if you had any idea how tough he made Cat's life." He got up and fetched a block of dope and a sharp knife to cut off a fresh slice.

"But I don't, do I? Because I never got the chance to find out, thanks to you two and the choices you made for me." Gabriel slammed the flat of his hand down on the table. "Well, I'm going to make up for lost time. I'm going back to Scotland. I'm going to find my grandfather and get to know him for myself. Maybe he's the ogre you and Daniel make him out to be. Or maybe he's just someone who wanted the best for his daughter. And judging by this"—he batted the letter, making the papers flutter in the dim light—"he wasn't so far off the mark, was he? I mean, my dad wasn't exactly a model citizen, was he?"

Matthias dropped the knife and stared at Gabriel. "I don't think going back is that great an idea."

"Why not? It's time I got to know my family, don't you think?"

"That's not the issue."

"Well, what is?"

Matthias made a small helpless gesture with his hands. "They're going to want to know where you've been for the last twenty-odd years. And that's kind of a problem for me."

"What's it got to do with you?"

"Think about it, Gabe. There's no statute of limitations for murder or kidnap. They're going to come after me and put me away for the rest of my life."

Toby had a gun too. "I won't tell them anything that implicates you," Gabriel said, contempt in the curl of his mouth. "You don't have to worry about your own skin. I'll take care of that."

Matthias laughed. "You really have no fucking idea who your grandfather is. You think you can just refuse Brodie Grant? He'll chase down your history, he'll backtrack and find out every move you've made all these years. He won't stop till he's nailed me to a fucking cross. This isn't just about you."

"This is my life." They were both shouting now, outrage and fear stoking the paranoia of dope and the abandon of alcohol. "If he gets me back, why the hell would my grandfather care about you?"

"Because he'll never give up the chance for revenge so he doesn't have to take responsibility."

"Responsibility? Responsibility for what?"

"For killing Cat." Even as he spoke, Matthias's face stretched in horror. He knew the enormity of what he'd said as soon as the words were out of his mouth.

Gabriel stared at him in disbelief. "You're crazy. You're saying my grandfather shot his own daughter?"

"That's exactly what I'm saying. I don't think he meant—"

Gabriel jumped to his feet, sending the chair crashing to the floor. "I can't believe—You lying piece of—You'd say anything," he shouted incoherently. "You brought a gun. You're the one who shot her, aren't you? That's what really happened. Not my grandfather. You. That's why you don't want me to go back, because you'll finally have to face what you did."

Matthias stood up, walking round the table towards Gabriel, hands outstretched. "You've got it so wrong," he said. "Please, Gabe."

Gabriel's face was a mask of rage and shock. He reached down for the knife on the table and rushed Matthias. Nothing in his mind but anger and pain, nothing as coherent as intent. But the result was as incontrovertible as if it had been the result of a meticulous plan. Matthias crumpled and fell backwards, a dark red blemish quickly spreading to a stain across the front of his T-shirt. Gabriel stood above him, panting and sobbing, not caring to make any effort to staunch the blood. *Toby had a gun too.*

Matthias clutched at his failing heart as it slowly ran out of blood to pump round his body. His heaving chest gradually subsided till it grew motionless. Gabriel had no idea how long it took Matthias to die, only that, by the end, his legs were so tired they could scarcely hold him up. He slumped to the floor where he

stood, just beyond the margin of the slowly congealing pool of blood that had spread beyond Matthias's body.

Time drifted past. Finally, what roused him was footsteps and lively chatter approaching along the loggia. Max and Luka swaggered in, full of the success of the evening's performance. When they saw the gory tableau in front of them, they stopped short. Max cursed, Luka crossed himself. Then Rado walked in with Ursula. She caught sight of Matthias and opened her mouth in a soundless scream, falling to her knees and crawling towards him.

"He killed my mother," Gabriel said, his voice flat and cold.

Ursula swung her head round to him, her lips curled back in a snarl. "You killed him?"

"I'm sorry," he whispered. "He killed my mother."

Ursula whimpered. "No. No, it's not true. He couldn't hurt a fly." She stretched out her hand tentatively, her fingertips brushing Matthias's dead hand.

"He had a gun. It's in the letter. Daniel left me a letter."

"What the fuck are we going to do?" Max yelped, breaking the macabre intimacy between them. "We can't call the cops."

"He's right," Rado said. "They'll pin it on one of us. One of the illegals, not the painter's son."

Ursula pressed her hands to her face, fingers splayed, as if she was going to claw her features apart. Her body heaved in a spasm of dry retching. Then somehow she visibly drew her strength together. Her face smeared with Matthias's blood like a terrible parody of night camouflage, she launched herself at Gabriel with a harrowing scream.

Max and Luka instinctively threw themselves between her and Gabriel, dragging her back, keeping her clawing fingers from his eyes. Panting, she spat on the floor. "We loved you like a son," she wailed. Then something in German that sounded like a curse.

"He killed my mother," Gabriel insisted. "Did you know that?"

"I wish he'd killed you," she screamed.

"Get her out of here," Rado shouted.

Max and Luka hauled her to her feet and half-carried her towards the door. "Pray I never see you again," Ursula screamed as she disappeared.

Rado crouched beside Gabriel. "What happened, man?"

"My dad left me a letter." He shook his head, dazed with shock and drink. "It's all over now, isn't it? He killed my mother, but I'm the one who's going to jail."

"Fuck, no," Rado said. "No way is Ursula going to the cops. It goes against everything she believes in." He put his arm round Gabriel. "Besides, we can't let her drag us all into this shit. No way I'm going back where I came from. Matthias is dead, there's nothing we can do to help him. No need to make things worse."

"She's not going to let me get away with this," Gabriel said, leaning into Rado. "You heard her. She's going to want to hurt me."

"We'll help her," Rado said. "We love you, man. And eventually she'll remember she does too."

Gabriel dropped his head into his hands and let the tears come. "What am I going to do?" he wailed.

Once his sobbing had subsided, Rado pulled him to his feet. "I hate to sound like a cold-hearted bastard, but the first thing you need to do is help me get rid of Matthias's body."

"What?"

Rado spread his hands. "No body, no murder. Even if we can't keep Ursula away from the cops, they're not going to sweat it if there's no body."

"You want me to help you bury him?" Gabriel sounded faint, as if this was one step more than he could manage.

"Bury him? No. Buried bodies have a way of turning up. We're going to carry him down to the field. Maurizio's pigs will eat anything."

By morning, Gabriel knew Rado had been right.

Thursday, 5th July 2007; Celadoria, near Greve in Chianti

Remembering that night now, Gabriel felt as though Bel Richmond was hollowing his stomach out with a spoon. Losing his father had been bad enough. But Daniel's letter and what it had led to had been devastating. It was as if his life was a piece of fabric that had been ripped from top to bottom and tossed in a heap. If the letter had plummeted him into a state of turmoil, killing Matthias had made matters infinitely worse. His father had not been the man he thought he was. His lies had poisoned so much. But Gabriel himself was worse than a liar. He was a killer. He'd committed an act that he would never have believed himself capable of. With such fundamental elements of his life exposed as a fantasy, how could he cling to any of it with confidence?

He'd grown up thinking his mother was an art teacher called Catherine. That she'd died giving birth to him. Gabriel had struggled with that guilt for as long as he could remember. He'd seen his father's isolation and sadness and had shouldered the blame for that too. He'd grown up carrying a weight that was completely bogus.

He didn't know who he was any more. His history had been just a story, made up to protect Daniel and Matthias from the consequences of the terrible thing they'd been part of. For their sake, he'd been wrenched out of the country where he belonged and brought up on alien soil. Who knew what his life would have been if he'd grown up in Scotland instead of Italy? He felt cast adrift, rootless and deliberately cheated out of his birthright.

His torment was made worse by constant fear, shivering behind him like the backdrop in a puppet booth. Every time he heard the sound of a car, he was on his feet, back to the wall, convinced that this time it was the carabinieri come for him at Ursula's insistence. He'd tried to cover his tracks, but he didn't have his father's experience, and he was afraid he hadn't succeeded.

But time had crawled past and after a few weeks of being holed

up like a sick animal, he had started to put himself back together. Gradually, he'd managed to find a way to distance the guilt, telling himself Matthias had lived free and clear for over twenty years, never paying a penny of the debt that was owed for Catriona's death. All Gabriel had done was force him to make amends for the life he'd stolen from all of them—Catriona, Daniel, and Gabriel himself. It wasn't entirely satisfactory from the perspective of the morality Daniel had instilled in him, but holding fast to this conviction made it possible for Gabriel to attempt to move forward, accommodating his remorse and assimilating his pain.

One overwhelming imperative drove him forward. He wanted to find the family that was his by rights, the clan he'd always craved, the tribe he belonged to. He wanted the home he'd been denied, a land where people looked like him rather than escapees from medieval paintings. But he'd known he wasn't ready yet. He had to get his head straight before he attempted to take on Sir Broderick Maclennan Grant. The little he had been able to glean from his father's letter, from Matthias, and from the Internet had left him certain that Grant would not give any claimant an easy time. Gabriel knew he needed to be able to hold his own and to keep his story straight in case that terrible April night ever came back to haunt him.

And now it looked as if it had. Fucking Bel Richmond with her digging and her determination was going to destroy the one hope he'd been clinging to for the past weeks. She knew she was on to something. Gabriel hadn't had much to do with the media, but he knew enough to realize that now she had the threads of her story, she wouldn't give up till she had nailed him. And when she published her scoop, any hope he had of making a new life with his mother's family would be dead in the water. Brodie Grant wouldn't be happy to embrace a murderer. Gabriel couldn't let it happen. He couldn't lose everything for a second time. It wasn't fair. It so wasn't fair.

Somehow, he remained composed, meeting her long, level

stare. He had to find out exactly what she knew. "What do you think happened?" he said, a sneer on his face. "Or should I say, what are you planning to tell the world happened?"

"I think you killed Matthias. I don't know whether you planned it or it was a spur-of-the-moment thing. But, like I said, there's a witness who can put you two together earlier that day. The only reason he hasn't told the police is that he doesn't understand the significance of what he saw. Of course, if I was to explain that to him . . . Well, it's not rocket science, is it, Adam? It took me three days to find you. I know the carabinieri have a reputation for being a bit slow on the uptake, so it might take them a bit longer. Time enough to get yourself under the protective wing of your grandfather, I'd have thought. Oh, but he's not your grandfather, is he? That's just my little fantasy."

"You can't prove any of this," he said. He poured the last of the wine into her glass, then went over to the wine rack to fetch another. He felt cornered. He'd come through a terrible ordeal. And now this fucking woman was going to steal the one hope that had held him together. His challenge was his way of giving her a chance to prevent his having to do whatever it took to stop her.

He glanced over his shoulder. Bel wasn't really paying attention to him now; she was absorbed in the chase, focused on turning the interview in the direction she sought. Absently, she said, "There are ways. And I know all of them."

He'd given her the chance and she'd deflected it. His past was corrupt beyond redemption. All he had left was the future. He couldn't let her take that from him. "I don't think so," he said, coming up behind her.

At the last minute, some primitive warning signal hit her brain and she swung round just in time to catch a flash of the blade as it headed unwaveringly for her.

Kirkcaldy

After Phil had made the first move, things had progressed at breakneck speed. Clothes stripped. Skin to feverish skin. Him on top. Her on top. Then to the bedroom. Face down, his hands cupping her breasts, her hands clinging to the struts of the bed-head. When they finally needed to pause for a second wind, they lay on their sides, grinning stupidly at each other.

"Whatever happened to foreplay?" Karen said, a giggle in her voice.

"That's what working together all these years has been," Phil said. "Foreplay. You getting me all het up. Your mind's as sexy as your body, you know that?"

She slid a hand down between them and let her fingertips caress the soft skin below his belly button. "I have wanted to do this for so long."

"Me too. But I really didn't want to fuck things up between us at work. We're a good team. I didn't want to chance spoiling that. We both love our work too much to risk it. Plus it's against the rules."

"So what's changed?" Karen said, a hollow feeling in her stomach.

"There's an inspector's job coming up in Dunfermline and I've been told unofficially that it's mine for the asking."

Karen pulled away, leaning on one elbow. "You're leaving CCRT?"

He sighed. "I've got to. I need to move up and there's not room for another inspector in the CCRT. Besides, this way I get to have you too." His face screwed up in anxiety. "If that's what you want. Obviously."

She knew how much he loved working cold cases. She also knew he was ambitious. After she'd blocked his career path with her promotion, she'd expected him to go sooner or later. What she hadn't bargained for was that she might figure in his calculations. "It's the right move for you," she said. "Better get out quick before

the Macaroon realizes he should hate you as much as he hates me. I'll miss working with you, though."

He wriggled close to her, gently rubbing the palms of his hands against her nipples. "There will be compensations," he said.

She let her hand drift downwards. "Apparently," she said. "But it's going to take a lot to make it up to me."

Boscolata, Tuscany

Carabiniere Nico Gallo crushed the cigarette under the heel of his highly polished boot and pushed himself off the olive tree he was leaning against. He brushed off the back of his shirt and his tightly fitting breeches and set off again along the path that bordered Boscolata's olive grove.

He was fed up. Hundreds of miles from his home in Calabria, living in a barracks only marginally better than a fisherman's shack, and still getting the shitty end of every assignment, he could hardly get through a day without regretting choosing a career in the carabinieri. His grandfather, who had encouraged him in his choice, had told him how women fell for men in uniform. That might have been the case in the old man's day, but it was the polar opposite now. All the women his age he seemed to meet were feminists, environmentalists, or anarchists. To them, his uniform was a provocation of a very different kind.

And to him, Boscolata was just another hippy commune inhabited by people with no respect for society. He bet they didn't pay their taxes. And he bet that the killer who had claimed the unknown victim at the Villa Totti wasn't far from where he was walking now. It was a waste of time, having a night patrol out here. If the killer had wanted to cover his tracks, he'd had months to do it. And even now, Nico reckoned everybody in Boscolata knew how to get inside the ruined villa without his having a clue they were in there. If this were his village back in the south, that's exactly how it would be.

Another round of the olive grove and he was going back to his car for a cup from the flask of espresso he'd thoughtfully brought with him. These were the milestones that made it possible to stay awake and alert: coffee, cigarettes, and chewing gum. When he got to the corner closest to the Villa Totti, he could have another cigarette.

As the sound of his match died away, Gallo realized there was another noise on the night air. This far up the hill, the night was silent but for the crickets, the odd night bird, and the occasional dog barking. But now the silence had been invaded by the straining sound of an engine climbing the steep dirt road to Boscolata and beyond. But curiously, it wasn't matched with the brilliance of headlights on full beam. He could make out pale glimmers through the trees and hedgerows, as if the vehicle was travelling on sidelights. Only one reason for that, in his books. The driver was up to something he didn't want to draw attention to.

Gallo glanced ruefully at his cigarette. He'd made sure he had enough for the night's duty, but that didn't mean he wanted to waste one. So he cupped it in his hand and moved closer to the villa to cut off anyone attempting to enter the crime scene.

It soon became clear he'd made the wrong choice. Instead of heading towards Boscolata and the villa, the lights swerved off to the right at the far end of the olive trees. Cursing, Gallo took a last drag on his cigarette, then started down the side of the grove as quickly and as quietly as he could.

He could just about make out the shape of a small hatchback. It stopped at the end of the trees, where the Totti property butted up against the substantial acreage farmed by the guy with the pigs. Maurizio, wasn't that the old man's name? Something like that. Gallo, about twenty metres away, edged closer, trying not to make a sound.

The car's interior light came on as the driver's door swung open. Gallo saw a tallish guy wearing dark sweats and a baseball cap get out and open the tailgate. He seemed to be dragging out a rolled-up carpet or something similar, bending down to get his

back underneath to take the weight. As he straightened up, staggering a little under the weight of his burden and approaching the sturdy wire fence that kept the pigs penned in, Gallo realized with a horrible lurch of his stomach that this wasn't an instance of midnight trash-dumping but something much more serious. The evil fucker was about to feed a body to the pigs. Everyone knew pigs would eat bloody anything and everything. And this was indisputably a body.

He grabbed his flashlight and turned it on. "Police! Freeze!" he shouted in the most melodramatic style he could muster. The man stumbled, tripped, and fell forward, his burden landing athwart the fence. He regained his feet and raced back to the car, reaching it seconds before Gallo. He jumped in and started the engine, throwing it into reverse just as Gallo hurled himself at the hood. The carabiniere tried to hang on, but the car was speeding backwards towards the track, jouncing and jittering every metre of the way, and he finally slid off in an ignominious heap as the car disappeared into the night.

"Oh God," he groaned, rolling over so he could reach his radio. "Control? This is Gallo, on guard at the Villa Totti."

"Roger that, Gallo. What's your ten?"

"Control, I don't know the ten-code for this. But some guy just tried to dump a body in a pig field."

Friday, 6th July 2007; Kirkcaldy

The phone penetrated Karen's light sleep on the first ring. Dazed and disorientated, she groped for it, thrilled into full consciousness by the mumble of "Phone," next to her ear. He was still here. No hit and run. He was still here. She grabbed the phone, forcing sticky eyelids apart. The clock read 05:47. She was CCRT. She didn't get calls at this time of the morning any more. "DI Pirie," she grunted.

"Morning, DI Pirie," a disgustingly bright voice said. "This is

Linda from Force Control. I've just had a Capitano di Stefano on from the carabinieri in Siena. I wouldn't usually have woken you, but he said it was urgent."

"It's OK, Linda," Karen said, rolling away from Phil and trying to get her head into work mode. What the hell could be quarter-to-six-in-the-morning urgent on a three-month-old maybe murder? "Fire away."

"There's not much to fire, Inspector. He said to tell you he's e-mailed you a photo to see if you can ID it. And it's urgent. He said it three times, so I think he meant it."

"I'll get right on to it. Thanks, Linda." She replaced the phone and Phil immediately pulled her to him with a different kind of urgency.

She squirmed round, trying to free herself from his grip. "I need to get up," she protested.

"So do I." He covered her mouth with his and started kissing her.

Karen pulled away, gasping. "Can you do quickies?"

He laughed. "I thought women didn't like quickies."

"Better learn how if you're going back on front-line policing," she said, drawing him into her.

Feeling only mildly guilty, Karen logged on to her e-mail. The promised message from di Stefano was the latest addition to her inbox. She clicked it open and set the attachment to download while she read the brief note. Someone tryed to feed a body to Maurizio Rossi's Cinta di Siena pigs. Maybe this is where the other victim went. Here is a picture of the face. Maybe you know who it is? God, that was a nasty thought. She'd heard that pigs had been known to eat everything but the belt buckle when unfortunate farmers had had accidents inside their pens, but it would never have occurred to her to consider it as a means of body disposal.

And then an even nastier thought occurred to her. *Pig eats victim. Pig incorporates human into its own meat. Pig gets turned*

into salami. And people end up eating people. Somehow, she didn't think the farmer was going to have much of a business left once this got out.

Karen hesitated, wondering why di Stefano thought she might recognize the victim. Could this be Adam Maclennan Grant, his future with his grandfather snatched from him at the last moment? Or the mysteriously disappeared Matthias, aka Toby Inglis? Anxiety dried her mouth, but she clicked on the attachment.

The face that filled her screen was definitely dead. The spark that animated even coma patients was entirely absent. But it was still shockingly unmistakable. The day before, Karen had interviewed Bel Richmond. And now she was dead.

A1, *Firenze–Milano*

There had been no reason to ditch Bel's hire car, Gabriel had decided. Not at this point. That mad bastard cop had shaken the living daylights out of him, but he couldn't have seen the licence plate. Nobody would be connecting a car hired by an English journalist with what had happened on the Boscolata hillside. Putting distance between himself and Tuscany was the most important thing now. Leave the past and its terrible necessities behind. Make a clean break and drive straight into the future.

It had been horrible, but he'd stripped the body, partly to make it easier for the pigs to do his dirty work for him, and partly to make it harder to identify her in the unlikely event that she was found soon enough to make identification a possibility. As it turned out, that had been a great decision. It had been bad enough when that crazy cop appeared out of nowhere. It would have been a million times worse if he'd left anything on the body that could make it easier to work out who she was.

And so the car would be safe for now. He'd park it in the long-stay at Zurich airport and pick up a flight. Thanks to Daniel's insistence that there was nothing for him there but pain and ghosts,

he'd never been to the UK before, had no idea what the security would be like. But there was no reason for them to look twice at him and his British passport.

He wished he hadn't had to kill Bel. It wasn't like he was some stone-cold killing machine. But he'd already lost everything once. He knew what that felt like, and he couldn't bear it to happen again. Even mice fight when they're cornered, and he definitely had more bottle than a mouse. She'd left him no choice. Like Matthias, she'd pushed him too far. OK, it had been different with Matthias. That time, he'd lost control. Realizing that someone he'd loved since childhood had been his mother's killer had cracked open some well of pain in his head, and he'd stabbed him before he even knew he had a knife in his hand.

With Bel, he'd known what he was doing. But he'd acted in self-preservation. He'd been on the very point of contacting his grandfather when Bel barged into his life, threatening everything. The last thing he needed was her spilling the beans, linking him to Matthias's murder. He wanted to arrive at his grandfather's house with a clean sheet, not have the life he'd been denied fucked up by some muck-raking journalist.

He kept telling himself that he'd done what he had to do. And that it was good that he felt bad about it. It showed he was basically a decent person. He'd been ambushed by events. It didn't mean he was a bad person. He desperately needed to believe that. He was on his way to start a new life. Within days, Gabriel Porteous would be dead and Adam Maclennan Grant would be safely under the wing of his rich and powerful grandfather.

There would be time to feel remorse later.

Rotheswell Castle

Susan Charleson clearly didn't like the police turning up without prior invitation. The few minutes' notice between Karen's arrival at the gate and her presence on the front doorstep hadn't been

quite long enough for Grant's right-hand woman to disguise her affront. "We weren't expecting you" replaced the welcome that had been uttered previously.

"Where is he?" Karen swept in, forcing Susan to take a couple of quick steps to the side.

"If you mean Sir Broderick, he is not yet available."

Karen made an ostentatious study of her watch. "Twenty-seven minutes past seven. I'm betting he's still at his breakfast. Are you going to take me to him, or am I going to have to find him myself?"

"This is outrageous," Susan said. "Does Assistant Chief Constable Lees know you're here, behaving in this high-handed manner?"

"I'm sure he soon will," Karen said over her shoulder as she set off down the hall. She threw open the first door she came to: a cloakroom. The next door: an office.

"Stop that," Susan said sharply. "You are exceeding your authority, Inspector." The next door: a small drawing room. Karen could hear Susan's running feet behind her. "Fine," Susan snapped as she overtook Karen. She stopped in front of her, spreading her arms wide, apparently under the illusion that would stop Karen if she was seriously minded to continue. "I'll take you to him."

Karen followed her through to the rear of the building. Susan opened the door on to a bright breakfast room that looked over to the lake and woods beyond. Karen had no eyes for the view or for the buffet laid out on the long sideboard. All she was interested in was the couple sitting at the table, their son perched between them. Grant immediately stood up and glowered at her. "What's going on?" he said.

"It's time for Lady Grant to get Alec ready for school," Karen said, realizing she was sounding like a bad script but not caring how foolish that felt.

"How dare you barge into my home shouting the odds." His was the first raised voice, but he appeared not to notice.

"I'm not shouting, sir. What I have to say, it's not appropriate

for me to say in front of a child." Karen met his glare, not backing down. Somehow, this morning she had lost what little fear of consequences she possessed.

Grant gave a quick, nonplussed look at his son and wife. "Then we'll go elsewhere, Inspector." He led the charge to the door. "Susan, coffee. In my office."

Karen struggled to keep up with his long stride, barely catching up as he stormed into a spartan room with a glass desk which held a large spiral-bound notebook and a slim laptop. Behind the desk was a functional, ergonomically designed office chair. Filing drawers lined one wall. Against the other were two chairs Karen recognized from a trip to Barcelona, where she'd mistakenly got off the city tour bus at the Mies van der Rohe pavilion and been surprisingly captivated by its calm and simplicity. Seeing them here grounded her somehow. She could hold her own against any big shot, she told herself.

Grant threw himself into his chair like a petulant child. "What the hell is all this in aid of?"

Karen dropped her heavy satchel on the floor and leaned against a filing cabinet, arms folded across her chest. She was dressed to impress in her smartest suit, one she'd bought from Hobbs in Edinburgh at the sales. She felt absolutely in control and to hell with Brodie Grant. "She's dead," she said succinctly.

Grant's head jerked back. "Who's dead?" He sounded indignant.

"Bel Richmond. Are you going to tell me what she was chasing?"

He attempted a nonchalant half-shrug. "I've no idea. She was a freelance journalist, not a member of my staff."

"She was working for you."

He waved a hand at her. The brush-off. "I was employing her to act as press liaison should anything come of this cold case inquiry." He actually curled his lip. "Which doesn't seem very likely at this point."

"She was working for you," Karen repeated. "She was doing a

lot more than press liaison. She wasn't a publicist. She was an investigative journalist, and that's precisely what she was doing for you. Investigating."

"I don't know where you get your ideas from, but I can assure you, you won't be having any more of them about this case after I've spoken to Simon Lees."

"Be my guest. I'll enjoy telling him how Bel Richmond flew out to Italy on your private jet yesterday. How she picked up a hire car on your company account at Florence airport. And how her killer was disturbed by the police while trying to feed her naked body to the pigs a couple of hundred yards from the house where Bel herself found the poster that kick-started this whole inquiry." Karen straightened up and crossed to the desk, leaning on it with her fists. "I am not the fucking numpty you take me for." She gave him glare for glare.

Before he could work out how to respond, a young woman in a black dress arrived with a tray of coffee. She looked around uncertainly. "On the desk, lassie," Grant said. Somehow Karen didn't think she was going to be offered a cup.

She waited till she heard the door close behind her, then she said, "I think you'd better tell me why Bel went to Italy. It's likely what got her killed."

Grant tilted his head back, thrusting his strong chin towards her. "As far as I am aware, Inspector, Fife police's jurisdiction doesn't stretch to Italy. This is nothing to do with you. So why don't you fuck off?"

Karen laughed out loud. "I've been told to fuck off by better men than you, Brodie," she said. "But you should know, I am here at the request of the Italian police."

"If the Italian police want to talk to me, they can come here and talk to me. Organ grinders, not monkeys. That's my way. Besides, if this was in any way official, you'd have your wee boy with you, taking notes. I do know my Scots law, Inspector. And now, as I previously requested, fuck off."

"Don't worry, I'm going. But for the record, I don't need

corroboration for a witness statement for the Italian police. I'll tell you something else for nothing. If I was your wife, I'd be seriously unhappy about all these women's bodies in your slipstream. Your daughter. Your wife. And now your hired gun."

His lips stretched back in a reptilian rictus. "How dare you!"

In spite of her determination, Grant had got under her skin. Karen reached for her bag and drew out the scale map of the ransom handover scene. "This is how I dare," she said, spreading it out on Grant's desk. "You think your money and your influence can buy anything. You think you can bury the truth like you've buried your wife and your daughter. Well, sir, I'm here to prove you wrong."

"I don't know what the hell you think you're talking about." Grant had to force out his words between stretched lips.

"The received account," she said, stabbing the map with her finger. "Cat takes the bag from your wife, the kidnappers fire a shot that hits her in the back and kills her. The police fire a shot that goes high and wide." She glanced up at him. His face was motionless, frozen in a mask of rage. She hoped her expression was giving as good as it got. "And then there's the truth: Cat takes the bag from your wife, she turns to take it back to the kidnappers. You start waving your gun around, the kidnappers plunge the beach into darkness, you fire." She looked him straight in the eye. "And you kill your daughter."

"This is a sick fantasy," Grant hissed.

"I know you've been in denial all these years, but that's the truth. And Jimmy Lawson is ready to tell it."

Grant slammed his hand down on the desk. "A convicted murderer? Who's going to believe him?" His lip quivered in a sneer.

"There's others who know you had a gun that night. They're retired now. There's nothing you can hold over their heads any more. You can maybe get Simon Lees to shut me up, but the genie's out of the bottle now. It wouldn't hurt you to start cooperating with me over Bel Richmond's murder."

"Get out of my house," Grant said. "Next time you come back, you'd better have a warrant."

Karen gave him a tight little smile. "You can count on it." She still had plenty of shots in reserve, but this wasn't the time to fire them. Mick Prentice and Gabriel Porteous could wait for another day. "It's not over, Brodie. It's not over till I say so."

The about-to-be-former Gabriel Porteous had no problem entering the UK. The immigration official at Edinburgh Airport swiped his passport, compared his image to the photograph, and nodded him through. He had to stick with his old ID for the car hire too. This collision of past and future was hard to balance. He wanted to let go of Gabriel and all he had done. He wanted to enter his new life clean and unencumbered. Emotionally, psychologically, and practically, he wanted no connection to his past life. No possibilities of awkward questions from the Italian authorities. Please God, his grandfather would accept that he wanted a clean break with his past. One thing was certain—he wouldn't have to exaggerate the shock and pain his father's letter had inflicted on him.

He had to stop at a petrol station and ask directions to Rotheswell Castle, but it was still only mid-morning when he approached the impressive front gate. He pulled up and got out, grinning at the CCTV. When the intercom asked who he was and what his business might be, he said, "I'm Adam Maclennan Grant. That's my business."

They kept him waiting almost five minutes before they opened the outer gate. At first, it pissed him off. His anxiety had reached an intolerable level. Then it dawned on him that you took precautions like this only when there was something serious to protect. So he waited, then he drove into the pen between the two sets of gates. He tolerated the security pat-down. He didn't complain when they searched his vehicle and asked him to open his hold-all and his backpack so they could rummage around. When they

finally let him through the inner gate and he caught his first glimpse of what he'd lost, his breath caught in his throat.

He drove slowly, making sure he had his emotions under control. He wanted this fresh start so badly. No more fuck-ups. He parked on the gravel near the front door and climbed out of the car, stretching luxuriously. He'd been folded into seats for too long. He squared his shoulders, straightened his spine, and walked up to the door. As he approached, it swung open. A woman in a tweed skirt and a woollen sweater stood in the doorway. Her hand flew up to her mouth involuntarily and she gasped, "Oh my God."

He gave her his best smile. "Hello. I'm Adam." He extended a hand. One look at this woman and he knew the kind of uptight manners expected in this house.

"Yes," the woman said. Training overcame emotion, and she took his hand in a firm grip and held on tight. "I'm Susan Charleson. I'm your gran—, I mean, I'm Sir Broderick's personal assistant. This is the most extraordinary shock. Surprise. Bolt from the blue." She burst out laughing. "Listen to me. I'm not usually like this. It's just that—well, I never imagined I'd see this day."

"I appreciate that. It's all been a bit of a shock for me too." He gently freed his hand. "Is my grandfather at home?"

"Come this way." She closed the door and ushered him down a hallway.

He'd been in some fine houses in Italy thanks to his father's business, but this place was utterly foreign. With its stone walls and its spare décor, it felt cold and naked. But it didn't hurt to make nice. "This is a beautiful house," he said. "I've never seen anything like it."

"Where do you live?" Susan asked as they turned into a long corridor.

"I grew up in Italy. But I'm planning on returning to my roots."

Susan stopped in front of a heavy studded oak door. She knocked and entered, beckoning Adam to follow. The room, a

book-lined refuge, was a blur to him. His total focus was on the white-haired man standing by the window, deep-set eyes unreadable, face immobile.

"Hello, sir," Adam said. To his surprise, he found it hard to speak. Emotion that he hadn't expected welled up, and he had to swallow hard to avoid tears.

The old man's face seemed to disintegrate before his eyes. An expression somewhere between smiling and sorrow engulfed him. He took a step towards Adam, then stopped. "Hello," he said, his voice choked too. He looked beyond Adam and waved Susan from the room.

The two men stared hungrily at each other. Adam managed to get himself under control, clearing his throat. "Sir, I'm sure you've had people claiming to be Catriona's son before. I just want to say that I don't want anything from you and I'm happy to undergo any tests—DNA, whatever—that you want. Until my father died three months ago, I had no idea who I really was. I've spent those three months wondering whether I should contact you or not . . . And, well, here I am." He took Daniel's letter from the inside pocket of his one good suit. "This is the letter he left me." He stretched out his arm to Grant, who took the creased sheets of paper. "I'll happily wait outside while you read it."

"There's no need for that," Grant said gruffly. "Sit down there, where I can see you." He took a chair opposite the one he had indicated and began to read. Several times he paused and scrutinized Adam, who forced himself to stay still and calm. At one point, Grant covered his mouth with his hand, the fingers visibly trembling. He came to the end and gazed hungrily at Adam. "If you're a fake, you're a bloody good one."

"There's also this—" Adam took a photograph from his pocket. Catriona sat on a kitchen chair, hands folded over the high curve of a heavily pregnant belly. Behind her, Mick leaned over her shoulder, one hand on the bump. They were both grinning. It had the slightly awkward look of something posed for the timer. "My mum and dad."

This time, Grant couldn't hold back his tears. Wordlessly, he held out his arms to his grandson. Adam, his eyes wet, got up and accepted the embrace.

It felt as if it went on forever and lasted no time at all. Finally they drew apart, each wiping their eyes with their hands. "You look like I did fifty years ago," Grant said heavily.

"You should still have the DNA test done," Adam said. "There are some bad people out there."

Grant gave him a long, measured look. "I don't think they're all on the outside," he said with an air of melancholy. "Bel Richmond was working for me."

Adam struggled not to show he recognized the name, but he could tell from his grandfather's face that he'd failed. "She came to see me," he said. "She never mentioned that you were her boss."

Grant gave a thin smile. "I wouldn't say I was her boss. But I did hire her to do a job for me. She did it so well it killed her."

Adam shook his head. "That can't be right. It was only last night that I spoke to her."

"It's right enough. I've had the police here earlier. Apparently her killer tried to feed her to the pigs right next door to the villa where your pal Matthias was squatting until round about the time when your father died," Grant continued grimly. "And the police are also investigating a presumed murder there. That one happened round about the time Matthias and his little troupe of puppeteers disappeared."

Adam raised his eyebrows. "That's bizarre," he said. "Who else is supposed to be dead?"

"They're not sure. The puppeteers scattered to the four winds. Bel was planning to track them down next. But she never got the chance. She was a good journalist. Good at sniffing things out."

"It sounds like it."

"So where is Matthias?" Grant asked.

"I don't know. The last time I saw him was the day I buried my father. I went back to the villa so he could give me the letter. I was upset when I realized he had known my real identity all along. I

was angry and upset that he and my dad had conspired to keep me from you all those years. When I left, I said I didn't want to hear from him again. I didn't even know they'd left Boscolata." He gave a delicate little shrug. "They must have fallen out with each other. I know the others sometimes got restive because Matthias took a bigger cut of the take. It must have got out of hand. Somebody got killed." He shook his head. "That's harsh."

"And Bel? What's your theory there?"

Adam had had a night drive and a flight to figure out the answer to that one. He hesitated for a moment, as if thinking about the possibilities. "If Bel was asking questions around Boscolata, word might have got back to the killer. I know at least one of the group was having sex with someone who lived there. Maybe his girlfriend told him about Bel and they were keeping tabs on her. If they found out she was coming to see me, they might have thought she was digging too deep and needed to be got rid of. I don't know. I've no idea how people like that think."

Grant's expression was as unreadable as it had been when Adam had first seen him. "You're very plausible," he said. "Some might say you're a chip off the old block." His face twisted momentarily in pain. "You're right about the DNA. We should have that done as soon as possible. Meanwhile, I think you should stay with us. Let us start getting to know you." His smile was disturbingly ambivalent. "The world's going to be very interested in you, Adam. We need to prepare for that. We don't necessarily need to be entirely frank. I've always been a great believer in privacy."

That had been a shaky moment when the old man revealed Bel was in his pocket. His questions were tougher than Adam had expected. But now he understood that a decision had been taken, a decision to opt for complicity. For the first time since Bel had walked through his door, the unbearable tension began to dissipate.

Friday, 13th July 2007; Glenrothes

The latest summons to the Macaroon's office wasn't entirely unexpected. Karen had been refusing to take no for an answer from him since she'd had a terse e-mail from Susan Charleson revealing the return of the prodigal. She badly wanted to talk to Brodie Grant and his murderous grandson, but of course she'd been warned off before she could even make her case to Lees. She'd known confronting Grant about his actions on the beach all those years ago would bring repercussions. Unsurprisingly, Grant had got his retaliation in first, accusing her of desperately looking for somebody to charge with something in a case where all the criminals were dead. Karen had had to listen to the Macaroon lecturing her on the importance of good relationships with the public. He reminded her that she had resolved three cold cases even though nobody would be tried for any of them. She had made the CCRT look good, and it would be extremely unhelpful if she pushed Sir Broderick Maclennan Grant into making them look bad.

When she'd raised the issue of Adam Maclennan Grant's possible involvement in two murders in Italy, the Macaroon had turned green and told her to back off a case that was none of her business.

Di Stefano had been in regular contact with Karen via phone and e-mail over the previous weeks. There was, he said, plenty of DNA on Bel's body. One of the teenagers who lived at Boscolata had identified Gabriel aka Adam as the man he'd seen with Matthias on the presumed day of the assumed murder at the Villa Totti. They'd found the house near Greve where a man answering to that description had been living. They'd found DNA there that matched what was on Bel's body. All they needed to bring a case before an investigating magistrate was a sample of DNA from the former Gabriel Porteous. Could Karen oblige?

Only when hell froze over.

Now, finally, the Macaroon had summoned her. Marshalling her thoughts, she walked into his office without knocking. This time, she was the one who got the shock. Sitting to one side of the desk, at an angle to the Macaroon but facing the visitor's chair, was Brodie Grant. He smiled at her discomfiture. Friday the thirteenth, right enough.

Without waiting to be asked, Karen sat down. "You wanted to see me, sir," she said, ignoring Grant.

"Karen, Sir Broderick has very kindly brought us his grandson's notarized statement about the recent events in Italy. He thought, and I agree with him, that this would be the most satisfactory way to proceed." He brandished a couple of sheets of paper at her.

Karen stared at him in disbelief. "Sir, a simple DNA test is the way to proceed."

Grant leaned forward. "I think you'll find that once you've read the statement, it's clear that a DNA test would be a waste of time and resources. No point in testing someone who's manifestly a witness, not a suspect. Whoever the Italian police are looking for, it's not my grandson."

"But—"

"And another thing, Inspector; my grandson and I will not be discussing with the media where he's been for the past twenty-two years. Obviously, we will be making public the fact that we have had this extraordinary reunion after all this time. But no details. I expect you and your team to respect that. If information leaks into the public domain, you can rest assured that I will pursue the person responsible and make sure they are held accountable."

"There will be no leaks from this office, I can assure you," the Macaroon said. "Will there, Karen?"

"No, sir," she said. No leaks. Nothing to contaminate Phil's imminent promotion or her own team.

Lees waved the papers at Karen. "There you go, Inspector. You can forward this on to your opposite number in Italy and then we

can draw a line under our own solved cases." He smiled winningly at Grant. "I'm glad we've been able to clear this up so satisfactorily."

"Me too," Grant said. "Such a pity we won't be seeing each other again, Inspector."

"Indeed. You take care, sir," she said, getting to her feet. "You want to take very good care of yourself. And your grandson. It would be tragic if Adam had to endure any more losses." Seething, Karen stalked out of the room. She steamed back to her own office, ready to rant. But Phil was away from his desk and nobody else would do. "Fuck, fuck, fuck," she muttered, slamming into her office just as the phone rang. For once, she ignored it. But the Mint stuck his head round the door. "It's some woman called Gibson looking for you."

"Put her on." She sighed. "Hello, Misha. What can I do for you?"

"I just wondered if there was any news. When your sergeant came round a couple of weeks ago to tell me you were pretty sure my father died earlier this year, he said there was a possibility he might have had children that we could test for a match. But then I didn't hear from you . . . "

Fuck, fuck, fuck, and fuck again. "It's not looking hopeful," Karen said. "The person in question is refusing to give any samples for testing."

"What do you mean, refusing? Doesn't he understand a child's life is at stake here?"

Karen could feel the emotional intensity down the phone line. "I think he's more concerned about keeping his own nose clean."

"You mean he's a criminal? I don't care about that. Does he not get it? I'm not going to give his DNA to anybody else. We can do it confidentially."

"I'll pass the request on," Karen said wearily.

"Can you not put me in direct touch with him? I'm begging you. This is my wee boy's life at stake. Every week that goes past, he's got less and less chance."

"I do understand that. But my hands are tied. I'm sorry. I will pass the request on, I promise."

As if she sensed Karen's frustration, Misha changed her approach. "I'm sorry. I appreciate how hard you've tried to help. I'm just desperate."

After the call, Karen sat staring into space. She couldn't bear the thought that Grant was protecting a murderer for his own selfish emotional ends. It wasn't exactly a surprise, given the way he'd covered up his own culpability in his daughter's death. But there had to be a way to get round this barrier. She and Phil had gone over their options so often during the past couple of weeks that it felt as if they'd worn a groove in her brain. They'd talked about stalking Adam, going for the publicly discarded Coke can or water bottle. They'd discussed stealing the rubbish from Rotheswell and having River go through it till she found a match for the Italian DNA. But they'd had to concede they were not so much clutching at straws as at shadows.

Karen leaned back in her chair and thought about the place where all this had started. Misha Gibson desperate for hope, prepared to do anything for her child. Just as Brodie Grant was for his grandson. The bonds between parents and children . . . And then, suddenly, it was there in front of her. Beautiful and cunning and deliciously ironic.

Almost tipping herself on to the floor, Karen shot up straight and grabbed the phone. She keyed in River Wilde's number and drummed her fingers on the desk. When River answered, Karen could hardly speak in sentences. "Listen, I just thought of something. If you've got half-siblings, you'd be able to see the connection in the DNA, right?"

"Yes. It wouldn't be as strong as with full siblings, but you'd see a correlation."

"If you had some DNA, and you got a sample that showed that degree of correlation, and you knew that person had a half-sibling, do you think that would be enough to get a warrant to take samples from the half-sibling?"

River hummed for a moment. "I could make the case," she said. "I think it would be enough."

Karen took a deep breath. "You know when we got Misha Gibson's DNA to check against the cave skeleton?"

"Yes," River said cautiously.

"Have you still got that?"

"Is your case still open?"

"If I was to say yes, what would your answer be?"

"If your case is still open, I'm legally entitled still to have possession of the DNA. If it's closed, the DNA should be destroyed."

"It's still open," Karen said. Which, technically, it was, since the only evidence against Mick Prentice in the death of Andy Kerr was circumstantial. Enough to close the file, certainly. But Karen hadn't actually returned it to the registry, so it wasn't closed as such.

"Then I still have the DNA."

"I need you to e-mail me a copy ASAP," Karen said, punching the air. She got to her feet and did a little dance round the office.

Fifteen minutes later, she was e-mailing a copy of Misha Gibson's DNA to di Stefano in Siena with a covering note. Please ask your DNA expert to compare these. I believe this to be the half-sibling of the man known as Gabriel Porteous. Let me know how you get on.

The next hours were a form of torture. By the end of the working day, there was still no word from Italy. When she got home, Karen couldn't leave the computer alone. Every ten minutes she was jumping up and checking her e-mail. "How quickly it fades," Phil teased her from the sofa.

"Yeah, right. If I wasn't doing it, you would be. You're as keen as I am to nail Brodie's grandson."

"You got me bang to rights, guv."

It was just after nine when the anticipated reply from di Stefano hit her inbox. Holding her breath, Karen opened the message. At first, she couldn't believe it. "No correlation?" she said. "No fucking correlation? How can that be? I was so sure . . ."

She threw herself down on the sofa, allowing Phil to cuddle her close. "I can't believe it either," he said. "We were all so sure that Adam was the killer." He flicked a finger at the anodyne statement Karen had brought home to show him. "Maybe he's telling the truth, bizarre though it sounds."

"No way," she said. "Murderous puppeteers following Bel through Italy? I've seen more credible episodes of *Scooby Doo*." She curled up, disconsolate, head tucked under Phil's chin. When the new idea hit, her head jerked so suddenly he nearly bit through his tongue. While he was moaning, Karen kept repeating, "It's a wise child that knows its father."

"What?" Phil finally said.

"What if Fergus is right?"

"Karen, what are you talking about?"

"Everybody thought Adam was Fergus's kid. Fergus thinks so. He shagged Cat around the right time, just a one-off. Maybe she'd had a row with Mick. Or maybe she was just pissed off because it was a Saturday night and he was with his wife and kid and not her. Whatever the reason, it happened." Karen was bouncing on her knees on the sofa, excitement making her a child again. "What if Mick was wrong all these years? What if Fergus really is Adam's dad?"

Phil grabbed her and gave her a resounding kiss on the forehead. "I told you right at the start I love your mind."

"No, you said it was sexy. Not quite the same thing." Karen nuzzled his cheek.

"Whatever. You are so smart, it turns me on."

"Do you think it's too late to ring him?"

Phil groaned. "Yes, Karen. It's an hour later where he lives. Leave it till the morning."

"Only if you promise to take my mind off it."

He flipped her over on to her back. "I'll do my best, boss."

Karen stretched out in the bath, enjoying the dual sensations of foam and water against her skin. Phil was playing cricket, which she now understood meant a quick game followed by a long drink with his mates. He'd stay at his own house tonight, rolling home at closing time after a skinful of lager. She didn't mind. Usually she met up with the girls for a curry and a gossip. But tonight she wanted her own company. She was expecting a phone call, and she didn't want to take it in a crowded pub or a noisy restaurant. She wanted to be sure of what she was hearing.

Fergus Sinclair had been suspicious when she'd called him out of the blue to ask for a DNA sample. Her pitch had been simple— a man had turned up claiming to be Adam, and Karen was determined to make every possible check on his bona fides. Sinclair had been cynical and excited by turns. In both states, he'd been convinced that he was the best litmus test available. "I'll know," he kept insisting. "It's an instinct. You know your own kids."

It wasn't the right time to share River's statistic that somewhere between 10 and 20 per cent of children were not actually the offspring of their attributed fathers and, in most of those cases, the fathers had no idea they weren't the dad. Karen kept falling back on appropriateness. Finally, he'd agreed to go to his local police station and give a DNA sample.

Karen had managed to persuade the German police duty officer to have the sample taken and couriered directly to River. The Macaroon would lose his mind when he saw the bill, but she was past caring. To speed things up, she'd persuaded di Stefano to e-mail a copy of the Italian killer's DNA to River.

And tonight, she would know. If the DNA said Fergus was the Italian killer's father, she'd be able to get a warrant to take a sample from Adam. Under Scots law, she could have detained him and taken a DNA sample without arresting or charging him. But she knew her career would be over if she attempted to treat Adam Maclennan Grant like any other suspect. She wouldn't go

near him without a sheriff's warrant. But once his DNA was in the system, even Brodie Grant's power couldn't keep him out of the clutches of the law. He'd have to pay for the lives he'd cut short.

Her thoughts stuttered to a halt when the phone rang. River had said nine o'clock, but it was barely half past seven. Probably her mother or one of the girls trying to persuade her to change her mind and join them. With a sigh, Karen stretched to pick up the phone from the stool by the bath.

"I've got Fergus Sinclair's DNA analysis in front of me," River said. "And I've also got one from Capitano di Stefano."

"And?" Karen could hardly breathe.

"A close correlation. Probably father and son."

Thursday, 19th July 2007; Newton of Wemyss

The voice is soft, like the sunlight that streams in at the window. "Say that again?"

"John's cousin's ex-wife. She moved to Australia. Outside Perth. Her second husband, he's a mining engineer or something." Words tumbling now, tripping over each other, a single stumble of sounds.

"And she's back?"

"That's what I'm telling you." Exasperated words, exasperated tone. "A twenty-fifth school reunion. Her daughter, Laurel, she's sixteen, she's come with her for a holiday. John met them at his mother's a couple of weeks ago. He didn't say anything because he didn't want to raise my hopes." A spurt of laughter. "This from Mr. Optimism."

"And it's right? It's going to work?"

"They're a match, Mum. Luke and Laurel. It's the best possible chance."

And this is how it ends.

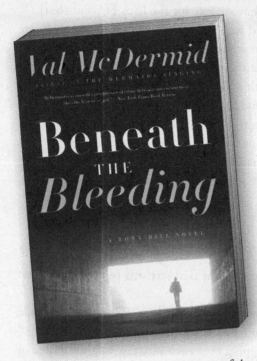